MECHANICAL ENGINEERING
(OBJECTIVE TYPE)

MECHANICAL ENGINEERING

(OBJECTIVE TYPE)

[With Multiple Choice Questions and Answers]

(Including Brief Theory)

[For Gate Test, U.P.S.C. Engineering Services Examinations (I.E.S.), I.A.S. Exams, (Engg. Group), Objective Type Tests of Public Undertaking and Private Limited Companies, Interviews, Degree Examinations, Quick Review of the Subject and other Companies Tests]

By

Dr. R.K. Bansal

B. Sc. Engg. (Mech.), M.Tech., Hons. (I.I.T., Delhi),
Ph.D., M.I.E. (India)

Formerly
Professor of Mechanical Engineering
Department of Mechanical Engineering
Delhi College of Engineering
Delhi

Presently

Dean (Academics)
Northern India Engineering College
Delhi

LAXMI PUBLICATIONS (P) LTD

(An ISO 9001:2008 Company)

BENGALURU • CHENNAI • COCHIN • GUWAHATI • HYDERABAD
JALANDHAR • KOLKATA • LUCKNOW • MUMBAI • RANCHI • NEW DELHI
BOSTON (USA) • ACCRA (GHANA) • NAIROBI (KENYA)

MECHANICAL ENGINEERING (OBJECTIVE TYPE)

Printed and bound in India
Typeset at Goswami Associates, Delhi
First Edition : 1984; Second Edition : 1987; Third Edition : 1991; Fourth Edition : 1993; Fifth Edition : 1993;
Sixth Edition : 2001; Seventh Edition : 2016
ISBN 978-81-318-0788-0

Published in India by

Laxmi Publications (P) Ltd.
(An ISO 9001:2008 Company)
113, GOLDEN HOUSE, DARYAGANJ,
NEW DELHI - 110002, INDIA
Telephone : 91-11-4353 2500, 4353 2501
Fax : 91-11-2325 2572, 4353 2528
www.laxmipublications.com info@laxmipublications.com

Branches

Bengaluru	080-26 75 69 30	
Chennai	044-24 34 47 26,	24 35 95 07
Cochin	0484-237 70 04,	405 13 03
Guwahati	0361-254 36 69,	251 38 81
Hyderabad	040-27 55 53 83,	27 55 53 93
Jalandhar	0181-222 12 72	
Kolkata	033-22 27 43 84	
Lucknow	0522-220 99 16	
Mumbai	022-24 93 12 61	
Ranchi	0651-220 44 64	

C—11420/016/02
Printed at: Jay Ambey Enterprises, Delhi

Dedicated

to

the loving memory of

my daughter, Babli

CONTENTS

PREFACE TO THE SEVENTH EDITION

The seventh edition has been thoroughly revised and enlarged. A new chapter on General Engineering that consists of subjects Physics, Chemistry, Mathematics and Electrical sciences has been included. The questions of 'Fill in the blanks' and 'Mark the true and false statements' have been added. These types of questions are generally asked in the latest test papers of most of the organisations. The text of each chapter has been thoroughly revised and enlarged.

Two typical objective type test papers, each containing 224 objective type questions from all chapters are given in the last chapter (chapter-15) for the sake of confidence and revision.

The book contains following 15 chapters:

1. Fluid Mechanics and Hydraulic Machines
2. Engineering Mechanics
3. Thermodynamics
4. Internal Combustion Engines and Nuclear Power Plants
5. Steam Boilers, Engines, Nozzle and Turbines
6. Compressor, Gas Turbines and Jet Engines
7. Heat Transfer, Refrigeration and Air Conditioning
8. Strength of Materials
9. Theory of Machines
10. Machine Design
11. Engineering Materials
12. Production Engineering
13. Industrial Engineering and Production Management
14. General Engineering and General Aptitude
15. Typical Objective Type Test Papers.

At the end of each chapter additional objective type questions from competitive examinations such as Gate, I.E.S. (Indian Engineering Service) Examination in Mech. Engg., I.A.S. (Indian Administrative Service) Examination and objective type tests of public undertakings and private limited companies have been included. To make the book more useful to the students and for quick review of the subject brief theory is given at the beginning of each chapter.

Though every care has been taken in checking the manuscripts and proofreading, yet claiming perfection is very difficult. I shall be very tankful to the readers and users of this book for pointing out any mistakes that might have crept in. Suggestions for improvement are most welcome and would be incorporated in the next edition with a view to make the book more useful.

—Author

Preface to the First Edition

I am glad to present the book entitled, 'Mechanical Engineering (Objective Type)' to the Engineering students preparing for the Graduate Aptitude Test in Engineering (GATE), U.P.S.C. Indian Engineering Service (I.E.S.) Examination in Mechanical Engineering, U.P.S.C. Indian Administrative Service (I.A.S.). Examination in Mechanical Engineering, objective type tests of public undertakings and private limited companies and interviews. The course-contents have been planned in such a way that the general requirements of all the above examinations are fulfilled.

The trend for objective type examination is on the increase. Since 1978, U.P.S.C. has started conducting objective type examinations for Indian Engineering Service examination on and Indian Administrative Service examination. Also the students, seeking admission to all Post Graduate degree course in Engineering/Technology in the country, will have to qualify the Graduate Aptitude Test in Engineering (GATE) with effect from July/August 1984 admission with Scholarship. These requirements have been kept in mind, while writing this book.

The book is written in a simple and easy-to-follow language, so that even an average student can grasp the subject by self study. To make the book more useful to the students and for quick review of this subject, brief conventional type theory is given at the beginning of each chapter. The book serves the purpose for both the papers I and II of Preliminary and Main examination of I.E.S. (Indian Engg. Service). At the end of each chapter, answers to all questions have been given. The typical objective type test papers each consisting 140 objective type questions from all chapters, are given in the end for the sake of revision.

To write a book of this type, a large number of standard books have been consulted. I am thankful to these authorities whose works have been consulted and give me a great help in preparing the book.

I express my appreciation and gratefulness to my publisher Shri R.K. Gupta (a Mechanical Engineer) for his most co-operative, painstaking attitude and untiring efforts for bringing out the book in a short period.

Mrs. Nirmal Bansal deserve special credit as she not only provided an ideal atmosphere at home for book writing but also gave inspiration and valuable suggestions.

Though every care has been taken in checking the manuscripts and proofreading, yet claiming perfection is very difficult. I shall be very thankful to the readers and users of this book for pointing out any mistakes that might have crept in. Suggestions for improvement are most welcome and would be incorporated in the next edition with a view to make the book more useful.

—Author

Chapter 1

FLUID MECHANICS AND HYDRAULIC MACHINES

I. THEORY

1.1. DEFINITIONS AND FLUID PROPERTIES

Fluid mechanics is that branch of science which deals with the behaviour of the fluid (*i.e.*, liquids or gases) when they are at rest or in motion. When the fluids are at rest, there will be no relative motion between adjacent fluid layers and hence velocity gradient $\left(\frac{du}{dy}\right)$, which is defined as the change of velocity between two adjacent fluid layers divided by the distance between the layers, will be zero. Also the shear stress $\tau = \mu \frac{du}{dy}$ will be zero in which $\frac{du}{dy}$ is the velocity gradient or **rate of shear strain.**

The law, which states that the shear stress (τ) is directly proportional to the rate of shear strain $\left(\frac{du}{dy}\right)$, is called **Newton's Law of viscosity.** Fluids which obey Newton's law of viscosity are known as **Newtonian fluids** and the fluids which do not obey this law are called **Non-Newtonian fluids.**

(*i*) **Density** or **mass density.** It is defined as the mass per unit volume of a fluid and is denoted by the symbol ρ (rho).

(*ii*) **Weight density** or **specific weight**. It is defined as the weight per unit volume of a fluid and is denoted by the symbol w.

Mathematically, $\rho = \dfrac{\text{Mass of fluid}}{\text{Volume of fluid}}$

and $$w = \frac{\text{Weight of fluid}}{\text{Volume of fluid}} = \frac{\text{Mass of fluid} \times g}{\text{Volume}} = \rho \times g.$$

The value of density (ρ) for water is 1000 kg/m^3 and of specific weight or weight density (w) is 1000 × 9.81 N/m^3 or 9810 N/m^3 in S.I. units.

(*iii*) **Specific volume.** It is defined as volume per unit mass and hence it is the reciprocal of mass density. **Specific gravity** is the ratio of weight density or mass density of the fluid to the weight density or mass density of a standard fluid at a standard temperature. For liquids, water is taken as a standard fluid at 4°C and for gases, air is taken as standard fluid.

(*iv*) **Viscosity.** It is defined as the property of a fluid which offers resistance to the movement of one layer of fluid over another adjacent layer of the fluid. Unit of viscosity in MKS is expressed as $\dfrac{\text{kgf-sec}}{\text{m}^2}$, in SI system as $\dfrac{\text{Ns}}{\text{m}^2}$ and in CGS as $\dfrac{\text{dyne-sec}}{\text{m}^2}$. The unit of viscosity in CGS is also called Poise.

The equivalent numerical value of one poise in MKS units is obtained by dividing 98.1 and in SI units is obtained by dividing 10.

Kinematic viscosity is defined as the ratio of dynamic viscosity to density of fluid. It is denoted by the Greek symbol (ν) called 'nu'. Unit of kinematic viscosity in MKS is m^2/sec and in CGS is cm^2/sec which is also called stoke. The viscosity of a liquid decreases with the increase of temperature while the viscosity of the gas increases.

(*v*) **Compressibility.** It is the reciprocal of the bulk modulus of elasticity, which is defined as the ratio of compressive stress to volumetric strain. Mathematically,

$$\text{Bulk Modulus} = \frac{\text{Increase of pressure}}{\text{Volumetric strain}} = \frac{dp}{-\left(\dfrac{dV}{V}\right)}$$

$$\therefore \quad \text{Compressibility} = \frac{1}{\text{Bulk modulus}} = \frac{-\left(\dfrac{dV}{V}\right)}{dp}$$

(*vi*) **Surface tension.** It is defined as the tensile force acting on the surface of a liquid in contact with a gas such that the contact surface behaves like a membrane under tension. It is expressed as force per unit length and is denoted by σ (called sigma). Hence unit of surface tension in MKS is kgf/m while in SI is N/m.

The relation between surface tension (σ) and difference of pressure (p) between inside and outside of a liquid drop is given by $p = \dfrac{4\sigma}{d}$

For a soap bubble, $p = \dfrac{8\sigma}{d}$

For a liquid jet, $p = \frac{2\sigma}{d}$.

(*vii*) **Capillarity.** It is defined as a phenomenon of rise or fall of a liquid surface in a small vertical tube held in a liquid relative to general level of the liquid. The rise or fall of liquid is given by

$$h = \frac{4\sigma \cos \theta}{wd}$$

where d = Dia. of tube

θ = Angle of contact between liquid and glass tube.

(*viii*) **Ideal fluid** is a fluid which offers no resistance to flow and is incompressible. Hence for ideal fluid viscosity (μ) is zero and density (ρ) is constant.

(*ix*) **Real fluid** is a fluid which offers resistance to flow. Hence viscosity for real fluid is not zero.

1.2. PRESSURE AND ITS MEASUREMENT

Pressure at a point is defined as the force per unit area. The Pascal's law states that intensity of pressure for a fluid at rest is equal in all directions. The pressure at any point in a incompressible fluid (*i.e.*, liquid) at rest is equal to the product of weight density of fluid and vertical height from free surface of the liquid.

Mathematically, $p = wz = \rho gz$.

(*i*) **Hydrostatic law** states that the rate of increase of pressure in the vertically downward direction is equal to the specific weight of the fluid *i.e.*, $\frac{dp}{dz} = w = \rho g$.

(*ii*) **Absolute pressure** is the pressure measured with reference to absolute zero pressure while gauge pressure is the pressure measured with reference to atmospheric pressure. Thus the pressure above the atmospheric pressure is called gauge pressure. Vacuum pressure is the pressure below the atmospheric pressure.

Mathematically,

Gauge pressure = Absolute pressure – Atmospheric pressure

Vacuum pressure = Atmospheric pressure – Absolute pressure.

(*iii*) **Manometers** are defined as the devices used for measuring the pressure at a point in a fluid. They are classified as:

1. Simple Manometers, and
2. Differential Manometers.

Simple manometers are used for measuring pressure at a point while differential manometers are used for measuring the difference of pressures between the two points in a pipe or two different pipes.

(*iv*) The pressure at a point in a static compressible fluid is obtained by combining two equations *i.e.*, equation of state for a gas $\left(\frac{p}{\rho} = RT\right)$ and the equation given by hydrostatic law

$\left(\frac{dp}{dz} = -\rho g\right)$. For isothermal process, the pressure at a height Z in a static compressible fluid is given as $p = p_o e^{-gZ/RT}$

(*v*) For adiabatic process the pressure and temperature at a height Z are

$$p = p_o \left[1 - \frac{\gamma - 1}{\gamma} \frac{gZ}{RT_o}\right]^{\frac{\gamma-1}{\gamma}} \text{ and } T = T_o \left[1 - \frac{\gamma - 1}{\gamma} \frac{gZ}{RT_o}\right]$$

where p_o = Absolute pressure at ground or sea-level

R = Gas constant, γ = Ratio of specific heats

T_o = Temperature at ground or sea-level.

1.3. HYDROSTATIC FORCES ON PLANE SURFACES

The **force** exerted by a static liquid on a vertical, horizontal and inclined surface immersed in the liquid is given by

$$F = \rho g A \bar{h}$$

where ρ = Density of the liquid

A = Area of the immersed surface

$\bar{h}$ = Depth of the centre of gravity of the immersed surface from free surface of the liquid.

(*i*) **Centre of pressure** is defined as the point of application of the resultant pressure on the surface. The depth of centre of pressure (h^*) from free surface of the liquid is given by

$$h^* = \frac{I_G \sin^2 \theta}{A\bar{h}} + \bar{h} \quad \text{for inclined surface}$$

$$= \frac{I_G}{A\bar{h}} + \bar{h} \quad \text{for vertical surface}$$

The centre of pressure for a plane vertical surface lies at a depth of two-third the total height of the immersed surface from free surface.

(*ii*) **The total force on a curved surface** is given by $F = \sqrt{F_x^2 + F_y^2}$

where F_x = Horizontal force on a curved surface and is equal to total pressure force on the projected area of the curved surface on the vertical plane

and F_y = Vertical force on the curved surface and is equal to the weight of the liquid actually or virtually supported by the curved surface.

The inclination of the resultant force on curved surface with horizontal is given by $\tan \theta = \frac{F_y}{F_x}$.

(iii) **The resultant force on a sluice gate** is given by

$$F = F_1 - F_2$$

where F_1 = Pressure force on the upstream side of the sluice gate

F_2 = Pressure force on the downstream side of the sluice gate.

(iv) **Lock-gates.** For a lock-gate, the reaction between the two gates (P) is equal to the reaction at the hinge (R), *i.e.*, $R = P$ and the reaction between the two gates (P) is given by $P = \dfrac{F}{2 \sin \theta}$

where F = Resultant water pressure on the lock-gate $= F_1 - F_2$

and θ = Inclination of the gate with the normal to the side of the lock.

1.4. BUOYANCY AND FLOATATION

Buoyant force is the upward force or thrust exerted by a liquid on body when the body is immersed in the liquid. The point through which the buoyant force is supposed to act is called **centre of buoyancy.** It is denoted by B. The point, about which a floating body starts oscillating when the body is given a small angular displacement, is known as **Metacentre.** It is denoted by M. The distance between the meta-centre (M) and centre of gravity (G) of a floating body is known as **metacentric height**. This is denoted by GM and mathematically it is given as

$$GM = \frac{I}{V} - BG$$

where I = Moment of Inertia of the plan of the floating body at the water surface

V = Volume of the body submerged in water

BG = Distance between the centre of gravity (G) and centre of buoyancy (B).

(i) **Conditions of equilibrium** of a floating and submerged body are:

Equilibrium	*Floating body*	*Submerged body*
(i) Stable	M should be above G	B should be above G
(ii) Unstable	M should be below G	B should be below G
(iii) Neutral	M and G coincide	B and G coincide

(ii) **The metacentric height (GM)** experimentally is given by

$$GM = \frac{wx}{W \tan \theta}$$

where w = Movable weight

x = Distance through which w is moved

W = Weight of floating body including w

θ = Angle through which floating body is tilted

(iii) **The time period of oscillation** of a floating body is given by $T = 2\pi \sqrt{\dfrac{k^2}{GM \times g}}$

where k = Radius of gyration, GM = Metacentric height.

1.5. KINEMATICS OF FLUID

Kinematics is defined as that branch of science which deals with the study of fluid in motion without considering the forces causing the motion. The fluids flow may be compressible or incompressible; steady or unsteady; uniform or non-uniform; laminar or turbulent; rotational or irrotational; one, two or three dimensional.

(*i*) If the density (ρ) changes from point to point during fluid flow, it is known **compressible flow.** But if density (ρ) is constant during fluid flow, it is called **incompressible flow.** Mathematically,

$$\rho \neq \text{Constant for compressible flow}$$

$$\rho = \text{Constant for incompressible flow.}$$

(*ii*) If the fluid characteristic like velocity, pressure, density, etc. do not change at a point with respect to time, the fluid flow is known as **steady flow.** If these fluid characteristic change with respect to time, the fluid flow is known as unsteady flow. Mathematically,

$$\left(\frac{\partial v}{\partial t}\right) = 0, \left(\frac{\partial p}{\partial t}\right) = 0 \text{ or } \left(\frac{\partial \rho}{\partial t}\right) = 0 \text{ for steady flow, and}$$

$$\left(\frac{\partial v}{\partial t}\right) \neq 0, \left(\frac{\partial p}{\partial t}\right) \neq 0 \text{ or } \left(\frac{\partial \rho}{\partial t}\right) \neq 0 \text{ for unsteady flow.}$$

(*iii*) If the velocity in a fluid flow does not change with respect to the length of direction of flow, the flow is said **uniform** and if the velocity change it is known **non-uniform** flow. Mathematically,

$$\left(\frac{\partial v}{\partial s}\right) = 0 \text{ for uniform, and } \left(\frac{\partial v}{\partial s}\right) \neq 0 \text{ for non-uniform flow.}$$

(*iv*) If the Reynold number (R_e) in a pipe is less than 2000, the flow is said to be **laminar** and if the Reynold number is more than 4000, the flow is said to be **turbulent.**

$$\text{Reynolds number } (R_e) \text{ is given by } R_e = \frac{\rho VD}{\mu} \text{ or } \frac{VD}{\nu}$$

where V = Velocity of fluid, D = Dia. of pipe

μ = Viscosity of fluid, ν = Kinematic viscosity of fluid.

(*v*) If the fluid particles while flowing along stream lines also rotate about their own axis, that flow is known as **rotational flow** and if the fluid particles, while flowing along stream lines, do not rotate about their own axis, that type of flow is called **irrotational flow.**

(*vi*) The rate of discharge for incompressible fluid is given by

$$Q = A \times V$$

(*vii*) **Continuity equation** is written is general form as

$$\rho AV = \text{constant}$$

and in differential form as $\dfrac{\partial u}{\partial x} + \dfrac{\partial v}{\partial y} + \dfrac{\partial w}{\partial z} = 0$ for three-dimensional flow and $$\frac{\partial u}{\partial x} + \frac{\partial v}{\partial y} = 0 \quad \text{for two-dimensional flow}$$

(*viii*) The components of acceleration in x, y and z direction are

$$a_x = u\frac{\partial u}{\partial x} + v\frac{\partial v}{\partial y} + w\frac{\partial u}{\partial z} + \frac{\partial u}{\partial t}$$

$$a_y = u\frac{\partial v}{\partial x} + v\frac{\partial v}{\partial y} + w\frac{\partial v}{\partial z} + \frac{\partial v}{\partial t}$$

$$a_z = u\frac{\partial w}{\partial x} + v\frac{\partial w}{\partial y} + w\frac{\partial w}{\partial z} + \frac{\partial w}{\partial t}$$

(*ix*) **Local acceleration** is defined as the rate of change of velocity at a given point. In the above components of acceleration the expressions $\frac{\partial u}{\partial t}$, $\frac{\partial v}{\partial x}$ and $\frac{\partial w}{\partial t}$ are called local acceleration.

(*x*) **Convective acceleration** is defined as the rate of change of velocity due to change of position of fluid particles in a fluid flow.

(*xi*) **Velocity potential function** (ϕ) is defined as the scalar function of space and time such that its negative derivative with respect to any direction gives the fluid velocity in that direction. Hence the components of velocity in x, y and z direction in terms of velocity potential are

$$u = -\frac{\partial \phi}{\partial x},\ v = -\frac{\partial \phi}{\partial y} \text{ and } w = -\frac{\partial \phi}{\partial z}.$$

(*xii*) **Stream function** (ψ) is defined as the scalar function of space and time, such that its partial derivative with respect to any direction gives the velocity component at right angles to that direction. It is defined only for two-dimensional flow. The velocity components in x and y directions in terms of stream function are

$$u = -\frac{\partial \psi}{\partial y} \quad \text{and} \quad v = \frac{\partial \psi}{\partial x}.$$

(*xiii*) **Equipotential line** is a line along which the velocity potential (ϕ) is constant. A grid obtained by drawing a series of equipotential lines and stream lines is called a flow net.

(*xiv*) **Angular deformation** or shear deformation is defined as the average change in the angle contained by two adjacent sides. It is also called shear strain rate and is given by

$$\text{Shear strain rate} = \frac{1}{2}\left[\frac{\partial v}{\partial x} + \frac{\partial u}{\partial y}\right]$$

Rotational components of a fluid particle are given as

$$\omega_z = \frac{1}{2}\left[\frac{\partial v}{\partial x} - \frac{\partial u}{\partial y}\right],\ \omega_x = \frac{1}{2}\left[\frac{\partial w}{\partial y} - \frac{\partial v}{\partial z}\right],\ \omega_y = \frac{1}{2}\left[\frac{\partial u}{\partial z} - \frac{\partial w}{\partial x}\right]$$

Vorticity is equal to two times the value of rotation.

(*xv*) **Vortex flow** is defined as the flow of a fluid along a curved path. It is of two types namely (*i*) Forced vortex flow and (*ii*) Free vortex flow. If the fluid particles are moving round a curved path with the help of some external torque the flow is called **forced vortex flow.** And if no external torque is acquired to rotate the fluid particles, the flow is called **free-vortex flow.** The relation between tangential velocity and radius for vortex flow is given by

$$r = \omega \times r \qquad \text{for forced vortex}$$

$$v \times r = \text{constant} \qquad \text{for free vortex.}$$

The pressure variation along the radial direction for vortex flow along a horizontal plane,

$$\frac{\partial p}{\partial r} = \frac{\rho v^2}{r}$$

For forced vortex flow, $z = \dfrac{v^2}{2g}$

For free vortex flow the equation is $\dfrac{p_1}{\rho g} + \dfrac{v_1^2}{2g} + z_1 = \dfrac{p_2}{\rho g} + \dfrac{v_2^2}{2g} + z_2.$

1.6. DYNAMICS OF FLUID

Dynamics of fluid flow is defined that branch of science which deals with the study of fluids in motion considering the forces which cause the flow.

(*i*) **Euler's equation** of motion is obtained by considering forces due to pressure and gravity. **Navier-Strokes equations** are obtained by considering pressure force, gravity force and viscous force. **Reynold's equation of motion** are obtained by considering pressure force, gravity force, viscous force and force due to turbulence.

(*ii*) **Bernoulli's equation** is obtained by integrating the Euler's equation of motion. It states that, "For a steady, ideal flow of an incompressible fluid, the total energy which consists of pressure energy $\left(\dfrac{p}{\rho g}\right)$, kinetic energy $\left(\dfrac{v^2}{2g}\right)$ and datum energy (*z*) at any point of the fluid is constant."

Mathematically, it is written as

$$\frac{p}{\rho g} + \frac{v^2}{2g} + z = \text{constant}$$

or

$$\frac{p_1}{\rho g} + \frac{v_1^2}{2g} + z_1 = \frac{p_2}{\rho g} + \frac{v_2^2}{2g} + z_2$$

Bernoulli's equation for **real fluids** is written as

$$\frac{p_1}{\rho g} + \frac{v_1^2}{2g} + z_1 = \frac{p_2}{\rho g} + \frac{v_2^2}{2g} + z_2 + h_L$$

where h_L = Loss of energy between section 1 and 2.

(*iii*) **Applications of Bernoulli's equation are:**

(*a*) Venturimeter, (*b*) Orificemeter, and (*c*) Pitot-tube.

The discharge through a venturimeter is given by

$$Q = C_d \frac{A_1 A_2}{\sqrt{A_1^2 - A_2^2}} \times \sqrt{2gh}$$

where h = Difference of pressure head in terms of fluid head flowing through venturimeter for a horizontal venturimeter

C_d = Coefficient of venturimeter

A_1 = Area at the inlet of venturimeter

A_2 = Area at the throat of the venturimeter.

The value of 'h' is given by the differential U-tube manometer. For a horizontal venturimeter or inclined venturimeter

$$h = x\left[\frac{S_h}{S_f} - 1\right] \quad \text{for manometer with heavier liquid}$$

$$= x\left[1 - \frac{S_l}{S_f}\right] \quad \text{for manometer with lighter liquid}$$

where x = Difference in the readings of the differential manometer

S_h = Specific gravity of heavier liquid in manometer

S_f = Specific gravity of liquid flowing through venturimeter

S_l = Specific gravity of lighter liquid in manometer.

(*iv*) **Pitot tube** is used to find the velocity of a flowing fluid at any point in a pipe or a channel. The velocity is given by the relation,

$$V = C_v\sqrt{2gh}$$

where C_v = Coefficient of pitot tube,

and h = Difference of the pressure head.

(*v*) **Momentum equation** states that the net force acting on a fluid mass is equal to the change in momentum per second (or rate of change of momentum) in that direction. Mathematically, it is written as

$$F = \frac{d}{dt}(mv)$$

where mv = Momentum.

(*vi*) **The impulse-momentum equation** is given by

$$F \times dt = d\,(mv)$$

and it states that the impulse of a force (F) acting on a fluid mass (m) in a short interval of time (dt) is equal to the change of momentum $d(mv)$ in the direction of force.

(*vii*) **Force on a bend,** exerted by a flowing fluid in the direction of x and y, are given by

$$F_x = \rho Q[v_{1x} - v_{2x}] + (p_1A_1)_x + (p_2A_2)_x$$

$$F_y = \rho Q[v_{1y} - v_{2y}] + (p_1A_1)_y + (p_2A_2)_y$$

where v_{1x} = Initial velocity in the x-direction

v_{2x} = Final velocity in the x-direction

$(p_1A_1)_x$ = Initial pressure force in x-direction

$(p_2A_2)_x$ = Final pressure force in x-direction and so on

Resultant force on the bend is $F_R = \sqrt{F_x^2 + F_y^2}$.

1.7. ORIFICE AND MOUTHPIECE

Orifice is a small opening of any cross-section on the side or at the bottom of a tank, through which a fluid is flowing. **A mouthpiece** is a short length of a pipe which is two or three times its diameter in length, fitted in a tank or vessel containing the fluid.

Hydraulic Coefficients

(*i*) There are three *hydraulic coefficients* namely,

(*a*) Coefficient of velocity, C_v (*b*) Coefficient of contraction, C_c

(*c*) Coefficient of discharge, C_d.

(*ii*) The expression for coefficient of velocity in terms of x, y coordinates from vena-contracta is

$$C_v = \frac{x}{\sqrt{4yH}}$$

where H = Height of water from the centre of orifice.

(*iii*) Coefficient of discharge for different types of mouthpieces are

(*a*) $C_d = 0.855$ for external mouthpiece

(*b*) $C_d = 0.707$ for internal mouthpiece running full

(*c*) $C_d = 0.50$ for internal mouthpiece running free

(*d*) $C_d = 1.0$ for convergent or convergent divergent.

1.8. NOTCH AND WEIR

Notch is a device used for measuring the rate of flow of a liquid through a small channel. A **weir** is a concrete or masonry structure placed in the open channel over which the flow occurs.

(*i*) The **discharge** through the following notches or weirs is given by

$$Q = \tfrac{2}{3} C_d \times L \times H^{3/2} \quad \text{for rectangular notch or weir}$$

$$= \frac{8}{15} C_d \times \tan\frac{\theta}{2} \times \sqrt{2g} \times H^{5/2} \quad \text{for a triangle notch or weir.}$$

(*ii*) **The discharge** through a trapezoidal notch or weir is equal to the sum of discharge through a rectangular notch and the discharge through a triangular notch.

(*iii*) **The error** in discharge due to error in measurement of head over a notch is given by

$$\frac{dQ}{Q} = \frac{3}{2}\frac{dH}{H} \qquad \text{for a rectangular notch}$$

$$= \frac{5}{2}\frac{dH}{H} \qquad \text{for a triangular notch}$$

where Q = Discharge through notch, and H = Head over the notch.

An error of 1% in measuring H will produce 1.5% error in discharge over a rectangular notch and 2.5% error in discharge over a triangular notch.

(*iv*) **Velocity of approach** (V_a) is defined as the velocity with which the water approaches the notch or weir. This is given by

$$V_a = \frac{\text{Discharge over the notch}}{\text{Cross-sectional area of channel}}$$

The head due to velocity of approach is given by $h_a = \frac{V_a^2}{2g}$

Discharge over a rectangular weir, considering velocity of approach is given by

$$Q = \tfrac{2}{3} C_d L \sqrt{2g}\,[(H_1 + h_a)^{3/2} - h_a^{3/2}]$$

1.9. VISCOUS FLOW

Viscous flow is the flow for which Reynold number is less than 2000 or the fluid flows in layers.

(*i*) For the viscous flow through **circular pipes,** the shear stress distribution, velocity distribution, ratio of maximum velocity to average velocity and difference of pressure head are given by

(*a*) $\tau = -\frac{\partial p}{\partial x}\frac{r}{2}$ (Linear)

(*b*) $u = -\frac{1}{4\mu}\frac{\partial p}{\partial x}[R^2 - r^2] = U_{\max}\left[1 - \left(\frac{r}{R}\right)^2\right]$ (Parabolic)

(*c*) $\frac{U_{\max}}{\bar{u}} = 2.0$

(*d*) $h_f = \frac{32\mu\bar{u}L}{\rho g D^2}$

where $\frac{\partial p}{\partial x}$ = Pressure gradient, τ = Shear stress

r = Radius at any point, R = Radius of the pipe

$U_{\max}$ = Maximum velocity, h_f = Loss of pressure head

$\bar{u}$ = Average velocity, D = Diameter of pipe.

(*ii*) For viscous flow between **two parallel plates,** the shear stress distribution, velocity distribution, ratio of maximum velocity to average velocity and difference of pressure head are given by

(*a*) $\tau = -\dfrac{1}{2}\dfrac{\partial p}{\partial x}[t - 2y]$ (Linear) (*b*) $u = -\dfrac{1}{2\mu}\dfrac{\partial p}{\partial x}[ty - y^2]$ (Parabolic)

(*c*) $\dfrac{U_{max}}{\overline{u}} = 1.5$ (*d*) $h_f = \dfrac{12\mu\overline{u}L}{\rho g t^2}$.

where t = Distance between two plates,

y = Distance from the plates.

(*iii*) **Kinetic energy correction factor** (α) is defined as the ratio of kinetic energy per second based on actual velocity to kinetic energy per second based on average velocity. For a circular pipe through which viscous flow is taking place, $\alpha = 2.0$.

(*iv*) **Momentum correction factor** (β) is defined as the ratio of momentum of a fluid based on actual velocity to the momentum of the fluid based on average velocity. For a circular pipe, having viscous flow, $\beta = \dfrac{4}{3}$.

(*v*) The **coefficient of friction** (f) which is a function of Reynold number is given by

$$f = \frac{16}{R_e} \quad \text{for viscous flow or for } R_e < 2000$$

$$= \frac{0.079}{R_e^{1/4}} \quad \text{for } R_e \text{ varying from 4000 to } 10^5$$

$$= 0.0008 + \frac{0.05525}{R_e^{0.237}} \quad \text{for } R_e \geq 10^5 \text{ but } \leq 4 \times 10^7.$$

1.10. TURBULENT FLOW

Smooth and rough boundaries. If the average height (k) of the irregularities projecting from the surface of the boundary is small compared with the thickness of the laminar sub-layer (δ'), the boundary is known as **smooth.** But if k is large in comparison to δ', the boundary is known as **rough.** Mathematically,

$$\frac{k}{\delta'} < 0.25 \quad \text{for smooth boundary}$$

$$> 6.0 \quad \text{for rough boundary.}$$

And if $\dfrac{k}{\delta'}$ lies between 0.25 to 6.0, the boundary is in transition.

Darcy formula is given by $h_f = \dfrac{4fLV^2}{d \times 2g}$

where h_f = Head loss due to friction and is known as **Major,** Head loss.

Chezy's formula is given by $V = C\sqrt{m \times i}$

where C = Chezy's constant,

m = Hydraulic mean depth = $\frac{d}{4}$ for circular pipe (running full),

i = Loss of head per unit length = $\frac{h_f}{L}$.

1.11. FLOW THROUGH PIPES

(*i*) **Minor losses**

(*a*) Loss of head due to *sudden expansion* (h_e) is given by $h_e = \frac{(V_1 - V_2)^2}{2g}$.

(*b*) Loss of head due to *sudden contraction* (h_c) is given by

$$h_c = \left(\frac{1}{C_c} - 1\right)^2 \frac{V_2^2}{2g},$$ where C_c = Coefficient of contraction

(*c*) Loss of head at the inlet of a pipe, $h_i = 0.5\,\frac{V^2}{2g}$.

(*d*) Loss of head at the outlet of a pipe, $h_o = \frac{V^2}{2g}$.

(*ii*) **Hydraulic gradient and total energy lines.** The line representing the sum of pressure head and datum head with respect to some reference line is called hydraulic gradient line (H.G.L.) while the line representing the sum of pressure head, datum head and velocity head with respect to some reference line is known as total energy line (T.E.L.).

(*iii*) The **equivalent size** of the pipes connected in series is given by

$$\frac{L}{d^5} = \frac{L_1}{d_1^5} + \frac{L_2}{d_2^5} + \frac{L_3}{d_3^5} + \ldots\ldots$$

where L = Equivalent length of pipe = $L_1 + L_2 + L_3$

d = Equivalent size of the pipe

d_1, d_2, d_3 = Diameters of pipes connected in series.

(*iv*) For **parallel pipes,** the loss of head in each pipe is same and rate of flow in main pipe is equal to the sum of the rate of flow in each pipe, connected in parallel *i.e.,*

(*a*) $Q = Q_1 + Q_2 + \ldots\ldots$ (*b*) $h_{f_1} = h_{f_2} = \ldots\ldots$

(*v*) **Power transmitted** through a pipe is given by,

$$\text{H.P.} = \frac{w \times Q \times [H - h_f]}{75}$$

where H = Total head at the inlet of pipe

h_f = Head lost due to friction

Efficiency of power transmission through pipes, $\eta = \frac{[H - h_f]}{H}$

This efficiency will be maximum when $h_f = \frac{H}{3}$.

Diameter of **nozzle** for maximum power transmission through nozzle is

$$d = \left(\frac{D^5}{8fL}\right)^{1/4}$$

where d = Diameter of nozzle at outlet, D = Diameter of pipe

L = Length of pipe, f = Coefficient of friction.

(*vi*) **Water hammer.** When a liquid is flowing through a long pipe fitted with a valve at the end of the pipe and the valve is closed suddenly, a pressure wave of high intensity is produced behind the valve. This pressure wave of high intensity is having the effect of hammering action of the walls of the pipe. This phenomenon is known as water hammer. The intensity of pressure rise (p_i) due to water hammer is given by

$$p_i = \frac{\rho L V}{t} \quad \text{If valve is closed gradually}$$

$$= V\sqrt{K\rho} \quad \text{If valve is closed suddenly}$$

$$= V\sqrt{\frac{\rho}{\frac{1}{K} + \frac{D}{Et}}} \quad \text{If valve is closed suddenly and pipe is elastic.}$$

where L = Length of pipe, V = Velocity of flow

K = Bulk modulus of fluid, D = Diameter of pipe

t = Time for closing valve

E = Modulus of elasticity for pipe material

The valve of closure is said to be gradual if $t > \frac{2L}{C}$

The valve closure is said to be sudden if $t < \frac{2L}{C}$

where C = Velocity of pressure wave produced due to water hammer $= \sqrt{\frac{K}{\rho}}$.

1.12. DIMENSIONAL AND MODEL ANALYSIS

(*i*) **Hydraulic similarities.** There are three types of similarities that must exist between the model and prototype. They are: (*a*) Geometric similarity, (*b*) Kinematic similarity, and (*c*) Dynamic similarity.

Geometric similarity means the similarity of all linear dimensions of model and prototype. **Kinematic** similarity means the similarity of motion between model and prototype. **Dynamic similarity** means the similarity of forces between the model and prototype.

(*ii*) **Dimensionless parameters.** They are five dimensionless parameters, namely: (*a*) Reynold's number, (*b*) Froude number, (*c*) Euler number, (*d*) Weber numbers, and (*e*) Mach number.

(*a*) **Reynold's number** is the ratio of inertia force to viscous force and is given by

$$R_e = \frac{\rho V D}{\mu} \text{ or } \frac{V \times D}{\nu} \text{ for pipe flow.}$$

(*b*) **Froude number** is the ratio of square root of the inertia force to gravity force and is given by

$$F_e = \sqrt{\frac{F_i}{F_g}} = \frac{V}{\sqrt{Lg}}.$$

(*c*) **Euler's number** is the ratio of square root of inertia force to pressure force and is given by

$$E = \sqrt{\frac{F_e}{F_p}} = \frac{V}{\sqrt{p/\rho}}$$

(*d*) **Weber number** is the ratio of square root of inertia force to surface tension force and is given by

$$W = \sqrt{\frac{F_i}{F_s}} = \frac{W}{\sqrt{\sigma/L\rho}}$$

(*e*) **Mach number** (M) is the ratio of square root of inertia force to elastic force and is given by

$$M = \sqrt{\frac{F_i}{F_e}} = \frac{V}{\sqrt{K/\rho}} = \frac{V}{C}.$$

where C = Velocity of sound wave in air.

(*iii*) **Models** are of two types namely (*a*) Undistorted and (*b*) Distorted model. If the models are geometrically similar to its proto-type the models are known as **undistorted model.** And if the models are having different scale ratios for horizontal and vertical dimensions, the models are known as **distorted model.**

1.13. BOUNDARY LAYERS

(*i*) **Boundary layer.** When a solid body is immersed in a flowing fluid, there is a narrow region to the fluid in the neighbourhood of the solid body, where the velocity of the fluid varies from zero to free-stream velocity. This narrow region of fluid is called boundary layer.

(*ii*) Boundary layer may be laminar or turbulent. If the Reynold number of the flow defined as

$$R_e = \frac{U \times x}{\nu}$$

is less than 5×10^5, the boundary layer is called laminary boundary layer. And if R_e is more than 5×10^5, the boundary layer is called turbulent boundary layer.

In the above expression, U = Free stream velocity,

x = Distance from leading edge,

ν = Kinematic viscosity.

(*iii*) **Displacement thickness (δ^*)** is given by $\delta^* = \int_0^\delta \left(1 - \frac{u}{U}\right) dy$

(*iv*) **Momentum thickness (θ)** is given by, $\theta = \int_0^\delta \frac{u}{U}\left(1 - \frac{u}{U}\right) dy$

(*v*) **Energy thickness** $= \int_0^\delta \frac{u}{U}\left[1 - \frac{u^2}{U^2}\right] dy$

(*vi*) **Von Karman momentum** integral equation is given

$$\frac{\tau_0}{\rho U^2} = \frac{\partial \theta}{\partial x}, \text{ where } \theta = \int_0^\delta \frac{u}{U}\left(1 - \frac{u}{U}\right) dy$$

(*vii*) Velocity profile for turbulent boundary is given by $\frac{u}{U} = \left(\frac{y}{\delta}\right)^{1/7}$

where u = Velocity within boundary layer

U = Free stream velocity

δ = Boundary layer thickness.

1.14. DRAG AND LIFT FORCES

(*i*) **Drag and lift forces.** The force, exerted by a flowing fluid on a solid body in the direction of motion, is called drag force while the force perpendicular to the direction of motion is called lift force. Mathematically, they are given as

$$F_D = C_D A \frac{\rho U^2}{2}, F_L = C_L A \frac{\rho U^2}{2}$$

where F_D = Drag force, F_L = Lift force

A = Projected area of body, U = Free-stream velocity

C_D = Coefficient of drag, C_L = Coefficient of lift.

(*ii*) The drag force on a sphere for $R_e \angle 0.2$ is given by

$$F_D = 3\pi\mu DU.$$

1.15. COMPRESSIBLE FLOW

(*i*) Bernoulli's equation for **compressible fluid** is given by

$$\frac{p}{\rho g} \log_e p + \frac{V^2}{2g} + Z = \text{constant for iso-thermal process}$$

$$\frac{\gamma}{\gamma-1}\frac{p}{\rho}+\frac{V^2}{2g}+Z = \text{constant for adiabatic process.}$$

(*ii*) **Velocity of sound wave** is given by

$$C = \sqrt{\frac{dp}{d\rho}} = \sqrt{\frac{k}{\rho}} \qquad \text{In terms of Bulk modulus } k$$

$$= \sqrt{\frac{p}{\rho}} = \sqrt{RT} \qquad \text{For isothermal process}$$

$$= \sqrt{\frac{\gamma p}{\rho}} = \sqrt{\gamma RT} \qquad \text{For adiabatic process}$$

(*iii*) If $M < 1$ flow is called **sub-sonic flow**

$M > 1$ flow is called **super-sonic flow**

$M = 1$ flow is called **sonic flow.**

In sub-sonic flow, the disturbance always moves ahead of the projectile. In sonic flow, the disturbance moves along the projectile while in super-sonic flow that disturbance lags behind the projectile.

(*iv*) **The mach angle** (α) is given by $\sin \alpha = \frac{C}{V} = \frac{1}{M}$

(*v*) **Area velocity relationship** for compressible fluid is given as $\frac{dA}{A} = \frac{dV}{V}\,[M^2 - 1]$

For $M < 1$, the velocity decreases with increase of area.
For $M > 1$, the velocity decreases with the decrease of area.

(*vi*) For maximum flow through an orifice of nozzle fitted to the tank the pressure ratio $\frac{p_2}{p_1}$ $= n = 0.528$.

(*vii*) The **compressibility correction factor** is given by

$$\text{C.C.F.} = \left[1 + \frac{M_1^2}{4} + \frac{2-\gamma}{4} M_1^4 + \ldots\ldots\right].$$

1.16. CHANNEL FLOW

(*i*) **Channel flow.** Channel flow means the flow of a liquid through a passage at atmospheric pressure. The flow of liquid in a channel takes place under the force of gravity which means the flow takes place due to the slope of the bed of the channel. The flow in a channel is classified as:

(*a*) Steady and unsteady flow

(*b*) Uniform and non-uniform flow

(*c*) Laminar and turbulent flow

(*d*) Sub-critical, critical and super-critical flow.

For the *steady flow*, the velocity at a point in a channel with respect to time should be constant. And for *unsteady flow*, the velocity at a point with respect to time should be variable.

The flow in a channel will be *uniform* if the velocity with respect to direction of flow is constant. The velocity will be constant if area of flow is constant. Area of flow will be constant if depth of flow is constant. Hence the flow in a channel will be uniform if the depth of flow is constant. But if the depth of flow is variable then the flow in a channel is known as *non-uniform.*

The non-uniform flow is divided into gradually varied flow (G.V.F.) and rapidly varied flow (R.V.E.). If the depth of flow varies gradually, then the non-uniform flow is known as gradually varied flow. And if the depth of flow varies rapidly, then non-uniform flow is called rapidly varied flow.

The flow in a channel is said to be *laminar* if the Reynolds number (R_e) is less than 500 or 600. Reynolds number is case of open channel flow is given by

$$R_e = \frac{\rho VR}{\mu}$$

where ρ = Density of liquid, V = Velocity of liquid, μ = Viscosity

R = Hydraulic radius or hydraulic mean depth

$$= \frac{A}{P} = \frac{\text{Area of flow}}{\text{Wetted perimeter}}.$$

If the Reynolds number is more than 2000, then the flow is open channel is known as *turbulent.*

(*ii*) The flow in open channel is said to be **sub-critical** if the Froude Number (F_e) is less than 1.0.

The Froude number is given by, $F_e = \dfrac{V}{\sqrt{gD}}$

where V = Velocity of flow

D = Hydraulic depth of channel

$$= \frac{A}{T}$$

where A = Wetted area, T = Top width of the channel.

(*iii*) The flow in a channel is known as **critical flow** if the Froude number is equal to 1.0. But if the Froude number is more than 1.0, then flow in the channel is known as **super-critical flow.**

(*iv*) **The velocity through a channel** is given by

(*a*) Chezy's Formula

(*b*) Manning's Formula.

Chezy's Formula

The velocity by Chezy's formula is given by, $V = C\sqrt{mi}$

where C = Chezy's constant and depends upon the surface of the channel

m = Hydraulic mean depth

$$= \frac{A}{P} = \frac{\text{Area of flow}}{\text{Wetted perimeter}}$$

i = Slope of the bed of the channel.

Manning's Formula

The velocity through a channel according to Manning's Formula is given by, $V = \frac{1}{N} m^{2/3} i^{1/2}$

where N = Manning's constant and depends upon the surface of the channel

m = Hydraulic mean depth

i = Slope of the bed of the channel.

(*v*) **The relation between Chezy's constant** (C) and Manning's constant (N) is given by

$$C = \frac{1}{N} m^{1/6}.$$

The Chezy's constant is not a dimensionless coefficients. The dimension of C is given by $L^{1/2} T^{-1}$.

(*vi*) **Different types of channels**

The different types of channels are

(*a*) Rectangular channel

(*b*) Trapezoidal channel

(*c*) Circular channel

(*d*) Triangular channel.

(*vii*) **Values of A (area of flow) and P (Wetted perimeter) for different channels.**

(*a*) *Rectangular channel* [See Fig. 1.1 (*a*)]

Let b = Width of channel, d = Depth of flow

then $A = b \times d$

$P = b + 2d$

(*b*) *Trapezoidal channel* [See Fig. 1.1 (*b*)]

Let b = Width of flow, d = Depth of flow

n = Side slope which is expressed as 1 vertical to n horizontal

then $A = (b + nd) \times d$

$P = b + 2d\sqrt{n^2 + 1}$

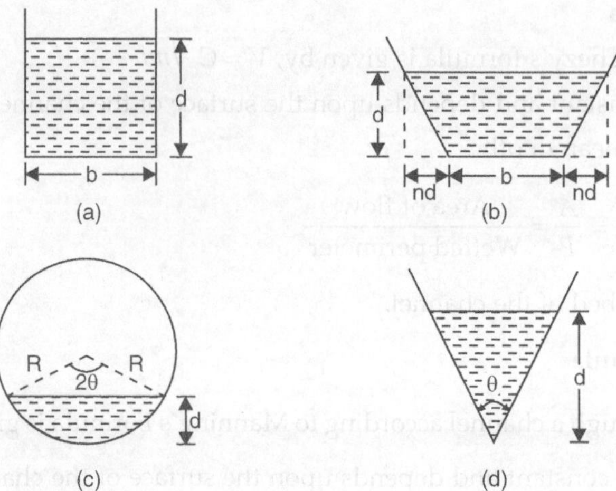

FIGURE 1.1

(*c*) *Circular channel* [Refer Fig. 1.1 (*c*)]

Let R = Radius of the channel, d = Depth of water in circular channel

2θ = Angle subtended at the centre of the channel by the free surface of water

then $$A = R^2\left(\theta - \frac{\sin 2\theta}{2}\right)$$

and $$P = 2R\theta$$

(*d*) *Triangular channel* [Refer Fig. 1.1 (*d*)]

Let d = Depth of flow, θ = Angle of triangular channel

then $$A = d^2 \tan\left(\frac{\theta}{2}\right)$$

and $$P = 2d \sec\left(\frac{\theta}{2}\right)$$

(*viii*) **Most efficient section of a channel.** The section of a channel is said to be most efficient if the discharge through the channel is maximum for a given area of flow, given surface resistance and given slope of the bed of the channel. The most efficient section is also known as most economical section of the channel.

(*a*) **A rectangular channel** will be most efficient, if

1. Depth of flow = Half of width of channel *i.e.*, $d = \frac{1}{2} \times b$

2. Hydraulic mean depth = Half of depth of flow *i.e.*, $m = \frac{1}{2} \times d$.

(*b*) **A trapezoidal channel** will be most efficient if

1. Length of sloping side = Half of top width

i.e., $$d\sqrt{n^2+1} = \frac{b+2nd}{2}.$$

2. Hydraulic mean depth = Half of depth of flow

i.e., $$m = \frac{d}{2}$$

3. The three sides of the trapezoidal section are the tangential to the semi-circle described on the water-line.

(*c*) **A circular channel** will be most efficient for maximum velocity, if

1. Depth of flow = 0.81 times the dia. of the circular channel.

2. Hydraulic mean depth = 0.3 times the dia. of the channel.

The flow through a circular channel will be maximum if

1. Depth of flow = 0.95 D.

(*ix*) **Specific energy.** The total energy of a flowing liquid in a channel with respect to the bed of channel is known as specific energy. It is given by

$$E = y + \frac{V^2}{2g}$$

where y = Depth of flow, V = Velocity of flow.

The curve which represents the variation of specific energy with the depth of flow is known as specific energy curve.

(*x*) **Critical depth.** The depth of flow of water at which the specific energy is minimum, is known as critical depth. For a rectangular channel, the critical depth (y_c) is given by

$$y_c = \left(\frac{q^2}{g}\right)^{1/3}$$

where q = Rate of flow per unit width of channel.

The velocity of flow, corresponding to critical depth is known as critical velocity.

(*xi*) **Hydraulic jump.** The sudden rise of water level which takes place due to the transformation of super-critical flow to the subcritical flow, is known as hydraulic jump. And when hydraulic jump takes place, a loss of energy due to eddy formation and turbulence occurs. The depth of water after hydraulic jump is given by

$$y_2 = -\frac{y_1}{2} + \sqrt{\left(\frac{y_1}{2}\right)^2 + \frac{2q^2}{gy_1}}$$

where y_2 = Depth of water after hydraulic jump

y_1 = Depth of water before hydraulic jump

q = Discharge per unit width.

The loss of energy (h_L) during hydraulic jump is given by, $h_L = \dfrac{(y_2 - y_1)^3}{4y_1y_2}$.

1.17. HYDRAULIC TURBINES

(*i*) **Hydraulic turbines.** The hydraulic machines which convert the hydraulic energy into the mechanical energy, are called turbines.

(*ii*) **The force exerted by a jet of water** on a stationary plate in the direction of jet is given by

$F_x = \rho A V^2$ for a vertical plate

$= \rho A V^2 \sin^2 \theta$ for an inclined plate

$= \rho A V^2 [1 + \cos \theta]$ for a curved plate

(*iii*) For force exerted by the jet of water having velocity V on a plate moving with a velocity u is given by

$F_x = \rho A (V - u)^2$ for a vertical plate

$= \rho A (V - u)^2 \sin^2 \theta$ for an inclined plate

$= \rho A (V - u)^2 [1 + \cos \theta]$ for a curved plate

(*iv*) **Efficiency of a series** of vanes is given as $\eta = \dfrac{2u(V - u)}{V^2}$

Efficiency will be maximum, when $u = \dfrac{V}{2}$. Maximum efficiency = 50%.

(*v*) **For a series of curved** radial vanes, the work done per unit weight per second

$$= \frac{1}{g}[V_{w_1 u_1} \pm V_{w_2 u_2}]$$

where V_{w_1}, V_{w_2} = Velocity of whirl at inlet and outlet.

(*vi*) **The net head** on the turbine is given by $H = H_g - h_f$

where H = Net head, H_g = Gross head, h_f = Head loss due to friction.

(*vii*) **The efficiencies** of a turbine are: (*a*) Hydraulic efficiency (η_h), (*b*) Mechanical efficiency (η_m), and (*c*) Overall efficiency (η_0).

(*a*) Hydraulic efficiency (η_h) is given by

$$\eta_h = \frac{\text{Power given by water of the runner}}{\text{Power supplied at inlet}} = \frac{\text{R.P.}}{\text{W.P.}}$$

$$= \rho \times Q \frac{[V_{w_1 u_1} \pm V_{w_2 u_2}]}{75} \Bigg/ \left(\frac{\rho \times Q \times g \times H}{75}\right) = \frac{(V_{w_1 u_1} \pm V_{w_2 u_2})}{gH}$$

(*b*) Mechanical efficiency (η_m) is given by $\eta_n = \dfrac{\text{S.P.}}{\text{R.P.}}$

(*c*) Overall efficiency is given by $\eta_0 = \dfrac{\text{S.P.}}{\text{W.P.}} = \eta_n \times \eta_h$.

(*viii*) **Impulse turbine.** If at the inlet of a turbine, total energy is only kinetic energy, the turbine is called impulse turbine. Pelton wheel is an impulse turbine.

(*ix*) **Reaction turbine.** If at the inlet of a turbine, the total energy is kinetic energy as well as pressure energy, the turbine is called reaction turbine. Francis and Kaplan turbines are reaction turbines.

(*x*) **Jet ratio (m)** is defined as the ratio of diameter (D) of Pelton wheel to the diameter (d) of the jet

or Jet ratio $m = \frac{D}{d}$

(*xi*) Pelton wheel is a tangential flow impulse turbine, Francis is an inward flow reaction turbine and Kaplan is an axial flow reaction turbine. The rate of flow (Q) through the turbine is given by

$$Q = \frac{\pi}{4} d^2 \times \sqrt{2gH} \quad \text{for Pelton Turbine}$$

$$= \pi D_1 B_1 V_{f1} \quad \text{for Francis Turbine}$$

$$= \frac{\pi}{4}[D_1^2 - D_0^2] \times V_{f1} \quad \text{for Kaplan Turbine}$$

where H = Net head, V_{f1} = Velocity of flow at inlet

D_0 = Dia. of Kaplan turbine, D_1 = Hub diameter.

(*xii*) **Draft-tube.** It is a pipe of gradually increasing area used for discharging water from the exit of a reaction turbine.

(*xiii*) **Specific speed** of a turbine is defined as the speed at which a turbine runs when it is working under a unit head and develops unit (*i.e.*, 1 kW power). It is given by $N_s = \frac{N\sqrt{P}}{H^{5/4}}$, where P = Shaft power, H = Net head on turbine.

(*a*) **Unit speed (N_u)** is the speed of a turbine, when the head on the turbine is one metre. It is given by $N_u = \frac{N}{\sqrt{H}}$.

(*b*) **Unit discharge (Q_u)** is the discharge through a turbine when the head (H) on the turbine is unity. It is given by $Q_u = \frac{Q}{\sqrt{H}}$.

(*c*) **Unit power (P_u)** is the power developed by a turbine when the head on the turbine is unity. It is given by $P_u = \frac{P}{H^{3/2}}$.

(*xiv*) **Characteristic curves.** The following three are the important characteristic curves of a turbine:

(*a*) Main characteristic curve, (*b*) Operating characteristic curves, and (*c*) Constant efficiency curves.

(*xv*) **Governing of a turbine** is defined as the operation by which the speed of the turbine is kept constant under all conditions of working.

1.18. CENTRIFUGAL PUMPS

(*i*) The hydraulic machine which converts the mechanical energy into pressure energy by means of centrifugal force is called **centrifugal pump**. The work done by impeller on water per second per unit weight of water

$$= \frac{1}{g} V_{w_2} \times u_2$$

where V_{w_2} = Velocity of whirl at outlet, u_2 = Tangential velocity of wheel at outlet.

(*ii*) **The manometric head (H_m)** is the head against which a centrifugal pump has to work. It is given as

$$H_m = \frac{V_{w_2} \times u_2}{g} - \text{Loss of head in impeller and casing}$$

$$= \frac{V_{w_2} \times u_2}{g} \quad \text{If losses in pump are zero}$$

$$= \text{Total head at outlet} - \text{Total head at inlet}$$

$$= \left(\frac{p_0}{w} + \frac{V_0^2}{2g} + Z_0\right) - \left(\frac{p_i}{w} + \frac{V_i^2}{w} + Z_i\right)$$

$$= h_s + h_d + h_{fd} + \frac{V_d^2}{2g}.$$

(*iii*) **The efficiencies of a pump** are: (*a*) Manometric efficiency (η_{man}), (*b*) Mechanical (η_m), and (*c*) overall efficiency (η_0). They are expressed as

$$\eta_{man} = \frac{gH_m}{V_{w_2} \times u_2}$$

$$\eta_m = \frac{\text{I.P.}}{\text{S.P.}} = \left(\rho \times Q \times \frac{V_{w_2} \times u_2}{75}\right) \Big/ \text{S.P.}$$

$$\eta_0 = \eta_{man} \times \eta_m$$

where I.P. = Power on impeller, S.P. = Shaft power.

(*iv*) **Multistage centrifugal** pumps are used to produce a high head or to discharge a large quantity of water. The produce a high head, the impellers are connected in series while to discharge a large quantity of liquid, the impellers are connected in parallel.

(*v*) **Specific speed of a pump** is defined as the speed at which a pump runs when the head developed is one metre and discharge is one cubic metre. It is given as $N_s = \dfrac{N\sqrt{Q}}{H_m^{3/4}}$, where H_m = Manometric head.

(*vi*) **Cavitation** is defined as the phenomenon of formation of vapour bubbles and sudden collapsing of the vapour bubbles.

1.19. RECIPROCATING PUMP

(*i*) The discharge through a reciprocating pump per second is given by

$$Q = \frac{ALN}{60} \quad \text{for a single acting}$$

$$= \frac{2ALN}{60} \quad \text{for a double acting}$$

where A = Area of position, L = Length of stroke.

(*ii*) **Work done by reciprocating** pump per second

$$= \frac{wALN}{60}(h_s + h_d) \quad \text{for a single acting}$$

$$= \frac{2wALN}{60}(h_s + h_d) \quad \text{for a double acting.}$$

(*iii*) **The pressure head due** to acceleration is given by

$$h_a = \frac{l}{g} \times \frac{A}{a} \times \omega^2 r \cos\theta$$

where l = Length of suction or delivery pipe

a = Area of suction or delivery pipe

r = Radius of crank = $L/2$

ω = Angular speed = $\dfrac{2\pi N}{60}$.

(*iv*) **Indicator diagram** is a graph between the pressure head in the cylinder and the distance travelled by the piston from inner dead centre for one complete revolution of the crank. Work done by the pump is proportional to the area of indicator diagram. Area of ideal indicator diagram is the same as the area of indicator diagram due to acceleration in suction and delivery pipes.

(*v*) **Air vessel** is a device used: (*a*) to obtain a continuous supply of water at uniform rate, (*b*) to save a considerable amount of work, and (*c*) to run the reciprocating pump at a high speed without separation.

The mean velocity ($\overline{V}$) for a single acting pump is $\overline{V} = \dfrac{A}{a}\dfrac{\omega r}{\pi}$.

The work saved by fitting air vessels in a single acting reciprocating pump is 84.8% while in a double acting, the work saved is 39.2%.

1.20. MISCELLANEOUS HYDRAULIC DEVICES

The miscellaneous hydraulic devices are hydraulic press, hydraulic accumulator, hydraulic intensifier, hydraulic ram, hydraulic lift, hydraulic cranes, hydraulic coupling and hydraulic torque converter.

(*i*) **Hydraulic press** is a device used for lifting heavy weights by the application of a much smaller force. **Hydraulic accumulator** is a device used for storing the energy of a fluid in the form of pressure energy. **Hydraulic intensifier** is a device used for increasing the pressure intensity of a liquid. **Hydraulic ram** is a pump which raises water without any external power (such as electricity) for its operation. **Hydraulic lift** is a device used for carrying persons or goods from one floor to another floor in a multi-storeyed building. **Hydraulic crane** is a device used for raising or transferring heavy weights. **Hydraulic coupling** is a device, in which power is transmitted from driving shaft to driven shaft without any change of torque while **torque convertor** is a device in which arrangement is provided for getting increased or decreased torque at the driven shaft.

(*ii*) **Capacity of a hydraulic accumulator** = $p \times A \times L$

where p = Liquid pressure supplied by pump

A = Area of sliding ram, and L = Stroke of the ram.

(*iii*) **Hydraulic ram** has two efficiencies namely D' Aubuisson's efficiency and Rankine efficiency. They are given by

$$\text{D' Aubuisson's, } \eta = \frac{wH}{Wh} \quad \text{and Rankine's, } \eta = \frac{w(H-h)}{(W-w)\times h}$$

where w = Weight of water raised/sec, W = Weight of water supplied raised/sec,

h = Height of water in supply tank, and H = Height of water raised.

II. OBJECTIVE TYPE QUESTIONS

Tick mark the most appropriate statement of the multiple choice answers:

Fluid Properties

1. An ideal fluid is defined as the fluid which

(*a*) is compressible ☐

(*b*) is incompressible ☐

(*c*) is incompressible and non-viscous (inviscid) ☐

(*d*) has negligible surface tension. ☐

2. Newton's law of viscosity states that

(*a*) shear stress is directly proportional to the velocity ☐

(*b*) shear stress is directly proportional to velocity gradient ☐

(*c*) shear stress is directly proportional to shear strain ☐

(*d*) shear stress is directly proportional to the viscosity. ☐

3. A Newtonian fluid is defined as the fluid which

(*a*) is incompressible and non-viscous ☐ (*b*) obeys Newton's law of viscosity ☐

(*c*) is highly viscous ☐ (*d*) is compressible and non-viscous. ☐

4. Kinematic viscosity is defined as equal to

(*a*) dynamic viscosity × density ☐ (*b*) dynamic viscosity/density ☐

(*c*) dynamic viscosity × pressure ☐ (*d*) pressure × density. ☐

5. Dynamic viscosity (μ) has the dimensions as

(*a*) MLT^{-2} ☐ (*b*) $ML^{-1}T^{-1}$ ☐

(*c*) $ML^{-1}T^{-2}$ ☐ (*d*) $M^{-1}L^{-1}T^{-1}$. ☐

6. Poise is the unit of

(*a*) mass density ☐ (*b*) kinematic viscosity ☐

(*c*) viscosity ☐ (*d*) velocity gradient. ☐

7. The increase of temperature

(*a*) increases the viscosity of a liquid ☐ (*b*) decreases the viscosity of a liquid ☐

(*c*) decreases the viscosity of a gas ☐ (*d*) increases the viscosity of a gas. ☐

8. Stoke is the unit of

(*a*) surface tension ☐ (*b*) viscosity ☐

(*c*) kinematic viscosity ☐ (*d*) none of the above. ☐

9. The multiplying factor for converting one poise into MKS unit of dynamic viscosity is

(*a*) 9.81 ☐ (*b*) 98.1 ☐

(*c*) 981 ☐ (*d*) 0.981. ☐

10. Surface tension has the units of

(*a*) force per unit area ☐ (*b*) force per unit length ☐

(*c*) force per unit volume ☐ (*d*) none of the above. ☐

11. The gases are considered incompressible when Mach number is

(*a*) equal to 1.0 ☐ (*b*) equal to 0.50 ☐

(*c*) more than 0.3 ☐ (*d*) less than 0.2. ☐

12. Kinematic viscosity (ν) is equal to

(*a*) $\mu \times \rho$ ☐ (*b*) $\frac{\mu}{\rho}$ ☐

(*c*) $\frac{\rho}{\mu}$ ☐ (*d*) none of the above. ☐

13. Compressibility is equal to

(*a*) $\frac{-\left(\frac{dV}{V}\right)}{dp}$ ☐ (*b*) $\frac{dp}{-\left(\frac{dV}{V}\right)}$ ☐

(*c*) $\frac{dp}{d\rho}$ ☐ (*d*) $\sqrt{\frac{dp}{d\rho}}$. ☐

14. Hydrostatic law of pressure is given as

(a) $\frac{\partial p}{\partial z} = \rho g$ ☐ (b) $\frac{\partial p}{\partial z} = 0$ ☐

(c) $\frac{\partial p}{\partial z} = z$ ☐ (d) $\frac{\partial p}{\partial z}$ = constant. ☐

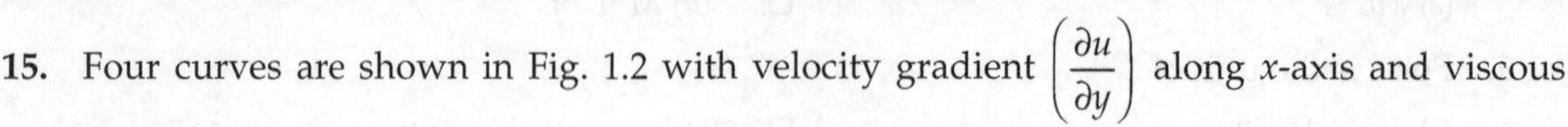

15. Four curves are shown in Fig. 1.2 with velocity gradient $\left(\frac{\partial u}{\partial y}\right)$ along x-axis and viscous shear stress (τ) along y-axis. Curve A corresponds to

(a) ideal fluid ☐

(b) newtonian fluid ☐

(c) non-newtonian fluid ☐

(d) ideal solid. ☐

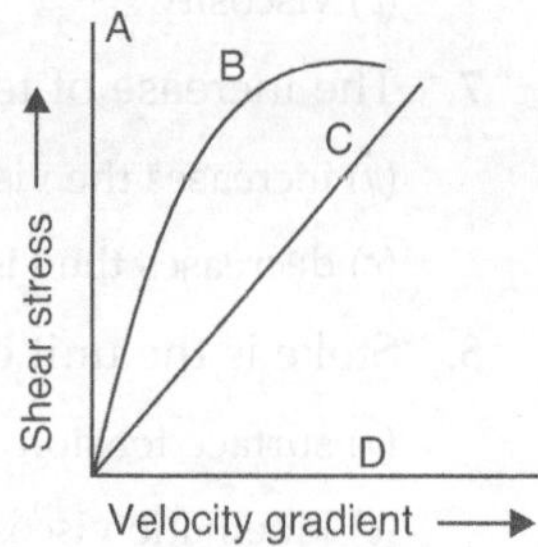

FIGURE 1.2

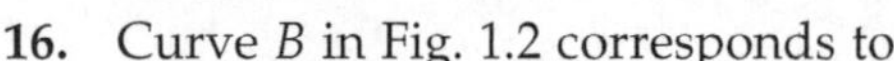

16. Curve B in Fig. 1.2 corresponds to

(a) ideal fluid ☐

(b) newtonian fluid ☐

(c) non-newtonian fluid ☐

(d) idealsolid. ☐

17. Curve C in Fig. 1.2 corresponds to

(a) ideal fluid ☐ (b) newtonian fluid ☐

(c) non-newtonian fluid ☐ (d) ideal solid. ☐

18. Curve D in Fig. 1.2 corresponds to

(a) ideal fluid ☐ (b) newtonian fluid ☐

(c) non-newtonian fluid ☐ (d) ideal solid. ☐

19. The relation between surface tension (σ) and difference of pressure (Δp) between the inside and outside of a liquid droplet is given as

(a) $\Delta p = \frac{\sigma}{4d}$ ☐ (b) $\Delta p = \frac{\sigma}{2d}$ ☐

(c) $\Delta p = \frac{4\sigma}{d}$ ☐ (d) $\Delta p = \frac{\sigma}{d}$. ☐

20. For a soap bubble, the surface tension (σ) and difference of pressure (Δp) are related as

(a) $\Delta p = \frac{\sigma}{4d}$ ☐ (b) $\Delta p = \frac{\sigma}{2d}$ ☐

(c) $\Delta p = \frac{4\sigma}{d}$ ☐ (d) $\Delta p = \frac{8\sigma}{d}$. ☐

21. For a liquid jet, the surface tension (σ) and difference of pressure (Δp) are related as

(a) $\Delta p = \frac{\sigma}{4d}$ ☐ (b) $\Delta p = \frac{\sigma}{2d}$ ☐

(c) $\Delta p = \frac{4\sigma}{d}$ ☐ (d) $\Delta p = \frac{2\sigma}{d}$. ☐

22. The capillary rise or fall of a liquid is given by

(a) $h = \dfrac{\sigma \cos \theta}{4\rho g d}$ ☐ (b) $h = \dfrac{4\sigma \cos \theta}{\rho g d}$ ☐

(c) $h = \dfrac{8\sigma \cos \theta}{\rho g d}$ ☐ (d) none of the above. ☐

Pressure and Hydrostatic Forces on Surfaces

23. Pascal's law states that pressure at a point is equal in all direction in

(a) a liquid at rest ☐ (b) a fluid at rest ☐

(c) a laminar flow ☐ (d) a turbulent flow. ☐

24. The hydrostatic law states that rate of increase of pressure in a vertical direction

(a) is equal to density of the fluid ☐ (b) is equal to specific weight of the fluid ☐

(c) is equal to weight of the fluid ☐ (d) none of the above. ☐

25. Fluid statics deals with the following forces

(a) viscous and gravity forces ☐ (b) viscous and gravity forces ☐

(c) gravity and pressure forces ☐ (d) surface tension and gravity forces. ☐

26. Gauge pressure at a point is equal to

(a) absolute pressure plus atmospheric pressure ☐

(b) absolute pressure minus atmospheric pressure ☐

(c) vacuum pressure plus absolute pressure ☐

(d) none of the above. ☐

27. Atmospheric pressure head in terms of water column is

(a) 7.5 m ☐ (b) 8.5 m ☐

(c) 9.81 m ☐ (d) 10.30 m. ☐

28. The hydrostatic pressure on a plane surface is equal to

(a) $\rho g A\bar{h}$ ☐ (b) $\rho g A\bar{h} \sin^2 \theta$ ☐

(c) $\frac{1}{4}\rho g A\bar{h}$ ☐ (d) $\rho g A\bar{h} \sin \theta$ ☐

where A = Area of plane surface and $\bar{h}$ = Depth of centroid of the plane area below the liquid free surface.

29. Centre of pressure of a plane surface immersed in a liquid is

(a) above the centre of gravity of the plane surface ☐

(b) at the centre of gravity of the plane surface ☐

(c) below the centre of gravity of the plane surface ☐

(d) none of the above. ☐

30. The resultant hydrostatic force acts through a point known as

(a) centre of gravity ☐ (b) centre of buoyancy ☐

(c) centre of pressure ☐ (d) none of the above. ☐

31. For submerged curved surface, the vertical component of the hydrostatic force is

(*a*) mass of the liquid supported by the curved surface ☐
(*b*) weight of the liquid supported by the curved surface ☐
(*c*) the force of the projected area of the curved surface on vertical plane ☐
(*d*) none of the above. ☐

32. Manometer is a device used for measuring

(*a*) velocity at a point in a fluid ☐ (*b*) pressure at a point in a fluid ☐
(*c*) discharge of a fluid ☐ (*d*) none of the above. ☐

33. Differential manometers are used for measuring

(*a*) velocity at a point in a fluid ☐
(*b*) pressure at a point in a fluid ☐
(*c*) difference of pressure between two points ☐
(*d*) none of the above. ☐

34. The pressure at a height Z in a static compressible fluid undergoing isothermal compression is given as

(*a*) $p = p_0\, e^{-\frac{gR}{ZT}}$ ☐ (*b*) $p = p_0\, e^{-\frac{gT}{RZ}}$ ☐
(*c*) $p = p_0 e^{-\frac{RT}{gZ}}$ ☐ (*d*) $p = p_0 e^{-\frac{gZ}{RT}}$ ☐

where p_0 = Pressure at ground level, R = Gas constant, T = Absolute temperature.

35. The pressure at a height Z in a static compressible fluid undergoing adiabatic compression is given by

(*a*) $p = p_0\left[1-\frac{\gamma-1}{\gamma}\frac{RT_0}{gZ}\right]^{\frac{\gamma}{\gamma-1}}$ ☐ (*b*) $p = p_0\left[1-\frac{\gamma}{\gamma-1}\frac{RT_0}{gZ}\right]^{\frac{\gamma}{\gamma-1}}$ ☐
(*c*) $p = p_0\left[1-\frac{\gamma-1}{\gamma}\frac{gZ}{RT_0}\right]^{\frac{\gamma}{\gamma-1}}$ ☐ (*d*) none of the above. ☐

36. The temperature at a height Z in a static compressible fluid undergoing adiabatic compression is given as

(*a*) $T = T_0\left[1-\frac{\gamma-1}{\gamma}\frac{RT_0}{gZ}\right]$ ☐ (*b*) $T = T_0\left[1-\frac{\gamma-1}{\gamma}\frac{gZ}{RT_0}\right]$ ☐
(*c*) $T = T_0\left[1-\frac{\gamma}{\gamma-1}\frac{RT_0}{gZ}\right]$ ☐ (*d*) none of the above. ☐

37. Temperature lapse-rate is given by

(*a*) $L = -\frac{R}{g}\left[\frac{\gamma-1}{\gamma}\right]$ ☐ (*b*) $L = -\frac{R}{g}\left[\frac{\gamma}{\gamma-1}\right]$ ☐
(*c*) $L = -\frac{g}{R}\left[\frac{\gamma-1}{\gamma}\right]$ ☐ (*d*) none of the above. ☐

38. When the fluid is at rest, the shear stress is

(a) maximum ☐ (b) zero ☐

(c) unpredictable ☐ (d) none of the above. ☐

39. The depth of centre of pressure of an inclined immersed surface from free surface of liquid is equal to

(a) $\frac{I_G}{A\bar{h}} + \bar{h}$ ☐ (b) $\frac{I_G A \sin^2 \theta}{\bar{h}} + \bar{h}$ ☐

(c) $\frac{I_G \sin^2 \theta}{A\bar{h}} + \bar{h}$ ☐ (d) $\frac{I_G \bar{h}}{A \sin^2 \theta} + \bar{h}$. ☐

40. The depth of centre of pressure of a vertical immersed surface from free surface of liquid is equal to

(a) $\frac{I_G}{A\bar{h}} + \bar{h}$ ☐ (b) $\frac{I_G A}{\bar{h}} + \bar{h}$ ☐

(c) $\frac{I_G \bar{h}}{A} + \bar{h}$ ☐ (d) $\frac{A\bar{h}}{I_G} + \bar{h}$. ☐

41. The centre of pressure for a plane vertical surface lies at a depth of

(a) half the height of the immersed surface ☐

(b) one-third the height of the immersed surface ☐

(c) two-third the height of the immersed surface ☐

(d) none of the above. ☐

42. The inlet length of a venturimeter

(a) is equal to the outlet length ☐ (b) is more than the outlet length ☐

(c) is less than the outlet length ☐ (d) none of the above. ☐

43. Flow of a fluid in a pipe takes place from

(a) higher level to lower level ☐ (b) higher pressure to lower pressure ☐

(c) higher energy to lower energy ☐ (d) none of the above. ☐

Buoyancy and Floatation

44. For a floating body, the buoyant force passes through the

(a) centre of gravity of the body ☐

(b) centre of gravity of the submerged part of the body ☐

(c) metacentre of the body ☐

(d) centroid of the liquid displaced by the body. ☐

45. The condition of stable equilibrium for a floating body is

(a) the metacentre M coincides with the centre of gravity G ☐

(b) the metacentre M is below centre of gravity G ☐

(c) the metacentre M is above centre of gravity G ☐

(d) the centre of buoyancy B is above centre of gravity G. ☐

46. A submerged body will be in stable equilibrium if the
(a) centre of buoyancy B is below the centre of gravity G ☐
(b) centre of buoyancy B coincides with G ☐
(c) centre of buoyancy B is above the metacentre M ☐
(d) centre of buoyancy B is above G. ☐

47. The metacentric height of a floating body is
(a) the distance between metacentre and centre of buoyancy ☐
(b) the distance between the centre of buoyancy and centre of gravity ☐
(c) the distance between metacentre and centre of gravity ☐
(d) none of the above. ☐

48. The point, through which the buoyant force is acting, is called
(a) centre of pressure ☐ (b) centre of gravity ☐
(c) centre of buoyancy ☐ (d) none of the above. ☐

49. The point, through which the weight is acting, is called
(a) centre of pressure ☐ (b) centre of gravity ☐
(c) centre of buoyancy ☐ (d) none of the above. ☐

50. The point, about which a floating body, starts oscillating when the body is tilted is called
(a) centre of pressure ☐ (b) centre of buoyancy ☐
(c) centre of gravity ☐ (d) metacentre. ☐

51. The metacentric height (GM) is given by
(a) $GM = BG - \frac{I}{V}$ ☐ (b) $GM = \frac{V}{I} - BG$ ☐
(c) $GM = \frac{I}{V} - BG$ ☐ (d) none of the above. ☐

52. For floating body, if the metacentre is above the centre of gravity, the equilibrium is called
(a) stable ☐ (b) unstable ☐
(c) neutral ☐ (d) none of the above. ☐

53. For a floating body, if the metacentre is below the centre of gravity, the equilibrium is called
(a) stable ☐ (b) unstable ☐
(c) neutral ☐ (d) none of the above. ☐

54. For a floating body, if the metacentre coincides with the centre of gravity, the equilibrium is called
(a) stable ☐ (b) unstable ☐
(c) neutral ☐ (d) none of the above. ☐

55. For a floating body, if centre of buoyancy is above the centre of gravity, the equilibrium is called
(a) stable ☐ (b) unstable ☐
(c) neutral ☐ (d) none of the above. ☐

56. For a submerged body, if the centre of buoyancy is above the centre of gravity, the equilibrium is called

(a) stable ☐ (b) unstable ☐
(c) neutral ☐ (d) none of the above. ☐

57. For a submerged body, if the centre of buoyancy is below the centre of gravity, the equilibrium is called

(a) stable ☐ (b) unstable ☐
(c) neutral ☐ (d) none of the above. ☐

58. For a submerged body, if the centre of buoyancy coincides with the centre of gravity, the equilibrium is called

(a) stable ☐ (b) unstable ☐
(c) neutral ☐ (d) none of the above. ☐

59. For a submerged body, if the metacentre is below the centre of gravity, the equilibrium is called

(a) stable ☐ (b) unstable ☐
(c) neutral ☐ (d) none of the above. ☐

60. The metacentric height (GM) experimentally is given as

(a) $GM = \dfrac{W \tan \theta}{wx}$ ☐ (b) $GM = \dfrac{w \tan \theta}{W \times x}$ ☐

(c) $GM = \dfrac{wx}{W \tan \theta}$ ☐ (d) $GM = \dfrac{Wx}{w \tan \theta}$ ☐

where w = Movable weight, W = Weight of floating body including w, θ = Angle of tilt.

61. The time period of oscillation of a floating body is given by

(a) $T = 2\pi \sqrt{\dfrac{GM \times g}{k^2}}$ ☐ (b) $T = 2\pi \sqrt{\dfrac{k^2}{GM \times g}}$ ☐

(c) $T = 2\pi \sqrt{\dfrac{GM}{gk^2}}$ ☐ (d) $T = 2\pi \sqrt{\dfrac{gk^2}{GM}}$ ☐

where k = Radius of gyration, GM = Metacentric height and T = Time period.

Kinematics and Dynamics of Flow

62. The necessary condition for the flow to be steady is that

(a) the velocity does not change from place to place ☐
(b) the velocity is constant at a point with respect to time ☐
(c) the velocity changes at a point with respect to time ☐
(d) none of the above. ☐

63. The necessary condition for the flow to be uniform is that

(a) the velocity is constant at a point with respect to time ☐
(b) the velocity is constant in the flow field with respect to space ☐

(c) the velocity changes at a point with respect to time ☐
(d) none of the above. ☐

64. The flow in the pipe is laminar if
(a) Reynold number is equal to 2500 ☐ (b) Reynold number is equal to 4000 ☐
(c) Reynold number is more than 2500 ☐ (d) None of the above. ☐

65. A stream line is a line
(a) which is along the path of a particle ☐
(b) which is always parallel to the main direction of flow ☐
(c) across which there is no flow ☐
(d) on which tangent drawn at any point gives the direction of velocity. ☐

66. Continuity equation can take the form
(a) $A_1V_1 = A_2V_2$ ☐ (b) $\rho_1A_1 = \rho_2A_2$ ☐
(c) $\rho_1A_1V_1 = \rho_2A_2V_2$ ☐ (d) $p_1A_1V_1 = p_2A_2V_2$. ☐

67. Pitot-tube is used for measurement of
(a) pressure ☐ (b) flow ☐
(c) velocity at a point ☐ (d) discharge. ☐

68. Bernoulli's theorem deals with the law of conservation of
(a) mass ☐ (b) momentum ☐
(c) energy ☐ (d) none of the above. ☐

69. Continuity equation deals with the law of conservation of
(a) mass ☐ (b) momentum ☐
(c) energy ☐ (d) none of the above. ☐

70. Irrotational flow means
(a) the fluid does not rotate while moving ☐
(b) the fluid moves in straight lines ☐
(c) the net rotation of fluid-particles about their mass centres is zero ☐
(d) none of the above. ☐

71. The velocity components in x and y-directions in terms of velocity potential (ϕ) are
(a) $u = -\dfrac{\partial\phi}{\partial x}, v = \dfrac{\partial\phi}{\partial y}$ ☐ (b) $u = \dfrac{\partial\phi}{\partial y}, v = \dfrac{\partial\phi}{\partial x}$ ☐
(c) $u = -\dfrac{\partial\phi}{\partial y}, v = -\dfrac{\partial\phi}{\partial x}$ ☐ (d) $u = -\dfrac{\partial\phi}{\partial x}, v = -\dfrac{\partial\phi}{\partial y}$. ☐

72. The velocity components in x and y-directions in terms of stream function (ψ) are
(a) $u = \dfrac{\partial\psi}{\partial x}, v = \dfrac{\partial\psi}{\partial y}$ ☐ (b) $u = -\dfrac{\partial\psi}{\partial x}, v = \dfrac{\partial\psi}{\partial y}$ ☐
(c) $u = \dfrac{\partial\psi}{\partial y}, v = \dfrac{\partial\psi}{\partial x}$ ☐ (d) $u = -\dfrac{\partial\psi}{\partial y}, v = \dfrac{\partial\psi}{\partial x}$. ☐

73. The relation between tangential velocity (v) and radius (r) is given by

(a) $V \times r$ = constant for forced vortex ☐ (b) V/r = constant for forced vortex ☐

(c) $V \times r$ = constant for free vortex ☐ (d) V/r = constant for free vortex. ☐

74. The pressure variation along the radial direction for vortex flow along a horizontal plane is given as

(a) $\frac{\partial p}{\partial r} = -\rho \frac{V^2}{r}$ ☐ (b) $\frac{\partial p}{\partial r} = \rho \frac{V}{r^2}$ ☐

(c) $\frac{\partial p}{\partial r} = \rho \frac{V^2}{r}$ ☐ (d) none of the above. ☐

75. For a forced vortex flow the height of paraboloid formed is equal to

(a) $\frac{p}{w} + \frac{V^2}{2g}$ ☐ (b) $\frac{V^2}{2g}$ ☐

(c) $\frac{V^2}{r^2 \times 2g}$ ☐ (d) $\frac{\omega r^2}{2g}$. ☐

76. Bernoulli's equation is derived making assumptions that

(a) the flow is uniform, steady and incompressible ☐

(b) the flow is non-viscous, uniform and steady ☐

(c) the flow is steady, non-viscous, incompressible and irrotational ☐

(d) none of the above. ☐

77. The Bernoulli's equation can take the form

(a) $\frac{p_1}{\rho_1} + \frac{V_1^2}{2g} + Z_1 = \frac{p_2}{\rho_2} + \frac{V_2^2}{2g} + Z_2$ ☐ (b) $\frac{p_1}{\rho_2 q} + \frac{V_1^2}{2} + Z_1 = \frac{p_2}{\rho_2 q} + \frac{V_2^2}{2} + Z_2$ ☐

(c) $\frac{p_1}{\rho_1 q} + \frac{V_1^2}{2g} + Z_1 = \frac{p_2}{\rho_2 q} + \frac{V_2^2}{2g} + gZ_2$ ☐ (d) $\frac{p_1}{\rho_1 g} + \frac{V_1^2}{2g} + Z_1 = \frac{p_2}{\rho_2 g} + \frac{V_2^2}{2g} + Z_2$. ☐

78. The flow rate through a circular pipe is measured by

(a) Pitot-tube ☐ (b) Venturi-meter ☐

(c) Orifice-meter ☐ (d) None of the above. ☐

79. If the velocity, pressure, density etc., do not change at a point with respect to time, the flow is called

(a) uniform ☐ (b) incompressible ☐

(c) non-uniform ☐ (d) steady. ☐

80. If the velocity, pressure, density etc., change at a point with respect to time, the flow is called

(a) uniform ☐ (b) compressible ☐

(c) unsteady ☐ (d) incompressible. ☐

81. If the velocity in a fluid flow does not change with respect to length of direction of flow, it is called

(a) steady flow ☐ (b) uniform flow ☐
(c) incompressible flow ☐ (d) rotational flow. ☐

82. If the velocity in a fluid flow changes with respect to length of direction of flow, it is called

(a) unsteady flow ☐ (b) compressible flow ☐
(c) irrotational flow ☐ (d) none of the above. ☐

83. If the density of a fluid is constant from point to point in a flow region, it is called

(a) steady flow ☐ (b) incompressible flow ☐
(c) uniform flow ☐ (d) rotational flow. ☐

84. If the density of a fluid changes from point to point in a flow region, it is called

(a) steady flow ☐ (b) unsteady flow ☐
(c) non-uniform flow ☐ (d) compressible flow. ☐

85. If the fluid particles move in straight lines and all the lines are parallel to the surface, the flow is called

(a) steady ☐ (b) uniform ☐
(c) compressible ☐ (d) laminar. ☐

86. If the fluid particles move in a zigzag way, the flow is called

(a) unsteady ☐ (b) non-uniform ☐
(c) turbulent ☐ (d) incompressible. ☐

87. The acceleration of a fluid particle in the direction of x is given by

(a) $A_x = u\frac{\partial}{\partial x} + v\frac{\partial v}{\partial y} + w\frac{\partial w}{\partial z} + \frac{\partial u}{\partial t}$ ☐ (b) $A_x = u\frac{\partial u}{\partial x} + u\frac{\partial v}{\partial y} + v\frac{\partial w}{\partial z} + \frac{\partial u}{\partial t}$ ☐

(c) $A_x = u\frac{\partial u}{\partial x} + v\frac{\partial u}{\partial y} + w\frac{\partial u}{\partial z} + \frac{\partial u}{\partial t}$ ☐ (d) none of the above. ☐

88. The local acceleration in the direction of x is given by

(a) $u\frac{\partial u}{\partial x} + \frac{\partial u}{\partial t}$ ☐ (b) $\frac{\partial u}{\partial t}$ ☐

(c) $u\frac{\partial u}{\partial x}$ ☐ (d) none of the above. ☐

89. The convective acceleration in the direction of x is given by

(a) $u\frac{\partial u}{\partial x} + v\frac{\partial v}{\partial y} + w\frac{\partial w}{\partial z}$ ☐ (b) $u\frac{\partial u}{\partial x} + u\frac{\partial u}{\partial y} + u\frac{\partial u}{\partial z}$ ☐

(c) $u\frac{\partial u}{\partial x} + u\frac{\partial v}{\partial y} + u\frac{\partial w}{\partial z}$ ☐ (d) $u\frac{\partial u}{\partial x} + v\frac{\partial u}{\partial y} + w\frac{\partial u}{\partial z}$. ☐

90. Shear strain rate is given by

(a) $\frac{1}{2}\left(\frac{\partial u}{\partial x}+\frac{\partial v}{\partial y}\right)$ ☐ (b) $\frac{1}{2}\frac{\partial u}{\partial x}+\frac{\partial v}{\partial y}$ ☐

(c) $\frac{1}{2}\left(\frac{\partial v}{\partial x}+\frac{\partial u}{\partial y}\right)$ ☐ (d) $\frac{1}{2}\frac{\partial v}{\partial x}+\frac{\partial u}{\partial y}$. ☐

91. For a two-dimensional fluid element in x-y plane, the rotational component is given as

(a) $\omega_z=\frac{1}{2}\left(\frac{\partial v}{\partial x}+\frac{\partial u}{\partial y}\right)$ ☐ (b) $\omega_z=\frac{1}{2}\left(\frac{\partial u}{\partial x}-\frac{\partial v}{\partial y}\right)$ ☐

(c) $\omega_z=\frac{1}{2}\left(\frac{\partial u}{\partial x}+\frac{\partial v}{\partial y}\right)$ ☐ (d) $\omega_z=\frac{1}{2}\left(\frac{\partial v}{\partial x}-\frac{\partial u}{\partial y}\right)$. ☐

92. Vorticity is given by

(a) two times the rotation ☐ (b) 1.5 times the rotation ☐

(c) three times the rotation ☐ (d) equal to the rotation. ☐

93. Study of fluid motion with the forces causing the flow is known as

(a) kinematics of fluid flow ☐ (b) dynamics of fluid flow ☐

(c) statics of fluid flow ☐ (d) none of the above. ☐

94. Study of fluid motion without considering the forces causing the flow is known as

(a) kinematics of fluid flow ☐ (b) dynamics of fluid flow ☐

(c) statics of fluid flow ☐ (d) none of the above. ☐

95. Study of fluid at rest, is known as

(a) kinematics ☐ (b) dynamics ☐

(c) statics ☐ (d) none of the above. ☐

96. The term $V^2/2g$ is known as

(a) kinetic energy ☐ (b) pressure energy ☐

(c) kinetic energy per unit weight ☐ (d) none of the above. ☐

97. The term $p/\rho g$ is known as

(a) kinetic energy per unit weight ☐ (b) pressure energy ☐

(c) pressure energy per unit weight ☐ (d) none of the above. ☐

98. The term Z is known as

(a) potential energy ☐ (b) pressure energy ☐

(c) potential energy per unit weight ☐ (d) none of the above. ☐

99. The discharge through a venturimeter is given as

(a) $Q=\frac{A_1^2A_2^2}{\sqrt{A_1^2-A_2^2}}\times\sqrt{2gh}$ ☐ (b) $Q=\frac{A_1A_2}{\sqrt{2A_1^2-A_2^2}}\times\sqrt{2gh}$ ☐

(c) $Q=\frac{A_1A_2}{\sqrt{A_1^2-A_2^2}}\times\sqrt{2gh}$ ☐ (d) none of the above. ☐

100. The difference of pressure head (h) measured by a mercury-oil differential manometer is given as

(*a*) $h = x\left[1 - \frac{S_g}{S_0}\right]$ ☐ (*b*) $h = x\,[S_g - S_0]$ ☐

(*c*) $h = x\,[S_0 - S_g]$ ☐ (*d*) $h = x\left[\frac{S_g}{S_0} - 1\right]$ ☐

where x = Difference of mercury level, S_g = Specific gravity of mercury, and S_0 = Specific gravity of oil.

101. The difference of pressure head (h) measured by a differential manometer containing lighter liquid is

(*a*) $h = x\left[1 - \frac{S_l}{S_0}\right]$ ☐ (*b*) $h = x\left[\frac{S_l}{S_0} - 1\right]$ ☐

(*c*) $h = x\,[S_0 - S_l]$ ☐ (*d*) none of the above ☐

where S_l = Specific gravity of lighter liquid in manometer

S_0 = Specific gravity of fluid flowing

x = Difference of lighter liquid levels in differential manometer.

102. Pitot-tube is used to measure

(*a*) discharge ☐ (*b*) average velocity ☐

(*c*) velocity at a point ☐ (*d*) pressure at a point. ☐

103. Venturimeter is used to measure

(*a*) discharge ☐ (*b*) average velocity ☐

(*c*) velocity at a point ☐ (*d*) pressure at a point. ☐

104. Orifice-meter is used to measure

(*a*) discharge ☐ (*b*) average velocity ☐

(*c*) velocity at a point ☐ (*d*) pressure at a point. ☐

105. For a sub-merged curved surface, the horizontal component of force due to static liquid is equal to

(*a*) weight of liquid supported by the curved surface ☐

(*b*) force on a projection of the curved surface on a vertical plane ☐

(*c*) area of curved surface × pressure at the centroid of the submerged area ☐

(*d*) none of the above. ☐

106. For a sub-merged curved surface, the component of force due to static liquid is equal to

(*a*) weight of the liquid supported by curved surface ☐

(*b*) force on a projection of the curved surface on a vertical plane ☐

(*c*) area of curved surface × pressure at the centroid of the sub-merged area ☐

(*d*) none of the above. ☐

107. An oil of specific gravity 0.7 and pressure 0.14 kgf/cm^2 will have the height of oil as

(*a*) 70 cm of oil ☐ (*b*) 2 m of oil ☐

(*c*) 20 cm of oil ☐ (*d*) 10 cm of oil. ☐

108. The difference in pressure head, measured by a mercury water differential manometer for a 20 m difference of mercury level will be

(*a*) 2.72 m ☐ (*b*) 2.52 m ☐
(*c*) 2.0 m ☐ (*d*) 0.2 m. ☐

109. The difference in pressure head, measured by a mercury-oil differential manometer for a 20 cm difference of mercury level will be (sp. gr. of oil = 0.8)

(*a*) 2.72 m of oil ☐ (*b*) 2.52 m of oil ☐
(*c*) 3.20 m of oil ☐ (*d*) 2.0 m of oil. ☐

110. The rate of flow through a venturimeter varies as

(*a*) H ☐ (*b*) $\sqrt{H}$ ☐
(*c*) $H^{3/2}$ ☐ (*d*) $H^{5/2}$. ☐

111. The rate of flow through a *V*-notch varies as

(*a*) H ☐ (*b*) $\sqrt{H}$ ☐
(*c*) $H^{3/2}$ ☐ (*d*) $H^{5/2}$. ☐

Orifices and Mouthpieces

112. The range for coefficient of discharge (C_d) for a venturimeter is

(*a*) 0.6 to 0.7 ☐ (*b*) 0.7 to 0.8 ☐
(*c*) 0.8 to 0.9 ☐ (*d*) 0.95 to 0.99. ☐

113. The coefficient of velocity (C_v) for an orifice is

(*a*) $C_v = \sqrt{\dfrac{4x^2}{yH}}$ ☐ (*b*) $C_v = \sqrt{\dfrac{2x}{4yH}}$ ☐
(*c*) $C_v = \sqrt{\dfrac{x^2}{4yH}}$ ☐ (*d*) none of the above. ☐

114. The coefficient of discharge (C_d) in terms of C_v and C_c is

(*a*) $C_d = \dfrac{C_v}{C_c}$ ☐ (*b*) $C_d = C_v \times C_c$ ☐
(*c*) $C_d = \dfrac{C_c}{C_v}$ ☐ (*d*) none of the above. ☐

115. An orifice is known as large orifice when the head of liquid from the centre of orifice is

(*a*) more than 10 times the depth of orifice ☐ (*b*) less than 10 times the depth of orifice ☐
(*c*) less than 5 times the depth of orifice ☐ (*d*) none of the above. ☐

116. Which mouthpiece is having maximum coefficient of discharge?

(*a*) external mouthpiece ☐ (*b*) convergent divergent mouthpiece ☐
(*c*) internal mouthpiece ☐ (*d*) none of the above. ☐

117. The coefficient of discharge (C_d)

(*a*) for an orifice is more than that for a mouthpiece ☐
(*b*) for internal mouthpiece is more than that external mouthpiece ☐

(c) for a mouthpiece is more than that for an orifice ☐
(d) none of the above. ☐

118. Orifices are used to measure
(a) velocity ☐ (b) pressure ☐
(c) rate of flow ☐ (d) none of the above. ☐

119. Mouthpieces are used to measure
(a) velocity ☐ (b) pressure ☐
(c) viscosity ☐ (d) rate of flow. ☐

120. The ratio of actual velocity of a jet of water at veena-contracta to the theoretical velocity is known as
(a) coefficient of discharge ☐ (b) coefficient of velocity ☐
(c) coefficient of contraction ☐ (d) coefficient of viscosity. ☐

121. The ratio of actual discharge of a jet of water to its theoretical discharge is known as
(a) coefficient of discharge ☐ (b) coefficient of velocity ☐
(c) coefficient of contraction ☐ (d) coefficient of viscosity. ☐

122. The ratio of the area of the jet of water at veena-contracta to the area of orifice is known as
(a) coefficient of discharge ☐ (b) coefficient of velocity ☐
(c) coefficient of contraction ☐ (d) coefficient of viscosity. ☐

123. The discharge through a large rectangular orifice is
(a) $\frac{2}{3} C_d \times b \times \sqrt{2g}(\sqrt{H_2} - \sqrt{H_1})$ ☐ (b) $\frac{8}{15} C_d \times b \times \sqrt{2g}\,(H_2^{3/2} - H_1^{3/2})$ ☐
(c) $\frac{2}{3} C_d \times b \times \sqrt{2g}\,(H_2^{3/2} - H_1^{3/2})$ ☐ (d) none of the above ☐
where b = Width of orifice, H_1 = Height of liquid above top edge of the orifice, H_2 = Height of liquid above bottom edge of orifice.

124. The discharge through fully submerged orifice is
(a) $C_d \times b \times (H_2 - H_1) \times \sqrt{2g} \times H^{3/2}$ ☐ (b) $C_d \times b \times (H_2 - H_1) \times \sqrt{2gH}$ ☐
(c) $C_d \times b \times (H_2^{3/2} - H_1^{3/2}) \times \sqrt{2gH}$ ☐ (d) none of the above ☐
where H = Difference of liquid levels on both sides of the orifice
H_1 = Height of liquid above top edge orifice of upstream side
H_2 = Height of liquid above bottom edge of orifice on upstream side.

Notches and Weirs

125. Notch is a device used for measuring
(a) rate of flow through pipes ☐ (b) rate of flow through a small channel ☐
(c) velocity through a pipe ☐ (d) velocity through a small channel. ☐

126. The discharge through a rectangular notch is given by

(*a*) $Q = \frac{2}{3} C_d \times L \times H^{5/2}$ ☐ (*b*) $Q = 2/3\, C_d \times L \times H^{3/2}$ ☐

(*c*) $Q = \frac{2}{3} C_d \times L \times H^{5/2}$ ☐ (*d*) $Q = 8/15\, C_d \times L \times H^{3/2}$. ☐

127. The discharge through a triangular notch is given by

(*a*) $Q = 2/3\, C_d \times \tan \frac{\theta}{2} \times \sqrt{2gH}$ ☐ (*b*) $Q = 2/3\, C_d \times \tan \frac{\theta}{2} \times \sqrt{2g} \times H^{3/2}$ ☐

(*c*) $Q = 8/15\, C_d \times \tan \frac{\theta}{2} \times \sqrt{2g}\; H^{5/2}$ ☐ (*d*) none of the above. ☐

where θ = Total angle of triangular notch, H = Head over notch.

128. The discharge through a trapezoidal notch is given as

(*a*) $Q = 2/3\, C_{d1} \times L \times H^{3/2} + 8/15\, C_{d2} \times \tan \theta/2 \times \sqrt{2g} \times H^{3/2}$ ☐

(*b*) $Q = 2/3\, C_{d1} \times L \times H^{5/2} + 8/15\, C_{d2} \times \tan \theta/2 \times \sqrt{2g}\; H^{3/2}$ ☐

(*c*) $Q = 2/3\, C_{d1} \times L \times H^{3/2} + 8/15\, C_{d2} \times \tan \theta/2 \times \sqrt{2g}\; H^{5/2}$ ☐

(*d*) none of the above ☐

where $\theta/2$ = Slope of the side of the trapezoidal notch.

129. The error in discharge due to the error in the measurement of head over a rectangular notch is given by

(*a*) $\frac{dQ}{Q} = \frac{5}{2} \frac{dH}{H}$ ☐ (*b*) $\frac{dQ}{Q} = \frac{3}{2} \frac{dH}{H}$ ☐

(*c*) $\frac{dQ}{Q} = \frac{7}{2} \frac{dH}{H}$ ☐ (*d*) $\frac{dQ}{Q} = \frac{1}{2} \frac{dH}{H}$. ☐

130. The error in discharge due to the error in the measurement of head over a triangular notch is given by

(*a*) $\frac{dQ}{Q} = \frac{5}{2} \frac{dH}{H}$ ☐ (*b*) $\frac{dQ}{Q} = \frac{3}{2} \frac{dH}{H}$ ☐

(*c*) $\frac{dQ}{Q} = \frac{7}{2} \frac{dH}{H}$ ☐ (*d*) $\frac{dQ}{Q} = \frac{1}{2} \frac{dH}{H}$. ☐

131. The velocity with which the water approaches a notch is called

(*a*) velocity of flow ☐ (*b*) velocity of approach ☐

(*c*) velocity of whirl ☐ (*d*) none of the above. ☐

132. The discharge over a rectangular notch considering velocity of approach is given as

(*a*) $Q = \frac{3}{2} C_d L \sqrt{2g}\, (H^{3/2} - h_a^{3/2})$ ☐ (*b*) $Q = \frac{2}{3} C_d L \sqrt{2g}\, (H - h_a)^{3/2}$ ☐

(*c*) $Q = \frac{2}{3} C_d L \sqrt{2g}\, [(H + h_a)^{3/2} - h_a^{3/2}]$ ☐ (*d*) none of the above ☐

where H = Head over notch, and h_a = Head due to velocity of approach.

133. The velocity of approach (V_a) is given by

(a) $V_a = \dfrac{\text{Discharged over notch}}{\text{Area of notch}}$ ☐ (b) $V_a = \dfrac{\text{Discharged over notch}}{\text{Area of channel}}$ ☐

(c) $V_a = \dfrac{\text{Discharged over notch}}{\text{Heat over notch} \times \text{Width of channel}}$ ☐ (d) none of the above. ☐

134. Francis's formula for a rectangular weir with end contraction suppressed is given as

(a) $Q = 1.84\ LH^{5/2}$ ☐ (b) $Q = 2/3\ L \times H^{3/2}$ ☐

(c) $Q = 1.84\ LH^{3/2}$ ☐ (d) $Q = 2/3\ L \times H^{5/2}$. ☐

135. Francis's formula for a rectangular weir for two end contractions is given by

(a) $Q = 1.84[L - 0.2 \times 2H]H^{5/2}$ ☐ (b) $Q = 1.84[L - 0.2H]H^{3/2}$ ☐

(c) $Q = 1.84[L - 0.2H]H^{5/2}$ ☐ (d) none of the above. ☐

136. Bazin's formula for discharge over a rectangular weir without velocity of approach is given by

(a) $Q = mL \times \sqrt{2gH^{5/2}}$ ☐ (b) $Q = mL \times \sqrt{2g} \times H^{3/2}$ ☐

(c) $Q = m \times L \times \sqrt{2gH}$ ☐ (d) none of the above ☐

where $m = 0.405 + \dfrac{0.003}{H}$ and H = Head over weir.

137. Cipolletti weir is a trapezoidal weir having side slope of

(a) 1 horizontal to 2 vertical ☐ (b) 4 horizontal to 1 vertical ☐

(c) 1 horizontal to 4 vertical ☐ (d) 1 horizontal to 3 vertical. ☐

Laminar and Turbulent Flow Through Pipes

138. A flow is said to be laminar when

(a) the fluid particles move in a zigzag way ☐

(b) the Reynold number is high ☐

(c) the fluid particles move in layers parallel to the boundary ☐

(d) none of the above. ☐

139. For the laminar flow through a circular pipe

(a) the maximum velocity = 1.5 times the average velocity ☐

(b) the maximum velocity = 2.0 times the average velocity ☐

(c) the maximum velocity = 2.5 times the average velocity ☐

(d) none of the above. ☐

140. The loss of pressure head for the laminar flow through pipes varies

(a) as the square of velocity ☐ (b) directly as the velocity ☐

(c) as the inverse of the velocity ☐ (d) none of the above. ☐

141. For the laminar flow through a pipe, the shear stress over the cross-section

(a) varies inversely as the distance from the centre of the pipe ☐

(b) varies directly as the distance from the surface of the pipe ☐

(*c*) varies directly as the distance from the centre of the pipe ☐
(*d*) remains constant over the cross-section. ☐

142. For the laminar flow between two parallel plates
(*a*) the maximum velocity = 2.0 times the average velocity ☐
(*b*) the maximum velocity = 2.5 times of the average velocity ☐
(*c*) the maximum velocity = 1.33 times the average velocity ☐
(*d*) none of the above. ☐

143. The valve of the kinetic energy correction factor (α) for the viscous flow through a circular pipe is
(*a*) 1.33 ☐ (*b*) 1.50 ☐
(*c*) 2.0 ☐ (*d*) 1.25. ☐

144. The valve of the momentum correction factor (β) for the viscous flow through a circular pipe is
(*a*) 1.33 ☐ (*b*) 1.50 ☐
(*c*) 2.0 ☐ (*d*) 1.25. ☐

145. The pressure drop per unit length of a pipe for laminar flow is
(*a*) equal to $\frac{12\mu\overline{U}L}{\rho g D^2}$ ☐ (*b*) equal to $\frac{12\mu\overline{U}}{\rho g D^2}$ ☐
(*c*) equal to $\frac{32\mu\overline{U}L}{\rho g D^2}$ ☐ (*d*) none of the above. ☐

146. For viscous flow between two parallel plates, the pressure drop per unit length is equal to
(*a*) $\frac{12\mu\overline{U}L}{\rho g D^2}$ ☐ (*b*) $\frac{12\mu\overline{U}L}{D^2}$ ☐
(*c*) $\frac{32\mu\overline{U}L}{D^2}$ ☐ (*d*) $\frac{12\mu\overline{U}}{D^2}$. ☐

147. The velocity distribution in laminar flow through a circular pipe follows the
(*a*) parabolic law ☐ (*b*) linear law ☐
(*c*) logarithmic law ☐ (*d*) none of the above. ☐

148. A boundary is known as hydrodynamically smooth, if
(*a*) $\frac{k}{\delta'} = 0.3$ ☐ (*b*) $\frac{k}{\delta'} > 0.3$ ☐
(*c*) $\frac{k}{\delta'} < 0.25$ ☐ (*d*) $\frac{k}{\delta'} = 6.0$ ☐
where k = Average height of the irregularities from the boundary and δ' = Thickness of laminar sub-layer.

149. The coefficient of friction for laminar flow through a circular pipe is given by
(*a*) $f = \frac{0.0791}{(R_e)^{1/4}}$ ☐ (*b*) $f = \frac{16}{R_e}$ ☐
(*c*) $f = \frac{64}{R_e}$ ☐ (*d*) none of the above. ☐

150. The loss of head due to sudden expansion of a pipe is given by

(a) $h_L = \frac{V_1^2 - V_2^2}{2g}$ ☐ (b) $h_L = \frac{0.5\, V_1^2}{2g}$ ☐

(c) $h_L = \frac{(V_1 - V_2)^2}{2g}$ ☐ (d) none of the above. ☐

151. The loss of head due to sudden contraction of a pipe is equal to

(a) $\left(\frac{1}{C_c} - 1\right)^2 \frac{V_2}{2g}$ ☐ (b) $\left(1 - \frac{1}{C_c}\right)^2 \frac{V_2}{2g}$ ☐

(c) $\frac{1}{C_c}\left(1 - \frac{V_2^{\,2}}{2g}\right)$ ☐ (d) none of the above. ☐

152. Hydraulic gradient line (H.G.L.) represents the sum of

(a) pressure head and kinetic head ☐

(b) kinetic head and datum head ☐

(c) pressure head, kinetic head and datum head ☐

(d) pressure head and datum head. ☐

153. Total Energy Line (T.E.L.) represents the sum of

(a) pressure head and kinetic head ☐

(b) kinetic head and datum head ☐

(c) pressure head and datum head ☐

(d) pressure head, kinetic head and datum head. ☐

154. When the pipes are connected in series, the total rate of flow

(a) is equal to the sum of the rate of flow in each pipe ☐

(b) is equal to the reciprocal of the sum of rate of flow in each pipe ☐

(c) is the same as flowing through each pipe ☐

(d) none of the above. ☐

155. Power transmitted through pipes will be maximum when

(a) head lost due to friction = $\frac{1}{2}$ total head at inlet of the pipe ☐

(b) head lost due to friction = $\frac{1}{4}$ total head at inlet of the pipe ☐

(c) head lost due to friction = total head at the inlet of the pipe ☐

(d) head lost due to friction = $\frac{1}{3}$ total head at the inlet of the pipe. ☐

156. The valve closure is said to be gradual if the time required to close the valve

(a) $t = \frac{2L}{C}$ ☐ (b) $t \le \frac{2L}{C}$ ☐

(c) $t < \frac{4L}{C}$ ☐ (d) $t > \frac{2L}{C}$. ☐

where L = Length of pipe, C = Velocity of pressure wave.

157. The velocity of pressure wave in terms of bulk modulus (K) and density (ρ) is given by

(*a*) $C = \sqrt{\frac{\rho}{K}}$ ☐ (*b*) $C = \sqrt{K\rho}$ ☐

(*c*) $C = \sqrt{\frac{K}{\rho}}$ ☐ (*d*) none of the above. ☐

158. The coefficient of friction in terms of shear stress is given by

(*a*) $f = \frac{2\tau V^2}{\tau_0}$ ☐ (*b*) $f = \frac{2\tau_0}{\rho V^2}$ ☐

(*c*) $f = \frac{\tau_0}{2\rho V^2}$ ☐ (*d*) $f = \frac{\rho V^2}{2\tau_0}$. ☐

159. Reynold shear stress for turbulent flow is given by

(*a*) $\tau = \overline{\rho u v'}$ ☐ (*b*) $\overline{\tau} = \mu \frac{\partial u}{\partial y}$ ☐

(*c*) $\overline{\tau} = \eta \frac{\overline{du}}{\partial y}$ ☐ (*d*) none of the above. ☐

where u', v' = Fluctuating component of velocity in the direction x and y and η = Eddy viscosity.

160. The shear stress in turbulent flow due to Prandtl is given by

(*a*) $\overline{\tau} = \rho l^2 \left(\frac{du}{dy}\right)^2$ ☐ (*b*) $\overline{\tau} = \rho^2 l \left(\frac{du}{dy}\right)^2$ ☐

(*c*) $\overline{\tau} = \rho^2 l^2 \left(\frac{du}{dy}\right)$ ☐ (*d*) none of the above. ☐

where l = Mixing length.

161. Shear velocity (u_*) is equal to

(*a*) $\sqrt{\rho\tau_0}$ ☐ (*b*) $\sqrt{\frac{\tau_0}{\rho}}$ ☐

(*c*) $\sqrt{\frac{\rho}{\tau_0}}$ ☐ (*d*) $\frac{1}{\sqrt{\rho\tau_0}}$ ☐

where τ_0 = Shear stress at the surface.

162. The velocity distribution in turbulent flow for pipes is given by

(*a*) $u = U_{max} + 5.5\, u_* \log_e (y/R)$ ☐ (*b*) $u = 2.5\, u_* \log_e (y/R)$ ☐

(*c*) $u = U_{max} + 2.5\, u_* \log_e (y/R)$ ☐ (*d*) none of the above. ☐

where u_* = Shear velocity, R = Radius of pipe, y = Distance from pipe wall, U_{max} = Centre-line velocity.

163. When the pipes are connected in parallel, the total loss of head

(*a*) is equal to the sum of the loss of head in each pipe ☐

(*b*) is same as in each pipe ☐

(*c*) is equal to the reciprocal of the sum of loss of head in each pipe ☐
(*d*) none of the above. ☐

164. L_1, L_2, L_3 are the length of three pipes, connected in series. If d_1, d_2 and d_3 are their diameters, then the equivalent size of the pipe is given by

(*a*) $\frac{L}{d^5} = \frac{L_1}{d_1^5} + \frac{L_2}{d_2^5} + \frac{L_3}{d_3^5}$ ☐ (*b*) $\frac{d^5}{L} = \frac{d_1^5}{L_1} + \frac{d_2^5}{L_2} + \frac{d_3^5}{L_3}$ ☐

(*c*) $Ld^5 = L_1 d_1^5 + L_2 d_2^5 + L_3 d_3^5$ ☐ (*d*) none of the above. ☐

where $L = L_1 + L_2 + L_3$.

165. The power transmitted in kW through pipe is given by

(*a*) $\frac{\rho \times g \times Q \times H}{75}$ ☐ (*b*) $\frac{\rho \times g \times Q \times h_f}{1000}$ ☐

(*c*) $\frac{\rho \times g \times Q \times (H - h_f)}{4500}$ ☐ (*d*) $\frac{\rho \times g \times Q \times (H - h_f)}{1000}$ ☐

where H = Total head at the inlet of pipe, h_f = Head lost due to friction in pipe and Q = Discharge per second.

166. Efficiency of power transmission through pipe is given by

(*a*) $\frac{H - h_f}{H}$ ☐ (*b*) $\frac{H}{H + h_f}$ ☐

(*c*) $\frac{H - h_f}{H + h_f}$ ☐ (*d*) none of the above. ☐

where H = Total head at inlet, h_f = Head lost due to friction.

167. Maximum efficiency of power transmission through pipe is

(*a*) 50% ☐ (*b*) 66.67% ☐
(*c*) 75% ☐ (*d*) 100%. ☐

168. Diameter of nozzle (d) for maximum power transmission is given by

(*a*) $d = \left(\frac{D^4}{8fL}\right)^{1/5}$ ☐ (*b*) $d = \left(\frac{D^5}{8fL}\right)^{1/5}$ ☐

(*c*) $d = \left(\frac{D^5}{8fL}\right)^{1/4}$ ☐ (*d*) none of the above. ☐

where D = Dia. of pipe, L = Length of pipe.

169. Water-hammer in pipes takes place when

(*a*) fluid is flowing with high velocity ☐
(*b*) fluid is flowing with high pressure ☐
(*c*) flowing fluid is suddenly brought to rest by closing the valve ☐
(*d*) flowing fluid is gradually brought to rest. ☐

170. The pressure rise (p_i) due to water hammer, when the valve is closed suddenly and pipe is assumed rigid, is equal to

(*a*) $V\sqrt{\frac{k}{\rho}}$ ☐ (*b*) $V\sqrt{k\rho}$ ☐

(*c*) $V\sqrt{\frac{\rho}{k}}$ ☐ (*d*) $Vk\rho$ ☐

where V = Velocity of flow, k = Bulk modulus of water, and ρ = Density of fluid.

171. The pressure rise (p_i) due to water hammer, when valve is closed gradually is equal to

(*a*) ρLV ☐ (*b*) $\frac{\rho LV}{t}$ ☐

(*c*) $\frac{\rho t}{VL}$ ☐ (*d*) $\frac{\rho}{LVt}$ ☐

where t = Time required to close the valve.

172. The pressure rise (p_i) due to water hammer, when valve is closed suddenly and pipe is elastic, is equal to

(*a*) $V \times \sqrt{\frac{kEt}{\rho D}}$ ☐ (*b*) $V \times \sqrt{\frac{\frac{1}{k}+\frac{D}{Et}}{\rho}}$ ☐

(*c*) $V \times \sqrt{\frac{\rho}{\frac{1}{k}+\frac{D}{Et}}}$ ☐ (*d*) none of the above. ☐

where E = Modulus of elasticity for pipe material, D = Diameter of pipe, t = Time required to close valve, and k = Bulk modulus of water.

173. The pressure rise (p_i) due to water hammer depends on

(*a*) the diameter of pipe only ☐
(*b*) the length of pipe only ☐
(*c*) the required to close the valve only ☐
(*d*) elastic properties of the pipe material only ☐
(*e*) elastic properties of liquid flowing through pipe only ☐
(*f*) all of the above. ☐

174. The valve closure is said to be sudden if the time required to close the valve

(*a*) $t = \frac{2L}{C}$ ☐ (*b*) $t < \frac{2L}{C}$ ☐

(*c*) $t > \frac{2L}{C}$ ☐ (*d*) none of the above ☐

where C = Velocity of pressure wave produced, and L = Length of pipe.

175. For a viscous flow through circular pipes, certain curves are shown in Fig. 1.3. Curve *A* is for

(*a*) shear stress distribution ☐
(*b*) velocity distribution ☐
(*c*) pressure distribution ☐
(*d*) none of the above. ☐

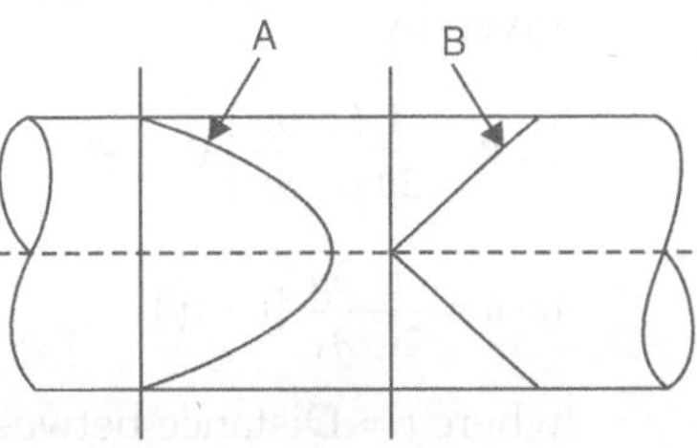

FIGURE 1.3

176. Curve *B* in Fig. 1.3 is for

(*a*) shear stress distribution ☐ (*b*) velocity distribution ☐

(*c*) pressure distribution ☐ (*d*) none of the above. ☐

177. Figure 1.4 shows four curves for velocity distribution across a section for Reynolds number equal to 1000, 4000, 6000 and 10000. Curve *A* corresponds to Reynold number equal to

(*a*) 1000 ☐

(*b*) 4000 ☐

(*c*) 6000 ☐

(*d*) 10000. ☐

A B C D

FIGURE 1.4

178. Curve *B* in Fig. 1.4 corresponds to Reynolds number

(*a*) 1000 ☐ (*b*) 4000 ☐

(*c*) 6000 ☐ (*d*) 10000. ☐

179. Curve *C* in Fig. 1.4 corresponds to the Reynold number

(*a*) 1000 ☐ (*b*) 4000 ☐

(*c*) 6000 ☐ (*d*) 10000. ☐

180. Curve *D* in Fig. 1.4 corresponds to the Reynold number

(*a*) 1000 ☐ (*b*) 4000 ☐

(*c*) 6000 ☐ (*d*) 10000. ☐

181. The shear stress distribution across a section of a circular pipe having viscous flow is given by

(*a*) $\tau = \frac{\partial p}{\partial x} r^2$ ☐ (*b*) $\tau = \frac{\partial p}{\partial x}\frac{r}{2}$ ☐

(*c*) $\tau = -\frac{\partial p}{\partial x}\frac{r}{2}$ ☐ (*d*) $\tau = -\frac{\partial p}{\partial x} \times 2r$. ☐

182. The velocity distribution across a section of a circular pipe having viscous flow is given by

(*a*) $u = U_{\max}\left[1-\left(\frac{r}{R}\right)^2\right]$ ☐ (*b*) $u = U_{\max}[R^2 - r^2]$ ☐

(*c*) $u = U_{\max}\left[1-\frac{r}{R}\right]^2$ ☐ (*d*) none of the above. ☐

183. The velocity distribution across a section of two fixed parallel plates having viscous flow is given by

(*a*) $u = \frac{1}{2\mu}\left(-\frac{\partial p}{\partial x}\right)(t^2 - y^2)$ ☐ (*b*) $u = -\frac{1}{2\mu}\frac{\partial p}{\partial x}[ty - y^2]$ ☐

(*c*) $u = \frac{1}{2\mu}\frac{\partial p}{\partial x}[y - ty]$ ☐ (*d*) $u = -\frac{1}{2\mu}\frac{\partial p}{\partial x}[t - y^2]$ ☐

where t = Distance between two plates and y is measured from the lower plate.

184. The shear stress distribution across a section of two fixed parallel plates having viscous flow is given by

(a) $\tau = -\frac{1}{2}\frac{\partial p}{\partial x}[t^2 - y^2]$ ☐ (b) $\tau = -\frac{1}{2}\frac{\partial p}{\partial x}[t - 2y]$ ☐

(c) $\tau = -\frac{1}{2}\frac{\partial p}{\partial x}[ty - y^2]$ ☐ (d) $\tau = \frac{1}{2}\frac{\partial p}{\partial x}[y - ty]$ ☐

where t = Distance between two parallel plates and y is measured from the plate.

Dimensional and Model Analysis

185. Reynold's number is defined as the

(a) ratio of inertia force to gravity force ☐ (b) ratio of viscous force to gravity force ☐

(c) ratio of viscous force to elastic force ☐ (d) ratio of inertia force to viscous force. ☐

186. Froude's number is defined as the ratio of

(a) inertia force to viscous force ☐ (b) inertia force to gravity force ☐

(c) inertia force to elastic force ☐ (d) inertia force to pressure force. ☐

187. Mach number is defined as the ratio of

(a) inertia force to viscous force ☐ (b) viscous force to surface tension force ☐

(c) viscous force to elastic force ☐ (d) inertia force to elastic force. ☐

188. Euler's number is the ratio of

(a) inertia force to pressure force ☐ (b) inertia force to elastic force ☐

(c) inertia force to gravity force ☐ (d) none of the above. ☐

189. Models are known undistorted model, if

(a) the prototype and model are having different scale ratios ☐

(b) the prototype and model are having same scale ratio ☐

(c) model and prototype are kinematically similar ☐

(d) none of the above. ☐

190. Geometric similarity between model and prototype means

(a) the similarity of discharge ☐ (b) the similarity of linear dimensions ☐

(c) the similarity of motion ☐ (d) the similarity of forces. ☐

191. Kinematic similarity between model and prototype means

(a) the similarity of forces ☐ (b) the similarity of shape ☐

(c) the similarity of motion ☐ (d) the similarity of discharge. ☐

192. Dynamic similarity between model and prototype means

(a) the similarity of forces ☐ (b) the similarity of motion ☐

(c) the similarity of shape ☐ (d) none of the above. ☐

193. Reynold number is expressed as

(a) $R_e = \frac{\rho\mu L}{V}$ ☐ (b) $R_e = \frac{V\mu L}{\rho}$ ☐

(c) $R_e = \frac{\rho V L}{\mu}$ ☐ (d) $R_e = \frac{V \times L}{\nu}$. ☐

194. Froude's number (F_e) is given by

(*a*) $F_e = V\sqrt{\frac{L}{g}}$ ☐ (*b*) $F_e = V\sqrt{\frac{g}{L}}$ ☐

(*c*) $F_e = \frac{V}{\sqrt{L \cdot g}}$ ☐ (*d*) none of the above. ☐

195. Mach number (M) is given by

(*a*) $M = \frac{C}{V}$ ☐ (*b*) $M = V \times C$ ☐

(*c*) $M = \frac{V}{C}$ ☐ (*d*) none of the above. ☐

196. The ratio of inertia force to viscous force is known as

(*a*) Reynold number ☐ (*b*) Froude number ☐

(*c*) Mach number ☐ (*d*) Euler number. ☐

197. The square root of the ratio of inertia force to gravity force is called

(*a*) Reynold number ☐ (*b*) Froude number ☐

(*c*) Mach number ☐ (*d*) Euler number. ☐

198. The square root of the ratio of inertia force to force due to compressibility is known as

(*a*) Reynold number ☐ (*b*) Froude number ☐

(*c*) Mach number ☐ (*d*) Euler number. ☐

199. The square root of the ratio of inertia force to pressure force is known as

(*a*) Reynold number ☐ (*b*) Froude number ☐

(*c*) Mach number ☐ (*d*) Euler number. ☐

200. Model analysis of pipes flow are based on

(*a*) Reynold number ☐ (*b*) Froude number ☐

(*c*) Mach number ☐ (*d*) Euler number. ☐

201. Model analysis of free surface flows are based on

(*a*) Reynolds number ☐ (*b*) Froude number ☐

(*c*) Mach number ☐ (*d*) Euler number. ☐

202. Model analysis of aeroplanes and projectile moving at super-sonic speed are based on

(*a*) Reynold number ☐ (*b*) Froude number ☐

(*c*) Mach number ☐ (*d*) Euler number. ☐

Boundary Layer Flow

203. Boundary layer on a flat plate is called laminar boundary layer if

(*a*) Reynold number is less than 2000 ☐ (*b*) Reynold number is less than 4000 ☐

(*c*) Reynold number is less than 5×10^5 ☐ (*d*) None of the above. ☐

204. Boundary layer thickness (δ) is the distance from the surface of the solid body in the direction perpendicular to flow, where the velocity of fluid is equal to

(*a*) Free stream velocity ☐ (*b*) 0.9 times the free stream velocity ☐

(*c*) 0.99 times the free stream velocity ☐ (*d*) None of the above. ☐

205. Displacement thickness (δ^*) is given by

(*a*) $\delta^* = \int_0^\delta \left(1 - \frac{U}{u}\right) dy$ ☐ (*b*) $\delta^* = \int_0^\delta \frac{U}{u}\left(1 - \frac{U}{u}\right) dy$ ☐

(*c*) $\delta^* = \int_0^\delta \frac{u}{U}\left(1 - \frac{u^2}{U^2}\right) dy$ ☐ (*d*) none of the above. ☐

206. Momentum thickness (θ) is given by

(*a*) $\theta = \int_0^\delta \frac{u}{U}\left(1 - \frac{u}{U}\right) dy$ ☐ (*b*) $\theta = \int_0^\delta \left(1 - \frac{u}{U}\right) dy$ ☐

(*c*) $\theta = \int_0^\delta \frac{u}{U}\left(1 - \frac{u^2}{U^2}\right) dy$ ☐ (*d*) none of the above. ☐

207. Energy thickness (δ^{**}) is equal to

(*a*) $\int_0^\delta \frac{u}{U}\left[1 - \frac{u}{U}\right] dy$ ☐ (*b*) $\int_0^\delta \frac{u}{U}\left(1 - \frac{u^2}{U^2}\right) dy$ ☐

(*c*) $\int_0^\delta \frac{u}{U}\left(1 - \frac{u}{U}\right)^2 dy$ ☐ (*d*) none of the above. ☐

208. Von-Karman momentum integral equation is given as

(*a*) $\frac{\tau_0}{\frac{1}{2}\rho U^2} = \frac{\partial \theta}{\partial x}$ ☐ (*b*) $\frac{\tau_0}{\rho U^2} = \frac{\partial \theta}{\partial x}$ ☐

(*c*) $\frac{\tau_0}{2\rho U^2} = \frac{\partial \theta}{\partial x}$ ☐ (*d*) none of the above. ☐

209. The boundary layer separation takes place if

(*a*) pressure gradient is zero ☐ (*b*) pressure gradient is +ve ☐

(*c*) pressure gradient is negative ☐ (*d*) none of the above. ☐

210. The condition for boundary layer separation is

(*a*) $\left(\frac{\partial u}{\partial y}\right)_{y=0} = +ve$ ☐ (*b*) $\left(\frac{\partial u}{\partial y}\right)_{y=0} = -ve$ ☐

(*c*) $\left(\frac{\partial u}{\partial y}\right)_{y=0} = 0$ ☐ (*d*) none of the above. ☐

211. The boundary layer flow will be attached to the surface if

(*a*) $\left(\frac{\partial u}{\partial y}\right)_{y=0} = 0$ ☐ (*b*) $\left(\frac{\delta u}{\delta y}\right)_{y=0} = +ve$ ☐

(*c*) $\left(\frac{\partial u}{\partial y}\right)_{y=0} = -ve$ ☐ (*d*) none of the above. ☐

212. The condition for detached flow is

(a) $\left(\frac{\partial u}{\partial y}\right)_{y=0} = 0$ ☐ (b) $\left(\frac{\partial u}{\partial y}\right)_{y=0} = +ve$ ☐

(c) $\left(\frac{\partial u}{\partial y}\right)_{y=0} = -ve$ ☐ (d) none of the above. ☐

213. Drag is defined as the force exerted by a flowing fluid on a solid body

(a) in the direction of flow ☐

(b) perpendicular to the direction of flow ☐

(c) in the direction which is at an angle of 45° to the direction of flow ☐

(d) none of the above. ☐

214. Lift force is defined as the force exerted by a flowing fluid on a solid body

(a) in the direction of flow ☐ (b) perpendicular to the direction of flow ☐

(c) at an angle of 45° to the direction of flow ☐ (d) none of the above. ☐

215. Drag force is expressed mathematically as

(a) $F_D = \frac{1}{2}\rho U^2 \times C_D \times A$ ☐ (b) $F_D = \frac{1}{2}\rho U^2 \times C_D \times A$ ☐

(c) $F_D = 2\rho U^2 \times C_D \times A$ ☐ (d) none of the above. ☐

216. Lift force (F_L) is expressed mathematically as

(a) $F_L = \frac{1}{2}\rho U^2 \times C_L$ ☐ (b) $F_L = \frac{1}{2}\rho U^2 \times C_L \times A$ ☐

(c) $F_L = 2\rho U^2 \times C_L \times A$ ☐ (d) $F_L = \rho U^2 \times C_L \times A$. ☐

217. Total drag on a body is the sum of

(a) pressure drag and velocity drag ☐ (b) pressure drag and friction drag ☐

(c) friction drag and velocity drag ☐ (d) none of the above. ☐

218. A body is called stream-lined body when it is placed in a flow and the surface of the body

(a) coincides with the streamlines ☐ (b) does not coincide with the streamlines ☐

(c) is perpendicular to the streamlines ☐ (d) none of the above. ☐

219. A body is called bluff body if the surface of the body

(a) coincides with the streamlines ☐ (b) does not coincide with the streamlines ☐

(c) is very smooth ☐ (d) none of the above. ☐

220. The drag on a sphere (F_D) for Reynolds number less than 0.2 is given by

(a) $F_D = 5\pi\mu DU$ ☐ (b) $F_D = 3\pi\mu DU$ ☐

(c) $F_D = 2\pi\mu DU$ ☐ (d) $F_D = \pi\mu DU$. ☐

221. The skin friction drag on a sphere (for Reynolds number less than 0.2) is equal to

(a) one-third of the total drag ☐ (b) half of the total drag ☐

(c) two-third of the total drag ☐ (d) none of the above. ☐

222. The pressure drag on a sphere (for Reynolds number less than 0.2) is equal to

(a) one-third of the total drag ☐ (b) half of the total drag ☐

(c) two-third of the total drag ☐ (d) none of the above. ☐

223. Terminal velocity of a falling body is equal to

(*a*) the maximum velocity with which body will fall ☐
(*b*) the maximum constant velocity with which body will fall ☐
(*c*) half of the maximum velocity ☐
(*d*) none of the above. ☐

224. When a falling body has attained terminal velocity, the weight of the body is equal to

(*a*) drag force minus buoyant force ☐ (*b*) buoyant force minus drag force ☐
(*c*) drag force plus the buoyant force ☐ (*d*) none of the above. ☐

225. The tangential velocity of ideal fluid at any point on the surface of the cylinder is given by

(*a*) $u_\theta = \frac{1}{2} U \sin \theta$ ☐ (*b*) $u_\theta = U \sin \theta$ ☐
(*c*) $u_\theta = \frac{1}{2} U \sin \theta$ ☐ (*d*) none of the above. ☐

226. The lift force (F_L) produced on a rotating circular cylinder in a uniform flow is given by

(*a*) $F_L = \frac{LUT}{\rho}$ ☐ (*b*) $F_L = \rho LUT$ ☐
(*c*) $F_L = \frac{\rho UT}{\rho}$ ☐ (*d*) $F_L = \frac{\rho LU}{\Gamma}$ ☐

where L = Length of the cylinder, U = Free stream velocity, Γ = Circulation.

227. The lift coefficient (C_L) for a rotating cylinder in a uniform flow is given by

(*a*) $C_L = \frac{\Gamma U}{R}$ ☐ (*b*) $C_L = \frac{\Gamma R}{U}$ ☐
(*c*) $C_L = \frac{\Gamma}{RU}$ ☐ (*d*) $C_L = \frac{RU}{\Gamma}$. ☐

228. The circulation developed on an airfoil is given by

(*a*) $\Gamma = \frac{CU \sin \alpha}{\pi}$ ☐ (*b*) $\Gamma = \pi CU \sin \alpha$ ☐
(*c*) $\Gamma = \frac{\pi CU}{\sin \alpha}$ ☐ (*d*) $\Gamma = \frac{\pi \sin \alpha}{CU}$. ☐

where C = Chord length, U = Velocity of airfoil, α = Angle of attack.

Boundary Layer Flow

229. The boundary layer takes place

(*a*) for ideal fluids ☐ (*b*) for pipe flow only ☐
(*c*) for real fluids ☐ (*d*) for flow over flat plate only. ☐

230. The boundary layer is called turbulent boundary layer if

(*a*) Reynold number is more than 2000 ☐ (*b*) Reynold number is more than 4000 ☐
(*c*) Reynold number is more than 5×10^5 ☐ (*d*) None of the above. ☐

231. Laminar sub-layer exists in

(*a*) laminar boundary layer region ☐ (*b*) turbulent boundary layer region ☐
(*c*) transition zone ☐ (*d*) none of the above. ☐

232. The thickness of laminar boundary layer at a distance x from the leading edge over a flat plate varies as

(a) $x^{4/5}$ □ (b) $x^{1/2}$ □

(c) $x^{1/5}$ □ (d) $x^{3/5}$. □

233. The thickness of turbulent boundary layer at a distance x from the leading edge over a flat plate varies as

(a) $x^{4/5}$ □ (b) $x^{1/2}$ □

(c) $x^{1/5}$ □ (d) $x^{3/5}$. □

234. The separation of boundary layer takes place in case of

(a) negative pressure gradient □ (b) positive pressure gradient □

(c) zero pressure gradient □ (d) none of the above. □

235. The velocity profile for turbulent boundary layer is

(a) $\frac{u}{U} = \sin\left(\frac{\pi}{2}\frac{y}{\delta}\right)$ □ (b) $\frac{u}{U} = \left(\frac{y}{\delta}\right)^{1/7}$ □

(c) $\frac{u}{U} = 2\left(\frac{y}{\delta}\right) - \left(\frac{y}{\delta}\right)^2$ □ (d) $\frac{u}{U} = \frac{3}{2}\left(\frac{y}{\delta}\right) - \frac{1}{2}\left(\frac{y}{\delta}\right)^3$. □

236. The drag force exerted by a fluid on a body immersed in the fluid is due to

(a) pressure and viscous force □ (b) pressure and gravity forces □

(c) pressure and turbulence forces □ (d) none of the above. □

Compressible Flow

237. Equation of state is expressed as

(a) $p\rho = RT$ □ (b) $\frac{p}{\rho} = RT$ □

(c) $\frac{\rho}{p} = RT$ □ (d) $\frac{p}{\rho} = \frac{R}{T}$ □

where p = Absolute pressure, T = Absolute temperature, R = Gas constant, ρ = Density of gas.

238. The continuity equation in differential form is

(a) $\frac{dA}{A} + \frac{dV}{V} + \frac{d\rho}{\rho} = 0$ □ (b) $AdA + VdV + \rho d\rho = 0$ □

(c) $\frac{A}{dA} + \frac{V}{dV} + \frac{\rho}{d\rho}$ = Constant □ (d) $\frac{dA}{A} + \frac{dV}{V} + \frac{d\rho}{\rho}$ = Constant. □

239. Velocity of sound wave (C) for isothermal process is given by

(a) $C = \sqrt{\frac{p}{\rho}}$ □ (b) $C = \sqrt{\frac{kp}{\rho}}$ □

(c) $C = \sqrt{p\rho}$ □ (d) $C = \sqrt{\frac{p}{k\rho}}$ □

where k = Ratio of specific heats.

240. Super-sonic flow means

(*a*) Mach number = 1.0 ☐ (*b*) Mach number = 1.0 ☐

(*c*) Mach number > 1.0 ☐ (*d*) None of the above. ☐

241. Sonic-flow means

(*a*) Mach number < 1.0 ☐ (*b*) Mach number = 1.0 ☐

(*c*) Mach number > 1.0 ☐ (*d*) None of the above. ☐

242. In sonic-flow, the disturbances, created by a projectile, moves

(*a*) along the projectile ☐ (*b*) ahead of the projectile ☐

(*c*) behind the projectile ☐ (*d*) none of the above. ☐

243. In super-sonic flow, the projectile (which creates disturbances) moves

(*a*) ahead of the disturbances ☐ (*b*) along the disturbances ☐

(*c*) behind the disturbances ☐ (*d*) none of the above. ☐

244. Mach angle (α) is given by

(*a*) $\sin \alpha = \frac{V}{C}$ ☐ (*b*) $\sin \alpha = VC$ ☐

(*c*) $\sin \alpha = \frac{C}{V}$ ☐ (*d*) $\sin \alpha = \frac{1}{VC}$. ☐

245. For sub-sonic flow, if the area of flow increases

(*a*) velocity is constant ☐ (*b*) velocity increases ☐

(*c*) velocity decreases ☐ (*d*) none of the above. ☐

246. For super-sonic flow, if the area of flow increases, then

(*a*) velocity decreases ☐ (*b*) velocity increases ☐

(*c*) velocity is constant ☐ (*d*) none of the above. ☐

247. The area velocity relationship for compressible fluids is

(*a*) $\frac{dA}{A} = \frac{dV}{V}[1 - M^2]$ ☐ (*b*) $\frac{dA}{A} = \frac{dV}{V}[M^2 - 1]$ ☐

(*c*) $\frac{dA}{A} = \frac{dV}{V}[1 - V^2]$ ☐ (*d*) $\frac{dA}{A} = \frac{dV}{V}[C^2 - 1]$. ☐

Channel Flow I

248. The flow in open channel is laminar if the Reynold number is

(*a*) 2000 ☐ (*b*) less than 2000 ☐

(*c*) less than 500 ☐ (*d*) none of the above. ☐

249. The flow in open channel is turbulent if the Reynold number is

(*a*) 2000 ☐ (*b*) more than 2000 ☐

(*c*) more than 4000 ☐ (*d*) 4000. ☐

250. If the Froude number in open channel flow is less than 1.0, the flow is called

(*a*) critical flow ☐ (*b*) super-critical ☐

(*c*) sub-critical ☐ (*d*) none of the above. ☐

251. If the Froude number in open channel flow is equal to 1.0, the flow is called

(*a*) critical flow ☐ (*b*) streaming flow ☐

(*c*) shooting flow ☐ (*d*) none of the above. ☐

252. If the Froude number in open channel flow is more than 1.0, the flow is called

(*a*) critical flow ☐ (*b*) streaming flow ☐

(*c*) shooting flow ☐ (*d*) none of the above. ☐

253. Chezy's formula is given as

(*a*) $V = i\sqrt{mC}$ ☐ (*b*) $V = C\sqrt{mi}$ ☐

(*c*) $V = m\sqrt{Ci}$ ☐ (*d*) none of the above. ☐

254. The discharge through a rectangular channel is maximum when

(*a*) $m = \frac{d}{3}$ ☐ (*b*) $m = \frac{d}{2}$ ☐

(*c*) $m = 2d$ ☐ (*d*) $m = \frac{3d}{2}$ ☐

where m = Hydraulic mean depth, d = Depth of flow.

255. The discharge through a trapezoidal channel is maximum when

(*a*) half of top width = sloping side ☐ (*b*) top width = half of sloping side ☐

(*c*) top width = 1.5 × sloping side ☐ (*d*) none of the above. ☐

256. The maximum velocity through a circular channel takes place when depth of flow is equal to

(*a*) 0.95 times the diameter ☐ (*b*) 0.5 times the diameter ☐

(*c*) 0.81 times the diameter ☐ (*d*) 0.3 times the diameter. ☐

257. The maximum discharge through a circular channel takes place when depth of flow is equal to

(*a*) 0.95 times the diameter ☐ (*b*) 0.3 times the diameter ☐

(*c*) 0.81 times the diameter ☐ (*d*) 0.5 times the diameter. ☐

258. Specific energy of a flowing fluid per unit weight is equal to

(*a*) $\frac{p}{\rho \times g} + \frac{V^2}{2g}$ ☐ (*b*) $\frac{p}{\rho \times g} + h$ ☐

(*c*) $\frac{V^2}{2g} + h$ ☐ (*d*) $\frac{p}{\rho \times g} + \frac{V^2}{2g} + h$. ☐

259. The depth of flow after hydraulic jump is

(*a*) $d_2 = \frac{d_1}{2}[\sqrt{1 + 8(F_e)_1^2} - 1]$ ☐ (*b*) $d_2 = \frac{d_1}{2}[1 + \sqrt{1 + 8(F_e)_1^2 - 1}]$ ☐

(*c*) $d_2 = \frac{d_1}{2} + \sqrt{\frac{d_1^2}{4} + 8(F_e)_1}$ ☐ (*d*) none of the above. ☐

260. The depth of flow at which specific energy is minimum is called

(a) normal depth ☐ (b) critical depth ☐

(c) alternate depth ☐ (d) none of the above. ☐

261. The critical depth (h_e) is given by

(a) $\left(\frac{q^2}{g}\right)^{1/2}$ ☐ (b) $\left(\frac{q}{g}\right)^{1/3}$ ☐

(c) $\left(\frac{q^2}{g}\right)^{1/3}$ ☐ (d) $\left(\frac{q^2}{g}\right)^{2/3}$ ☐

where q = Rate of flow per unit width of channel.

262. For a circular channel, the wetted perimeter is given by

(a) $\frac{R\theta}{2}$ ☐ (b) $3R\theta$ ☐

(c) $2R\theta$ ☐ (d) $R\theta$ ☐

where R = Radius of circular channel and θ = Half the angle subtended by the water surface at the centre.

263. For a circular channel, the area of flow is given by

(a) $R^2\left(2\theta - \frac{\sin 2\theta}{2}\right)$ ☐ (b) $R^2\left(\theta - \frac{\sin 2\theta}{2}\right)$ ☐

(c) $R^2(\theta - \sin 2\theta)$ ☐ (d) none of the above ☐

where θ = Half the angle subtended by water surface at the centre and R = Radius of circular channel.

264. The hydraulic mean depth is given by

(a) $\frac{P}{A}$ ☐ (b) $\frac{P^2}{A}$ ☐

(c) $\frac{A}{P}$ ☐ (d) $\sqrt{\frac{A}{P}}$ ☐

where A = Area and P = Wetted perimeter.

265. A most economical section is one which for a given cross-sectional area, slope of bed (i) and co-efficient of resistance has

(a) maximum wetted perimeter ☐ (b) maximum discharge ☐

(c) maximum depth of flow ☐ (d) none of the above. ☐

Hydraulic Turbines

266. Specific speed of a turbine is defined as the speed of the turbine which

(a) produces unit power at unit head ☐ (b) produces unit power at unit discharge ☐

(c) delivers unit discharge at unit head ☐ (d) delivers unit discharge at unit power. ☐

267. A pump is defined as a device which converts
(*a*) hydraulic energy into mechanical energy ☐
(*b*) mechanical energy into hydraulic energy ☐
(*c*) kinetic energy into mechanical energy ☐
(*d*) none of the above. ☐

268. A turbine is a device which converts
(*a*) hydraulic energy into mechanical energy ☐
(*b*) mechanical energy into hydraulic energy ☐
(*c*) kinetic energy into mechanical energy ☐
(*d*) electrical energy into mechanical energy. ☐

269. The force exerted by a jet of water on a stationary vertical plate in the direction of jet is given by
(*a*) $F_x = \rho AV^2 \sin^2 \theta$ ☐ (*b*) $F_x = \rho AV^2 [1 + \cos \theta]$ ☐
(*c*) $F_x = \rho AV^2$ ☐ (*d*) none of the above. ☐

270. The force exerted by a jet of water on a stationary inclined plate in the direction of jet is given by
(*a*) $F_x = \rho AV^2$ ☐ (*b*) $F_x = \rho AV^2 \sin^2 \theta$ ☐
(*c*) $F_x = \rho AV^2[1 + \cos \theta]$ ☐ (*d*) $F_x = \rho AV^2 [1 + \sin \theta]$. ☐

271. The force exerted by a jet of water on a stationary curved plate in the direction of jet is equal to
(*a*) ρAV^2 ☐ (*b*) $\rho AV^2 \sin^2 \theta$ ☐
(*c*) $\rho AV^2 (1 + \cos \theta)$ ☐ (*d*) $\rho AV^2 [1 + \sin \theta]$. ☐

272. The force exerted by a jet of water having velocity V on a vertical plate, moving with a velocity u is given by
(*a*) $F_x = \rho A(V - u)^2 \sin^2 \theta$ ☐ (*b*) $F_x = \rho A(V - u)^2$ ☐
(*c*) $F_x = \rho A(V - u)^2 [1 + \cos \theta]$ ☐ (*d*) none of the above. ☐

273. The force exerted by a jet of water having velocity V on a series of vertical plates moving with velocity u is given by
(*a*) $P_x = \rho AV^2$ ☐ (*b*) $F_x = \rho A(V - u)^2$ ☐
(*c*) $F_x = \rho AVu$ ☐ (*d*) none of the above. ☐

274. Efficiency of the jet of water having velocity V striking a series of vertical plates moving with a velocity u is given by
(*a*) $\eta = \dfrac{2V(V-u)}{u^2}$ ☐ (*b*) $\eta = \dfrac{2u(V-u)}{V^2}$ ☐
(*c*) $\eta = \dfrac{u^2}{V^2(V-u)}$ ☐ (*d*) none of the above. ☐

275. Efficiency, of the jet of water having velocity V and striking a series of vertical plates moving with a velocity u, is maximum when
(*a*) $u = 2V$ ☐ (*b*) $u = \dfrac{V}{2}$ ☐
(*c*) $u = \dfrac{3V}{2}$ ☐ (*d*) $u = \dfrac{4V}{3}$. ☐

276. Maximum efficiency of a series of vertical plates is

(*a*) 66.67% ☐ (*b*) 33.33% ☐

(*c*) 50% ☐ (*d*) 80%. ☐

277. For a series of curved radial vanes, the work done per second per unit weight is equal to

(*a*) $\frac{1}{g}\ Vw_1u_2 + Vw_2u_2$ ☐ (*b*) $\frac{1}{g}\ [V_1u_1 + V_2u_2]$ ☐

(*c*) $\frac{1}{g}\ [Vw_1u_2 \pm Vw_2u_2]$ ☐ (*d*) none of the above. ☐

278. The net head (H) on the turbine is given by

(*a*) H = Gross Head + Head lost due to friction ☐

(*b*) H = Gross Head – Head lost due to friction ☐

(*c*) H = Gross Head + $\frac{V^2}{2g}$ – Head lost due to friction. ☐

(*d*) none of the above. ☐

279. Hydraulic efficiency of a turbine is defined as the ratio of

(*a*) power available at the inlet of turbine to power given by water to the runner ☐

(*b*) power at the shaft of the turbine to power given by water to the runner ☐

(*c*) power at the shaft of the turbine to the power at the inlet of turbine ☐

(*d*) none of the above. ☐

280. Mechanical efficiency of a turbine is the ratio of

(*a*) power at the inlet to the power at the shaft of turbine ☐

(*b*) power at the shaft to the power given to the runner ☐

(*c*) power at the shaft to power at the inlet of turbine ☐

(*d*) none of the above. ☐

281. The overall efficiency of a turbine is the ratio of

(*a*) power at the inlet of turbine to the power at the shaft ☐

(*b*) power at the shaft to the power given to the runner ☐

(*c*) power at the shaft to the power at the inlet of turbine ☐

(*d*) none of the above. ☐

282. The relation between hydraulic efficiency (η_h), mechanical efficiency (η_m) and overall efficiency (η_0) is

(*a*) $\eta_h = \eta_0 \times \eta_m$ ☐ (*b*) $\eta_0 = \eta_h \times \eta_m$ ☐

(*c*) $\eta_0 = \frac{\eta_m}{\eta_h}$ ☐ (*d*) none of the above. ☐

283. A turbine is called impulse if at the inlet of the turbine

(*a*) total energy is only kinetic energy ☐

(*b*) total energy is only pressure energy ☐

(*c*) total energy is the sum of kinetic energy and pressure energy ☐

(*d*) none of the above. ☐

284. A turbine is called reaction turbine if at the inlet of the turbine the total energy is

(*a*) kinematic energy only ☐ (*b*) kinetic energy and pressure energy ☐

(*c*) pressure energy only ☐ (*d*) none of the above. ☐

285. Tick mark the correct statement

(*a*) Pelton wheel is a reaction turbine ☐ (*b*) Pelton wheel is a radial flow turbine ☐

(*c*) Pelton wheel is an impulse turbine ☐ (*d*) None of the above. ☐

286. Francis turbine is

(*a*) an impulse turbine ☐ (*b*) a radial flow impulse turbine ☐

(*c*) an axial flow turbine ☐ (*d*) a reaction radial flow turbine. ☐

287. Kaplan turbine is

(*a*) an impulse turbine ☐ (*b*) a radial flow impulse turbine ☐

(*c*) an axial flow reaction turbine ☐ (*d*) a radial flow reaction turbine. ☐

288. Jet ratio (m) is defined as the ratio of

(*a*) diameter of jet of water to diameter of Pelton wheel ☐

(*b*) velocity of vane to the velocity of jet of water ☐

(*c*) velocity of flow of the jet of water ☐

(*d*) diameter of Pelton wheel to diameter of the jet of water. ☐

289. Flow ratio is defined as the ratio of

(*a*) velocity of flow at inlet to the velocity given by $\sqrt{2gH}$ ☐

(*b*) velocity of runner at inlet to the velocity of flow at inlet ☐

(*c*) velocity of runner to the velocity given by $\sqrt{2gH}$ ☐

(*d*) none of the above. ☐

290. Speed ratio is given by

(*a*) $\frac{u}{\sqrt{2gh}}$ ☐ (*b*) $\frac{V_f}{\sqrt{2gh}}$ ☐

(*c*) $\frac{\sqrt{2gH}}{V_f}$ ☐ (*d*) $\frac{V_w}{\sqrt{2gH}}$. ☐

291. The speed ratio for Pelton wheel varies from

(*a*) 0.45 to 0.50 ☐ (*b*) 0.6 to 0.7 ☐

(*c*) 0.3 to 0.4 ☐ (*d*) 0.8 to 0.9. ☐

292. The discharge through Pelton Turbine is given by

(*a*) $Q = \pi DBV_f$ ☐ (*b*) $Q = \frac{\pi}{4} d^2 \times \sqrt{2gH}$ ☐

(*c*) $Q = \frac{\pi}{4} [D_0^2 - D_6^2] \times V_f$ ☐ (*d*) none of the above. ☐

293. The discharge through Francis Turbine is given by

(a) $Q = \pi DBV_f$ ☐ (b) $Q = \frac{\pi}{4} d^2 \times \sqrt{2gh}$ ☐

(c) $Q = \frac{\pi}{4} [D_0^2 - D_6^2] \times V_f$ ☐ (d) none of the above. ☐

294. The discharge through Kaplan turbine is given by

(a) $Q = \pi DBV_f$ ☐ (b) $Q = \frac{\pi}{4} d^2 \times \sqrt{2gH}$ ☐

(c) $Q = \frac{\pi}{4} [D_0^2 - D_6^2] \times V_f$ ☐ (d) $Q = 0.9\ \pi DBV_f$. ☐

295. Draft tube is used for discharging water from the exit of

(a) an impulse turbine ☐ (b) a Francis turbine ☐

(c) a Kaplan turbine ☐ (d) a Pelton wheel. ☐

296. Specific speed of a turbine is defined as the speed at which the turbine runs when

(a) working under unit head and discharging one litre per second ☐

(b) working under unit head and develops unit horse power ☐

(c) develops unit horse power and discharges on litre per second ☐

(d) none of the above. ☐

297. The specific speed (N_s) of a turbine is given by

(a) $N_s = \frac{N\sqrt{P}}{H^{3/4}}$ ☐ (b) $N_s = \frac{N\sqrt{Q}}{H^{3/4}}$ ☐

(c) $N_s = \frac{N\sqrt{P}}{H^{5/4}}$ ☐ (d) $N_s = \frac{NP^{5/4}}{\sqrt{H}}$. ☐

298. Unit speed is the speed of a turbine when it is working

(a) under unit head and develops unit power ☐

(b) under unit head and discharge one m^3/sec ☐

(c) under unit head ☐

(d) none of the above. ☐

299. Unit discharge is the discharge of a turbine when

(a) the head on turbine is unity and it develops unit power ☐

(b) the head on turbine is unity and it moves at unit speed ☐

(c) the head on the turbine is unity ☐

(d) none of the above. ☐

300. Unit power is the power developed by a turbine when

(a) head on turbine is unity and discharge is also unity ☐

(b) head = one metre and speed is unity ☐

(c) head on turbine is unity ☐

(d) none of the above. ☐

301. The unit speed (N_u) is given by the expression

(a) $N_u = \frac{N}{H^{3/2}}$ ☐ (b) $N_u = \frac{N}{H^{3/4}}$ ☐

(c) $N_u = \frac{N}{\sqrt{H}}$ ☐ (d) $N_u = \frac{N}{H^{5/4}}$. ☐

302. The unit discharge (Q_u) is given by the expression

(a) $Q_u = \frac{Q}{\sqrt{H}}$ ☐ (b) $Q_u = \frac{Q}{H^{3/2}}$ ☐

(c) $Q_u = \frac{Q}{H^{3/4}}$ ☐ (d) $Q_u = \frac{Q}{H^{5/4}}$. ☐

303. Unit power (P_u) is given by the expression

(a) $P_u = \frac{P}{\sqrt{H}}$ ☐ (b) $P_u = \frac{P}{H^{3/2}}$ ☐

(c) $P_u = \frac{P}{H^{3/4}}$ ☐ (d) $P_u = \frac{P}{H^{5/4}}$. ☐

304. The unit discharge (Q_u) and unit speed (N_u) curves for different turbines are shown in Fig. 1.5. Curve *A* is for

(a) Francis Turbine ☐

(b) Kaplan Turbine ☐

(c) Pelton Turbine ☐

(d) Propeller Turbine. ☐

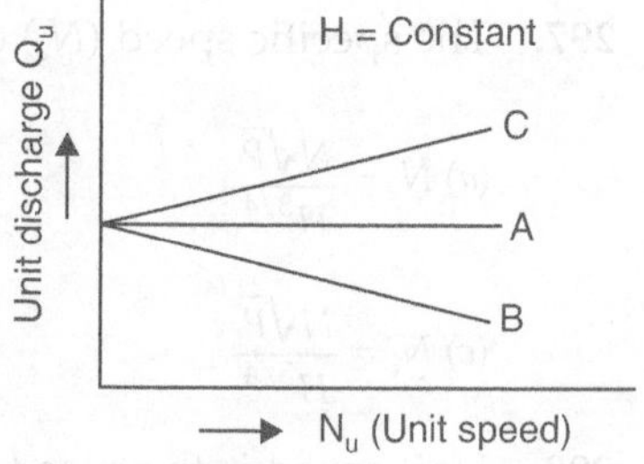

FIGURE 1.5

305. Curve *B* in Fig. 1.5 is for

(a) Francis Turbine ☐

(b) Kaplan Turbine ☐

(c) Pelton Turbine ☐

(d) Propeller Turbine. ☐

306. Curve *C* in Fig. 1.5 is for

(a) Francis Turbine ☐ (b) Kaplan Turbine ☐

(c) Pelton Turbine ☐ (d) Propeller Turbine. ☐

307. Tick mark the correct statement

(a) Curves at constant speed are called main characteristic curves. ☐

(b) Curves at constant head are called main characteristic curves. ☐

(c) Curves at constant efficiency are called operating characteristic curves. ☐

(d) Curves at constant efficiency are called main characteristic curves. ☐

308. Main characteristic curve of a turbine means

(a) curves at constant speed ☐ (b) curves at constant efficiency ☐

(c) curves at constant head ☐ (d) none of the above. ☐

309. Operating characteristic curves of a turbine means

(*a*) curves drawn at constant speed ☐ (*b*) curves drawn at constant efficiency ☐

(*c*) curves drawn at constant head ☐ (*d*) none of the above. ☐

310. Muschel curves means

(*a*) curves at constant head ☐ (*b*) curves at constant speed ☐

(*c*) curves at constant efficiency ☐ (*d*) none of the above. ☐

311. Governing of a turbine means

(*a*) the head is kept constant under all condition of working ☐

(*b*) the speed is kept constant under all conditions ☐

(*c*) the discharge is kept constant under all conditions ☐

(*d*) none of the above. ☐

Centrifugal and Reciprocating Pumps

312. The work done by impeller of a centrifugal pump on water per second per unit weight of water is given by

(*a*) $\frac{1}{g} Vw_1u_1$ ☐ (*b*) $\frac{1}{g} Vw_2u_2$ ☐

(*c*) $\frac{1}{g} (Vw_1u_2 - Vw_1u_1)$ ☐ (*d*) none of the above. ☐

313. The manometer head (H_m) of a centrifugal pump is given by

(*a*) pressure head at outlet of pump—pressure head at inlet ☐

(*b*) total head at inlet—total head at outlet ☐

(*c*) total head at outlet—total head at inlet ☐

(*d*) none of the above. ☐

314. The manometric efficiency (η_{man}) of a centrifugal pump is given by

(*a*) $\frac{H_m}{gVw_2u_2}$ ☐ (*b*) $\frac{gH_m}{Vw_2u_2}$ ☐

(*c*) $\frac{Vw_2u_2}{gH_m}$ ☐ (*d*) $\frac{g \times Vw_2u_2}{H_m}$. ☐

315. Mechanical efficiency ($\eta_{mech.}$) of a centrifugal pump is given by

(*a*) (power at the impeller)/shaft power ☐

(*b*) shaft power/power at the impeller ☐

(*c*) power possessed by water/power at the impeller ☐

(*d*) power possessed by water/shaft power. ☐

316. To produce a high head by multistage centrifugal pumps, the impellers are connected

(*a*) in parallel ☐ (*b*) in series ☐

(*c*) in parallel and in series both ☐ (*d*) none of the above. ☐

317. To discharge a large quantity of liquid by multistage centrifugal pump, the impellers are connected

(*a*) in parallel ☐ (*b*) in series ☐

(*c*) in parallel and in series ☐ (*d*) none of the above. ☐

318. Specific speed of a pump is the speed at which a pump runs when
(*a*) head developed is unity and discharge is one cubic metre ☐
(*b*) head developed is unity and shaft horse power is also unity ☐
(*c*) discharge is one cubic metre and shaft horse power is unity ☐
(*d*) none of the above. ☐

319. The specific speed (N_s) of a pump is given by the expression
(*a*) $N_s = \frac{N\sqrt{Q}}{H_m^{5/4}}$ ☐ (*b*) $N_s = \frac{N\sqrt{P}}{H_m^{3/4}}$ ☐
(*c*) $N_s = \frac{N\sqrt{Q}}{H_m^{3/4}}$ ☐ (*d*) $N_s = \frac{N\sqrt{P}}{H_m^{5/4}}$. ☐

320. The operating characteristic curves of a centrifugal pump are shown in Fig. 1.6.
Curve *A* is for
(*a*) head ☐
(*b*) efficiency ☐
(*c*) power ☐
(*d*) none of the above. ☐

FIGURE 1.6

321. Curve *B* in Fig. 1.6 is for
(*a*) head ☐ (*b*) efficiency ☐
(*c*) power ☐ (*d*) none of the above. ☐

322. Curve C in Fig. 1.6 is for
(*a*) head ☐ (*b*) efficiency ☐
(*c*) power ☐ (*d*) none of the above. ☐

323. Cavitation will take place if the pressure of the flowing fluid at any point is
(*a*) more than vapour pressure of the fluid ☐ (*b*) equal to vapour pressure of the fluid ☐
(*c*) is less than vapour pressure of the fluid ☐ (*d*) none of the above. ☐

324. Cavitation can take place in case of
(*a*) Pelton Wheel ☐ (*b*) Francis Turbine ☐
(*c*) Reciprocating pump ☐ (*d*) Centrifugal pump. ☐

325. Tick mark the correct statement
(*a*) Centrifugal pump convert mechanical energy into hydraulic energy by sucking liquid into chamber. ☐
(*b*) Reciprocating pumps convert mechanical energy into hydraulic energy by means of centrifugal force. ☐
(*c*) Centrifugal pumps convert mechanical energy into hydraulic energy by means of centrifugal force. ☐
(*d*) Reciprocating pumps convert hydraulic energy into mechanical energy. ☐

326. The discharge through a single acting reciprocating pump is

(a) $Q = \frac{ALN}{60}$ ☐ (b) $Q = \frac{2ALN}{60}$ ☐

(c) $Q = ALN$ ☐ (d) $Q = 2ALN$. ☐

327. The pressure head due to acceleration (h_a) is reciprocating pump is given by

(a) $h_a = \frac{l}{g} \times \frac{a}{A} \times \omega^2 r \sin\theta$ ☐ (b) $h_a = \frac{l}{g} \times \frac{A}{a} \times \omega^2 r \sin\theta$ ☐

(c) $h_a = \frac{l}{g} \times \frac{A}{a} \times \omega^2 r \cos\theta$ ☐ (d) $h_a = \frac{A}{a} \omega^2 r \sin\theta$ ☐

where A = Area of cylinder, a = Area of pipe and r = Radius of crank.

328. Indicator diagram shows for one complete revolution of crank the

(a) variation of kinetic head in the cylinder ☐

(b) variation of pressure head in the cylinder ☐

(c) variation of kinetic and pressure head in the cylinder ☐

(d) none of the above. ☐

329. Air vessel in a reciprocating pump is used

(a) to obtain a continuous supply of water at uniform rate ☐

(b) to reduce suction head ☐

(c) to increase the delivery head ☐

(d) none of the above. ☐

330. The work saved by fitting as air vessel to a single acting reciprocating pump is

(a) 39.2% ☐ (b) 84.8% ☐

(c) 48.8% ☐ (d) 92.3%. ☐

331. The work saved by fitting an air vessel to a double acting reciprocating pump is

(a) 39.2% ☐ (b) 84.8% ☐

(c) 48.8% ☐ (d) 92.3%. ☐

332. The pressure, at which separation takes place, is known separation pressure or separation pressure head. For water, the limiting value of separation pressure head is

(a) 2.5 m (abs.) ☐ (b) 7.5 m (abs.) ☐

(c) 10.3 m (abs.) ☐ (d) 5 m (abs.). ☐

333. During suction stroke of a reciprocating pump, the separation may take place

(a) at the end of suction stroke ☐ (b) in the middle of suction stroke ☐

(c) in the beginning of suction stroke ☐ (d) none of the above. ☐

334. During delivery stroke of a reciprocating pump, the separation may take place

(a) at the end of delivery stroke ☐ (b) in the middle of delivery stroke ☐

(c) in the beginning of the delivery stroke ☐ (d) none of the above. ☐

Miscellaneous Hydraulic Devices

335. Hydraulic accumulator is a device used for

(*a*) lifting heavy weights ☐

(*b*) storing the energy of a fluid in the form of pressure energy ☐

(*c*) increasing the pressure intensity of a fluid ☐

(*d*) none of the above. ☐

336. Hydraulic intensifier is a device used for

(*a*) storing energy of a fluid in the form of pressure energy ☐

(*b*) increasing pressure intensity of a fluid ☐

(*c*) transmitting power from one shaft to another ☐

(*d*) none of the above. ☐

337. Hydraulic ram is a pump which works

(*a*) on the principle of water-hammer ☐ (*b*) on the principle of centrifugal action ☐

(*c*) on the principle of reciprocating action ☐ (*d*) none of the above. ☐

338. Hydraulic coupling is a device used for

(*a*) transmitting same torque to the driven shaft ☐

(*b*) transmitting increased torque to the driven shaft ☐

(*c*) transmitting decreased torque to the driven shaft ☐

(*d*) none of the above. ☐

339. Torque converter is a device used for

(*a*) transmitting same torque to the driven shaft ☐

(*b*) transmitting increased torque to the driven shaft ☐

(*c*) transmitting decreased torque to the driven shaft ☐

(*d*) transmitting increased or decreased torque to the driven shaft. ☐

340. Capacity of a hydraulic accumulator is given as equal to

(*a*) pressure of water supplied by pump × volume of accumulator ☐

(*b*) pressure of water × area of accumulator ☐

(*c*) pressure of water × stroke of the ram of accumulator ☐

(*d*) none of the above. ☐

341. Kaplan turbine is a propeller turbine in which the vanes fixed on the bub are

(*a*) non-adjustable ☐ (*b*) adjustable ☐

(*c*) fixed ☐ (*d*) none of the above. ☐

342. If the head on the turbine is more than 300 m, the type of turbine used should be

(*a*) Kaplan ☐ (*b*) Francis ☐

(*c*) Pelton ☐ (*d*) Propeller. ☐

343. If the specific speed of a turbine is more than 300, the type of turbine is

(*a*) Pelton ☐ (*b*) Kaplan ☐

(*c*) Francis ☐ (*d*) Pelton with more jets. ☐

344. Run-away speed of a Pelton wheel means

(*a*) full load speed ☐

(*b*) no load speed ☐

(*c*) no load speed with no governor mechanism ☐

(*d*) none of the above. ☐

345. Spouting velocity means

(*a*) actual velocity of jet ☐ (*b*) ideal velocity of jet ☐

(*c*) half of ideal velocity of jet ☐ (*d*) none of the above. ☐

346. Surge tank in a pipe line is used to

(*a*) reduce the loss of head due to friction in pipe ☐

(*b*) make the flow uniform in pipe ☐

(*c*) relieve the pressure due to water hammer ☐

(*d*) none of the above. ☐

347. Hydraulic ram is a device used for

(*a*) storing energy of a water in the form of pressure energy ☐

(*b*) increasing pressure intensity of water ☐

(*c*) lifting small quantity of water to a greater height by means of large quantity of water falling through small height ☐

(*d*) none of the above. ☐

348. For low head and high discharge, the suitable turbine is

(*a*) Pelton ☐ (*b*) Francis ☐

(*c*) Kaplan ☐ (*d*) None of the above. ☐

349. For high head and low discharge, the suitable turbine is

(*a*) Pelton ☐ (*b*) Francis ☐

(*c*) Kaplan ☐ (*d*) None of the above. ☐

350. The flow of water, leaving the impeller, in a centrifugal pump casing is

(*a*) forced vortex flow ☐ (*b*) free vortex flow ☐

(*c*) centrifugal flow ☐ (*d*) none of the above. ☐

351. Rotameter is used for measuring

(*a*) density of fluids ☐ (*b*) velocity of fluids in pipes ☐

(*c*) discharge of fluids ☐ (*d*) viscosity of fluids. ☐

352. A current meter is a device used for measuring

(*a*) velocity ☐ (*b*) viscosity ☐

(*c*) current ☐ (*d*) pressure. ☐

353. A hot wire anemometer is a device used for measuring

(*a*) viscosity ☐ (*b*) velocity of gases ☐

(*c*) pressure of gases ☐ (*d*) none of the above. ☐

354. D′ Aubuissons efficiency of a Hydraulic Ram as compared to Rankine's efficiency is

(*a*) less ☐ (*b*) more ☐

(*c*) equal ☐ (*d*) none of the above. ☐

355. The value of specific weight for water in S.I. units is equal to

(*a*) 981 N/m³ ☐ (*b*) 98.1 N/m³ ☐
(*c*) 9810 N/m³ ☐ (*d*) 1000 N/m³. ☐

356. The angle of contact (θ) between water and glass tube in case of capillary rise is equal to

(*a*) 0° ☐ (*b*) 90° ☐
(*c*) 128° ☐ (*d*) 150°. ☐

357. The angle of contact (θ) between mercury and glass tube in case of capillary depression is

(*a*) 0° ☐ (*b*) 90° ☐
(*c*) 128° ☐ (*d*) 150°. ☐

358. Numerical value of gauge pressure is

(*a*) more than absolute pressure ☐ (*b*) less than absolute pressure ☐
(*c*) equal to the absolute pressure ☐ (*d*) none of the above. ☐

359. Hydraulic mean depth is given by

(*a*) $\frac{P}{A}$ ☐ (*b*) $\frac{A}{P}$ ☐
(*c*) $A \times P$ ☐ (*d*) $\frac{1}{AP}$ ☐

where A = Area and P = Wetted perimeter.

360. For a superonic flow, velocity increases

(*a*) with the decrease of area of flow ☐ (*b*) with the increase of area of flow ☐
(*c*) when area of flow is constant ☐ (*d*) none of the above. ☐

Channel Flow II

361. A flow in a channel will be laminar, if Reynolds number of the flow

(*a*) is less than 2000 ☐ (*b*) is less than 1500 ☐
(*c*) is less than 1000 ☐ (*d*) is less than 500. ☐

362. The flow in the channel will be turbulent if Reynolds number is more than

(*a*) 500 ☐ (*b*) 1000 ☐
(*c*) 1500 ☐ (*d*) 2000. ☐

363. If the depth of flow along the length of the channel is constant, then flow is known as

(*a*) laminar ☐ (*b*) uniform ☐
(*c*) non-uniform ☐ (*d*) steady. ☐

364. If the depth of flow along the length of the channel is variable, then flow is known as

(*a*) laminar ☐ (*b*) uniform ☐
(*c*) non-uniform ☐ (*d*) steady. ☐

365. If the velocity at a section of a channel is constant with respect to time, then the flow in the channel is known as

(*a*) uniform ☐ (*b*) laminar ☐
(*c*) steady ☐ (*d*) unsteady. ☐

366. If the velocity at a section of a channel is variable with respect to time, then the flow in the channel is known as

(*a*) uniform ☐ (*b*) laminar ☐

(*c*) steady ☐ (*d*) unsteady. ☐

367. If the depth of flow in a channel is varying slowly, then the flow is known as

(*a*) G.V.F. ☐ (*b*) R.V.F. ☐

(*c*) P.V.F. ☐ (*d*) M.V.F. ☐

368. If the depth of flow in a channel is varying rapidly, then the flow is known as

(*a*) G.V.F. ☐ (*b*) R.V.F. ☐

(*c*) P.V.F. ☐ (*d*) M.V.F. ☐

369. For sub-critical flow in a channel

(*a*) $F_e = 1$ ☐ (*b*) $F_e > 1$ ☐

(*c*) $F_e < 1$ ☐ (*d*) none of the above. ☐

where F_e = Froude number.

370. Hydraulic mean depth is given by

(*a*) $m = \frac{P}{A}$ ☐ (*b*) $\frac{A}{P}$ ☐

(*c*) $m = A \times P$ ☐ (*d*) $\frac{1}{AP}$ ☐

where A = Area of flow, P = Wetted perimeter, and m = Hydraulic mean depth.

371. The Froude number is given by

(*a*) $F_e = \sqrt{\frac{V}{gD}}$ ☐ (*b*) $F_e = \frac{V}{\sqrt{gD}}$ ☐

(*c*) $F_e = \frac{V \times g}{\sqrt{D}}$ ☐ (*d*) $F_e = \frac{g}{\sqrt{VD}}$ ☐

where V = Velocity of flow, D = Hydraulic depth of channel, and

g = Acceleration due to gravity.

372. The hydraulic depth of a channel is equal to

(*a*) $D = A \times T$ ☐ (*b*) $D = \frac{A}{T}$ ☐

(*c*) $D = \frac{1}{AT}$ ☐ (*d*) $D = \frac{T}{A}$ ☐

where A = Area of flow and T = Top width.

373. Chezy's formula is used for finding ...... in a channel.

(*a*) velocity ☐ (*b*) discharge ☐

(*c*) slope ☐ (*d*) hydraulic mean depth. ☐

374. Chezy's formula is given by

(a) $V = \sqrt{C\,mi}$ ☐ (b) $Q = AC\sqrt{mi}$ ☐

(c) $V = C\sqrt{mi}$ ☐ (d) $m = \frac{A}{P}$ ☐

where V = Velocity of flow, C = Chezy's constant,

m = Hydraulic mean depth, and i = Slope of the bed of channel.

375. Manning's formula is used for finding ...... in a channel.

(a) rate of flow ☐ (b) velocity ☐

(c) hydraulic mean depth ☐ (d) slope. ☐

376. Manning's formula is given by

(a) $\frac{1}{N}\sqrt{mi}$ ☐ (b) $\frac{1}{N}\, m^{2/3}\, i^{1/2}$ ☐

(c) $\frac{1}{N}\, m^{1/6}\, i^{1/2}$ ☐ (d) $\frac{1}{N}\, m^{2/5}\, i^{3/4}$ ☐

where N = Manning's constant, m = Hydraulic mean depth, and

i = Slope of the bed of channel.

377. The relation between Chezy's constant and Manning's constant is given by

(a) $C = \frac{1}{N}\, m^{1/2}$ ☐ (b) $C = \frac{1}{N} m^{1/3}$ ☐

(c) $C = \frac{1}{N} m^{1/6}$ ☐ (d) $C = \frac{1}{N} m^{1/9}$. ☐

378. Which one of the following is a dimensionless constant?

(a) Chezy's constant ☐ (b) Manning's constant ☐

(c) Reynolds number ☐ (d) none of the above. ☐

379. The dimensions of the Chezy's constant is given by

(a) $L^{1/2}\, T^{1/2}$ ☐ (b) $L^{1/2}\, T^{3/4}$ ☐

(c) $L^{1/2}\, T^{-1}$ ☐ (d) $LT^{1/2}$. ☐

380. The expression $\left(\frac{b \times d}{b + 2d}\right)$ represents the hydraulic mean depth for

(a) triangular channel ☐ (b) rectangular channel ☐

(c) trapezoidal channel ☐ (d) circular channel ☐

where b = Width of channel and d = Depth of flow.

381. The area of flow of a trapezoidal channel is given by

(a) $A = (b + 2nd) \times d$ ☐ (b) $A = (b + nd) \times d$ ☐

(c) $A = (2b + nd) \times d$ ☐ (d) $A = (b + nd) \times 2d$ ☐

where b = Width, d = Depth, and n = Side slope of the channel.

382. The wetted perimeter in a trapezoidal is given by

(*a*) $P = b + d\sqrt{n^2+1}$ ☐ (*b*) $P = 2b + d\sqrt{n^2+1}$ ☐

(*c*) $P = b + 2d\sqrt{n^2+1}$ ☐ (*d*) $P = 2b + d\sqrt{n^2+1}$. ☐

383. If angle subtended at the centre of the circular channel by free surface of water is '2θ', then wetted perimeter is given by,

(*a*) $P = R \times \theta$ ☐ (*b*) $P = 2R\theta$ ☐

(*c*) $P = 4R\theta$ ☐ (*d*) $P = 3R\theta$. ☐

384. The area of flow for a circular channel is given by

(*a*) $A = R^2 (\theta - 2 \sin 2\theta)$ ☐ (*b*) $A = 2R^2 (2\theta - \sin 2\theta)$ ☐

(*c*) $A = R^2\left(\theta - \frac{\sin 2\theta}{2}\right)$ ☐ (*d*) $A = R^2\left(\frac{\theta}{2} - \sin 2\theta\right)$ ☐

where R = Radius and 2θ = Angle at the centre.

385. If the angle of the triangular channel is θ, then area of flow is given by

(*a*) $A = 2d^2 \tan 2\theta$ ☐ (*b*) $A = d^2 \tan \theta$ ☐

(*c*) $A = 3d^2 \tan \frac{\theta}{2}$ ☐ (*d*) $A = d^2 \tan\left(\frac{\theta}{2}\right)$ ☐

where d = Depth of flow.

386. The wetted perimeter P in the above question is given by

(*a*) $P = 2d \sec\left(\frac{\theta}{2}\right)$ ☐ (*b*) $P = \frac{d}{2} \sec \theta$ ☐

(*c*) $P = d \sec 2\theta$ ☐ (*d*) $P = 2d \sec \theta$. ☐

387. A section of a channel is most efficient if the discharge through the channel is maximum for a given

(*a*) area of flow ☐ (*b*) surface resistance ☐

(*c*) slope of the bed of channel ☐ (*d*) all of the above. ☐

388. A rectangular channel will be most efficient if depth of flow is equal to

(*a*) width of the channel ☐ (*b*) twice the width of the channel ☐

(*c*) half the width of channel ☐ (*d*) one-quarter width of channel. ☐

389. For the most efficient rectangular channel, the hydraulic mean depth should be equal to

(*a*) depth of flow ☐ (*b*) half of depth of flow ☐

(*c*) twice the depth of flow ☐ (*d*) one-quarter depth of flow. ☐

390. A trapezoidal channel will be most efficient if length of sloping side is equal to

(*a*) twice the top width ☐ (*b*) top width ☐

(*c*) half of the top width ☐ (*d*) one-quarter of top width. ☐

391. If d = depth of flow, b = width of trapezoidal channel and n = slope of the sides of the channel, then for most efficient trapezoidal channel, the condition is

(a) $d\sqrt{n^2+1} = b + 2nd$ ☐ (b) $2d\sqrt{n^2+1} = b + nd$ ☐

(c) $d\sqrt{n^2+1} = b + 2nd$ ☐ (d) $2d\sqrt{n^2+1} = b + 2nd$. ☐

392. For the most efficient trapezoidal channel, the hydraulic mean depth should be

(a) twice the depth of flow ☐ (b) 1.5 times the depth of flow ☐

(c) equal to the depth of flow ☐ (d) half the depth of flow. ☐

393. A trapezoidal channel will be most efficient if

(a) three sides of trapezoidal are equal ☐

(b) three sides of trapezoidal are unequal ☐

(c) three sides of trapezoidal section are tangential to the semi-circle described on the water line ☐

(d) none of the above. ☐

394. A circular channel will have maximum velocity if depth of flow is equal to

(a) diameter of circular channel ☐

(b) 0.81 times the diameter of circular channel ☐

(c) 0.75 times the diameter of circular channel ☐

(d) 0.625 times the diameter of circular channel. ☐

395. The rate of flow through a circular channel will be maximum if depth of flow is equal to

(a) diameter of circular channel ☐ (b) 0.95 times the dia. of circular channel ☐

(c) 0.81 times the dia. of circular channel ☐ (d) 0.75 times the dia. of circular channel. ☐

396. The total energy of flowing water in a channel with respect to bed of channel is known as

(a) total energy ☐ (b) specific energy ☐

(c) hydraulic gradient ☐ (d) mechanical energy. ☐

397. The specific energy is given by

(a) $y + \dfrac{V^2}{2g}$ ☐ (b) $2y + \dfrac{V^2}{2g}$ ☐

(c) $y + \dfrac{V^2}{g}$ ☐ (d) $\dfrac{P}{w} + z + \dfrac{V^2}{2g}$ ☐

where y = Depth of flow and V = Velocity of flow.

398. Critical depth is the depth of water at which

(a) specific energy is maximum ☐

(b) specific energy is minimum ☐

(c) specific energy is zero ☐

(d) specific energy is one plus minimum specific energy. ☐

399. For a rectangular channel, the critical depth (y_c) is given by

(a) $y_c = \frac{q^2}{g}$ ☐ (b) $y_c = \left(\frac{3q^2}{g}\right)^{1/3}$ ☐

(c) $y_c = \left(\frac{q^2}{g}\right)^{1/3}$ ☐ (d) $y_c = \left(\frac{q^2}{g}\right)^{1/4}$ ☐

where q = Rate of flow per unit width.

400. The velocity of flow, corresponding to critical depth is known as

(a) maximum velocity ☐ (b) minimum velocity ☐

(c) critical velocity ☐ (d) average velocity. ☐

401. Hydraulic jump will take place if depth of flow is

(a) more than critical depth ☐ (b) equal to critical depth ☐

(c) less than critical depth ☐ (d) two times the critical depth. ☐

402. The loss of energy during hydraulic jump is given by

(a) $h_L = \frac{(y_1 + y_2)}{4y_1y_2}$ ☐ (b) $h_L = \frac{(y_2 - y_2)^2}{2y_1y_2}$ ☐

(c) $h_L = \frac{(y_2 - y_1)^2}{2y_1y_2}$ ☐ (d) $h_L = \frac{(y_2 - y_1)^3}{4y_1y_2}$ ☐

where y_1 = Depth of water before hydraulic jump

y_2 = Depth of water after hydraulic jump.

Tick mark the true or false statement:

403. Newton's law of viscosity states that the shear stress is directly proportional to rate of shear strain.

(a) True ☐ (b) False. ☐

404. Kinematic viscosity is equal to viscosity multiplied by density.

(a) True ☐ (b) False. ☐

405. Surface tension has the unit of force per unit length.

(a) True ☐ (b) False. ☐

406. Gauge pressure at a point is equal to absolute pressure minus atmospheric pressure.

(a) True ☐ (b) False. ☐

407. The increase of temperature decreases the viscosity of a gas.

(a) True ☐ (b) False. ☐

408. The gases are considered incompressible when Mach number is less than 0.2.

(a) True ☐ (b) False. ☐

409. Atmospheric pressure head in terms of meter of water is 7.5 m.

(a) True ☐ (b) False. ☐

410. The centre of pressure of a plane horizontal surface immersed in a liquid coincides with the centre of gravity.

(*a*) True ☐ (*b*) False. ☐

411. For a floating body, the buoyant force passes through the centre of gravity of the body.

(*a*) True ☐ (*b*) False. ☐

412. A submerged body will be in stable equilibrium if the centre of buoyancy is above the centre of gravity.

(*a*) True ☐ (*b*) False. ☐

413. The necessary condition for the flow to be steady is that the velocity changes at a point with respect to time.

(*a*) True ☐ (*b*) False. ☐

414. The flow in the pipe is laminar if the Reynolds number is equal or less than 2000.

(*a*) True ☐ (*b*) False. ☐

415. Continuity equation is based on law of conservation of energy.

(*a*) True ☐ (*b*) False. ☐

416. The pressure variation along the radial direction for vertex flow in a horizontal plane is expressed as $\frac{dp}{dr} = \rho \frac{V^2}{r}$.

(*a*) True ☐ (*b*) False. ☐

417. The coefficient of discharge for an orifice is more than that of for a mouthpiece.

(*a*) True ☐ (*b*) False. ☐

418. For the laminar flow through a circular pipe, the variation of shear stress across the cross-section of the pipe is parabolic.

(*a*) True ☐ (*b*) False. ☐

419. The value of momentum correction factor for the viscous flow through a circular pipe is equal to two.

(*a*) True ☐ (*b*) False. ☐

420. For a hydrodynamically smooth boundary, the term $\frac{k}{\delta'}$ should be less than 0.25 (where k = average height of the irregularities from the boundary and δ' is the thickness of laminar sub-layer).

(*a*) True ☐ (*b*) False. ☐

421. The relation between coefficient of friction (f) and Reynolds number (R_e) for laminar flow through a pipe is given by $f = \frac{16}{R_e}$.

(*a*) True ☐ (*b*) False. ☐

422. Power transmitted through a pipe will be maximum when the head lost due to friction in the pipe is one-third of the total at the inlet of the pipe.

(*a*) True ☐ (*b*) False. ☐

423. The ratio of inertia force to viscous force is known as Reynolds number.

(*a*) True ☐ (*b*) False. ☐

424. Boundary layer on a flat plate is known as laminar boundary layer if Reynolds number is less than 2000.

(*a*) True ☐ (*b*) False. ☐

425. The boundary layer separation takes place if pressure gradient is positive.

(*a*) True ☐ (*b*) False. ☐

426. The force exerted by a flowing fluid on a solid body in the direction of flow is known as drag force.

(*a*) True ☐ (*b*) False. ☐

427. A body is called streamlined body when it is placed in a flow and the surface of the body coincides with the streamlines.

(*a*) True ☐ (*b*) False. ☐

428. The terminal velocity of a falling body is equal to the maximum constant velocity with which body will fall.

(*a*) True ☐ (*b*) False. ☐

429. For isothermal process, the velocity of sound wave is equal to $\sqrt{\frac{p}{\rho}}$, where p = pressure and ρ = density.

(*a*) True ☐ (*b*) False. ☐

430. Flow of a fluid in a pipe takes place from higher pressure to lower pressure and not from higher energy to lower energy.

(*a*) True ☐ (*b*) False. ☐

431. The shear velocity (u_*) is equal to $\sqrt{\frac{\tau_0}{\rho}}$, where τ_0 = shear stress at the surface and ρ = density of fluid.

(*a*) True ☐ (*b*) False. ☐

432. When pipes are connected in parallel, the total loss of head is equal to the sum of the loss of head in each pipe.

(*a*) True ☐ (*b*) False. ☐

433. The maximum efficiency of power transmission through pipe is 50%.

(*a*) True ☐ (*b*) False. ☐

434. Water-hammer in pipes takes place when the flowing fluid is suddenly brought to rest by closing the valve.

(*a*) True ☐ (*b*) False. ☐

435. For super-sonic flow, if the area of flow increases, then the velocity also increases.

(*a*) True ☐ (*b*) False. ☐

436. The velocity through a channel for uniform flow by Chezy's formula is given by, $V = C\sqrt{m \times i}$, where C = Chezy's constant, m = hydraulic mean depth and i = slope of the bed of channel.

(*a*) True ☐ (*b*) False. ☐

437. Specific speed of a turbine is defined as the speed of the turbine which produces unit power at unit discharge.

(*a*) True ☐ (*b*) False. ☐

438. A turbine converts hydraulic energy into mechanical energy whereas a pump converts mechanical energy into hydraulic energy.

(*a*) True ☐ (*b*) False. ☐

439. A jet of water is having a velocity V and strikes a series of vertical plates which are moving with velocity u. The efficiency of the system will be maximum if $u = \frac{V}{2}$.

(*a*) True ☐ (*b*) False. ☐

440. Jet ratio is defined as the ratio of the diameter of the jet to the diameter of Pelton turbine.

(*a*) True ☐ (*b*) False. ☐

Fill in the blanks:

441. Kaplan tubine is ...... turbine.

(*a*) a reaction ☐ (*b*) an impulse ☐

442. The term $\left(\frac{N}{\sqrt{H}}\right)$ represents ...... of a turbine.

(*a*) specific speed ☐ (*b*) unit speed. ☐

where N = Speed of the turbine and H = Net head on the turbine.

443. Stoke is the unit of ....

(*a*) viscosity ☐ (*b*) kinematic viscosity. ☐

444. Absolute pressure ...... atmospheric pressure is equal to gauge pressure.

(*a*) plus ☐ (*b*) minus. ☐

445. If the Reynolds number is equal to or less than ...... , the flow in pipe is known as laminar.

(*a*) 4000 ☐ (*b*) 2000. ☐

446. The pitot-tube is used to measure ...... at a point in a flowing liquid.

(*a*) discharge ☐ (*b*) velocity. ☐

447. The velocity profile, across the section of a pipe for laminar flow, follows ...... law.

(*a*) linear ☐ (*b*) parabolic. ☐

448. The ratio of inertia force to ...... force is known as Reynolds number.

(*a*) gravity ☐ (*b*) viscous. ☐

449. Kinematic similarity means similarity of ...... between model and prototype.

(*a*) motion ☐ (*b*) forces. ☐

450. The force exerted by a flowing fluid on a solid body ...... is known as drag force.

(*a*) in the direction of flow ☐ (*b*) perpendicular to the direction of flow. ☐

III. OBJECTIVE TYPE QUESTIONS FROM COMPETITIVE EXAMINATIONS

Tick mark the most appropriate answer:

1. For a Newtonian fluid

(*a*) shear stress is proportional to shear strain ☐
(*b*) rate of shear stress is proportional to shear strain ☐
(*c*) shear stress is proportional to rate of shear strain ☐
(*d*) rate of shear stress is proportional to rate of shear strain. ☐

(GATE-ME-2006)

2. Stoke is the unit of

(*a*) surface tension ☐ (*b*) viscosity ☐
(*c*) kinematic viscosity ☐ (*d*) none of the above. ☐

3. In a two-dimensional velocity field with velocities u and v along x- and y-directions respectively, the convective acceleration along the x-direction is given by

(*a*) $u\dfrac{\partial u}{\partial x}+v\dfrac{\partial u}{\partial y}$ ☐ (*b*) $u\dfrac{\partial u}{\partial x}+v\dfrac{\partial v}{\partial y}$ ☐

(*c*) $u\dfrac{\partial v}{\partial x}+v\dfrac{\partial u}{\partial y}$ ☐ (*d*) $v\dfrac{\partial u}{\partial x}+u\dfrac{\partial u}{\partial y}$. ☐

(GATE-ME-2006)

4. Centre of pressure of a plane vertical surface immersed in a liquid is

(*a*) above the C.G. of the plane surface ☐
(*b*) at the C.G. of the plane surface ☐
(*c*) below the C.G. of the plane surface ☐
(*d*) none of the above. ☐

5. Maximum velocity of a one-dimensional incompressible fully developed viscous flow, between two fixed parallel plates, is 6 ms^{-1}. The mean velocity (in ms^{-1}) of the flow is

(*a*) 2 ☐ (*b*) 3 ☐
(*c*) 4 ☐ (*d*) 5. ☐

(GATE-ME-2010)

6. If 'x' is the distance measured from the leading edge of a flat plate, then laminar boundary layer thickness varies as

(*a*) $1/x$ ☐ (*b*) $x^{4/5}$ ☐
(*c*) x^2 ☐ (*d*) $x^{1/2}$. ☐

(GATE-ME-2002)

7. Continuity equation deals with the law of conservation of

(a) mass ☐ (b) momentum ☐

(c) energy ☐ (d) none of the above. ☐

8. Consider an incompressible laminar boundary layer flow over a flat plate of length L, aligned with the direction of an incoming uniform free stream. If F is the ratio of the drag force on the front half of the plate to the drag force on the rear half, then

(a) $F < 1/2$ ☐ (b) $F = 1/2$ ☐

(c) $F = 1$ ☐ (d) $F > 1$. ☐

(GATE-ME-2007)

9. For a floating body, if the metacentre is above the centre of gravity, the equilibrium is said to be

(a) stable ☐ (b) unstable ☐

(c) neutral ☐ (d) none of the above. ☐

10. A Kaplan turbine is

(a) a high head mixed flow turbine ☐ (b) a low head axial flow turbine ☐

(c) an outward flow reaction turbine ☐ (d) an impulse inward flow turbine. ☐

(GATE-ME-1997)

11. A jet of water issues from a Nozzle with a velocity 20 m/s and it impinges normally on a flat plate moving away from it at 10 m/s. The cross-sectional area of the jet is 0.01 m^2, and the density of water = 1000 kg/m^3. The force developed on the plate is

FIGURE 1.7

(a) 1000 N ☐ (b) 100 N ☐

(c) 10 N ☐ (d) 2000 N. ☐

(GATE-ME-2010)

12. The ratio of inertia force to viscous force is known as

(a) Reynolds number ☐ (b) Froude number ☐

(c) Mach number ☐ (d) Euler number. ☐

13. Water flows through a vertical contraction from a pipe of diameter d to another of diameter $d/2$ (see Fig. 1.8). The flow velocity at the inlet to the contraction is 2 m/s and pressure 200 kN/m^2. If the height of the contraction measures 2 m, then pressure at the exit of the contraction will be very nearly

(a) 168 kN/m^2 ☐

(b) 192 kN/m^2 ☐

(c) 150 kN/m^2 ☐

(d) 174 kN/m^2. ☐

(GATE-ME-1999)

FIGURE 1.8

14. The power transmitted by a flowing fluid through a pipe will be maximum if head loss due to friction in pipe is

(a) equal to total head at inlet ☐ (b) half the total head at inlet ☐

(c) one-third of the total head at inlet ☐ (d) one-fourth of the total head at inlet. ☐

15. A water container is kept on a weighing balance. Water from a tap is falling vertically into the container with a volume flow rate of 'Q'; the velocity of the water when it hits the water surface is 'U'. At a particular instant of time the total mass of the container and water is 'm'. The force registered by the weighing balance at this instant of time is

(a) $mg + \rho QU$ ☐ (b) $mg + 2\rho QU$ ☐

(c) $mg + \rho QU^2/2$ ☐ (d) $\rho QU^2/2$. ☐

(GATE-ME-2003)

16. The thickness of turbulent boundary layer at a distance x from the leading edge of the flat plate varies as

(a) $x^{4/5}$ ☐ (b) $x^{1/2}$ ☐

(c) $x^{1/5}$ ☐ (d) $x^{2/5}$. ☐

17. If there are 'm' physical quantity and 'n' fundamental dimensions in a particular process, the number of non-dimensional parameters is

(a) $m + n$ ☐ (b) $m \times n$ ☐

(c) $m - n$ ☐ (d) m/n. ☐

(GATE-ME-2002)

18. The velocity profile of a fully developed laminar flow in a straight circular pipe, as shown in Fig. 1.9 is given by the expression

$$u(r) = \frac{-R^2}{4\mu}\left(\frac{dp}{dx}\right)\left(1-\frac{r^2}{R^2}\right)$$

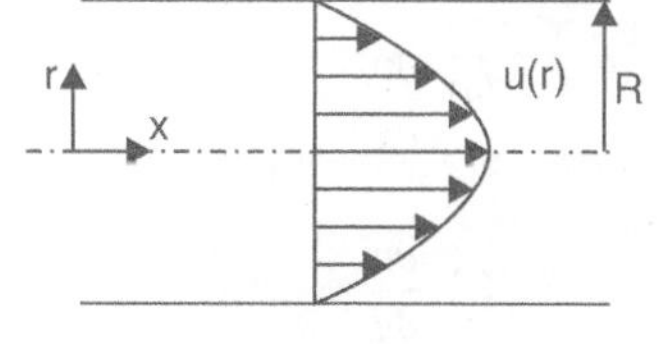

FIGURE 1.9

where $\frac{dp}{dx}$ is a constant.

The average velocity of fluid in the pipe is

(a) $\frac{-R^2}{8\mu}\left(\frac{dp}{dx}\right)$ ☐ (b) $\frac{-R^2}{4\mu}\left(\frac{dp}{dx}\right)$ ☐

(c) $\frac{-R^2}{2\mu}\left(\frac{dp}{dx}\right)$ ☐ (d) $\frac{-R^2}{\mu}\left(\frac{dp}{dx}\right)$. ☐

(GATE-ME-2009)

19. An incompressible fluid flows over a flat plate with zero pressure gradient. The boundary layer thickness is 1 mm at a location where the Reynolds number is 1000. If the velocity of the fluid alone is increased by a factor of 4, then the boundary layer thickness sat the same location, in mm will be

(a) 4 ☐ (b) 2 ☐
(c) 0.5 ☐ (d) 0.25. ☐

(GATE-ME-2012)

20. Existence of velocity potential implies that
(a) Fluid is in continuum ☐ (b) Fluid is irrotational ☐
(c) Fluid is ideal ☐ (d) Fluid is compressible. ☐

(GATE-ME-1994)

21. A streamline and an equipotential line in a flow field
(a) are parallel to each other ☐ (b) are perpendicular to each other ☐
(c) intersect at an acute angle ☐ (d) are identical. ☐

(GATE-ME-2011)

22. A hydraulic turbine develops 1000 kW power for a head of 40 m. If the head is reduced to 20 m, the power developed (in kW) is
(a) 177 ☐ (b) 354 ☐
(c) 500 ☐ (d) 707. ☐

(GATE-ME-2010)

23. The velocity components in the x- and y-directions of a two dimensional potential flow are u and v, respectively. Then $\frac{\partial u}{\partial y}$ is equal to

(a) $\frac{\partial v}{\partial x}$ ☐ (b) $-\frac{\partial v}{\partial x}$ ☐

(c) $\frac{\partial v}{\partial y}$ ☐ (d) $-\frac{\partial v}{\partial y}$. ☐

(GATE-ME-2005)

24. In a Pelton wheel, the bucket peripheral speed is 10 m/s, the water jet velocity is 25 m/s and volumetric flow rate of the jet is 0.1 m^3/s. If the jet deflection angle is 120°C and the flow is ideal, the power developed is
(a) 7.5 kW ☐ (b) 15.0 kW ☐
(c) 22.5 kW ☐ (d) 37.5 kW. ☐

(GATE-ME-2006)

25. The SI unit of kinematic viscosity (v) is
(a) m^2/sec ☐ (b) kg/m-sec ☐
(c) m/sec^2 ☐ (d) m^3/sec^2. ☐

(GATE-ME-2001)

26. The stream function in a two dimensional flow field is given by $\psi = x^2 - y^2$. The magnitude of the velocity at point (1, 1) is

(*a*) 2 ☐ (*b*) $2\sqrt{2}$ ☐
(*c*) 4 ☐ (*d*) 8. ☐

(GATE-ME-1989)

27. Flow separation in flow past in a solid object is caused by

(*a*) a reduction of pressure to vapour pressure ☐
(*b*) a negative pressure gradient ☐
(*c*) a positive pressure gradient ☐
(*d*) the boundary layer thickness reducing to zero. ☐

(GATE-ME-2002)

28. The discharge in m^3/s for laminar flow through a pipe of diameter 0.04 m having a centre line velocity of 1.5 m/s is

(*a*) $3\pi/50$ ☐ (*b*) $3\pi/2500$ ☐
(*c*) $3\pi/5000$ ☐ (*d*) $3\pi/10000$. ☐

(GATE-ME-1988)

29. The dimension of surface tension is

(*a*) ML^{-1} ☐ (*b*) L^2T^{-1} ☐
(*c*) $ML^{-1}T^{-1}$ ☐ (*d*) MT^{-2}. ☐

(GATE-ME-1996)

30. Refer to Fig. 1.10, the absolute pressure of gas *A* in the bulb is

(*a*) 771.2 mm Hg ☐
(*b*) 752.65 mm Hg ☐
(*c*) 767.35 mm Hg ☐
(*d*) 748.8 mm Hg. ☐

(GATE-ME-1997)

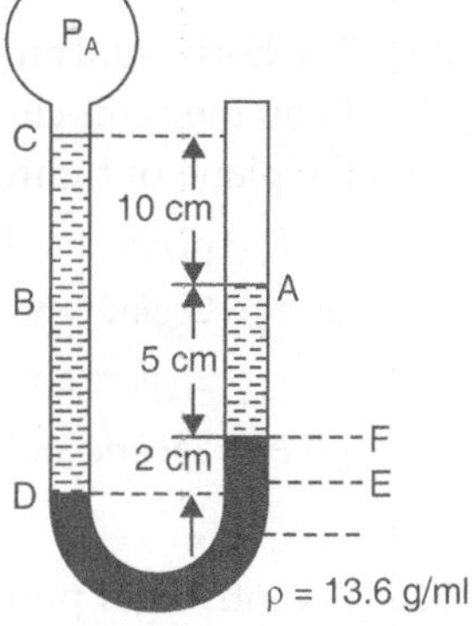

P_A
C
B
A
D
F
E
10 cm
5 cm
2 cm
ρ = 13.6 g/ml

FIGURE 1.10

31. For air flow over a flat plate, velocity (*U*) and boundary layer thickness (δ) can be expressed respectively, as

$$\frac{U}{U_\infty} = \frac{3}{2}\frac{y}{\delta} - \frac{1}{2}\left(\frac{y}{\delta}\right)^3 ; \delta = \frac{4.64x}{\sqrt{Re_x}}$$

If the free stream velocity is 2 m/s, and air has kinematic viscosity of 1.5×10^{-5} m²/s and density of 1.23 kg/m³, then wall shear stress at $x = 1$ m, is

(*a*) 2.36×10^2 N/m² ☐ (*b*) 43.6×10^{-3} N/m² ☐
(*c*) 4.36×10^{-3} N/m² ☐ (*d*) 2.18×10^{-3} N/m². ☐

(GATE-ME-2004)

32. For laminar flow through a long pipe, the pressure drop per unit length increases

(a) in linear proportion to the cross-sectional area ☐

(b) in proportion to the diameter of the pipe ☐

(c) in inverse proportion to the cross-sectional area ☐

(d) in inverse proportion to the square of cross-sectional area. ☐

(GATE-ME-1996)

33. In order to have maximum power from a Pelton turbine, the bucket speed must be

(a) equal to the jet speed ☐ (b) equal to half of the jet speed ☐

(c) equal to twice the jet speed ☐ (d) independent of the jet speed. ☐

(GATE-ME-2013)

34. A journal bearing has a shaft diameter of 40 mm and a length of 40 mm. The shaft is rotating at 20 rad/s and the viscosity of the lubricant is 20 mPa-s. The clearance is 0.020 mm. The loss of torque due to the viscosity of the lubricant is approximately

(a) 0.040 Nm ☐ (b) 0.252 Nm ☐

(c) 0.400 Nm ☐ (d) 0.652 Nm. ☐

(GATE-ME-2008)

35. A circular plate 1 m in diameter is submerged vertically in water such that its upper edge is 8 m below the free surface of water. The total hydrostatic pressure force on one side of plate is

(a) 6.7 kN ☐ (b) 65.4 kN ☐

(c) 45.0 kN ☐ (d) 77.0 kN. ☐

(GATE-ME-1988)

36. The horizontal and vertical hydrostatic forces F_x and F_y on the semi-circular gate, having a width 'w' into the plane of figure, are

(a) $F_x = \rho ghrw$ and $F_y = 0$ ☐

(b) $F_x = 2\rho ghrw$ and $F_y = 0$ ☐

(c) $F_x = 2\rho ghrw$ and $F_y = \rho gwr^2/2$ ☐

(d) $F_x = 2\rho ghrw$ and $F_y = \pi\rho gwr^2/2$. ☐

(GATE-ME-2001)

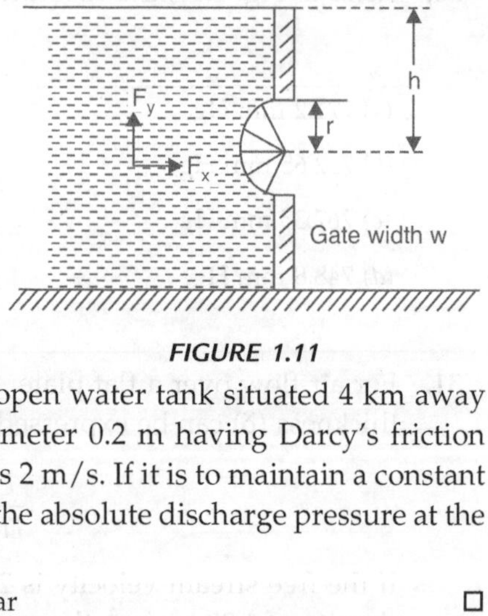

FIGURE 1.11

37. A centrifugal pump is required to pump water to an open water tank situated 4 km away from the location of the pump through a pipe of diameter 0.2 m having Darcy's friction factor for 0.01. The average speed of water in the pipe is 2 m/s. If it is to maintain a constant head of 5 m in the tank, neglecting other minor losses, the absolute discharge pressure at the pump exist is

(a) 0.449 bar ☐ (b) 5.503 bar ☐

(c) 44.911 bar ☐ (d) 55.203 bar. ☐

(GATE-ME-2004)

38. At a hydro electric power plant site, available head and flow rates are 24.5 m and 10.1 m^3/s respectively. If the turbine to be installed is required to run at 4.0 revolution per second (rps) with an overall efficiency of 90%, then suitable type of turbine for this site is

(*a*) Francis ☐ (*b*) Kaplan ☐

(*c*) Pelton ☐ (*d*) Propeller. ☐

(GATE-ME-2004)

39. A large hydraulic turbine is to generate 300 kW at 100 rpm under a head of 40 m. For initial testing, a 1 : 4 scale model of the turbine operates under a head of 10 m. The power generated by the model (in kW) will be

(*a*) 2.34 ☐ (*b*) 4.68 ☐

(*c*) 9.38 ☐ (*d*) 18.75. ☐

(GATE-ME-2006)

40. Match the items in columns I and II.

Column I	Column II
P. Centrifugal compressor	1. Axial flow
Q. Centrifugal pump	2. Surging
R. Pelton wheel	3. Priming
S. Kaplan turbine	4. Pure impulse

(*a*) *P*-2, *Q*-3, *R*-4, *S*-1 ☐ (*b*) *P*-2, *Q*-3, *R*-1, *S*-4 ☐

(*c*) *P*-3, *Q*-4, *R*-1, *S*-2 ☐ (*d*) *P*-1, *Q*-2, *R*-3, *S*-4. ☐

(GATE-ME-2007)

41. The velocity profile in fully developed laminar flow in a pipe of diameter D is given by $u = u_0(1 - 4r^2/D^2)$, where r is the radial distance from the centre. If the viscosity of the fluid is μ, the pressure drop across a length L of the pipe is

(*a*) $\frac{\mu u_0 L}{D^2}$ ☐ (*b*) $\frac{4\mu u_0 L}{D^2}$ ☐

(*c*) $\frac{8\mu u_0 L}{D^2}$ ☐ (*d*) $\frac{16\mu u_0 L}{D^2}$. ☐

(GATE-ME-2006)

42. The inlet angle of runner blades of a Francis turbine is 90°. The blades are so shaped that the tangential component of velocity at blade outlet is zero. The flow velocity remains constant throughout the blade passage and is equal to half of the blade velocity at runner inlet. The blade efficiency of the runner is

(*a*) 25% ☐ (*b*) 50% ☐

(*c*) 80% ☐ (*d*) 89%. ☐

(GATE-ME-2007)

43. A turbine is called impulse if at the inlet of the turbine

(*a*) total energy is only kinetic energy ☐
(*b*) total energy is only pressure energy ☐
(*c*) total energy is the sum of kinetic energy and pressure energy ☐
(*d*) none of the above. ☐

44. The velocity components in the x- and y-directions are given by $u = \lambda xy^3 - x^2y$, $v = xy^2 - \frac{3}{4}y^4$. The value of λ for a possible flow field involving an incompressible fluid is

(*a*) $-\frac{3}{4}$ ☐ (*b*) $-\frac{4}{3}$ ☐
(*c*) $\frac{4}{3}$ ☐ (*d*) 3. ☐

(GATE-ME-1995)

45. The specific speed of a turbine is given by

(*a*) $N_s = \frac{N\sqrt{P}}{H^{3/4}}$ ☐ (*b*) $N_s = \frac{N\sqrt{Q}}{H^{3/4}}$ ☐
(*c*) $N_s = \frac{N\sqrt{P}}{H^{5/4}}$ ☐ (*d*) $N_s = \frac{N\sqrt{P}}{H^{3/2}}$. ☐

46. A fluid flow is represented by the velocity field $\vec{V} = ax\vec{i} + ay\vec{j}$, where a is a constant. The equation of stream line passing through a point (1, 2) is

(*a*) $x - 2y = 0$ ☐ (*b*) $2x + y = 0$ ☐
(*c*) $2x - y = 0$ ☐ (*d*) $x + 2y = 0$. ☐

(GATE-ME-2004)

47. For high head and low discharge, the suitable turbine is

(*a*) Pelton ☐ (*b*) Francis ☐
(*c*) Kaplan ☐ (*d*) None of the above. ☐

48. Streamlines, path lines and streak lines are virtually identical for

(*a*) uniform flow ☐ (*b*) flow of ideal fluids ☐
(*c*) steady flow ☐ (*d*) non uniform flow. ☐

(GATE-ME-1994)

49. The value of kinetic energy correction factor for the viscous flow through a circular pipe is

(*a*) 1.33 ☐ (*b*) 1.5 ☐
(*c*) 2.0 ☐ (*d*) 2.25. ☐

50. In a flow field the stream lines and equipotential lines

(*a*) are parallel ☐
(*b*) cut at any angle ☐
(*c*) are orthogonal every where in the field ☐
(*d*) cut orthogonal except at the stagnation points. ☐

(GATE-ME-1994)

51. The velocity distribution in laminar flow through circular pipe, follows the

(*a*) parabolic law ☐ (*b*) linear law ☐

(*c*) logarithmic law ☐ (*d*) none of the above. ☐

52. For a fluid element in a two dimensional flow field (x-y plane), it will undergo

(*a*) translation only ☐ (*b*) translation and rotation ☐

(*c*) translation and deformation ☐ (*d*) deformation only. ☐

(GATE-ME-1994)

53. Two pipes of uniform section but different diameters carry water at the same volumetric flow rate. Water properties are the same in the two pipes. The Reynolds number, based on the pipe diameter

(*a*) is the same in the both pipes ☐ (*b*) is large in the narrow pipe ☐

(*c*) is smaller in the narrower pipe ☐ (*d*) depends on the pipe material. ☐

(GATE-PI-2008)

54. An incompressible fluid (kinematic viscosity = 7.4×10^{-7} m^2/s, specific gravity, 0.88) is held between two parallel plates. If the top plate is moved with a velocity of 0.5 m/s while the bottom one is held stationary, the fluid attains a linear velocity profile in the gap of 0.5 mm between these plates; the shear stress in **Pascal** on the surface of bottom plate is

(*a*) 0.651×10^{-3} ☐ (*b*) 0.651 ☐

(*c*) 6.51 ☐ (*d*) 0.651×10^{3}. ☐

(GATE-ME-2004)

55. In the given Fig. 1.12, if the pressure of gas in bulb A is 50 cm Hg vacuum and p_{atm} = 76 cm Hg, then height of column H is equal to

(*a*) 26 cm ☐

(*b*) 50 cm ☐

(*c*) 76 cm ☐

(*d*) 126 cm. ☐

(GATE-ME-2000)

FIGURE 1.12

ANSWERS

Answers to Objective Type Questions

1. (*c*)	**2.** (*b*)	**3.** (*b*)	**4.** (*b*)	**5.** (*b*)	**6.** (*c*)
7. (*b*), (*d*)	**8.** (*c*)	**9.** (*b*)	**10.** (*b*)	**11.** (*d*)	**12.** (*b*)
13. (*a*)	**14.** (*a*)	**15.** (*d*)	**16.** (*c*)	**17.** (*b*)	**18.** (*a*)
19. (*c*)	**20.** (*d*)	**21.** (*d*)	**22.** (*b*)	**23.** (*b*)	**24.** (*b*)
25. (*c*)	**26.** (*b*)	**27.** (*d*)	**28.** (*a*)	**29.** (*c*)	**30.** (*c*)
31. (*b*)	**32.** (*b*)	**33.** (*c*)	**34.** (*d*)	**35.** (*c*)	**36.** (*b*)
37. (*c*)	**38.** (*b*)	**39.** (*c*)	**40.** (*a*)	**41.** (*c*)	**42.** (*c*)
43. (*c*)	**44.** (*c*)	**45.** (*b*)	**46.** (*d*)	**47.** (*c*)	**48.** (*c*)
49. (*b*)	**50.** (*d*)	**51.** (*c*)	**52.** (*a*)	**53.** (*b*)	**54.** (*c*)

55. (*d*)	**56.** (*a*)	**57.** (*b*)	**58.** (*c*)	**59.** (*d*)	**60.** (*c*)
61. (*b*)	**62.** (*b*)	**63.** (*b*)	**64.** (*d*)	**65.** (*c*)	**66.** (*c*)
67. (*c*)	**68.** (*c*)	**69.** (*a*)	**70.** (*c*)	**71.** (*d*)	**72.** (*d*)
73. (*b*), (*c*)	**74.** (*c*)	**75.** (*b*)	**76.** (*c*)	**77.** (*d*)	**78.** (*c*), (*b*)
79. (*d*)	**80.** (*c*)	**81.** (*c*)	**82.** (*d*)	**83.** (*b*)	**84.** (*d*)
85. (*d*)	**86.** (*c*)	**87.** (*c*)	**88.** (*b*)	**89.** (*d*)	**90.** (*c*)
91. (*d*)	**92.** (*a*)	**93.** (*a*)	**94.** (*a*)	**95.** (*c*)	**96.** (*c*)
97. (*c*)	**98.** (*c*)	**99.** (*c*)	**100.** (*d*)	**101.** (*a*)	**102.** (*c*)
103. (*a*)	**104.** (*a*)	**105.** (*b*)	**106.** (*a*)	**107.** (*b*)	**108.** (*c*)
109. (*c*)	**110.** (*b*)	**111.** (*d*)	**112.** (*d*)	**113.** (*c*)	**114.** (*b*)
115. (*b*)	**116.** (*b*)	**117.** (*c*)	**118.** (*c*)	**119.** (*d*)	**120.** (*b*)
121. (*a*)	**122.** (*c*)	**123.** (*c*)	**124.** (*b*)	**125.** (*b*)	**126.** (*b*)
127. (*c*)	**128.** (*c*)	**129.** (*b*)	**130.** (*a*)	**131.** (*b*)	**132.** (*c*)
133. (*b*)	**134.** (*c*)	**135.** (*b*)	**136.** (*b*)	**137.** (*c*)	**138.** (*c*)
139. (*b*)	**140.** (*b*)	**141.** (*c*)	**142.** (*d*)	**143.** (*c*)	**144.** (*a*)
145. (*d*)	**146.** (*d*)	**147.** (*a*)	**148.** (*c*)	**149.** (*b*)	**150.** (*c*)
151. (*d*)	**152.** (*d*)	**153.** (*d*)	**154.** (*c*)	**155.** (*d*)	**156.** (*d*)
157. (*c*)	**158.** (*b*)	**159.** (*a*)	**160.** (*a*)	**161.** (*b*)	**162.** (*c*)
163. (*b*)	**164.** (*a*)	**165.** (*d*)	**166.** (*a*)	**167.** (*b*)	**168.** (*c*)
169. (*c*)	**170.** (*b*)	**171.** (*b*)	**172.** (*c*)	**173.** (*f*)	**174.** (*b*)
175. (*b*)	**176.** (*a*)	**177.** (*a*)	**178.** (*b*)	**179.** (*c*)	**180.** (*d*)
181. (*c*)	**182.** (*a*)	**183.** (*b*)	**184.** (*b*)	**185.** (*d*)	**186.** (*b*)
187. (*d*)	**188.** (*a*)	**189.** (*b*)	**190.** (*b*)	**191.** (*c*)	**192.** (*c*)
193. (*c*)	**194.** (*c*)	**195.** (*c*)	**196.** (*a*)	**197.** (*b*)	**198.** (*c*)
199. (*d*)	**200.** (*a*)	**201.** (*b*)	**202.** (*c*)	**203.** (*c*)	**204.** (*c*)
205. (*d*)	**206.** (*a*)	**207.** (*b*)	**208.** (*b*)	**209.** (*b*)	**210.** (*c*)
211. (*b*)	**212.** (*c*)	**213.** (*a*)	**214.** (*b*)	**215.** (*a*)	**216.** (*b*)
217. (*b*)	**218.** (*a*)	**219.** (*b*)	**220.** (*b*)	**221.** (*c*)	**222.** (*a*)
223. (*b*)	**224.** (*c*)	**225.** (*c*)	**226.** (*b*)	**227.** (*c*)	**228.** (*b*)
229. (*c*)	**230.** (*c*)	**231.** (*b*)	**232.** (*b*)	**233.** (*a*)	**234.** (*b*)
235. (*b*)	**236.** (*a*)	**237.** (*b*)	**238.** (*a*)	**239.** (*a*)	**240.** (*c*)
241. (*b*)	**242.** (*a*)	**243.** (*a*)	**244.** (*c*)	**245.** (*c*)	**246.** (*b*)
247. (*b*)	**248.** (*c*)	**249.** (*b*)	**250.** (*c*)	**251.** (*a*)	**252.** (*c*)
253. (*b*)	**254.** (*b*)	**255.** (*a*)	**256.** (*c*)	**257.** (*a*)	**258.** (*c*)
259. (*a*)	**260.** (*b*)	**261.** (*c*)	**262.** (*c*)	**263.** (*b*)	**264.** (*c*)
265. (*b*)	**266.** (*a*)	**267.** (*b*)	**268.** (*a*)	**269.** (*c*)	**270.** (*d*)
271. (*c*)	**272.** (*b*)	**273.** (*a*)	**274.** (*b*)	**275.** (*b*)	**276.** (*c*)
277. (*c*)	**278.** (*b*)	**279.** (*d*)	**280.** (*b*)	**281.** (*c*)	**282.** (*b*)
283. (*a*)	**284.** (*b*)	**285.** (*c*)	**286.** (*d*)	**287.** (*c*)	**288.** (*d*)
289. (*a*)	**290.** (*a*)	**291.** (*a*)	**292.** (*b*)	**293.** (*a*)	**294.** (*c*)
295. (*b*), (*c*)	**296.** (*b*)	**297.** (*c*)	**298.** (*c*)	**299.** (*c*)	**300.** (*c*)
301. (*c*)	**302.** (*a*)	**303.** (*b*)	**304.** (*c*)	**305.** (*a*)	**306.** (*b*)

307. (*b*)	308. (*c*)	309. (*a*)	310. (*c*)	311. (*b*)	312. (*b*)
313. (*c*)	314. (*b*)	315. (*a*)	316. (*b*)	317. (*a*)	318. (*a*)
319. (*c*)	320. (*a*)	321. (*b*)	322. (*c*)	323. (*c*)	324. (*b*), (*d*)
325. (*c*)	326. (*a*)	327. (*c*)	328. (*b*)	329. (*a*)	330. (*b*)
331. (*a*)	332. (*a*)	333. (*c*)	334. (*a*)	335. (*b*)	336. (*b*)
337. (*a*)	338. (*a*)	339. (*d*)	340. (*a*)	341. (*b*)	342. (*c*)
343. (*b*)	344. (*c*)	345. (*b*)	346. (*c*)	347. (*c*)	348. (*c*)
349. (*a*)	350. (*b*)	351. (*c*)	352. (*a*)	353. (*b*)	354. (*b*)
355. (*c*)	356. (*a*)	357. (*c*)	358. (*b*)	359. (*b*)	360. (*b*)
361. (*d*)	362. (*d*)	363. (*b*)	364. (*c*)	365. (*c*)	366. (*d*)
367. (*a*)	368. (*b*)	369. (*c*)	370. (*b*)	371. (*b*)	372. (*b*)
373. (*a*)	374. (*c*)	375. (*b*)	376. (*b*)	377. (*c*)	378. (*c*)
379. (*c*)	380. (*b*)	381. (*b*)	382. (*c*)	383. (*b*)	384. (*c*)
385. (*d*)	386. (*a*)	387. (*d*)	388. (*c*)	389. (*b*)	390. (*c*)
391. (*d*)	392. (*d*)	393. (*c*)	394. (*b*)	395. (*b*)	396. (*b*)
397. (*a*)	398. (*b*)	399. (*c*)	400. (*c*)	401. (*c*)	402. (*d*).

True/False

403. (*a*)	404. (*b*)	405. (*a*)	406. (*a*)	407. (*b*)	408. (*a*)
409. (*b*)	410. (*a*)	411. (*b*)	412. (*a*)	413. (*b*)	414. (*a*)
415. (*b*)	416. (*a*)	417. (*b*)	418. (*b*)	419. (*b*)	420. (*a*)
421. (*a*)	422. (*a*)	423. (*a*)	424. (*b*)	425. (*a*)	426. (*a*)
427. (*a*)	428. (*a*)	429. (*a*)	430. (*b*)	431. (*a*)	432. (*b*)
433. (*b*)	434. (*a*)	435. (*a*)	436. (*a*)	437. (*b*)	438. (*a*)
439. (*a*)	440. (*b*).				

Fill in the Blanks

441. (*a*)	442. (*b*)	443. (*b*)	444. (*b*)	445. (*b*)	446. (*b*)
447. (*b*)	448. (*b*)	449. (*a*)	450. (*a*).		

Answers to Objective Type Questions from Competitive Examinations

1. (*c*)	2. (*c*)	3. (*a*)	4. (*c*)	5. (*c*)	6. (*d*)
7. (*a*)	8. (*d*)	9. (*a*)	10. (*b*)	11. (*a*)	12. (*a*)
13. (*c*)	14. (*c*)	15. (*a*)	16. (*a*)	17. (*c*)	18. (*a*)
19. (*c*)	20. (*b*)	21. (*b*)	22. (*b*)	23. (*a*)	24. (*c*)
25. (*a*)	26. (*b*)	27. (*c*)	28. (*d*)	29. (*d*)	30. (*a*)
31. (*c*)	32. (*c*)	33. (*b*)	34. (*a*)	35. (*b*)	36. (*d*)
37. (*b*)	38. (*a*)	39. (*a*)	40. (*a*)	41. (*d*)	42. (*c*)
43. (*a*)	44. (*d*)	45. (*c*)	46. (*c*)	47. (*a*)	48. (*c*)
49. (*c*)	50. (*d*)	51. (*a*)	52. (*b*)	53. (*b*)	54. (*b*)
55. (*b*).					

Chapter 2 ENGINEERING MECHANICS

I. THEORY

2.1. DEFINITIONS

Mechanics is that branch of science which deals with the bodies when they are at rest or in motion. When the bodies are at rest, the branch of mechanics is known as *Statics* and if the bodies are in motion, the branch of mechanics is known as *'Dynamics*. Dynamics is further divided into two parts namely (*i*) Kinematics and (*ii*) Kinetics. *Kinematics* is the branch of mechanics which deals with the study of rigid bodies in motion without considering the forces, which cause motion. *Kinetics* is the branch of mechanics which deals with the study of rigid bodies in motion, taking into consideration the forces.

1. Force is that action which moves or tends to move a body. The units of force are (*i*) newton (N) in S.I. units, and (*ii*) dyne in C.G.S. units.

2. Newton is a force which acts on a mass of one kilogram and produces an acceleration of one metre per second square. **Dyne** is a force which acts on a mass of one gram and produces an acceleration of one centimetre per second square. The relation between newton (N) and dyne is given by

$$1 \text{ N} = 10^5 \text{ dyne}$$

Force is a vector quantity which means it is having magnitude and direction. A single force which produces the same effect as a number of forces acting together is called the *resultant* of these forces. If the forces are acting in a straight line, their resultant is equal to the algebraical sum of the forces. If the forces are acting in different directions, their resultant is obtained by: (*a*) Law of triangle of forces, (*b*) Law of parallelogram of forces, and (*c*) Law of polygon of forces.

(*a*) Law of triangle of forces states that if two forces acting on a body are represented in magnitude and direction by the two sides of a triangle taken in order, then their resultant is given by the third side of the triangle taken in the opposite order.

(b) **Law of parallelogram of forces** states that if two forces, acting at a point of a body, be represented in magnitude and direction by the two adjacent sides of a parallelogram, their resultant may be represented in magnitude and direction by the diagonal of the parallelogram, which passes through their point of intersection.

(c) **Law of polygon of forces** states that if a number of forces acting on a point of a body are represented in magnitude and direction by the sides of a polygon, taken in order, then their resultant is represented in magnitude and direction by the closing side of the polygon taken in the opposite direction. Conversely, if any number of forces acting at a point can be represented in magnitude and direction by the sides of a polygon taken in order, the forces are in equilibrium.

3. The forces acting on a body may be: (*a*) Coplanar, (*b*) Non-coplanar, (*c*) Concurrent, (*d*) Non-concurrent, (*e*) Coplanar concurrent, and (*f*) Collinear etc.

4. Coplanar forces are those forces, whose lines of action lie on the same plane. *Non-coplanar* forces are those forces whose lines of action do not lie in the same plane. *Concurrent forces* are those forces, which meet at a point and if the forces do not meet at a point, the forces are called *non-concurrent.* If the lines of action of the forces lie in the same plane and they meet at a point, those forces are called *coplanar concurrent forces. Collinear forces* are those forces, whose lines of action lie on the same line.

5. Lami's theorem states that if three coplanar forces acting at a point be in equilibrium, then each force is proportional to the sine of the angle between the other two.

6. Moment of a force about a point is the product of the magnitude of the force and perpendicular distance of its line of action from the point.

7. When a number of forces acting on a rigid body are in equilibrium, then the sum of moments of the forces which tend to turn the body in one direction about any given axis is equal to the sum of the moments of the forces which tend to turn the body in the opposite direction about the same axis. This is known as *Principle of Moments.*

8. When a number of coplanar forces are acting on a particle, the algebraic sum of the moments of all the forces about any point is equal to the moment of their resultant force about the same point. This is known as **Varignon's theorem of moments.**

9. A system of coplanar forces will be in equilibrium if the sum of the resolved components of the forces of the system in any two perpendicular directions is zero separately and the sum of the moments of the forces about a point in their plane is zero. Conversely, if a system of coplanar forces is in equilibrium, the sum of the resolved components of the forces of the system in any two perpendicular directions must be separately zero and also the algebraic sum of their moments about any point in their plane must be zero.

10. A couple consists of two equal, opposite and parallel forces acting on a body. The perpendicular distance between the two parallel forces is called the *arm* of the couple. The *moment of a couple* is equal to the product of the magnitude of one of the forces and the sum of the couple. The couple tends to rotate a body. If two couples are acting on a body, the body will be in equilibrium if both the couples have equal moments, are acting in the same plane and their directions of rotation are opposite.

2.2. EQUATIONS OF MOTIONS

The rate of change of displacement of a body is called velocity or linear velocity (V) while the rate of change of angular displacement is called angular velocity (ω). The relation between linear velocity and angular velocity is given by

$$V = \omega r$$

The rate of change of velocity is called acceleration (a) while the rate of change of angular velocity is called angular acceleration (α). The relation between acceleration (a) and angular acceleration is given by

$$a = \alpha r$$

The acceleration (a) in differential form is represented by $a = \dfrac{dv}{dt}$ in terms of velocity

$= \dfrac{d^2s}{dt^2}$ in terms of displacement

$= \dfrac{Vdv}{ds}$ in terms of velocity and displacement.

***(a)* Equations for Linear Motions.** For a body moving with a uniform acceleration (a) the equations are

(*i*) $V = u + at$ (*ii*) $S = ut + \frac{1}{2}at^2$

(*iii*) $V^2 = u^2 + 2aS$

where u = Initial velocity, V = Final velocity

t = Time taken in sec, S = Distance travelled.

The distance travelled by a body in nth second is given by

$$S_n = u + \frac{a}{2}(2n - 1)$$

***(b)* Equations for Angular Motions.** For a body rotating with a uniform angular acceleration (α), the equations are

(*i*) $\omega = \omega_0 + \alpha t$ (*ii*) $\theta = \omega_0 t + \frac{1}{2}\alpha t^2$

(*iii*) $\omega^2 = \omega_0^2 + 2\alpha\theta$

where ω_0 = Initial angular velocity in rad/sec

ω = Final velocity

t = Time in sec

θ = Angle traversed in radians.

The angle traversed by a body in nth second, $\theta_n = \omega_0 + \dfrac{\alpha}{2}(2n - 1)$.

2.3. KINETICS

Kinetics deals with the study of rigid bodies in motion, where the forces which couse motion are considered.

When two bodies are connected by a light inextensible string and pass over a smooth pulley, both the bodies will have equal acceleration (a) and tension (T) in both sides of the string will be equal.

(*i*) Let W_1 and W_2 are the weights attached at the two ends of the string as shown in Fig. 2.1. If $W_1 > W_2$ or W_1 is moving downwards, then

$$\text{Acceleration,} \quad a = \left(\frac{W_1 - W_2}{W_1 + W_2}\right) g \text{ m/s}^2,$$

$$\text{Tension,} \quad T = \frac{2W_1W_2}{W_1 + W_2} \text{ and}$$

FIGURE 2.1

$$\text{Pressure on the pulley, } 2T = \frac{4W_1W_2}{W_1 + W_2}.$$

(*ii*) If the weight W_2 is placed on a smooth horizontal plane and weight W_1 is hanging free, as shown in Fig. 2.2 ($W_1 > W_2$), then

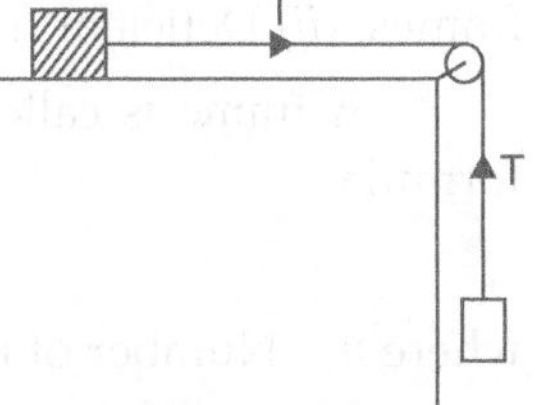

FIGURE 2.2

$$\text{Acceleration,} \quad a = \frac{W_1 \times g}{W_1 + W_2}$$

$$\text{and Tension,} \quad T = \frac{W_1W_2}{W_1 + W_2}.$$

(*iii*) If the weight W_2 is placed on a rough horizontal plane (having co-efficient of friction, μ) and weight W_1 hangs freely. When W_1 moves downward.

$$\text{Acceleration,} \quad a = \frac{(W_1 - \mu W_2)g}{W_1 + W_2}$$

$$\text{and Tension,} \quad T = \frac{(1 + \mu)W_1W_2}{(W_1 + W_2)}.$$

(*iv*) If the weight W_2 is lying on a smooth inclined plane (having inclination α) and W_1 hangs freely as shown in Fig. 2.3. When W_1 moves downward, then

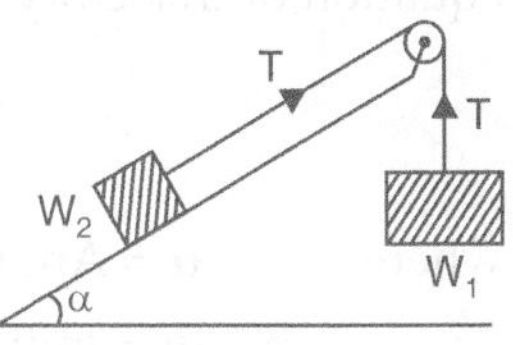

FIGURE 2.3

$$\text{Acceleration,} \quad a = \frac{(W_1 - W_2 \sin\alpha)g}{(W_1 + W_2)}$$

$$\text{and Tension,} \quad T = \frac{W_1W_2(1 + \sin\alpha)}{(W_1 + W_2)}.$$

(*v*) Weight W_2 is lying on a rough inclined plane (having inclination α and co-efficient of friction μ) and W_1 hangs freely. When W_1 moves downwards, then

$$\text{Acceleration,} \quad a = \frac{(W_1 - W_2 \sin\alpha - \mu\, W_2 \cos\alpha)}{(W_1 + W_2)} \times g$$

$$\text{and Tension,} \quad T = \frac{W_1W_2(1 + \sin\alpha + \mu \cos\alpha)}{(W_1 + W_2)}$$

The acceleration of a body moving down an inclined plane is given by

$$a = g \sin\theta \quad \text{when inclined plane is smooth}$$

$$= g[\sin\theta - \mu\cos\theta] \quad \text{when inclined plane is rough}$$

where θ = Inclination of plane with horizontal and

μ = Co-efficient of friction between body and inclined plane.

The tension (T) in the cables supporting a lift is given by

$$T = W\left(1 + \frac{a}{g}\right) \quad \text{when lift is moving up}$$

$$= W\left(1 + \frac{a}{g}\right) \quad \text{when lift is moving down.}$$

2.4. FRAMES

A frame may be defined as a structure, made up of several bars, rivetted or welded together. The bars are also known as the members of the frame. There are three types of frames namely (*i*) Perfect frames, (*ii*) Deficient frame, and (*iii*) Redundant frames.

A frame is called a *perfect frame,* which has got the number of members as given by the formula

$$n = (2j - 3)$$

where n = Number of members and j = Number of joints.

A frame is called *deficient frame* if it has got less number of members than given by $n = 2j - 3$. A frame is called *redundant frame* if it has got more number of members than given by $n = 2j - 3$.

2.5. PROJECTILES

Projectile is defined as an object which is projected in air with certain initial velocity at a certain angle. The object traces some path in air and falls on the ground at a point, other than the point of projection. The path traced by the projectile is known as *trajectory,* which is parabolic in shape. The equation of trajectory is

$$y = x\tan\alpha - \frac{gx^2}{2u^2\cos^2 x}$$

where α = Angle of projection

u = Initial velocity of projectile

x and y = Horizontal and vertical distances covered by the projectile during the time 't'.

The velocity with which a projectile is thrown in air is known as *velocity of projection.* And the angle to the horizontal at which a projectile is thrown in air is called *angle of projection.*

The horizontal range (R), which is the distance between the point of projection and the point where the projectile strikes the ground, is given as

$$R = \frac{u^2 \sin 2\alpha}{g}$$

R will be maximum for a given value of u, when $\sin 2\alpha = 1 = \sin 90°$ or $\alpha = 45°$. Then maximum value of R will be as

$$R_{max} = \frac{u^2}{g}$$

Time of flight (t), which is the time taken by the projectile to reach maximum height and to return back to the ground is given as,

$$t = \frac{2u \sin \alpha}{g}$$

Maximum height (H), reached by projectile is given as

$$H = \frac{u^2 \sin^2 \alpha}{2g}$$

where u = Velocity of projection, and α = Angle of projection.

2.6. FRICTION

The property of bodies, by virtue of which a force is exerted by a stationary body on the other moving body to resist its motion, is called friction and the force exerted is called force of friction.

***(i)* Static friction.** Static friction is the force called into play, when one body rests upon another and a force is applied to make one slide over the other but the body is not sliding. If the body is moving (or sliding) then the force of friction that exists between the moving body and stationary body is called *dynamic friction*. The force of friction, which exists when the body is just on the point of moving (or sliding), is called *limiting friction*. Limiting friction is the maximum amount of friction that comes into play. Dynamic friction is always less than the limiting friction.

***(ii)* The coefficient of friction (μ).** It is the ratio of force of limiting friction to normal reaction between two bodies. And angle of friction (ϕ) is the angle between the normal reaction and resultant of limiting friction and normal reaction. The relation between coefficient of friction (μ) and angle of friction (ϕ) is given by

$$\mu = \tan \phi.$$

***(iii)* Angle of repose.** It is the inclination of a plane with the horizontal at which the body placed on the inclined plane is just on the point of moving down.

Or

The angle of repose is defined as the maximum inclination of a plane at which a body remains in equilibrium over the inclined plane by the assistance of friction only.

***(iv)* Friction on a horizontal plane.** The force P required to move the body of weight W on a rough horizontal plane as shown in Fig. 2.4 is given by

$$P = W \tan \phi$$

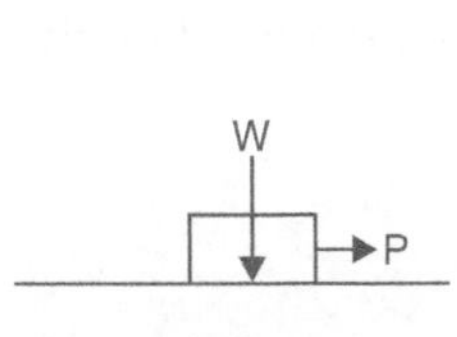

FIGURE 2.4

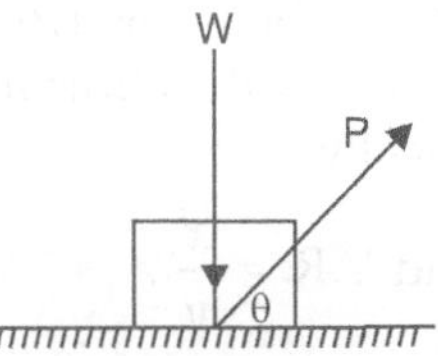

FIGURE 2.5

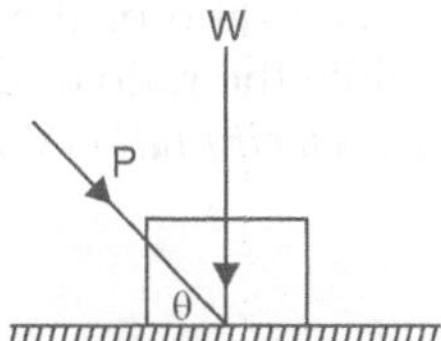

FIGURE 2.6

If P is inclined at an angle θ with the horizontal plane as shown in Fig. 2.5, then the value of P necessary to move the body is given by

$$P = \frac{W \sin \phi}{\cos (\theta - \phi)}$$

where ϕ = Angle of friction and P will be minimum when $\theta = \phi$

$P_{min} = W \sin \phi$.

For Fig. 2.6, the necessary value of P to move that body is given by

$$P = \frac{W \sin \phi}{\cos (\theta + \phi)}$$

(v) Friction on an inclined plane. Figure 2.7 shows a body of weight W, placed on an inclined plane of inclination α and force P is applied at an angle θ with the plane. The necessary value of P to move the body up is given by

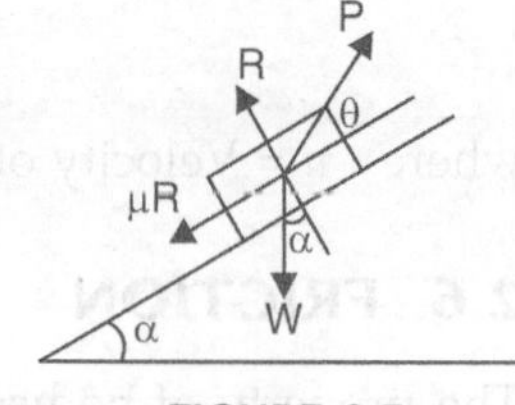

FIGURE 2.7

$$P = W \frac{\sin (\alpha - \phi)}{\cos (\theta - \phi)}$$

where ϕ = Angle of friction.

If the body is on the point of moving down, then the value of P is given by

$$P = W \frac{\sin (\alpha - \phi)}{\cos (\theta + \phi)}.$$

In Fig. 2.8, force P is applied horizontally and the body is on the point of moving up the plane. The necessary value of P is given by

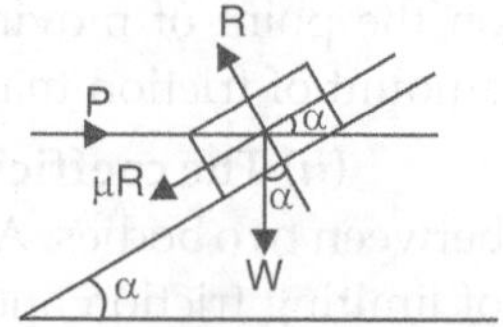

FIGURE 2.8

$$P = W \tan (\alpha + \phi).$$

If the body (Fig. 2.8) is on the point of moving down the plane, then the necessary horizontal force P is given by

$$P = W \tan (\alpha - \phi).$$

2.7. LIFTING MACHINES

Lifting machines are the machines used for lifting heavy loads or weights by applying comparatively small force. The important lifting machines are:

(*i*) Wheel and axle—simple and differential.
(*ii*) Worm and worm-wheel.
(*iii*) Purchase crab-winch-single and double.
(*iv*) Pulleys.
(*v*) Screw jack—simple and differential.

(*i*) The ratio of the load lifted (W) to the effort (P) applied is called *mechanical advantage* (M.A.), while the ratio of distance (x) moved by the effort to the distance (y) moved by the load is known as *velocity ratio* (V.R.). Mathematically

$$\text{M.A.} = \frac{W}{P} \text{ and V.R.} = \frac{x}{y}.$$

(ii) **Efficiency of a machine** is given by,

$$\eta = \frac{\text{Output}}{\text{Input}} = \frac{W \times \text{Distance moved by } W}{P \times \text{Distance moved by } P} = \frac{W \times y}{P \times x}$$

$$= \frac{\frac{W}{P}}{\frac{x}{y}} = \frac{\text{M.A.}}{\text{V.R.}}$$

If the efficiency of a machine is 100%, the machine is known as **ideal machine.** If the efficiency of a machine is less than 50%, it is known as **self-locking machine**. Self-locking machines are also called irreversible machine. If the efficiency of a machine is more than 50%, the machine is known as reversible machine.

(iii) **Law of machine** is expressed as

$$P = mW + c$$

where P = Effort applied, W = Load lifted and m and c = Constants.

The amount of friction, present in a machine, may be expressed in terms of effort or load. Additional effort is required to overcome this friction. Additional load is to be lifted if friction is not there.

Additional effort $= P - \dfrac{W}{\text{V.R.}}$ and

Additional load $= P \times \text{V.R.} - W$

where V.R. = Velocity ratio, P = Actual effort applied

W = Actual load lifted.

(iv) **Screw jack**

It is used for lifting heavy weights. The angle of screw (α) in terms of pitch (p) is given by

$$\tan \alpha = \frac{p}{\pi d}$$

where d = Mean diameter of screw.

(a) The effort applied horizontally at the mean radius of the screw jack to lift a load of W is given by

$$P = W \tan (\alpha + \phi)$$

and to lower the load W, the effort is given by $P = W \tan (\alpha - \phi)$

where α = Angle of screw, and ϕ = Angle of friction.

(b) The effort applied at the end of the handle of a screw jack is given by

$$P = \frac{Wd}{2L} \tan (\alpha + \phi) \text{ to lift a load } W$$

$$= \frac{Wd}{2L} \tan (\alpha - \phi) \text{ to lower a load } W$$

where d = Mean diameter of the screw, L = Length of the handle

α = Angle of screw, ϕ = Angle of friction.

(*c*) **Efficiency of a screw-jack** is independent of the weight lifted or effort applied. Efficiency of a screw-jack for raising a load W is given by

$$\eta = \frac{\tan \alpha}{\tan (\alpha + \phi)}$$

The efficiency of a screw-jack will be maximum if

$$\alpha = 45° - \frac{\phi}{2}.$$

The maximum efficiency of a screw-jack is given by

$$\eta_{max} = \frac{1 - \sin \phi}{1 + \sin \phi}$$

where ϕ = Angle of friction.

(*v*) Simple wheel and axle

Let W = Load lifted.

P = Effort applied,

D = Diameter of wheel, and

d = Diameter of axle.

Then velocity ratio (V.R.) of this system is given by,

$$\text{V.R.} = \frac{D}{d} \text{ and } \eta = \frac{\text{M.A.}}{\text{V.R.}}.$$

(*vi*) Differential wheel and axle

There are two axles and one wheel in this system.

Let D = Diameter of wheel,

d_1 = Diameter of axle B, and

d_2 = Diameter of axle C.

The diameter of axle B is more than the diameter of axle C.

Then, $$\text{V.R.} = \frac{2D}{(d_1 - d_2)} \text{ and } \eta = \frac{\text{M.A.}}{\text{V.R.}}.$$

(*vii*) Worm and worm wheel

Let L = Radius of the wheel (or length of handle)

r = Radius of load drum

T = Number of teeth on the worm wheel.

(*a*) *For a single-threaded worm*

If the effort wheel completes one revolution, then the worm will also complete one revolution. But the worm wheel will move through one teeth. Hence for one revolution of the effort wheel:

Distance moved by effort = $2\pi L$

Distance moved by load = Distance of one teeth

$$= \frac{\text{Distance of one revolution of load drum}}{\text{Number of teeth on worm wheel}}$$

$$= \frac{2\pi r}{T}$$

$$\therefore \qquad \text{V.R.} = \frac{\text{Distance moved by effort}}{\text{Distance moved by load}}$$

$$= \frac{2\pi L}{\left(\dfrac{2\pi r}{T}\right)} = \frac{L \times T}{r}.$$

(b) *For a double-threaded worm*

Here the worm-wheel will move by two teeths and load will move by $\dfrac{2\pi r \times 2}{T}$

$$\therefore \qquad \text{V.R.} = \frac{2\pi L}{\left(\dfrac{2\pi r \times 2}{T}\right)} = \frac{L \times T}{2r}$$

If worm is n-threaded, then V.R. $= \dfrac{L \times T}{n \times r}$.

(*viii*) Single purchase crab winch

It consists of an effort axle and a load axle. On effort axle, a small toothed wheel known as pinion is mounted whereas on load axle, a large toothed wheel known as spur wheel is mounted. The pinion meshes with spur wheel.

Let T_1 = Number of teeth on the pinion,

T_2 = Number of teeth on the spur wheel,

L = Length of lever arm, and

D = Diameter of the load axle.

When lever arm makes one revolution, the pinion also makes one revolution. The spur wheel makes $\left(\dfrac{T_1}{T_2}\right)$ revolution. Also the load axle makes $\left(\dfrac{T_1}{T_2}\right)$ revolution.

Distance moved by effort $= 2\pi L$

Distance moved by load $= (\pi D) \times \dfrac{T_1}{T_2}$

$$\therefore \qquad \text{V.R.} = \frac{2\pi L}{\pi D \times \dfrac{T_1}{T_2}} = 2 \times \frac{L}{D} \times \frac{T_2}{T_1}.$$

(*ix*) Double purchase crab winch

It consists of effort axle, load axle and intermediate axle. On the intermediate axle a pinion and a spur wheel is mounted. The pinion of intermediate axle gears with the spur wheel of the load axle. And the spur wheel of the intermediate axle gears with the pinion of the effort axle.

Let T_1 = Number of teeth on pinion of effort axle

T_2 = Number of teeth on the spur wheel of intermediate axle

T_3 = Number of teeth on the pinion of intermediate axle

T_4 = Number of teeth on the spur wheel of load axle

L = Length of lever, and

D = Diameter of load axle.

Distance moved by load $= \pi D \times \frac{T_1}{T_2} \times \frac{T_3}{T_4}$.

Distance moved by effort $= 2\pi L$

$$\therefore \quad \text{V.R.} = \frac{2\pi L}{\pi D \times \frac{T_1}{T_2} \times \frac{T_3}{T_4}} = \frac{2L}{D} \times \frac{T_2}{T_1} \times \frac{T_4}{T_3}.$$

(x) Pulleys

A pulley may be a fixed pulley or a movable pulley. For a single fixed pulley, mechanical advantage is equal to one, whereas in case of a single movable pulley, the mechanical advantage is more than one.

The pulleys are generally used in certain combinations to obtain a higher mechanical advantage and efficiency. The important system of pulleys are:

(*a*) First system of pulley,

(*b*) Second system of pulley, and

(*c*) Third system of pulley.

For these system of pulleys, the velocity ratio is given by:

(*a*) *For first system*

$$\text{V.R.} = 2^n$$

where n = Number of movable pulleys.

(*b*) *For second system*

$$\text{V.R.} = n$$

where n = Number of segments supporting the load

= Total number of pulleys in two blocks.

If the weight of lower block is considered, than also

$$\text{V.R.} = n.$$

(*c*) *For third system*

$$\text{V.R.} = 2^n - 1$$

where n = Number of pulleys.

For Weston's Differential Pulley Block

$$\text{V.R.} = \frac{2D}{(D - d)}$$

where D = Diameter of large pulley of upper block

d = Diameter of smaller pulley of upper block.

Sometimes, this velocity ratio is also expressed as

$$\text{V.R.} = \frac{2T_1}{T_1 - T_2}$$

where T_1 = Number of teeth of larger pulley and

T_2 = Number of teeth of smaller pulley.

2.8. VIRTUAL WORK

Virtual work is the work done by a force on a body due to small virtual (*i.e.* imaginary) displacement. Principle of Virtual work states, "If a system of forces acting on a body or system of bodies be in equilibrium and if the system is imagined to undergo a small displacement consistent with the geometrical conditions, then the algebraic sum of the virtual work done by the forces of the system is zero".

For most of the problems solved by virtual work method, it is assumed that there is no friction in hinges, bearing or along sliding surfaces. And if all the forces act in the same direction, the system is given virtual displacement in the direction of forces.

2.9. CENTRE OF GRAVITY (C.G.)

The point, through which the whole weight of a body is supposed to act, is called the centre of gravity of the body.

The location of the C.G. of the following cases are:

(*i*) The C.G. of a rectangle or parallelogram is at the point of intersection of its diagonals.

(*ii*) The C.G. of a semi-circular arc [Fig. 2.9 (*a*)] is at $2r/\pi$ above base *AB*.

(*iii*) The C.G. of a semi-circle [Fig. 2.9 (*b*)] is at $4r/3\pi$ above base *AB*.

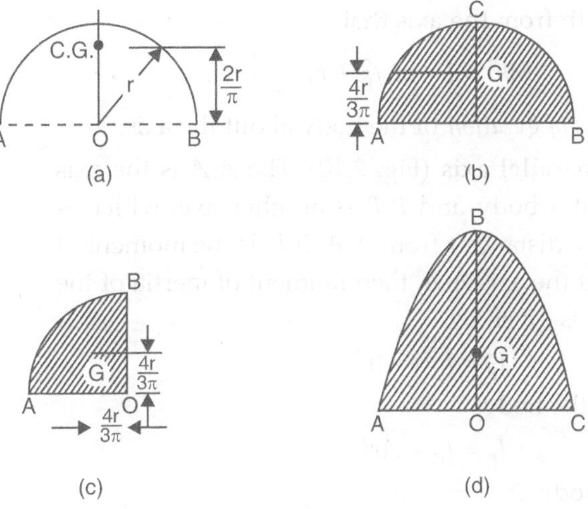

FIGURE 2.9

(*iv*) The C.G. of a quadrant of a circle [Fig. 2.9 (*c*)] is at $4r/3\pi$ from *AO* and *OB*.

(*v*) The C.G. of a triangle lies at the point of concurrence of the medians of the triangle.

(*vi*) C.G. of a solid right circular cone is at $\frac{1}{4}$ of the total height above the base on axis.

(*vii*) C.G. of a thin hollow right circular cone is at $\frac{1}{3}$ of the total height above the base on axis.

(*viii*) C.G. of a solid hemisphere is at $3r/8$ from base.

(*ix*) C.G. of a thin hollow hemisphere is at $r/2$ from base.

(*x*) C.G. of solid pyramid or cone is at $\frac{1}{4}$ of the total height above base.

(*xi*) C.G. of hollow pyramid or cone is at $\frac{1}{3}$ of the total height above base.

(*xii*) C.G. of prism or cylinder hollow or solid is at $\frac{1}{2}$ of the total height.

(*xiii*) C.G. of a parabola [Fig. 2.9 (*d*)] is at $\frac{2}{5}$ of the axis *OB* above *AC*.

The C.G. of a body consisting of different areas is given by

$$\bar{x} = \frac{a_1x_1 + a_2x_2 + a_3x_3 + \ldots\ldots}{a_1 + a_2 + a_3 + \ldots\ldots}$$

and

$$\bar{y} = \frac{a_1y_1 + a_2y_2 + a_3y_3 + \ldots\ldots}{a_1 + a_2 + a_3 + \ldots\ldots}$$

If a given section is symmetrical about X-X-axis or Y-Y-axis, the C.G. of the section will lie on the axis of symmetry.

2.10. MOMENT OF INERTIA (M.O.I.)

The moment of inertia of a body of mass *M* about an axis is given by

$$m_1r_1^2 + m_2r_2^2 + \ldots\ldots = \Sigma m_1r_1^2 = I$$

where m_1, m_2, ... are the masses of very small portions of the body and r_1, r_2, ... are their distances from the axis.

If *k* is such a length from the axis that

$$Mk^2 = I = m_1r_1^2 + m_2r_2^2 + \ldots..$$

then *k* is called the *radius of gyration* of the body about the axis.

***(a)* Theorem of parallel axis** (Fig. 2.10). The *A-A* is the axis passing through C.G. of a body and *B-B* is another axis which is parallel to *A-A* and is at a distance *r* from *A-A*. If I_G is the moment of inertia of the body about the axis *A-A*, then moment of inertia of the body about axis *B-B* (I_B) is given by

$$I_B = I_G + Mr^2$$

where *M* is the mass of the body

$$I_B = I_G + Ar^2$$

where *A* = Area of the body.

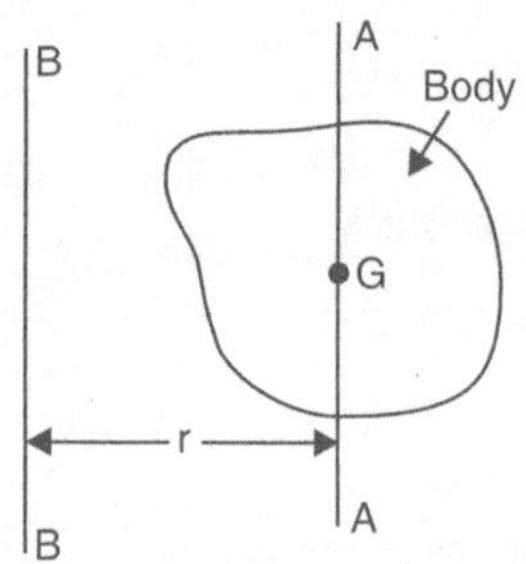

FIGURE 2.10

***(b)* Theorem of perpendicular axis** (Fig. 2.11). Let *A* is the lamina, lying in the plane *x-y*, where *OX* and *OY* are two axes at right angles to one another. If *OZ* is the axis at right angles to the plane *x-y*, then moment of inertia of the lamina about *OZ* is equal to the sum of moments of inertia about axes *OX* and *OY*. Mathematically,

$$I_z = I_x + I_y$$

where I_z = M.O.I. about axis *OZ*

I_x = M.O.I. about axis *OX*

I_y = M.O.I. about axis *OY*.

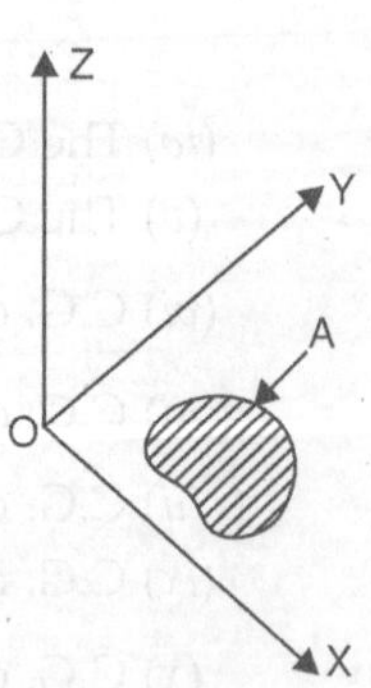

FIGURE 2.11

(c) M.O.I. of various sections

(*i*) Rectangular section $I = \frac{1}{12} BD^3$.

(*ii*) Hollow rectangular section $I = \frac{BD^3 - bd^3}{12}$.

(*iii*) $I = \frac{B}{12}(D^3 - d^3)$.

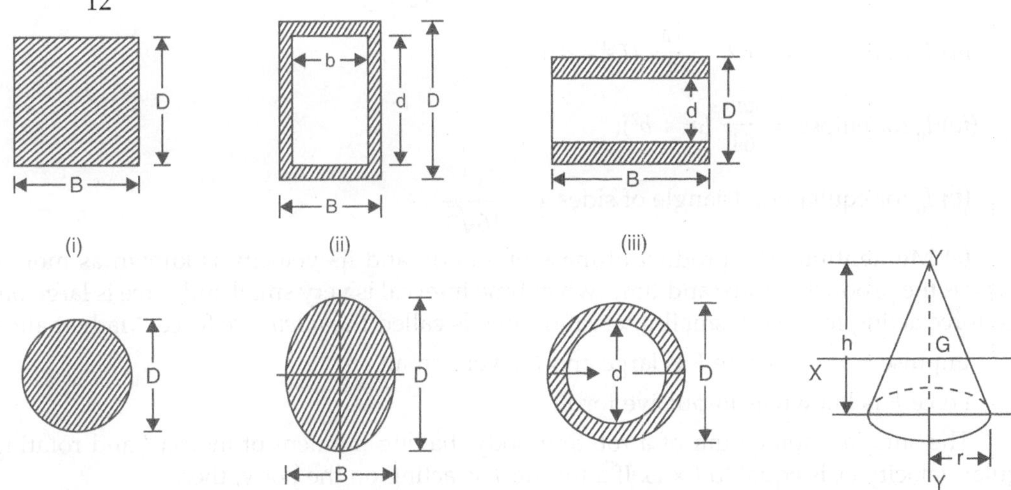

FIGURE 2.12

(*iv*) Solid circular section $I = \frac{\pi}{64} D^4$.

(*v*) Elliptical section $I = \frac{\pi}{64} BD^3$.

(*vi*) Hollow circular section $I = \frac{\pi}{64}(D^4 - d^4)$.

(*vii*) M.O.I. of a triangle about base $= \frac{bh^3}{12}$, where b = Base and h = Height.

(*viii*) M.O.I. of a triangle about an axis passing through its C.G. $= \frac{bh^3}{36}$.

(*ix*) M.O.I. of a solid cone about Y-$Y = \frac{3}{10} Mr^2$.

M.O.I. of a solid cone about axis X-X passing through C.G. of the cone

$$= \frac{3}{20} M\left(r^2 + \frac{h^2}{4}\right).$$

(*x*) M.O.I. of solid sphere $= \frac{2}{5} Mr^2$.

(*xi*) M.O.I. of solid circular cylinder about vertical axis $= \frac{Mr^2}{2}$.

(*d*) Polar moment of inertia, (I_0)

It is the M.O.I. of a plane figure about an axis at right angle to its plane and passing through the centre of gravity of the figure.

(*i*) Polar moment of inertia of a circle = $\frac{\pi d^4}{32}$.

(*ii*) Polar moment of inertia of rectangle = $\frac{bh(b^2+h^2)}{12}$.

(*iii*) I_0 of a hollow circle = $\frac{\pi}{32}(D^4 - d^4)$.

(*iv*) I_0 for ellipse = $\frac{\pi ab}{64}(a^2 + b^2)$.

(*v*) I_0 for equilateral triangle of sides $x = \frac{x^4}{16\sqrt{3}}$.

(*e*) Momentum. The product of mass of a body and its velocity is known as momentum. *Impulse* is the product of force and time, when time interval is very small and force is large one. The large force acting for a very small interval of time is called the *impulsive* force. Mathematically,

Impulse = $F \times t$, where F is large and t is very small.

Force F is known as impulsive force.

The angular momentum of a rotating body, having moment of inertia I and rotating with angular velocity ω, is equal to $I \times \omega$. If a torque T is acting on the body, then

$$T = I\alpha, \text{ where } \alpha = \text{Angular acceleration.}$$

The unit of torque will be in Newton-metre as I is in kg m^2. If torque is to be obtained in kgf-m, then it should be divided by 'g' (*i.e.*, 9.81).

(*f*) Centripetal and centrifugal forces. Centripetal force is the force which acts on a body, rotating along a circular path, along the radius of the circular path and is always directed towards the centre of the path.

The force acting on a body, rotating in a circular or curved path along the radius and away from the centre, is called *centrifugal force.* This is given by

$$\text{Centrifugal force} = m\frac{V^2}{r} \text{ or } m\omega^2 r$$

where m = Mass of the body,

r = Radius of the circular path,

ω = Angular velocity of the body, and

V = Linear velocity of body.

(*g*) Banking of roads. Banking or super-elevation of roads is the process of raising the outer edge of the roads above the inner edge. The angle of super-elevation or banking (θ) is given by

$$\theta = \tan^{-1}\left(\frac{V^2}{gr}\right)$$

where V = Velocity of vehicle and r = Radius of circular path.

The maximum velocity (V_{max}) of a vehicle on a level circular path to avoid

(*a*) Skidding is $V_{max} = \sqrt{\mu g r}$

and (*b*) Over turning is $V_{max} = \sqrt{\dfrac{d\,gr}{2h}}$

where h = Height of C.G. of vehicle from ground level,

d = Distance between the centre lines of the wheels, and

r = Radius of circular path.

2.11. SIMPLE HARMONIC MOTION (S.H.M.)

A body is said to describe simple harmonic motion (S.H.M.) if it moves in a straight line such that its acceleration is proportional to its distance from a fixed point and is always directed towards the fixed point.

(*i*) For a simple harmonic motion, *periodic time (T)* is given by

$$T = \frac{2\pi}{\omega}, \text{ where } \omega = \text{Angular velocity of the particle.}$$

(*ii*) Number of cycles per second is called *frequency* and it is equal to $\frac{1}{T}$, where T is the periodic time.

$$\therefore \quad \text{Frequency} = \frac{1}{T} = \frac{\omega}{2\pi}.$$

(*iii*) *The velocity of* the particle moving with S.H.M. at a distance y from the mean position is

$$V = \omega\sqrt{r^2 - y^2}, \text{ where } r = \text{Amplitude of the motion}$$

and acceleration is given by $a = -\omega^2 y$.

(*iv*) The velocity of the particle moving with S.H.M. is maximum, when it is passing through mean position. Hence when $y = 0$, velocity is maximum and is given by, $V_{max} = \omega r$.

(*v*) The acceleration of the particle moving with S.H.M. is maximum, when it is at its extreme position, *i.e.*, when $y = r$ = amplitude.

$\therefore$ Maximum acceleration, $a_{max} = -\omega^2 r$.

When particle is at the mean position, acceleration is zero and when particle is at its extreme position, velocity is zero.

(*vi*) Time period of a **simple pendulum** is given by

$$T = 2\pi\sqrt{\frac{L}{g}}, \text{ where } L = \text{Length of pendulum.}$$

When the bob of the pendulum moves from one extremity to the other, half an oscillation is said to be completed. This constitutes **one Beat**. A pendulum, which executes one beat per second, is called **seconds pendulum.** Hence for a seconds pendulum, the time period is equal to two seconds. Length of a second pendulum is obtained from

$$T = 2\pi\sqrt{\frac{L}{g}} = 2 \qquad \therefore \quad L = \left(\frac{2}{2\pi} \times \sqrt{g}\right)^2 = \left(\frac{1}{\pi}\sqrt{9.81}\right)^2$$

$$= 0.994 \text{ m or } 99.40 \text{ cm.}$$

(*vii*) The time period of an **elastic string or spring** is given by

$$T = 2\pi\sqrt{\frac{\delta}{g}}, \qquad \text{where } \delta = \text{Static extension.}$$

(*viii*) **Gain or loss of oscillations due to** change in length or g of a simple pendulum is given by

$$\frac{dn}{n} = \frac{dg}{2g} - \frac{dl}{2l}$$

where $\begin{cases} \dfrac{dn}{n} = \text{Change in number of oscillation} \\ \dfrac{dg}{g} = \text{Change of gravity} \\ \dfrac{dl}{l} = \text{Change of length of pendulum.} \end{cases}$

(*a*) If L is constant, then $\dfrac{dn}{n} = \dfrac{dg}{2g}$

(*b*) If g is constant, then $\dfrac{dn}{n} = -\dfrac{dl}{2l}$

–ve sign means that $\dfrac{dl}{l}$ decreases, as $\dfrac{dn}{n}$ increases. Or with the increase of length, the clock goes slow.

(*ix*) **Compound pendulum.** A body of any shape or size which is made to oscillate about a fixed axis is called a compound pendulum. The body shown in Fig. 2.13 oscillates about an axis *Y-Y* or through point of suspension *O*. *G* is the centre of gravity of the body and $OG = h$. $OP = L$, the length of a simple pendulum which has the same time period as compound pendulum. The time period of compound pendulum is

$$T = 2\pi\sqrt{\frac{k^2 + h^2}{gh}}$$

where k = Radius of gyration and h = OG or length between the point of suspension (*O*) and centre of gravity (*G*).

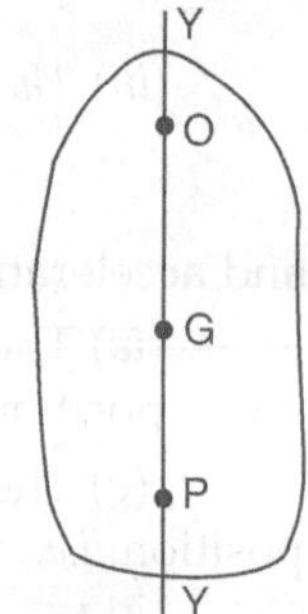

FIGURE 2.13

The point *P* is called *centre of oscillation*. If *P* be made centre of suspension, *O* will become centre of oscillation. Hence centre of oscillation and suspension are interchangeable.

The point *P* is also called *centre of percussion* which is the point on the body at which if a blow is given, no reaction will be felt at the point of suspension.

2.12. BELTS AND ROPES DRIVES

The belts and ropes are used for transmitting power from one shaft to another shaft. The belts are of the following important types:

(*i*) Open belt drive,

(*ii*) Cross belt drive, and

(*iii*) Compound belt drive.

(*i*) **The velocity ratio** of a belt is given by

$$\frac{N_2}{N_1} = \frac{d_1}{d_2} \quad \text{if thickness of belt is neglected}$$

$$= \frac{d_1 + t}{d_2 + t} \quad \text{if thickness of belt is considered}$$

$$= \frac{d_1}{d_2} \times \left[\frac{E \times \sqrt{f_2}}{E + \sqrt{f_1}} \right] \quad \text{if creep of belt is considered}$$

where N_1 = Speed of driver pulley,
N_2 = Speed of driven pulley,
d_1 = Diameter of driver pulley,
d_2 = Diameter of driven pulley,
t = Thickness of belt,
E = Young's modulus,
f_1 = Stress in the belt on tight side, and
f_2 = Stress in the belt on slack side.

(*ii*) **The velocity ratio of a compound belt drive is given by**

$$\frac{\text{Speed of last follower}}{\text{Speed of first driver}} = \frac{\text{Product of diameter of drivers}}{\text{Product of diameter of follower}}.$$

(*iii*) **Length of a belt**

The length of a belt is given by,

$$L = \pi (r_1 + r_2) + \frac{(r_1 - r_2)^2}{x} + 2x \quad \text{for an open belt}$$

$$= \pi (r_1 + r_2) + \frac{(r_1 + r_2)^2}{x} + 2x \quad \text{for a crossed belt}$$

where r_1 = Radius of larger pulley,
r_2 = Radius of smaller pulley,
x = Distance between the centres of two pulleys, and
L = Length of the belt.

(*iv*) **Ratio of tensions**

The ratio of the tensions on the two sides of a belt is given by,

$$\frac{T_1}{T_2} = e^{\mu \times \theta}$$

where T_1 = Tension on the tight side,
T_2 = Tension on the slack side,
μ = Co-efficient of friction between belt and pulley,
θ = Angle of contact in radians
= $(180 - 2\alpha)$ for an open belt
= $(180 + 2\alpha)$ for a crossed belt

where $\alpha = \sin^{-1}\dfrac{(r_1 - r_2)}{x}$ for an open belt

$= \sin^{-1}\dfrac{(r_1 + r_2)}{x}$ for a crossed belt.

(v) Power transmitted by a belt

Power transmitted is given by

$$P = \frac{(T_1 - T_2) \times v}{75},$$ where T_1 and T_2 are in kgf

$$= (T_1 - T_2) \times v \text{ watt},$$ where T_1 and T_2 are in newton.

(A) The centrifugal tension (T_c) is given by

$$T_c = \frac{w}{g} \times v^2 \text{ kgf}$$ in M.K.S. units

$$= m \times v^2 \text{ Newton}$$ in S.I. units

where w = Weight of belt per metre length in kgf,

m = Mass of belt per metre length in kg.

(B) The maximum tension (T_{max}) is given by

$$T_{max} = f \times \text{Area of belt}$$

$$= f \times (b \times t)$$

$$= T_1 + T_c$$ if centrifugal tension is considered

$$= T_1$$ if centrifugal tension is neglected.

(C) For maximum power transmission:

(*a*) Velocity of belt (v) is given by

$$v = \sqrt{\frac{g \times T_{max}}{3w}}$$ in M.K.S. units

$$= \sqrt{\frac{T_{max}}{3 \times m}}$$ in S.I. units

(*b*) $T_{max} = 3T_c$ or $T_c = \frac{1}{3}\, T_{max}$

(*c*) $T_1 = \frac{2}{3}\, T_{max}$.

(D) Initial tension (T_0) in a belt is given by

$$T_0 = \frac{T_1 + T_2}{2}$$ if T_c is neglected

$$= \frac{T_1 + T_2 + 2T_e}{2}$$ if T_c is considered.

2.13. WORK, POWER AND ENERGY

Work. Work is the product of force and distance. Mathematically the work done is given by

Work done $= F \times S$ When force and distance are in the same direction

$= F \cos \theta \times S$ When force acts at an angle θ with the direction of displacement.

(*i*) **The unit of work** is Nm (or Joule). Hence one **Joule** is the work done by a force of 1 N when the displacement is 1 m.

(*ii*) The work done is also represented by the **area of force-displacement curve**.

(*iii*) The work done on a **rotating body** by a torque (T) is given by

$$\text{Work done} = T \times \theta$$

where θ = Angular displacement is radians.

Power. The rate of doing work is known as power. Hence the power is the work done per second. The unit of power is Nm/s or Watt (W).

Hence,

$$\text{Power} = \text{Work done per second}$$

$$= \frac{\text{Work done}}{\text{Time}} = \frac{\text{Force} \times \text{Distance}}{\text{Time}}$$

$$= \text{Force} \times \frac{\text{Distance}}{\text{Time}} = \text{Force} \times \text{Velocity}$$

The force and velocity should be in the same direction.

The power required to rotate a body is given by,

$$\text{Power} = \text{Torque} \times \text{Angular velocity}$$

$$= T \times \omega$$

$$= T \times \frac{2\pi N}{60} = \frac{2\pi NT}{60} \text{ watt.}$$

II. OBJECTIVE TYPE QUESTIONS

Tick mark the most appropriate of the multiple choice answers:

Basic Definitions

1. Triangle law of forces states that if two forces acting at a point are represented in magnitude and direction by the two sides of the triangle taken in order, then their resultant is given by the

(*a*) third side of the triangle taken in the same order ☐

(*b*) third side of the triangle taken in the opposite order ☐

(*c*) sum of the two forces acting ☐

(*d*) none of the above. ☐

2. Law of polygon of forces states that

(*a*) if a number of forces acting at a point are represented by the sides of a polygon taken in order, then their resultant is represented in magnitude and direction by the closing side of the polygon, taken in the same order. ☐

(*b*) if a number of forces acting at a point are represented in magnitude and direction by the sides of a polygon taken in order, then their resultant is represented in magnitude and direction by the closing side of the polygon, taken in the opposite order. ☐

(*c*) the resultant of a number of forces acting on a point is the sum of all forces. ☐

(*d*) none of the above. ☐

3. Two forces P and Q are acting at an angle θ, their resultant (R) is given by

(a) $R = \sqrt{P^2 + Q^2 + 2PQ \sin 2\theta}$ ☐ (b) $R = \sqrt{P^2 + Q^2 + 2PQ \cos \theta}$ ☐

(c) $R = \sqrt{P^2 + Q^2 - 2PQ \cos \theta}$ ☐ (d) $R = \sqrt{P^2 + Q^2 + 2PQ \cos 2\theta}$. ☐

4. Two forces A and B are acting at an angle θ and their resultant B makes an angle α with the force A, then

(a) $\tan \alpha = \dfrac{B \sin \theta}{B + A \cos \theta}$ ☐ (b) $\tan \alpha = \dfrac{A \sin \theta}{A + B \cos \theta}$ ☐

(c) $\tan \alpha = \dfrac{B \sin \theta}{A + B \cos \theta}$ ☐ (d) $\tan \alpha = \dfrac{A \cos \theta}{B + A \sin \theta}$. ☐

5. Two forces A and B are acting at an angle θ and their resultant R makes an angle α with the force A, then

(a) $\cos \alpha = \dfrac{A + B \sin \theta}{\sqrt{A^2 + B^2 - 2AB \cos \theta}}$ ☐ (b) $\cos \alpha = \dfrac{A + B \sin \theta}{\sqrt{A^2 + B^2 + 2AB \cos \theta}}$ ☐

(c) $\cos \alpha = \dfrac{B \sin \theta}{\sqrt{A^2 + B^2 + 2AB \cos \theta}}$ ☐ (d) none of the above. ☐

6. Lami's theorem states that if

(a) three forces acting at a point are in equilibrium, they can be represented by the three sides of a triangle. ☐

(b) the three forces acting at a point can be represented in magnitude and direction by the sides of a triangle, the forces are in equilibrium. ☐

(c) three forces acting at a point are in equilibrium, each force is proportional to the sine of the angle between the other two. ☐

(d) none of the above. ☐

7. The forces which do not meet at a point are called

(a) non-coplanar forces ☐ (b) coplanar forces ☐

(c) non-concurrent forces ☐ (d) concurrent forces. ☐

8. The forces whose lines of action do not lie in the same plane, are called

(a) non-coplanar forces ☐ (b) coplanar forces ☐

(c) non-concurrent forces ☐ (d) none of the above. ☐

9. The forces, whose line of action lie on the same line, are known as

(a) coplanar forces ☐ (b) concurrent forces ☐

(c) collinear forces ☐ (d) none of the above. ☐

10. The forces, whose lines of action does not lie in the same plane but are meeting at one point, are known as

(a) coplanar concurrent forces ☐ (b) non-coplanar concurrent forces ☐

(c) non-coplanar non-concurrent forces ☐ (d) none of the above. ☐

11. The forces, whose lines of action lie in the same plane and are meeting at one point, are known as

(a) coplanar concurrent forces ☐ (b) coplanar non-concurrent forces ☐

(c) non-coplanar concurrent forces ☐ (d) none of the above. ☐

12. Coplanar concurrent forces means the lines of action of forces

(*a*) lie in the same plane ☐

(*b*) lie in the same plane but the forces are not meeting at one point ☐

(*c*) lie in the same plane and forces are meeting at one point ☐

(*d*) none of the above. ☐

13. The forces, which meet at a point, are known as

(*a*) collinear forces ☐ (*b*) coplanar forces ☐

(*c*) concurrent forces ☐ (*d*) none of the above. ☐

14. The forces, which lie in the same plane, are called

(*a*) collinear forces ☐ (*b*) coplanar forces ☐

(*c*) concurrent forces ☐ (*d*) none of the above. ☐

15. Tick mark the correct statement

(*a*) The algebraic sum of the resolved parts of a number of force in a given direction is equal to their resultant. ☐

(*b*) The algebraic sum of the resolved parts of a number of forces in a given direction is equal to two times their resultant. ☐

(*c*) The algebraic sum of the resolved parts of a number of forces in a given direction is equal to the resolved component of the resultant in that direction. ☐

(*d*) None of the above. ☐

16. The statement—if three forces acting at a point can be represented in magnitude and direction by the sides of a triangle taken in order, the forces are in equilibrium—is known as

(*a*) Lami's theorem ☐ (*b*) Law of polygon of forces ☐

(*c*) Law of triangle of forces ☐ (*d*) Newton's law of forces. ☐

17. The statement—the algebraic sum of the moments taken about any point in the plane of forces is zero—is known as

(*a*) Law of polygon of forces ☐ (*b*) Lami's theorem ☐

(*c*) Newton's law of forces ☐ (*d*) Law of moments. ☐

18. Two couples will balance one another when they are in the same plane and

(*a*) have equal moments and their direction of rotation is same ☐

(*b*) have unequal moments and their direction of rotation is opposite ☐

(*c*) have equal moments and their direction of rotation is opposite ☐

(*d*) none of the above. ☐

19. The number of members (n) and number of joints (j) in a perfect frame is given by

(*a*) $n = (3j - 2)$ ☐ (*b*) $n = (2j - 3)$ ☐

(*c*) $j = (2n - 3)$ ☐ (*d*) $j = (3n - 2)$. ☐

20. A frame, which has got less number of members than given by the formula $n = (2j - 3)$, is called a

(*a*) perfect frame ☐ (*b*) deficient frame ☐

(*c*) redundant frame ☐ (*d*) none of the above. ☐

21. A frame, which has got more number of members than given by the formula $n = (2j - 3)$, is called a

(*a*) perfect frame ☐ (*b*) deficient frame ☐
(*c*) redundant frame ☐ (*d*) none of the above. ☐

22. A frame, which has got the number of members equal to the number of members given by $n = (2j - 3)$, is called a

(*a*) perfect frame ☐ (*b*) deficient frame ☐
(*c*) redundant frame ☐ (*d*) none of the above. ☐

23. The resultant of two forces each equal to $\frac{P}{4}$ and acting at right angles is

(*a*) $\frac{P}{2}$ ☐ (*b*) $\frac{P}{2\sqrt{2}}$ ☐
(*c*) $\sqrt{2P}$ ☐ (*d*) $\frac{P}{\sqrt{2}}$. ☐

24. Two forces of magnitudes 4 and 5 N act at an angle of 60°, the resultant force is equal to

(*a*) 6 N ☐ (*b*) $\sqrt{61}$ ☐
(*c*) 7 N ☐ (*d*) 9 N. ☐

25. A body will be in equilibrium when

(*a*) the algebraic sum of vertical components of all forces is zero ☐
(*b*) the algebraic sum of horizontal components of all forces is zero ☐
(*c*) the algebraic sum of moments of all forces about a point is zero ☐
(*d*) all the above. ☐

Equations of Motion

26. Rate of change of displacement of a body is called

(*a*) acceleration ☐ (*b*) velocity ☐
(*c*) momentum ☐ (*d*) none of the above. ☐

27. Rate of change of velocity of body is called

(*a*) acceleration ☐ (*b*) velocity ☐
(*c*) momentum ☐ (*d*) none of the above. ☐

28. The product of mass and velocity of a body is called

(*a*) acceleration ☐ (*b*) velocity ☐
(*c*) momentum ☐ (*d*) none of the above. ☐

29. If a body is moving with a uniform acceleration (a), then final velocity (V) of the body after time 't' is equal to

(*a*) $ut + \frac{1}{2}at^2$ ☐ (*b*) $u + at$ ☐
(*c*) $u^2 + 2aS$ ☐ (*d*) none of the above. ☐

where u = Initial velocity, S = Distance travelled in t seconds.

30. If a body is moving with a uniform acceleration (a), then the distance travelled by a body after time 't' is equal to

(a) $ut + \frac{1}{2}at^2$ ☐ (b) $u + at$ ☐

(c) $u^2 + 2aV$ ☐ (d) none of the above. ☐

31. If a body is moving with a uniform acceleration (a), then the distance travelled by the body in nth second is given by

(a) $\frac{u+a}{2}(1 - 2n)$ ☐ (b) $\frac{u+a}{2}(n - 2)$ ☐

(c) $u + \frac{a}{2}(2n - 1)$ ☐ (d) none of the above. ☐

32. If a body is moving in a curved path, the motion of the body is called

(a) rectilinear ☐ (b) rotational ☐

(c) curvilinear ☐ (d) none of the above. ☐

33. If a body is moving in a straight line, the motion of the body is called

(a) rectilinear ☐ (b) rotational ☐

(c) curvilinear ☐ (d) none of the above. ☐

34. If a body is moving in a circular path, the motion of the body is called

(a) rectilinear ☐ (b) rotational ☐

(c) curvilinear ☐ (d) none of the above. ☐

35. Rate of change of angular velocity is called

(a) acceleration ☐ (b) angular acceleration ☐

(c) kinetic energy ☐ (d) none of the above. ☐

36. The relation between linear acceleration (a) and angular acceleration (α) is given by

(a) $\alpha = a \times r$ ☐ (b) $\alpha = \frac{a}{r}$ ☐

(c) $\alpha = \frac{1}{a \times r}$ ☐ (d) $\alpha = \frac{r}{a}$. ☐

37. The angular displacement by a rotating body in the nth second is equal to

(a) $\omega_0 + \left(\frac{n-2}{2}\right)\alpha$ ☐ (b) $\left(\frac{\omega_0 \times n}{2}\right) \times \alpha$ ☐

(c) $\omega_0 + \left(\frac{2n-1}{2}\right)\alpha$ ☐ (d) $\omega_0 + \left(\frac{1-2n}{2}\right)\alpha$. ☐

where ω_0 = Initial angular velocity of the body moving in a circle

α = Uniform angular acceleration.

38. The linear velocity (V) of a rotating body is given by

(a) $V = \frac{\omega}{r}$ ☐ (b) $V = \omega \times r$ ☐

(c) $V = \frac{1}{\omega r}$ ☐ (d) none of the above. ☐

39. The expression ($\frac{1}{2} mV^2$) denotes

(a) centrifugal force ☐ (b) kinetic energy ☐

(c) potential energy ☐ (d) none of the above. ☐

40. The expression $\left(\frac{mV^2}{r}\right)$ represents

(*a*) centrifugal force ☐ (*b*) kinetic energy ☐
(*c*) potential energy ☐ (*d*) none of the above. ☐

41. The expression ($\frac{1}{2}$ $I\omega^2$) represents

(*a*) centrifugal force ☐ (*b*) kinetic energy ☐
(*c*) kinetic energy of rotation ☐ (*d*) potential energy. ☐

42. A force P of high magnitude acts on a body for a small interval of time (Δt). The product of P and Δt is called

(*a*) impulsive force ☐ (*b*) kinetic energy of the body ☐
(*c*) impulse ☐ (*d*) none of the above. ☐

43. The force P in question 42 is called

(*a*) impluse ☐ (*b*) impulsive force ☐
(*c*) propelling force ☐ (*d*) none of the above. ☐

44. Energy lost by a body (of mass m and moving with a velocity V) when it strikes another body (of mass M at rest) due to impact is equal to

(*a*) $\frac{mV^2}{2g}\left(1+\frac{m}{m+M}\right)$ ☐ (*b*) $\frac{mV^2}{2g}\ (m+M-1)$ ☐
(*c*) $\frac{mV^2}{2g}\left(1-\frac{m}{m+M}\right)$ ☐ (*d*) none of the above. ☐

45. Tension in a cable supporting a lift, when left is going up is equal to

(*a*) $W\left(1-\frac{a}{g}\right)$ ☐ (*b*) $W\left(1+\frac{a}{g}\right)$ ☐
(*c*) $W\left(W-\frac{W}{g}\right)$ ☐ (*d*) $W\left(g+\frac{a}{g}\right)$ ☐

where a = Uniform acceleration of lift, and W = Weight carried by lift.

46. Tension a cable supporting a lift, when lift is going down is equal to

(*a*) $W\left(1-\frac{a}{g}\right)$ ☐ (*b*) $W\left(1+\frac{a}{g}\right)$ ☐
(*c*) $a\left(W-\frac{W}{g}\right)$ ☐ (*d*) $W\left(g+\frac{a}{g}\right)$ ☐

where a = Uniform acceleration of lift, and W = Weight carried by lift.

47. When two bodies of mass (m and $2m$) are connected by a light inextensible string and pass over a smooth pulley, then acceleration of one body is

(*a*) equal to the acceleration of the other body ☐
(*b*) two time the acceleration of the other body ☐
(*c*) half the acceleration of the other body ☐
(*d*) none of the above. ☐

48. When two bodies of mass (M and $2M$) are connected by a light inextensible string and pass over a smooth pulley, then

(a) tension in both sides of the string will be equal ☐
(b) tension in one side of the string is two times the tension in the other side of the string ☐
(c) tension in one side of the string is half the tension in the other side of the string ☐
(d) none of the above. ☐

49. Two bodies of masses m_1 and m_2 are connected by a light inextensible string and pass over a smooth pulley. If the mass m_1 is coming down, then the acceleration of both the bodies is equal to

(a) $\dfrac{g(m_1 + m_2)}{(m_1 - m_2)}$ ☐ (b) $\dfrac{g(m_1 - m_2)}{(m_1 + m_2)}$ ☐

(c) $\dfrac{g(m_1 m_2)}{m_1 - m_2}$ ☐ (d) $\dfrac{g(m_1 m_2)}{m_1 + m_2}$. ☐

50. For question 49, the tension in the string will be equal to

(a) $\dfrac{2(m_1 + m_2)}{m_1 - m_2}$ ☐ (b) $\dfrac{2m_1 m_2}{m_1 - m_2}$ ☐

(c) $\dfrac{2m_1 m_2}{m_1 + m_2}$ ☐ (d) $\dfrac{2(m_1 + m_2)}{m_1 - m_2}$. ☐

51. Figure 2.14 shows the two bodies of masses m_1 and m_2 connected by a light inextensible string and a passing over a smooth pulley. Mass m_2 lies on a smooth horizontal plane. When mass m_1 is moving downward the acceleration of the two bodies is equal to

(a) $\dfrac{m_1 g}{m_1 - m_2}$ m/s² ☐

(b) $\dfrac{m_1 g}{m_1 + m_2}$ m/s² ☐

(c) $\dfrac{m_2 g}{m_1 + m_2}$ ☐

(d) $\dfrac{m_2 g}{m_1 - m_2}$. ☐

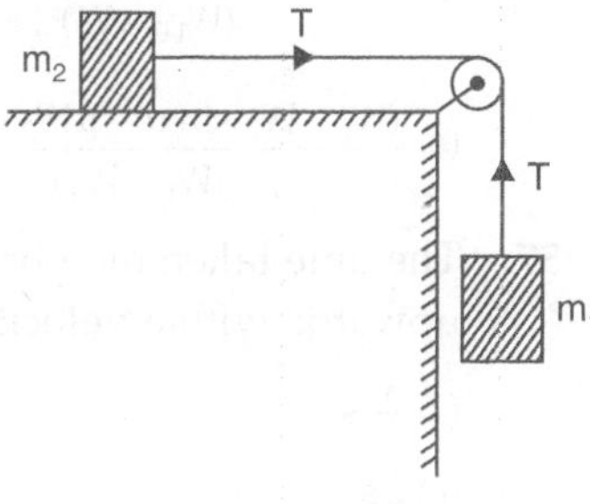

m₂ T T m₁

FIGURE 2.14

52. Refer to Fig. 2.14, the tension (T) will be equal to

(a) $\dfrac{m_1 m_2}{m_1 - m_2}$ N ☐ (b) $\dfrac{m_1 m_2}{m_1 + m_2}$ ☐

(c) $\dfrac{m_1 - m_2}{m_1 m_2}$ ☐ (d) $\dfrac{m_1 + m_2}{m_1 m_2}$. ☐

53. If the weight W_2 in Fig. 2.14 is resting on a rough horizontal plane (having co-efficient of friction as μ), then the acceleration is equal to

(a) $\dfrac{(W_1 + \mu W_2)g}{(W_1 + W_2)}$ ☐ (b) $\dfrac{(\mu W_1 + W_2)g}{(W_1 + W_2)}$ ☐

(c) $\dfrac{(W_1 - \mu W_2)g}{(W_1 + W_2)}$ ☐ (d) $\dfrac{(\mu W_1 + W_2)g}{W_1 - W_2}$. ☐

54. If the weight W_2 in Fig. 2.14 is resting on a rough horizontal plane (having co-efficient of friction as μ), then tension in the string is equal to

(a) $\frac{(1+\mu)W_1W_2}{W_1+W_2}$ ☐ (b) $\frac{(1+\mu)W_1W_2}{W_1-W_2}$ ☐

(c) $\frac{\mu W_1W_2}{W_1+W_2}$ ☐ (d) $\frac{\mu W_1W_2}{W_1-W_2}$. ☐

55. Two weights W_1 and W_2 are connected by a light inextensible string. Weight W_2 is placed on a smooth inclined plane of inclination α and W_1 hangs freely as shown in Fig. 2.15. If W_1 moves downwards then acceleration is equal to

(a) $\frac{W_1-W_2\sin\alpha}{(W_1+W_2)}$ ☐

(b) $\frac{(W_1-W_2\sin\alpha)g}{W_1+W_2}$ ☐

(c) $\frac{W_1+W_2}{W_1-W_2\sin\alpha}$ ☐

(d) $\frac{(W_1+W_2)g}{W_1-W_2\sin\alpha}$. ☐

FIGURE 2.15

56. If in Fig. 2.15, weight W_2 is placed on a rough inclined plane of inclination α and co-efficient of friction μ, then the acceleration is equal to

(a) $\frac{(W_1-W_2\sin\alpha-\mu W_2\cos\alpha)}{(W_1+W_2)}$ ☐ (b) $\frac{(W_1+W_2\sin\alpha-\mu W_2\cos\alpha)}{(W_1+W_2)}$ ☐

(c) $\frac{(W_1-W_2\sin\alpha-\mu W_2\cos\alpha)g}{(W_1+W_2)}$ ☐ (d) $\frac{(W_1-W_2\sin\alpha)g}{(W_1+W_2)}$. ☐

57. The time taken by a ball (of weight 500 N) to return back to earth, if it is thrown vertically upwards with a velocity 4.9 m/s is equal to

(a) $\frac{1}{2}$ s ☐ (b) 1 s ☐

(c) 2 s ☐ (d) 3 s. ☐

58. The maximum height attained by a ball (of weight 500 N) which is thrown vertically upwards with a velocity 4.9 m/s is equal to

(a) 100 cm ☐ (b) 245 cm ☐

(c) 122.5 cm ☐ (d) 980 cm. ☐

59. A tower is of height 100 m. A stone is thrown up from the foot of water with a velocity of 20 m/s and at the same time another stone is dropped from the top of the tower. The two stones will meet after

(a) 10 s ☐ (b) 5 s ☐

(c) 2 s ☐ (d) 7.5 s. ☐

60. The maximum height reached by a stone (of weight 50 N) which is thrown vertically upward with an initial velocity 19.6 m/s would be

(a) 20 m ☐ (b) 19.6 m ☐

(c) 30 m ☐ (d) 25 m. ☐

61. A body is moving with a velocity of 2 m/s. If the velocity of body becomes 5 m/s after 4 seconds, the acceleration of the body would be

(*a*) 1 m/s^2 (*b*) 0.75 m/s^2
(*c*) 1.5 m/s^2 (*d*) 0.375 m/s^2.

62. A body is rotating with an angular velocity of 5 radians/s. After 4 seconds, the angular velocity of the body becomes 1.3 radians per sec. The angular acceleration of the body would be

(*a*) 3 rad/s^2 (*b*) 2 rad/s^2
(*c*) 1 rad/s^2 (*d*) 1.5 rad/s^2.

63. A flywheel starting from rest and accelerating uniformly performs 20 revolution in 4 seconds. The angular velocity of flywheel after 8 seconds would be

(*a*) 30 rad/s (*b*) 35 rad/s
(*c*) 40 rad/s (*d*) 55 rad/s.

64. A body is moving in a straight line with an initial velocity of 4 m/s. After 5 seconds the velocity of the body becomes 9 m/s. The distance travelled by the body in third second would be

(*a*) 6 m (*b*) 5.5 m
(*c*) 6.5 m (*d*) 4 m.

65. A body is rotating with an angular velocity of 5 radians/s. After 4 seconds, the angular acceleration of the body becomes 13 radians/s. If the body is rotating with uniform acceleration, the angle covered by the body in the third second would be

(*a*) 20 radians (*b*) 25 radians
(*c*) 15 radians (*d*) 10 radians.

66. A body is moving with a velocity of 10 m/s. The time required, to stop the body within a distance of 5 m, is equal to

(*a*) 3 seconds (*b*) 5 seconds
(*c*) 1 second (*d*) 0.5 second

67. A stone dropped into a well is heard to strike the water after 4 seconds. If the velocity of sound is 350 m/s, the depth of well would be

(*a*) 150 m (*b*) 70.75 m
(*c*) 100 m (*d*) 35.375 m.

68. A light string passes over a smooth, weightless pulley and has weights 40 N and 60 N attached to its end as shown in Fig. 2.16. The tension in string will be

(*a*) 60 N
(*b*) 50 N
(*c*) 48 N
(*d*) 20 N.

40 N
60 N

FIGURE 2.16

69. Refer to Fig. 2.16, the acceleration, with which the weight 60 N descends, is

(*a*) $\frac{g}{5}$ (*b*) $\frac{g}{4}$
(*c*) $2g$ (*d*) $5g$.

70. Refer to Fig. 2.16, the pressure on pulley would be

(*a*) 100 N ☐ (*b*) 96 N ☐

(*c*) 20 N ☐ (*d*) 50 N. ☐

71. Figure 2.17 shows two weights 40 N and 60 N connected by a light inextensible string and passes over a smooth weightless pulley. The weight 40 N is resting on a rough horizontal plane with $\mu = 0.3$ and weight 60 N hangs freely and is moving downward. The tension in the string would be

(*a*) 20 N ☐

(*b*) 100 N ☐

(*c*) 31.2 N ☐

(*d*) 50 N. ☐

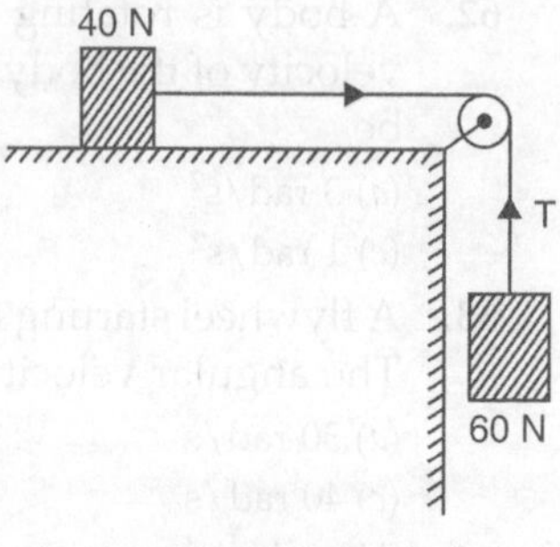

FIGURE 2.17

72. Refer to Fig. 2.17, the acceleration with which weight 60 N descends would be

(*a*) 0.08 g ☐

(*b*) 20 g ☐

(*c*) 100 g ☐

(*d*) 50 g. ☐

73. Two weights of 50 N and 150 N (of two blocks *A* and *B* respectively) are connected by a string and frictionless and weightless pulleys as shown in Fig. 2.18. The tension in the string would be

(*a*) 100 N ☐

(*b*) 200 N ☐

(*c*) 64.3 N ☐

(*d*) 50 N. ☐

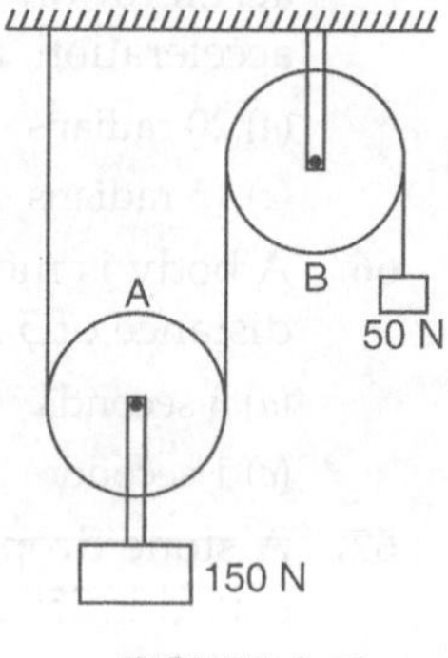

FIGURE 2.18

74. Refer to Fig. 2.18, the acceleration to block *A* would be

(*a*) $\frac{g}{7}$ ☐ (*b*) $\frac{g}{5}$ ☐

(*c*) $\frac{2g}{5}$ ☐ (*d*) $\frac{2g}{7}$. ☐

75. Refer to Fig. 2.18, the acceleration of block *B* would be

(*a*) $\frac{g}{7}$ ☐ (*b*) $\frac{g}{5}$ ☐

(*c*) $\frac{2g}{5}$ ☐ (*d*) $\frac{2g}{7}$. ☐

Projectiles

76. The path traced by a projectile in the space is

(*a*) hyperbolic ☐ (*b*) parabolic ☐

(*c*) linear ☐ (*d*) none of the above. ☐

77. The equation of the path travelled by a projectile in space is given as

(a) $y = x \tan \alpha - \dfrac{2gx^2}{u^2 \cos^2 \alpha}$ ☐ (b) $y = x \cos \alpha - \dfrac{2gx^2}{u^2 \tan^2 \alpha}$ ☐

(c) $y = x \tan \alpha - \dfrac{gx^2}{2u^2 \cos^2 \alpha}$ ☐ (d) $y = x \sin \alpha - \dfrac{gx^2}{2u^2 \cos^2 \alpha}$. ☐

where x, y are the co-ordinates of any point on the path, from point of projection and u = velocity of projection.

78. The range of the projectile is

(a) horizontal distance between the point of projection and the point where the projectile strikes the ground ☐

(b) maximum vertical height attained by the projectile ☐

(c) half of the maximum vertical height attained by the projectile ☐

(d) none of the above. ☐

79. The range (R) is given by

(a) $\dfrac{u^2 \sin^2 2\alpha}{2g}$ ☐ (b) $\dfrac{u^2 \sin^2 2x}{g}$ ☐

(c) $\dfrac{u^2 \sin^2 2\alpha}{g}$ ☐ (d) none of the above. ☐

where α = Angle of projection, and u = Velocity of projection.

80. Time of flight of projectile is defined as the time

(a) taken by the projectile to reach maximum height ☐

(b) taken by the projectile to reach maximum height and to return back to the ground ☐

(c) taken by the projectile to return from maximum height to the ground ☐

(d) none of the above. ☐

81. The time of flight (t) is equal to

(a) $\dfrac{u \sin \alpha}{2g}$ ☐ (b) $\dfrac{2u \sin \alpha}{g}$ ☐

(c) $\dfrac{2g}{u \sin \alpha}$ ☐ (d) $\dfrac{g}{2u \sin \alpha}$. ☐

82. Maximum height attained by a projectile is equal to

(a) $\dfrac{u^2 \sin \alpha}{2g}$ ☐ (b) $\dfrac{u \sin^2 \alpha}{2g}$ ☐

(c) $\dfrac{u^2 \sin^2 \alpha}{2g}$ ☐ (d) $\dfrac{2u^2 \sin^2 \alpha}{g}$. ☐

83. The velocity of projectile at a height k is equal to

(a) $\sqrt{u^2 + 2gh}$ ☐ (b) $\sqrt{u - 2gh}$ ☐

(c) $\sqrt{2u^2 - gh}$ ☐ (d) $\sqrt{gh - 2u^2}$. ☐

84. Velocity of projectile after an interval 't' is given by

(a) $\sqrt{u^2 + g^2t^2 - (2 \sin \alpha) \times gt}$ ☐ (b) $\sqrt{u^2 - (2 \sin \alpha) \times gt}$ ☐

(c) $\sqrt{u^2 + g^2t^2 - (2u \sin \alpha) \times gt}$ ☐ (d) none of the above. ☐

85. Maximum horizontal range of a projectile, having velocity of projection u and angle of projection as α, is equal to

(a) $\dfrac{u^2 \sin 2x}{2g}$ ☐ (b) $\dfrac{u^2}{2g}$ ☐

(c) $\dfrac{2g}{u^2}$ ☐ (d) $\dfrac{g}{2u^2}$. ☐

86. If R_{max} is the maximum range of a projectile, then the range (R) of the projectile when fixed at an angle of 30°, with the same initial velocity would be

(a) $\sqrt{2}\, R_{max}$ ☐ (b) $\dfrac{\sqrt{3}}{2} R_{max}$ ☐

(c) $\frac{1}{8} R_{max}$ ☐ (d) $\frac{1}{4} R_{max}$. ☐

87. If R_{max} is the maximum range of a projectile, then the maximum height of the projectile when fired at angle 30°, with the same velocity would be

(a) $\frac{1}{4} R_{max}$ ☐ (b) $\frac{1}{2} R_{max}$ ☐

(c) $\frac{1}{8} R_{max}$ ☐ (d) none of the above. ☐

88. The angle of projection for maximum range of a projectile is

(a) 90° ☐ (b) 30° ☐

(c) 60° ☐ (d) 45°. ☐

89. One newton is equal to

(a) 10^3 dyne ☐ (b) 10^2 dyne ☐

(c) 10^5 dyne ☐ (d) 10^4 dyne. ☐

90. Dyne is the force acting on a mass of

(a) one kg to produce an acceleration of one m/cm² ☐

(b) one g to produce an acceleration of one m/sec² ☐

(c) one kg to produce an acceleration of one cm/sec² ☐

(d) one g to produce an acceleration of one cm/sec². ☐

91. One newton is a force acting on a mass of

(a) one g to produce an acceleration of one m/sec² ☐

(b) one kg to produce an acceleration of one m/sec² ☐

(c) one kg to produce an acceleration of one cm/sec². ☐

(d) none of the above. ☐

92. Joule is the unit of

(a) velocity ☐ (b) force ☐

(c) work ☐ (d) acceleration. ☐

93. Joule is expressed in S.I. units as

(*a*) N m/s ☐ (*b*) N m^2 ☐

(*c*) N m ☐ (*d*) m N. ☐

94. Watt is the unit of

(*a*) force ☐ (*b*) velocity ☐

(*c*) work ☐ (*d*) power. ☐

95. Watt is expressed in S.I. units as

(*a*) N m/s ☐ (*b*) N m^2/s ☐

(*c*) J/s ☐ (*d*) N m. ☐

96. Pressure in S.I. units is expressed as

(*a*) N/m ☐ (*b*) N/m^2 ☐

(*c*) N m^2 ☐ (*d*) N m. ☐

97. One metric horse power is equal to

(*a*) 746 watts ☐ (*b*) 736 watts ☐

(*c*) 550 watts ☐ (*d*) 75 watts. ☐

98. Momentum of a body is given by

(*a*) mass × velocity ☐ (*b*) mass × change of velocity ☐

(*c*) moment × distance ☐ (*d*) velocity × acceleration. ☐

99. The expression $I \times \omega$ (where I = moment of inertia and ω = angular velocity) represents

(*a*) power ☐ (*b*) angular momentum ☐

(*c*) moment (angular) ☐ (*d*) none of the above. ☐

100. The torque (T) acting on a rotating body is given as $T = I\alpha$, where I = M.O.I. in kg-m^2 and α = angular acceleration in rad/sec^2. The units of T will be as

(*a*) kgf-m ☐ (*b*) N m ☐

(*c*) dyne metre ☐ (*d*) none of the above. ☐

Friction

101. Limiting force of friction is defined as the frictional force which exists when a body

(*a*) is moving with maximum velocity ☐ (*b*) is stationary ☐

(*c*) just begins to slide over the surface ☐ (*d*) none of the above. ☐

102. Co-efficient of friction is the ratio of

(*a*) force of friction to reaction between two bodies ☐

(*b*) force of friction to normal reaction between two bodies ☐

(*c*) force of limiting friction to reaction between two bodies ☐

(*d*) force of limiting friction to normal reaction between two bodies. ☐

103. Angle of friction (ϕ) is the angle between the

(*a*) limiting friction and normal reaction ☐

(*b*) limiting friction and the resultant of limiting friction and normal reaction ☐

(*c*) normal reaction and the resultant of limiting friction and normal reaction ☐

(*d*) none of the above. ☐

104. The co-efficient of friction (*a*) in terms of angle of friction (ϕ) is given by

(*a*) $\phi = \tan \mu$ ☐ (*b*) $\mu = \sin \phi$ ☐

(*c*) $\mu = \tan \phi$ ☐ (*d*) $u = \dfrac{1}{\tan \phi}$. ☐

105. The force of friction which exists when the body is in motion is called

(*a*) static friction ☐ (*b*) limiting friction ☐

(*c*) dynamic friction ☐ (*d*) none of the above. ☐

106. Dynamic friction is always

(*a*) more than static friction ☐ (*b*) more than limiting friction ☐

(*c*) less than limiting friction ☐ (*d*) none of the above. ☐

107. A body of weight *W* is resting on a horizontal plane. A force *P* is applied parallel to the plane to move the body. The value of *P*, necessary to move the body against the resistance of friction is

(*a*) $W/\tan \phi$ ☐ (*b*) $W \sin \phi$ ☐

(*c*) $W \tan \phi$ ☐ (*d*) $W \cos \phi$ ☐

where ϕ = Angle of friction.

108. The force *P* is applied at an angle θ with the horizontal plane on which a body of weight *W* is placed as shown in Fig. 2.19. The value of *P*, necessary to move the body is equal to

(*a*) $\dfrac{W \cos \phi}{\cos (\theta - \phi)}$ ☐

(*b*) $W \dfrac{\sin \phi}{\cos (\theta - \phi)}$ ☐

(*c*) $\dfrac{W \tan \phi}{\sin (\theta - \phi)}$ ☐

(*d*) none of the above ☐

where ϕ = Angle of friction.

FIGURE 2.19

109. Refer to Fig. 2.19, the necessary force *P* to move the body will be minimum, when

(*a*) $\theta = 2\phi$ ☐ (*b*) $\theta = \phi$ ☐

(*c*) $\theta = \phi/2$ ☐ (*d*) none of the above ☐

where ϕ = Angle of friction.

110. Refer to Fig. 2.19, the minimum force *P* to move the body is

(*a*) $W \tan \phi$ ☐ (*b*) $W \cos \phi$ ☐

(*c*) $W \sin \phi$ ☐ (*d*) $W/\sin \phi$ ☐

where ϕ = Angle of friction.

111. The necessary force *P*, applied at an angle θ with the horizontal plane, on which a body of weight *W* is placed, to move the body is (see Fig. 2.20)

(*a*) $\dfrac{W \cos \phi}{\cos (\theta - \phi)}$ ☐

(*b*) $\dfrac{W \sin \phi}{\cos (\theta - \phi)}$ ☐

FIGURE 2.20

(*c*) $\frac{W \tan \phi}{\cos(\theta + \phi)}$ ☐ (*d*) $\frac{W \tan \phi}{\cos(\theta - \phi)}$ ☐

where ϕ = Angle of friction.

112. Refer to Fig. 2.20, the minimum force P to move the body is equal to

(*a*) $W \tan \phi$ ☐ (*b*) $W \sin \phi$ ☐

(*c*) $W \cos \phi$ ☐ (*d*) $W/\tan \phi$ ☐

where ϕ = Angle of friction.

113. Refer to Fig. 2.20, the force P will be minimum when

(*a*) $\theta = 2\phi$ ☐ (*b*) $\theta = \phi$ ☐

(*c*) $\theta = (90 - \phi)$ ☐ (*d*) $\theta = (90 - 2\phi)$. ☐

114. The necessary force P, applied parallel to an inclined plane having inclination α with horizontal to move a body of weight W, up the plane is equal to (Refer to Fig. 2.21).

(*a*) $\frac{W \sin \alpha}{\cos \phi}$ ☐

(*b*) $\frac{W \sin(\alpha + \phi)}{\cos \phi}$ ☐

(*c*) $\frac{W \cos(\alpha + \phi)}{\cos \phi}$ ☐

(*d*) $\frac{W \tan(\alpha + \phi)}{\cos \phi}$ ☐

FIGURE 2.21

where ϕ = Angle of friction.

115. Refer to Fig. 2.21, if the body is on the point of moving down the plane, then necessary force P will be equal to

(*a*) $\frac{W \sin(\alpha + \phi)}{\cos \phi}$ ☐ (*b*) $\frac{W \sin(\alpha - \phi)}{\tan \phi}$ ☐

(*c*) $\frac{W \sin(\alpha - \phi)}{\cos \phi}$ ☐ (*d*) $\frac{W \sin(\alpha - \phi)}{\tan \phi}$ ☐

116. Figure 2.22 shows a body of weight W placed on an inclined plane having inclination α with the horizontal. Force P is applied horizontally. When the body is on the point of moving up the plane, the necessary value of P is

(*a*) $W \sin(\alpha + \phi)$ ☐ (*b*) $W \tan(\alpha + \phi)$ ☐

(*c*) $W \cos(\alpha + \phi)$ ☐ (*d*) $W \sin(\alpha - \phi)$. ☐

117. Refer to Fig. 2.22, if the body is on the point of moving down the plane, the necessary force P would be

(*a*) $W \sin (\alpha - \phi)$ ☐

(*b*) $W \tan (\alpha + \phi)$ ☐

(*c*) $W \tan (\alpha - \phi)$ ☐

(*d*) $W \sin (\alpha + \phi)$. ☐

FIGURE 2.22

118. A ladder of weight 250 N is placed against a smooth vertical wall and a rough horizontal floor ($\mu = 0.3$) as shown in Fig. 2.23. If the ladder is on the point of sliding, the reaction at A will be

(*a*) 250 N ☐

(*b*) 261 N ☐

(*c*) 125 N ☐

(*d*) 500 N. ☐

FIGURE 2.23

119. Refer to Fig. 2.23, the reaction at B will be

(*a*) 250 N ☐ (*b*) 125 N ☐

(*c*) 75 N ☐ (*d*) 500 N. ☐

120. If in question 118, a man of weight 500 N stands on the ladder at its middle point. The reaction at A will be (when ladder is on the point of sliding)

(*a*) 750 N ☐ (*b*) 783 N ☐

(*c*) 250 N ☐ (*d*) 1000 N. ☐

121. For question 120, the reaction at B will be

(*a*) 750 N ☐ (*b*) 783 N ☐

(*c*) 225 N ☐ (*d*) 1000 N. ☐

122. A horizontal force of 400 N is applied on a body of weight 1200 N, placed on a horizontal plane. If the body is just on the point of motion, the angle of friction would be

(*a*) 20° ☐ (*b*) 18° 26′ ☐

(*c*) 10° ☐ (*d*) 25°. ☐

123. A body of weight W is placed on an inclined plane. The angle made by the inclined plane with horizontal, when the body is on the point of moving down is called

(*a*) angle of inclination ☐ (*b*) angle of repose ☐

(*c*) angle of friction ☐ (*d*) angle of limiting friction. ☐

124. If angle of friction is zero

(*a*) force of friction will act normal to the plane ☐

(*b*) force of friction will act opposite to the direction of motion ☐

(*c*) force of friction will be zero ☐

(*d*) force of friction will be infinite. ☐

125. If P is the effort required to lift a load (W), then mechanical advantage (M.A.) is given by

(a) $\frac{P}{W}$ ☐ (b) $P \times W$ ☐

(c) $\frac{W}{P}$ ☐ (d) $\frac{1}{P \times W}$. ☐

126. If D is the distance moved by the effort and d is the distance moved by the load, then velocity ratio (V.R.) is given by

(a) $\frac{d}{D}$ ☐ (b) $\frac{D}{d}$ ☐

(c) $d \times D$ ☐ (d) $\frac{1}{D \times d}$. ☐

127. Efficiency of a machine in terms of mechanical advantage (M.A.) and velocity ratio (V.R.) is given by

(a) $\frac{\text{V.R.}}{\text{M.A.}}$ ☐ (b) $\frac{1}{(\text{V.R.}) \times (\text{M.A.})}$ ☐

(c) $\frac{\text{M.R.}}{\text{V.R.}}$ ☐ (d) (V.R.) × (M.A.). ☐

128. Self-locking machine is one which has efficiency

(a) 100% ☐ (b) less than 50% ☐

(c) more than 50% ☐ (d) none of the above. ☐

129. A reversible machine is one which have efficiency

(a) 100% ☐ (b) less than 50% ☐

(c) more than 50% ☐ (d) none of the above. ☐

130. Non-reversible machine is also called

(a) ideal machine ☐ (b) self-locking machine ☐

(c) actual machine ☐ (d) none of the above. ☐

131. The law of the machine is given by

(a) $W = mP + C$ ☐ (b) $P = mW + C$ ☐

(c) $C = W + mP$ ☐ (d) $C = mW + P$ ☐

where P = Effort applied, W = Load lifted,

m = Constant and equal to the slope of the line,

C = Another constant.

132. For an ideal machine

(a) M.A. is less than V.R. ☐ (b) M.A. is greater than V.R. ☐

(c) M.A. is equal to V.R. ☐ (d) None of the above. ☐

133. In an actual machine, the amount of friction present may be expressed in terms of effort. Additional effort is required to overcome this friction. The value of additional effort required is equal to

(a) $\frac{W}{\text{V.R.}} P$ ☐ (b) $P - \frac{W}{\text{V.R.}}$ ☐

(c) $P + \frac{W}{\text{V.R.}}$ ☐ (d) $P + \frac{\text{V.R.}}{W}$ ☐

where V.R. = Velocity ratio, and W = Load applied.

134. The amount of friction, present in a machine, may be expressed in terms of load. Additional load is to be lifted if the friction is not present. The value of this additional load is equal to

(a) $W \times P \times$ V.R. ☐ (b) $\frac{P}{\text{V.R.}} + W$ ☐

(c) $P \times$ V.R. $- W$ ☐ (d) $\frac{\text{V.R.}}{P} + W.$ ☐

135. The efficiency of a screw-jack for raising a load W is equal to

(a) $\frac{\cos \alpha}{\tan (\alpha + \phi)}$ ☐ (b) $\frac{\sin \alpha}{\tan (\alpha + \phi)}$ ☐

(c) $\frac{\tan (\alpha + \phi)}{\tan \alpha}$ ☐ (d) $\frac{\tan \alpha}{\tan (\alpha + \phi)}$ ☐

where α = Angle of screw (Helix angle), and ϕ = Angle of friction.

136. The efficiency of a screw-jack will be maximum, if angle of screw or helix angle (α) is equal to

(a) $\frac{\pi}{2} - \phi$ ☐ (b) $\frac{\pi}{4} - \phi$ ☐

(c) $\frac{\pi}{4} + \phi$ ☐ (d) $\frac{\pi}{4} - \frac{\phi}{2}$ ☐

where ϕ = Angle of friction.

137. The maximum value of efficiency of a screw-jack for raising a load W is equal to

(a) $\frac{1 - \tan \phi}{1 + \sin \phi}$ ☐ (b) $\frac{\sin \phi}{1 + \cos \phi}$ ☐

(c) $\frac{1 - \sin \phi}{1 + \sin \phi}$ ☐ (d) $\frac{\cos \phi}{1 + \sin \phi}$. ☐

138. The efficiency of a screw-jack for a given value of angle of friction

(a) depends upon the weight lifted only ☐

(b) depends upon the effort applied only ☐

(c) depends upon weight and effort only ☐

(d) independent of weight lifted or effort applied. ☐

139. For a differential wheel and axle having diameter of effort wheel as D and diameters of two axles as d_1 and d_2 ($d_1 > d_2$), the velocity ratio (V.R.) is equal to

(a) $\frac{2D}{d_1 + d_2}$ ☐ (b) $\frac{2D}{2d_1 - d_2}$ ☐

(c) $\frac{2D}{d_1 - d_2}$ ☐ (d) $\frac{D}{d_1 - d_2}$. ☐

Centre of Gravity

140. The C.G. of a triangle lies at the point of concurrence of

(a) the right bisectors of the angle of the triangle ☐

(b) the medians of the triangle ☐

(c) the altitudes from the vertices on the opposite side ☐

(d) none of the above. ☐

141. The C.G. of solid hemisphere lies on the central radius at a distance

(a) $\frac{3r}{4}$ from the plane base ☐ (b) $\frac{3r}{8}$ from the plane base ☐

(c) $\frac{8r}{3}$ from the plane base ☐ (d) none of the above. ☐

142. The C.G. of a semi-circular lamina lies on the central radius at a distance of

(a) $\frac{4r}{3\pi}$ from base diameter ☐ (b) $\frac{3r}{8}$ from base diameter ☐

(c) $\frac{8r}{3}$ from base diameter ☐ (d) none of the above. ☐

143. The C.G. of a solid right circular cone lies on the axis at a height

(a) half of the total height above the base ☐

(b) one-third of the total height above the base ☐

(c) one-fourth of the total height above the base ☐

(d) none of the above. ☐

144. The C.G. of a thin hollow right circular cone lies on the axis at a height

(a) half of the total height ☐ (b) one-third of the total height ☐

(c) one-fourth of the total height ☐ (d) none of the above. ☐

145. The C.G. of a semi-circular arc is at the central radius at a distance of

(a) $\frac{3r}{4}$ from base diameter ☐ (b) $\frac{3r}{8}$ from base diameter ☐

(c) $\frac{2r}{\pi}$ above base diameter ☐ (d) $\frac{r}{2\pi}$ above base diameter. ☐

146. The C.G. of a quadrant of a circle is at a distance of

(a) $\frac{3r}{4\pi}$ from the axis ☐ (b) $\frac{4r}{3\pi}$ from the axis ☐

(c) $\frac{3r}{8}$ from the axis ☐ (d) $\frac{8r}{3}$ from the axis. ☐

147. The C.G. of a thin hollow hemisphere is at a distance of

(a) $\frac{r}{3}$ from base ☐ (b) $\frac{r}{2}$ from base ☐

(c) $\frac{r}{4}$ from base ☐ (d) none of the above. ☐

148. The C.G. of a solid cone lies on the central axis at a distance of

(a) $\frac{1}{3}$ of the total height above base ☐ (b) half the total height above base ☐

(c) $\frac{1}{4}$ of the total height above base ☐ (d) 2/5 of total height above base. ☐

149. The C.G. of a hollow cone lies on the central axis above base at a distance of

(a) $\frac{1}{3}$ of the total height ☐ (b) half of the total height ☐

(c) $\frac{1}{4}$ of the total height ☐ (d) 2/5 of the total height. ☐

150. The C.G. of a hollow pyramid lies on the central axis above base at a distance of

(a) $\frac{1}{4}$ of the total height ☐ (b) $\frac{1}{3}$ of the total height ☐

(c) $\frac{1}{2}$ of the total height ☐ (d) 2/5 of the total height. ☐

151. The C.G. of a hollow cylinder lies on the vertical axis above base at a distance of

(a) $\frac{1}{4}$th of the total height ☐ (b) $\frac{1}{3}$rd of the total height ☐

(c) half of the total height ☐ (d) 2/5th of the total height. ☐

152. The C.G. of a parabola as shown in Fig. 2.24 lies on the axis OB at a distance of

(a) $\frac{1}{5}$th of OB ☐

(b) $\frac{1}{4}$th of OB ☐

(c) 2/5th of OB ☐

(d) $\frac{1}{3}$rd of OB. ☐

FIGURE 2.24

Moment of Inertia

153. If A = area of the body, I_G = moment of inertia of body about an axis passing through its C.G. and I_0 = moment of inertia of the body about an axis parallel to the axis passing through C.G. and at a distance x, then according to the theorem of parallel axis

(a) $I_G = I_0 + Ax^2$ ☐ (b) $I_0 = I_G + Ax^2$ ☐

(c) $I_0 = I_G - Ax^2$ ☐ (d) none of the above. ☐

154. If I_x = M.O.I. about x-axis and I_y = M.O.I. about y-axis, then moment of inertia about z-axis is given by

(a) $I_z = I_x - I_y$ ☐ (b) $I_z = I_y - I_x$ ☐

(c) $I_z = I_x \times I_y$ ☐ (d) none of the above. ☐

155. A thin rod of length L and mass M will have moment of inertia about an axis passing through one of its edge and perpendicular to the rod,

(a) $\frac{Ml^2}{12}$ ☐ (b) $\frac{ML^2}{6}$ ☐

(c) $\frac{ML^3}{3}$ ☐ (d) $\frac{ML^2}{3}$. ☐

156. Moment of inertia of a rectangular section having b = Width and d = Depth about x-axis is given by

(a) $I_x = \frac{bd^3}{12}$ ☐ (b) $I_x = \frac{b^3d}{12}$ ☐

(c) $I_x = \frac{b^2d^2}{6}$ ☐ (d) none of the above. ☐

157. M.O.I. of a circular section of diameter d about an axis passing through its C.G. lying in the plane of the section is given by

(a) $I_x = \frac{\pi d^4}{32}$ ☐ (b) $I_x = \frac{\pi d^4}{64}$ ☐

(c) $I_x = \frac{\pi d^4}{16}$ ☐ (d) none of the above. ☐

158. M.O.I. of a triangular section about an axis passing through its base is given by

(a) $I = \frac{bh^3}{12}$ ☐ (b) $I = \frac{bh^3}{32}$ ☐

(c) $I = \frac{bh^2}{36}$ ☐ (d) none of the above ☐

where b = Width at a base, and h = Height of triangle.

159. M.O.I. of a triangular section, about an axis passing through its C.G. is

(a) $I = \frac{bh^3}{12}$ ☐ (b) $I = \frac{bh^3}{32}$ ☐

(c) $I = \frac{bh^2}{36}$ ☐ (d) none of the above ☐

160. M.O.I. of a solid sphere of mass M and radius R is given by

(a) $I = \frac{MR^2}{12}$ ☐ (b) $I = \frac{2}{5} MR^2$ ☐

(c) $I = \frac{MR^2}{36}$ ☐ (d) $I = \frac{3}{5} MR^2$. ☐

161. M.O.I. of a thin spherical shell of mass M and radius r is given as

(a) $\frac{2}{5} Mr^2$ ☐ (b) $\frac{2}{3} Mr^2$ ☐

(c) $\frac{3}{2} Mr^2$ ☐ (d) $\frac{4}{3} Mr^2$. ☐

162. M.O.I. of a solid cone about its vertical axis is

(a) $\frac{10}{3} Mr^2$ ☐ (b) $\frac{5}{3} Mr^2$ ☐

(c) $\frac{3}{10} Mr^2$ ☐ (d) $\frac{3}{5} Mr^2$. ☐

163. The units of moment of inertia of mass are

(a) kg-m^3 ☐ (b) kg-m^2 ☐

(c) kg-m ☐ (d) kg-m^4. ☐

164. The units of moment of inertia of area are

(a) kg-m^3 ☐ (b) kg-m^2 ☐

(c) m^4 ☐ (d) none of the above. ☐

165. M.O.I. of a solid cone about an axis passing through its C.G. and parallel to base is

(a) $\frac{3}{10} Mr^2$ ☐ (b) $\frac{3}{5} Mr^2$ ☐

(c) $\frac{3}{20} M\left(r^2 + \frac{h^2}{4}\right)$ ☐ (d) $\frac{10}{3} Mr^2$. ☐

166. M.O.I. of a solid circular cylinder about vertical axis is

(a) $\frac{3}{10} Mr^2$ ☐ (b) $\frac{3}{5} Mr^2$ ☐

(c) $\frac{3}{2} Mr^2$ ☐ (d) $\frac{Mr^2}{2}$. ☐

167. M.O.I. of elliptical section, of major axis = D and minor axis = B, is

(a) $\frac{\pi}{64} D^4$ ☐ (b) $\frac{\pi}{64} d^4$ ☐

(c) $\frac{\pi}{64} BD^3$ ☐ (d) $\frac{\pi}{64} B^2D^2$. ☐

168. Polar moment of inertia of a circle (I_0) is given by

(a) $\frac{\pi}{64} d^4$ ☐ (b) $\frac{\pi}{32} d^4$ ☐

(c) $\frac{\pi}{16} d^4$ ☐ (d) $\frac{\pi}{32} d^3$. ☐

169. Polar moment of inertia of a rectangle ($b \times h$) is given by

(a) $\frac{bh^3}{12}$ ☐ (b) $\frac{hb^3}{12}$ ☐

(c) $\frac{bh(b^2+h^2)}{12}$ ☐ (d) $\frac{b^2h^2(b+h)}{12}$. ☐

170. Polar moment of inertia of an equilateral triangle of sides x is given by

(a) $\frac{x^4}{16}$ ☐ (b) $\frac{x^4}{16\sqrt{3}}$ ☐

(c) $\frac{x^4}{32}$ ☐ (d) $\frac{x^4}{64}$. ☐

171. Polar moment of inertia is

(a) the moment of inertia of an area about an axis parallel to centroidal axis ☐

(b) equal to moment of inertia ☐

(c) the moment of an area about an axis which is not lying in the plane of the area ☐

(d) the moment of inertia of an area about a line or axis perpendicular to the plane of the area. ☐

172. The moment of inertia of a triangle (having base = b and height = h) with respect to an axis through the apex and parallel to the base is

(a) $\frac{bh^3}{12}$ ☐ (b) $\frac{bh^3}{36}$ ☐

(c) $\frac{bh^3}{4}$ ☐ (d) $\frac{bh^3}{10}$. ☐

173. Newton's second law for rotary motion states that

(a) rate of change of rotation of a body about a fixed axis is directly proportional to the impressed external force and takes place in the direction of force. ☐

(b) rate of change of momentum is directly proportional to the impressed force and takes plane in the direction of force. ☐

(c) rate of change of rotation (angular momentum) is directly proportional to the impressed external torque and takes place in the direction of force. ☐

(d) none of the above. ☐

174. The relation between external torque (T) acting on a body and the angular acceleration (α) is given by

(a) $T = \frac{I}{\alpha}$ ☐ (b) $T = \frac{\alpha}{I}$ ☐

(c) $T = I\alpha$ ☐ (d) none of the above ☐

where T is in newton-metres.

Circular Motion

175. If a body is moving in a circular path, a force comes into play which acts along the radius of circular path and is directed towards the centre of the path. This force is called

(a) centrifugal force ☐ (b) centripetal force ☐

(c) shear force ☐ (d) none of the above. ☐

176. The force, acting on a body moving in a circular path along the radius away from the centre of the path, is called

(a) centrifugal force ☐ (b) centripetal force ☐

(c) shear force ☐ (d) none of the above. ☐

177. Centrifugal force is given by

(a) $\frac{W}{g}\frac{\omega^2}{r}$ ☐ (b) $\frac{W}{g}\,\omega^2 r^2$ ☐

(c) $\frac{W}{g}\,V^2 r$ ☐ (d) $\frac{W}{g}\frac{V^2}{r}$ ☐

where W = Weight of body in kgf, V = Linear velocity,

ω = Angular velocity, and r = Radius of circular path.

178. The expression $T \times \omega$ (where T = torque in N m and ω angular velocity) gives

(a) work done due to rotation ☐ (b) horse power ☐

(c) power in watts ☐ (d) force. ☐

179. Kinetic energy due to rotation is expressed as

(a) K.E. = $\frac{I\omega}{2g}$ ☐ (b) K.E. = $\frac{I^2 \times \omega}{2g}$ ☐

(c) K.E. = $\frac{I\omega^2}{2g}$ ☐ (d) none of the above. ☐

180. Radius of gyration is expressed as

(a) $k = \sqrt{\frac{m}{I}}$ ☐ (b) $\sqrt{\frac{I}{m}}$ ☐

(c) $k = \sqrt{mI}$ ☐ (d) $\frac{I}{\sqrt{mI}}$. ☐

181. Super-elevation (or banking) of roads is the process of

(a) raising the inner edge of the road above the outer edge ☐

(b) raising the outer edge of the roads above the inner edge ☐

(c) keeping both the edges at the same level ☐

(d) none of the above. ☐

182. The angle of super-elevation (or banking) is given by

(*a*) $\theta = \tan^{-1}\frac{V^2}{g}$ ☐ (*b*) $\theta = \tan^{-1}\frac{V^2}{gr}$ ☐

(*c*) $\theta = \tan^{-1}\frac{gr}{V^2}$ ☐ (*d*) $\theta = \tan^{-1}\frac{r}{gV^2}$ ☐

where V = Velocity of the vehicle, r = Radius of circular path.

183. The maximum velocity of a vehicle on a level circular path to avoid skidding is given by

(*a*) $\sqrt{\frac{\mu}{gr}}$ ☐ (*b*) $\sqrt{\mu gr}$ ☐

(*c*) $\sqrt{\frac{\mu r}{g}}$ ☐ (*d*) $\frac{1}{\sqrt{\mu gr}}$ ☐

where μ = Co-efficient of friction between the wheels of the vehicle and the ground

r = Radius of circular path.

184. The maximum velocity (V_{max}) of a vehicle on a level circular path to avoid outer-turning is given by

(*a*) $\sqrt{\frac{dgr}{2h}}$ ☐ (*b*) $\sqrt{\frac{gr}{2hd}}$ ☐

(*c*) $\sqrt{\frac{2h}{dgr}}$ ☐ (*d*) $\sqrt{\frac{dg}{2hr}}$ ☐

where h = Height of C.G. of the vehicle from the ground level

d = Distance between the centre lines of the wheels

r = Radius of circular path.

185. The gravitational acceleration at a place is 6 times the value of gravitational acceleration at earth, the weight of the body at that place will be

(*a*) one-sixth of the weight at earth ☐ (*b*) same as at earth ☐

(*c*) 6 times the weight at earth ☐ (*d*) none of the above. ☐

186. A flywheel, mounted on a shaft, weighs 500 kg and has a radius of gyration 50 cm about its axis of rotation. If the flywheel starts from rest and shaft is subjected to a moment of 625 Nm, the speed of shaft after 4 seconds is equal to

(*a*) 30 rad per sec ☐ (*b*) 20 rad/sec ☐

(*c*) 10 rad per sec ☐ (*d*) 40 rad/sec. ☐

187. A curved road is generally provided a slope, which is known as

(*a*) angle of banking ☐ (*b*) angle of repose ☐

(*c*) angle of friction ☐ (*d*) angle of reaction. ☐

188. The angle of banking provided on the curved roads depends upon

(*a*) the velocity of vehicle only ☐

(*b*) the square of velocity of vehicle only ☐

(*c*) the square or velocity of vehicle and radius of circular path ☐

(*d*) co-efficient of friction between the vehicle and road contact point. ☐

189. To prevent side thrust on the wheel flanges of a train, moving round a curve

(*a*) outside rails are raised ☐ (*b*) inner rails are raised ☐
(*c*) rails are kept at the same level ☐ (*d*) thrust eliminators are provided. ☐

Simple Harmonic Motion

190. A body is said to describe simple harmonic motion if it moves in a straight line such that

(*a*) its acceleration is proportional to its distance from a fixed point on the straight line and is directed away from the fixed point ☐
(*b*) its velocity is proportional to its distance from a fixed point and is directed towards the fixed point ☐
(*c*) its velocity is proportional to its distance from a fixed point and is directed away from the fixed point ☐
(*d*) none of the above. ☐

191. The time period of one oscillation of a simple pendulum is given by

(*a*) $\pi\sqrt{\frac{L}{g}}$ ☐ (*b*) $2\pi\sqrt{\frac{L}{g}}$ ☐
(*c*) $2\pi\sqrt{\frac{L}{2g}}$ ☐ (*d*) $\pi\sqrt{\frac{2L}{g}}$. ☐

192. Beat is equal to

(*a*) one oscillation ☐ (*b*) twice the oscillation ☐
(*c*) half the oscillation ☐ (*d*) none of the above. ☐

193. If a pendulum executes one beat per second, it is called

(*a*) simple pendulum ☐ (*b*) second's pendulum ☐
(*c*) compound pendulum ☐ (*d*) none of the above. ☐

194. Length of a second's pendulum is equal to

(*a*) 50 cm ☐ (*b*) 99.4 cm ☐
(*c*) 150 cm ☐ (*d*) 80 cm. ☐

195. The periodic time (T) of a S.H.M. is given by

(*a*) $T = \frac{\omega}{2\pi}$ ☐ (*b*) $T = \frac{\pi}{2\omega}$ ☐
(*c*) $T = \frac{2\pi}{\omega}$ ☐ (*d*) $T = \frac{2\omega}{\pi}$. ☐

196. The velocity of a particle moving with S.H.M. is maximum, when it is at

(*a*) extreme position ☐
(*b*) its mean position ☐
(*c*) a point between its mean and extreme position ☐
(*d*) none of the above. ☐

197. The acceleration of a particle moving with S.H.M. is maximum, when it is at

(*a*) its extreme position ☐
(*b*) its mean position ☐
(*c*) a point between its mean and extreme positions ☐
(*d*) none of the above. ☐

198. The velocity of a particle moving with S.H.M. is zero, when it is at

(*a*) its extreme position ☐

(*b*) its mean position ☐

(*c*) a point between its mean and extreme position ☐

(*d*) none of the above. ☐

199. The acceleration of a particle moving with S.H.M. is zero, when it is at

(*a*) its extreme position ☐

(*b*) its mean position ☐

(*c*) a point between its mean and extreme position ☐

(*d*) none of the above. ☐

200. The period of oscillation of a helical spring is given by

(*a*) $T = 2\pi\sqrt{\frac{g}{\delta}}$ ☐ (*b*) $T = 2\pi\sqrt{\frac{\delta}{g}}$ ☐

(*c*) $T = \sqrt{\frac{g}{2\delta}}$ ☐ (*d*) $T = \pi\sqrt{\frac{2\delta}{g}}$ ☐

where δ = Static extension.

201. For a compound pendulum, the time period is given as

(*a*) $T = 2\pi\sqrt{\frac{K_G^2}{gh}}$ ☐ (*b*) $T = 2\pi\sqrt{\frac{K_G^2 + h^2}{gh}}$ ☐

(*c*) $T = 2\pi\sqrt{\frac{gh}{K_G^2}}$ ☐ (*d*) $T = 2\pi\sqrt{\frac{gh}{K_G^2 + h^2}}$ ☐

where K_G = Radius of gyration

h = Distance from the point of suspension and centre of the gravity of the body.

202. Centre of percussion is a point at which

(*a*) resultant pressure force is supposed to act ☐

(*b*) weight of the body is acting ☐

(*c*) if a blow is given, no reaction will be felt at the point of suspension of a body ☐

(*d*) resultant force of buoyancy is supposed to act. ☐

203. Centre of oscillation and centre of suspension for a compound pendulum are

(*a*) acting at the same point ☐ (*b*) interchangable ☐

(*c*) not-interchangable ☐ (*d*) none of the above. ☐

204. The velocity of a particle moving with S.H.M. at a distance from mean position is given by

(*a*) $\sqrt{\omega(r^2 - y^2)}$ ☐ (*b*) $\omega\sqrt{r^2 - y^2}$ ☐

(*c*) $\omega\sqrt{1 - \frac{y^2}{r^2}}$ ☐ (*d*) $\sqrt{\omega\left(1 - \frac{y^2}{r^2}\right)}$ ☐

where r = Amplitude of the motion, and ω = Angular velocity.

205. The acceleration of a particle moving with S.H.M. at a distance y from mean position is given by

(*a*) $-\omega y^2$ ☐ (*b*) ωy^2 ☐

(*c*) $-\omega^2 y$ ☐ (*d*) $-\omega^2/y$. ☐

206. The gain or loss of number of oscillations are given by

(*a*) $\frac{dn}{n}=\frac{dg}{g}-\frac{dl}{2l}$ ☐ (*b*) $\frac{dn}{n}=\frac{dl}{l}-\frac{dg}{2g}$ ☐

(*c*) $\frac{dn}{n}=\frac{dg}{g}-\frac{dl}{l}$ ☐ (*d*) $\frac{dn}{n}=\frac{dg}{2g}-\frac{dl}{2l}$. ☐

207. At a place, the clock will go slow if

(*a*) length of pendulum is increased ☐ (*b*) length of pendulum is decreased ☐

(*c*) mass of its bob is increased ☐ (*d*) none of the above. ☐

208. The length of a simple pendulum which has the same time period as compound pendulum is given by

(*a*) $h+\frac{k^2}{h^2}$ ☐ (*b*) $h^2+\frac{k^2}{h}$ ☐

(*c*) $h+\frac{k^2}{h}$ ☐ (*d*) $h+\frac{k}{h^2}$ ☐

where h = Length between the point of suspension and centre of gravity, and
k = Radius of gyration.

209. The time period of a simple pendulum will be doubled if

(*a*) its length is doubled ☐ (*b*) its length is halved ☐

(*c*) its length is increased four times ☐ (*d*) its length is increased eight times. ☐

210. At a place, a clock will go fast if

(*a*) its length is increased ☐ (*b*) its length is decreased ☐

(*c*) mass of its bob is increased ☐ (*d*) mass of its bob is decreased. ☐

211. The maximum velocity of a particle moving with S.H.M. is 2 m/sec and maximum acceleration is 20 m/sec^2. The frequency of the motion is equal to

(*a*) $\frac{20}{\pi}$ ☐ (*b*) $\frac{10}{\pi}$ ☐

(*c*) $\frac{5}{\pi}$ ☐ (*d*) $\frac{1}{\pi}$. ☐

212. A body moving with S.H.M. is having frequency as three vibrations per second and amplitude as 20 cm. The maximum velocity of the body is equal to

(*a*) 0.377 m ☐ (*b*) 3.77 m ☐

(*c*) 37.7 m ☐ (*d*) 0.00377 m. ☐

213. The periodic time (T) of a particle with S.H.M. is

(*a*) directly proportional to the mass of the particle ☐

(*b*) directly proportional to the angular velocity ☐

(*c*) directly proportional to the square of angular velocity ☐

(*d*) inversely proportional to the angular velocity. ☐

Basic Definitions

214. A force is acting on a mass of one kilogram and produces an acceleration of one metre per second square. Then the force is known as

(*a*) dyne ☐ (*b*) newton ☐
(*c*) kg-weight ☐ (*d*) kgf. ☐

215. A force is acting on a mass of one gram and produces an acceleration of one cm per second square. Then the force is known as

(*a*) dyne ☐ (*b*) newton ☐
(*c*) gm-weight ☐ (*d*) gmf. ☐

216. One newton is equal to

(*a*) 10^3 dyne ☐ (*b*) 10^4 dyne ☐
(*c*) 10^5 dyne ☐ (*d*) 10^6 dyne. ☐

217. The linear acceleration a is equal to

(*a*) $\frac{dv}{dt}$ ☐ (*b*) $v\frac{dv}{ds}$ ☐
(*c*) $\frac{d^2s}{dt^2}$ ☐ (*d*) any one of the above ☐
(*e*) none of the above. ☐

218. The acceleration of a body moving down an inclined smooth plane is equal to

(*a*) $g \cos \theta$ ☐ (*b*) $-g \sin \theta$ ☐
(*c*) $g \sin \theta$ ☐ (*d*) $g \tan \theta$ ☐

where θ = Inclination of plane with horizontal.

219. The acceleration of a body moving up an inclined smooth plane is equal to

(*a*) $g \cos \theta$ ☐ (*b*) $-g \sin \theta$ ☐
(*c*) $g \sin \theta$ ☐ (*d*) $g \tan \theta$ ☐

where θ = Inclination of plane with horizontal.

220. The acceleration of a body moving down a rough inclined plane is equal to

(*a*) $g \sin \theta$ ☐ (*b*) $g [\sin \theta - \mu \cos \theta]$ ☐
(*c*) $g \tan \theta$ ☐ (*d*) $g [\sin \theta - \mu \sin \theta]$ ☐

where θ = Inclination of plane with the horizontal.

221. The radius of gyration (k) for a circular lamina is equal to

(*a*) $\sqrt{2}\,R$ ☐ (*b*) $\frac{R}{\sqrt{2}}$ ☐
(*c*) $0.6324\,R$ ☐ (*d*) $0.5\,R$ ☐

where R = Radius.

222. The radius of gyration (k) for a solid cylinder is equal to

(*a*) $\sqrt{2}\,R$ ☐ (*b*) $\frac{R}{\sqrt{2}}$ ☐
(*c*) $0.6324\,R$ ☐ (*d*) $0.5\,R$. ☐

223. The radius of gyration (k) for a solid sphere is equal to

(*a*) $\sqrt{2}\ R$ ☐ (*b*) $\frac{R}{\sqrt{2}}$ ☐

(*c*) 0.6324 R ☐ (*d*) 0.5 R ☐

224. The kinetic energy due to rotation of a body is equal to

(*a*) $\frac{1}{2} I\omega^2$ ☐ (*b*) $\frac{1}{2} mV^2$ ☐

(*c*) $2I\omega^3$ ☐ (*d*) $\frac{1}{2} I^2\omega$. ☐

225. If a body is having motion of translation as well as motion of rotation, then total kinetic energy is equal to

(*a*) $\frac{1}{2} mV^2$ ☐ (*b*) $\frac{1}{2} mV^2 + \frac{1}{2} I\omega^2$ ☐

(*c*) $\frac{1}{2} mV^2 + \frac{1}{2} \omega I^2$ ☐ (*d*) none of the above. ☐

226. The angular acceleration (α) of a rotating body is equal to

(*a*) $\frac{d\omega}{dt}$ ☐ (*b*) $\frac{d^2\theta}{dt^2}$ ☐

(*c*) $\omega \frac{d\omega}{d\theta}$ ☐ (*d*) any one of the above. ☐

227. The radius of gyration (k) is equal to

(*a*) $\sqrt{\frac{A}{I}}$ ☐ (*b*) $\sqrt{\frac{I}{A}}$ ☐

(*c*) $\sqrt{AI}$ ☐ (*d*) $\sqrt{\frac{I}{AI}}$ ☐

where I = Moment of inertia, and A = Area.

228. When a projectile is projected on an inclined plane, the range on the inclined plane will be maximum if

(*a*) $2\alpha - \beta = 90°$ ☐ (*b*) $2\alpha + \beta = 90°$ ☐

(*c*) $\alpha + 2\beta = 90°$ ☐ (*d*) $\alpha - 2\beta = 90°$ ☐

where α = Angle of projection with the horizontal, and

β = Inclination of the inclined plane with the horizontal.

229. If the projectile is projected down an inclined plane, the range on the inclined plane will be maximum if

(*a*) $2\alpha - \beta = 90°$ ☐ (*b*) $2\alpha + \beta = 90°$ ☐

(*c*) $\alpha + 2\beta = 90°$ ☐ (*d*) $\alpha - 2\beta = 90°$ ☐

where α = Angle of projection with the horizontal, and

β = Inclination of the inclined plane with the horizontal.

230. Angle of repose is

(*a*) less than angle of friction ☐ (*b*) more than angle of friction ☐

(*c*) equal to angle of friction ☐ (*d*) none of the above. ☐

Lifting Machines

231. For a simple wheel and axle, the velocity ratio is given by

(a) $D/(d + 1)$ ☐ (b) d/D ☐
(c) D/d ☐ (d) $(D + 1)/d$ ☐

where D = Diameter of wheel, and d = Diameter of axle.

232. A weight of 48 N is to be raised by a wheel and axle. The diameters of wheel and axle are 400 mm and 100 mm respectively. The force required at the wheel is 16 N. Then the velocity ratio of this machine will be

(a) $\frac{1}{4}$ ☐ (b) $\frac{2}{4}$ ☐
(c) $\frac{3}{4}$ ☐ (d) 4. ☐

233. In the above question, the mechanical advantage will be

(a) $\frac{1}{3}$ ☐ (b) 3 ☐
(c) 4 ☐ (d) 5. ☐

234. In question 232, the efficiency of the machine will be

(a) 90% ☐ (b) 80% ☐
(c) 75% ☐ (d) 50%. ☐

235. For a differential wheel and axle, the velocity ratio is given by

(a) $D/(d_1 - d_2)$ ☐ (b) $2D/(d_1 - d_2)$ ☐
(c) $D/2(d_1 - d_2)$ ☐ (d) $2D/(2d_1 - d_2)$ ☐

where D = Dia. of wheel, d_1 = Dia. of bigger axle, d_2 = Dia. of smaller axle.

236. For a differential wheel and axle, the diameter of wheel is 25 cm. The diameters of differential axles are 10 cm and 9 cm. An effort of 30 N is applied to lift a load of 900 N. Then the velocity ratio of the machine will be

(a) 10 ☐ (b) 25 ☐
(c) 50 ☐ (d) 60. ☐

237. In question 236, the mechanical advantage will be

(a) 10 ☐ (b) 20 ☐
(c) 30 ☐ (d) 40. ☐

238. In question 236, the efficiency of the machine will be

(a) 40% ☐ (b) 60% ☐
(c) 75% ☐ (d) 80%. ☐

239. The differential wheel and axle is having large value of .... as compared to simple wheel and axle.

(a) mechanical advantage ☐ (b) velocity ratio ☐
(c) efficiency ☐ (d) all of the above. ☐

240. The velocity ratio of worm and worm-wheel, when the worm is single threaded, is given by

(a) $\frac{2LT}{r}$ ☐ (b) $\frac{LT}{r}$ ☐
(c) $\frac{L \times T}{2r}$ ☐ (d) $\frac{L \times T}{3r}$ ☐

where L = Length of the handle, T = Number of teeth on the worm wheel, and
r = Radius of load drum.

241. When the worm is double-threaded, then the velocity ratio of worm and worm wheel is

(*a*) $\frac{2LT}{r}$ ☐ (*b*) $\frac{LT}{r}$ ☐

(*c*) $\frac{L \times T}{2r}$ ☐ (*d*) $\frac{L \times T}{3r}$ ☐

242. The number of teeth on the worm-wheel of a single-threaded worm and worm wheel is 60. Calculate the velocity ratio if the diameter of the effort wheel is 25 cm and that of load drum is 12.5 cm. The load lifted is 600 N when a force of 20 N is applied to this machine

(*a*) 60 ☐ (*b*) 120 ☐

(*c*) 180 ☐ (*d*) 240. ☐

243. In question 242, the mechanical advantage would be

(*a*) 30 ☐ (*b*) 45 ☐

(*c*) 60 ☐ (*d*) 75. ☐

244. In question 242, the efficiency of the machine would be

(*a*) 75% ☐ (*b*) 60% ☐

(*c*) 50% ☐ (*d*) 25%. ☐

245. The velocity ratio of single purchase crab winch is given by

(*a*) $\frac{L}{D} \times \frac{T_2}{T_1}$ ☐ (*b*) $\frac{2L}{D} \times \frac{T_2}{T_1}$ ☐

(*c*) $\frac{L}{2D} \times \frac{T_2}{T_1}$ ☐ (*d*) $\frac{L}{3D} \times \frac{T_2}{T_1}$ ☐

where L = Length of lever arm, D = Diameter of the load axle,

T_1 = Number of teeth on the pinion, and T_2 = Number of teeth on the spur wheel.

246. The number of teeth on pinion and spur wheel of a single purchase crab winch are 10 and 100 respectively. The diameter of load axle is 30 cm and length of lever arm is also 30 cm. The velocity ratio of this machine will be

(*a*) 10 ☐ (*b*) 20 ☐

(*c*) 30 ☐ (*d*) 40. ☐

247. For a single fixed pulley, the mechanical advantage is

(*a*) less than one ☐ (*b*) more than one ☐

(*c*) equal to one ☐ (*d*) none of the above. ☐

248. For a single movable pulley, the mechanical advantage is

(*a*) less than one ☐ (*b*) more than one ☐

(*c*) equal to one ☐ (*d*) none of the above. ☐

249. The velocity ratio for the first system of pulleys is equal to

(*a*) $2 \times n$ ☐ (*b*) 2^n ☐

(*c*) $2/n$ ☐ (*d*) $\frac{1}{2^n}$ ☐

where n = No. of movable pulleys.

250. For the first system of pulleys, there are four movable pulleys. A load of 1440 N is lifted by an effort of 100 N. Then the velocity ratio of this pulley will be

(*a*) 4 ☐ (*b*) 8 ☐
(*c*) 16 ☐ (*d*) 32. ☐

251. In question 250, the efficiency of the pulley will be

(*a*) 90% ☐ (*b*) 80% ☐
(*c*) 60% ☐ (*d*) 50%. ☐

252. In case of second system of pulley, the velocity ratio is equal to

(*a*) $n + 1$ ☐ (*b*) $n - 1$ ☐
(*c*) n ☐ (*d*) $n + 2$ ☐

where n = No. of segments supporting the movable block or load.

253. A weight of 200 N is to be lifted by an effort of 60 N, by second system of pulleys having three pulleys in the upper block and two pulleys in the lower block. The velocity ratio of the system will be

(*a*) 2 ☐ (*b*) 3 ☐
(*c*) 1 ☐ (*d*) 5. ☐

254. In question 253, the efficiency of the system will be

(*a*) 80% ☐ (*b*) 66.67% ☐
(*c*) 50% ☐ (*d*) 33.33%. ☐

255. In third system of pulley, the velocity ratio is equal to

(*a*) 2^n ☐ (*b*) $2^n - 1$ ☐
(*c*) $2^n + 1$ ☐ (*d*) $2^n + 2$ ☐

where n = No. of pulleys in the system.

256. There are four pulleys in a third system of pulley. The load lifted is 1800 N by an effort of 160 N. Then the velocity ratio of the pulley will be

(*a*) 16 ☐ (*b*) 15 ☐
(*c*) 17 ☐ (*d*) 18. ☐

257. In question 256, the efficiency of the machine will be

(*a*) 50% ☐ (*b*) 60% ☐
(*c*) 70% ☐ (*d*) 75%. ☐

258. In a Weston differential pulley block, the upper block has two pulleys of diameters 25 cm and 20 cm. A load of 100 N is lifted by an effort of 20 N, by this machine. The velocity ratio of this machine is

(*a*) 40 ☐ (*b*) 30 ☐
(*c*) 20 ☐ (*d*) 10. ☐

259. The efficiency of the machine given in question 258, is

(*a*) 50% ☐ (*b*) 60% ☐
(*c*) 70% ☐ (*d*) 75%. ☐

Belt Drive

260. The velocity ratio of a belt is given by (if thickness of belt is neglected)

(a) $\frac{N_2}{N_1} = \frac{d_2}{d_1}$ ☐ (b) $\frac{N_2}{N_1} = \frac{d_1}{d_2}$ ☐

(c) $\frac{N_2}{N_1} = \frac{d_2+1}{d_1+1}$ ☐ (d) $\frac{N_2}{N_1} = \frac{d_1+1}{d_2+1}$ ☐

where N_1 = Speed of driver pulley, N_2 = Speed of driven pulley,
d_1 = Diameter of driver pulley, and d_2 = Diameter of driven pulley.

261. The velocity ratio of a belt (if thickness of belt is considered) is given by

(a) $\frac{N_2}{N_1} = \frac{d_2+t}{d_1+t}$ ☐ (b) $\frac{N_2}{N_1} = \frac{d_1+2}{d_2+2}$ ☐

(c) $\frac{N_2}{N_1} = \frac{d_1+t}{d_2+t}$ ☐ (d) $\frac{N_2}{N_1} = \frac{d_2+1}{d_1+1}$. ☐

262. The velocity ratio of a compound belt drive is given by

(a) $\frac{\text{Speed of last follower}}{\text{Speed of first driver}} = \frac{\text{Product diameter of follower}}{\text{Product of diameter driver}}$ ☐

(b) $\frac{\text{Speed of last follower}}{\text{Speed of first driver}} = \frac{\text{Diameter of last follower}}{\text{Diameter of first driver}}$ ☐

(c) $\frac{\text{Speed of last follower}}{\text{Speed of first driver}} = \frac{\text{Product of diameter of drivers}}{\text{Product of diameter of follower}}$ ☐

(d) None of the above. ☐

263. A shaft is driven with the help of a belt which is passing over the engine and shaft. The engine is running at 200 r.p.m. The diameters of engine pulley is 51 cm and that of shaft is 30 cm. The speed of the shaft will be

(a) 200 r.p.m. ☐ (b) 300 r.p.m. ☐

(c) 340 r.p.m. ☐ (d) 400 r.p.m. ☐

264. In the question 263, the speed ratio would be

(a) 2 ☐ (b) 1.7 ☐

(c) 1.5 ☐ (d) 1.3. ☐

265. In question 263, what should be diameter of the pulley of engine, so that the speed of shaft is 400 r.p.m.?

(a) 100 cm ☐ (b) 75 cm ☐

(c) 60 cm ☐ (d) 40 cm. ☐

266. A shaft is driven by a belt of thickness 1 cm. The belt is passing over the engine and the shaft. The engine is running at 310 r.p.m. The diameter of engine pulley is 50 cm and that of shaft is 30 cm. The speed of the shaft will be

(a) 400 r.p.m. ☐ (b) 500 r.p.m. ☐

(c) 510 r.p.m. ☐ (d) 520 r.p.m. ☐

267. The speed ratio in question 266, would be

(a) 2 ☐ (b) $\frac{50}{30}$ ☐

(c) $\frac{51}{31}$ ☐ (d) $\frac{52}{32}$. ☐

268. An engine is running at 200 r.p.m. With the help of belt, the engine drives a line shaft. The diameters of pulleys on engine and on line shafts are 80 cm and 40 cm respectively. A 100 cm diameter pulley one line shaft drives a 20 cm diameter pulley keyed to a dynamo shaft. The speed of dynamo shaft would be

(a) 1000 r.p.m. ☐ (b) 2000 r.p.m. ☐
(c) 1500 r.p.m. ☐ (d) 4000 r.p.m. ☐

269. In question 268, the speed ratio of the compound belt drive would be

(a) 10 ☐ (b) 15 ☐
(c) 20 ☐ (d) 25. ☐

270. For an open belt drive, the length (L) of the belt is equal to

(a) $\pi(r_1 + r_2) + \dfrac{(r_1 + r_2)^2}{2x} + x$ ☐ (b) $\pi(r_1 - r_2) + \dfrac{(r_1 - r_2)^2}{x} + 2x$ ☐

(c) $\pi(r_1 + r_2)^2 + \dfrac{(r_1 - r_2)^2}{2x} + 2x$ ☐ (d) $\pi(r_1 + r_2) + \dfrac{(r_1 + r_2)^2}{x} + 2x$ ☐

where r_1 = Radius of larger pulley, r_2 = Radius of smaller pulley, and

x = Distance between the centres of two pulleys.

271. For a crossed belt-drive, the length (L) of the belt is equal to

(a) $\pi(r_1 + r_2) + \dfrac{(r_1 + r_2)^2}{2x} + x$ ☐ (b) $\pi(r_1 - r_2) + \dfrac{(r_1 - r_2)^2}{x} + 2x$ ☐

(c) $\pi(r_1 + r_2)^2 + \dfrac{(r_1 - r_2)^2}{x} + 2x$ ☐ (d) $\pi(r_1 + r_2) + \dfrac{(r_1 + r_2)^2}{x} + 2x$ ☐

272. The length of the open belt drive depends upon

(a) sum of the radii ☐ (b) difference of the radii ☐
(c) both the sum and difference of the radii ☐ (d) none of the above. ☐

273. The length of the crossed belt drive depends upon

(a) sum of radii ☐ (b) difference of radii ☐
(c) both sum and difference of the radii ☐ (d) none of the above. ☐

274. The ratio of tensions on the two sides of a belt is given by

(a) $\dfrac{T_1}{T_2} = e \times \mu \times \theta$ ☐ (b) $\dfrac{T_1}{T_2} = e^{\mu \times \theta}$ ☐

(c) $\dfrac{T_1}{T_2} = \mu^{e \times \theta}$ ☐ (d) $\dfrac{T_1}{T_2} = e^{(\mu/\theta)}$ ☐

where T_1 = Tension on the tight side, T_2 = Tension on the slack side

μ = Co-efficient of friction between belt and pulley

θ = Angle of contact in radians.

275. The angle of contact for an open belt-drive is given by

(a) $\theta = (180 + 2\alpha)$ ☐ (b) $\theta = (180 - 2\alpha)$ ☐

(c) $\theta = \dfrac{180 + \alpha}{2}$ ☐ (d) $\theta = \dfrac{180 - \alpha}{2}$. ☐

276. The angle of contact for a crossed belt is given by

(*a*) $\theta = (180 + 2\alpha)$ ☐ (*b*) $\theta = (180 - 2\alpha)$ ☐

(*c*) $\theta = \frac{180 + \alpha}{2}$ ☐ (*d*) $\theta = \frac{180 - \alpha}{2}$. ☐

277. The value of α for an open-belt drive is given by

(*a*) $\alpha = \sin^{-1}\left(\frac{r_1 - r_2}{2x}\right)$ ☐ (*b*) $\alpha = \sin^{-1}\left(\frac{r_1 + r_2}{2x}\right)$ ☐

(*c*) $\alpha = \sin^{-1}\left(\frac{r_1 - r_2}{x}\right)$ ☐ (*d*) $\alpha = \sin^{-1}\left(\frac{r_1 + r_2}{x}\right)$. ☐

278. The value of α for a crossed-belt drive is given by

(*a*) $\alpha = \sin^{-1}\left(\frac{r_1 - r_2}{2x}\right)$ ☐ (*b*) $\alpha = \sin^{-1}\left(\frac{r_1 + r_2}{2x}\right)$ ☐

(*c*) $\alpha = \sin^{-1}\left(\frac{r_1 - r_2}{x}\right)$ ☐ (*d*) $\alpha = \sin^{-1}\left(\frac{r_1 + r_2}{x}\right)$. ☐

279. The power transmitted by a belt is given by

(*a*) $P = (T_1 - T_2) \times v$ ☐ (*b*) $P = \frac{(T_1 - T_2) \times v}{4500}$ ☐

(*c*) $P = \frac{(T_1 - T_2) \times v}{75}$ ☐ (*d*) $P = \frac{(T_1 + T_2) \times v}{75}$ ☐

where T_1 and T_2 are tension in N.

280. The power in kilo-watt transmitted by a belt is given by

(*a*) $P = (T_1 - T_2) \times v$ ☐ (*b*) $P = \frac{(T_1 - T_2) \times v}{1000}$ ☐

(*c*) $P = \frac{(T_1 - T_2) \times v}{75}$ ☐ (*d*) $P = \frac{(T_1 + t_2) \times v}{75}$ ☐

where T_1 and T_2 are in Newton.

281. The centrifugal tension (T_c) in belt is given by

(*a*) $T_c = \frac{1}{2} mv^2$ Newton ☐ (*b*) $T_c = m \times v^2$ Newton ☐

(*c*) $T_c = 2 \times m \times v^2$ Newton ☐ (*d*) $T_c = 3 \times m \times v^2$ Newton. ☐

282. For maximum power transmission, the velocity of belt is given by

(*a*) $v = \sqrt{\frac{T_{max}}{m}}$ ☐ (*b*) $v = \sqrt{\frac{3T_{max}}{m}}$ ☐

(*c*) $v = \sqrt{\frac{T_{max}}{3m}}$ ☐ (*d*) $v = \sqrt{\frac{2T_{max}}{m}}$ ☐

where T_{max} = Maximum tension in the belt, m = Mass of belt per metre length in kg.

283. For maximum power transmission, maximum tension in the belt is given by

(*a*) $T_{max} = 2T_c$ ☐ (*b*) $T_{max} = 3T_c$ ☐

(*c*) $T_{max} = 4T_c$ ☐ (*d*) $T_{max} = \frac{1}{3} T_c$ ☐

where T_c = Centrifugal tension in the belt.

284. For maximum power transmission, the tension on the tight side should be

(*a*) one-third of maximum tension ☐ (*b*) two-third of maximum tension ☐

(*c*) one-quarter of maximum tension ☐ (*d*) three-fourth of maximum tension. ☐

285. The initial tension in the belt is given by (if centrifugal tension is neglected)

(*a*) $T_0 = \frac{T_1 + T_2}{3}$ ☐ (*b*) $T_0 = \frac{T_1 + T_2}{2}$ ☐

(*c*) $T_0 = \frac{T_1 + T_2}{4}$ ☐ (*d*) $T_0 = \frac{2T_1 + T_2}{3}$. ☐

286. If centrifugal tension is considered, then initial tension in the belt is given by

(*a*) $T_0 = \frac{T_1 + T_2 + T_c}{3}$ ☐ (*b*) $T_0 = \frac{T_1 + T_2 + 2T_c}{3}$ ☐

(*c*) $T_0 = \frac{T_1 + T_2 + 2T_c}{3}$ ☐ (*d*) $T_0 = \frac{T_1 + T_2 + T_c}{2}$. ☐

287. The tensions on the tight and slack sides of a belt are 200 N and 100 N. If centrifugal tension is neglected, then initial tension in the belt will be

(*a*) 25 N ☐ (*b*) 300 N ☐

(*c*) 150 N ☐ (*d*) 175 N. ☐

288. If in question 287, the belt is running at 4 m/sec, then the power transmitted by the belt will be

(*a*) 1200 W ☐ (*b*) 600 W ☐

(*c*) 400 W ☐ (*d*) 200 W. ☐

289. If in question 287, the centrifugal tension is considered and mass of belt is 0.5 kg per metre length when belt is running at 4 m/sec then centrifugal tension will be equal to

(*a*) 16 N ☐ (*b*) 8 N ☐

(*c*) 4 N ☐ (*d*) 2 N. ☐

290. For the above question, the initial tension will be equal to

(*a*) 150 N ☐ (*b*) 158 N ☐

(*c*) 164 N ☐ (*d*) 172 N. ☐

Tick mark the true or false statements:

291. The resultant of two forces *A* and *B* (which are coplaner and are acting at an angle θ) is given by $R = \sqrt{A^2 + B^2 + 2AB \sin \theta}$

(*a*) True ☐ (*b*) False. ☐

292. For a perfect frame, the number of members (*n*) and number of joints (*j*) is given by $n = 2j - 3$

(*a*) True ☐ (*b*) False. ☐

III. OBJECTIVE TYPE QUESTIONS FROM COMPETITIVE EXAMINATIONS

Tick mark the most appropriate answer:

1. The resultant of two forces each equal to $\frac{P}{4}$ and acting at right angles is

(*a*) $\frac{P}{2}$ ☐ (*b*) $\frac{P}{2\sqrt{2}}$ ☐

(*c*) $\sqrt{2}P$ ☐ (*d*) $\frac{P}{\sqrt{2}}$. ☐

2. Two forces of magnitudes 4 and 5 N act at an angle of 60°, the resultant force is equal to

(*a*) $\sqrt{41}$ N ☐ (*b*) $\sqrt{61}$ N ☐

(*c*) 7 N ☐ (*d*) 9 N. ☐

3. For the motion of a rigid link in any mechanism,
   (*a*) the velocity of one of the ends should be zero ☐
   (*b*) the velocity of one end with respect to that of the other should be perpendicular to the link ☐
   (*c*) the acceleration of one end with respect to that of the other should be perpendicular to the links ☐
   (*d*) the two ends of the link may have different components of velocity along the link. ☐
4. The linear momentum of a particle
   (*a*) must be directed along the velocity of the particle ☐
   (*b*) is the dot product of the mass with its velocity ☐
   (*c*) is the cross product of the mass with its velocity ☐
   (*d*) is a scalar quantity. ☐
5. The angular momentum of a particle is the
   (*a*) linear momentum per unit angle ☐
   (*b*) product of the mass with its angular velocity ☐
   (*c*) moment of the product of the mass and the angular velocity about and origin ☐
   (*d*) cross product of the position vector and the linear momentum. ☐
6. In all engineering problems a frame of reference at rest with respect to the earth is taken as an inertial frame. The assumption is valid because
   (*a*) the centrifugal force on the earth and the force of attraction between the earth and the sun balance each other ☐
   (*b*) the acceleration and angular velocity of the earth **is** so small, that the error caused is the negligible ☐
   (*c*) the error due to the acceleration of the earth **is** taken care of by the experimental calculation of the value of g ☐
   (*d*) the earth does not have any acceleration ☐
   (*e*) of some reason other than any of the above. ☐
7. Zero work done by a system of forces acting on a body implies that
   (*a*) the resultant of the system of forces is zero ☐
   (*b*) the cross product of the resultant of the system of forces and the vector in the direction of motion of the body is zero ☐

(c) the body does not have any motion ☐

(d) the motion of the body is in a direction perpendicular to the direction of the simplest resultant of the system of forces. ☐

8. The force of gravitation between two bodies will be inversely proportional to the square of the distance between their centres of masses if the

(a) bodies are of constant densities ☐

(b) bodies are spherical ☐

(c) bodies are of any arbitrary shape ☐

(d) bodies are of the same shape and size. ☐

9. A force F acting on a rigid body at a point *P* can be replaced by a force of equal magnitude and in the same direction at a point *Q* on the body, together with a moment

(a) equal in magnitude to *PQ* times *F*, acting normal to the plane of *F* and *PQ* ☐

(b) equal in magnitude to *F* times the distance moved in the lines of the lines of actions of the force, acting in the plane of *PQ* and *F* ☐

(c) given by $F \times QP$ ☐

(d) given by $F \times PQ$. ☐

10. The Lami's theorem

(a) relates the forces with the sines of angles ☐

(b) states that, for equilibrium under the action of three concurrent forces, there is a unique constant of proportionality between a force and the angle between the other two forces ☐

(c) may be applied to consider a relationship between forces and angles of a polygon representation of forces ☐

(d) may be applied for concurrent of non-current forces keeping a body in equilibrium. ☐

11. If the sum of all the forces acting on a body is zero, it may be concluded that the

(a) body must be in equilibrium ☐

(b) body cannot be in equilibrium ☐

(c) body may be in equilibrium provided the forces are concurrent ☐

(d) body may be in equilibrium provided the forces are parallel. ☐

12. The maximum bending moment in a simply supported beam of length *L* loaded by a concentrated load *W* at the mid-point is given by

(a) WL ☐ (b) $WL/2$ ☐

(c) $WL/4$ ☐ (d) $WL/8$ ☐

(e) $WL/16$. ☐

13. The condition for a screw jack to be self-locking is that

(a) its efficiency should be the maximum possible. ☐

(b) its efficiency should be the minimum possible. ☐

(c) its efficiency should be more than 50% ☐

(d) it should not unwind to lower the load if left to itself. ☐

14. The maximum efficiency of a machine

(*a*) should be 100% under ideal conditions ☐

(*b*) is directly proportional to the velocity ratio ☐

(*c*) is given by mechanical advantage divided by velocity ratio ☐

(*d*) should occur when the load is 50% of maximum permissible. ☐

15. The first moment of a triangular area of base b and height h taken about an axis coincident with the base is given by

(*a*) $bh^3/12$ ☐ (*b*) $b^2h/6$ ☐

(*c*) $bh^2/6$ ☐ (*d*) $h/3$. ☐

16. Given that there is a rectangle and a triangle, each of base b and area A, the first moment of the area to the rectangle about its base

(*a*) equals the first moment of the triangular area about its base b ☐

(*b*) is more than the first moment of the triangular area about its base b ☐

(*c*) is less than the first moment of the triangular area about its base b ☐

(*d*) equals twice that of the triangular area about its base b. ☐

17. The unit vector 'normal' to a curve

(*a*) is directed towards the local centre of curvature ☐

(*b*) is directed outward along the join of the centre of curvature and the point ☐

(*c*) is the same as the radial unit vector ☐

(*d*) must only be perpendicular to the path of the point. ☐

18. The Coriolis acceleration may not vanish

(*a*) if the relative velocity of the moving point becomes zero ☐

(*b*) if the rotational velocity of the moving frame becomes zero ☐

(*c*) if the rotational velocity of the moving frame and the relative velocity become collinear ☐

(*d*) if the angular acceleration of the point becomes zero. ☐

19. The impulse-momentum principle is applicable

(*a*) if there is no external force acting on the body ☐

(*b*) only if a force acts for an infinitesimally short time on the body ☐

(*c*) when the momentum is conserved ☐

(*d*) wherever Newton's law is applicable. ☐

20. The principle of conservation of mechanical energy requires that

(*a*) the acceleration should be zero ☐

(*b*) there should be no external forces ☐

(*c*) the motion should be restricted to the gravitational field only ☐

(*d*) the force-field should be conservative. ☐

21. For a perfectly plastic central impact

(*a*) the entire kinetic energy of the two bodies must be lost ☐

(*b*) the two bodies must move stuck together, whether the impact is direct or indirect ☐

(*c*) the two bodies must move stuck together only if the impact is direct ☐

(*d*) a body, initially at rest, should stay at rest. ☐

22. When two bodies collide without the presence of any other forces or force fields

(*a*) their total momentum must be conserved ☐
(*b*) their total kinetic energy must be conserved ☐
(*c*) the collision must be direct ☐
(*d*) the collision must be central. ☐

23. The moments of inertia of a solid circular section of radius *r* and of a hollow circular section of radii *r* and *R*, each about the diametrical lines, are equal. Then,

(*a*) $R = 2r$ ☐ (*b*) $R = \sqrt{2r}$ ☐
(*c*) $R^2 = \sqrt{2r^2}$ ☐ (*d*) $R = 4r$. ☐

24. A rectangle is divided into two smaller rectangles by drawing the neutral axis parallel to its smaller sides. If the moments of inertia of the smaller rectangles of areas A_1 and A_2 about the neutral axis are I_1 and I_2, the moment of inertia of the total area of the given rectangle about the neutral axis would be

(*a*) $I_1 + I_2$ ☐ (*b*) $I_1 - I_2$ ☐
(*c*) $\sqrt{I_1^2 + I_2^2}$ ☐ (*d*) $| I_1 | + | I_2 |$. ☐

25. A solid cylinder, solid sphere and a loop, each of radius *R* and mass *m*, are released simultaneously from rest on an incline at the same elevation. The order in which they will pass through a common mark at the same level will be

(*a*) cylinder, sphere and hoop ☐ (*b*) sphere, cylinder and hoop ☐
(*c*) hoop, sphere and cylinder ☐ (*d*) hoop, cylinder and sphere. ☐

26. The principle of conservation of angular momentum for a rigid body is applicable when

(*a*) the external forces add to zero ☐
(*b*) the internal forces add to zero ☐
(*c*) the external forces and moments are absent ☐
(*d*) the resultant external moment is zero. ☐

27. The time taken by a small frictionless bead to slide on a thin wire in the gravitational field is the minimum if the shape of the wire is

(*a*) a straight line ☐ (*b*) a cycloid ☐
(*c*) a parabola ☐ (*d*) an involute. ☐

28. Resonance occurs when

(*a*) a freely vibrating system is made to vibrate at increasingly higher frequencies ☐
(*b*) the forcing frequency equals the natural frequency of the system ☐
(*c*) the system vibrates at its natural frequency ☐
(*d*) the amplitude of vibrations exceeds twice the amplitude of free vibrations. ☐

29. A rigid body is acted upon by a force system. It can in general be brought to equilibrium by the application of a

(*a*) force acting on a suitable point on the body ☐
(*b*) force acting anywhere along a suitable line ☐
(*c*) force acting along a suitable line and a moment along the direction of the force ☐
(*d*) force acting along a suitable line and a moment in the direction perpendicular to the direction of force. ☐

30. The simplest resultant of a spatial parallel force system is always

(*a*) a wrench ☐ (*b*) a force ☐

(*c*) a moment ☐ (*d*) a force and moment. ☐

31. The point of contraflexure in a loaded beam is one where

(*a*) the bending moment is maximum ☐ (*b*) the shear force is maximum ☐

(*c*) the bending moment changes sign ☐ (*d*) the shear force changes sign. ☐

32. In a symmetrically overhung simply supported beam, the maximum bending moment will be the least possible when

(*a*) the supports are near the ends ☐

(*b*) the supports coincide to becomes a single support at the centre ☐

(*c*) the distance between the supports becomes one-third of the length of the beam ☐

(*d*) the numerical value of bending moment at either supports and at the centre of the beam are equal. ☐

33. A rigid body is in equilibrium. Given that the moment of all the forces acting on the body about some axis is zero and also given that forces are concurrent, implies that

(*a*) the resultant force is zero ☐

(*b*) the forces has a line of action passing through the axis ☐

(*c*) the resultant forces has a line of action parallel to the axis ☐

(*d*) any of (*a*), (*b*), (*c*) can be true. ☐

34. A body is acted upon by a force system. It can in general be brought to equilibrium by the application of

(*a*) a force acting on a suitable point on the body ☐

(*b*) a force acting anywhere along a suitable line ☐

(*c*) a force acting along a suitable line a moment along the direction of the force ☐

(*d*) a force acting along a suitable line and a moment in the direction perpendicular to the direction of force. ☐

35. The coefficient of friction between two surfaces is the constant of proportionality between the applied tangential force and the normal reaction

(*a*) at the instant of application of the force ☐ (*b*) at any instant when the body is at rest ☐

(*c*) at the instant of impending motion ☐ (*d*) at an instant after the motion takes place. ☐

36. The ratio between the tensions in the right side and slack side of a flat belt drive increases

(*a*) in direct proportion to the angle of lap ☐

(*b*) exponentially as the angle of lap increases ☐

(*c*) in direct proportion to the coefficient of friction ☐

(*d*) proportional to the width of the belt. ☐

37. Tension in a cable supporting a lift, when lift is going up with a uniform acceleration '*a*' is equal to

(*a*) $W\left(1-\frac{a}{g}\right]$ ☐ (*b*) $W(1 + a/g)$ ☐

(*c*) $W\left(\frac{a}{g}-1\right]$ ☐ (*d*) $W(a/g + 1)$. ☐

where W = Weight carried by lift, and

g = Acceleration due to gravity.

38. Simple harmonic motion is

(*a*) the motion of a point in a circle ☐
(*b*) another name for oscillatory motion ☐
(*c*) a projection of the circular motion of a particle ☐
(*d*) a projection of the circular motion of a particle at constant speed on a diameter of the circle. ☐

39. One of the following assumption is not necessary in obtaining the equation for parabolic trajectory of a particle

(*a*) Air resistance is negligible ☐
(*b*) The gravitational acceleration '*g*' is constant ☐
(*c*) The body can be represented by a particle ☐
(*d*) The body must not change its mass during the motion. ☐

40. The centroid of a body

(*a*) must be a point on that body ☐
(*b*) is a point which can be made to lie on or outside the body by changing the coordinate system ☐
(*c*) is a fixed point in space regardless of the orientation of the body ☐
(*d*) is a unique point fixed with respect to the body. ☐

41. The first moment of area of a semicircular area about its diameter '*d*' is given by

(*a*) $d^3/12$ ☐ (*b*) $d^3/24$ ☐
(*c*) $d^3/6$ ☐ (*d*) $d^3/36$. ☐

42. A rigid body, in translation

(*a*) can only move in a straight line ☐ (*b*) may move along a straight or curved path ☐
(*c*) cannot move on a circular path ☐ (*d*) must undergo plane motion only. ☐

43. The instantaneous centre of rotation

(*a*) should also be the instantaneous centre of acceleration ☐
(*b*) is a hypothetical concept ☐
(*c*) can exist for any space motion ☐
(*d*) must exist for any plane motion. ☐

44. The D'Alembert principle

(*a*) is a hypothetical principle ☐
(*b*) provides no special advantage over Newton's law ☐
(*c*) is based upon the existence of inertia forces ☐
(*d*) allows a dynamical problem to be treated as a statical problem. ☐

45. The gravitational acceleration at a place is 6 times the value of gravitational acceleration at earth, the weight of body at that place will be

(*a*) one-sixth of weight at earth ☐ (*b*) same as at earth ☐
(*c*) six times the weight at earth ☐ (*d*) none of the above. ☐

46. Central impact of two bodies

(*a*) also implies direct impact ☐
(*b*) requires that the bodies should not rotate at all ☐
(*c*) must always be elastic impact ☐
(*d*) can only be direct impact. ☐

47. The coefficient of restitution is defined on the basis of
(*a*) velocity components along the line of impact only ☐
(*b*) velocity component normal to the line of impact ☐
(*c*) velocity vectors before and after the collision ☐
(*d*) energies of bodies before and after the impact. ☐

48. Euler's equation of motion should not be applied about
(*a*) the centre of mass of the body ☐
(*b*) the fixed point about which the body may rotate ☐
(*c*) a point accelerating towards the centre of mass ☐
(*d*) a point fixed in space ☐
(*e*) a point other than the above points. ☐

49. The kinetic energy of a rigid body of mass m moving such that the velocity of its centre of mass is V_c and the rotational velocity is ω, is given by
(*a*) $1/2\, m\, V_c^2$ ☐ (*b*) $1/2\, I_c\, \omega^2$ ☐
(*c*) $1/2\, mV_c^2 + 1/2\, I_c\, \omega^2$ ☐ (*d*) none of the above expressions. ☐

50. A linear response implies that
(*a*) the elements are in one line ☐ (*b*) the response is along the given line ☐
(*c*) the sensitivity is constant ☐ (*d*) the response is not exponential. ☐

51. A vibrating system
(*a*) does not pass through an equilibrium position ☐
(*b*) passes through the equilibrium position once every cycle ☐
(*c*) passes through the equilibrium position twice every cycle ☐
(*d*) starts from an equilibrium position and does not returnal to it. ☐

52. The existence of potential energy implies that
(*a*) a general force field exists ☐
(*b*) a conservative force field exists ☐
(*c*) there must either be a spring or gravitational field ☐
(*d*) the body should be in equilibrium. ☐

53. Stability of equilibrium of a body requires that
(*a*) $\dfrac{d\,PE}{ds} = 0$ ☐ (*b*) $\dfrac{d\,PE}{ds} = 0$ and $\dfrac{d^2\,PE}{ds^2} < 0$ ☐
(*c*) $\dfrac{d\,PE}{ds} = 0$ and $\dfrac{d^2\,PE}{ds^2} > 0$ ☐ (*d*) $\dfrac{d^2\,PE}{ds^2} = 0$. ☐

54. The momentum of a system of two bodies is conserved
(*a*) if either body does not exert a force on the other ☐
(*b*) under all circumstances ☐
(*c*) when there is no external force acting on either body ☐
(*d*) when the external forces act only on one body. ☐

55. The moment of inertia of a rectangular lamina of sides b and h about its neutral axis parallel to the side b is given by
(*a*) $bh^3/12$ ☐ (*b*) $bh^3/36$ ☐
(*c*) $hb^3/12$ ☐ (*d*) $hb^3/36$. ☐

56. The second moment of a plane area about any axis as compared to its second moment about the neutral axis is

(*a*) always more ☐ (*b*) always less ☐

(*c*) sometimes more ☐ (*d*) equal. ☐

57. A plane system of forces has a single force resultant if the sum of the moments and

(*a*) moments of all the forces about the origin is zero ☐

(*b*) moments of all the forces about any point on the plane is zero ☐

(*c*) moments of all the forces about any point on a particular line in that plane is zero ☐

(*d*) moments of all the forces about any point on or outside the plane is zero. ☐

58. For a plane system of forces to have the simplest resultant as a single moment the

(*a*) forces must be parallel ☐

(*b*) force system must constitute of moments and/or couples only ☐

(*c*) force system cannot have the forces as concurrent ☐

(*d*) force system cannot have an odd number of forces. ☐

59. A rigid body is in equilibrium under the action of three forces. It implies that the

(*a*) forces must be concurrent ☐

(*b*) forces must be coplanar ☐

(*c*) forces must either be concurrent or coplanar ☐

(*d*) forces must pass through the centre of mass. ☐

60. A rigid body is in equilibrium when the

(*a*) rate of change of linear momentum is zero ☐

(*b*) sum of all the external forces is zero ☐

(*c*) sum of moments of all the forces acting on the body about some point on the body is zero ☐

(*d*) rate of change of angular momentum is zero. ☐

61. A thin rigid beam hinged at one end and roller-supported at its mid-point is said to be

(*a*) a symmetrical simply supported beam ☐ (*b*) an overhanging simply supported beam ☐

(*c*) a cantilever beam ☐ (*d*) a fixed beam. ☐

62. The shear force at a section in a beam is given by the

(*a*) external force at that section ☐

(*b*) transverse component of the external force at that section ☐

(*c*) transverse force from the part of the beam on one side of the section to that on the other side of the section ☐

(*d*) addition of the forces at (*b*) and (*c*) above. ☐

63. The bending moment at a section in a beam is given by the

(*a*) external moment at that section ☐

(*b*) summation of all the moments about that section ☐

(*c*) summation of moments of all the forces about the section ☐

(*d*) net moment exerted by the part of the beam on one side of the section to that on the other side of the section. ☐

64. The force of friction between two bodies in contact is

(*a*) a function of the relative velocity between them ☐

(*b*) dependant on the areas of contact ☐

(*c*) always normal to the surface of contact ☐

(*d*) never shown in the free-body diagram of the system of these two bodies. ☐

65. The frictional force on a body acted upon by a force on a rough horizontal surface is

(*a*) always equal and opposite to the horizontal component of the force ☐

(*b*) equal and opposite to the applied force ☐

(*c*) equal and opposite to the horizontal component of the applied force if the body is at rest or moving with a constant velocity ☐

(*d*) independent of the vertical component of the force. ☐

66. The displacement of a point

(*a*) implies the distance moved by the point ☐

(*b*) is a vector, from the initial to the final position of the point ☐

(*c*) is always less than the distance traversed by the point ☐

(*d*) is independent of the distance and the direction of movement of the point. ☐

67. The relationship $V^2 - U^2 = 2as$, with conventional notation, is applicable

(*a*) for all possible motions of a point ☐ (*b*) for constant velocity of a point ☐

(*c*) for constant acceleration of a point ☐ (*d*) for vertical motion only. ☐

68. The length of a line can be considered to be concentrated at the centroid of that line for the purpose of calculating

(*a*) the area of the surface of revolution generated by revolving the line about any axis outside it ☐

(*b*) the average distance of the line from any axis outside it ☐

(*c*) the volume of the body generated by revolving the line about any axis outside it ☐

(*d*) none of the above. ☐

69. The centre of volume and centre of mass of a body coincide

(*a*) if and only if the body is of uniform density ☐

(*b*) if the body is geometrically symmetrical about the centre of mass ☐

(*c*) the density variation is symmetrical about the centroid ☐

(*d*) if the body is made of homogeneous material. ☐

70. The motion of a particle, in general, is described by

(*a*) Newton's law and not the work-energy equation ☐

(*b*) the Impulse-momentum principle alone if there is no external force ☐

(*c*) the Newton's law, the work-energy equation, impulse momentum principle ☐

(*d*) the principles of conservation of energy and momentum. ☐

71. Virtual work refers to

(*a*) work by virtue of actual forces ☐

(*b*) work by virtue of actual displacements ☐

(*c*) work in overcoming the constraints ☐

(*d*) work associated with a possible displacement. ☐

72. The principle of virtual work states that the virtual work should be zero for
   (*a*) a body to be in equilibrium, in general ☐
   (*b*) a body in equilibrium if it does not rotate ☐
   (*c*) a body moving with a constant acceleration ☐
   (*d*) none of the above. ☐

73. Degrees of freedom of a rigid body imply
   (*a*) the angles that it may turn through ☐
   (*b*) the angular motions the body can have ☐
   (*c*) the constraints to its motion ☐
   (*d*) the total number of modes of displacement. ☐

74. The tensions on the tight and slack sides of a belt are 200 N and 100 N. If centrifugal tension is neglected then initial tension in the belt will be
   (*a*) 25 N ☐ (*b*) 300 N ☐
   (*c*) 150 N ☐ (*d*) 175 N. ☐

75. The centre of percussion is a point on a rigid body
   (*a*) through which the resultant of the applied forces acts ☐
   (*b*) where the impact is made ☐
   (*c*) where the largest external force acts ☐
   (*d*) about which the body may rotate. ☐

ANSWERS

Answers to Objective Type Questions

1. (*b*)	2. (*b*)	3. (*b*)	4. (*c*)	5. (*b*)	6. (*c*)
7. (*c*)	8. (*a*)	9. (*c*)	10. (*b*)	11. (*a*)	12. (*c*)
13. (*c*)	14. (*b*)	15. (*c*)	16. (*c*)	17. (*d*)	18. (*c*)
19. (*b*)	20. (*b*)	21. (*c*)	22. (*a*)	23. (*b*)	24. (*b*)
25. (*d*)	26. (*b*)	27. (*a*)	28. (*c*)	29. (*b*)	30. (*a*)
31. (*c*)	32. (*c*)	33. (*a*)	34. (*b*)	35. (*b*)	36. (*b*)
37. (*c*)	38. (*b*)	39. (*b*)	40. (*a*)	41. (*c*)	42. (*c*)
43. (*b*)	44. (*c*)	45. (*b*)	46. (*a*)	47. (*a*)	48. (*e*)
49. (*b*)	50. (*c*)	51. (*b*)	52. (*b*)	53. (*c*)	54. (*a*)
55. (*b*)	56. (*c*)	57. (*b*)	58. (*c*)	59. (*b*)	60. (*b*)
61. (*b*)	62. (*b*)	63. (*c*)	64. (*c*)	65. (*d*)	66. (*c*)
67. (*b*)	68. (*c*)	69. (*a*)	70. (*b*)	71. (*c*)	72. (*a*)
73. (*c*)	74. (*a*)	75. (*d*)	76. (*b*)	77. (*c*)	78. (*a*)
79. (*c*)	80. (*b*)	81. (*b*)	82. (*c*)	83. (*b*)	84. (*c*)
85. (*b*)	86. (*b*)	87. (*c*)	88. (*d*)	89. (*c*)	90. (*d*)
91. (*b*)	92. (*c*)	93. (*c*)	94. (*d*)	95. (*a*), (*c*)	96. (*b*)

97. (*b*)	**98.** (*a*)	**99.** (*b*)	**100.** (*b*)	**101.** (*c*)	**102.** (*d*)
103. (*c*)	**104.** (*c*)	**105.** (*c*)	**106.** (*c*)	**107.** (*c*)	**108.** (*b*)
109. (*b*)	**110.** (*c*)	**111.** (*c*)	**112.** (*b*)	**113.** (*c*)	**114.** (*b*)
115. (*c*)	**116.** (*b*)	**117.** (*c*)	**118.** (*b*)	**119.** (*c*)	**120.** (*b*)
121. (*c*)	**122.** (*b*)	**123.** (*b*)	**124.** (*c*)	**125.** (*c*)	**126.** (*b*)
127. (*c*)	**128** (*b*)	**129.** (*c*)	**130.** (*b*)	**131.** (*b*)	**132.** (*c*)
133. (*b*)	**134.** (*c*)	**135.** (*d*)	**136.** (*d*)	**137.** (*c*)	**138.** (*d*)
139. (*c*)	**140.** (*b*)	**141.** (*b*)	**142.** (*a*)	**143.** (*c*)	**144.** (*b*)
145. (*c*)	**146.** (*b*)	**147.** (*b*)	**148.** (*c*)	**149.** (*a*)	**150.** (*b*)
151. (*c*)	**152.** (*c*)	**153.** (*b*)	**154.** (*c*)	**155.** (*d*)	**156.** (*a*)
157. (*b*)	**158.** (*c*)	**159.** (*a*)	**160.** (*b*)	**161.** (*b*)	**162.** (*c*)
163. (*b*)	**164.** (*c*)	**165.** (*c*)	**166.** (*d*)	**167.** (*c*)	**168.** (*b*)
169. (*c*)	**170.** (*b*)	**171.** (*d*)	**172.** (*c*)	**173.** (*c*)	**174.** (*c*)
175. (*b*)	**176.** (*a*)	**177.** (*d*)	**178.** (*c*)	**179.** (*c*)	**180.** (*b*)
181. (*b*)	**182.** (*b*)	**183.** (*b*)	**184.** (*a*)	**185.** (*c*)	**186.** (*c*)
187. (*a*)	**188.** (*c*)	**189.** (*a*)	**190.** (*d*)	**191.** (*b*)	**192.** (*c*)
193. (*b*)	**194.** (*b*)	**195.** (*c*)	**196.** (*b*)	**197.** (*a*)	**198.** (*a*)
199. (*b*)	**200.** (*b*)	**201.** (*b*)	**202.** (*c*)	**203.** (*b*)	**204.** (*b*)
205. (*c*)	**206.** (*d*)	**207.** (*a*)	**208.** (*c*)	**209.** (*c*)	**210.** (*b*)
211. (*c*)	**212.** (*b*)	**213.** (*d*)	**214.** (*b*)	**215.** (*a*)	**216.** (*c*)
217. (*d*)	**218.** (*c*)	**219.** (*b*)	**220.** (*b*)	**221.** (*b*)	**222.** (*b*)
223. (*c*)	**224.** (*a*)	**225.** (*b*)	**226.** (*d*)	**227.** (*b*)	**228** (*a*)
229. (*b*)	**230.** (*c*)	**231.** (*c*)	**232.** (*d*)	**233.** (*b*)	**234.** (*c*)
235. (*b*)	**236.** (*c*)	**237.** (*c*)	**238.** (*b*)	**239.** (*d*)	**240.** (*b*)
241. (*c*)	**242.** (*b*)	**243.** (*a*)	**244.** (*d*)	**245.** (*b*)	**246.** (*b*)
247. (*c*)	**248.** (*b*)	**249.** (*b*)	**250.** (*c*)	**251.** (*a*)	**252.** (*c*)
253. (*d*)	**254.** (*b*)	**255.** (*b*)	**256.** (*b*)	**257.** (*d*)	**258.** (*d*)
259. (*a*)	**260.** (*b*)	**261.** (*c*)	**262.** (*c*)	**263.** (*c*)	**264.** (*b*)
265. (*c*)	**266.** (*c*)	**267.** (*c*)	**268.** (*b*)	**269.** (*a*)	**270.** (*c*)
271. (*d*)	**272.** (*c*)	**273.** (*a*)	**274.** (*b*)	**275.** (*b*)	**276.** (*a*)
277. (*c*)	**278.** (*d*)	**279.** (*a*)	**280.** (*b*)	**281.** (*b*)	**282.** (*c*)
283. (*b*)	**284.** (*b*)	**285.** (*b*)	**286.** (*c*)	**287.** (*c*)	**288.** (*c*)
289. (*b*)	**290.** (*b*).				

True/False

291. (*b*) **292.** (*a*).

Answers to Objective Type Questions from Competitive Examinations

1. (*b*)	**2.** (*b*)	**3.** (*b*)	**4.** (*b*)	**5.** (*d*)	**6.** (*b*)
7. (*c*)	**8.** (*b*)	**9.** (*c*)	**10.** (*b*)	**11.** (*c*)	**12.** (*c*)
13. (*d*)	**14.** (*c*)	**15.** (*c*)	**16.** (*c*)	**17.** (*a*)	**18.** (*d*)
19. (*d*)	**20.** (*d*)	**21.** (*c*)	**22.** (*a*)	**23.** (*c*)	**24.** (*a*)
25. (*b*)	**26.** (*d*)	**27.** (*b*)	**28.** (*b*)	**29.** (*c*)	**30.** (*c*)
31. (*c*)	**32.** (*d*)	**33.** (*d*)	**34.** (*c*)	**35.** (*c*)	**36.** (*b*)
37. (*b*)	**38.** (*d*)	**39.** (*d*)	**40.** (*d*)	**41.** (*a*)	**42.** (*a*)
43. (*d*)	**44.** (*d*)	**45.** (*c*)	**46.** (*c*)	**47.** (*a*)	**48.** (*e*)
49. (*c*)	**50.** (*c*)	**51.** (*a*)	**52.** (*b*)	**53.** (*c*)	**54.** (*c*)
55. (*a*)	**56.** (*a*)	**57.** (*c*)	**58.** (*b*)	**59.** (*b*)	**60.** (*b*)
61. (*b*)	**62.** (*c*)	**63.** (*d*)	**64.** (*d*)	**65.** (*c*)	**66.** (*b*)
67. (*c*)	**68.** (*a*)	**69.** (*c*)	**70.** (*c*)	**71.** (*d*)	**72.** (*a*)
73. (*d*)	**74.** (*c*)	**75.** (*a*).			

Chapter 3 THERMODYNAMICS

I. THEORY

3.1. DEFINITIONS

The branch of science, which deals with the energies possessed by gases and vapours, their conversion in terms of heat and work and their relationship with properties of the system is called thermodynamics. A *property* is a measurable or observable characteristic of a system. Properties are classified as *intensive property* and *extensive property.*

The properties, which are independent of the mass, are called *intensive properties.* Examples of intensive properties are : temperature, pressure, density etc. The properties, which are dependent on mass of the system, are called *extensive properties.* Examples of extensive properties are : total volume, total energy, etc. A *system* in thermodynamics is defined as a region in space or a definite quantity of matter bounded by a closed surface. The space external to a system is known as *surrounding* as shown in Fig. 3.1.

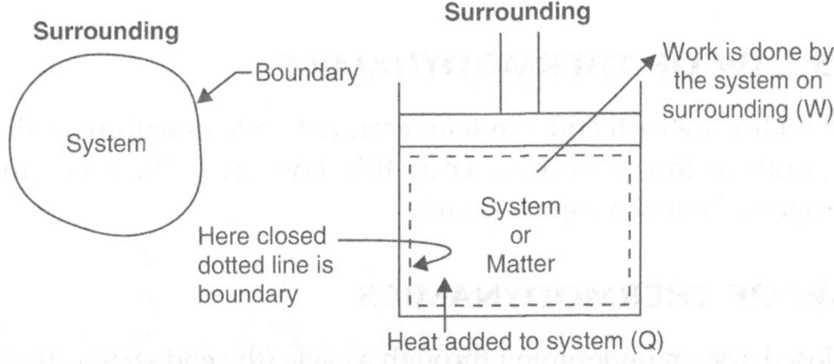

FIGURE 3.1

1. Thermodynamic systems. They are classified as (*i*) closed system, (*ii*) open system, and (*iii*) isolated system. A system, in which there is no transfer of mass across its boundaries, is called closed system. In closed system, energy may cross the boundaries of the system. A system, in which there is transfer of mass and energy across its boundaries, is called *an open system.* A system, in which there is no transfer of mass and energy across its boundaries, is called an *isolated system.*

2. Work. It is energy in transition which flows from a system to the surroundings during a given process if the sole effect external to the system can be reduced to the raising of a weight. Work done by the system is considered positive and work done on the system is considered negative.

3. Process. It is defined as any change that the system may undergo. Any process or series of processes in which the system returns to its original condition or state is called a **cyclic process.** A process in which system departs from equilibrium state only infinitesimally at every instant, is called **quasi-static process.**

A process may be divided as:

(*i*) Reversible process, and

(*ii*) Irreversible process.

(*i*) Reversible process. A process is said to be reversible if it can be performed in the reverse direction **exactly** in the same manner as in the forward direction. Examples of reversible process are: isothermal process and adiabatic process.

(*ii*) Irreversible process. A process is said to be irreversible if it cannot be performed in the reverse direction exactly in the same manner as in the forward direction. Examples of irreversible process are: constant volume process, constant pressure process, throtling process and polytropic process.

4. Conditions of reversibility. A process will be reversible if it satisfies the following two conditions :

(*i*) A substance undergoing a reversible change must at all instants be in thermodynamic equilibrium with its surroundings. In other words, the changes in pressure and volume of the working substance must take place at an extremely slow rate so that when the substance is taking heat, its temperature differs by an extremely small amount dT from hotter body and when it is loosing heat, the temperature should again differ by an extremely small amount from the cold body. In fact as dT approaches zero the process approaches complete reversibility.

(*ii*) There should be no loss of heat by conduction, convection and radiation etc. Also the process should be frictionless.

3.2. ZEROTH'S LAW OF THERMODYNAMICS

It states that if two bodies are each equal in temperature to a temperature of the third body, they are equal in temperature to each other. Thus this law gives the basis for measuring the thermodynamic property known as temperature.

3.3. FIRST LAW OF THERMODYNAMICS

It states that in a closed system undergoing through a cycle (the end state is the same as the initial state) the net work delivered to the surroundings is proportional to the net heat taken from the surroundings or the heat and work are mutually convertible.

Mathematically, it is written as

$$J\oint dQ = \oint dW$$

where $\oint$ = Integration along cyclic path, and

$\oint dQ$ = Net heat taken from surroundings during the cyclic process

$\oint dW$ = Net work delivered to the surrounding during the cyclic process

J = Mechanical equivalent of heat or Joule's constant

= 427 if heat is expressed in kilo calories and work in kgf-m.

3.4. SECOND LAW OF THERMODYNAMICS

It can be stated in two different ways :

1. According to *Kelvin Planck* statement it is impossible to construct an engine, which while operating in a cycle produces no other effect except to extract heat from a single reservoir and do equivalent amount of work. This implies that perpetual motion machine of the second kind (defind as 100% efficient engine) is impossible.

2. According to *Clausius statement,* heat energy cannot flow from lower temperature to higher temperature without the aid of an external energy. This implies that a heat pump cannot operate without the input of work.

1. Heat. It is energy *in transition* which flows by virtue of temperature difference from one system to another. If the heat is transferred by the surroundings to the system, it is said to be positive and if the heat is transferred by the system to the surrounding, it is said to the negative, kilo-calorie (kcal) is the unit of heat.

2. Specific heat. Specific heat of a substance is defined as the heat required to raise unit degree of temperature of unit mass of the substance. Solids and liquids have one specific heat while gases have two specific heats, namely, the *specific heat at constant pressure,* C_p and the *specific heat at constant volume,* C_v, C_p is always greater than C_v by the amount of work done in constant pressure expansion of the gas. The ratio of the specific heats of a gas is a constant and is denoted by the symbol γ or mathematically,

$$\frac{C_p}{C_v} = \gamma$$

For atmospheric air,

$$C_p = 0.24,\ C_v = 0.171 \text{ and } \gamma = 1.4.$$

Specific heat at constant volume (C_v). The amount of heat required to heat one kilogram of a gas through 1°C at constant volume is known as specific heat at constant volume. This is represented by C_v. The unit of C_v is kcal/kg K.

Specific heat at constant pressure (C_p). The amount of heat required to heat one kilogram of a gas through 1°C at constant pressure is known as specific heat at constant pressure. This is represented by C_p. The unit of C_p is also kcal/kg K.

The heat supplied to a gas when it is heated at constant volume is given by,

$$Q = m \times C_v \times dT$$

where m = Mass of gas, dT = Rise of temperature.

The heat supplied to a gas, when it is heated at constant pressure is given by,

$$Q = m \times C_p \times dT.$$

Specific heats in terms of gas constant. The two specific heats, *(i.e.,* C_p and C_v*)* in term of gas constant are related by

$$C_p - C_v = \frac{R}{J}$$

where R = Gas constant in kgf-m/kg K.

The expressions for C_p and C_v in term of gas constant are also given by

$$C_p = \frac{\gamma R}{J(\gamma - 1)}$$

and $$C_v = \frac{R}{J(g-1)}.$$

3. Energy. There exists a property of a system called energy. E, such that change in its value is the algebraic sum of the heat supplied and the work done during any change in state.

Mathematically, it is written as

Change in energy = Algebraic sum of heat supplied and work done

or $$dE = Q - W$$

where Q = Heat supplied, W = Work done

and dE = Change in energy.

The energy E is the sum of kinetic energy (K.E.), potential energy (P.E.) and internal energy *(U)* which is due to the motions of the molecules. The internal energy *(U)* changes with the rise of temperature.

Hence, $$dE = dU + d(\text{K.E.}) + d(\text{P.E.}).$$

For a closed or non-flow system, kinetic energy and potential energy terms are zero and hence energy (E) is equal to internal energy (U). Hence the equation $dE = Q - W$ becomes as

$$dU = Q - W \text{ or } Q = W + dU.$$

3.5. GAS LAWS

A perfect gas it defined as a gas which strictly obeys the Boyle's and Charle's laws. Air is considered as a perfect gas. The following are the gas laws :

(i) **Boyle's Law**. It states that volume of a given mass of a perfect gas varies inversely as the absolute pressure, when temperature is constant, *i.e.,*

$$V \propto \frac{1}{p} \text{ or } pV = \text{constant if } T \text{ is constant.}$$

(*ii*) **Charle's Law.** It states that the volume of a given mass of a perfect gas varies directly as its absolute temperature if pressure remains constant, *i.e.*,

$$V \propto T \quad \text{or} \quad \frac{V}{T} = \text{Constant if } p \text{ is constant.}$$

(*iii*) **Characteristic Equation of Gas.** It is obtained by combining Boyle's law and Charle's law. The characteristic equation of gas is given as

$$pV = mRT$$

where p = Absolute pressure in N/m^2 (or kgf/m^2)

V = Volume of gas in m^3

m = Mass of the gas

T = Absolute temperature and R = Constant for the gas in J/kg K (or kgf-m/kg K).

(*iv*) **Joule's Law.** It states that the change of internal energy (dU) is proportional to the change of temperature (dT), *i.e.*,

$$dU \propto dT$$

or

$$dU = \text{mass} \times \text{constant} \times dT$$

$$(U_2 - U_1) = m \times C_v \times (T_2 - T_1)$$

where constant = C_v.

(*v*) **Avogadro's Law.** It states that the molecular weight of all the perfect gases occupy the same volume under the same conditions of pressure and temperature.

(*vi*) **Universal Gas Constant.** It is defined as the product of the characteristic gas constant and molecular weight of the gas. If M is the molecular weight of a gas, then universal gas constant is equal to MR. Its value is equal to 848 kgf-m/kg/mol K or 8314 Nm/kg mol K.

(*vii*) **Dalton's Law of Partial Pressures.** It states that the pressure of a mixture of gases is the sum of the pressure which each gas would exert if it existed alone in the space occupied by the mixture at the same temperature.

Work done in a fully resisted expansion = $\int p dV$, which is equal to area *under p-V* diagram whereas the work done in unresisted or free expansion is zero.

3.6. NON-FLOW PROCESSES OR CLOSED SYSTEM

The following are the non-flow processes :

(*i*) **Constant volume process (Isometric).** Figure 3.1 (*a*) shows the constant volume process on *p-V* diagram. Here work done is zero as area of curve 1-2 on *p-V* diagram is zero.

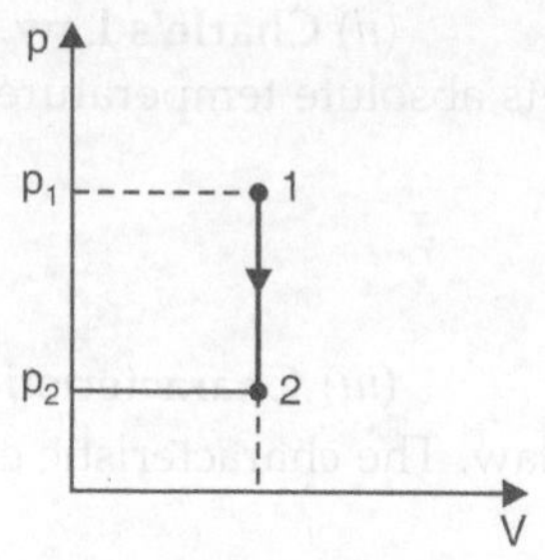

FIGURE 3.1 (a)

The first law of thermodynamics gives

$$dQ = dW + dU = 0 + dU = dU$$

or $$Q = U_2 - U_1.$$

Hence the addition of heat at constant volume process increases the temperature and consequently the internal energy of the system. The ratio of increase of internal energy to the corresponding increase of temperature is called the specific heat at constant volume. For a unit mass,

$$\text{Specific heat of constant volume } (C_v) = \frac{\text{Increase of internal energy}}{\text{Increase of temperature}} = \frac{U_2 - U_1}{T_2 - T_1}$$

$$\therefore \quad U_2 - U_1 = C_v(T_2 - T_1)$$

$$\therefore \quad Q = U_2 - U_1 = C_v(T_2 - T_0).$$

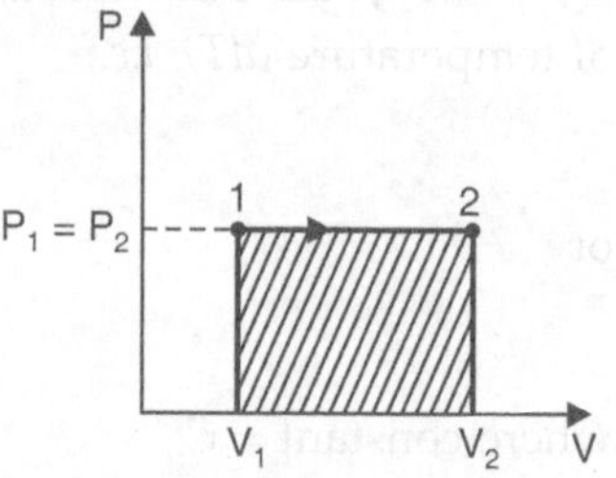

FIGURE 3.1 (b)

(ii) **Constant pressure process.** The constant pressure process on p-V diagram is represented by a horizontal line (1-2) as shown in Fig. 3.1 (*b*). The heating causes increase in volume and temperature and external work is done. From first law of thermodynamics,

$$dQ = dW + dU = pdV + dU \quad (\because \quad dW = pdV)$$

$$= d(pV) + dU$$

($\because$ pressure is constant and can be taken inside the differential)

$$= d(pV + U).$$

The term $(pV + U)$ is the sum of pressure volume product and internal energy and is known as enthalpy (H).

$$\therefore \quad dQ = dH \quad \text{or} \quad Q = H_2 - H_1.$$

The addition of heat at constant pressure increase the temperature and consequently enthalpy of the system. The ratio of increase in enthalpy to the corresponding increase in temperature is called the specific heat at constant pressure (C_p). For a unit mass,

$$C_p = \frac{H_2 - H_1}{T_2 - T_1} \quad \text{or} \quad H_2 - H_1 = C_p(T_2 - T_1)$$

$$\therefore \quad Q = H_2 - H_1 = C_p[T_2 - T_1]$$

Work done, $$W = \int_1^2 pdV = p[V_2 - V_1] = mR(T_2 - T_1)$$

$$= \frac{mR}{J}(T_2 - T_1) \text{ heat units}$$

$$= \frac{R}{J}(T_2 - T_1) \text{ for a unit mass}$$

Now, $$dQ = dW + dU \quad \text{or} \quad Q = W + (U_2 + U_1)$$

Substituting the values of Q, W and $(U_2 - U_1)$, we get

$$C_p(T_2 - T_1) = \frac{R}{J}(T_2 - T_1) + C_v(T_2 - T_1) \quad \text{or} \quad C_p = \frac{R}{J} + C_v$$

$$\therefore \quad (C_p - C_v) = \frac{R}{J}$$

As $\quad \dfrac{C_p}{C_v} = \gamma, C_v = \dfrac{R}{J(\gamma - 1)}$ and $C_p = \dfrac{\gamma R}{J(\gamma - 1)}$.

(*iii*) **Constant temperature (Isothermal).** For a perfect gas, constant temperature expansion is same as hyperbolic expansion ($pV = C$) as shown in Fig. 3.1 (*c*). Work done,

$$W = mRT \log_e \frac{V_2}{V_1}$$

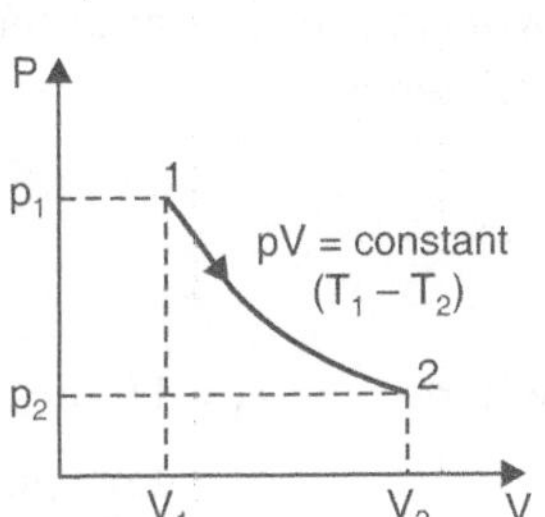

FIGURE 3.1 (c)

Change of internal energy,

$$U_2 - U_1 = m \times C_v \times (T_2 - T_1) = 0$$

Heat supplied, $\quad Q = W + (U_2 - U_1)$

$$= mRT \log_e \frac{V_2}{V_1}$$

(*iv*) **Polytropic expansion ($pV^n = C$) [Fig. 3.1 (*d*)].**

$$\text{Work done} = \frac{p_1V_1 - p_2V_2}{n - 1}$$

(*v*) **Adiabatic expansion ($pV^\gamma = C$).** Hence work done is obtained by changing n to γ. Hence work done

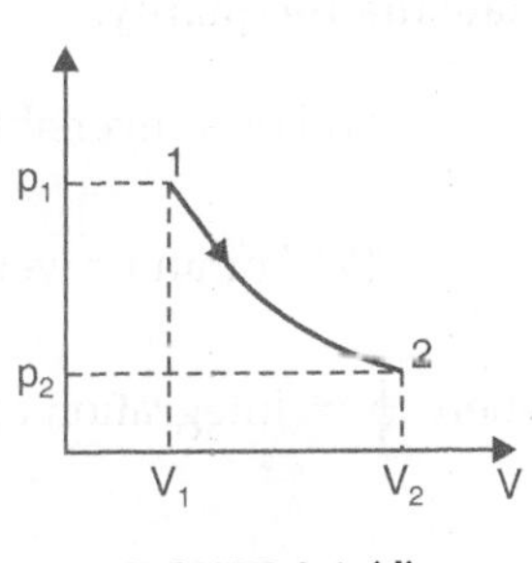

FIGURE 3.1 (d)

$$= \frac{p_1V_1 - p_2V_2}{\gamma - 1} = \frac{mR(T_1 - T_2)}{(\gamma - 1)}$$

$$= \frac{mR(T_1 - T_2)}{J(\gamma - 1)} \text{ heat units}$$

$$= mC_v[T_1 - T_2] \quad \left[\because \; C_v = \frac{R}{J(\gamma - 1)}\right] \text{ heat units}$$

(*vi*) **Free expansion.** If a fluid expands suddenly into a vacuum chamber through an orifice of large size, the expansion of fluid is called free expansion. In this case no external heat is supplied and also there is no work done and hence there is no *charge in internal energy*. Then temperature of the fluid before and after expansion is same. Enthalpy of the fluid is constant during the process. It is different from isothermal process where trmperature is constant but external work is done in that process.

(*vii*) **Throttle process.** It is defined as the process which take place in such a ways that the gas expands from high pressure to low pressure without doing any work and there is no change in kinetic energy and potental energy of the gas and also there is no heat transfer. Hence the net enthalpy of the gas remains constant.

(*viii*) **Reversible process.** A process is reversible if it can be carried out in reverse direction exatly in the same manner as in the forward direction and the surroundings before and after the process are exactly in the same condition. If a process involves (*i*) friction, (*ii*) unresisted expansion, and (*iii*) heat transfer with a finit temperature differnce then it will not be reversible. Isothermal and adiabatic processes (without internal friction, *i.e.*, friction between molecules and molecules or between molecules and cylinder suface) are the only reversible pocesses. In actual practice, reversibility can be approached but cannot be achived.

Reversible engines are most efficient and all reversible engnes have the same efficiency working in the same temperature range. The efficiency is given by (usually called Carnot cycle efficiency).

$$\eta = \frac{T_1 - T_2}{T_1} \quad \text{or} \quad \frac{Q_1 - Q_2}{Q_1}$$

where T_1 = Highest absolute temperature,

T_2 = Lowest absolute temperature,

Q_1 = Heat supplied, Q_2 = Heat rejected.

(*ix*) **Clausius inequality.** The mathematical statement that cyclic $\int \frac{dQ}{T} \leq 0$ is termed as **Clausius Inequality.**

(*a*) For a reversible cycle, $\oint \frac{dQ}{T} = 0$, and

(*b*) For an irreversible cycle, $\oint \frac{dQ}{T} < 0$

where $\oint$ = Integration along cyclic path.

3.7. ENTROPY

It is the property of a system which increases with the addition of heat. Entropy itself cannot be defined but change of entropy can be defined. For a reversible cycle, the increse of entropy multiplied by the absolute temperature gives the heat received by the gas. Mathematically, for reversible cycle

$$\text{Cyclic} \int \frac{dQ}{T} = 0 \quad \text{or} \quad \int \frac{dQ}{T} = ds = \text{chance of entropy}$$

$$\therefore \quad dQ = Tds \quad \text{and} \quad Q = \int Tds.$$

For any reversible process (heating and expansion), the area under the *T-S* diagram will give the total heat absorbed or rejected.

Change of entropy in irreversible processes is found out by assuming any reversible path between the end states and calculating

$$\int_1^2 \left(\frac{dQ}{T}\right)_{\text{rever}}$$

1. Isentropic and polytropic processes. Isentropic process is a reversible adiabatic process. For an adiabatic process, heat supplied or rejected is zero. Thus for a reversible adiabatic process,

$$ds = \frac{dQ}{T} = 0.$$

Hence the entropy for an isentropic process is constant or change of entropy is zero. *Polytropic processes* are processes where entropy is not constant.

2. Entropy change in any thermodynamic process. Figure 3.2 shows *T-S* diagram for reversible adiabatic (isentropic), thermal ($T = C$), constant volume and constant pressure expansion of a gas. The entropy change per unit mass ($s_2 - s_1$) for various thermodynamic processes change per unit mass ($s_2 - s_1$) for various thermodynamic processes are:

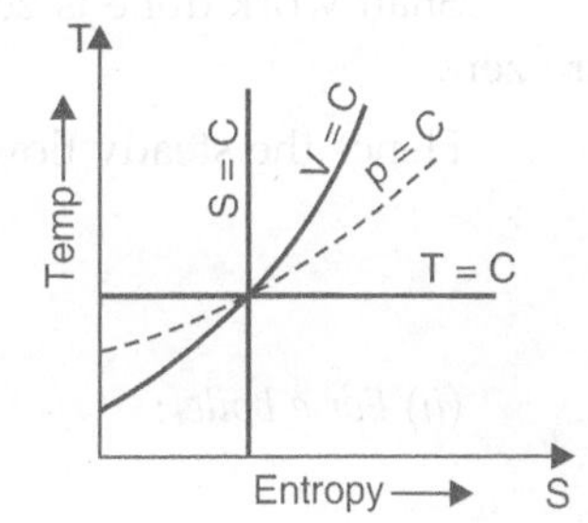

FIGURE 3.2

$$s_2 - s_1 = C_p \log_e \left(\frac{T_2}{T_1}\right) \quad \text{for constant pressure } (p = C)$$

$$(s_2 - s_1) = C_v \log_e \left(\frac{T_2}{T_1}\right) \quad \text{for constant volume } (V = C).$$

$$(s_2 - s_1) = \frac{R}{J} \log_e \left(\frac{v_2}{v_1}\right)$$

$$= \frac{R}{J} \log_e \left(\frac{p_1}{p_2}\right) \quad \text{for isothermal process } (T = C)$$

$$(s_2 - s_1) = \left(\frac{\gamma - n}{\gamma - 1}\right) \frac{R}{J} \log_e \left(\frac{v_2}{v_1}\right) \quad \text{for polytropic process.}$$

3. Steady flow energy equation. The steady flow energy equation is expressed as :

Total energy entering the system = Total energy leaving the system + losses

But, total energy = Internal energy + kinetic energy + pressure energy + potential energy + heat supplied

$$= u_1 + \frac{v_1^2}{2gJ} + \frac{p_1 v_1}{J} + \frac{Z_1}{J} + Q$$

where u_1 = Internal energy in heat units per unit weight

$\frac{v_1^2}{2gJ}$ = K.E. in heat units per unit weight.

$\frac{p_1 v_1}{J}$ = Pressure energy in heat units per unit weight.

$\frac{Z_1}{J}$ = Potential energy is heat units per unit weight.

Also, $u_1 + \frac{p_1 v_1}{J} = h_1.$

∴ Steady flow energy equation is written as

$$h_1 + \frac{V_1^2}{2gJ} + \frac{Z_1}{J} + Q = h_2 + \frac{V_2^2}{2gJ} + \frac{Z_2}{J} + W_s + \text{Losses}$$

where W_s = Shaft work done.

(i) *For a nozzle:*

Shaft work done is zero. Heat supplied is approximately zero and $Z_2 - Z_1 = 0$. Also losses are zero.

Hence the steady flow equation becomes as

$$h_1 + \frac{V_1^2}{2gJ} = h_2 + \frac{V_2^2}{2gJ}$$

(ii) *For a boiler:*

$$W_s = 0,\ Z_2 - Z_1 = 0 \quad \text{and} \quad \frac{V_2^2}{2g} - \frac{V_1^2}{2g} = 0$$

Hence steady flow energy equation becomes as

$$h_1 + Q = h_2 \quad \text{or} \quad Q = h_2 - h_1.$$

3.8. AIR CYCLES

Air cycles are the cycles, which use the prefect gas as a medium that is having the properties of air at room temperature. The air cycles are: (*i*) Carnot cycle, (*ii*) Otto cycle, (*iii*) Diesel cycle, (*iv*) Stirling cycle, (*v*) Brayton cycle, (*vi*) Ericsson cycle, and (*vii*) Dual cycle.

3.9. MEAN EFFECTIVE PRESSURE

It is defined as the ratio of work done per cycle to the stroke volume. Mean effective pressure is also equal to the area of pressure volume (*p-V*) diagram divided by length of the diagram.

3.10. EFFICIENCIES OF THE ENGINES

The various engine efficiencies are:

(*i*) Air standard efficiency, (*ii*) Thermal efficiency, (*iii*) Relative efficiency, (*iv*) Mechanical efficiency, and (*v*) Overall efficiency.

(*i*) Air standard efficiency is a hypothetical efficiency, which is used for comparing the efficiencies of various cycles. It is defined as the ratio of ideal work done to heat supplied. Mathematically, it is written as

$$\text{Air standard efficiency} = \frac{\text{Heat supplied} - \text{Heat rejected}}{\text{Heat supplied}}.$$

(*ii*) Thermal efficiency of an engine is the ratio of indicated work done to heat supplied.

(*iii*) Relative efficiency is the ratio of thermal efficiency to air standard efficiency.

(iv) **Mechanical efficiency** is the ratio of brake or actual work done to the indicated work done.

(v) **Overall efficiency** is the ratio of brake or actual work done to the heat supplied. This is also called brake thermal efficiency.

3.11. CARNOT CYCLE

Figure 3.3 shows the Carnot cycle on *p-V* and *T-S* diagrams. It consists of two isothermal and two reversible adiabatic processes.

Heat is supplied isothermally during process 2-3 at a temperature T_1 and heat is rejected isothermally during process 4-1 at a temperature T_2. During processes 3-4 and 1-2, no heat is supplied. For a closed cycle, the ratio of expansion and compression should be same (*i.e.*, both equal to *r*). Hence $\frac{V_3}{V_2} = \frac{V_4}{V_1}$.

For one kg mass of air,

Heat supplied $= p_2V_2 \log_e \left(\frac{V_3}{V_2}\right) = p_2V_2 \log_e r = RT_1 \log_e r$

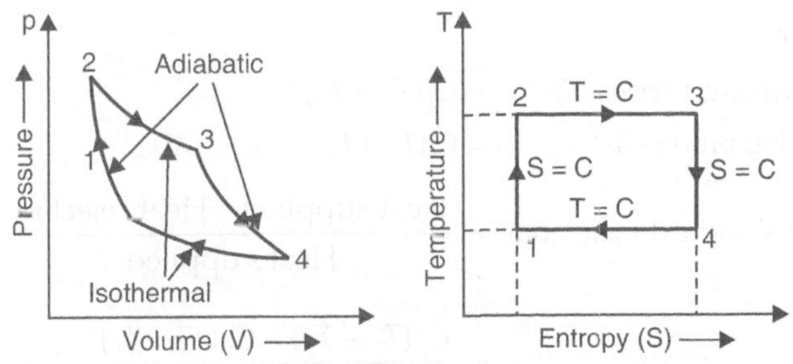

FIGURE 3.3. Carnot cycle on *p-V* and *T-S* diagrams.

Heat rejected $= p_4V_4 \log_e \left(\frac{V_4}{V_1}\right) = p_4V_4 \log_e r = RT_2 \log_e r$

$\therefore$ Efficiency of the cycle $= \frac{\text{Heat supplied} - \text{Heat rejected}}{\text{Heat supplied}}$

$$= \frac{RT_1 \log_e r - RT_2 \log_e r}{RT_1 \log_e r} = \frac{T_1 - T_2}{T_1}$$

where T_1 = Highest absolute temperature,

T_2 = Lowest absolute temperature.

The efficiency will be maximum, if heat is rejected at the lowest possible temperature.

3.12. OTTO CYCLE

Figure 3.4 shows the Otto cycle on *p-V* and *T-S* diagrams. It consists of two reversible adiabatic and two constant volume processes. Heat is supplied during the process 2-3 at constant volume and is rejected during the process 4-1 at constant volume. During processes 3-4 and 1-2, no heat is supplied. Let the clearance is unity and ratio of compression and expansion are equal and is *r*.

$$\therefore \qquad r = \frac{V_1}{V_2} = \frac{V_4}{V_3}$$

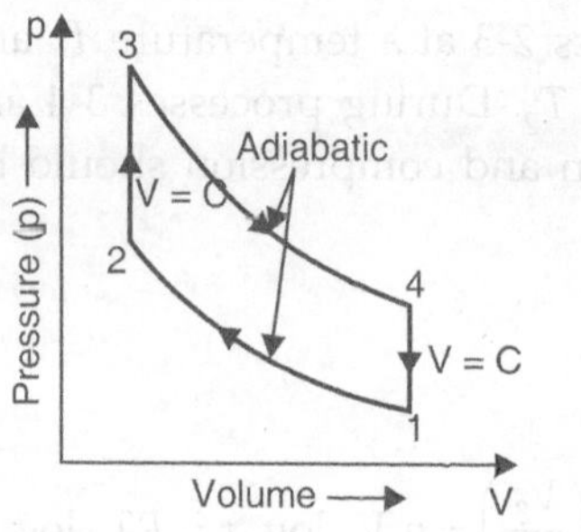

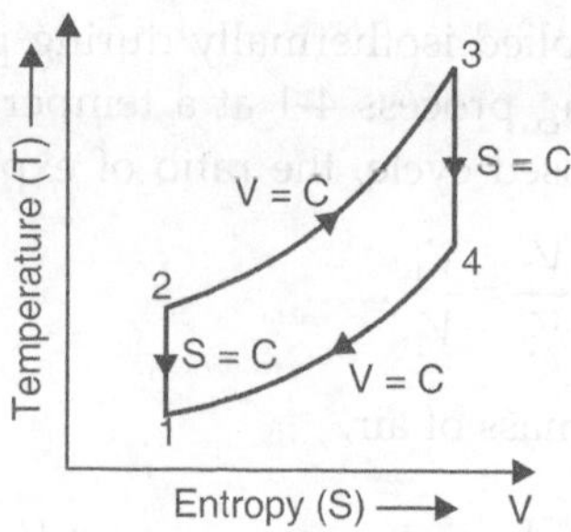

FIGURE 3.4

For one kg of air,

Heat supplied during process 2-3 $= C_v(T_3 - T_2)$

Heat rejected during process 4-1 $= C_v(T_4 - T_1)$

$$\therefore \qquad \text{Air standard efficiency} = \frac{\text{Heat supplied} - \text{Heat rejected}}{\text{Heat supplied}}$$

$$= \frac{C_v(T_3 - T_2) - C_v(T_4 - T_1)}{C_v(T_3 - T_2)} = 1 - \frac{(T_4 - T_1)}{(T_3 - T_2)} \qquad \ldots(A)$$

Also we have

$$\frac{T_2}{T_1} = \left(\frac{V_1}{V_2}\right)^{\gamma-1} = r^{\gamma-1} \quad \text{and} \quad \frac{T_3}{T_4} = \left(\frac{V_4}{V_3}\right)^{g^{\gamma-1}} = r^{\gamma-1}$$

$$\therefore \qquad T_1 = \frac{T_1}{r^{\gamma-1}} \quad \text{and} \quad T_4 = \frac{T_3}{r^{\gamma-1}}$$

Substituting the values of T_1 and T_4 in equation (*A*), we get

$$\text{Air standard efficiency} = 1 - \frac{\left(\frac{T_3}{r^{\gamma-1}} - \frac{T_2}{r^{\gamma-1}}\right)}{(T_3 - T_2)} = 1 - \frac{1}{r^{\gamma-1}}$$

Petrol and gas engines run on otto cycle.

3.13. DIESEL CYCLE

Figure 3.5 shows the Diesel cycle on *p-V* and *T-S* diagrams, it consists of a constant pressure, a constant volume and two adiabatic processes. Heat is supplied during constant pressure process 2-3 and is rejected during the constant volume process 4-1. During adiabatic process 3-4 and 1-2, no heat is supplied.

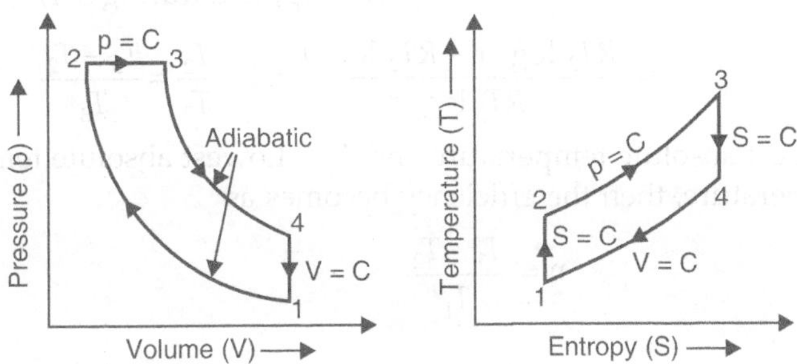

FIGURE 3.5

For one kg mass of air

$$\text{Air standard efficiency} = \frac{\text{Heat supplied} - \text{Heat rejected}}{\text{Heat supplied}}$$

$$= \frac{C_p(T_3 - T_2) - C_v(T_4 - T_1)}{C_p(T_3 - T_2)} = 1 - \frac{C_v(T_4 - T_1)}{C_p(T_3 - T_2)}$$

$$= 1 - \frac{1}{\gamma}\frac{(T_4 - T_1)}{(T_3 - T_2)} \qquad \left(\because \frac{C_p}{C_v} = \gamma\right)$$

Let $r = \text{Compression ratio} = \dfrac{V_1}{V_2}$

$r_c = \text{Heat addition cut-off ratio} = \dfrac{V_3}{V_2}$

Then $\dfrac{T_2}{T_1} = r^{\gamma-1}$ and $\dfrac{T_3}{T_2} = \dfrac{V_3}{V_2} = r_c$

Now, $T_3 = T_2 r_c = T_1 r^{\gamma-1} r_c$

Similarly, $T_4 = T_1 r_c^{\gamma}$. Hence air standard efficiency becomes as $= 1 - \dfrac{1}{\gamma r^{\gamma-1}}\left[\dfrac{r_c^{\gamma} - 1}{(r_c - 1)}\right]$.

3.14. STIRLING CYCLE

Figure 3.6 shows stirling cycle on *p-V* diagram. It consists of two isothermal and two constant volume processes. The constant volume processes are performed with the help of regenerator to make the cycle reversible. If the efficiency of the regenerator is 100% then the heat supplied during constant volume process 2-3 and heat rejected in constant volume process 4-1 will be equal.

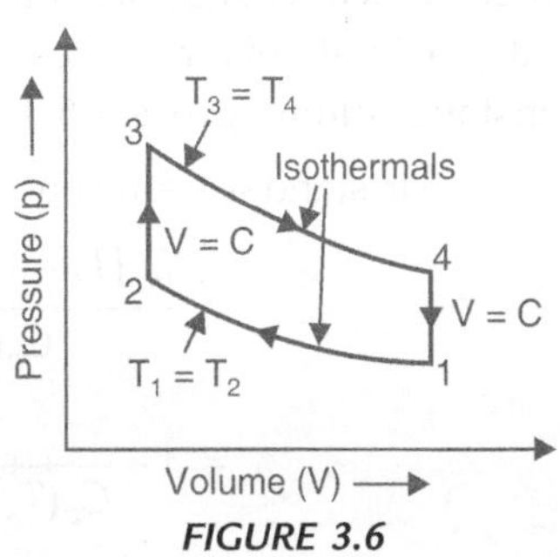

FIGURE 3.6

The work done will be equal to the difference of heat supplied during isothermal process 3-4 and heat rejected during isothermal process 1-2. For a unit mass,

Air standard efficiency

$$= \frac{\text{(Heat supplied during 3-4)} - \text{(heat rejected during 1-2)}}{\text{(Heat supplied during 3-4)}}$$

$$= \frac{RT_3 \log_e r - RT_2 \log_e r}{RT_3 \log_e r} = 1 - \frac{T_2}{T_3} = \frac{T_3 - T_2}{T_3}$$

But T_3 = Highest absolute temperature and T_2 = Lowest absolute temperature. If T_1 is the highest absolute temperature, then the efficiency becomes as

$$\eta = \frac{T_1 - T_2}{T_1}.$$

3.15. BRAYTON CYCLE

Figure 3.7 shows Brayton cycle on *p-V* diagram. It consists of two constant pressure and two adiabatic processes. Heat is supplied during constant pressure process 2-3 and is rejected during constant pressure process 4-1.

$$\text{Air standard efficiency} = \frac{\text{Heat supplied (process 2-3)} - \text{Heat rejected (process 4-1)}}{\text{Heat supplied (process 2-3)}}$$

$$= \frac{C_p(T_3 - T_2) - C_p[T_4 - T_1]}{C_p[T_3 - T_2]}$$

$$= 1 - \frac{(T_4 - T_1)}{(T_3 - T_2)}$$

But $$T_4 = \frac{T_3}{r^{\gamma-1}} \quad \text{and} \quad T_1 = \frac{T_2}{r^{\gamma-1}}$$

$\therefore$ Air standard efficiency $= 1 - \dfrac{1}{r^{\gamma-1}}$.

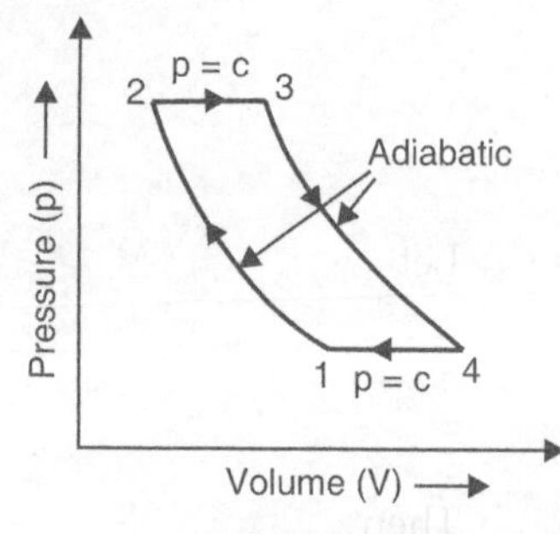

FIGURE 3.7

3.16. DUAL CYCLE

Figure 3.8 shows the Dual cycle. It consists of two constant volume processes, two adiabatic processes and one constant pressure process. Heat is supplied partly at constant volume process 2-3 and partly at constant pressure process 3-4. Heat is rejected at constant volume process 5-1.

Air standard efficiency

$$= \frac{C_v[T_3 - T_2] + C_p[T_4 - T_3] - C_v[T_5 - T_1]}{C_v(T_3 - T_2) + C_p(T_4 - T_3)}$$

$$= 1 - \frac{C_v(T_5 - T_1)}{C_v(T_3 - T_2) + C_p\,(T_4 - T_3)}$$

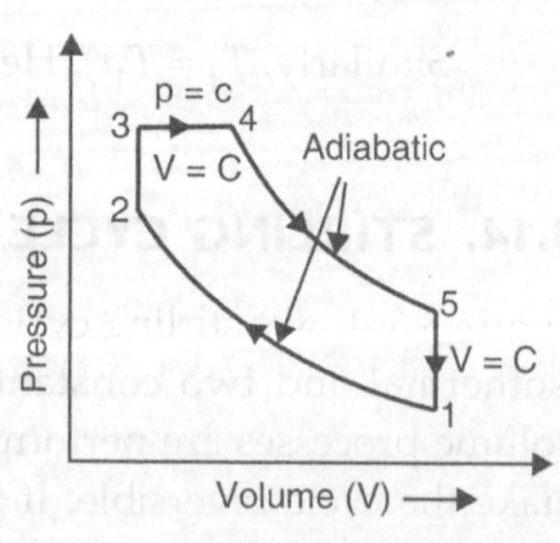

FIGURE 3.8

3.17. FUEL AND COMBUSTION

Fuel is a substance which produces a large amount of heat when burnt with oxygen in atmospheric air. Combustion is the chemical combination of a fuel with oxygen to give light and heat. Slow combustion gives light whereas rapid combustion of fuels gives heat and light as in case of boilers, furnaces and I.C. engines.

Fuels are classified as : (*i*) solid fuels, (*ii*) liquid fuels and (*iii*) gaseous fuels. *Solid fuels* are : Wood, peat, lignite or brown coal, Bituminous coal and Anthracite. *Liquid fuels* are : Petrol, kerosene, diesel oil and fuel oil. *Gaseous fuels* are : Coal, gas, producer gas, water gas, coke oven gas and blast furnace gas.

A **good fuel** is one which has low ignition point, has high calorific value, produces minimum quantity of smoke and gases, should be easy to store and convenient for transportation and is economical.

The heat liberated in kcal by complete combustion of 1 kg of fuel (solid or liquid) is known as *calorific value of fuels.* For gaseous fuels, the caloric value is defined as the heat liberated in kcal by complete combustion of 1 m^3 of the gas at a specific temperature and pressure. Calorific value may be divided into (*i*) higher calorific value, and (*ii*) lower calorific value. The total heat liberated in kcal per kg or per nr is known as *higher calorific value* (H.C.V.) of a fuel. If the heat carried away by the products of the combustion is not recovered, then the amount of heat obtained per kg or per m^3 of fuel is called *lower calorific value* (L.C.V.). The relation between lower calorific value (L.C.V.) and higher calorific value (H.C.V.) is given by

$$\text{L.C.V.} = (\text{H.C.V.} - \text{Mass of water} \times 588.76) \text{ kcal}$$

The calorific value of a fuel is determined by the two calorimeters namely (*i*) Bomb calorimeter, and (*ii*) Junker's gas calorimeter. Bomb calorimeter is generally used for determining the higher calorific value of a solid fuel and Junker's gas calorimeter is used for determining the calorific value of gaseous fuel.

3.18. COMBUSTION EQUATIONS

The following are the equations of stoichiometric or chemically correct mixtures

1. Burning of carbon (C) to carbon dioxide (CO_2)

$$C + O_2 = CO_2$$

(*a*) *By weight:*

$$12 \text{ kg of C} + 32 \text{ kg of } O_2 = 44 \text{ kg of } CO_2$$

$$\therefore \quad 1 \text{ kg of C} + \frac{32}{12} \text{ kg of } O_2 = \frac{44}{12} \text{ kg of } CO_2$$

or

$$1 \text{ kg of C} + \frac{8}{3} \text{ kg of } O_2 = \frac{11}{3} \text{ kg of } CO_2$$

(*b*) *By volume:*

$$1 \text{ mol of C} + 1 \text{ mol of } O_2 = 1 \text{ mol of } CO_2$$

2. Burning of carbon to carbon monoxide

$$2C + O_2 = 2CO$$

(a) *By weight:*

$$24 \text{ kg C} + 32 \text{ kg } O_2 = 56 \text{ kg CO}$$

$$\therefore \quad 1 \text{ kg C} + \frac{32}{24} \text{ kg } O_2 = \frac{56}{24} \text{ kg CO}$$

or $$1 \text{ kg C} + \frac{4}{3} \text{ kg } O_2 = \frac{7}{3} \text{ kg CO}$$

(b) *By volume:*

$$2 \text{ mol C} + 1 \text{ mol } O_2 = 2 \text{ mol CO}$$

3. Burning of carbon monoxide to carbon dioxide

$$2CO + O_2 = 2CO_2$$

(a) *By weight:*

$$56 \text{ kg CO} + 32 \text{ kg } O_2 = 88 \text{ kg } CO_2$$

$$\therefore \quad 1 \text{ kg CO} + \frac{32}{56} \text{ kg } O_2 = \frac{88}{56} \text{ kg } CO_2$$

or $$1 \text{ kg CO} + \frac{4}{7} \text{ kg } O_2 = \frac{11}{7} \text{ kg } CO_2$$

4. Burning of hydrogen of water or steam

$$2H_2 + O_2 = 2H_2O$$

(a) *By weight:*

$$4 \text{ kg } H_2 + 32 \text{ kg } O_2 = 36 \text{ kg of } H_2O$$

or $$1 \text{ kg } H_2 + 8 \text{ kg } O_2 = 9 \text{ kg of } H_2O$$

(b) *By volume:*

$$2 \text{ mol } H_2 + 1 \text{ mol } O_2 = 2 \text{ mol } H_2O$$

5. Burning of sulphur to sulphur dioxide

$$S + O_2 = SO_2$$

(a) *By weight:*

$$32 \text{ kg S} + 32 \text{ kg } O_2 = 64 \text{ kg } SO_2$$

or $$1 \text{ kg S} + 1 \text{ kg } O_2 = 2 \text{ kg } SO_2$$

(b) *By volume:*

$$1 \text{ mol S} + 1 \text{ mol } O_2 = 1 \text{ mol } SO_2$$

6. Burning of marsh gas or methane

$$CH_4 + 2O_2 = CO_2 + 2H_2O$$

(a) *By weight:*

$$16 \text{ kg } CH_4 + 64 \text{ kg } O_2 = 44 \text{ kg } CO_2 + 36 \text{ kg } H_2O$$

or $$1 \text{ kg } CH_4 + 4 \text{ kg } O_2 = \frac{11}{4} \text{ kg } CO_2 + \frac{9}{4} \text{ kg } H_2O$$

(b) *By volume:*

$$1 \text{ mol } CH_4 + 2 \text{ mol } O_2 = 1 \text{ mol } CO_2 + 2 \text{ mol } H_2O.$$

(*a*) **Minimum air required for complete combustion of a fuel.** It is obtained from the total weight or volume of oxygen required for complete combustion. For a liquid and solid fuel, the total weight of oxygen is determined and for a gaseous fuel the total volume of oxygen is determined. Air contains 23 parts of O_2 by weight and 21 parts by volume. If a fuel contains C parts by weight of carbon, H_2 parts by weight of hydrogen. O_2 parts by weight of oxygen and S parts by weight of sulphur, then total weight of oxygen required for complete combustion of 1 kg of fuel is

$$= \frac{8}{3}C + 8\left(H_2 + \frac{O_2}{8}\right) + S$$

$$\therefore \quad \text{Minimum air required per kg of fuel} = \frac{100}{23}\left[\frac{8}{3}C + 8\left(H_2 + \frac{O_2}{8}\right) + S\right].$$

(*b*) **Conversion of volumetric analysis to gravimetric analysis.** The volumetric analysis of a gas can be converted to gravimetric analysis by multiplying the volume of each constituent of the gas by its molecular weight. The product of volume and molecular weight gives of proportional weights of the constituents of the gas. The analysis by weight is obtained by adding all the proportional weights of the constituents and dividing each by the total weight.

(*c*) **Conversion of gravimetric analysis to volumetric analysis.** If the weight of each constituent of a fuel is known, then the volumetric analysis is obtained by dividing the weight of each constituent by its molecular weight. This gives the proportional volumes of the constituents. By adding these proportional volumes and dividing each proportional volume by the total, the volumetric analysis is obtained.

II. OBJECTIVE TYPE QUESTIONS

Tick mark the most appropriate statement of the multiple choice answers:

Definitions and Laws of Thermodynamics

1. Tick mark the correct statement
   (*a*) Intensive properties are dependent on mass of the system ☐
   (*b*) Extensive properties are independent of the mass ☐
   (*c*) Intensive properties are independent of the mass ☐
   (*d*) Temperature and pressure are extensive properties. ☐
2. Zeroth's law of thermodynamics states that
   (*a*) if two bodies are each equal in temperature to the temperature of a third body, they are equal in temperature to each other ☐
   (*b*) if two bodies are each equal in temperature to the temperature of a third body, they are having different temperature ☐
   (*c*) if two bodies are each different in temperature to the temperature of a third body, they are equal in temperature to each other ☐
   (*d*) none of the above. ☐

3. A closed system is one which
(*a*) permits the passage of energy and matter across the boundaries ☐
(*b*) permits the passage of energy across the boundary but does not permit the passage of matter ☐
(*c*) does not permit the passage of energy and matter across the boundary ☐
(*d*) does not permit the passage of energy but permits the passage of matter across it. ☐

4. An open system is one which
(*a*) permits the passage of energy and matter across the boundaries ☐
(*b*) permits the passage of energy across the boundary but does not permit the passage of matter ☐
(*c*) does not permit the passage of energy and matter across the boundary ☐
(*d*) none of the above. ☐

5. An isolated system is one which
(*a*) permits the passage of energy and matter across the boundaries ☐
(*b*) permits the passage of energy only ☐
(*c*) does not permit the passage of energy and matter across it ☐
(*d*) permits the passage or mass only. ☐

6. Specific heat is defined as the amount of heat required
(*a*) to raise unit degree of temperature of a substance ☐
(*b*) to raise unit mass of substance through unit degree of temperature ☐
(*c*) to raise unit mass of substance through 10 degrees of temperature ☐
(*d*) none of the above. ☐

7. The specific heat of a gas at constant pressure
(*a*) is equal to the specific heat at constant volume ☐
(*b*) is two times the specific heat at constant volume ☐
(*c*) is always greater than the specific heat at constant volume ☐
(*d*) none of the above. ☐

8. The ratio of specific heats of a gas at constant pressure and at constant volume is
(*a*) always a constant ☐
(*b*) always varies with the pressure ☐
(*c*) always varies with the temperature ☐
(*d*) none of the above. ☐

9. A perfect gas obeys
(*a*) Boyle's law only ☐ (*b*) Charle's law only ☐
(*c*) Both Boyle's and Charle's law ☐ (*d*) None of the above. ☐

10. According to Boyle's law
(*a*) $p_1V_1 = p_2V_2$ if p is constant ☐ (*b*) $\frac{p_1}{T_1} = \frac{p_2}{T_2}$ if volume is constant ☐
(*c*) $\frac{V_1}{T_1} = \frac{V_2}{T_2}$ is constant ☐ (*d*) none ot the above. ☐

11. According to Charle's law

(*a*) $p_1V_1 = p_2V_2$ if T is constant □ (*b*) $\frac{p_1}{T_1} = \frac{p_2}{T_2}$ if V is constant □

(*c*) $\frac{V_1}{T_1} = \frac{V_2}{T_2}$ if p is constant □ (*d*) none of the above. □

12. Characteristic equation of a gas is given by

(*a*) $pV = RT$, where V = Volume of gas □ (*b*) $pV = mRT$, where m = Mass of the gas □

(*c*) $mpV = RT$ □ (*d*) none of the above. □

13. Joule's law states that

(*a*) change of pressure is proportional to the change of temperature □

(*b*) change of volume is proportional to the change of temperature □

(*c*) change of internal energy is proportional to the change of temperature □

(*d*) none of the above. □

14. Displacement work is equal to

(*a*) $\int pdV$ □ (*b*) $\int Vdp$ □

(*c*) $\int Fdx$ □ (*d*) none of the above. □

where p = Pressure, V = Volume, and F = Force.

15. Change of internal energy is equal to heat interchange Q in case of

(*a*) non-flow constant volume process □ (*b*) non-flow constant pressure process □

(*c*) non-flow isothermal process □ (*d*) none of the above. □

16. The change of enthalpy (dH) is equal to heat in interchange Q in case of

(*a*) non-flow constant volume process □

(*b*) non-flow constant pressure process □

(*c*) non-flow isothermal process □

(*d*) none of the above. □

17. Work done in expansion is equal to heat supplied in case of

(*a*) non-flow constant volume process □ (*b*) non-flow constant pressure process □

(*c*) non-flow isothermal process □ (*d*) none of the above. □

Non-Flow Processes

18. For a polytropic expansion (pV^n = Constant), the temperature and pressure are related as

(*a*) $\frac{T_1}{T_2} = \left(\frac{p_1}{p_2}\right)^{\frac{n}{n-1}}$ □ (*b*) $\frac{T_1}{T_2} = \left(\frac{p_1}{p_2}\right)^{n-1}$ □

(*c*) $\frac{T_1}{T_2} = \left(\frac{p_1}{p_2}\right)^{\frac{n}{n-1}}$ □ (*d*) $\frac{T_1}{T_2} = \left(\frac{p_1}{p_2}\right)^{\frac{n-1}{n}}$. □

where n = Index of expansion.

19. For a polytropic process, the temperature and volume are related as

(a) $\frac{T_1}{T_2} = \left(\frac{V_1}{V_2}\right)^{\frac{n}{n-1}}$ ☐ (b) $\frac{T_1}{T_2} = \left(\frac{V_1}{V_2}\right)^{n-1}$ ☐

(c) $\frac{T_1}{T_2} = \left(\frac{V_2}{V_1}\right)^{n-1}$ ☐ (d) $\frac{T_1}{T_2} = \left(\frac{V_1}{V_2}\right)^{\frac{n-1}{n}}$. ☐

where n = Index of expansion.

20. Choose the wrong statement

(a) Displacement work is equal to $\int pdV$ ☐

(b) Intensive properties are independent of mass ☐

(c) An isolated system permits the passage of energy only ☐

(d) The specific heat of a gas at constant pressure is always less than the specific heat at constant volume. ☐

21. Choose the correct statement

(a) In free expansion, the work done = $\int pdV$ ☐

(b) In free expansion, enthalpy of the fluid changes during the process ☐

(c) In adiabatic expansion, the heat is supplied or rejected to the surrounding ☐

(d) In free expansion, the external heat supplied is zero and external work done is also zero. ☐

22. Figure 3.9 shows the various non-flow processes on *p-V* diagram. Curve *A* represents

(a) constant pressure process ☐

(b) constant volume process ☐

(c) adiabatic process ☐

(d) isothermal process. ☐

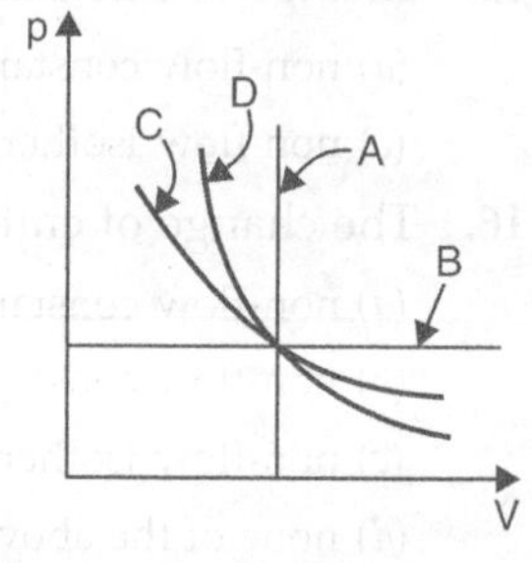

FIGURE 3.9

23. Curve *B* in Fig. 3.9 represents

(a) constant pressure process ☐

(b) constant volume process ☐

(c) adiabatic process ☐

(d) isothermal process. ☐

24. Curve C in Fig. 3.9 represents

(a) constant pressure process ☐ (b) constant volume process ☐

(c) adiabatic process ☐ (d) Isothermal process. ☐

25. Curve *D* in Fig. 3.9 represents

(a) constant pressure process ☐ (b) constant volume process ☐

(c) adiabatic process ☐ (d) isothermal process. ☐

26. During expansion, for the same increase in volume, the pressure drop is

(a) less in adiabatic process as compared to isothermal process ☐

(b) less in adiabatic process as compared to constant pressure process ☐

(c) greater in adiabatic process as compared to isothermal process ☐

(d) none of the above. ☐

27. For the same expansion ratio, work done by the gas in case of adiabatic process is
    (a) more than the work done in case of isothermal process ☐
    (b) less than the work done in case of isothermal process ☐
    (c) same as in case of isothermal process ☐
    (d) none of the above. ☐

28. The first law of thermodynamics states
    (a) if two bodies are each equal in temperature to a temperature of the third body, they are equal in temperature to each other ☐
    (b) that in a closed system undergoing through a cycle the net work delivered to the surroundings is proportional to the net heat taken from the surroundings ☐
    (c) heat cannot flow from lower temperature to higher temperature without the aid of an external agency ☐
    (d) that the net work can be produced during a cycle without some supply of heat. ☐

29. The statement—heat cannot flow from lower temperature to higher temperature without the aid of an external agency—is
    (a) Zeroth's law of thermodynamics ☐ (b) first law of thermodynamics ☐
    (c) second law of thermodynamics ☐ (d) none of the above. ☐

Air Cycles

30. A Carnot cycle consists of
    (a) two constant pressure and two adiabatic processes ☐
    (b) two constant volume and two adiabatic processes ☐
    (c) two isothermal and two adiabatic processes ☐
    (d) one constant pressure, one constant volume and two adiabatic processes. ☐

31. The efficiency of the Carnot cycle is equal to
    (a) $\frac{T_1 + T_2}{T_1}$ ☐ (b) $\frac{T_1}{T_1 + T_2}$ ☐
    (c) $\frac{T_1 - T_2}{T_1}$ ☐ (d) $\frac{T_1}{T_1 - T_2}$ ☐

32. The sum of internal energy and pressure volume product is called
    (a) entropy ☐ (b) enthalpy ☐
    (c) heat supplied ☐ (d) none of the above. ☐

33. The energy associated with molecular motions is called
    (a) K.E. of gases ☐ (b) internal energy ☐
    (c) entropy ☐ (d) enthalpy. ☐

34. Isentropic process means
    (a) a process in which no heat is supplied or rejected ☐
    (b) a process in which no heat is supplied or rejected and entropy is constant ☐
    (c) a process in which temperature is constant ☐
    (d) none of the above. ☐

35. A process, in which no heat is supplied or rejected and entropy is not constant, is called

(a) isentropic ☐ (b) isothermal ☐

(c) polytropic ☐ (d) hyperbolic. ☐

36. The change of entropy, when a unit mass of a perfect gas is heated from T_1 to T_2 at constant pressure is equal to

(a) $C_p \log_e \frac{T_1 + T_2}{T_1}$ ☐ (b) $C_p \log_e \frac{T_1}{T_1 + T_2}$ ☐

(c) $C_p \log_e \frac{T_2}{T_1}$ ☐ (d) $C_p \log_e \frac{T_1}{T_2}$ ☐

37. The change of entropy, when a unit mass of a perfect gas is heated from T_1 to T_2 at constant volume is equal to

(a) $C_v \log_e \frac{T_1 + T_2}{T_1}$ ☐ (b) $C_v \log_e \frac{T_1}{T_1 + T_2}$ ☐

(c) $C_p \log_e \frac{T_2}{T_1}$ ☐ (d) $C_v \log_e \frac{T_2}{T_1}$. ☐

38. The change of entropy for a unit mass of a perfect gas at constant temperature is equal to

(a) $\frac{R}{J} \log_e \frac{p_1 + p_2}{p_1}$ ☐ (b) $\frac{R}{J} \log_e \frac{p_1}{p_2}$ ☐

(c) $\frac{R}{J} \log_e \frac{p_1}{p_1 + p_2}$ ☐ (d) $\frac{R}{J} \log_e \frac{p_2 - p_1}{p_1}$ ☐

Efficiencies of Engines

39. Air standard efficiency is equat to

(a) $\frac{\text{heat supplied} + \text{heat rejected}}{\text{heat supplied}}$ ☐ (b) $\frac{\text{heat supplied}}{\text{heat supplied} + \text{heat rejected}}$ ☐

(c) $\frac{\text{heat rejected}}{\text{heat supplied} + \text{heat rejected}}$ ☐ (d) $\frac{\text{heat supplied} - \text{heat rejected}}{\text{heat supplied}}$ ☐

40. Thermal efficiency of an engine is equal to

(a) $\frac{\text{ideal work done}}{\text{heat supplied}}$ ☐ (b) $\frac{\text{indicated work done}}{\text{heat supplied}}$ ☐

(c) $\frac{\text{actual work done}}{\text{heat supplied}}$ ☐ (d) none of the above. ☐

41. Relative efficiency is equal to

(a) $\frac{\text{air standard efficiency}}{\text{thermal efficiency}}$ ☐ (b) $\frac{\text{thermal efficency}}{\text{air standard efficiency}}$ ☐

(c) $\frac{\text{overall efficiency}}{\text{air standard efficiency}}$ ☐ (d) none of the above. ☐

42. Overal efficiency is equal to

(*a*) $\frac{\text{indicated work done}}{\text{heat supplied}}$ ☐ (*b*) $\frac{\text{actual work done}}{\text{heat supplied}}$ ☐

(*c*) $\frac{\text{ideal work done}}{\text{heat supplied}}$ ☐ (*d*) none of the above. ☐

43. Mean effective pressure (m.e.p.) is equal to

(*a*) $\frac{p_1 + p_2}{2}$ ☐ (*b*) $\frac{\text{work done per cycle}}{\text{length}}$ ☐

(*c*) $\frac{\text{work done per cycle}}{\text{stroke length}}$ ☐ (*d*) none of the above. ☐

44. Stirling cycle consists of

(*a*) two isothermal, and two adiabatic processes ☐

(*b*) two isothermal, one constant volume and one constant pressure processes ☐

(*c*) two isothermal and two constant volume processes ☐

(*d*) done isothermal, one adiabatic and two constant volume processes. ☐

45. Air standard efficiency of Stirling cycle is

(*a*) more than that of Carnot cycle ☐ (*b*) less than that of Carnot cycle ☐

(*c*) same as of Carnot cycle ☐ (*d*) none of the above. ☐

46. Stirling cycle consists of

(*a*) two isothermal and two constant volume processes ☐

(*b*) two isothermal and two constant pressure processes ☐

(*c*) two constant volume and two adiabatic processes ☐

(*d*) two isothermal and two adiabatic processes. ☐

47. Otto cycle consists of

(*a*) two constant volume and two adiabatic pocesses ☐

(*b*) two constant volume and two isothermal processes ☐

(*c*) tow adiabatic and two isothermal processes ☐

(*d*) two constant pressure and two adiabatic processes ☐

48. Diesel cycle consists of

(*a*) two constant volume and two adiabatic processes ☐

(*b*) two constant volume and two isothermal processes ☐

(*c*) one constant pressure, one constant volume and two adiabatic processes ☐

(*d*) one constant pressure, one constant volume and two isothermal processes. ☐

49. Air standard efficiency of an Otto cycle is equal to

(*a*) $1 - \frac{1}{r^{\gamma+1}}$ ☐ (*b*) $1 - \frac{r}{r^{\gamma-1}}$ ☐

(*c*) $1 - \frac{1}{r^{\gamma-1}}$ ☐ (*d*) $1 + \frac{r}{r^{\gamma-1}}$ ☐

where r = Ratio of compression and expansion,

γ = Ratio of two specific heats.

50. Mean effective pressure of Carnot cycle is equal to

(*a*) $\dfrac{R(T_2 - T_1)\log_e r}{r}$ ☐ (*b*) $\dfrac{(T_2 - T_1)\log_e r}{\text{swept volume}}$ ☐

(*c*) $\dfrac{R\log_e r}{r(T_2 - T_1)}$ ☐ (*d*) $\dfrac{R(T_2 - T_1)\log_e r}{\text{swept volume}}$ ☐

where T_2 = Highest absolute temperature,
T_1 = Lowest absolute temperature, and
r = Ratio of compression and expansion.

51. For the same compression ratio and same heat input, the thermal efficiency of Otto cycle is

(*a*) less than that of Diesel cycle ☐ (*b*) more than that of Diesel cycle ☐
(*c*) equal to Diesel engine ☐ (*d*) none of the above. ☐

52. For constant maximum pressure and heat input, the thermal efficiency of Otto cycle is

(*a*) less than that of Diesel engine ☐ (*b*) more that that of Diesel engine ☐
(*c*) equal to Diesel engine ☐ (*d*) none of the above. ☐

53. Compression ratio is defined as the ratio of

(*a*) $\dfrac{\text{total volume}}{\text{swept volume}}$ ☐ (*b*) $\dfrac{\text{swept volume}}{\text{total volume}}$ ☐

(*c*) $\dfrac{\text{total volume}}{\text{clearance volume}}$ ☐ (*d*) $\dfrac{\text{swept volume}}{\text{clearance volume}}$ ☐

54. Mean effective pressure is obtained if the work done is divided by

(*a*) total volume ☐ (*b*) swept volume ☐
(*c*) clearance volume ☐ (*d*) none of them. ☐

55. Otto cycle is a theoretical cycle on which

(*a*) diesel engines run ☐ (*b*) only petrol engines run ☐
(*c*) petrol and gas engine runs ☐ (*d*) none of the above. ☐

56. The compression ratios for Diesel engines lie in the range of

(*a*) 5 to 8 ☐ (*b*) 15 to 20 ☐
(*c*) 3 to 6 ☐ (*d*) 30 to 40. ☐

57. The compression ratios for petrol engines lie in the range of

(*a*) 5 to 8 ☐ (*b*) 15 to 20 ☐
(*c*) 3 to 6 ☐ (*d*) 30 to 40. ☐

58. The first law of thermodynamics is expressed by the equation

(*a*) $W = JH$ ☐ (*b*) $H = JW$ ☐
(*c*) $W = J + H$ ☐ (*d*) $H = J + W$. ☐

where J = Mechanical equivalent of heat, W = Work, and H = Heat.

59. Unit of energy is

(*a*) Watt (W) ☐ (*b*) Joule (J) ☐
(*c*) Joule-metre (J m) ☐ (*d*) none of the above. ☐

60. Unit of power is

(*a*) Watt (W) ☐ (*b*) Joule (J) ☐

(*c*) Joule-metre (J m) ☐ (*d*) none of the above. ☐

61. Unit of universal gas constant is

(*a*) Nm kg/K ☐ (*b*) Nm/kg K ☐

(*c*) m/K ☐ (*d*) kg-m/K. ☐

62. The unit N m/s is called

(*a*) Joule (J) ☐ (*b*) Watt (W) ☐

(*c*) Calorie ☐ (*d*) none of the above. ☐

63. The unit N m is called

(*a*) Joule (J) ☐ (*b*) Watt (W) ☐

(*c*) Calorie ☐ (*d*) none of the above. ☐

64. Bar is the unit of

(*a*) power ☐ (*b*) energy ☐

(*c*) pressure ☐ (*d*) entropy. ☐

65. One bar in S.I. units is equal to

(*a*) 1×10^4 N/m^2 ☐ (*b*) 1×10^5 N/m^2 ☐

(*c*) 1×10^3 N/m^2 ☐ (*d*) 1 N/m^2. ☐

66. One bar is also equal to

(*a*) 1.033 kgf/cm^2 ☐ (*b*) 14.7 kgf/cm^2 ☐

(*c*) 1.0197 kgf/cm^2 ☐ (*d*) 1 kgf/cm^2. ☐

67. One atmospheric pressure is equal to

(*a*) 1.033 kgf/cm^2 ☐ (*b*) 14.7 kgf/cm^2 ☐

(*c*) 1.0197 kgf/cm^2 ☐ (*d*) 1 kgf/cm^2. ☐

68. If the work done is given in kgf-m, then to convert kgf-m into equivalent heat units of kilo-calorie (kcal), kgf-m should be divided by

(*a*) 1000 ☐ (*b*) 427 ☐

(*c*) 746 ☐ (*d*) 735. ☐

69. The value of gas constant for air, which has specific heat at constant volume equal to 715 N m/kg K and specific heat at constant pressure equal to 998 N m/kg K, is

(*a*) $\frac{998}{715}$ N m/kg K ☐ (*b*) 283 Nm/kg K ☐

(*c*) 1713 N m/kg K ☐ (*d*) none of the above. ☐

70. Choose the correct statement

(*a*) Entropy is an intensive property ☐ (*b*) Density is an intensive property ☐

(*c*) Density is an intensive property ☐ (*d*) Pressure is an extensive property. ☐

71. Thermal power plants use the cycle known as

(*a*) Otto ☐ (*b*) Carnot ☐
(*c*) Brayton ☐ (*d*) Rankine. ☐

72. When two dissimilar metals are heated at one end and cooled at the other, an e.m.f. is developed which is proportional to

(*a*) product of temperatures at two ends ☐
(*b*) difference of temperature at two ends ☐
(*c*) ratio of temperature at two ends ☐
(*d*) none of the above. ☐

73. Number of processes in a Rankine cycle are

(*a*) two ☐ (*b*) three ☐
(*c*) four ☐ (*d*) five. ☐

74. Kelvin Planck's law deals with

(*a*) conservation of mass ☐ (*b*) conservation of momentum ☐
(*c*) conservation of heat ☐ (*d*) conversion of heat into work. ☐

75. Gas turbine cycle consists of

(*a*) two isentropic and two constant pressure ☐
(*b*) tow isentropic and two constant volume ☐
(*c*) two isentropic and two isothermal ☐
(*d*) none of the above. ☐

76. Choose a wrong statement

(*a*) When a process undergoes a complete cycle, then the change of entropy is zero ☐
(*b*) A vertical line represents an isentropic process on *T-S* diagram ☐
(*c*) Second law of thermodynamics defines entropy ☐
(*d*) Carnot is the cycle used in thermal power plants. ☐

77. The thermal efficiency of an engine, which is supplied heat at the rate of 15,000 N m/s and gives an output of 4500 W, is

(*a*) 45% ☐ (*b*) 50% ☐
(*c*) 30% ☐ (*d*) 33%. ☐

78. Which one is not the property of a thermodynamic system?

(*a*) pressure ☐ (*b*) enthalpy ☐
(*c*) entropy ☐ (*d*) heat. ☐

79. Which one is not the intensive property of a thermodynamic system?

(*a*) temperature ☐ (*b*) total volume ☐
(*c*) pressure ☐ (*d*) density. ☐

80. Which one is not the extensive property of a thermodynamic system?

(*a*) mass ☐ (*b*) total volume ☐
(*c*) total internal energy ☐ (*d*) density. ☐

81. Which one is a correct statement?

(*a*) The change of property depends upon the path followed. ☐

(*b*) The change of property is fixed by the end states and is independent of path followed. ☐

(*c*) The change of property is not fixed by the end states. ☐

(*d*) The change of property is not a property of thermodynamic system. ☐

82. Thermal equilibrium between two or more bodies exists

(*a*) when there is no change in temperature of each body when they are brought together ☐

(*b*) when there is not change in pressure when they are brought together ☐

(*c*) when there is no change in density when they are brought together ☐

(*d*) when there is no change in any observable properties when they are brought together. ☐

83. The statement that when any two bodies are in thermal equilibrium with third, they are also equal in thermal equilibrium with each other is

(*a*) first law of thermodynamics ☐ (*b*) Zeroth law of thermodynamics ☐

(*c*) second law of thermodynamics ☐ (*d*) none of the above. ☐

84. The law, which gives the basis for measuring the thermodynamic property known as temperature, is called

(*a*) first law of thermodynamics ☐ (*b*) second law of thermodynamics ☐

(*c*) third law of thermodynamics ☐ (*d*) Zeroth law of thermodynamics. ☐

85. The law, which states that heat and work are mutually convertible, is known as

(*a*) Zeroth law of thermodynamics ☐ (*b*) first law of thermodynamics ☐

(*c*) second law of thermodynamics ☐ (*d*) none of the above. ☐

86. In metric system of units, one kilogram-calorie (kcal) heat unit is equivalent to

(*a*) 500 kgf-m ☐ (*b*) 427 kgf-m ☐

(*c*) 4.186 kgf-m ☐ (*d*) none of the above. ☐

87. The first law of thermodynamic was given by

(*a*) Joule ☐ (*b*) Charles ☐

(*c*) Wilson ☐ (*d*) Watts. ☐

88. According to first law of thermodynamics

(*a*) net work can be produced during a cycle without some supply of heat ☐

(*b*) net work cannot be produced during a cycle with some supply of heat ☐

(*c*) net work cannot be produced during a cycle without some supply of heat ☐

(*d*) none of the above. ☐

89. Choose the wrong statement

(*a*) Otto cycle consists of two constant volume and two adiabatic processes. ☐

(*b*) Stirling cycle consists of two constant volume and two isothermal processes. ☐

(*c*) Carnot cycle consists of two adiabatic and two isothermal processes. ☐

(*d*) Diesel cycle consists of two constant pressure and two adiabatic processes. ☐

90. A perpetual motion machine of the first kind (defined as a machine which produces power without consuming any energy) is

(*a*) impossible according to second law of thermodynamics ☐

(*b*) impossible according to first law of thermodynamics ☐

(*c*) possible according to first law of thermodynamics ☐

(*d*) possible according to second law of thermodynamics. ☐

91. A perpetual motion machine of the second kind (defined as 100% efficient engine) is

(*a*) impossible according to second law of thermodynamics ☐

(*b*) impossible according to first law of thermodynamics ☐

(*c*) possible according to first law of thermodynamics ☐

(*d*) possible according to second law of thermodynamics. ☐

92. According to Kelvin Planck statement a perpetual motion machine of

(*a*) second kind is impossible ☐ (*b*) first kind is possible ☐

(*c*) second kind is impossible ☐ (*d*) first kind is impossible. ☐

93. If C_p is the specific heat at constant pressure and C_v is the specific heat at constant volume, then

(*a*) $C_p < C_v$ ☐ (*b*) $C_p = C_v$ ☐

(*c*) $C_p > C_v$ ☐ (*d*) none of the above. ☐

94. For atmospheric air, the specific heat at constant pressure is equal to

(*a*) 0.17 ☐ (*b*) 1.4 ☐

(*c*) 0.24 ☐ (*d*) 0.42. ☐

95. For atmospheric air, the specific heat at constant volume is equal to

(*a*) 0.171 ☐ (*b*) 1.4 ☐

(*c*) 0.24 ☐ (*d*) 0.42. ☐

96. For atmospheric air, ratio C_p/C_v is equal to

(*a*) 0.171 ☐ (*b*) 1.4 ☐

(*c*) 1.20 ☐ (*d*) 1.3. ☐

97. Liquids have

(*a*) one specific heat ☐ (*b*) two specific heats ☐

(*c*) three specific heats ☐ (*d*) none of the above. ☐

98. Gases have

(*a*) one specific heat ☐ (*b*) two specific heats ☐

(*c*) three specific heats ☐ (*d*) none of the above. ☐

99. Choose the correct statement

(*a*) Free expansion in a non-flow process is same as isothermal process. ☐

(*b*) Enthalpy of the gas is constant is isothermal process. ☐

(*c*) Enthalpy of the gas is constant in free expansion. ☐

(*d*) The temperature and enthalpy in case of free expansion and isothermal process have constant values. ☐

100. Figure 3.10 shows the four different curves obtained by compressing a perfect gas. Curve *EA* shows the compression at

(*a*) constant pressure ☐
(*b*) adiabatic ☐
(*c*) isothermal ☐
(*d*) according to law $pV^{1.1} = C$. ☐

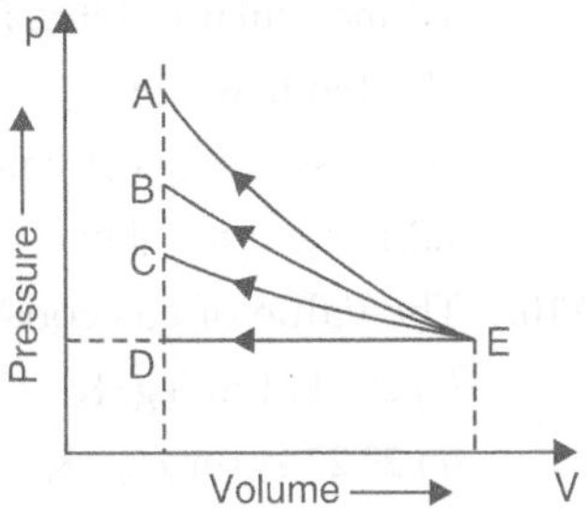

FIGURE 3.10

101. Refer to Fig. 3.10, the curve *EB* shows the compresssion at

(*a*) constant pressure ☐
(*b*) adiabatic ☐
(*c*) isothermal ☐
(*d*) according to law $pV^{1.1} = C$. ☐

102. Refer to Fig. 3.10, the curve *EC* shows the compression at

(*a*) constant pressure ☐ (*b*) adiabatic ☐
(*c*) isothermal ☐ (*d*) according to law $pV^{1.1} = C$. ☐

103. Refer to Fig. 3.10, the curve *ED* shows the compression at

(*a*) constant pressure ☐ (*b*) adiabatic ☐
(*c*) isothermal ☐ (*d*) according to law $pV^{1.1} = C$. ☐

104. Figure 3.11 shows the four different curves obtained by expanding a perfect gas. Curve *EA* shows the expansion at

(*a*) constant pressure ☐
(*b*) adiabatic ☐
(*c*) isothermal ☐
(*d*) according to law $pV^{1.1} = C$. ☐

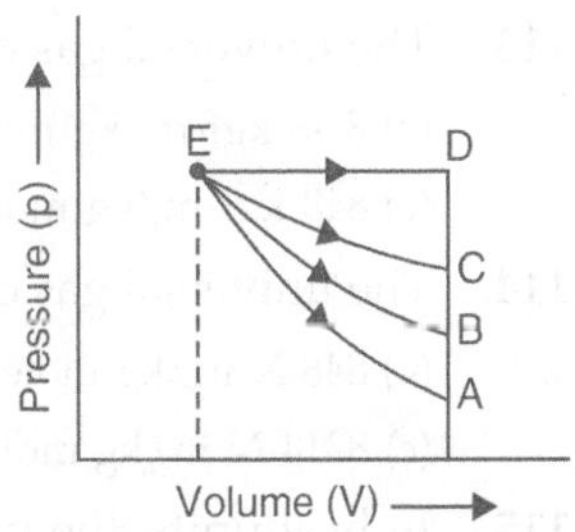

FIGURE 3.11

105. Refer to Fig. 3.11, the curve *EB* shows the compression at

(*a*) constant pressure ☐
(*b*) adiabatic ☐
(*c*) isothermal ☐
(*d*) according to law $pV^{1.1} = C$. ☐

106. Refer to Fig. 3.11, the curve *EC* shows the exponsion at

(*a*) constant pressure ☐ (*b*) adiabatic ☐
(*c*) isothermal ☐ (*d*) according to the law $pV^{1.1} = C$. ☐

107. Refer to Fig. 3.11, the curve *ED* shows the expansion at

(*a*) constant pressure ☐ (*b*) adiabatic ☐
(*c*) isothermal ☐ (*d*) according to the law $pV^{1.1} = C$. ☐

108. One mole of a gas is equal to

(*a*) atomic weight of the gas ☐ (*b*) 2.241 m^3 ☐
(*c*) 22.41 m^3 ☐ (*d*) 224.1 m^3. ☐

109. One kg mole of a gas is equal to

(*a*) molecular weight of the gas

(*b*) atomic weight

(*c*) mass of a gas of volume 22.41 m^3 at N.T.P

(*d*) none of the above.

110. The value of gas constant for air in M.K.S. units is equal to

(*a*) 287 kgf-m/kg/K (*b*) 28.7 kgf-m/kg/K

(*c*) 29.27 kgf-m/kg/K (*d*) 87.2 kgf-m/kg/K

111. The value of gas constant for air in S.I. units is equal to

(*a*) 287 N m/kg/K (*b*) 28.7 Nm/kg/K

(*c*) 29.27 kgf m/kg/K (*d*) 0.287 J/kg/K

112. Universal gas constant is equal to

(*a*) $\frac{R}{M}$ (*b*) MR

(*c*) $\frac{M}{R}$ (*d*) $\frac{1}{MR}$.

where R = Characteristic gas constant, and

M = Molecular weight.

113. The universal gas constant in M.K.S. units is equal to

(*a*) 8.48 kgf-m/kg mole/K (*b*) 84.8 kgf-m/kg mole/K

(*c*) 848 kgf-m/kg mole/K (*d*) 8314 kgf-m/kg mole/K

114. The universal gas constant in S.I. unit is equal to

(*a*) 848 N m/kg mole/K (*b*) 84.8 N m/kg mole/K

(*c*) 8314 N m/kg mole/K (*d*) 8314 J/kg mole/K.

115. In heat units, the value of universal gas constant is equal to

(*a*) 19.86 kcal/kg mole/K (*b*) 1.986 kcal/kg mole/K

(*c*) 198.6 kcal/kg mole/K (*d*) none of the above.

116. The heat supplied at constant volume for a non-flow process is

(*a*) $mC_p [T_2 - T_1]$ (*b*) $mC_v(T_2 - T_1)$

(*c*) $mC_p[T_2 + T_1]$ (*d*) $mC_v (T_2 + T_1)$

117. The heat supplied to a gas at constant pressure for a non-flow process is

(*a*) $mC_p[T_2 - T_1]$ (*b*) $mC_v [T_2 - T_1]$

(*c*) $mC_p [T_2 + T_1]$ (*d*) $mC_v [T_2 + T_1]$

118. The relation between two specific heats of a gas (*i.e.*, C_p and C_v) and gas constant is given by

(*a*) $C_p - C_v = \frac{R+1}{J}$ (*b*) $C_v - C_p = \frac{R}{J}$

(*c*) $C_p - C_v = \frac{R}{J}$ (*d*) $C_p - C_v = \frac{J}{R}$.

119. The value of specific heat of a gas at constant pressure in terms of gas constant is given by

(a) $C_p = \dfrac{R}{J(\gamma - 1)}$ ☐ (b) $C_p = \dfrac{\gamma R}{J(\gamma - 1)}$ ☐

(c) $C_p = \dfrac{J(\gamma - 1)}{R}$ ☐ (d) $C_p = \dfrac{J(\gamma - 1)}{\gamma R}$ ☐

where $\gamma = C_p/C_v$, J = Mechanical equivalent of heat.

120. The value of specific heat of a gas at constant volume in terms of a gas constant is given by

(a) $C_v = \dfrac{R}{J(\gamma - 1)}$ ☐ (b) $C_v = \dfrac{\gamma R}{J(\gamma - 1)}$ ☐

(c) $C_v = \dfrac{J(\gamma - 1)}{R}$ ☐ (d) $C_v = \dfrac{J(\gamma - 1)}{\gamma R}$. ☐

121. In case of hyperbolic expansion of a gas, the work done is equal to

(a) $pV_1 \log_e \dfrac{V_1 + V_2}{V_2}$ ☐ (b) $p_1V_1 \log_e \dfrac{V_2}{V_1}$ ☐

(c) $(p_1V_1 - p_2V_2)$ ☐ (d) $p_1V_1 \log_e \dfrac{V_1}{V_2}$. ☐

where V_1 = Volume of gas before expression, and

V_2 = Volume after expansion.

122. In case of hyperbolic expansion of a gas, heat supplied is

(a) more than the work done ☐ (b) less than the work done ☐

(c) equal to the work done ☐ (d) none of the above. ☐

123. In case of adiabatic expansion of a gas, the work done is equal to

(a) $(p_1V_1 - p_2V_2)$ ☐ (b) $\dfrac{(p_1V_1 - p_2V_2)}{(\gamma - 1)}$ ☐

(c) $\dfrac{(p_1V_1 - p_2V_2)}{1 + \gamma}$ ☐ (d) $\dfrac{mR(T_1 - T_2)}{(\gamma - 1)}$ ☐

124. Adiabatic expansion of a gas is given by the law

(a) pV = constant ☐ (b) pV^{γ} = constant ☐

(c) pV^0 = constant ☐ (d) pV^{∞} = constant. ☐

125. The difference between free expansion and throttling is

(a) enthalpy is constant in free expansion but not in throttling ☐

(b) temperature is constant in free expansion but not in throttling ☐

(c) that is case of free expansion the work done is zero but in case of throttling work done is not zero. ☐

(d) that in case of free expension the gas leaves with large velocity where as in throttling the gas has negligible velocity. ☐

126. A perfect gas is heated at constant pressure. The final volume of the gas becomes 1.5 times the initial volume. If its initial temperature is 30°C, the final temperature will be

(a) 45°C ☐ (b) 100°C ☐

(c) 177°C ☐ (d) 330°C. ☐

127. The work done, in a non-flow reversible process when volume (V)changes from 1 m^3 to 3 m^3, and $p = (-3V + 12)$ kgf/cm^2, is

(*a*) 1×10^4 kgf-m (*b*) 12×10^4 kgf-m

(*c*) 12 kgf-m (*d*) 20 kgf-m.

128. A process will be reversible if it involves

(*a*) friction

(*b*) unresisted expansion

(*c*) heat transfer with a finite temperature difference

(*d*) none of the above.

129. The condition for a cycle to be irreversible is

(*a*) cyclic $\int \frac{dQ}{T} < 0$ (*b*) cyclic $\int \frac{dQ}{T} = 0$

(*c*) cyclic $\int \frac{dQ}{T} > 0$ (*d*) none of the above.

130. The condition for a cycle to be irreversible is

(*a*) cyclic $\int \frac{dQ}{T} < 0$ (*b*) cyclic $\int \frac{dQ}{T} = 0$

(*c*) cyclic $\int \frac{dQ}{T} > 0$ (*d*) all of the above.

131. If cyclic $\int \frac{dQ}{T} > 0$, the cycle is

(*a*) reversible (*b*) irreversible

(*c*) impossible (*d*) none of the above.

132. A heat engine is supplied with 300 kcal/sec of heat at constant fixed temperature of 250°C. If 200 kcal/sec are rejected at 10°C, the cycle is

(*a*) reversible (*b*) irreversible

(*c*) impossible (*d*) none of the above.

133. Entropy itself cannot be defined but change of entropy is defined for a reversible process as

(*a*) $\frac{\text{heat added system}}{\text{absolute temperture}}$ (*b*) (heat added) × (absolute temperature)

(*c*) $\frac{\text{absolute temperature}}{\text{heat added}}$ (*d*) none of the above.

134. Figure 3.12 shows the reversible process of *T-S* diagrams. Curve *A* is for

(*a*) adiabatic process

(*b*) isothermal process

(*c*) constant volume process

(*d*) constant pressure process.

FIGURE 3.12

135. Curve *B* in Fig. 3.12 is for

(*a*) adiabatic process ☐ (*b*) isothermal process ☐

(*c*) constant volume process ☐ (*d*) constant pressure process. ☐

136. Curve *C* in Fig. 3.12 is for

(*a*) adiabatic process ☐ (*b*) isothermal process ☐

(*c*) constant volume process ☐ (*d*) constant pressure process. ☐

137. Curve *D* in Fig. 3.12 is for

(*a*) adiabatic process ☐ (*b*) isothermal process ☐

(*c*) constant volume process ☐ (*d*) constant pressure process. ☐

138. A process in which no heat is supplied or rejected and entropy is constant is known as

(*a*) adiabatic process ☐ (*b*) isentropic process ☐

(*c*) polytropic process ☐ (*d*) isothermal process. ☐

139. The change of entropy of a unit mass of a perfect gas for polytropic process, is equal to

(*a*) $\left(\frac{\gamma-n}{1-n}\right)\log_e\frac{T_2}{T_1}$ ☐ (*b*) $\left(\frac{\gamma-n}{1-n}\right)C_v\log_e\frac{T_2}{T_1}$ ☐

(*c*) $\left(\frac{1-n}{\gamma-n}\right)C_v\log_e\frac{T_2}{T_1}$ ☐ (*d*) $\left(\frac{1+n}{\gamma-n}\right)C_v\log_e\frac{T_2}{T_1}$. ☐

140. At a temperature of 227°C the heat supplied to an engine is 250 kcal per second. Heat rejected takes place at a constant temperature of 27°C. If heat rejected 200 kcal/sec. Then applying Clausius inequality, determine whether the process would be

(*a*) irreversible ☐ (*b*) reversible ☐

(*c*) impossible ☐ (*d*) none of the above. ☐

141. If in the above question, the heat rejected is 100 kcal/ sec, the process would be

(*a*) irreversible ☐ (*b*) reversible ☐

(*c*) impossible ☐ (*d*) none of the above. ☐

142. If in the above question, the heat rejected is 150 kcal/sec, the process would be

(*a*) irreversible ☐ (*b*) reversible ☐

(*c*) impossible ☐ (*d*) none of the above. ☐

143. For a non-flow process, the work done is given by

(*a*) $\int pdV$ ☐ (*b*) $\int Vdp$ ☐

(*c*) $-\int pdV$ ☐ (*d*) $-\int Vdp$. ☐

144. For a flow process the work done is equal to

(*a*) $\int pdV$ ☐ (*b*) $\int Vdp$ ☐

(*c*) $-\int pdV$ ☐ (*d*) $-\int Vdp$. ☐

Air Cycles

145. Figure 3.13 shows different air cycles on p-V diagram.

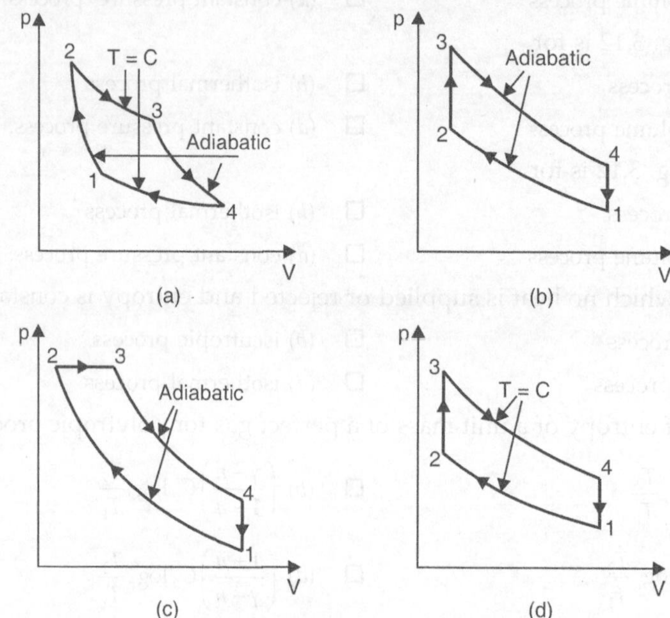

FIGURE 3.13

Figure 3.13 (*a*) shows

(*a*) Otto cycle ☐ (*b*) Carnot cycle ☐
(*c*) Stirling cycle ☐ (*d*) Diesel cycle. ☐

146. Figure 3.13 (*b*) shows

(*a*) Otto cycle ☐ (*b*) Carnot cycle ☐
(*c*) Stirling cycle ☐ (*d*) Diesel cycle. ☐

147. Figure 3.13 (*c*) shows

(*a*) Otto cycle ☐ (*b*) Carnot cycle ☐
(*c*) Stirling cycle ☐ (*d*) Diesel cycle. ☐

148. Figure 3.13 (*d*) shows

(*a*) Otto cycle ☐ (*b*) Carnot cycle ☐
(*c*) Stirling cycle ☐ (*d*) Diesel cycle. ☐

149. Figure 3.14 shows the different air cycles on T-S diagrams. Figure 3.14(*a*) shows

(*a*) Otto cycle ☐ (*b*) Carnot cycle ☐
(*c*) Stirling cycle ☐ (*d*) Diesel cycle. ☐

150. Figure 3.14 (*b*) shows

(*a*) Otto cycle ☐ (*b*) Carnot cycle ☐
(*c*) Stirling cycle ☐ (*d*) Diesel cycle. ☐

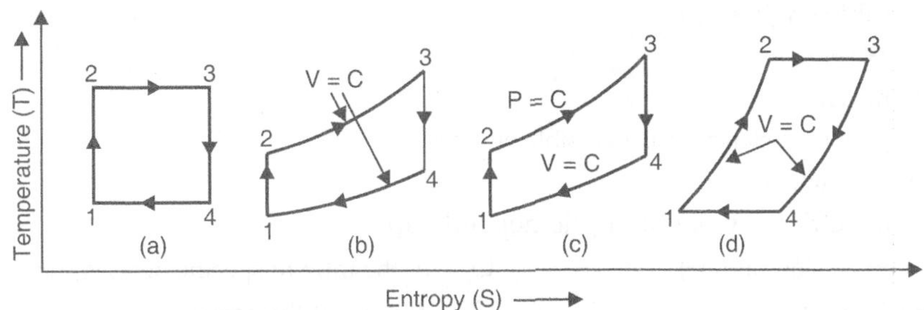

FIGURE 3.14

151. Figure 3.14 (*c*) shows

(*a*) Otto cycle ☐ (*b*) Carnot cycle ☐

(*c*) Stirling cycle ☐ (*d*) Diesel cycle. ☐

152. Figure 3.14 (*d*) shows

(*a*) Otto cycle ☐ (*b*) Carnot cycle ☐

(*c*) Stirling cycle ☐ (*d*) Diesel cycle. ☐

153. Mean effective pressure in case of Carnot cycle is given by

(*a*) $\dfrac{R(T_2 - T_1)\log_e \dfrac{V_3}{V_2}}{(V_3 - V_2)}$ ☐ (*b*) $\dfrac{R[T_2 - T_1]\log_e \dfrac{V_3}{V_2}}{(V_3 - V_2)}$ ☐

(*c*) $\dfrac{(T_2 - T_1)\log_e \dfrac{V_3}{V_2}}{(V_4 - V_2)}$ ☐ (*d*) $\dfrac{(T_2 - T_1)\log_e \dfrac{V_3}{V_2}}{(V_3 - V_2)}$ ☐

where $(V_4 - V_2)$ = Stroke length, and $\dfrac{V_3}{V_2}$ = Expansion ratio.

154. The efficiency of the Carnot cycle may be increased by

(*a*) increasing the highest temperature ☐

(*b*) decreasing the lowest temperature ☐

(*c*) increasing the lowest temperature ☐

(*d*) both (*a*) and (*b*). ☐

155. In a Carnot cycle, heat is supplied at 227°C and heat is rejected at 27°C, the efficiency of the cycle will be

(*a*) 80% ☐ (*b*) 70% ☐

(*c*) 66.67% ☐ (*d*) 50%. ☐

156. The Carnot cycle cannot be realised in actual practice due to

(*a*) very high pressure developed in the cylinder ☐

(*b*) high volume ratios in the cylinder ☐

(*c*) both (*a*) and (*b*) ☐

(*d*) none of the above. ☐

157. The reversible engines are

(*a*) most efficient ☐
(*b*) least efficient ☐
(*c*) having same efficiency as irreversible engines ☐
(*d*) none of the above. ☐

158. The thermal efficiency of Otto cycle depends upon

(*a*) the compression ratio (r) ☐ (*b*) the ratio fo specific heats (γ) ☐
(*c*) both (*a*) and (*b*) ☐ (*d*) none of the above. ☐

159. The ideal cycle for comparing the performance of steam plants is

(*a*) Carnot cycle ☐ (*b*) Rankine cycle ☐
(*c*) Otto cycle ☐ (*d*) Stirling cycle. ☐

160. The relationship between entropy, enthalpy and work is given by

(*a*) $Tds = dH + Vdp$ ☐ (*b*) $dH = Vdp - Tds$ ☐
(*c*) $Tds = dH - Vdp$ ☐ (*d*) $Vdp = \frac{dH}{Tds}$. ☐

161. A gas follows the law $pV^n = C$. If the value of $n = 1$, the process is known as

(*a*) adiabatic ☐ (*b*) isothermal ☐
(*c*) isentropic ☐ (*d*) polytropic. ☐

162. For any thermodynamic process, the area under T-S curve (temperature-entropy curve) represents

(*a*) work done by the system ☐ (*b*) work done on the system ☐
(*c*) head absorbed or rejected ☐ (*d*) none of the above. ☐

163. Entropy depends upon

(*a*) heat and work ☐ (*b*) volume and temperature ☐
(*c*) temperature and pressure ☐ (*d*) all of the above. ☐

164. When heat is absorbed by a gas, change of entropy is considered to be

(*a*) zero ☐ (*b*) positive ☐
(*c*) negative ☐ (*d*) none of the above. ☐

165. Internal energy of a gas is a function of

(*a*) pressure ☐ (*b*) pressure and volume ☐
(*c*) temperature ☐ (*d*) entropy. ☐

166. First law of thermodynamics deals with

(*a*) conservation of mass ☐ (*b*) conservation of heat ☐
(*c*) conservation of momentum ☐ (*d*) conservation of energy. ☐

167. The law which states that the absolute pressure of a given mass of a perfect gas varies inversely at its volume when the temperature remains constant is known as

(*a*) Charle's law ☐ (*b*) Boyle's law ☐
(*c*) Gay-Lussac law ☐ (*d*) Avogadro's law. ☐

168. The law which states that the volume of a given mass of a perfect gas varies directly as its absolute temperature, when the absolute pressure remains constant is known as

(*a*) Charle's law ☐ (*b*) Boyle's law ☐
(*c*) Gay-Lussac law ☐ (*d*) Avogadro's law. ☐

169. The law which states that the equal volumes of all gases, at the same temperature and pressure, contain equal number of molecules is known as

(*a*) Charle's law ☐ (*b*) Boyle's law ☐
(*c*) Gay-Lussac law ☐ (*d*) Avogadro's law. ☐

170. The law which states that the absolute pressure of a perfect gas varies directly with the absolute temperature volume is kept constant is known as

(*a*) Charle's law ☐ (*b*) Boyle's law ☐
(*c*) Gay-Lussac law ☐ (*d*) Avogadro's law. ☐

171. The law which states that change of internal energy of a perfect gas is directly proportional to the change of temperature, is known as

(*a*) Boyle's law ☐ (*b*) Charle's law ☐
(*c*) Joule's law ☐ (*d*) Avogadro's law. ☐

172. The absolute zero pressure corresponds to the temperature of

(*a*) 273°C ☐ (*b*) –273°C ☐
(*c*) 0°C ☐ (*d*) none of the above. ☐

173. The absolute zero temperature corresponds to

(*a*) 273°C ☐ (*b*) –273°C ☐
(*c*) 0°C ☐ (*d*) –237°C. ☐

174. Calorie is the amount of heat required to raise the temperature through 1°C of a gas of mass

(*a*) one kg ☐ (*b*) half kg ☐
(*c*) one gm ☐ (*d*) 100 gm. ☐

175. The heat supplied to a gas at constant volume

(*a*) increases the temperature of the gas ☐ (*b*) increases the internal energy of the gas ☐
(*c*) external work is done ☐ (*d*) none of the above ☐

176. The heat supplied to a gas at constant pressure

(*a*) increases the temperature of the gas ☐ (*b*) external work is done during expansion ☐
(*c*) increases the internal energy of the gas ☐ (*d*) both (*a*) and (*b*). ☐

177. The gas constant (R) is equal to

(*a*) ratio of two specific heats ☐ (*b*) sum of two specific heats ☐
(*c*) difference of two specific heats ☐ (*d*) product of two specific heats. ☐

178. Choose the correct statement.

(*a*) The hyperbolic process is governed by Charle's law. ☐
(*b*) The change of internal energy in isothermal process is maximum. ☐
(*c*) The heating of a gas at constant pressure is governed by Boyle's law. ☐
(*d*) The ratio of specific heat at constant pressure to the specific heat at constant volume is always greater than one. ☐

179. In a reversible adiabatic process, the term $\left(\frac{p_1}{p_2}\right)^{\frac{\gamma-1}{\gamma}}$ is equal to

(*a*) heat absorbed or rejected ☐ (*b*) polytropic index (n) ☐

(*c*) ratio of $\frac{T_1}{T_2}$ ☐ (*d*) work done during adiabatic expansion. ☐

180. In a reversible adiabatic process, the term $\left(\frac{p_1V_1 - p_2V_2}{\gamma-1}\right)$ is equal to

(*a*) heat absorbed or rejected ☐ (*b*) polytropic index (n) ☐

(*c*) ratio of $\frac{T_1}{T_2}$ ☐ (*d*) work done during adiabatic expansion. ☐

181. In a polytropic process, the term $\left(\frac{\gamma-n}{\gamma-1}\right)\frac{p_1V_1 - p_2V_2}{J(n-1)}$ is equal to

(*a*) heat absorbed or rejected ☐ (*b*) polytropic index (n) ☐

(*c*) ratio of $\frac{T_1}{T_2}$ ☐ (*d*) work done during adiabatic expansion. ☐

182. The term $\left[\log\left(\frac{p_2}{p_1}\right)\Big/\log\left(\frac{V_1}{V_2}\right)\right]$ is equal to

(*a*) work done during isothermal process ☐ (*b*) work done during adiabatic expansion ☐

(*c*) ratio of $\frac{T_1}{T_2}$ ☐ (*d*) the polytropic index (n). ☐

183. The area under the temperature-entropy curve (T-S curve) of any thermodynamic process represents

(*a*) work done during the process ☐ (*b*) heat absorbed only ☐

(*c*) heat rejected only ☐ (*d*) heat absorbed or rejected. ☐

184. Choose the wrong statement

(*a*) The change of entropy is positive when heat is absorbed by a gas. ☐

(*b*) Loss of heat takes place in an irreversible process. ☐

(*c*) The isothermal and adiabatic processes are reversible processes. ☐

(*d*) The absolute zero pressure can be attained at a temperature of 0°C. ☐

185. The heat transfer during polytropic process is

(*a*) $\frac{\gamma-n}{\gamma-1}\times\frac{p_1V_1 - p_2V_2}{J(n-1)}$ ☐ (*b*) $\frac{p_1V_1 - p_2V_2}{n-1}$ ☐

(*c*) $\left(\frac{1}{n-1}\right)(p_1V_1 - p_2V_2)$ ☐ (*d*) $\frac{n-1}{1}(p_1V_1 - p_2V_2)$ ☐

186. The term $\left(\frac{\gamma - n}{\gamma - 1}\right) \times m\, C_v\, (T_1 - T_2)$ is equal to

(*a*) heat transferred during polytropic process ☐
(*b*) work done ☐
(*c*) ratio of two specific heats ☐
(*d*) none of the above. ☐

187. If the cut-off is decreased, the efficiency of Diesel cycle

(*a*) increases ☐ (*b*) decreases ☐
(*c*) is same ☐ (*d*) none of the above. ☐

188. If cut-off is zero, the efficiency of Diesel cycle

(*a*) is equal to Otto cycle ☐ (*b*) is more than Otto cycle ☐
(*c*) approaches to Otto cycle ☐ (*d*) none of the above. ☐

189. Select the correct statement.

(*a*) The efficiency of Diesel cycle increases with increase in cut-off. ☐
(*b*) The efficiency of Diesel cycle is equal to Otto cycle for the same compression ratio. ☐
(*c*) For the same compression ratio, the efficiency of dual combustion cycle is less than Diesel cycle. ☐
(*d*) An Otto cycle consists of two constant volumes and two adiabatic processes. ☐

190. The efficiency of diesel cycle compared to dual combustion cycle for the same compression ratio is

(*a*) more ☐ (*b*) less ☐
(*c*) equal ☐ (*d*) none of the above. ☐

191. Choose the wrong statement

(*a*) Thermodynamic system is a definite area or space where some thermodynamic process takes place. ☐
(*b*) For a perfect gas, according to Boyle's law, $p \times V$= constant. ☐
(*c*) Volume is an executive property of a thermodynamic system. ☐
(*d*) Temperature is not intensive property of a thermodynamic system. ☐

192. Choose the correct statement

(*a*) Open system permits the passage of energy only across the boundaries. ☐
(*b*) First law of thermodynamics deals with conservation of mass. ☐
(*c*) One bar in S.I. units is equal to 1×10^5 N/m^2. ☐
(*d*) What is equal to 10 Nm/s? ☐

Fuel and Combustion

193. Calorific value of a liquid or solid fuel is the amount of heat liberated

(*a*) in kcal by complete combustion of 1 m^3 of fuel ☐
(*b*) in kcal when the temperature of the fuel is raised by 1°C ☐
(*c*) in kcal by complete combustion of 1 kg of fuel ☐
(*d*) none of the above. ☐

194. Calorfic value of a gaseous fuel is the amount of heat liberated in kcal

(*a*) by complete combustion of 1 m^3 of fuel at S.T.P. ☐
(*b*) when the temperature of the fuel is raised by 1°C ☐
(*c*) by complete combustion of 1 kg of fuel ☐
(*d*) none of the above. ☐

195. A good fuel is one which gas

(*a*) low ignition point and low calorific value ☐
(*b*) high ignition point and low calorific value ☐
(*c*) high ignition point and high calorific value ☐
(*d*) low ignition point and high calorific value. ☐

196. Bomb calorimeter is used to determine

(*a*) the viscosity of a liquid ☐ (*b*) calorific value of solid or liquid fuels ☐
(*c*) calorific value of a gaseous fuel ☐ (*d*) none of the above. ☐

197. Gas producers are used to

(*a*) determine calorific value of a gaseous fuel ☐
(*b*) determine viscosity of a gas ☐
(*c*) determine calorific value of a solid fuel ☐
(*d*) convert solid fuels into combustible fuels. ☐

198. Gas producers in which the steam-air mixture is forced by external means are known as

(*a*) forced gas producers ☐ (*b*) free gas producers ☐
(*c*) pressure producers ☐ (*d*) suction producers. ☐

199. Gas producers in which the steam-air mixture is drawn by the suction stroke of the engine are known as

(*a*) forced producers ☐ (*b*) free producers ☐
(*c*) pressure producers ☐ (*d*) suction producers. ☐

200. Avogadro's hypothesis states that

(*a*) the pressure of a mixture of gases is the sum of the partial pressure of each gas ☐
(*b*) equal volumes of different perfect gases contain equal number of molecules at the same condition of temperature and pressure ☐
(*c*) equal volumes of all gases, at the same temperature and pressure, contain different number of molecules ☐
(*d*) none of the above. ☐

201. When carbon (C) is burnt to carbon dioxide (CO_2), one kg of carbon requires

(*a*) $\frac{8}{3}$ kg O_2 to produce 11 kg CO_2 ☐ (*b*) 8 kg O_2 to produce $\frac{11}{3}$ kg CO_2 ☐
(*c*) $\frac{8}{3}$ kg O_2 to produce $\frac{11}{3}$ kg CO_2 ☐ (*d*) 8 kg O_2 to produce 11 kg CO_2. ☐

202. When carbon is burnt to carbon monoxide, one kg of carbon requires

(*a*) $\frac{4}{3}$ kg O_2 to produce 7 kg CO ☐ (*b*) 4 kg O_2 to produce $\frac{7}{3}$ kg CO ☐

(*c*) 4 kg O_2 produce 7 kg CO ☐ (*d*) $\frac{4}{3}$ kg O_2 to produce $\frac{7}{3}$ kg CO. ☐

203. When carbon is burnt to carbon monoxide, there is

(*a*) molecular expansion ☐ (*b*) molecular contraction ☐

(*c*) no molecular expansion ☐ (*d*) no molecular contraction. ☐

204. When carbon monoxide is burnt to carbon dioxide, there is

(*a*) molecular expansion ☐ (*b*) molecular contraction ☐

(*c*) no molecular expansion ☐ (*d*) no molecular contraction. ☐

205. When carbon monoxide is burnt to *carbon dioxide*, one kg of carbon monoxide requires

(*a*) $\frac{4}{7}$ kg O_2 to produce $\frac{11}{7}$ kg CO_2 ☐ (*b*) 4 kg O_2 to produce $\frac{11}{7}$ kg CO_2 ☐

(*c*) 4 kg O_2 to produce $\frac{11}{7}$ kg CO_2 ☐ (*d*) $\frac{4}{7}$ kg O_2 to produce 11 kg CO_2. ☐

206. When sulphur is burnt to sulphur dioxide, one kg of sulphur requires

(*a*) one kg O_2 to produce one kg S ☐ (*b*) one kg O_2 to produce 2 kg S ☐

(*c*) 2 kg O_2 to produce one kg S ☐ (*d*) 2 kg O_2 to produce 2 kg S. ☐

207. When sulphur is burnt to sulphur dioxide, there is

(*a*) molecular expansion ☐ (*b*) molecular contraction ☐

(*c*) no molecular expansion or contraction ☐ (*d*) none of the above. ☐

208. When one kg of solid or liquid fuel, containing C, H_2, O_2 and S, is completely burnt, minimum weight of air required is

(*a*) $\left(\frac{8}{3}C + 8H_2 + S - O_2\right)$ kg ☐ (*b*) $\frac{100}{23}\left[\frac{8}{3}C + 8H_2 + S + O_2\right]$ kg ☐

(*c*) $\frac{100}{23}\left[\frac{8C}{3} + 8\left(H_2 - \frac{O_2}{8}\right) + S\right]$ ☐ (*d*) $\frac{23}{100}\left[\frac{8C}{3} + 8\left(H_2 + \frac{O_2}{8}\right) + S\right]$ kg. ☐

209. Air contains by weight

(*a*) 23 parts O_2 and 77 parts N_2 ☐ (*b*) 21 parts O_2 and 79 parts N_2 ☐

(*c*) 77 parts O_2 and 23 parts N_2 ☐ (*d*) 79 parts O_2 and 21 parts N_2. ☐

210. Air contains by volume

(*a*) 23 parts O_2 and 77 parts N_2 ☐ (*b*) 21 parts O_2 and 79 parts N_2 ☐

(*c*) 77 parts O_2 and 23 parts N_2 ☐ (*d*) 79 parts O_2 and 21 parts N_2. ☐

211. Select the wrong statement

(*a*) The air supplied should be in excess of the theoretical air to ensure complete and rapid combustion of a fuel. ☐

(*b*) One kg of carbon produces $\frac{11}{3}$ kg of carbon dioxide. ☐

(*c*) Hydrogen is the lightest substance. ☐

(*d*) Coal gas is obtained by mixing coal and gas at ambient conditions. ☐

212. Select the correct statement

(*a*) Coke oven gas is produced by burning coke in an oven. ☐
(*b*) A good fuel should have high ignition point. ☐
(*c*) Gasoline is the lightest and most volatile liquid fuel as compared to kerosene and fuel oil. ☐
(*d*) Producer gas is obtained by carbonisation of coal. ☐

Miscellaneous Questions

213. One atmospheric pressure is equal to

(*a*) 14.7 kgf/cm^2 ☐ (*b*) 1 kgf/m^2 ☐
(c) 1.033 kgf/cm^2 ☐ (*d*) 1.033 kgf/m^2. ☐

214. The boiling and melting points, for water, coincide at

(*a*) 273 K ☐ (*b*) 200 K ☐
(c) 100 K ☐ (*d*) 0 K. ☐

215. The entropy, in a steam turbine when steam expands

(*a*) decreases linearly ☐ (*b*) increases linearly ☐
(*c*) increases exponentially ☐ (*d*) does not change. ☐

216. The statement that energy can neither be created nor destroyed but only converted from one form to another, is known as

(*a*) second law of thermodynamics ☐ (*b*) first law of thermodynamics ☐
(*c*) Zeroth law of thermodynamics ☐ (*d*) none of the above. ☐

217. For steady flow

(*a*) $\left(\frac{\partial v}{\partial t}\right)$ is constant ☐ (*b*) $\left(\frac{\partial v}{\partial t}\right)$ is zero ☐
(c) $\left(\frac{\partial v}{\partial s}\right)$ is constant ☐ (*d*) $\left(\frac{\partial v}{\partial s}\right)$ is zero. ☐

218. For a boiler, the steady flow energy equation is given by

(*a*) $h_1 + \frac{V_1^2}{2gJ} = h_2 + \frac{V_2^2}{2gJ}$ ☐ (*b*) $Q = (h_2 - h_1)$ ☐
(c) $h_1 + \frac{V_1^2}{2gJ} + Q = h_2 + \frac{V_2^2}{2gJ}$ ☐ (*d*) $W_s = (h_2 - h_1) + Q$. ☐

219. The device, which converts heat energy into useful work when operating in a cyclic process, is known as

(*a*) heat pump ☐ (*b*) heat engine ☐
(c) compressor ☐ (*d*) none of the above. ☐

220. The device, which delivers the heat from low temperature to a high temperature in a cyclic process, by utilizing mechanical work, is known as

(*a*) heat pump ☐ (*b*) heat engine ☐
(c) compressor ☐ (*d*) none of the above. ☐

221. One bar is equal to

(a) 1 N/m^2 ☐ (b) 100 N/m^2 ☐

(c) 1000 N/m^2 ☐ (d) 100000 N/m^2. ☐

222. One technical atmosphere has a pressure equal to

(a) 1 N/m^2 ☐ (b) 1 kgf/m^2 ☐

(c) 1 kgf/cm^2 ☐ (d) 1 lb/in^2. ☐

223. Standard temperature and pressure (S.T.P.) refers to a temperature of

(a) 20°C and pressure of 76 cm of mercury ☐ (b) 15°C and pressure of 76 cm of mercury ☐

(c) 0°C and pressure of 76 cm of mercury ☐ (d) 0°C and pressure of 75 cm of mercury. ☐

224. Normal temperature and pressure (N.T.P.) refer to a temperature of

(a) 20°C and pressure of 76 cm of mercury ☐ (b) 15°C and pressure of 76 cm of mercury ☐

(c) 0°C and pressure of 76 cm of mercury ☐ (d) 0°C and pressure of 75 cm of mercury. ☐

225. The root mean square velocity (*i.e.*, R.M.S. velocity) of the moving molecules is given by

(a) $\dfrac{(C_1 + C_2 + C_3 + ... + C_N)^2}{N}$ ☐ (b) $\left(\dfrac{C_1^2\, C_2^2 + C_3^2 + ...}{N^2}\right)^{1/2}$ ☐

(c) $\left(\dfrac{C_1^2 + C_2^2 + C_3^2 + ... + C_n^2}{N}\right)^{1/2}$ ☐ (d) $\left(\dfrac{C_1 \times C_2 + C_2 \times C_3 + ...}{N}\right)^{1/2}$ ☐

where C_1, C_2, C_3 are the velocities of individual molecules

N = No. of molecules.

226. The R.M.S. velocity of N molecules in a gas, if the temperature is constant

(a) goes on increasing ☐ (b) goes on decreasing ☐

(c) remains constant ☐ (d) none of the above. ☐

227. A gas of mass 10 kg and at a temperature of 20°C is heated to a temperature of 50°C. The specific heats at constant pressure and at constant volume are 0.2 and 0.15 respectively. The heat supplied to the gas when heated at constant pressure, will be

(a) 30 kcal ☐ (b) 45 kcal ☐

(c) 60 kcal ☐ (d) 75 kcal. ☐

228. In question 227, the heat supplied to the gas when heated at constant volume, will be

(a) 30 kcal ☐ (b) 45 kcal ☐

(c) 60 kcal ☐ (d) 75 kcal. ☐

229. The velocity, with which greatest percentage of molecules is found to move at any temperature, is known as

(a) root mean square velocity ☐ (b) mean velocity ☐

(c) most probable velocity ☐ (d) shear velocity. ☐

230. The internal energy of a gas depends upon

(a) pressure ☐ (b) volume ☐

(c) velocity ☐ (d) temperature. ☐

231. The enthalpy of vaporization, at critical point is

(*a*) maximum ☐ (*b*) minimum ☐

(*c*) zero ☐ (*d*) none of the above. ☐

232. The change in enthalpy, for a reversible adiabatic process, is

(*a*) maximum ☐ (*b*) minimum ☐

(*c*) unity ☐ (*d*) zero. ☐

233. The area under a curve on T-ϕ diagram represents

(*a*) heat transfer during reversible process ☐ (*b*) heat transfer during irreversible process ☐

(*c*) change of entropy ☐ (*d*) change of enthalpy. ☐

234. Specific fuel consumption is defined as

(*a*) fuel consumption per B.H.P. ☐ (*b*) fuel consumption per hour ☐

(*c*) fuel consumption per hour per B.H.P. ☐ (*d*) fuel consumption per I.H.P. ☐

235. If hot water and cold water are mixed, then entropy of the system will

(*a*) increase ☐ (*b*) decrease ☐

(*c*) remain constant ☐ (*d*) none of the above. ☐

236. Temperature of the order of — 160°C can be measured accurately by

(*a*) thermocouple ☐ (*b*) mercury thermometer ☐

(*c*) gas thermometer ☐ (*d*) pyrometer. ☐

237. Second law of thermodynamics defines

(*a*) heat ☐ (*b*) entropy ☐

(*c*) enthalpy ☐ (*d*) work. ☐

238. The ideal efficiency of a simple gas turbine depends upon

(*a*) temperature ratio ☐ (*b*) volume ratio ☐

(*c*) pressure ratio ☐ (*d*) none of the above. ☐

239. The amount of heat generated per kg of fuel is known as

(*a*) heat energy ☐ (*b*) calorific value ☐

(*c*) specific heat ☐ (*d*) internal energy. ☐

240. In an isothermal process, internal energy

(*a*) decreases ☐ (*b*) increases ☐

(*c*) has no change ☐ (*d*) none of the above. ☐

Tick mark the true or false statements:

241. In an isolated system, heat and work crosses the boundary of the system, but mass of the substance does not.

(*a*) True ☐ (*b*) False. ☐

242. In closed system, energy may cross the boundaries of the system.

(*a*) True ☐ (*b*) False. ☐

243. The properties, which are dependent on mass of the system are called intensive properties.

(*a*) True ☐ (*b*) False. ☐

244. The space and matter external to a system are known as surrounding.
(*a*) True ☐ (*b*) False. ☐

245. Temperature is a extensive property.
(*a*) True ☐ (*b*) False. ☐

246. Total volume is an intensive property.
(*a*) True ☐ (*b*) False. ☐

247. The unit of power in S.I. units is horse power.
(*a*) True ☐ (*b*) False. ☐

248. According to Gay-Lussac law for a perfect gas, p/T = constant, if v is kept constant.
(*a*) True ☐ (*b*) False. ☐

249. Equal volume of all gases, at the same temperature and pressure, contain equal number of molecules. This statement is called Avogadro's law.
(*a*) True ☐ (*b*) False. ☐

250. First law of thermodynamics deals with conservation of energy.
(*a*) True ☐ (*b*) False. ☐

251. When two dissimilar metals are heated at one end and cooled at the other, the e.m.f. developed is proportional to the difference of temperature at the two ends.
(*a*) True ☐ (*b*) False. ☐

252. It is impossible to transfer heat from a body at a lower temperature to a body at a higher temperature, without the aid of an external source.
(*a*) True ☐ (*b*) False. ☐

253. According to Boyle's law, the absolute pressure of a given mass of a perfect gas varies inversely as its volume, when the temperature remains constant.
(*a*) True ☐ (*b*) False. ☐

254. The total energy of the system remains constant according to first law of thermodynamics.
(*a*) True ☐ (*b*) False. ☐

255. The amount of heat required to raise the temperature of unit mass of a gas through 1°C at constant pressure is known as specific heat at constant pressure.
(*a*) True ☐ (*b*) False. ☐

256. The product of the molecular weight of a gas and the gas constant is known as universal gas constant.
(*a*) True ☐ (*b*) False. ☐

257. The heat supplied increases the internal energy of a gas, if the gas is heated at constant volume.
(*a*) True ☐ (*b*) False. ☐

258. The heating of a gas at constant volume is governed by Charle's law.
(*a*) True ☐ (*b*) False. ☐

259. According to isothermal process, the gas should be heated or expanded in such a way that product of its pressure and volume remains constant.
(*a*) True ☐ (*b*) False. ☐

260. A process, in which working substance neither receives nor gives out heat to its surroundings during its expansion and compression, is known as adiabatic process.

(*a*) True ☐ (*b*) False. ☐

261. In an isothermal change, there is no change of internal energy.

(*a*) True ☐ (*b*) False. ☐

262. The value of one bar in S.I. unit is equal to 1×10^4 N/m^2.

(*a*) True ☐ (*b*) False. ☐

263. The absolute zero temperature is taken as 273°C.

(*a*) True ☐ (*b*) False. ☐

264. Absolute pressure is equal to gauge pressure plus atmospheric pressure.

(*a*) True ☐ (*b*) False. ☐

265. Watt is the unit of energy in S.I. units.

(*a*) True ☐ (*b*) False. ☐

266. One newton-metre (1 Nm) is the S.I. unit of watt.

(*a*) True ☐ (*b*) False. ☐

267. Change of internal energy of a perfect gas is directly proportional to the change of temperature according to Boyle's law.

(*a*) True ☐ (*b*) False. ☐

268. Some quantity of air, in excess of the theoretical or minimum air is supplied, to ensure complete and rapid combustion of fuel.

(*a*) True ☐ (*b*) False. ☐

269. Free expansion process is a process in which s gas is allowed to expand suddenly into a vacuum chamber through an orifice of large dimensions.

(*a*) True ☐ (*b*) False. ☐

270. Throttling process is a process in which a perfect gas is expanded through an aperture of minute dimensions.

(*a*) True ☐ (*b*) False. ☐

271. Work done in a free expansion is maximum.

(*a*) True ☐ (*b*) False. ☐

272. In the equation $pV^n = C$, if $n = 0$, the process is known as isothermal.

(*a*) True ☐ (*b*) False. ☐

273. In the equation $pV^n = C$, if $n = 1$, the process is known as hyperbolic.

(*a*) True ☐ (*b*) False. ☐

274. There is a gain of heat in an irreversible process.

(*a*) True ☐ (*b*) False. ☐

275. The isothermal and adiabatic processes arc reversible processes.

(*a*) True ☐ (*b*) False. ☐

276. For reversible engine, the Carnot cycle has maximum efficiency.

(*a*) True ☐ (*b*) False. ☐

277. The efficiency of Carnot cycle increases, if the highest temperature is decreased.

(*a*) True ☐ (*b*) False. ☐

278. The efficiency of Joule cycle is less than Carnot cycle.

(*a*) True ☐ (*b*) False. ☐

279. Otto cycle is known as constant volume cycle.

(*a*) True ☐ (*b*) False. ☐

280. With the increase in cut off, the efficiency of Diesel cycle increases

(*a*) True ☐ (*b*) False. ☐

281. A closed cycle gas turbine works of Joule's cycle.

(*a*) True ☐ (*b*) False. ☐

282. For atmospheric air, the value of specific heat at constant pressure is 0.24.

(*a*) True ☐ (*b*) False. ☐

283. For a non-flow process, the work done is given by $\int pdV$.

(*a*) True ☐ (*b*) False. ☐

284. In a Carnot cycle, heat is supplied at 227°C and heat is rejected at 27°C, the efficiency of the cycle will be 66.67%.

(*a*) True ☐ (*b*) False. ☐

Fill in the blanks :

285. A ...... system is one in which there is no transfer of mass across its boundaries.

(*a*) open ☐ (*b*) closed. ☐

286. Extensive properties depend on ...... of the system.

(*a*) pressure ☐ (*b*) mass. ☐

287. The volume of a given mass of a perfect gas varies ...... as its absolute temperature, when the absolute pressure remains constant, according to Charle's law.

(*a*) directly ☐ (*b*) indirectly. ☐

288. According to Gay-Lussac law, the absolute pressure of a given mass of a perfect gas varies ...... as its absolute temperature, when the volume remains constant.

(*a*) indirectly ☐ (*b*) directly. ☐

289. The Joule's law states that change of internal energy of a perfect gas is ...... proportional to the change of temperature.

(*a*) directly ☐ (*b*) indirectly. ☐

290. It is ...... to construct an engine working on a cyclic process, whose sole purpose is to convert heat energy into work.

(*a*) possible ☐ (*b*) impossible. ☐

291. The gas constant (R) is equal to the ...... of two specific heats.

(*a*) ratio ☐ (*b*) difference. ☐

292. The amount of heat required to raise the temperature of ...... water through 1°C is called calorie.

(*a*) 1 kg ☐ (*b*) 1 gm. ☐

293. The ratio of specific heat at constant pressure and specific heat at constant volume is always ...... one.
(*a*) less than ☐ (*b*) greater than. ☐

294. The value of gas constant (*R*) in S.I. units is ......
(*a*) 28.7 J/kg/K ☐ (*b*) 287 J/kg/K. ☐

295. The value of specific heat at constant pressure is ...... that of at constant volume.
(*a*) less than ☐ (*b*) more than. ☐

296. One kg of carbon produces ...... kg of carbon dioxide.
(*a*) 3/11 ☐ (*b*) 11/3. ☐

297. The oxygen atom is ...... times heavier than the hydrogen atom.
(*a*) 16 ☐ (*b*) 12. ☐

298. The natural solid fuel is ......
(*a*) coke ☐ (*b*) wood. ☐

299. All the commercial liquid fuels are derived from ......
(*a*) gasoline ☐ (*b*) crude oil. ☐

300. The lightest and most Volatile liquid fuel is ....
(*a*) gasoline ☐ (*b*) kerosene. ☐

301. The atomic weight of nitrogen is ...... oxygen.
(*a*) more than ☐ (*b*) less than. ☐

302. A monoatomic molecule consists of ...... atoms.
(*a*) one ☐ (*b*) two. ☐

303. The ......calorific value of solid and liquid fuels are determined by bomb calorimeter.
(*a*) lower ☐ (*b*) higher. ☐

304. The amount of heat obtained by the complete combustion of 1 kg of a fuel when the products of its combustion are cooled down to the temperature of supplied air is called ...... calorific value of fuel.
(*a*) lower ☐ (*b*) higher. ☐

305. A good fuel should have ...... ignition point.
(*a*) low ☐ (*b*) high. ☐

306. Hydrogen is the ...... substance.
(*a*) heaviest ☐ (*b*) lightest. ☐

307. When carbonisation of coal is carried out at 500°C to 700°C, the ...... is obtained.
(*a*) hard coke ☐ (*b*) soft coke. ☐

308. The efficiency of a dual combustion cycle ...... upon the cut off ratio.
(*a*) does not depend ☐ (*b*) depends. ☐

309. Diesel cycle consists of ...... processes.
(*a*) two constant pressure and two adiabatic ☐
(*b*) one constant pressure, one constant volume and two adiabatic. ☐

310. The efficiency of Diesel cycle, for the same compression ratio, is ...... Otto cycle.

(*a*) more than ☐ (*b*) less than. ☐

311. Stirling cycle consists of ...... and two isothermal processes.

(*a*) two adiabatic ☐ (*b*) two constant volume. ☐

312. Otto cycle consists of ...... and two adiabatic processes.

(*a*) two isothermal ☐ (*b*) two constant volume. ☐

313. The throttling process is a ...... process

(*a*) reversible ☐ (*b*) irreversible. ☐

III. OBJECTIVE TYPE QUESTIONS FROM COMPETITIVE EXAMINATIONS

Tick mark the most appropriate answer:

1. For an ideal gas, which one is a function of temperature only

(*a*) specific heat at constant volume ☐ (*b*) specific heat at constant temperature ☐

(*c*) internal energy ☐ (*d*) entropy. ☐

2. Enthalpy of an ideal gas depends on

(*a*) pressure ☐ (*b*) temperature ☐

(*c*) volume ☐ (*d*) molecular weight. ☐

3. In isothermal process

(*a*) temperature increases gradually ☐ (*b*) volume remains constant ☐

(*c*) pressure remains constant ☐ (*d*) enthalopy change is maximum ☐

(*e*) change internal energy is zero. ☐

4. During throttling process

(*a*) internal energy does not change ☐ (*b*) pressure does not change ☐

(*c*) entropy does not change ☐ (*d*) enthalpy does not change ☐

(*e*) volume change is negligible. ☐

5. Which one of the following identity is correct?

(*a*) $G = MR$ ☐ (*b*) $G = M/R$ ☐

(*c*) $G = R/M$ ☐ (*d*) $G = M^2R$. ☐

where G is universal gas constant,

R is characteristic gas constant, and M is the molecular weight

6. Which one of the following gases will have the maximum value of gas constant R?

(*a*) nitrogen ☐ (*b*) carbon dioxide ☐

(*c*) sulphur dioxide ☐ (*d*) oxygen. ☐

7. Which law of thermodynamics the basis of temperature measurement?

(*a*) Zeroth law of thermodynamics ☐ (*b*) First law of thermodynamics ☐

(*c*) Second law of thermodynamics ☐ (*d*) Law of stable equilibrium. ☐

8. The Kelvin temperature of a system can be measured by a
   (*a*) mercury- in- glass thermometer ☐ (*b*) thermocouple ☐
   (*c*) constant-volume gas thermometer ☐ (*d*) resistance thermometer. ☐
9. The process or system that do not involve heat are called
   (*a*) isothermal processes ☐ (*b*) equilibrium processes ☐
   (*c*) thermal processes ☐ (*d*) adiabatic processes. ☐
10. In a reversible adiabatic process the ratio (T_1/T_2) is equal to
   (*a*) $\left(\frac{p_1}{p_2}\right)^{\frac{\gamma-1}{\gamma}}$ ☐ (*b*) $\left(\frac{v_1}{v_2}\right)^{\frac{\gamma-1}{\gamma}}$ ☐
   (*c*) $(v_1 - v_2)^{\frac{\gamma-1}{2\gamma}}$ ☐ (*d*) $\left(\frac{v_2}{v_1}\right)^{\gamma}$. ☐
   where γ is the adiabatic exponent.
11. Indicate the correct relation between the gas constant R, the adiabatic exponent γ and the specific heat at constant volume. The specific heat at constant volume is equal to
   (*a*) $\frac{R}{\gamma-1}$ ☐ (*b*) $\frac{\gamma-1}{\gamma R}$ ☐
   (*c*) $\frac{(\gamma-1)R}{\gamma}$ ☐ (*d*) $\frac{\gamma R}{\gamma-1}$. ☐
12. The specific heat of gas remains constant at all pressures and temperatures. This statement pertains to
   (*a*) Joule's law ☐ (*b*) Regnault's law ☐
   (*c*) Avogadro's law ☐ (*d*) Maxwell law. ☐
13. The efficiency of a thermodynamic cycle cannot be infinite since it
   (*a*) violates the first law of thermodynamics ☐
   (*b*) violates the third law of thermodynamics ☐
   (*c*) violates the third law of thermodynamics ☐
   (*d*) rejects no heat. ☐
14. A thermodynamic cycle is impossible if
   (*a*) $\oint \frac{dQ}{T} < 0$ ☐ (*b*) $\oint \frac{dQ}{T} = 0$ ☐
   (*c*) $\oint \frac{dQ}{T} > 0$ ☐ (*d*) $\oint ds > 0$. ☐
15. When two bodies are in thermal equilibrium with a third body they are also in thermal equilibrium with each other. This statement is called
   (*a*) Zeroth law of thermodynamics ☐ (*b*) first law of thermodynamics ☐
   (*c*) second law of thermodynamics ☐ (*d*) Kelvin Planck's law. ☐
16. The temperature at which the volume of a gas becomes zero is called
   (*a*) absolute scale of temperature ☐ (*b*) absolute zero temperature ☐
   (*c*) absolute temperature ☐ (*d*) none of the above. ☐

17. The value of one bar (in SI units) is equal to

(*a*) 100 N/m^2 ☐ (*b*) 1000 N/m^2 ☐

(*c*) 1×10^4 N/m^2 ☐ (*d*) 1×10^5 N/m^2 ☐

(*e*) 1×10^6 N/m^2. ☐

18. The relation, $T\,ds = du + p\,dv$, is true for

(*a*) any system, any process ☐ (*b*) any system, reversible process ☐

(*c*) closed system any process ☐ (*d*) closed system and reversible processes only. ☐

19. Which of the following is a correct statement?

(*a*) A reversible adiabatic process is an isentropic process ☐

(*b*) An irreversible adiabatic process is a constant entropy process ☐

(*c*) Entropy decreases during an irreversible adiabatic process ☐

(*d*) An isentropic process is an adiabatic process. ☐

20. For an inviscid incompressible fluid flowing through a duct, the steady flow energy equation reduces to

(*a*) Euler equation ☐ (*b*) Bernoulli equation ☐

(*c*) Stoke equation ☐ (*d*) Navier-stoke equation. ☐

21. Reversible steady flow work interaction is equal to

(*a*) $\int_1^2 p\,dv$ ☐ (*b*) $-\int_1^2 v\,dp$ ☐

(*c*) $u_1 - u_2$ ☐ (*d*) $p_1v_1 - p_2v_2$. ☐

22. In SI units, the value of the universal as constant is

(*a*) 0.8314 J/mole/K ☐ (*b*) 8.314 J/mole/K ☐

(*c*) 83.14 J/mole/K ☐ (*d*) 831.4 J/mole/K ☐

(*e*) 8314 J/mole/K. ☐

23. When the gas is heated at constant pressure, the heat supplied

(*a*) increases the internal energy of the gas ☐

(*b*) increases the temperature of the gas ☐

(*c*) does some external work during expansion ☐

(*d*) both (*b*) and (*c*). ☐

24. The gas constant (R) is equal to the

(*a*) sum of two specific heats ☐ (*b*) difference of two specific heats ☐

(*c*) product of two specific heats ☐ (*d*) ratio of two specific heats. ☐

25. A perfect gas enclosed in a piston-cylinder arrangement executes a reversible adiabatic expansion process. Consider the following statements:

1. entropy will increase
2. entropy change will be zero
3. entropy change of the surroundings will be zero of these statements.

(*a*) 1 and 3 are correct ☐ (*b*) 2 alone is correct ☐

(*c*) 2 and 3 are correct ☐ (*d*) 1 alone is correct. ☐

26. The standard fixed point of thermometry is the
(a) ice point ☐ (b) sulphur point ☐
(c) triple point of water ☐ (d) normal boiling point of water. ☐

27. The first law of thermodynamic was given by
(a) Joule ☐ (b) Charles ☐
(c) Wilson ☐ (d) Watts. ☐

28. If the thermal efficiency of a Carnot engine is 1/5, the COP of a Carnot refrigerator is
(a) 5 ☐ (b) 4 ☐
(c) 6 ☐ (d) 3. ☐

29. Two insulated tanks containing ideal gases at different pressures and temperatures are connected to each other and gases are allowed to mix. The process that occurs can be called
(a) free expansion ☐ (b) constant enthalpy ☐
(c) constant internal energy ☐ (d) reversible adiabatic. ☐

30. In comparison with the slopes of constant pressure lines for an ideal gas on a *T-S* plot, the slopes of constant volume lines are
(a) less ☐ (b) more ☐
(c) equal ☐ (d) unpredictable. ☐

31. A Carnot cycle operates between two temperatures T_1 and $T_2 (T_1 > T_2)$. If T_1 is increased by ΔT and T_2 is decreased by ΔT, the efficiency η_2 in the second case and the efficiency η_1 in the first case are related by
(a) $\eta_1 > \eta_2$ ☐ (b) $\eta_2 > \eta_2$ ☐
(c) $\eta_1 < \eta_2$ ☐ (d) unpredictable. ☐

32. The absolute zero pressure will be
(a) when molecular momentum of the system becomes zero ☐
(b) at sea level ☐
(c) at the temperature of –273 K ☐
(d) under vacuum conditions. ☐
(e) at the centre of the earth. ☐

33. Absolute zero temperature is taken as
(a) –273°C ☐ (b) 273°C ☐
(c) 237°C ☐ (d) –373°C. ☐

34. Which of the following is *correct* ?
(a) Absolute pressure = gauge pressure + atmospheric pressure ☐
(b) Gauge pressure = absolute pressure + atmospheric pressure ☐
(c) Atmospheric pressure = absolute pressure + gauges pressure ☐
(d) Absolute pressure = gauge pressure – atmospheric pressure. ☐

35. The internal energy of an ideal gas is a function of

(*a*) pressure only ☐ (*b*) absolute temperature only ☐

(*c*) pressure and volume ☐ (*d*) pressure, volume and temperature. ☐

36. For a closed system, the difference between heat added to and work done by the system is equal to

(*a*) enthalpy ☐ (*b*) enternal energy ☐

(*c*) Gibbs function ☐ (*d*) flow work. ☐

37. Heat transferred to a closed stationary system at constant volume is equal to

(*a*) work transfer ☐ (*b*) increase in internal energy ☐

(*c*) increases in enthalpy ☐ (*d*) increase in Gibbs function ☐

38. For a system undergoing phase change like melting or vaporization, which one remains constant?

(*a*) Enthalpy ☐ (*b*) Entropy ☐

(*c*) Specific volume ☐ (*d*) Gibbs function? ☐

39. Which one does not change during a throttling process?

(*a*) Enthalpy ☐ (*b*) Entropy ☐

(*c*) Volume ☐ (*d*) Pressure. ☐

40. The heat absorbed or rejected during a polytropic process is

(*a*) $\left(\frac{\gamma - n}{\gamma - 1}\right)$ × work done ☐ (*b*) $\left(\frac{\gamma - n}{\gamma - 1}\right)^2$ × work done ☐

(*c*) $\left(\frac{\gamma - n}{\gamma - 1}\right)^{1/2}$ × work done ☐ (*d*) $\left(\frac{\gamma - n}{\gamma - 1}\right)^3$ × work done. ☐

where γ = Adiabatic exponent, n = Polytropic index.

41. The internal energy of an ideal gas is a function of its absolute temperature only. This statement refers to

(*a*) Avogadro's law ☐ (*b*) Maxwell law ☐

(*c*) Joule's law ☐ (*d*) Regnault's law. ☐

42. The identify $\delta Q = p\, dv + du$ is true for

(*a*) any process and open system ☐ (*b*) any process and closed system ☐

(*c*) any process and any system ☐ (*d*) reversible process any closed system. ☐

43. Work done in a free expansion process is

(*a*) minimum ☐ (*b*) maximum ☐

(*c*) zero ☐ (*d*) none of the above. ☐

44. Match List I with List II and select the answer from the code given below:

List I (Equipment in a refrigeration system)	**List II (Purpose)**
A. Compressor	1. Enthalpy remains constant
B. Evaporator	2. Enthalpy increases
C. Throttle valve	3. Enthalpy increases but pressure remains constant
D. Condenser	4. Enthalpy decreases but pressure remains constant

Code:

	A	*B*	*C*	*D*	
(*a*)	3	2	1	4	☐
(*b*)	2	3	4	1	☐
(*c*)	2	3	1	4	☐
(*d*)	4	2	3	1	☐

45. In an aircraft refrigeration system, the pressure at the cooling turbine outlet is equal to

(*a*) ambient pressure ☐ (*b*) cabin pressure ☐

(*c*) pressure at compressor inlet ☐ (*d*) none of the above. ☐

46. If the temperature of the source is increased, the efficiency of the Carnot engine

(*a*) decreases ☐

(*b*) increases ☐

(*c*) does not change ☐

(*d*) will be equal to the efficiency of a practical engine ☐

(*e*) depends on other factors. ☐

47. The efficiency of an ideal Carnot engine depends on

(*a*) working substance ☐

(*b*) on the temperature of the source only ☐

(*c*) on the temperature of the sink only ☐

(*d*) on the temperatures of both the source and the sink. ☐

48. The efficiency of a Carnot engine using an ideal gas as the working substance is

(*a*) $\frac{T_1 - T_2}{T_1}$ ☐ (*b*) $\frac{T_1}{T_1 - T_2}$ ☐

(*c*) $\frac{T_1 T_2}{T_1 - T_2}$ ☐ (*d*) $\frac{T_1 - T_2}{T_1 T_2}$ ☐

(*e*) $\frac{T_2(T_1 - T_2)}{T_1(T_1 + T_2)}$ ☐

49. When a system is in equilibrium any conceivable change in entropy would be

(*a*) maximum ☐ (*b*) zero ☐

(*c*) positive ☐ (*d*) negative ☐

50. An isentropic process

(*a*) is always reversible ☐ (*b*) is always adiabatic ☐

(*c*) need not be adiabatic or reversible ☐ (*d*) is always frictionless. ☐

51. The work done by an ideal gas undergoing polytropic expansion from State 1 to State 2 is

(*a*) $\frac{n(p_1v_1 - p_2v_2)}{n-1}$ ☐ (*b*) $\frac{p_2v_2 - p_1v_1}{n-1}$ ☐

(*c*) $\frac{p_1v_1 - p_2v_2}{n-1}$ ☐ (*d*) $\frac{p_1v_1 - p_2v_2}{\gamma - 1}$. ☐

52. Match List I with List II and choose the correct answer from the code

List I (Law of thermodynamics)	**List II (Defines)**
A. First	1. Absolute zero temperature
B. Second	2. Internal energy
C. Zeroth	3. Temperature
D. Third	4. Entropy

Code:

	A	*B*	*C*	*D*	
(*a*)	1	2	3	4	☐
(*b*)	3	4	2	1	☐
(*c*)	4	2	1	3	☐
(*d*)	2	4	3	1	☐

53. An isolated system

(*a*) is a specified region where transfer of energy and/or mass take place ☐

(*b*) is a region of constant mass and only energy is allowed to cross the boundaries ☐

(*c*) cannot transfer either energy or mass to or from the surroundings ☐

(*d*) is one in which mass within the system is not necessarily constant ☐

(*e*) none of the above ☐

54. In an extensive property of a thermodynamic system

(*a*) extensive heat is transferred ☐ (*b*) extensive work is done ☐

(*c*) extensive energy is utilised ☐ (*d*) all of the above ☐

(*e*) none of the above ☐

55. Which of the following is an intensive property of a thermodynamic system ?

(*a*) Volume ☐ (*b*) Temperature ☐

(*c*) Mass ☐ (*d*) Energy. ☐

56. In a reversible cycle, the entropy of the system

(*a*) increases ☐ (*b*) decreases ☐

(*c*) does not change ☐ (*d*) first increases and then decreases ☐

(*e*) depends on the properties of working substance. ☐

57. A frictionless heat engine can be 100% efficient only if its exhaust temperature is

(*a*) equal to its input temperature ☐ (*b*) less than its input temperature ☐
(*c*) 0°C ☐ (*d*) 0 K. ☐

58. Kelvin-Planck's law deals with

(*a*) conservation of energy ☐ (*b*) conservation of heat ☐
(*c*) conservation of mass ☐ (*d*) conversion of heat into work ☐
(*e*) conversion of work into heat. ☐

59. When a liquid boils at constant pressure, which of the following parameter increases

(*a*) temperature ☐ (*b*) latent heat of vaporization ☐
(*c*) kinetic energy ☐ (*d*) entropy. ☐

60. The work done by a closed system will increase when the value of polytropic index n

(*a*) increases ☐ (*b*) decreases ☐
(*c*) first decreases and then increases ☐ (*d*) first increases and then decreases. ☐

61. The cycle in which heat is supplied at constant volume and rejected at constant pressure is known as

(*a*) Otto cycle ☐ (*b*) Dual cycle ☐
(*c*) Atkinson cycle ☐ (*d*) Stirling cycle. ☐

62. A definite area or space where some thermodynamic process takes place is known as

(*a*) thermodynamic system ☐ (*b*) thermodynamic cycle ☐
(*c*) thermodynamic process ☐ (*d*) thermodynamic law. ☐

63. An open system is one in which

(*a*) heat and work cross the boundary of the system, but the mass of the working substance does not ☐
(*b*) mass of working substance crosses the boundary of the system but the heat and work do not ☐
(*c*) both the heat and work as well as mass of the working substances cross the boundary of the system ☐
(*d*) neither the heat and work nor the mass of the working substances cross the boundary of the system. ☐

64. The characteristic equation of gases $pV = mRT$ holds good for

(*a*) monoatomic gases ☐ (*b*) diatomic gas ☐
(*c*) real gases ☐ (*d*) ideal gases. ☐

65. A gas which obeys kinetic theory perfectly is known as

(*a*) monoatomic gas ☐ (*b*) diatomic gas ☐
(*c*) real gas ☐ (*d*) pure gas ☐
(*e*) perfect gas. ☐

ANSWERS

Answers to Objective Type Questions

1. (*c*) 2. (*a*) 3. (*b*) 4. (*a*) 5. (*b*) 6. (*b*)
7. (*c*) 8. (*a*) 9. (*c*) 10. (*a*) 11. (*c*) 12. (*b*)
13. (*c*) 14. (*a*) 15. (*a*) 16. (*b*) 17. (*c*) 18. (*d*)
19. (*c*) 20. (*d*) 21. (*d*) 22. (*b*) 23. (*a*) 24. (*d*)
25. (*c*) 26. (*c*) 27. (*b*) 28. (*b*) 29. (*c*) 30. (*c*)
31. (*c*) 32. (*b*) 33. (*b*) 34. (*b*) 35. (*c*) 36. (*c*)
37. (*d*) 38. (*b*) 39. (*d*) 40. (*b*) 41. (*b*) 42. (*b*)
43. (*c*) 44. (*c*) 45. (*b*) 46. (*a*) 47. (*a*) 48. (*c*)
49. (*c*) 50. (*d*) 51. (*b*) 52. (*a*) 53. (*c*) 54. (*b*)
55. (*c*) 56. (*b*) 57. (*a*) 58. (*a*) 59. (*b*) 60. (*a*)
61. (*b*) 62. (*b*) 63. (*a*) 64. (*c*) 65. (*b*) 66. (*c*)
67. (*a*) 68. (*b*) 69. (*b*) 70. (*e*) 71. (*d*) 72. (*b*)
73. (*d*) 74. (*d*) 75. (*a*) 76. (*d*) 77. (*c*) 78. (*d*)
79. (*b*) 80. (*d*) 81. (*b*) 82. (*c*) 83. (*b*) 84. (*d*)
85. (*b*) 86. (*b*) 87. (*a*) 88. (*c*) 89. (*d*) 90. (*b*)
91. (*a*) 92. (*a*) 93. (*c*) 94. (*c*) 95. (*a*) 96. (*b*)
97. (*a*) 98. (*b*) 99. (*c*) 100. (*b*) 101. (*d*) 102. (*c*)
103. (*a*) 104. (*b*) 105. (*d*) 106. (*c*) 107. (*a*) 108. (*b*)
109. (*a*) 110. (*c*) 111. (*a*) 112. (*b*) 113. (*c*) 114. (*c*)
115. (*b*) 116. (*b*) 117. (*a*) 118. (*c*) 119. (*b*) 120. (*a*)
121. (*b*) 122. (*c*) 123. (*b*), (*d*) 124. (*b*) 125. (*d*) 126. (*c*)
127. (*b*) 128. (*d*) 129. (*b*) 130. (*a*) 131. (*c*) 132. (*b*)
133. (*a*) 134. (*b*) 135. (*d*) 136. (*c*) 137. (*a*) 138. (*b*)
139. (*b*) 140. (*a*) 141. (*c*) 142. (*b*) 143. (*a*) 144. (*d*)
145. (*b*) 146. (*a*) 147. (*d*) 148. (*c*) 149. (*b*) 150. (*a*)
151. (*d*) 152. (*c*) 153. (*b*) 154. (*d*) 155. (*c*) 156. (*c*)
157. (*a*) 158. (*c*) 159. (*b*) 160. (*c*) 161. (*b*) 162. (*c*)
163. (*c*) 164. (*b*) 165. (*c*) 166. (*d*) 167. (*b*) 168. (*a*)
169. (*d*) 170. (*c*) 171. (*c*) 172. (*d*) 173. (*b*) 174. (*c*)
175. (*a*) 176. (*d*) 177. (*c*) 178. (*d*) 179. (*c*) 180. (*d*)
181. (*a*) 182. (*d*) 183. (*d*) 184. (*d*) 185. (*a*) 186. (*a*)
187. (*a*) 188. (*c*) 189. (*d*) 190. (*b*) 191. (*d*) 192. (*c*)
193. (*c*) 194. (*a*) 195. (*d*) 196. (*b*) 197. (*c*) 198. (*c*)
199. (*d*) 200. (*b*) 201. (*c*) 202. (*d*) 203. (*a*) 204. (*b*)
205. (*a*) 206. (*b*) 207. (*c*) 208. (*c*) 209. (*a*) 210. (*b*)
211. (*d*) 212. (*c*) 213. (*c*) 214. (*a*) 215. (*a*) 216. (*b*)
217. (*b*) 218. (*b*) 219. (*b*) 220. (*a*) 221. (*d*) 222. (*c*)

223. (*b*)	224. (*a*)	225. (*c*)	226. (*c*)	227. (*c*)	228. (*b*)
229. (*c*)	230. (*d*)	231. (*c*)	232. (*d*)	233. (*a*)	234. (*c*)
235. (*a*)	236. (*c*)	237. (*b*)	238. (*c*)	239. (*b*)	240. (*c*).

True/False

241. (*b*)	242. (*a*)	243. (*b*)	244. (*a*)	245. (*b*)	246. (*b*)
247. (*b*)	248. (*a*)	249. (*a*)	250. (*a*)	251. (*a*)	252. (*a*)
253. (*a*)	254. (*a*)	255. (*a*)	256. (*a*)	257. (*b*)	258. (*b*)
259. (*b*)	260. (*a*)	261. (*a*)	262. (*b*)	263. (*b*)	264. (*a*)
265. (*b*)	266. (*b*)	267. (*a*)	268. (*a*)	269. (*b*)	270. (*a*)
271. (*b*)	272. (*b*)	273. (*a*)	274. (*b*)	275. (*a*)	276. (*a*)
277. (*b*)	278. (*a*)	279. (*a*)	280. (*b*)	281. (*a*)	282. (*a*)
283. (*a*)	284. (*a*).				

Fill in the Blanks

285. (*b*)	286. (*b*)	287. (*a*)	288. (*b*)	289. (*a*)	290. (*b*)
291. (*b*)	292. (*b*)	293. (*b*)	294. (*b*)	295. (*b*)	296. (*b*)
297. (*a*)	298. (*b*)	299. (*b*)	300. (*a*)	301. (*b*)	302. (*a*)
303. (*b*)	304. (*b*)	305. (*a*)	306. (*b*)	307. (*b*)	308. (*b*)
309. (*b*)	310. (*b*)	311. (*b*)	312. (*b*)	313. (*b*).	

Answers to Objective Type Questions from Competitive Examinations

1. (*c*)	2. (*b*)	3. (*e*)	4. (*d*)	5. (*a*)	6. (*a*)
7. (*a*)	8. (*c*)	9. (*d*)	10. (*a*)	11. (*a*)	12. (*b*)
13. (*b*)	14. (*c*)	15. (*a*)	16. (*b*)	17. (*d*)	18. (*d*)
19. (*a*)	20. (*b*)	21. (*b*)	22. (*c*)	23. (*d*)	24. (*b*)
25. (*b*)	26. (*c*)	27. (*a*)	28. (*b*)	29. (*c*)	30. (*b*)
31. (*c*)	32. (*a*)	33. (*a*)	34. (*a*)	35. (*b*)	36. (*b*)
37. (*b*)	38. (*d*)	39. (*a*)	40. (*a*)	41. (*c*)	42. (*d*)
43. (*c*)	44. (*c*)	45. (*b*)	46. (*b*)	47. (*d*)	48. (*a*)
49. (*b*)	50. (*c*)	51. (*c*)	52. (*d*)	53. (*c*)	54. (*e*)
55. (*b*)	56. (*c*)	57. (*d*)	58. (*d*)	59. (*b*)	60. (*d*)
61. (*b*)	62. (*a*)	63. (*c*)	64. (*c*)	65. (*c*).	

Chapter 4 INTERNAL COMBUSTION ENGINES AND NUCLEAR POWER PLANTS

I. THEORY

4.1. INTRODUCTION

The engine, in which the combustion of fuel takes place inside the cylinder of the engine, is known as *internal combustion (i.e.,* I.C.) *engine.* If the combustion of fuel takes place outside the engine *(i.e.,* in boiler), then that engine is called *steam engine.* Internal combustion engines (as compared to steam engines) are more compact, occupy less space, more efficient and portable. Hence they are used in motor cycles, motor cars, aircraft and ships.

The power produced inside the cylinder of an I.C. engine is known as indicated power (I.P.). This power is measured by taking indicator diagram by an instrument known as indicator.

In case of I.C. engines, a very high temperature can be achieved and hence the efficiency of I.C. engines is more than that of steam engines. The efficiency of I.C. engines is of the order of 35% and that of steam engine is about 15%.

4.2. CLASSIFICATION OF I.C. ENGINES

The I.C. engines are **classified** according to:

1. *Type of fuel used, i.e.,* Petrol engine, diesel engine or gas engine.
2. *Cycle of operation, i.e.,* Otto cycle engine, diesel cycle engine or dual combustion cycle engine.
3. *Method of ignition, i.e.,* spark ignition engine (S.I.) and compression ignition engine (C.I.).
4. *Number of strokes, i.e.,* two-stroke cycle engine or four-stroke cycle engine.
5. *Method of fuel injection, i.e.,* air injection or airless or solid injection engines.
6. *Method of cooling, i.e.,* air cooled engine or water cooled engine.

4.3. SOME DEFINITIONS

The following terms are mostly used in an I.C. Engine :

1. Top or inner dead centre (T.D.C.)
2. Bottom or outer dead centre (B.D.C.)
3. Clearance volume (C.V. or V_c)
4. Swept volume (S.V. or V_s)
5. Stroke (L)
6. Compression ratio (r)
7. Four-stroke or two-stroke engine
8. Piston speed (N).

1. Top or Inner Dead Centre

The top most position of the piston towards cover end side of the cylinder is known as top dead centre (*i.e.*, T.D.C.) or inner dead centre [*i.e.*, I.D.C., as shown in Fig. 4.1 (*a*)].

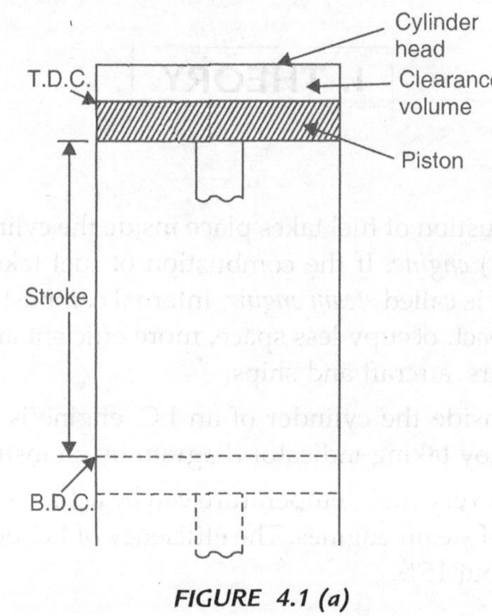

FIGURE 4.1 (a)

2. Bottom or Outer Dead Centre

The lowest position of the piston towards the crank end side of the cylinder is known as bottom dead centre (*i.e.*, B.D.C.) or outer dead centre (O.D.C.).

3. Clearance Volume

When the piston is at top dead centre there is some volume between the piston and cylinder head. This volume is known as clearance volume which is represented by V_c.

4. Swept Volume

The volume corresponding to the piston displacement from T.D.C. (or B.D.C.) is known as swept volume. This is represented by V_s. Mathematically, swept volume is given by,

$$V_s = \frac{\pi}{4} D^2 \times L$$

where D = Inside diameter of cylinder or Bore of the cylinder, and

L = Stroke length.

5. Stroke

It is the distance travelled by the piston from one of its dead centre position to the other dead centre position. This is shown in Fig. 4.1 (*a*) and is represented by L.

6. Compression Ratio (*r*)

Compression ratio is the ratio of the total volume of the cylinder to clearance volume. It is represented by r.

If V_c = Clearance volume,

V_s = Swept volume,

r = Compression ratio.

Then total volume of the cylinder = Swept volume + Clearance volume

$$= V_s + V_c$$

$$\therefore \quad \text{Compression ratio,} \quad r = \frac{V_s + V_c}{V_c} = \frac{V_s}{V_c} + 1$$

Generally, the clearance volume is expressed as the percentage of the swept volume. Hence if clearance volume is 10% of the swept volume, then compression ratio will be given as

$$r = \frac{V_s}{\left(\frac{10}{100} \text{ of } V_s\right)} + 1 = \frac{100}{10} + 1 = 11.$$

7. Four Stroke and Two-Stroke Engines

A four-stroke engine is one which requires four-strokes of the piston to complete the cycle or the engine which requires two revolutions of the crank shaft to complete the cycle is known as four-stroke engine.

A two-stroke engine is one which requires two-stroke of the piston or one revolution of the crank shaft to complete the cycle.

8. Piston Speed (*N*)

Piston speed is the distance travelled by the piston in unit time. Generally, the time is taken as one minute. Hence distance travelled by piston in one minute is known as piston speed.

If N = r.p.m. of the engine

L = Stroke length in metre

Then piston speed = (Distance moved by piston in one revolution) × No. of revolutions per minute = $2L \times$ Nm/sec

The distance moved by piston in one revolution of the crank is $2L$.

4.4. OTTO CYCLE

Figure 4.1 (*b*) shows the Otto cycle on *p-V* and *T-S* diagrams. It consists of two reversible adiabatic and two constant volume processes. Heat is supplied during the process 2-3 at constant volume and is rejected during the process 4-1 at constant volume. During processes 3-4 and 1-2, no heat is supplied. Let the clearance is unity and ratio of compression and expansion are equal and is *r*.

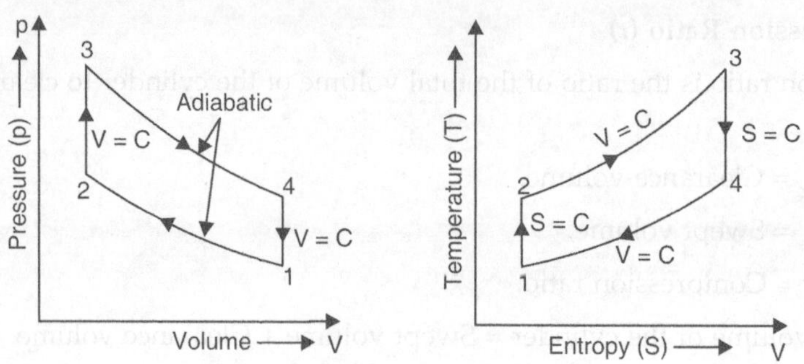

FIGURE 4.1 (b)

$$\therefore \qquad r = \frac{V_1}{V_2} = \frac{V_4}{V_3}$$

For one kg of air,

Heat supplied during process 2-3 $= C_v (T_3 - T_2)$

Heat rejected during process 4-1 $= C_v (T_4 - T_1)$

$$\therefore \text{ Air standard efficiency} = \frac{\text{Heat supplied} - \text{Heat rejected}}{\text{Heat supplied}}$$

$$= \frac{C_v(T_3 - T_2) - C_v(T_4 - T_1)}{C_v(T_3 - T_2)} = 1 - \frac{(T_4 - T_1)}{(T_3 - T_2)} \qquad \text{...}(A)$$

Also, we have $\quad \dfrac{T_2}{T_1} = \left(\dfrac{V_1}{V_2}\right)^{\gamma-1} = r^{\gamma-1}$

and $\quad \dfrac{T_3}{T_4} = \left(\dfrac{V_4}{V_3}\right)^{\gamma-1} = r^{\gamma-1}$

$$\therefore \qquad T_1 = \frac{T_2}{r^{\gamma-1}} \text{ and } T_4 = \frac{T_3}{r^{\gamma-1}}$$

Substituting the values of T_1 and T_4 in equation (*A*), we get

$$\text{Air standard efficiency} = 1 - \frac{\left(\dfrac{T_3}{r^{\gamma-1}} - \dfrac{T_2}{r^{\gamma-1}}\right)}{(T_3 - T_2)} = 1 - \frac{1}{r^{\gamma-1}}$$

Petrol and gas engine run on Otto cycle.

4.5. DIESEL CYCLE

Figure 4.1 (*c*) shows the Diesel cycle on *p-V* and *T-S* diagrams. It consists of a constant pressure, a constant volume and two adiabatic processes. Heat is supplied during constant pressure process 2-3 and is rejected during the constant volume process 4-1. During adiabatic process 3-4 and 1-2, no heat is supplied.

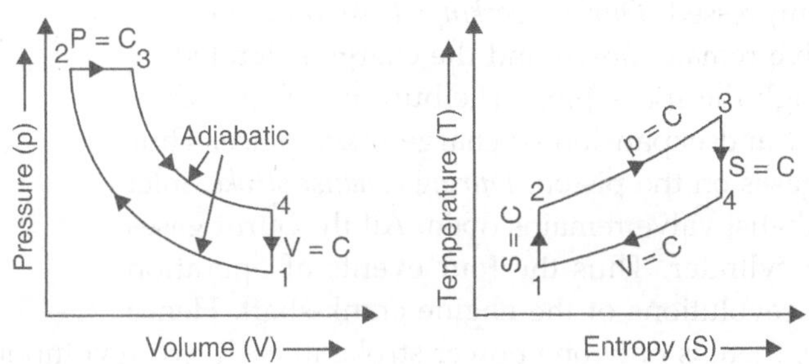

FIGURE 4.1 (c)

For one kg mass of air

Air standard efficiency $= \dfrac{\text{Heat supplied} - \text{Heat rejected}}{\text{Heat supplied}}$

$$= \frac{C_p(T_3 - T_2) - C_v(T_4 - T_1)}{C_p(T_3 - T_2)} = 1 - \frac{C_v(T_4 - T_1)}{C_p(T_3 - T_2)}$$

$$= 1 - \frac{1}{\gamma}\frac{(T_4 - T_1)}{(T_3 - T_2)} \qquad \left(\because \frac{C_p}{C_v} = \gamma\right)$$

Let $r = \text{Compression ratio} = \dfrac{V_1}{V_2}$

$r_c = \text{Heat addition cut-off ratio} = \dfrac{V_3}{V_2}$

Then $\dfrac{T_2}{T_1} = r^{\gamma-1}$ and $\dfrac{T_3}{T_2} = \dfrac{V_3}{V_2} = r_c$

Now, $T_3 = T_2 r_c = T_1 r^{\gamma-1} r_c$

Similarly, $T_4 = T_1 r_c^{\gamma}$. Hence air standard efficiency becomes as

$$= 1 - \frac{1}{\gamma r^{\gamma-1}}\left[\frac{r_c^{\gamma} - 1}{(r_c - 1)}\right].$$

4.6. TWO-STROKE OR FOUR-STROKE CYCLE ENGINE

If the basic four events of operation (*i.e.*, suction stroke, compression stroke, working or expansion stroke and exhaust stroke) are completed in two-strokes, the engine is known as two-stroke cycle engine. And if the above four events of operations are completed in four-strokes, the engine is known as four-stroke cycle engine.

1. Four-stroke Petrol Engine (S.I. Engine)

For a four-stroke petrol engine, when the piston moves outward, suction stroke starts. *During suction stroke,* the mixture of air and petrol in a required proportions enters the cylinder through inlet valve. The exhaust valve remains closed. *During compression stroke* the inlet valve and exhaust valve remain closed. The mixture of air and petrol is compressed. *During working stroke* both the inlet valve and exhaust valve remain closed and the charge is ignited by an electric spark through the spark-plug. The burning of fuel starts and energy is released and expansion of charge takes place. Thus work is done by the gases on the piston. *During exhaust stroke,* inlet valve is closed and exhaust valve remains open. All the burnt gases are driven out of the cylinder. Thus the four events of operation require two complete revolutions of the engine crank-shaft. Hence for a four-stroke cycle, there is only one power stroke in each two revolutions of the crank-shaft. The cycle on which the above engine works is known as **Otto cycle**.

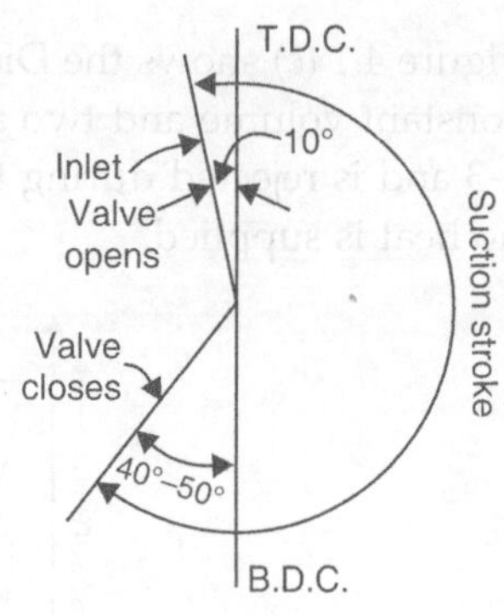

FIGURE 4.2

2. Valve Timing Diagrams of Four-Stroke Petrol Engine (or S.I. Engine)

A valve (inlet or exhaust) should open at the start of a stroke and should close at the end of a stroke. But if a valve opens before the beginning of a stroke and closes after the end of the stroke, greater efficiency will be obtained. Hence *the inlet valve* during suction stroke, opens 10° before the top dead centre (T.D.C.) or (O.D.C.) and closes about 40 to 50° after the bottom dead centre (B.D.C.) as shown in Fig. 4.2. As the inlet valve remains open for more than 180°, maximum charge is admitted into the cylinder.

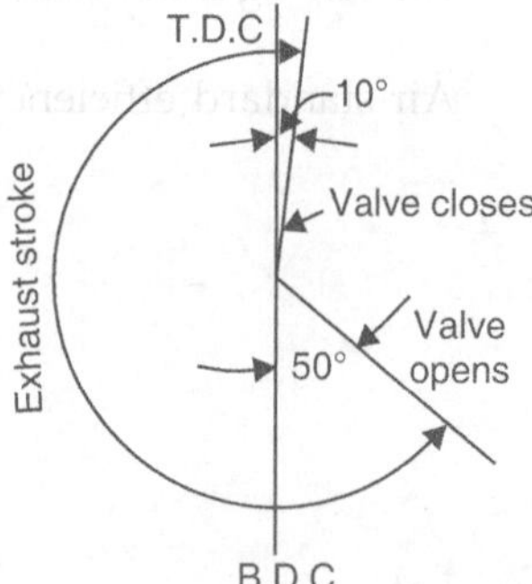

FIGURE 4.3

The exhaust valve during exhaust stroke opens about 50° before the B.D.C. and closest about 10° after the T.D.C. as shown in Fig. 4.3. Thus the period of exhaust is about 240° and hence the complete and effective scavenging is ensured.

The spark is ignited, during compression stroke about 10° before T.D.C. as whole fuel will not burn instantaneously. Also the fuel has some delay period and spark requires some time to travel and spread throughout the cylinder.

3. Two-stroke Petrol Engine (S.I. Engine)

In case of two-stroke engine, there are three parts, *i.e.,* inlet, exhaust and transfer ports instead of inlet and exhaust valves. Also in two-stroke engines, the suction and exhaust strokes are eliminated.

The mixture of air and petrol vapour from carburettor enters the casing through inlet. This charge is get compressed during power stroke in the casing. Near the end of power stroke, the exhaust port which is slightly higher than the transfer port, opens and burnt gases are released to the atmosphere through exhaust port. Immediately afterwards, the transfer port is

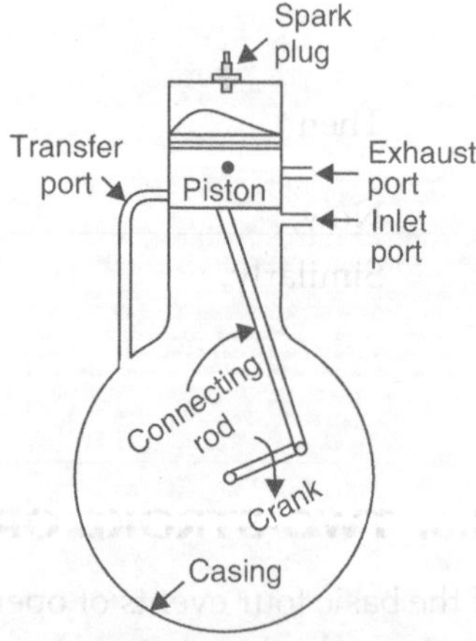

FIGURE 4.4. Two-stroke petrol engine.

uncovered by the piston and the fresh compressed charge from the casing enters the cylinder. This helps in driving out the burnt gases. In case of two-stroke engine, there is only one power stroke during one revolution of the crank.

4. Comparison of Two-stroke and Four-stroke Cycle Engines

(*i*) For the same speed, power developed by a two-stroke engine is two times the power developed by a four-stroke engine.

(*ii*) For the same power, a two-stroke engine is lighter and occupies less floor area as compared to four-stroke engine.

(*iii*) In case of a two-stroke engine, a lighter flywheel and a lighter foundation of the engine is required.

(*iv*) In case of two-stroke engine, there are no valves and hence mechanism is very simple.

(*v*) In two-stroke engines, scavenging (the process of removing exhaust gases from the cylinder) is very poor and hence the fresh charge is polluted.

(*vi*) In case of two-stroke engines, inlet and exhaust ports open simultaneously and some fresh charge is lost. The compression ratio is also lower. Therefore the thermal efficiency is lower.

(*vii*) In case of two-stroke engines, there is more wear and tear and there is more consumption of lubricating oil.

4.7. CARBURETTOR AND CARBURETION

The carburettor is a device used for atomising and vaporising the fuel and mixing it with air in the correct proportions to suit the charging operation conditions in the engine. *Atomization* means only reducing the fuel to fine particles whereas *vaporisation* means the change of phase of fuel from liquid phase to vapour phase. *Carburetion* is the process of breaking up and mixing the petrol with air.

1. Air-Fuel mixtures. The correct mixture of air and fuel (petrol) theoretically is about 15 : 1. But in actual practice, the air fuel ratio for average speeds varies from 15 : 1 to 17:1. For maximum power, a richer mixture of the ratio of approximately 12 : 1 is used.

2. Electric spark. For igniting the air petrol mixture in case of spark ignition engine (S.I.), spark plug is used. A high voltage of the order of 20,000 volts is produced by an ignition coil and this high voltage is directed to the spark plug by the distributor. The spark jumps across the gap of the spark plug and fuel is ignited. The spark gas is approximately 1 mm.

3. Four-stroke diesel engine (compression ignition, *i.e.*, C.I. engine). The four-stroke diesel engine is similar to four-strokes petrol engine except that there is no spark plug and no carburettor in case of diesel engine. But the diesel engine has three values, *i.e.*, inlet valve, exhaust valve and fuel injection valve. During *suction stroke*, only air is drawn in the cylinder through inlet valve. The exhaust and fuel injection valves remain closed. During *compression stroke*, all the valves are closed. The air gets compressed to a pressure about $3.5 \times 10^6 \ \frac{N}{m^2}$ and attains a very high temperature, which is sufficient to ignite the fuel. The fuel (diesel) is injected into the cylinder just before the beginning of the *working stroke*. The fuel gets ignited by the high temperature of the air, attained during compression stroke. Burning of the fuel starts and energy is released and expansion of the fuel takes place. The exhaust valve and inlet valves remain closed. During *exhaust stroke,* the inlet and fuel injection valves remain closed and exhaust valve remain open. All the burnt gases are driven out of the cylinder. Then the cycle again repeats.

4.8. ENGINE POWER

The engine has got mainly the two powers :

1. Indicated power (I.P.) 2. Brake power (B.P.)

1. Indicated power (I.P.) is the power produced inside the cylinder. Mathematically, it is given by the relation,

$$\text{I.P.} = \frac{p_m LAN}{60{,}000} \text{ kW}$$

where p_m = Mean effective pressure in N/m^2

L = Stroke length in m

A = Area in m^2

N = Number of power strokes/minute.

For a two-stroke engine number of power strokes/minute is equal to the r.p.m. of the engine while for a four-stroke engine, number of power strokes per minute is equal to half the r.p.m. of the engine.

2. Brake power (B.P.) is the power available at the shaft. Some of the power produced inside the cylinder is used to overcome the frictional resistance between the cylinder surface and piston rings and to drive fuel pumps, governor, lubricating oil and water circulating pump, etc. This used power is known as frictional power (F.P.). Then brake power is given by

$$\text{B.P.} = \text{I.P.} - \text{F.P.}$$

4.9. ENGINE EFFICIENCIES

The various engine efficiencies are :

(i) Mechanical efficiency (η_m), *(ii)* Indicated thermal efficiency (η_i), *(iii)* Brake thermal efficiency (η_b), *(iv)* Relative efficiency (η_r), and *(v)* Volumetric efficiency (η_v).

(i) **Mechanical efficiency** (η_m) is the ratio of brake power to indicated power or mathematically it is given as

$$\eta_m = \frac{\text{B.P.}}{\text{I.P.}}$$

(ii) **Indicated thermal efficiency** (η_i) is ratio of indicated power to the heat energy of the fuel supplied during the same interval of time, Mathematically, it is given as

$$\eta_i = \frac{(\text{I.P.}) \times 60{,}000}{427 \times W \times C}$$

where W = Weight of fuel supplied/min., and

C = Calorific valve of fuel in kcal/kg.

(*iii*) **Brake thermal efficiency** (η_b) is the ratio of brake power to the heat energy of the fuel supplied during the same interval of time. Mathematically, it is given as

$$\eta_b = \frac{(\text{B.P.}) \times 60{,}000}{427 \times W \times C}$$

where W = Weight of fuel burnt/min., and

C = Calorific valve of fuel.

Brake thermal efficiency for an I.C. engine is also equal to overall efficiency (η_o).

(*iv*) **Relative efficiency** (η_r) is the ratio of indicated thermal efficiency to the corresponding air standard cycle efficiency. For petrol engine, the relative efficiency (η_r) is given as

$$\eta_r = \frac{\eta_i}{1 - \left(\frac{1}{r}\right)^{\gamma}}.$$

(*v*) **Volumetric efficiency** (η_v) is the ratio of actual volume admitted to the engine cylinder during suction stroke referred to normal pressure and temperature to the swept volume of the piston. This is also defined as the ratio of mass of air admitted to the cylinder during suction stroke to the mass of free air equivalent to the piston displacement at intake temperature and pressure condition.

1. Measurement of indicated power. The power developed in the cylinder is called indicated power. It is obtained by taking indicator diagram from an instrument called indicator.

Indicated power for multi-cylinder engines is determined from *Morse test.*

2. Specific fuel consumption is the mass of fuel consumed per B.P. per hour.

4.10. GOVERNING OF I.C. ENGINES

The I.C. engines are governed by the following methods :

1. Hit and Miss governing, 2. Quantitative governing, and 3. Qualitative governing.

1. Hit and Miss governing. This method is commonly used in small gas engine. In this type of governing, the inlet valve remain closed during a cycle so that no fuel is admitted.

2. Quantitative governing. In this type of governing, the quantity of the mixture (fuel and air) entering the cylinder is regulated.

3. Qualitative governing. In this type of governing, the quality of the mixture (fuel and air) is altered by the action of the governor. The air supply remains constant, but the amount of fuel entering the cylinder is regulated.

4.11. IMPORTANT TERMS USED IN I.C. ENGINES

1. Supercharging is the process of supplying the intake air to the engine cylinder at a pressure greater than the pressure of the surrounding atmosphere. This is done by blower or compressor. With the increase of pressure, the density and temperature also increases. Supercharging is necessary to increase the power output or to maintain power at altitudes. Due to higher intake temperature, supercharging increases the possibility of detonation in petrol engines and to decreases the possibility of knocking the diesel engines.

2. **Scavenging** is the process of removing the burnt gases in I.C. engines from the combustion chamber of the engine cylinder.

3. **Auto-ignition** is the phenomenon by which a fuel catches fire without external flame. Iso-octane helps to resist auto-ignition and normal heptane accelerates auto-ignition.

4. **Pre-ignition** is the ignition of the charge in spark ignition engine before the spark occurs in the spark plug. Or, pre-ignition is the spontaneous combustion of the mixture before the end of the compression stroke. This is due to the cylinder walls being too hot, spark plug electrodes are over heated or very hot carbon deposits.

5. **Ignition Lag** is defined as the time taken by the fuel after injection to reach upto auto-ignition temperature.

6. **Injection Lag** is defined as the time before actual fuel injection and pump plunger starts to pump fuel.

7. **Detonation.** The rapid auto-ignition of a portion of a fuel causes a pressure wave of high intensity to be set up in the cylinder of an I.C. engine. This pressure wave of high intensity propagates rapidly through the gas gives a violet blow to the walls of the cylinder, to the combustion chamber and to the piston. A loud pulsating noise known as knocking is produced which gives voilent vibration to the engine. The term detonation is used to indicate presence of gas vibration and the term knock is used to include all phenomenon that arise from auto-ignition. The knock in S.I. engines is characterised by sudden auto-ignition of the charge farthest away from the spark plug while in case of C.I. engines it is characterised by the sudden auto-ignition of the mixture at the very beginning of the combustion process.

8. **Knocking in S.I. Engines.** The knocking in the S.I. engine can be prevented if the end mixture (the charge away from the spark plug) is having :

(*a*) Low density, (*b*) Low temperature, (*c*) Long ignition delay, and (*d*) Lean or rich mixture.

(*a*) **Density.** The possibility of knocking in S.I. engines will increase if the density of the unburnt mixture is increased by any one of the following methods :

(*i*) Super-charging the engine,

(*ii*) Opening throttle,

(*iii*) Increasing the compression ratio,

(*iv*) Advancing the spark timing.

(*b*) **Temperature.** The possibility of knocking in S.I. engines will increase if the temperature of the unburnt mixture is increased by any of the following methods :

(*i*) Supercharging,

(*ii*) Increasing the inlet air temperature,

(*iii*) Increasing the coolant temperature, and

(*iv*) Raising the compression ratio,

(*v*) Increasing the temperature of the cylinder and combustion chamber walls.

(*c*) **Ignition delay.** The possibility of knocking in S.I. engines will increase if the time of the exposure of the unburnt mixture of auto-ignition conditions by any of the following factors is increased :

(*i*) decreasing the speed of the engine,

(*ii*) decreasing the turbulance of the mixture,

(*iii*) increasing the distance of the flame travel in combustion chamber.

(*d*) **Composition.** The possibility of knocking is increased by :

(*i*) Low-self ignition temperature,

(*ii*) Chemically correct mixture, and

(*iii*) Short ignition delay.

A rich mixture is sometimes used to suppress knocking.

9. Knocking in C.I. Engines. The knocking, in the C.I. engines, can be prevented if the first element of the fuel and air is having

(*a*) high density, (*b*) high temperature

(*c*) a short ignition delay, and (*d*) reactive mixture.

(*a*) **Density.** The possibility of knocking in C.I. engines will increase if the density of the initially formed mixture is decreased by the following methods :

(*i*) Decreasing the inlet air pressure, and

(*ii*) Decreasing the compression ratio.

(*b*) **Temperature.** The possibility of knocking in C.I. engines will increase if the temperature of the initially formed mixture is decreased by the following methods:

(*i*) Decreasing the air inlet temperature,

(*ii*) Decreasing the compression ratio,

(*iii*) Decreasing the coolant temperature,

(*iv*) Decreasing the cylinder and combustion chamber walls temperature.

(*c*) **Ignition delay.** The possibility of knocking in C.I. engine will increase if the time for forming a homogeneous mixture by any of the following methods, is increased :

(*i*) Decreasing the injection pressure,

(*ii*) Increasing the speed of the engine, and

(*iii*) Increasing the rate of injection.

(*d*) **Composition.**The possibility of knocking in C.I. engine will increase if the composition of the initially formed mixture is having :

(*i*) High viscosity,

(*ii*) Low volatility.

(*iii*) High self-ignition temperature, and

(*iv*) Low ignition delay.

10. Octane Number. The percentage by volume of iso-octane, in a mixture of iso-octance and normal-heptane (*n*-heptance) which shows the same tendency to knock the given fuel when tested in a specified test engine under specified conditions, is known as octane number of the fuel. If a fuel has octane number as 85, it means that fuel is having same knocking tendency as a mixture of 85% iso-octance and 15% *n*-heptane by volume.

11. Cetane number. The percentage by volume of cetane in a mixture of cetane and α-methyl-naphthalene which has the same ignition delay as the given fuel when tested in a specified test engine under specified conditions, is known as cetane number. If a fuel has cetane number 45, it means that the fuel is having the same ignition delay as a mixture of 45% cetane and 55% α-methyl-naphthalene by volume.

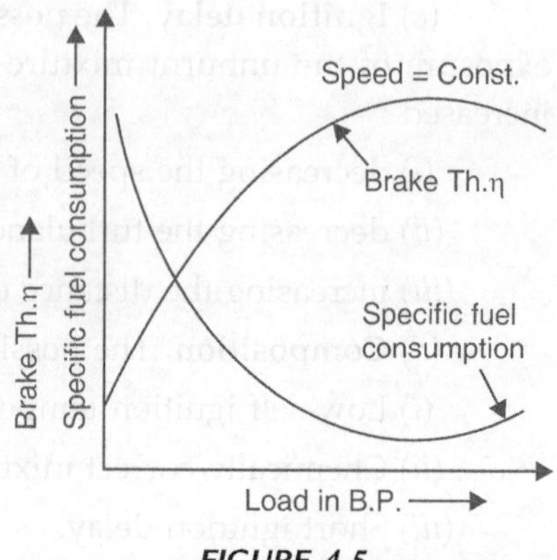

FIGURE 4.5

4.12. PERFORMANCE TEST ON I.C. ENGINE

The performance test on an I.C. engine is carried out at constant speed and at variable speed. The variation of specific fuel consumption and brake thermal efficiency at constant speed with the variation of load in B.P. on engine is shown in Fig. 4.5. It is seen that when brake thermal efficiency is maximum, specific fuel consumption (*i.e.*, fuel consumption per B.P. per hour) is minimum.

II. OBJECTIVE TYPE QUESTIONS

Tick mark the most appropriate statement of the multiple choice answers:

1. In an I.C. engine fuel is
   (*a*) burnt outside the cylinder ☐ (*b*) inside the cylinder ☐
   (*c*) not burnt anywhere ☐ (*d*) none of the above. ☐
2. The power produced inside the cylinder of an I.C. engine is known as
   (*a*) break power ☐ (*b*) indicated power ☐
   (*c*) frictional power ☐ (*d*) none of the above. ☐
3. The net indicated power (I.P.) of an I.C. engine given by indicator diagram is equal to
   (*a*) total power produced + pumping power ☐
   (*b*) total power produced – pumping power ☐
   (*c*) $\frac{\text{total power produced}}{\text{pumping power}}$ ☐
   (*d*) none of the above. ☐
4. Pumping power is equal to
   (*a*) total power produced ☐ (*b*) power spent in suction and exhaust stroke ☐
   (*c*) power spent in a complete cycle ☐ (*d*) none of the above. ☐
5. The power available at the shaft of an I.C. engine is known as break power and is equal to
   (*a*) total power produced—frictional power ☐ (*b*) net I.P. – frictional power ☐
   (*c*) net I.P. + frictional power ☐ (*d*) $\frac{\text{net I.P.}}{\text{frictional power}}$ ☐

6. The four operation *i.e.*, suction, compression, expansion and exhaust in a four-stroke cycle engine, are completed in the number of revolutions of the crankshaft equal to

(*a*) one ☐ (*b*) two ☐
(*c*) three ☐ (*d*) four. ☐

7. In a two-stroke engine, the working cycle is completed in the number of revolutions of the crank-shaft equal to

(*a*) one ☐ (*b*) two ☐
(*c*) three ☐ (*d*) four. ☐

8. Tick the correct statement

(*a*) The four-stroke engine has twice as many power strokes as a two-stroke cycle engine, at the same engine speed. ☐
(*b*) For the same power, a four-stroke cycle engine is lighter and occupies less floor area. ☐
(*c*) A four-stroke engine requires a lighter flywheel as compared to two-stroke engine. ☐
(*d*) A four-stroke cycle engine, theoretically develops half power as that of two-stroke cycle engine. ☐

9. Choose the wrong statement

(*a*) The two-stroke cycle engines have lighter flywheel. ☐
(*b*) The thermal efficiency of a two-stroke cycle engine is less than that of a four-stroke engine. ☐
(*c*) Diesel engine is an internal combustion engine. ☐
(*d*) Compression ratio of I.C. engine is the volume displaced by the piston per stroke and clearance volume in cylinder. ☐

10. In petrol engine, suction consists of

(*a*) air only ☐ (*b*) a mixture of air and fuel ☐
(*c*) fuel only ☐ (*d*) none of the above. ☐

11. In diesel engine, the suction consists of

(*a*) air only ☐ (*b*) a mixture of air and fuel ☐
(*c*) fuel only ☐ (*d*) none of the above. ☐

12. In petrol engine, ignition takes place

(*a*) automatically due to high temperature of compressed air ☐
(*b*) automatically due to high temperature of compressed fuel ☐
(*c*) by means of a spark ☐
(*d*) none of the above. ☐

13. In diesel engine, the ignition takes place

(*a*) automatically due to high temperature of compressed air ☐
(*b*) automatically due to high temperature of compressed fuel ☐
(*c*) by means of a spark ☐
(*d*) none of the above. ☐

14. Which one is a spark ignition engine?

(*a*) petrol engines ☐ (*b*) diesel engines ☐
(*c*) steam engines ☐ (*d*) none of the above. ☐

15. Which one is a compression engine?
(*a*) petrol engines ☐ (*b*) diesel engines ☐
(*c*) steam engines ☐ (*d*) none of the above. ☐

16. For the same power, a two-stroke cycle engine occupies the floor area which as compared to four-stroke cycle is
(*a*) more ☐ (*b*) less ☐
(*c*) equal ☐ (*d*) none of the above. ☐

17. The flywheel used in two-stroke cycle engine as compared to four-stroke cycle engine is
(*a*) heavy in weight ☐ (*b*) same in weight ☐
(*c*) light in weight ☐ (*d*) none of the above. ☐

18. The thermal efficiency of a two-stroke cycle engine as compared to four-stroke cycle engine is
(*a*) more ☐ (*b*) equal ☐
(*c*) less ☐ (*d*) none of the above. ☐

19. Two-stroke cycle engines use the fuel
(*a*) petrol and air only ☐ (*b*) diesel and air only ☐
(*c*) steam only ☐ (*d*) both (*a*) and (*b*). ☐

20. The process of removing the burnt gases in I.C. engines from the combustion chamber of the engine cylinder is known as
(*a*) superchanging ☐ (*b*) scavenging ☐
(*c*) polymerisation ☐ (*d*) detonation. ☐

21. The process of suppling the intake air to the engine cylinder at a density greater than the density of the surrounding atmosphere is known as
(*a*) supercharging ☐ (*b*) scavenging ☐
(*c*) polymerisation ☐ (*d*) detonation. ☐

22. The process in which the molecules of a compound becomes larger is known as
(*a*) supercharging ☐ (*b*) scavenging ☐
(*c*) polymerisation ☐ (*d*) detonation. ☐

23. The process, in which the violent sound pulsations within the cylinder of an I.C. engine are produced, is known as
(*a*) supercharging ☐ (*b*) scavenging ☐
(*c*) polymerisation ☐ (*d*) detonation. ☐

24. The ratio of the shaft power to the indicated power is called
(*a*) volumetric efficiency ☐ (*b*) relative efficiency ☐
(*c*) mechanical efficiency ☐ (*d*) indicated thermal efficiency. ☐

25. The ratio of shaft power to the energy supplied by the fuel is called
(*a*) volumetric efficiency ☐ (*b*) relative efficiency ☐
(*c*) mechanical efficiency ☐ (*d*) indicated thermal efficiency. ☐

26. Indicated power of a four-stroke engine is given by

(*a*) $\frac{p_m LAN}{1000 \times 60}$ ☐ (*b*) $\frac{p_m LAN}{2 \times 1000 \times 60}$ ☐

(*c*) $\frac{2p_m LAN}{1000 \times 60}$ ☐ (*d*) $\frac{4p_m LAN}{1000 \times 60}$. ☐

where p_m = Mean effective pressure, L = Stroke, A = Area, and N = r.p.m. of engine.

27. Morse test is used to determine

(*a*) indicated power for multicylinder engines ☐

(*b*) shaft power ☐

(*c*) mean effective pressure ☐

(*d*) temperature of the exhaust gases. ☐

28. Choose the wrong statement about a four-stroke cycle diesel engine

(*a*) The inlet valve opens at 20° before the top dead centre and closes at 40° after bottom dead centre ☐

(*b*) The compression start at 40° after bottom dead centre and ends at 10° before the top dead centre ☐

(*c*) The exhaust valve opens at 30° before bottom dead centre and closes at 10° after top dead centre ☐

(*d*) The fuel valve opens at top dead centre and closes at bottom dead centre. ☐

29. Choose the correct statement about a four-stroke diesel engine

(*a*) A diesel engine has only two valves ☐

(*b*) A diesel engine is known as compression ignition engine ☐

(*c*) A diesel engine is known as spark ignition engine ☐

(*d*) The fuel injection starts at top dead centre and ends at 20° after top dead centre. ☐

30. In a four-stroke cycle petrol engine, the thing which opens at 50° before bottom dead centre and closes at 15° after top dead centre is known as

(*a*) expansion ☐ (*b*) compression ☐

(*c*) exhaust valve ☐ (*d*) inlet valve. ☐

31. In a four-stroke cycle petrol engine, the thing which starts at 40° after bottom dead centre and ends at 30° before top dead centre is known as

(*a*) expansion ☐ (*b*) compression ☐

(*c*) exhaust valve ☐ (*d*) inlet valve. ☐

32. In a four-strokes cycle petrol engine, the thing which opens at 20° before top dead centre and closes at 40° after bottom dead centre is known as

(*a*) expansion ☐ (*b*) compression ☐

(*c*) exhaust valve ☐ (*d*) inlet valve. ☐

33. In a four-stroke cycle petrol engine, the thing which starts at 30° before top dead centre and ends at 50° before bottom dead centre is known as

(*a*) expansion ☐ (*b*) compression ☐

(*c*) exhaust valve ☐ (*d*) inlet valve. ☐

Engine Efficiencies

34. The mechanical efficiency (η_m) of an I.C. engine is equal to

(a) $\frac{\text{I.P.}}{\text{B.P.}}$ ☐ (b) $\frac{\text{B.P.}}{\text{I.P.}}$ ☐

(c) $\frac{\text{B.P.}}{\text{F.P.}}$ ☐ (d) $\frac{\text{F.P.}}{\text{B.P.}}$. ☐

35. The relation between frictional power (F.P.), brake power (B.P.) and indicated power (I.P.) is given by

(a) F.P. = I.P. – B.P. ☐ (b) F.P. = $\frac{\text{B.P.}}{\text{I.P.}}$ ☐

(c) F.P. = 1 – (I.P. × B.P.) ☐ (d) F.P. = (I.P. × B.P.) – 1. ☐

36. The ratio of the B.P. to the heat energy of the fuel supplied during the same interval of time is called

(a) brake thermal efficiency ☐ (b) indicated thermal efficiency ☐

(c) volumetric efficiency ☐ (d) relative efficiency. ☐

37. The ratio of the I.P. to the heat energy of the fuel supplied during the same interval of time is called

(a) brake thermal efficiency ☐ (b) indicated thermal efficiency ☐

(c) volumetric efficiency ☐ (d) relative efficiency. ☐

38. For a four-stroke engine, the ratio of the volume of the charge admitted to the engine cylinder during suction stroke at N.T.P. to the swept volume of the piston is called

(a) brake thermal efficiency ☐ (b) indicated thermal efficiency ☐

(c) volumetric efficiency ☐ (d) relative efficiency. ☐

39. The ratio of the indicated thermal efficiency to the corresponding ideal air standard efficiency is called

(a) brake thermal efficiency ☐ (b) indicated thermal efficiency ☐

(c) volumetric efficiency ☐ (d) relative efficiency. ☐

40. The ratio of the work available at the shaft to the energy supplied during the same interval of time is known as

(a) brake thermal efficiency ☐ (b) indicated thermal efficiency ☐

(c) mechanical efficiency ☐ (d) overall efficiency. ☐

41. Choose the correct statement

(a) Specific fuel consumption is defined as fuel consumed per B.P. ☐

(b) Combustion in compression ignition engine is homogeneous ☐

(c) The brake power is always greater than indicated power ☐

(d) For the same compression ratio Otto cycle is more efficient than the Diesel cycle. ☐

42. Choose the wrong statement

(a) The ratio of the indicated thermal efficiency to the air standard efficiency is called relative efficiency ☐

(b) Most high speed compression engines operate on dual combustion cycle ☐

(c) Diesel fuel is more difficult to ignite as compared to petrol ☐

(d) The mean effective pressure, obtained from engine indicator, gives the maximum pressure developed in the cylinder. ☐

Specific Fuel Consumption

43. The specific fuel consumption per B.P. hour for a diesel engine as compared to that of petrol engine is

(*a*) same ☐ (*b*) higher ☐
(*c*) lower ☐ (*d*) none of the above. ☐

44. If the specific fuel consumption per B.P. hour is approximately 0.25 kg, the engine is

(*a*) diesel engine ☐ (*b*) petrol engine ☐
(*c*) steam engine ☐ (*d*) none of the above. ☐

45. If the specific fuel consumption per B.P. hour is approximately 0.2 kg, the engine is

(*a*) diesel engine ☐ (*b*) petrol engine ☐
(*c*) steam engine ☐ (*d*) none of the above. ☐

46. The efficiency of an I.C. engine will increase if the intake air temperature

(*a*) increases ☐ (*b*) decreases ☐
(*c*) remain same ☐ (*d*) none of the above. ☐

47. The indicator on an engine is used to determine

(*a*) temperature ☐ (*b*) B.P. ☐
(*c*) speed ☐ (*d*) I.P. and mean effective pressure. ☐

48. If the thermal efficiency of an I.C. engine is approximately 70%, then the engine would be

(*a*) diesel engine ☐ (*b*) petrol engine ☐
(*c*) gas engine ☐ (*d*) none of the above. ☐

49. If the thermal efficiency of an I.C. engine is approximately 30%, then the engine would be

(*a*) diesel engine ☐ (*b*) petrol engine ☐
(*c*) gas engine ☐ (*d*) none of the above. ☐

50. The expression $1 - \left(\frac{1}{r}\right)^{\gamma-1}$ denotes the following efficiencies of an I.C. engine

(*a*) mechanical efficiency ☐ (*b*) relative efficiency ☐
(*c*) air standard efficiency ☐ (*d*) volumetric efficiency. ☐

Petrol and Diesel Engines

51. The compression ratio for a diesel engine as compared to petrol engine is

(*a*) higher ☐ (*b*) lower ☐
(*c*) same ☐ (*d*) none of the above. ☐

52. The compression ratio for a petrol engine varies from

(*a*) 25 to 40 ☐ (*b*) 10 to 15 ☐
(*c*) 6 to 10 ☐ (*d*) 15 to 25. ☐

53. The compression ratio for a diesel engine varies from

(*a*) 25 to 40 ☐ (*b*) 10 to 15 ☐
(*c*) 6 to 10 ☐ (*d*) 15 to 25. ☐

54. If the compression ratio in petrol engine is higher than that of diesel engine, then

(*a*) ignition of fuel will be delayed ☐ (*b*) pre-ignition of fuel will take place ☐
(*c*) knocking will take place ☐ (*d*) detonation will take place. ☐

55. Air fuel ratio theoretically for a petrol engine is approximately

(*a*) 25 : 1 ☐ (*b*) 20 : 1 ☐
(*c*) 10 : 1 ☐ (*d*) 15 : 1. ☐

56. The pressure at the end of compression in petrol engine as compared to that of diesel engine, would be

(*a*) higher ☐ (*b*) lower ☐
(*c*) same ☐ (*d*) none of the above. ☐

57. The approximate pressure at the end of the compression in case of diesel engine is

(*a*) 1.5×10^6 N/m^2 ☐ (*b*) 3.5×10^6 N/m^2 ☐
(*c*) 2.0×10^6 N/m^2 ☐ (*d*) 2.5×10^6 N/m^2. ☐

58. The approximate pressure at the end of the compression in petrol engine is

(*a*) 1.5×10^6 N/m^2 ☐ (*b*) 1.0×10^6 N/m^2 ☐
(*c*) 2.0×10^6 N/m^2 ☐ (*d*) 3.5×10^6 N/m^2. ☐

59. The mechanism used for controlling air-fuel ratio in petrol engine is known as

(*a*) injector ☐ (*b*) governor ☐
(*c*) carburettor ☐ (*d*) none of them. ☐

60. Choose the correct statement

(*a*) The spark plug gap is normally kept from 0.8 mm to 1.0 mm. ☐
(*b*) Piston rings are generally made of aluminium. ☐
(*c*) A fuel will detonate less if it has higher self-ignition temperature. ☐
(*d*) The delay period in petrol engine is of the order of .04 sec. ☐

61. Choose the wrong statement

(*a*) The delay period is petrol engine is of the order of 0.002 sec. ☐
(*b*) The spark play gap is normally kept from 0.4 mm to 0.6 mm. ☐
(*c*) Piston rings are generally made of cast iron. ☐
(*d*) A fuel will detonate less if it has lower self-ignition temperature. ☐

62. The relative fuel-air ratio (F_R) is equal to

(*a*) actual fuel-air-ratio × chemically correct fuel-air ratio ☐
(*b*) (chemically correct fuel air ratio)/actual fuel-air ratio ☐
(*c*) (actual fuel-air ratio)/chemically correct fuel-air ratio ☐
(*d*) none of the above. ☐

63. For the same compression ratio, Diesel cycle as compared to Otto cycle is thermally

(*a*) more efficient ☐ (*b*) less efficient ☐
(*c*) having same efficiency ☐ (*d*) none of the above. ☐

64. For the same compression ratio, the mean effective pressure for the Diesel cycle as compared to Otto cycle is

(*a*) more ☐ (*b*) less ☐
(*c*) same ☐ (*d*) none of the above. ☐

Knocking and Detonation

65. In spark ignition engines, the possibility of knocking can be reduced by
    (*a*) advancing the spark timing ☐ (*b*) by increasing the coolant temperature ☐
    (*c*) reducing compression ratio ☐ (*d*) none of the above. ☐

66. Advancing the spark timing in spark ignition engine, the possibility of knock will
    (*a*) increase ☐ (*b*) decrease ☐
    (*c*) not take place ☐ (*d*) none of the above. ☐

67. In spark ignition engines, decreasing the coolant temperature the possibility of knock will
    (*a*) increase ☐ (*b*) decrease ☐
    (*c*) not take place ☐ (*d*) none of the above. ☐

68. In spark ignition engines, reducing the compression ratio, the possibility of knock will
    (*a*) increase ☐ (*b*) decrease ☐
    (*c*) not take place ☐ (*d*) none of the above. ☐

69. Increasing the speed of the engine, in a spark ignition engine, the possibility of knock will (when a fixed octane rating fuel is used)
    (*a*) increase ☐ (*b*) decrease ☐
    (*c*) not take place ☐ (*d*) none of the above. ☐

70. The fuels in order of decreasing knock tendency for spark ignition engines are
    (*a*) paraffins, aromatics, naphthenes ☐ (*b*) naphthenes, paraffins, aromatics ☐
    (*c*) paraffins, naphthenes, aromatics ☐ (*d*) none of the above. ☐

71. If a fuel has higher self-ignition temperature, the tendency for detonation would
    (*a*) be more ☐ (*b*) be less ☐
    (*c*) not take place ☐ (*d*) none of the above. ☐

72. For spark ignition engine fuels, the anti-knock agent is
    (*a*) naphthene ☐ (*b*) amyl nitrate ☐
    (*c*) tetra-ethyl lead ☐ (*d*) none of the above. ☐

73. For compression ignition engine fuels, the anti-knock agent is
    (*a*) naphthene ☐ (*b*) paraffins ☐
    (*c*) amyl nitrate ☐ (*d*) none of the above. ☐

74. By reducing compression ratio, the knocking tendency in compression ignition engine will
    (*a*) increase ☐ (*b*) decrease ☐
    (*c*) not take place ☐ (*d*) none of the above. ☐

75. By increasing the wall temperature of a combustion chamber of a compression ignition engines, the knocking tendency will
    (*a*) increase ☐ (*b*) decrease ☐
    (*c*) not take place ☐ (*d*) none of the above. ☐

76. In petrol engines, the detonating tendency increases when
    (*a*) engine speed is increased ☐ (*b*) engine speed is decreased ☐
    (*c*) compression ratio is increased ☐ (*d*) compression ratio is decreased. ☐

77. The ignition quality of petrol is measured by

(*a*) calorific value ☐ (*b*) specific fuel consumption ☐

(*c*) octane number ☐ (*d*) cetane number. ☐

78. The ignition quality of diesel is measured by

(*a*) calorific value ☐ (*b*) specific fuel consumption ☐

(*c*) octane number ☐ (*d*) cetane number. ☐

79. Choose the wrong statement

(*a*) The fuels in order of decreasing knock tendency for spark ignition engines are paraffins, naphthenes and aromatics. ☐

(*b*) Cetane number is the measure of ignition quality of diesel oil ☐

(*c*) The detonating tendency in petrol engine increases with decrease of compression ratio. ☐

(*d*) Anti-knock for compression ignition engine is amyl nitrate. ☐

80. Choose the correct statement

(*a*) The bottom ring on the piston of an I.C. engine is called groove ring. ☐

(*b*) Iso-octane is the fuel which has maximum resistance to detonation. ☐

(*c*) Iso-octane is the fuel which detonates easily. ☐

(*d*) Highest useful compression ratio is the compression ratio at which an engine operates smoothly. ☐

Octane and Cetane Numbers

81. If a mixture contains iso-octane 65% and normal heptane (*n*-Haptane) 35°, it means the mixture has

(*a*) cetane number = 35 ☐ (*b*) cetane number = 65 ☐

(*c*) octane number = 35 ☐ (*d*) octane number = 65. ☐

82. The petrol generally used in petrol engines has the octane number of the order of

(*a*) 100–120 ☐ (*b*) 80–90 ☐

(*c*) 60–70 ☐ (*d*) 40–50. ☐

83. The diesel generally used in diesel engines, has the cetane number of the order of

(*a*) 60–80 ☐ (*b*) 40–60 ☐

(*c*) 30–40 ☐ (*d*) 10–20. ☐

84. For a four-stroke diesel engine, the air combustion per rated B.P. per hour would be

(*a*) 16 kg ☐ (*b*) 22 kg ☐

(*c*) 2 kg ☐ (*d*) 5 to 7 kg. ☐

85. The octane number of cetane is

(*a*) 40 ☐ (*b*) 30 ☐

(*c*) 10 ☐ (*d*) none of the above. ☐

86. The octane number of iso-octane is

(*a*) 40 ☐ (*b*) 30 ☐

(*c*) 10 ☐ (*d*) 100. ☐

87. The fuel which, has maximum resistance to detonation, is

(*a*) *n*-Haptane ☐ (*b*) iso-octane ☐

(*c*) benzene ☐ (*d*) alcohol. ☐

88. The fuel, which detonates easily, is

(*a*) *n*-Haptane ☐ (*b*) iso-octane ☐

(*c*) benzene ☐ (*d*) alcohol. ☐

89. The efficiency is determined by Morse-test in multi-cylinder, is

(*a*) relative efficiency ☐ (*b*) volumetric efficiency ☐

(*c*) mechanical efficiency ☐ (*d*) thermal efficiency. ☐

90. The time before actual fuel injection and pump plunger starts to pump fuel is known as

(*a*) ignition lag ☐ (*b*) injection lag ☐

(*c*) pre-ignition ☐ (*d*) knocking. ☐

91. The time taken by fuel after injection to reach up to auto-ignition temperature is known as

(*a*) ignition lag ☐ (*b*) injection lag ☐

(*c*) pre-ignition ☐ (*d*) knocking. ☐

92. The ignition of the charge in spark ignition engine before the spark in the spark plug occurs is known as

(*a*) ignition lag ☐ (*b*) injection lag ☐

(*c*) pre-ignition ☐ (*d*) knocking. ☐

93. The phenomenon by which a fuel catches fire without external flame is known as

(*a*) ignition lag ☐ (*b*) injection lag ☐

(*c*) pre-ignition ☐ (*d*) auto-ignition. ☐

94. Morse test is performed on I.C. engines to determine mechanical efficiency of

(*a*) single cylinder spark ignition engine ☐ (*b*) single cylinder compression ignition engine ☐

(*c*) multi-cylinder engines ☐ (*d*) none of the above. ☐

95. For a four cylinder in-line internal combustion engine, the most popular firing order is

(*a*) 1–4–3–2 ☐ (*b*) 1–3–4–2 ☐

(*c*) 1–2–3–4 ☐ (*d*) 1–2–4–3. ☐

96. The increase in cut-off ratio, of a diesel cycle with fixed compression ratio, would

(*a*) decreased m.e.p. ☐ (*b*) increase m.e.p. ☐

(*c*) keep same m.e.p. ☐ (*d*) none of the above. ☐

Carburettor and Carburetion

97. Carburettor is used for a

(*a*) spark ignition engine ☐ (*b*) compression ignition engine ☐

(*c*) steam engine ☐ (*d*) gas engine. ☐

98. Fuel injector is used for a

(*a*) spark ignition engine ☐ (*b*) compression ignition engine ☐

(*c*) steam engine ☐ (*d*) gas engine. ☐

99. To supply a high voltage to the spark plug in a spark ignition engine

(a) distributor is used ☐ (b) ignition coil is used ☐
(c) carburettor is used ☐ (d) none of the above. ☐

100. For providing correct firing order in the spark ignition engines,

(a) distributor is used ☐ (b) ignition coil is used ☐
(c) carburettor is used ☐ (d) none of the above. ☐

101. For breaking up and mixing the petrol with air in the spark ignition engine

(a) distributor is used ☐ (b) ignition coil is used ☐
(c) carburettor is used ☐ (d) none of the above. ☐

102. Over engine cylinders in scooters, the fins are provided for

(a) higher efficiency ☐ (b) higher strength of cylinder is ☐
(c) better cooling ☐ (d) none of the above. ☐

103. To produce a spark across the gap, the voltage required is

(a) 1000 to 2000 volts ☐ (b) 2000 to 4000 volts ☐
(c) 4000 to 6000 volts ☐ (d) 6000 to 1000 volts. ☐

104. A spark plug is used in

(a) petrol engine ☐ (b) diesel engine ☐
(c) gas engine ☐ (d) steam only. ☐

105. During suction stroke, a petrol engine draws a mixture of

(a) petrol and diesel ☐ (b) diesel and air ☐
(c) air and petrol ☐ (d) air only. ☐

106. During suction stroke, a diesel engine draws a mixture of

(a) petrol engine ☐ (b) diesel engine ☐
(c) air and petrol ☐ (d) air only. ☐

107. Diesel engines have higher compression ratio as compared to petrol engines. Higher compression ratio in petrol engines would result in

(a) auto-ignition of fuel ☐ (b) pre-ignition of fuel ☐
(c) detonation ☐ (d) none of the above. ☐

108. The running cost of diesel engine as compared to petrol engine is

(a) higher ☐ (b) lower ☐
(c) same ☐ (d) none of the above. ☐

109. There is no need of injecting fuel by external source to the engines working on

(a) diesel cycle ☐ (b) otto cycle ☐
(c) dual combustion cycle ☐ (d) none of the above. ☐

110. High speed C.I. engines mostly operate on

(a) otto cycle ☐ (b) diesel cycle ☐
(c) dual combustion cycle ☐ (d) none of the above. ☐

111. In C.I. engines, the combustion is
(*a*) hetrogeneous ☐ (*b*) homogeneous ☐
(*c*) both (*a*) and (*b*) ☐ (*d*) none of the above. ☐

112. The function of a carburettor is to supply
(*a*) air and diesel ☐ (*b*) air only ☐
(*c*) air and petrol ☐ (*d*) petrol only. ☐

113. The function of a distributor in a coil ignition system of an internal combustion engine is
(*a*) to supply a high voltage ☐ (*b*) to provide correct firing order ☐
(*c*) to supply air and petrol ☐ (*d*) to distribute current. ☐

114. The function of ignition coil in spark ignition current is
(*a*) to supply a high voltage ☐ (*b*) to provide correct firing order ☐
(*c*) to supply air and petrol ☐ (*d*) to ignite the fuel. ☐

115. By supercharging, the power developed by the engine
(*a*) increases ☐ (*b*) decreases ☐
(*c*) is same ☐ (*d*) none of he above. ☐

116. The function of lubrication in I.C. engine is
(*a*) to reduce friction ☐ (*b*) to dissipate the heat due to friction ☐
(*c*) to cool the cylinder ☐ (*d*) none of the above. ☐

117. The size of the flywheel for two-stroke engine as compared to four-stroke engine should be
(*a*) smaller ☐ (*b*) bigger ☐
(*c*) same ☐ (*d*) none of the above. ☐

118. The size of the flywheel for the diesel engines as compared to petrol engine should be
(*a*) smaller ☐ (*b*) bigger ☐
(*c*) same ☐ (*d*) none of the above. ☐

Governing of I.C. Engines

119. For the petrol engines, the type of governing used is
(*a*) quality governing ☐ (*b*) hit and miss governing ☐
(*c*) quantity governing ☐ (*d*) none of the above. ☐

120. For the diesel engines, the type of governing used is
(*a*) quality governing ☐ (*b*) hit and miss governing ☐
(*c*) quantity governing ☐ (*d*) none of the above. ☐

121. Choose the correct statement
(*a*) The injection pressure developed by injector in a diesel engine is of the order of 10×10^5 N/m^2. ☐
(*b*) Thermal efficiency of I.C. engine on weak mixture is lower. ☐
(*c*) The air-fuel ratio in petrol engines is controlled by governor. ☐
(*d*) In hit and miss governing, the fuel supply is completely cut-off during one or more number of cycles. ☐

122. Choose the wrong statement

(*a*) The air-fuel ratio in petrol engines is controlled by carburettor. ☐

(*b*) During idling stage gasoline does not flow through the carburettor tube because of venturi vacuum. ☐

(*c*) Compression ratio for S.I. engines usually varies from 6 to 12. ☐

(*d*) Tendency to detonation is S.I. engines increases with decrease of compression ratio. ☐

123. The diesel oil is having cetane number about

(*a*) 100 ☐ (*b*) 10 ☐

(*c*) 20 to 40 ☐ (*d*) 40 to 60. ☐

124. The idling engine requires

(*a*) lean fuel air mixture ☐ (*b*) rich fuel air mixture ☐

(*c*) air without fuel ☐ (*d*) none of the above. ☐

125. When all the four wheels are powered, it is known as

(*a*) two-wheel drive ☐ (*b*) four-wheel drive ☐

(*c*) one-wheel drive ☐ (*d*) none of the above. ☐

126. The function of differential in automobiles is

(*a*) to enable the automobile to turn through right angle ☐

(*b*) to allow the rear movement of the automobile ☐

(*c*) to permit two rear wheels to run independently ☐

(*d*) to permit two rear wheels to have flexibility of relative speed. ☐

127. For speed travel, the suitable propulsion device is

(*a*) Turbojet ☐ (*b*) Propeller ☐

(*c*) Rocket ☐ (*d*) Turboprop. ☐

128. The type of engine used for tighter bombers is

(*a*) Turbojet ☐ (*b*) Propeller ☐

(*c*) Rocket ☐ (*d*) Turboprop. ☐

129. The ambient air for propulsion is not used in the following device

(*a*) Turbojet ☐ (*b*) Propelle ☐

(*c*) Rocket ☐ (*d*) Turboprop. ☐

130. The turbine of the turbojet engine as compared to that of turboprop engine is

(*a*) smaller ☐ (*b*) bigger ☐

(*c*) same ☐ (*d*) none of the above. ☐

131. The propulsion efficiency used in practice is approximately

(*a*) 90% ☐ (*b*) 80% ☐

(*c*) 60% ☐ (*d*) 30%. ☐

132. The efficiency of a turbojet engine will be maximum at a speed of

(*a*) 8000 km/hr ☐ (*b*) 5000 km/hr ☐

(*c*) 1000 km/hr ☐ (*d*) 2400 km/hr. ☐

133. In an I.C. engine, firing order depends upon

(*a*) crank shaft design ☐ (*b*) arrangement of cylinder ☐

(*c*) number of cylinders ☐ (*d*) all of the above ☐

(*e*) none of the above. ☐

134. In hit and miss governing

(*a*) mixture strength is maintained constant ☐

(*b*) quantity of fuel is varied to suit the load on engine ☐

(*c*) the fuel supply is cut-off completely during one or more number of cycles ☐

(*d*) none of the above. ☐

135. In quantity governing

(*a*) mixture strength is maintained constant and quantity of charge drawn into the cylinder is changed to suit the load ☐

(*b*) quantity of fuel is varied to suit the load on engine ☐

(*c*) the fuel supply is cut-off completely during one or more number of cycles ☐

(*d*) none of the above. ☐

136. In quality governing

(*a*) mixture strength is maintained constant and quantity of charge drawn into the cylinder is changed ☐

(*b*) quantity to fuel entering the cylinder is varied, keeping air supply constant ☐

(*c*) the fuel supply is cut-off completely during one or more number of cycles ☐

(*d*) none of the above. ☐

137. The thermal efficiency of I.C. engines, as compared to steam engines, is

(*a*) higher ☐ (*b*) lower ☐

(*c*) same ☐ (*d*) none of the above. ☐

138. The cycle, on which a gas engine works, is called

(*a*) Diesel cycle ☐ (*b*) Otto cycle ☐

(*c*) Dual cycle ☐ (*d*) None of the above. ☐

139. When the load on C.I. engine decreases, its efficiency, increases. This is due to

(*a*) hit and miss governing ☐ (*b*) higher maximum temperature ☐

(*c*) quantity governing ☐ (*d*) quality governing. ☐

140. Knocking of the S.I. engine can be prevented if the end gas has

(*a*) higher density ☐ (*b*) high temperature ☐

(*c*) long ignition delay ☐ (*d*) none of the above. ☐

141. Knocking in the C.I. engine can be prevented if the first element of fuel and air has

(*a*) a short delay ☐ (*b*) a low density ☐

(*c*) a low temperature ☐ (*d*) all of the above. ☐

142. The increase of the delay period (time lag) is due to increase of

(*a*) temperature ☐ (*b*) speed of engine ☐

(*c*) load ☐ (*d*) density induction air. ☐

143. The name of investigator for delay period in I.C. engines is

(*a*) Tizard and Pye ☐ (*b*) Diesel ☐

(*c*) Watts ☐ (*d*) Ricardo. ☐

144. The phenomenon responsible for violent sound pulsations within the cylinder of an I.C. engine is known as

(*a*) detonation ☐ (*b*) pre-ignition ☐

(*c*) auto-ignition ☐ (*d*) turbulence. ☐

145. Detonation can be controlled by

(*a*) increasing inlet pressure ☐ (*b*) retarding spark timing ☐

(*c*) reducing speed of engine ☐ (*d*) none of the above. ☐

146. In spark ignition engines, before the passage of spark, the ignition of the charge by some hot surface is known as

(*a*) detonation ☐ (*b*) pre-ignition ☐

(*c*) auto-ignition ☐ (*d*) all the above. ☐

147. The torque given by an engine will be maximum when the engine

(*a*) runs at lowest speed ☐ (*b*) runs at maximum speed ☐

(*c*) develop maximum power ☐ (*d*) none of the above. ☐

148. For cooling the pistons of diesel engines, the commonly fluid used is

(*a*) air ☐ (*b*) water ☐

(*c*) fuel oil ☐ (*d*) lubricating oil. ☐

149. Decrease in engine speed from no load to full load is called

(*a*) hunting ☐ (*b*) speed drop ☐

(*c*) performance number ☐ (*d*) carburetion. ☐

150. Over-control by the governor causes

(*a*) hunting ☐ (*b*) speed drop ☐

(*c*) carburetion ☐ (*d*) none of the above. ☐

151. The breaking up and mixing the petrol with air is known as

(*a*) hunting ☐ (*b*) speed drop ☐

(*c*) carburetion ☐ (*d*) none of the above. ☐

152. The percentage of CO in exhaust gases for C.I. engines is

(*a*) 5–10% ☐ (*b*) 10–20% ☐

(*c*) zero ☐ (*d*) none of the above. ☐

153. The percentage of CO in exhaust gases for petrol engines for maximum output is

(*a*) 5–10% ☐ (*b*) 10–20% ☐

(*c*) zero ☐ (*d*) considerable per cent. ☐

154. By supercharging the petrol engines, the possibility of detonation

(*a*) decreases ☐ (*b*) increases ☐

(*c*) remains constant ☐ (*d*) none of the above. ☐

155. By supercharging the diesel engine, the possibility of knocking

(*a*) decreases (*b*) increases

(*c*) remains constant (*d*) none of the above.

156. Possibility of knock in S.I. engine will increase if

(*a*) spark timing is retarded (*b*) spark timing is advanced

(*c*) spark timing is normal (*d*) none of the above.

157. If a fuel is having octane number as 85, it means that fuel is having same knocking tendency as a mixture of

(*a*) 15% *n*-heptane and 85% iso-octane by weight

(*b*) 15% *n*-heptane and 85% iso-octane by volume

(*c*) 85% *n*-heptane and 15% iso-octane by weight

(*d*) 85% *n*-heptane and 15% iso-octane by volume.

158. If a fuel is having cetane number as 45, it means that the fuel is having same ignition delay as a mixture of

(*a*) 45% *n*-haptane and 55% iso-octane

(*b*) 55% iso-octane and 45% *n*-heptane

(*c*) 45% cetane and 55% α-methyl naphthalene

(*d*) 55% cetane and 45% α-methyl naphthalene.

159. In Fig. 4.6 four curves are shown. Which is correct for the variation of brake thermal efficiency against load in B.P. for constant speed?

(*a*) curve *A*

(*b*) curve *B*

(*c*) curve *C*

(*d*) curve *D*.

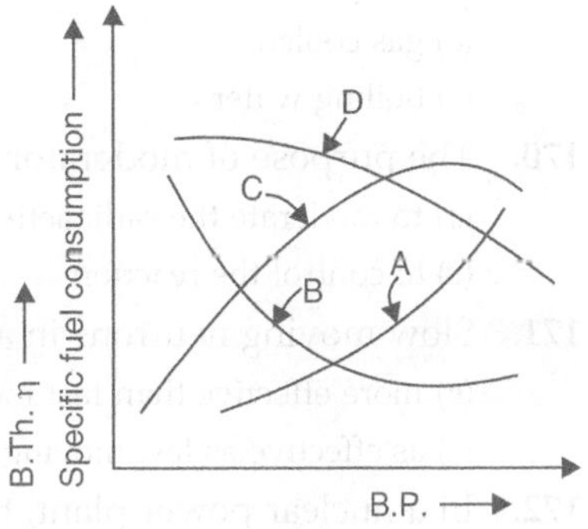

FIGURE 4.6

160. Which of the curves in Fig. 4.6 gives the variation of specific fuel consumption against load in B.P. for constant speed?

(*a*) curve *A* (*b*) curve *B*

(*c*) curve *C* (*d*) curve *D*.

Nuclear Power Plants

161. Uranium is used as a primary fuel in

(*a*) steam power plant (*b*) hydroelectric power plant

(*c*) nuclear power plant (*d*) none of the above.

162. The energy produced by 4.5 tonnes of high grade coal is equivalent to the energy produced by

(*a*) one kg of uranium (*b*) one gram of uranium

(*c*) 100 gm of uranium (*d*) 10 gm of uranium.

163. Tarapur is the place in India where the first following power plant was located

(*a*) steam power plant ☐ (*b*) hydroelectric power plant ☐
(*c*) nuclear power plant ☐ (*d*) none of the above. ☐

164. Isotopes of uranium are

(*a*) U_{235} ☐ (*b*) U_{234} ☐
(*c*) U_{238} ☐ (*d*) all of the above. ☐

165. The isotope of the uranium, which is mostly used is

(*a*) U_{235} ☐ (*b*) U_{234} ☐
(*c*) U_{238} ☐ (*d*) U_{240}. ☐

166. The fissionable material, produced in a reactor core by the action of a neutron on thorium, is

(*a*) U_{235} ☐ (*b*) U_{233} ☐
(*c*) U_{238} ☐ (*d*) U_{240}. ☐

167. The fissionable material, produced in a reactor core by the action of neutron on U_{238}, is

(*a*) Pu_{238} ☐ (*b*) Pu_{239} ☐
(*c*) U_{233} ☐ (*d*) U_{234}. ☐

168. The fissionable materials are

(*a*) U_{233} and Pu_{238} ☐ (*b*) U_{233} and Pu_{239} ☐
(*c*) U_{238} and Pu_{238} ☐ (*d*) U_{235} and Pu_{235}. ☐

169. The type of reactor for the nuclear power plant at Tarapur is of

(*a*) gas cooled ☐ (*b*) pressurized water ☐
(*c*) boiling water ☐ (*d*) none of the above. ☐

170. The purpose of moderator in a nuclear power plant is

(*a*) to moderate the radioactive pollution ☐ (*b*) to reduce the temperature ☐
(*c*) to control the reaction ☐ (*d*) to reduce the speed of fast moving neutrons. ☐

171. Slow moving neutrons in a nuclear power plant are

(*a*) more effective than fast moving neutron ☐ (*b*) less effective than fast moving neutron ☐
(*c*) as effective as fast moving neutron ☐ (*d*) none of the above. ☐

172. In a nuclear power plant, the most commonly used moderator is

(*a*) concrete ☐ (*b*) graphite ☐
(*c*) concrete and bricks ☐ (*d*) concrete and graphite. ☐

173. Tick mark the correct statement

(*a*) Molten lead is used as a coolant in nuclear plant. ☐
(*b*) The function of control rods in nuclear plants is to control fuel consumption ☐
(*c*) Control rods are made of lead ☐
(*d*) Uranium-233 (U_{233}) is produced by neutron irradiation of thorium. ☐

174. Pick up the wrong statement

(*a*) Light water is used as a coolant in nuclear plant. ☐
(*b*) The first nuclear power plant in India is located at Tarapur. ☐
(*c*) Enriched uranium is one in which percentage of U_{235} has been artificially increased. ☐
(*d*) Plutonium is produced artificially. ☐

175. Fast breeder reactor uses

(*a*) water as coolant ☐ (*b*) moderator ☐

(*c*) 90% of U_{235} ☐ (*d*) 100% of U_{235}. ☐

176. Plastics of thin layers of metals are used to stop

(*a*) electrons ☐ (*b*) neutrons ☐

(*c*) alpha and beta particles ☐ (*d*) none. ☐

177. To return the neutrons back into the core of a nuclear reactor

(*a*) shielding is done ☐ (*b*) reflector is used ☐

(*c*) moderator is used ☐ (*d*) control rod is used. ☐

178. To protect against neutron and gamma rays

(*a*) reflector is used ☐ (*b*) moderator is used ☐

(*c*) shielding is done ☐ (*d*) control rod is used. ☐

179. To absorb excess neutrons in nuclear reactors

(*a*) reflector is used ☐ (*b*) moderator is used ☐

(*c*) shielding is done ☐ (*d*) control rod is used. ☐

180. To slow down the speed of fast moving neutrons

(*a*) reflector is used ☐ (*b*) moderator is used ☐

(*c*) shielding is done ☐ (*d*) control rod is used. ☐

181. By neutron irradiation of thorium

(*a*) plutonium is produced ☐ (*b*) uranium-233 is produced ☐

(*c*) U_{225} is produced ☐ (*d*) U_{238} is produced. ☐

182. By neutron irradiation of U_{235}

(*a*) plutonium is produced ☐ (*b*) uranium-233 is produced ☐

(*c*) U_{238} produced ☐ (*d*) U_{239} is produced. ☐

183. 235 is

(*a*) primary fuel ☐

(*b*) secondary fuel ☐

(*c*) produced by the action of a neutron on thorium ☐

(*d*) ferrite materials. ☐

184. U_{238} and Pu_{239} are

(*a*) primary fuels ☐ (*b*) secondary fuels ☐

(*c*) ferrite materials ☐ (*d*) none of the above. ☐

185. U_{238} and Th_{232} are

(*a*) primary fuels ☐ (*b*) secondary fuels ☐

(*c*) ferrite materials ☐ (*d*) none of the above. ☐

186. The nuclear energy is commonly measured in

(*a*) neutron volts ☐ (*b*) electron volts ☐

(*c*) million electron volts ☐ (*d*) watts. ☐

187. The material, which absorbs neutrons and undergoes spontaneous changes leading to the formation of fissionable materials, is called
(*a*) primary fuel ☐ (*b*) secondary fuel ☐
(*c*) ferrite material ☐ (*d*) none of the above. ☐

188. The percentage composition of neutral uranium is
(*a*) 0.712% U_{238}, 0.006% U_{235}, 99.282% U_{234} ☐ (*b*) 0.712% U_{235}, .006% U_{238}, 99.282% U_{234} ☐
(*c*) 99.282% U_{238}, .006% U_{234} .712% U_{235} ☐ (*d*) 99.282% U_{238}, 0.712% U_{235}, 006% U_{234}. ☐

189. The energy in million electron volts released from uranium fission is approximately
(*a*) 800 ☐ (*b*) 400 ☐
(*c*) 200 ☐ (*d*) 20. ☐

190. Pick up the correct statement
(*a*) The nuclear power plant at Tarapur has fast breeder type reactor. ☐
(*b*) A moderator generally used in nuclear power plant is graphite and concrete. ☐
(*c*) The size of the reactor is said to be critical when it produces tremendous power. ☐
(*d*) Reactors designed for propulsion applications arc designed for pure uranium. ☐

191. Pick up the wrong statement
(*a*) Boiling water reactor uses ordinary water as moderator, coolant and working fluid. ☐
(*b*) The fast breeder reactor uses graphite as moderator. ☐
(*c*) Enriched uranium is one in which percentage of U_{235} has been artificially increased. ☐
(*d*) A moderator does not absorb neutrons. ☐

192. Shielding in nuclear reactor is generally done for protection against
(*a*) electrons ☐ (*b*) neutron and gamma rays ☐
(*c*) α and β rays ☐ (*d*) X-rays. ☐

193. Fast breeder reactors produces
(*a*) less fuel than they consume ☐ (*b*) more fuel than they consume ☐
(*c*) same fuel as they consume ☐ (*d*) no fuel. ☐

194. Fast breeder reactors use
(*a*) water as moderator ☐ (*b*) carbon dioxide as moderator ☐
(*c*) graphite as moderator ☐ (*d*) no moderator. ☐

195. Fast breeder reactors
(*a*) use water as coolant ☐ (*b*) are liquid-metal cooled ☐
(*c*) use no coolant ☐ (*d*) carbon dioxide as coolant. ☐

196. The production of neutron for a critical reactor, is
(*a*) zero ☐
(*b*) infinite ☐
(*c*) equal to the number of neutrons lost by leakage ☐
(*d*) none of the above. ☐

197. When reflector is used in a nuclear power plant, the leakage of neutron
(*a*) is zero ☐ (*b*) increases ☐
(*c*) decreases ☐ (*d*) none of the above. ☐

198. Reflector in nuclear power plant is used

(*a*) to protect against neutron and gamma rays ☐
(*b*) to absorb excess neutrons ☐
(*c*) to slow down the speed of fast moving neutron ☐
(*d*) to return the neutrons back into the core of the reactor. ☐

199. Control rod in a nuclear plant is used

(*a*) to protect against neutron and gamma rays ☐
(*b*) to absorb excess neutrons ☐
(*c*) to slow down the speed of fast moving neutron ☐
(*d*) to return the neutrons back into the core of the reactor. ☐

200. Moderator in a nuclear power plant is used

(*a*) to protect against neutron and gamma rays ☐
(*b*) to absorb excess neutrons ☐
(*c*) to slow down the speed of fast moving neutron ☐
(*d*) to return the neutrons back into the core of the reactor. ☐

201. Shielding in a nuclear power plant is done

(*a*) to protect against neutron and gamma rays ☐
(*b*) to absorb excess neutrons ☐
(*c*) to slow down the speed of fast moving neutron ☐
(*d*) to return the neutrons back into the core of the reactor. ☐

202. Fast neutrons have energies about

(*a*) 100 eV ☐ (*b*) 500 eV ☐
(*c*) 10 eV ☐ (*d*) more than 1000 eV. ☐

203. The various small shapes into which a solid fuel for nuclear power plant may be made, are

(*a*) plates only ☐ (*b*) pins only ☐
(*c*) pallets only ☐ (*d*) any one of the above. ☐

204. By high energy (fast) neutrons

(*a*) U_{235} will undergo fission ☐ (*b*) U_{238} will undergo fission ☐
(*c*) U_{234} will undergo fission ☐ (*d*) none of the above. ☐

205. By fast or slow neutrons

(*a*) U_{235} will undergo fission ☐ (*b*) U_{238} will undergo fission ☐
(*c*) U_{234} will undergo fission ☐ (*d*) none of the above. ☐

206. A nuclear power plant is considered to be economical if

(*a*) reflector is used ☐ (*b*) conlrol rod is used ☐
(*c*) used fuel is reprocessed ☐ (*d*) moderator is used. ☐

207. Pick up the wrong statement

(*a*) A fast breeder reactor uses 90% U_{235}. ☐
(*b*) A fast breeder reactor has no moderator. ☐
(*c*) Pressurised water reactor is designed to prevent the water coolant from boiling in the core. ☐
(*d*) Reactors designed for propulsion applications are designed for natural uranium. ☐

208. Pick up correct statement.
(a) A reactor is critical when it produces no power. ☐
(b) U_{235} and Pu_{238} are secondary fuel. ☐
(c) U_{235} is a primary fuel. ☐
(d) U_{233} is produced artificially. ☐

209. The desirable type of neutrons is triggering fission, is
(a) slow moving ☐ (b) fast moving ☐
(c) stationary ☐ (d) none of the above. ☐

210. For shielding, the most commonly used material is
(a) graphite ☐ (b) carbon ☐
(c) lead or concrete ☐ (d) none of the above. ☐

211. The function of nuclear reactor is
(a) to propel aircrafts, ships and submarine only ☐
(b) to produce fissionable materials only ☐
(c) to produce heat tor thermoelectric power only ☐
(d) all of the above. ☐

212. Because of a reflector, the fuel required in a nuclear power plant to generate sufficient neutrons to sustain a chain reaction is
(a) more ☐ (b) less ☐
(c) zero ☐ (d) none of the above. ☐

213. The material for control rod is
(a) graphite ☐ (b) lead ☐
(c) boron or cadmium ☐ (d) zinc. ☐

214. The production of the neutron is exactly balanced by the loss of neutron through leakage in a reactor when it becomes
(a) above critical ☐ (b) critical ☐
(c) below critical ☐ (d) none of the above. ☐

215. For propulsion applications, reactors are designed for
(a) enriched uranium ☐ (b) natural uranium ☐
(c) any form of uranium ☐ (d) pure uranium. ☐

Tick mark the true or false statement:

216. Special reactors, called breeder reactors, supply a net gain of fissionable material.
(a) True ☐ (b) False. ☐

217. For shielding most commonly used material is carbon.
(a) True ☐ (b) False. ☐

218. In India, the first nuclear power plant was located at Tarapur.
(a) True ☐ (b) False. ☐

219. A nuclear power plant differs from a conventional steam power station only in the steam generating part.
(a) True ☐ (b) False. ☐

220. To measure the nuclear energy, the unit million electron volts is commonly used.

(*a*) True ☐ (*b*) False. ☐

221. The nuclear power station at Tarapur has the reactor of gas cooled lype.

(*a*) True ☐ (*b*) False. ☐

222. Internal combustion engines are the engines in which the combustion of fuel takes place inside the engine cylinder.

(*a*) True ☐ (*b*) False. ☐

223. The working pressure inside the cylinder of an I.C. engine is very low as compared to steam engine.

(*a*) True ☐ (*b*) False. ☐

224. To measure the calorific value of gaseous fuels, Bomb calorimeter is used,

(*a*) True ☐ (*b*) False. ☐

225. The area under a curve on T-ϕ diagram represents change of entropy.

(*a*) True ☐ (*b*) False. ☐

226. The working cycle in a two-stroke engine, is completed in two revolutions of the crank.

(*a*) True ☐ (*b*) False. ☐

227. The petrol engine works on Rankine cycle.

(*a*) True ☐ (*b*) False. ☐

228. An engine indicator is used to determine B.H.P.

(*a*) True ☐ (*b*) False. ☐

229. In a diesel engine, the fuel is ignited by spark.

(*a*) True ☐ (*b*) False. ☐

230. The efficiency of an I.C. engine will decreases if the intake air temperature increases.

(*a*) True ☐ (*b*) False. ☐

231. Iso-octane has octane number of 100.

(*a*) True ☐ (*b*) False. ☐

232. Octane number represents the ignition quality of petrol.

(*a*) True ☐ (*b*) False. ☐

233. The theoretical correct mixture of air and petrol is 20 : 1.

(*a*) True ☐ (*b*) False. ☐

234. The self-ignition temperature of diesel oil compared to petrol is higher.

(*a*) True ☐ (*b*) False. ☐

235. With increase in cut-off ratio, the m.e.p of diesel cycle having fixed compression ratio, will increase.

(*a*) True ☐ (*b*) False. ☐

236. If p = mean effective pressure, L = stroke, A = area of piston and N = r.p.m. of engine, then indicated power of a four-stroke engine will be equal to PLAN/4.

(*a*) True ☐ (*b*) False. ☐

237. The pressure inside the engine cylinder during suction stroke of a four-stroke petrol engine is above the atmospheric pressure.

(*a*) True ☐ (*b*) False. ☐

238. The diesel engines are known as compression ignition engine.

(*a*) True ☐ (*b*) False. ☐

239. In a petrol engine, the charge is ignited with the help of a spark plug.

(*a*) True ☐ (*b*) False. ☐

240. The spark plug gap is normally maintained at 1 mm to 2 mm.

(*a*) True ☐ (*b*) False. ☐

241. The knocking in diesel engines for given fuel, will be enhanced by increasing compression ratio.

(*a*) True ☐ (*b*) False. ☐

242. Otto cycle is more efficient than the Diesel, for the same compression ratio.

(*a*) True ☐ (*b*) False. ☐

243. The running cost of a petrol engine is lower than of a diesel engine.

(*a*) True ☐ (*b*) False. ☐

244. A four-stroke cycle engine develops twice the power as that of a two-stroke cycle engine.

(*a*) True ☐ (*b*) False. ☐

245. The compression ratio for a petrol engine varies from 15 to 25.

(*a*) True ☐ (*b*) False. ☐

246. The compression ratio for a diesel engine varies from 15 to 25.

(*a*) True ☐ (*b*) False. ☐

247. The process of removing the burnt gases from the combustion chamber of an internal combustion engine, is known as scavenging.

(*a*) True ☐ (*b*) False. ☐

248. A supercharger receives air from the atmosphere surrounding the engine, compresses it to a higher pressure and feeds it into the inlet valve of the engine.

(*a*) True ☐ (*b*) False. ☐

249. Diesel fuel is more difficult to ignite compared to petrol.

(*a*) True ☐ (*b*) False. ☐

250. The rating of diesel engine increases linearly, with increase in air-inlet temperature.

(*a*) True ☐ (*b*) False. ☐

251. A diesel engine as compared to petrol engine (both running at rated load) is more efficient.

(*a*) True ☐ (*b*) False. ☐

252. Piston rings are usually made of cast iron.

(*a*) True ☐ (*b*) False. ☐

253. The first elements of fuel and air should have low temperature to reduce the possibility of knock in the C.I. engines.

(*a*) True ☐ (*b*) False. ☐

254. If a fuel is having higher self ignition temperature, then it will detonate less.
(*a*) True ☐ (*b*) False. ☐

255. The compression ratio in petrol engine is kept low as compared to that of a diesel engine, because higher compression ratio in petrol engine would lead to pre-ignition of fuel.
(*a*) True ☐ (*b*) False. ☐

256. An engine will generate maximum torque when it runs at lowest speed.
(*a*) True ☐ (*b*) False. ☐

257. When an engine is idling, it requires lean fuel air mixture.
(*a*) True ☐ (*b*) False. ☐

258. Violet sound pulsations within the cylinder of an I.C. engine, are due to detonation.
(*a*) True ☐ (*b*) False. ☐

259. The delay period or time lag in I.C. engines was investigated by Ricardo.
(*a*) True ☐ (*b*) False. ☐

260. In petrol engine, the air fuel ratio is controlled by injector.
(*a*) True ☐ (*b*) False. ☐

261. A diesel engine has air inlet valve, exhaust valve and fuel injection valve.
(*a*) True ☐ (*b*) False. ☐

262. A carburettor is required in petrol engine.
(*a*) True ☐ (*b*) False. ☐

263. Detonation in petrol engines can be suppressed or reduced by the addition of small amount of lead ethide.
(*a*) True ☐ (*b*) False. ☐

264. Higher compression ratio in diesel engines results in higher pressure.
(*a*) True ☐ (*b*) False. ☐

265. The ratio of I.P. to B.P. of an I.C. engine is known as mechanical efficiency.
(*a*) True ☐ (*b*) False. ☐

266. The brake power is always greater than indicated power in case of an I.C. engine.
(*a*) True ☐ (*b*) False. ☐

267. The ratio of the indicated thermal efficiency to air standard efficiency is known as relative efficiency of an I.C. engine.
(*a*) True ☐ (*b*) False. ☐

268. If r = compression ratio and γ = ratio of two specific heats, then air standard efficiency of an I.C. engine is given by $1 - (r)^{(1-\gamma)}$.
(*a*) True ☐ (*b*) False. ☐

269. A single cylinder engine will have the same indicated power as multi-cylinder engines.
(*a*) True ☐ (*b*) False. ☐

270. If the speed of the engine is increased, then indicated power will decrease.
(*a*) True ☐ (*b*) False. ☐

Fill in the blanks:

271. A four-stroke engine, theoretically, should develop ...... power as that of a two-stroke cycle engine.

(*a*) same. ☐ (*b*) half. ☐

272. The mean effective pressure of a diesel engine having fixed compression ratio will ...... if cut off ratio decreases.

(*a*) decrease ☐ (*b*) increase. ☐

273. A four-stroke cycle engine gives ...... mechanical efficiency than a two-stroke cycle engine.

(*a*) higher ☐ (*b*) lower. ☐

274. Thermal efficiency of a two-stroke cycle engine is ...... a four-stroke cycle engine.

(*a*) greater than ☐ (*b*) less than. ☐

275. Number of working strokes per minute for a two-stroke cycle engine are ...... the speed of the engine in r.p.m.

(*a*) equal to ☐ (*b*) one-half. ☐

276. The pressure inside the cylinder is ......the atmospheric pressure during the exhaust stroke.

(*a*) below ☐ (*b*) above. ☐

277. The sensing element in the control system of nuclear reactors measure the ...... of the neutron flux in the reactor.

(*a*) temperature ☐ (*b*) density. ☐

278. Reflector in nuclear power plants ...... neutron leakage.

(*a*) increases ☐ (*b*) decreases. ☐

279. A moderator ...... neutrons.

(*a*) absorbs ☐ (*b*) does not absorb. ☐

280. Generally used moderator is ...... in nuclear plants.

(*a*) heavy water ☐ (*b*) graphite and concrete. ☐

281. The nuclear power station at Tarapur has the reactor of the ...... type.

(*a*) gas cooled ☐ (*b*) boiling water. ☐

282. The thermal efficiency of diesel engines is ...... on weak mixtures.

(*a*) lower ☐ (*b*) higher. ☐

283. The first nuclear power plant was located at ...... in India.

(*a*) Kota ☐ (*b*) Tarapur. ☐

284. From uranium fission, the energy released is about ...... million electron volts.

(*a*) 200 ☐ (*b*) 400. ☐

285. With the help of a ......, the charge is ignited in a petrol engine.

(*a*) spark plug ☐ (*b*) fuel injector. ☐

286. The diesel engines are also known as ...... engines.

(*a*) compression ignition ☐ (*b*) spark ignition. ☐

287. Supercharging is the process of supplying the intake of an engine with air at a density ...... than the density of the surrounding atmosphere.

(*a*) lesser ☐ (*b*) greater. ☐

288. Specific fuel consumption is defined as the fuel consumption ......

(*a*) per hour ☐ (*b*) per hour per B.H.P. ☐

289. The function of providing the correct tiring order in the engine in spark ignition engines is performed by ......

(*a*) spark plug ☐ (*b*) distributor. ☐

290. The primary fuel is ...... in nuclear power plants.

(*a*) U^{238} ☐ (*b*) U^{235} ☐

291. Control rod in nuclear power plants are made of ......

(*a*) graphite ☐ (*b*) cadmium. ☐

292. In the spark ignition engines, an ignition coil supplies ...... voltage to the spark plug.

(*a*) low ☐ (*b*) high. ☐

293. In hit and miss governing, the fuel supply is cut-off ...... during one or more number of cycles.

(*a*) partially ☐ (*b*) completely. ☐

294. In spark ignition engines, the advancing of spark timing will ...... knocking tendency.

(*a*) decrease ☐ (*b*) increase. ☐

295. To prevent knocking in spark ignition engines, the charge away from the spark plug should have ......

(*a*) low temperature ☐ (*b*) high temperature ☐

296. The ...... is the power available at the crank shaft.

(*a*) indicated horse power ☐ (*b*) brake horse power. ☐

297. The ratio of the indicated thermal efficiency to the air standard efficiency is known as ...... efficiency.

(*a*) mechanical ☐ (*b*) relative. ☐

298. The thermal efficiency of petrol engine is about ......

(*a*) 70% ☐ (*b*) 30%. ☐

299. The thermal efficiency of diesel engine is about ......

(*a*) 70% ☐ (*b*) 30%. ☐

300. The compression ratio in petrol engine is kept low as compared to a diesel engine because higher compression ratio in petrol engine would lead to ...... of fuel.

(*a*) detonation ☐ (*b*) pre-ignition. ☐

301. The relation ...... gives the mean effective pressure of an engine determined by indicator diagram.

(*a*) $\frac{A \times L}{S}$ ☐ (*b*) $\frac{A \times S}{L}$ ☐

where A = Area of indicator diagram, S = Spring constant

L = Length of indicator diagram.

302. Speed drop is the decrease in engine speed from ......

(*a*) 25% load to 75% load ☐ (*b*) no load to full load. ☐

303. Four-stroke petrol engines as compared to two-stroke petrol engine having same output rating and same compression ratio have ...... thermal efficiency.

(*a*) lower ☐ (*b*) higher. ☐

304. For the diesel engine, ... governing is used.

(*a*) quality ☐ (*b*) quantity. ☐

305. For the petrol engine, ... governing is used.

(*a*) quality ☐ (*b*) quantity. ☐

III. OBJECTIVE TYPE QUESTIONS FROM COMPETITIVE EXAMINATIONS

Tick mark the most appropriate answer :

1. Octane number of the fuel used commercially for diesel engine in India is in the range

(*a*) 80 to 90 ☐ (*b*) 60 to 80 ☐

(*c*) 60 to 70 ☐ (*d*) 40 to 45. ☐

2. The knocking tendency in C.I. engines increases with

(*a*) decrease of compression ratio ☐ (*b*) increase of compression ratio ☐

(*c*) increasing the temperature of inlet air ☐ (*d*) increasing cooling water temperature. ☐

3. Desirable characteristic of combustion chamber for S.I. engines to avoid knock is

(*a*) small bore ☐

(*b*) short ratio of flame path to bore ☐

(*c*) absence of hot surfaces in the end region of gas ☐

(*d*) all of the above. ☐

4. For the same compression ratio and heat supplied, the air standard efficiency of an Otto cycle compared to that of a Diesel cycle is

(*a*) less ☐ (*b*) more ☐

(*c*) equal ☐ (*d*) unpredictable. ☐

5. For the same compression ratio and heat input, the cycles in decreasing order of thermal efficiency are

(*a*) Otto, Dual, Diesel ☐ (*b*) Diesel, Otto, Dual ☐

(*c*) Dual, Diesel, Otto ☐ (*d*) Otto, Diesel, Dual. ☐

6. Which standard cycle consists of two isothermals connected by two constant volume processes ?

(*a*) Brayton cycle ☐ (*b*) Ericsson cycle ☐

(*c*) Stirling cycle ☐ (*d*) Atkinson cycle. ☐

7. The thermal efficiency of good I.C. engine at the rated load is in the range of

(*a*) 80 to 90% ☐ (*b*) 60 to 70% ☐

(*c*) 30 to 35% ☐ (*d*) 10 to 20%. ☐

8. In case of S.I. engine, to have best thermal efficiency the fuel-air mixture ratio should be
   (a) lean ☐ (b) rich ☐
   (c) may be lean or rich ☐ (d) chemically correct. ☐
9. The fuel-air ratio, for maximum power of S.I. engines, should be
   (a) lean ☐ (b) rich ☐
   (c) may be lean or rich ☐ (d) chemically correct. ☐
10. Which is not a part of petrol engine ?
    (a) valve mechanism ☐ (b) fuel injector ☐
    (c) induction coil ☐ (d) air filter. ☐
11. Indicate the number of valves in a single cylinder two-stroke petrol engine
    (a) one valve ☐ (b) two valves ☐
    (c) four valves ☐ (d) no valve. ☐
12. The number of ports in the cylinder of a two-stroke engine crankcase compression and no crankcase valve are
    (a) one ☐ (b) two ☐
    (c) three ☐ (d) four. ☐
13. In a four-stroke cycle S.I. engine the camshaft runs at
    (a) the same speed as crankshaft ☐
    (b) half the speed of crankshaft ☐
    (c) twice the speed of crankshaft ☐
    (d) any speed irrespective of crankshaft speed. ☐
14. Which one of the following an S.I. engine?
    (a) Diesel engine ☐ (b) Petrol engine ☐
    (c) Gas engine ☐ (d) none of the above. ☐
15. Which one of the following is C.I. engine?
    (a) Diesel engine ☐ (b) Petrol engine ☐
    (c) Gas engine ☐ (d) none of the above. ☐
16. Which of the following processes is not associated with Diesel cycle?
    (a) constant volume ☐ (b) constant pressure ☐
    (c) isothermal ☐ (d) adiabatic. ☐
17. For same output, same speed and same compression ratio the thermal efficiency of a two-stroke cycle petrol engine as compared to that for four-stroke cycle petrol engine is
    (a) more ☐
    (b) less ☐
    (c) same as long as compression ratio is same ☐
    (d) same as long as output is same. ☐
18. The ratio of brake power to indicated power of an I.C. engine is called
    (a) mechanical efficiency ☐ (b) thermal efficiency ☐
    (c) volumetric efficiency ☐ (d) relative efficiency. ☐

19. The specific fuel consumption of a diesel engine as compared to that for petrol engines is

(*a*) lower ☐ (*b*) higher ☐

(*c*) same for same output ☐ (*d*) none of the above. ☐

20. Which aspect is not true for a two-stroke cycle engine when compared with a four-strok cycle engine ?

(*a*) cycle is completed in one stroke of the piston ☐

(*b*) uniform turning moment and hence a lighter flywheel ☐

(*c*) theoretically develops twice the power ☐

(*d*) noisy exhaust ☐

(*e*) more consumption of lubricating oil. ☐

21. A two-stroke engine has a speed of 750 rpm. A four-stroke engine having an identical cylinder size runs at 1500 rpm. The theoretical output of the two-stroke engine will

(*a*) be twice that of the four-stroke engine ☐

(*b*) be half that of the four-stroke engine ☐

(*c*) be the same as that of the four-stroke engine ☐

(*d*) depend upon whether it is SI or CI engine. ☐

22. The top of the piston in two-stroke engine is

(*a*) flat ☐ (*b*) slanted ☐

(*c*) crown shaped ☐ (*d*) convex shaped. ☐

23. The thermal efficiency of petrol engine as compared to diesel engine is

(*a*) lower ☐ (*b*) higher ☐

(*c*) same for same power output ☐ (*d*) same for same speed. ☐

24. Compression ratio of petrol engines is in the range of

(*a*) 2 to 3 ☐ (*b*) 7 to 10 ☐

(*c*) 16 to 20 ☐ (*d*) none of the above ☐

25. Compression ratio of diesel engines may have a range

(*a*) 8 to 10 ☐ (*b*) 10 to 15 ☐

(*c*) 16 to 20 ☐ (*d*) none of the above. ☐

26. Which one of the following gas has the maximum value of specific heat ratio (γ)?

(*a*) Oxygen ☐ (*b*) Helium ☐

(*c*) Methane ☐ (*d*) Carbon dioxide. ☐

27. For the same compression ratio and heat rejection, which of the following cycle is most efficient

(*a*) Otto cycle ☐ (*b*) Diesel cycle ☐

(*c*) Dual cycle ☐ (*d*) Brayton cycle. ☐

28. For petrol engines, the method of governing employed is

(*a*) quantity governing ☐ (*b*) quality governing ☐

(*c*) hit and miss governing ☐ (*d*) none of the above. ☐

29. For diesel engines, the method of governing employed is

(*a*) quantity governing ☐ (*b*) quality governing ☐
(*c*) hit and miss governing ☐ (*d*) none of the above. ☐

30. Voltage developed to strike spark in the spark plug is in the range

(*a*) 6 to 12 volts ☐ (*b*) 1000 to 2000 volts ☐
(*c*) 20000 to 25000 volts ☐ (*d*) none of the above. ☐

31. In a 4-cylinder petrol engine the standard firing order is

(*a*) 1-2-3-4 ☐ (*b*) 1-4-3-2 ☐
(*c*) 1-3-2-4 ☐ (*d*) 1-3-4-2. ☐

32. A cycle consisting of two reversible isothermal processes and two-reversible isobaric processes is known as

(*a*) Atkinson cycle ☐ (*b*) Ericsson cycle ☐
(*c*) Stirling cycle ☐ (*d*) Brayton cycle. ☐

33. A Stirling cycle and a Carnot cycle operate between 50°C and 350°C, and their thermal efficiencies are η_s and η_c respectively. Which of the following identities is true ?

(*a*) $\eta_s > \eta_c$ ☐ (*b*) $\eta_s = \eta_c$ ☐
(*c*) $\eta_s < \eta_c$ ☐ (*d*) none of the above. ☐

34. Which is not the common component between a petrol and a diesel engine ?

(*a*) camshaft ☐ (*b*) exhaust silencer ☐
(*c*) spray nozzle ☐ (*d*) dynamo. ☐

35. In a four-stroke cycle engine, the four operations namely suction, compression, expansion and exhaust are completed in the number of revolutions of crankshaft equal to

(*a*) four ☐ (*b*) three ☐
(*c*) two ☐ (*d*) one. ☐

36. In a two-stroke cycle engine, the operations namely suction, compression, expansion and exhaust are completed in the number of revolutions of crankshaft equal to

(*a*) four ☐ (*b*) three ☐
(*c*) two ☐ (*d*) one. ☐

37. Which of the following constituents of a fuel does not contribute to its calorific value on combustion?

(*a*) carbon ☐ (*b*) hydrogen ☐
(*c*) sulphur ☐ (*d*) nitrogen. ☐

38. The torque developed by the engine is maximum at

(*a*) minimum speed of engine ☐
(*b*) maximum powder speed of engine ☐
(*c*) minimum volumetric efficiency speed of engine ☐
(*d*) maximum power speed of engine. ☐

39. Iso-octane content in a fuel for S.I. engines

(*a*) retards auto-ignition ☐ (*b*) accelerates auto-ignition ☐
(*c*) does not affect auto-ignition ☐ (*d*) none of the above. ☐

40. Normal heptane content in fuel for S.I. engines
(*a*) retards auto-ignition ☐ (*b*) accelerates auto-ignition ☐
(*c*) does not affect auto-ignition ☐ (*d*) none of the above. ☐

41. The knocking in S.I. engines increases with
(*a*) increase in inlet air temperature ☐ (*b*) increase in compression ratio ☐
(*c*) increase in cooling water temperature ☐ (*d*) all of the above. ☐

42. Scavenging air means
(*a*) air used under compression ☐
(*b*) air used for forcing the burnt gases out of the cylinder during the exhaust period ☐
(*c*) forced air for cooling the engine cylinder ☐
(*d*) burnt air containing combustion products. ☐

43. In a four-stroke cycle petrol engine, during suction stroke
(*a*) only air is sucked in ☐ (*b*) only petrol is sucked in ☐
(*c*) mixture of petrol and air is sucked in ☐ (*d*) none of the above. ☐

44. In a four-stroke cycle diesel engine, during suction stroke
(*a*) only air is sucked in ☐ (*b*) only fuel is sucked in ☐
(*c*) mixture of fuel and air is sucked in ☐ (*d*) none of the above. ☐

45. The two-stroke cycle engine has
(*a*) one suction valve and one exhaust valve operated by one cam ☐
(*b*) one suction valve and one exhaust valve operated by two cams ☐
(*c*) only ports covered and uncovered by piston to effect charging and exhausting ☐
(*d*) none of the above. ☐

46. For the same maximum pressure and temperature of the cycle and for the same heat rejection which air standard cycle has the maximum efficiency
(*a*) Otto cycle ☐ (*b*) Diesel cycle ☐
(*c*) Dual cycle ☐ (*d*) Brayton cycle. ☐

47. For the same compression ratio, the efficiency of the Brayton cycle is
(*a*) equal to Diesel cycle ☐ (*b*) equal to Otto cycle ☐
(*c*) equal to Dual cycle ☐ (*d*) greater than Diesel cycle. ☐

48. Which is the correct statement in context of two-stroke engines ?
(*a*) compression ratio is always lower than that of a four-stroke cycle engine ☐
(*b*) there is only one valve for inlet and exhaust ☐
(*c*) charge enters the engine cylinder through ports only ☐
(*d*) a Diesel engine cannot operate on two-stroke cycle. ☐

49. Compared to four-stroke cycle engine, a two-stroke cycle engine
(*a*) can be easily started ☐ (*b*) has lesser shocks and vibrations ☐
(*c*) has lower fuel consumption ☐ (*d*) is smaller in size for the same output. ☐

50. Choose the false statement

(*a*) liquid fuels have higher calorific value than solid fuels ☐

(*b*) a good fuel has a high ignition point ☐

(*c*) coal gas consists mainly of hydrogen, carbon monoxide and various hydrocarbons ☐

(*d*) calorific value of a gaseous fuel is expressed in kcal/m^3 ☐

51. In a gas turbine unit with regenerator, a perfect regeneration means

(*a*) $T_1 < T_4$ ☐ (*b*) $T_3 > T_4$ ☐

(*c*) $T_3 = T_4$ ☐ (*d*) none of the above. ☐

where T_4 is the temperature of gases leaving the turbine and T_3 is the temperature of air coming out of the regenerator.

52. Consider the following statements. In an, irreversible process

1. entropy always increases
2. the sum of the entropy of all bodies taking part in a process always increases
3. once created, entropy can not be destroyed.

Identify the correct statements using the codes given below:

(*a*) 1 and 2 ☐ (*b*) 1 and 3 ☐

(*c*) 2 and 3 ☐ (*d*) 1, 2 and 3. ☐

53. Consider the following statements

The definition of

1. temperature is due to Zeroth law of thermodynamics.
2. entropy is due to first law of thermodynamics
3. internal energy is due to second law of thermodynamics
4. reversibility is due to Kelvin-Planck's statement.

Identify the correct statements using the codes given below :

(*a*) 1, 2 and 3 ☐ (*b*) 1, 3 and 4 ☐

(*c*) 1 alone ☐ (*d*) 2 alone. ☐

(ENGG SERVICES-1993)

54. The knocking in S.I. engines gets reduced by

(*a*) increasing the compression ratio ☐ (*b*) retarding the spark advance ☐

(*c*) increasing inlet air temperature ☐ (*d*) increasing the cooling water temperature. ☐

55. The knocking tendency in petrol engines will increase when

(*a*) speed is decreased ☐ (*b*) speed is increased ☐

(*c*) fuel-air ratio is made rich ☐ (*d*) fuel-air ratio is made lean. ☐

56. Petrol commercially available in India for Indian passenger cars has octane number in the range

(*a*) 40 to 50 ☐ (*b*) 60 to 70 ☐

(*c*) 80 to 85 ☐ (*d*) 95 to 100. ☐

57. Which one of the followings is not a true statement ?
(*a*) lower the sink temperature, more is the heat rejected for a given heat supply ☐
(*b*) a system is said to be in its dead state when its available energy is zero ☐
(*c*) availability refers to the maximum possible useful work that could be obtained in constant environmental conditions ☐
(*d*) entropy increase is proportional to energy degradation. ☐

58. Which parameter remains constant during Joule- Thomson expansion ?
(*a*) temperature ☐ (*b*) pressure ☐
(*c*) enthalpy ☐ (*d*) volume. ☐

59. Very high speed engines are generally
(*a*) Gas engines ☐ (*b*) S.I. engines ☐
(*c*) C.I. engines ☐ (*d*) Steam engines. ☐

60. In S.I. engine, to develop high voltage for spark plug
(*a*) battery is installed ☐ (*b*) distributor is installed ☐
(*c*) carburettor is installed ☐ (*d*) ignition coil is installed. ☐

ANSWERS

Answers to Objective Type Questions

1. (*b*)	**2.** (*b*)	**3.** (*b*)	**4.** (*b*)	**5.** (*b*)	**6.** (*b*)
7. (*a*)	**8.** (*d*)	**9.** (*d*)	**10.** (*b*)	**11.** (*a*)	**12.** (*c*)
13. (*a*)	**14.** (*a*)	**15.** (*b*)	**16.** (*b*)	**17.** (*c*)	**18.** (*c*)
19. (*d*)	**20.** (*b*)	**21.** (*a*)	**22.** (*c*)	**23.** (*d*)	**24.** (*c*)
25. (*d*)	**26.** (*b*)	**27.** (*a*)	**28.** (*d*)	**29.** (*b*)	**30.** (*c*)
31. (*b*)	**32.** (*d*)	**33.** (*a*)	**34.** (*b*)	**35.** (*a*)	**36.** (*a*)
37. (*b*)	**38.** (*c*)	**39.** (*d*)	**40.** (*d*)	**41.** (*d*)	**42.** (*d*)
43. (*c*)	**44.** (*b*)	**45.** (*a*)	**46.** (*b*)	**47.** (*d*)	**48.** (*a*)
49. (*b*)	**50.** (*c*)	**51.** (*a*)	**52.** (*c*)	**53.** (*d*)	**54.** (*b*)
55. (*d*)	**56.** (*b*)	**57.** (*b*)	**58.** (*b*)	**59.** (*c*)	**60.** (*c*)
61. (*d*)	**62.** (*c*)	**63.** (*b*)	**64.** (*a*)	**65.** (*c*)	**66.** (*a*)
67. (*b*)	**68.** (*b*)	**69.** (*b*)	**70.** (*c*)	**71.** (*b*)	**72.** (*b*)
73. (*c*)	**74.** (*a*)	**75.** (*b*)	**76.** (*c*)	**77.** (*c*)	**78.** (*d*)
79. (*c*)	**80.** (*b*)	**81.** (*d*)	**82.** (*b*)	**83.** (*b*)	**84.** (*d*)
85. (*d*)	**86.** (*d*)	**87.** (*b*)	**88.** (*a*)	**89.** (*c*)	**90.** (*b*)
91. (*a*)	**92.** (*c*)	**93.** (*d*)	**94.** (*c*)	**95.** (*b*)	**96.** (*b*)
97. (*a*)	**98.** (*b*)	**99.** (*b*)	**100.** (*a*)	**101.** (*c*)	**102.** (*c*)
103. (*d*)	**104.** (*a*)	**105.** (*c*)	**106.** (*d*)	**107.** (*b*)	**108.** (*b*)
109. (*b*)	**110.** (*c*)	**111.** (*a*)	**112.** (*c*)	**113.** (*c*)	**114.** (*a*)
115. (*a*)	**116.** (*b*)	**117.** (*a*)	**118.** (*b*)	**119.** (*a*)	**120.** (*c*)

121. (*d*)	**122.** (*d*)	**123.** (*d*)	**124.** (*b*)	**125.** (*b*)	**126.** (*d*)
127. (*c*)	**128.** (*a*)	**129.** (*c*)	**130.** (*b*)	**131.** (*c*)	**132.** (*d*)
133. (*d*)	**134.** (*c*)	**135.** (*a*)	**136.** (*b*)	**137.** (*a*)	**138.** (*b*)
139. (*d*)	**140.** (*c*)	**141.** (*a*)	**142.** (*c*)	**143.** (*a*)	**144.** (*a*)
145. (*b*)	**146.** (*b*)	**147.** (*a*)	**148.** (*d*)	**149.** (*c*)	**150.** (*a*)
151. (*c*)	**152.** (*c*)	**153.** (*d*)	**154.** (*b*)	**155.** (*a*)	**156.** (*b*)
157. (*b*)	**158.** (*c*)	**159.** (*c*)	**160.** (*b*)	**161.** (*c*)	**162.** (*b*)
163. (*c*)	**164.** (*d*)	**165.** (*a*)	**166.** (*b*)	**167.** (*a*)	**168.** (*b*)
169. (*c*)	**170.** (*d*)	**171.** (*a*)	**172.** (*d*)	**173.** (*d*)	**174.** (*d*)
175. (*c*)	**176.** (*c*)	**177.** (*b*)	**178.** (*c*)	**179.** (*d*)	**180.** (*b*)
181. (*b*)	**182.** (*a*)	**183.** (*a*)	**184.** (*b*)	**185.** (*c*)	**186.** (*c*)
187. (*b*)	**188.** (*d*)	**189.** (*c*)	**190.** (*b*)	**191.** (*b*)	**192.** (*b*)
193. (*b*)	**194.** (*d*)	**195.** (*b*)	**196.** (*c*)	**197.** (*b*)	**198.** (*c*)
199. (*b*)	**200.** (*c*)	**201.** (*a*)	**202.** (*d*)	**203.** (*d*)	**204.** (*b*)
205. (*a*)	**206.** (*c*)	**207.** (*d*)	**208.** (*c*)	**209.** (*a*)	**210.** (*c*)
211. (*d*)	**212.** (*b*)	**213.** (*c*)	**214.** (*b*)	**215.** (*a*).	

True/False

216. (*a*)	**217.** (*b*)	**218.** (*a*)	**219.** (*a*)	**220.** (*a*)	**221.** (*b*)
222. (*a*)	**223.** (*a*)	**224.** (*b*)	**225.** (*b*)	**226.** (*b*)	**227.** (*b*)
228. (*b*)	**229.** (*b*)	**230.** (*a*)	**231.** (*a*)	**232.** (*a*)	**233.** (*b*)
234. (*a*)	**235.** (*a*)	**236.** (*b*)	**237.** (*b*)	**238.** (*a*)	**239.** (*a*)
240. (*b*)	**241.** (*b*)	**242.** (*a*)	**243.** (*b*)	**244.** (*b*)	**245.** (*b*)
246. (*a*)	**247.** (*a*)	**248.** (*a*)	**249.** (*a*)	**250.** (*b*)	**251.** (*a*)
252. (*a*)	**253.** (*b*)	**254.** (*a*)	**255.** (*a*)	**256.** (*a*)	**257.** (*b*)
258. (*a*)	**259.** (*b*)	**260.** (*b*)	**261.** (*a*)	**262.** (*a*)	**263.** (*a*)
264. (*a*)	**265.** (*b*)	**266.** (*b*)	**267.** (*a*)	**268.** (*a*)	**269.** (*b*)
270. (*b*).					

Fill in the Blanks

271. (*b*)	**272.** (*a*)	**273.** (*b*)	**274.** (*b*)	**275.** (*b*)	**276.** (*b*)
277. (*b*)	**278.** (*b*)	**279.** (*b*)	**280.** (*b*)	**281.** (*b*)	**282.** (*b*)
283. (*b*)	**284.** (*a*)	**285.** (*a*)	**286.** (*a*)	**287.** (*b*)	**288.** (*b*)
289. (*b*)	**290.** (*b*)	**291.** (*b*)	**292.** (*b*)	**293.** (*b*)	**294.** (*b*)
295. (*a*)	**296.** (*b*)	**297.** (*b*)	**298.** (*b*)	**299.** (*a*)	**300.** (*b*)
301. (*b*)	**302.** (*b*)	**303.** (*b*)	**304.** (*b*)	**305.** (*a*).	

Answers to Objective Type Questions from Competitive Examinations

1. (*d*)	**2.** (*a*)	**3.** (*d*)	**4.** (*b*)	**5.** (*a*)	**6.** (*c*)
7. (*c*)	**8.** (*c*)	**9.** (*b*)	**10.** (*b*)	**11.** (*d*)	**12.** (*c*)
13. (*b*)	**14.** (*b*)	**15.** (*a*)	**16.** (*c*)	**17.** (*b*)	**18.** (*a*)
19. (*a*)	**20.** (*a*)	**21.** (*c*)	**22.** (*c*)	**23.** (*a*)	**24.** (*b*)
25. (*c*)	**26.** (*b*)	**27.** (*a*)	**28.** (*a*)	**29.** (*b*)	**30.** (*c*)
31. (*d*)	**32.** (*b*)	**33.** (*b*)	**34.** (*c*)	**35.** (*c*)	**36.** (*d*)
37. (*d*)	**38.** (*c*)	**39.** (*a*)	**40.** (*b*)	**41.** (*d*)	**42.** (*b*)
43. (*c*)	**44.** (*a*)	**45.** (*c*)	**46.** (*b*)	**47.** (*b*)	**48.** (*d*)
49. (*a*)	**50.** (*b*)	**51.** (*c*)	**52.** (*a*)	**53.** (*c*)	**54.** (*b*)
55. (*a*)	**56.** (*c*)	**57.** (*a*)	**58.** (*c*)	**59.** (*b*)	**60.** (*d*).

Chapter 5

STEAM BOILERS, ENGINES, NOZZLES AND TURBINES

I. THEORY

5.1. BOILERS

Boiler is a device which is used for generating steam. This generated steam is supplied to steam engines or turbines for power generation. The generated steam may also be used for process work in cotton mills, sugar factories etc. and for heating purposes. Boilers are also called 'steam generators'.

5.2. CLASSIFICATION OF BOILERS

The main classifications of boilers are :

(*i*) Fire tube boilers or water tube boilers ;

(*ii*) Stationary or mobile boilers ;

(*iii*) Internally fired or externally fired boilers ;

(*iv*) Horizontal or vertical boilers ;

(*v*) According to heat source ; and

(*vi*) Natural and forced circulation boilers.

(*i*) **Fire tube or water tube boilers.** If the hot flue gases, from the boiler furnace, flow through the tubes and water surrounds these tubes then the boiler is known as fire tube boiler. Examples of fire-tube boilers are : Cochran, Cornish, Lancashire, Locomotive boilers etc. If the water flows through the tubes, which are surrounded by the fire or hot flue gases from the boiler furnace, then the boiler is known as water-tube boiler. Examples of water-tube boilers are : Babcock and Wilcox, Stirling etc.

(*ii*) **Stationary or mobile boilers.** If the boilers are used at one place only, they are called stationary boilers. And if the boilers are portable, they are called mobile boilers. Locomotive and marine boilers are mobile boilers.

(*iii*) **Internally fired or externally fired boilers.** If the furnace is placed in the region of boiling water, the boiler is called internally fired boiler and if the furnace is placed outside the region of boiling water, the boiler is known as externally fired boiler. Lancashire boiler is an internally fired boiler.

(*iv*) **Horizontal or vertical boilers.** If the tubes of heating surface are horizontal, the boiler is known as horizontal boiler. And if the tubes of heating surface are vertical, the boiler is known as vertical boiler.

(*v*) **According to heat sources.** The sources of heat are obtained :

(*a*) by the combustion of fuel in solid, liquid or gaseous form ;

(*b*) by electrical energy ;

(*c*) by nuclear energy ; and

(*d*) by hot waste gases. The boilers are classified according to the type of fuel used by the boiler.

(*vi*) **Natural and forced circulation boilers.** If the circulation of water in the boilers is due to convection currents produced by heat, the boilers are known as natural circulation boilers. If the circulation of water is by pumps, the boilers are known as forced circulation boilers. Examples of natural circulation boilers are : Babcock and Wilcox boiler, Lancashire boiler etc. Examples of forced circulation boilers are : La Mont boiler, Benson boiler, Velox boiler etc.

5.2.1. Lancashire Boiler

It is a horizontal stationary, internally fired, fire tube type and natural circulation boiler. It is approximately 7 to 9 m in length and 2 to 3 m in diameter. It has two parallel furnace flue tubes. The steam pressure is upto 16 kgf/cm^2 and evaporative capacity is upto 8000 kg/hr.

5.2.2. Cornish Boiler

It is similar in construction to Lancashire boiler. But it has only one furnace flue tube, while Lancashire boiler has got two. The length and diameter of its shell is less than that of Lancashire boiler. The length of the shell varies from 4 to 8 m and diameter varies from 1.25 to 1.75 m. The working pressure range and capacity of Cornish boiler as compared to Lancashire boiler is low.

5.2.3. Cochran Boiler

It is a fire tube type vertical multi-tubular, portable boiler. It is approximately 2.75 m in diameter and 5.5 m in height. It has got a number of horizontal fire tubes. The working pressure range and evaporative capacity are approximately 20 kgf/cm^2 and 3500 kg/hr respectively.

5.2.4. Locomotive Boiler

It is a portable internally fired, horizontal multi-tubular fire tube boiler. The fire box is separate and the shell is horizontal. The main drum consists of fire box at one end and smoke box at the other end. The hot flue gases from fire box flow to smoke box through the tubes, which are surrounded with water in shell. The normal pressure range is upto 20 kgf/cm^2.

5.2.5. High Pressure Boilers

Benson boiler, Loeffler boiler, La Mont boiler and Velox boiler are high pressure boilers.

5.2.6. Benson Boiler

It is a very light weight high pressure boiler. It has no drum. The steam can be generated within 20 minutes. The normal pressure range is from 30 kgf/cm^2 to 250 kgf/cm^2 and evaporative capacity is upto 135 tons of steam per hour.

5.2.7. La-Mont Boiler

It is a high pressure, water tube type, forced circulation boiler. It produces 100 tons of superheated steam per hour at a temperature of 500°C and at a pressure of about 135 kgf/cm^2.

5.3. COMPARISON OF WATER-TUBE AND FIRE-TUBE BOILERS

1. The initial cost of water-tube boiler is higher compared to fire-tube boiler for the same capacity.
2. Water-tube boilers are mostly externally fired and fire-tube boilers are mostly internally fired.
3. The water-tube boiler occupies less space than fire-tube boiler for a given power.
4. As compared to fire-tube boilers, all parts of water-tube boiler are readily accessible for cleaning and inspection.
5. A water-tube boiler will tend to produce wet steam.
6. A water-tube boilers are eminently suitable for the highest pressure without excessive thickness of metal.

5.4. BOILER MOUNTINGS AND ACCESSORIES

The boiler *mountings* are water level indicator, stop valve, safety valve, pressure gauge, blow off clock, feed check valve and fusible plug. The boiler *accessories* are economiser superheater feed pump, injector, air preheater, steam trap, steam separator and pressure reducing valve.

5.4.1. Boiler Mountings

1. **Water level indicator** indicates the exact level of water at any time in the boiler.
2. **Stop valve** is also called junction valve. The function of stop valve is to shut off steam or control the flow of steam from boiler to main steam pipe to requirements.
3. **Safety valve.** The function of safety valve is to prevent the steam pressure in the boiler exceeding a pre-determined maximum pressure. The safety valves may be spring load safety valve, dead weight safety valve, lever safety valve and high steam and low water safety valve. Dead weight safety valve is usually used on Lancashire boilers and lever safety valve

is used for stationary boilers. The high steam and low water safety valve blows out if the steam pressure is higher than the working pressure or when the water level in the boiler is low.

4. **Pressure gauge** shows the pressure of the steam formed at any time.
5. **Blow off cock** is used to remove mud, scale or sediments collected at the bottom of the boiler. Also it is used to empty the boiler when required.
6. **Feed check valve** is used to control the supply of water into the boiler. It does not allow any water to flow back from the boiler.
7. **Fusible plug** is used to put off the fire in the furnace of the boiler when the level of water falls to an unsafe limit.

5.4.2. Boiler Accessories

1. **Economiser** is a device used for heating feed water which is supplied to the boiler by utilising the heat in the exhaust flue gases before leaving through the chimney.
2. **Super Heater** is a device used for drying wet steam and then increasing the temperature of the dry steam without increasing its pressure by utilising the heat of combustion products.
3. **Feed Pump** is a device used for pumping water to the boiler.
4. **Injector** is used for pumping water to a vertical or a locomotive boiler and also to heat the feed water.
5. **Air Preheater** is a device used to heat the air before it flows into the furnace for combustion, by utilising the heat in the exhaust flue gases.
6. **Steam Trap** is used to drain off water resulting from the partial condensation of steam from steam pipes.
7. **Steam Separator** is a device used for separating suspended water particles carried by the steam on its way from the boiler to the engine.
8. **Pressure reducing valve** is used for maintaining constant pressure in the steam supply pipe line.

5.5. TERMS USED IN BOILERS

1. **Evaporative capacity of a boiler.** It is expressed as the amount of water evaporated in kg/hr or steam produced in kg/hr or steam produced in kg per kg of fuel burnt.
2. **Equivalent evaporation.** It is the amount of water evaporated from and at 100°C into dry and saturated steam at normal atmospheric pressure. The equivalent evaporation in kg/kg of fuel burnt is given by,

   Equivalent evaporation = $W (H_S - h)/L$

   where W = Weight of water evaporated into steam,

   H_S = Total heat of steam,

   h = Sensible heat of feed water, and

   L = Latent heat of steam.
3. **Factor of evaporation.** It is always more than one and is equal to the term $(H_S - h)/L$.

5.6. EFFICIENCY OF A BOILER

It is ratio of the heat absorbed by feed water to the heat supplied by fuel in a given time. It is given as

$$\text{Boiler, } \eta = \frac{W_S\,(H_S - h)}{W \times C}$$

where W_S = Weight of steam produced/hr,

W = Weight of fuel burnt/hr,

C = Calorific value of fuel,

H_S = Total heat of steam produced, and

h = Sensible heat of feed water.

5.7. HEAT LOSSES IN A BOILER

The heat losses which occur in a boiler are :

(*i*) Heat carried away by dry flue gases,

(*ii*) Heat lost due to unburnt carbon in ashpit,

(*iii*) Heat lost due to incomplete combustion of carbon to carbon monoxide,

(*iv*) Heat carried away by moisture in fuel,

(*v*) Heat lost due to radiations, and

(*vi*) Heat carried away by steam formed by combustion per kg of fuel.

5.8. BOILER DRAUGHT

It is defined as the small pressure difference which causes flow of gases to take place inside the boiler. This pressure difference is very small and is generally measured in mm of water. Draught is of two types namely natural draught and artificial draught. Draught produced by a chimney is known as *natural draught* and draught produced by fan, steam jet etc. is known as *artificial draught.*

5.9. A CHIMNEY

It is made of masonry, steel for concrete. The term chimney relates to a masonry structure and stack refers to a metallic chimney.

5.9.1. Chimney Calculations

(*i*) Density of the chimney gases in terms of mass of air used in kg/kg of fuel at T K

$$= 1.293 \left(\frac{m+1}{m}\right) \times \frac{273}{T}$$

where m = Mass of air used in kg/kg of fuel.

(*ii*) Pressure causing draught in the chimney

$$= 353\,H\left[\frac{1}{T_1} - \left(\frac{m+1}{m}\right)\times\frac{1}{T}\right] \text{ kgf/m}^2 \text{ or mm of } H_2O$$

where H = Height of chimney in m,

T_1 = Absolute atmosphere temperature, and

T = Absolute temperature of flue gases in the chimney.

(*iii*) Velocity of flue gases in the chimney

$$V = \sqrt{2g(H' - h_f)}$$

where H' = Height of hot gas column in m, and

h_f = Loss due to friction.

(*iv*) For maximum discharge through the chimney, the height of chimney should be equal to the height of the hot gas column producing draught.

(*v*) For maximum discharge through chimney, the draught (h) in mm of water

$$= 176.5\,\frac{H}{T_1}$$

where H = Height of chimney in m, and

T_1 = Absolute atmospheric temperature of air.

5.10. BOILER EFFICIENCY

It is defined as the ratio of the heat usefully employed in generation of steam to the heat supplied in the fuel in the same period.

5.11. STEAM NOZZLE

The device, which converts heat energy of the steam into kinetic energy, is called steam nozzle. It consists of a passage of varying cross-section. The steam enters the nozzle at a high pressure and a low velocity and leaves the nozzle at a low pressure and a high velocity.

5.11.1. Classification of Nozzles

The nozzles arc classified into three following types:

1. Convergent nozzle,
2. Divergent nozzle, and
3. Convergent-Divergent nozzle.

If the cross-section of the nozzle decreases continuously from the entrance to the exit, the nozzle is known as convergent nozzle. If the cross-section of the nozzle increases from entrance to exit, it is known as divergent nozzle and if the cross-section first decreases and then increases continuously, it is called convergent-divergent nozzle.

5.11.2. Expansion of Steam in Nozzle

Expansion of steam in nozzle follows Rankine cycle. As the expansion of steam in nozzle is very rapid it is assumed that the expansion is adiabatic. Velocity of steam leaving the nozzle is given by

$$V_2 = \sqrt{2gJ\,\Delta h + V_1^2}$$

where J = 427 kgf-m/kcal,

Δh = Heat drop (theoretical), and

V_1 = Velocity at inlet.

If the velocity at inlet (*i.e.*, V_j) is negligible, then velocity at outlet of nozzle is given by

$$V_2 = \sqrt{2gJ\,\Delta h} = \sqrt{2 \times 9.81 \times 427\,\Delta h} = 91.53\sqrt{\Delta h} \text{ in MKS units}$$

$$= 44.72\sqrt{\Delta h} \text{ – in SI units}$$

The isentropic expansion of steam through nozzle is approximately represented by the equation

$$pV^n = \text{constant}$$

where n □ 1.30 for steam initially superheated

□ 1.135 for steam initially dry saturated.

The velocity at the outlet of the nozzle (when velocity at inlet is negligible) in terms of pressures is given by

$$V_2 = \sqrt{2g\left(\frac{n}{n-1}\right)p_1 v_1\left[1 - \left(\frac{p_2}{p_1}\right)^{(n-1)/n}\right]}$$

5.11.3. Condition for Maximum Discharge

Nozzle is designed for maximum discharge. There is only one value of the ratio $\left(\frac{p_2}{p_1}\right)$, which will give maximum discharge for the nozzle. This ratio is called critical pressure ratio. It is given by

$$\left(\frac{p_2}{p_1}\right) = \left(\frac{2}{n+1}\right)^{\frac{n}{n-1}}$$

where p_2 = Pressure at throat, and p_1 = Initial pressure of steam.

The pressure p_2 is also called **critical pressure.**

For a saturated dry steam, n = 1.135 and critical pressure ratio becomes as

$$\frac{p_2}{p_1} = \left(\frac{2}{1.135 + 1}\right)^{\frac{1.135}{1.135 - 1.0}} = 0.58.$$

For steam which initially superheated, n = 1.3 and critical pressure ratio becomes as

$$\frac{p_2}{p_1} = \left(\frac{1}{1.3+1}\right)^{\frac{1.3}{1.3-1.0}} = 0.545.$$

The maximum discharge is given by

$$m = \frac{A_2}{v_1}\sqrt{2g\left(\frac{n}{n+1}\right)p_1 v_1\left(\frac{2}{n+1}\right)^{(2/n-1)}}$$

where A_2 = Throat area.

Hence *maximum discharge* through a convergent-divergent nozzle depends on the initial condition of steam and throat area only. It is *independent of the pressure* at outlet of the nozzle.

5.11.4. Effect of Friction

The frictional losses in the nozzle :

(*i*) reduces the exit velocity,

(*ii*) reduces the heat drop by 10 to 15%, and

(*iii*) increases the final dryness fraction of steam.

The effect of friction from inlet to the throat of a convergent-divergent nozzle is negligible and the whole of friction loss is assumed between throat and exit. The exit velocity of steam when friction is considered is given by

$$V_2 = 91.53\sqrt{K\,\Delta h} \quad \text{in MKS units}$$

where K is the *nozzle efficiency* which is defined as the ratio of useful heat drop to isentropic heat drop.

5.12. STEAM ENGINE

Steam engine is a device, which converts the heat energy of the steam into mechanical work by the reciprocating motion of the piston. It works on Rankine cycle. The steam engine may be a simple steam engine or a compounded steam engine. If the expansion of the steam takes place in a single cylinder, it is known as simple steam engine. And if the expansion of the steam takes place in two or more cylinders, the engine is known as compound engine.

5.12.1. Classification of Steam Engines

The main classification of the steam engines according to their constructions and operating features are :

1. *According to the position of cylinders as:*

(*a*) Vertical engine,

(*b*) Horizontal engine, and

(*c*) Inclined engine.

2. *According to the number of working strokes per revolution as* :

   (*a*) Single acting engine, and

   (*b*) Double acting engine.

   If the steam acts on one side of the piston, if is known as *single acting steam engine.* In case of single acting steam engine, there is only one working stroke per revolution. But if the steam acts on both sides of the piston, then the engine is known as double acting steam engine. In case of double acting, there are two working strokes per revolution.

3. *According to the expansion of steam as*:

   (*a*) Simple steam engine,

   (*b*) Compound steam engine,

   (*c*) Triple expansion engine, and

   (*d*) Quadruple expansion engine.

   In simple steam engine, steam expands in one cylinder only where as in compound steam engine, it expands in two cylinders. The first cylinder is known as high pressure cylinder and second cylinder is known as low pressure cylinder. If the steam expands successively in three cylinders, then the steam engine is known as triple expansion engine.

4. *According to the way, the exhaust takes place as:*

   (*a*) Condensing engine, and

   (*b*) Non-condensing engine.

   If the exhaust takes place into a condenser (in which pressure is lower than the atmospheric pressure), the steam engine is known as condensing engine. But if the exhaust takes place directly to the atmosphere, the steam engine is known as non-condensing engine.

5. *According to the speed of engine as:*

   (*a*) High speed engine (if speed is more than 200 r.p.m.),

   (*b*) Medium speed engine (if speed is between 100 to 200 r.p.m.), and

   (*c*) Low speed engine (if speed is less than 100 r.p.m.).

6. *According to length of stroke as:*

   (*a*) Long stroke engine (if stroke > 1.5 dia, of cylinder)

   (*b*) Short stroke engine (if stroke ≤ 1.5 dia. of cylinder).

7. *According to the governing method as:*

   (*a*) Throttle governed engines, and

   (*b*) Automatic cut-off governed engines.

If the speed of engine is controlled by means of a value in steam pipe which regulates the pressure of the steam entering the engine, then the engine is known as throttle governed engine. In automatic cut-off governed engine, the pressure of entering steam remains constant and the governor controls the amount of the steam admitted to the cylinder.

5.12.2. Theoretical Indicator Diagram

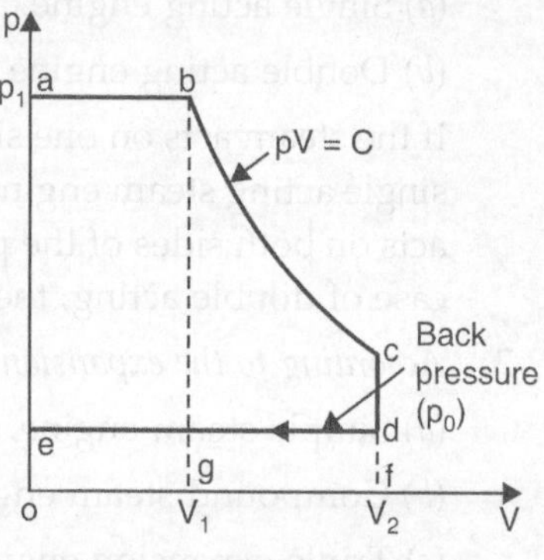

FIGURE 5.1

Theoretical indicator diagram for a reciprocating steam engine is shown in Fig. 5.1 without clearance,. The steam is admitted at point *a* at constant pressure p_1 upto *b*. At point *b*, the steam is cut-off. The high pressure steam expands from *b* to *c* with hyperbolic expansion (*i.e.*, $pV = C$). At C, the exhaust port opens and steam pressure falls to back pressure p_s, which may be equal to atmospheric pressure if exhaust goes to atmosphere otherwise it is less than atmospheric pressure. The point C is also called the point of release. The exhaust of steam at constant pressure is represented by the line *d-e*. The fresh steam is admitted at *e* and pressure immediately rise to point *a*.

Hypothetical (theoretical) work done per cycle

= Area of diagram *abcde*

= Area of *abgo* + area *of bcfg* – area of *oedf*

$$= p_1V_1 + p_1V_1 \log_e \left(\frac{V_2}{V_1}\right) p_bV_2$$

$$= p_1V_1(1 + \log_e r) - p_bV_2 \qquad \left(\because \frac{V_2}{V_1} = r = \text{ratio of expansion}\right)$$

The theoretical mean effective pressure (p_m)

$$= \frac{\text{Theoretical work done per cycle}}{V_2}$$

$$= \frac{p_1V_1(1 + \log_e r) - p_bV_2}{V_2}$$

$$= p_1\frac{V_1}{V_2}(1 + \log_e r) - p_b = \frac{p_1}{r}(1 + \log_e r) - p_b$$

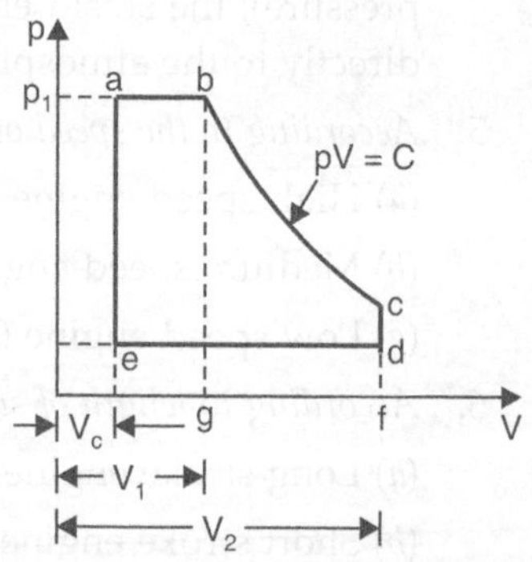

FIGURE 5.2

The theoretical indicator diagram for a reciprocating steam engine *with clearance* is shown in Fig. 5.2.

Hypothetical work done per cycle

$$= p_1V_1 (1 + \log_e r) - p_b V_2 - (p_1 + p_b) V_c$$

where V_c = Clearance volume

Theoretical mean effective pressure

$$p_m = \frac{p_1V_1 (1 + \log_e r) - p_2V_2 - (p_1 - p_b)V_c}{(V_2 - V_c)}$$

5.12.3. Hypothetical (theoretical) Indicator Diagram, Actual Indicator Diagram and Diagram Factor

Figure 5.3 shows the hypothetical indicator diagram by dotted lines and actual indicator diagram by thick lines, which is obtained by an engine indicator during a test. The various points on the actual indicator diagram represents as given below:

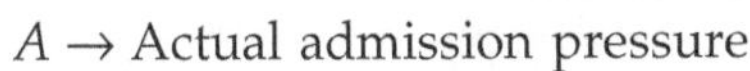

$A \rightarrow$ Actual admission pressure

$B \rightarrow$ Point of cut-off

$C \rightarrow$ Point of release

$D \rightarrow$ The exhaust port closes and entrapped steam is compressed to point E.

$E \rightarrow$ Admission port opens and the entrapped steam (or the cushion steam) mixes with the admission steam.

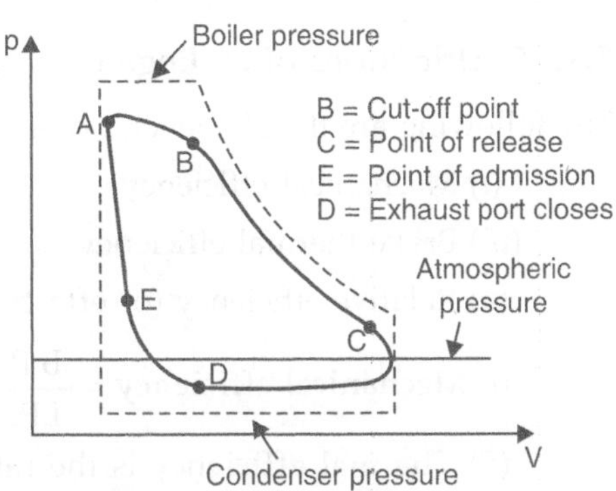

FIGURE 5.3

The area of actual indicator diagram is less than the area of theoretical indicator diagram as shown in Fig. 5.3. The ratio of the area of actual indicator diagram to the area of hypothetical indicator diagram is called *Diagram factor* (D.F.). Mathematically,

$$\text{D.F.} = \frac{\text{Area of actual indicator diagram}}{\text{Arera of hypothetical indicator diagram}}$$

$$= \frac{\text{Mean effective presure from actual indicator diagram}}{\text{Mean effective pressure from hypothetical indicator diagram}}$$

The value of D.F. is always less than 1.0.

5.12.4. Power a Steam Engines

1. Indicated Power (I.P.) for a single acting reciprocating steam engine is given by

$$\text{I.P.} = \frac{p_m \times L \times A \times N}{60{,}000} \text{ kW}$$

where, p_m = Actual mean effective pressure in N/mm^2,

A = Area of piston in mm^2, and

L = Length of stroke in m and N = r.p.m.

For a double acting $\quad \text{I.P.} = \dfrac{2p_m \times L \times A \times N}{60{,}000}$ kW.

2. Brake Power (B.P.) is the power available at the crank shaft and is measured by dynamometer. It is given as

$$\text{B.P.} = \frac{2\pi N(W - S)r}{60{,}000} \text{ for a rope brake dynamometer}$$

where W = Weight suspended in N, N = r.p.m.,

S = Spring balance reading in N, and

r = Radius of flywheel + Radius of rope in m.

3. Frictional Power (F.P.) is the difference of I.P. and B.P.

∴ F.P. = I.P. – B.P.

5.12.5. Efficiencies of an Engine

The following are the efficiencies of a steam engine:

(*i*) Mechanical efficiency, (*ii*) Thermal efficiency,

(*iii*) Brake thermal efficiency, (*iv*) Overall efficiency, and

(*v*) Relative efficiency (or effective ratio).

(*i*) **Mechanical efficiency** $= \frac{\text{B.P}}{\text{I.P.}}$

(*ii*) **Thermal efficiency** is the ratio of the indicated work done to the energy supplied in steam. If m is the mass of steam used per minute, h_1 is the enthalpy of steam supplied in kcal/kg and h_f is the enthalpy of the water returned to the hot well in kcal/kg, then the energy supplied in steam minute

$$= m \times (h_1 - h_f) \text{ kcal/minute}$$

$$= m \times (h_1 - h_f) \times J \text{ Nm/minute}$$

Indicated work done per minute = I.P × 1000 × 60 Nm/minute

$$\therefore \quad \text{Thermal efficiency} = \frac{\text{Indicated work done per minute}}{\text{Energy supplied per minute in steam}}$$

$$= \frac{\text{I.P} \times 60{,}000}{m \times (h_1 - h_f) \times J}.$$

(*iii*) **Brake thermal efficiency** is the ratio of the brake or shaft work obtained to the energy supplied in steam. It is given by

$$\text{Brake thermal efficiency} = \frac{\text{B.P.} \times 60{,}000}{m \times (h_1 - h_f) \times J}.$$

(*iv*) **Overall efficiency** is the ratio of shaft work to the energy supplied by fuel for the generation of steam in boiler. If m is the mass of fuel burnt per minute and C is the calorific value of the fuel, then overall efficiency is given by

$$\text{Overall efficiency} = \frac{\text{B.P.} \times 60{,}000}{m \times C \times J}$$

(*v*) **Relative efficiency** is the ratio of thermal efficiency to the corresponding Rankine efficiency. Mathematically,

$$\text{Relative efficiency} = \frac{\text{Thermal efficiency}}{\text{Rankine efficiency}}.$$

5.12.6. Specific Steam Consumption

It is the quantity of steam consumed per hour per B.P. or per I.P.

5.13. SATURATION CURVE AND MISSING QUANTITY

Saturation curve is the curve showing the volume that the saturated and dry steam, in the cylinder of the steam engine during expansion stroke, would occupy. Due to condensation of the steam in the cylinder, the actual volume occupied by steam is less than the dry and saturated steam. Curve *AB* in Fig. 5.4 represents saturation curve. *The missing quantity* is the difference in the theoretical volume (given by saturation curve) and the actual volume at any point. In Fig. 5.4, at pressure *OM*,

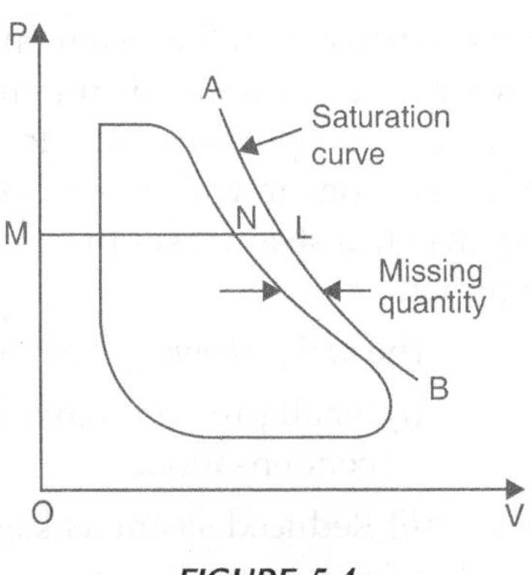

FIGURE 5.4

$$MN = \text{Actual volume in the cylinder,}$$
$$ML = \text{Volume given by saturation curve,}$$
$$NL = \text{Missing quantity, and}$$
$$\frac{MN}{ML} = \text{Dryness fraction at } N.$$

The missing quantity is due to condensation of steam in the cylinder and due to leakage past the piston. The missing quantity may be reduced by proper steam jacketing of the cylinder walls, using super heated steam and by reducing the temperature range of the working steam.

5.14. GOVERNING OF SIMPLE STEAM ENGINE

It is done by the following two methods:

(*i*) By throttle governing, and

(*ii*) By cut-off governing.

(*i*) Throttle Governing is done by means of a throttle valve, which alters the pressure of the steam at admission, keeping the cut-off ratio constant. In this type of governing, the steam consumption is directly proportional to the indicated power. Hence if a graph between consumption of steam and indicated power is plotted, the graph will be a straight line as shown in Fig. 5.5. This line is called *Willan's line* and the principle is known as *Willan's law.* It states that the rate of steam consumption is directly proportional to the indicated power, if the steam engine is throttle governed. Mathematically,

FIGURE 5.5

$$W = A + B \times (\text{I.P.})$$

(*ii*) Cut-off Governing is done by the alternation of the ratio of cut-off, keeping the admission pressure constant. It is more efficient and economical. The steam consumption is about 75% of throttle governed engine. The curve for cut-off governing is shown in Fig. 5.5.

5.15. COMPOUND STEAM ENGINE

If the expansion of the steam takes place in two or more than two cylinders in series, the engine is known as compound steam engine. The cylinders are known as high pressure cylinder (H.P. cylinder), low pressure cylinder (L.P. cylinder) and intermediate pressure cylinder. High pressure cylinder is one in which expansion of steam first occurs and low pressure cylinder is one in which expansion of steam takes place in the last. The remaining cylinders are called intermediate pressure cylinder.

The *main advantages* of compound engines are:

(*i*) Small pressure range and small temperature range per cylinder and hence reduction in condensation,

(*ii*) Reduced steam leakage past piston and valves,

(*iii*) Reduced stroke length, less weight and more uniform torque and hence lighter flywheel,

(*iv*) Can be started in any position,

(*v*) Reheating between high-pressure cylinder and low pressure cylinder is possible, and

(*vi*) Saving in steam consumption and hence increase in thermal efficiency.

5.15.1. Methods of Compounding Steam Engines

The methods are:

(*i*) Tandem Compound Engine,

(*ii*) Woolfe Compound Engine, and

(*iii*) Receive Compound Engine.

(*i*) Tandem Compound Engine consists of a common piston rod for the high pressure cylinder and low pressure cylinder, working on the same crank. The exhaust steam from the high pressure cylinder is allowed to pass directly to the low pressure cylinder. The maximum turning moment, due to each cylinder on the crank-shaft, acts at the same instant and hence it requires a large flywheel.

(*ii*) Woolfe Compound Engine of two cylinders consists of two cranks which are at 180° to each other. The two cylinders are arranged side by side and the exhaust steam from the H.P. cylinder is allowed to pass directly into the L.P. cylinder. This type of compound engine also requires large flywheel.

(*iii*) Receiver Compound Engine of two cylinders consists of two cranks which are placed at 90° to each other. The exhaust from the H.P. cylinder is passed into a receiver which is placed between the two cylinders. The L.P. cylinder draws the steam from this cylinder. As the two cycles are out of phase by 90°, the turning moment on the crank shaft is more uniform and hence a smaller flywheel is required.

5.15.2. Method of Governing of Compound Engines

The methods of governing of the compound steam engines are:

(*i*) Throttle governing,

(*ii*) Cut-off governing in H.P. cylinder, and

(*iii*) Cut-off governing in L.P. cylinder.

(*i*) **Throttle governing** is done by throttling the steam before it enters the H.P. cylinder. In this method there is more steam consumption and less amount of work done per kg of steam.

(*ii*) **Cut-off governing in H.P. cylinder** is done by varying the cut-off in the high pressure cylinder with the help of Mayer's expansion valve. The amount of work done per kg of steam is more and hence this method is more economical. In this method, the work done in H.P. cylinder is approximately the same but the work done in L.P. cylinder reduces and hence work done in two cylinders are different.

(*iii*) **Cut-off governing in L.P. cylinder** is done by varying the cut-off in L.P. cylinder only. This does not change the total work done by the engine but only changes the ratio of work done in two cylinders.

5.16. STEAM TURBINES

The reciprocating steam engines utilise the pressure energy of the steam whereas the steam turbines utilise the dynamic action of the steam. When the steam is flowing through a nozzle or fixed blades of a steam turbine, there is a heat drop which increases the velocity of the steam. This high velocity steam strikes the curved vanes and change of momentum takes place. Due to change of momentum, force is exerted on the moving blades and power is obtained.

5.17. CLASSIFICATION OF STEAM TURBINES

The steam turbines are classified as:

1. *According to the action of steam as:*
   (*a*) Impulse (*b*) Reaction
   (*c*) Combination of impulse and reaction.
2. *According to the direction of steam flow as:*
   (*a*) Axial (*b*) Radial
   (*c*) Tangential (*d*) Mixed.
3. *According to the pressure of steam as:*
   (*a*) High pressure (*b*) Medium pressure
   (*c*) Low pressure.
4. *According to terminal pressure as:*
   (*a*) Condensing (*b*) Non-condensing.
5. According to the step reduction as:
   (*a*) Single stage (*b*) Multi-stage.

5.17.1. Impulse Turbine

In an impulse turbine, steam is expanded in the nozzle thereby there is pressure drop and heat drop in the nozzle only. The pressure at inlet and outlet of the moving blades remain constant as shown in Fig. 5.6.

The simple impulse turbine is De-Laval Turbine which has a very high speed of the order of 30,000 r.p.m.

Other impulse are the Curtis, Rateau and Zoelly.

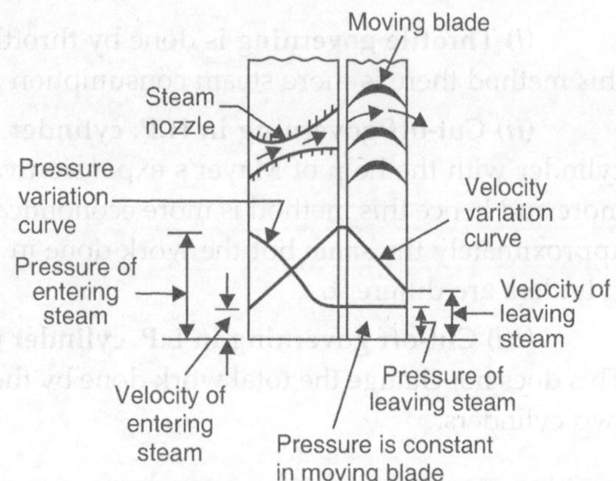

FIGURE 5.6

To reduce the speed of the impulse turbines to practical limits, the three following methods are used :

(*a*) Pressure compounding,

(*b*) Velocity compounding, and

(*c*) Pressure-velocity compounding.

(*a*) **Pressure Compounding** turbines consists of rings of fixed nozzles followed by the rings of moving blades as shown in Fig. 5.7. The rotors with moving blades are keyed to the shaft in series. The total pressure drop is divided equally between all the nozzle rings. The pressure remains constant over each ring of moving blades. Each stage consists of one ring of fixed nozzles and one ring of moving blades. The steam from boiler enters the first ring of nozzles. The pressure of the steam decreases and velocity increases in the nozzles. From the nozzles, the steam passes over the first ring of moving blades. In the moving blades, the kinetic energy of the steam is used and thus velocity decreases. The exhaust from the first moving blade enters the next nozzle ring, where the velocity of steam again increases.

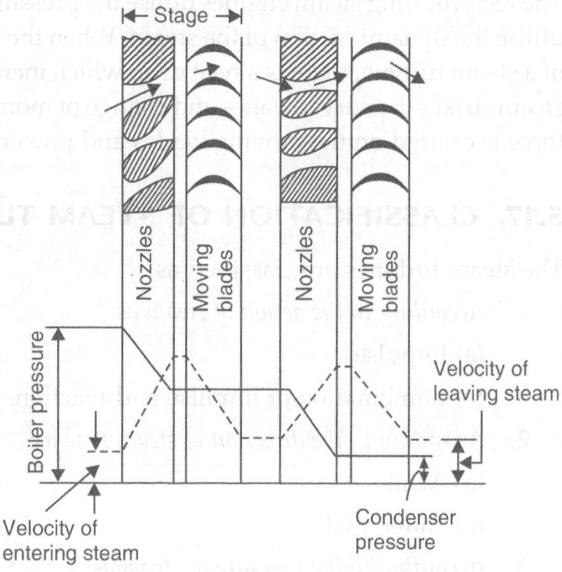

FIGURE 5.7. Pressure compounding.

This steam again passes over the second ring of moving blades and so on. The ratio of blade velocity to steam velocity in this type of compounding is constant and hence this is the most efficient type of turbine. But it requires a large number of stages. These turbines are also called **Rateau** and **Zoelly turbines.**

(*b*) **Velocity Compounding** turbines consist of a set of nozzles, rows of moving blades fixed to the shaft and rows of fixed blades mounted on the casing as shown in Fig. 5.8. The total heat drop takes place in the nozzles and high velocity of steam is

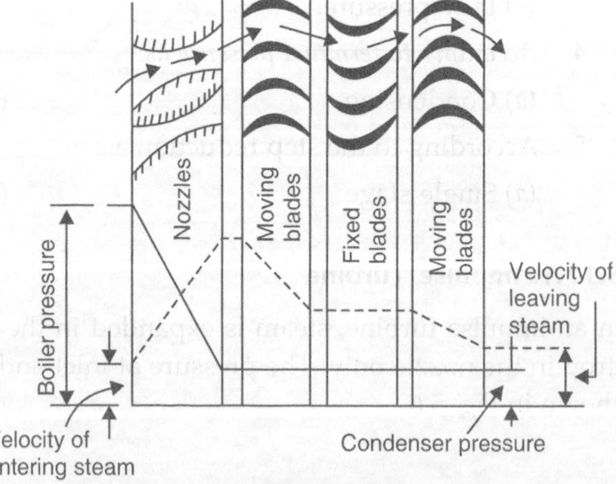

FIGURE 5.8

generated in the nozzle, which passes over the first set of moving blades where only a part of the kinetic energy is absorbed. This steam again passes over a set of fixed blades, which only changes the direction of the steam. The steam from the fixed blades passes over the second set of moving blades, where again a part of kinetic energy is absorbed. All the pressure drop takes place in the nozzles only. The pressure is constant over each set of moving and fixed blades. This type of turbine is also called *Curtis turbine.* This type of turbine requires less number of stages. But it has got low efficiency.

(*c*) **Pressure-Velocity Compounding** turbine has the advantages of the pressure compounding and velocity-compounding turbines. The total pressure drop of steam is divided into stages and velocity obtained in each stage is also compounded as shown in Fig. 5.9. It requires less number of stages as bigger pressure drop takes place in each stage. But the efficiency is lower than the pressure compounded turbines.

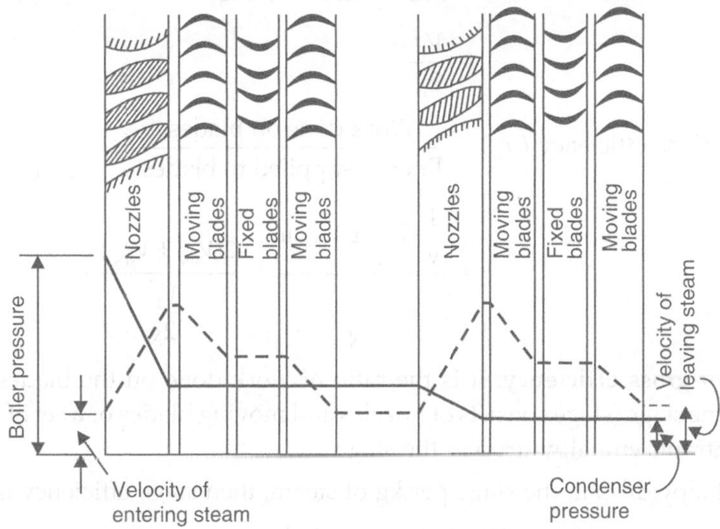

FIGURE 5.9. Pressure-velocity compounding.

5.17.2. Reaction Turbines

Reaction turbines consist of large number of stages, each stage is having fixed and moving blades as shown in Fig. 5.10. The heat drop takes place in both fixed and moving blades. The steam is admitted for the whole of circumference first through the fixed blades. There is a small drop in pressure and the velocity of steam increases. Then the steam passes over the first row of moving blades where change of momentum takes place and hence force acts on the blades.

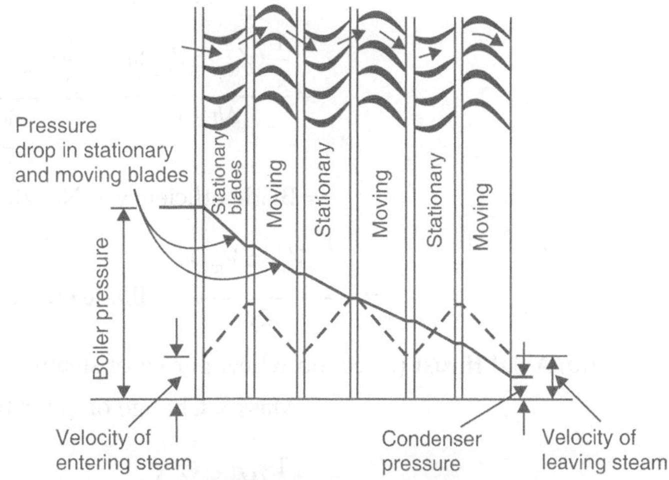

FIGURE 5.10. Reaction turbine.

Also there is small drop of pressure, when steam is passing over the moving blades. This increases the kinetic energy of the steam. This type of turbine is also called **Parson's Reaction Turbine.**

5.17.3. Work done and Efficiencies for Impulse Turbines

(*i*) **Work done** by steam on blades per sec per kg of steam striking

$$= \frac{1}{g}(V_{w_1} \pm V_{w_2}) \times u$$

where u = Blade velocity.

(*ii*) **Blade or diagram efficiency.** It is defined as the ratio of work done on the blades to the energy supplied to the blades. The energy supplied to the blades per kg of steam

$$= \text{K.E. of steam at inlet}$$

$$= \frac{V_1^2}{2g}$$

$$\therefore \quad \text{Blade efficiency } (\eta_b) = \frac{\text{Work done on blades}}{\text{Energy supplied to blades}}$$

$$= \frac{\frac{1}{g}(V_{w_1} \pm V_{w_2})u}{\frac{V_1^2}{2g}} = \frac{2(V_{w_1} \pm V_{w_2})u}{\frac{V_1^2}{2g}}$$

(*iii*) **Stage or gross efficiency.** It is the ratio of work done on the blades per kg of steam (flowing through the stage (stage consists of nozzles and moving blades or a set of fixed and moving blades) to the isentropic enthalpy drop in the stage.

If Δh = Enthalpy drop in the stage per kg of steam, then stage efficiency is given by

$$\text{Stage efficiency} = \frac{\text{Work done on blades per kg of steam}}{\text{Enthalpy drop in the stage per kg of steam}}$$

$$= \frac{\frac{1}{g}(V_{w_2} \pm V_{w_2})u}{\Delta h \times J} = \frac{\frac{1}{g}(V_{w_2} \pm V_{w_2})u}{\frac{V_1^2}{2g}} \times \frac{\frac{V_1^2}{2g}}{\Delta h \times J}$$

$$= \text{Blade efficiency} \times \text{Nozzle efficiency}$$

$$\left(\because \frac{\frac{1}{g}(V_{w_1} \pm V_{w_2})u}{2g} = \text{Blade efficiency and } \frac{V_1^2}{2g\,\Delta h \times J} = \text{Nozzle efficiency} \right)$$

(*iv*) **Axial thrust** (force) on wheel per kg of steam

$$= \text{Mass} \times \text{Change of velocity in axial direction}$$

$$= \frac{1}{g}(V_{fi} - V_{fo})$$

(*v*) **Energy lost** due to friction per kg of steam in blades

$$= \frac{(V_{r_i}^2 - V_{r_0}^2)}{2g}.$$

5.17.4. Work Done and Efficiencies for Reaction Turbine (Parson's Turbine)

The work done by steam, blade efficiency, gross efficiency and axial thrust for reaction turbines are given by the same expression as that of impulse turbines. But in case of reaction turbines, the term, degree of reaction is more important. *Degree of reaction* is defined as the ratio of heat drop in the moving blades to the sum of heat drop in fixed and moving blades.

(*i*) $\therefore$ Degree of reaction $= \dfrac{\text{Heat drop in moving blades}}{\text{(Total heat drop in fixed and moving blades)}}$

The degree of reaction for Parson's turbine with symmetrical blades is 50%.

(*ii*) Heat drop through fixed blades per kg of steam $= \dfrac{V_1^2 - V_2^2}{2gJ}$.

(*iii*) Heat drop through moving blades per kg of steam $= \dfrac{(V_{r_1}^2 - V_{r_2}^2)}{2gJ}$.

5.17.5. Condition for Maximum Efficiency

(*i*) **For Impulse Turbine**

Let ρ = Blade speed ratio, *i.e.*, ratio of blade speed to steam speed

$$= \frac{u}{V_1}.$$

The efficiency of an impulse turbine is maximum when blade speed (u) is given by

$$u = \frac{V_1 \cos\alpha}{2}$$

where V_1 = Absolute velocity of steam entering the blades, and

α = Nozzle angle

or $$\frac{u}{V_1} = \frac{\cos\alpha}{2} \quad \text{or} \quad \rho = \frac{\cos\alpha}{2} \qquad \left(\because \frac{u}{V_1} = \rho\right)$$

Maximum efficiency $= (1 + KC)\dfrac{\cos^2\alpha}{2}$

where K = Friction factor $= \dfrac{V_{r_2}}{V_{r_1}}$, and

$$C = \frac{\text{Cosine of angle made by } V_{r_2}}{\text{Cosine of angle made by } V_{r_1}}$$

(*ii*) **For Reaction Turbine** the condition of maximum efficiency is derived, making the following assumptions :

(*a*) The moving and fixed blades are symmetrical, and

(*b*) Degree of reaction is 50%.

The efficiency is maximum when $u = V_1 \cos \alpha$ or $\rho = \dfrac{u}{V_1} = \cos \alpha$

and maximum efficiency $= \dfrac{2\cos^2 \alpha}{1+\cos^2 \alpha}$.

5.17.6. Reheat Factor

Reheat factor is defined as the ratio of cumulative enthalpy drop to isentropic enthalpy drop. Figure 5.11 shows the expansion of steam in multistage turbine (here three stage turbine is considered). For the first stage A_1B_1 is the isentropic enthalpy drop out of which A_1C_1 is the useful enthalpy drop.

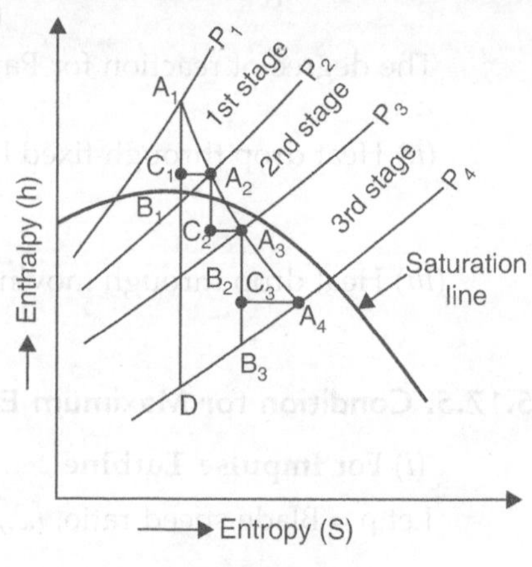

FIGURE 5.11

The ratio $\dfrac{A_1C_1}{A_1B_1}$ is called stage efficiency. At the exit from the first stage, the steam is at A_2 and not at B_1. Thus the quality of the steam at exit from first stage is improved.

For the 2nd stage,

$$A_2C_2 = \text{Useful enthalpy drop}$$
$$A_2B_2 = \text{Isentropic enthalpy drop.}$$

$$\text{2nd stage efficiency} = \frac{A_2C_2}{A_2B_2}$$

For the 3rd stage,

$$A_3C_3 = \text{Useful enthalpy drop}$$
$$A_3B_3 = \text{Isentropic enthalpy drop}$$

$$\therefore \quad \text{3rd stage efficiency} = \frac{A_3C_3}{A_3B_3}.$$

For the three stage turbine,

Cumulative heat drop $= (A_1B_1 + A_2B_2 + A_3B_3)$

Isentropic enthalpy drop $= A_1D$

$$\therefore \quad \text{Reheat factor (R.F.)} = \frac{A_1B_1 + A_2B_2 + A_3B_3}{A_1D} \quad \ldots(i)$$

The reheat factor depends on

(*i*) turbine stage efficiency

(*ii*) initial pressure and temperature

(*iii*) exit pressure, and

(*iv*) number of stages. The reheat factor is more if there are large number of stages. Its value varies from 1.02 to 1.06.

Due to reheat factor the efficiency of a turbine as a whole is more than its individual stages. The efficiency of the complete turbine is known as **internal efficiency** of the turbine. Hence internal efficiency

$$= \frac{\text{Total useful heat drop}}{\text{Isentropic heat drop}}$$

With reference to Fig. 5.11,

Total useful heat drop $= A_1C_1 + A_2B_2 + A_3B_3$

Isentropic heat drop considering the complete turbine as a single stage $= A_1 D$.

$$\therefore \quad \text{Internal efficiency} = \frac{(A_1C_1 + A_2C_2 + A_3C_3)}{A_1D}$$

But from equation (*i*), $\text{R.F.} = \dfrac{A_1B_1 + A_2B_2 + A_3B_3}{A_1D}$ or $A_1D = \dfrac{A_1B_1 + A_2B_2 + A_3B_3}{\text{R.F.}}$

$$\therefore \quad \text{Internal efficiency} = \left(\frac{A_1C_1 + A_2C_2 + A_3C_3}{A_1B_1 + A_2B_2 + A_3B_3}\right) \times \text{R.F.}$$

$$= \text{Stage efficiency} \times \text{R.F.}$$

assuming stage efficiency constant for all stages.

5.17.7. Steam Turbine Governing

The function of a governor is to regulate the supply of steam to the turbine in such a way as to keep its speed fairly constant from no-load to full-load.

Governing of steam turbine are done by the following methods :

(*i*) Throttle governing

(*ii*) Nozzle control governing

(*iii*) Bypass governing

(*i*) Throttle Governing. In throttle governing, the steam pressure at which steam is admitted to the turbine is reduced at part loads. As throttling is an irreversible process, hence thermodynamically the throttle governing is not efficient. The mechanism of throttle governing is simple. Throttle governing is used on small turbines. The steam flow to the turbine is throttle by balanced throttle valve actuated by a centrifugal governor.

(*ii*) Nozzle Control Governing. In nozzle control governing, various arrangements of valves and groups of nozzles are employed. The principle of nozzle control is accomplished by uncovering as many steam passages as are necessary to meet the load by poppet valves.

(*iii*) Bypass Governing. In bypass governing, all the steam entering the turbine passes through the main throttle valve which is under the control of speed governor and enters the nozzle box or the steam chest. In certain cases, this would suffice for all loads upto economical load, the governing being effected by the throttling. For loads greater than the economic load a bypass valve is opened. The bypass valve is not opened until the lift of throttle valve exceeds a certain amount, also as the load diminishes the bypass valve closes first. The bypass valve is under the control of speed governor for all loads within this range.

In reaction turbines, because of pressure drop required in moving blades, nozzle control governing is not possible, and throttle governing plus bypass governing is used.

II. OBJECTIVE TYPE QUESTIONS

Tick mark the most appropriate statement of the multiple choice answers:

Boilers

1. A boiler is said to be fire-tube boiler if
   - (*a*) water passes through tubes and hot products of combustion from furnace are around the tubes ☐
   - (*b*) the hot products of combustion passes through the tubes and water around it ☐
   - (*c*) forced circulation takes place ☐
   - (*d*) none of the above. ☐
2. A boiler is said to be water-tube boiler if
   - (*a*) water passes through tubes and hot products of combustion are around the tubes ☐
   - (*b*) the hot products of combustion passes through the tubes and water around it ☐
   - (*c*) forced circulation takes place ☐
   - (*d*) none of the above. ☐
3. Which of the following is a fire-tube boiler?
   - (*a*) Lancashire boiler ☐ (*b*) Locomotive boiler ☐
   - (*c*) Cochran boiler ☐ (*d*) all of the above. ☐
4. Which of the following is a water-tube boiler?
   - (*a*) Babcock and Wilcox boiler ☐ (*b*) Stirling boiler ☐
   - (*c*) Yarrow boiler ☐ (*d*) all of the above. ☐
5. The diameter of the cylindrical shell of a Lancashire boiler varies from
   - (*a*) 2 to 3 m ☐ (*b*) 1 to 2 m ☐
   - (*c*) 3.5 to 1.0 m ☐ (*d*) 4 to 6 m. ☐
6. The diameter of the cylindrical shell of a Cornish boiler varies from
   - (*a*) 2 to 3 m ☐ (*b*) 1.25 to 1.75 m ☐
   - (*c*) 0.5 to 1.0 m ☐ (*d*) 4 to 6 m. ☐
7. The diameter of the cylindrical shell of a Cornish boiler as compared to that of Lancashire boiler is
   - (*a*) more ☐ (*b*) less ☐
   - (*c*) same ☐ (*d*) none of the above. ☐

8. The Lancashire boiler has length from
   (*a*) 2 to 4 m ☐ (*b*) 4 to 6 m ☐
   (*c*) 7 to 9 m ☐ (*d*) 10 to 12 m. ☐
9. The normal working pressure range for Lancashire boiler is upto
   (*a*) 16×10^5 N/m^2 ☐ (*b*) 303×10^5 N/m^2 ☐
   (*c*) 40×10^5 N/m^2 ☐ (*d*) 53×10^5 N/m^2. ☐
10. The normal working pressure range for Cornish boiler as compared to that of Lancashire boiler is
   (*a*) same ☐ (*b*) high ☐
   (*c*) low ☐ (*d*) none. ☐
11. Choose the correct statement
   (*a*) The Cornish boiler contains two fire tubes and Lancashire boiler contains one. ☐
   (*b*) In a fire-tube boiler, the water is contained inside the tubes and hot gases are around the tubes. ☐
   (*c*) Lancashire boiler is a fire-tube boiler. ☐
   (*d*) Locomotive boiler is a water-tube boiler. ☐
12. A device, which is used to generate and supply steam at a high pressure and temperature, is called
   (*a*) steam turbine ☐ (*b*) steam boiler ☐
   (*c*) steam engine ☐ (*d*) none of the above. ☐
13. If in a boiler, the circulation of water is by convection currents which are set up during the heating of water, then the boiler is known as
   (*a*) forced circulation boiler ☐ (*b*) natural circulation boiler ☐
   (*c*) internally fired boiler ☐ (*d*) externally fired boiler. ☐
14. If in a boiler, the circulation of water is by pumps, then the boiler is known as
   (*a*) forced circulation boiler ☐ (*b*) natural circulation boiler ☐
   (*c*) internally fired boiler ☐ (*d*) externally fired boiler. ☐
15. If in a boiler, the combustion takes place outside the region of boiling water, the boiler is known as
   (*a*) forced circulation boiler ☐ (*b*) natural circulation boiler ☐
   (*c*) internally fired boiler ☐ (*d*) externally fired boiler. ☐
16. If in a boiler, the combustion takes place inside the region of boiling water, the boiler is known as
   (*a*) forced circulation boiler ☐ (*b*) natural circulation boiler ☐
   (*c*) internally fired boiler ☐ (*d*) externally fired boiler. ☐
17. La-Mont boiler is a
   (*a*) low pressure water-tube boiler ☐ (*b*) forced circulation boiler ☐
   (*c*) natural circulation boiler ☐ (*d*) none of the above. ☐

18. Which of the following is a high pressure boiler?

(*a*) La-Mont boiler ☐ (*b*) Benson boiler ☐
(*c*) Loeffler boiler ☐ (*d*) all of the above ☐
(*e*) none of the above. ☐

19. La-Mont boiler produces steam at a pressure which varies from

(*a*) 5×10^5 to 10×10^5 N/m^2 ☐ (*b*) 10×10^5 to 15×10^5 N/m^2 ☐
(*c*) 1×10^5 to 5×10^5 N/m^2 ☐ (*d*) 35×10^5 to 200×10^5 N/m^2. ☐

20. Benson boiler produces steam upto a rate of

(*a*) 100 kg/hr only ☐ (*b*) 10^4 kg/hr only ☐
(*c*) 135 ton/hr only ☐ (*d*) 100 kg/hr only. ☐

21. The rate of flow of steam in case of water-tube boilers as compared to fire-tube boilers is

(*a*) less ☐ (*b*) more ☐
(*c*) same ☐ (*d*) none of the above. ☐

22. A Benson boiler requires

(*a*) two drums ☐ (*b*) one drum ☐
(*c*) no drum ☐ (*d*) three drum. ☐

23. Choose the wrong statement

(*a*) Steam boiler is a device used for generating and supplying steam at a high pressure and temperature ☐
(*b*) La-Mont boiler is a forced circulation boiler ☐
(*c*) Benson boiler is a high pressure boiler ☐
(*d*) A Benson boiler requires two drums. ☐

Boiler Mountings

24. The fitting mounted on the boiler, whose function is to put off the fire in the furnace, when level of water falls to an unsafe limit is called

(*a*) safety valve ☐ (*b*) stop valve ☐
(*c*) fusible plug ☐ (*d*) blow off cock. ☐

25. The fitting mounted on the boiler, whose function is to prevent the steam pressure in a boiler exceeding a fixed maximum pressure, is called

(*a*) safety valve ☐ (*b*) stop valve ☐
(*c*) fusible valve ☐ (*d*) blow off cock. ☐

26. The fitting mounted on the boiler, whose function is to control the flow of steam from boiler to the main steam pipe and to shut off steam completely when required, is called

(*a*) safety valve ☐ (*b*) stop valve ☐
(*c*) fusible valve ☐ (*d*) blow off cock. ☐

27. The fitting mounted on the boiler, whose function is to empty the boiler when required and to discharge mud and scale which are accumulated at the bottom of boiler is called

(*a*) safety valve ☐ (*b*) stop valve ☐
(*c*) fusible valve ☐ (*d*) blow off cock. ☐

28. The fitting mounted on the boiler, whose function is to regulate the supply of water pumped into the boiler by the feed pump, is called

(*a*) water level indicator ☐ (*b*) feed check valve ☐
(*c*) blow off water ☐ (*d*) stop valve. ☐

29. The device, which indicates the exact level or water in the boiler at any instant, is called

(*a*) water level indicator ☐ (*b*) feed check valve ☐
(*c*) blow off cock ☐ (*d*) stop valve. ☐

Boiler Accessories

30. A device, whose function is to heat feed water by utilising the heat in the exhaust flue gases before leaving through the chimney, is called

(*a*) superheater ☐ (*b*) economiser ☐
(*c*) air preheater ☐ (*d*) none of the above. ☐

31. A device in which the heat of the combustion products is utilised first dry wet steam and then to rise its temperature without raising its pressure, is called

(*a*) superheater ☐ (*b*) economiser ☐
(*c*) air preheater ☐ (*d*) feed pump. ☐

32. A device, in which some portion of waste heat of fluge gases is recovered to heat the air before it passes into the furnace for combustion purposes, is called

(*a*) superheater ☐ (*b*) economiser ☐
(*c*) air preheater ☐ (*d*) feed pump. ☐

33. A device, which is used for pumping water into the boiler and also for heating the feed water, is called

(*a*) economiser ☐ (*b*) feed pump ☐
(*c*) injector ☐ (*d*) air preheater. ☐

34. A device used for raising the pressure of feed water, is called

(*a*) economiser ☐ (*b*) feed pump ☐
(*c*) injector ☐ (*d*) air preheater. ☐

35. Which of the following are boiler mountings ?

(*a*) economiser ☐ (*b*) fusible plug ☐
(*c*) superheater ☐ (*d*) injector. ☐

36. Which of the following are boiler accessories?

(*a*) safety valve ☐ (*b*) stop valve ☐
(*c*) economiser ☐ (*d*) blow off cock. ☐

37. Choose the correct statement

(*a*) Size of the boiler tubes is specified by inside diameter. ☐
(*b*) An economiser decreases the steam raising capacity of a boiler. ☐
(*c*) The basic purpose of a drum in boiler is to serve as storage of steam. ☐
(*d*) Injector is used for pumping into the boiler and also for heating the feed water. ☐

38. Choose the wrong statement
   (*a*) Safety valve is a boiler mounting. ☐
   (*b*) Economiser is a boiler accessories. ☐
   (*c*) The supply of water pumped into the boiler is regulated by feed check valve. ☐
   (*d*) Super heater heats the feed water by utilizing the heat in the exhaust the gases before having through the chimney. ☐

39. The function of high steam and low water safety valve is
   (*a*) to blow out if the steam pressure is higher than working pressure ☐
   (*b*) to blow out steam when the water level in the boiler is low ☐
   (*c*) both (*a*) and (*b*) ☐
   (*d*) none of the above. ☐

40. Which safety valve of the following should be used for portable boiler?
   (*a*) dead weight safety valve ☐
   (*b*) spring loaded safety valve ☐
   (*c*) lever safety valve ☐
   (*d*) high steam and low water safety valve. ☐

41. The function of junction valve is to
   (*a*) regulate the supply of water ☐
   (*b*) regulate the flow of steam ☐
   (*c*) put off the fire in the furnace ☐
   (*d*) indicate the level of water. ☐

42. A device, which is connected to the steam supply pipe line to maintain constant pressure, is called
   (*a*) steam separator ☐ (*b*) steam trap ☐
   (*c*) pressure reducing valve ☐ (*d*) injector. ☐

43. A device, which drains of water from steam pipes and jackets without allowing steam of escape through it, is called
   (*a*) steam separator ☐ (*b*) steam trap ☐
   (*c*) injector ☐ (*d*) economiser. ☐

44. The device, which separates the suspended water particles from steam which is on its way from the boiler to the engine, is called
   (*a*) steam separator ☐ (*b*) steam trap ☐
   (*c*) injector ☐ (*d*) economiser. ☐

45. Choose the correct statement
   (*a*) Fusible plug controls the flow of steam ☐
   (*b*) Stop valve puts off the fire in the furnace when level of water falls to an unsafe limit. ☐
   (*c*) Blow off cock blows out steam if the steam pressure is higher than the working pressure. ☐
   (*d*) Feed check valve regulate the supply of water in the boiler. ☐

46. Choose the wrong statement

(a) Water level indicator indicates the exact level of water in the boiler. ☐

(b) Economiser heats the feed water by utilising the heat in the exhaust flue gases before leaving through the chimney. ☐

(c) Injector is used for pumping water and also for heating the feed water. ☐

(d) Fusible plug is a boiler accessory. ☐

47. The performance of a boiler is measured by

(a) amount of water evaporated per hour ☐ (b) steam produced in kg/hr ☐

(c) steam produced in kg/kg of fuel burnt ☐ (d) all of the above ☐

(e) none of the above. ☐

48. To compare the capacity of boilers, the feed water temperature and working pressure adopted are

(a) 60°C and normal atmospheric pressure ☐

(b) 60°C and 15 kgf/cm^2 pressure ☐

(c) 100°C and normal atmospheric pressure ☐

(d) 100°C and 15 kgf/cm^2 pressure. ☐

49. Equivalent evaporation means

(a) amount of water evaporated per kg of fuel burnt ☐

(b) amount of steam produced in kg/hr ☐

(c) amount of water evaporated from and at 100° into dry and satured steam at normal atmospheric pressure ☐

(d) the evaporation of 15.653 kg of water per hour from and at 100°C. ☐

50. If H_S = total heat of steam at the working pressure, h = sensible heat of feed water and L = latent heat of steam, then the factor of evaporation is given by

(a) $\frac{H_S + h}{L}$ ☐ (b) $\frac{L}{(H_S + h)}$ ☐

(c) $\frac{H_S - h}{L}$ ☐ (d) $\frac{L}{H_S - h}$. ☐

51. The equivalent evaporation is given by

(a) $\frac{W(H_S + h)}{L}$ ☐ (b) $\frac{W(H_S - h)}{L}$ ☐

(c) $\frac{(W \times L)}{(H_S - h)}$ ☐ (d) $\frac{(W \times L)}{(H_S + h)}$. ☐

where W = Weight of water actually evaporated into steam at the working pressure,

H_S = Total heat of steam,

h = Sensible heat of feed water, and

L = Latent heat of steam.

Efficiency and Heat Losses of a Boiler

52. The efficiency of a boiler is defined as

(*a*) ratio of heat supplied by the fuel to the heat absorbed by feed water ☐

(*b*) ratio of heat absorbed by feed water to the heat supplied by fuel in a given time ☐

(*c*) ratio of the weight of water evaporated to the total water supplied in the boiler for a given time ☐

(*d*) none of the above. ☐

53. Boiler efficiency (η) is expressed as equal to

(*a*) $\dfrac{W_S(H_S + h)}{WC}$ ☐ (*b*) $\dfrac{W_S(H_S - h)}{WC}$ ☐

(*c*) $\dfrac{W \times C}{W_S(h \times H_S)}$ ☐ (*d*) $\dfrac{WC}{W_S(H_S - h)}$. ☐

where, W = Weight of fuel burnt/hr, H_S = Total heat of steel produced

h = Sensible heat of feed water, W_S = Weight of steam produced/hr

C = Calorific value of fuel per kg.

54. The main object of a boiler trial is

(*a*) to determine the thermal efficiency of the boiler when working at a definite pressure ☐

(*b*) to draw up heat balance sheet for the boiler ☐

(*c*) both (*a*) and (*b*) ☐

(*d*) none of the above. ☐

55. The amount of heat lost in a boiler is equal to

(*a*) the sum of heat supplied by fuel and heat utilised in raising steam from feed water ☐

(*b*) difference of heat supplied and heat utilised in raising steam from feed water ☐

(*c*) the ratio of heat supplied by fuel of heat utilised by feed water. ☐

(*d*) none of the above. ☐

56. Which of the following losses occur in a boiler?

(*a*) heat carried away by dry flue gases ☐

(*b*) heat carried away by moisture in fuel ☐

(*c*) heat lost due to unburnt carbon ☐

(*d*) heat carried away by steam formed by combustion per kg of fuel ☐

(*e*) all of the above ☐

57. Maximum heat loss in a boiler occur due to

(*a*) flue gases ☐ (*b*) incomplete combustion ☐

(*c*) unburnt carbon ☐ (*d*) moisture in fuel ☐

58. The maximum design working pressure for a fire-tube boiler is

(*a*) 80×10^5 N/m^2 ☐ (*b*) 200×10^5 N/m^2 ☐

(*c*) 16×10^5 N/m^2 ☐ (*d*) 1×10^5 N/m^2. ☐

59. Orsatmeter is a device used for
(*a*) mass flow of the flue gas ☐ (*b*) gravimetric analysis of the flue gas ☐
(*c*) volumetric analysis of the flue gas ☐ (*d*) none of the above. ☐

60. Locomotive type boiler is
(*a*) vertical tubular fire-tube type ☐ (*b*) horizontal tubular water tube-boiler ☐
(*c*) horizontal multi-tubular fire-tube type ☐ (*d*) none of the above. ☐

61. Choose the correct statement
(*a*) The increase in pressure reduces the boiling point of liquid. ☐
(*b*) Cochran boiler is a horizontal water-tube boiler. ☐
(*c*) Hygrometry deals with temperature of air. ☐
(*d*) The evaporation of 15.653 kg of water per hour from and at 100°C is known as one boiler power. ☐

62. Choose the wrong statement
(*a*) Maximum heat loss in a boiler is due to flue gases. ☐
(*b*) Orsatmeter is used for volumetric analysis of the flue gas. ☐
(*c*) La-Mont boiler works on formed circulation. ☐
(*d*) Water-tube boilers are internally fired. ☐

Boiler Draught and Chimney

63. Boiler draught is defined as
(*a*) small pressure difference which causes flow of gases to take place inside the boiler ☐
(*b*) sum of pressure above and below the fire grate ☐
(*c*) the device which produces vacuum in the boiler ☐
(*d*) none of the above. ☐

64. Draught produced by a chimney is known as
(*a*) induced draught ☐ (*b*) forced draught ☐
(*c*) natural draught ☐ (*d*) none of the above. ☐

65. Stack refers to chimney which is made of
(*a*) masonry structure ☐ (*b*) concrete structure ☐
(*c*) metal ☐ (*d*) none of the above. ☐

66. The amount of draught produced by a brick chimney as compared to the draught produced by a steel chimney is
(*a*) less ☐ (*b*) more ☐
(*c*) same ☐ (*d*) none of the above. ☐

67. The condition for maximum discharge through a chimney is that height of chimney should be
(*a*) two times the height of column of hot gases producing draught ☐
(*b*) equal to the height of column of hot gases producing draught ☐
(*c*) half the height of column of hot gases producing drought ☐
(*d*) none of the above. ☐

68. If H' is the equivalent height of the draught produced in a chimney and h_f is the head lost due to friction, the velocity of flue gases in the chimney is given by

(a) $\sqrt{2gH'}$ ☐ (b) $2g\sqrt{H'+h_f}$ ☐

(c) $\sqrt{2g(H'-h_f)}$ ☐ (d) $\sqrt{2g(H'+h_f)}$. ☐

69. For maximum discharge through a chimney of given height and cross-section, the ratio of absolute temperature of chimney gases to the absolute temperature of the outside air is equal to

(a) $\frac{2m}{m+1}$ ☐ (b) $\frac{m+1}{2m}$ ☐

(c) $\frac{2(m+1)}{m}$ ☐ (d) $\frac{m}{2m+1}$ ☐

where m = Mass of air used in kg per kg of fuel.

70. Density of chimney gas at an absolute temperature of T K is equal to

(a) $\frac{273}{T}\times\left(\frac{m}{m+1}\right)\times 1.293$ ☐ (b) $1.293\left(\frac{m+1}{m}\right)\times\frac{273}{T}$ ☐

(c) $\left(\frac{m+1}{m}\right)\times\frac{1}{1.293}\times\frac{2.73}{T}$ ☐ (d) $\left(\frac{m}{m+1}\right)\times\frac{273}{T}$. ☐

where m = Mass of air used in kg per kg of fuel.

71. The draught in terms of height of column of water in mm for the hot gases flowing through the chimney is equal to

(a) $353\left[\frac{1}{T_1}-\frac{m+1}{m}\frac{1}{T}\right]$ ☐ (b) $353\,H\left[\frac{m+1}{m}-\frac{1}{T_1}\right]$ ☐

(c) $353\left[\frac{1}{T_1}+\frac{m+1}{m}\frac{1}{T}\right]$ ☐ (d) $353\,H\left[\frac{1}{T_1}-\frac{m+1}{m}\frac{1}{T}\right]$ ☐

where H = Height of chimney in metre,

T = Absolute temperature of flue gasses inside the chimney,

T_1 = Absolute temperature of outside air, and

m = Mass of air per kg of fuel.

72. The density of air at 0°C is equal to

(a) 1.293 kg/m^3 ☐ (b) 12.93 kg/m^3 ☐

(c) 129.3 kg/m^3 ☐ (d) 0.1293 kg/m^3. ☐

73. The volume per unit mass of air at 0°C is equal to

(a) 7.734 m^3/kg ☐ (b) 0.7734 m^3/kg ☐

(c) 0.07734 m^3/kg ☐ (d) 1.293 m^3/kg. ☐

74. For maximum discharge of flue gases through a chimney, the drought (h) in mm of water is equal to

(a) $353\frac{H}{T}$ ☐ (b) $\frac{176.5\,H}{T_1}$ ☐

(c) $\frac{T_1}{353\, H}$ ☐ (d) $\frac{T_1}{176.5\, H}$. ☐

where H = Height of chimney in metre, and

T_1 = Absolute temperature of outside air.

75. Artificial draught is produced by a

(a) steam jet ☐ (b) fan ☐

(c) chimney ☐ (d) both (a) and (b). ☐

76. The type of draught, which is more economical when drought required is more than 40 mm, is

(a) natural draught ☐ (b) artificial draught ☐

(c) both (a) and (b) ☐ (d) none of the above. ☐

77. The artificial draught produces

(a) less smoke ☐ (b) higher furnace temperature ☐

(c) more draught ☐ (d) all of the above. ☐

78. Choose the correct statement

(a) Draught produced by a chimney is known as forced draught. ☐

(b) Steel chimney produces less draught than that of brick chimney. ☐

(c) For maximum discharge, the height of chimney should be equal to the height of the hot gas column producing draught. ☐

(d) Natural draught produces more draught than artificial draught. ☐

79. Choose the wrong statement

(a) For maximum discharge through a chimney, the draught in mm of water is equal to (176.5 × height of chimney)/absolute temperature of outside air. ☐

(b) Artificial draught as compared to natural draught, produces less smoke. ☐

(c) Draught is the small pressure difference which causes flow of the flue gases. ☐

(d) Draught produced by a chimney is known as artificial draught. ☐

80. In case of locomotive boilers, draught is produced by

(a) fan ☐ (b) steam jet ☐

(c) chimney ☐ (d) all of the above. ☐

81. The draught produced by a fan which is placed at the bottom of the chimney and sucks the gas from boiler side, is known as

(a) forced draught ☐ (b) induced draught ☐

(c) natural draught ☐ (d) none of the above. ☐

82. If a fan or blower is installed near as at the base of a boiler to force air through fuel on stoker or grate either direct or through an air preheater, the type of draught produced is called

(a) forced draught ☐ (b) induced draught ☐

(c) balanced draught ☐ (d) none of he above. ☐

83. If a fan is installed near the base of a boiler to force air through fuel and another fan is installed near the base of the chimney to such the gas from boiler, the type of draught produced is called

(*a*) forced draught ☐ (*b*) induced draught ☐

(*c*) balanced draught ☐ (*d*) none of the above. ☐

84. The power required to run a fan, which produces draught is equal to

(*a*) $\frac{h \times V \times \eta}{60,000}$ ☐ (*b*) $\frac{h \times V}{\eta \times 60,000}$ ☐

(*c*) $\frac{h \times V}{60,000}$ ☐ (*d*) $\frac{h \times \eta}{V \times 60,000}$. ☐

where V = Volume of gas flowing through fan in m^3/min,

η = Efficiency of fan

h = Draught produced by fan in mm of water column.

85. For the same draught developed, the forced draught fan as compared to induced draught fan requires

(*a*) more power ☐ (*b*) less power ☐

(*c*) same power ☐ (*d*) none of the above. ☐

86. Choose the correct statement

(*a*) The heating value of coal increases during storage. ☐

(*b*) A water-tube boiler compared to fire-tube boiler has more heating surface for the same diameter and thickness of the tube. ☐

(*c*) Balanced draught is produced by one induced draught fan and a chimney. ☐

(*d*) In case of locomotive boilers, draught is produced by a fan. ☐

Basic Questions

87. On kcal/kg in S.I. unit is equal to

(*a*) 427 J/kg ☐ (*b*) 4.72 J/kg ☐

(*c*) 4.1868 kJ/kg ☐ (*d*) 42.7 J/kg. ☐

88. Mechanical equivalent (J) of heat is equal to

(*a*) 427 J/kcal ☐ (*b*) 427 kgf-m/kcal ☐

(*c*) 427 J/kg ☐ (*d*) 4.1868 kJ/kg. ☐

89. The critical point is the point at which

(*a*) melting and boiling temperature became equal ☐

(*b*) the change of volume accompanying evaporation is maximum ☐

(*c*) the change of volume accompanying evaporation is zero ☐

(*d*) none of the above. ☐

90. At the critical point, the latent heat of vaporisation is

(*a*) maximum ☐ (*b*) minimum ☐

(*c*) zero ☐ (*d*) none of the above. ☐

91. The temperature, at which the melting and boiling temperature become equal, is equal to

(*a*) 273.16 K ☐ (*b*) 647.3 K ☐

(*c*) 373.16 K ☐ (*d*) 173.16 K. ☐

92. At the critical point, the temperature for water is equal to

(*a*) 273.16 K ☐ (*b*) 647.3 K ☐

(*c*) 373.16 K ☐ (*d*) 173.16 K. ☐

93. At the critical point, the pressure for water is equal to

(*a*) 1 kgf/cm^2 ☐ (*b*) 200 kgf/cm^2 ☐

(*c*) 225.65 kgf/cm^2 ☐ (*d*) zero. ☐

94. At the critical point, the specific volume of water is equal to

(*a*) 1 m^3/kg ☐ (*b*) 0.00317 m^3/kg ☐

(*c*) 0 ☐ (*d*) 0.00622m^3/kg. ☐

95. The pressure, at which the melting and boiling temperature become equal, is equal to

(*a*) 225.65 kgf/cm^2 ☐ (*b*) 1 kgf/cm^2 ☐

(*c*) 0.00622 kgf/cm^2 ☐ (*d*) 0.5 kgf/cm^2 ☐

96. Sublimation of ice to steam takes place if the pressure is

(*a*) 225.65 kgf/cm^2 ☐ (*b*) 1 kgf/cm^2 ☐

(*c*) less than 0.0062 kgf/cm^2 ☐ (*d*) 1 kgf/cm^2. ☐

97. At a low pressure

(*a*) boiling point rises markedly and melting point drops slightly ☐

(*b*) boiling point rises slightly and melting point drops markedly ☐

(*c*) boiling point drops markedly and melting point rises slightly ☐

(*d*) boiling point drops slightly and metaling point rises markedly. ☐

98. Dryness fraction of wet steam is given by

(*a*) ration of mass of dry steam to the mass of suspended liquid water ☐

(*b*) ratio of mass of suspended liquid water to mass of dry steam ☐

(*c*) ratio of mass of dry stream to the sum of mass of suspended liquid water and dry steam ☐

(*d*) ratio of sum of mass of suspended liquid water and dry steam to the mass of dry steam. ☐

99. Which of the curves shown in Fig. 5.12 represents the variation of saturation temperature with pressure?

(*a*) curve *A* ☐

(*b*) curve *B* ☐

(*c*) curve *C* ☐

(*d*) curve *D* ☐

100. Which of the curves shown in Fig. 5.12 represents the variation of specific volume of steam with pressure?

(*a*) curve *A* ☐

(*b*) curve *B* ☐

(*c*) curve *C* ☐

(*d*) curve *D* ☐

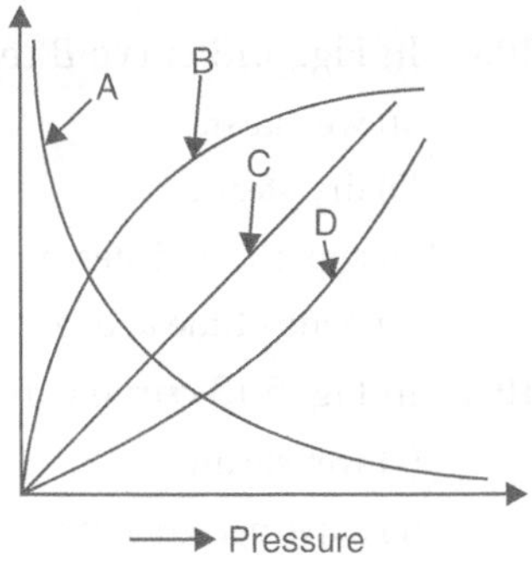

FIGURE 5.12

101. If the mass of dry steam is equal to 10 kg and mass of suspended liquid water is 5 kg in a mixture of wet steam, then dryness fraction is equal to

(*a*) 0.50 ☐ (*b*) 0.33 ☐

(*c*) 0.667 ☐ (*d*) 1.0 ☐

102. Sublimation region is the region where

(*a*) liquid and vapour phases are in equilibrium ☐

(*b*) solid and liquid phases are in equilibrium ☐

(*c*) solid, liquid and vapour phases are in equilibrium ☐

(*d*) sollid and vapour phases are in equilibrium. ☐

103. The equivalent evaporation of boiler is measure to compare

(*a*) any type of boilers operating under and conditions ☐

(*b*) the two different boilers of the same make ☐

(*c*) the given boiler with the model ☐

(*d*) none of the above. ☐

Steam Engine

104. The standard cycle for comparing the performance of steam plants is

(*a*) Carnot cycle ☐ (*b*) Rankine cycle ☐

(*c*) Otto cycle ☐ (*d*) Joule cycle. ☐

105. Figure 5.13 show the *p-V* diagram representing the Rankine cycle (neglecting feed pump work) for wet, dry and super heated steam. Curve *A* represents the diagram for

(*a*) wet steam ☐

(*b*) dry steam ☐

(*c*) superheated steam ☐

(*d*) none of the above. ☐

FIGURE 5.13

106. In Fig. 5.13 curve *B* represents the diagram for

(*a*) wet steam ☐

(*b*) dry steam ☐

(*c*) superheated steam ☐

(*d*) none of the above. ☐

107. In Fig. 5.13 curve *C* represents the diagram for

(*a*) wet steam ☐ (*b*) dry steam ☐

(*c*) superheated steam ☐ (*d*) none of the above. ☐

108. Figure 5.14 shows the *T-S* diagram representing the Rankine cycle (neglecting feed pump work) for wet, dry and superheated steam. Curve *A* represents the diagram for

(*a*) superheated steam ☐ (*b*) dry steam ☐

(*c*) wet steam ☐ (*d*) none of the above. ☐

109. In Fig. 5.14 curve *B* represents the diagram for

(*a*) superheated steam ☐

(*b*) dry steam ☐

(*c*) wet steam ☐

(*d*) none of the above. ☐

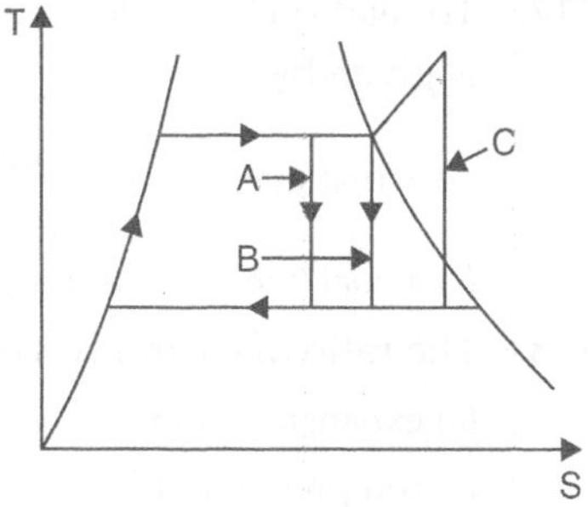

110. In Fig. 5.14, curve *C* represents the diagram for

(*a*) superheated steam ☐

(*b*) dry steam ☐

(*c*) wet steam ☐

(*d*) none of the above. ☐

FIGURE 5.14

111. Figure 5.15 shows the Rankine cycle and the Carnot cycle on *T-S* diagram. The Carnot cycle is represented by the area

(*a*) *abcd* ☐

(*b*) *ebcd* ☐

(*c*) *habcf* ☐

(*d*) *gebcdf.* ☐

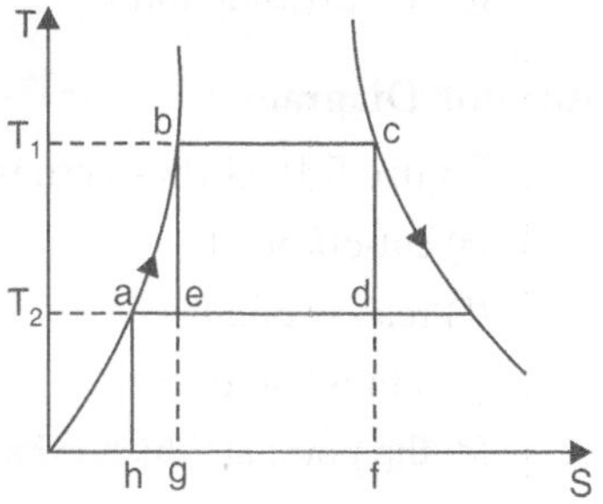

FIGURE 5.15

112. In Fig. 5.15, the Rankine cycle is represented by the area

(*a*) *abcd* ☐

(*b*) *ebcd* ☐

(*c*) *habcf* ☐

(*d*) *gebcdf* ☐

113. In Fig. 5.15, the Rankine efficiency is equal to the ratio of

(*a*) area *abcd* to area *ebcd* ☐ (*b*) area *abcd* to area *habcdf* ☐

(*c*) area *ebcd* to area *habcdf* ☐ (*d*) area *ebcd* to area *gebcdf.* ☐

114. In Fig. 5.15, the Carnot cycle efficiency is equal to the ratio of

(*a*) area *abcd* to area *ebcd* ☐ (*b*) area *abcd* to area *habcdf* ☐

(*c*) area *ebcd* to area *habcdf* ☐ (*d*) area *ebcd* to area gebcdf. ☐

115. Diagram factor (D.F.) is defined as

(*a*) product of area of actual indicator diagram and area of theoretical indicator diagram ☐

(*b*) ratio of area of actual indicator diagram and area of theoretical indicator diagram ☐

(*c*) ratio of area of theoreticcal indicator diagram to area of actual indicator diagram ☐

(*d*) none of the above. ☐

116. The diagram factor is always

(*a*) more than 1.0 ☐ (*b*) equal to 1.0 ☐

(*c*) equal to zero ☐ (*d*) less than 1.0. ☐

117. To relation between actual mean effective pressure and theoretical **mean effective pressure** is given by

(*a*) actual m.e.p. = $\frac{\text{Theoretical m.e.p.}}{\text{Diagram factor}}$ ☐ (*b*) actual m.e.p = $\frac{\text{Diagram factor}}{\text{Theoretical m.e.p.}}$ ☐

(*c*) actual m.e.p = Theoretical m.e.p. × D.F. ☐ (*d*) none of the above. ☐

118. The ratio of clearance volume to the swept volume is known as

(*a*) expansion ratio ☐ (*b*) cut-off ratio ☐

(*c*) compression ratio ☐ (*d*) clearance ratio. ☐

119. The ratio of the volume at cut-off to the swept volume is known as

(*a*) expansion ratio ☐ (*b*) cut off ratio ☐

(*c*) compression ratio ☐ (*d*) clearance ratio. ☐

Indicator Diagram

120. Figure 5.16 shows actual indicator diagram of steam engine. The point *A* is called

(*a*) cut-off point ☐

(*b*) release point ☐

(*c*) admission point ☐

(*d*) the point at which exhaust port closes. ☐

121. In Fig. 5.16, the point *C* represents

(*a*) cut-off point ☐

(*b*) release point ☐

(*c*) admission point ☐

(*d*) the point at which exhaust port closes. ☐

FIGURE 5.16

122. In Fig. 5.16, the point *D* represents

(*a*) cutt-off point ☐ (*b*) release point ☐

(*c*) admission point ☐ (*d*) the point at which exhaust port closes. ☐

123. In Fig. 5.16, the point *E* represents

(*a*) cut-off point ☐ (*b*) release point ☐

(*c*) admission point ☐ (*d*) the point at which exhaust port closes. ☐

124. Choose the wrong statement. The area of actual indicator diagram is less than that of the theoretical indicator diagram due to

(*a*) pressure drop because of wire drawing ☐

(*b*) gradual cut-off and release ☐

(*c*) expansion is different from hyperbolic ☐

(*d*) release and admission takes place at the end of the stroke. ☐

125. The saturation curve is the curve showing the volume of the steam in the cylinder during expansion stroke (*i.e.*, from the cut-off point to the point at release) in a steam engine if the condition of the steam is

(*a*) 80% dry ☐ (*b*) 90% dry ☐

(*c*) 95% dry ☐ (*d*) 100% dry. ☐

126. Missing quantity is defined as

(*a*) The difference in the volume at any point between the saturation curve and actual expansion curve. ☐

(*b*) The product of volume given by saturation curve and actual expansion curve. ☐

(*c*) The difference in volume given by the saturation curve and clearance volume. ☐

(*d*) None of the above. ☐

127. The steam, in a steam engine, entrapped in the cylinder after the exhaust valve closes, is known as

(*a*) missing quantity ☐ (*b*) cushion steam ☐

(*c*) clearance volume steam ☐ (*d*) none of the above. ☐

Governing of Steam Engine

128. Which one method is adopted for governing the steam engine?

(*a*) by-pass governing ☐ (*b*) nozzle control governing ☐

(*c*) cut-off governing ☐ (*d*) combination of (*a*) and (*b*). ☐

129. Which one method is common for governing the steam engine and steam turbines?

(*a*) by-pass governing ☐ (*b*) nozzle control governing ☐

(*c*) cut-off governing ☐ (*d*) throttle governing. ☐

130. Cut-off governing in steam engine as compared to throttle governing is

(*a*) more efficient and economical ☐ (*b*) less efficient and less economical ☐

(*c*) more efficient and less economical ☐ (*d*) less efficient but more economical. ☐

131. Willan's law states that if a steam engine is governed by throttling then

(*a*) rate of steam consumption is inversely proportional to the indicated power ☐

(*b*) rate of steam consumption is directly proportional to the indicated power. ☐

(*c*) rate of steam consumption varies with the square of indicated power. ☐

(*d*) rate of steam consumption varies with the square root of I.P. ☐

132. Willan's line is a straight line graph between the rate of steam consumption and

(*a*) pressure of steam ☐ (*b*) temperature of steam ☐

(*c*) indicated power ☐ (*d*) none of the above. ☐

133. The law followed by Willan's line is expressed as

(*a*) $W = C + mP$ ☐ (*b*) $P = mW + C$ ☐

(*c*) $W = C - mP$ ☐ (*d*) $P = mW + C$. ☐

where P = I.P, and W = Steam consumption per hour m and C are constant of proportionality.

134. Choose the correct statement

(*a*) In throttle governing of steam engine, the volume of intake steam is varied. ☐

(*b*) In cut-off governing of steam engine, the pressure of intake steam is varied. ☐

(*c*) In throttle governing of steam engine, the temperature of intake steam is varried. ☐

(*d*) In throttle governing of steam engine, the pressure of intake steam is varied. ☐

135. Choose the wrong statement

(*a*) Actual mean effective pressure is equal to the theoretical mean effective pressure divided by diagram factor. ☐

(*b*) Cushion steam is the steam left in the clearance space from the previous stroke. ☐

(*c*) The ratio of volume at cut-off to the swept volume is known as cut-off ratio. ☐

(*d*) The average volume of diagram factor lies between 1.0 to 1.1. ☐

136. During the stroke of steam engine, the sequence of events are

(*a*) cut-off, compression, admission, release ☐ (*b*) cut-off, admission, compression, release ☐

(*c*) admission, cut-off release, compression ☐ (*d*) admission, release, compression, cut-off. ☐

137. The part, which provides simple harmonic motion to the *D*- slide valve of a steam engine is called

(*a*) eccentric rod ☐ (*b*) valve rod ☐

(*c*) piston rod ☐ (*d*) flywheel. ☐

138. The part, which converts rotary motion of the crank-shaft into reciprocating motion of the valve rod, is called

(*a*) eccentric rod ☐ (*b*) *D*-slide valve ☐

(*c*) piston rod ☐ (*d*) flywheel. ☐

139. The part, which is used to exhaust steam from the cylinder at proper moment is called

(*a*) eccentric rod ☐ (*b*) D-slide valve ☐

(*c*) piston rod ☐ (*d*) flywheel. ☐

140. Choose the wrong statement

(*a*) The governor keeps the engine speed constant at all load conditions. ☐

(*b*) The flywheel prevents the fluctuation of speed. ☐

(*c*) Eccentric provides reciprocating motion to the slide valve. ☐

(*d*) A single acting steam engine produces twice the power produced by a double acting steam engine. ☐

141. A steam engine develops 10 I.P. when 200 kg of steam per hour is used and 30 I.P. when 400 kg of steam per hour is used. If the engine is throttle-governed the steam consumption in kg/hr, when the engine develops 20 I.P., would be

(*a*) 250 ☐ (*b*) 300 ☐

(*c*) 350 ☐ (*d*) 275. ☐

142. Figure 5.17 shows the variation of I.P. with the rate of steam consumed in kg per hour for a steam engine. For cut-off governing, the curve is

(*a*) *A* ☐

(*b*) *B* ☐

(*c*) *C* ☐

(*d*) *D*. ☐

143. Refer to Fig. 5.17, for throttle-governing, the curve is

(*a*) *A* ☐

(*b*) *C* ☐

(*c*) *B* ☐

(*d*) *D* ☐

FIGURE 5.17

144. By compounding the steam engines, condensation is

(*a*) increased ☐ (*b*) reduced ☐

(*c*) constant ☐ (*d*) none of the above. ☐

145. By compounding the steam engines, the stroke length is

(*a*) increased ☐ (*b*) reduced ☐

(*c*) constant ☐ (*d*) none of the above. ☐

146. A compound steam engine as compared to simple steam engine, requires

(*a*) heavier flywheel ☐ (*b*) lighter flywheel ☐

(*c*) same flywheel ☐ (*d*) none of the above ☐

147. Tandom type of two cylinder compound engine has

(*a*) two cranks at 180° to each other ☐ (*b*) one crank only ☐

(*c*) two cranks at 90° to each other ☐ (*d*) none of the above. ☐

148. Woolfe's type of two cylinder compound engine has

(*a*) two cranks at 180° to each other ☐ (*b*) one crank only ☐

(*c*) two cranks at 90° to each other ☐ (*d*) none of the above. ☐

149. Receiver type of two cylinder compound engine has

(*a*) two cranks at 180° to each other ☐ (*b*) one crank only ☐

(*c*) two cranks at 90° to each other ☐ (*d*) none of the above. ☐

150. Choose the wrong statement

(*a*) The high pressure and low pressure cylinder of a Tandem type compound engine have common piston rod. ☐

(*b*) The high pressure and pressure cylinders of Woolfe type compound engine have separate piston rods. ☐

(*c*) The low pressure cylinder of a Tandem type compound engine is placed nearest the crank shaft. ☐

(*d*) A tandem type compound engine requires a lighter flywheel. ☐

151. Choose the correct statement

(*a*) A cross-compounding of steam engines, cylinders are arranged at 180° and each cylinder has common piston, connecting rod and piston. ☐

(*b*) A receiver type compound engine requires a heavier flywheel. ☐

(*c*) A compound steam engine requires heavier flywheel than that of simple steam engine. ☐

(*d*) By compounding the steam engines, the temperature range per cylinder is reduced, with corresponding reduction in condensation. ☐

152. If the motion of D-slide valve is assumed simple harmonic, the positions of crank at admission and cut-off is given by

(*a*) $\theta = \sin^{-1}\left(\frac{2e}{L}\right) - \alpha$ ☐ (*b*) $\theta = \sin^{-1}\left(\frac{2e}{L}\right) + \alpha$ ☐

(*c*) $\theta = \sin^{-1}\left(\frac{2s}{L}\right) - \alpha$ ☐ (*d*) $\theta = \sin^{-1}\left(\frac{-2e}{L}\right) - \alpha$. ☐

where θ = Crank angle, α = Stroke of advance of eccentric,

L = Stroke of valve, s = Steam lap.

153. If the motion of D-slide valve is assumed simple harmonic, the position of crank at release and compression is given by

(*a*) $\theta = \sin^{-1}\left(\frac{2e}{L}\right) - \alpha$ ☐ (*b*) $\theta = \sin^{-1}\left(\frac{2s}{L}\right) + \alpha$ ☐

(*c*) $\theta = \sin^{-1}\left(\frac{2s}{L}\right) - \alpha$ ☐ (*d*) $\theta = \sin^{-1}\left(\frac{-2e}{L}\right) - \alpha$. ☐

where s = Steam lap, e = Exhaust lap, θ = Crank angle,

α = Stroke of advance of eccentric, and

L = Stroke of valve.

154. The D-slide valve of a steam engine has steam lap = 14 mm, valve stroke = 56 mm and angle of advance = 40°. The crank angle at admission is equal to

(*a*) 10° ☐ (*b*) 75° ☐

(*c*) –10° ☐ (*d*) –75°. ☐

155. For the question 154, the crank angle for cut-off will be equal to

(*a*) 10° ☐ (*b*) 110° ☐

(*c*) –10° ☐ (*d*) 75°. ☐

156. If the angle of advance is 150° and steam engine is to be reversed then eccentric should be shifted by

(*a*) 165° ☐ (*b*) 180° ☐

(*c*) 205° ☐ (*d*) 15°. ☐

157. Various curves for steam consumption in kg/hr, specific steam consumption in kg/B.P./hr, thermal efficiency and mechanical efficiency versus B.P. are shown in Fig. 5.18. Steam consumption/hr is given by

(*a*) curve *A* ☐

(*b*) curve *B* ☐

(*c*) curve *C* ☐

(*d*) curve *D*. ☐

158. In Fig. 5.18, specific steam consumption in kg/B.P./hr is given by

(*a*) curve *A* ☐

(*b*) curve *B* ☐

(*c*) curve *C* ☐

(*d*) curve *D*. ☐

Steam consumption (kg/hr)
Specific steam consumption
Thermal efficiency
Mech. efficiency
A
B
C
D
B.P.

FIGURE 5.18

159. In Fig. 5.18, thermal efficiency is given by

(*a*) curve *A* ☐ (*b*) curve *B* ☐

(*c*) curve *C* ☐ (*d*) curve *D*. ☐

160. In Fig. 5.18, mechanical efficiency is given by

(*a*) curve *A* ☐ (*b*) curve *B* ☐

(*c*) curve *C* ☐ (*d*) curve *D*. ☐

161. The hypothetical work done per cycle for a reciprocating steam engine without clearance is equal to

(a) p_1V_1 $(1 + \log_e r) + p_b V_2$ ☐ (b) $p_1V_1(1 - \log_e r) - b_b V_2$ ☐

(c) $p_1V_1(1 + \log_e r) - p_bV_2$ ☐ (d) p_1V_1 $(1 - \log_e r) + p_b V_2$ ☐

where p_1 = Steam pressure, p_b = Back pressure, V_1 = Volume at cut-off,

V_2 = Volume at release, and r = Ratio of expansion.

162. The theoretical mean effective pressure for a reciprocating steam engine without clearance is equal to

(a) $\frac{p_1}{r}(1 - \log_e r) - p_b$ ☐ (b) $\frac{p_1}{r}(1 + \log_e r) + p_b$ ☐

(c) $\frac{p_1}{r}(1 - \log_e r) + p_b$ ☐ (d) $\frac{p_1}{r}(1 + \log_e r) - p_b$. ☐

163. The hypothetical work done per cycle for a reciprocating steam engine, having clearance volume as V_c, is equal to

(a) p_1V_1 $(1 + \log_e r) - p_bV_2$ ☐ (b) $p_1V_1(1 + \log_e r) - p_hV_2 - (p_1 - p_b)$ V_c ☐

(c) $p_1V_1(1 + \log_e r) + p_bV_2$ ☐ (d) $p_1V_1(1 + \log_e r) + p_hV_2 - (p_1 - p_b)$ V_c. ☐

where p_1 = Steam pressure, p_b = Back pressure, and

r = Ratio of expansion $\left(\frac{V_1}{V_2}\right)$.

164. If A = area of indicator diagram in cm^2, L = base width of indicator diagram in cm and S = pressure scale in kgf/cm^2 per cm, then mean effective pressure in kgf/cm^2 is equal to

(a) $\frac{AL}{S}$ ☐ (b) ALS ☐

(c) $\frac{AS}{L}$ ☐ (d) $\frac{S}{AL}$. ☐

165. If the theoretical area of an indicator diagram of simple steam engine is 24 cm^2, base width = 6 cm and pressure scale = 1 kgf/cm^2 per cm, then mean effective pressure is equal to

(a) 6 kgf/cm^2 ☐ (b) 4 kgf/cm^2 ☐

(c) 124 kgf/cm^2 ☐ (d) 2 kgf/cm^2. ☐

166. The diagram factor for a reciprocating steam engine can be increased by

(a) using Corliss of drop valves ☐ (b) steam jacketing ☐

(c) both (a) and (b) ☐ (d) none of the above. ☐

167. In a reciprocating steam engine, the clearance volume is 10% of the stroke volume and cut-off takes place when piston has travelled 0.3 of the stroke. The expansion ratio is equal to

(a) 2.5 ☐ (b) 2.75 ☐

(c) 4.0 ☐ (d) 2.0. ☐

168. Missing quantity is due to

(a) condensation of steam ☐ (b) leakage past the piston ☐

(c) both (a) and (b) ☐ (d) none of the above. ☐

169. The missing quantity may be reduced by

(*a*) steam jacketing of cylinder wells ☐
(*b*) using superheated steam ☐
(*c*) reducing temperature range of working steam ☐
(*d*) all of the above ☐
(*e*) none of the above. ☐

170. During expansion stroke of a simple reciprocating steam engine, the ratio of actual volume in the cylinder to the volume given by the saturation curve at any point represents

(*a*) missing quantity ☐ (*b*) dryness fraction at that point ☐
(*c*) cut-off ratio ☐ (*d*) expansion ratio. ☐

171. Cut-off governing as compared to throttle-governing is

(*a*) less efficient but more economical ☐
(*b*) more efficient but less efficient ☐
(*c*) more efficient and more economical ☐
(*d*) less efficient and less economical. ☐

172. Which one of the following steam engines requires a smaller flywheel?

(*a*) Trandem type ☐ (*b*) Woolfe type ☐
(*c*) Receiver type ☐ (*d*) Simple steam engine. ☐

173. Which of the following are the methods of governing of compound engine ?

(*a*) throttle governing ☐ (*b*) cut-off governing in L.P. cylinder ☐
(*c*) cut-off governing in H.P. cylinder ☐ (*d*) all of the above. ☐
(*e*) none of the above. ☐

174. In reciprocating steam engines

(*a*) pressure energy and dynamic action of steam is utilised ☐
(*b*) pressure energy of steam is utilised and dynamic action of steam is negligible ☐
(*c*) dynamic action of steam is utilised and pressure energy is negligible ☐
(*d*) none of the above ☐

175. Steam engine is governed by throttle-governing, in which intake steam

(*a*) temperature is varied ☐ (*b*) volume is varied ☐
(*c*) pressure is varied ☐ (*d*) all of the above are varied. ☐

176. Cut-off governing of steam engine is done by varying

(*a*) temperature of intake steam ☐ (*b*) volume of intake steam ☐
(*c*) pressure of intake steam ☐ (*d*) none of the above. ☐

177. The material, used for piston rings, is

(*a*) alloy steel ☐ (*b*) cast iron ☐
(*c*) carbon steel ☐ (*d*) copper. ☐

178. The cooling of a 250 MW generator is done by

(*a*) water ☐ (*b*) air ☐
(*c*) nitrogen ☐ (*d*) hydrogen. ☐

Steam Nozzle

179. A steam nozzle is a device used for converting

(a) heat energy of steam into pressure energy ☐
(b) heat energy of steam into kinetic energy ☐
(c) pressure energy of steam into kinetic energy ☐
(d) pressure energy of steam into heat energy. ☐

180. If the exit pressure equal to or more than critical pressure, the nozzle used should be

(a) divergent ☐ (b) convergent ☐
(c) convergent-divergent ☐ (d) none of the above. ☐

181. If the exit pressure is less than the critical pressure, the nozzle used should be

(a) divergent ☐ (b) convergent ☐
(c) convergent-divergent ☐ (d) none of the above. ☐

182. Expansion of steam in nozzle is a flow process. This expansion is assumed as

(a) iso-thermal ☐ (b) adiabatic ☐
(c) polytropic ☐ (d) none of the above. ☐

183. The ideal expansion of steam in a nozzle follows

(a) Otto cycle ☐ (b) Diesel cycle ☐
(c) Rankine cycle minus its feed pump term ☐ (d) None of the above. ☐

184. The velocity at the outlet of the nozzle (if initial velocity is neglected) is equal to

(a) $9.81\sqrt{\Delta h}$ ☐ (b) $91.53\sqrt{\Delta h}$ ☐
(c) $190\sqrt{\Delta h}$ ☐ (d) $50\sqrt{\Delta h}$. ☐

where Δh = Theoretical heat drop.

185. The isentropic expansion of steam (initially dry saturated) through the nozzle may be approximately represented by the equation

(a) $PV^{1.4}$ = constant ☐ (b) $PV^{1.3}$ = constant ☐
(c) $PV^{1.135}$ = constant ☐ (d) $PV^{1.2}$ = constant. ☐

186. The isentropic expansion of steam (initially superheated) through the nozzle may be represented by the equation

(a) $PV^{1.4}$ = constant ☐ (b) $PV^{1.3}$ = constant ☐
(c) $PV^{1.135}$ = constant ☐ (d) $PV^{1.2}$ = constant. ☐

187. The ratio of the pressures at the exit and at inlet of the nozzle for maximum discharge is given by

(a) $\frac{p_2}{p_1} = \left(\frac{2}{n+1}\right)^{\frac{n-1}{n}}$ ☐ (b) $\frac{p_2}{p_1} = \left(\frac{2}{n+1}\right)^{\frac{n}{n-1}}$ ☐

(c) $\frac{p_2}{p_1} = \left(\frac{2}{n+1}\right)^{\frac{n}{n-1}}$ ☐ (d) $\frac{p_2}{p_1} = \left(\frac{n+1}{2}\right)^{\frac{n}{n+1}}$. ☐

188. For the steam which is initially saturated and flowing through a nozzle, the ratio of pressures at exit and at inlet for maximum discharge is equal to

(*a*) 0.545 ☐ (*b*) 0.582 ☐
(*c*) 50 ☐ (*d*) 0.60. ☐

189. For the steam which is initially superheated and flowing through a nozzle, the ratio of pressures at exit and at inlet for maximum discharge is equal to

(*a*) 0.545 ☐ (*b*) 0.582 ☐
(*c*) 0.50 ☐ (*d*) 0.60. ☐

190. Nozzle is designed for

(*a*) maximum perssure at outlet ☐
(*b*) maximum discharge ☐
(*c*) maximum pressure and maximum discharge ☐
(*d*) none of the above. ☐

191. The fractional losses in the nozzle

(*a*) reduce the exit velocity of nozzle ☐ (*b*) increase the exit velocity ☐
(*c*) have no effect on exit velocity ☐ (*d*) none of the above. ☐

192. The frictional losses in the nozzle

(*a*) decrease the final dryness fraction of steam ☐
(*b*) increase the final dryness fraction of steam ☐
(*c*) has no effect on final dryness fraction of the steam ☐
(*d*) none of the above. ☐

193. The exit velocity of nozzle considering friction and neglecting initial velocity is given by

(*a*) $V_2 = 91.53\ \sqrt{k\,\Delta h}$ ☐ (*b*) $V_2 = 9.81\ \sqrt{k\,\Delta h}$ ☐
(*c*) $V_2 = 100\ \sqrt{k\,\Delta h}$ ☐ (*d*) none of the above. ☐

194. For the super-saturated flow in the nozzle, the heat drop

(*a*) increases ☐ (*b*) decreases ☐
(*c*) remains constant ☐ (*d*) none of the above. ☐

195. For the super-saturated flow in the nozzle the discharge

(*a*) increases ☐ (*b*) decreases ☐
(*c*) remains constant ☐ (*d*) none of the above. ☐

196. The problems on super-saturated flow

(*a*) are solved by Mollier chart ☐
(*b*) cannot be solved by Mollier chart ☐
(*c*) can be solved by Mollier chart if Wilson line is drawn on it ☐
(*d*) none of the above. ☐

197. The steam injector is a device used for

(*a*) converting water into steam ☐
(*b*) converting steam into water ☐

(*c*) feeding water into a boiler, using either live boiler steam or exhaust steam ☐
(*d*) none of the above. ☐

198. In case of nozzle, the whole of friction loss is assumed

(*a*) between inlet and throat ☐ (*b*) between inlet and outlet ☐
(*c*) between throat and exit ☐ (*d*) none of the above. ☐

199. Critical pressure ratio for a convergent nozzle is

(*a*) the ratio of pressure at the outlet and inlet of nozzle ☐
(*b*) the ratio of pressure at the inlet and outlet of nozzle ☐
(*c*) the ratio of pressure at outlet and inlet of nozzle when mass flow is maximum ☐
(*d*) the ratio of pressure at outlet and inlet of nozzle when mass flow is minimum. ☐

200. If the exit pressure for a convergent nozzle is equal to inlet pressure of the nozzle, then mass rate of flow through nozzle is

(*a*) minimum ☐ (*b*) zero ☐
(*c*) maximum ☐ (*d*) constant. ☐

201. If the exit pressure for a convergent nozzle is decreased to such a valve that the ratio of exit pressure to inlet pressure is equal to critical pressure ratio, then the mass rate of flow through the nozzle will be

(*a*) minimum ☐ (*b*) zero ☐
(*c*) maximum ☐ (*d*) constant. ☐

202. If the exit pressure for a convergent nozzle is less than critical pressure, the mass rate of flow will be

(*a*) decreasing ☐ (*b*) increasing ☐
(*c*) constant ☐ (*d*) zero. ☐

203. Figure 5.19 shows the variation of mass rate of flow per unit area of a gas with pressure ratio at exit to inlet (p_2/p_1) for a convergent nozzle. The actual curve is given by

(*a*) curve *OBC* ☐
(*b*) curve *DBC* ☐
(*c*) curve *ABC* ☐
(*d*) curve *OBD*. ☐

204. In Fig. 5.19, the theoretical curve is given by

(*a*) curve *OBC* ☐
(*b*) curve *DBC* ☐
(*c*) curve *ABC* ☐
(*d*) curve *OBD*. ☐

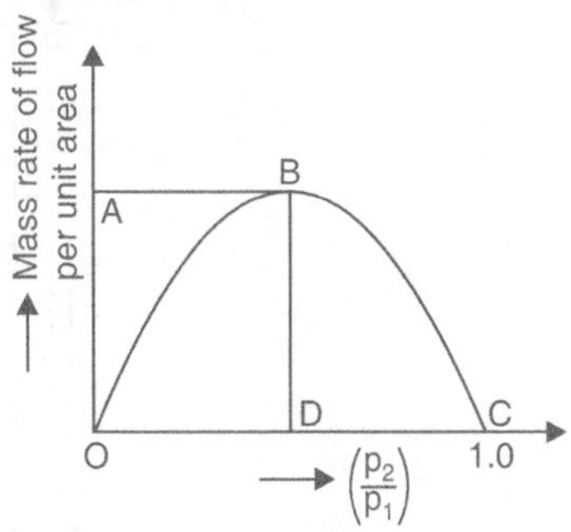

FIGURE 5.19

205. The mass rate of flow of a gas through a convergent nozzle depends upon

(*a*) pressure ratio at exit to inlet only ☐ (*b*) pressure and density at the inlet only ☐
(*c*) adiabatic exponent γ only ☐ (*d*) all the above. ☐

206. The maximum mass rate of flow of a gas through a convergent nozzle is equal to

(a) $A_2\sqrt{\gamma\frac{p_1}{\rho_1}}\left(\frac{2}{\gamma+1}\right)^{\frac{\gamma+1}{2(\gamma-1)}}$ ☐ (b) $A_2\sqrt{\gamma p_1\rho_1}\left(\frac{2}{\gamma+1}\right)^{\frac{\gamma+1}{2(\gamma-1)}}$ ☐

(c) $A_2\sqrt{\gamma p_1\rho_1}\left(\frac{2}{\gamma+1}\right)^{\frac{\gamma+1}{2(\gamma-1)}}$ ☐ (d) $A_2\sqrt{\frac{\gamma p_1}{p_1}}\left(\frac{2}{\gamma+1}\right)^{\frac{\gamma+1}{2(\gamma-1)}}$ ☐

207. The maximum mass rate of flow of a gas through a convergent nozzle depend upon

(a) pressure and density at the inlet only ☐ (b) adiabatic exponent γ only ☐
(c) pressure ratio at the exit and inlet only ☐ (d) all of the above ☐
(e) both (a) and (b) only. ☐

208. At the critical pressure ratio for a convergent nozzle, the velocity at outlet will be

(a) more that sonic velocity ☐ (b) less than the sonic velocity ☐
(c) equal to the sonic velocity ☐ (d) none of the above. ☐

209. Figure 5.20 shows the pressure distributions along the axis of a convergent-divergent nozzle, whose inlet is connected to a large reservoir, in which velocity is negligible and pressure at the outlet of the nozzle (p) is varied.

It the pressure at the outlet of nozzle is equal to inlet pressure (p_1), then the pressure distribution is given by

(a) curve *AB* ☐ (b) curve *ACE* ☐
(c) curve *ADF* ☐ (d) curve *ADKLM*. ☐

210. Refer to Fig. 5.20, the pressure at the outlet, corresponding to point *M* is called

(a) critical pressure ☐ (b) design pressure ☐
(c) normal pressure ☐ (d) none of the above. ☐

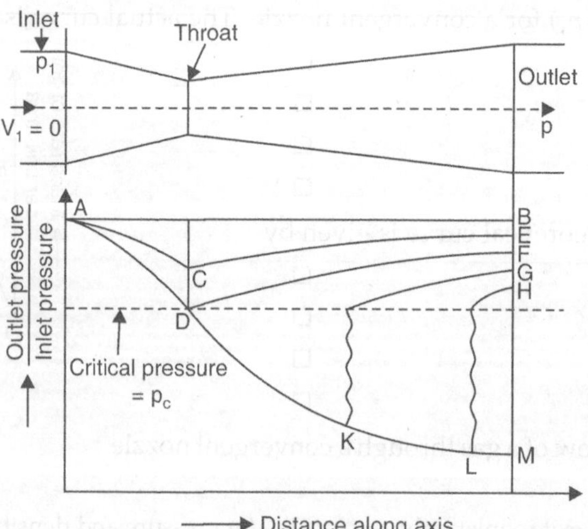

FIGURE 5.20

211. Refer to Fig. 5.20, if the pressure at the outlet is reduced to such a valve that critical pressure exit at the throat, then pressure distribution is given by

(*a*) curve *AB* ☐ (*b*) curve *ACE* ☐
(*c*) curve *ADF* ☐ (*d*) curve *ADM*. ☐

212. Refer to Fig. 5.20, The pressure at the throat has become critical corresponding to pressure at outlet. If the pressure is further reduced at the outlet, the pressure distribution is given by

(*a*) curve *AB* ☐ (*b*) curve *ACE* ☐
(*c*) curve *ADF* ☐ (*d*) curve *ADKG*. ☐

213. Refer to Fig. 5.20, the pressure at the throat has become critical corresponding to some pressure at outlet. If the pressure at outlet is more than this pressure but less than inlet pressure, the pressure distribution is given by

(*a*) curve *AB* ☐ (*b*) curve *ACE* ☐
(*c*) curve *ADF* ☐ (*d*) curve *ADKG*. ☐

214. Refer to Fig. 5.20, the design pressure ratio is given by

(*a*) $\frac{p_1}{p_M}$ ☐ (*b*) $\frac{p_1}{p_C}$ ☐
(*c*) $\frac{p_M}{p_1}$ ☐ (*d*) $p_1 \times p_M$. ☐

215. Formation of shock wave front in case of convergent-divergent nozzle takes place

(*a*) at throat ☐ (*b*) in convergent portion ☐
(*c*) in divergent portion ☐ (*d*) none of the above. ☐

216. Figure 5.21 shows the velocity distributions for a convergent-divergent nozzle when the outlet pressure is varied from the inlet pressure to design pressure. If outlet pressure is same as inlet pressure, then velocity distribution is given by

(*a*) curve *D* ☐ (*b*) curve *C* ☐
(*c*) curve *B* ☐ (*d*) curve *A*. ☐

217. Refer to Fig. 5.21, if the pressure at outlet is such that the velocity at the throat is less than sonic velocity, then velocity distribution is given by

(*a*) curve *D* ☐ (*b*) curve *C* ☐
(*c*) curve *B* ☐ (*d*) curve *A*. ☐

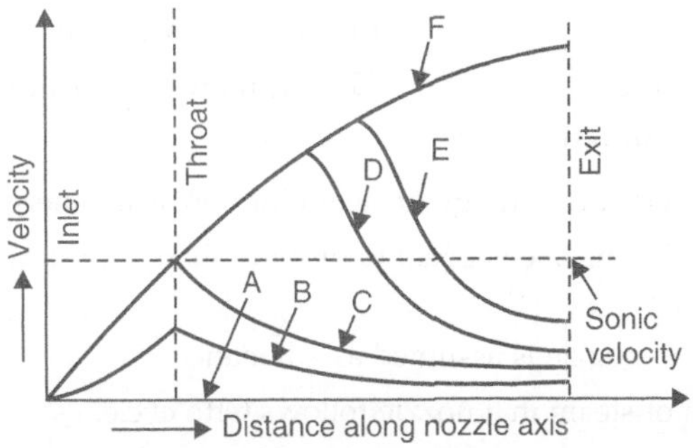

FIGURE 5.21

218. Refer to Fig. 5.21, if the pressure at outlet is such that the velocity at the throat has just become equal to sonic velocity then velocity distribution is given by

(*a*) curve *D* ☐ (*b*) curve *C* ☐
(*c*) curve *B* ☐ (*d*) curve *A*. ☐

219. Refer to Fig. 5.21, if the pressure at outlet is equal to the design pressure, then the velocity distribution is given by

(*a*) curve *A* ☐ (*b*) curve *C* ☐
(*c*) curve *D* ☐ (*d*) curve *F*. ☐

220. Refer to Fig. 5.21 if the pressure at outlet is slightly more than the design pressure, then the velocity distribution is given by

(*a*) curve *A* ☐ (*b*) curve *C* ☐
(*c*) curve *E* ☐ (*d*) curve *F*. ☐

221. Figure 5.22 shows various pressure distribution curves for a convergent nozzle. The inlet of the nozzle is connected to a larger reservoir in which velocity is negligible and pressure is maintained constant. The pressure at the outlet of the nozzle is varied from inlet pressure to pressure less than critical.

Curve *A* holds good when pressure at exit is equal to

(*a*) critical pressure ☐
(*b*) inlet pressure ☐
(*c*) more than critical pressure ☐
(*d*) less than critical pressure. ☐

FIGURE 5.22

222. In Fig. 5.22 curve *B* hold good when exit pressure is equal to

(*a*) critical pressure ☐ (*b*) inlet pressure ☐
(*c*) less than critical pressure ☐ (*d*) more than critical pressure. ☐

223. In Fig. 5.22, curve *C* holds good when exit pressure is equal to

(*a*) critical pressure ☐ (*b*) inlet pressure ☐
(*c*) less than critical pressure ☐ (*d*) more than critical pressure. ☐

224. In Fig. 5.22, curve *D* holds good when exit pressure is equal to

(*a*) critical pressure ☐ (*b*) inlet pressure ☐
(*c*) less than critical pressure ☐ (*d*) more than critical pressure. ☐

225. Choose the correct statement

(*a*) A steam nozzle converts heat energy of steam into pressure energy. ☐
(*b*) If the exit pressure for a nozzle is equal to or more than critical pressure, the divergent nozzle should be used. ☐
(*c*) Expansion of steam in nozzle is assumed as adiabatic. ☐
(*d*) The ideal expansion of steam in a nozzle follows Otto cycle. ☐

226. Choose the wrong statement

(*a*) For maximum discharge through a nozzle of initial saturated steam, the ratio of pressures at exit and at inlet is equal to 0.582. ☐

(*b*) Nozzle is designed for maximum discharge. ☐

(*c*) The frictional losses in the nozzle increases the final dryness fraction of steam. ☐

(*d*) Steam injector converts water into steam. ☐

227. Nozzle efficiency is defined as

(*a*) ratio of useful heat drop to isentropic heat drop ☐

(*b*) product of useful heat drop to isentropic heat drop ☐

(*c*) ratio of isentropic heat drop to useful heat drop ☐

(*d*) none of the above. ☐

228. The frictional losses in the nozzle

(*a*) reduces to heat drop ☐ (*b*) increases the heat drop ☐

(*c*) keep the heat drop constant ☐ (*d*) none of the above. ☐

229. The velocity of steam at the throat of a nozzle corresponding to critical pressure is

(*a*) more than the velocity of sound ☐ (*b*) less than the velocity of sound ☐

(*c*) equal to the velocity of sound ☐ (*d*) none of the above. ☐

230. Choose the correct statement

(*a*) In case of nozzle, whole of friction loss is assumed between inlet and throat. ☐

(*b*) For a convergent nozzle, mass rate of flow is maximum when exit pressure is equal to inlet pressure. ☐

(*c*) Shock wave front is formed in the convergent portion of a convergent-divergent nozzle. ☐

(*d*) The super saturation reduces the exit velocity of the nozzle. ☐

231. Choose the wrong statement

(*a*) The discharge is maximum at critical pressure. ☐

(*b*) The critical pressure ratio for initially wet steam is 0.582. ☐

(*c*) The velocity of steam at the throat is equal to sonic velocity when pressure is critical. ☐

(*d*) The mach number is more than one for the convergent portion of the nozzle. ☐

232. The maximum discharge of steam through a convergent-divergent nozzle depends on

(*a*) initial condition of steam ☐ (*b*) throat area ☐

(*c*) pressure at exit of nozzle ☐ (*d*) both (*a*) and (*b*) only ☐

(*e*) none of the above. ☐

233. When a nozzle operates with the maximum mass flow, it is said to be

(*a*) choked ☐ (*b*) under-expanding ☐

(*c*) over-expanding ☐ (*d*) none of the above. ☐

234. When the back pressure of a nozzle is below the designed valve of pressure at the exit of the nozzle, the nozzle is said to be

(*a*) choked ☐ (*b*) under-expanding ☐

(*c*) over-expanding ☐ (*d*) none of the above. ☐

235. At the outlet of a nozzle, the steam is at a

(*a*) low pressure and low velocity ☐ (*b*) low pressure and high velocity ☐

(*c*) high pressure and low velocity ☐ (*d*) high pressure and high velocity ☐

236. Choose the correct statement

(*a*) The friction in the nozzle increases the exit velocity of steam. ☐

(*b*) The friction in the nozzle decreases dryness friction of steam. ☐

(*c*) The pressure of steam increases while flowing through a nozzle. ☐

(*d*) The critical pressure ratio for initially superheated steam is less than that of initially dry saturated steam. ☐

237. At the inlet of the nozzle, the steam is at a

(*a*) low pressure and low velocity ☐ (*b*) low pressure and high velocity ☐

(*c*) high pressure and low velocity ☐ (*d*) high pressure and high velocity. ☐

238. In reciprocating steam engines

(*a*) the kinetic energy of the steam is utilized ☐

(*b*) the pressure energy of the steam is utilized ☐

(*c*) dynamic action of the steam is utilized ☐

(*d*) none of the above. ☐

Steam Turbines

239. In steam turbines

(*a*) the static action of steam is used ☐ (*b*) pressure energy of steam is used ☐

(*c*) dynamic action of steam is used ☐ (*d*) none of the above. ☐

240. The thermodynamic efficiency of a steam turbine is

(*a*) lower than that of steam engine ☐ (*b*) higher than that of steam engine ☐

(*c*) same as of steam engine ☐ (*d*) none of the above. ☐

241. The pressure on the two sides of the moving blades in case of impulse turbine

(*a*) increases ☐ (*b*) decreases ☐

(*c*) remains constant ☐ (*d*) none of the above. ☐

242. In case of impulse steam turbine

(*a*) the steam is expanded in nozzles causing pressure and heat drop in nozzles only ☐

(*b*) the steam is expanded both in moving and fixed blades ☐

(*c*) the steam is expanded in moving blades only ☐

(*d*) none of the above. ☐

243. The pressure drop on the sides of the moving blades in case of reaction turbine

(*a*) increases ☐ (*b*) decreases ☐

(*c*) remains constant ☐ (*d*) none of the above. ☐

244. In case of reaction steam turbine

(*a*) the steam is expanded in nozzles only ☐

(*b*) the steam is expanded in moving blades only ☐

(*c*) the steam is expanded both in fixed and moving blades continuously ☐

(*d*) none of the above. ☐

245. De-laval turbine is a

(*a*) single wheel impulse turbine ☐ (*b*) pressure-compounded turbine ☐
(*c*) velocity-compounded turbine ☐ (*d*) reaction turbine. ☐

246. Curtis turbine is a

(*a*) pressure compounded turbine ☐ (*b*) velocity-compounded turbine ☐
(*c*) pressure-velocity compounded turbine ☐ (*d*) reaction turbine. ☐

247. Rateau and Zoelly turbines are

(*a*) pressure-compounded turbines ☐ (*b*) velocity-compounded turbines ☐
(*c*) pressure-velocity compounded turbine ☐ (*d*) reaction turbines. ☐

248. Parson's turbine is a

(*a*) single wheel impulse turbine ☐ (*b*) pressure-compound turbine ☐
(*c*) velocity-compound turbine ☐ (*d*) reaction turbine. ☐

249. The work done by steam on impulse turbine per kg of steam is equal to

(*a*) $\frac{(V_{w_1} \pm V_{w_2})u}{g}$ ☐ (*b*) $\frac{(V_{w_1} \pm V_{w_2})V_{w_1}}{g}$ ☐

(*c*) $\frac{(V_{w_1} \pm V_{w_2} + 1)u}{g}$ ☐ (*d*) $\frac{u}{g(V_{w_1} + V_{w_2})}$. ☐

where V_{w_1} = Velocity of whirl at inlet of blade,

V_{w_2} = Velocity of whirl at outlet of blade, and

u = Tangential velocity of blade.

250. Blade efficiency (η_b) is equal to

(*a*) $\frac{\text{Work done on blades}}{\text{Energy supplied per stage}}$ ☐ (*b*) $\frac{\text{Work done on blades}}{\text{Energy supplied to blades}}$ ☐

(*c*) $\frac{\text{Energy supplied per stage}}{\text{Work done on blades}}$ ☐ (*d*) none of the above. ☐

251. The steam consumption in case of steam turbines as compared to reciprocating steam engine is

(*a*) more ☐ (*b*) same ☐
(*c*) less ☐ (*d*) can be more or less. ☐

252. Select the wrong statement

(*a*) In an impulse turbine, steam expands wholly in the nozzle. ☐
(*b*) The action of steam in a steam turbine is dynamic. ☐
(*c*) De-laval turbine is a simple reaction turbine. ☐
(*d*) The efficiency of a steam turbine is higher than that of steam engine. ☐

Efficiency of a Turbine

253. Stage efficiency of a turbine is equal to

(a) $\frac{\text{Work done on blades}}{\text{Total energy supplied}}$ ☐ (b) $\frac{\text{Work done on blades}}{\text{Energy supplied per stage}}$ ☐

(c) $\frac{\text{Energy supplied per stage}}{\text{Work done on blades}}$ ☐ (d) $\frac{\text{Total energy supplied}}{\text{Work done on blades}}$. ☐

254. Stage efficiency is also called

(a) diagram efficiency ☐ (b) blade efficiency ☐

(c) gross efficiency ☐ (d) none of the above. ☐

255. Stage efficiency (η_s) of a steam turbine is equal to

(a) nozzle efficiency × blade efficiency ☐ (b) blade efficiency nozzle efficiency ☐

(c) nozzle efficiency/blade efficiency ☐ (d) none of the above. ☐

256. The blade efficiency of a steam turbine is equal to

(a) $\frac{(V_{w_1} \pm V_{w_2}) \times u}{2V_1^2}$ ☐ (b) $\frac{2(V_{w_1} \pm V_{w_2}) \times u}{V_1^2}$ ☐

(c) $\frac{2V_1(V_{w_1} \pm V_{w_2})}{2u^2}$ ☐ (d) $\frac{V_1(V_{w_1} \pm V_{w_2})}{2u^2}$. ☐

where V_{w_1} = Velocity of whirl at inlet, V_{w_2} = Velocity of whirl at outlet

V_1 = Absolute velocity of steam at inlet, and u = Velocity of blade.

257. The maximum blade efficiency in a single stage impulse turbine is equal to

(a) $(1 + KC)\cos^2\alpha$ ☐ (b) $(1 + KC)\frac{\cos^2\alpha}{2}$ ☐

(c) $2(1 + KC)\cos^2\alpha$ ☐ (d) $(1 + KC)\cos^2 2\alpha$. ☐

where α = Angle made by absolute velocity at inlet,

K = Ratio of relative velocities at outlet to inlet, and

C = Ratio of the cosine of the angle made by relative velocities at outlet to inlet.

258. The condition for maximum blade efficiency for a single stage impulse turbine is that the blade speed ratio should be equal to

(a) $2\cos\alpha$ ☐ (b) $\cos 2\alpha$ ☐

(c) $\frac{\cos\alpha}{2}$ ☐ (d) $\frac{1}{2}\cos 2\alpha$. ☐

where α = Angle made by absolute velocity at inlet.

259. In case of reaction turbine, the heat drop takes place on

(a) fixed blade only ☐ (b) moving blades only ☐

(c) both fixed and moving blades ☐ (d) none of the above. ☐

260. Degree of reaction is defined as the ratio of

(*a*) enthalpy drop in the moving blades/enthalpy drop in fixed blades ☐
(*b*) enthalpy drop in moving bladed/sum of enthalpy drop in moving and fixed blades ☐
(*c*) enthalpy drop in fixed blades/sum of enthalpy drop in moving and the blades ☐
(*d*) none of the above. ☐

261. For Parson's reaction turbine, degree of reaction is equal to

(*a*) 80% ☐ (*b*) 50% ☐
(*c*) 75% ☐ (*d*) 100%. ☐

262. For Parson's reaction turbine, the condition for maximum efficiency is given by

(*a*) $\rho = 2 \cos \alpha$ ☐ (*b*) $\rho = \cos^2 \alpha$ ☐
(*c*) $p = \frac{\cos \alpha}{2}$ ☐ (*d*) $\rho = \cos \alpha$. ☐

where p = Speed ratio, and α = Angle made by absolute velocity at inlet.

263. For Parson's reaction turbine, the maximum efficiency is equal to

(*a*) $\frac{1+\cos^2 \alpha}{2 \cos^2 \alpha}$ ☐ (*b*) $\frac{\cos^2 \alpha}{1+\cos^2 \alpha}$ ☐
(*c*) $\frac{2 \cos^2 \alpha}{1+\cos^2 \alpha}$ ☐ (*d*) $\frac{\cos^2 \alpha}{2(1+\cos^2 \alpha)}$. ☐

264. Reheat factor (R.F.) is defined as the

(*a*) ratio of cumulative enthalpy drop to isentropic enthalpy drop ☐
(*b*) product of cumulative enthalpy drop and isentropic enthalpy drop ☐
(*c*) ratio of isentropic enthalpy drop to cumulative enthalpy drop ☐
(*d*) none of the above ☐

265. Reheat factor depends upon

(*a*) stage efficiency only ☐ (*b*) exit pressure only ☐
(*c*) initial pressure and temperature only ☐ (*d*) all of the above ☐
(*e*) none of the above. ☐

266. The degree of reaction, for a turbine in which heat drop in moving blade is 8 kcal/kg and in fixed blade 12 kcal/kg would be

(*a*) 80% ☐ (*b*) 60% ☐
(*c*) 40% ☐ (*d*) 20%. ☐

267. The value of reheat factor increases if the number of stages

(*a*) are less ☐ (*b*) are large ☐
(*c*) are constant ☐ (*d*) none of the above. ☐

268. The value of reheat factor varies normally between

(*a*) 1.0 to 1.5 ☐ (*b*) 1.5 to 2.0 ☐
(*c*) 1.02 to 1.06 ☐ (*d*) none of the above. ☐

269. Internal turbine efficiency is defined as the ratio of

(a) $\frac{\text{Total useful heat drop}}{\text{Adiabatic heat drop}}$ ☐ (b) $\frac{\text{Adiabatic heat drop}}{\text{Heat supplied}}$ ☐

(c) $\frac{\text{Useful heat drop}}{\text{Heat supplied}}$ ☐ (d) none of the above. ☐

270. If the stage efficiency (η_s) is constant for all stages, then internal turbine efficiency is equal to

(a) η_s/R.F. ☐ (b) $\eta_s \times$ R.F. ☐

(c) $\frac{\text{R.F.}}{\eta_s}$ ☐ (d) $\frac{1}{\eta_s \times \text{R.F.}}$ ☐

where R.F. = Reheat factor.

271. Rankine efficiency is defined as the ratio of

(a) $\frac{\text{Total useful heat drop}}{\text{Adiabatic heat drop}}$ ☐ (b) $\frac{\text{Adiabatic heat drop}}{\text{Heat supplied}}$ ☐

(c) $\frac{\text{Useful heat drop}}{\text{Heat supplied}}$ ☐ (d) none of the above. ☐

272. The ratio, useful heat drop to heat supplied, gives

(a) Rankine efficiency ☐ (b) stage efficiency ☐

(c) overall efficiency ☐ (d) internal turbine efficiency. ☐

273. The ratio of actual work done per kg to Rankine work done per kg is equal to

(a) Rankine efficiency ☐ (b) stage efficiency ☐

(c) overall efficiency ☐ (d) relative efficiency. ☐

274. The enthalpy-entropy diagram for a three stage turbine is shown in Fig. 5.23. The efficiency for the 1st stage is equal to

(a) 40% ☐ (b) 60% ☐

(c) 80% ☐ (d) 83.33% ☐

275. The efficiency for the 2nd stage (refer to Fig. 5.23) is equal to

(a) 83.33% ☐

(b) 80% ☐

(c) 60% ☐

(d) 40%. ☐

276. The efficiency for the 3rd stage (refer to Fig. 5.23) is equal to

(a) 80% ☐

(b) 88.33% ☐

(c) 60% ☐

(d) 40%. ☐

FIGURE 5.23

277. The internal efficiency of the turbine shown in Fig. 5.23 is

(*a*) 80% ☐ (*b*) 88.23% ☐
(*c*) 40% ☐ (*d*) 60%. ☐

278. For the three stage turbine shown in Fig. 5.23, the reheat factor is equal to

(*a*) $\frac{170}{180}$ ☐ (*b*) $\frac{180}{170}$ ☐
(*c*) $\frac{180}{150}$ ☐ (*d*) $\frac{170}{150}$. ☐

279. Choose the correct statement

(*a*) Reheat factor is always less than 1.0 ☐
(*b*) The ratio of adiabatic heat drop to the heat supplied is equal to stage efficiency. ☐
(*c*) In reaction turbine if the degree of reaction is zero, then heat drop in moving blades is zero. ☐
(*d*) Curtis turbine is a simple reaction type turbine. ☐

280. Choose the wrong statement

(*a*) The Parson's reaction turbine has identical, fixed and moving blades. ☐
(*b*) The Parson's reaction turbine has degree of reaction equal to 50%. ☐
(*c*) Rateau turbine is a pressure-compounded turbine. ☐
(*d*) The maximum efficiency of a reaction turbine is equal to $\frac{2\sin^2\alpha}{1+\sin^2\alpha}$. ☐

281. In reaction turbine, the expansion of steam over the blades represents

(*a*) free-expansion process ☐ (*b*) isothermal process ☐
(*c*) adiabatic process ☐ (*d*) throttling process. ☐

282. The process of draining steam for heating the feed water is known as

(*a*) reheating of steam ☐ (*b*) bleeding ☐
(*c*) governing ☐ (*d*) cooling. ☐

283. The process of maintaining the speed of a steam turbine constant for various load conditions is known as

(*a*) reheating ☐ (*b*) bleeding ☐
(*c*) governing ☐ (*d*) cooling. ☐

284. The steam turbines are governed commonly by the method of

(*a*) by-pass governing ☐ (*b*) throttle governing ☐
(*c*) nozzle control governing ☐ (*d*) none of the above. ☐

285. Choose the wrong statement

(*a*) Bleeding increases the thermodynamic efficiency of a turbine. ☐
(*b*) Bleeding increases the power developed by the turbine. ☐
(*c*) Bleeding decreases the power developed by the turbine. ☐
(*d*) Boiler is supplied with hot water due to bleeding. ☐

286. Choose the correct statement

(*a*) If the pressure at the inlet of a turbine increases, the turbine is tripped. ☐
(*b*) Rateau turbine is a velocity compounded turbine. ☐

(c) Turbine is tripped when the speed shoots up. ☐
(d) The air ejector injects are in turbine for cooling purposes. ☐

287. In pressure compounded impulse turbine the pressure drop for each stage (starting from the 1st stage to the last stage) is
(a) equal ☐ (b) increasing ☐
(c) decreasing ☐ (d) first increases then decreases. ☐

288. In pressure compounded impulse turbine, the pressure over each ring of moving blades is
(a) increasing ☐ (b) decreasing ☐
(c) constant ☐ (d) first increases then decreases. ☐

289. In pressure compounded impulse turbine all the pressure drop takes place in
(a) moving blades ☐ (b) nozzles ☐
(c) both moving blades and nozzles ☐ (d) none of them. ☐

290. In case of pressure compounded impulse turbine, the ratio of blade velocity to steam velocity
(a) increasing ☐ (b) decreasing ☐
(c) first increases then decreases ☐ (d) remains constant. ☐

291. The number of stages required in case of pressure compounded turbine as compared to velocity-compound turbine is
(a) less ☐ (b) more ☐
(c) same ☐ (d) none of the above. ☐

292. The efficiency of velocity-compounded steam turbine as compared to pressure-compounded turbine is
(a) less ☐ (b) more ☐
(c) same ☐ (d) may be more or less. ☐

293. To reduce the speed of impulse turbines to practical limits, the method used is
(a) velocity-compounding ☐ (b) pressure-compounding ☐
(c) pressure-velocity compounding ☐ (d) all of the above ☐
(e) none of the above. ☐

294. The pressure-velocity compounded turbine is having efficiency
(a) more than pressure compounded turbine ☐
(b) less than velocity compounded turbine ☐
(c) less than pressure compounded turbine ☐
(d) equal to velocity compounded turbine. ☐

295. The axial force on the wheel of an impulse turbine per kg of steam is equal to
(a) $\frac{1}{g}(V_{w_1}+V_{w_2})$ ☐ (b) $\frac{1}{g}(V_{r_1}-V_{r_2})$ ☐
(c) $\frac{1}{g}(V_{f_1}-V_{f_2})$ ☐ (d) $\frac{1}{g}(V_1-V_2)$. ☐

where V_{w_1}, V_f, V_r and V are velocity of whirl, velocity of flow, relative velocity and absolute velocity of steam respectively.

296. Energy lost due to friction per kg of steam in blades is equal to

(*a*) $\frac{(V_{w_1}^2 - V_{w_2}^2)}{2g}$ ☐ (*b*) $\frac{(V_{r_1}^2 - V_{r_2}^2)}{2g}$ ☐

(*c*) $\frac{(V_{f_1}^2 - V_{f_2}^2)}{2g}$ ☐ (*d*) $\frac{(V_1^2 - V_2^2)}{2g}$. ☐

297. Heat drop through fixed blades per kg of steam through Parson's reaction turbine is equal to

(*a*) $\frac{1}{g}(V_{w_1}^2 - V_{w_2}^2)$ ☐ (*b*) $\frac{1}{2g}(V_{r_1}^2 - V_{r_2}^2)$ ☐

(*c*) $\frac{1}{g}(V_{f_1}^2 - V_{f_2}^2)$ ☐ (*d*) $\frac{1}{2g}(V_1^2 - V_2^2)$. ☐

298. Choose the correct statement

(*a*) The efficiency of an impulse turbine is maximum when speed ratio = (cos α). ☐

(*b*) The efficiency of a reaction turbine (with degree of reaction half and fixed and moving blades symmetrical) is maximum when speed ratio = (cos α)/2.0. ☐

(*c*) In pressure-compounded steam turbine the pressure drop per stage is equal. ☐

(*d*) The efficiency of a velocity-compounded steam turbine is more than the efficiency of a pressure-compounded steam turbine. ☐

299. The nozzle angle of an impulse turbine is 20°. The efficiency of the turbine will be maximum when speed ratio is equal to

(*a*) 0.50 ☐ (*b*) 0.47 ☐

(*c*) 0.25 ☐ (*d*) 1.0. ☐

300. For Parson's turbine, having moving and fixed blades symmetrical and 50% degree of reaction, the outlet angle of the moving blade as compared to guide blade angle (*i.e.*, angle of fixed blades) is

(*a*) more ☐ (*b*) less ☐

(*c*) same ☐ (*d*) may be more or less. ☐

301. The outlet angle of the moving blade of a Parson's turbine (having moving and fixed blades symmetrical and 50% degree of reaction) is 20°, the efficiency of the turbine will be maximum when the speed ratio is equal to

(*a*) 0.50 ☐ (*b*) 0.469 ☐

(*c*) 0.938 ☐ (*d*) 1.0. ☐

302. The expression $V_{r_o} = KV_{r_i}$ (where V_{r_o} and V_{r_i} are the relative velocities at outlet and inlet and K is a friction factor) is applicable to

(*a*) reaction turbines ☐ (*b*) impulse turbine ☐

(*c*) both (*a*) and (*b*) ☐ (*d*) none of the above. ☐

303. Bleeding

(*a*) increases the thermodynamic efficiency ☐

(*b*) reduces the temperature stresses in the boiler ☐

(c) prevents the condensation of SO_2 gases on economiser ☐
(d) all of the above. ☐
(e) none of the above. ☐

304. Bleeding
(a) increase the cost of plant ☐ (b) reduces the work done per kg of steam ☐
(c) both (a) and (b) ☐ (d) none of the above. ☐

305. Reheating
(a) increases thermal efficiency ☐ (b) increases work done per kg of steam ☐
(c) increases dryness fraction of steam ☐ (d) reduces blade erosion ☐
(e) all of the above ☐ (f) none of the above. ☐

306. The efficiency of steam turbines may be improved by
(a) reheating of steam ☐ (b) regenerative feed heating ☐
(c) binary vapour plants ☐ (d) any of the above methods ☐
(e) none of the above. ☐

307. The ratio of the cumulative enthalpy drop in a multi-stage turbine to isentropic enthalpy drop is called
(a) internal efficiency of turbine ☐ (b) reheat factor ☐
(c) Rankine efficiency ☐ (d) none of the above. ☐

308. A binary vapour plant consists of
(a) mercury boiler ☐ (b) steam turbine ☐
(c) steam condenser ☐ (d) economiser ☐
(e) super heater ☐ (f) all of the above ☐
(g) none of the above. ☐

309. Choose the wrong statement
(a) In reaction steam turbines, the pressure is reduced in the fixed blades as well as in moving blades. ☐
(b) A reaction turbine as compared to an impulse turbine, for a given power has less rows of blades. ☐
(c) The steam expands wholly in the nozzle in an impulse turbine. ☐
(d) In reaction turbines, the velocity of steam is increased in the fixed blades as well as in moving blades. ☐

310. In a steam power plant, a condenser
(a) reduces back pressure of steam ☐ (b) increases expansion ratio of steam ☐
(c) reduces cost of power plant ☐ (d) reduces the temperature of exhaust steam ☐
(e) all of the above. ☐

311. The circulating water flows through tubes which are surrounded by steam, this type of condenser is called
(a) jet condenser ☐ (b) evaporative condenser ☐
(c) surface condenser ☐ (d) none of the above. ☐

312. The function of steam separator is
(a) to heat the air entering the boiler furnace, by the heat carried away by flue gases going as a waste through the chimney. ☐
(b) to drain away automatically the water collected in the steam pipe as a result of partial condensation ☐
(c) to separate water particles from steam before it is supplied to a steam engine or turbine. ☐
(d) to reduce the pressure to a constant value on the delivery side. ☐

313. The device used to heat the air before it is supplied to the furnace by the heat carried away by flue gases is known as

(*a*) super heater ☐ (*b*) steam separator ☐

(*c*) steam trap ☐ (*d*) air pre-heater. ☐

314. The device used to drain away automatically the water collected in the steam pipe as result of partial condensation, is known as

(*a*) super heater ☐ (*b*) steam separator ☐

(*c*) steam trap ☐ (*d*) air pre-heater. ☐

315. The weight of water evaporated from water at 100°C to dry and saturated steam at 100°C by utilizing the same amount of heat as would have been used under the actual working conditions, is known as

(*a*) factor of evaporation ☐ (*b*) equivalent evaporation ☐

(*c*) boiler efficiency ☐ (*d*) boiler power. ☐

316. The ratio of the heat actually utilized in generation of steam to the heat supplied by the fuel in the same period is known as

(*a*) efficiency of turbine ☐ (*b*) efficiency of economiser ☐

(*c*) boiler efficiency ☐ (*d*) chimney efficiency. ☐

317. The ratio of the heat absorbed by the feed water in the economiser to the heat in flue gases entering the economiser reckoned above the temperature of air supplied to the boiler, is known as

(*a*) turbine efficiency ☐ (*b*) economiser efficiency ☐

(*c*) boiler efficiency ☐ (*d*) chimney efficiency. ☐

318. One boiler power means evaporation of certain amount of water per hour from and at 100°C into dry saturated steam. This amount of water should be

(*a*) 10 kg ☐ (*b*) 15.653 kg ☐

(*c*) 18.568 kg ☐ (*d*) 22.635 kg. ☐

319. In a compound engine, the diameter of the low pressure cylinder as compared to the diameter of the high pressure cylinder, is made

(*a*) smaller ☐ (*b*) larger ☐

(*c*) equal ☐ (*d*) none of the above. ☐

320. Willian's law states that the rate of steam consumption of a steam engine provided with throttled governor is

(*a*) proportional to the square of indicated power ☐

(*b*) proportional to the indicated power ☐

(*c*) proportional to the reciprocal of I.P. ☐

(*d*) inversely proportional to the square of I.P. ☐

321. In case of single acting steam engine, the number of working stroke per revolution are

(*a*) one ☐ (*b*) two ☐

(*c*) three ☐ (*d*) four. ☐

322. If the steam acts on one side of the piston, then steam engine is known as

(*a*) single acting ☐ (*b*) double acting ☐
(*c*) triple acting ☐ (*d*) quadruple acting. ☐

323. In double acting steam engine, the number of working strokes are

(*a*) one ☐ (*b*) two ☐
(*c*) three ☐ (*d*) four. ☐

324. For double acting steam engine, the steam acts on

(*a*) one side of the piston ☐ (*b*) both sides of the piston ☐
(*c*) three sides of the piston ☐ (*d*) none of the above. ☐

325. In compound steam engine, the steam expands in

(*a*) one cylinder ☐ (*b*) two cylinders ☐
(*c*) three cylinders ☐ (*d*) four cylinders. ☐

326. In simple steam engine, the steam expands in

(*a*) one cylinder ☐ (*b*) two cylinders ☐
(*c*) three cylinders ☐ (*d*) four cylinders. ☐

327. The steam expands successively in three cylinders in ...... expansion steam engine.

(*a*) simple ☐ (*b*) compound ☐
(*c*) triple ☐ (*d*) quadruple. ☐

328. In compound steam engine, the cylinders are known as

(*a*) high pressure cylinders ☐
(*b*) low pressure cylinders ☐
(*c*) first is high pressure cylinder and second is low pressure cylinder ☐
(*d*) none of the above. ☐

329. If the exhaust of a steam engine passes through a condenser in which pressure is less than the atmospheric pressure, then the steam engine is known as

(*a*) simple engine ☐ (*b*) compound engine ☐
(*c*) condensing engine ☐ (*d*) non-condensing engine. ☐

330. If the exhaust of a steam engine takes place directly to the atmosphere, then the steam engine is known as

(*a*) simple ☐ (*b*) compound ☐
(*c*) condensing ☐ (*d*) non-condensing. ☐

331. For a long stroke steam engine, stroke should be

(*a*) equal to dia. of the cylinder ☐ (*b*) less than the dia. of the cylinder ☐
(*c*) 1.25 times the dia. of the cylinder ☐ (*d*) more than 1.5 times the dia. of the cylinder. ☐

332. If the speed of the steam engine is controlled by means of a valve in steam pipe which regulates the pressure of the steam entering the engine, then the engine is known as

(*a*) by-pass governing ☐ (*b*) nozzle governing ☐
(*c*) throttle governing ☐ (*d*) automatic cut-off governing. ☐

333. If the pressure of the entering steam remains constant and the governor controls the amount of the steam admitted to the cylinder, this type of governing is known as

(*a*) by-pass governing ☐ (*b*) nozzle governing ☐

(*c*) throttle governing ☐ (*d*) automatic cut-off governing. ☐

For the following questions, choose the correct word out of is/is not.

334. If the hot products of combustion or flue gases passes through the tubes of a boiler and the outside surfaces of these tubes is exposed to water, then the boiler is/is not a fire-tube boiler.

335. Cochran boiler is/is not a water-tube boiler.

336. The device, which converts heat energy of the steam into kinetic energy, is/is not a steam nozzle.

337. If the velocity at the inlet of the nozzle is negligible, then the velocity at the outlet of nozzle is/is not equal to 91.53 $\sqrt{\Delta h}$, where Δh = heat drop.

338. Thermal efficiency of a steam turbine is/is not higher than that of a reciprocating steam engine.

339. Heavy foundation is/is not required in case of steam turbine as compared to reciprocating steam engine.

340. The steam turbine develops power at a uniform rate and hence a flywheel is/is not required.

341. The discharge through a nozzle is/is not maximum when the ratio of pressure at inlet to throat reaches the critical value.

342. The effect of friction is/is not to decrease the heat drop.

343. The maximum discharge through a convergent-divergent nozzle is/is not independent of the pressure at outlet of nozzle.

344. To force water into boiler under pressure, a steam injector is/is not used.

345. For a convergent-divergent nozzle, if friction loss is given, the whole of friction loss is/is not assumed to occur between throat and exist.

346. The critical pressure ratio for a nozzle, when initial condition of steam is dry and saturated, is/is not equal to 0.545.

347. The velocity of a simple turbine is/is not very high.

348. The simplest and easiest way of producing artificial draught is/is not by steam jet.

349. In forced draught system, a fan or blower is/is not placed at the base of the chimney.

Fill in the blanks:

350. The device used for generating steam is known as ....

351. Cochran is .... boiler.

352. Steam turbines are classified as impulse turbine and .... turbine.

353. The pressure at inlet and outlet of the moving blade of .... turbine remain constant.

354. In a single acting steam engine, the steam acts on .... side of the piston.

355. There are .... working stroke per revolution of the crank shaft in case of double acting steam engine.

356. If the conversion of heat energy of steam into mechanical work occurs in one cylinder only, then this steam engine is known as .... steam engine.

357. If the steam expands successively in three cylinders, then the steam engine is known as .... expansion engine.

358. If the exhaust steam, in a steam engine, passes directly to the atmosphere, then this engine is known as .... engine.

359. In case of condensing steam engines, the exhaust steam passes into .... where pressure is lower than that of atmosphere.

360. For high speed steam engines, the r.p.m. is approximately more than ....

361. If the length of the stroke in a steam engine is more than 15 times the cylinder diameter, then the steam engine is known as .... stroke engine.

362. There is only one value of the ratio of pressures at throat to inlet corresponding to which the discharge of the steam through a nozzle is maximum. That pressure ratio is known as ....

363. Mechanical draught is produced by ....

364. Steam turbines are classified as impulse and .... turbines.

365. In an impulse turbine, ... of the steam remains constant while passing over the blades.

366. In a simple impulse turbine, the speed of the turbine is very ....

367. The effect of blade friction in case of impulse turbine is to .... V_{w_2}.

368. Due to blade friction, the work done per kg of steam becomes ....

369. The mass flow rate in a simple impulse turbine is 20 kg/sec. The velocities of flow at inlet and outlet are 240 m/sec and 210 m/sec. Then axial thrust will be .... N.

370. In case of reaction turbines, the steam expands continuously in both the fixed and moving blades, so its relative velocity .... due to the expansion of the steam.

371. The ratio of isentropic heat drop in moving blades to isentropic heat drop in the entire stage of the reaction turbine is known as ....

372. The process of abstracting steam at a certain section of the turbine and subsequently using if for heating feed water supplied to the boiler is known as ....

373. Nozzle governing is .... efficient than throttle-governing.

III. OBJECTIVE TYPE QUESTIONS FROM COMPETITIVE EXAMINATIONS

Tick mark the most appropriate answer:

1. The material, used for piston rings, is
   (*a*) alloy steel ☐ (*b*) cast iron ☐
   (*c*) carbon steel ☐ (*d*) copper. ☐

2. The cooling of a 250 MW generator is done by
   (*a*) water ☐ (*b*) air ☐
   (*c*) nitrogen ☐ (*d*) hydrogen. ☐

3. A steam nozzle is a device used for converting
   (*a*) heat energy of steam into pressure energy ☐
   (*b*) heat energy of steam into kinetic energy ☐
   (*c*) pressure energy of steam into kinetic energy ☐
   (*d*) pressure energy of steam into heat energy. ☐
4. The effect of considering friction losses in steam nozzle for the same pressure ratio leads to
   (*a*) increase in exit velocity from the nozzle ☐
   (*b*) decrease in exit velocity from nozzle ☐
   (*c*) no change in exit velocity from nozzle ☐
   (*d*) increase or decrease depending upon the exit quality of steam. ☐
5. Reheat factor in steam turbine depends on
   (*a*) exit pressure only ☐ (*b*) stage efficiency only ☐
   (*c*) initial pressure and temperature only ☐ (*d*) all of the above. ☐
6. The effect of considering friction in steam nozzles for the same pressure ratio leads to
   (*a*) increase in dryness fraction of exit steam ☐
   (*b*) decrease in dryness fraction of exit steam ☐
   (*c*) no change in the quality of exit steam ☐
   (*d*) decrease or increase of dryness fraction of exit steam depending upon inlet quality. ☐
7. Select the wrong statement
   (*a*) In the impules turbine, steam expands wholly in the nozzle. ☐
   (*b*) The action of steam in a steam turbine is dynamic. ☐
   (*c*) De-laval turbine is a simple reaction turbine. ☐
   (*d*) The efficiency of a steam turbine is higher than that of steam engine. ☐
8. Stage efficiency of a turbine is equal to
   (*a*) $\dfrac{\text{Work done on blades}}{\text{Total energy supplied}}$ ☐ (*b*) $\dfrac{\text{Work done on blades}}{\text{Energy supplied per stage}}$ ☐
   (*c*) $\dfrac{\text{Energy supplied per stage}}{\text{Work done on blades}}$ ☐ (*d*) $\dfrac{\text{Total energy supplied}}{\text{Work done on blades}}$. ☐
9. In case of impules steam turbine
   (*a*) there is enthalpy drop in fixed and moving blades ☐
   (*b*) there is enthalpy drop only in moving blades ☐
   (*c*) there is enthalpy drop in nozzles ☐
   (*d*) none of the above. ☐
10. De-Laval turbine is
   (*a*) pressure compounded impulse turbine ☐ (*b*) velocity compounded impulse turbine ☐
   (*c*) simple single wheel impulse turbine ☐ (*d*) simple single wheel reaction turbine. ☐

11. Choose the wrong statement
   (*a*) Steam boiler is a device used for generating and supplying steam at a high pressure and temperature. ☐
   (*b*) La-Mont boiler is a forced circulation boiler. ☐
   (*c*) Benson boiler is a high pressure boiler. ☐
   (*d*) A Benson boiler requires two drums. ☐
12. Which of the following is a high pressure boiler?
   (*a*) La-Mont boiler ☐ (*b*) Benson boiler ☐
   (*c*) Loeffler boiler ☐ (*d*) all of the above ☐
   (*e*) none of the above. ☐
13. The ratio of exit pressure to inlet pressure of maximum mass flow rate per unit area of steam through nozzle when steam is initially superheated is
   (*a*) 0.555 ☐ (*b*) 0.578 ☐
   (*c*) 0.5457 ☐ (*d*) 0.6. ☐
14. The critical pressure ratio of a convergent nozzle is defined as
   (*a*) the ratio of outlet pressure to inlet pressure of nozzle ☐
   (*b*) the ratio of inlet pressure to outlet pressure of nozzle ☐
   (*c*) the ratio of outlet pressure to inlet pressure only when mass flow rate per unit area is minimum ☐
   (*d*) the ratio of outlet pressure only when mass flow rate per unit area is maximum. ☐
15. The insentropic expansion of steam through nozzle for the steam initially dry saturated at inlet is approximated by equation
   (*a*) $pv = C$ ☐ (*b*) $pv^{1.4} = C$ ☐
   (*c*) $pv^{1.3} = C$ ☐ (*d*) $pv^{1.135} = C$. ☐
16. The simple steam engine as compared to compound steam engine for the same output has
   (*a*) smaller flywheel ☐ (*b*) large flywheel ☐
   (*c*) same size flywheel as output is same ☐ (*d*) same size flywheel as speed is same. ☐
17. The leakage past the piston initial condensation by compounding the steam engine is
   (*a*) increased ☐ (*b*) decreased ☐
   (*c*) unaffected ☐ (*d*) depends on methods of compounding. ☐
18. Woolfe type compound steam engines have
   (*a*) two cranks at 180° phase difference ☐ (*b*) two cranks at 90° phase difference ☐
   (*c*) only one crank with no phase difference ☐ (*d*) two cranks with no phase difference. ☐
19. The ratio of clearance volume to the swept volume is known as
   (*a*) expansion ratio ☐ (*b*) cut-off ratio ☐
   (*c*) compression ratio ☐ (*d*) clearance ratio. ☐
20. The ratio of the volume at cut-off to the swept volume is known as
   (*a*) expansion ratio ☐ (*b*) cut-off ratio ☐
   (*c*) compression ratio ☐ (*d*) clearance ratio. ☐

21. For Parson's reaction turbine, degree of reaction is equal to
(a) 80% ☐ (b) 50% ☐
(c) 75% ☐ (d) 100%. ☐

22. For Parson's reaction turbine, the condition for maximum efficiency is given by
(a) $\rho = 2\cos\alpha$ ☐ (b) $\rho = \cos^2\alpha$ ☐
(c) $\rho = \frac{\cos\alpha}{2}$ ☐ (d) $\rho = \cos\alpha$. ☐
where ρ = Speed ratio, and α = Angle made by absolute velocity at inlet.

23. For Parson's reaction turbine, the maximum efficiency is equal to
(a) $\frac{1+\cos^2\alpha}{2\cos^2\alpha}$ ☐ (b) $\frac{\cos^2\alpha}{1+\cos^2\alpha}$ ☐
(c) $\frac{2\cos^2\alpha}{1+\cos^2\alpha}$ ☐ (d) $\frac{\cos^2\alpha}{2(1+\cos^2\alpha)}$. ☐

24. For multistage steam turbine reheat factor is defined as
(a) stage efficiency × nozzle efficiency ☐ (b) commulative enthalpy drop × η_{nozzle} ☐
(c) $\frac{\text{commulative enthalpy drop}}{\text{isentropic enthalpy drop}}$ ☐ (d) $\frac{\text{isentropic enthalpy drop}}{\text{cumulative actual enthalpy drop}}$. ☐

25. The value of reheat factor normally varies from
(a) 0.5 to 0.6 ☐ (b) 0.9 to 0.95 ☐
(c) 0.9 to 1.06 ☐ (d) 1.2 to 1.6. ☐

26. The isentropic expansion of steam (initially saturated) through the nozzle may be approximately represented by the equation
(a) $PV^{1.4}$ = constant ☐ (b) $PV^{1.3}$ = constant ☐
(c) $PV^{1.135}$ = constant ☐ (d) $PV^{1.2}$ = constant. ☐

27. The isentropic expansion of steam (initially superheated at inlet) through the nozzle may be represented by the equation
(a) $PV^{1.4}$ = constant ☐ (b) $PV^{1.3}$ = constant ☐
(c) $PV^{1.135}$ = constant ☐ (d) $PV^{1.2}$ = constant. ☐

28. The ratio of the pressure at the exit and at inlet of the steam nozzle for maximum discharge is given by
(a) $\frac{p_2}{p_1} = \left(\frac{2}{n+1}\right)^{\frac{n-1}{n}}$ ☐ (b) $\frac{p_2}{p_1} = \left(\frac{2}{n+1}\right)^{\frac{n}{n-1}}$ ☐
(c) $\frac{p_2}{p_1} = \left(\frac{2}{n+1}\right)^{\frac{n}{n-1}}$ ☐ (d) $\frac{p_2}{p_1} = \left(\frac{n+1}{2}\right)^{\frac{n}{n+1}}$. ☐
where p_2 = Exit pressure, p_1 = Inlet pressure, and n = Index of esentropic expansion.

29. Diagram factor (D.F.) is defined as
(a) product of area of actual indicator diagram and area of theoretical indicator diagram ☐
(b) ratio of area of actual indicator diagram and area of theoretical indicator diagram ☐

(*c*) ratio of area of theoretical indicator diagram to area of actual indicator diagram ☐
(*d*) none of the above. ☐

30. The diagram factor is always
(*a*) more than 1.0 ☐ (*b*) equal to 1.0 ☐
(*c*) equal to zero ☐ (*d*) less than 1.0. ☐

31. The pressure on the two sides of the impulse wheel of steam turbine
(*a*) is same ☐ (*b*) is different ☐
(*c*) increases from one side to the other side ☐ (*d*) decreases from one side to the other side. ☐

32. In De-Laval steam turbine
(*a*) the pressure in the turbine rotor is approximately same as in condenser ☐
(*b*) the pressure in the turbine rotor is higher than pressure in the condenser ☐
(*c*) the pressure in the turbine rotor gradually decrease from inlet to exit to condenser ☐
(*d*) none of the above. ☐

33. In case of reaction steam turbine
(*a*) there is enthalpy drop both fixed and moving blades ☐
(*b*) there is enthalpy drop only in fixed blades ☐
(*c*) there is enthalpy drop only in moving blades ☐
(*d*) none of the above. ☐

34. For throttle governing of steam engine
(*a*) rate of steam consumption varies inversely as the speed of the engine ☐
(*b*) rate of steam consumption varies directly as the speed of the engine ☐
(*c*) rate of steam consumption is directly proportional to I.P. of the engine ☐
(*d*) rate of steam consumption is inversely proportional to I.P. of the engine. ☐

35. If the angle of advance is 150° and a steam engine is to be reversed then eccentric should be shifted by
(*a*) 165° ☐ (*b*) 180° ☐
(*c*) 205° ☐ (*d*) 15°. ☐

36. For the steam which is initially saturated and flowing through a nozzle, the ratio of pressures at exit and at inlet for maximum discharge is equal to
(*a*) 0.545 ☐ (*b*) 0.582 ☐
(*c*) 0.5 ☐ (*d*) 0.60. ☐

37. For the steam which is initially superheated and flowing through a nozzle, the ratio of pressures at exit and at inlet for maximum discharge is equal to
(*a*) 0.545 ☐ (*b*) 0.582 ☐
(*c*) 0.50 ☐ (*d*) 0.60. ☐

38. Nozzle is designed for
(*a*) maximum pressure at outlet ☐
(*b*) maximum discharge ☐
(*c*) maximum pressure and maximum discharge ☐
(*d*) none of the above. ☐

39. The condition for maximum blade efficiency for a single stage impulse turbine is that the blade speed ratio should be equal to

(*a*) $2 \cos \alpha$ ☐ (*b*) $\cos 2\alpha$ ☐

(*c*) $\frac{\cos \alpha}{2}$ ☐ (*d*) $\frac{1}{2} \cos 2\alpha$ ☐

where α = Angle made by absolute velocity at inlet.

40. In case of reaction turbines, the heat drop takes place on

(*a*) fixed blade only ☐ (*b*) moving blades only ☐

(*c*) both fixed and moving blades ☐ (*d*) none of the above. ☐

41. Degree of reaction is defined as the ratio of

(*a*) enthalpy drop in the moving blades/enthalpy drop in fixed blades ☐

(*b*) enthalpy drop in moving blades/sum of enthalpy drop in moving and fixed blades ☐

(*c*) enthalpy drop in fixed blades/sum of enthalpy drop in moving and the blades ☐

(*d*) none of the above. ☐

42. For Parson's reaction steam turbine, degree of reaction is

(*a*) 75% ☐ (*b*) 100% ☐

(*c*) 50% ☐ (*d*) 60%. ☐

43. The maximum efficiency for Parson's reaction turbine is given by

(*a*) $\eta_{max} = \frac{\cos\alpha}{1+\cos\alpha}$ ☐ (*b*) $\eta_{max} = \frac{2\cos\alpha}{1+\cos\alpha}$ ☐

(*c*) $\eta_{max} = \frac{2\cos^2\alpha}{1+\cos^2\alpha}$ ☐ (*d*) $\eta_{max} = \frac{1+\cos^2\alpha}{2\cos^2\alpha}$. ☐

44. For a steam engine the 'diagram factor' is defined as

(*a*) area of hypothetical indicator diagram × area of actual indicator diagram ☐

(*b*) $\frac{\text{area of actual indicator diagram}}{\text{area of hypothetical indicator diagram}}$ ☐

(*c*) $\frac{\text{area of hypothetical indicator diagram}}{\text{area of actual indicator diagram}}$ ☐

(*d*) none of the above. ☐

45. The function of D-slide valve in steam engine is

(*a*) only to exhaust steam from the cylinder ☐

(*b*) only to admit steam in the cylinder ☐

(*c*) to admit steam and also exhaust steam from the cylinder ☐

(*d*) none of the above. ☐

46. Stage efficiency is also called

(*a*) diagram efficiency ☐ (*b*) blade efficiency ☐

(*c*) gross efficiency ☐ (*d*) none of the above. ☐

47. Stage efficiency (η_s) of a steam turbine is equal to

(*a*) nozzle efficiency × blade efficiency ☐ (*b*) blade efficiency/nozzle efficiency ☐

(*c*) nozzle efficiency/blade efficiency ☐ (*d*) none of the above. ☐

48. Benson boiler produces steam upto a rate of
(*a*) 100 kg/hr only ☐ (*b*) 10^4 kg/hr ☐
(*c*) 135 ton/hr only ☐ (*d*) 100 kg/hr. ☐

49. The rate of flow of steam in case of water-tube boilers as compared to fire-tube boilers is
(*a*) less ☐ (*b*) more ☐
(*c*) same ☐ (*d*) none of the above. ☐

50. A Benson boiler requires
(*a*) two drums ☐ (*b*) one drum ☐
(*c*) no drum ☐ (*d*) three drums. ☐

51. Steam turbines are governed by the following methods
(*a*) Throttle governing ☐ (*b*) Nozzle control governing ☐
(*c*) By-pass governing ☐ (*d*) All of the above. ☐

52. In steam turbines the reheat factor
(*a*) increases with the increases in number of stages ☐
(*b*) decreases with the increase in number of stages ☐
(*c*) remains same irrespective of number of stages ☐
(*d*) none of the above. ☐

53. Which of the following are boiler mountings?
(*a*) economiser ☐ (*b*) fusible plug ☐
(*c*) superheater ☐ (*d*) injector. ☐

54. Which of the following are boiler accessories?
(*a*) safety valve ☐ (*b*) stop valve ☐
(*c*) economiser ☐ (*d*) blow off cock. ☐

55. Choose the wrong statement. The area of actual indicator diagram is less than that of the theoretical indicator diagram due to
(*a*) pressure drop because of wire drawing ☐
(*b*) gradural cut-off and release ☐
(*c*) expansion is different from hyperbolic ☐
(*d*) release and admission takes place at the end of the stroke. ☐

56. The saturation curve is the curve showing the volume of the steam in the cylinder during expansion stroke (*i.e.*, from the cut-off point to the point at release) in a steam engine if the condition of the steam is
(*a*) 80% dry ☐ (*b*) 90% dry ☐
(*c*) 95% dry ☐ (*d*) 100% dry. ☐

57. Missing quantity is defined as
(*a*) the difference in the volume at any point between the saturation curve and actual expansion curve ☐
(*b*) the product of volume given by saturation curve and actual expansion curve ☐
(*c*) the difference in volume given by the saturation curve and clearance volume ☐
(*d*) none of the above. ☐

58. The steam, in a steam engine, entrapped in the cylinder after the exhaust valve closes, is known as
(*a*) missing quantity ☐ (*b*) cushion steam ☐
(*c*) clearance volume steam ☐ (*d*) none of the above. ☐

59. Governing of compound steam engines is done by
(*a*) throttling steam to high pressure cylinder ☐
(*b*) cut-off variation in high pressure cylinder ☐
(*c*) cut-off variations simultaneously in high pressure and low pressure cylinders ☐
(*d*) all of the above methods. ☐

60. The equation used for expansion in compound steam engines namely $pV = C$ signifies
(*a*) isothermal expansion ☐ (*b*) adiabatic expansion ☐
(*c*) parabolic expansion ☐ (*d*) hyperbolic expansion. ☐

61. In a two-cylinder compound steam engine with continuous expansion from initial pressure to back pressure is geometric mean of the initial and back pressure if
(*a*) work is shared equally by high pressure and low pressure cylinder ☐
(*b*) the initial loads on the pistons of high pressure cylinder and low pressure cylinder are same ☐
(*c*) both the cases (*a*) and (*b*) ☐
(*d*) none of the above. ☐

62. In reciprocating steam engines
(*a*) pressure energy and dynamic action of steam is utilised ☐
(*b*) pressure energy of steam is utilised and dynamic action of steam is negligible ☐
(*c*) dynamic action of steam is utilised and pressure energy is negligible ☐
(*d*) none of the above. ☐

63. Steam engine is governed by throttle governing, in which intake steam
(*a*) temperature is varied ☐ (*b*) volume is varied ☐
(*c*) pressure is varied ☐ (*d*) all of the above are varied. ☐

64. Cut-off governing of steam engine is done by varying
(*a*) temperature of intake steam ☐ (*b*) volume of intake steam ☐
(*c*) pressure of intake steam ☐ (*d*) none of the above. ☐

65. If the exit pressure equal to or more than critical pressure, the nozzle used should be
(*a*) divergent ☐ (*b*) convergent ☐
(*c*) convergent-divergent ☐ (*d*) none of the above. ☐

66. Choose the correct statement
(*a*) Size of the boiler tubes is specified by inside diameter. ☐
(*b*) An economiser decreases the steam raising capacity of a boiler. ☐
(*c*) The basic purpose of a drum in boiler is to serve as storage of steam. ☐
(*d*) Injector is used for pumping into the boiler and also for heating the feed water. ☐

67. Choose the wrong statement
   (*a*) Safety valve is a boiler mounting. ☐
   (*b*) Economiser is a boiler accessories. ☐
   (*c*) The supply of water pumped into the boiler is regulated by feed check valve. ☐
   (*d*) Super heater heats the feed water by utilizing the heat in the exhaust the gases before having through the chimney. ☐
68. The function of high steam and low water safety valve is
   (*a*) to blow out if the steam pressure is higher than working pressure ☐
   (*b*) to blow out steam when the water level in the boiler is low ☐
   (*c*) both (*a*) and (*b*) ☐
   (*d*) none of the above. ☐
69. Willan's law states that if a steam engine is governed by throttling then
   (*a*) rate of steam consumption is inversely proportional to the indicated power ☐
   (*b*) rate of steam consumption is directly proportional to the indicated power ☐
   (*c*) rate of steam consumption varies with the square of indicated power ☐
   (*d*) rate of steam consumption varies with the square root of I.P. ☐
70. Willan's line is a straight line graph between the rate of steam consumption and
   (*a*) pressure of steam ☐ (*b*) temperature of steam ☐
   (*c*) indicated power ☐ (*d*) none of the above. ☐

ANSWERS

Answers to Objective Type Questions

1. (*b*)	**2.** (*a*)	**3.** (*d*)	**4.** (*d*)	**5.** (*a*)	**6.** (*b*)
7. (*b*)	**8.** (*c*)	**9.** (*a*)	**10.** (*c*)	**11.** (*c*)	**12.** (*b*)
13. (*b*)	**14.** (*a*)	**15.** (*d*)	**16.** (*c*)	**17.** (*b*)	**18.** (*d*)
19. (*d*)	**20.** (*c*)	**21.** (*b*)	**22.** (*c*)	**23.** (*d*)	**24.** (*c*)
25. (*a*)	**26.** (*b*)	**27.** (*d*)	**28.** (*b*)	**29.** (*a*)	**30.** (*b*)
31. (*a*)	**32.** (*c*)	**33.** (*c*)	**34.** (*b*)	**35.** (*b*)	**36.** (*c*)
37. (*d*)	**38.** (*d*)	**39.** (*c*)	**40.** (*d*)	**41.** (*b*)	**42.** (*c*)
43. (*b*)	**44.** (*a*)	**45.** (*d*)	**46.** (*d*)	**47.** (*d*)	**48.** (*c*)
49. (*c*)	**50.** (*c*)	**51.** (*b*)	**52.** (*b*)	**53.** (*b*)	**54.** (*c*)
55. (*b*)	**56.** (*e*)	**57.** (*a*)	**58.** (*c*)	**59.** (*c*)	**60.** (*c*)
61. (*d*)	**62.** (*d*)	**63.** (*a*)	**64.** (*c*)	**65.** (*c*)	**66.** (*a*)
67. (*b*)	**68.** (*c*)	**69.** (*c*)	**70.** (*b*)	**71.** (*d*)	**72.** (*a*)
73. (*b*)	**74.** (*b*)	**75.** (*d*)	**76.** (*b*)	**77.** (*d*)	**78.** (*c*)
79. (*d*)	**80.** (*b*)	**81.** (*b*)	**82.** (*a*)	**83.** (*c*)	**84.** (*b*)
85. (*b*)	**86.** (*b*)	**87.** (*c*)	**88.** (*b*)	**89.** (*c*)	**90.** (*c*)
91. (*a*)	**92.** (*b*)	**93.** (*c*)	**94.** (*b*)	**95.** (*b*)	**96.** (*c*)
97. (*c*)	**98.** (*c*)	**99.** (*b*)	**100.** (*a*)	**101.** (*c*)	**102.** (*d*)
103. (*a*)	**104.** (*b*)	**105.** (*c*)	**106.** (*b*)	**107.** (*a*)	**108.** (*c*)
109. (*b*)	**110.** (*a*)	**111.** (*b*)	**112.** (*a*)	**113.** (*b*)	**114.** (*d*)
115. (*b*)	**116.** (*d*)	**117.** (*c*)	**118.** (*d*)	**119.** (*b*)	**120.** (*c*)
121. (*a*)	**122.** (*b*)	**123.** (*d*)	**124.** (*d*)	**125.** (*d*)	**126.** (*a*)
127. (*b*)	**128.** (*c*)	**129.** (*d*)	**130.** (*a*)	**131.** (*b*)	**132.** (*c*)

133. (*b*)	**134.** (*d*)	**135.** (*d*)	**136.** (*c*)	**137.** (*b*)	**138.** (*a*)
139. (*b*)	**140.** (*d*)	**141.** (*b*)	**142.** (*b*)	**143.** (*a*)	**144.** (*b*)
145. (*b*)	**146.** (*b*)	**147.** (*b*)	**148.** (*a*)	**149.** (*c*)	**150.** (*d*)
151. (*d*)	**152.** (*c*)	**153.** (*d*)	**154.** (*c*)	**155.** (*b*)	**156.** (*c*)
157. (*c*)	**158.** (*d*)	**159.** (*b*)	**160.** (*a*)	**161.** (*c*)	**162.** (*d*)
163. (*b*)	**164.** (*c*)	**165.** (*b*)	**166.** (*c*)	**167.** (*b*)	**168.** (*c*)
169. (*d*)	**170.** (*b*)	**171.** (*c*)	**172.** (*c*)	**173.** (*d*)	**174.** (*b*)
175. (*c*)	**176.** (*b*)	**177.** (*b*)	**178.** (*d*)	**179.** (*b*)	**180.** (*b*)
181. (*c*)	**182.** (*b*)	**183.** (*c*)	**184.** (*b*)	**185.** (*c*)	**186.** (*b*)
187. (*b*)	**188.** (*b*)	**189.** (*a*)	**190.** (*b*)	**191.** (*a*)	**192.** (*b*)
193. (*a*)	**194.** (*b*)	**195.** (*a*)	**196.** (*c*)	**197.** (*c*)	**198.** (*c*)
199. (*c*)	**200.** (*b*)	**201.** (*c*)	**202.** (*c*)	**203.** (*c*)	**204.** (*a*)
205. (*d*)	**206.** (*c*)	**207.** (*c*)	**208.** (*c*)	**209.** (*a*)	**210.** (*b*)
211. (*c*)	**212.** (*d*)	**213.** (*b*)	**214.** (*a*)	**215.** (*c*)	**216.** (*d*)
217. (*c*)	**218.** (*b*)	**219.** (*d*)	**220.** (*c*)	**221.** (*c*)	**222.** (*a*)
223. (*d*)	**224.** (*b*)	**225.** (*c*)	**226.** (*d*)	**227.** (*a*)	**228.** (*a*)
229. (*c*)	**230.** (*d*)	**231.** (*d*)	**232.** (*d*)	**233.** (*a*)	**234.** (*b*)
235. (*b*)	**236.** (*d*)	**237.** (*c*)	**238.** (*b*)	**239.** (*c*)	**240.** (*b*)
241. (*c*)	**242.** (*a*)	**243.** (*a*)	**244.** (*c*)	**245.** (*a*)	**246.** (*b*)
247. (*a*)	**248.** (*d*)	**249.** (*a*)	**250.** (*b*)	**251.** (*c*)	**252.** (*c*)
253. (*b*)	**254.** (*c*)	**255.** (*a*)	**256.** (*b*)	**257.** (*b*)	**258.** (*c*)
259. (*c*)	**260.** (*b*)	**261.** (*b*)	**262.** (*d*)	**263.** (*c*)	**264.** (*a*)
265. (*d*)	**266.** (*c*)	**267.** (*b*)	**268.** (*c*)	**269.** (*a*)	**270.** (*b*)
171. (*b*)	**272.** (*c*)	**273.** (*d*)	**274.** (*d*)	**275.** (*a*)	**276.** (*b*)
277. (*b*)	**278.** (*b*)	**279.** (*c*)	**280.** (*d*)	**281.** (*c*)	**282.** (*b*)
283. (*c*)	**284.** (*c*)	**285.** (*b*)	**286.** (*c*)	**287.** (*a*)	**288.** (*c*)
289. (*b*)	**290.** (*d*)	**291.** (*b*)	**292.** (*a*)	**293.** (*d*)	**294.** (*c*)
295. (*c*)	**296.** (*b*)	**297.** (*d*)	**298.** (*c*)	**299.** (*b*)	**300.** (*c*)
301. (*c*)	**302.** (*b*)	**303.** (*d*)	**304.** (*c*)	**305.** (*e*)	**306.** (*d*)
307. (*b*)	**308.** (*f*)	**309.** (*d*)	**310.** (*e*)	**311.** (*c*)	**312.** (*c*)
313. (*d*)	**314.** (*c*)	**315.** (*b*)	**316.** (*c*)	**317.** (*b*)	**318.** (*b*)
319. (*b*)	**320.** (*b*)	**321.** (*a*)	**322.** (*a*)	**323.** (*b*)	**324.** (*b*)
325. (*b*)	**326.** (*a*)	**327.** (*c*)	**328** (*c*)	**329.** (*c*)	**330.** (*d*)
331. (*d*)	**332.** (*c*)	**333.** (*d*).			

For the following questions, choose the correct word out of is/is not

334. is	**335.** is not	**336.** is	**337.** is	**338.** is	**339.** is not
340. is not	**341.** is	**342.** is	**343.** is	**344.** is	**345.** is
346. is not	**347.** is	**348.** is	**349.** is not.		

Fill in the Blanks

350. boiler	**351.** fire-tube	**352.** reaction	**353.** impulse	**354.** one	**355.** two
356. simple	**357.** triple	**358.** non-condensing		**359.** condenser	
360. 200 r.p.m.	**361.** long	**362.** critical pressure ratio			
363. fans or blower		**364.** reaction	**365.** pressure	**366.** high	**367.** reduce
368. less	**369.** (600)	**370.** increases	**371.** degree of reduction		**372.** bleeding
373. more.					

Answers to Objective Type Questions from Competitive Examinations

1. (*b*)	2. (*d*)	3. (*b*)	4. (*b*)	5. (*d*)	6. (*a*)
7. (*c*)	8. (*b*)	9. (*c*)	10. (*c*)	11. (*d*)	12. (*d*)
13. (*c*)	14. (*d*)	15. (*d*)	16. (*b*)	17. (*b*)	18. (*a*)
19. (*d*)	20. (*b*)	21. (*b*)	22. (*d*)	23. (*c*)	24. (*c*)
25. (*c*)	26. (*c*)	27. (*b*)	28. (*b*)	29. (*b*)	30. (*d*)
31. (*a*)	32. (*a*)	33. (*a*)	34. (*b*)	35. (*c*)	36. (*b*)
37. (*a*)	38. (*b*)	39. (*c*)	40. (*c*)	41. (*b*)	42. (*c*)
43. (*c*)	44. (*b*)	45. (*b*)	46. (*c*)	47. (*a*)	48. (*c*)
49. (*b*)	50. (*c*)	51. (*d*)	52. (*a*)	53. (*b*)	54. (*c*)
55. (*d*)	56. (*d*)	57. (*a*)	58. (*b*)	59. (*d*)	60. (*d*)
61. (*c*)	62. (*b*)	63. (*c*)	64. (*b*)	65. (*b*)	66. (*d*)
67. (*d*)	68. (*c*)	69. (*b*)	70. (*c*).		

Chapter 6 COMPRESSORS, GAS TURBINES AND JET ENGINES

I. THEORY

6.1. AIR COMPRESSOR

Air compressor is a machine which provides gas or air at high pressure by compression. The compressed air is used for driving pneumatic tools, for cooling large buildings, for conveying solid and powdered materials in pipe lines, for paint spraying etc.

6.2. CLASSIFICATION OF COMPRESSORS

The air compressors are mainly classified as:

1. *According to moving parts as:*
   (*a*) Reciprocating compressor (*b*) Centrifugal compressor
   (*c*) Rotary compressor.
2. *According to number of stages as:*
   (*a*) Single stage compressor (*b*) Multi-stage compressor.
3. *According to number of power cylinders as:*
   (*a*) Single cylinder compressor (*b*) Multi-cylinder compressor.
4. *According to number of air cylinders as:*
   (*a*) Simple compressor (*b*) Duplex compressor
   (*c*) Triplex compressor.
5. *According to power drive as:*
   (*a*) Direct drive compressor (*b*) Belt drive compressor
   (*c*) Chain drive compressor.

6. *According to method of cooling as:*
   (*a*) Air cooled compressor (*b*) Water cooled compressor.
7. *According to the principle of operation as:*
   (*a*) Positive displacement compressor (*b*) Dynamic compressors.

The reciprocating compressor and rotary compressors are positive displacement compressor. But centrifugal compressors and axial flow compressors are dynamic compressors.

The air compressors are *classified* as positive displacement compressors and non-positive displacement compressors. They are also classified as reciprocating compressors or rotary compressors. The *examples* of positive displacement compressors arc: reciprocating compressors, root blower and vane pump. The examples of non-positive displacement compressors are: centrifugal and axial compressors.

6.3. WORK DONE IN SINGLE STAGE RECIPROCATING AIR COMPRESSOR NEGLECTING CLEARANCE VOLUME

Figure 6.1 shows the theoretical air compression cycle without clearance on *p-V* diagram. The different operations are : suction, compression and discharge. The air is sucked at constant pressure P_1 and is delivered at constant pressure p_2. The work done in the cycle is given by the area of the *p-V* diagram.

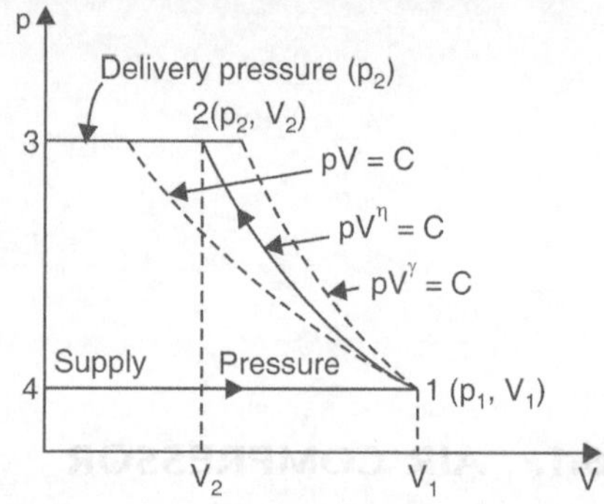

FIGURE 6.1

6.3.1. Work Done When Compression Follows the Law $pV^n = C$

Work done per cycle = Area of *p-V* diagram 1-2-3-4

= (Area under 1-2) + (Area under 2-3) – (Area under 1-4)

$$= \left(\frac{p_2V_2 - p_1V_1}{n-1}\right) + p_2V_2 - p_1V_1$$

$$= (p_2V_2 - p_1V_1)\left(\frac{1}{n-1} + 1\right) = \left(\frac{n}{n-1}\right)(p_2V_2 - p_1V_1)$$

$$= \left(\frac{n}{n-1}\right)p_1V_1\left[\left(\frac{p_2}{p_1}\right)^{\frac{n-1}{n}} - 1\right]$$

$$= \left(\frac{n}{n-1}\right)mRT_1\left[\left(\frac{p_2}{p_1}\right)^{\frac{n-1}{n}} - 1\right] \qquad (\because\ p_1V_1 = mRT_1)$$

6.3.2. Work Done When Compression Follows the Law $pV^\gamma = C$

The work done per cycle is obtained by changing n to γ.

$$\therefore \text{ Work done per cycle} = \left(\frac{\gamma}{\gamma-1}\right)mRT_1\left[\left(\frac{p_2}{p_1}\right)^{\frac{\gamma-1}{\gamma}} - 1\right] \text{ or } \left(\frac{\gamma}{\gamma-1}\right)p_1V_1\left[\left(\frac{p_2}{p_1}\right)^{\frac{n-1}{n}} - 1\right].$$

6.3.3. Work Done When Compression Follows the Isothermal Law (*i.e.*, pV = C)

Work done per cycle $= p_1V_1 \log_e\left(\frac{V_1}{V_2}\right) + p_2V_2 - p_1V_1$

$$= p_1V_1 \log_e\left(\frac{V_1}{V_2}\right) \qquad (\because\ p_2V_2 = p_1V_1)$$

$$= p_1V_1 \log_e\left(\frac{p_2}{p_1}\right) = p_1V_1 \log_e r$$

where r = Compression ratio $= \frac{V_1}{V_2} = \frac{p_2}{p_1}$.

The work done is minimum when compression follows isothermal law (*i.e.*, $pV = C$ or $n = 1$) and is maximum when compression is adiabatic (*i.e.*, $n = \gamma$). Isothermal compression is not possible in practice as the compressor would need to run at very low speed. In practice, the value of n varies from 1.1 to 1.3.

The performance of a reciprocating air compressor is given by *isothermal efficiency* which is the ratio of isothermal work and actual indicator work.

6.4. EFFICIENCIES OF A COMPRESSOR

The efficiencies of a compressor are:

(*i*) isothermal efficiency, (*ii*) adiabatic efficiency,

(*iii*) mechanical efficiency, and (*iv*) volumetric efficiency.

(*i*) **Isothermal efficiency.** It is the ratio of isothermal work to actual indicator work. Mathematically,

$$\text{Isothermal efficiency} = \frac{p_1V_1 \log_e\left(\frac{p_2}{p_1}\right)}{\left(\frac{n}{n-1}\right)p_1V_1\left[\left(\frac{p_2}{p_1}\right)^{\frac{n-1}{n}} - 1\right]} = \frac{\log_e\left(\frac{p_2}{p_1}\right)}{\left(\frac{n}{n-1}\right)\left[\left(\frac{p_2}{p_1}\right)^{\frac{n-1}{n}} - 1\right]}$$

(*ii*) **Adiabatic efficiency.** It is the ratio of adiabatic work to actual work of a compressor. Mathematically,

$$\text{Adiabatic efficiency} = \frac{p_1V_1\left(\frac{\gamma}{\gamma-1}\right)\left[\left(\frac{p_2}{p_1}\right)^{\frac{\gamma-1}{\gamma}} - 1\right]}{p_1V_1\left(\frac{n}{n-1}\right)\left[\left(\frac{p_2}{p_1}\right)^{\frac{n-1}{n}} - 1\right]}$$

$$= \frac{\left(\frac{\gamma}{\gamma-1}\right)\left[\left(\frac{p_2}{p_1}\right)^{\frac{\gamma-1}{\gamma}} - 1\right]}{\left(\frac{n}{n-1}\right)\left[\left(\frac{p_2}{p_1}\right)^{\frac{n-1}{n}} - 1\right]}.$$

(*iii*) **Mechanical efficiency.** It is the ratio of I.H.P. to B.H.P. of the motor.

(*iv*) **Volumetric efficiency.** It is the ratio of the actual mass of air pumped by the compressor to the mass of air which the compressor would pump if it handled a volume of air equal to swept volume at the suction stroke at free air condition (*i.e.*, intake condition).

1. If the expansion and compression follows the same law, then volumetric efficiency in terms of clearance ratio and pressure ratio is given by

$$\text{Volumetric efficiency} = 1 + C - C\left(\frac{p_2}{p_1}\right)^{\frac{1}{n}} \quad \text{or} \quad 1 + C - C\left(\frac{V_1}{V_2}\right)$$

where C = Clearance ratio, and $\left(\frac{p_2}{p_1}\right)$ = Pressure ratio.

2. The volumetric efficiency decreases with the increase of pressure ratio. Also the volumetric efficiency decreases with the increase of clearance ratio.

3. Thus the *effect of clearance* in compressor is to reduce the volumetric efficiency but the clearance is not having any effect on the work done per kg of air delivered. The clearance volume in single stage compressors is more than that of multi-stage compressors. It is about 4 to 10% of stroke volume in case of single stage compressors and 1.5 to 3% of stroke volume in multi-stage compressors.

6.5. WORK DONE IN SINGLE STAGE COMPRESSOR WITH CLEARANCE VOLUME

Figure 6.2 shows the *p-V* diagram with clearance. The air is sucked at constant pressure p_1 and is delivered at constant pressure p_2. At the end of delivery stroke, the volume V_3 having pressure p_2 is left inside the cylinder. This high pressure air will expand to V_4 at the beginning of suction stroke. The fresh air inhaled in the suction stroke is only $(V_1 - V_4)$. Thus the clearance reduces the volume of gas actually sucked in per working cycle. The work done is equal to the area of curve 1-2-3-4-1.

FIGURE 6.2

$\therefore$ Work done per cycle = (Area of 1-2-6-5-1) − (Area of 3-4-5-6-3)

$$= \left(\frac{n}{n-1}\right) p_1 V_1 \left[\left(\frac{p_2}{p_1}\right)^{\frac{n-1}{n}} - 1\right] - \left(\frac{n}{n-1}\right) p_4 V_4 \left[\left(\frac{p_3}{p_4}\right)^{\frac{n-1}{n}} - 1\right]$$

But $p_4 = p_1$ and $p_3 = p_2$,

$$\text{Hence work done per cycle} = \left(\frac{n}{n-1}\right) p_1 V_1 \left[\left(\frac{p_2}{p_1}\right)^{\frac{n-1}{n}} - 1\right] - \left(\frac{n}{n-1}\right) p_1 V_4 \left[\left(\frac{p_2}{p_1}\right)^{\frac{n-1}{n}} - 1\right]$$

$$= \left(\frac{n}{n-1}\right) p_1 (V_1 - V_4) \left[\left(\frac{p_2}{p_1}\right)^{\frac{n-1}{n}} - 1\right]$$

$$= \left(\frac{n}{n-1}\right) m\, RT_1 \left[\left(\frac{p_2}{p_1}\right)^{\frac{n-1}{n}} - 1\right]$$

where m is the mass of air corresponding to $(V_1 - V_4)$ which is the volume of air inhaled during suction stroke.

6.6. MULTI-STAGE COMPRESSOR

When the pressure ratio is high, multi-stage compressors are used. Multi-stage compression is very efficient and improves the volumetric efficiency. The other advantages of multi-stage compression are smaller leakage loss, lighter cylinders more uniform torque and smaller size of flywheel.

6.6.1. Work Done in Two-stage Compressor Neglecting Clearance

Figure 6.3 shows the p-V diagram and schematic diagram for a two-stage compressor neglecting clearance. The air after compression in the first stage (*i.e.*, L.P. cylinder) is cooled in an intercooler at constant pressure. Then this cooled air is sent to second stage called high pressure (H.P. cylinder). Cooling in the inter-cooler is said to be *perfect* if the air is cooled to its original temperature. For perfect cooling, the point 2' will lie on the curve pV = C. The *assumptions* made for calculating the work done in multi-stage compressor are:

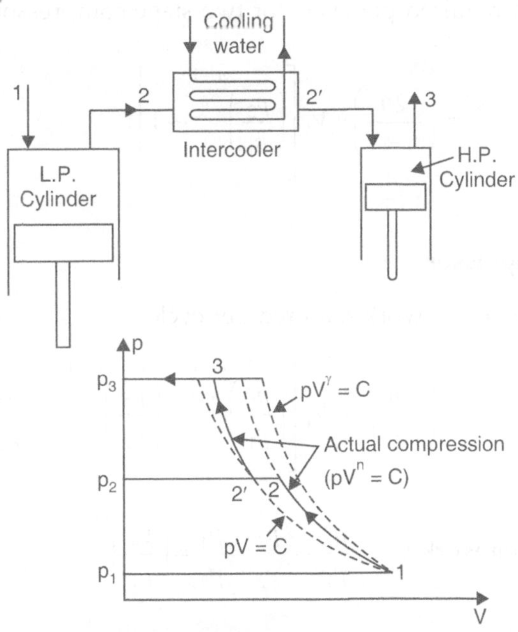

FIGURE 6.3

(*i*) intercooler is perfect,

(*ii*) the index of compression in each stage is same,

(*iii*) suction and delivery pressures remain constant during each stage.

∴ Work required per cycle = Work required in L.P. cylinder + Work required in H.P. cylinder

$$= \left(\frac{n}{n-1}\right) p_1 V_1 \left[\left(\frac{p_2}{p_1}\right)^{\frac{n-1}{n}} - 1\right] + \left(\frac{n}{n-1}\right) p_2' V_2' \left[\left(\frac{p_3}{p_2'}\right)^{\frac{n-1}{n}} - 1\right].$$

For a perfect intercooling, the point 2′ lies on the isothermal line (*i.e.*, $pV = C$).

$$\therefore \quad p_1 V_1 = p_2' V_2'$$

$$\therefore \quad \text{Work required per cycle} = \left(\frac{n}{n-1}\right) p_1 V_1 \left[\left(\frac{p_2}{p_1}\right)^{\frac{n-1}{n}} - 1\right] + \left(\frac{n}{n-1}\right) p_1 V_1 \left[\left(\frac{p_3}{p_2}\right)^{\frac{n-1}{n}} - 1\right]$$

$$= \left(\frac{n}{n-1}\right) p_1 V_1 \left[\left(\frac{p_2}{p_1}\right)^{\frac{n-1}{n}} + \left(\frac{p_3}{p_2}\right)^{\frac{n-1}{n}} - 2\right].$$

The work done will be *minimum*, when $p_2 = \sqrt{p_1 \times p_3}$ or $\dfrac{p_2}{p_1} = \dfrac{p_3}{p_2}$

or pressure ratio is same in each stage.

∴ Minimum work required per cycle for two stage compressor

$$= \left(\frac{2n}{n-1}\right) p_1 V_1 \left[\left(\frac{p_3}{p_1}\right)^{\frac{n-1}{2n}} - 1\right].$$

6.6.2. Three-stage Compressor

For a three-stage compressor, the work required per cycle

$$= \left(\frac{n}{n-1}\right) p_1 V_1 \left[\left(\frac{p_2}{p_1}\right)^{\frac{n-1}{n}} + \left(\frac{p_3}{p_2}\right)^{\frac{n-1}{n}} + \left(\frac{p_4}{p_2}\right)^{\frac{n+1}{n}} - 3\right]$$

and condition for minimum work is $\dfrac{p_2}{p_1} = \dfrac{p_3}{p_2} = \dfrac{p_4}{p_3} = \left(\dfrac{p_4}{p_1}\right)^{1/3}$

or
$$p_2 = p_1\left(\frac{p_4}{p_1}\right)^{1/3}$$

and
$$p_3 = p_1 = \left(\frac{p_4}{p_1}\right)^{2/3}$$

∴ Minimum work required for three-stage compressor per cycle

$$= \left(\frac{3n}{n-1}\right)p_1V_1\left[\left(\frac{p_4}{p_1}\right)^{\frac{n-1}{3n}} - 1\right].$$

6.6.3. Control of Compressors

Control of compressors means the controlling of the amount of air delivered by the compressor. There are three main following methods of controlling the air delivered :

1. Throttle control,
2. Clearance control, and
3. Blow-off control.

1. Throttle control. The pressure in the receiver controls the opening of the suction valve of the compressor. If the pressure in the receiver exceeds the normal pressure, the quantity of air during suction stroke is reduced by the partial closure of the suction valve by the excess pressure.

2. Clearance control. When the pressure in the receiver exceeds the normal pressure, the clearance pockets are brought into communication with the cylinder by automatically operated valves. The increase of clearance volume decreases the volumetric efficiency.

3. Blow-off control. When the pressure in the receiver exceeds the normal pressure by-pass valve opens to the atmosphere.

6.7. ROTARY COMPRESSORS

The machines, which provide gas or air at high pressure due to rotation, are called rotary compressors. They may be positive displacement rotary compressors (such as roots blower and Vane type blower) or non-positive steady-flow rotary compressors (such as centrifugal compressors and axial flow compressors).

6.7.1. Roots Efficiency

It is the ratio of the work required per minute for ideal isentropic compression to the work required per minute to drive the roots blower compressor. Mathematically, it is given by

$$\eta_{\text{roots}} = \frac{\frac{\gamma}{\gamma-1}\left[r_p^{\frac{\gamma-1}{\gamma}} - 1\right]}{(r_p - 1)}$$

where r_p = Pressure ratio = $\frac{p_2}{p_1}$.

$$= \frac{C_p J}{R}\left[\frac{r_p^{\frac{\gamma-1}{\gamma}} - 1}{r_p - 1}\right]$$

where C_p = Specific heat at constant pressure.

6.8. CENTRIFUGAL COMPRESSORS

They are like centrifugal pumps. They consist of an impeller rotating at high speeds which may be upto 30000 r.p.m. The air enters the impeller at the centre called the eye and flows radially outward with increasing pressure and temperature. A gradually increasing area of diffuser converts the kinetic energy of the air into pressure energy. A pressure ratio of 4 : 1 can be obtained in a single-stage centrifugal compressor.

6.9. AXIAL FLOW COMPRESSOR

It consists of a central drum on which moving blades are mounted. The fixed blades are on outer casing. A stage, consists of one ring of moving blades and one ring of fixed blades. A pressure ratio of 10 : 1 can be obtained in multi-stage compressor. The number of stages used vary from 5 to 14.

6.10. STATIC TEMPERATURE

Static temperature of a moving gas is the actual temperature recorded by a thermometer moving with the gas with the speed equal to that of the gas. The **stagnation temperature** is the temperature recorded by a stationary thermometer when the moving gas is brought to rest isentropically. The **stagnation pressure** is the pressure of the gas when the moving gas is brought to rest isentropically.

In the absence of heat and work transfer, the stagnation temperature (T_0), stagnation enthalpy (h_0) and stagnation pressure (p_0) are given as

$$T_0 = T + \frac{V^2}{2gJC_p} \quad h_0 = h + \frac{V^2}{2gJ} \quad \text{and} \quad p_0 = p\left(\frac{T_0}{T}\right)^{\frac{\gamma-1}{\gamma}}$$

where T, h, p and V are the temperature, enthalpy, pressure and velocity of the moving gas respectively.

6.11. WORK DONE BY THE IMPELLER

The work done by impeller of a centrifugal compressor/kg of air in terms of stagnation temperature, stagnation enthalpy and stagnation pressure when the impeller is radial vaned and there is no heat transfer and also the *slip factor* (which is the ratio of velocity of whirl at outlet to velocity of blade at outlet) is unity, is given as,

$$\text{Work done/kg} = C_p J\,[T_{01} - T_{02}] = h_{01} - h_{02}$$

$$= C_p T_{01}\left[\left(\frac{p_{02}}{p_{01}}\right)^{\frac{\gamma-1}{\gamma}} - 1\right]$$

where T_{01} and T_{02} are stagnation temperature at inlet and outlet

h_{01} and h_{02} are stagnation enthalpy at inlet and outlet

p_{01} and p_{02} are stagnation pressures at inlet and outlet.

For an ideal centrifugal-compressor, the ratio of total pressures at outlet to that of inlet is given by

$$\frac{p_{02}}{p_{01}} = \left(1 + \frac{V_b^2}{gJC_p T_{01}}\right)^{\frac{\gamma}{\gamma-1}}$$

where V_b = Tip velocity of impeller, T_{01} = Total temperature at inlet

p_{02} = Total pressure at outlet, p_{01} = Total pressure at inlet.

6.12. EFFICIENCY OF A ROTARY COMPRESSOR

The basis for efficiency of a reciprocating compressor is taken as isothermal compression whereas the basis for efficiency of a rotary compressor is taken as isentropic compression. Isentropic efficiency is the ratio of isentropic work to actual work. In case of rotary compressor, the index of compression is always more than γ. If h_1 is the enthalpy of air at inlet to rotary compressor, h_2 is the enthalpy at outlet for actual compression and h_2' is the enthalpy at outlet for isentropic compression, then isentropic efficiency (η_{isen}) is given by

$$\eta_{isen} = \frac{h_2' - h_1}{h_2 - h_1} = \frac{T_2' - T_1}{T_2 - T_1}$$

where T_1 = Inlet temperature,

T_2 = Actual delivery temperature, and

T_2' = Delivery temperature after isentropic compression.

6.13. DEGREE OF REACTION

It is defined as the ratio of the pressure rise in rotor to the total pressure rise in a stage. A stage consists of one row of rotor and one row of moving blades. The degree of reaction is also equal to the ratio of enthalpy rise in rotor to the enthalpy rise in stage.

1. For minimum fluid friction and blade tip clearance losses, the blades of an axial flow compressor are generally designed for 50% degree of reaction. Also when degree of reaction is 50%, the blades are symmetrical.

2. The *blades* of turbines may have profiles consisting of circular arcs and straight lines, the compressor blades are of aerofoil section.

3. **Stalling** is the phenomenon in which the air stream in not able to follow the blade contour of an axial flow compressor.

6.14. GAS TURBINE

Gas turbines are similar to steam turbines. In gas turbine plant, the air from rotary compressor is passed into combustion chamber in which fuel is burnt, the product of combustion then strikes the blades of the turbine with high velocity and work is produced.

The gas turbines are **classified** as:

1. Open-cycle gas turbines,
2. Closed-cycle gas turbine, and
3. Semi-closed cycle gas turbine.

A gas turbine is said to work on **closed cycle** if the working fluid in the plant does not come in contact with the atmospheric air, and the fluid is used over and over again as shown in Fig. 6.4. The gas turbine is said to work on *open-cycle* if the entire working fluid in the plant is taken from atmosphere and is again returned to the atmosphere after work is produced on the turbine as shown in Fig. 6.5. The gas turbine is said to work on *semi-closed cycle* if a part of working fluid is retained in a plant while the remaining part flows from and to the atmosphere.

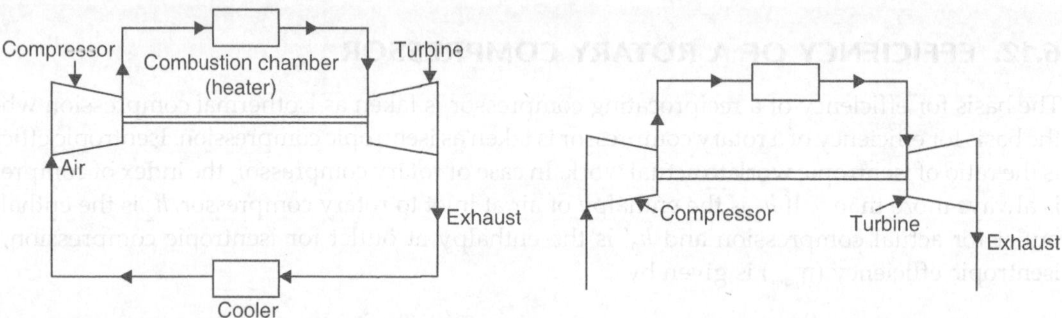

FIGURE 6.4. Closed-cycle gas turbine.

FIGURE 6.5. Open-cycle gas turbine.

Closed-cycle and open-cycle gas turbines work on Brayton or Joule cycle, which is shown in Fig. 6.6 on *p-V* and *T-S* diagram.

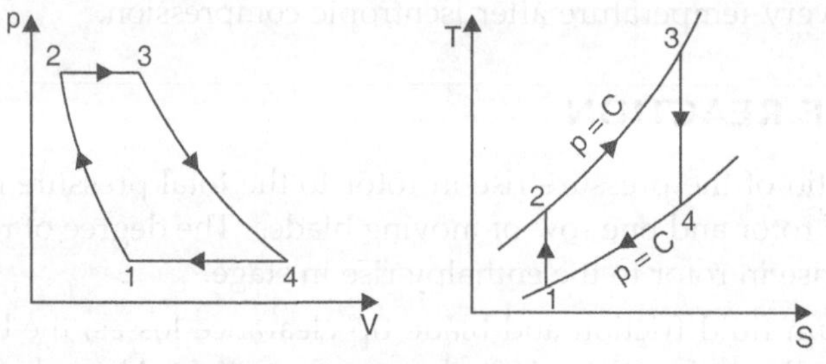

FIGURE 6.6

Different operations are:

Curve 1-2 shows the work done by compressor on the air.

Curve 2-3 shows the addition of heat at constant pressure in heater.

Curve 3-4 shows the work done by air on turbine.

Curve 4-1 shows the heat rejection in the cooler at constant pressure.

Let one kg of air is following through the cycle

Then thermal efficiency $= \dfrac{\text{Net work output}}{\text{Heat supplied}}$

$$= \frac{\text{(Work done by air on turbine – Work done by compressor on air)}}{\text{(Heat supplied)}}$$

$$= \frac{C_p[T_3 - T_4] - C_p[T_2 - T_1]}{C_p[T_3 - T_2]} = \frac{(T_3 - T_4) - (T_2 - T_1)}{(T_3 - T_2)}$$

But $\quad T_2 = T_1 \left(\dfrac{p_2}{p_1}\right)^{\frac{\gamma-1}{\gamma}}$

and $\quad T_3 = T_4 \left(\dfrac{p_3}{p_4}\right)^{\frac{\gamma-1}{\gamma}} = T_4 \left(\dfrac{p_2}{p_1}\right)^{\frac{\gamma-1}{\gamma}}$

Substituting these values in the above equation, we get

$$\text{Thermal efficiency} = 1 - \frac{1}{\left(\dfrac{p_2}{p_1}\right)^{\frac{\gamma-1}{\gamma}}} = 1 - \left(\frac{1}{r_p}\right)^{\frac{\gamma-1}{\gamma}}$$

where r_p = Pressure ratio $= \dfrac{p_2}{p_1}$.

6.14.1. Work Ratio

Work ratio of the turbine plant is defined as the ratio of net work output from the plant to the work obtained from turbine.

$$\therefore \qquad \text{Work ratio} = \frac{C_p[T_3 - T_4] - C_p[T_2 - T_1]}{C_p[T_3 - T_4]} = 1 - \frac{T_2 - T_1}{T_3 - T_4}$$

Substituting the values of T_2 and T_3 from above, we get

$$\text{Work ratio} = 1 - \frac{T_1}{T_4} = 1 - \frac{T_1}{T_3}(r_p)^{\frac{\gamma-1}{\gamma}}$$

where T_1 = Absolute air inlet temperature to compressor, and

T_3 = Absolute air inlet temperature to turbine.

6.14.2. The Ideal Air Standard Efficiency

The ideal air standard efficiency (*i.e.*, thermal efficiency when the compression and expansion are isentropic) is independent of the maximum temperature. It is a function of pressure ratio only. But the work ratio is a function of pressure ratio and temperature ratio. If T_1 (the air inlet temperature to compressor) is very low and T_3 (the air inlet temperature to turbine) is very high, the work ratio will be maximum. In actual practice, the lowest possible temperature is atmospheric temperature and highest possible temperature is about 1000°C.

6.14.3. Maximum Specific Work Output

For *maximum specific work* output, *i.e.*, maximum work output per unit mass of air flow, the pressure ratio (r_p) is given by

$$r_p = \left(\frac{T_3}{T_1}\right)^{\frac{\gamma}{2(\gamma-1)}}$$

If the temperature T_2 and T_1 are fixed, then the value of maximum pressure ratio is given by

$$r_{p(\max)} = \left(\frac{T_3}{T_1}\right)^{\frac{\gamma}{\gamma-1}}$$

Hence pressure ratio in terms of maximum pressure ratio is given by

$$r_p = \sqrt{r_{p(\max)}}\ .$$

6.15. TURBINE EFFICIENCY

It is defined as the ratio of actual work done on the turbine to the work due to isentropic expansion. Figure 6.7 shows on *T-S* diagram actual expansion and actual compression along with isentropic expansion and compression in the turbine and compressor of a closed-cycle gas turbine plant.

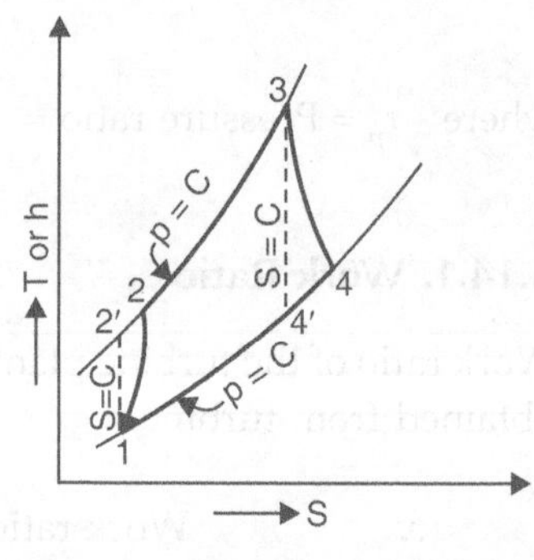

FIGURE 6.7

The turbine efficiency (η_t) is given by

$$\eta_t = \frac{C_p[T_3 - T_4]}{C_p[T_3 - T_4']} = \frac{T_3 - T_4}{T_3' - T_4'}$$

$$= \frac{\text{Actual temperature drop}}{\text{Isentropic temp. drop}}.$$

6.16. COMPRESSOR EFFICIENCY (η_c)

It is defined as the ratio of the work due to isentropic compression to the work required for actual compression. Mathematically,

$$\therefore \qquad \eta_c = \frac{C_p[T_2' - T_1]}{C_p[T_2 - T_1]} = \frac{T_2' - T_1}{T_2 - T_1}$$

$$= \frac{\text{Isentropic increase in temperature}}{\text{Actual increase in temperature}}.$$

6.16.1. Modification of Simple Gas-turbine Cycle for Higher Efficiency

The efficiency of a simple gas turbine cycle can be increased by the following methods:

(*i*) Gas turbine cycles with heat exchanger or regenerator.

(*ii*) Gas turbine cycle with reheating.

(*iii*) Gas turbine cycle with intercooling.

(*iv*) Gas turbine cycle with reheating and heat exchanger.

6.17. JET PROPULSION AND ROCKET PROPULSION

The jet produced by the combustion of fuel, is used for the propulsion of aircrafts and missiles. Jet propulsion is the term used when the oxygen for combustion is taken from the surroundings and rocket propulsion is the term used when the aircraft uses its own oxygen for combustion.

The propulsion power is given by

$$\text{Propulsion power} = \frac{W(V-u)u}{g \times 1000} \text{ kW}$$

where W = Weight of gases discharged per second,

V = Relative velocity of jet to aircraft, and

u = Velocity of aircraft.

The propulsion efficiency is given by,

$$\text{Propulsion, } \eta = \frac{2u}{V+u}.$$

The propulsion efficiency of a screw propeller and a jet engine are shown in Fig. 6.8. For a screw propeller aircraft, the efficiency increases with the aircraft speed initially. But when the speed of the propeller tip becomes equal to sonic velocity, the efficiency decreases. The efficiency of a jet engine increases with the speed of aircraft. The jet engines are more efficient than a screw propeller when the speed of aircraft exceeds 800 km/hr. Also jet engines are more efficient for high altitude flying.

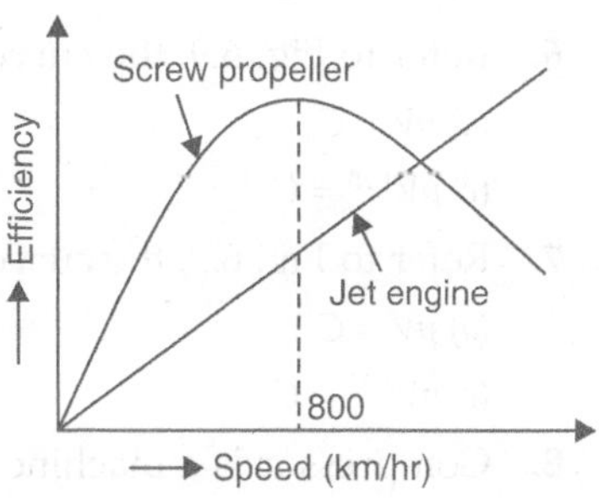

FIGURE 6.8

II. OBJECTIVE TYPE QUESTIONS

Tick mark the most appropriate statement of the multiple choice answers:

Compressor

1. The work done by the prime mover to run the compressor is minimum if compression follows

(*a*) adiabatic law ($pV^r = C$) ☐ (*b*) isothermal law ($pV = C$) ☐

(*c*) $pV^{1.1}$ = constant ☐ (*d*) $pV^{1.25}$ = constant. ☐

2. From work point of view, isothermal compression should be used in case of a compressor, but isothermal compression necessitates
   (a) high running of machine ☐ (b) very slow running of machine ☐
   (c) running of machine at constant speed ☐ (d) none of the above. ☐
3. The practical values of index of compression (*i.e.*, n) for a compressor lie in the range of
   (a) 1.4 to 1.6 ☐ (b) 1.3 to 1.4 ☐
   (c) 1.2 to 1.4 ☐ (d) 1.1 to 1.3. ☐
4. Figure 6.9 shows the theoretical air compression cycle neglecting clearance volume on p-V diagram.
   Four curves of compression following the laws $pV = C$, $pV^{1.15} = C$, $pV^{1.25} = C$ and $pV^{1.4} = C$ are shown. Curve A follows the law of compression given by
   (a) $pV = C$ ☐
   (b) $pV^{1.15} = C$ ☐
   (c) $pV^{1.25} = C$ ☐
   (d) $pV^{1.4} = C$. ☐

FIGURE 6.9

5. Refer to Fig. 6.9, the curve B follows the law of compression given by
   (a) $pV = C$ ☐ (b) $pV^{1.15} = C$ ☐
   (c) $pV^{1.25} = C$ ☐ (d) $pV^{1.4} = C$. ☐
6. Refer to Fig. 6.9, the curve C follows the law of compression
   (a) $pV = C$ ☐ (b) $pV^{1.15} = C$ ☐
   (c) $pV^{1.25} = C$ ☐ (d) $pV^{1.4} = C$. ☐
7. Refer to Fig. 6.9, the curve D follows the law of compression
   (a) $pV = C$ ☐ (b) $pV^{1.15} = C$ ☐
   (c) $pV^{1.25} = C$ ☐ (d) $pV^{1.4} = C$. ☐
8. Compressor is a machine which
   (a) provides gas at high pressure and is driven by prime mover ☐
   (b) provides gas at high pressure and is not driven by prime mover ☐
   (c) provides gas at high temperature and is not driven by prime mover ☐
   (d) none of the above. ☐
9. In a single stage, single acting reciprocating air compressor without clearance, the expression
   $p_1 V_1 \log_e \frac{p_2}{p_1}$ represents the work done on the air per cycle for
   (a) isothermal compression ☐ (b) adiabatic compression ☐
   (c) polytropic compression ☐ (d) none of the above. ☐
10. In a single stage, single acting reciprocating air compressor without clearance, the expression
   $\frac{\gamma}{\gamma - 1} R\,(T_2 - T_1)$ represents the work done on the air per cycle for

(a) isothermal compression ☐ (b) adiabatic compression ☐
(c) polytropic compression ☐ (d) none of the above. ☐

11. The work done in a single stage compressor neglecting clearance volume, when compression follows the law $pV^n = C$, is equal to

(a) $\frac{n}{n-1} R(T_2 - T_1)$ ☐ (b) $\frac{n}{n-1} (p_2V_2 - p_1V_1)$ ☐

(c) $\left(\frac{n}{n-1}\right) p_1V_1 \left[\left(\frac{p_2}{p_1}\right)^{\frac{n-1}{n}} - 1\right]$ ☐ (d) all of the above ☐

(e) none of the above. ☐

12. The clearance volume is provided in reciprocating compressors to

(a) reduce the work done ☐
(b) to increase the volumetric efficiency ☐
(c) provide space for valves and also to be sure that the piston done not strike the cylinder at the end of the stroke ☐
(d) none of the above. ☐

13. The clearance ratio is defined as the ratio of

(a) clearance volume to swept volume ☐ (b) clearance volume to cylinder volume ☐
(c) swept volume to clearance volume ☐ (d) none of the above. ☐

14. The clearance ratio for a single stage compressor lies between

(a) 15 to 20% ☐ (b) 20 to 30% ☐
(c) 4 to 10% ☐ (d) 1 to 2%. ☐

15. The clearance ratio for multi-stage compressor lies between

(a) 10 to 15% in L.P. and H.P. cylinder ☐ (b) 4 to 10% in both cylinders ☐
(c) 1 to 2% in both cylinders ☐ (d) 1.5 to 3% in both cylinders. ☐

16. The mechanical efficiency for an air compressor is equal to

(a) $\frac{\text{Shaft power}}{\text{Indicated power}}$ ☐ (b) $\frac{\text{Indicated power}}{\text{Shaft power}}$ ☐

(c) Shaft power × I.P. ☐ (d) none of the above. ☐

17. Free air condition means the air at

(a) 15°C and 1 kgf/cm^2 with a relative humidity 60% ☐
(b) standard atmospheric conditions and 0°C ☐
(c) atmospheric conditions at any specific location ☐
(d) 20° and 1 kgf/cm^2 and a relative humidity of 36%. ☐

18. Standard air condition means the air at

(a) 15°C and 1 kgf/cm^2 with a relative humidity of 60% ☐
(b) standard atmospheric conditions and 0°C ☐
(c) atmospheric conditions at any specific location ☐
(d) 20°C and 1 kgf/cm^2 and a relative humidity of 36%. ☐

19. The compressor capacity is equal to

(a) volume of the air sucked ☐ (b) volume of air delivered ☐

(c) sum of (a) and (b) ☐ (d) none of the above. ☐

20. Compressor capacity is having the unit

(a) m^3/kg ☐ (b) m^3/min ☐

(c) kg/m^3 ☐ (d) none of the above. ☐

Efficiency of a Compressor

21. Volumetric efficiency is defined as the ratio of

(a) volume of free air actually delivered by the machine per minute to the total swept volume by the piston per minute ☐

(b) swept volume per minute to volume delivered per minute ☐

(c) clearance volume to swept volume ☐

(d) none of the above. ☐

22. If for a single-stage compressor, V_1 = volume of air before compression, V_p = stroke volume and V_c = clearance volume, then the clearance ratio (C) is equal to

(a) $\frac{V_1 + V_p}{V_p}$ ☐ (b) $\frac{V_1 - V_p}{V_p}$ ☐

(c) $\frac{V_1}{V_p}$ ☐ (d) $\frac{V_p}{V_1}$ ☐

23. With the increase of pressure ratio, the volumetric efficiency of an air compressor

(a) increases ☐ (b) decreases ☐

(c) constant ☐ (d) first increases then decreases. ☐

24. In multi-stage compression, the leakage of air past the piston of a cylinder as compared to single stage compression, is

(a) more ☐ (b) less ☐

(c) constant ☐ (d) none of the above. ☐

25. In multi-stage compression as compared to single stage compression, the volumetric efficiency is

(a) more ☐ (b) less ☐

(c) constant ☐ (d) none of the above. ☐

26. Choose the correct statement

(a) In a double-acting compressor, the air is compressed in two cylinders. ☐

(b) When air is to be compressed at a high pressure, then it is advantageous to use single-stage compression. ☐

(c) The compressor capacity is the volume of air sucked by the compressor during its suction stroke. ☐

(d) Compressor air is used in driving pneumatic tools, paint spraying etc. ☐

27. Choose the wrong statement

(a) Compressor is a machine used to raise the pressure of air ☐

(*b*) The volume of air delivered by the compressor is known as compressor capacity. ☐

(*c*) The capacity of a compressor is expressed in m^3/min. ☐

(*d*) Standard air is the air at 0°C and 1 kgf/cm^2 with relative humidity of 36 percent. ☐

Multi-stage Compressor

28. In multi-stage compressors, the air after compression in the first stage is cooled in an intercooler at constant pressure before passing into the second stage. The cooling is said to be perfect if

(*a*) air is cooled to a temperature more than its original temperature ☐

(*b*) air is cooled to a temperature less than its original temperature ☐

(*c*) air is cooled to its original temperature ☐

(*d*) none of the above. ☐

29. The assumptions made for calculating the work done in multi-stage compressor are

(*a*) the index of compression is same in each stage ☐

(*b*) intercooling in each stage is at constant pressure ☐

(*c*) suction and delivery pressures remain constant during each stage ☐

(*d*) all of the above. ☐

30. The work done in case of two-stage reciprocating air compressor with incomplete intercooling is equal to

(*a*) $\frac{n}{n-1} p_1V_1 \left[\left(\frac{p_2}{p_1} \right)^{\frac{n-1}{n}} + \left(\frac{p_3}{p_2} \right)^{\frac{n-1}{n}} - 2 \right]$ ☐

(*b*) $\frac{2n}{n-1} (p_1V_1 + p_2V_2)$ ☐

(*c*) $\left(\frac{n}{n-1} \right) p_1V_1 \left[\left(\frac{p_2}{p_1} \right)^{\frac{n-1}{n}} - 1 \right] + \frac{n}{n-1} p_2V_2 \left[\left(\frac{p_3}{p_2} \right)^{\frac{n-1}{n}} - 1 \right]$ ☐

(*d*) none of the above. ☐

31. If the intercooling is perfect, the work done in case of two stage reciprocating air compressor is equal to

(*a*) $\frac{n}{n-1} p_1V_1 \left[\left(\frac{p_2}{p_1} \right)^{\frac{n-1}{n}} + \left(\frac{p_3}{p_2} \right)^{\frac{n-1}{n}} - 2 \right]$ ☐

(*b*) $\frac{2n}{n-1} (p_1V_1 + p_2V_2)$ ☐

(*c*) $\left(\frac{n}{n-1} \right) p_1V_1 \left[\left(\frac{p_2}{p_1} \right)^{\frac{n-1}{n}} - 1 \right] + \frac{n}{n-1} p_2V_2 \left[\left(\frac{p_3}{p_2} \right)^{\frac{n-1}{n}} - 1 \right]$ ☐

(*d*) none of the above. ☐

32. For a two-stage reciprocating air compressor, for minimum work required the intercooler pressure is given by

(a) $p_2 = \sqrt{p_1 p_3}$ ☐ (b) $p_2 = p_1 \times p_3$ ☐

(c) $p_2 = \dfrac{1}{\sqrt{p_1 p_3}}$ ☐ (d) $p_2 = \dfrac{1}{p_1 p_3}$. ☐

33. For a two-stage reciprocating air compressor, the minimum work required is equal to

(a) $\dfrac{2n}{n-1} p_1 V_1 \left[1 + \left(\dfrac{p_3}{p_1}\right)^{\frac{n-1}{2n}} \right]$ ☐ (b) $\dfrac{2n}{n-1} p_1 V_1 \left[1 \left(\dfrac{p_3}{p_1}\right)^{\frac{n-1}{2n}} \right]$ ☐

(c) $\dfrac{2n}{n-1} p_1 V_1 \left[\left(\dfrac{p_3}{p_1}\right)^{\frac{n-1}{n}} - 1 \right]$ ☐ (d) $\dfrac{2n}{n-1} p_1 V_1 \left[\left(\dfrac{p_3}{p_1}\right)^{\frac{n-1}{2n}} - 1 \right]$. ☐

34. For a three stage reciprocating air compressor, for perfect intercooling and minimum work required, the condition is

(a) $\dfrac{p_2}{p_1} = \dfrac{p_3}{p_2} = \dfrac{p_4}{p_3}$ ☐ (b) $p_1 p_2 = p_3 p_4$ ☐

(c) $p_1 p_2 = \sqrt{p_3 p_4}$ ☐ (d) $p_1 p_2 = \dfrac{1}{\sqrt{p_3 p_4}}$. ☐

35. For a three-stage reciprocating air compressor, for perfect intercooling the minimum work required is

(a) $\dfrac{3n}{n-1} p_1 V_1 \left[1 + \left(\dfrac{p_4}{p_1}\right)^{\frac{n-1}{n}} \right]$ ☐ (b) $\dfrac{3n}{n-1} p_1 V_1 \left[\left(\dfrac{p_4}{p_1}\right)^{\frac{n-1}{n}} - 1 \right]$ ☐

(c) $\dfrac{3n}{n-1} p_1 V_1 \left[\left(\dfrac{p_4}{p_1}\right)^{\frac{n-1}{3n}} - 1 \right]$ ☐ (d) $\dfrac{n}{n-1} p_4 V_1 \left[\left(\dfrac{p_4}{p_1}\right)^{\frac{n-1}{n}} - 1 \right]$. ☐

36. The phenomenon of unsteady, periodic and reversed flow is known as

(a) choking ☐ (b) surging ☐

(c) stalling ☐ (d) none of the above. ☐

37. The phenomenon of fixed mass flow rate in compressor regardless of pressure ratio is known as

(a) choking ☐ (b) surging ☐

(c) stalling ☐ (d) none of the above. ☐

38. The phenomenon of air steam not able to follow the blade contour of an axial compressor is known as

(a) choking ☐ (b) surging ☐

(c) stalling ☐ (d) none of the above. ☐

39. The common method of controlling the air delivered by the compressor is

(*a*) blow off control ☐ (*b*) clearance control ☐

(*c*) throttle control ☐ (*d*) all of the above. ☐

40. The work done in a single-stage single-acting reciprocating compressor without clearance, is

(*a*) maximum during isothermal compression ☐

(*b*) minimum during adiabatic compression ☐

(*c*) minimum during isothermal compression ☐

(*d*) none of the above. ☐

41. Volumetric efficiency of a compressor

(*a*) increases with increase in compression ratio ☐

(*b*) decreases with increase in compression ratio ☐

(*c*) is constant with increase or decrease in compression ratio ☐

(*d*) none of the above. ☐

42. The ratio of minimum work done or perfect cooling for a two-stage reciprocating compressor having compression indices as ***m*** and ***n*** for first and second stage, is given by

(*a*) $\frac{m(n-1)}{n(m+1)}$ ☐ (*b*) $\frac{m(n-1)}{n(m-1)}$ ☐

(*c*) $\frac{m(n+1)}{n(m+1)}$ ☐ (*d*) $\frac{m(n+1)}{n(m-1)}$. ☐

43. For a two-stage compressor, for maximum efficiency the condition is that the intermediate pressure should be

(*a*) average of the initial and final pressure ☐ (*b*) geometric mean of initial and final pressure ☐

(*c*) product of two pressures ☐ (*d*) the ratio of two pressures. ☐

44. For a two-stage compressor, the ratio of diameters of L.P. cylinder to H.P. cylinder is equal to

(*a*) square of the ratio of final pressure to initial pressure ☐

(*b*) the ratio of final pressure to initial pressure ☐

(*c*) the square root of the ratio of final pressure to initial pressure ☐

(*d*) cube root of the ratio of final pressure to initial pressure. ☐

45. For a three-stage compressor for maximum efficiency, the compression ratio is equal to

(*a*) square of the ratio of final pressure to initial pressure ☐

(*b*) the ratio of final pressure to initial pressure ☐

(*c*) square root of the ratio of final pressure to initial pressure ☐

(*d*) cube root of the ratio of final pressure to initial pressure. ☐

46. For a three-stage compressor for perfect intercooling, the minimum work required is equal to

(a) 3 × work required in one stage ☐ (b) $\frac{3n}{n-1} p_1 V_1 \left[\left(\frac{p_2}{p_1} \right)^{\frac{n-1}{n}} - 1 \right]$ ☐

(c) $\frac{3n}{n-1} p_1 V_1 \left[\left(\frac{p_4}{p_1} \right)^{\frac{n-1}{3n}} - 1 \right]$ ☐ (d) all of the above. ☐

47. In case of air engines of air motors

(a) intake is at higher pressure and discharge at lower pressure ☐

(b) intake is at lower pressure and discharge at higher pressure ☐

(c) intake and discharge is at the same pressure ☐

(d) none of the above. ☐

48. The work obtained in case of an air engine with zero clearance is equal to

(a) $\frac{n}{n-1} p_1 V_1 \left[\left(\frac{p_2}{p_1} \right)^{\frac{n-1}{n}} - 1 \right]$ ☐ (b) $\frac{n}{n-1} \left[1 - \left(\frac{p_2}{p_1} \right)^{\frac{n-1}{n}} \right]$ ☐

(c) $\frac{n}{n-1} p_1 V_1 \left[\left(\frac{p_2}{p_1} \right)^{\frac{n-1}{n}} + 1 \right]$ ☐ (d) none of the above. ☐

49. For the minimum work required to compress air in multi-stage compression

(a) the compression ratio in each stage is same ☐

(b) the work done is the same in each stage ☐

(c) the air is cooled to the initial temperature after each stage of compression ☐

(d) all of the above. ☐

50. The volumetric efficiency of a compressor in terms of clearance ratio (K) is equal to

(a) $1 + K + K \left(\frac{p_2}{p_1} \right)^{1/n}$ ☐ (b) $1 - K - K \left(\frac{p_2}{p_1} \right)^{1/n}$ ☐

(c) $1 + K - K \left(\frac{p_2}{p_1} \right)^{1/n}$ ☐ (d) $1 + K - K \left(\frac{p_2}{p_1} \right)^{\frac{n-1}{n}}$. ☐

51. The volumetric efficiency of a compressor, for the same compression ratio, will increase if the index of compression (n) is

(a) increased ☐ (b) decreased ☐

(c) constant ☐ (d) none of the above. ☐

52. For the least work to be done on a compressor, the compression should be

(a) adiabatic ☐ (b) isothermal ☐

(c) polytropic ☐ (d) none of the above. ☐

53. For the highest capacity of a compressor, the intake temperature should be

(*a*) highest ☐ (*b*) lowest ☐

(*c*) atmospheric ☐ (*d*) none of the above. ☐

54. By empolying intercooler in multi-stage compressor, compression obtained is

(*a*) adiabatic ☐ (*b*) isothermal ☐

(*c*) polytropic ☐ (*d*) none of the above. ☐

55. Compression ratio is defined as the ratio of

(*a*) suction volume and clearance volume ☐

(*b*) intake pressure to discharge pressure ☐

(*c*) absolute discharge pressure to absolute intake pressure ☐

(*d*) none of the above. ☐

56. Choose the correct statement

(*a*) Volumetric efficiency is the reciprocal of the compression ratio. ☐

(*b*) Volumetric efficiency of air compressor is of the order of 35 to 50%. ☐

(*c*) Optimum intermediate pressure in two stage compressor is geometric mean of the two. ☐

(*d*) Volumetric efficiency decreases with decrease in compression ratio. ☐

57. Choose the wrong statement

(*a*) In case of air engine, intake is at higher pressure and discharge at lower pressure. ☐

(*b*) In case of compressor, intake is at lower pressure and discharge at higher pressure. ☐

(*c*) Cylinder clearance in a compressor should be as small as possible. ☐

(*d*) The volumetric efficiency increases as the value of index n is decreased. ☐

58. For a two-stage compressor, the safe delivery pressure lies in the range of

(*a*) 1 to 2 kgf/cm^2 ☐ (*b*) 2 to 5 kgf/cm^2 ☐

(*c*) 5 to 35 kgf/cm^2 ☐ (*d*) 35 to 90 kgf/cm^2. ☐

59. For a single stage compressor, the safe delivery pressure is

(*a*) 6 to 8 kgf/cm^2 ☐ (*b*) 8 to 20 kgf/cm^2 ☐

(*c*) upto 1 kgf/cm^2 ☐ (*d*) more than 20 kgf/cm^2. ☐

60. For delivering small quantities of air at high pressure

(*a*) rotary compressor are used ☐ (*b*) reciprocating air compressor are used ☐

(*c*) air engines are used ☐ (*d*) none of the above. ☐

61. For delivering large quantities of air at low pressure

(*a*) rotary compressors are used ☐ (*b*) reciprocating air compressors are used ☐

(*c*) air engines are used ☐ (*d*) none of the above. ☐

62. The suction pressure and delivery pressure of a two-stage reciprocating compressors are 1 kgf/cm^2 and 16 kgf/cm^2. If the cooling is perfect, then intermediate pressure is equal to

(*a*) 8 kgf/cm^2 ☐ (*b*) 4 kgf/cm^2 ☐

(*c*) 2 kgf/cm^2 ☐ (*d*) 8.5 kgf/cm^2. ☐

63. The suction pressure and delivery pressure of a three-stage compressor are 1 kgf/cm^2 and 27 kgf/cm^2 respectively. For maximum efficiency, the compression ratio is equal to

(*a*) 9 ☐ (*b*) 9.33 ☐

(*c*) 3 ☐ (*d*) 4. ☐

64. The pressure at the beginning of first and third stage of a four-stage compressor are 1 kgf/cm^2 and 125 kgf/cm^2, then the delivery pressure at the fourth-stage will be

(*a*) 625 kgf/cm^2 ☐ (*b*) 125 kgf/cm^2 ☐

(*c*) 25 kgf/cm^2 ☐ (*d*) 1 kgf/cm^2. ☐

65. Which of the following are positive displacement compressors ?

(*a*) Reciprocating compressor ☐ (*b*) Root blower ☐

(*c*) Vane pump compressor ☐ (*d*) all of the above ☐

(*e*) none of the above. ☐

66. Which of the following are non-positive displacement compressors ?

(*a*) reciprocating compressor ☐ (*b*) centrifugal compressor ☐

(*c*) axial compressor ☐ (*d*) both (*b*) and (*c*) ☐

(*e*) none of the above. ☐

67. Figure 6.10 shows the typical performance curves of a water-cooled single-stage reciprocating air compressor. Which curve shows the variation of shaft power versus delivery pressure ?

(*a*) curve *A* ☐ (*b*) curve *B* ☐

(*c*) curve *C* ☐ (*d*) curve *D*. ☐

68. In Fig. 6.10, which curve shows the variation of actual capacity versus delivery pressure ?

(*a*) curve *A* ☐

(*b*) curve *B* ☐

(*c*) curve *C* ☐

(*d*) curve *D*. ☐

69. In Fig. 6.10, which curve shows the variation of volumetric efficiency versus delivery pressure

(*a*) curve *A* ☐

(*b*) curve *B* ☐

(*c*) curve *C* ☐

(*d*) curve *D*. ☐

D
C
B
A
Delivery pressure (kgf/cm^2)

FIGURE 6.10

70. In Fig. 6.10, curve *A* shows the variation of

(*a*) shaft power ☐ (*b*) volumetric efficiency ☐

(*c*) actual capacity ☐ (*d*) none of the above. ☐

71. Blower is a compressor which produces air pressure upto

(*a*) 10 kgf/cm^2 ☐ (*b*) 5 kgf/cm^2 ☐
(*c*) 2.5 kgf/cm^2 ☐ (*d*) 20 kgf/cm. ☐

72. Reciprocating air compressors are suitable for delivering air of

(*a*) small quantity at low pressure ☐ (*b*) large quantity at low pressure ☐
(*c*) small quantity at high pressure ☐ (*d*) large quantity at high pressure. ☐

73. Rotary compressors are suitable for delivering air of

(*a*) small quantity at low pressure ☐ (*b*) large quantity at low pressure ☐
(*c*) small quantity at high pressure ☐ (*d*) large quantity at high pressure. ☐

Rotary Compressor

74. In a rotary air compressor, the maximum delivery pressure is limited to

(*a*) 100 kgf/cm^2 ☐ (*b*) 50 kgf/cm^2 ☐
(*c*) 25 kgf/cm^2 ☐ (*d*) 10 kgf/cm^2. ☐

75. The efficiency of a rotary air compressor as compared to that of reciprocating air compressor is

(*a*) more ☐ (*b*) less ☐
(*c*) nearly the same ☐ (*d*) none of the above. ☐

76. Choose the wrong statement

(*a*) The rotary compressors are used for delivering large quantity of air at low pressure. ☐
(*b*) The speed of rotary compressor is high as compared to reciprocating air compressor. ☐
(*c*) For a single stage compressor, the safe delivery pressure is 6 to 8 kgf/cm^2. ☐
(*d*) A compressor at high altitude will required more power. ☐

77. The efficiency of a roots blower is defined as

(*a*) the ratio of work required for ideal isentropic compressor to the work done by the roots blower on the air during the same interval of time. ☐
(*b*) the ratio of work done by the roots blower to the work required for ideal isentropic compressor per minute. ☐
(*c*) product of the ideal work required for isentropic compressor to the work done by the roots blower ☐
(*d*) none of the above. ☐

78. The efficiency of a roots blower is equal to

(*a*) $\dfrac{r_p^{\frac{\gamma-1}{\gamma}}-1}{(r_p-1)}$ ☐ (*b*) $\left(\dfrac{\gamma}{\gamma-1}\right)\times\dfrac{r_p^{\frac{\gamma-1}{\gamma}}-1}{r_p+1}$ ☐

(*c*) $\left(\dfrac{\gamma}{\gamma-1}\right)\times\left(\dfrac{r_p^{\frac{\gamma-1}{\gamma}}-1}{r_p-1}\right)$ ☐ (*d*) $\left(\dfrac{-1}{\gamma}\right)\times\left(1-r_p^{\frac{\gamma-1}{\gamma}}\right)$ ☐

where r_p = Pressure ratio = Ratio of delivery pressure to intake pressure of air.

79. The criterion of thermodynamic efficiency is
(a) isothermal compression for reciprocating and rotary compressors ☐
(b) isentropic compression for reciprocating and rotary compressors ☐
(c) isothermal compression for reciprocating compressor and isentropic compression for rotary compressor ☐
(d) isentropic compression for reciprocating compressor and isothermal compression for rotary compressor. ☐

80. The static temperature of a moving gas is the actual temperature recorded by a
(a) thermometer which is stationary ☐
(b) thermometer which is moving with the gas with the speed more than the speed of the gas ☐
(c) thermometer which is moving with the gas with the speed less than the speed of the gas ☐
(d) thermometer which is moving with the gas with the speed equal to that of the gas. ☐

81. The stagnation temperature of a moving gas is the temperature recorded by a
(a) thermometer which is stationary ☐
(b) thermometer which is moving with the gas with the speed equal to that of the gas ☐
(c) thermometer which is stationary and the moving gas is brought to rest under reversible adiabatic conditions ☐
(d) none of the above. ☐

82. The stagnation temperature is equal to
(a) $T + \frac{V^2}{gJC_p}$ ☐ (b) $T + \frac{V^2}{2gJC_p}$ ☐
(c) $T - \frac{V^2}{gJC_p}$ ☐ (d) $T - \frac{C_pV^2}{gJ}$ ☐
where T = Static temperature, and V = Velocity of the gas.

83. The static temperature of a moving gas will decrease if the velocity of the gas
(a) decreases ☐ (b) increases ☐
(c) is constant ☐ (d) none of the above. ☐

84. In the absence of heat and work transfer (*i.e.*, $q = 0$ and $W = 0$) the total head temperature
(a) is increasing ☐ (b) is decreasing ☐
(c) remains constant ☐ (d) none of the above. ☐

85. The stagnation enthalpy of a moving gas is equal to
(a) $h - \frac{V^2}{2gJ}$ ☐ (b) $h + \frac{V^2}{2gJ}$ ☐
(c) $h - \frac{2V^2}{gJ}$ ☐ (d) $h + \frac{2V^2}{gJ}$. ☐
where h = Static enthalpy.

86. The stagnation pressure of a moving gas is equal to

(a) $p + \frac{V^2}{2gJ}$ ☐ (b) $p + \left(\frac{T_0}{T}\right)^{\gamma}$ ☐

(c) $p\left(\frac{T_0}{T}\right)^{\frac{\gamma-1}{\gamma}}$ ☐ (d) $p\left(\frac{T}{T_0}\right)^{\frac{\gamma-1}{\gamma}}$. ☐

where p = Static pressure, T_0 = Stagnation temperature, and T = Static temperature.

Centrifugal Compressor

87. The slip factor for a centrifugal compressor is defined as

(a) the ratio of blade velocity at outlet to outlet whirl velocity ☐
(b) ratio of outlet whirl velocity to blade velocity at outlet ☐
(c) the ratio of blade velocity at inlet to blade velocity at outlet ☐
(d) the ratio of blade velocity at inlet to outlet whirl velocity. ☐

88. In case of centrifugal compressor, the work done per kg by an ideal radial vaned impeller (with no heat transfer and having slip factor equal to one) is given by

(a) $C_pJ\,(T_1 - T_2)$ ☐ (b) $C_pJ(T_{01} + T_{02})$ ☐
(c) $C_pJ\,(T_{01} - T_{02})$ ☐ (d) $C_pJ\,(T_1 - T_{02})$. ☐

where T_1 and T_2 = Static temperatures at inlet and outlet, and
T_{01} and T_{02} = Stagnation temperatures at inlet and outlet.

89. In aeroplanes, the type of rotary compressor used is

(a) radial flow type ☐ (b) centrifugal type ☐
(c) axial flow type ☐ (d) none of the above. ☐

90. In gas turbines, the type of rotary compressor used is

(a) radial flow type ☐ (b) centrifugal type ☐
(c) axial flow type ☐ (d) none of the above. ☐

91. For an ideal centrifugal compressor, the power given to the compressed air depends upon

(a) total inlet temperature ☐ (b) total pressure ratios of the compressor ☐
(c) mass of flow of air ☐ (d) all of the above ☐
(e) none of the above. ☐

92. For an ideal centrifugal compressor, the ratio of total pressures at outlet to inlet is equal to

(a) $\left(1+\frac{V_b^{\,2}}{gJC_pT_{01}}\right)^{\frac{\gamma-1}{\gamma}}$ ☐ (b) $\left(1\times\frac{V_b^{\,2}}{gJC_pT_{01}}\right)^{\frac{\gamma}{\gamma-1}}$ ☐

(c) $\left(1+\frac{V_b^{\,2}}{gJC_pT_{01}}\right)^{\frac{\gamma}{\gamma-1}}$ ☐ (d) none of the above ☐

where V_b = Tip velocity of the impeller, and
T_{01} = Total temperature at inlet.

93. For an ideal centrifugal compressor, the pressure produced depends on

(*a*) total inlet temperature ☐ (*b*) square of impeller velocity ☐
(*c*) impeller diameter ☐ (*d*) all of the above ☐
(*e*) both (*a*) and (*b*). ☐

94. Adiabatic efficiency of a compressor is equal to

(*a*) ratio of actual work required to isentropic work required for compressing air for the same pressure ratio ☐
(*b*) ratio of the isentropic work to actual work required for compressing air for the same pressure ratio ☐
(*c*) product of isentropic work and actual work required ☐
(*d*) none of the above. ☐

95. The temperature rise ratio (*i.e.*, adiabatic efficiency) of a centrifugal compressor is equal to

(*a*) $\frac{T_2' + T_1}{T_2 + T_1}$ ☐ (*b*) $\frac{T_2 - T_1}{T_2' - T_1}$ ☐
(*c*) $\frac{T_2' - T_1}{T_2 - T_1}$ ☐ (*d*) $\frac{T_2 + T_1}{T_2' + T_1}$. ☐

where T_1 = Absolute temperature at inlet,
T_2 = Actual absolute delivery temperature, and
T_2' = Absolute delivery, temperature after isentropic compression.

96. The power input factor varies from

(*a*) 0.8 to 0.9 ☐ (*b*) 0.5 to 0.8 ☐
(*c*) 1.035 to 1.04 ☐ (*d*) 1.1 to 1.5. ☐

97. The ratio of enthalpy rise in rotor to enthalpy rise in stage is called

(*a*) adiabatic efficiency ☐ (*b*) degree of reaction ☐
(*c*) static pressure ratio ☐ (*d*) none of the above. ☐

98. For minimum fluid friction and blade tip clearance losses, the blades of an axial flow compressor are designed for

(*a*) 80% reaction ☐ (*b*) 60% reaction ☐
(*c*) 50% reaction ☐ (*d*) 40% reaction. ☐

99. The blades of compressor have profiles consisting of

(*a*) circular arcs ☐ (*b*) straight lines ☐
(*c*) aerofoil section ☐ (*d*) none of the above. ☐

100. Degree of reaction for an axial flow compressor is also equal to

(*a*) ratio of increase in pressure in rotor to total increase in pressure in the stage ☐
(*b*) ratio of total increase in pressure in the stage to increase in pressure in rotor ☐
(*c*) product of pressure in rotor and pressure in the stage ☐
(*d*) none of the above. ☐

101. For an axial flow compressor, if degree of reaction is 50%

(*a*) the blades are symmetrical ☐ (*b*) blades are unsymmetrical ☐

(*c*) blades are curved ☐ (*d*) none of the above. ☐

102. Choose the correct statement

(*a*) The ratio of inlet whirl velocity to the blade velocity at outlet is called slip factor for a centrifugal compressor. ☐

(*b*) An axial flow compressor is mostly suitable for low volume flow rate. ☐

(*c*) The air enters the impeller axially in a centrifugal compressor and leaves the vane radially. ☐

(*d*) By lowering the inlet temperature and increasing the tip speed of a centrifugal compressor, the pressure ratio can be increased. ☐

103. Choose the wrong statement

(*a*) If degree of reaction is 50%, the blades are symmetrical. ☐

(*b*) Degree of reaction is the ratio of increase in pressure in rotor to total increase in pressure in the stage. ☐

(*c*) The blades of an axial flow compressor are designed for 50% reaction for minimum fluid friction and blade tip clearance losses. ☐

(*d*) The power input factor varies from 0.8 to 0.9. ☐

104. The frontal area for a centrifugal compressor as compared to axial flow compressor is

(*a*) large ☐ (*b*) same ☐

(*c*) small ☐ (*d*) none of the above. ☐

105. In axial flow compressor, stalling of blades is the phenomenon of

(*a*) air steam not able to follow the blade contour ☐

(*b*) air stream blocking the passage ☐

(*c*) air stream flowing with sonic velocity ☐

(*d*) none of the above. ☐

106. For minimum work on a compressor, the air should be taken from

(*a*) source of high temperature air ☐ (*b*) source of low temperature air ☐

(*c*) atmosphere ☐ (*d*) none of the above. ☐

107. In multi-stage compressor for minimum work, the work done in

(*a*) all stages should be different ☐ (*b*) all stages should be equal ☐

(*c*) first stage should be least ☐ (*d*) last stage should be least. ☐

108. In case of reciprocating compressor, the performance is expressed as

(*a*) $\dfrac{\text{Indicated work}}{\text{Isothermal work}}$ ☐ (*b*) $\dfrac{\text{Isothermal work}}{\text{Indicated work}}$ ☐

(*c*) $\dfrac{\text{Adiabatic work}}{\text{Indicated work}}$ ☐ (*d*) none of the above. ☐

109. At high altitudes the performance of air compressor as compared to that at sea-level would be

(*a*) inferior ☐ (*b*) better ☐

(*c*) same ☐ (*d*) none of the above. ☐

110. In multi-stage compressors, intercooling is done to

(*a*) minimise the work of compression (*b*) cool the air
(*c*) reduce the volume of air (*d*) none of the above.

111. The use of after cooler is to

(*a*) minimise the work of compression (*b*) cool the air
(*c*) reduce the volume of air (*d*) none of the above.

112. Centrifugal blowers are used to supply

(*a*) small volume of air at low pressure
(*b*) small volume of air at high pressure
(*c*) large volume of air at low pressure
(*d*) large volume of air at high pressure.

113. In a compressor, the cylinder clearance should be

(*a*) minimum (*b*) maximum
(*c*) about 100% of swept volume (*d*) none of the above.

114. A centrifugal compressor

(*a*) converts kinetic energy into pressure energy
(*b*) converts pressure energy into kinetic energy
(*c*) converts heat energy into kinetic energy
(*d*) none of the above.

115. The clearance in compressors

(*a*) increase the work done on air/kg (*b*) decreases the work done on air/kg
(*c*) has not effect on work done/kg of air (*d*) none of the above.

116. The clearance volume in single stage compressor as compared to multi-stage compressor is

(*a*) more (*b*) less
(*c*) same (*d*) none of the above.

117. Figure 6.11 shows the compression curves in a reciprocating air compressor on *T-S* diagram. The isothermal compression is followed by

(*a*) curve *A*
(*b*) curve *B*
(*c*) curve *C*
(*d*) curve *D*.

FIGURE 6.11

118. In Fig. 6.11, the adiabatic compression is followed by

(*a*) curve *A*
(*b*) curve *B*
(*c*) curve *C*
(*d*) curve *D*.

119. In Fig. 6.11, the compression given by the law $pV^n = C$ (where $n < \gamma$) is followed by

(a) curve A ☐ (b) curve B ☐

(c) curve C ☐ (d) curve D ☐

120. In Fig. 6.11, the compression given by the law $pV^n = C$ (where $n < \gamma$) is followed by

(a) curve A ☐ (b) curve B ☐

(c) curve C ☐ (d) curve D. ☐

121. The isothermal efficiency of a single stage compressor is equal to

(a) $\left[\log_e\left(\frac{p_2}{p_1}\right)\right] \Big/ \left(\frac{n}{n-1}\right)\left[\left(\frac{p_2}{p_1}\right)^{\frac{n-1}{1}} - 1\right]$ ☐ (b) $\log_e\left(\frac{p_1}{p_2}\right) \Big/ \left(\frac{n-1}{n}\right)\left[\left(\frac{p_2}{p_1}\right)^{\frac{n-1}{n}} - 1\right]$ ☐

(c) $\left(\frac{n}{n-1}\right)\left[\left(\frac{p_2}{p_1}\right)^{\frac{n-1}{n}} - 1\right] \Big/ \log_e\left(\frac{p_2}{p_1}\right)$ ☐ (d) $\left(\frac{n-1}{n}\right)\left[\left(\frac{p_2}{p_1}\right)^{\frac{n-1}{n}} - 1\right] \Big/ \log_e\left(\frac{p_2}{p_1}\right)$. ☐

where p_2 = Delivery pressure, p_1 = Suction pressure, and

n = Actual index of compression.

122. The adiabatic efficiency of a single-stage compressor is equal to

(a) $\log_e\left(\frac{p_2}{p_1}\right) \Big/ \left(\frac{n}{n-1}\right)\left[\left(\frac{p_2}{p_1}\right)^{\frac{n-1}{n}} - 1\right]$ ☐

(b) $\log_e\left(\frac{p_1}{p_2}\right) \Big/ \left(\frac{n-1}{n}\right)\left[\left(\frac{p_2}{p_1}\right)^{\frac{n-1}{n}} - 1\right]$ ☐

(c) $\left(\frac{\gamma}{\gamma-1}\right)\left[\left(\frac{p_2}{p_1}\right)^{\frac{\gamma-1}{\gamma}} - 1\right] \Big/ \left(\frac{n}{n-1}\right)\left[\left(\frac{p_2}{p_1}\right)^{\frac{n-1}{n}} - 1\right]$ ☐

(d) none of the above. ☐

where p_2 = Delivery pressure, p_1 = Suction pressure, and

n = Actual index of compression.

123. The work done per cycle for a single stage compressor with clearance volume having the same index of compression and expansion is equal to

(a) $\left(\frac{n}{n-1}\right)p_1V_1\left[\left(\frac{p_2}{p_1}\right)^{\frac{n-1}{n}} - 1\right]$ ☐ (b) $\left(\frac{n}{n-1}\right)p_1V_4\left[\left(\frac{p_2}{p_1}\right)^{\frac{n-1}{n}} - 1\right]$ ☐

(c) $\left(\frac{n}{n-1}\right)p_1(V_1 - V_4)\left[\left(\frac{p_2}{p_1}\right)^{\frac{n-1}{1}} - 1\right]$ ☐ (d) $\left(\frac{n-1}{n}\right)p_1(V_1 - V_4)\left[\left(\frac{p_2}{p_1}\right)^{\frac{n-1}{n}} - 1\right]$. ☐

where V_1 = Volume of air at the beginning of compression stroke,

V_4 = Volume of air at the end of expansion,

p_1 = Suction pressure, and p_2 = Delivery pressure.

124. For a three-stage compressor, the suction pressure and delivery pressure are 1 kgf/cm^2 and 64 kgf/cm^2. For maximum efficiency the pressure at the end of 1st stage will be

(a) 2 kgf/cm^2 ☐ (b) 4 kgf/cm^2 ☐
(c) 8 kgf/cm^2 ☐ (d) 16 kgf/cm^2. ☐

125. In question 124, the pressure at the end of 2nd stage will be

(a) 2 kgf/cm^2 ☐ (b) 4 kgf/cm^2 ☐
(c) 8 kgf/cm^2 ☐ (d) 16 kgf/cm^2. ☐

Gas Turbine

126. If the working fluid in a plant does not come in contact with the atmospheric air, and is used over and over again, the gas turbine is said to work on

(a) semi-closed cycle ☐ (b) open cycle ☐
(c) closed cycle ☐ (d) none of the above. ☐

127. If the entire working fluid in a plant is taken from atmosphere and is again returned to the atmosphere, the gas turbine is said to work on

(a) semi-closed cycle ☐ (b) open cycle ☐
(c) closed cycle ☐ (d) none of the above. ☐

128. If a part of the working fluid is retained in a plant while the remaining part flows from and to the atmosphere, the gas turbine is said to work on

(a) semi-closed cycle ☐ (b) open cycle ☐
(c) closed cycle ☐ (d) none of the above. ☐

129. Choose the correct statement

(a) For minimum work on a compressor, the air should be taken from a source of high temperature air. ☐
(b) The performance of air compressor at high altitudes as compared to that at sea-level would be better. ☐
(c) Intercooling in multi-stage compressors is done to minimise the work of compression. ☐
(d) In a compressor, the cylinder clearance should be maximum. ☐

130. Closed-cycle gas turbine works on

(a) Rankine cycle ☐ (b) Carnot cycle ☐
(c) Brayton or Joule cycle ☐ (d) Otto cycle. ☐

131. An open-cycle gas turbine works on

(a) Rankine cycle ☐ (b) Carnot cycle ☐
(c) Brayton or Joule cycle ☐ (d) Otto cycle. ☐

132. A closed-cycle gas turbine consists of

(a) compressor and turbine ☐ (b) heating chamber and turbine ☐
(c) cooling chamber and turbine ☐ (d) compressor, heater, cooler and turbine. ☐

133. The ideal air standard efficiency (with isentropic expansion and compression) of a Joule cycle is equal to

(a) $1 + r_p^{\frac{\gamma-1}{\gamma}}$ ☐ (b) $1 - r_p^{\frac{\gamma-1}{\gamma}}$ ☐

(c) $1 - \left(\frac{1}{r_p}\right)^{\frac{\gamma-1}{\gamma}}$ ☐ (d) $r_p^{\frac{\gamma-1}{\gamma}} - 1$. ☐

where r_p = Pressure ratio.

134. The work ratio which is the ratio of the net work obtained from the plant to the turbine work is equal to

(*a*) $1 - r_p^{\frac{\gamma-1}{\gamma}}$ ☐ (*b*) $1 - \frac{T_1}{T_3} r_p^{\frac{\gamma-1}{\gamma}}$ ☐

(*c*) $1 + \frac{T_1}{T_2} r_p^{\frac{\gamma-1}{\gamma}}$ ☐ (*d*) $1 + r_p^{\frac{\gamma-1}{\gamma}}$. ☐

where r_p = Pressure ratio, T_1 = Air inlet temperature to compressor, and T_3 = Air inlet temperature to turbine.

135. The work ratio of a closed-cycle gas turbine plant is the function of

(*a*) pressure ratio of the cycle only ☐

(*b*) temperature ratio of the cycle only ☐

(*c*) pressure ratio and temperature ratio both ☐

(*d*) none of the above. ☐

136. The work ratio of a closed-cycle gas turbine plant can be increased by

(*a*) decreasing the air inlet temperature to compressor ☐

(*b*) increasing the air inlet temperature to turbine ☐

(*c*) both (*a*) and (*b*) ☐

(*d*) none of the above. ☐

137. For the maximum work output per unit mass flow through a closed-cycle gas turbine plant, the optimum pressure ratio is given by

(*a*) $r_p = \left(\frac{T_3}{T_1}\right)^{\frac{\gamma-1}{\gamma}}$ ☐ (*b*) $r_p = \left(\frac{T_3}{T_1}\right)^{\frac{2\gamma}{\gamma-1}}$ ☐

(*c*) $r_p = \left(\frac{T_3}{T_1}\right)^{\frac{\gamma}{2\gamma-1}}$ ☐ (*d*) $r_p = \left(\frac{T_3}{T_1}\right)^{\frac{\gamma}{2(\gamma-1)}}$. ☐

where T_3 = Air inlet temperature to turbine, and T_1 = Air inlet temperature to compressor.

138. The specific work output for a closed-cycle gas turbine plant will be maximum if the temperature at the end of expansion is

(*a*) less than the temperature at the end of compression ☐

(*b*) more than the temperature at the end of compression ☐

(*c*) equal to that at the end of compression ☐

(*d*) none of the above. ☐

139. Choose the correct statement

(*a*) Gas turbine as compared to steam turbine has more efficiency. ☐

(*b*) Gas turbine as compared to I.C. engine has less efficiency. ☐

(*c*) The specific weight per power developed of a gas turbine is less than that of an I.C. engine ☐

(*d*) A closed-cycle gas turbine does not work on the same cycle as that of an open-cycle gas turbine. ☐

140. Choose the wrong statement

(a) The thermal efficiency of a diesel plant is higher than that of a gas turbine. ☐

(b) The mechanical efficiency of a gas turbine is higher than that of an I.C. engine. ☐

(c) A closed-cycle gas turbine has higher efficiency than that of an open-cycle gas turbine. ☐

(d) Reheating in a gas turbine decreases the thermal efficiency. ☐

141. Figure 6.12 shows the T-S or h-S diagram for a closed-gas turbine plant when turbine and compressor efficiencies are considered. The turbine efficiency is given by

(a) $\frac{h_3 - h_4'}{h_3 - h_4}$ ☐

(b) $\frac{h_3 - h_4}{h_3 - h_4'}$ ☐

(c) $\frac{h_2' - h_1}{h_2 - h_1}$ ☐

(d) $\frac{h_2 - h_1}{h_2' - h_1}$. ☐

FIGURE 6.12

142. Refer to Fig. 6.12, the compressor efficiency is equal to

(a) $\frac{h_3 - h_4'}{h_3 - h_4}$ ☐ (b) $\frac{h_3 - h_4}{h_3 - h_4'}$ ☐

(c) $\frac{h_2' - h_1}{h_2 - h_1}$ ☐ (d) $\frac{h_2 - h_1}{h_2' - h_1}$. ☐

143. For the maximum work output per unit mass flow through a closed-cycle gas turbine plant, the optimum pressure ratio (when the machine efficiencies, *i.e.*, turbine and compressor efficiencies are considered) is equal to

(a) $\left(\frac{T_3}{T_1}\right)^{\frac{\gamma}{\gamma-1}} \times \eta_t \times \eta_c$ ☐ (b) $\left(\frac{T_3}{T_2}\right)^{\frac{\gamma}{2(\gamma-1)}} \times \eta_t \times \eta_c$ ☐

(c) $\left(\eta_t \times \eta_c \times \frac{T_3}{T_1}\right)^{\frac{\gamma}{2(\gamma-1)}}$ ☐ (d) $\left(\eta_t \times \eta_c \times \frac{T_3}{T_1}\right)^{\frac{\gamma}{2\gamma-1}}$. ☐

where η_t = Turbine efficiency, η_c = Compressor efficiency,

T_1 = Air inlet temperature to compressor, and

T_3 = Air inlet temperature to turbine.

144. Figure 6.13 shows the various curves with respect to pressure ratios for a closed-gas turbine plant having $\eta_t = 0.9$, $\eta_c = 0.85$ and a constant temperature ratio. The ideal cycle thermal efficiency is given by

(a) curve A ☐ (b) curve B ☐

(c) curve C ☐ (d) curve D. ☐

145. In Fig. 6.13, the specific work output per kg per sec is given by

(*a*) curve *A* ☐
(*b*) curve *B* ☐
(*c*) curve *C* ☐
(*d*) curve *D*. ☐

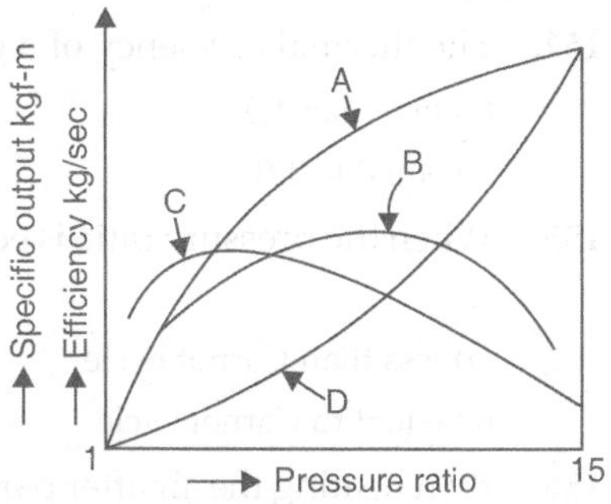

146. In Fig. 6.13, the thermal efficiency of the practical cycle is given by

(*a*) curve *A* ☐
(*b*) curve *B* ☐
(*c*) curve *C* ☐
(*d*) curve *D*. ☐

FIGURE 6.13

147. In Fig. 6.13, the curve *D* represents

(*a*) ideal cycle thermal efficiency ☐ (*b*) specific work output per kg/sec ☐
(*c*) thermal efficiency of practical cycle ☐ (*d*) none of the above. ☐

148. In a gas turbine, the maximum temperature is

(*a*) 2000°C ☐ (*b*) 1000°C ☐
(*c*) 700°C ☐ (*d*) 200°C. ☐

149. In the air after practical expansion in the turbine is reheated, the efficiency of the gas turbine cycle

(*a*) decreases ☐ (*b*) increases ☐
(*c*) remains constant ☐ (*d*) first increases then decreases. ☐

150. By having multi-stage compressor with intercoolers, the efficiency of a gas turbine cycle

(*a*) increases ☐ (*b*) decreases ☐
(*c*) remains constant ☐ (*d*) first increases then decreases. ☐

151. If the exhaust from the gas turbine is utilised in heating the compressed air, the efficiency of the gas turbine cycle will

(*a*) decreases ☐ (*b*) increase ☐
(*c*) remain constant ☐ (*d*) first increases then decreases. ☐

152. The thermal efficiency of a gas turbine cycle with heat exchanger (or generator) considering compression and expansion as isentropic, is equal to

(*a*) $1 + \frac{T_1}{T_3} r_p^{\frac{\gamma-1}{\gamma}}$ ☐ (*b*) $1 - \frac{T_1}{T_3} r_p^{\frac{\gamma-1}{\gamma}}$ ☐

(*c*) $1 + \left(\frac{T_1}{T_3} r_p\right)^{\frac{\gamma-1}{\gamma}}$ ☐ (*d*) $1 - \left(\frac{T_1}{T_3} r_p\right)^{\frac{\gamma-1}{\gamma}}$. ☐

153. The thermal efficiency of a gas turbine cycle with regenerator increases with

(*a*) the increase of pressure ratio ☐ (*b*) the decrease of pressure ratio ☐
(*c*) constant pressure ratio ☐ (*d*) none of the above. ☐

154. The thermal efficiency of a gas turbine with regenerator is maximum when pressure ratio is

(*a*) less than 1.3 ☐ (*b*) more than 1.0 ☐

(*c*) equal to 1.0 ☐ (*d*) zero. ☐

155. When the pressure ratio is equal to 1.0, the thermal efficiency of a gas turbine with regenerator is

(*a*) less than Carnot cycle ☐ (*b*) more than Carnot cycle ☐

(*c*) equal to Carnot cycle ☐ (*d*) none of the above. ☐

156. By reheating the air after partial expansion in the turbine, the maximum work will be obtained when

(*a*) $r_p^{\frac{\gamma-1}{\gamma}} = \left(\frac{T_3}{T_1}\right)^2$ ☐ (*b*) $r_p^{\frac{\gamma-1}{\gamma}} = \left(\frac{T_3}{T_1}\right)^{3/2}$ ☐

(*c*) $r_p^{\frac{\gamma-1}{\gamma}} = \left(\frac{T_3}{T_1}\right)^{2/3}$ ☐ (*d*) $r_p^{\frac{\gamma-1}{\gamma}} = \left(\frac{T_1}{T_3}\right)^{3/2}$. ☐

157. The plant, which is smaller in size and lower in weight for the same power, is

(*a*) steam plant ☐ (*b*) gas turbine plant ☐

(*c*) I.C. engine plant ☐ (*d*) diesel plant. ☐

158. The power produced in kW/ton for a gas turbine is

(*a*) 140 ☐ (*b*) 27 ☐

(*c*) 10 ☐ (*d*) 50. ☐

159. The power produced in kW/ton for a diesel plant is

(*a*) 140 ☐ (*b*) 27 ☐

(*c*) 10 ☐ (*d*) 50. ☐

160. The power produced in kW/ton for a steam plant is

(*a*) 140 ☐ (*b*) 27 ☐

(*c*) 10 ☐ (*d*) 50. ☐

161. For standby and peak load, the most useful plant is

(*a*) steam plant ☐ (*b*) I.C. engine plant ☐

(*c*) gas turbine plant ☐ (*d*) diesel plant. ☐

Jet Propulsion and Rocket Propulsion

162. The propulsive power of a jet engine is equal to

(*a*) $\frac{W(V-u)u}{75}$ ☐ (*b*) $\frac{W(V+u)u}{g \times 75}$ ☐

(*c*) $\frac{W(V-u)u}{g \times 75}$ ☐ (*d*) $\frac{W(V+u)u}{75}$. ☐

where W = Weight of gases discharged/second, u = Velocity of aircraft, and

V = Relative velocity of jet to aircraft = Velocity of jet + Velocity of aircraft.

163. The efficiency of propulsion of a jet engine is equal to

(*a*) $\frac{2u}{V - u}$ ☐ (*b*) $\frac{2u}{V + u}$ ☐

(*c*) $\frac{V - u}{2u}$ ☐ (*d*) $\frac{V + u}{2u}$. ☐

164. The air fuel ratio in a jet engine is

(*a*) 10 : 1 ☐ (*b*) 40 : 1 ☐

(*c*) 60 : 1 ☐ (*d*) 90 : 1. ☐

165. In rocket propulsion, the oxygen for combustion of its fuel is

(*a*) taken from surrounding air ☐

(*b*) taken from the rocket itself ☐

(*c*) taken from compressed atmospheric air ☐

(*d*) none of the above. ☐

166. In jet propulsion, the oxygen for combustion of the fuel is

(*a*) taken from surrounding air ☐ (*b*) taken from the aircraft itself ☐

(*c*) taken from compressed atmospheric air ☐ (*d*) none of the above. ☐

Additional Questions

167. Gas turbine can be applied to an aircraft for

(*a*) screw propulsion only ☐ (*b*) jet propulsion only ☐

(*c*) both (*a*) and (*b*) ☐ (*d*) none of the above. ☐

168. At high altitudes, the efficiency of a screw propeller as compared to jet engine is

(*a*) more ☐ (*b*) less ☐

(*c*) same ☐ (*d*) none of the above. ☐

169. A simple gas turbine cycle will have more efficiency and more work if

(*a*) multi-stage compressor with intercoolers are used ☐

(*b*) heat exchangers are used ☐

(*c*) the air partial expansion in the turbine is reheated ☐

(*d*) all of the above ☐

(*e*) none of the above. ☐

170. The ratio of specific weight per h.p. of I.C. engine and gas turbines is of the order of

(*a*) 6 : 1 ☐ (*b*) 1 : 4 ☐

(*c*) 2 : 1 ☐ (*d*) 3 : 1. ☐

171. For a given inlet temperature, the thermal efficiency of a simple gas turbine with increase in pressure ratio

(*a*) decreases ☐ (*b*) increases ☐

(*c*) first increases then decreases ☐ (*d*) remains constant. ☐

172. Figure 6.14 shows various curves for propulsion efficiency versus speed. The propulsion efficiency of a jet engine is given by

(*a*) curve *A* ☐
(*b*) curve *B* ☐
(*c*) curve *C* ☐
(*d*) curve *D*. ☐

FIGURE 6.14

173. In Fig. 6.14, the propulsion efficiency of a screw propeller engine is given by

(*a*) curve *A* ☐
(*b*) curve *B* ☐
(*c*) curve *C* ☐
(*d*) curve *D*. ☐

174. Choose the correct statement

(*a*) In gas turbine, the maximum combustion pressure is more than that of an I.C. engine. ☐
(*b*) Low weight and small frontal area are the requirements for a gas turbine to be used in aircrafts. ☐
(*c*) Gas turbine blades are given a rake opposite to the direction of motion of blades. ☐
(*d*) The pressure ratio in gas turbine is of the order of 10 : 1. ☐

175. Work ratio of a turbine plant is the ratio of

(*a*) $\frac{\text{net work output}}{\text{heat supplied}}$ ☐ (*b*) $\frac{\text{heat supplied}}{\text{net work output}}$ ☐

(*c*) $\frac{\text{net work output}}{\text{work from turbine}}$ ☐ (*d*) (net work output) × heat supplied. ☐

176. To increase the work ratio of a closed-cycle gas turbine plant, the inlet air temperature to compressor should be as low as possible and air inlet temperature to turbine should be as high as possible. In actual practice the lower temperature and higher temperature are

(*a*) 0°C and 2000°C ☐ (*b*) 20°C and 1000°C ☐
(*c*) atmospheric temperature and 1000°C ☐ (*d*) – 10°C and 1000°C. ☐

177. In a closed-cycle gas turbine, the air inlet temperature to the compressor is 20°C. After compression, the actual temperature is 207.5°C whereas the temperature after isentropic compression is 170°C. The efficiency of the compressor would be

(*a*) 50% ☐ (*b*) 80% ☐
(*c*) 60% ☐ (*d*) 90%. ☐

178. The compression ratio of a compressor is the ratio of

(*a*) volume at outlet to volume at inlet ☐
(*b*) absolute pressure at outlet to absolute pressure at inlet ☐
(*c*) temperature at outlet to temperature at inlet ☐
(*d*) none of the above. ☐

179. In case of air compressor, FAD stands for

(*a*) forced adiabatic diffuser ☐ (*b*) free automatic delivery of compressed air ☐
(*c*) free air delivered ☐ (*d*) free automatic datum. ☐

180. Free air delivered is the volume of air delivered under the conditions of temperature and pressure existing at

(*a*) the compressor outlet ☐ (*b*) the compressor inlet ☐
(*c*) the condenser outlet ☐ (*d*) the condenser inlet. ☐

181. The machine which compresses air in only one end of a cylinder, is known as

(*a*) single cylinder compressor ☐ (*b*) single stage compressor ☐
(*c*) single acting compressor ☐ (*d*) rotary compressor. ☐

182. The machine which compresses air from atmospheric pressure to desired discharge pressure in a single operation, is known as

(*a*) single cylinder compressor ☐ (*b*) single stage compressor ☐
(*c*) single acting compressor ☐ (*d*) rotary compressor. ☐

183. The machine in which compression is effected by a rotating vane or impeller to give the air the desired pressure, is known as

(*a*) single cylinder compressor ☐ (*b*) single stage compressor ☐
(*c*) single acting compressor ☐ (*d*) rotary compressor. ☐

184. The ratio of the capacity of a compressor to the piston displacement of the compressor, is known as

(*a*) compressor efficiency ☐ (*b*) brake power ☐
(*c*) volumetric efficiency ☐ (*d*) theoretical power. ☐

185. The ratio of the theoretical power to the brake power of a compressor is known as

(*a*) compressor efficiency ☐ (*b*) volumetric efficiency ☐
(*c*) power factor ☐ (*d*) none of the above. ☐

186. A four stage compressor should be used if pressure developed is above

(*a*) 20 kgf/cm^2 ☐ (*b*) 40 kgf/cm^2 ☐
(*c*) 60 kgf/cm^2 ☐ (*d*) 90 kgf/cm^2. ☐

187. A three stage compressor should be used if pressure developed is in the range of

(*a*) 0 to 5.5 kgf/cm^2 ☐ (*b*) 5.5 to 35 kgf/cm^2 ☐
(*c*) 35 to 84 kgf/cm^2 ☐ (*d*) 84 to 200 kgf/cm^2. ☐

188. If a gas is compressed isothermally instead of adiabatically under same initial condition and final pressure, then the work required to compress the given quantity of the gas will be

(*a*) more ☐ (*b*) same ☐
(*c*) less ☐ (*d*) none of the above. ☐

189. For the maximum efficiency of a two stage compressor, the intermediate pressure should be

(*a*) arithmetic mean of initial and final pressures ☐
(*b*) geometric mean of initial and final pressures ☐
(*c*) trigonometric mean of initial and final pressures ☐
(*d*) none of the above. ☐

190. For a two stage air compressor, the initial and final pressures are 1 kgf/cm^2 and 9 kgf/cm^2 respectively. Then the intermediate pressure will be

(*a*) 2 kgf/cm^2 ☐ (*b*) 3 kgf/cm^2 ☐

(*c*) 4.5 kgf/cm^2 ☐ (*d*) 6 kgf/cm^2. ☐

191. The volume within the cylinder when the piston is at the end of its inward travel plus the volume within the passage leading to the valves is known as

(*a*) swept volume ☐ (*b*) total volume ☐

(*c*) clearance volume ☐ (*d*) none of the above. ☐

192. A three stage reciprocating air compressor, compresses air from 1 bar to 27 bar. For a perfect intercooling, the pressure ratio in each stage will be

(*a*) 2 ☐ (*b*) 3 ☐

(*c*) 4 ☐ (*d*) 5. ☐

193. In question 192, initial pressure $p_1 = 1$ bar and final pressure $p_4 = 27$ bar, then the intermediate pressure p_2 will be

(*a*) 3 bar ☐ (*b*) 5 bar ☐

(*c*) 8 bar ☐ (*d*) 9 bar. ☐

194. In question 192, the intermediate pressure p_3 will be

(*a*) 3 bar ☐ (*b*) 5 bar ☐

(*c*) 8 bar ☐ (*d*) 9 bar. ☐

195. The criterion of thermodynamic efficiency of reciprocating compressor is

(*a*) adiabatic ☐ (*b*) isothermal ☐

(*c*) isentropic ☐ (*d*) parabolic. ☐

196. The criterion of thermodynamic efficiency of a rotary compressor is

(*a*) adiabatic ☐ (*b*) isothermal ☐

(*c*) isentropic ☐ (*d*) parabolic. ☐

197. Rotary compressor are used for supplying

(*a*) small quantity of air at low pressure rise ☐

(*b*) large quantity of air at high pressure rise ☐

(*c*) large quantity of air at low pressure rise ☐

(*d*) small quantity of air at high pressure rise. ☐

198. For a centrifugal compressor, when the impeller has backward curved vanes, then with an increase in flow rate the Euler head *H* will be

(*a*) unaffected ☐ (*b*) decreasing ☐

(*c*) increasing ☐ (*d*) none of the above. ☐

199. For a centrifugal compressor, when the impeller has radial vanes, then with an increase in flow rate the Euler head *H* will be

(*a*) unaffected ☐ (*b*) decreasing ☐

(*c*) increasing ☐ (*d*) none of the above. ☐

200. For a centrifugal compressor, when the impeller has forward curved vanes, then with an increase in flow rate, the Euler head H will be

(*a*) unaffected ☐ (*b*) decreasing ☐
(*c*) increasing ☐ (*d*) none of the above. ☐

Tick mark the true and false statement:

201. In compressor, the phenomenon of chocking means the fixed mass flow rate regardless of pressure ratio.

(*a*) True ☐ (*b*) False. ☐

202. The volumetric efficiency of compressor increases with increase in compression ratio.

(*a*) True ☐ (*b*) False. ☐

203. Centrifugal compressor is best suited for supplying intermittent small quantity of air at high pressure.

(*a*) True ☐ (*b*) False. ☐

204. The thermodynamic efficiency of a rotary compressor is based on adiabatic compression.

(*a*) True ☐ (*b*) False. ☐

205. The indicated work per unit mass of air delivered is directly proportional to clearance volume.

(*a*) True ☐ (*b*) False. ☐

206. For minimum work in compressor, operating between limits p_1 and p_3, the best intercooler pressure p_2 is equal to $\sqrt{p_1 \times p_3}$

(*a*) True ☐ (*b*) False. ☐

207. If the compression in the compressor is isothermal, then least work will be done.

(*a*) True ☐ (*b*) False. ☐

208. The ratio of the discharge pressure to the inlet pressure of air is called compression ratio.

(*a*) True ☐ (*b*) False. ☐

209. The volume of air sucked by the compressor during its suction stroke is known capacity of compressor.

(*a*) True ☐ (*b*) False. ☐

210. Intercooling in multi-stage compressors is done to minimise the work of compression.

(*a*) True ☐ (*b*) False. ☐

211. The work done on all stages should be equal for minimum work in multi-stage compression.

(*a*) True ☐ (*b*) False. ☐

212. The compressor ratio for the compressor is always greater than unity.

(*a*) True ☐ (*b*) False. ☐

213. In double acting compressor, the air is compressed in two cylinders.

(*a*) True ☐ (*b*) False. ☐

214. If the temperature of intake air is lowest, the compressor capacity will be lowest.

(*a*) True ☐ (*b*) False. ☐

215. For maximum efficiency in multi-stage compressor, the pressure ratio and work done for each stage should be same.

(*a*) True ☐ (*b*) False. ☐

216. Compressor capacity is the actual volume of air delivered by a compressor, when reduced to normal temperature and pressure conditions.

(*a*) True ☐ (*b*) False. ☐

217. The mean effective pressure is the ratio of work done per cycle to the stroke volume of the compressor.

(*a*) True ☐ (*b*) False. ☐

218. A compressor at high altitude will draw more power.

(*a*) True ☐ (*b*) False. ☐

219. A centrifugal compressor works on the principle of conversion of kinetic energy into pressure energy.

(*a*) True ☐ (*b*) False. ☐

220. Actual compression curve for the compressor is in between isothermal and adiabatic.

(*a*) True ☐ (*b*) False. ☐

221. The thermal efficiency, of a gas turbine for optimum pressure ratio, increases as the turbine inlet temperature increases.

(*a*) True ☐ (*b*) False. ☐

222. Number of stages in any gas turbine is equal to total heat drop in turbine divided by heat drop in a stage.

(*a*) True ☐ (*b*) False. ☐

223. Intercooling in gas turbine results in increase in net output but decrease in thermal efficiency.

(*a*) True ☐ (*b*) False. ☐

224. In the cross compounding of the gas turbine plant, h.p. compressor is connected to h.p. turbine and l.p. compressor is connected to l.p. turbine.

(*a*) True ☐ (*b*) False. ☐

225. If the aircraft velocity is equal to twice the jet velocity, then rocket works with maximum overall efficiency.

(*a*) True ☐ (*b*) False. ☐

226. In gas turbines, the high air fuel ratio is used to reduce the exit temperature.

(*a*) True ☐ (*b*) False. ☐

227. The compression ratio in jet engines varies as the square of the speed.

(*a*) True ☐ (*b*) False. ☐

228. By decreasing the compression work and by increasing the turbine work, the work ratio of a gas turbine may be improved.

(*a*) True ☐ (*b*) False. ☐

229. The ratio of highest pressure in a cycle to lowest pressure in gas turbines is known as the pressure ratio.
(*a*) True ☐ (*b*) False. ☐

230. If p_1 is the maximum pressure of a cycle and p_2 minimum pressure, then for maximum work output in a two stage expansion gas turbine with perfect reheating, the intermediate pressure should be $\sqrt{p_1 \times p_2}$.
(*a*) True ☐ (*b*) False. ☐

231. Rotary compressors are used for delivering small quantities of air at low pressures.
(*a*) True ☐ (*b*) False. ☐

232. The efficiency of a jet engine is higher at high altitudes.
(*a*) True ☐ (*b*) False. ☐

233. A gas turbine used in aircraft should have low weight and small frontal area.
(*a*) True ☐ (*b*) False. ☐

234. The air is obtained from the atmosphere and compressed in an air compressor, in a gas turbine.
(*a*) True ☐ (*b*) False. ☐

235. An increase in speed at a given pressure ratio, in a centrifugal compressor causes increases in flow and decrease in efficiency.
(*a*) True ☐ (*b*) False. ☐

Fill in the blanks:

236. Compression ratio in case of air pressure is the ratio of the absolute discharge pressure to the absolute ....
(*a*) inlet pressure ☐ (*b*) outlet pressure. ☐

237. Single acting compressor compresses air at .... of a cylinder.
(*a*) one end ☐ (*b*) both ends. ☐

238. Multistage compressor produces the desired final pressure through .... stages.
(*a*) one ☐ (*b*) two or more. ☐

239. The ratio of the theoretical power to the brake power is known as ....
(*a*) compressor efficiency ☐ (*b*) mechanical efficiency. ☐

240. The ratio of the capacity of a compressor to the piston displacement of the compressor is known as .... efficiency.
(*a*) mechanical ☐ (*b*) volumetric. ☐

241. The work required to compress a given quantity of gas is, ... if it is compressed isothermally instead of adiabatically under same initial condition and final pressure.
(*a*) more ☐ (*b*) less. ☐

242. For the maximum efficiency of a two stage compressor, the intermediate pressure is the .... mean of initial and final pressures.
(*a*) arithmetic ☐ (*b*) geometric. ☐

243. Clearance volume is generally expressed as percentage of piston ....

(*a*) velocity ☐ (*b*) displacement. ☐

244. Rotary compressors run at a very .... speed.

(*a*) slow ☐ (*b*) high. ☐

245. Rotary compressors are suitable for supplying .... quantity of air.

(*a*) small ☐ (*b*) large. ☐

246. The maximum combustion pressure in gas turbine is .... as compared to I.C. engine.

(*a*) less ☐ (*b*) more. ☐

247. In a gas turbine, the compression ratio is of the order of ....

(*a*) 20 : 1 ☐ (*b*) 8 : 1. ☐

248. In jet engines, the compression ratio varies as .... of the speed.

(*a*) square ☐ (*b*) cube. ☐

249. Aircraft units employ .... type of gas turbine.

(*a*) closed ☐ (*b*) open. ☐

250. A gas turbine used in aircraft should have .... weight and small frontal area.

(*a*) high ☐ (*b*) low. ☐

251. Gas turbine blades are given a rake .... direction of motion of blades.

(*a*) opposite to the ☐ (*b*) in the. ☐

252. The thermal efficiency of a gas turbine is .... as compared to a diesel plant.

(*a*) more ☐ (*b*) less. ☐

253. The minimum work required for a three stage reciprocating air compressor is .... the work required for each stage.

(*a*) two times ☐ (*b*) three times. ☐

254. Volumetric efficiency of a compressor decreases with .... in compression ratio.

(*a*) increase ☐ (*b*) decrease. ☐

255. The work done on a compressor will be .... if air is taken from a source of low temperature air.

(*a*) maximum ☐ (*b*) minimum. ☐

256. In a centrifugal compressor, the pressure ratio can be increased by .... the tip speed and lowering inlet temperature.

(*a*) decreasing ☐ (*b*) increasing. ☐

257. The efficiency of a jet engine is .... at high altitude.

(*a*) lower ☐ (*b*) higher. ☐

258. A jet engine has .... propeller.

(*a*) one ☐ (*b*) no. ☐

III. OBJECTIVE TYPE QUESTIONS FROM COMPETITIVE EXAMINATIONS

Tick mark the most appropriate answer:

1. The law of compression desired for reciprocating air compressor is isothermal and that may be possible by
   (*a*) very low speeds ☐
   (*b*) very high speeds ☐
   (*c*) any speed as speed does not affect the compression law ☐
   (*d*) none of the above. ☐
2. In jet propulsion, the oxygen for combustion of the fuel is
   (*a*) taken from surrounding air ☐ (*b*) taken from the aircraft itself ☐
   (*c*) taken from compressed atmospheric air ☐ (*d*) none of the above. ☐
3. Work input to the air compressor with 'n' as index of compression
   (*a*) increases with increase in value of n ☐
   (*b*) decreases with increase in value of n ☐
   (*c*) remains same whatever the value of n ☐
   (*d*) first increases and then decreases with increase of value of n. ☐
4. The air fuel ratio in a jet engine is
   (*a*) 10 : 1 ☐ (*b*) 40 : 1 ☐
   (*c*) 60 : 1 ☐ (*d*) 90 : 1. ☐
5. Work done in a single-stage, single-acting air compressor without clearance per kg of air delivered when the compression process is isothermal is given by
   (*a*) $\frac{n}{n-1} p_1 v_1 \left[\left(\frac{p_2}{p_1} \right)^{\frac{n-1}{n}} - 1 \right]$ ☐ (*b*) $\frac{\gamma}{\gamma - 1} p_1 v_1 \left[\left(\frac{p_2}{p_1} \right)^{\frac{\gamma-1}{\gamma}} - 1 \right]$ ☐
   (*c*) $p_1 v_1 \log_e \frac{p_2}{p_1}$ ☐ (*d*) $\frac{n}{n-1} p_1 v_1 \log_e \frac{p_2}{p_1}$. ☐
6. In rocket propulsion, the oxygen for combustion of its fuel is
   (*a*) taken from surrounding air ☐ (*b*) taken from the rocket itself ☐
   (*c*) taken from compressed atmospheric air ☐ (*d*) none of the above. ☐
7. The thermal efficiency of a gas turbine cycle with ideal regenerative heat exchanger is
   (*a*) equal to work ratio ☐ (*b*) less than work ratio ☐
   (*c*) more than work ratio ☐ (*d*) unpredictable. ☐
8. The volumetric efficiency of a compressor, for the same compression ratio, will increase if the index of compression (n) is
   (*a*) increased ☐ (*b*) decreased ☐
   (*c*) constant ☐ (*d*) none of the above. ☐

9. In a two-stage gas turbine plant reheating after first stage
   (*a*) decreases thermal efficiency ☐ (*b*) increases thermal efficiency ☐
   (*c*) does not effect thermal efficiency ☐ (*d*) none of the above. ☐
10. In a two-stage gas turbine plant, reheating after first stage
   (*a*) increases work ratio ☐ (*b*) decreases work ratio ☐
   (*c*) does not affect work ratio ☐ (*d*) none of the above. ☐
11. For the least work to be done on a compressor, the compression should be
   (*a*) adiabatic ☐ (*b*) isothermal ☐
   (*c*) polytropic ☐ (*d*) none of the above. ☐
12. Work done in a two-stage reciprocating air compressor with perfect intercooling is given by
   (*a*) $\frac{n}{n-1} p_1 V_1 \left[\left(\frac{p_2}{p_1} \right)^{\frac{n-1}{n}} + \left(\frac{p_3}{p_1} \right)^{\frac{n-1}{n}} - 2 \right]$ ☐ (*b*) $\frac{2n}{n-1} p_1 V_1 \left[\left(\frac{p_3}{p_1} \right)^{\frac{n-1}{n}} - 1 \right]$ ☐
   (*c*) $\frac{n}{n-1} p_1 V_1 \left[\left(\frac{p_3}{p_1} \right)^{\frac{n-1}{n}} - 1 \right]$ ☐ (*d*) $\frac{n}{n-1} p_1 V_1 \left[\left(\frac{p_2}{p_1} \right)^{\frac{n-1}{n}} \left(\frac{p_3}{p_2} \right)^{\frac{n-1}{n}} - 2 \right]$. ☐
13. In reciprocating air compressor the method of controlling the quantity of air delivered is done by
   (*a*) throttle control ☐ (*b*) blow-off control ☐
   (*c*) clearance control ☐ (*d*) all of the above. ☐
14. In a single stage, single acting reciprocating air compressor without clearance, the expression $p_1 V_1 \log_e \frac{p_2}{p_1}$ represents the work done on the air per cycle for
   (*a*) isothermal compression ☐ (*b*) adiabatic compression ☐
   (*c*) polytropic compression ☐ (*d*) none of the above. ☐
15. The efficiency of Vane type air compressor as compared to roots air compressor for the same pressure ratio is
   (*a*) more ☐ (*b*) less ☐
   (*c*) same ☐ (*d*) may be more or less. ☐
16. Choose the correct statement
   (*a*) In gas turbine, the maximum combustion pressure is more than that of an I.C. engine. ☐
   (*b*) Low weight and small frontal area are the requirements for a gas turbine to be used in aircrafts. ☐
   (*c*) Gas turbine blades are given a rake opposite to the direction of motion of blades. ☐
   (*d*) The pressure ratio in gas turbine is of the order of 10 : 1. ☐
17. In centrifugal air compressor the pressure developed depends on
   (*a*) impeller tip velocity ☐ (*b*) inlet-temperature ☐
   (*c*) compression index ☐ (*d*) all of the above. ☐

18. With the increase of pressure ratio, the volumetric efficiency of an air compressor
(*a*) increases ☐ (*b*) decreases ☐
(*c*) constant ☐ (*d*) first increases then decreases. ☐

19. In a centrifugal air compressor the pressure ratio is increased by
(*a*) increasing the speed of impeller keeping its diameter fixed ☐
(*b*) increasing the diameter of the impeller keeping its speed constant ☐
(*c*) reducing inlet temperature, keeping impeller diameter and speed fixed ☐
(*d*) all of the above. ☐

20. Work ratio of a turbine plant is the ratio of
(*a*) $\dfrac{\text{net work output}}{\text{heat supplied}}$ ☐ (*b*) $\dfrac{\text{heat supplied}}{\text{net work output}}$ ☐
(*c*) $\dfrac{\text{net work output}}{\text{work from turbine}}$ ☐ (*d*) (net work output) × heat supplied. ☐

21. In multi-stage compression, the leakage of air past the piston of a cylinder as compared to single stage compression, is
(*a*) more ☐ (*b*) less ☐
(*c*) constant ☐ (*d*) none of the above. ☐

22. In multi-stage compression as compared to single stage compression, the volumetric efficiency is
(*a*) more ☐ (*b*) less ☐
(*c*) constant ☐ (*d*) none of the above. ☐

23. The ratio of specific weight per h.p. of I.C. engine and gas turbines is of the order of
(*a*) 6 : 1 ☐ (*b*) 1 : 4 ☐
(*c*) 2 : 1 ☐ (*d*) 3 : 1. ☐

24. For a given inlet temperature, the thermal efficiency of a simple gas turbine with increase in pressure ratio
(*a*) decreases ☐ (*b*) increases ☐
(*c*) first increases then decreases ☐ (*d*) remains constant. ☐

25. Work done in a two stage reciprocating air compressor with imperfect cooling is given by

(*a*) $\dfrac{n}{n-1}p_1V_1\left[\left(\dfrac{p_2}{p_1}\right)^{\frac{n-1}{n}}-1\right]+\dfrac{n}{n-1}p_2V_2\left[\left(\dfrac{p_3}{p_2}\right)^{\frac{n-1}{n}}-1\right]$ ☐

(*b*) $\dfrac{2n}{n-1}p_1V_1\left[\left(\dfrac{p_2}{p_1}\right)^{\frac{n-1}{n}}-1\right]$ ☐

(*c*) $\dfrac{2n}{n-1}p_1V_1\left[\left(\dfrac{p_3}{p_1}\right)^{\frac{n-1}{2n}}-1\right]$ ☐

(*d*) $\dfrac{n-1}{2n}p_1V_1\left[\left(\dfrac{p_3}{p_1}\right)^{\frac{n-1}{2n}}-1\right]$ ☐

26. Gas turbine can be applied to an aircraft for

(*a*) screw propulsion only ☐ (*b*) jet propulsion only ☐

(*c*) both (*a*) and (*b*) ☐ (*d*) none of the above. ☐

27. At high altitudes, the efficiency of a screw propeller as compared to jet engine is

(*a*) more ☐ (*b*) less ☐

(*c*) same ☐ (*d*) none of the above. ☐

28. A simple gas turbine cycle will have more efficiency and more work if

(*a*) multi-stage compressor with intercoolers are used ☐

(*b*) heat exchangers are used ☐

(*c*) the air partial expansion in the turbine is reheated ☐

(*d*) all of the above. ☐

29. In a two-stage gas turbine plant, with intercooling and reheating

(*a*) both work ratio and thermal efficiency improve ☐

(*b*) work ratio improves but thermal efficiency decreases ☐

(*c*) thermal efficiency improves but work ratio decreases ☐

(*d*) both work ratio and thermal efficiency decrease. ☐

30. Choose the correct statement

(*a*) Volumetric efficiency is the reciprocal of the compression ratio. ☐

(*b*) Volumetric efficiency of air compressor is of the order of 35 to 50%. ☐

(*c*) Optimum intermediate pressure in two stage compressor is geometric mean of the two. ☐

(*d*) Volumetric efficiency decreases with decrease in compression ratio. ☐

31. For a jet-propulsion unit, ideally the compressor work and turbine work are

(*a*) equal ☐ (*b*) unequal ☐

(*c*) not related to each other ☐ (*d*) unpredictable. ☐

32. For a two-stage compressor, the safe delivery pressure lies in the range of

(*a*) 1 to 2 × 10^3 N/m^2 ☐ (*b*) 2 to 5 × 10^3 N/m^2 ☐

(*c*) 5 to 35 × 10^3 N/m^2 ☐ (*d*) 35 to 90 × 10^3 N/m^2. ☐

33. Greater the difference between jet velocity and aeroplane velocity

(*a*) greater the propulsive efficiency ☐ (*b*) less the propulsive efficiency ☐

(*c*) unaffected is the propulsive efficiency ☐ (*d*) none of the above. ☐

34. Choose the wrong statement

(*a*) In case of air engine, intake is at higher pressure and discharge at lower pressure. ☐

(*b*) In case of compressor, intake is at lower pressure and discharge at higher pressure. ☐

(*c*) Cylinder clearance in a compressor should be as small as possible. ☐

(*d*) The volumetric efficiency increases as the value of index *n* is decreased. ☐

35. With suction pressure being atmospheric, increase in delivery pressure with fixed clearance volume

(*a*) increase volumetric efficiency ☐

(*b*) decreases volumetric efficiency ☐

(*c*) does not change volumetric efficiency ☐

(*d*) first increases volumetric efficiency and then decreases it. ☐

36. In a single stage, single acting reciprocating air compressor without clearance, the expression $\frac{\gamma}{\gamma-1} R(T_2 - T_1)$ represents the work done on the air per cycle for

(*a*) isothermal compression ☐ (*b*) adiabatic compression ☐

(*c*) polytropic compression ☐ (*d*) none of the above. ☐

37. Mechanical efficiency of reciprocating air compressor is expressed as

(*a*) $\frac{\text{B.P.}}{\text{I.P.}}$ ☐ (*b*) $\frac{\text{I.P.}}{\text{B.P.}}$ ☐

(*c*) $\frac{\text{F.P.}}{\text{B.P.}}$ ☐ (*d*) $\frac{\text{F.P.}}{\text{I.P.}}$. ☐

38. For the same overall pressure ratio, the leakage of air past the piston for multi-stage compression as compared to single-stage compression, is

(*a*) more ☐ (*b*) less ☐

(*c*) constant ☐ (*d*) may be more or less. ☐

39. The work done in a single stage compressor neglecting clearance volume, when compression follows the law $pV^n = C$, is equal to

(*a*) $\frac{n}{n-1} R(T_2 - T_1)$ ☐ (*b*) $\frac{n}{n-1} (p_2V_2 - p_1V_1)$ ☐

(*c*) $\left(\frac{n}{n-1}\right) p_1V_1 \left[\left(\frac{p_2}{p_1}\right)^{\frac{n-1}{n}} - 1\right]$ ☐ (*d*) all of the above ☐

(*e*) none of the above ☐

40. The work ratio of closed cycle gas turbine plant depends upon

(*a*) pressure ratio of the cycle and specific heat ratio ☐

(*b*) temperature ratio of the cycle and specific heat ratio ☐

(*c*) pressure ratio, temperature ratio and specific heat ratio ☐

(*d*) only on pressure ratio. ☐

41. Thermal efficiency of closed cycle gas turbine plant increases by

(*a*) reheating ☐ (*b*) intercooling ☐

(*c*) regenerator ☐ (*d*) all of the above. ☐

42. with the increase in pressure ratio thermal efficiency of a simple gas turbine plant with fixed turbine inlet temperature

(*a*) decreases ☐ (*b*) increases ☐

(*c*) first increases and then decreases ☐ (*d*) first decreases and then increases. ☐

43. The clearance volume is provided in reciprocating compressors to

(*a*) reduce the work done ☐

(*b*) to increase the volumetric efficiency ☐

(c) provide space for valves and also to be sure that the piston do not strike the cylinder at the end of the stroke ☐

(d) none of the above. ☐

44. The clearance ratio for a single stage compressor lies between

(a) 15 to 20% ☐ (b) 20 to 30% ☐

(c) 4 to 10%. ☐ (d) 1 to 2%. ☐

45. Mechanical efficiency of a gas turbine as compared to internal combustion reciprocating engine is

(a) higher ☐ (b) lower ☐

(c) same ☐ (d) unpredictable. ☐

46. For a two-stage compressor, for maximum efficiency the condition is that the intermediate pressure should be

(a) average of the initial and final pressure ☐ (b) geometric mean of initial and final pressure ☐

(c) product of two pressures ☐ (d) the ratio of two pressures. ☐

47. For a gas turbine the pressure ratio may be in the range

(a) 2 to 3 ☐ (b) 3 to 5 ☐

(c) 16 to 18 ☐ (d) 18 to 22. ☐

48. The clearance ratio is defined as the ratio of

(a) clearance volume to swept volume ☐ (b) clearance volume to cylinder volume ☐

(c) swept volume to clearance volume ☐ (d) none of the above. ☐

49. The air standard efficiency of closed gas turbine cycle is given by (r_p = pressure ratio for the compressor and turbine)

(a) $\eta = 1 - \frac{1}{(r_p)^{\gamma-1}}$ ☐ (b) $\eta = 1 - (r_p)^{\gamma-1}$ ☐

(c) $\eta = 1 - \left(\frac{1}{r_p}\right)^{\frac{\gamma-1}{\gamma}}$ ☐ (d) $\eta = (r_p)^{\frac{\gamma-1}{\gamma}} - 1$. ☐

50. For a two-stage compressor, the ratio of diameters of L.P. cylinder to H.P. cylinder is equal to

(a) square of the ratio of final pressure to initial pressure ☐

(b) the ratio of final pressure to initial pressure ☐

(c) the square root of the ratio of final pressure to initial pressure ☐

(d) cube root of the ratio of final pressure to initial pressure. ☐

51. The clearance volume in reciprocating all compressor

(a) to reduce the work done for per kg of air delivered ☐

(b) to increase the volumetric efficiency of the compressor ☐

(c) to accommodate valves in the head of the compressor ☐

(d) to create turbulence in the air to be delivered. ☐

52. With increase in clearance volume, the ideal work of compressing 1 kg of air

(*a*) increases ☐ (*b*) decreases ☐

(*c*) remains same ☐ (*d*) first increases and then decreases. ☐

53. In case of air engines of air motors

(*a*) intake is at higher pressure and discharge at lower pressure ☐

(*b*) intake is at lower pressure and discharge at higher pressure ☐

(*c*) intake and discharge is at the same pressure ☐

(*d*) none of the above. ☐

54. In reciprocating air compressors the clearance ratio is given by

(*a*) $\frac{\text{Total volume of cylinder}}{\text{Clearance volume}}$ ☐ (*b*) $\frac{\text{Swept volume of cylinder}}{\text{Clearance volume}}$ ☐

(*c*) $\frac{\text{Clearance volume}}{\text{Swept volume of cylinder}}$ ☐ (*d*) $\frac{\text{Clearance volume}}{\text{Total volume of cylinder}}$. ☐

55. The work obtained in case of an air engine with zero clearance is equal to

(*a*) $\frac{n}{n-1} p_1 V_1 \left[\left(\frac{p_2}{p_1} \right)^{\frac{n-1}{n}} - 1 \right]$ ☐ (*b*) $\frac{n}{n-1} \left[1 - \left(\frac{p_2}{p_1} \right)^{\frac{n-1}{n}} \right]$ ☐

(*c*) $\frac{n}{n-1} p_1 V_1 \left[\left(\frac{p_2}{p_1} \right)^{\frac{n-1}{n}} + 1 \right]$ ☐ (*d*) none of the above. ☐

56. The work input to air compressor is minimum if the compression law followed is

(*a*) $pV^{1.35} = C$ ☐ (*b*) isothermal $pV = C$ ☐

(*c*) isentropic $pV^{\gamma} = C$ ☐ (*d*) none of the above. ☐

57. The propulsive power of a jet engine is equal to

(*a*) $\frac{W(V-u)u}{60{,}000}$ ☐ (*b*) $\frac{W(V+u)u}{g \times 60{,}000}$ ☐

(*c*) $\frac{W(V-u)u}{g \times 60{,}000}$ ☐ (*d*) $\frac{W(V+u)u}{60{,}000}$. ☐

where W = Weight of gases discharged/second, u = Velocity of aircraft, and

V = Relative velocity of jet to aircraft = Velocity of jet + Velocity of aircraft.

58. The efficiency of propulsion of a jet engine is equal to

(*a*) $\frac{2u}{V-u}$ ☐ (*b*) $\frac{2u}{V+u}$ ☐

(*c*) $\frac{V-u}{2u}$ ☐ (*d*) $\frac{V+u}{2u}$. ☐

59. Choose the correct statement

(*a*) For minimum work on a compressor, the air should be taken from a source of high temperature air. ☐

(*b*) The performance of air compressor at high altitudes as compared to that at sea-level would be better. ☐

(*c*) Intercooling in multi-stage compressors is done to minimise the work of compression. ☐
(*d*) In a compressor, the cylinder clearance should be maximum. ☐

60. Closed-cycle gas turbine works on
(*a*) Rankine cycle ☐ (*b*) Carnot cycle ☐
(*c*) Brayton or Joule cycle ☐ (*d*) Otto cycle. ☐

61. An open-cycle gas turbine works on
(*a*) Rankine cycle ☐ (*b*) Carnot cycle ☐
(*c*) Brayton or Joule cycle ☐ (*d*) Otto cycle. ☐

62. In a two stage compressor, efficiency will be maximum if
(*a*) $p_1 = \frac{p_2 + p_3}{2}$ ☐ (*b*) $p_1 = \sqrt{p_2 \times p_3}$ ☐
(*c*) $p_2 = \sqrt{p_1 \times p_3}$ ☐ (*d*) $p_2 = \frac{p_1 + p_3}{2}$. ☐

63. Choose the correct statement for an air turbine
(*a*) it is valveless ☐ (*b*) it requires no internal lubrication ☐
(*c*) it is small in size ☐ (*d*) it is light in weight ☐
(*e*) all of the above ☐ (*f*) none of the above ☐

ANSWERS

Answers to Objective Type Questions

1. (*b*)	**2.** (*b*)	**3.** (*d*)	**4.** (*d*)	**5.** (*c*)	**6.** (*b*)
7. (*a*)	**8.** (*a*)	**9.** (*a*)	**10.** (*b*)	**11.** (*d*)	**12.** (*c*)
13. (*a*)	**14.** (*c*)	**15.** (*d*)	**16.** (*b*)	**17.** (*c*)	**18.** (*d*)
19. (*b*)	**20.** (*b*)	**21.** (*a*)	**22.** (*b*)	**23.** (*b*)	**24.** (*b*)
25. (*a*)	**26.** (*d*)	**27.** (*a*)	**28.** (*c*)	**29.** (*d*)	**30.** (*c*)
31. (*a*)	**32.** (*a*)	**33.** (*d*)	**34.** (*a*)	**35.** (*c*)	**36.** (*b*)
37. (*a*)	**38.** (*c*)	**39.** (*d*)	**40.** (*c*)	**41.** (*b*)	**42.** (*b*)
43. (*b*)	**44.** (*c*)	**45.** (*d*)	**46.** (*d*)	**47.** (*a*)	**48.** (*b*)
49. (*d*)	**50.** (*c*)	**51.** (*d*)	**52.** (*b*)	**53.** (*b*)	**54.** (*b*)
55. (*c*)	**56.** (*c*)	**57.** (*d*)	**58.** (*c*)	**59.** (*a*)	**60.** (*b*)
61. (*a*)	**62.** (*b*)	**63.** (*e*)	**64.** (*a*)	**65.** (*d*)	**66.** (*d*)
67. (*c*)	**68.** (*b*)	**69.** (*d*)	**70.** (*d*)	**71.** (*c*)	**72.** (*c*)
73. (*b*)	**74.** (*d*)	**75.** (*c*)	**76.** (*d*)	**77.** (*a*)	**78.** (*c*)
79. (*c*)	**80.** (*d*)	**81.** (*c*)	**82.** (*b*)	**83.** (*b*)	**84.** (*c*)
85. (*b*)	**86.** (*c*)	**87.** (*b*)	**88.** (*c*)	**89.** (*c*)	**90.** (*c*)
91. (*d*)	**92.** (*c*)	**93.** (*e*)	**94.** (*b*)	**95.** (*a*)	**96.** (*c*)
97. (*b*)	**98.** (*c*)	**99.** (*c*)	**100.** (*a*)	**101.** (*b*)	**102.** (*d*)
103. (*d*)	**104.** (*a*)	**105.** (*a*)	**106.** (*b*)	**107.** (*b*)	**108.** (*b*)
109. (*a*)	**110.** (*a*)	**111.** (*b*)	**112.** (*c*)	**113.** (*a*)	**114.** (*a*)
115. (*c*)	**116.** (*a*)	**117.** (*d*)	**118.** (*b*)	**119.** (*c*)	**120.** (*a*)

121. (*a*)	122. (*c*)	123. (*c*)	124. (*b*)	125. (*d*)	126. (*b*)
127. (*b*)	128. (*a*)	129. (*c*)	130. (*c*)	131. (*c*)	132. (*d*)
133. (*c*)	134. (*b*)	135. (*c*)	136. (*c*)	137. (*d*)	138. (*c*)
139. (*c*)	140. (*d*)	141. (*b*)	142. (*c*)	143. (*c*)	144. (*a*)
145. (*c*)	146. (*b*)	147. (*d*)	148. (*c*)	149. (*b*)	150. (*a*)
151. (*b*)	152. (*b*)	153. (*b*)	154. (*c*)	155. (*c*)	156. (*c*)
157. (*b*)	158. (*a*)	159. (*b*)	160. (*c*)	161. (*c*)	162. (*c*)
163. (*b*)	164. (*c*)	165. (*b*)	166. (*a*)	167. (*c*)	168. (*b*)
169. (*d*)	170. (*a*)	171. (*c*)	172. (*c*)	173. (*a*)	174. (*b*)
175. (*c*)	176. (*c*)	177. (*b*)	178. (*b*)	179. (*c*)	180. (*b*)
181. (*c*)	182. (*b*)	183. (*d*)	184. (*c*)	185. (*a*)	186. (*d*)
187. (*c*)	188. (*c*)	189. (*b*)	190. (*b*)	191. (*c*)	192. (*b*)
193. (*a*)	194. (*d*)	195. (*b*)	196. (*a*)	197. (*c*)	198. (*b*)
199. (*a*)	200. (*c*)				

True/False

201. (*a*)	202. (*b*)	203. (*b*)	204. (*a*)	205. (*b*)	206. (*a*)
207. (*a*)	208. (*a*)	209. (*b*)	210. (*a*)	211. (*a*)	212. (*a*)
213. (*b*)	214. (*b*)	215. (*a*)	216. (*b*)	217. (*a*)	218. (*b*)
219. (*a*)	220. (*a*)	221. (*b*)	222. (*a*)	223. (*a*)	224. (*a*)
225. (*b*)	226. (*a*)	227. (*a*)	228. (*a*)	229. (*a*)	230. (*a*)
231. (*a*)	232. (*a*)	233. (*a*)	234. (*a*)	235. (*a*)	

Fill in the Blanks

236. (*a*)	237. (*a*)	238. (*b*)	239. (*a*)	240. (*b*)	241. (*b*)
242. (*b*)	243. (*b*)	244. (*b*)	245. (*b*)	246. (*a*)	247. (*b*)
248. (*a*)	249. (*b*)	250. (*b*)	251. (*b*)	252. (*b*)	253. (*b*)
254. (*a*)	255. (*b*)	256. (*b*)	257. (*b*)	258. (*b*)	

Answers to Objective Type Questions from Competitive Examinations

1. (*a*)	2. (*a*)	3. (*a*)	4. (*c*)	5. (*c*)	6. (*b*)
7. (*a*)	8. (*d*)	9. (*a*)	10. (*a*)	11. (*b*)	12. (*b*)
13. (*d*)	14. (*a*)	15. (*a*)	16. (*b*)	17. (*d*)	18. (*b*)
19. (*d*)	20. (*c*)	21. (*b*)	22. (*a*)	23. (*a*)	24. (*c*)
25. (*a*)	26. (*c*)	27. (*b*)	28. (*d*)	29. (*b*)	30. (*c*)
31. (*a*)	32. (*c*)	33. (*a*)	34. (*d*)	35. (*b*)	36. (*b*)
37. (*b*)	38. (*b*)	39. (*d*)	40. (*c*)	41. (*d*)	42. (*c*)
43. (*c*)	44. (*c*)	45. (*a*)	46. (*b*)	47. (*c*)	48. (*a*)
49. (*c*)	50. (*c*)	51. (*c*)	52. (*c*)	53. (*a*)	54. (*c*)
55. (*b*)	56. (*b*)	57. (*c*)	58. (*b*)	59. (*c*)	60. (*c*)
61. (*c*)	62. (*c*)	63. (*e*)			

Chapter 7

Heat Transfer, Refrigeration and Air Conditioning

I. THEORY

7.1. HEAT TRANSFER

Heat transfer or heat flow, according to the second law of thermodynamics, takes place from a body at higher temperature to a body at lower temperature. There are three methods of heat transfer: (*i*) Conduction, (*ii*) Convection, and (*iii*) Radiation.

The heat transfer per unit time is called rate of heat transfer. The rate of heat transfer is considered constant if the temperature with respect to time is constant. The rate of heat transfer is considered variable if the temperature changes with respect to time. If the rate of heat transfer is constant it is known as *steady state heat transfer* and if the rate of heat transfer is variable, it is known as *unsteady-heat transfer.*

7.1.1. Conduction

Conduction is the process of heat transfer from one molecule of the body to another molecule without actual motion of the molecule. The heat transfer in a metal rod is by conduction. The *Fourier's law* is the basic law of heat transfer by conduction.

7.1.2. Convection

It is the process of heat transfer from one particle of the fluid to another by the actual motion of the heated particle. Heating of rooms by room heaters is due to convection.

7.1.3. Radiation

It is the process of heat transfer from a hot body to a cold body in a straight line, without affecting the intervening medium. Heat transfer by radiation is due to electromagnetic waves which require no medium for its propagation. The amount of heat transfer by radiation depends on (*i*) The nature of the body,

(*ii*) Temperature of the body, and

(*iii*) Kind and extent of the surface of the body.

7.2. FOURIER'S LAW OF HEAT CONDUCTION

It states that the one-dimensional steady state rate of heat flow (Q) due to conduction per unit area is directly proportional to the temperature gradient.

or $$\frac{Q}{A} \propto \frac{dt}{dx} = -k\frac{dt}{dx}$$

where k = Constant of proportionality.

$$\therefore \quad Q = -kA\frac{dt}{dx} \qquad \text{(–ve sign is due to decrease of temperature as } x \text{ increases)}$$

where Q = Rate of heat flow,

A = Area of heat flow normal to direction of heat flow,

k = Coefficient of thermal conductivity or thermal conductivity, and

$\frac{dt}{dx}$ = Temperature gradient in the direction of heat flow.

The following **assumptions** are made to find the flow of heat due to conduction:

(*i*) The flow of heat takes place at steady state.

(*ii*) The heat flow takes place only along x-direction. No heat flow takes place along y and z-directions.

(*iii*) Thermal conductivity remains constant and does not vary with temperature.

7.2.1. Thermal Conductivity (k)

It is defined as the heat flow per unit time across unit area when temperature gradient is unity. It is written as

$$k = -\frac{Q}{A} \times \frac{dx}{dt}$$

The unit of thermal conductivity in MKS system is kcal/hr m°C whereas in SI system W/m K, *i.e.*, Watt/metre K.

$$1\frac{\text{kcal}}{\text{hr m°C}} = 1.163\frac{\text{W}}{\text{m K}}$$

and $$1\frac{\text{W}}{\text{m K}} = 0.86 \text{ kcal/hr m°C}$$

The heat flow through a body by conduction is also written as

$$Q = -kA\frac{dt}{dx} = \frac{-kA(t_2 - t_1)}{(x_2 - x_1)} = \frac{kA(t_1 - t_2)}{(x_2 - x_1)} = \frac{kA(t_1 - t_2)}{\Delta x}$$

The term $\frac{kA}{\Delta x}$ is known as *thermal conductance* whereas the term $\frac{\Delta x}{kA}$ is known as *thermal resistance.*

The term $\frac{Q}{A}$ is known as *thermal loading.*

7.2.2. Conduction Through a Flat Composite Wall

Figure 7.1 shows a composite wall of three different materials of thickness x_1, x_2, x_3 and thermal conductivities k_1, k_2 and k_3 respectively. The surface temperatures are t_1, t_2, t_3 and t_4. For the steady-state, the heat flow Q through an area A (normal to the direction of heat flow) is given as

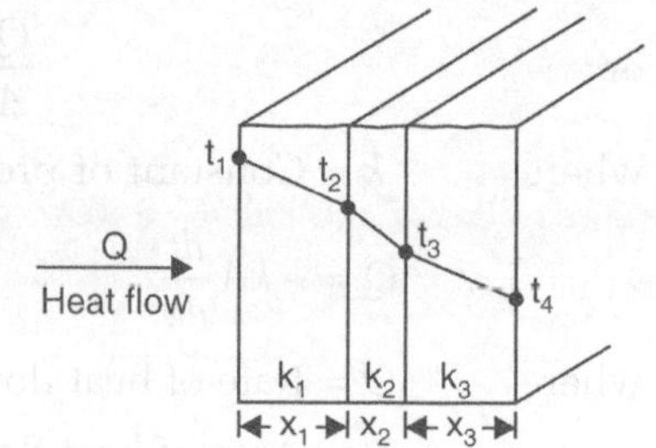

FIGURE 7.1

$$Q = \frac{k_1 A(t_2 - t_1)}{x_1}$$

$$= \frac{k_2 A(t_3 - t_2)}{x_2} = \frac{k_3 A(t_4 - t_3)}{x_3}$$

or $$-(t_2 - t_1) = \frac{Qx_1}{k_1 A} \text{ hence } (t_1 - t_2) = \frac{Qx_1}{k_1 A}$$

Similarly, $$(t_2 - t_3) = \frac{Qx_2}{k_2 A} \text{ and } (t_3 - t_4) = \frac{Qx_3}{k_3 A}$$

Adding , we get $(t_1 - t_2) + (t_2 - t_3) + (t_3 - t_4) = \frac{Qx_1}{k_1 A} + \frac{Qx_2}{k_2 A} + \frac{Qx_3}{k_3 A}$

or $$(t_1 - t_4) = Q\left[\frac{x_1}{k_1 A} + \frac{x_2}{k_2 A} + \frac{x_3}{k_3 A}\right]$$

or $$Q = \frac{(t_1 - t_4)}{\left(\frac{x_1}{k_1 A} + \frac{x_2}{k_2 A} + \frac{x_3}{k_3 A}\right)} = \frac{(t_1 - t_4)}{\sum \frac{x}{kA}}.$$

7.2.3. Conduction Through Hollow Cylinder

Conduction through hollow cylinder of internal radius = r_1, of external radius = r_2 and of length = L, in the radial direction is given by the equation

$$Q = \frac{2\pi Lk(t_1 - t_2)}{\log_e(r_2 / r_1)}$$

where t_1 = Internal surface temperature,

t_2 = External surface temperature, and

Q = Rate of heat flow in radial direction.

Mean area of heat transfer (A_m) is given by $A_m = \dfrac{(A_2 - A_1)}{\log_e (r_2 / r_1)}$

where A_2 = External surface area, and A_1 = Internal surface area.

The mean radius (r_m) for the hollow cylinder is given by $r_m = \dfrac{(r_2 - r_1)}{\log_e (r_2 / r_1)}$.

7.2.4. Conduction Through Composite Cylinder

The rate of heat flow in the radial direction through composite cylinder of layers of three different material is given by (Refer to Fig. 7.2).

$$Q = \frac{(t_1 - t_4)}{\dfrac{1}{2\pi k_1 L} \log_e \left(\dfrac{r_2}{r_1}\right) + \dfrac{1}{2\pi k_2 L} \log_e \left(\dfrac{r_3}{r_2}\right) + \dfrac{1}{2\pi k_3 L} \log_e \left(\dfrac{r_4}{r_3}\right)}.$$

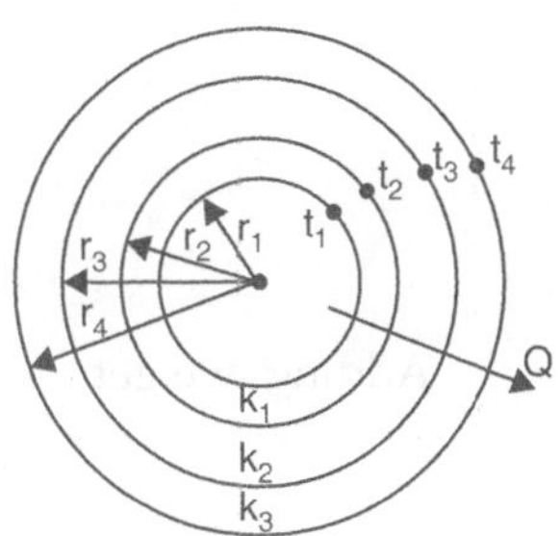

FIGURE 7.2

7.2.5. Conduction Through Sphere

The rate of heat flow in the radial outwards direction through thick sphere of inside radius = r_1 and of outside radius = r_2 is given by

$$Q = \frac{(t_1 - t_2)}{\left(\dfrac{(r_2 - r_1)}{4\pi k r_1 r_2}\right)} = \frac{4\pi k r_1 r_2 (t_1 - t_2)}{(r_2 - r_1)}$$

Mean area of heat transfer (A_m) is given by

$$A_m = 4\pi r_1 r_2$$

and *mean radius* (r_m) is given by $r_m = \sqrt{r_1 r_2}$.

7.3. NEWTON'S LAW OF COOLING

It states that the rate of heat transfer from a solid surface of area A, at a temperature t_w to a fluid at a temperature t_f, is given by

$$Q = hA(t_w - t_f)$$

where h = Film heat transfer coefficient and is having unit as kcal/m^2 hr°C.

7.4. HEAT TRANSFER FROM A HOT FLUID TO A COLD FLUID THROUGH A WALL

The rate of flow of heat (Q) from a hot fluid at temperature t_i to cold fluid at temperature t_o through a wall of thickness x_1 and thermal conductivity k is given by (Refer to Fig. 7.3)

$$Q = h_i A(t_i - t_1)$$

$$= \frac{kA}{x_1}(t_1 - t_2) = h_o A(t_2 - t_o)$$

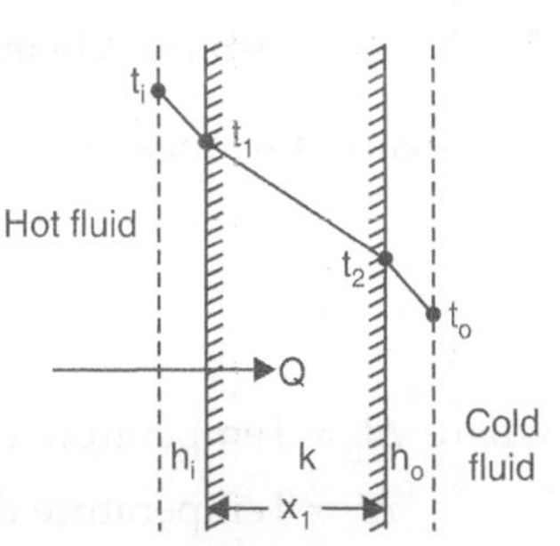

FIGURE 7.3

where h_i = Film heat transfer coefficient for hot fluid,

h_o = Film heat transfer coefficient for cold fluid, and

A = Area of heat transfer.

$$\therefore \quad t_i - t_1 = \frac{Q}{h_i A}$$

$$t_1 - t_2 = \frac{Q}{\left(\frac{kA}{x_1}\right)}$$

$$t_2 - t_o = \frac{Q}{h_o A}$$

Adding, we get

$$(t_i - t_o) = Q\left[\frac{1}{h_i A} + \frac{1}{\left(\frac{kA}{x_1}\right)} + \frac{1}{h_o A}\right]$$

$$\therefore \quad Q = \frac{(t_i - t_o)}{\left(\frac{1}{h_i A} + \frac{x_1}{kA} + \frac{1}{h_o A}\right)} \quad \text{or} \quad \frac{A(t_i - t_o)}{\left(\frac{1}{h_i} + \frac{x_1}{k} + \frac{1}{h_o}\right)} \qquad ...(i)$$

The rate of heat flow can also be expressed in terms of overall coefficient of heat transfer (U) as

$$Q = UA(t_i - t_o) \qquad ...(ii)$$

Comparing equations (*i*) and (*ii*), we get

$$\frac{1}{U} = \frac{1}{h_i} + \frac{x_1}{k} + \frac{1}{h_o}.$$

7.4.1. Critical Insulation

The critical radius of the insulation is defined as the outer radius of insulation which gives maximum heat flow. The critical radius of insulation for a hollow cylinder is equal to (k/h) and for a sphere is equal to $\frac{2k}{h}$, where k is the thermal conductivity of insulation and h is the film coefficient of heat transfer at outer radius.

7.4.2. Logarithmic Mean Temperature Difference (LMTD)

The logarithmic mean temperature difference for a heat exchanger is given by

$$\Delta t_m = \frac{\Delta t_o - \Delta t_i}{\log_e\left(\frac{\Delta t_o}{\Delta t}\right)}$$

where Δt_o = Temperature difference between hot and cold fluids at outlet,

Δt_i = Temperature difference between hot and cold fluids at inlet, and

Δt_m = Logarithmic mean temperature difference.

7.4.3. Forced Convection

The heat transfer co-efficient (h) in case of forced convection depends on viscosity (μ), density (ρ), thermal conductivity (k), specific heat (C_p), temperature difference between surface and fluid (θ), fluid velocity (V) and characteristic linear dimension (l). From dimensional analysis, we can write them in dimensionless group as

$$\frac{hl}{k} = f\left[\left(\frac{C_p\mu}{k}\right)\cdot\left(\frac{\rho Vl}{\mu}\right)\right]$$

where the dimensionless group,

$\frac{hl}{k}$ is known as Nusselt number and written as N_u,

$\frac{C_p\mu}{k}$ is known as Prandtl number and written as P_r, *and*

$\frac{\rho Vl}{\mu}$ is known as Reynolds number and written as R_e

$$\therefore \qquad N_u = f(P_r, R_e)$$

The N_u, P_r and R_e are calculated at mean film temperature t_f given by

$$t_f = \frac{t_b + t_w}{2}$$

where t_b = Mean bulk temperature, and t_w = Wall surface temperature.

7.4.4. Free Convection

In free convection, the buoyancy force (which is caused by the difference in density of a fluid due to temperature) causes the motion of fluid particles. The density (ρ) of a fluid element which is at a temperature (θ) above the surrounding fluid of density ρ_s is given by

$$\rho = \frac{\rho_s}{(1+\beta\theta)}$$

where β = Coefficient of cubical expansion.

For a unit volume, the force of buoyancy causing the fluid motion is given by

$$F = (\rho_s - \rho) \times g = \rho\beta\theta g$$

The dimension analysis of free convection gives the following dimensionless groups

$$\frac{hl}{k} = f\left[\left(\frac{C_p\mu}{k}\right),\left(\frac{\beta g\rho^2 l^3\theta}{\mu^2}\right)\right]$$

where the dimensionless group $\frac{\beta g\rho^2 l^3\theta}{\mu^2}$ is called Grashof number (G_r).

$$\therefore \qquad N_u = f(P_r, G_r).$$

The above relation is usually expressed as

$$N_u = C(G_r, P_r)^m$$

For laminar flow $10^4 < G_r \times P_r < 10^9$

For tubulent flow $G_r \times P_r > 10^9$

The values of C and m are:

$C = 0.53$, $m = 0.25$ for horizontal cylinders and laminar flow,

$C = 0.59$, $m = 0.25$ for vertical plates and cylinders and turbulent flow,

$C = 0.1$, $m = 0.33$ for vertical plates and cylinders and turbulent flow, and

$C = 0.13$, $m = 0.33$ for horizontal cylinders and turbulent flow.

7.5. RADIATION FALLING ON A BODY

When the radiations are falling on a body, the body may absorb some of the radiations, may reflect some of the radiation and may transmit the remaining radiations. If Q_i is the total incident radiations, then mathematically,

$$C_i = Q_a + Q_r + Q_t$$

where Q_a = Radiation energy absorbed, Q_r = Radiation energy reflected, and
Q_t = Radiation energy transmitted.

The above equation can also be written as $1 = \frac{Q_a}{Q_i} + \frac{Q_r}{Q_i} + \frac{Q_t}{Q_i} = \alpha + \rho + \tau$

where α = The ratio of energy absorbed to the total incident energy and is known as *absorptivity*,
ρ = The ratio of energy reflected to the total incident energy and is known as *reflectivity*, and
τ = The ratio of energy transmitted to the total incident energy and is known as *transmissivity*.

For solids mostly $\tau = 0$ and hence, we have for solids $\alpha + \rho = 1$.

The bodies are classified as black, white, transparent and opaque depending upon the radiating properties.

(*i*) **Black body** absorbs all radiations $\therefore \alpha = 1$

(*ii*) **White body** reflects all radiations $\therefore \rho = 1$

(*iii*) **Transparent body** transmits all radiations $\therefore \tau = 1$

(*iv*) **Opaque body** does not transmit any radiation

$$\therefore \quad \alpha + \rho = 1 \ (\tau = 0).$$

7.5.1. Total Emissive Power

It is defined as the total amount of radiation emitted by a body per unit time.

7.5.2. Monochromatic Emission Power

If the energy emitted per unit area per unit time from a black body in the spectral range λ to $\lambda + d\lambda$, is E_b then, we have

$$E_b = \int_0^\infty E_\lambda d\lambda$$

where E_λ is known as *monochromatic emissive power.*

For maximum emissive power, the value of wavelength is given by *Wien's Law* which is given as

$$\lambda_{max} \times T = \text{constant.}$$

The value of constant is equal to 2884 μ K.

7.5.3. Stefan Boltzman Law

It states that the total emission from a black body per unit area per unit time is proportional to the fourth power of the absolute temperature of the body. Mathematical expression for this law is

$E_b = \sigma T^4$, where σ = Stefan-Boltzman constant = 4.97×10^{-8}.

7.5.4. Kirchhoff's Law

It states that the total emissive power for any body at a given temperature is equal to its absorptivity multiplied by the total emissive power of a perfect black body at the same temperature.

7.5.5. Emissivity

It is defined as the ratio of the total emissive power of a body to the total emissive power of a black body. It is denoted by $\in$.

Mathematically, $$\in = \frac{E}{E_b}$$

where E = Total emissive power, and

E_b = Total emissive power of a black body.

7.5.6. Grey Body

If the ratio of emission of a body to that of the black body at a given temperature is constant for all wavelengths, the body is called grey body.

7.6. REFRIGERATION

Refrigeration is the cooling of a system (or a region) below the temperature of its surroundings. Refrigeration is a process of removing heat from a low temperature region and transferring it to a high temperature region. This flow of heat against a reverse temperature gradient is not allowed by second law of thermodynamics unless necessary extra-energy is supplied by an external source.

A machine which cools the region (or a body) below the temperature of its surrounding is known as **a refrigerator**. The amount of heat removed from the region (or from a body) is known as refrigerating effect.

The various applications of refrigeration are manufacturing of ice, cooling of buildings and cooling of space so that food, drinks and medicines are stored.

7.6.1. Coefficient of Performance (C.O.P.)

Coefficient of performance represents the performance of a refrigerator. It is defined as the ratio of heat abstracted or absorbed from the system to the work done on the system. Mathematically, it is written as

$$\text{C.O.P.} = \frac{\text{Heat abstracted or absorbed from the system}}{\text{Work done on the system}} = \frac{N}{W}$$

where N is also equal to net refrigerating effect.

Relative coefficient of performance is the ratio of actual and theoretical coefficient of performance.

7.6.2. Units of Refrigeration

Units of refrigeration is commonly expressed as ton of refrigeration. One ton of refrigeration is defined as the capacity to freeze one ton of water from and at 0°C in 24 hours. In actual practice, one ton of refrigeration

$$N = 3000 \text{ kcal/hr} \quad \text{or} \quad 50 \text{ kcal/min}$$
$$= 210 \text{ kJ/min} \quad \text{or} \quad 3.5 \text{ kW}.$$

7.6.3. Ideal Refrigerator (or Carnot Refrigerator)

It is one which works on reversed Carnot cycle. It is most efficient theoretical air refrigerator. Figure 7.4 shows the p-V and T-s diagram for an air refrigerator working on reversed Carnot cycle.

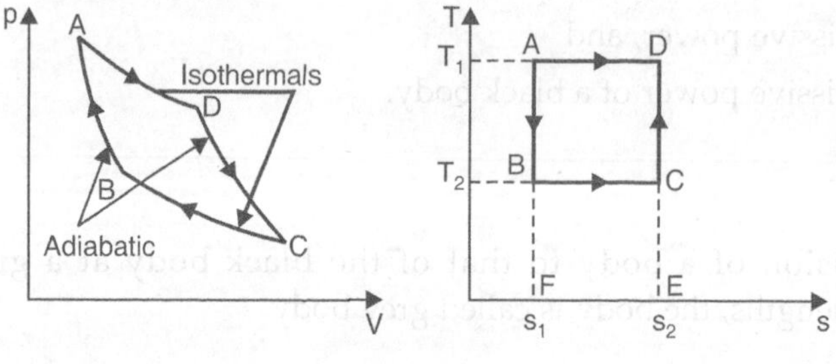

FIGURE 7.4

Starting from point A, the air in the clearance space of the cylinder expands adiabatically to point B, causing the temperature to fall from T_1 to T_2. From B, the air is expanded isothermally to point C. During process B-C, the heat is absorbed from the body to be cooled. From C, the air is compressed adiabatically to D and the temperature of air rises from T_2 to T_1. From D, the air is compressed isothermally to A. During process D-A, heat is rejected to the atmosphere. Consider 1 kg of air from T-s diagram, we get

Heat absorbed from body to be cooled $= \text{area } BCEF = T_2(s_2 - s_1)$

Work done per cycle $= \text{area } ABCD = (T_1 - T_2)(s_2 - s_1)$

$$\therefore \quad \text{C.O.P.} = \frac{\text{Heat absorbed}}{\text{Work done per cycle}} = \frac{T_2(s_2 - s_1)}{(T_1 - T_2)(s_2 - s_1)} = \frac{T_2}{T_1 - T_2}$$

where T_1 = Highest absolute temperature, and

T_2 = Lowest absolute temperature.

Now the coefficient of performance can be improved either by lowering the higher temperature (T_1) or by raising the lower temperature (T_2).

7.6.4. Bell-Coleman Refrigerator

It consists of a compression cylinder (air compressor) and an expansion cylinder (air motor). Figure 7.5 shows the *p-V* diagram for Bell-Coleman cycle. The processes are represented as:

(*i*) 6-1 represents the suction of air during suction stroke of the compressor cylinder.

(*ii*) 1-2 represents the adiabatic compression of the air in compressor cylinder.

(*iii*) 2-3 represents the cooling of the hot compressed air at constant pressure. During this process heat is rejected.

(*iv*) 3-4 represents the adiabatic expansion of the high pressure air in the expansion cylinder.

(*v*) 4-1 represents the heating of the cold expanded air at constant pressure. During this process heat is absorbed by the air.

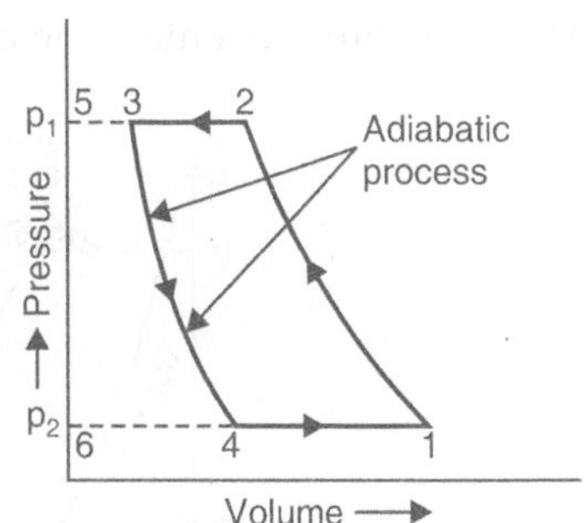

FIGURE 7.5

For one kg of air.

Heat absorbed by air $= C_p[T_1 - T_4]$

Heat rejected by air $= C_p[T_2 - T_3]$

∴ Work done on the air = Heat rejected by air – heat absorbed by air

$$= C_p[T_2 - T_3] - C_p[T_1 - T_4]$$

$$\text{C.O.P.} = \frac{\text{Heat absorbed}}{\text{Work done}}$$

$$= \frac{C_p[T_1 - T_4]}{C_p[T_2 - T_3] - C_p[T_1 - T_4]} = \frac{(T_1 - T_4)}{(T_2 - T_3) - (T_1 - T_4)}.$$

The ratio of expansion in the expansion cylinder is the same as the compression ratio in the compression cylinder so that the upper and lower limits of pressure can be maintained same. Then

$$\frac{p_2}{p_1} = \frac{p_3}{p_4} = r_p = \text{compression ratio of expansion ratio}$$

$$\therefore \qquad \text{C.O.P} = \left(r_p^{\frac{\gamma-1}{\gamma}} - 1\right).$$

7.6.5. Vapour Compression Refrigerator

It consists of a compressor, condenser, throttle valve and an evaporator as shown in Fig. 7.6. Most of the refrigerators use ammonia, carbon dioxide, Freon-12 or other vapours as the working fluid.

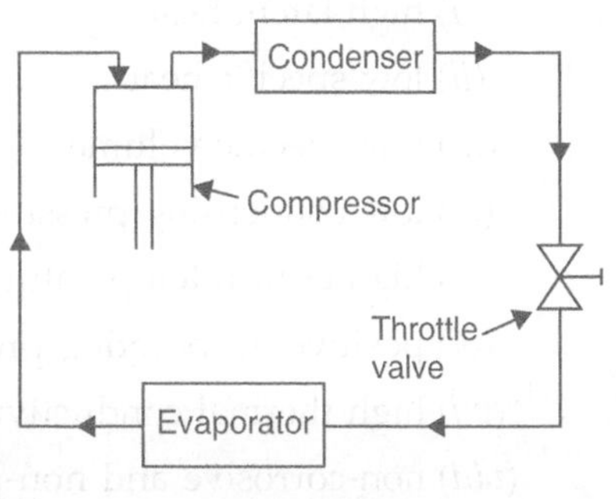

FIGURE 7.6

Figure 7.7 shows *p-V* and *T-s* diagrams for vapour compression refrigeration cycle. During the suction stroke of the compressor, the refrigerant in the condition of wet vapour is drawn from the evaporator. In compressor, the vapours are compressed adiabatically and the temperature and pressure of vapours are raised. This process is shown by *BC*. The high pressure vapour is cooled in the condenser at the same pressure. This process is shown by *CD*. The high pressure liquid is now expanded through a throttle valve, lowering its temperature and pressure. This process is shown by *DA*. The low temperature vapour now absorbs heat from the evaporator at constant pressure. Then the vapours enters the compressor during suction stroke and cycle is repeated.

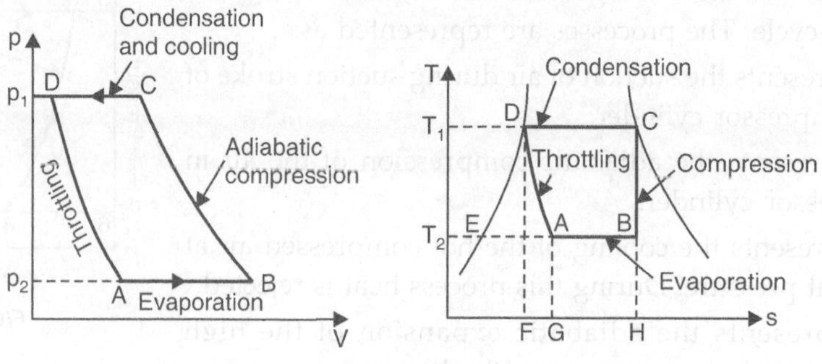

FIGURE 7.7

Heat absorbed in evaporator = Area *ABHG*

Work done by compressor = Area *BCDE*

$$\therefore \qquad \text{C.O.P.} = \frac{\text{Heat absorbed}}{\text{Work done by compressor}} = \frac{\text{Area } ABHG}{\text{Area } BCDE}.$$

7.6.6. Under-cooling

It is the process of reducing the temperature of the liquid refrigerant below saturation temperature before throttling. This increases the coefficient of performance of the cycle. This is done by inserting a liquid cooler between condenser and throttle valve.

7.7. PROPERTIES OF A REFRIGERANT

A good refrigerant should have the following properties:

(*i*) high latent heat
(*ii*) low specific heat
(*iii*) low specific volume
(*iv*) low condensing pressure
(*v*) high critical temperature
(*vi*) positive evaporating pressure
(*vii*) high thermal conductivity
(*viii*) non-corrosive and non-poisonous
(*ix*) cheap.

The following refrigerants are mostly used:

(*i*) Ammonia (NH_3)

(*ii*) Carbon dioxide (CO_2)

(*iii*) Sulphur dioxide (SO_2)

(*iv*) Freon-12(CCl_2F_2)

(*v*) Freon-22 ($CHClF_2$).

(*i*) **Ammonia (NH_3).** It has high latent heat, moderate working pressure and low freezing temperature (–77.8°C). It is used mostly for large commercial installations. Its boiling temperature is –33.3°C and critical temperature 132.6°C. It is highly toxic and has a strong smell. It attacks brass and bronze but is non-corrosive to iron and steel.

(*ii*) **Carbon dioxide (CO_2).** The boiling point of CO_2 is –78.5°C and critical temperature 31°C. It has low specific volume (0.016 m^3/kg at –15°C) and hence plant is very compact. For ships where space is a vital consideration, CO_2 is mostly used. It is non-toxic, non-corrosive and non-inflammable. It is cheap but it has high working pressure (about 70 kgf/cm^2).

(*iii*) **Sulphur dioxide (SO_2).** It has boiling temperature –25.6°C and critical temperature 157.2°C. It has low working pressure (4.5 kgf/cm^2). It is highly toxic, and is corrosive when in contact with moisture. It is used in small and domestic plants.

(*iv*) **Freon-12 (CCl_2F_2).** It is colourless, odourless and non-toxic. It is mostly used for domestic plants. It has operating pressure of about 8 kgf/cm^2. As compared to ammonia it has small specific volume but has a high cost.

(*v*) **Freon-22 ($CHClF_2$).** It has boiling temperature as –40.8°C and critical temperature as 96.2°C. It is used in small to medium commercial plants. As compared to Freon-12, it has low specific volume but has high cost.

7.8. VAPOUR ABSORPTION CYCLE

In case of vapour compression cycle, a large amount of work is required to compress the vapour of the refrigerant for increasing its pressure. But in case of vapour absorption cycle, the refrigerant vapour is dissolved in an inert liquid at the same pressure as the evaporator. The pressure of this solution is raised by the pump and very little work is required in increasing the pressure of the liquid solution. After raising the pressure, the refrigerant is separated from the solution by heating. The vapour of the refrigerant is passed through condenser, throttle valve and evaporator as shown in Fig. 7.8. Then again the cycle is repeated.

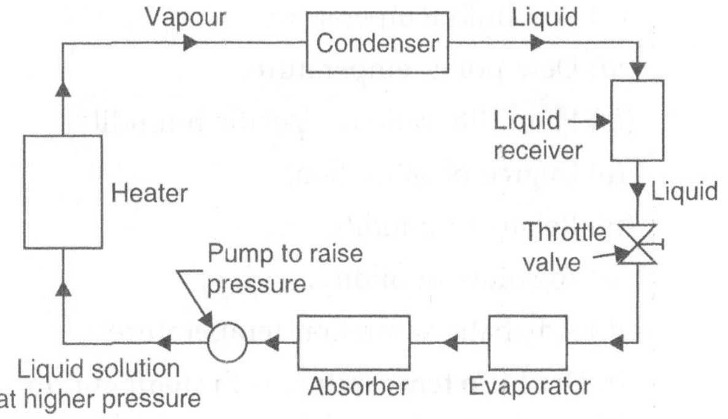

FIGURE 7.8

7.8.1. Electrolux Refrigerator

It is an absorption type refrigerator which has no moving machinery, no noise and can be used where no electricity is available.

7.9. AIR CONDITIONING

Air conditioning may be defined as the process of treating air so as to simultaneously control its temperature, humidity, clearliness and distribution to meet the requirements of the conditioned space. Hence the important actions involved in the operation of an air conditioning system are:

1. Temperature control,
2. Humidity control,
3. Air filtering, cleaning and purification, and
4. Air movement circulation.

A complete air conditioning system provides automatic control of these conditions for both winter and summer.

7.9.1. Psychrometry

Psychrometry is the branch of science which deals with the behaviour of mixture of air and water vapour. Atmospheric air is considered to be a mixture of dry air and water vapour. The control of moisture (or water vapour) contents in atmosphere is essential. This branch of science, with the properties of air-vapour mixture and their measurement and control, is important in air conditioning.

7.9.2. Psychrometric Terms

The followings are the psychrometric terms:

(*i*) Dry bulb temperature,

(*ii*) Wet bulb temperature,

(*iii*) Dew point temperature,

(*iv*) Humidity ratio or specific humidity,

(*v*) Degree of saturation,

(*vi*) Relative humidity,

(*vii*) Absolute humidity, and

(*viii*) Adiabatic saturation temperature.

(*i*) **Dry bulb temperature** is the temperature of the air recorded by an ordinary thermometer which is not affected by moisture or radiation.

(*ii*) **Wet bulb temperature** is the temperature indicated by a wet bulb thermometer. Wet bulb temperature is an indirect measure of the moisture content of air.

(*iii*) **Dew point temperature** is the temperature of air at which condensation of moisture starts when the air is cooled. This temperature is equal to the saturation temperature of water vapour at its partial pressure.

(*iv*) **Humidity ratio or specific humidity** is defined as the ratio of mass of water vapour to the mass of dry air in a given volume of air-water vapour mixture. If for a given mixture of volume *V*.

m_s = Mass of water vapour, and m_a = Mass of dry air,

then humidity ratio = $\frac{m_s}{m_a}$.

Humidity ratio, mathematically, is also expressed as

$= \frac{v_a}{v_s}$, where v_a and v_s are specific volumes of dry air and water vapour respectively

$= 0.622 \frac{p_s}{p_a}$, where p_s and p_a are the pressure of water vapour and dry air respectively

$= 0.622 \frac{p_s}{(p_b - p_s)}$, where $p_a = p_b - p_s$

p_b = Barometric pressure or total pressure of moist air.

(*v*) **Degree of saturation** (μ) is defined as the ratio of mass of water vapour in a unit mass of dry air to mass of water vapour in a unit mass of dry air saturated at the same temperature. It is also the ratio of specific humidity of moist air to the specific humidity of the saturated air at the same temperature. It is denoted by μ.

$$\therefore \quad \mu = \frac{\text{Specific humidity of moist air}}{\text{Specific humidity of saturated air at the same temp.}}$$

$$= \frac{0.622\left(\frac{p_s}{p_b - p_s}\right)}{0.622\left(\frac{p_g}{p_b - p_g}\right)} = \frac{p_s}{p_g}\left[\frac{1-\frac{p_g}{p_b}}{1-\frac{p_s}{p_b}}\right]$$

where p_b = Barometric pressure

p_g = Pressure at saturation read from steam tables corresponding to dry bulb temperature

p_s = Pressure of water vapours in a mixture of air-water vapour.

(*vi*) **Relative humidity** (ϕ) is the ratio of actual mass of water vapour in a given volume to the mass of water vapour saturated at the same temperature in the same volume. It is also defined as the ratio of the actual partial pressure of vapour in a given volume of mixture to the partial pressure of vapour when the same volume of mixture is saturated at the same temperature. Thus mathematically,

$$\phi = \frac{\text{Mass of water vapour in given volume}}{\text{Mass of water vapour if saturated in the same volume at the same temperature}}$$

$$= \frac{p_s}{p_g} = \frac{v_g}{v_s}.$$

Thus when actual partial pressure of vapour is equal to partial pressure of the saturated vapour, relative humidity is one.

Relation between relative humidity (ϕ) and degree of saturation (μ).

$$\mu = \frac{p_s}{p_g}\left[\frac{\left(1-\frac{p_g}{p_b}\right)}{\left(1-\frac{p_g}{p_b}\right)}\right] = \phi\frac{\left(1-\frac{p_g}{p_b}\right)}{\left(1-\frac{p_s}{p_g}\times\frac{p_g}{p_b}\right)}$$

$$= \phi\frac{\left(1-\frac{p_g}{p_b}\right)}{\left(1-\phi\frac{p_g}{p_b}\right)} = \phi\frac{(p_b - p_g)}{(p_b - \phi p_g)}$$

or

$$\phi = \frac{\mu}{1-(1-\mu)\frac{p_g}{p}}$$

(*vii*) **Absolute humidity** is the actual mass of water vapour in a unit volume of air.

(*viii*) **Adiabatic saturation temperature** is the temperature at which water by evaporating into air can bring the air to saturation adiabatically at the same temperature.

7.9.3. Psychrometric Chart

Psychrometric chart is a nomogram which shows the graphical representation of the properties of moist air. This chart is used for solving the psychrometric problems, as the properties of moist air can be read directly from this chart. This chart is generally constructed for standard atmospheric pressure of 760 mm of Hg or 1.033 kgf/cm^2, taking dry bulb temperature as abscissa and humidity ratio as the ordinate. The other lines, that the chart contains, are as follows:

(*i*) **Wet bulb temperature lines.** These lines are inclined, straight and non-uniformly speed.

(*ii*) **Relative humidity** lines are curved.

(*iii*) **Volume lines** are straight inclined lines.

(*iv*) **Humidity ratio lines** are straight, parallel, horizontal and uniformly spaced.

(*v*) **Dry bulb temperature lines** are straight, parallel, vertical and uniformly spaced.

7.9.4. Psychrometric Processes

The following are the main psychrometric processes:

(*i*) Sensible heating or cooling.

(*ii*) Humidification as dehumidification.

(*iii*) Cooling and dehumidification.

(*iv*) Heating and humidification.

(*v*) Adiabatic or evaporative cooling.

(*vi*) Chemical dehumidification.

(*i*) **Sensible heating or cooling.** The process of adding heat to moist air without changing the moisture content (or humidity ratio) is called sensible heating, whereas the process of removing heat from moist air without changing the moisture content (or humidity ratio) is called sensible cooling. In both these processes, the humidity ratio remains constant. In Fig. 7.9, the line 1-2 shows sensible heating and line 3-4 shows sensible cooling. In sensible heating process the dry bulb temperature and enthalpy of moist air increases whereas in sensible cooling process, the dry bulb temperature end enthalpy decreases.

(*ii*) **Humidification or dehumidification.** The process of adding moisture at constant dry bulb temperature to moist air is called humidification, whereas the process of removing moisture at constant dry bulb temperature from the moist air is called dehumidification. In Fig. 7.10, the line 1-2 shows the process of humidification whereas the line 3-4 shows the process of dehumidification. During humidification process relative humidity increases whereas during dehumidification process relative humidity decreases.

(*iii*) **Cooling and dehumidification** process of moist air is shown in Fig. 7.11. When the moist air is cooled below its dew point temperature, the condensation takes place. Line 1-2 represents cooling of moist air at constant humidity ratio upto dew point temperature. Line 2-3 represents further cooling of the air. The line 1-3 represents the cooling and dehumidification process.

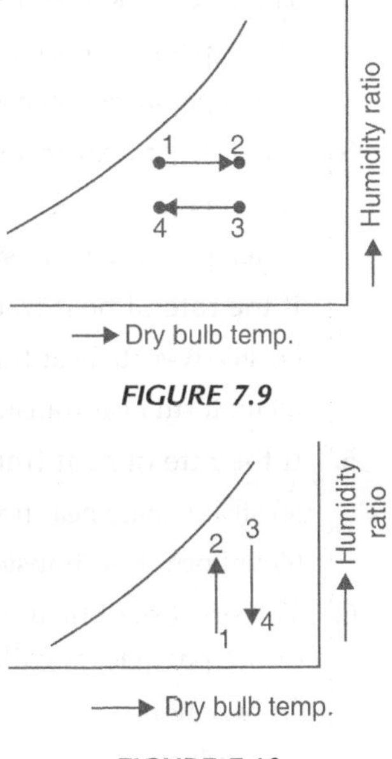

FIGURE 7.9

FIGURE 7.10

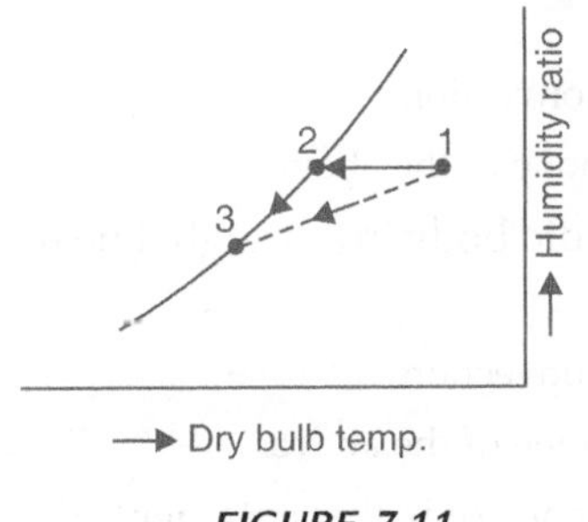

FIGURE 7.11

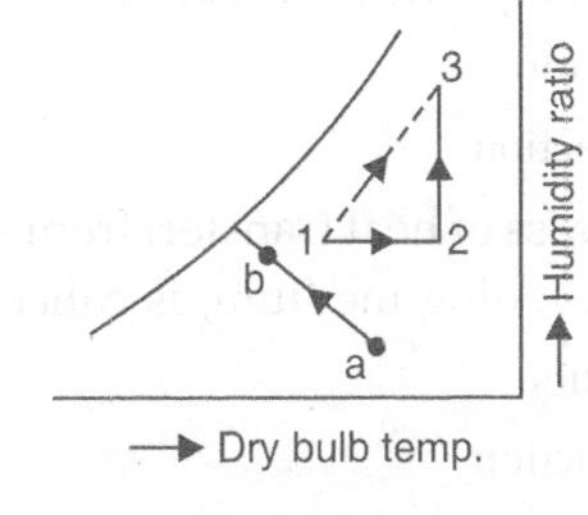

FIGURE 7.12

(*iv*) **Heating and humidification** process of moist air is shown in Fig. 7.12. Line 1-2 shows heating at constant humidity ratio and line 2-3 shows the humidification process at constant dry bulb temperature. The line 1-3 shows the heating and humidification process.

(*v*) **Adiabatic or evaporative cooling** process is shown on psychrometric chart by a constant wet bulb temperature line as shown in Fig. 7.12, by line *a-b*. During adiabatic cooling process, the relative humidity increases.

(*vi*) **Chemical dehumidification** is done by certain chemical substances which have affinity for moisture. During this process, the wet bulb temperature remains constant.

II. OBJECTIVE TYPE QUESTIONS

Tick mark the most appropriate statement of the multiple choice answers:

Heat Transfer

1. Heat transfer takes place by the process of

(*a*) conduction ☐ (*b*) convection ☐

(*c*) radiation ☐ (*d*) all of the above. ☐

2. The rate of heat transfer is constant if
   (*a*) temperature decreases with time ☐ (*b*) temperature increases with time ☐
   (*c*) temperature is constant with time ☐ (*d*) none of the above. ☐
3. The rate of heat transfer is variable if
   (*a*) temperature decreases with time ☐ (*b*) temperature increases with time ☐
   (*c*) temperature is constant with time ☐ (*d*) temperature changes with time. ☐
4. If the rate of heat transfer is constant, it is known as
   (*a*) steady-state heat transfer ☐ (*b*) unsteady-state heat transfer ☐
   (*c*) uniform heat transfer ☐ (*d*) non-uniform heat transfer. ☐
5. If the rate of heat transfer is variable, it is known as
   (*a*) steady-state heat transfer ☐ (*b*) unsteady-state heat transfer ☐
   (*c*) uniform heat transfer ☐ (*d*) non-uniform heat transfer. ☐
6. The process of heat transfer, from one particle of the body to another without actual motion of the particle, is called
   (*a*) radiation ☐ (*b*) convection ☐
   (*c*) conduction ☐ (*d*) none of the above. ☐
7. The process of heat transfer, from one particle of the body to another by the actual motion of the heated particles, is called
   (*a*) radiation ☐ (*b*) convection ☐
   (*c*) conduction ☐ (*d*) none of the above. ☐
8. The process of heat transfer, from a hot body to a cold body in a straight line without affecting the intervening medium, is called
   (*a*) radiation ☐ (*b*) convection ☐
   (*c*) conduction ☐ (*d*) none of the above. ☐
9. Heat energy can be obtained from other type of energy is the statement of
   (*a*) Zeroth law of thermodynamics ☐ (*b*) first law of thermodynamics ☐
   (*c*) second law of thermodynamics ☐ (*d*) Fourier's law. ☐
10. The heat transfer takes place according to
    (*a*) first law of thermodynamics ☐ (*b*) Zeroth law of thermodynamics ☐
    (*c*) second law of thermodynamics ☐ (*d*) Fourier's law. ☐
11. The basic law of heat conduction is called
    (*a*) Newton's law of cooling ☐ (*b*) Fourier's law ☐
    (*c*) Kirchhoff's law ☐ (*d*) Stefan's law. ☐
12. The rate of heat transfer from a solid surface to a fluid is obtained from
    (*a*) Newton's law of cooling ☐ (*b*) Fourier's law ☐
    (*c*) Kirchhoff's law ☐ (*d*) Stefan's law. ☐
13. Fourier's law is based on assumption that
    (*a*) heat flow through a solid is one-dimensional ☐
    (*b*) heat flow is in steady-state ☐

(c) both (a) and (b) ☐
(d) none of the above. ☐

14. According to Fourier's law, heat flow (Q) due to conduction is given by

(a) $Q = -K\dfrac{dt}{dx}$ ☐ (b) $Q = -KA\dfrac{dt}{dx}$ ☐
(c) $Q = KA\dfrac{dx}{dt}$ ☐ (d) $Q = -KA\dfrac{dx}{dt}$ ☐

where K = Coefficient of thermal conductivity,
A = Area of heat flow perpendicular to the direction of flow, and
$\dfrac{dt}{dx}$ = Temperature gradient in the direction of heat flow.

15. The coefficient of thermal conductivity is defined as the heat flow per unit time

(a) through unit thickness ☐
(b) when temperature difference of unity is maintained between opposite faces ☐
(c) when temperature gradient is unity ☐
(d) across unit area when temperature gradient is unity. ☐

16. The thermal conductivity in MKS units is expressed as

(a) kcal/hr m°C ☐ (b) kcal/hr m^2 °C ☐
(c) kcal/m°C ☐ (d) kcal/hr m^3°C. ☐

17. The thermal conductivity in S.I. units is expressed as

(a) J/m^2 K ☐ (b) W/m K ☐
(c) W/m K sec ☐ (d) Wm/m K. ☐

18. The heat flow through a body is given as $Q = -KA\dfrac{dt}{dx} = -\dfrac{KA(t_2 - t_1)}{(x_2 - x_1)} = \dfrac{KA(t_1 - t_2)}{\Delta x}$.

Then term $\dfrac{KA}{\Delta x}$ is known as

(a) thermal resistance ☐ (b) thermal conductance ☐
(c) thermal loading ☐ (d) none of the above. ☐

19. The term $\dfrac{\Delta x}{AK}\left(\text{in heat flow equation } Q = \dfrac{KA(t_1 - t_2)}{\Delta x}\right)$ is known as

(a) thermal resistance ☐ (b) thermal conductance ☐
(c) thermal loading ☐ (d) none of the above. ☐

20. The term $\dfrac{Q}{A}\left(\text{in heat flow equation } Q = \dfrac{KA(t_1 - t_2)}{\Delta x}\right)$ is known as

(a) thermal resistance ☐ (b) thermal conductance ☐
(c) thermal loading ☐ (d) none of the above. ☐

Heat Transfer Through a Composite Wall

21. A furnace wall of thickness 1 m and of surface area 2 m^2, is made of a material whose thermal conductivity is $\frac{1}{4}$ kcal/hr m°C. The temperature of inner surface of the wall is 1000°C and of outer surface 200°C. The heat flow through the wall is

(*a*) 500 kcal/hr ☐ (*b*) 400 kcal/hr ☐
(*c*) 300 kcal/hr ☐ (*d*) 200 kcal/hr. ☐

22. In question 21, the thermal resistance to heat flow is equal to

(*a*) 3°C hr/kcal ☐ (*b*) 4°C hr/kcal ☐
(*c*) 2°C hr/kcal ☐ (*d*) 5°C hr/kcal. ☐

23. In question 21, if the thickness of wall is $\frac{1}{2}$ m and surface area is 4 m^2 then heat flow will be

(*a*) 250 kcal/hr ☐ (*b*) 400 kcal/hr ☐
(*c*) 1600 kcal/hr ☐ (*d*) 800 kcal/hr. ☐

24. In question 21, the thermal conductance of the wall is equal to

(*a*) 3 kcal/hr°C ☐ (*b*) 4 kcal/hr°C ☐
(*c*) 2 kcal/hr°C ☐ (*d*) 0.5 kcal/hr°C. ☐

25. In question 21, the thermal loading of the heating surface or heat flux density is equal to

(*a*) 200 kcal/hr m^2 ☐ (*b*) 300 kcal/hr m^2 ☐
(*c*) 400 kcal/hr m^2 ☐ (*d*) 500 kcal/hr m^2. ☐

26. A composite wall of two layers of thickness Δx_1, Δx_2 and of thermal conductivities k_1 and k_2, is having cross-sectional area A normal to the path of heat flow. If the wall surface temperatures are t_1 and t_3 then rate of heat flow (Q) is equal to

(*a*) $\dfrac{A(t_1 - t_3)}{\dfrac{\Delta x_1}{k_1} + \dfrac{\Delta x_2}{k_2}}$ ☐ (*b*) $\dfrac{Ak_1k_2(t_1 - t_3)}{\Delta x_1 + \Delta x_2}$ ☐

(*c*) $\dfrac{(Ak_1 + Ak_2)(t_1 - t_3)}{(\Delta x_1 + \Delta x_2)}$ ☐ (*d*) $\dfrac{(t_1 - t_3)}{\left(\dfrac{\Delta x_1}{Ak_1} + \dfrac{x_2}{Ak_2}\right)}$. ☐

27. A composite wall of three hetrogeneous layers of thickness 0.3 m, 0.2 m, 0.15 m and of thermal conductivities 0.3, 0.2, 0.15 kcal/hr m°C is having surface are 1.0 m^2. If the inner and outer temperatures of the composite wall are 1000°C and 40°C, the rate of heat flow will be equal to

(*a*) 300 kcal/hr ☐ (*b*) 200 kcal/hr ☐
(*c*) 320 kcal/hr ☐ (*d*) 500 kcal/hr. ☐

28. According to Newton's law of cooling, the amount of heat transfer from a solid surface of area A at a temperature t_s to a fluid at a temperature t_f is given by

(*a*) $Q = \dfrac{hA}{(t_s - t_f)}$ ☐ (*b*) $Q = \dfrac{h}{A(t_s - t_f)}$ ☐

(*c*) $Q = hA(t_s - t_f)$ ☐ (*d*) $Q = (t_s - t_f)/hA$. ☐

where h = Film heat transfer coefficient.

29. The unit of film heat transfer coefficient (h) in MKS is

(*a*) kcal/m^2 hr°C ☐ (*b*) kcal/m hr°C ☐

(*c*) kcal/hr°C ☐ (*d*) kcal m/hr°C. ☐

30. A wall of surface area A (of thickness Δx and of conductivity k) contains hot fluid at temperature t_1 to one side and cold fluid at temperature t_2 to other side. The rate of heat transfer from hot fluid to cold fluid is equal to

(*a*) $(t_1 - t_2)/\left(\frac{1}{h_1 A} + \frac{\Delta x}{kA} + \frac{1}{h_2 A}\right)$ ☐ (*b*) $\frac{(t_1 - t_2)(h_1 + h_2)}{A(\Delta x)}$ ☐

(*c*) $\frac{(t_1 - t_2)\left(\frac{1}{h_1} + \frac{1}{h_2}\right)A}{\Delta x}$ ☐ (*d*) $\left(\frac{1}{h_1 A} + \frac{\Delta x}{kA} + \frac{1}{h_2 A}\right)(t_1 - t_2)$. ☐

where h_1, h_2 = Film coefficient of heat transfer of hot and cold fluids.

Conduction Through Hollow Cylinder

31. The rate of radial heat flow per unit length through the walls of a hollow cylinder of inner radius r_1 and outer radius r_2 with t_1 and t_2 temperatures at radii r_1 and r_2 is equal to

(*a*) $\frac{2\pi k(t_1 + t_2)}{\log_e(r_2 / r_1)}$ ☐ (*b*) $\frac{2\pi(t_1 + t_2)}{k \log_e(r_2 / r_1)}$ ☐

(*c*) $\frac{2\pi k(t_1 - t_2)}{\log_e(r_2 / r_1)}$ ☐ (*d*) $\frac{\log_e(r_2 / r_1)}{2\pi k(t_1 - t_2)}$ ☐

32. The mean radius (r_m) for a hollow cylinder of inner and outer radii r_1 and r_2 for the same rate of radial heat flow is equal to

(*a*) $\frac{(r_2 + r_1)}{2}$ ☐ (*b*) $\frac{(r_2 + r_1)}{2 \log_e(r_2 / r_1)}$ ☐

(*c*) $\frac{(r_2 - r_1)}{2 \log_e(r_2 / r_1)}$ ☐ (*d*) $\frac{(r_2 - r_1)}{\log_e(r_2 / r_1)}$. ☐

33. For steady flow and constant value of conductivity, the temperature distribution for a plane wall is

(*a*) parabolic ☐ (*b*) linear ☐

(*c*) logarithmic ☐ (*d*) cubic. ☐

34. For steady flow and constant value of conductivity, the temperature distribution for a hollow cylinder of radii r_1 and r_2 is

(*a*) parabolic ☐ (*b*) linear ☐

(*c*) logarithmic function of radii ☐ (*d*) cubic. ☐

35. The rate of heat transfer through a hollow cylinder of radii r_1 and r_2 depends on

(*a*) difference of radii ($r_2 - r_1$) ☐ (*b*) ratio of radii $\left(\frac{r_2}{r_1}\right)$ ☐

(*c*) product of radii ($r_2 \times r_1$) ☐ (*d*) sum of radii ($r_2 + r_1$). ☐

36. For the same temperature difference, the rate of heat transfer through a hollow cylinder of inner and outer radii r_1 and r_2 respectively, increases as the ratio $\left(\frac{r_2}{r_1}\right)$

(*a*) increases ☐ (*b*) decreases ☐
(*c*) constant ☐ (*d*) none of above. ☐

37. The rate of heat flow through the walls of a *hollow* sphere of inner and outer radii r_1 and r_2 maintained at temperatures t_1 and t_2 respectively is equal to

(*a*) $\frac{4\pi k r_1 r_2 (t_1 - t_2)}{(r_2 + r_1)}$ ☐ (*b*) $\frac{k r_1 r_2 (t_1 - t_2)}{(r_2 - r_1)}$ ☐

(*c*) $\frac{k r_1 r_2 (t_1 - t_2)}{4\pi (r_2 - r_1)}$ ☐ (*d*) $\frac{4\pi k r_1 r_2 (t_1 - t_2)}{(r_2 - r_1)}$. ☐

38. The mean radius of heat transfer for a hollow sphere of inner and outer radii r_1 and r_2 is equal to

(*a*) $(r_2 - r_1)$ ☐ (*b*) $r_2 \times r_1$ ☐

(*c*) $\sqrt{r_1 r_2}$ ☐ (*d*) $\frac{r_2}{r_1}$. ☐

39. Critical radius of a hollow cylinder is defined as

(*a*) inner radius which would give maximum heat flow ☐
(*b*) outer radius which would give minimum heat flow ☐
(*c*) outer radius which would give maximum heat flow ☐
(*d*) none of the above. ☐

40. The critical radius of insulation for a hollow cylinder is equal to

(*a*) k/h ☐ (*b*) $k \times h$ ☐

(*c*) $\sqrt{k \times h}$ ☐ (*d*) h/k. ☐

where k = Thermal conductivity of insulation, and
h = Film coefficient of heat transfer at outer radius.

41. The critical radius of insulation for sphere is equal to

(*a*) $2k \times h$ ☐ (*b*) $2k/h$ ☐

(*c*) $h/2k$ ☐ (*d*) $\sqrt{2k \times h}$. ☐

where k = Thermal conductivity of insulation, and
h = Film coefficient of heat transfer at outer radius.

42. The average temperature difference between the two fluids in case of counter flow heat exchanger as compared to parallel flow heat exchanger is

(*a*) more ☐ (*b*) less ☐
(*c*) constant ☐ (*d*) none of the above. ☐

43. The logarithmic mean temperature difference for a heat exchanger is equal to

(a) $\dfrac{\Delta t_o + \Delta t_i}{\log_e \Delta t_o / \Delta t_i}$ ☐ (b) $\dfrac{(\Delta t_o - \Delta t_i)}{\log_e (\Delta t_o / \Delta t_i)}$ ☐

(c) $\log_e(\Delta t_o - \Delta t_i)$ ☐ (d) $\dfrac{1}{2}\log_e (\Delta t_o - \Delta t_i)$. ☐

where Δt_o = Temperature difference between hot and cold fluids at outlet, and

Δt_i = Temperature difference between hot and cold fluids at inlet.

44. In case of heat exchanger, the value of logarithmic mean temperature difference should be

(a) as small as possible ☐ (b) as large as possible ☐

(c) constant ☐ (d) none of the above. ☐

45. The value of logarithmic mean temperature for counter flow heat exchanger as compared to parallel flow exchanger is always

(a) less ☐ (b) more ☐

(c) constant ☐ (d) none of the above. ☐

46. Choose the correct statement.

(a) Fourier's law of heat conduction gives the heat flow for ☐

(b) Thermal conductivity of air increases with decrease in temperature. ☐

(c) Thermal conductivity of solids increases with rise in temperature. ☐

(d) The unit of thermal conductivity in S.I. units is W/m K. ☐

47. Choose the wrong statement

(a) Fourier's law of heat conduction is $Q = -KA\dfrac{dt}{dx}$. ☐

(b) In case of solids, the heat transfer takes place due to conduction. ☐

(c) In case of liquids, the heat transfer takes place due to convection. ☐

(d) The unit of thermal conductivity in M.K.S. unit is kcal/hr m^2°C. ☐

48. For the counter flow heat exchanger

(a) both the fluids at exit are in their hottest state ☐

(b) both the fluids at inlet are in their coldest sate ☐

(c) both the fluids at inlet are in their hottest state ☐

(d) one fluid is hottest and other is coldest at inlet. ☐

49. If h = coefficient of heat transfer, k = thermal conductivity, and l is characteristic linear dimension then the term $\dfrac{hl}{k}$ is called

(a) Reynolds number ☐ (b) Nusselt number ☐

(c) Prandtl number ☐ (d) Froude number. ☐

50. If μ = coefficient of viscosity, C_p = specific heat at constant pressure and k is thermal conductivity then the term $\frac{\mu C_p}{k}$ is called

(a) Reynold number ☐ (b) Nusselt number ☐
(c) Prandtl number ☐ (d) Froude number. ☐

51. The term $\frac{\rho VL}{\mu}$ is called

(a) Reynold number ☐ (b) Nusselt number ☐
(c) Prandtl number ☐ (d) Froude number. ☐

52. The film temperature at which the fluid properties are to be evaluated (as Reynold, Prandtl and Nusselt's number are function of fluid properties) is given by

(a) $2(t_b + t_w)$ ☐ (b) $\frac{t_b + t_w}{2}$ ☐

(c) $2(t_b - t_w)$ ☐ (d) $\frac{t_b + t_w}{2}$. ☐

where t_b = Mean bulk temperature, and t_w = Wall surface temperature.

53. In free convection, motion of the fluid is caused

(a) by the weight of the fluid element ☐
(b) by the hydrostatic force of the element ☐
(c) by the buoyancy forces arising from variation in density of the fluid with the temperature ☐
(d) none of the above. ☐

54. The density of a fluid element which is at a temperature θ above the surrounding fluid is equal to

(a) $\frac{\rho_s}{(1-\beta\theta)}$ ☐ (b) $\rho_s (1 + \beta\theta)$ ☐

(c) $\rho_s (1 - \beta\theta)$ ☐ (d) $\frac{\rho_s}{(1+\beta\theta)}$. ☐

where ρ_s = Density of surrounding fluid, and
β = Coefficient of cubical expansion.

55. For a unit volume, the buoyancy force causing the fluid motion is equal to

(a) $\rho\beta\theta g$ ☐ (b) $\frac{\rho\beta\theta}{g}$ ☐

(c) $(\rho_s - \rho) \times g$ ☐ (d) both (a) and (c). ☐

56. The term $\left(\frac{\beta g \rho^2 l^2 \theta}{\mu^2}\right)$ is a dimensionless number. It is called

(a) Reynold number ☐ (b) Nusselt number ☐
(c) Grashof number ☐ (d) Prandtl number. ☐

57. The process which transmits energy by means of electromagnetic waves is called
(a) convection ☐ (b) radiation ☐
(c) conduction ☐ (d) none of the above. ☐

Radiation

58. The amount of radiation depends upon
(a) temperature of the body only ☐
(b) surface area of the body only ☐
(c) nature of the body only ☐
(d) all of the above. ☐

59. Radiant energy, being electromagnetic in character
(a) needs medium for its propagation ☐
(b) needs no medium ☐
(c) needs heating before propagation ☐
(d) none of the above. ☐

60. The heating effect is caused by the wavelengths when they are in the range of
(a) 100 μ to 150 μ ☐ (b) 0.1 μ to 100 μ ☐
(c) 150 μ to 200 μ ☐ (d) none of the above. ☐

61. The ratio of the energy absorbed by a body to the total incident radiation energy falling on the body, is called
(a) transmissivity ☐ (b) reflectivity ☐
(c) absorptivity ☐ (d) none of the above. ☐

62. The ratio of the energy reflected by a body to the total incident radiation energy falling on the body, is called
(a) transmissivity ☐ (b) reflectivity ☐
(c) absorptivity ☐ (d) none of the above. ☐

63. The ratio of the energy transmitted by a body to the total incident radiation energy falling on the body, is called
(a) transmissivity ☐ (b) reflectivity ☐
(c) absorptivity ☐ (d) none of the above. ☐

64. Depending on the radiating properties, the bodies are classified as
(a) black body ☐ (b) white body ☐
(c) transparent body ☐ (d) opaque body ☐
(e) all of the above. ☐

65. A body, which absorbs all the radiations falling on it, is called
(a) opaque body ☐ (b) white body ☐
(c) black body ☐ (d) transparent body. ☐

66. A body, which reflects all the radiations falling on it, is called
(a) opaque body ☐ (b) white body ☐
(c) black body ☐ (d) transparent body. ☐

67. A body, which transmits all the radiations falling on it, is called

(*a*) opaque body ☐ (*b*) white body ☐

(*c*) black body ☐ (*d*) transparent body. ☐

68. A body, which does not transmit any radiations but partly reflects and party absorbs the radiations, is called

(*a*) opaque body ☐ (*b*) white body ☐

(*c*) black body ☐ (*d*) transparent body. ☐

69. A hollow sphere with uniform interior temperature and a small hole behaves very nearly as a

(*a*) opaque body ☐ (*b*) white body ☐

(*c*) black body ☐ (*d*) transparent body. ☐

70. Choose the correct statement

(*a*) The total amount of radiation absorbed by a body per unit time is known as total emissive power. ☐

(*b*) The value of the wavelength for minimum emissive power is given by Kirchhoff' s law. ☐

(*c*) Mathematically Wien's law is expressed as $\lambda'_{max} \times T$ = constant. ☐

(*d*) A perfect black body is one which is black in colour. ☐

71. Stefan Boltzman Law states that the total emission from a black body per unit area per unit time

(*a*) varies directly with the absolute temperature ☐

(*b*) varies inversely with the absolute temperature ☐

(*c*) varies directly with the square of the absolute temperature ☐

(*d*) varies directly with the fourth power of the absolute temperature. ☐

72. 'The total emissive power for any body at a given temperature is equal to its absorptivity multiplied by the total emissive power of a perfect black body at the same temperature'— this statement is known as

(*a*) Wien's law ☐ (*b*) Planck's law ☐

(*c*) Kirchhoff's law ☐ (*d*) Stefan Boltzman law. ☐

73. The ratio of total emissive power of a body to the total emissive power of a black body is called

(*a*) absorptivity ☐ (*b*) reflectivity ☐

(*c*) transmissivity ☐ (*d*) emissivity. ☐

74. If the ratio of emission of a body to that of a black body at a given temperature is constant for all wavelengths, the body is called

(*a*) black body ☐ (*b*) grey body ☐

(*c*) white body ☐ (*d*) opaque body. ☐

Additional Questions on Heat Transfer

75. The heat transfer, from one particle of a fluid to another by the actual motion (which is caused by the difference in density due to temperature) of the particle known as

(*a*) forced convection ☐ (*b*) free convection ☐
(*c*) artificial convection ☐ (*d*) none of the above. ☐

76. The heat transfer, from one particle of a fluid to another by the actual motion (which is caused by some mechanical means) of the fluid is known as

(*a*) forced convection ☐ (*b*) free convection ☐
(*c*) artificial convection ☐ (*d*) none of the above. ☐

77. The heat transfer by radiation takes place by means of

(*a*) electromagnetic waves ☐
(*b*) molecular energy interchanges ☐
(*c*) flow of electrons ☐
(*d*) flow of fluid. ☐

78. The amount of radiation depends on

(*a*) nature of the body ☐ (*b*) temperature of the body ☐
(*c*) type of surface of the body ☐ (*d*) all of the above ☐
(*e*) none of the above. ☐

79. Choose the correct statement

(*a*) The reciprocal of thermal resistance is known as thermal loading. ☐
(*b*) The reciprocal thermal conductance is known as thermal loading. ☐
(*c*) The reciprocal of thermal resistance is known as thermal conductance. ☐
(*d*) The unit of thermal conductivity in S.I. system is W/m^2 K. ☐

80. The rate of heat flow through a composite wall of three layers of thickness 0.3 m, 0.2 m, 0.15 m and of corresponding thermal conductivities 0.3, 0.2 and 0.15 kcal/hr m°C is 320 kcal/hr. If the surface area normal to the direction of flow of heat is 1 m^2 and inner surface temperature is 1000°C,then the interface temperature at the end of 1st layer will be

(*a*) 700°C ☐ (*b*) 680°C ☐
(*c*) 500°C ☐ (*d*) 360°C. ☐

81. In the above question, the interface temperature at the end of the 2nd layer would be

(*a*) 700°C ☐ (*b*) 680°C ☐
(*c*) 500°C ☐ (*d*) 360°C. ☐

82. In question 80, the temperature at the outer surface of the wall would be

(*a*) 700°C ☐ (*b*) 680°C ☐
(*c*) 40°C ☐ (*d*) 360°C. ☐

83. Choose the correct statement

(*a*) The film heat transfer co-efficient is property of the fluid. ☐
(*b*) The thermal conductivity is not a property of material. ☐
(*c*) A body which is black in colour is called black body. ☐
(*d*) A body which reflects all radiations is called white body. ☐

84. The overall heat transfer co-efficient (U) for a composite wall of thickness x_1, x_2, x_3 and of corresponding thermal conductivities k_1, k_2, k_3 is given by equation

(a) $\frac{1}{U} = \frac{k_1}{x_1} + \frac{k_2}{x_2} + \frac{k_3}{x_3}$ ☐ (b) $U = \frac{k_1}{x_1} + \frac{k_2}{x_2} + \frac{k_3}{x_3}$ ☐

(c) $\frac{1}{U} = \frac{x_1}{k_1} + \frac{x_2}{k_2} + \frac{x_3}{k_3}$ ☐ (d) $U = \frac{x_1}{k_1} + \frac{x_2}{k_2} + \frac{x_3}{k_3}$. ☐

85. The overall heat transfer co-efficient (U), when the heat is transferred from hot fluid (of film heat transfer co-efficient h_i) to a cold fluid (of film heat transfer coefficient h_o) through a wall of thickness (Δx) and thermal conductivity (k) is given by

(a) $\frac{1}{U} = h_i + \frac{k}{\Delta x} + h_o$ ☐ (b) $U = h_i + \frac{k}{\Delta x} + h_o$ ☐

(c) $\frac{1}{U} = \frac{1}{h_i} + \frac{\Delta x}{k} + \frac{1}{h_o}$ ☐ (d) $U = \frac{1}{h_i} + \frac{\Delta x}{k} + h_o$. ☐

86. For a forced convection, Nusselt number is a function of

(a) Prandtl and Grashof numbers ☐ (b) Grashof number only ☐

(c) Reynold and Grashof numbers ☐ (d) Reynold and Prandtl numbers. ☐

87. For a free convection, Nusselt number is a function of

(a) Prandtl and Grash of number ☐

(b) Reynold and Grashof numbers ☐

(c) Reynold number only ☐

(d) Reynold and Prandtl number. ☐

88. Planck's law holds good for

(a) black bodies ☐ (b) white bodies ☐

(c) grey bodies ☐ (d) all coloured bodies. ☐

Refrigeration

89. The vapour compression refrigerators are based on the principle that fluids absorb heat while changing from a

(a) vapour phase to liquid phase and give up heat in changing from a liquid phase to vapour phase ☐

(b) liquid phase to vapour phase and give up heat in changing from a vapour phase to liquid phase ☐

(c) vapour phase to liquid phase only ☐

(d) none of the above. ☐

90. The difference between vapour compression and vapour absorption refrigeration cycle is that

(a) absorption cycle requires more work input ☐

(b) absorption cycle requires less work input ☐

(c) vapour compression cycle requires a pump ☐

(d) none of the above. ☐

91. The refrigerant used in vapour absorption refrigerator is

(*a*) Freon-12 ☐ (*b*) ammonia ☐

(*c*) CO_2 ☐ (*d*) aqua-ammonia. ☐

92. Heat is absorbed by a refrigerant, during a refrigeration cycle in a

(*a*) condenser ☐ (*b*) evaporator ☐

(*c*) compressor ☐ (*d*) throttle valve. ☐

93. Heat is rejected by a refrigerant, during a refrigeration cycle in a

(*a*) condenser ☐ (*b*) evaporator ☐

(*c*) compressor ☐ (*d*) throttle valve. ☐

94. In a vapour compression cycle, the refrigerant is in the form of superheated vapour before passing through

(*a*) condenser ☐ (*b*) evaporator ☐

(*c*) compressor ☐ (*d*) throttle valve. ☐

95. In a vapour compression cycle, the refrigerant is the form of dry saturated vapour before entering

(*a*) condenser ☐ (*b*) evaporator ☐

(*c*) compressor ☐ (*d*) throttle valve. ☐

96. A domestic refrigerator has the co-efficient of performance

(*a*) equal to 1.0 ☐ (*b*) less than 1.0 ☐

(*c*) more than 1.0 ☐ (*d*) any value. ☐

97. Choose the wrong statement

(*a*) A refrigerant should have high latent heat. ☐

(*b*) A refrigerant should have positive evaporating pressure. ☐

(*c*) A refrigerant should be cheap. ☐

(*d*) A refrigerant should have low thermal conductivity. ☐

98. Choose the correct statement

(*a*) Super-heating in a refrigeration cycle increases C.O.P. ☐

(*b*) Sub-cooling in a refrigeration cycle decreases C.O.P. ☐

(*c*) The refrigerant used in aeroplanes is Freon-II. ☐

(*d*) Ammonia-absorption refrigeration cycle requires less work input as compared to vapour compression refrigeration cycle. ☐

99. In a vapour compression refrigeration cycle, the lowest temperature occurs in

(*a*) condenser ☐ (*b*) evaporator ☐

(*c*) compressor ☐ (*d*) throttle valve. ☐

100. The pipe material for the refrigerators using ammonia as refrigerant should be of

(*a*) copper ☐ (*b*) cast steel or wrought iron ☐

(*c*) brass ☐ (*d*) aluminium. ☐

101. A refrigerating machine

(*a*) removes heat from a high temperature region to transfer it to a low temperature region ☐

(*b*) rejects energy to a low temperature region ☐

(*c*) removes heat from a low temperature region to transfer it to a high temperature region ☐

(*d*) none of the above. ☐

102. The coefficient of performance (C.O.P.) is defined as

(*a*) (Work done on the system)/(Heat absorbed from the system) ☐

(*b*) (Heat absorbed from system)/(Work done on the system) ☐

(*c*) (Heat absorbed—Work done)/(Work done on the system) ☐

(*d*) (Work done on the system)/(Heat absorbed—Work done). ☐

103. The coefficient of performance (C.O.P.) of a refrigerator working on a reserved Carnot cycle is mathematically equal to

(*a*) $\frac{(T_1 - T_2)}{T_1}$ ☐ (*b*) $\left(\frac{T_1}{T_2 - T_1}\right)$ ☐

(*c*) $\left(\frac{T_1}{T_1 - T_2}\right)$ ☐ (*d*) $\frac{(T_2 - T_1)}{T_1}$. ☐

where T_2 = Highest absolute temperature, and T_1 = Lowest absolute temperature.

104. Relative co-efficient of performance is equal to

(*a*) ratio of actual and theoretical C.O.P. ☐

(*b*) product of actual and theoretical C.O.P. ☐

(*c*) difference of theoretical and actual C.O.P. ☐

(*d*) ratio of theoretical and actual C.O.P. ☐

105. The C.O.P. of a refrigerator working on a reversed Carnot cycle using perfect gas is equal to

(*a*) $(1 - r^{\gamma-1})$ ☐ (*b*) $\frac{1}{(r^{\gamma-1} - 1)}$ ☐

(*c*) $\frac{1}{1 - r^{\gamma-1}}$ ☐ (*d*) $\frac{1}{1 + r^{\gamma-1}}$. ☐

where r = Volume ratio for isentropic compression or expansion.

106. The capacity of a refrigerator is one tonne. This means that the heat removing capacity (or refrigeration effect) of the refrigerator is

(*a*) 1000 kcal/hr ☐ (*b*) 2000 kcal/hr ☐

(*c*) 3000 kcal/hr ☐ (*d*) 4000 kcal/hr. ☐

107. The highest and lowest temperatures of the refrigerant in a refrigeration system which operates on the reversed Carnot cycle, at 27°C and –33°C respectively. The C.O.P. of the cycle is equal to

(*a*) 3.0 ☐ (*b*) 2.0 ☐

(*c*) 6.0 ☐ (*d*) 1.5. ☐

108. If in question 107, the capacity of the refrigerator is 8 tons, the work done on the refrigerant per hour will be

(*a*) 3000 kcal/hr ☐ (*b*) 4000 kcal/hr ☐
(*c*) 2000 kcal/hr ☐ (*d*) 1000 kcal/hr. ☐

109. The C.O.P. of a refrigerating machine, when lower temperature is fixed, can be improved by

(*a*) raising the higher temperature ☐ (*b*) lowering the higher temperature ☐
(*c*) keeping higher temperature constant ☐ (*d*) first increasing then decreasing. ☐

110. The C.O.P. of a refrigerating machine, when higher temperature is fixed, can be improved by

(*a*) raising the lower temperature ☐
(*b*) lowering the lower temperature ☐
(*c*) keeping the lower temperature constant ☐
(*d*) none of the above. ☐

111. The refrigerating effect means

(*a*) the heat rejected from the system per hour ☐
(*b*) the work done on the system ☐
(*c*) heat absorbed in the cycle from the body or space to be cooled ☐
(*d*) none of the above. ☐

112. One tonne refrigeration in S.I. units is equivalent to

(*a*) 50 kJ/min ☐ (*b*) 3.5 kW ☐
(*c*) 3000 J/min ☐ (*d*) 200 kJ/min. ☐

113. The reversed Carnot cycle can be used as a heat pump. The C.O.P. for Carnot heat pump is equal to

(*a*) $\frac{T_2}{T_1 - T_2}$ ☐ (*b*) $\frac{T_2}{(T_2 - T_1)}$ ☐

(*c*) $\frac{T_1 - T_2}{T_2}$ ☐ (*d*) $\frac{T_2 - T_1}{T_2}$. ☐

where T_2 = Highest absolute temperature, and T_1 = Lowest absolute temperature.

114. The co-efficient of performance (C.O.P.) of a a refrigerator working on a reversed Carnot cycle is 5. The ratio of highest absolute temperature to lowest absolute temperature (*i.e.*, T_2/T_1) would be

(*a*) 1.25 ☐ (*b*) 1.3 ☐
(*c*) 1.4 ☐ (*d*) 1.2. ☐

115. If the ratio of highest absolute temperature to lowest absolute temperature is the same as in question 114 and the cycle is used as heat pump the co-efficient of performance for the Carnot heat pump would be

(*a*) 5 ☐ (*b*) 4 ☐
(*c*) 6 ☐ (*d*) 3. ☐

116. Choose the correct statement

(*a*) The co-efficient of performance is always less than one. ☐

(*b*) The horse power per ton of refrigeration is 4.75 × (C.O.P.). ☐

(*c*) In S.I. units, one tonne of refrigeration is equal to 210 kJ/min. ☐

(*d*) In a refrigerating machine heat rejected is less than heat absorbed. ☐

117. The cycle, on which air refrigerator works, is

(*a*) Rankine cycle ☐ (*b*) Bell-Coleman cycle ☐

(*c*) Carnot cycle ☐ (*d*) Reversed Carnot cycle ☐

(*e*) both (*b*) and (*d*). ☐

118. The C.O.P. of a refrigerator working on Bell-Coleman cycle having the same compression and expansion ratio, is equal to

(*a*) $\left(\frac{1}{r_p}\right)^{\frac{\gamma-1}{\gamma}}$ ☐ (*b*) $\frac{1}{r_p^{\frac{\gamma-1}{\gamma}}-1}$ ☐

(*c*) $r_p^{\frac{\gamma-1}{\gamma}}-1$ ☐ (*d*) $(r_p-1)^{\frac{\gamma-1}{\gamma}}$ ☐

where r_p = Compression ratio or expansion ratio.

119. Which of the following is used as working fluid vapour in vapour compression refrigerator?

(*a*) carbon dioxide ☐ (*b*) ammonia ☐

(*c*) Freon-12 ☐ (*d*) all of the above ☐

(*e*) none of the above. ☐

120. The function of oil separator in vapour compression refrigerator is to

(*a*) lubricate the compressor ☐

(*b*) reduce the heat transfer efficiency ☐

(*c*) prevent the oil being carried to condenser ☐

(*d*) none of the above. ☐

121. Wet compression cycle for a vapour compression refrigerator means

(*a*) the entire compression of the vapour should be in wet region ☐

(*b*) the entire compression of the vapour should be in superheated region ☐

(*c*) the vapour enters the compressor in dry and saturated state but leaves the compressor in wet state ☐

(*d*) none of the above. ☐

122. Dry compression cycle for a vapour compression refrigerator means

(*a*) the entire compression of the vapour should be in wet region ☐

(*b*) the entire compression of the vapour should be in superheated region ☐

(*c*) the vapour enters the compressor in dry and saturated state but leaves the compressor in wet state ☐

(*d*) none of the above. ☐

123. In a vapour compression refrigerator, the oil separator is installed between

(*a*) evaporator and compressor ☐ (*b*) compressor and condenser ☐

(*c*) condenser and throttle valve ☐ (*d*) throttle valve and evaporator. ☐

124. The coefficient of performance in wet compression as compared to dry compression is

(*a*) lower ☐ (*b*) higher ☐
(*c*) same ☐ (*d*) none of the above. ☐

125. In actual practice, dry compression is preferred as in dry compression

(*a*) there is higher volumetric efficiency ☐
(*b*) there is higher mechanical efficiency ☐
(*c*) there is less chances of damage of the compressor ☐
(*d*) there is low first cost of motor and compressor ☐
(*e*) all of the above ☐
(*f*) none of the above. ☐

126. Undercooling in a refrigeration cycle

(*a*) decreases the net refrigerating effect ☐
(*b*) increases the net refrigerating effect ☐
(*c*) has no effect on net refrigerating effect ☐
(*d*) first increases then decreases the net refrigerating effect. ☐

127. The coefficient of performance of a vapour compression refrigerating cycle using expansion cylinder instead of throttle valve is

(*a*) higher ☐ (*b*) lower ☐
(*c*) same ☐ (*d*) none of the above. ☐

128. Undercooling is the process of cooling the refrigerant

(*a*) after compression ☐
(*b*) before compression ☐
(*c*) after condensation before throttling ☐
(*d*) none of the above. ☐

129. The main parts of a vapour compression refrigerating in sequence are

(*a*) compressor, condenser, evaporator and throttle valve ☐
(*b*) compressor, throttle valve, condenser and evaporator ☐
(*c*) compressor, evaporator, condenser and throttle valve ☐
(*d*) compressor, condenser, throttle valve and evaporator. ☐

130. If a refrigerant is having low specific heat, then coefficient of performance will be

(*a*) lower ☐ (*b*) higher ☐
(*c*) same ☐ (*d*) none of the above. ☐

131. The net refrigerating effect for a refrigerant is

(*a*) inversely proportional to latent heat of refrigerant ☐
(*b*) directly proportional to latent heat of refrigerant ☐
(*c*) directly proportional to the square of latent heat ☐
(*d*) none of the above. ☐

132. A good refrigerant should be of

(*a*) low specific heat ☐ (*b*) high latent heat ☐

(*c*) high critical pressure ☐ (*d*) high thermal conductivity ☐

(*e*) all of the above ☐ (*f*) none of the above. ☐

133. The evaporating pressure of a refrigerant should be

(*a*) equal to atmospheric pressure ☐

(*b*) lower than atmospheric pressure ☐

(*c*) higher than atmospheric pressure ☐

(*d*) none of the above. ☐

134. The specific volume of a refrigerant should be

(*a*) high ☐ (*b*) low ☐

(*c*) normal ☐ (*d*) none of the above. ☐

135. Which of the following refrigerant has the lowest specific volume ?

(*a*) ammonia ☐ (*b*) carbon dioxide ☐

(*c*) Freon-12 ☐ (*d*) Freon-22. ☐

136. Which of the following refrigerant has the highest latent heat ?

(*a*) ammonia ☐ (*b*) carbon dioxide ☐

(*c*) Freon-12 ☐ (*d*) Freon-22. ☐

137. The refrigerant, which has the lowest boiling point, is

(*a*) Freon-12 ☐ (*b*) Freon-22 ☐

(*c*) carbon dioxide ☐ (*d*) ammonia. ☐

138. In vapour absorption refrigerant

(*a*) a vapour compressor is used ☐

(*b*) a liquid pump is used ☐

(*c*) both a vapour compressor and a liquid pump is used ☐

(*d*) none of the above. ☐

139. The C.O.P. of the vapour absorption cycle as compared to vapour compression cycle is

(*a*) higher ☐ (*b*) lower ☐

(*c*) same ☐ (*d*) none of the above. ☐

140. The C.O.P. of the vapour absorption cycle can be increased by using

(*a*) heat exchangers ☐ (*b*) vapour compressor ☐

(*c*) both (*a*) and (*b*) ☐ (*d*) none of the above. ☐

141. Electrolux refrigerator is a

(*a*) vapour compression refrigerator ☐

(*b*) vapour absorption refrigerator without a liquid pump ☐

(*c*) vapour absorption refrigerator without any pump ☐

(*d*) none of the above. ☐

142. In electrolux refrigerator

(*a*) hydrogen evaporates in ammonia ☐ (*b*) ammonia evaporates in hydrogen ☐
(*c*) ammonia is absorbed in hydrogen ☐ (*d*) ammonia is absorbed in water. ☐

143. In electrolux refrigerator

(*a*) there is no noise ☐ (*b*) there is no moving machinery ☐
(*c*) no electricity is required ☐ (*d*) all of the above ☐
(*e*) none of the above. ☐

144. The refrigerating system, used for most of the domestic refrigerators, is of the type of

(*a*) vapour absorption ☐ (*b*) electrolux refrigerator ☐
(*c*) vapour compression ☐ (*d*) air refrigeration cycle. ☐

Questions on Refrigerant

145. The widely used refrigerant in domestic refrigerators, is

(*a*) carbon dioxide ☐ (*b*) ammonia ☐
(*c*) Freon-12 ☐ (*d*) sulphur dioxide. ☐

146. The refrigerant usually used for refrigeration in aeroplanes is

(*a*) carbon dioxide ☐ (*b*) ammonia ☐
(*c*) Freon-12 ☐ (*d*) air. ☐

147. The refrigerant used for steam jet refrigeration is

(*a*) carbon dioxide ☐ (*b*) ammonia ☐
(*c*) Freon-12 ☐ (*d*) water. ☐

148. The refrigerants, in present practice, are identified by numbers. The refrigerant number R-717 means

(*a*) carbon dioxide ☐ (*b*) ammonia ☐
(*c*) sulphur dioxide ☐ (*d*) methyl chloride. ☐

149. The refrigerant number R-744 means

(*a*) carbon dioxide ☐ (*b*) ammonia ☐
(*c*) sulphur dioxide ☐ (*d*) methyl chloride. ☐

150. The refrigerant number R-764 means

(*a*) carbon dioxide ☐ (*b*) ammonia ☐
(*c*) sulphur dioxide ☐ (*d*) methyl chloride. ☐

151. The refrigerant number R-40 means

(*a*) carbon dioxide ☐ (*b*) ammonia ☐
(*c*) sulphur dioxide ☐ (*d*) methyl chloride. ☐

152. The refrigerants (ammonia, carbon dioxide, Freon-22 and Freon-12) in order of decreasing boiling point are

(*a*) ammonia, CO_2, Freon-12, Freon-22 ☐ (*b*) ammonia, Freon-12, Freon-22, CO_2 ☐
(*c*) Freon-12, ammonia, Freon-22, CO_2 ☐ (*d*) Freon-12, Freon-22, ammonia, CO_2. ☐

153. The boiling point of Freon-12 is

(*a*) –40.8°C ☐ (*b*) –29.8°C ☐
(*c*) –33.3°C ☐ (*d*) –78.5°C. ☐

154. The boiling point of Freon-22 is

(*a*) –40.8°C ☐ (*b*) –29.8°C ☐
(*c*) –33.3°C ☐ (*d*) –78.5°C. ☐

155. The boiling point of carbon dioxide as compared to ammonia is

(*a*) higher ☐ (*b*) lower ☐
(*c*) same ☐ (*d*) none of the above. ☐

156. The refrigerants (NH_3, CO_2, F-22, F-12) in order of decreasing freezing point are

(*a*) NH_3, CO_2, F-22, F-12 ☐ (*b*) NH_3, CO_2, F-12, F-22 ☐
(*c*) CO_2, NH_3, F-22, F-12 ☐ (*d*) F-12, F-22, NH_3, CO_2. ☐

157. The freezing point of ammonia is

(*a*) –77.8°C ☐ (*b*) –157.8°C ☐
(*c*) –160°C ☐ (*d*) –110°C. ☐

158. The freezing point of carbon dioxide is

(*a*) –77.8°C ☐ (*b*) –157.8°C ☐
(*c*) –160°C ☐ (*d*) –110°C. ☐

159. For large commercial installations, ammonia is used as a refrigerant because

(*a*) it has large, latent heat ☐ (*b*) it is relatively cheap ☐
(*c*) it has moderate working pressure ☐ (*d*) all of the above ☐
(*e*) none of the above. ☐

160. For domestic refrigerators, Freon-12 is mostly used as a refrigerators because

(*a*) it has low specific volume ☐ (*b*) it is non-toxic ☐
(*c*) it is colourless and odourless ☐ (*d*) all of the above ☐
(*e*) none of the above. ☐

161. Freon-12, as compared to ammonia, is having

(*a*) less cost ☐ (*b*) high cost ☐
(*c*) same cost ☐ (*d*) none of the above. ☐

162. The refrigerants (NH_3, CO_2, F-22, F-12) in order of decreasing specific volume are

(*a*) NH_3, CO_2, F-22, F-12 ☐ (*b*) F-12, F-22, CO_2, NH_3 ☐
(*c*) NH_3, F-12, F-22, CO_2 ☐ (*d*) CO_2, NH_3, F-12, F-22. ☐

163. The refrigerants with lowest specific volume is

(*a*) NH_3 ☐ (*b*) CO_2 ☐
(*c*) F-12 ☐ (*d*) F-22. ☐

164. For ships refrigerations where space is a vital consideration, the refrigerant used is

(*a*) NH_3 ☐ (*b*) CO_2 ☐
(*c*) F-22 ☐ (*d*) F-12. ☐

165. Which of the refrigerants is having lowest working pressure?

(*a*) CO_2 ☐ (*b*) SO_2 ☐
(*c*) NH_3 ☐ (*d*) F-12. ☐

166. Which of the refrigerants is having highest working pressure?

(*a*) CO_2 ☐ (*b*) SO_2 ☐
(*c*) NH_3 ☐ (*d*) F-12. ☐

167. Which of the refrigerant is highly toxic?

(*a*) CO_2 ☐ (*b*) SO_2 ☐
(*c*) NH_3 ☐ (*d*) F-12. ☐

168. The evaporator and condenser pressure should be

(*a*) below atmospheric pressure ☐ (*b*) above atmospheric pressure ☐
(*c*) at the atmospheric pressure ☐ (*d*) none of the above. ☐

169. Freon-11 has the evaporator pressure

(*a*) less than atmospheric pressure ☐ (*b*) more than atmospheric pressure ☐
(*c*) equal to atmospheric pressure ☐ (*d*) none of the above. ☐

170. Critical temperature is the temperature above which

(*a*) a gas will immediately liquefy ☐ (*b*) a gas will never liquefy ☐
(*c*) water will evaporate ☐ (*d*) none of the above. ☐

171. If the critical temperature of a refrigerant is very near the condensing temperature, the power requirements are

(*a*) small ☐ (*b*) large ☐
(*c*) same ☐ (*d*) none of the above. ☐

172. The critical temperature of ammonia as compared to that of Freon-22 is

(*a*) low ☐ (*b*) high ☐
(*c*) same ☐ (*d*) none of the above. ☐

173. A refrigerant should have the freezing temperature

(*a*) below –35°C ☐ (*b*) above –35°C ☐
(*c*) at –35°C ☐ (*d*) none of the above. ☐

174. Power requirements of a refrigerator are

(*a*) directly proportional to C.O.P. ☐
(*b*) inversely proportional to C.O.P. ☐
(*c*) directly proportional to the square root of C.O.P. ☐
(*d*) inversely proportional to the square root of C.O.P. ☐

175. Carbon dioxide, as compared to other refrigerants, has

(*a*) lowest C.O.P. ☐ (*b*) highest C.O.P. ☐
(*c*) same C.O.P. ☐ (*d*) none of the above. ☐

176. The latent heat of vaporization is maximum in case of

(*a*) CO_2 ☐ (*b*) NH_3 ☐
(*c*) Freon-22 ☐ (*d*) Freon-12. ☐

177. The liquid and vapour refrigerants should have viscosity

(*a*) low ☐ (*b*) high ☐
(*c*) normal ☐ (*d*) none of the above. ☐

178. Choose the wrong statement

(*a*) Ammonia attacks copper and copper alloys. ☐
(*b*) Ammonia is non-corrosive to iron and steel. ☐
(*c*) Freon groups do not attack any metals. ☐
(*d*) CO_2 is having high specific volume as compared to ammonia. ☐

179. Choose the correct statement

(*a*) Vapour compression cycle requires a liquid pump. ☐
(*b*) Undercooling decreases the net refrigerating effect. ☐
(*c*) A good refrigerant should have low latent heat. ☐
(*d*) Electrolux refrigerator requires no electricity for its operation. ☐

180. The refrigerant, which is non-toxic and non-flammable, is

(*a*) Sulphur dioxide ☐ (*b*) Freon-12 ☐
(*c*) Ammonia ☐ (*d*) CO_2 ☐
(*e*) both (*b*) and (*d*) ☐ (*f*) both (*a*) and (*c*). ☐

181. The refrigerant, which is extremely toxic and non-flammable is

(*a*) Sulphur dioxide ☐ (*b*) Freon-12 ☐
(*c*) CO_2 ☐ (*d*) none of the above. ☐

182. The refrigerant which is colourless and odourless is

(*a*) Sulphur dioxide ☐ (*b*) CO_2 ☐
(*c*) Freon-12 ☐ (*d*) Ammonia. ☐

183. The refrigerant carbon dioxide is not used widely because of its

(*a*) high power requirements per tonne of refrigeration and high operating pressures. ☐
(*b*) low power requirements and high operating pressures. ☐
(*c*) low power requirements and low operating pressures. ☐
(*d*) high power requirements and low operating pressures. ☐

184. Freon-12 has operating pressures of about

(*a*) 15 kg/cm^2 ☐ (*b*) 30 kg/cm^2 ☐
(*c*) 2 kg/cm^2 ☐ (*d*) 8 kg/cm^2. ☐

185. The temperature of air at which condensation of moisture begins when the air is cooled is known as

(*a*) dry bulb temperature ☐ (*b*) dew point temperature ☐
(*c*) wet bulb temperature ☐ (*d*) saturation temperature. ☐

186. The temperature of air recorded by an ordinary thermometer is known as

(*a*) dry bulb temperature ☐ (*b*) dew point temperature ☐
(*c*) wet bulb temperature ☐ (*d*) saturation temperature. ☐

187. Humidity ratio is defined as the ratio of

(*a*) mass or water vapour to the mass of air-water vapour mixture □

(*b*) mass of water vapour to the mass of dry air in a given mixture of air-water vapour mixture □

(*c*) mass of dry air to the mass of air water vapour mixture □

(*d*) mass of dry air to mass of water vapour in an air-water vapour mixture. □

188. Humidity ratio is also called

(*a*) relative humidity □ (*b*) specific humidity □

(*c*) absolute humidity □ (*d*) none of the above. □

189. If m_a = mass of dry air, and m_s = mass of water vapour in a air-water vapour mixture, then humidity ratio is given by

(*a*) $\frac{m_s}{m_a + m_s}$ □ (*b*) $\frac{m_a}{m_a + m_s}$ □

(*c*) $\frac{m_s}{m_a}$ □ (*d*) $\frac{m_a}{m_s}$. □

190. Degree of saturation is defined as the ratio of

(*a*) specific humidity of moist air to the specific humidity of the saturated air at the same temperature □

(*b*) mass of water vapour associated with unit mass of dry air to mass of water vapour associated with unit mass of dry air saturated at the same temperature □

(*c*) any one of (*a*) and (*b*) □

(*d*) none of the above. □

191. The humidity ratio in terms of specific volume of dry air (v_a) and specific volume of water vapour (v_s) is given by

(*a*) $\frac{v_s}{v_a}$ □ (*b*) $\frac{v_s}{(v_a + v_s)}$ □

(*c*) $\frac{v_a}{v_s}$ □ (*d*) $\frac{v_a}{(v_a + v_s)}$. □

Psychrometry

192. The science which deals with the behaviour of mixture of air and water vapour is called

(*a*) psychrometry □ (*b*) airconditioning □

(*c*) hygrometry □ (*d*) any one of (*a*) and (*c*) □

(*e*) none of the above. □

193. The humidity ratio, mathematically, is equal to

(*a*) $0.622 \frac{p_s}{(p - p_s)}$ □ (*b*) $\frac{0.622\,(p - p_s)}{p_s}$ □

(*c*) $0.622 \frac{p}{p_s}$ □ (*d*) $\frac{0.622 \times p}{(p - p_s)}$. □

where p_s = Partial pressure of superheated water vapour, and

p = Barometric pressure of mixture.

194. The relative humidity of air is defined as the ratio of

(a) mass of water vapour in a given volume to the total mass of the mixture of air and water vapour ☐

(b) mass of water vapour in a given volume to the mass of water vapour if saturated at the same temperature ☐

(c) mass of water vapour in a given volume to the mass of air ☐

(d) mass of air to the mass of water vapour in the mixture of air and water vapour. ☐

195. The relative humidity, mathematically, is equal to

(a) $\frac{p_s}{p_g}$ ☐ (b) $\frac{v_g}{v_s}$ ☐

(c) $\frac{p_g}{p_s}$ ☐ (d) any one of (a) and (b). ☐

where p_s = Actual partial pressure of superheated water vapour,

p_g = Pressure at saturation corresponding to dry bulb temperature,

v_g = Specific volume of vapour corresponding to saturated temperature, and

v_s = Specific volume of vapour corresponding to pressure p_s.

196. The relation between relative humidity (ϕ) and degree of saturation (μ) is given by

(a) $\mu = \left(\frac{p - p_g}{p - p_s}\right)\phi$ ☐ (b) $\mu = \phi\left(\frac{p - p_s}{p - p_g}\right)$ ☐

(c) $\mu = \frac{\phi \times p_g}{(p - p_s)}$ ☐ (d) $\mu = \phi\left[\frac{p + p_s}{p - p_g}\right]$ ☐

197. The pressure at saturation corresponding to dry bulb temperature is 0.08 kgf/cm^2 and the vapour pressure of water in the mixture of air and water vapour is 0.032 kgf/cm^3. If the barometric pressure of mixture is 1.032 kgf/cm^2, then relative humidity will be

(a) 50% ☐ (b) 40% ☐

(c) 30% ☐ (d) 20%. ☐

198. In question 197, the humidity ratio would be

(a) 0.02 ☐ (b) 0.03 ☐

(c) 0.035 ☐ (d) 0.04. ☐

199. In question 197, the degree of saturation will be

(a) 0.50 ☐ (b) 0.4 ☐

(c) 0.38 ☐ (d) 0.32. ☐

200. Absolute humidity is defined as

(a) actual mass of air in a unit volume of air and water vapour mixture ☐

(b) actual mass of water vapour in a unit volume of air ☐

(c) actual mass of water vapour in a unit volume of air and water vapour mixture ☐

(d) actual mass of air in a unit volume of water vapour. ☐

201. Wet bulb depression is equal to
(*a*) sum of dry bulb temperature and wet bulb temperature ☐
(*b*) average of dry bulb temperature and wet bulb temperature ☐
(*c*) difference of dry bulb temperature and wet bulb temperature ☐
(*d*) product of dry bulb and wet bulb temperatures. ☐

202. If the relative humidity is high, the rate of evaporation of water will be
(*a*) high ☐ (*b*) low ☐
(*c*) zero ☐ (*d*) none of the above. ☐

203. The wet bulb depression is zero when relative humidity is equal to
(*a*) zero ☐ (*b*) 0.5 ☐
(*c*) 0.75 ☐ (*d*) 1.0. ☐

Psychrometric Chart

204. In a psychrometric chart, the horizontal scale shows
(*a*) wet-bulb temperature ☐ (*b*) dry bulb temperature ☐
(*c*) adiabatic saturation temperature ☐ (*d*) specific humidity. ☐

205. In a psychrometric chart, the vertical scale shows
(*a*) wet-bulb temperature ☐ (*b*) dry bulb temperature ☐
(*c*) adiabatic saturation temperature ☐ (*d*) specific humidity. ☐

206. In a psychrometric chart, wet bulb temperature lines are
(*a*) curved ☐
(*b*) inclined and straight but non-uniformly spaced ☐
(*c*) horizontal and non-uniformly spaced ☐
(*d*) horizontal and uniformly spaced. ☐

207. In a psychrometric chart, relative humidity lines are
(*a*) curved ☐
(*b*) inclined and straight but non-uniformly spaced ☐
(*c*) horizontal and non-uniformly spaced ☐
(*d*) horizontal and uniformly spaced. ☐

208. In a psychrometric chart, dew point temperature lines are
(*a*) curved ☐
(*b*) inclined and straight but non-uniformly spaced ☐
(*c*) horizontal and non-uniformly spaced ☐
(*d*) horizontal and uniformly spaced. ☐

Psychrometric Processes

209. Sensible heat factor (SHF) is defined as the ratio of
(*a*) latent heat to sensible heat ☐ (*b*) sensible heat to latent heat ☐
(*c*) latent heat to total heat ☐ (*d*) sensible heat to total heat. ☐

210. Which of the following is a psychrometric process?

(*a*) humidification ☐ (*b*) dehumidification ☐
(*c*) sensible heating ☐ (*d*) adiabatic mixing of streams ☐
(*e*) all of the above ☐ (*f*) none of the above. ☐

211. The process of adding heat to moist air at the same humidity ratio is known as

(*a*) sensible heating ☐ (*b*) sensible cooling ☐
(*c*) humidification ☐ (*d*) dehumidification. ☐

212. The process of cooling of air at the same humidity ratio is known as

(*a*) sensible heating ☐ (*b*) sensible cooling ☐
(*c*) humidification ☐ (*d*) dehumidification. ☐

213. The process of adding moisture to the air at constant dry bulb temperature is known as

(*a*) sensible heating ☐ (*b*) sensible cooling ☐
(*c*) humidification ☐ (*d*) dehumidification. ☐

214. The process of removing moisture from air at constant dry bulb temperature is known as

(*a*) sensible heating ☐ (*b*) sensible cooling ☐
(*c*) humidification ☐ (*d*) dehumidification. ☐

215. For sensible heating, the air is passed over a heating coils, which may be

(*a*) electric resistance coils ☐ (*b*) hot water passed through coils ☐
(*c*) steam passed through coils ☐ (*d*) any one of the above. ☐

216. In case of sensible heating of air, the bypass factor (BF) is equal to

(*a*) $\dfrac{t_{d_2} - t_{d_1}}{t_{d_3} - t_{d_1}}$ ☐ (*b*) $\dfrac{t_{d_3} - t_{d_2}}{t_{d_3} - t_{d_1}}$ ☐

(*c*) $\dfrac{t_{d_3} - t_{d_1}}{t_{d_2} - t_{d_1}}$ ☐ (*d*) $\dfrac{t_{d_3} - t_{d_2}}{t_{d_2} - t_{d_1}}$. ☐

where t_{d_1} = Dry bulb temperature of air entering the heating coil,

t_{d_2} = Dry bulb temperature of air leaving the heating coil, and

t_{d_3} = Temperature of heating coil.

217. In case of sensible cooling of air, the bypass factor is equal to

(*a*) $\dfrac{t_{d_2} - t_{d_1}}{t_{d_1} - t_{d_3}}$ ☐ (*b*) $\dfrac{t_{d_1} - t_{d_2}}{t_{d_2} - t_{d_3}}$ ☐

(*c*) $\dfrac{t_{d_3} - t_{d_1}}{t_{d_2} - t_{d_3}}$ ☐ (*d*) $\dfrac{t_{d_2} - t_{d_3}}{t_{d_1} - t_{d_2}}$. ☐

where t_{d_1} = Dry bulb temperature of air entering the cooling coil,

t_{d_2} = Dry bulb temperature of air leaving the cooling coil, and

t_{d_3} = Temperature of cooling coil.

218. Figure 7.13 shows some psychrometric process. Curve 1-2 represents

(*a*) sensible cooling ☐ (*b*) humidification ☐
(*c*) dehumidification ☐ (*d*) sensible heating. ☐

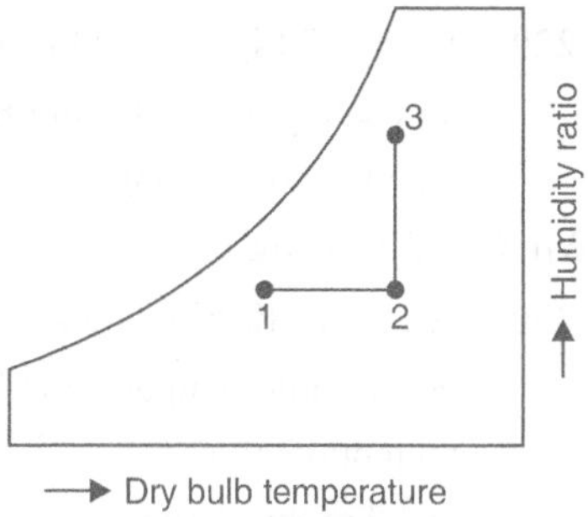

219. In Fig. 7.13 curve 2-3 represents

(*a*) sensible cooling ☐
(*b*) humidification ☐
(*c*) dehumidification ☐
(*d*) sensible heating. ☐

FIGURE 7.13

220. In Fig. 7.13, curve 2-1 represents

(*a*) sensible cooling ☐
(*b*) humidification ☐
(*c*) dehumidification ☐
(*d*) sensible heating. ☐

221. In Fig. 7.13 curve 3-2 represents

(*a*) sensible cooling ☐ (*b*) humidification ☐
(*c*) dehumidification ☐ (*d*) sensible heating. ☐

222. In sensible cooling of air, the relative humidity

(*a*) is constant ☐ (*b*) decreases ☐
(*c*) increases ☐ (*d*) first increases then decreases. ☐

223. In sensible cooling of air, the wet bulb temperature

(*a*) is constant ☐ (*b*) increases ☐
(*c*) decreases ☐ (*d*) first increases then decreases. ☐

224. During humidification process, humidity ratio

(*a*) is constant ☐ (*b*) increases ☐
(*c*) decreases ☐ (*d*) first increases then decreases. ☐

225. During humidification process, relative humidity

(*a*) is constant ☐ (*b*) increases ☐
(*c*) decreases ☐ (*d*) first increases then decreases. ☐

226. Figure 7.14 shows four curves. Curve *A* represents

(*a*) cooling and dehumidification ☐ (*b*) heating and humidification ☐
(*c*) adiabatic cooling ☐ (*d*) chemical dehumidification. ☐

227. In Fig. 7.14, curve *B* represents

(*a*) cooling and dehumidification ☐
(*b*) heating and dehumidification ☐
(*c*) adiabatic cooling ☐
(*d*) chemical dehumidification. ☐

FIGURE 7.14

228. in Fig. 7.14, curve *C* represents

(*a*) cooling and dehumidification ☐
(*b*) heating and humidification ☐
(*c*) adiabatic cooling ☐
(*d*) chemical dehumidification. ☐

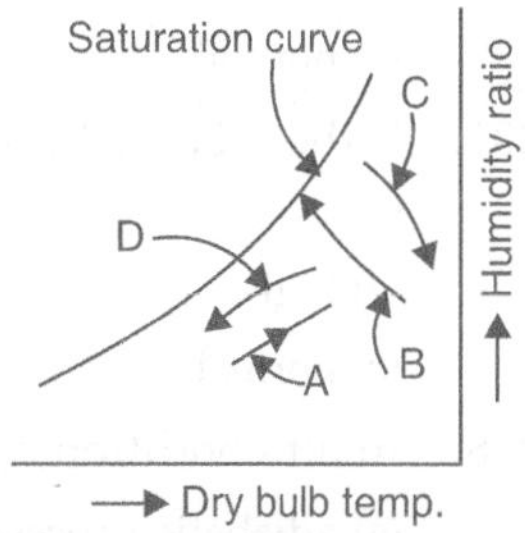

229. In Fig. 7.14, curve *D* represents

(*a*) cooling and dehumidification ☐ (*b*) heating and humidification ☐
(*c*) adiabatic cooling ☐ (*d*) chemical dehumidification. ☐

Air Conditioning

230. Air conditioning means control of

(*a*) dry bulb temperature ☐ (*b*) humidity ☐
(*c*) motion of air ☐ (*d*) purity of air ☐
(*e*) all of the above ☐ (*f*) none of the above. ☐

231. Choose the correct statement

(*a*) During heating and humidification process, humidity ratio decreases. ☐
(*b*) During cooling and dehumidification process, humidity ratio increases. ☐
(*c*) During cooling and dehumidification process, dry bulb temperature increases. ☐
(*d*) During heating and humidification process, dry bulb temperature increases. ☐

232. Choose the wrong statement

(*a*) The humidity ratio is constant during sensible heating of air. ☐
(*b*) The relative humidity increases during sensible cooling of air. ☐
(*c*) Dry bulb temperature is constant during humidification process. ☐
(*d*) As air gets wet, the relative humidity increases. ☐

233. In summer air conditioning, the process used is known as

(*a*) heating and humidification ☐ (*b*) dehumidification ☐
(*c*) cooling and dehumidification ☐ (*d*) humidification. ☐

234. In winter air conditioning, the process used is known as

(*a*) heating and humidification ☐ (*b*) dehumidification ☐
(*c*) cooling and dehumidification ☐ (*d*) humidification. ☐

235. The atmospheric air at dry bulb temperature 14°C enters a heating coil maintained at 38°C. If the air leaves the heating coil at 26°C, the by-pass factor of the heating coil would be

(*a*) 0.4 ☐ (*b*) 0.5 ☐
(*c*) 0.6 ☐ (*d*) 0.7. ☐

236. If the relative humidity is low, the rate of evaporation of water will be

(*a*) high ☐ (*b*) low ☐
(*c*) normal ☐ (*d*) zero. ☐

237. If the relative humidity of atmosphere is equal to one then rate of evaporation of water will be

(*a*) high ☐ (*b*) low ☐
(*c*) normal ☐ (*d*) zero. ☐

238. In air conditioning, the mixing of two or more steam of moist air follows

(*a*) adiabatic process ☐ (*b*) isothermal process ☐
(*c*) polytropic process ☐ (*d*) constant pressure process. ☐

239. The coil efficiency, in case of sensible cooling of air, is equal to

(*a*) B.F. + 1 ☐ (*b*) B.F. – 1 ☐

(*c*) 1 – B.F. ☐ (*d*) 1/B.F. ☐

where B.F. = Bypass factor.

240. If sensible heat added is 10 kcal/sec and latent heat added is 5 kcal/sec, then sensible heat factor is equal to

(*a*) 0.5 ☐ (*b*) 0.75 ☐

(*c*) 0.67 ☐ (*d*) 0.40. ☐

241. The heat that is added or removed from a substance without a change of state is known as

(*a*) sensible heat ☐ (*b*) latent heat ☐

(*c*) total heat ☐ (*d*) none of the above. ☐

242. The moisture content of the air is indicated by

(*a*) dry bulb temperature ☐ (*b*) wet bulb temperature ☐

(*c*) dew point temperature ☐ (*d*) adiabatic saturation temperature ☐

(*e*) both (*b*) and (*c*). ☐

243. Air is dehumidified by

(*a*) cooling ☐ (*b*) heating ☐

(*c*) injecting water ☐ (*d*) injecting steam. ☐

244. During adiabatic saturation process on unsaturated air, the parameter which remains constant is

(*a*) dry bulb temperature ☐ (*b*) dew point temperature ☐

(*c*) wet bulb temperature ☐ (*d*) relative humidity. ☐

245. If the relative humidity is 100%, it means

(*a*) dry bulb temperature, wet bulb temperature, dew point temperature and saturation temperature are equal. ☐

(*b*) dry bulb temperature equals wet bulb temperature. ☐

(*c*) saturation temperature equals wet bulb temperature. ☐

(*d*) wet bulb temperature equals dew point temperature. ☐

246. Which of the following is a correct statement ?

(*a*) Wet bulb temperature and adiabatic saturation temperature are thermodynamic properties. ☐

(*b*) Wet bulb temperature and adiabatic saturation temperature are not thermodynamic properties. ☐

(*c*) Wet bulb temperature is a thermodynamic property but adiabatic saturation temperature is not. ☐

(*d*) Adiabatic saturation temperature is a thermodynamic property but wet bulb temperature is not. ☐

247. Dew point temperature is equal to

(*a*) wet bulb temperature ☐

(*b*) dry bulb temperature ☐

(*c*) difference of dry bulb temperature and wet bulb temperature ☐

(*d*) the saturation temperature of water at vapour at its partial pressure. ☐

248. If pressure at saturation is small in comparison to barometric pressure, then relative humidity is

(*a*) less than degree of saturation ☐ (*b*) equal to degree of saturation ☐
(*c*) more than degree of saturation ☐ (*d*) none of the above. ☐

249. Which one of the following is a correct statement for sensible heating and sensible cooling ?

(*a*) In sensible heating process the dry bulb temperature increases but enthalpy of moist air decreases. ☐
(*b*) In sensible process dry bulb temperature decreases but enthalpy of moist air increases. ☐
(*c*) In sensible heating process both dry bulb temperature and enthalpy of moist air increases. ☐
(*d*) In sensible heating process both dry bulb temperature and enthalpy decreases. ☐

250. Choose the wrong statement

(*a*) Horizontal line on psychromatic chart represents sensible heating or cooling process. ☐
(*b*) Boiling temperature of CO_2 is –78.5°C. ☐
(*c*) Curved lines on psychromatic chart represents relative humidity lines. ☐
(*d*) Relative humidity is the ratio of actual humidity to absolute humidity. ☐

Tick mark the true or false statement:

251. If T_1 = lowest absolute temperature, and T_2 = highest absolute temperature, then C.O.P., of a refrigerator working on a reversed Carnot cycle is $\pi/(T_2 - T_1)$.

(*a*) True ☐ (*b*) false. ☐

252. One tonne of refrigeration means that the heat removing capacity is 100 kcal/minute.

(*a*) True ☐ (*b*) false. ☐

253. The pressure above which a liquid will remain a liquid, is known as critical pressure of the liquid.

(*a*) True ☐ (*b*) false. ☐

254. The reciprocal of the efficiency of a heat engine is known as the co-efficient of performance.

(*a*) True ☐ (*b*) False. ☐

255. The expression $\rho VL/\mu$, is known as Reynolds number (where ρ = density, V = velocity, L = linear dimension, and μ = viscosity).

(*a*) True ☐ (*b*) False. ☐

256. The heat transfer in case of solids, takes place according to convection.

(*a*) True ☐ (*b*) False. ☐

257. The heat transfer takes place according to first law of thermodynamics.

(*a*) True ☐ (*b*) False. ☐

258. With the rise in temperature, thermal conductivity of air decreases.

(*a*) True ☐ (*b*) False. ☐

259. Superheating in a refrigeration cycle decreases C.O.P.

(*a*) True ☐ (*b*) False. ☐

260. Vertical lines on pressure-enthalpy chart show the constant total heat lines.

(*a*) True ☐ (*b*) False. ☐

261. Critical temperature is the temperature above which a gas will never liquify.
(*a*) True ☐ (*b*) False. ☐

262. If p_v = partial pressure of water vapour in the air and p_s is the saturation pressure of water vapour at same temperature, then the ratio p_s/p_v is known as relative humidity.
(*a*) True ☐ (*b*) False. ☐

263. The ratio of the thermal conductivity of insulating material to heat transfer coefficient at outer surface of a cylindrical pipe is known as critical radius of insulation for the cylindrical pipe.
(*a*) True ☐ (*b*) False. ☐

264. The thermal conductivity is expressed as kcal-m/hr.
(*a*) True ☐ (*b*) False. ☐

265. With the rise in temperature normally, the thermal conductivity of solids increases.
(*a*) True ☐ (*b*) False. ☐

266. With the presence of moisture, the insulation ability of an insulator would decrease.
(*a*) True ☐ (*b*) False. ☐

267. Radiation pyrometer are used for measuring the temperature of steam at around 550°C.
(*a*) True ☐ (*b*) False. ☐

268. If the absorptivity of a body does not vary with temperature and wavelength of the incident ray, the body is known as grey.
(*a*) True ☐ (*b*) False. ☐

269. Two times the thermal conductivity of insulating material divided by heat transfer coefficient at outer surface of a spherical shell gives the critical radius of insulation for the spherical shell.
(*a*) True ☐ (*b*) False. ☐

270. Clapeyron equation is a relation between temperature and enthalpy.
(*a*) True ☐ (*b*) False. ☐

271. The latent heat of vaporisation of a refrigerant increases at lower temperature and pressures.
(*a*) True ☐ (*b*) False. ☐

272. In a refrigeration cycle, C.O.P. increases if under-cooling is done.
(*a*) True ☐ (*b*) False. ☐

273. If the vapour removes sensible heat from the refrigerant, then sub-cooling occurs.
(*a*) True ☐ (*b*) False. ☐

274. The liquid refrigerant is changed into the compressor suction or discharge valve ports.
(*a*) True ☐ (*b*) False. ☐

275. The heat required to increase the temperature of a liquid or vapour is known as latent heat.
(*a*) True ☐ (*b*) False. ☐

276. With rise in temperature, the thermal conductivity of air increases.
(*a*) True ☐ (*b*) False. ☐

277. The word L.M.T.D. in case of heat transfer stands for Log mean temperature difference.
(*a*) True ☐ (*b*) False. ☐

278. In case of counter flow heat exchanger, the L.M.T.D. is lower as compared to parallel flow heat exchanger.
(*a*) True ☐ (*b*) False. ☐

279. In heat transfer problems, the concept of overall coefficient of heat transfer is used.
(*a*) True ☐ (*b*) False. ☐

280. Air refrigerator operates on Rankine cycle.
(*a*) True ☐ (*b*) False. ☐

281. In absorption type of refrigeration system, aqua ammonia is used.
(*a*) True ☐ (*b*) False. ☐

282. The ratio of work done by compressor to the refrigerant effect is known C.O.P.
(*a*) True ☐ (*b*) False. ☐

283. The heat flow for one-dimensional cases, is given by Fourier's law of heat conduction.
(*a*) True ☐ (*b*) False. ☐

284. The heat conducted in unit time across unit area through unit thickness when a temperature difference of unity is maintained is known as thermal conductivity.
(*a*) True ☐ (*b*) False. ☐

285. The heat transfer by conduction through a thick sphere and through a thick cylinder is same.
(*a*) True ☐ (*b*) False. ☐

286. Heat exchanger is a device used for transferring heat from one fluid to another fluid.
(*a*) True ☐ (*b*) False. ☐

287. Both the fluids at inlet are in their hottest state in counter-current flow heat exchanger.
(*a*) True ☐ (*b*) False. ☐

288. The ratio of the thickness of film of fluid to the thermal conductivity is known as film coefficient,
(*a*) True ☐ (*b*) False. ☐

289. The emissivity and absorptivity are same, at thermal equilibrium.
(*a*) True ☐ (*d*) False. ☐

290. Emissivity is the ratio of emissive power of a body to the emissive power of a perfectly black body,
(*a*) True ☐ (*b*) False. ☐

291. If h = thermal diffusivity, k = thermal conductivity, ρ = density and S = specific heat, then $k/\rho S$ represents the thermal diffusivity.
(*a*) True ☐ (*b*) False. ☐

292. The heat of sun reaches to us according to convection.
(*a*) True ☐ (*b*) False. ☐

293. The amount of heat flow through a body by conduction is inversely proportional to the thickness of the body.

(*a*) True ☐ (*b*) False. ☐

294. The product of actual C.O.P. and theoretical C.O.P. is known as relative co-efficient of performance,

(*a*) True ☐ (*b*) False. ☐

295. The coefficient of performance is always less than one.

(*a*) True ☐ (*b*) False. ☐

296. Heat is rejected by a refrigerant in a condenser during a refrigeration cycle.

(*a*) True ☐ (*b*) False. ☐

297. A refrigerant should have high critical temperature.

(*a*) True ☐ (*b*) False. ☐

298. Freon-II has the highest critical point temperature.

(*a*) True ☐ (*b*) False. ☐

299. During gas charging in an ammonia vapour compression system, the connection is made at the compressor outlet.

(*a*) True ☐ (*b*) False. ☐

300. Ammonia evaporates in hydrogen in electrolux refrigerator.

(*a*) True ☐ (*b*) False. ☐

301. If p_v = partial pressure of water vapour, p_a = partial pressure of dry air, and p_b = barometric pressure, then according to Dalton's law of partial pressures, $p_a = p_b + p_v$.

(*a*) True ☐ (*b*) False. ☐

302. Wet bulb temperature is the temperature of air recorded by a thermometer, when its bulb is surrounded by a wet cloth exposed to the air.

(*a*) True ☐ (*b*) False. ☐

303. The process of removing water vapour from the surrounding air is known as dehydration.

(*a*) True ☐ (*b*) False. ☐

304. The dry bulb temperature increases, during heating and dehumidification process.

(*a*) True ☐ (*b*) False. ☐

305. If S.H. = Sensible heat, and L.H. = Latent heat, then the ratio of sensible heat to the sum of sensible heat and latent heat [*i.e.*, S.H./(S.H. + L.H.)] gives sensible heat factor.

(*a*) True ☐ (*b*) False. ☐

306. For a given dry bulb temperature, as the relative humidity decreases, the wet bulb temperature will increase.

(*a*) True ☐ (*b*) False. ☐

307. The dry bulb temperature, wet bulb temperature and dew point temperature will be different at 100% relative humidity.

(*a*) True ☐ (*b*) False. ☐

308. The ratio of mass of water vapour in air in a given volume at a given temperature to the mass of water vapour contained in the same volume at the same temperature when air is saturated, is known as relative humidity.

(*a*) True ☐ (*b*) False. ☐

309. Wet bulb temperature is the indication of amount of moisture in air.

(*a*) True ☐ (*b*) False. ☐

310. The process of changing vapour into liquid at constant temperature, is known as condensation.

(*a*) True ☐ (*b*) False. ☐

311. In a refrigeration cycle, the oil separator is installed between evaporator and compressor.

(*a*) True ☐ (*b*) False. ☐

312. In a refrigeration system, using ammonia, the leaks are detected by sulphur sticks which on detection gave white smoke.

(*a*) True ☐ (*b*) False. ☐

Fill in the blanks:

313. Critical temperature is the temperature .... which a gas will never liquid.

(*a*) below ☐ (*b*) above. ☐

314. If the heat removing capacity of a refrigerator is ...., then it is known as one tonne refrigerator.

(*a*) 50 kcal/min ☐ (*b*) 60 kcal/min. ☐

315. The unit of thermal conductivity is .... in S.I. units.

(*a*) W/m/K/s ☐ (*b*) W/m K. ☐

316. According to Stefan's law, the total radiation from a black body per second per unit area is directly proportional to .... of the absolute temperature.

(*a*) third power ☐ (*b*) fourth power. ☐

317. A body which absorbs heat radians of all wave lengths falling on it, it called ....

(*a*) a grey body ☐ (*b*) a black body. ☐

318. The rate of heat flow through a thick wall due to conduction is .... with the thickness of the wall.

(*a*) directly proportional ☐ (*b*) inversely proportional. ☐

319. In free convection the movement of the fluid particles are caused by .... only.

(*a*) pumps ☐ (*b*) temperature difference. ☐

320. Thermal conductivity of solid metals .... with the rise in temperature.

(*a*) increases ☐ (*b*) decreases. ☐

321. Logarithmic mean temperature difference is .... to the arithmetic mean temperature difference.

(*a*) equal ☐ (*b*) not equal. ☐

322. The term $\rho VL/\mu$ is known .... number.

(*a*) Prandtl ☐ (*b*) Reynold. ☐

323. If h = film co-efficient, L = linear dimension, and k = thermal conductivity then the term $h \times L/k$ is known .... number.

(*a*) Prandtl ☐ (*b*) Nusselt. ☐

324. The emissivity of a polished silver body is .... as compared to black body.

(*a*) very low ☐ (*b*) very high. ☐

325. A .... is one whose absorptivity does not vary with temperature and wavelength of the incident ray.

(*a*) black body ☐ (*b*) grey body. ☐

326. Emissivity is the ratio of emissive power of a body to emissive power of a ....

(*a*) white body ☐ (*b*) perfectly black body. ☐

327. The wet bulb indicates .... humidity of air.

(*a*) specific ☐ (*b*) relative. ☐

328. The relative humidity .... as air gets drier.

(*a*) decreases ☐ (*b*) increases. ☐

329. If *RH* = relative humidity, and *DBT* = dry bulb temperature then comfort conditions are .... in air conditioning.

(*a*) 25°C *DBT* and 80% *RH* ☐ (*b*) 22°C *DBT* and 60% *RH*. ☐

330. The sensible heat .... during humidification process.

(*a*) decreases ☐ (*b*) increases. ☐

331. During dehumidification process .... remains constant.

(*a*) relative humidity ☐ (*b*) dry bulb temperature. ☐

332. If H_1 = total heat of air entering the heating coil, H_2 = total heat of air leaving the heating coil, and H_3 = total heat of air at the end of humidification process, then the term $(H_2 - H_1)/(H_3 - H_1)$ represents .... during the heating and humidification process.

(*a*) latent heat factor ☐ (*b*) sensible heat factor. ☐

333. The specific humidity .... during humidification process.

(*a*) decreases ☐ (*b*) increases. ☐

334. Freon-12 has the .... freezing point.

(*a*) highest ☐ (*b*) lowest. ☐

335. The performance of a refrigerating machine is measured in terms of ...., which is the ratio of refrigerating effect to work input.

(*a*) efficiency ☐ (*b*) co-efficient of performance. ☐

336. The C.O.P. of a Carnot heat pump is .... than the C.O.P. of Carnot refrigerator by unity.

(*a*) less ☐ (*b*) more. ☐

337. The wet bulb temperature is same as dew point at .... relative humidity.

(*a*) 50% ☐ (*b*) 100%. ☐

338. As relative humidity decreases, the dew point will be .... than wet bulb temperature

(*a*) lower ☐ (*b*) higher. ☐

339. The lowest temperature occur in .... in a vapour compression cycle.

(*a*) evaporator ☐ (*b*) condenser. ☐

340. The temperature at which refrigerant gas becomes liquid is .... in a refrigerator.

(*a*) condensing temperature ☐ (*b*) critical temperature. ☐

341. The vapour pressure of refrigerant should be .... than atmospheric pressure

(*a*) higher ☐ (*b*) lower. ☐

342. If Δt_1 = temperature difference between the hot and cold fluids at inlet, and Δt_2 = temperature difference between hot and cold fluids at exit, then the term .... gives the L.M.T.D. for a heat exchanger.

(*a*) $\dfrac{\log(\Delta t_2 - \Delta t_1)}{\Delta t_2 - \Delta t_1}$ ☐ (*b*) $\dfrac{\Delta t_2 - \Delta t_1}{\log\left(\dfrac{\Delta t_2}{\Delta t_1}\right)}$. ☐

343. If A_1 = surface area of inner wall of a hollow sphere, A_2 = surface area of outer wall, r_1 = inner radius of sphere, r_2 = outer radius, t_1 = temperature of inner wall, and t_2 = temperature of outer wall, then the term .... represents the rate of heat flow.

(*a*) $K\sqrt{A_1A_2}\,\dfrac{t_1 - t_2}{r_2 - r_1}$ ☐ (*b*) $K\sqrt{r_1r_2} \times \dfrac{A_1A_2}{\sqrt{t_1t_2}}$. ☐

344. Nusselt number is a function of .... in free convection heat transfer.

(*a*) Reynold and Prandtl numbers ☐ (*b*) Grashof and Reynold number. ☐

345. According to ...., emissive power and absorptivity are constant for all bodies.

(*a*) Stefan's law ☐ (*b*) Kirchhoff's law. ☐

346. Thermal diffusivity of a substance is .... to thermal conductivity.

(*a*) inversely proportional ☐ (*b*) directly proportional. ☐

347. The ratio of .... is known as film co-efficient.

(*a*) equivalent thickness of film to thermal conductivity ☐

(*b*) thermal conductivity to equivalent thickness of film. ☐

348. According to ...., $Q = \rho AT^4$, where A = area, and T = absolute temperature

(*a*) Stefan's law ☐ (*b*) Newton's law. ☐

349. According to .... the wavelength corresponding to maximum energy is proportional to the absolute temperature.

(*a*) Stefan's law ☐ (*b*) Wein's law. ☐

350. If T_1 = lowest absolute temperature, and T_2 = Highest absolute temperature, then term ...... represents the C.O.P. of a refrigerator working on a reversed Carnot cycle.

(*a*) $\dfrac{T_1}{T_2 - T_1}$ ☐ (*b*) $\dfrac{T_2 - T_1}{T_1}$. ☐

III. OBJECTIVE TYPE QUESTIONS FROM COMPETITIVE EXAMINATIONS

Tick mark the most appropriate answer:

1. The Fourier's law of heat transfer by conduction is expressed as

(a) $Q = kA^2 \frac{dt}{dx}$ ☐ (b) $Q = -kA \frac{dt}{dx}$ ☐

(c) $Q = k^2 A \frac{dx}{dt}$ ☐ (d) $Q = k^3 A \frac{dx}{dt}$. ☐

where Q = Heat transfer rate, A = Surface area perpendicular to the direction of heat flow, dt = Temperature difference for short perpendicular distance dx, and d_2 = Thermal conductivity.

2. The ratio of heat transfer by convection to that by conduction is called

(a) Stanton number ☐ (b) Nusselt number ☐

(c) Biot number ☐ (d) Peclet number. ☐

3. The temperature of air-water mixture recorded by a thermometer when its bulb is covered with a wick saturated with water and placed in the air-water stream is called

(a) wet bulb temperature ☐ (b) dew point temperature ☐

(c) saturation temperature ☐ (d) critical temperature. ☐

4. What happens when the thickness of insulation on a pipe exceeds the critical value?

(a) there is decrease in the heat flow rate ☐

(b) there is increase in the heat flow rate ☐

(c) the heat flow rate remains constant ☐

(d) the temperature rises at the junction between pipe and insulation. ☐

5. Wet bulb depression represents the difference between

(a) dry bulb temperature and wet bulb temperature ☐

(b) dry bulb temperature and dew point temperature ☐

(c) dew point temperature and saturation temperature. ☐

6. In a refrigerator plant, if the condenser temperature increases, the power input to the compressor will

(a) decrease ☐ (b) increase ☐

(c) remain the same ☐ (d) be unpredictable. ☐

7. For the given conditions, the logarithmic mean temperature difference for parallel flow for evaporators and condensers is .... that for counter flow

(a) equal to ☐ (b) greater than ☐

(c) smaller than ☐ (d) any one of the above. ☐

8. When air is adiabatically saturated, the temperature attained is

(a) dew point temperature ☐ (b) dry bulb temperature ☐

(c) wet bulb temperature ☐ (d) triple point temperature. ☐

9. Consider the following statements pertaining to heat transfer through fins:

(1) Fins must be arranged at right angles to the direction of flow of the working fluid

(2) The temperature along the fin is variable and accordingly the heat transfer rate varies along the fin elements.

(3) Fins are equally effective irrespective whether they are on the hot side or cold side of the fluid

(4) Fins are made of materials that have thermal conductivity higher than that of the wall.

Identify the correct statements

(*a*) 1 and 2 ☐ (*b*) 1 and 3 ☐
(*c*) 2 and 3 ☐ (*d*) 3 and 4. ☐

10. Free convection heat flow depends on all of the followings, except
(*a*) density ☐ (*b*) coefficient of viscosity ☐
(*c*) gravitational force ☐ (*d*) velocity. ☐

11. The heat transfer is constant when
(*a*) temperature remains constant with time ☐ (*b*) temperature decreases with time ☐
(*c*) temperature increases with time ☐ (*d*) any of these. ☐

12. The coefficient of thermal conductivity is defined as
(*a*) quantity of heat transfer per unit area per one degree drop in temperature ☐
(*b*) quantity of heat transfer per one degree temperature drop per unit area ☐
(*c*) quantity of heat transfer per unit time per unit area ☐
(*d*) quantity of heat transfer per unit time per unit area per one degree temperature drop per unit length. ☐

13. In a vapour compression system, the working fluid is superheated vapour at entrance to
(*a*) evaporator ☐ (*b*) condenser ☐
(*c*) compressor ☐ (*d*) expansion valve. ☐

14. The more effective way of increasing the efficiency of a Carnot engine is to
(*a*) increase higher temperature ☐ (*b*) increase lower temperature ☐
(*c*) decrease lower temperature ☐ (*d*) decrease higher temperature. ☐

15. The free convection heat transfer is significantly affected by
(*a*) Reynolds number ☐ (*b*) Grashof number ☐
(*c*) Prandtl number ☐ (*d*) Stanton number. ☐

16. Which of the following is not the necessary condition for the Fourier's conduction heat equation $Q = -kA\dfrac{dt}{dx}$?
(*a*) Steady state ☐
(*b*) One dimensional heat flow ☐
(*c*) Constant value of thermal conductivity ☐
(*d*) Constant and uniform temperatures at the wall surfaces. ☐

17. Thermal boundary layer is a region where
(*a*) heat dissipation is negligible ☐
(*b*) inertia and convection terms are of the same order of magnitude ☐
(*c*) convection and dissipation terms are of the same order of magnitude ☐
(*d*) convection and conduction terms are of the same order of magnitude. ☐

18. Consider the following statements:

The Fourier heat conduction equation $Q = -kA\frac{dt}{dx}$ presumes

1. Steady state conditions
2. Constant value of thermal conductivity
3. Uniform temperature at the wall surfaces
4. One-dimensional heat flow.

Which of these statements are correct?

(*a*) 1, 2 and 3 ☐ (*b*) 1, 2 and 4 ☐
(*c*) 2, 3 and 4 ☐ (*d*) 1, 3 and 4. ☐

19. Which of the following aspects is not true in the context of dry compression in a vapour compression refrigeration cycle?

(*a*) Increase in refrigerating effect ☐
(*b*) Increase in power consumption ☐
(*c*) Improvement in coefficient of performance ☐
(*d*) Complete evaporation in the evaporator. ☐

20. The thermal resistance for heat conduction through a hollow sphere of inner radius r_1 and outer radius r_2 is

(*a*) $\frac{r_2 - r_1}{4\pi k r_1 r_2}$ ☐ (*b*) $\frac{(r_2 - r_1) r_1 r_2}{4\pi k}$ ☐

(*c*) $\frac{4\pi k(r_2 - r_1)}{r_1 r_2}$ ☐ (*d*) $\frac{k(r_2 - r_1)}{4\pi r_1 r_2}$. ☐

where k = Thermal conductivity of the material of sphere.

21. Which of the following components in a vapour absorption plant is a substitute for the compressor of the vapour compression system?

(*a*) Absorber ☐ (*b*) Heat exchanger and generator ☐
(*c*) Aqua pump and generator ☐ (*d*) Absorber, equa pump and generator. ☐

22. The thermal conductivity is expressed as

(*a*) W/m K ☐ (*b*) W/m^2 K ☐
(*c*) W/hm K ☐ (*d*) W/h^{2}m^2 K. ☐

23. Heat transfer from higher temperature to low temperature takes place according to

(*a*) Fourier law ☐ (*b*) First law of thermodynamics ☐
(*c*) Second law of thermodynamics ☐ (*d*) Zeroth law of thermodynamics. ☐

24. Consider the following properties of air-water vapour mixture:

1. dry bulb temperature.
2. wet bulb temperature.
3. specific humidity.
4. relative humidity.

The psychometric chart shows the relationships between

(*a*) 1, 2 only ☐ (*b*) 2, 3 only ☐
(*c*) 1, 2, 4 only ☐ (*d*) all of the above. ☐

25. For a given dry bulb temperature as the relative humidity decreases, the wet bulb temperature will

(*a*) increase ☐ (*b*) decrease ☐
(*c*) be the same ☐ (*d*) depend on other factors. ☐

26. The heat flow equation through a sphere of inner radius r_1 and outer radius r_2 is to be written in the same form as that for heat flow through a plane wall. For wall thickness $(r_2 - r_1)$, the equivalent mean radius for the spherical shell is

(*a*) $\frac{r_1 + r_1}{2}$ ☐ (*b*) $r_1 r_2$ ☐
(*c*) $\sqrt{r_1 r_2}$ ☐ (*d*) $\frac{r_1 + r_2}{\log_e(r_2/r_1)}$. ☐

27. The saturation temperature at the partial pressure of water vapour in the air-water vapour mixture is called

(*a*) dry bulb temperature ☐ (*b*) wet bulb temperature ☐
(*c*) dew point temperature ☐ (*d*) saturation temperature. ☐

28. A satellite in space exchanges heat with the surroundings essentially by

(*a*) conduction ☐ (*b*) convection ☐
(*c*) radiation ☐ (*d*) conduction and convection put together. ☐

29. A body cooling from 80 to 70°C takes 10 minutes when left exposed to environmental conditions. If the body is to cool further from 70 to 60°C under the same external conditions, it will take

(*a*) same time of 10 minutes ☐
(*b*) more that 10 minutes ☐
(*c*) less than 10 minutes ☐
(*d*) time will depend upon the environmental conditions (Leet). ☐

30. A pipe carrying steam at 215°C traverses as room and heat is lost to the surrounding air at 27°C. The major fraction of heat loss to the surroundings will be essentially due to

(*a*) conduction ☐ (*b*) convection ☐
(*c*) radiation ☐ (*d*) conduction and radiation put together. ☐

31. A hot metal piece kept in air cools from 80 to 70°C in t_1 seconds, from 70 to 60°C in t_2 seconds and from 60 to 50°C in t_3 seconds. Then

(*a*) $t_1 = t_2 = t_3$ ☐ (*b*) $t_1 < t_2 < t_3$ ☐
(*c*) $t_1 > t_2 > t_3$ ☐ (*d*) none of the above. ☐

32. A humidification process means

(*a*) decrease in relative humidity ☐ (*b*) an increase in specific humidity ☐
(*c*) a decrease in temperature ☐ (*d*) an increase in temperature. ☐

33. For steady state heat flow and constant value of thermal conductivity, the temperature distribution for a plane wall is a

(*a*) linear ☐ (*b*) parabolic ☐
(*c*) logarithmic ☐ (*d*) exponential curve. ☐

34. In an adiabatic saturation process

(*a*) the enthalpy remains constant ☐ (*b*) the temperature remains constant ☐
(*c*) the absolute humidity remains constant ☐ (*d*) the relative humidity remains constant. ☐

35. Critical thickness of insulation for spheres is given by

(*a*) k/h ☐ (*b*) $k/4\pi h$ ☐
(*c*) $h/2k$ ☐ (*d*) $2k/h$. ☐

where k is the coefficient of thermal conductivity and h is convective heat transfer coefficient.

36. Electrolux refrigerator essentially works on vapour

(*a*) absorption system ☐ (*b*) vapour compression system ☐
(*c*) thermo-electric system ☐ (*d*) vortex tube system. ☐

37. The intensity of solar radiation on earth is

(*a*) 1 kW/m^2 ☐ (*b*) 2 kW/m^2 ☐
(*c*) 5 kW/m^2 ☐ (*d*) 10 kW/m^2. ☐

38. Hydrogen is essential in an electrolux refrigeration system, because

(*a*) it acts as a catalyst in the evaporator ☐
(*b*) it helps in maintaining a low partial pressure for the evaporating ammonia ☐
(*c*) the cooled hydrogen leaving the heat exchanges tools the refrigerant entering the evaporator ☐
(*d*) the reaction between hydrogen and ammonia is endothermic in evaporators and exothermic in adsorbate. ☐

39. A furnace is made of a brick wall of thickness 0.5 m and conductivity 0.75 W/m K. For the same heat loss and temperature drop, what should be the thickness of diatomite earth of thermal conductivity 0.15 W/m K?

(*a*) 0.05 m ☐ (*b*) 0.1 m ☐
(*c*) 0.2 m ☐ (*d*) 0.5 m. ☐

40. Two walls of same thickness and cross-sectional area have thermal conductivities in the ratio 1 : 2. If same temperature difference is maintained across the wall faces, the ratio of heat flow Q_1/Q_2 will be

(*a*) 1/2 ☐ (*b*) 1 ☐
(*c*) 2 ☐ (*d*) 4. ☐

41. An ideal gas at 27°C is heated at constant pressure till the volume becomes three times. The temperature of the gas will then be

(*a*) 81°C ☐ (*b*) 900°C ☐
(*c*) 627°C ☐ (*d*) 927°C. ☐

42. Gas turbines work on

(*a*) constant volume cycle ☐ (*b*) constant pressure cycle ☐
(*c*) constant temperature cycle ☐ (*d*) constant enthalpy cycle. ☐

43. The radial heat transfer rate through hollow cylinder increases as the ratio of outer radius to inner radius

(*a*) decreases ☐ (*b*) increases ☐
(*c*) constant ☐ (*d*) none of the above. ☐

44. Stefan-Boltzmann law is expressed as

(*a*) $Q = \sigma AT^4$ ☐ (*b*) $Q = \sigma A^2T^4$ ☐
(*c*) $Q = \sigma AT^2$ ☐ (*d*) $Q = AT^4$. ☐

45. When the dry bulb and wet bulb temperature of air are same, then relative humidity of air will be

(*a*) zero per cent ☐ (*b*) 50 per cent ☐
(*c*) 66.7 per cent ☐ (*d*) 100 per cent. ☐

46. The quantity of heat radiation is dependent on

(*a*) area of the body only ☐ (*b*) shape of the body only ☐
(*c*) temperature of the body only ☐ (*d*) on all (*a*), (*b*) and (*c*). ☐

47. The wet bulb depression is zero when relative humidity equals

(*a*) zero ☐ (*b*) 0.5 ☐
(*c*) 0.75 ☐ (*d*) 1.0. ☐

48. The law of thermodynamics according to which heat transfer takes place is

(*a*) Zeroth ☐ (*b*) first ☐
(*c*) second ☐ (*d*) third. ☐

49. The convective heat transfer coefficient in laminar flow over a flat plate

(*a*) increases with distance ☐ (*b*) increases if a higher viscosity fluid is used ☐
(*c*) increases if a denser fluid is used ☐ (*d*) decreases with increase in free stream velocity. ☐

50. The refrigerant Freon-12 is a compound consisting of

(*a*) carbon, hydrogen and fluorine ☐ (*b*) carbon, hydrogen and chlorine ☐
(*c*) carbon, chlorine and fluorine ☐ (*d*) carbon, hydrogen, chlorine and fluorine. ☐

51. The refrigerant commonly used for commercial ice plants is

(*a*) Freon-12 ☐ (*b*) NH_3 ☐
(*c*) CO_2 ☐ (*d*) Air. ☐

52. Ammonia is preferred as a refrigerant in large commercial installations because

(*a*) it is non-toxic ☐ (*b*) it has low latent heat ☐
(*c*) it has low working pressure ☐ (*d*) it is relatively cheap. ☐

53. For a cylindrical rod with uniformly distributed heat sources, the thermal gradient dt/dr at half the radius location will be

(*a*) one-fourth of that at the surface ☐ (*b*) one-half of that at the surface ☐
(*c*) twice of that at the surface ☐ (*d*) four times of that at the surface. ☐

54. The constant wet bulb lines are non-uniformly spaced and inclined sloping down towards the right of psychrometric diagram and follow very closely the lines of

(a) constant volume ☐ (b) constant relative humidity ☐
(c) constant enthalpy ☐ (d) constant moisture content. ☐

55. The wall thickness and the temperatures at the various wall faces of a composite wall made of three layers of different materials are as indicated in Fig. 7.15. These parameters indicate that

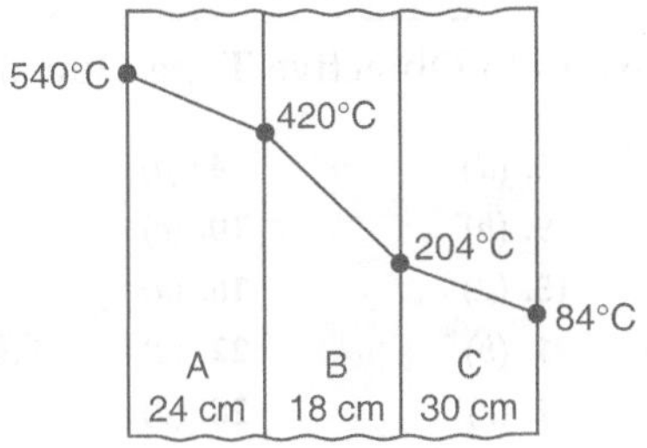

540°C
420°C
204°C
84°C
A 24 cm
B 18 cm
C 30 cm

FIGURE 7.15

(a) layer A and C are made of the same material ☐
(b) there is generation of heat at the interface of walls A and B ☐
(c) free surfaces of materials A and C are being heated ☐
(d) of all the materials, the material B has the lowest value of thermal conductivity. ☐

56. A convergent duct produces

(a) acceleration in both subsonic and supersonic flow ☐
(b) deceleration in both subsonic and supersonic flow ☐
(c) acceleration in subsonic and deceleration in supersonic flow ☐
(d) deceleration in subsonic and acceleration in supersonic flow. ☐

57. Which one of the following statements is not true in the context of thermal radiations?

(a) thermal radiations are electromagnetic waves ☐
(b) thermal radiations travel in free space with a velocity 3×10^8 m/s ☐
(c) all bodies emit thermal radiations at all temperatures ☐
(d) thermal radiations are not reflected from a mirror. ☐

58. The ratio of the thickness of thermal boundary layer δ_t to thickness of hydrodynamic boundary δ_h is given by $\frac{\delta_t}{\delta_h}$ = (Prandtl number)^n

The index n is

(a) −1/3 ☐ (b) −2/3 ☐
(c) 1 ☐ (d) −1. ☐

59. The value of Prandtl number for air is about

(a) 0.1 ☐ (b) 0.4 ☐
(c) 0.7 ☐ (d) 1.1. ☐

60. Heat transfer by radiation is encountered least in

(a) boiler furnace ☐ (b) insulated steam pipe ☐
(c) electric bulb ☐ (d) nuclear reactor. ☐

61. The essential condition for the transfer of heat from one body to another is

(*a*) both bodies must be in physical contact ☐

(*b*) heat content of one body must be more than that of the other ☐

(*c*) one of the bodies must have a high value of thermal conductivity ☐

(*d*) there must exist a temperature difference between the bodies. ☐

ANSWERS

Answers to Objective Type Questions

1. (*d*)	**2.** (*c*)	**3.** (*d*)	**4.** (*a*)	**5.** (*b*)	**6.** (*c*)
7. (*b*)	**8.** (*a*)	**9.** (*b*)	**10.** (*c*)	**11.** (*b*)	**12.** (*a*)
13. (*c*)	**14.** (*b*)	**15.** (*d*)	**16.** (*a*)	**17.** (*b*)	**18.** (*b*)
19. (*a*)	**20.** (*c*)	**21.** (*b*)	**22.** (*c*)	**23.** (*c*)	**24.** (*d*)
25. (*a*)	**26.** (*d*)	**27.** (*c*)	**28.** (*c*)	**29.** (*a*)	**30.** (*a*)
31. (*c*)	**32.** (*d*)	**33.** (*b*)	**34.** (*c*)	**35.** (*b*)	**36.** (*b*)
37. (*d*)	**38.** (*c*)	**39.** (*c*)	**40.** (*a*)	**41.** (*b*)	**42.** (*a*)
43. (*b*)	**44.** (*b*)	**45.** (*b*)	**46.** (*d*)	**47.** (*d*)	**48.** (*c*)
49. (*b*)	**50.** (*c*)	**51.** (*a*)	**52.** (*b*)	**53.** (*c*)	**54.** (*d*)
55. (*d*)	**56.** (*c*)	**57.** (*b*)	**58.** (*c*)	**59.** (*b*)	**60.** (*b*)
61. (*a*)	**62.** (*b*)	**63.** (*a*)	**64.** (*e*)	**65.** (*c*)	**66.** (*b*)
67. (*d*)	**68.** (*a*)	**69.** (*c*)	**70.** (*c*)	**71.** (*d*)	**72.** (*c*)
73. (*d*)	**74.** (*b*)	**75.** (*b*)	**76.** (*a*)	**77.** (*a*)	**78.** (*d*)
79. (*c*)	**80.** (*b*)	**81.** (*d*)	**82.** (*c*)	**83.** (*d*)	**84.** (*c*)
85. (*c*)	**86.** (*d*)	**87.** (*a*)	**88.** (*a*)	**89.** (*b*)	**90.** (*b*)
91. (*d*)	**92.** (*b*)	**93.** (*a*)	**94.** (*a*)	**95.** (*c*)	**96.** (*c*)
97. (*d*)	**98.** (*d*)	**99.** (*b*)	**100.** (*b*)	**101.** (*c*)	**102.** (*b*)
103. (*b*)	**104.** (*a*)	**105.** (*c*)	**106.** (*c*)	**107.** (*c*)	**108.** (*b*)
109. (*b*)	**110.** (*a*)	**111.** (*c*)	**112.** (*b*)	**113.** (*b*)	**114.** (*d*)
115. (*c*)	**116.** (*c*)	**117.** (*e*)	**118.** (*b*)	**119.** (*d*)	**120.** (*c*)
121. (*a*)	**122.** (*b*)	**123.** (*b*)	**124.** (*b*)	**125.** (*e*)	**126.** (*b*)
127. (*a*)	**128.** (*c*)	**129.** (*d*)	**130.** (*b*)	**131.** (*b*)	**132.** (*e*)
133. (*c*)	**134.** (*b*)	**135.** (*b*)	**136.** (*a*)	**137.** (*c*)	**138.** (*b*)
139. (*b*)	**140.** (*a*)	**141.** (*c*)	**142.** (*b*)	**143.** (*d*)	**144.** (*a*)
145. (*c*)	**146.** (*d*)	**147.** (*d*)	**148.** (*b*)	**149.** (*a*)	**150.** (*c*)
151. (*d*)	**152.** (*c*)	**153.** (*b*)	**154.** (*a*)	**155.** (*b*)	**156.** (*b*)
157. (*a*)	**158.** (*d*)	**159.** (*d*)	**160.** (*d*)	**161.** (*b*)	**162.** (*c*)
163. (*b*)	**164.** (*b*)	**165.** (*b*)	**166.** (*a*)	**167.** (*b*)	**168.** (*b*)
169. (*a*)	**170.** (*b*)	**171.** (*b*)	**172.** (*b*)	**173.** (*a*)	**174.** (*b*)
175. (*a*)	**176.** (*b*)	**177.** (*a*)	**178.** (*d*)	**179.** (*d*)	**180.** (*e*)
181. (*a*)	**182.** (*e*)	**183.** (*a*)	**184.** (*d*)	**185.** (*b*)	**186.** (*a*)
187. (*b*)	**188.** (*b*)	**189.** (*c*)	**190.** (*c*)	**191.** (*c*)	**192.** (*d*)
193. (*a*)	**194.** (*b*)	**195.** (*d*)	**196.** (*a*)	**197.** (*b*)	**198.** (*a*)
199. (*c*)	**200.** (*b*)	**201.** (*c*)	**202.** (*b*)	**203.** (*d*)	**204.** (*b*)
205. (*d*)	**206.** (*b*)	**207.** (*a*)	**208.** (*c*)	**209.** (*d*)	**210.** (*e*)

211. (*a*)	**212.** (*b*)	**213.** (*c*)	**214.** (*d*)	**215.** (*d*)	**216.** (*b*)
217. (*a*)	**218.** (*d*)	**219.** (*b*)	**220.** (*a*)	**221.** (*c*)	**222.** (*c*)
223. (*c*)	**224.** (*b*)	**225.** (*b*)	**226.** (*b*)	**227.** (*c*)	**228.** (*d*)
229. (*a*)	**230.** (*e*)	**231.** (*d*)	**232.** (*d*)	**233.** (*c*)	**234.** (*a*)
235. (*b*)	**236.** (*a*)	**237.** (*d*)	**238.** (*a*)	**239.** (*c*)	**240.** (*c*)
241. (*a*)	**242.** (*e*)	**243.** (*a*)	**244.** (*c*)	**245.** (*a*)	**246.** (*d*)
247. (*d*)	**248.** (*b*)	**249.** (*c*)	**250.** (*d*).		

True/False

251. (*a*)	**252.** (*b*)	**253.** (*a*)	**254.** (*a*)	**255.** (*a*)	**256.** (*b*)
257. (*b*)	**258.** (*b*)	**259.** (*a*)	**260.** (*a*)	**261.** (*a*)	**262.** (*b*)
263. (*a*)	**264.** (*b*)	**265.** (*b*)	**266.** (*a*)	**267.** (*b*)	**268.** (*a*)
269. (*a*)	**270.** (*b*)	**271.** (*a*)	**272.** (*a*)	**273.** (*a*)	**274.** (*b*)
275. (*b*)	**276.** (*a*)	**277.** (*a*)	**278.** (*b*)	**279.** (*a*)	**280.** (*b*)
281. (*a*)	**282.** (*b*)	**283.** (*a*)	**284.** (*a*)	**285.** (*b*)	**286.** (*a*)
287. (*a*)	**288.** (*b*)	**289.** (*a*)	**290.** (*a*)	**291.** (*a*)	**292.** (*b*)
293. (*a*)	**294.** (*b*)	**295.** (*b*)	**296.** (*a*)	**297.** (*a*)	**298.** (*a*)
299. (*a*)	**300.** (*a*)	**301.** (*b*)	**302.** (*a*)	**303.** (*a*)	**304.** (*b*)
305. (*a*)	**306.** (*b*)	**307.** (*b*)	**308.** (*a*)	**309.** (*a*)	**310.** (*a*)
311. (*b*)	**312.** (*a*).				

Fill in the Blanks

313. (*b*)	**314.** (*a*)	**315.** (*b*)	**316.** (*b*)	**317.** (*b*)	**318.** (*b*)
319. (*b*)	**320.** (*b*)	**321.** (*b*)	**322.** (*b*)	**323.** (*b*)	**324.** (*a*)
325. (*b*)	**326.** (*b*)	**327.** (*b*)	**328.** (*b*)	**329.** (*b*)	**330.** (*b*)
331. (*b*)	**332.** (*b*)	**333.** (*b*)	**334.** (*b*)	**335.** (*b*)	**336.** (*b*)
337. (*b*)	**338.** (*a*)	**339.** (*a*)	**340.** (*a*)	**341.** (*a*)	**342.** (*b*)
343. (*a*)	**344.** (*a*)	**345.** (*b*)	**346.** (*b*)	**347.** (*b*)	**348.** (*a*)
349. (*b*)	**350.** (*a*).				

Answers to Objective Type Questions from Competitive Examinations

1. (*b*)	**2.** (*b*)	**3.** (*a*)	**4.** (*b*)	**5.** (*a*)	**6.** (*b*)
7. (*a*)	**8.** (*c*)	**9.** (*a*)	**10.** (*d*)	**11.** (*a*)	**12.** (*d*)
13. (*b*)	**14.** (*c*)	**15.** (*b*)	**16.** (*c*)	**17.** (*c*)	**18.** (*d*)
19. (*a*)	**20.** (*a*)	**21.** (*c*)	**22.** (*a*)	**23.** (*c*)	**24.** (*d*)
25. (*b*)	**26.** (*c*)	**27.** (*c*)	**28.** (*c*)	**29.** (*b*)	**30.** (*b*)
31. (*b*)	**32.** (*b*)	**33.** (*a*)	**34.** (*a*)	**35.** (*d*)	**36.** (*a*)
37. (*a*)	**38.** (*b*)	**39.** (*b*)	**40.** (*a*)	**41.** (*c*)	**42.** (*b*)
43. (*a*)	**44.** (*a*)	**45.** (*d*)	**46.** (*c*)	**47.** (*d*)	**48.** (*b*)
49. (*c*)	**50.** (*c*)	**51.** (*b*)	**52.** (*c*)	**53.** (*b*)	**54.** (*c*)
55. (*d*)	**56.** (*d*)	**57.** (*d*)	**58.** (*a*)	**59.** (*c*)	**60.** (*b*)
61. (*d*).					

Chapter 8 STRENGTH OF MATERIALS

I. THEORY

8.1. BASIC DEFINITIONS

When an external force acts on a body, the body tends to undergo some deformation. Due to cohesion between the molecules, the body resists deformation. This resistance by which material of the body oppose the deformation is known as **Strength of Materials**. The resistance or force per unit area of deformation is called **stress**. The measure of deformation in the body is known as **strain**. The external force acting on the body is called **load**. Mathematically, stress is written as,

$$\sigma = \frac{P}{A}$$

where σ = Stress, P = External force or load, and
A = Cross-sectional area of the body.

8.2. TYPES OF STRESSES

There are three types of simple stresses:

1. Tensile stress, 2. Compressive stress, and 3. Shear stress.

8.2.1. Types of Strains

There are four types of strains:

1. Tensile strain,
2. Compressive strain,
3. Shear strain, and
4. Volumetric strain.

Strain is defined as the ratio of change of dimension of a body (subjected to some external force) to the original dimension. If the change in dimension is due to tensile stress which means increase in length then the ratio of increase of length to the original length is known as *tensile strain.* If the change in dimension is due to compressive stress which means decrease in length then the ratio of decrease in length to the original length is known as *compressive strain*. If the change in dimension is due to shear stress which means the angle through which a body is distorted then this angle is known as *shear strain*. Volumetric strain is the ratio of change of volume of the body to its original volume.

8.2.2. Elastic Body and Elastic Strains

Elastic strains are the strains which disappear with the removal of external load acting on the body. Elastic body is the body which resumes its original dimensions on the removal of the external load acting on the body.

8.2.3. Plastic Body and Plastic Strains

Plastic strains are the strains which remain even after the removal of the external load acting on the body and the body behaving in this manner is known as plastic body.

8.3. HOOKE'S LAW

It states that for a material loaded within elastic-limits the stress is proportional to the strain, produced by the stress. Mathematically, it is written as,

$$\text{Stress} \propto \text{Corresponding strain}$$

or The ratio of stress to the corresponding strain is a constant.

This constant is known as Modulus of elasticity or Modulus or Rigidity.

8.3.1. Modulus of Elasticity or Young's Modulus (E)

It is defined as the ratio of tensile or compressive stress to the corresponding strain. This is denoted by *E*.

$$\therefore \quad E = \frac{\text{Tensile stress}}{\text{Tensile strain}} \text{ or } \frac{\text{Compressive stress}}{\text{Compressive strain}} \text{ or } \frac{\text{Stress}}{\text{Corresponding strain}}.$$

8.3.2. Modulus of Rigidity or Shear Modulus of Elasticity

It is defined as the ratio of shear stress to the corresponding shear strain within elastic limit. This is denoted by *C* or *G* or *N*.

$$\therefore \quad C \text{ (or } G \text{ or } N) = \frac{\text{Shear stress}}{\text{Shear strain}}.$$

8.3.3. Bulk Modulus (K)

It is defined as the ratio of normal stress on each face of a solid cube to the volumetric strain. It is denoted by *K*.

$$\therefore \quad K = \frac{\text{Normal stress on each face of a cube}}{\text{Volumetric strain}}$$

8.4. EXTENSION OF A UNIFORM ROD OR UNIFORMLY TAPERING ROD

When a uniform rod of diameter (D) is subjected to an axial tensile load (P), the extension of the rod (δl) is given by

$$\delta l = \frac{4Pl}{\pi D^2 E}$$

where E = Modulus of elasticity of the rod, and l = Length of the body.

The extension of the rod which tapers uniformly from a diameter D_1 to a diameter D_2 and carries an axial tensile load (P) is given by

$$\delta l = \frac{4PL}{\pi E D_1 D_2}.$$

8.5. STRESS AND EXTENSION OF A UNIFORM ROD OR CONICAL ROD DUE TO ITS OWN WEIGHT

(*a*) A rod of length (l) and diameter (D) is rigidly fixed at the upper end and is hanging vertically. The stress due to rod's own weight at any section at a distance 'x' from the lower end

$$= wx$$

where w = Weight per unit volume of the rod.

The **total extension** produced in the rod due to its own weight $= \dfrac{wl^2}{2E}$.

(*b*) **A conical rod** of length (l) and base diameter (D) is rigidly fixed with its base diameter at the upper end and is hanging vertically. The stress due to its own weight at any section at a distance 'x' from the lower end

$$= \frac{wx}{3}$$

where w = Weight per unit volume of the conical rod.

The **total extension** produced in the conical rod due to its own weight $= \dfrac{wl^2}{6E}$.

8.6. BARS OF COMPOSITE SECTIONS

A bar made up of two or more different materials, rigidly fixed with each other and behaving as one unit for extension or compression when subjected to an axial tensile or compressive loads, is called a composite bar. For the composite bars the following **two points** are important:

(*i*) The extension or compression (*i.e.*, deformation) in each bar is equal. Hence deformation per unit length, *i.e.*, strain in each bar is equal.

(*ii*) The total external load on the composite bar is equal to the total sum of the loads carried by each different materials.

8.7. THERMAL STRESSES

Thermal stresses are set up in a body, when the temperature of the body is raised or lowered and the body is not allowed to expand or contract freely. But if the body is allowed to expand or contact freely, no stresses will be set up in the body.

A rod of length (l) is heated and let the temperature rise in the rod be T. If the rod is free to expand then the extension of the rod

$$= \alpha T l$$

where α = Coefficient of linear expansion.

If the rod is prevented from expanding, compressive thermal stress and strain will be set up in the rod.

$$\text{Thermal strain} = \frac{\text{Extension prevented}}{\text{Original length}} = \frac{\alpha T l}{l} = \alpha T$$

$$\therefore \quad \text{Thermal stress} = \text{Thermal strain} \times E = \alpha T E.$$

8.8. POISSON'S RATIO AND LATERAL STRAIN

When a body is subjected to an axial tensile load, there is an increase in the length of the body. But at the same time there is a decrease in other dimensions of the body at right angles to the line of action of the applied load. The strain in the direction of applied load is called *longitudinal strain.* And the strains at right angles to the direction of applied load is known as *lateral strains.* The ratio of lateral strain to longitudinal strain is called the Poisson's ratio and it is generally denoted by $1/m$. Hence mathematically,

$$\text{Poisson's ratio} \left(\frac{1}{m}\right) = \frac{\text{Lateral strain}}{\text{Longitudinal strain}}.$$

The value of Poisson's ratio for most of the materials lies in the range of 0.25 to 0.35.

8.9. RESILIENCE

Resilience is defined as the work done in producing strain per unit volume of a material, whereas, the **proof resilience** is the maximum strain energy stored in a unit volume of a material without permanent deformation (*i.e.*, upto elastic limit).

$$\text{Resilience} = \frac{\sigma^2}{2E}, \text{ where } \sigma \text{ is the stress induced in the body}$$

$$\text{Proof resilience} = \frac{\sigma_{EL}^2}{2E}, \text{ where } \sigma_{EL} = \text{stress at the elastic limit.}$$

8.9.1. Shear Resilience

Shear resilience is the amount of energy stored in a material per unit volume without permanent deformation (*i.e.*, upto elastic limit) when it is subjected to shear stress, Mathematically,

$$\text{Shear resilience} = \frac{\tau^2}{2C}$$

where τ = Shear stress, and C = Modulus of rigidity.

8.9.2. Relations between Modulus of Elasticity (E), Modulus of Rigidity (C) and Bulk Modulus (K) in Terms of Poisson's Ratio

The followings are the important relations:

(*i*) Between Modulus of Elasticity (*E*) and Modulus of Rigidity (C),

$$E = 2C\left(1+\frac{1}{m}\right) \quad \text{or} \quad C = \frac{mE}{2(m+1)}$$

(*ii*) Between Modulus of Elasticity (*E*) and the Bulk Modulus (*K*),

$$E = 3K\left(1-\frac{2}{m}\right) \quad \text{or} \quad K = \frac{mE}{3(m-2)}.$$

(*iii*) Between Modulus of Elasticity (*E*), Modulus of Rigidity (C) and Bulk Modulus (*K*),

$$E = \frac{9KC}{C+3K}.$$

8.10. STRESSES ON AN INCLINED SECTION OF A BAR CARRYING AXIAL LOAD

8.10.1. A Bar Subjected to One Tensile Stress

A bar of cross-sectional area '*A*' carries an axial tensile load, which produces a tensile stress σ across the sections *AB* and *CD* as shown in Fig. 8.1. On a section *AE*, which is inclined at an angle θ with section *AB*, normal and tangential stresses will be induced.

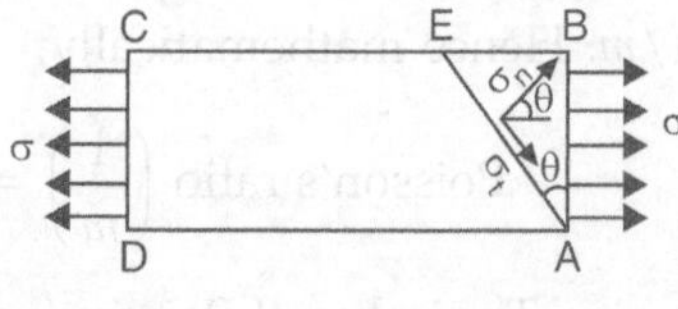

FIGURE 8.1

Normal stress (σ_n) on plane *AE*

$$= \frac{\text{Total load normal to plane } AE}{\text{Area of plane } AE}$$

$$= \frac{\sigma A \cos\theta}{(A/\cos\theta)} = \sigma\cos^2\theta$$

Tangential or shear stress (σ_t) on plane *AE*

$$= \frac{\text{Total load along the plane } AE}{\text{Area of plane } AE}$$

$$= \frac{\sigma A \sin\theta}{(A/\cos\theta)} = \sigma\sin\theta\cos\theta = \frac{\sigma}{2}\sin 2\theta$$

Resultant stress on plane *AE* $= \sqrt{\sigma_n^2+\sigma_t^2} = \sigma\cos\theta$

(*i*) When $\theta = 0$, Normal stress (σ_n) is maximum and equal to p.

(*ii*) When $\theta = 90°$, Normal stress (σ_n) is minimum and equal to zero and also shear stress is equal to zero.

(*iii*) When $\theta = 45°$, shear stress or tangential stress is maximum and equal to $\frac{\sigma}{2}$.

8.10.2. A Bar Subjected to Two Like Principal Stresses

A bar subjected to two like principal stresses is shown in Fig. 8.2.

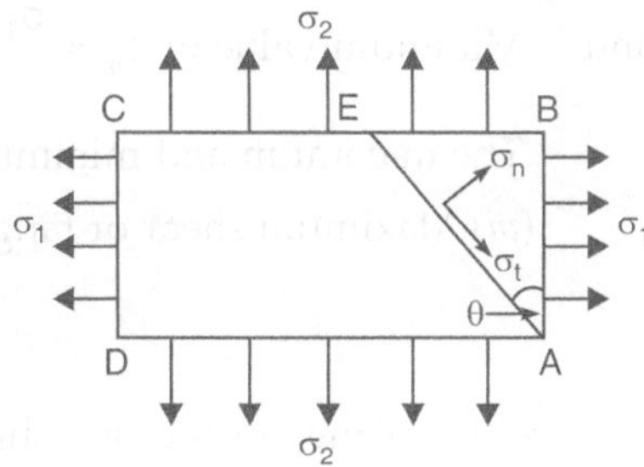

FIGURE 8.2

(*i*) The normal stress on plane *AE*,

$$\sigma_n = \frac{\sigma_1 + \sigma_2}{2} + \frac{\sigma_1 - \sigma_2}{2} \cos 2\theta .$$

(*ii*) The tangential or shear stress on the plane *AE*,

$$\sigma_t = \frac{\sigma_1 - \sigma_2}{2} \sin 2\theta.$$

(*iii*) The resultant stress (σ_r) on the plane *AE*

$$= \sqrt{\sigma_n^2 + \sigma_t^2} = \sqrt{\sigma_1^2 \cos^2 \theta + \sigma_2^2 \sin^2 \theta} .$$

Obliquity. It is the angle made by the line of action of the resultant stress with the normal to the plane. It is denoted by ϕ. Then $\tan \phi = \frac{\sigma_t}{\sigma_n}$.

(*iv*) When $\theta = 45°$ or $135°$, the shear stress on the inclined plane is maximum and equal to $\left(\frac{\sigma_1 - \sigma_2}{2}\right)$.

(*v*) When $\theta = 0$, then normal stress is equal to σ_1.

(*vi*) When $\theta = 90°$, the normal stress is equal to σ_2.

8.10.3. A Bar Subjected to Two Direct Stresses in Two Mutually Perpendicular Direction Accompanied by a Shear Stress (τ)

A bar subjected to two direct stresses in two mutually perpendicular direction accompanied by a shear stress (τ) is shown in Fig. 8.3.

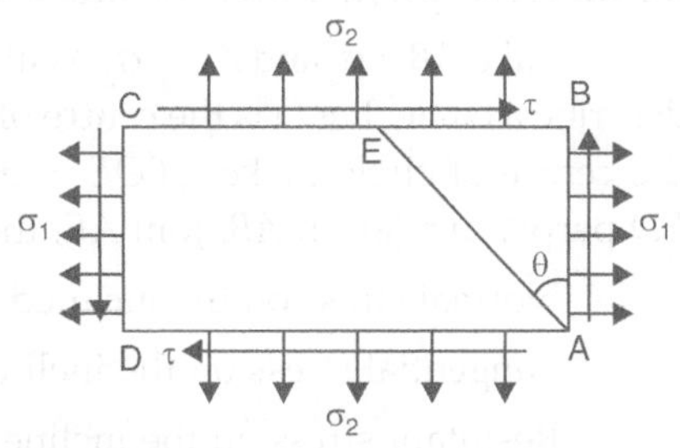

FIGURE 8.3

(*i*) The normal stress on the plane *AE*,

$$\sigma_n = \frac{\sigma_1 + \sigma_2}{2} + \frac{\sigma_1 - \sigma_2}{2} \cos 2\theta + \tau \sin 2\theta.$$

(*ii*) The tangential or shear stress on the plane *AE*,

$$\sigma_t = \frac{\sigma_1 + \sigma_2}{2} \sin 2\theta - \tau \cos 2\theta.$$

(*iii*) **Principal plane.** It is a plane which is not subjected to shear stress.

(*iv*) **Principal stresses.** They are the stresses acting normal to principal planes.

(*v*) The maximum value of the normal stress for the case shown in Fig. 8.3 is given by

$$\text{Maximum value of } \sigma_n = \frac{\sigma_1 + \sigma_2}{2} + \sqrt{\left(\frac{\sigma_1 - \sigma_2}{2}\right)^2 + \tau^2}$$

and $$\text{Minimum value of } \sigma_n = \frac{\sigma_1 + \sigma_2}{2} - \sqrt{\left(\frac{\sigma_1 - \sigma_2}{2}\right)^2 + \tau^2}.$$

The maximum and minimum values of σ_n, are also called principal stresses

(*vi*) Maximum shear or tangential stress is given by

$$\tau_{max} = \frac{(\text{Max. value of } \sigma_n - \text{Min. value of } \sigma_n)}{2}$$

(*vii*) A plane, for the case shown in Fig. 8.3, will be a principal plane, if

$$\tan 2\theta = \frac{2\tau}{\sigma_1 - \sigma_2}.$$

8.11. MOHR'S CIRCLE OF STRESSES

Mohr's circle of stresses is used for finding the normal, tangential and resultant stresses on an inclined plane. The Mohr's circle of stresses is drawn for the following cases:

(*i*) A bar subjected to two unequal like principal stresses

(*ii*) A bar subjected to two unequal unlike principal stresses.

8.11.1. A Bar Subjected to Two Unequal Like Principal Stresses

Let σ_1 and σ_2 be two unequal tensile (or compressive) principal stresses acting on a bar. It is required to find the normal, tangential and resultant stresses on a plane inclined at an angle θ with the major principal plane. Figure 8.4 shows the Mohr's circle of stresses for this case. Mohr's circle of stresses is drawn as:

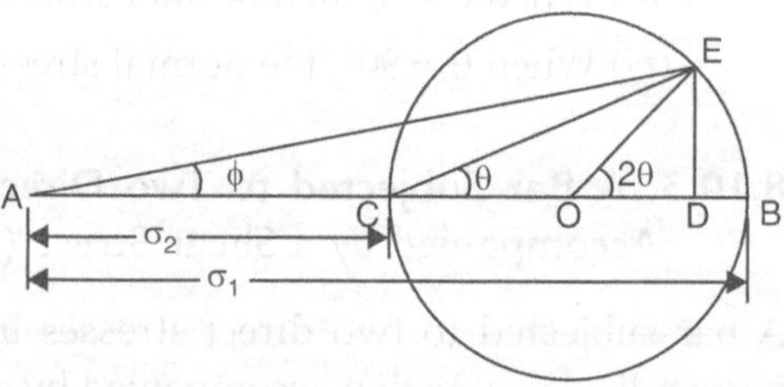

FIGURE 8.4

Take $AB = \sigma_1$ and $AC = \sigma_2$. With BC as diameter describe a circle. Let O is the centre of the circle. With the centre of circle make $\angle BOE = 2\theta$. From E draw ED perpendicular on AB. Join AE, then

Normal stress on the inclined plane $= AD$

Tangential stress on the inclined plane $= ED$

Resultant stress on the inclined plane $= AE$

Radius of Mohr's circle $= \dfrac{\sigma_1 - \sigma_2}{2}$

Obliquity is represented by angle ϕ.

8.11.2. A Bar is Subjected to Two Unequal Unlike Principal Stresses

Let σ_1 be the tensile principal stress and σ_2 be the compressive principal stress. The Mohr's circle is shown in Fig. 8.5.

Take any point *A*. From *A*, take $AB = \sigma_1$ (+) and $AC = \sigma_2$ (–). Draw a circle with *BC* as diameter. Let *O* be the centre of the circle. From *O*, draw an angle $BOE = 2\theta$. From *E* draw *ED* normal on *AB*. Join *AE* and *CE*. Then

Normal stress on inclined plane $= AD$

Shear stress on inclined plane $= ED$

Resultant stress on inclined plane $= AE$

Radius of Mohr's circle $= \dfrac{\sigma_1 + \sigma_2}{2}$

Angle ϕ represents the obliquity.

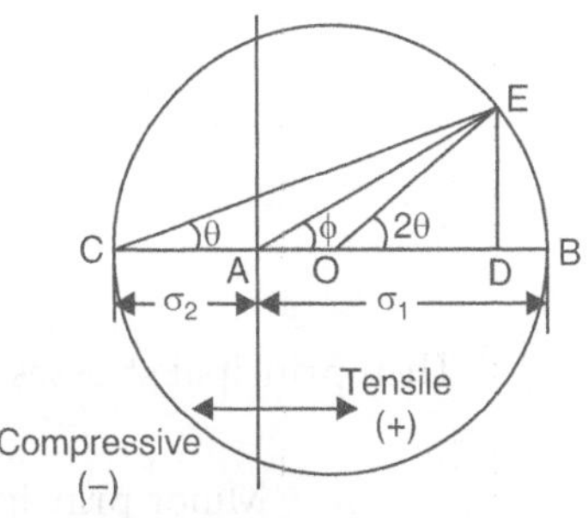

FIGURE 8.5

8.12. PRINCIPAL STRESSES ON A POINT ON A SHAFT SUBJECTED TO BENDING AND TORSION

When a shaft of diameter '*d*' is subjected to bending as well as torsion, the principal stresses are calculated first finding the bending stress and shear stress at any point. The bending stress (σ) at a distance *y* from the neutral axis is given by

$$\frac{\sigma}{y} = \frac{M}{I} \quad \text{or} \quad \sigma = \frac{M}{I}y$$

The shear stress (τ) at a radius '*r*' is given by

$$\frac{\tau}{r} = \frac{T}{J} \quad \text{or} \quad \tau = \frac{T}{J} \times r$$

where J = Polar moment of inertia of the shaft.

The principal stress are given by $\sigma_1 = \dfrac{\sigma}{2} + \sqrt{\dfrac{\sigma^2}{4} + \tau^2}$

and
$$\sigma_2 = \frac{\sigma}{2} - \sqrt{\frac{\sigma^2}{4} + \tau^2}$$

The location of the principal planes through the point is given by $\tan 2\theta = \dfrac{2\tau}{\sigma}$.

The maximum shear stress (τ_{max}) is given by $\tau_{max} = \dfrac{\sigma_1 - \sigma_2}{2}$.

The maximum bending stress and shear stress occur at a point on a shaft when the point lies on the surface of the shaft. Then the maximum bending stress σ will be

$$\sigma = \frac{M}{I} \times y = \frac{M}{I} \times \frac{d}{2} \qquad \left(\because\ y = \frac{d}{2}\right)$$

$$= \frac{M}{\frac{\pi}{64}d^4} \times \frac{d}{2} = \frac{32M}{\pi d^3} \qquad \left(\because \; I = \frac{\pi}{64}d^4\right)$$

and shear stress on the surface of the shaft,

$$\tau = \frac{T}{J} \times r = \frac{T}{\frac{\pi}{32}d^4} \times \frac{d}{2} \qquad \left(\because \; r = \frac{d}{2} \text{ and } J = \frac{\pi}{32}d^4\right)$$

$$= \frac{16T}{\pi d^3}.$$

Then principal stresses on a point lying on the surface of the shaft are given by

$$\text{Minor principal stress} = \frac{\sigma}{2} + \sqrt{\frac{\sigma^2}{4} + \tau^2} = \frac{16}{\pi d^3}[M + \sqrt{M^2 + T^2}]$$

and $$\text{Major principal stress} = \frac{\sigma}{2} - \sqrt{\frac{\sigma^2}{2} + \tau^2} = \frac{16}{\pi d^3}[M - \sqrt{M^2 + T^2}].$$

The position of principal planes through a point lying on the surface of the shaft is given by

$$\tan 2\theta = \frac{2\tau}{\sigma} = \frac{2 \times 16T}{\pi d^3} \times \frac{1}{\frac{32M}{\pi d^3}} = \frac{T}{M}.$$

8.13. STRAIN ENERGY STORED IN A STRAINED MATERIAL

Figure 8.6 shows the stress-strain curve for a material upto elastic limit when loaded in one direction. The strain energy stored in the material per unit volume

$$= \text{area of triangle } OAB$$

$$= \frac{1}{2} \times BO \times AB = \frac{1}{2} \times e \times \sigma = \frac{1}{2}\sigma e.$$

(*a*) If two principal stresses σ_1 and σ_2 exist at a point in a strained material, there will be two principal strains e_1 and e_2. They are given by

$$e_1 = \frac{\sigma_1}{E} - \frac{\sigma_2}{mE} \quad \text{and } e_2 = \frac{\sigma_2}{E} - \frac{\sigma_1}{mE}$$

Then strain energy stored per unit volume

$$= \frac{1}{2}\sigma_1 e_1 + \frac{1}{2}\sigma_2 e_2$$

$$= \frac{1}{2}\sigma_1\left(\frac{\sigma_1}{E} - \frac{\sigma_2}{mE}\right) + \frac{1}{2}\sigma_2\left(\frac{\sigma_2}{E} - \frac{\sigma_1}{mE}\right)$$

$$= \frac{1}{2E}\left(\sigma_1^2 + \sigma_2^2 - \frac{2\sigma_1\sigma_2}{m}\right).$$

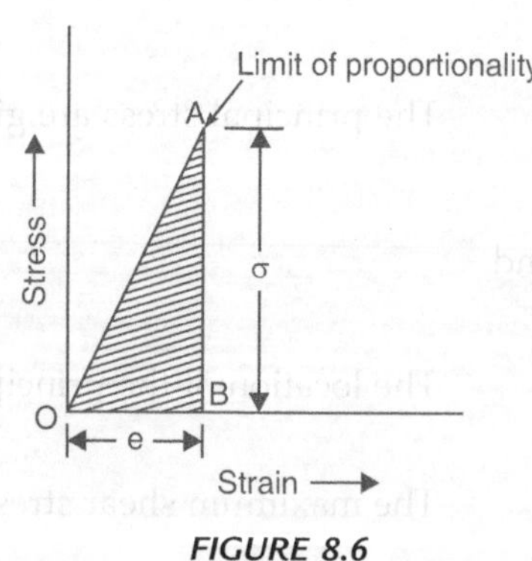

FIGURE 8.6

(*b*) If three principal stresses σ_1, σ_2 and σ_3 exist at a point, then strain energy stored per unit volume will be

$$= \frac{1}{2E}\left[\sigma_1^2 + \sigma_2^2 + \sigma_3^2 - \frac{2\sigma_1\sigma_2 + 2\sigma_2\sigma_3 + 2\sigma_3\sigma_1}{m}\right].$$

8.14. SHEAR FORCE DIAGRAM AND BENDING MOMENT DIAGRAM

8.14.1. Shear Force Diagram

It is a diagram which shows the values of shear force at various sections of a structural member.

8.14.2. Bending Moment Diagram

It is a diagram which shows the values of bending moment at various sections of a structural member.

8.14.3. Beam

It is a structural member subjected to a system of external forces at right angles to its axis. They are classified as:

(*i*) Cantilever, (*ii*) Simply supported,
(*iii*) Built-in or fixed, (*iv*) Continuous, and
(*v*) Overhanging.

(*i*) **Cantilever beam is** a structural member which is fixed at one end and free at the other end.

(*ii*) **Simply supported beam** is a beam whose both ends are freely resting on supports.

(*iii*) **Built-in or fixed beam** is a beam whose both ends are fixed. They are also called encastred beam or fixed beam.

(*iv*) **Continuous beam** is a beam which is having more than two supports.

(*v*) **Overhanging beam** is a beam whose one or both ends project beyond the supports.

8.14.4. Types of Loading

There are three types of loading:

(*i*) Concentrated loads or point loads,
(*ii*) Uniformly distributed loads, and
(*iii*) Uniformly varying loads.

8.14.5. Relation Between Shear Force (S.F.) and Bending Moment (B.M.) at a Section

(*i*) The shear force at any section is equal to rate of change of bending moment, *i.e.*,

$$\text{S.F.} = \frac{d}{dx}\,(\text{B.M.})$$

(*ii*) Intensity of loading (w) is equal to rate of change of shear force, *i.e.*, $w = \frac{d}{dx}$ (S.F.).

(*iii*) Where shear force changes sign, the bending moment will be minimum.

8.14.6. Point of Contra-flexure or Point of Inflexion

It is the point where the bending moment changes sign from +ve to –ve or *vice versa*.

8.14.7. Bending Equation

It is given by $\frac{M}{I} = \frac{E}{R} = \frac{\sigma}{y}$

where M = Bending moment,

I = Moment of inertia,

R = Radius of curvature, and

y = Distance from neutral axis.

8.14.8. Section Modulus

It is defined as the ratio of moment of inertia about the neutral axis to the distance from the neutral axis of the extreme fibre of the section. It is denoted by Z.

$\therefore$ $Z = \frac{I}{y}$, where y is the distance of the extreme fibre from neutral axis.

8.14.9. Differential Equations of Flexure

The followings are the differential equations of flexure:

(*a*) Bending moment, $M = EI\frac{d^2y}{dx^2}$

(*b*) Shear force, $F = \frac{dM}{dx} = \frac{EId^3y}{dx^3}$

(*c*) Rate of loading, $w = \frac{dF}{dx} = \frac{EId^4y}{dx^4}$

(*d*) Slope = $\frac{dy}{dx}$

(*e*) Deflection = y.

8.15. DEFLECTION AND SLOPE

The shear force diagram, bending moment diagram, deflection and slope for the following cases will be considered:

1. Uniform beam simply supported at the ends which consists of a
   (*a*) Point load at the centre,
   (*b*) Uniformly distributed load over the whole span, and
   (*c*) Point load at a distance '*a*' from one end and '*b*' from the other.
2. Cantilevers which consists of a
   (*a*) Point load at the free end,
   (*b*) Uniformly distributed load over the whole length, and
   (*c*) Uniformly varying load.
3. Fixed beam which consists of a
   (*a*) Point load at the centre, and
   (*b*) Uniformly distributed load over the whole span.

8.15.1. Uniform Beam Simply Supported at the Ends

(*a*) **Beam with a point load at the centre.** A beam *AB* of span *l*, simply supported at the ends *A* and *B* and carrying a point load (*W*) at the centre is shown in Fig. 8.7 (*a*).

(*i*) The shear force and bending moment diagrams are shown in Figs. 8.7 (*b*) and (*c*) respectively.

(*ii*) The bending moments are zero at the ends and maximum at the centre where shear force changes sign.

The maximum bending moment is equal to $\left(\frac{Wl}{4}\right)$.

(*iii*) **Deflection** at the ends *A* and *B* is zero. Maximum deflection occurs at the centre (*i.e.*, at point C). The maximum deflection at the centre

$$= \frac{Wl^3}{48EI}.$$

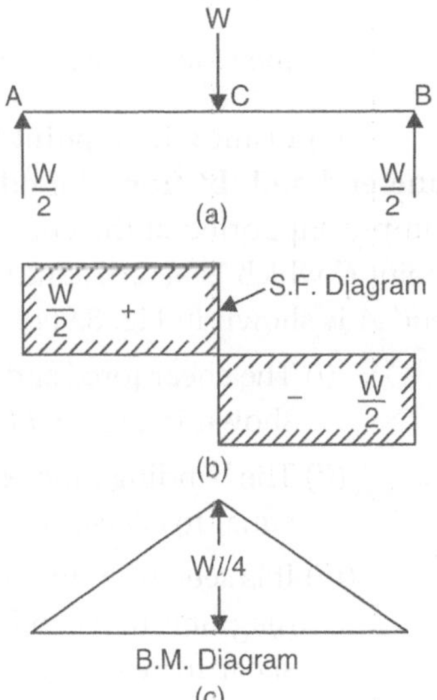

FIGURE 8.7

(*iv*) **The slope** at C is zero.

The slope at the end *A* is equal to $\frac{Wl^2}{16EI}$.

(*v*) The shear force diagram consists of two rectangles whereas the bending moment diagram consists of a triangle.

(*b*) **Beam with uniformly distributed load.** A beam *AB* of span *l*. Simply supported at the ends and carrying a uniformly distributed load of *w* per unit length over the whole span is shown in Fig. 8.8 (*a*).

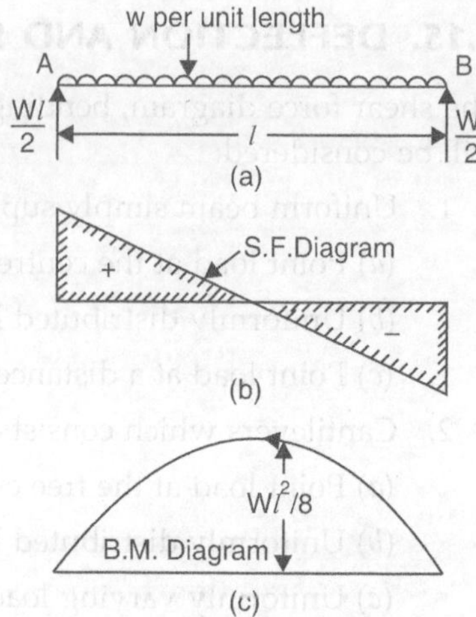

FIGURE 8.8

(*i*) The shear force and bending moment diagrams are shown in Figs. 8.8 (*b*) and (*c*) respectively.

(*ii*) The shear force diagram consists of two triangles and bending moment diagram is a parabola.

(*iii*) The shear force is zero at the centre, whereas the bending moment is maximum at the centre where shear force changes sign. The value of maximum bending moment is equal to ($wl^2/8$).

(*iv*) **The deflection** at the ends is zero and maximum deflection occurs at the centre. The value of maximum deflection at the centre is equal to

$$\frac{5wl^4}{384\,El} \quad \text{or} \quad \frac{5Wl^3}{384\,EI}$$

where W = Total load = $w \times l$.

(*v*) **The slope** at the centre is zero, whereas at the ends the slope is equal to $\frac{wl^2}{24EI}$ or $\frac{Wl^2}{24EI}$.

(*c*) **Beam with a point load (W) at a distance 'a' from one end and 'b' from the other end.** A beam *AB* of span *l*, simply supported at the ends and carrying a point load *W* at point *C* which is at a distance *'a'* from end *B* and *'b'* from the end *A* is shown in Fig. 8.9(*a*).

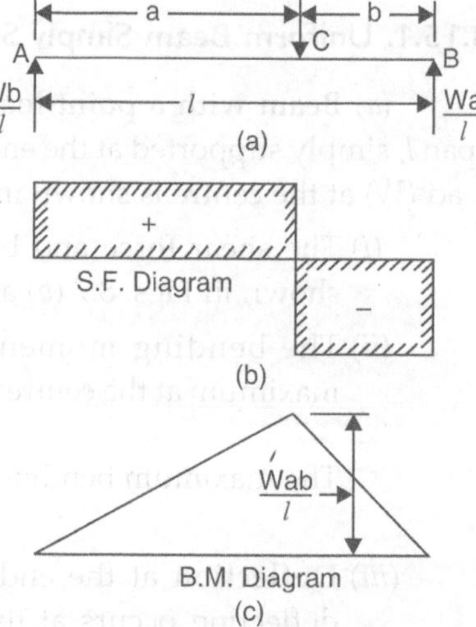

FIGURE 8.9

(*i*) The shear force and bending moment diagrams are shown in Figs. 8.9 (*b*) and (*c*) respectively.

(*ii*) The bending moment is maximum at point *C* and is equal to ($W\,ab/l$)

(*iii*) It is seen from the shear force and bending moment diagrams that bending moment is maximum where shear force changes sign.

(*iv*) **The deflection** at the ends *A* and *B* is zero whereas under the load *W* at point *C* it is equal to [$Wa^2b^2/3EI\,(a+b)$].

(*v*) If $a > b$, then maximum deflection occurs between *A* and *C*. The value of maximum deflection = $Wb(l^2 - b^2)^{3/2}/9\sqrt{3}\,EI\,(a+b)$.

(*vi*) The distance of the point, at which maximum deflection occurs, from *A* is equal to

$$\sqrt{\frac{(l^2 - b^2)}{3}} \quad \text{or} \quad \sqrt{\frac{a^2 + 2ab}{3}}.$$

8.15.2. Cantilevers

(*a*) **A cantilever with a point load (W) at the free end.** A cantilever *AB* of length *l*, carrying a point load (*W*) at the free end *B* is shown in Fig. 8.10(*a*). The shear force and bending moment diagrams are shown in Figs. 8.10(*b*) and (*c*) respectively. The shear force diagram is a rectangle, whereas bending moment diagram is a triangle. The bending moment is zero at the free end and maximum at the fixed end *A*.

(*i*) **The deflection** is zero at fixed end *A* and maximum at the free end *B*. The deflection at *B* is equal to $Wl^3/3EI$.

(*ii*) **The slope** at the free end $B = Wl^2/2EI$.

(*b*) **A cantilever carrying a uniformly distributed load of w per unit length over the whole length (*l*).** It is shown in Fig. 8.11 (*a*). The shear force and bending moment diagrams are shown in Figs. 8.11 (*b*) and (*c*) respectively. The shear force diagram is a triangle whereas the bending moment diagram is a parabola. The shear force and bending moments are zero at the free end *B* and they are maximum at the fixed end *A*. The maximum bending moment is equal to $\frac{wl^2}{2}$.

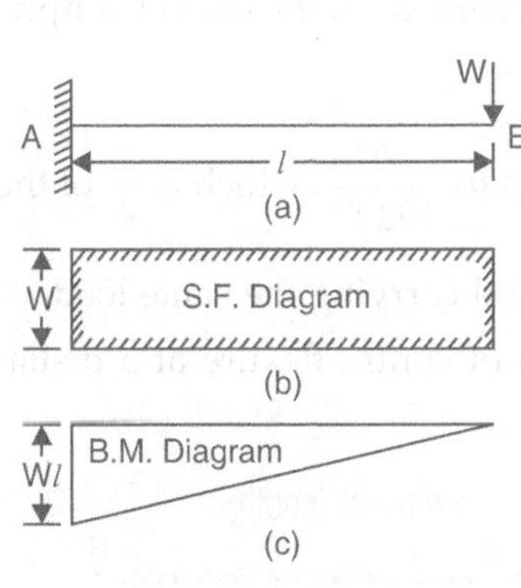

FIGURE 8.10

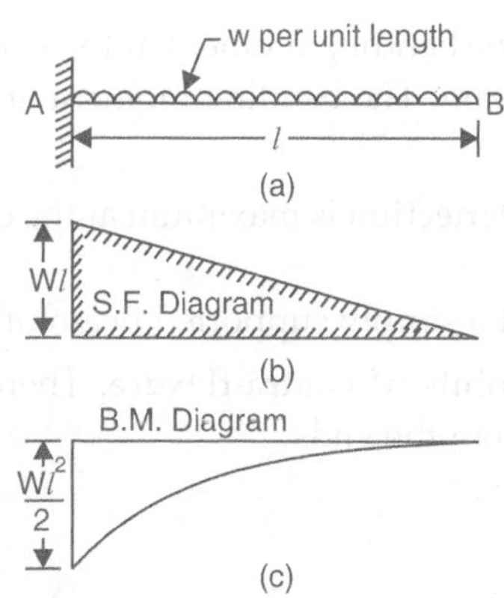

FIGURE 8.11

Deflection is maximum at the free end *B* and the magnitude of maximum deflection

$$= \frac{wl^4}{8EI} \quad \text{or} \quad \frac{Wl^4}{8EI}, \qquad \text{where } W = wl.$$

Slope at the free end *B* $= \frac{wl^5}{6EI}$ or $\frac{Wl^2}{6EI}$.

(*c*) **A cantilever of length (*l*) carrying a load whose intensity varies uniformly from zero at free end to w per unit length at the fixed end.** It is shown in Fig. 8.12(*a*). The shear force diagram and bending moment diagrams are shown in Figs. 8.12(*b*) and (*c*). The shear force diagram follows a parabolic curve whereas the bending moment diagram follows a cubic curve. The shear force and bending moments are maximum at the fixed ends and their values are :

Maximum shear force $= \dfrac{wl}{2}$

Maximum bending moment $= \dfrac{wl^2}{2}$

Deflection at the free end $= \dfrac{wl^4}{30\,EI}$

Slope at the free end $= \dfrac{wl^3}{24\,EI}$.

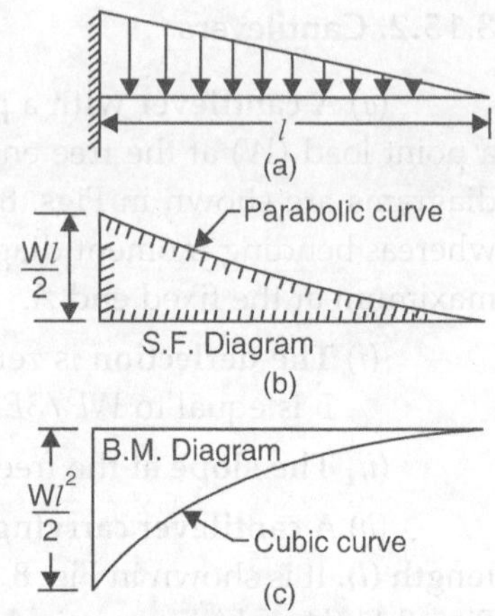

FIGURE 8.12

8.15.3. Fixed Beam

(*a*) **Fixed beam carrying a point load at the centre.** Figure 8.13 (*a*) shows a beam of span *l* carrying a point load (*W*) at the centre. The shear force and bending moment diagrams are shown in Figs. 8.13 (*b*) and (*c*) respectively.

(*i*) The bending moment at the ends is not equal to zero, as in the case of simply supported beam. The bending moment at the centre = $Wl/8$.

(*ii*) **Deflection** is maximum at the centre and is equal to $\dfrac{Wl^3}{192\,EI}$, which is $\dfrac{1}{4}$ of the deflection of a simply supported beam of the same span and carrying the same load at the centre.

(*iii*) **Points of contra-flexure.** There are two points of contra-flexure at a distance of $(l/4)$ from the ends.

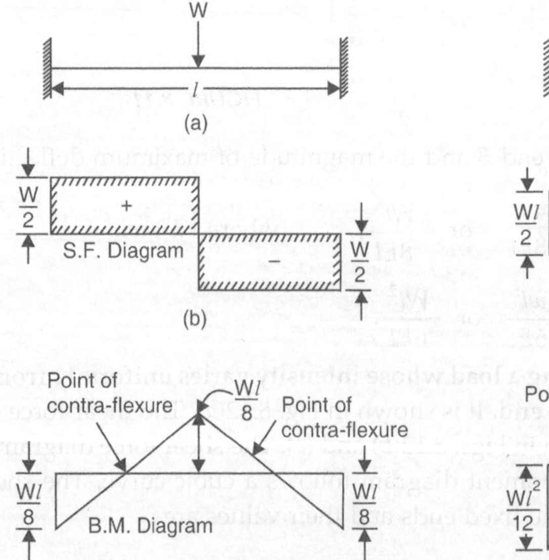

FIGURE 8.13

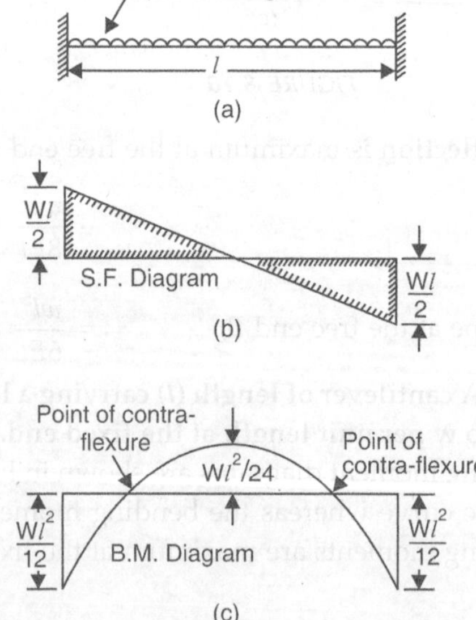

FIGURE 8.14

(*b*) **Fixed beam carrying a uniformly distributed load of w per unit length over the whole span.** This beam is shown in Fig. 8.14(*a*). The shear force and bending moment diagrams are shown in Figs. 8.14(*b*) and (*c*) respectively.

(*i*) The bending moments at the ends = $\frac{wl^2}{12}$ whereas at the centre it is equal to $\frac{wl^2}{24}$.

(*ii*) **The deflection** at the centre is equal to $\frac{wl^4}{384}$ which is $\frac{1}{5}$ of the central deflection of the simply supported beam of the same span and carrying the same uniform distributed load.

(*iii*) There are *two points of* contra-flexure, which are equidistant from the centre of the span at a distance of $(l/2\sqrt{3})$ from the centre of the span.

8.16. TORSION OF SHAFTS

When a circular shaft is subjected to equal and opposite twisting moments at its two ends, the shaft is said to be under pure torsion. The effect of torsion is to produce shear stress in the material of the shaft. The shear stress at any point of the shaft is proportional to its distance from the central axis of the shaft. The *torsion equation* for a shaft subjected to a torque is

$$\frac{\tau}{R} = \frac{q}{r} = \frac{C\theta}{l} = \frac{T}{J}$$

where τ = Shear stress at a distance R from the axis of shaft,

q = Shear stress at a distance r from the axis of the shaft,

l = Length of shaft,

θ = Angle of twist in radians,

C = Modulus of rigidity of the material of the shaft,

T = Torque applied, and

J = Polar moment of inertia of the shaft.

Torque transmitted. The torque transmitted by a rotating shaft is given by

$$T = \frac{\pi}{16}\tau D^3 \quad \text{For a solid shaft}$$

$$= \frac{\pi}{16}\tau\frac{(D^4 - d^4)}{D} \quad \text{For a hollow shaft}$$

where τ = Maximum allowable shear stress,

D = Outer diameter of the shaft, and

d = Inner diameter of the shaft.

The polar moment of inertia of a shaft is given by

$$J = \frac{\pi D^4}{32} \qquad \text{For a solid shaft}$$

$$= \frac{\pi}{32}(D^4 - d^4) \qquad \text{For a hollow shaft.}$$

8.16.1. Polar Modulus

Polar modulus of a shaft section is the ratio of polar moment of intertia of the shaft section to the maximum radius. The greatest twisting moment which a given shaft section can resist

= Maximum allowable shear stress × Polar modulus.

18.16.2. Torsional Rigidity

Torsional rigidity of a shaft is the product of modulus of rigidity (C) and polar moment of inertia (J). Torsional rigidity is also defined as the torque required to produce a twist of one radian per unit length of the shaft.

8.16.3. Power Transmitted by a Rotating Shaft

Power transmitted by a rotating shaft subjected to a torque (T) is given by

$$P = 2\pi NT/60{,}000$$

The strength of a shaft is measured by the amount of torque or power that the shaft can transmit. A *hollow shaft* of the same sectional area as that of solid shaft can transmit more torque. Hence a hollow shaft as compared to solid shaft is more strong.

8.16.4. Torsional Resilience

It is the amount of energy stored in a material of a shaft due to shear stress when the shaft is subjected to torsion. When the shear stress is uniform, the energy stored = $\frac{q^3}{2C}$ × volume. But in case of the shaft subjected to torsion, shear stress varies from zero at the axis to maximum value (τ) at the surface. The energy stored by the shaft subjected to torsion (or torsional resilience) is given by

$$\text{Torsional resilience} = \frac{\tau^2}{4C} \times \text{Volume of shaft}$$

where C = Modulus of rigidity.

For a hollow shaft of internal diameter d and external diameter D, the torsional resilience

$$= \frac{\tau^2}{4C}\left(\frac{D^2 + d^2}{D^2}\right) \times \text{Volume of the shaft.}$$

8.17. SPRINGS

Springs are the elastic bodies which absorb energy due to resilience. The absorbed energy may be released as and when required. They are of two types, *i.e.*, (*i*) Laminated or leaf springs, and (*ii*) Helical springs.

8.17.1. Laminated Springs

They consist of number of flat plates as shown in Fig. 8.15, connected in such a way that they can slide over the adjacent plates when load is applied on the spring. The plates are of equal width and thickness but are having different lengths so that the section modulus (Z) varies directly with the bending moment. These springs which are used in cars, lorries etc., get loaded at the ends and are supported in the middle.

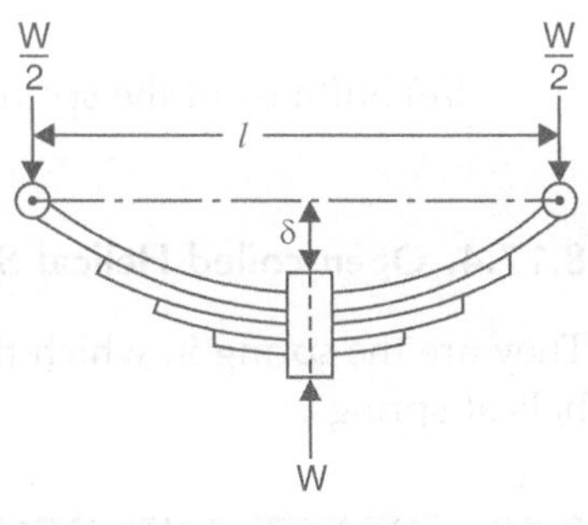

FIGURE 8.15

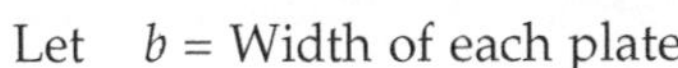

Let b = Width of each plate

n = Number of plates

l = Span of spring

σ_b = Maximum bending stress developed in the plates

t = Thickness of each plate

W = Point load acting at the centre of the spring.

The maximum bending stress,

$$\sigma_b = \frac{3}{2} \times \frac{Ml}{nbt^2}$$

and central deflection,
$$\delta = \frac{3}{8}\frac{Wl^3}{Enbt^3} = \frac{fl^2}{4Et}.$$

8.17.2. Helical Springs

They are thick spring wires coiled into a helix. They are of two types, *i.e.*, (*i*) Close-coiled, and (*ii*) Open-coiled.

8.17.3. Close-coiled Helical Springs

They are the springs in which helix angle is very small or in other words the pitch between two adjacent turns is small. A closed-coiled helical spring carrying an axial load W is shown in Fig. 8.16.

Let d = Diameter of spring wire

p = Pitch of the helical spring

n = Number of coils

R = Mean radius of spring coil

W = Axial load on spring

C = Modulus of rigidity

(*i*) Then maximum shear stress, $\tau = \dfrac{16\,WR}{nd^3}$

(*ii*) Strain energy stored in the spring $= \dfrac{32\,W^2 R^3 n}{Cd^4}$

(*iii*) Deflection in the spring, $\delta = \dfrac{64\,WR^3 n}{Cd^4}$

(*iv*) Stiffness of the spring, $s = \dfrac{Cd^4}{64\,R^3 n}$.

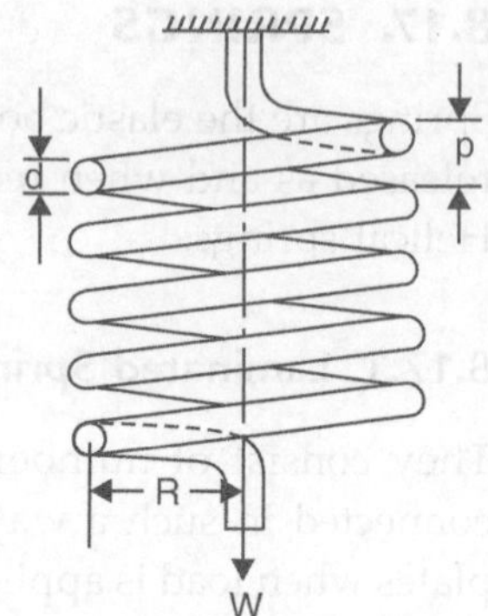

FIGURE 8.16

8.17.4. Open-coiled Helical Springs

They are the spring in which the pitch is large or helix angle is large as compared to close-coiled helical springs.

8.18. DIRECT AND BENDING STRESSES

If a short column carries an axial vertical point load, the intensity of stress on the section of the column will be uniform. This stress is called direct stress. But if the vertical point load is acting eccentrically (with an eccentricity *e*), the section will be subjected to direct and bending stresses. The direct stress is compressive whereas the bending stress at one end of the section will be compressive and at the other end it will be tensile. The resultant stress at one end will be compressive while at the other end the resultant stress may be compressive, zero or tensile depending upon the eccentricity. *The condition* for the stresses to remain wholly compressive is

(*i*) For rectangular section ($a \times b$) of the column, the load should be applied within the middle third of the base, *i.e.*, within the middle $\dfrac{a}{3}$ and $\dfrac{b}{3}$. The figure within which the load may be placed so as not to produce tensile stresses is called the *core* or kernel of the section.

(*ii*) For a solid circular section, they should be applied within concentric circle of diameter $d/4$.

8.19. THIN CYLINDERS AND SPHERES

A cylindrical vessel is said to be thin if the ratio of its internal diameter to the thickness of the wall is more than 20. When the thin cylinders are subjected to an internal fluid pressure. The longitudinal stress (*i.e.*, stress acting lengthwise) and circumferential stress (*i.e.*, stress acting along the circumference) are induced in the material of the vessel. These stresses are given as:

Longitudinal stress $= \dfrac{p \times D}{4t}$

Circumferential or hoop stress $= \dfrac{p \times D}{2t}$

where p = Internal fluid pressure,

D = Internal diameter of cylindrical vessel, and

t = Thickness of the vessel.

Both these stresses are tensile in nature. Also hoop stress is two times the longitudinal stress. The greatest shear stress = $\frac{p \times D}{8t}$.

8.19.1. Strains in Cylindrical Shells

The strains in a thin cylindrical shell are:

(*i*) Longitudinal strain = $\frac{p \times D}{4tE} - \frac{pD}{2tEm} = \frac{PD}{2tE}\left(\frac{1}{2} - \frac{1}{m}\right)$

(*ii*) Circumferential strain = $\frac{pD}{2tE} - \frac{pD}{4tEm} = \frac{pD}{2tE}\left(1 - \frac{1}{2m}\right)$

(*iii*) Volumetric strain = longitudinal strain + 2 × circumferential strain.

8.19.2. Riveted Cylindrical Shells

In case of riveted cylindrical shells, the efficiency of the joints is to be taken into account while determining the longitudinal and circumferential stresses. If

η_e = Efficiency of the circumferential joint, and

η_l= Efficiency of the longitudinal joint, then

Hoop or circumferential stress = $\frac{p \times D}{2t\eta \cdot t}$ and

Longitudinal stress = $\frac{p \times D}{4t\eta_e}$.

8.19.3. Thin Spherical Shells

In case of thin spherical shell, the hoop stress and longitudinal stress at any point are equal to $\frac{p \times D}{4t}$. There is no shear stress anywhere in the shell. The strain in any direction

$$= \frac{p \times D}{4tE} - \frac{p \times D}{4tEm} = \frac{pD}{4tE}\left(1 - \frac{1}{m}\right)$$

Volumetric strain = $\frac{3pD}{4tE}\left(1 - \frac{1}{m}\right)$.

8.20. THICK CYLINDERS AND SPHERES

In case of thick cylinders at any point the three stresses (namely radial stress, hoop stress and longitudinal stress) are acting. The hoop stress was assumed uniform across the thickness of thin cylinder. But in case of thick cylinder, the hoop stress varies from maximum value at the inner circumference to a minimum value at the outer circumference. The longitudinal stress is uniformly distributed. These stresses are given as

$$\left.\begin{aligned}\text{Radial stress} &= \frac{b}{x^2} - a \\ \text{Hoop stress} &= \frac{b}{x^2} + a\end{aligned}\right\}$$, where a and b are constants and are determined from initial known conditions and x is any radius.

$$\text{Longitudinal stress} = \frac{p \times d}{4t}.$$

8.20.1. Thick Spherical Shells

In case of thick spherical shells;

$$\text{Radial stress} = \frac{2b}{x^3} - a$$

$$\text{Hoop stress} = \frac{b}{x^3} + a$$

where a and b are constants and are evaluated from the known initial conditions, x is any radius.

8.21. COLUMNS AND STRUTS

Column are the vertical members of a structure which carry axial compressive loads. The struts are any members of a structure which carry axial compressive loads. The maximum axial compressive load which a column can take without failure is known as *critical load* or *buckling load or crippling load.*

8.21.1. Slenderness Ratio

It is the ratio of the length of the column to the least radius of gyration of the cross-sectional area of the column. The greater the slenderness ratio the lesser will be the critical load.

8.21.2. Buckling Factor

It is the ratio of the equivalent length of the column to the least radius of gyration of the cross-sectional area of the column.

8.21.3. Equivalent Length

Equivalent length of a column with given end condition is the length of an equivalent column of the same material and same section with both ends hinged having the value of crippling load equal to that of the given column.

8.21.4. End Condition

End conditions of a loaded column are

(*i*) one end fixed and the other end is free
(*ii*) both ends hinged

(*iii*) both ends fixed

(*iv*) one end is fixed and other end is hinged.

8.21.5. Euler's Theory of Long Columns

In case of long columns, the stress due to direct load is very small in the comparison with the stress due to bending. Hence in Euler's theory of long columns, the failure is due to bending only. The crippling load (P), according to this theory for various end conditions of long columns, along with their equivalent lengths are given below:

l = *Actual length of column* l_e = *Effective length*			
S.No.	*End conditions*	*Crippling load (P)*	*Equivalent length in terms of actual length*
1.	One end fixed, other end free	$P = \dfrac{\pi^2 EI}{4l^2}$	$l_e = 2l$
2.	Both ends hinged	$P = \dfrac{\pi^2 EI}{l^2}$	$l_e = l$
3.	Both ends fixed	$P = \dfrac{4\pi^2 EI}{l^2}$	$l_e = \dfrac{l}{2}$
4.	One end fixed other end hinge	$p = \dfrac{2\pi^2 EI}{l^2}$	$l_e = \dfrac{l}{\sqrt{2}}$

The crippling load by Euler's formula in general can be written as

$$P = \frac{\pi^2 EI}{l_e^2}$$

where l_e = Equivalent length and I = Least moment of inertia.

8.21.6. Rankine's Formula

It is an empirical formula and the crippling load (P) according to this formula is given by

$$P = \frac{\sigma_c A}{1 + a\left(\frac{l_e}{k}\right)^2}$$

where $a = \dfrac{\sigma_c}{\pi^2 E} = \dfrac{1}{7500}$ for mild steel

σ_c = Crushing stress of the material of the column

$\dfrac{l_e}{k}$ = Buckling factor = $\dfrac{\text{Equivalent length}}{\text{Least radius of gyration}}$

k = Least radius of gyration.

8.21.7. Gordon's Formula

The crippling load by this formula is given by

$$P = \frac{\sigma_c A}{1 + b\left(\frac{l_e}{d}\right)^2}$$

where b is a constant and d is the least diameter or breadth of the section, σ_c = Crushing stress of column material.

8.21.8. Johnson's Parabolic Formula

It is given by $P = \sigma_c A - eA\left(\frac{l_e}{k}\right)^2$

where e = A constant, A = Area of cross-section of column

l_e = Equivalent length, k = Least radius of gyration.

8.22. STRESS-STRAIN CURVE FOR A MILD STEEL BAR

It is obtained by performing a tensile test on a piece of mild steel of uniform section and of given lengths by increasing the load from zero to the load till the piece breaks. The strain (which is the ratio of increase of length to the original length) and stress (which is the ratio of load applied to the cross-section area of the bar) are calculated for all loads and a graph between stress and strain is plotted as shown in Fig. 8.17.

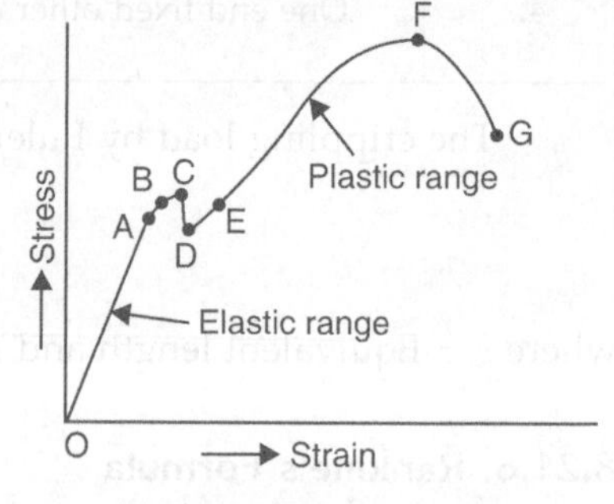

FIGURE 8.17

The strain is proportional to the stress upto point A, which is known as *limit of proportionality*. Hooke's Law holds good upto point A. When the load of stress is increased beyond A and upto point B, the material is elastic and can regain its shape and size when the load is removed. But Hooke's Law is not applicable. The point B upto which material is elastic is called the *elastic limit*. Beyond the point B, if material is stressed, it will not be elastic but it will be plastic, *i.e.*, the material will not regain its shape and size. When the load is removed. The point C on the curve represents upper yield point whereas point D represents lower yield point. At point E *yielding*, the process by which strain increasing without increasing stress, ceases. After point E, strain can only be increased if load (or stress) is increased. The stress corresponding to point F is called *maximum stress* or *ultimate stress*. From point F to G, it is observed that the piece continues to extend even if the load is no longer increased till it breaks. The stress corresponding to point G is known as *breaking stress*.

II. OBJECTIVE TYPE QUESTIONS

Tick mark the most appropriate statement of the multiple choice answers:

Basic Definitions

1. With elastic limit in a loaded material, stress is
   (*a*) inversely proportional to strain ☐ (*b*) directly proportional to strain ☐
   (*c*) equal to strain ☐ (*d*) none of the above. ☐
2. The ratio of linear stress to linear strain is known as
   (*a*) Poisson's ratio ☐ (*b*) bulk modulus ☐
   (*c*) modulus of rigidity ☐ (*d*) modulus of elasticity. ☐
3. The ratio of lateral strain to longitudinal strain is called
   (*a*) Poisson's ratio ☐ (*b*) bulk modulus ☐
   (*c*) modulus of rigidity ☐ (*d*) modulus of elasticity. ☐
4. The ratio of shear stress to shear strain is called
   (*a*) Poisson's ratio ☐ (*b*) bulk modulus ☐
   (*c*) modulus of rigidity ☐ (*d*) modulus of elasticity. ☐
5. The ratio of normal stress of each face of a solid cube to volumetric strain is called
   (*a*) Poisson's ratio ☐ (*b*) bulk modulus ☐
   (*c*) modulus of rigidity ☐ (*d*) modulus of elasticity. ☐
6. Hooke's law holds good upto
   (*a*) proportional ☐ (*b*) yield point ☐
   (*c*) elastic limit ☐ (*d*) plastic limit. ☐
7. The property of a material by virtue of which a body returns to its original shape after removal of the load is known as
   (*a*) ductility ☐ (*b*) plasticity ☐
   (*c*) elasticity ☐ (*d*) resilience. ☐
8. A tensile force (*P*) is acting on a body of length (*l*) and area of cross-section (*A*). The change in length would be
   (*a*) $\frac{P}{lAE}$ ☐ (*b*) $\frac{PE}{Al}$ ☐
   (*c*) $\frac{Pl}{AE}$ ☐ (*d*) $\frac{Al}{PE}$ ☐
9. The modulus of elasticity (*E*) and modulus of rigidity (*C*) are related by
   (*a*) $C = \frac{mE}{3(m-2)}$ ☐ (*b*) $C = \frac{mE}{2(m+1)}$ ☐
   (*c*) $C = \frac{3(m-2)}{mE}$ ☐ (*d*) $C = \frac{2(m+1)}{mE}$ ☐

   where $\frac{1}{m}$ = Poisson's ratio.

10. The modulus of elasticity (*E*) and bulk modulus (*K*) are related by

(*a*) $K = \frac{mE}{3(m-2)}$ ☐ (*b*) $K = \frac{mE}{2(m+1)}$ ☐

(*c*) $K = \frac{3(m-2)}{mE}$ ☐ (*d*) $K = \frac{2(m+1)}{mE}$ ☐

where $\frac{1}{m}$ = Poisson's ratio.

11. Figure 8.18 shows the stress-strain curve for mild steel. Hook's law holds good upto

(*a*) point *A* ☐

(*b*) point *B* ☐

(*c*) point *C* ☐

(*d*) point *D*. ☐

FIGURE 8.18

12. The point *A* in Fig. 8.18 represents

(*a*) proportional limit ☐

(*b*) upper yield point ☐

(*c*) lower yield point ☐

(*d*) breaking point. ☐

13. The point *C* in Fig. 8.18 represents

(*a*) elastic limit ☐ (*b*) upper yield point ☐

(*c*) lower yield point ☐ (*d*) breaking point. ☐

14. The point *D* in Fig. 8.18 represents

(*a*) elastic limit ☐ (*b*) upper yield point ☐

(*c*) lower yield point ☐ (*d*) breaking point. ☐

15. The point *F* in Fig. 8.18 represents

(*a*) elastic limit ☐ (*b*) upper yield point ☐

(*c*) lower yield point ☐ (*d*) breaking point. ☐

16. The stress corresponding to point *E* in Fig. 8.18 is called

(*a*) breaking stress ☐ (*b*) ultimate stress ☐

(*c*) yield point stress ☐ (*d*) nominal stress. ☐

17. The stress corresponding to point *F* in Fig. 8.18 is called

(*a*) breaking stress ☐ (*b*) ultimate stress ☐

(*c*) yield point stress ☐ (*d*) nominal stress. ☐

18. A rod of steel of 1 cm^2 in cross-sectional area and 100 cm long is subjected to an axial pull of 20000 N. If $E = 20 \times 10^6$ N/cm^2, the elongation will be

(*a*) 1 cm ☐ (*b*) 0.2 cm ☐

(*c*) 0.1 cm ☐ (*d*) 0.15 cm. ☐

19. The modulus of elasticity in S.I. units is expressed as

(*a*) N/m ☐ (*b*) N/m^2 ☐

(*c*) Nm/S ☐ (*d*) no unit. ☐

20. The stress at any section of a rod (due to its own weight) of length (l) and diameter (d) rigidly fixed at the upper end and hanging is

(*a*) directly proportional to x ☐ (*b*) inversely proportional to x ☐

(*c*) directly proportional to x^2 ☐ (*d*) inversely proportional to x^2 ☐

where x = Distance of section from the lower end.

21. The elongation produced in a rod (by its own weight) of length (l) and diameter (d) rigidly fixed at the upper end and hanging is equal to

(*a*) $\frac{wl}{2E}$ ☐ (*b*) $\frac{wl^2}{2E}$ ☐

(*c*) $\frac{wl^3}{2E}$ ☐ (*d*) $\frac{wl^4}{2E}$ ☐

where w = Weight per unit volume of the rod, E = Modulus of elasticity.

22. The ratio of modulus of rigidity to modulus of elasticity for a Poisson's ratio of 0.25 would be

(*a*) 0.5 ☐ (*b*) 0.4 ☐

(*c*) 0.3 ☐ (*d*) 1.0 ☐

23. The ratio of bulk modulus to modulus of elasticity for a Poisson's ratio of 0.25 would be

(*a*) 2/3 ☐ (*b*) 1/3 ☐

(*c*) 4/3 ☐ (*d*) 1.0. ☐

24. The relation between modulus of elasticity (E), modulus of rigidity (C) and bulk modulus (K) is given by

(*a*) $E = \frac{3KC}{C + 9K}$ ☐ (*b*) $E = \frac{9KC}{C + 3K}$ ☐

(*c*) $E = \frac{C + 9K}{3KC}$ ☐ (*d*) $E = \frac{C + 3K}{9KC}$ ☐

25. The ratio of modulus of rigidity of bulk modulus for a Poisson's ratio of 0.25 would be

(*a*) 2/3 ☐ (*b*) 2/5 ☐

(*c*) 3/5 ☐ (*d*) 4/5. ☐

26. The work done in producing strain on a material per unit volume is called

(*a*) resilience ☐ (*b*) ductility ☐

(*c*) elasticity ☐ (*d*) plasticity. ☐

27. The properly by virtue of which a metal can be beaten into plates is called

(*a*) ductility ☐ (*b*) malleability ☐

(*c*) resilience ☐ (*d*) plasticity. ☐

28. Choose the wrong statement

(*a*) Elongation produced in a rod (by its own weight) which is rigidly fixed at the upper end and hanging is equal to that produced by a load half its weight applied at the end. ☐

(*b*) The stress at any section of a rod on account of its own weight is directly proportional to the distance of the section from the lower end. ☐

(c) Modulus of elasticity is having the same unit as stress. □

(d) If a material expands freely due to heating, it will develop thermal stresses. □

29. A rod of length (l) tapers uniformly from a diameter D_1 to a diameter D_2 and carries an axial tensile load P. The extension of the rod would be

(a) $\frac{\pi Pl}{4ED_1D_2}$ □ (b) $\frac{4Pl}{\pi ED_1D_2}$ □

(c) $\frac{\pi EPl}{4D_1D_2}$ □ (d) $\frac{4PEl}{\pi D_1D_2}$. □

30. If in question 29, the rod is of uniform diameter (D), then extension would be

(a) $\frac{\pi Pl}{4ED^2}$ □ (b) $\frac{4Pl}{\pi ED^2}$ □

(c) $\frac{\pi EPl}{4D^2}$ □ (d) $\frac{4PEl}{\pi D^2}$. □

31. A rod of length l and cross-sectional area A rotates about an axis passing through one end of the rod. The extension produced in the bar due to centrifugal forces would be

(a) $\frac{w\omega l^2}{3E}$ □ (b) $\frac{w\omega^2 l^3}{gE}$ □

(c) $\frac{w\omega^2 l^3}{3gE}$ □ (d) $\frac{3gE}{w\omega^2 l^3}$. □

32. For the bars of composite section

(a) the load carried by different materials is the same as the total external load □

(b) the extension in different materials is different □

(c) the total external load is equal to the total sum of the loads carried by different materials □

(d) strain in all materials is equal □

(e) both (c) and (d). □

33. Two materials are having modulus of elasticities, modulus of rigidities and bulk modulus as (E_1, E_2), (C_1, C_2), and (K_1, K_2). The modular ratio is given by

(a) $\frac{E_1}{C_2}$ □ (b) $\frac{E_1}{K_2}$ □

(c) $\frac{K_1}{E_2}$ □ (d) $\frac{E_1}{K_2}$. □

34. For uniform strength, a bar which is fixed at the upper end and is subjected to an external load P at the lower end, the area (A) at any section at a distance x from the lower end is given by

(a) $A_2 e^{\frac{w}{px}}$ □ (b) $A_2 e^{\frac{p}{wx}}$ □

(c) $A_2 e^{\frac{wx}{p}}$ □ (d) $A_2 \times \frac{p}{wx}$ □

where A_2 = Area at the lower end, w = Weight per unit volume, p = Uniform stress intensity in the bar.

35. The thermal stress is given by

(a) $E\alpha T$ ☐ (b) $\frac{ET}{\alpha}$ ☐

(c) $\frac{E\alpha}{T}$ ☐ (d) $\frac{1}{E\alpha T}$ ☐

where α = Co-efficient of linear expansion, T = Rise in temperature.

36. A steel tyre of internal diameter d is heated so that it can be shrunk on a wheel of diameter D. If now the steel tyre be cooled, it will grip the wheel. The tensile stress induced circumferentially is called

(a) shear stress ☐ (b) hoop stress ☐

(c) longitudinal stress ☐ (d) ultimate stress. ☐

37. In question 36, the strain produced due to cooling is equal to

(a) $\frac{D+d}{d}$ ☐ (b) $\frac{D\times d}{D-d}$ ☐

(c) $\frac{D-d}{d}$ ☐ (d) $\frac{d}{D+d}$. ☐

38. The stress produced due to cooling in question 36, is equal to

(a) $\left(\frac{D+d}{d}\right)E$ ☐ (b) $\left(\frac{D\times d}{D-d}\right)E$ ☐

(c) $\left(\frac{D-d}{d}\right)E$ ☐ (d) $\left(\frac{d}{D+d}\right)E$. ☐

39. Choose the correct statement

(a) The strain produced per unit volume is called resilience ☐

(b) The maximum strain produced per unit volume is called proof resilience ☐

(c) The least strain energy stored in a unit volume is called proof resilience ☐

(d) The greatest strain energy stored in a unit volume of a material without permanent deformation is called proof resilience. ☐

40. Choose the wrong statement

(a) The thermal stresses are only set up in the bodies, when they are not allowed to expand or contract freely with the rise or fall of temperature. ☐

(b) If α is the co-efficient of linear expansion and T is rise in temperature, then thermal stress is equal to $E\alpha T$. ☐

(c) The thermal stress is independent of cross-sectional area of the bar. ☐

(d) Rubber is more elastic than steel. ☐

41. The proof resilience is given by

(a) $\frac{2\sigma^2}{E}$ ☐ (b) $\frac{2E}{\sigma^2}$ ☐

(c) $\frac{\sigma^2}{2E}$ ☐ (d) $\frac{E}{2\sigma^2}$. ☐

where σ = Stress at the elastic limit.

42. If D is the diameter of a sphere, then volumetric strain is equal to

(*a*) two times the strain of diameter ☐ (*b*) 1.5 times the strain of diameter ☐

(*c*) three times the strain of diameter ☐ (*d*) the strain of diameter. ☐

43. If l be the length and D be the diameter of a cylindrical rod, then volumetric strain of the rod is equal to

(*a*) strain of length plus strain of diameter ☐

(*b*) strain of diameter ☐

(*c*) strain of length + twice the strain of diameter ☐

(*d*) strain of length. ☐

44. The extension per unit length of the rod due to suddenly applied load as compared to the same load gradually applied to the same rod is

(*a*) same ☐ (*b*) double ☐

(*c*) three times ☐ (*d*) half. ☐

45. The stress due to suddenly applied load as compared to the stress due to the same load gradually applied to the same rod is

(*a*) half ☐ (*b*) same ☐

(*c*) double ☐ (*d*) three times. ☐

46. The instantaneous stress (f) induced in a bar of length (l) and of cross-sectional area (A) due to impact load P, falling from a height h, is equal to

(*a*) $\frac{P}{A}+\sqrt{\frac{P^2}{A^2}+\frac{2PhE}{Al}}$ ☐ (*b*) $\frac{P}{A}+\sqrt{\frac{P^2}{A^2}-\frac{2PhE}{Al}}$ ☐

(*c*) $\frac{P}{A}+\sqrt{1+\frac{2PhE}{Al}}$ ☐ (*d*) $\frac{P}{A}+\sqrt{1+\frac{2hAE}{Pl}}$ ☐

47. The cross-sectional area of a rod is 10 cm^2. A pull of 10 tonnes is suddenly applied to the rod. The maximum stress produced in the rod would be

(*a*) 1 tonne/cm^2 ☐ (*b*) 2 tonne/cm^2 ☐

(*c*) 0.5 tonne/cm^2 ☐ (*d*) 4 tonne/cm^2. ☐

48. A steel specimen 1.5 cm^2 in cross-section stretches by 0.005 cm over a 5 cm gauge length under an axial load of 3000 N. The strain energy stored in the specimen at this stage would be

(*a*) 15 N cm ☐ (*b*) 7.5 N cm ☐

(*c*) 30 N cm ☐ (*d*) 10 N cm. ☐

49. In question 48, if the load at the elastic limit for the specimen as 5000 kg, the elongation at the elastic limit would be

(*a*) $\frac{1}{100}$ cm ☐ (*b*) $\frac{1}{50}$ cm ☐

(*c*) $\frac{1}{120}$ cm ☐ (*d*) $\frac{1}{150}$ cm. ☐

50. In questions 48 and 49 the proof resilience would be

(a) 15 N cm ☐ (b) 7.5 N cm ☐

(c) 20.833 N cm ☐ (d) 10 N cm. ☐

51. The tensile force at a distance y from support in a vertical hanging bar of length l which carries a load P at the bottom is equal to

(a) P ☐ (b) $P + wl$ ☐

(c) $P + w(l - y)$ ☐ (d) $P + wy$ ☐

where w = Weight per unit length.

Stresses on Inclined Plane

52. The normal stress on an oblique plane at an angle θ to the cross-section of a body which is subjected to a direct tensile stress (p) is equal to

(a) $\frac{p}{2} \sin 2\theta$ ☐ (b) $p \cos \theta$ ☐

(c) $p \cos^2 \theta$ ☐ (d) $p \sin^2 \theta$. ☐

53. In question 52, the normal stress on the oblique plane will be maximum when θ is equal to

(a) 90° ☐ (b) 45° ☐

(c) 0° ☐ (d) 30°. ☐

54. The tangential or shear stress on an oblique plane at an angle θ to the cross-section of a body which is subjected to a direct tensile stress (p) is equal to

(a) $\frac{p}{2} \sin 2\theta$ ☐ (b) $p \cos \theta$ ☐

(c) $p \cos^2 \theta$ ☐ (d) $p \sin^2 \theta$. ☐

55. In question 54, the tangential or shear stress on the oblique plane will be maximum when θ is equal to

(a) 90° ☐ (b) 45° ☐

(c) 0° ☐ (d) 30°. ☐

56. The resultant stress on an oblique plane at an angle θ to the cross-section of a body which is subjected to a direct tensile stress is

(a) $p/2 \sin 2\theta$ ☐ (b) $p \cos \theta$ ☐

(c) $p \cos^2 \theta$ ☐ (d) $p \sin^2 \theta$. ☐

57. If a member is subjected to an axial tensile load, the plane normal to the axis of loading carries

(a) minimum normal stress ☐ (b) maximum normal stress ☐

(c) maximum shear stress ☐ (d) none of the above. ☐

58. If a member is subjected to an axial tensile load, the plane inclined at 45° to the axis of loading carries.

(a) minimum shear stress ☐ (b) maximum normal stress ☐

(c) maximum shear stress ☐ (d) none of the above. ☐

59. The maximum shear stress induced in a member which is subjected to an axial load is equal to

(*a*) maximum normal stress ☐ (*b*) half of maximum normal stress ☐
(*c*) twice the maximum normal stress ☐ (*d*) thrice the maximum normal stress. ☐

60. If a member, whose tensile strength is more than two times the shear strength, is subjected to an axial load upto failure, the failure of the member will occur by

(*a*) maximum normal stress ☐ (*b*) maximum shear stress ☐
(*c*) normal stress or shear stress ☐ (*d*) none of the above. ☐

61. Principal plane is a plane on which shear stress is

(*a*) maximum ☐ (*b*) minimum ☐
(*c*) average of maximum and minimum ☐ (*d*) zero. ☐

62. Principal stresses are the stresses acting normal to

(*a*) a plane ☐ (*b*) an oblique plane ☐
(*c*) a principal plane ☐ (*d*) a plane having minimum shear stress. ☐

63. A member of cross-sectional area 10 cm^2 is subjected to an axial load of 10,000 N. A plane makes an angle of 30° to the cross-section. The normal stress on this plane will be

(*a*) 1000 N ☐ (*b*) 750 N ☐
(*c*) 433 N ☐ (*d*) 0. ☐

64. For the question 63, the shear stress or tangential stress on the 30° plane will be

(*a*) 1000 N ☐ (*b*) 750 N ☐
(*c*) 433 N ☐ (*d*) 0. ☐

65. If in question 63, the plane makes an angle of 60° instead of 30° to the cross-section, the normal stress on 60° plane will be

(*a*) double of 30° plane ☐ (*b*) equal to 30° plane ☐
(*c*) half of 30° plane ☐ (*d*) one-third of 30° plane. ☐

66. For the question 65, the shear or tangential stress on 60° plane will be

(*a*) double of 30° plane ☐ (*b*) equal to 30° plane ☐
(*c*) half of 30° plane ☐ (*d*) one-third of 30° plane. ☐

67. A body is subjected to two principal stresses σ_1 and σ_2 in two mutually perpendicular directions, the normal stress, on an oblique plane at an angle θ with the principal plane on which σ_1 is acting, is equal to

(*a*) $\frac{\sigma_1+\sigma_2}{2}+\frac{\sigma_1-\sigma_2}{2}\sin 2\theta$ ☐ (*b*) $\frac{\sigma_1+\sigma_2}{2}+\frac{\sigma_1-\sigma_2}{2}\cos 2\theta$ ☐

(*c*) $\frac{\sigma_1-\sigma_2}{2}\sin 2\theta$ ☐ (*d*) $\sqrt{\sigma_1^2\cos^2\theta+\sigma_2^2\sin^2\theta}$. ☐

68. For the question 67, the tangential stress on the oblique plane is equal to

(*a*) $\frac{\sigma_1+\sigma_2}{2}+\frac{\sigma_1-\sigma_2}{2}\sin 2\theta$ ☐ (*b*) $\frac{\sigma_1+\sigma_2}{2}+\frac{\sigma_1-\sigma_2}{2}\cos 2\theta$ ☐

(*c*) $\frac{\sigma_1-\sigma_2}{2}\sin 2\theta$ ☐ (*d*) $\sqrt{\sigma_1^2\cos^2\theta+\sigma_2^2\sin^2\theta}$. ☐

69. For the question 67, the resultant stress on the oblique plane is equal to

(a) $\frac{\sigma_1+\sigma_2}{2}+\frac{\sigma_1-\sigma_2}{2}\sin 2\theta$ ☐ (b) $\frac{\sigma_1+\sigma_2}{2}+\frac{\sigma_1-\sigma_2}{2}\cos 2\theta$ ☐

(c) $\frac{\sigma_1-\sigma_2}{2}\sin 2\theta$ ☐ (d) $\sqrt{\sigma_1^2\cos^2\theta+\sigma_2^2\sin^2\theta}$. ☐

70. Choose the wrong statement

(a) To avoid the shear failure of a material (which is subjected to a tensile load) along a plane at an angle of 45° to the direction of tensile stress, the material should have its shear strength at least equal to half the tensile strength. ☐

(b) Key is made weaker link in the design of pulley, key and shaft. ☐

(c) The planes, which carry no shear stress, are known as principal planes. ☐

(d) The normal stress on an oblique plane will be maximum when the angle of oblique plane with cross-section is 45°. ☐

71. Choose the correct statement

(a) The circumferential stress induced in a thin-walled cylindrical vessel is $(pd/4t)$. ☐

(b) The longitudinal stress induced in a thin walled cylindrical vessel is $(pd/2t)$. ☐

(c) The strain in a cylindrical bar (of length L metres) which deforms by l cm is $0.01\ l/L$. ☐

(d) When a body is subjected to a direct tensile stress (p), the maximum tangential stress is equal to the direct tensile stress. ☐

72. A body is subjected to two principal stresses σ_1 and σ_2 in two mutually perpendicular directions. Obliquity is the angle made by

(a) oblique plane with the direction of σ_1 ☐

(b) oblique plane with the direction of σ_2 ☐

(c) the line of action of the resultant stress with the normal of the oblique plane ☐

(d) the line of action of the resultant stress with the oblique plane. ☐

73. The obliquity (ϕ) is expressed mathematically as

(a) $\tan\phi=\frac{\sigma_1}{\sigma_r}$ ☐ (b) $\tan\phi=\frac{\sigma_2}{\sigma_r}$ ☐

(c) $\tan\phi=\frac{\sigma_1}{\sigma_n}$ ☐ (d) $\tan\phi=\frac{\sigma_r}{\sigma_n}$ ☐

where σ_1 and σ_2 = Two principal stresses at right angles, σ_r = Resultant stress

σ_r and σ_n = Tangential and normal stresses on an oblique plane.

74. Two principal tensile stresses of magnitudes 400 N/cm² and 200 N/cm² are acting at a point across two perpendicular planes. An oblique plane makes an angle of 30° with the major principal plane. The normal stress on the oblique plane will be

(a) 600 N/cm² ☐ (b) 350 N/cm² ☐

(b) 86.6 N/cm² ☐ (d) 173.2 N/cm² ☐

75. In question 74, the tangential stress along the oblique plane will be

(a) 600 N/cm² ☐ (b) 350 N/cm² ☐

(b) 86.6 N/cm² ☐ (d) 173.2 N/cm² ☐

76. In question 74, the resultant stress acting on the oblique plane is

(*a*) $\sqrt{600^2 + 86.6^2}$ ☐ (*b*) $\sqrt{350^2 + 86.6^2}$ ☐

(*c*) $\sqrt{350^2 + 173.2^2}$ ☐ (*d*) $\sqrt{86.6^2 + 173.2^2}$ ☐

77. The obliquity is equal to

(*a*) $\tan^{-1} \dfrac{86.6}{350}$ ☐ (*b*) $\tan^{-1} \dfrac{86.6}{600}$ ☐

(*c*) $\tan^{-1} \dfrac{350}{86.6}$ ☐ (*d*) $\tan^{-1} \dfrac{600}{86.6}$ ☐

78. Choose the wrong statement

(*a*) The resultant, of the two equal and like principal stresses, on any plane is equal to either of the principal stresses. ☐

(*b*) The resultant stress of the two equal and like principal stresses on any plane, is normal to the plane. ☐

(*c*) The resultant, of the two equal and unlike principal stresses, on any plane is equal is to either of the principal stresses. ☐

(*d*) The resultant stress of two equal and unlike principal stresses on any plane, is normal to the plane. ☐

79. Two unequal like principal stresses are acting at a point across two perpendicular directions. An oblique plane makes an angle θ with the major principal plane. As the angle θ increases, obliquity (ϕ)

(*a*) decreases ☐ (*b*) increases ☐

(*c*) first increases then decreases ☐ (*d*) remains constant. ☐

80. In question 79, the maximum obliquity (ϕ_{max}) is given by

(*a*) $\phi_{max} = \tan^{-1} \dfrac{\sigma_1 - \sigma_2}{\sigma_1 + \sigma_2}$ ☐ (*b*) $\phi_{max} = \tan^{-1} \dfrac{\sigma_1 + \sigma_2}{\sigma_1 - \sigma_2}$ ☐

(*c*) $\phi_{max} = \sin^{-1} \dfrac{\sigma_1 - \sigma_2}{\sigma_1 + \sigma_2}$ ☐ (*d*) $\phi_{max} = \sin^{-1} \dfrac{\sigma_1 + \sigma_2}{\sigma_1 - \sigma_2}$ ☐

where σ_1 = Major principal stress and σ_2 = Minor principal stress.

Mohr's Circle of Stresses

81. Two unequal like principal stresses are acting at a point across two perpendicular planes. The resultant stress acting on a plane inclined at an angle θ with the major principal plane is to be determined by Mohr's circle of stress as shown in Fig. 8.19. Two principal stresses in Fig. 8.19 acting on the point are given by

(*a*) *AO* and *OB* ☐

(*b*) *AC* and *BC* ☐

(*c*) *AB* and *AC* ☐

(*d*) *AD* and *AB*. ☐

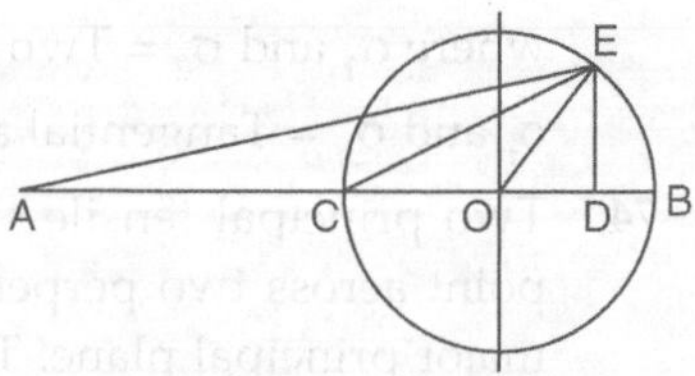

FIGURE 8.19. Mohr's circle.

82. In Fig. 8.19, the radius of the Mohr's circle is

(*a*) sum of two principal stress ☐ (*b*) difference of two principal stresses ☐

(*c*) half of difference of principal stresses ☐ (*d*) half of sum of two principal stresses. ☐

83. In Fig. 8.19, the normal stress is given by

(*a*) *AE* ☐ (*b*) *ED* ☐

(*c*) *CE* ☐ (*d*) *AD*. ☐

84. In Fig. 8.19, the tangential stress is given by

(*a*) *AE* ☐ (*b*) *ED* ☐

(*c*) *CE* ☐ (*d*) *AD*. ☐

85. In Fig. 8.19, the resultant stress is given by

(*a*) *AC* ☐ (*b*) *ED* ☐

(*c*) *CE* ☐ (*d*) *AD*. ☐

86. In Fig. 8.19, obliquity is represented by the

(*a*) angle *ECB* ☐ (*b*) angle *EOD* ☐

(*c*) angle *EAD* ☐ (*d*) angle *AED*. ☐

87. In Fig. 8.19, the angle of oblique plane is represented by

(*a*) angle *ECB* ☐ (*b*) angle *EOD* ☐

(*c*) angle *EAD* ☐ (*d*) angle *AED*. ☐

88. In Fig. 8.19, the maximum tangential stress is equal to

(*a*) radius of the Mohr's circle ☐ (*b*) diameter of Mohr's circle ☐

(*c*) circumference of Mohr's circle ☐ (*d*) half of radius of Mohr's circle. ☐

89. In which of the following cases, Mohr's circle is used to determine the stresses on an oblique plane?

(*a*) two unequal like principal stresses ☐

(*b*) two unequal unlike principal stresses ☐

(*c*) direct tensile stress in one plane accompanied by a shear stress ☐

(*d*) all of the above ☐

(*e*) none of the above. ☐

90. Figure 8.20 shows the Mohr's circle of stress for two unequal unlike principal stresses. The two unequal unlike principal stresses are given by

(*a*) *AC* and *AD* ☐

(*b*) *AC* and *AB* ☐

(*c*) *CE* and *CD* ☐

(*d*) *CO* and *OB*. ☐

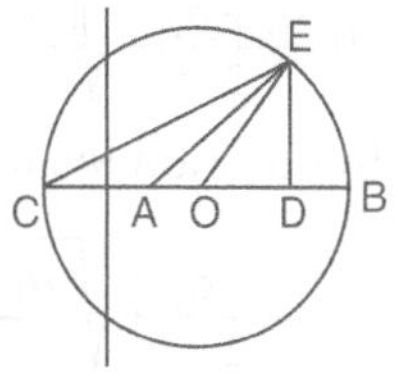

FIGURE 8.20. Mohr's circle.

91. In Fig. 8.20, radius of Mohr's circle is

(*a*) sum of two principle stresses ☐

(*b*) difference of two principle stresses ☐

(*c*) half of difference of principle stresses ☐

(*d*) half of sum of two stresses. ☐

92. In Fig. 8.20, the normal stress is given by

(a) ED ☐ (b) AE ☐

(c) AD ☐ (d) CE. ☐

93. In Fig. 8.20, the tangential stress is given by

(a) ED ☐ (b) AE ☐

(c) AD ☐ (d) CE. ☐

94. In Fig. 8.20, the resultant stress is given by

(a) ED ☐ (b) AE ☐

(c) AD ☐ (d) CE. ☐

95. In Fig. 8.20, obliquity is represented by

(a) angle ECD ☐ (b) angle EAD ☐

(c) angle EOD ☐ (d) angle AED ☐

96. In Fig. 8.20, the angle of oblique plane on which normal and tangential stresses are determined, is given by

(a) angle ECD ☐ (b) angle EAD ☐

(c) angle EOD ☐ (d) angle AED. ☐

97. In Fig. 8.20, the maximum shear stress is equal to

(a) radius of Mohr's circle ☐ (b) diameter to Mohr's circle ☐

(c) circumference of Mohr's circle ☐ (d) half of the radius of Mohr's circle. ☐

98. Figure 8.21 shows a body subjected to two unequal like direct stresses (σ_1 and σ_2) in two mutually perpendicular planes along with a simple shear stress (τ). The maximum normal stress will be

(a) $\frac{\sigma_1+\sigma_2}{2}+\sqrt{\left(\frac{\sigma_1-\sigma_2}{2}\right)^2+\tau^2}$ ☐

(b) $\frac{\sigma_1+\sigma_2}{2}+\sqrt{\left(\frac{(\sigma_1+\sigma_2)^2}{2}\right)+\tau^2}$ ☐

(c) $\frac{\sigma_1+\sigma_2}{2}-\sqrt{\left(\frac{(\sigma_1-\sigma_2)^2}{2}\right)^2+\tau^2}$ ☐

(d) $\frac{\sigma_1+\sigma_2}{2}-\sqrt{\left(\frac{\sigma_1+\sigma_2}{2}\right)^2+\tau^2}$ ☐

FIGURE 8.21. Mohr's circle.

99. In question 98, the minimum normal stress will be

(a) $\frac{\sigma_1+\sigma_2}{2}+\sqrt{\left(\frac{\sigma_1-\sigma_2}{2}\right)^2+\tau^2}$ ☐ (b) $\frac{\sigma_1+\sigma_2}{2}+\sqrt{\left(\frac{\sigma_1+\sigma_2}{2}\right)^2+\tau^2}$ ☐

(c) $\frac{\sigma_1+\sigma_2}{2}-\sqrt{\left(\frac{\sigma_1-\sigma_2}{2}\right)^2+\tau^2}$ ☐ (d) $\frac{\sigma_1+\sigma_2}{2}-\sqrt{\left(\frac{\sigma_1+\sigma_2}{2}\right)^2+\tau^2}$ ☐

100. In question 98, the maximum shear stress will be

(*a*) $\sqrt{\left(\frac{\sigma_1 - \sigma_2}{2}\right)^2 + \tau^2}$ ☐ (*b*) $\sqrt{\left(\frac{\sigma_1 + \sigma_2}{2}\right)^2 + \tau^2}$ ☐

(*c*) $\sqrt{(\sigma_1 - \sigma_2)^2 + 4\tau^2}$ ☐ (*d*) $\sqrt{(\sigma_1 + \sigma_2)^2 + 4\tau^2}$ ☐

101. A body is subjected to a direct tensile stress of 300 N/cm^2 in one plane accompanied by a simple shear stress of 200 N/cm^2. The maximum normal stress will be

(*a*) 250 N/cm^2 ☐ (*b*) 400 N/cm^2 ☐

(*c*) – 100 N/cm^2 ☐ (*d*) 300 N/cm^2. ☐

102. In question 101, the minimum normal stress will be

(*a*) 250 N/cm^2 ☐ (*b*) 400 N/cm^2 ☐

(*c*) – 100 N/cm^2 ☐ (*d*) 300 N/cm^2. ☐

103. In question 101, the maximum shear stress will be

(*a*) 250 N/cm^2 ☐ (*b*) 400 N/cm^2 ☐

(*c*) – 100 N/cm^2 ☐ (*d*) 300 N/cm^2. ☐

104. A body is subjected to a tensile stress of 1200 N/cm^2 on one plane and a tensile stress of 600 N/cm^2 on the other right-angled plane together with shear stresses of 400 N/cm^2 on the same planes. The maximum normal stress will be

(*a*) 900 N/cm^2 ☐ (*b*) 1400 N/cm^2 ☐

(*c*) 400 N/cm^2 ☐ (*d*) 500 N/cm^4. ☐

105. In question 104, the minimum normal stress would be

(*a*) 900 N/cm^2 ☐ (*b*) 1400 N/cm^2 ☐

(*c*) 400 N/cm^2 ☐ (*d*) 500 N/cm^4. ☐

106. In question 104, the greatest shear stress would be

(*a*) 900 N/cm^2 ☐ (*b*) 1400 N/cm^2 ☐

(*c*) 400 N/cm^2 ☐ (*d*) 500 N/cm^4. ☐

107. A strained element is subjected to principal stresses σ_1 and σ_2 in two mutually perpendicular direction. The strain (e_1) in the direction of principal stress (σ_1) is given by

(*a*) $e_1 = \frac{\sigma_1}{E} + \frac{\sigma_1}{mE}$ ☐ (*b*) $e_1 = \frac{\sigma_1}{E} - \frac{\sigma_2}{mE}$ ☐

(*c*) $e_1 = \frac{\sigma_2}{E} + \frac{\sigma_1}{mE}$ ☐ (*d*) $e_1 = \frac{\sigma_2}{E} - \frac{\sigma_1}{mE}$. ☐

where $\frac{1}{m}$ = Poisson's ratio.

108. In question 107, the strain (e_2) in the direction of principal stress (σ_2) is given by

(*a*) $e_2 = \frac{\sigma_1}{E} + \frac{\sigma_1}{mE}$ ☐ (*b*) $e_2 = \frac{\sigma_1}{E} - \frac{\sigma_2}{mE}$ ☐

(*c*) $e_2 = \frac{\sigma_2}{E} + \frac{\sigma_1}{mE}$ ☐ (*d*) $e_2 = \frac{\sigma_2}{E} - \frac{\sigma_1}{mE}$. ☐

109. In question 107, the strain energy stored per unit volume is equal to

(a) $\frac{1}{2}\sigma_1 e_1 + \frac{1}{2}\sigma_2 e_2$ ☐ (b) $\frac{1}{2E}\left(\sigma_1^2 + \sigma_2^2 - \frac{2\sigma_1\sigma_2}{m}\right)$ ☐

(c) both (a) and (b) ☐ (d) none of the above. ☐

110. A strained element is subjected to two principal stresses of 800 N/cm² and – 200 N/cm². If the Poisson ratio = 0.25 and $E = 2 \times 10^6$ N/cm², the strain in the direction of principal stress (800 N/cm²) will be

(a) 0.000425 ☐ (b) 0.0002 ☐

(c) 0.0004 ☐ (d) 0.0001. ☐

111. In question 110, the strain in the direction of principal stress (– 200 N/cm²) will be

(a) 0.000425 ☐ (b) 0.0002 ☐

(c) 0.0004 ☐ (d) 0.0001. ☐

112. In question 110, the strain energy stored per unit volume would be

(a) 0.125 N/cm² ☐ (b) 0.250 N/cm² ☐

(c) 0.38 N/cm² ☐ (d) 1.0 N/cm². ☐

113. If σ_1, σ_2 and σ_3 be the principal stresses at a point in a strained material, then strain energy stored per unit volume will be

(a) $\frac{1}{2E}\left(\sigma_1^2 + \sigma_2^2 + \sigma_3^2 - \frac{2\sigma_1\sigma_2}{m}\right)$ ☐

(b) $\frac{1}{2E}\left(\sigma_1^2 + \sigma_2^2 + \sigma_3^2 + \frac{2\sigma_1\sigma_2\sigma_3}{m}\right)$ ☐

(c) $\frac{1}{2E}\left(\sigma_1^2 + \sigma_2^2 + \sigma_3^2 - \frac{2\sigma_1\sigma_2 + 2\sigma_2\sigma_3 + 2\sigma_3\sigma_1}{m}\right)$ ☐

(d) none of the above. ☐

114. The stress and strain curves for four different materials, *i.e.*, mild steel, cast iron, brass and cast aluminium are shown in Fig. 8.22. The curve *A* is for

(a) mild steel ☐ (b) cast iron ☐

(c) brass ☐ (d) cast aluminium. ☐

115. Curve *B* in Fig. 8.22 is for

(a) mild steel ☐

(b) cast iron ☐

(c) brass ☐

(d) cast aluminium. ☐

116. Curve *C* in Fig. 8.22 is for

(a) mild steel ☐

(b) cast iron ☐

(c) brass ☐

(d) cast aluminium. ☐

FIGURE 8.22

117. Curve D in Fig. 8.22 is for

(*a*) mild steel ☐ (*b*) cast iron ☐
(*c*) brass ☐ (*d*) cast aluminium. ☐

118. A tensile test is performed on a specimen of 4 cm^2 cross-section. The maximum load on the specimen was 4000 N. The area of cross-section at neck was 2 cm^2. The specimen will have the ultimate tensile stress as

(*a*) 1000 N/cm^2 ☐ (*b*) 2000 N/cm^2 ☐
(*c*) 3000 N/cm^2 ☐ (*d*) 500 N/cm^2. ☐

119. If a support is non-yielding, it means

(*a*) slope of the beam at the support is zero ☐ (*b*) support is frictionless ☐
(*c*) support hold member firmly ☐ (*d*) none of the above. ☐

120. A conical rod of length l is subjected to a tensile force F. The diameter of the rod at bottom is D and at top d. The tensile stress at a distance x from the top end will be

(*a*) $\frac{4}{\pi[D-d]^2}$ ☐ (*b*) $\frac{4Fl^2}{\pi[(D-d)x+ld]^2}$ ☐

(*c*) $\frac{4F}{[\pi D-x]^2}$ ☐ (*d*) $\frac{4F}{\pi D^2}$. ☐

121. The elongation of a conical bar due to its own weight is equal to

(*a*) $\frac{wl}{2E}$ ☐ (*b*) $\frac{wl^2}{6E}$ ☐

(*c*) $\frac{wl^3}{6E}$ ☐ (*d*) $\frac{wl^4}{6E}$. ☐

where l = Length of bar and w = Weight per unit volume.

122. If a beam is fixed at both its ends, it is called a

(*a*) fixed beam ☐ (*b*) built-in beam ☐
(*c*) encastered beam ☐ (*d*) any one of the above ☐
(*e*) none of the above. ☐

123. If a beam is supported on more than two supports, it is called a

(*a*) built-in beam ☐ (*b*) continuous beam ☐
(*c*) simply supported beam ☐ (*d*) encastered beam. ☐

124. Choose the wrong statement

(*a*) The shear force at any section of a beam is equal to the total sum of the forces acting on the beam on any one side of the section. ☐

(*b*) The magnitude of the bending moment at any section of a beam is equal to the vector sum of the moments (about the section) due to the forces acting on the beam on any one side of the section. ☐

(*c*) A diagram which shows the values of shear forces at various sections of structural member is called a shear force diagram. ☐

(*d*) A simply supported beam is one which is supported on more than two supports. ☐

Shear Force and Bending Moment Diagrams

125. A simply supported beam of span (l) carries a point load (W) at the centre of the beam. The bending moment diagram will be a

(*a*) parabola with maximum ordinate at the centre of the beam ☐
(*b*) parabola with maximum ordinate at one end of the beam ☐
(*c*) triangle with maximum ordinate at the centre of the beam ☐
(*d*) triangle with maximum ordinate at one end of the beam. ☐

126. A simply supported beam of span (l) carries a uniformly distributed load (w N per unit length) over the whole span. The bending moment diagram will be a

(*a*) parabola with maximum ordinate at the centre of the beam ☐
(*b*) parabola with maximum ordinate at one end of the beam ☐
(*c*) triangle with maximum ordinate at the centre of the span ☐
(*d*) triangle with maximum ordinate at one end of the beam. ☐

127. A cantilever of length (l) carries a point load (W) at the free end. The bending moment diagram will be a

(*a*) parabola with maximum ordinate at the centre of the beam ☐
(*b*) parabola with maximum ordinate at the cantilever end ☐
(*c*) triangle with maximum ordinate at the free end ☐
(*d*) triangle with maximum ordinate at the cantilever end. ☐

128. A cantilever of length (l) carries a uniformly distributed load over the whole length. The bending moment diagram will be

(*a*) parabola with maximum ordinate at the centre ☐
(*b*) parabola with maximum ordinate at the cantilever end ☐
(*c*) triangle with maximum ordinate at the free end ☐
(*d*) triangle with maximum ordinate at the cantilever end. ☐

129. A simply supported beam of span (l) carries a point load (W) at the centre of the beam. The shear force diagram will be

(*a*) a rectangle ☐ (*b*) a triangle ☐
(*c*) two equal and opposite rectangles ☐ (*d*) two equal and opposite triangles. ☐

130. A simply supported beam of span (l) carries a uniformly distributed load over the whole span. The shear force diagram will be

(*a*) a rectangle ☐ (*b*) a triangle ☐
(*c*) two equal and opposite rectangles ☐ (*d*) two equal and opposite triangles. ☐

131. A cantilever of length (l) carries a point load (W) at the free end. The shear force diagram will be

(*a*) two equal and opposite rectangles ☐ (*b*) a rectangle ☐
(*c*) two equal and opposite triangles ☐ (*d*) a triangle. ☐

132. A cantilever of length (l) carries a uniformly distributed load over the whole length. The shear force diagram will be

(*a*) two equal and opposite rectangles ☐ (*b*) a rectangle ☐
(*c*) two equal and opposite triangles ☐ (*d*) a triangle. ☐

133. The bending moment on a section is maximum where shearing force is

(*a*) minimum ☐ (*b*) maximum ☐
(*c*) zero ☐ (*d*) equal ☐
(*e*) changing sing ☐ (*f*) none of the above. ☐

134. The point of zero bending moment, where the continuous curve of bending moment changes sign, is called

(*a*) the point of contra-flexure ☐ (*b*) the point of inflation ☐
(*c*) the point of virtual hinge ☐ (*d*) all of the above. ☐

135. The point of contra-flexure occurs only in

(*a*) continuous beams ☐ (*b*) cantilever beams ☐
(*c*) overhanging beams ☐ (*d*) simply supported beams ☐
(*e*) all of the above ☐ (*f*) none of the above. ☐

136. The bending moment at a section, where shear force is zero, will be

(*a*) zero ☐ (*b*) maximum ☐
(*c*) minimum ☐ (*d*) either minimum or maximum ☐
(*e*) none of the above. ☐

137. A beam of uniform strength is one which has same

(*a*) bending stress at every section ☐ (*b*) deflection throughout the beam ☐
(*c*) bending moment throughout the beam ☐ (*d*) shear force throughout the beam. ☐

138. The shear force and bending moment are zero at the free end of a cantilever, if it carries a

(*a*) point load at the free end ☐
(*b*) uniformly distributed load over the whole length ☐
(*c*) point load in the middle of its length ☐
(*d*) none of the above. ☐

139. A simply supported beam carries a uniformly distributed load of w kgf per unit length over the whole span (l). The shear force at the centre is

(*a*) $\frac{wl}{2}$ ☐ (*b*) $\frac{wl^2}{8}$ ☐
(*c*) $\frac{wl}{4}$ ☐ (*d*) zero. ☐

140. For question 139, the shear force at the supported ends will be

(*a*) $\pm \frac{wl}{2}$ ☐ (*b*) $\frac{wl^2}{8}$ ☐
(*c*) $\frac{wl}{8}$ ☐ (*d*) zero. ☐

141. For question 139, the bending moment at the supported ends will be

(*a*) $\pm \frac{wl}{2}$ ☐ (*b*) $\frac{wl^2}{8}$ ☐
(*c*) $\frac{wl}{4}$ ☐ (*d*) zero. ☐

142. For question 139, the bending moment at the centre will be

(a) $\pm \frac{wl}{2}$ ☐ (b) $\frac{wl^2}{8}$ ☐

(c) $\frac{wl}{4}$ ☐ (d) zero. ☐

143. For question 139, the point of contra-flexure is at

(a) the supported end ☐ (b) the middle of the beam ☐

(c) distance $\frac{l}{4}$ from the supported ends ☐ (d) none of the above. ☐

144. A cantilever of length (l) carries a uniformly distributed load w kgf per unit length for the whole length. The shear force at the free end will be

(a) wl ☐ (b) $\frac{wl^2}{2}$ ☐

(c) $\frac{wl}{2}$ ☐ (d) zero. ☐

145. In question 144, the shear force at the fixed end will be

(a) wl ☐ (b) $\frac{wl^2}{2}$ ☐

(c) $\frac{wl}{2}$ ☐ (d) zero. ☐

146. In question 144, the bending moment at the fixed end will be

(a) wl ☐ (b) $\frac{wl^2}{2}$ ☐

(c) $\frac{wl}{2}$ ☐ (d) zero. ☐

147. In question 144, the bending moment at the free end will be

(a) wl ☐ (b) $\frac{wl^2}{2}$ ☐

(c) $\frac{wl}{2}$ ☐ (d) zero. ☐

148. In question 144, point of contra-flexure is at

(a) the free end ☐ (b) the fixed end ☐

(c) the middle of the beam ☐ (d) nowhere. ☐

149. Figure 8.23 shows a simply supported beam of span 10 m. It carries a point load of 200 kgf at a distance 7 m from the end A. The reaction at A will be

(a) 140 kgf ☐

(b) 60 kgf ☐

(c) 420 kgf ☐

(d) 100 kgf. ☐

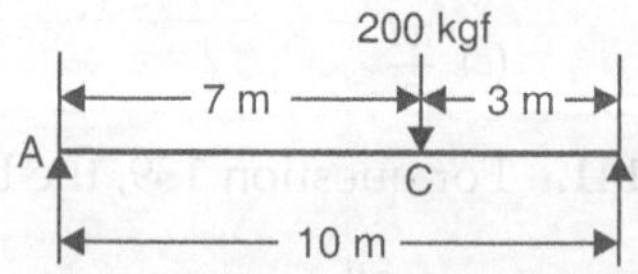

FIGURE 8.23

150. For the beam shown in Fig. 8.23, the reaction at B will be

(*a*) 140 kgf ☐ (*b*) 60 kgf ☐
(*c*) 420 kgf ☐ (*d*) 100 kgf. ☐

151. For the beam shown in Fig. 8.23, the maximum bending moment is at

(*a*) point A ☐ (*b*) point B ☐
(*c*) point C ☐ (*d*) point between A and C. ☐

152. For the beam shown in Fig. 8.23, the maximum bending moment

(*a*) 140 kgf-m ☐ (*b*) 60 kgf-m ☐
(*c*) 420 kgf-m ☐ (*d*) 100 kgf-m. ☐

153. For the beam shown in Fig. 8.23, the shear force diagram will consist of

(*a*) a rectangle ☐ (*b*) a triangle ☐
(*c*) two rectangles ☐ (*d*) two triangles. ☐

154. The ratio of volumetric strain to linear strain in any of the three axes of a solid cube which is subjected by equal normal forces of the similar nature on all the faces, will be

(*a*) 2 ☐ (*b*) 3 ☐
(*c*) 1 ☐ (*d*) 1.5. ☐

155. Ties are the structure numbers which carry

(*a*) axial compressive loads ☐ (*b*) axial tensile loads ☐
(*c*) torsional loads ☐ (*d*) none of the above. ☐

156. In a tensile test of a specimen, the ratio of maximum load to the original cross-sectional area of the test piece is called

(*a*) yield stress ☐ (*b*) ultimate stress ☐
(*c*) safe stress ☐ (*d*) breaking stress. ☐

157. A short column of rectangular section carries a vertical point load W axially. The stress on the section of the column will be

(*a*) uniform ☐
(*b*) zero at one end and maximum at the other end ☐
(*c*) zero at the axis and maximum on the outer ends ☐
(*d*) tensile on one end and compressive on the other. ☐

158. If in question 157, the short column carries a vertical point load eccentrically, then the stress on the section will be

(*a*) uniform ☐
(*b*) zero at the axis and maximum on the outer ends ☐
(*c*) zero at one end and maximum at the other end ☐
(*d*) tensile on one end and compressive on the other ☐
(*e*) non-uniform and will depend upon the amount of eccentricity. ☐

159. A solid circular shaft of diameter D carries an axial load W. If the same load is applied axially on a hollow circular shaft of inner diameter as $\frac{D}{2}$, the ratio of stresses in a solid shaft to that of hollow shaft would be

(a) $\frac{1}{2}$ ☐ (b) $\frac{1}{4}$ ☐

(c) $\frac{4}{3}$ ☐ (d) $\frac{3}{4}$. ☐

160. If in question 159, the inner diameter of hollow shaft is $D/3$, then the ratio of stress in a solid shaft to that of hollow shaft would be

(a) $\frac{1}{2}$ ☐ (b) $\frac{1}{4}$ ☐

(c) $\frac{8}{9}$ ☐ (d) $\frac{9}{8}$. ☐

161. Under tensile test, a test piece of a material is having 40% elongation whereas another test piece of the same dimensions but of different material is having 25% elongation. Then the ductility of the first material as compared to that of second material is

(a) less ☐ (b) same ☐

(c) more ☐ (d) none of the above. ☐

162. The diameter of a cast iron round bar, on which tensile test is performed, at fracture will

(a) increase ☐ (b) decrease ☐

(c) be approximately same ☐ (d) none of the above. ☐

163. The diameter of a mild steel round bar, on which tensile test is performed, at fracture will

(a) increase ☐ (b) decrease ☐

(c) be same ☐ (d) none of the above. ☐

Deflection of Beams

164. If a member is subjected to a uniform bending moment (M), the radius of curvature of the deflected form of the member is given by

(a) $\frac{M}{R} = \frac{E}{I}$ ☐ (b) $\frac{M}{I} = \frac{E}{R}$ ☐

(c) $\frac{M}{I} = \frac{R}{E}$ ☐ (d) $\frac{M}{E} = RI$. ☐

165. Which one of the following equations is correct?

(a) $\frac{1}{R} = \frac{d^2y}{dx^2} = \frac{EI}{M}$ ☐ (b) $\frac{1}{R} = \frac{d^2y}{dx^2} = \frac{M}{EI}$ ☐

(c) $R = \frac{d^2y}{dx^2} = \frac{M}{EI}$ ☐ (d) $R = \frac{d^2y}{dx^2} = \frac{EI}{M}$. ☐

where R = Radius of curvature and M = Bending moment.

166. The expression $EI\,\frac{d^2y}{dx^2}$ at a section of a member represents

(a) shearing force ☐ (b) rate of loading ☐

(c) bending moment ☐ (d) slope. ☐

167. The expression $EI \frac{d^3y}{dx^3}$ at a section of a member represents

(*a*) shearing force ☐ (*b*) rate of loading ☐

(*c*) bending moment ☐ (*d*) slope. ☐

168. The expression $EI \frac{d^4y}{dx^4}$ at a section of a member represents

(*a*) shearing force ☐ (*b*) rate of loading ☐

(*c*) bending moment ☐ (*d*) slope. ☐

169. A cantilever of length (*l*) carries a point load (*W*) at the free end. The downward deflection at the free end is equal to

(*a*) $\frac{Wl^2}{8EI}$ ☐ (*b*) $\frac{Wl^3}{3EI}$ ☐

(*c*) $\frac{5Wl^3}{384\ EI}$ ☐ (*d*) $\frac{Wl^3}{48\ EI}$. ☐

170. In question 169, the slope at the free end will be

(*a*) $\frac{Wl^2}{6EI}$ ☐ (*b*) $\frac{Wl^2}{2EI}$ ☐

(*c*) $\frac{Wl^2}{24\ EI}$ ☐ (*d*) $\frac{Wl^2}{16\ EI}$. ☐

171. A cantilever of length (*l*) carries a uniformly distributed load *w* per unit length over the whole length. The downward deflection at the free end will be

(*a*) $\frac{Wl^3}{8EI}$ ☐ (*b*) $\frac{Wl^3}{3EI}$ ☐

(*c*) $\frac{4Wl^3}{384\ EI}$ ☐ (*d*) $\frac{Wl^3}{48\ EI}$. ☐

where $W = w \times l$ = Total load.

172. In question 171, the slope at the free end will be

(*a*) $\frac{Wl^2}{6EI}$ ☐ (*b*) $\frac{Wl^2}{2EI}$ ☐

(*c*) $\frac{Wl^2}{24\ EI}$ ☐ (*d*) $\frac{Wl^2}{16\ EI}$. ☐

where W = Total load = $w \times l$.

173. A uniform simply supported beam of span (*l*) carries a point load (*W*) at the centre. The downward deflection at the centre will be

(*a*) $\frac{Wl^3}{8EI}$ ☐ (*b*) $\frac{Wl^3}{3EI}$ ☐

(*c*) $\frac{5Wl^3}{384\ EI}$ ☐ (*d*) $\frac{Wl^3}{48\ EI}$. ☐

174. In question 173, the slope at the support will be

(a) $\frac{Wl^2}{6EI}$ ☐ (b) $\frac{Wl^3}{2EI}$ ☐

(c) $\frac{Wl^2}{24\,EI}$ ☐ (d) $\frac{Wl^2}{16\,EI}$. ☐

175. A uniformly simply supported beam of span (l) carries a uniformly distributed load w per unit length over the whole span. The downward deflection at the centre will be

(a) $\frac{Wl^3}{8EI}$ ☐ (b) $\frac{Wl^3}{3EI}$ ☐

(c) $\frac{5Wl^3}{384\,EI}$ ☐ (d) $\frac{Wl^3}{48\,EI}$. ☐

where $W = w \times l$ = Total load.

176. In question 175, the slope at the supported will be

(a) $\frac{Wl^2}{6EI}$ ☐ (b) $\frac{Wl^2}{2EI}$ ☐

(c) $\frac{Wl^2}{24\,EI}$ ☐ (d) $\frac{Wl^2}{16\,EI}$. ☐

where $W = w \times l$ = Total load.

177. Figure 8.24, shows a simply supported beam of length L. It carries a point load W at a point C. The distance of point load W from left end is more than that of from right end, *i.e.*, $a > b$. The deflection at C is

(a) $\frac{Wab}{3EIL}$ ☐

(b) $\frac{3Wa^2b^2}{EIL}$ ☐

(c) $\frac{Wa^2b^2}{3EIL}$ ☐

(d) $\frac{Wa^2b}{3EIL}$ ☐

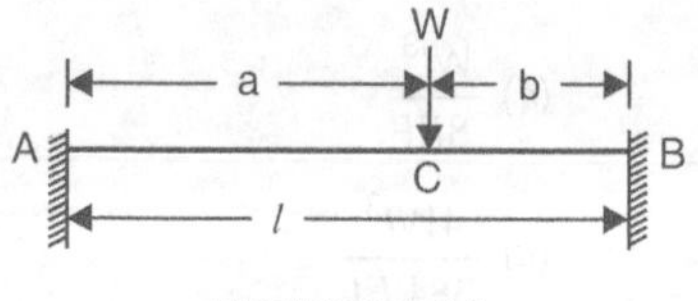

W
a
b
A
B
C
l

FIGURE 8.24

where E = Modulus of elasticity and I = Moment of inertia

178. In Fig. 8.24, the maximum deflection is

(a) $\frac{Wa(a^2+2ab)^{3/2}}{8EIL}$ ☐ (b) $\frac{Wb(a^2+2ab)^{3/2}}{9\sqrt{3}EIL}$ ☐

(c) $\frac{Wab(a^2+b^2)^{3/2}}{9\sqrt{3}EIL}$ ☐ (d) $\frac{Wa^2b^2}{3EIL}$. ☐

179. In Fig. 8.24, the maximum deflection lies at

(a) point C ☐ (b) point B ☐

(c) between points A and E ☐ (d) between points B and C. ☐

180. In Fig. 8.24, the distance of maximum deflection from left end will be

(a) $\sqrt{a^2 + 2ab}$ ☐ (b) $\sqrt{\frac{a^2 + 2ab}{3}}$ ☐

(c) $\sqrt{\frac{L^2 - b^2}{3}}$ ☐ (d) $\sqrt{\frac{L^2 - a^2}{3}}$ ☐

181. If in question 177, L = 6 m, a = 4 m, W = 60000 N, moment of inertia (I) = 8000 cm^4 and modulus of elasticity $E = 2 \times 10^7$ N/cm^2 then the deflection at C will be

(a) 1.33 cm ☐ (b) 0.75 cm ☐

(c) 1 cm ☐ (d) 1.5 cm. ☐

182. For the question 181, the distance of maximum deflection from the left end would be

(a) 3 m ☐ (b) 3.5 m ☐

(c) 3.27 m ☐ (d) 2 m. ☐

183. A simply supported beam of span (l) carries a uniformly distributed load w kgf per unit length over the whole span. The deflection at the centre is y. If the length of the beam is doubled, the deflection at the centre would be

(a) $2y$ ☐ (b) $4y$ ☐

(c) $8y$ ☐ (d) $16y$. ☐

184. If in question 183, instead of making the length of the beam double, the uniform distributed load is changed to $2w$ per unit length. Then the deflected at the centre would be

(a) $2y$ ☐ (b) $4y$ ☐

(c) $\frac{y}{2}$ ☐ (d) $\frac{y}{8}$. ☐

185. A simply supported beam is of rectangular section. It carries a uniformly distributed load over the whole span. The deflection at the centre is y. If the width of the beam is doubled, the deflection at the centre would be

(a) $2y$ ☐ (b) $4y$ ☐

(c) $\frac{y}{2}$ ☐ (d) $\frac{y}{8}$. ☐

186. If in question 185, instead of making the width of the beam as double, the depth of the beam is made double. Then the deflection at the centre would be

(a) $2y$ ☐ (b) $4y$ ☐

(c) $\frac{y}{2}$ ☐ (d) $\frac{y}{8}$. ☐

187. A simply supported beam carries a uniformly distributed load over the whole span. The deflection at the centre is y. If the distributed load per unit length is doubled and also depth of the beam is doubled, then the deflection at the centre would be

(a) $2y$ ☐ (b) $4y$ ☐

(c) $\frac{y}{2}$ ☐ (d) $\frac{y}{4}$. ☐

188. The slope at the free end of a cantilever of length 1 m is 1°. If the cantilever carries uniformly distributed load over the whole length, then the deflection at the free end will be

(*a*) 1 cm ☐ (*b*) 1.309 cm ☐

(*c*) 2.618 cm ☐ (*d*) 3.927 cm. ☐

189. A cantilever of length (l) carries a point load (W) at a distance x from the fixed end, then the deflection at the free end will be

(*a*) $\frac{Wx^3}{3EI} + \frac{Wx^2}{2EI} \times l$ ☐ (*b*) $\frac{Wl^3}{3EI}$ ☐

(*c*) $\frac{Wx^3}{3EI} + \frac{Wx^2}{2EI}(l - x)$ ☐ (*d*) $\frac{Wx^3}{2EI} + \frac{Wx^3}{3EI}(l - x)$. ☐

190. A cantilever of length (l) carries a uniformly distributed load of w per unit length for a distance x from the fixed end, then the deflection at the free end will be

(*a*) $\frac{Wx^4}{8EI} + \frac{Wx^3}{6EI} \times l$ ☐ (*b*) $\frac{Wx^4}{8EI}$ ☐

(*c*) $\frac{El^4}{8EI}$ ☐ (*d*) $\frac{Wx^4}{8EI} + \frac{W^3x}{8EI} \times (1 - x)$. ☐

191. A cantilever of length (l) carries a distributed load whose intensity varies from zero at the free end to w per unit length at the fixed end. The deflection at the free end will be

(*a*) $\frac{wl^4}{3EI}$ ☐ (*b*) $\frac{wl^4}{8EI}$ ☐

(*c*) $\frac{11}{120}\frac{wl^4}{EI}$ ☐ (*d*) $\frac{wl^4}{30EI}$. ☐

192. A cantilever of length (l) carries a distributed load whose intensity varies uniformly from zero at the fixed end to w per unit length at the free end. The deflection at the free end will be

(*a*) $\frac{wl^4}{3EI}$ ☐ (*b*) $\frac{wl^4}{8EI}$ ☐

(*c*) $\frac{11}{120}\frac{wl^4}{EI}$ ☐ (*d*) $\frac{wl^4}{30EI}$. ☐

193. The statement that 'the deflection at any point in a beam subjected to any load system is equal to the partial derivative of the total strain energy stored with respect to the load acting at the point in the direction in which deflection is desired' is called

(*a*) Bettle's law ☐ (*b*) The first theorem of Castigliano ☐

(*c*) Clapeyron's theorem ☐ (*d*) Maxwell's law. ☐

194. A laminated spring 1 m long carries a central point load of 200 kg. The spring is made up of plates each 5 cm wide and 1 cm thick. The bending stress in the plates is limited to 1000 kg/cm^2. The number of plates required, will be

(*a*) 3 ☐ (*b*) 5 ☐

(*c*) 6 ☐ (*d*) 8. ☐

195. In question 194, if $E = 2 \times 10^6$ kg/cm^2 the deflection under the given load of 200 kg will be

(a) 1 cm ☐ (b) 1.25 cm ☐

(c) 1.3 cm ☐ (d) 1.40 cm. ☐

Fixed Beam

196. A fixed beam is a beam whose end supports are such that the end slopes

(a) are maximum ☐ (b) are minimum ☐

(c) are zero ☐ (d) none of the above. ☐

197. A fixed beam of length (l) carries a point load (W) at the centre. The deflection at the centre is

(a) same as for a simply supported beam ☐

(b) half of the deflection for a simply supported beam ☐

(c) one-fourth of the deflection for a simply supported beam ☐

(d) double the deflection for a simply supported beam. ☐

198. For the question 197, the number of points of contra-flexure

(a) is one ☐ (b) are two ☐

(c) are three ☐ (d) is none. ☐

199. For the question 197, the points of contra-flexure lies at

(a) the fixed ends ☐ (b) the middle of the beam ☐

(c) $\frac{l}{4}$ from the ends ☐ (d) none of the above. ☐

200. For the question 197, the bending moment at the centre is

(a) same as for a simply supported beam ☐

(b) half of the bending moment for a simply supported beam ☐

(c) one-fourth of the bending moment for a simply supported beam ☐

(d) double the bending moment for a simply supported beam. ☐

201. For the question 197, the bending moment at the fixed ends is

(a) zero ☐ (b) $\frac{Wl}{4}$ ☐

(c) $\frac{Wl}{8}$ ☐ (d) $\frac{Wl}{2}$ ☐

202. A fixed beam of span (l) carries a uniformly distributed load of w per unit length over the whole span. The deflection at the centre is

(a) equal of the central deflection of a simply supported beam ☐

(b) half of the central deflection for a simply supported beam ☐

(c) one-fourth of the central deflection for a simply supported beam ☐

(d) one-fifth of the central deflection of the simply supported beam. ☐

203. For the question 202, the points of contra-flexure lies at

(a) the fixed ends ☐ (b) the middle of the beam ☐

(c) $\frac{l}{4}$ from the ends ☐ (d) $\frac{l}{2\sqrt{3}}$ from the centre of the span. ☐

204. For the solution of problems on fixed beams, the condition is
(*a*) area of free B.M. diagram = area of fixed B.M. diagram ☐
(*b*) the distance of the centroid of the free B.M. diagram from an end should be equal to the distance of the centroid of fixed B.M. diagram from the same end ☐
(*c*) both (*a*) and (*b*) ☐
(*d*) none of the above. ☐

205. In question 202, the end moments are each equal to
(*a*) $\frac{wl^2}{8}$ ☐ (*b*) $\frac{wl^6}{6}$ ☐
(*c*) $\frac{wl^2}{12}$ ☐ (*d*) $\frac{wl^2}{4}$. ☐

206. A fixed beam of span (l) carries a point load (W) at a distance a from one end as shown in Fig. 8.25. The fixing moment at the end A will be given by
(*a*) $\frac{Wa^2b}{l^2}$ ☐
(*b*) $\frac{Wab^2}{l^2}$ ☐
(*c*) $\frac{Wa^2b^2}{l^2}$ ☐
(*d*) $\frac{Wab^2}{l}$. ☐

FIGURE 8.25

207. In question 206, the fixing moments at B will be
(*a*) $\frac{Wa^2b}{l^2}$ ☐ (*b*) $\frac{Wab^2}{l^2}$ ☐
(*c*) $\frac{Wa^2b^2}{l^2}$ ☐ (*d*) $\frac{Wab^2}{l}$. ☐

208. For the fixed beam shown in Fig. 8.25, if $a > b$, then fixing moment at A will be
(*a*) greater than that of B ☐ (*b*) lesser than that of B ☐
(*c*) same as at A ☐ (*d*) none of the above. ☐

209. For the fixed beam shown in Fig. 8.25, the deflection at C will be
(*a*) $\frac{Wab^3}{3EIl^3}$ ☐ (*b*) $\frac{Wa^3b}{3EIl^3}$ ☐
(*c*) $\frac{Wa^3b^3}{3EIl^3}$ ☐ (*d*) $\frac{Wa^2b^2}{3EIl^2}$. ☐

210. For the fixed beam shown in Fig. 8.25, if $a > b$ then maximum deflection occurs
(*a*) at C ☐ (*b*) between C and B ☐
(*c*) between A and C ☐ (*d*) at A or at B. ☐

211. For the Fixed beam shown in Fig. 8.25, the maximum deflection is equal to

(a) $\frac{2}{3}\frac{Wa^2b^2}{(3a+b)^2 EI}$ ☐ (b) $\frac{2}{3}\frac{Wa^3b^2}{(3a+b)^2 EI}$ ☐

(c) $\frac{3}{2}\frac{Wa^2b^2}{(3a+b)^2 EI}$ ☐ (d) $\frac{3}{2}\frac{Wa^3b^2}{(3a+b)^2 EI}$. ☐

212. A fixed beam of length (l) carries a point load (W) at the centre. The deflection at the centre is equal to

(a) $\frac{Wl^3}{48\,EI}$ ☐ (b) $\frac{Wl^3}{96\,EI}$ ☐

(c) $\frac{Wl^3}{192\,EI}$ ☐ (d) $\frac{Wl^3}{384\,EI}$. ☐

213. A fixed beam of length (l) carries a uniformly distributed load of w per unit length over the whole span. The deflection at the centre is equal to

(a) $\frac{Wl^3}{48\,EI}$ ☐ (b) $\frac{Wl^3}{96\,EI}$ ☐

(c) $\frac{Wl^3}{192\,EI}$ ☐ (d) $\frac{Wl^3}{384\,EI}$. ☐

where $W = w \times l$.

214. The ratio of the maximum central deflection of a fixed beam of length (l) when it carries a uniformly distributed load (w) per unit length over the whole length to the central deflection of the fixed beam of length (l) when it carries a point load (W) at the centre such that $W = wl$, is equal to

(a) 1.0 ☐ (b) 2.0 ☐

(c) 3.0 ☐ (d) 4.0. ☐

215. Figure 8.26 shows four bending moment diagrams for a beam of length (l). Figure 8.26 (a) is the bending moment diagram for a

(a) fixed beam with a point load at centre ☐

(b) fixed beam with a uniformly distributed load over the whole span ☐

(c) simply supported beam with a point load at the centre ☐

(d) simply supported beam with a uniformly distributed load over the whole span. ☐

216. Figure 8.26 (b) represents the bending moment diagram for a

(a) fixed beam with a point load at centre ☐

(b) fixed beam with a uniformly distributed load over the whole span ☐

(c) simply supported beam with a point load at the centre ☐

(d) simply supported beam with a uniformly distributed load over the whole span. ☐

(a) (b) (c) (d)

FIGURE 8.26

217. Figure 8.26 (*c*) represents the bending moment diagram for a

(*a*) fixed beam with a point load at centre ☐
(*b*) Fixed beam with a uniformly distributed load over the whole span ☐
(*c*) simply supported beam with a point load at the centre ☐
(*d*) simply supported beam with a uniformly distributed load over the whole span. ☐

218. Figure 8.26 (*d*) represents the bending moment diagram for

(*a*) fixed beam with a point load at the centre ☐
(*b*) fixed beam with uniformly distributed load over the whole span ☐
(*c*) simply supported beam with a point load at the centre ☐
(*d*) simply supported beam with uniformly distributed load over the whole span. ☐

219. The Brinnell hardness number is equal to

(*a*) $\frac{2P}{\pi D[D+\sqrt{D^2-d^2}]}$ ☐ (*b*) $\frac{2P}{\pi D[D-\sqrt{D^2-d^2}]}$ ☐

(*c*) $\frac{P}{\pi D[D-\sqrt{D^2-d^2}]}$ ☐ (*d*) $\frac{\pi D[D+\sqrt{D^2-d^2}]}{2P}$. ☐

where P = Standard load in kg, D = Diameter in steel ball in mm

d = Diameter of indent in mm.

220. In a concrete beam the steel bars are embedded

(*a*) near top section ☐ (*b*) near bottom section ☐
(*c*) in the centre ☐ (*d*) anywhere. ☐

221. The load on a circular column of diameter (*d*) for keeping the stress wholly compressive may be applied anywhere within a concentric circle of diameter

(*a*) $\frac{d}{8}$ ☐ (*b*) $\frac{d}{4}$ ☐

(*c*) $\frac{d}{3}$ ☐ (*d*) $\frac{d}{2}$ ☐

222. Three beams of circular, square, rectangular (with depth twice the width) sections and of same length are subjected to the same maximum bending moment. If allowable stress is the same then the section which will require maximum weight of the same material will be

(*a*) rectangular ☐ (*b*) square ☐
(*c*) circular ☐ (*d*) none of the above. ☐

223. For the question 222, the section which will require minimum weight of the same material will be

(*a*) rectangular ☐ (*b*) square ☐
(*c*) circular ☐ (*d*) none of the above. ☐

224. For the question 222, the ratios of weight of circular beam, weight of rectangular beam and weight of square beam is

(*a*) 1 : 0.7938 : 1.118 ☐ (*b*) 07938 : 1 : 1.118 ☐
(*c*) 1.118 : 07938 : 1 ☐ (*d*) 07938 : 1118 : 1. ☐

225. Kernel or core of a section is the figure within which load may be placed so as

(*a*) to produce tensile stress at one end and compressive stress at the other end ☐

(*b*) to produce tensile stress at both the ends of the section ☐

(*c*) to produce tensile stress in the middle of the section ☐

(*d*) not to produce tensile stress anywhere in the section. ☐

226. The maximum eccentricity (*e*) of the vertical load on a short column of solid circular section of diameter (*D*), without producing any tensile stress along the cross-section is

(*a*) $\frac{D}{4}$ ☐ (*b*) $\frac{D}{8}$ ☐

(*c*) $\frac{D}{12}$ ☐ (*d*) $\frac{D}{16}$ ☐

227. The maximum eccentricity (*e*) of the vertical load on a short column of external diameter (*D*) an internal diameter (*d*), without producing any tensile stress along the cross-section is

(*a*) $\frac{D^2 - d^2}{8D}$ ☐ (*b*) $\frac{D^2 + d^2}{D}$ ☐

(*c*) $\frac{D^2 + d^2}{8D}$ ☐ (*d*) $\frac{8D}{D^2 + d^2}$ ☐

228. A short pillar of cross-section 100 cm × 100 cm carries a point load of 100000 N acting with an eccentricity of 10 cm. The stress due to direct load is equal to

(*a*) 10 N/cm² ☐ (*b*) 6 N/cm² ☐

(*c*) 16 N/cm2 ☐ (*d*) 4 N/cm². ☐

229. In question 228, the section modulus is equal to

(*a*) 10^6 cm³ ☐ (*b*) 16.67×10^4 cm³ ☐

(*c*) 10^5 cm³ ☐ (*d*) 10^4 cm³. ☐

230. In question 228, stress due to moment is equal to

(*a*) ± 10 N/cm² ☐ (*b*) ± 6 N/cm² ☐

(*c*) ± 16 N/cm² ☐ (*d*) ± 4 N/cm². ☐

231. In question 228, the maximum stress on the section is

(*a*) 10 N/cm² ☐ (*b*) 26 N/cm² ☐

(*c*) 16 N/cm² ☐ (*d*) 4 N/cm². ☐

232. In question 228, the minimum stress on the section is equal to

(*a*) 10 N/cm² ☐ (*b*) 2 N/cm² ☐

(*c*) 16 N/cm² ☐ (*d*) 4 N/cm². ☐

233. A short column of external diameter 40 cm and internal diameter 20 cm carries an eccentric load 10000 N. The maximum eccentricity which the load can have without producing tension on the cross-section is

(*a*) 10 cm ☐ (*b*) 8 cm ☐

(*c*) 7.5 cm ☐ (*d*) 6.25 cm. ☐

234. A steel plate 6 cm wide and 1 cm thick is to be bent into a circular arc of radius 10 m. If $E = 2 \times 10^6$ N/cm^2, then the maximum stress induced will be

(a) 2000 N/cm^2 ☐ (b) 1000 N/cm^2 ☐
(c) 500 N/cm^2 ☐ (d) 400 N/cm^2 ☐

235. In question 234, the bending moment which produces the maximum shear stress will be

(a) 2000 N-cm ☐ (b) 10000 N-cm ☐
(c) 1000 N-cm ☐ (d) 500 N-cm. ☐

236. The ratio of moment of inertia about the neutral axis to the distance of the most distant point of the section from the neutral axis is called

(a) moment of inertia ☐ (b) section modulus ☐
(c) polar moment of inertia ☐ (d) modulus of rigidity. ☐

237. The relation between maximum stress (f) offered by a section, moment of resistance (M) and section modulus (Z) is given by

(a) $M = \dfrac{f}{Z}$ ☐ (b) $M = \dfrac{Z}{f}$ ☐
(c) $M = f \times Z$ ☐ (d) $M = \dfrac{1}{f \times Z}$ ☐

238. Choose the correct statement

(a) Section modulus of a hollow circular section of external diameter D and internal diameter d is equal to $\dfrac{\pi(D^4 - d^4)}{64D}$. ☐

(b) Section modulus of a circular section on diameter D is $\dfrac{\pi D^4}{32}$. ☐

(c) section modulus of a rectangular section is $\dfrac{bd^2}{6}$. ☐

(d) Section modulus of a square section $b \times b$ is $\dfrac{b^4}{12}$. ☐

239. A beam of uniform strength can be designed by

(a) varying the depth of the beam but maintaining constant width ☐
(b) varying the width of the beam but maintaining constant depth ☐
(c) varying width and depth ☐
(d) any one of the above ☐
(e) none of the above. ☐

240. For a beam of uniform strength having constant depth, the width at a distance x from the support varies with

(a) $\sqrt{x}$ ☐ (b) x ☐
(c) $x^{1/4}$ ☐ (d) $x^{3/4}$ ☐

241. For a beam of uniform strength having constant width, the depth of the beam at a distance x from the support varies with

(a) $\sqrt{x}$ ☐ (b) x ☐
(c) $x^{1/4}$ ☐ (d) $x^{3/4}$ ☐

242. Flitched beam means a

(a) continuous beam ☐
(b) fixed beam ☐
(c) beam of composite section consisting of a wooden beam strengthened by mild steel plates ☐
(d) none of the above. ☐

243. The shear stress required to cause plastic deformation of solid metal is called

(a) proof stress ☐ (b) flow stress ☐
(c) rupture stress ☐ (d) ultimate stress. ☐

244. The stress which will cause a specified permanent deformation in a material (usually 0.01% or less) is called

(a) proof stress ☐ (b) flow stress ☐
(c) rupture stress ☐ (d) ultimate stress. ☐

245. The stress obtained by dividing the load at the moment of incipient fracture, by the area supporting that load is called

(a) proof stress ☐ (b) flow stress ☐
(c) rupture stress ☐ (d) ultimate stress. ☐

246. The statement that 'If unit loads rest upon a beam at the two points A and *B*, then the deflection at A due to unit load at *B* is equal to the deflection at *B* due to unit load at A' is given by

(a) Mohr ☐ (b) Castigliano ☐
(c) Max well ☐ (d) Rankine. ☐

247. A short column of rectangular section carries a point load (*W*) acting with an eccentricity (*e*). The shape of Kernel area would be

(a) square ☐ (b) rectangle ☐
(c) circle ☐ (d) rhombus. ☐

248. If in question 247, the column is of circular section then the shape of Kernel area would be

(a) square ☐ (b) rectangle ☐
(c) circle ☐ (d) rhombus. ☐

249. Every cross-section of a shaft, which is subjected to a twisting moment, is under

(a) compressive stress ☐ (b) shear stress ☐
(c) tensile stress ☐ (d) bending stress. ☐

250. The shear stress at any point of a shaft, subjected to twisting moment, is

(a) proportional to its distance from the central axis of the shaft ☐
(b) inversely proportional to its distance from the central axis of the shaft ☐
(c) proportional to the square of its distance from the central axis of the shaft ☐
(d) none of the above. ☐

Torsion of Shafts

251. When a shaft is subjected to tension, the relation between maximum shear stress (f), modulus of rigidity of the shaft (C) and angle of twist (θ) is given by

(a) $\frac{C\theta}{\tau} = \frac{R}{L}$ ☐ (b) $\frac{C\theta}{L} = \frac{\tau}{R}$ ☐

(c) $\frac{C\theta}{R} = \frac{\tau}{L}$ ☐ (d) $\frac{C}{L\theta} = \frac{\tau}{R}$. ☐

where L = Length of shaft and R = Radius of shaft.

252. A solid shaft of diameter D transmits the torque equal to

(a) $\frac{\pi}{32}\ \tau D^3$ ☐ (b) $\frac{\pi}{64}\ \tau D^3$ ☐

(c) $\frac{\pi}{16}\ \tau D^3$ ☐ (d) $\frac{\pi}{8}\ \tau D^3$. ☐

where τ = Maximum allowable shear stress.

253. The torque transmitted by a hollow shaft to external diameter (D) and internal diameter (d) is equal to

(a) $\frac{\pi}{32}\ \tau\,[D^3 - d^3]$ ☐ (b) $\frac{\pi}{16}\ \tau\,[D^3 - d^3]$ ☐

(c) $\frac{\pi}{16}\ \tau \left[\frac{D^4 - d^4}{D}\right]$ ☐ (d) $\frac{\pi}{32}\ \tau \left[\frac{D^4 - d^4}{D}\right]$. ☐

254. Polar moment of inertia of a solid circular shaft of diameter D is equal to

(a) $\frac{\pi D^3}{32}$ ☐ (b) $\frac{\pi D^4}{32}$ ☐

(c) $\frac{\pi D^3}{64}$ ☐ (d) $\frac{\pi D^4}{64}$. ☐

255. Polar moment of inertia of a hollow circular shaft is equal to

(a) $\frac{\pi}{32}\ [D^3 - d^3]$ ☐ (b) $\frac{\pi}{32}\ [D^4 - d^4]$ ☐

(c) $\frac{\pi}{64}\ [D^3 - d^3]$ ☐ (d) $\frac{\pi}{64}\ [D^4 - d^4]$. ☐

256. Torsional rigidity of a shaft is equal to

(a) product of modulus of rigidity and polar moment of inertia ☐

(b) sum of modulus of rigidity and polar moment of inertia ☐

(c) difference of modulus of rigidity and polar moment of inertia ☐

(d) ratio of modulus of rigidity and polar moment of inertia. ☐

257. Polar modulus of a shaft section is equal to

(a) product of polar moment of inertia and maximum radius of the shaft ☐

(b) ratio of polar moment of inertia to maximum radius of the shaft ☐

(c) sum of polar moment of inertia and maximum radius of the shaft ☐

(d) difference of polar moment of inertia and maximum radius of the shaft. ☐

258. The torsional rigidity of a shaft is defined as the torque required to produce

(*a*) maximum twist in the shaft ☐

(*b*) maximum shear stress in the shaft ☐

(*c*) minimum twist in the shaft ☐

(*d*) a twist of one radian per unit length of the shaft. ☐

259. The flexural rigidity of the deflection of beams is expressed as

(*a*) $\frac{I}{E}$ ☐ (*b*) $\frac{E}{I}$ ☐

(*c*) EI ☐ (*d*) $\frac{I}{EI}$. ☐

where E = Modulus of elasticity and I = Moment of inertia.

260. The greatest twisting moment which a given shaft section can resist is equal to

(*a*) polar modulus × τ ☐ (*b*) polar modulus/τ ☐

(*c*) τ/polar modulus ☐ (*d*) none of the above. ☐

261. Choose the correct statement

(*a*) Shafts of the same material and length having the same polar modulus have the same strength. ☐

(*b*) For a shaft of a given material, the magnitude of polar modulus is a measure of its strength in resisting torsion. ☐

(*c*) From a number of shafts of the same length and material the shaft with greatest polar modulus will resist the maximum twisting moment. ☐

(*d*) all of the above. ☐

(*e*) none of the above. ☐

262. Two shafts, one solid and the other hollow, are made of the same materials and are having same length and weight. The hollow shaft as compared to solid shaft is

(*a*) more strong ☐ (*b*) less strong ☐

(*c*) having same strength ☐ (*d*) none of the above. ☐

263. Two shafts, one solid and the other hollow, are of same length and material. They are subjected to the same torque and attain the same permissible maximum shear stress at the same time. The weight of hollow shaft will be

(*a*) more ☐ (*b*) less ☐

(*c*) same as of solid shaft ☐ (*d*) none of the above. ☐

264. The torque transmitted, by a solid shaft of diameter 40 mm if the shear stress is not to exceed 400 kg/cm^2, would be

(*a*) 1.6 × π kg-m ☐ (*b*) 16 π kg-m ☐

(*c*) 0.8 × π kg-m ☐ (*d*) 0.4 × π kg-m. ☐

265. If in question 264, the diameter of the solid shaft is doubled, then torque transmitted would be

(*a*) same ☐ (*b*) double ☐

(*c*) four times ☐ (*d*) eight times. ☐

266. If in question 264, the diameter of the solid shaft is made 20 mm, then torque transmitted would be

(*a*) same ☐ (*b*) one-half ☐

(*c*) one-eight ☐ (*d*) one-fourth. ☐

267. The torsion equation is given by

(*a*) $\frac{T}{J}=\frac{\tau}{R}=\frac{L}{C\theta}$ ☐ (*b*) $\frac{T}{R}=\frac{\tau}{J}=\frac{C\theta}{L}$ ☐

(*c*) $\frac{T}{J}=\frac{\tau}{R}=\frac{C\theta}{L}$ ☐ (*d*) $\frac{T}{\tau}=\frac{R}{J}=\frac{C\theta}{L}$. ☐

268. The assumption made, while determining the shear stress in a circular shaft subjected to torsion, is that

(*a*) the material of the shaft is uniform ☐

(*b*) the twist along the shaft is uniform ☐

(*c*) cross-sections of the shaft is plane and circular before and alter the twist ☐

(*d*) all of the above ☐

(*e*) none of the above. ☐

269. When a shaft of diameter (*d*) is subjected to combined twisting moment (*T*) and bending moment (*M*), the maximum shear stress (τ) is equal to

(*a*) $\frac{R}{J}\sqrt{M^2+T^2}$ ☐ (*b*) $\frac{J}{R}\sqrt{(M^2+T^2)}$ ☐

(*c*) $\frac{R}{J}(M^2+T^2)$ ☐ (*d*) $\frac{J}{R}(M^2+T^2)$. ☐

where *J* = Polar moment of inertia of the shaft.

270. In question 269, the maximum normal stress is given by

(*a*) $\frac{16}{\pi d^3}(M-\sqrt{M^2+T^2})$ ☐ (*b*) $\frac{16}{\pi d^3}(M+\sqrt{M^2+T^2})$ ☐

(*c*) $\frac{16}{\pi d^3}(M^2-\sqrt{M^2-T^2})$ ☐ (*d*) $\frac{16}{\pi d^3}(M+\sqrt{M^2-T^2})$. ☐

271. In question 269, the minimum normal stress is given by

(*a*) $\frac{16}{\pi d^3}(M-\sqrt{M^2+T^2})$ ☐ (*b*) $\frac{16}{\pi d^3}(M^2+\sqrt{M^2+T^2})$ ☐

(*c*) $\frac{16}{\pi d^3}(M-\sqrt{M^2-T^2})$ ☐ (*d*) $\frac{16}{\pi d^3}(M+\sqrt{M^2-T^2})$. ☐

272. In question 269, the position of principal planes is given by

(*a*) $\tan 2\theta = \frac{M}{T}$ ☐ (*b*) $\tan 2\theta = \frac{T}{M}$ ☐

(*c*) $\tan 2\theta = M \times T$ ☐ (*d*) $\tan 2\theta = \frac{2T}{M}$ ☐

Combination of Bending Moment and Twisting Moment

273. A shaft of diameter 8 cm is subjected to a bending moment of 30000 N-cm and a twisting moment of 40000 N-cm. The maximum normal stress induced in the shaft is equal to

(a) $\frac{2500}{\pi}$ N/cm^2 ☐ (b) $\frac{5000}{\pi}$ N/cm^2 ☐

(c) $\frac{1575}{\pi}$ N/cm^2 ☐ (d) $\frac{3150}{\pi}$ N/cm^2 ☐

274. In question 273, the minimum normal stress induced is equal to

(a) 2500 N/cm^2 ☐ (b) $\frac{5000}{\pi}$ N/cm^2 ☐

(c) $\frac{1575}{\pi}$ N/cm^2 ☐ (d) $-\frac{625}{\pi}$ N/cm^2. ☐

275. In question 273, the position of principal planes is given by

(a) tan 2θ = 1.33 ☐ (b) tan 2θ = 0.75 ☐

(c) tan 2θ = 2.0 ☐ (d) tan 2θ = 1.50. ☐

276. In question 273, the maximum shear stress induced in the shaft is equal to

(a) 2500 N/cm^2 ☐ (b) $\frac{5000}{\pi}$ N/cm^2 ☐

(c) $\frac{1575}{\pi}$ N/cm^2 ☐ (d) $\frac{3150}{\pi}$ N/cm^2 ☐

277. When a shaft of diameter (*d*) is subjected to a maximum bending moment (*M*) and a torque (*T*). The equivalent bending moment (M_e) is equal to

(a) $(\sqrt{M^2+T^2})$ ☐ (b) $\frac{1}{2}(M - \sqrt{M^2+T^2})$ ☐

(c) $\frac{1}{2}(M + \sqrt{M^2+T^2})$ ☐ (d) $(M + \sqrt{M^2-T^2})$. ☐

278. For the question 277, the equivalent torque (T_e) is equal to

(a) $(\sqrt{M^2+T^2})$ ☐ (b) $\frac{1}{2}(M - \sqrt{M^2+T^2})$ ☐

(c) $\frac{1}{2}(M + \sqrt{M^2+T^2})$ ☐ (d) $(M + \sqrt{M^2-T^2})$. ☐

279. Equivalent torque is the torque which alone produces

(a) maximum normal stress ☐ (b) minimum normal stress ☐

(c) maximum shear stress ☐ (d) minimum shear stress. ☐

280. Equivalent bending moment is the bending moment which alone produces

(a) maximum normal stress ☐ (b) minimum normal stress ☐

(c) maximum shear stress ☐ (d) minimum shear stress. ☐

281. A solid shaft of diameter 10 cm is subjected to a maximum bending moment of 40,000 N-cm and maximum torque of 30,000 N-cm. The equivalent bending moment is

(a) 50000 N-cm ☐ (b) 45000 N-cm ☐

(c) 40000 N-cm ☐ (d) 35000 N-cm. ☐

282. In question 281, the equivalent torque is equal to

(*a*) 50000 N-cm ☐ (*b*) 45000 N-cm ☐
(*c*) 40000 N-cm ☐ (*d*) 35000 N-cm. ☐

283. In question 281, the maximum principal stress will be

(*a*) $\frac{800}{\pi}$ N/cm^2 ☐ (*b*) $\frac{1600}{\pi}$ N/cm^2 ☐
(*c*) $\frac{1440}{\pi}$ N/cm^2 ☐ (*d*) $\frac{1000}{\pi}$ N/cm^2. ☐

284. In question 281, the maximum shear stress will be

(*a*) $\frac{800}{\pi}$ N/cm^2 ☐ (*b*) $\frac{1600}{\pi}$ N/cm^2 ☐
(*c*) $\frac{1440}{\pi}$ N/cm^2 ☐ (*d*) $\frac{1000}{\pi}$ N/cm^2. ☐

285. A solid shaft of diameter 10 cm transmits 140 horse power at 140 r.p.m. The torque on the shaft will be

(*a*) $\frac{4500}{\pi}$ kg-m ☐ (*b*) $\frac{2250}{\pi}$ kg-m ☐
(*c*) $\frac{3600}{\pi^2}$ kg-m ☐ (*d*) $\frac{1125}{\pi}$ kg-m. ☐

286. In question 285, the maximum shear stress induced will be

(*a*) $\frac{4500}{\pi}$ kg/cm^2 ☐ (*b*) $\frac{2250}{\pi}$ kg/cm^2 ☐
(*c*) $\frac{3600}{\pi^2}$ kg/cm^2 ☐ (*d*) $\frac{1125}{\pi}$ kg/cm^2. ☐

Strain Energy and Springs

287. The cross-section of a member is subjected to a uniform shear stress of intensity τ. Then the strain energy stored per unit volume is equal to

(*a*) $\frac{2\tau^2}{C}$ ☐ (*b*) $\frac{\tau^2}{2C}$ ☐
(*c*) $\frac{C}{2\tau^2}$ ☐ (*d*) $\frac{2C}{\tau^2}$. ☐

where C = Modulus of rigidity.

288. The strain energy stored by a solid shaft per unit volume, due to torsion is equal to

(*a*) $\frac{\tau^2}{2C}$ ☐ (*b*) $\frac{\tau^2}{4C}$ ☐
(*c*) $\frac{\tau^2}{3C}$ ☐ (*d*) $\frac{\tau^2}{5C}$. ☐

where τ = Maximum shear stress at the surface of the shaft.

289. The strain energy stored by a hollow shaft of external diameter *(D)* and internal diameter *(d)* per unit volume, due to torsion is

(*a*) $\frac{\tau^2}{2C}$ ☐ (*b*) $\frac{\tau^2}{4C}\left(\frac{D^2}{D^2+d^2}\right)$ ☐

(*c*) $\frac{\tau^2}{4C}\left(\frac{D^2+d^2}{D^2}\right)$ ☐ (*d*) $\frac{\tau^2}{2C}\left(\frac{D^2+d^2}{D^2}\right)$. ☐

290. The deflection of a close-coiled helical spring, subjected to an axial load W is equal to

(*a*) $\frac{64W^2R^5n}{Cd^3}$ ☐ (*b*) $\frac{64W^2R^3n}{Cd^3}$ ☐

(*c*) $\frac{64WR^3n}{Cd^4}$ ☐ (*d*) $\frac{32W^2R^3n}{Cd^4}$. ☐

where d = Diameter of spring, n = Number of coils,

R = Mean radius of coils, C = Modulus of rigidity.

291. The stiffness of the close-coiled helical spring, subjected to an axial load W, is equal to

(*a*) $\frac{Cd^4}{64R^4n}$ ☐ (*b*) $\frac{Cd^3}{64R^3n}$ ☐

(*c*) $\frac{Cd^4}{32R^3n}$ ☐ (*d*) $\frac{Cd^4}{64R^2n}$. ☐

292. For the question 291, the maximum shear stress at any section of the helical-spring coil would be

(*a*) $\frac{16WR}{\pi d^3}$ ☐ (*b*) $\frac{WR}{16\pi d^3}$ ☐

(*c*) $\frac{16WR}{\pi d^2}$ ☐ (*d*) $\frac{64\,WR}{\pi d^3}$. ☐

293. A closely coiled helical spring is made out of 1 cm diameter steel rod. The spring carries an axial pull of 20 N. The coil consists of 10 complete turns with a mean radius of 5 cm. If modulus of rigidity of spring material is 8×10^5 N/cm², the maximum shear stress induced in the section of the rod would be

(*a*) $\frac{3200}{\pi}$ N/cm² ☐ (*b*) $\frac{1600}{\pi}$ N/cm² ☐

(*c*) $\frac{800}{\pi}$ N/cm² ☐ (*d*) $\frac{400}{\pi}$ N/cm². ☐

294. In question 293, the deflection of the spring would be

(*a*) 2 cm ☐ (*b*) 4 cm ☐

(*c*) 1 cm ☐ (*d*) 1.5 cm. ☐

295. In question 293, stiffness of the spring would be

(*a*) 20 N per cm ☐ (*b*) 15 N per cm ☐

(*c*) 10 N per cm ☐ (*d*) 8 N per cm. ☐

296. In question 293, the strain energy stored by the spring is equal to

(*a*) 40 N-cm ☐ (*b*) 30 N-cm ☐
(*c*) 25 N-cm ☐ (*d*) 20 N-cm. ☐

297. For calculating stresses in a hollow shaft-subjected to torsion, the radius taken into consideration is

(*a*) outer radius ☐ (*b*) inner radius ☐
(*c*) mean radius ☐ (*d*) none of the above. ☐

298. The ratio of torque transmitted by hollow shaft (of inner diameter equal to half the outer diameter) to the torque transmitted by a solid shaft of the same material and of the same weight is equal to

(*a*) $\frac{1}{2}$ ☐ (*b*) $\frac{1}{4}$ ☐
(*c*) $\frac{3}{4}$ ☐ (*d*) $\frac{15}{16}$. ☐

299. The shear stress at a point in a shaft subjected to a torque is

(*a*) directly proportional to the polar moment of inertia and to the distance of the point from the axis ☐
(*b*) directly proportional to the applied torque and polar moment of inertia ☐
(*c*) directly proportional to the applied torque and inversely proportional to the polar moment of inertia ☐
(*d*) inversely proportional to the polar moment of inertia. ☐

300. The strips in the Laminated spring are in different lengths for

(*a*) improved appearance ☐ (*b*) material economy ☐
(*c*) equal distribution of stress ☐ (*d*) reducing the weight. ☐

Thin Cylinders and Spheres

301. A cylindrical vessel is said to be thin if the ratio of its internal diameter to the wall thickness is

(*a*) less than 20 ☐ (*b*) equal to 20 ☐
(*c*) more than 20 ☐ (*d*) none of the above. ☐

302. The hoop or circumferential stress in a thin cylindrical shell of diameter (D), length (L) and thickness (t), when subjected to an internal pressure (p) is equal to

(*a*) $\frac{pD}{4t}$ ☐ (*b*) $\frac{pD}{2t}$ ☐
(*c*) $\frac{2pD}{t}$ ☐ (*d*) $\frac{4pD}{t}$. ☐

303. The longitudinal or axial stress in a thin cylindrical shell of diameter (D). Length (L) and thickness (t), when subjected to an internal pressure (p) is equal to

(*a*) $\frac{pD}{4t}$ ☐ (*b*) $\frac{pD}{2t}$ ☐
(*c*) $\frac{2pD}{t}$ ☐ (*d*) $\frac{4pD}{t}$. ☐

304. The maximum shear stress in a thin cylindrical shell, when subjected to an internal pressure (*p*) is equal to

(*a*) $\frac{pD}{4t}$ ☐ (*b*) $\frac{pD}{8t}$ ☐

(*c*) $\frac{pD}{2t}$ ☐ (*d*) $\frac{pD}{t}$. ☐

305. The maximum shear stress in a thin spherical shell, when subjected to an internal pressure (*p*) is equal to

(*a*) $\frac{pD}{4t}$ ☐ (*b*) $\frac{pD}{8t}$ ☐

(*c*) $\frac{pD}{2t}$ ☐ (*d*) zero. ☐

306. The hoop or circumferential stress in a thin spherical shell, when subjected to an internal pressure (*p*) is equal to

(*a*) $\frac{pD}{4t}$ ☐ (*b*) $\frac{pD}{2t}$ ☐

(*c*) $\frac{pD}{8t}$ ☐ (*d*) $\frac{2pD}{t}$. ☐

307. The hoop or circumferential stress in a riveted cylindrical shell, when subjected to an internal pressure (*p*) is equal to

(*a*) $\frac{pD}{4t\eta_l}$ ☐ (*b*) $\frac{pD}{4t\eta_c}$ ☐

(*c*) $\frac{pD}{2t\eta_l}$ ☐ (*d*) $\frac{dD}{2t\eta_c}$. ☐

where D = Internal diameter, η_l = Efficiency of longitudinal joint, and

η_c = Efficiency of circumferential joint.

308. The longitudinal stress in a riveted cylindrical shell, when subjected to internal pressure (*p*) is equal to

(*a*) $\frac{pD}{4t\eta_l}$ ☐ (*b*) $\frac{pD}{4t\eta_c}$ ☐

(*c*) $\frac{pD}{2t\eta_l}$ ☐ (*d*) $\frac{dD}{2t\eta_c}$. ☐

where η_l = Efficiency of longitudinal joint, and

η_c = Efficiency of circumferential joint.

309. Choose the correct statement

(*a*) The hoop stress in a thin cylindrical shell in compressive stress. ☐

(*b*) The shear stress in a thin spherical shell is more than that of in a thin cylindrical shell. ☐

(*c*) The design of thin cylindrical shell is based on hoop stress. ☐

(*d*) The ratio of hoop stress to longitudinal stress for a thin cylindrical shell is 1/2. ☐

310. A water main 1 m in diameter contains a fluid having pressure 10 kg/cm^2. If the maximum permissible tensile stress in the metal is 200 kg/cm^2, the thickness of the metal required would be

(*a*) 2 cm ☐ (*b*) 2.5 cm ☐
(*c*) 1 cm ☐ (*d*) 0.5 cm. ☐

311. The circumferential strain in case of thin cylindrical shell, when subjected to internal pressure (*p*), is equal to

(*a*) $\frac{pd}{2tE}\left(\frac{1}{2}-\frac{1}{m}\right)$ ☐ (*b*) $\frac{pd}{2tE}\left(1-\frac{1}{2m}\right)$ ☐
(*c*) $\frac{pd}{4tE}\left(1-\frac{1}{m}\right)$ ☐ (*d*) $\frac{3pd}{4tE}\left(1-\frac{1}{m}\right)$. ☐

312. The longitudinal strain in case of thin cylindrical shell, when subjected to internal pressure (*p*), is equal to

(*a*) $\frac{pd}{2tE}\left(\frac{1}{2}-\frac{1}{m}\right)$ ☐ (*b*) $\frac{pd}{2tE}\left(1-\frac{1}{2m}\right)$ ☐
(*c*) $\frac{pd}{4tE}\left(1-\frac{1}{m}\right)$ ☐ (*d*) $\frac{3pd}{4tE}\left(1-\frac{1}{m}\right)$. ☐

313. The strain in any direction in case of a thin spherical shell, when subjected to internal pressure (*p*), is equal to

(*a*) $\frac{pd}{2tE}\left(\frac{1}{2}-\frac{1}{m}\right)$ ☐ (*b*) $\frac{pd}{2tE}\left(1-\frac{1}{2m}\right)$ ☐
(*c*) $\frac{pd}{4tE}\left(1-\frac{1}{m}\right)$ ☐ (*d*) $\frac{3pd}{2tE}\left(1-\frac{1}{m}\right)$. ☐

314. The volumetric strain in case of thin spherical shell, when subjected to internal pressure (*p*), is equal to

(*a*) $\frac{pd}{2tE}\left(\frac{1}{2}-\frac{1}{m}\right)$ ☐ (*b*) $\frac{pd}{2tE}\left(1-\frac{1}{2m}\right)$ ☐
(*c*) $\frac{pd}{4tE}\left(1-\frac{1}{m}\right)$ ☐ (*d*) $\frac{3pd}{2tE}\left(1-\frac{1}{m}\right)$. ☐

315. The circumferential strain in case of thin cylindrical shell, when subjected to internal pressure (*p*), is

(*a*) more than diametral strain ☐ (*b*) less than diametral strain ☐
(*c*) equal to diametral strain ☐ (*d*) none of the above. ☐

316. The volumetric strain in case of a thin cylindrical shell, when subjected to internal pressure (*p*), is equal to

(*a*) sum of circumferential and longitudinal strains ☐
(*b*) difference of circumferential and longitudinal strains ☐
(*c*) twice the circumferential strain plus longitudinal strain ☐
(*d*) twice the longitudinal strain plus circumferential strain. ☐

317. A thin cylindrical drum 100 cm in diameter and 10 m long has a shell thickness of 1 cm. The drum is subjected to an internal pressure of 400 N/cm². If $E = 2 \times 10^7$ N/cm² and Poisson's ratio = 0.3, then circumferential strain will be

(*a*) 0.085×10^{-3} ☐ (*b*) 0.20×10^{-3} ☐

(*c*) 0.85×10^{-3} ☐ (*d*) 1.9×10^{-3}. ☐

318. For the question 317, the longitudinal strain will be

(*a*) 0.085×10^{-3} ☐ (*b*) 0.20×10^{-3} ☐

(*c*) 0.85×10^{-3} ☐ (*d*) 1.9×10^{-3}. ☐

319. For the question 317, the volumetric strain will be

(*a*) 0.085×10^{-3} ☐ (*b*) 0.20×10^{-3} ☐

(*c*) 0.85×10^{-3} ☐ (*d*) 1.9×10^{-3}. ☐

320. For the question 317, the ratio of longitudinal strain to circumferential strain will be

(*a*) 8/17 ☐ (*b*) 4/17 ☐

(*c*) 2/17 ☐ (*d*) 1/17. ☐

321. If in question 317, Poisson's ratio is made equal to 1/4, then ratio of longitudinal strain to circumferential strain will be

(*a*) 3/7 ☐ (*b*) 5/7 ☐

(*c*) 4/7 ☐ (*d*) 2/7. ☐

322. Choose the correct statement

(*a*) The circumferential strain in case of thin cylindrical shell is equal to $\frac{pd}{2tE}\left(\frac{1}{2} - \frac{1}{m}\right)$. ☐

(*b*) The longitudinal strain in case of thin cylindrical shell is equal to $\frac{pd}{2tE}\left(1 - \frac{1}{2m}\right)$. ☐

(*c*) The longitudinal stress in a thin cylindrical shell is equal to $\frac{pd}{2t}$. ☐

(*d*) The ratio of longitudinal strain to circumferential strain in case of thin cylindrical shell is equal to

$\left(\frac{m-2}{2m-1}\right)$ ☐

where $\frac{1}{m}$ = Poisson's ratio.

323. Choose the wrong statement

(*a*) The hoop stress in a thin cylindrical is twice of longitudinal stress. ☐

(*b*) Wire winding of thin cylinders makes the cylinder to withstand a higher fluid pressure. ☐

(*c*) A cylindrical section having no joint is known as seamless section. ☐

(*d*) The ratio of longitudinal strain to volumetric strain in case of thin cylindrical shell is equal to

$\left(\frac{2m-1}{5m-4}\right)$. ☐

324. A cylindrical thin drum 80 cm in diameter and 3 m long has a shell thickness of 1 cm. If the drum is subjected to an internal pressure of 25 N/cm^2, then the hoop stress will be

(*a*) 500 N/cm^2 ☐ (*b*) 1000 N/cm^2 ☐
(*c*) 250 N/cm^2 ☐ (*d*) 1500 N/cm^2. ☐

325. In question 324, the longitudinal stress would be

(*a*) 500 N/cm^2 ☐ (*b*) 1000 N/cm^2 ☐
(*c*) 250 N/cm^2 ☐ (*d*) 1500 N/cm^2. ☐

326. In the assembly of pulley, key and shaft

(*a*) key is made strongest link ☐
(*b*) key is made weaker link ☐
(*c*) all the three are designed for the same strength ☐
(*d*) pulley is made weaker. ☐

327. A coil is having stiffness *k*. It is cut into two halves, then the stiffness of the cut coils will be

(*a*) same ☐ (*b*) half ☐
(*c*) double ☐ (*d*) one-fourth. ☐

Thick Cylinders and Spheres

328. The hoop stress in case of thick cylinders across the thickness

(*a*) is uniformly distributed ☐
(*b*) varies from maximum at the outer circumference to minimum at inner circumference ☐
(*c*) varies from maximum at the inner circumference to minimum at outer circumference ☐
(*d*) is zero. ☐

329. The longitudinal stress in case of thick cylinders

(*a*) is uniformly distributed ☐
(*b*) varies from maximum at the outer circumference to minimum at inner circumference ☐
(*c*) varies from maximum at the inner circumference to minimum at outer circumference ☐
(*d*) none of the above. ☐

330. In case of thick cylinders, at any point the three principal stresses, *i.e.*, radial, circumferential and longitudinal

(*a*) are all tensile stresses ☐ (*b*) are all compressive stresses ☐
(*c*) are all shear stresses ☐ (*d*) none of the above. ☐

331. In case of thick cylinders, the difference of circumferential stress and radial stress

(*a*) is constant ☐ (*b*) varies directly with radius ☐
(*c*) varies inversely with radius ☐ (*d*) none of the above. ☐

332. The radial stress at any radius x in case of thick cylinder subjected to internal pressure (p) is equal to

(a) $\frac{b^2}{x^2} + a$ ☐ (b) $\frac{b}{x^2} - a$ ☐

(c) $\frac{x^2}{b} + a$ ☐ (d) $\frac{x^2}{b} - a.$ ☐

where a and b are constants.

333. The circumferential stress at any radius r in case of thick cylinder subjected to internal pressure (p) is equal to

(a) $(b/x^2) + a$ ☐

(b) $(b/x^2) - a$ ☐

(c) $(x^2/b) + a$ ☐

(d) $(x^2/b) - a$ ☐

where a and b are constants.

FIGURE 8.27

334. Figure 8.27 shows the different curves for the stresses distribution across the thickness of a thick cylinder. The radial stress distribution is given by

(a) curve A ☐ (b) curve B ☐

(c) curve C ☐ (d) curve D. ☐

335. In Fig. 8.27, the circumferential stress (hoop stress) distribution is given by

(a) curve A ☐ (b) curve B ☐

(c) curve C ☐ (d) curve D. ☐

336. In Fig. 8.27, the longitudinal stress distribution is given by

(a) curve A ☐ (b) curve B ☐

(c) curve C ☐ (d) curve D. ☐

337. When a thick cylinder is subjected to internal fluid pressure (pi), the minimum value of radial stress is

(a) $\frac{2p_i R_i^2}{R_o^2 - R_i^2}$ ☐ (b) p_i ☐

(c) O ☐ (d) $\left(\frac{R_o^2 + R_i^2}{R_o^2 - R_i^2}\right) \times p_i.$ ☐

where R_o = Outer radius and R_i = Inner radius of the thick cylinder.

338. When a thick cylinder is subjected to internal fluid pressure (p_i), the minimum value of circumferential stress is

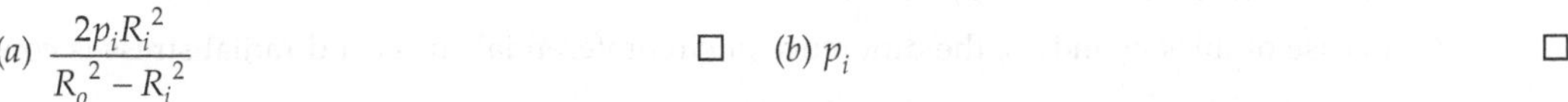

(a) $\frac{2p_i R_i^2}{R_o^2 - R_i^2}$ ☐ (b) p_i ☐

(c) O ☐ (d) $\left(\frac{R_o^2 + R_i^2}{R_o^2 - R_i^2}\right) \times p_i.$ ☐

339. When a thick cylinder is subjected to internal fluid pressure (p_i), the maximum value of radial stress is

(a) $\frac{2p_iR_i^2}{R_o^2 - R_i^2}$ ☐ (b) p_i ☐

(c) O ☐ (d) $\left(\frac{R_o^2 + R_i^2}{R_o^2 - R_i^2}\right) \times p_i$. ☐

340. When a thick cylinder is subjected to internal fluid pressure (p_i), the maximum value of circumferential stress is

(a) $\frac{2p_iR_i^2}{R_o^2 - R_i^2}$ ☐ (b) p_i ☐

(c) O ☐ (d) $\left(\frac{R_o^2 + R_i^2}{R_o^2 + R_i^2}\right) \times p_i$. ☐

341. In case of cylinders which have to carry high internal fluid pressure, the method adopted is to

(a) wind strong steel wire under tension on the cylinder ☐

(b) shrink one cylinder over the other ☐

(c) both (a) and (b) ☐

(d) none of the above. ☐

342. The radial stress, at any radius x in case of a thick spherical shell subjected to internal fluid pressure p, is equal to

(a) $\left(\frac{2b}{x^3}\right) - a$ ☐ (b) $\left(\frac{b}{x^3}\right) + a$ ☐

(c) $\left(\frac{x^3}{2b}\right) - a$ ☐ (d) $\left(\frac{x^3}{b}\right) + a$. ☐

where a and b are constants.

343. The circumferential stress, at any radius x in case of a thick spherical shell subjected to internal fluid pressure p, is equal to

(a) $\left(\frac{2b}{x^3}\right) - a$ ☐ (b) $\frac{b}{x^3} + a$ ☐

(c) $\left(\frac{x^3}{2b}\right) - a$ ☐ (d) $\left(\frac{x^3}{b}\right) + a$. ☐

344. Choose the correct statement

(a) When a thick cylinder is subjected to internal fluid pressure, the circumferential stress is maximum at the outer surface of the cylinder. ☐

(b) In case of thick cylinders, the difference of circumferential stress and radial stress is constant. ☐

(c) In case of thick cylinders, the minimum value of radial stress is equal to internal fluid pressure. ☐

(d) The single thick cylinder withstands high internal fluid pressure as compared to compound cylinder. ☐

345. A pipe of internal diameter as 40 cm and external diameter as 80 cm contains a fluid at a pressure of 750 N/cm^2. The maximum circumferential stress (hoop stress) would be

(*a*) 1250 N/cm^2 ☐ (*b*) 500 N/cm^2 ☐

(*c*) 250 N/cm^2 ☐ (*d*) 2000 N/cm^2. ☐

346. In question 345, the minimum circumferential stress would be

(*a*) 1250 N/cm^2 ☐ (*b*) 500 N/cm^2 ☐

(*c*) 250 N/cm^2 ☐ (*d*) 2000 N/cm^2. ☐

347. In question 345, the circumferential stress at a radius 30 cm would be

(*a*) 1250 N/cm^2 ☐ (*b*) 500 N/cm^2 ☐

(*c*) 694.4 N/cm^2 ☐ (*d*) 194.4 N/cm^2. ☐

348. In question 345, the radial stress at a radius 30 cm would be

(*a*) 125 kg/cm^2 ☐ (*b*) 50 kg/cm^2 ☐

(*c*) 69.44 kg/cm^2 ☐ (*d*) 19.44 kg/cm^2 ☐

Riveted and Welded Joints

349. The diameter of the rivet by Unwin's formula in terms of plate thickness is given by

(*a*) $d = 16\sqrt{t}$ ☐ (*b*) $d = 9\sqrt{t}$ ☐

(*c*) $d = 6\sqrt{t}$ ☐ (*d*) $d = 3\sqrt{t}$ ☐

where d and t are in mm.

350. The shearing strength per pitch length in ease of lap joint is equal to

(*a*) $n \times \frac{\pi}{4} d^2 \times \tau$ ☐ (*b*) $2n \times \frac{\pi}{4} d^2 \times \tau$ ☐

(*c*) $3n \times \frac{\pi}{4} d^2 \times \tau$ ☐ (*d*) $4n \times \frac{\pi}{4} d^2 \times \tau$ ☐

where n = Number of rivets per pitch length

d = Diameter of rivet and τ = Permissible shear stress.

351. The shearing strength per pitch length in case of butt joint is equal to

(*a*) $n \times \frac{\pi}{4} d^2 \times \tau$ ☐ (*b*) $2n \times \frac{\pi}{4} d^2 \times \tau$ ☐

(*c*) $3n \times \frac{\pi}{4} d^2 \times \tau$ ☐ (*d*) $4n \times \frac{\pi}{4} d^2 \times \tau$ ☐

where n = Number of rivets per pitch length.

352. The tearing strength per pitch length of a riveted joint is equal to

(*a*) $(p - d) \times t \times \sigma_c$ ☐ (*b*) $(p - d) \times t \times \sigma_t$ ☐

(*c*) $(p - d) \times t \times \tau$ ☐ (*d*) $(p - 2d) \times t \times \sigma_t$ ☐

where p = Pitch, p = Diameter of rivet, t = Thickness of plates and σ_c, σ_t and

τ = Safe crushing, tensile and shear stresses respectively.

353. The bearing or crushing strength per pitch length of a riveted joint is equal to

(*a*) $\frac{\pi}{4} d^2 \times \sigma_c \times n$ ☐ (*b*) $\pi d \times t \times \sigma_c \times n$ ☐

(*c*) $d \times t \times \sigma_c \times n$ ☐ (*d*) $p \times t \times \sigma_c \times n$ ☐

where n = Number of rivets per pitch length, σ_c = Safe crushing stress.

354. In case of riveted joint 'margin' is the distance between the

(*a*) centres of two consecutive rivets in a row ☐

(*b*) centre of rivet hole to the nearest edge of plate ☐

(*c*) centres of rivets in adjacent rows ☐

(*d*) none of the above. ☐

355. If the margin in case of riveted joint is at least 1.5 *d*, there will be

(*a*) tearing off the plate between the rivet hole and edge of the plate ☐

(*b*) tearing off the plates between rivets ☐

(*c*) no tearing off the plate between the rivet hole and edge of the plate ☐

(*d*) no crushing of the joint. ☐

356. The expression $p \times t \times \sigma_t$, in case of a riveted joint is called

(*a*) tearing strength ☐ (*b*) bearing strength ☐

(*c*) crushing strength ☐ (*d*) strength of solid plate. ☐

357. If P_t, P_s and P_c are the maximum load per pitch length for a riveted joint from tearing, shearing and crushing considerations respectively, then efficiency of the joint is equal to

(*a*) $\frac{P_t}{\text{strength of solid plate}}$ ☐ (*b*) $\frac{P_s}{\text{strength of solid plate}}$ ☐

(*c*) $\frac{P_c}{\text{strength of solid plate}}$ ☐ (*d*) $\frac{\text{least of } P_t, P_s \text{ and } P_c}{\text{strength of solid plate}}$ ☐

(*e*) $\frac{\text{maximum of } P_t, P_s \text{ and } P_c}{\text{strength of solid plate}}$ ☐ (*f*) none of the above. ☐

358. In riveted joint the tensile, shearing and crusing stresses are based on the

(*a*) diameter of drilled hole ☐

(*b*) diameter of rivet ☐

(*c*) mean of the diameter of rivet and drilled hole ☐

(*d*) none of the above. ☐

359. Generally a rivet joint fails by

(*a*) shearing of rivets ☐

(*b*) crushing of rivets ☐

(*c*) tearing of the plate across the row of rivets ☐

(*d*) all of the above ☐

(*e*) any one of the above. ☐

360. A welded joint as compared to riveted joint, has

(*a*) less strength ☐ (*b*) more strength ☐

(*c*) same strength ☐ (*d*) none of the above. ☐

361. In the diamond-riveted arrangement, the nine rivets have been arranged in four rows with one rivet in the first row, two rivets in the second row, three rivets in the third row and three rivets in the fourth row. The weakest section is the section passing through

(*a*) fourth row ☐ (*b*) third row ☐

(*c*) second row ☐ (*d*) one rivet hole in first row. ☐

362. The efficiency of a diamond-riveted joint, having width of the plate as b and diameter of hole as d, is equal to

(*a*) $\frac{d}{b}$ ☐ (*b*) $\frac{b+d}{b}$ ☐

(*c*) $\frac{b-d}{b}$ ☐ (*d*) $\frac{b-d}{d}$. ☐

363. The material, of which rivets are made, should be

(*a*) hard ☐ (*b*) malleable ☐

(*c*) ductile ☐ (*d*) touch. ☐

Column and Strut

364. Strut is defined is a

(*a*) member of a structure which carries a tensile load ☐

(*b*) member of a structure which carries an axial compressive load ☐

(*c*) vertical member of a structure which carries a tensile load ☐

(*d*) none of the above. ☐

365. Column is defined as a

(*a*) member of a structure which carries a tensile load ☐

(*b*) member of a structure which carries an axial compressive load ☐

(*c*) vertical member of a structure which carries a tensile load ☐

(*d*) vertical member of a structure which carries an axial compressive load. ☐

366. The maximum axial compressive load which a column can take without failure by lateral deflection is called

(*a*) critical load ☐ (*b*) buckling load ☐

(*c*) crippling load ☐ (*d*) any one of the above. ☐

367. Slenderness ratio is defined as the ratio of

(*a*) equivalent length of the column to the minimum radius of gyration ☐

(*b*) length of the column to the minimum radius of gyration ☐

(*c*) length of the column to the area of cross-section of the column ☐

(*d*) minimum radius of gyration to the area of cross-section of the column. ☐

368. Buckling factor is defined as the ratio of

(*a*) equivalent length of a column to the minimum radius of gyration ☐

(b) length of the column to the minimum radius of gyration ☐
(c) length of the column to the area of cross-section of the column ☐
(d) none of the above. ☐

369. A loaded column is having the tendency to deflect. On account of this tendency, the critical load
(a) decreases with the decrease in length ☐
(b) decreases with the increase in length ☐
(c) first decreases then increases with the decrease in length ☐
(d) first increases then decreases with the decrease in length. ☐

370. A loaded column fails due to
(a) stress due to direct load ☐ (b) stress due to bending ☐
(c) both (a) and (b) ☐ (d) none of the above. ☐

371. The crippling load, according to Euler's theory of long columns, when both ends of the column are hinged, is equal to
(a) $\frac{4\pi^2 EI}{l^2}$ ☐ (b) $\frac{\pi^2 EI}{l^2}$ ☐
(c) $\frac{\pi^2 EI}{4l^2}$ ☐ (d) $\frac{2\pi^2 EI}{l^2}$. ☐
where l = Length of column.

372. The crippling load, according to Euler's theory of long column when one end of the column is fixed and other end is free, is equal to
(a) $\frac{4\pi^2 EI}{l^2}$ ☐ (b) $\frac{\pi^2 EI}{l^2}$ ☐
(c) $\frac{\pi^2 EI}{4l^2}$ ☐ (d) $\frac{2\pi^2 EI}{l^2}$. ☐

373. The crippling load, according to Euler's theory of long column when both ends of the column are fixed, is equal to
(a) $\frac{4\pi^2 EI}{l^2}$ ☐ (b) $\frac{\pi^2 EI}{l^2}$ ☐
(c) $\frac{\pi^2 EI}{4l^2}$ ☐ (d) $\frac{2\pi^2 EI}{l^2}$. ☐

374. The crippling load, according to Euler's theory of long column when one end of the column is fixed and the other end is hinged, is equal to
(a) $\frac{4\pi^2 EI}{l^2}$ ☐ (b) $\frac{\pi^2 EI}{l^2}$ ☐
(c) $\frac{\pi^2 EI}{4l^2}$ ☐ (d) $\frac{2\pi^2 EI}{l^2}$. ☐

375. The ratio of crippling load, for a column of length (l) with both ends fixed to the crippling load of the same column with both ends hinged, is equal to

(*a*) 2.0 ☐ (*b*) 4.0 ☐
(*c*) 0.25 ☐ (*d*) 0.50. ☐

376. The ratio of crippling load, for a column of length (l) with both ends fixed to the crippling load of the same column with one end fixed and other end is free, is equal to

(*a*) 2.0 ☐ (*b*) 4.0 ☐
(*c*) 8.0 ☐ (*d*) 16.0. ☐

377. The ratio of crippling load, for a column of length (l) with both ends fixed to the crippling load, of the same column with one end fixed and other end is hinged, is equal to

(*a*) 2.0 ☐ (*b*) 4.0 ☐
(*c*) 8.0 ☐ (*d*) 16.0. ☐

378. The equivalent length of a given column with given end conditions is the length of a column of the same material and section with hinged ends having crippling load equal to

(*a*) two times that of the given column ☐ (*b*) half that of given column ☐
(*c*) four timest at of the given column ☐ (*d*) that of the given column. ☐

379. The equivalent length is equal to actual length of a column with

(*a*) one end fixed and other end free ☐ (*b*) both ends fixed ☐
(*c*) one end fixed and other end hinged ☐ (*d*) both ends hinged. ☐

380. The equivalent length is twice the actual length of a column with

(*a*) one end fixed and other end free ☐ (*b*) both ends fixed ☐
(*c*) one end fixed and other end hinged ☐ (*d*) both ends hinged. ☐

381. The equivalent length is equal to half of the actual length of a column with

(*a*) one end fixed and other end free ☐ (*b*) both ends fixed ☐
(*c*) one end fixed and other end hinged ☐ (*d*) both ends hinged. ☐

382. The equivalent length is equal to actual length divided by $\sqrt{2}$ for a column with

(*a*) one end fixed and other end free ☐ (*b*) both ends fixed ☐
(*c*) one end fixed and other end hinged ☐ (*d*) both ends hinged. ☐

383. The equivalent length is equal to four times the actual length of a column with

(*a*) one end fixed and other end free ☐ (*b*) both ends fixed ☐
(*c*) one end fixed and other end hinged ☐ (*d*) both ends hinged ☐
(*e*) none of the above. ☐

384. Choose the wrong statement

(*a*) Column is a vertical member of a structure which carries an axial compressive load. ☐
(*b*) The ratio of length of a column to its minimum radius of gyration is called slenderness ratio. ☐
(*c*) A column tends to buckle in the direction of the minimum moment of inertia. ☐
(*d*) The equivalent length of a column with one end fixed and other end is free is half of its actual length. ☐

385. Choose the correct statement

(*a*) Euler's formula holds good only for short columns. □

(*b*) A short column is one which has the ratio of its length to least radius of gyration more than 100. □

(*c*) A column with both ends fixed has minimum equivalent (or effective) length. □

(*d*) The equivalent length of a column with one end fixed and other end hinged is half of its actual length. □

386. The Euler's formula for the long columns can be written as

$$P = K\frac{\pi^2 EI}{l^2}$$

where P = Crippling load, l = Actual length and K = Constant.

The value of K for a column with both ends fixed should be

(*a*) 2.0 □ (*b*) 1.0 □

(*c*) 4.0 □ (*d*) $\frac{1}{4}$. □

387. A bar of length 4 m has flexural rigidity, *i.e.*, $EI = 1.6 \times 10^9$ N-cm^2. If the bar is used as a column with both ends hinged the crippling load will be approximately

(*a*) 49350 N □ (*b*) 98700 N □

(*c*) 197400 N □ (*d*) 394800 N. □

388. For the question 387, if the same bar is used as a column with one end fixed and other end hinged, the crippling load will be

(*a*) 49350 N □ (*b*) 98700 N □

(*c*) 197400 N □ (*d*) 394800 N. □

389. For the question 387, if the bar is used as a column with both ends fixed, the crippling load would be

(*a*) 49350 N □ (*b*) 98700 N □

(*c*) 197400 N □ (*d*) 394800 N. □

390. For the question 387, if the bar is used as a column with one end fixed and other end free, the crippling load would be

(*a*) 24675 N □ (*b*) 49350 N □

(*c*) 197400 N □ (*d*) 394800 N. □

391. Rankine's formula is an empirical formula which is used for

(*a*) long columns □ (*b*) short columns □

(*c*) both long and short columns □ (*d*) none of the above. □

392. The crippling load by Rankine's formula is

(*a*) $\dfrac{\sigma_c A}{1 + a\left(\dfrac{l}{k}\right)^2}$ □ (*b*) $\dfrac{\sigma_c A}{1 + a\left(\dfrac{l_c}{k}\right)^2}$ □

(*c*) $\dfrac{\sigma_c A}{1 - a\left(\dfrac{l}{k}\right)^2}$ □ (*d*) $\dfrac{\sigma_c A}{1 - a\left(\dfrac{l_c}{k}\right)^2}$. □

where A = Area of cross-section of the column, σ_c = Crushing stress
a = Rankine's constant, k = Least radius of gyration
l = Actual length of column, l_c = Equivalent length of column.

393. The Rankine's constant (a) in Rankine's formula is equal to

(*a*) $\frac{\pi^2 E}{\sigma_c}$ ☐ (*b*) $\frac{\pi^2}{E\sigma_c}$ ☐

(*c*) $\frac{E\sigma_c}{\pi^2}$ ☐ (*d*) $\frac{\sigma_c}{\pi^2 E}$. ☐

394. The Rankine's constant (a) for a given material of a column depends upon the

(*a*) length of the column ☐ (*b*) diameter of the column ☐

(*c*) length and diameter ☐ (*d*) none of the above. ☐

395. The expression $(\sigma_c A)\Big/\left(1+b\left(\frac{l_c}{d}\right)^2\right)$ is known as

(*a*) Rankine's formula ☐ (*b*) Gordon's formula ☐

(*c*) Straight line formula ☐ (*d*) Johnson's parabolic formula ☐

where d = Least diameter or width of the section
b = Constant and l_c = Equivalent length.

396. The expression $\left[\sigma_c A - eA\left(\frac{l_c}{k}\right)^2\right]$, where e is a constant, is known as

(*a*) Rankine's formula ☐ (*b*) Gordon's formula ☐

(*c*) Straight line formula ☐ (*d*) Johnson's parabolic formula ☐

397. The relation between crippling load, safe load and factor of safety (S.F.) is given by

(*a*) crippling load = safe load/S.F. ☐ (*b*) crippling load = safe load × S.F. ☐

(*c*) crippling load = S.F./safe load ☐ (*d*) crippling load – safe load = S.F. ☐

398. A hollow cast iron column of 3.5 m length, of 100 cm^2 cross-sectional area, of 4900 cm^4 moment of inertia and of 1 × 10^6 kgf/cm^2 modulus of elasticity is fixed at both ends. The radius of gyration of the column would be

(*a*) 49 cm ☐ (*b*) 98 cm ☐

(*c*) 7 cm ☐ (*d*) 14 cm. ☐

399. In question 398, the slenderness ratio for the column would be

(*a*) 100 ☐ (*b*) 50 ☐

(*c*) 25 ☐ (*d*) 20. ☐

400. In question 398, equivalent (or effective length) of the column would be

(*a*) 7.0 m ☐ (*b*) 14.0 m ☐

(*c*) 3.5 m ☐ (*d*) 1.75 m. ☐

401. In question 398, the buckling factor of the column would be

(*a*) 100 ☐ (*b*) 50 ☐

(*c*) 25 ☐ (*d*) 20. ☐

402. In question 398, Euler's crippling load is equal to

(*a*) 789500 kgf ☐ (*b*) 1579000 kgf ☐
(*c*) 394750 kgf ☐ (*d*) 3158000 kgf. ☐

403. In question 398, if Rankine's constant $a = \frac{1}{1600}$ and crush stress (σ_c) = 5500 kg/cm², then Rankine's critical load will be equal to

(*a*) 39550 kgf ☐ (*b*) 395500 kgf ☐
(*c*) 19775 kgf ☐ (*d*) 197750 kgf ☐

404. If in questions 398 and 403, factor of safety be four, then safe load by Rankines formula would be

(*a*) 197750 kgf ☐ (*b*) 19775 kgf ☐
(*c*) 98875 kgf ☐ (*d*) 49439 kgf. ☐

405. Choose the wrong statement

(*a*) Shafts of the same material and length having the same polar modulus have the same strength. ☐

(*b*) The measure of the strength of the shaft in resisting torsion is given by the magnitude of the polar modulus. ☐

(*c*) The polar modulus of a solid shaft = $\frac{\pi d^3}{16}$. ☐

(*d*) Polar modulus is the same as polar moment of inertia. ☐

Additional Questions

406. A cantilever of length (l) carries a load whose intensity varies uniformly from zero at the free end to w per unit length at the fixed end, the bending moment diagram will be a

(*a*) straight line curve ☐ (*b*) parabolic curve ☐
(*c*) cubic curve ☐ (*d*) combination of (*a*) and (*b*). ☐

407. A simply supported beam is overhanging equally on both sides and carries a uniformly distributed load of w per unit length over the whole length. The length between the supports is (l) and length of overhang to one side is 'a'. If $l > 2a$, then the number of points of contra-flexure will

(*a*) zero ☐ (*b*) one ☐
(*c*) two ☐ (*d*) three. ☐

408. If in question 407, $l = 2a$, the number of points of contra-flexure will be

(*a*) zero ☐ (*b*) one ☐
(*c*) two ☐ (*d*) three. ☐

409. If in question 407, $l < 2a$, the number of two points of contra-flexure will be

(*a*) zero ☐ (*b*) one ☐
(*c*) two ☐ (*d*) three. ☐

410. In question 407, the shear force diagram will consist of

(*a*) two triangles ☐ (*b*) two rectangles ☐
(*c*) four triangles ☐ (*d*) four rectangles. ☐

411. For the same loading, the maximum bending moment for a fixed beam as compared to simply supported beam is

(*a*) more ☐ (*b*) less ☐

(*c*) same ☐ (*d*) none of the above. ☐

412. For the same loading, the maximum deflection for a fixed beam as compared to simply supported beam is

(*a*) more ☐ (*b*) same ☐

(*c*) less ☐ (*d*) none of the above. ☐

413. In a fixed beam, temperature variation produces

(*a*) large stresses ☐ (*b*) small stresses ☐

(*c*) zero stress ☐ (*d*) none of the above. ☐

414. In a simply supported beam, the temperature variation produces

(*a*) large stresses ☐ (*b*) small stresses ☐

(*c*) zero stress ☐ (*d*) none of the above. ☐

Tick mark the true or false statement:

415. If a pressure vessel has $\frac{d}{t} \geq 10$, then it is known as thin shell

(where d = internal diameter and t = thickness of shall).

(*a*) True ☐ (*b*) False. ☐

416. For a thick shell, the ratio of wall thickness of the diameter of the shell should be less than 1/10.

(*a*) True ☐ (*b*) False. ☐

417. The circumferential stress in a thin cylinder is half of the longitudinal stress.

(*a*) True ☐ (*b*) False. ☐

418. The longitudinal stress in a thin cylinder in terms of internal diameter (d), thickness of cylinder (t) and internal pressure (p) is equal to $p \times d/2t$.

(*a*) True ☐ (*b*) False. ☐

419. The hoop stress in a thin cylinder is $p \times d/4t$.

(*a*) True ☐ (*b*) False. ☐

420. The design of thin cylinder is based on hoop stress.

(*a*) True ☐ (*b*) False. ☐

421. The circumferential strain in a thin cylinder, subjected to internal fluid pressure p is $\left[\frac{p \times d}{2tE}\right]\left(1 - \frac{1}{2m}\right)$.

(*a*) True ☐ (*b*) False. ☐

422. The volumetric strain in a thin cylinder subjected to internal pressure is equal to two times the circumferential strain plus longitudinal strain.

(*a*) True ☐ (*b*) False. ☐

423. In a thick cylinder, the circumferential stress is maximum at the inner radius.

(*a*) True ☐ (*b*) False. ☐

424. The volumetric strain in case of a thin cylinder in terms of internal diameter (d), thickness (t), internal fluid pressure (p) and Poisson's ratio $\left(\frac{1}{m}\right)$ is equal to $\left(\frac{pd}{2tE}\right)\left[2.5 - \left(\frac{2}{m}\right)\right]$

(*a*) True ☐ (*b*) False. ☐

425. The volumetric strain in case of a thin spherical shell is $\left(\frac{3pd}{4tE}\right)\left(1-\frac{1}{m}\right)$

(*a*) True ☐ (*b*) False. ☐

426. There will be an increase in length and decrease in diameter when a thin cylinder is subjected to an internal fluid pressure (p).

(*a*) True ☐ (*b*) False. ☐

427. The ratio of longitudinal strain to circumferential strain, when a thin cylinder is subjected to internal pressure p is in term of Poisson's ratio only is $(m-2)/(2m-1)$.

(*a*) True ☐ (*b*) False. ☐

428. The shear stress produced in a shaft which is subjected to torsion, is zero at the centre and maximum at the circumference.

(*a*) True ☐ (*b*) False. ☐

429. The horse power transmitted by a shaft is $2\pi NT/4500$ where T = torque in kgf-m and N = speed in r.p.m.

(*a*) True ☐ (*b*) False. ☐

430. The term $(T \times w/1000)$ represents the power transmitted by a shaft in kW where T = torque in Nm and w is in rad/s.

(*a*) True ☐ (*b*) False. ☐

431. The torque required to produce a twist of one radian in unit length of a shaft, is known as torsional rigidity of the shaft.

(*a*) True ☐ (*b*) False. ☐

432. The polar modulus of a shaft is expressed by J/R where J = polar moment of inertia and R is the radius of shaft.

(*a*) True ☐ (*b*) False. ☐

433. The term $\pi d^3/16$ represents the polar moment of inertia of a solid shaft.

(*a*) True ☐ (*b*) False. ☐

434. The polar modulus of a hollow shaft of outer diameter (D) and inner diameter (d) is given by $\pi(D^3 - d^3)/16$.

(*a*) True ☐ (*b*) False. ☐

435. The strength of a shaft will be 27 times than that of another shaft if both the shafts are made of the same material but the dia. of the first shaft should be three times the dia. of the second shaft.

(*a*) True ☐ (*b*) False. ☐

436. The polar moment of inertia of a hollow shaft is $\pi (D^4 - d^4)/64$, where D = outer diameter and d = inner diameter.

(*a*) True ☐ (*b*) False. ☐

437. For thin cylindrical shells, Lami's theory is used.

(*a*) True ☐ (*b*) False. ☐

438. The ratio of length of a column to least radius of gyration is known as slenderness ratio.

(*a*) True ☐ (*b*) False. ☐

439. For short columns, slenderness ratio is more than 80.

(*a*) True ☐ (*b*) False. ☐

440. A vertical column will tend to buckle in the direction of the minimum moment of inertia.

(*a*) True ☐ (*b*) False. ☐

441. The Rankine's formula holds good for both short and long columns.

(*a*) True ☐ (*b*) False. ☐

442. Euler's formula holds good only for short columns.

(*a*) True ☐ (*b*) False. ☐

443. If one end of a column is fixed and other end is free, then the column will have maximum equivalent length.

(*a*) True ☐ (*b*) False. ☐

444. Crippling load is the load at which the column will start bending.

(*a*) True ☐ (*b*) False. ☐

445. Short column fails due to bending stress.

(*a*) True ☐ (*b*) False. ☐

446. The bending stress in a long column are more than the direct stresses.

(*a*) True ☐ (*b*) False. ☐

447. The crippling load for a column hinged at both ends is four times the crippling load for the same column when fixed at both ends.

(*a*) True ☐ (*b*) False. ☐

448. The equivalent length for a column fixed at both ends is one-half of the actual length.

(*a*) True ☐ (*b*) False. ☐

449. The strain is having the unit of N/mm.

(*a*) True ☐ (*b*) False. ☐

450. The stress is having the unit of N/m^2.

(*a*) True ☐ (*b*) False. ☐

451. The ratio of tensile stress to tensile strain upto elastic limit is known as Young's modulus.

(*a*) True ☐ (*b*) False. ☐

452. The ratio of compressive stress to compressive strain upto elastic limit is known as bulk modulus.

(*a*) True ☐ (*b*) False. ☐

453. The ratio of shear stress to shear strain upto elastic limit is known as modulus of rigidity.

(*a*) True ☐ (*b*) False. ☐

454. The stress is directly proportional to the corresponding strain within elastic limit. This is known as Hook's law.

(*a*) True ☐ (*b*) False. ☐

455. The ratio of longitudinal strain of lateral strain is known as Poisson's ratio.

(*a*) True ☐ (*b*) False. ☐

456. If P is tensile force acting on a body of area of cross-section A, length L and of Young's modulus E, then change in length is equal to $P \times L/AE$.

(*a*) True ☐ (*b*) False. ☐

457. A body is subjected to a direct load equal to the weight of the body, then the deformation of the body under its own weight is double the deformation due to the direct load.

(*a*) True ☐ (*b*) False. ☐

458. A circular bar tapering uniformly from diameter D_1 at one end to diameter D_2 at the other end and subjected to an axial force P, will have the extension equal to the extension of the circular bar of diameter $\sqrt{D_1 \times D_2}$ when subjected to the same load P.

(*a*) True ☐ (*b*) False. ☐

459. The ratio of the modulus of elasticities of two materials is known as modular ratio.

(*a*) True ☐ (*b*) False. ☐

460. The relation between Young's modulus (E) and bulk modulus (K) is given by $K = mE/2(m-3)$.

(*a*) True ☐ (*b*) False. ☐

461. A shear stress across a plane, is always accompanied by a balancing shear stress across the plane and normal to it.

(*a*) True ☐ (*b*) False. ☐

462. Due to shear stress (τ), the strain energy stored in a body is equal to $\tau^2 \times V/2C$ where C = shear modulus and V = volume of the body.

(*a*) True ☐ (*b*) False. ☐

463. The strain energy stored in a body, when load is applied gradually is equal to $\sigma^2 \times V/2E$ where σ = stress in the body, V = volume of the body and E = Young's modulus.

(*a*) True ☐ (*b*) False. ☐

464. The stress produced in a body due to suddenly applied load is one-half of the stress produced when the same load is applied gradually.

(*a*) True ☐ (*b*) False. ☐

465. The radius of Mohr's circle is equal to the maximum shear stress.

(*a*) True ☐ (*b*) False. ☐

466. A body is subjected to two tensile stresses mutually ⊥ to each other, then the maximum shear stress is one-half of the algebraic difference of these two stresses.

(*a*) True ☐ (*b*) False. ☐

467. The B.M. is maximum when shear force along a section is zero.

(*a*) True ☐ (*b*) False. ☐

468. Co-efficient of cubical expansion is twice the coefficient of linear expansion.

(*a*) True ☐ (*b*) False. ☐

469. If the bending stress is same at every section along its longitudinal axis, then beam is said to be of uniform strength.

(*a*) True ☐ (*b*) False. ☐

470. The intensity of stress which causes unit strain is called modulus of elasticity.

(*a*) True ☐ (*b*) False. ☐

471. Elasticity is the property of a material by virtue of which a body does not return to its original shape after removal of the load.

(*a*) True ☐ (*b*) False. ☐

472. Ductility is the property of a material which allows it to be drawn into smaller section.

(*a*) True ☐ (*b*) False. ☐

473. The unit of stress and Young's modulus are same.

(*a*) True ☐ (*b*) False. ☐

474. The thermal stresses depend upon the cross-sectional area of the bar.

(*a*) True ☐ (*b*) False. ☐

475. The strain at right angles to longitudinal strain is known as lateral strain.

(*a*) True ☐ (*b*) False. ☐

476. The Poisson's ratio for most of materials varies from 3 to 4.

(*a*) True ☐ (*b*) False. ☐

477. The length of a bar will decrease but width and thickness will increase if the bar is subjected to an axial push.

(*a*) True ☐ (*b*) False. ☐

478. The strength of a shaft is judged by the torque transmitted by it.

(*a*) True ☐ (*b*) False. ☐

479. The width of all leaves in a leaf spring is equal.

(*a*) True ☐ (*b*) False. ☐

480. Margin is the distance between the centre of a rivet hole to the nearest edge of the plate.

(*a*) True ☐ (*b*) False. ☐

481. If d = diameter of rivet hole, then the value of margin should be equal to 'd'

(*a*) True ☐ (*b*) False. ☐

482. If P_t = pull required to tear of the plate, P_s = pull required to shear off the rivet per pitch length and P_c = pull required to crush the rivet per pitch length, then the strength of riveted joint is equal to minimum value of P_t, P_s or P_c.

(*a*) True ☐ (*b*) False. ☐

483. The ratio of strength to the joint of the strength of solid plate is known as the efficiency of riveted joint.

(*a*) True ☐ (*b*) False. ☐

484. The term $p \times t \times \sigma_t$ in which p = pitch, t = thickness of plate and σ_t = permissible tensile stress, is known as strength of solid plate.

(*a*) True ☐ (*b*) False. ☐

485. If d = dia. of rivet hole in mm and t = thickness of plate in mm, then according to Unwin's formula, $d = 1.4\sqrt{t}$.

(*a*) True ☐ (*b*) False. ☐

486. Rivets are generally specified by its shank diameter.

(*a*) True ☐ (*b*) False. ☐

487. Fatigue testing machine is used to determine the impact test.

(*a*) True ☐ (*b*) False. ☐

488. The ratio of ultimate stress to working stress is known as factor of safety.

(*a*) True ☐ (*b*) False. ☐

489. In case of cast iron, the percentage reduction in area under tensile test, is approximately 12%.

(*a*) True ☐ (*b*) False. ☐

490. The tensile test is performed generally on ductile materials.

(*a*) True ☐ (*b*) False. ☐

491. The compression test is performed generally on brittle materials.

(*a*) True ☐ (*b*) False. ☐

492. The stress normal to principal plane is known as principal stress.

(*a*) True ☐ (*b*) False. ☐

493. The plane on which shear stress is maximum is known as principal plane.

(*a*) True ☐ (*b*) False. ☐

494. A body is subjected to a tensile stress σ along x-axis only. The shear stress will be maximum along a section inclined at an angle of 45° to the normal of the section.

(*a*) True ☐ (*b*) False. ☐

495. In the above question, the value of maximum shear stress will be two-third of the tensile stress σ.

(*a*) True ☐ (*b*) False. ☐

496. A body is subjected to a direct tensile stress σ along x-axis accompanied by a simple shear stress τ, then maximum normal stress is $(\sigma/2) + \sqrt{\sigma^2 + 4\tau^2/2}$.

(*a*) True ☐ (*b*) False. ☐

497. In question 496, the maximum shear stress is $\sqrt{\sigma^2 + 4\tau^2}$.

(*a*) True ☐ (*b*) False. ☐

498. Cantilever is a beam which is fixed at one end and free at the other.

(*a*) True ☐ (*b*) False. ☐

499. A fixed beam is one whose both ends are encastered.

(*a*) True ☐ (*b*) False. ☐

500. If a beam is supported on more than two supports, then it is known as continuous beam,

(*a*) True ☐ (*b*) False. ☐

501. A cantilever carries a load at the free end. If this load is increased, then failure of the beam will take place at the free end.

(*a*) True ☐ (*b*) False. ☐

502. In case of simply supported beams, the bending moment is maximum at the supports.

(*a*) True ☐ (*b*) False. ☐

503. A simply supported beam of span L carries a point load W at the centre of the beam. The maximum between B.M. will be equal to $W \times L/2$.

(*a*) True ☐ (*b*) False. ☐

504. In question 503, the B.M. diagram will be rectangle.

(*a*) True ☐ (*b*) False. ☐

505. In question 503, the shear force diagram will be an isosceles triangle.

(*a*) True ☐ (*b*) False. ☐

506. In question 503, the maximum bending moment will be at the centre.

(*a*) True ☐ (*b*) False. ☐

507. In question 503, the shear force will change sign at the centre.

(*a*) True ☐ (*b*) False. ☐

508. The point where shear force changes sign, is known as point of contraflexure.

(*a*) True ☐ (*b*) False. ☐

509. The point of contraflexure lies at the centre of a simply supported beam carrying a uniformly distributed load.

(*a*) True ☐ (*b*) False. ☐

510. A continuous beam has only two supports at the ends.

(*a*) True ☐ (*b*) False. ☐

Fill in the blanks:

511. In .... beams, the point of contraflexure occurs.

(*a*) fixed ☐ (*b*) overhanging. ☐

512. The shear force .... is at the free end of a cantilever beam.

(*a*) maximum ☐ (*b*) zero. ☐

513. A beam of uniform strength has same .... at every section.

(*a*) bending moment ☐ (*b*) bending stress. ☐

514. The force of friction at the base of the dam should be .... the force exerted by water on the dam to avoid sliding of the dam at the base.

(*a*) less than ☐ (*b*) more than. ☐

515. There will be no .... of the dam at the base, if the resultant (of the force exerted by water and weight of dam) cut the base.

(*a*) sliding ☐ (*b*) overturning. ☐

516. The bending stress in the beam will .... if the section modulus of the beam is increased.

(*a*) decrease ☐ (*b*) increase. ☐

517. A beam of uniform strength is having depth constant. Its width will vary .... to bending moment.

(*a*) inversely ☐ (*b*) in proportional to. ☐

518. A neutral axis of a beam is subjected to .... stress.

(*a*) maximum ☐ (*b*) zero. ☐

519. For eccentrically loaded struts .... members are preferred.

(*a*) solid ☐ (*b*) hollow. ☐

520. A square beam and a circular beam of the same areas of cross-sections are subjected to equal bending moments, then .... will be more economical.

(*a*) circular beam ☐ (*b*) square beam. ☐

521. A beam is simply supported at both ends. The maximum deflection is equal to $PL^3/48EI$. This deflection is due to ......

(*a*) U.D.L. ☐ (*b*) central load P at the middle. ☐

522. For a, ... the ratio of dia. to thickness should be more than 15.

(*a*) thick cylinder ☐ (*b*) thin cylinder. ☐

523. In a thin cylinder, the circumferential stress is ...... of the longitudinal stress.

(*a*) two times ☐ (*b*) half. ☐

524. The ...... stress in a thin cylinder is $p \times d/4t$.

(*a*) hoop ☐ (*b*) longitudinal. ☐

525. On ...... stress, the design of this cylinder is based.

(*a*) hoop ☐ (*b*) longitudinal. ☐

526. The volumetric strain in a thin cylinder subjected to internal pressure is equal to ...... times the circumferential strain plus longitudinal strain.

(*a*) 1.5 ☐ (*b*) two. ☐

527. In a thick cylinder, the circumferential stress is ...... at the inner radius.

(*a*) minimum ☐ (*b*) maximum. ☐

528. The thickness of a thin cylinderical shell with hemispherical ends is ...... that of spherical ends.

(*a*) less than ☐ (*b*) more than. ☐

529. If p = internal fluid pressure, d = dia., t thickness and $1/m$ = Poisson's ratio, then the volumetric strain in case of a thin ...... shell will be $\left(\frac{3pd}{4E}\right)\left(1-\frac{1}{m}\right)$.

(*a*) cylindrical ☐ (*b*) spherical. ☐

530. The section modulus is ...... for a rectangular section.

(a) $\frac{bd^2}{12}$ ☐ (b) $\frac{bd^2}{6}$ ☐

where b = Width and d = Depth of the section.

531. The section modulus isfor a hollow rectangular section.

(a) $\frac{BD^3 - bd^3}{6D}$ ☐ (b) $\frac{BD^2 - bd^2}{6}$ ☐

where B, D = Outer width and depth, b, d = Inner width and depth.

532. The section modulus is .... for a circular section.

(a) $\frac{\pi d^3}{16}$ ☐ (b) $\frac{\pi d^3}{32}$ ☐

where d = Dia. of circular section.

533. The bending equation gives the relation between ...... and moment of inertia.

(a) bending moment ☐ (b) shear force. ☐

534. The netural axis of a symmetrical circular section lies at a distance of ...... from the outermost layer.

(a) $\frac{2d}{3}$ ☐ (b) $\frac{d}{2}$ ☐

where d = Dia. of the section.

535. The shear stress distribution is across a rectangular section.

(a) linear ☐ (b) parabolic. ☐

536. The maximum shear stress is .... for a rectangular section.

(a) at outermost layer ☐ (b) at the N.A. ☐

537. The maximum shear stress for a rectangular section is given by, τ_{max} = ... τ_{av}.

(a) 2.0 ☐ (b) 1.5 ☐

where τ_{av} = Average shear stress.

538. The shear stress is .... at the N.A. for a circular section.

(a) minimum ☐ (b) maximum. ☐

539. The relation between maximum shear stress and average shear stress for a circular section is given by, τ_{max} = ... τ_{av}.

(a) $\frac{3}{4}$ ☐ (b) $\frac{4}{3}$. ☐

540. The shear stress is maximum at .... for a triangular section.

(a) N.A. ☐ (b) a height of $h/2$. ☐

541. The shear stress changes at .... the function of web and flange of an I-section.

(a) gradually ☐ (b) abruptly. ☐

III. OBJECTIVE TYPE QUESTIONS FROM COMPETITIVE EXAMINATIONS

Tick mark the most appropriate answer:

1. For the state of plane stress shown in Fig. 8.28, the maximum and minimum principal stresses are:

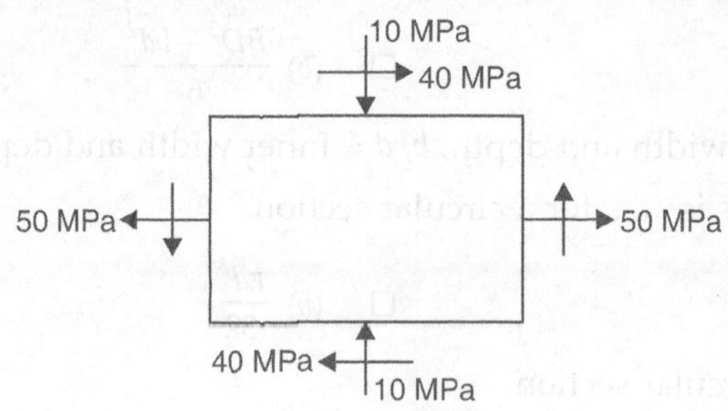

FIGURE 8.28

(*a*) 60 MPa and 30 MPa ☐ (*b*) 50 MPa and 10 MPa ☐
(*c*) 40 MPa and 20 MPa ☐ (*d*) 70 MPa and 30 MPa. ☐

2. Which of the following is true (μ = Poisson's ratio)

(*a*) $0 < \mu < -\frac{1}{2}$ ☐ (*b*) $1 < \mu < 0$ ☐
(*c*) $1 < \mu < -1$ ☐ (*d*) $\infty < \mu < -\infty$ ☐

3. For most brittle materials, the ultimate strength in compression is much large than the ultimate strength in tension. This is mainly due to

(*a*) presence of flaws and microscopic cracks or cavities ☐
(*b*) necking in tension ☐
(*c*) severity of tensile stress as compared to compressive stress ☐
(*d*) non-linearity of stress-strain diagram. ☐

4. Bending moment M and torque T is applied on a solid circular shaft. If the maximum bending stress equals to maximum shear stress developed, then M is equal to

(*a*) $\frac{T}{2}$ ☐ (*b*) T ☐
(*c*) $2T$ ☐ (*d*) $4T$. ☐

5. When bending moment M and torque T is applied on a shaft then equivalent torque is

(*a*) $M + T$ ☐ (*b*) $\sqrt{M^2 + T^2}$ ☐
(*c*) $\frac{1}{2}\sqrt{M^2 + T^2}$ ☐ (*d*) $\frac{1}{2}(M + \sqrt{M^2 + T^2})$. ☐

6. The bending moment diagram for the case shown in Fig. 8.29 below will be as shown in figure

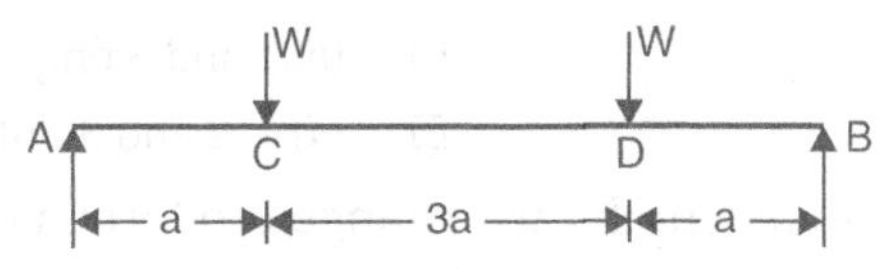

(a)

(b)

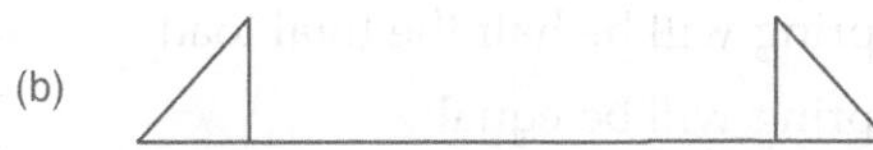

(c)

(d)

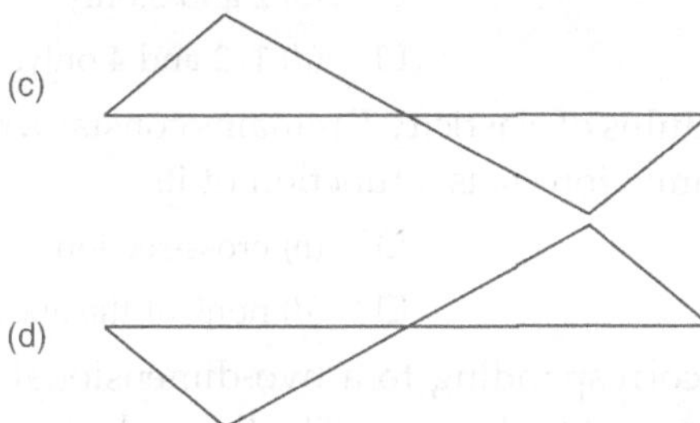

FIGURE 8.29

7. If a prismatic bar be subjected to an axial tensile stress σ, the shear stress induced at a plane inclined at θ with the axis will be

(*a*) $\frac{\sigma}{2} \sin 2\theta$ ☐ (*b*) $\frac{\sigma}{2} \cos 2\theta$ ☐

(*c*) $\frac{\sigma}{2} \cos^2 \theta$ ☐ (*d*) $\frac{\sigma}{2} \sin^2 \theta$. ☐

8. A vertical hanging bar of length *L* and weighing *w* N/unit length carries a load *W* at the bottom as shown in Fig. 8.30. The tensile force in the bar at a distance *y* from the support will be given by

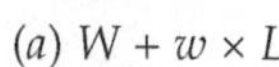
(*a*) $W + w \times L$ ☐

(*b*) $W + w(L - y)$ ☐

(*c*) $(W + w) \times \frac{y}{L}$ ☐

(*d*) $W + \frac{W}{w}(L - y)$. ☐

FIGURE 8.30

9. In case of bi-axial state of normal stresses, the normal stress on 45° plane is equal to

(*a*) sum of normal stresses ☐ (*b*) difference of normal stresses ☐

(*c*) half the sum of normal stresses ☐ (*d*) half the difference of normal stresses. ☐

10. The temperature stress is a function of

1. Co-efficient of linear expansion

2. Temperature rise

3. Modulus of elasticity. The correct answer is

(a) 1 and 2 only ☐ (b) 1 and 3 only ☐

(c) 2 and 3 only ☐ (d) 1, 2 and 3 only. ☐

11. When two springs of equal lengths are arranged to form a cluster spring then which of the following statements are true :

1. Angle of twist in both the springs will be equal
2. Deflection of both the springs will be equal
3. Load taken by each spring will be half the total load
4. Shear stress in each spring will be equal

(a) 1 and 2 only ☐ (b) 2 and 3 only ☐

(c) 3 and 4 only ☐ (d) 1, 2 and 4 only. ☐

12. When σ and Young's modulus of elasticity E remains constant, the energy absorbing capacity of part subjected to dynamic forces, is a function of its

(a) length ☐ (b) cross-section ☐

(c) volume ☐ (d) none of the above. ☐

13. If the principal stresses corresponding to a two-dimensional state of stress are σ_1 and σ_2 when σ_1 is greater than σ_2 and both are tensile, then which one of the following would be correct criterion for failure by yielding, according to maximum shear stress theory ?

(a) $\frac{\sigma_1 - \sigma_2}{2} = \pm \frac{\sigma_{yp}}{2}$ ☐ (b) $\frac{\sigma_1}{2} = \pm \frac{\sigma_{yp}}{2}$ ☐

(c) $\frac{\sigma_2}{2} = \pm \frac{\sigma_{yp}}{2}$ ☐ (d) $\sigma_1 = \pm 2\sigma_{yp}$. ☐

14. The buckling load will be maximum for a column, if

(a) one end of the column is clamped and other end is free ☐

(b) both ends of the column are clamped ☐

(c) both ends of the column are hinged ☐

(d) one end of the column is hinged and other end is free. ☐

15. A length of 10 mm diameter steel wire is coiled to a close-coiled helical spring having 8 coils of 75 mm mean diameter, and the spring has a stiffness k. If the same length of the wire is coiled to 10 coils of 60 mm mean diameter, then the spring stiffness will be

(a) k ☐ (b) 1.25 k ☐

(c) 1.56 k ☐ (d) 1.95 k. ☐

16. A metal pipe of 1 m diameter contains a fluid having a pressure of 100 N/cm^2. If the permissible tensile stress in the metal is 2 kN/cm^2, then the thickness of the metal required for making the pipe would be

(a) 5 mm ☐ (b) 10 mm ☐

(c) 20 mm ☐ (d) 25 mm. ☐

17. Circumferential and longitudinal strains in cylindrical boiler under internal steam pressure, are e_1 and e_2 respectively. Change in volume of the boiler cylinder per unit volume will be

(*a*) $e_1 + 2e_2$ ☐ (*b*) $e_1e_2^2$ ☐

(*c*) $2e_1 + e_2$ ☐ (*d*) $e_1^2\, e_2$. ☐

18. In the assembly of pulley, key and shaft

(*a*) pulley is made the weakest ☐

(*b*) key is made the weakest ☐

(*c*) key is made the strongest ☐

(*d*) all the three are designed for equal strength. ☐

19. Shear stress distribution diagram of a beam of rectangular cross-section, subject to transverse loading will be

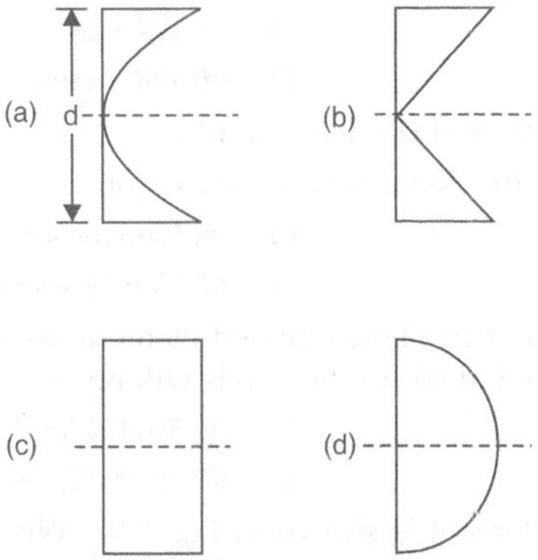

(a) d (b) (c) (d)

FIGURE 8.31

20. A horizontal beam with square cross-section is simply supported with sides of the square horizontal and vertical and carries a distributed loading that produces maximum bending stress σ in the beam. When the beam is placed with one of the diagonals horizontal, the maximum bending stress will be

(*a*) $\frac{\sigma}{\sqrt{2}}$ ☐ (*b*) σ ☐

(*c*) $\sqrt{2} \times \sigma$ ☐ (*d*) 2σ. ☐

21. A shaft was initially subjected to bending moment and then was subjected to torsion. If the magnitude of the bending moment is found to be the same as that of the torque, then the ratio of maximum bending stress to shear stress would be

(*a*) 0.25 ☐ (*b*) 0.50 ☐

(*c*) 2.0 ☐ (*d*) 4.0. ☐

22. A simply supported beam of rectangular section 4 cm by 6 cm carries a mid-span concentrated load such that 6 cm side lies parallel to the line of action of loading; deflection under the load is δ. If the beam is now supported with the 4 cm side parallel to the line of action of loading, the deflection under the load will be

(*a*) 0.44 × δ ☐ (*b*) 0.67 × δ ☐

(*c*) 1.50 × δ ☐ (*d*) 2.25 × δ. ☐

23. A beam *AB* is hinge-supported at its ends and is loaded by a couple *P* × *C* as shown in Fig. 8.32. The magnitude of shearing force at a section *x* of the beam is

(*a*) 0 ☐

(*b*) *P* ☐

(*c*) $\frac{P}{2L}$ ☐

(*d*) $\frac{P \times C}{2L}$ ☐

FIGURE 8.32

24. The frictional torque for square thread at mean radius (*R*) while raising load (*W*) is given by

(*a*) *WR* × tan (ϕ – α) ☐ (*b*) *WR* × tan (ϕ + α) ☐

(*c*) *WR* × tan α ☐ (*d*) *WR* × tan ϕ ☐

where ϕ = Angle of friction and α = Helix angle.

25. Design of shafts made of brittle materials is based on

(*a*) Guest's theory ☐ (*b*) Rankine's theory ☐

(*c*) St. Venant's theory ☐ (*d*) Von Mises theory. ☐

26. Principal stresses at a point in plane stressed element are $\sigma_x = \sigma_y$ = 5000 N/cm². Normal stress on the plane inclined at 45° to the *x*-axis will be

(*a*) 0 ☐ (*b*) 5000 N/cm² ☐

(*c*) 7070 N/cm² ☐ (*d*) 10000 N/cm². ☐

27. State of stress in plane element is shown in Fig. 8.33. Which one of the following figures shown in Fig. 8.34 is the correct sketch of Mohr's circle of the state of stress ?

FIGURE 8.33

FIGURE 8.34

28. A steel rod of 1 sq. cm cross-sectional area is 100 cm long and has a Young's modulus of elasticity 20 × 10⁶ N/cm². It is subjected to an axial pull of 20 kN. The elongation of the rod will be

(*a*) 0.05 cm ☐ (*b*) 0.1 cm ☐

(*c*) 0.15 cm ☐ (*d*) 0.20 cm. ☐

29. If the area of cross-section of a wire is circular and if the radius of this wire decreases to half its original value due to stretch to the wire by a load, then modulus of elasticity of the wire be

(*a*) one-fourth of its original value ☐ (*b*) halved ☐
(*c*) doubled ☐ (*d*) unaffected ☐

E depends upon the material. It is independent of area, load etc.

30. Match list I with list II and select the correct answer using codes given below the lists:

List I (Material properties)	**List II** (Test of determine material properties)
A. Ductility	1. Impact test
B. Toughness	2. Fatigue test
C. Endurance limit	3. Tension test
D. Resistance to penetration	4. Hardness test

Codes:

	A	*B*	*C*	*D*	
(*a*)	3	2	1	4	☐
(*b*)	4	2	1	3	☐
(*c*)	3	1	2	4	☐
(*d*)	4	1	2	3	☐

31. If a material had a modulus of elasticity of 21×10^6 N/cm^2 and a modulus of rigidity of 8×10^6 N/cm^2, then approximate value of the Poisson's ratio of the material would be

(*a*) 0.26 ☐ (*b*) 0.31 ☐
(*c*) 0.47 ☐ (*d*) 0.5. ☐

32. A shaft can safely transmit 90 kW while rotating at a given speed. If this shaft is replaced by a shaft of diameter double the previous one and rotated at half the speed of the previous, the power that can be transmitted by the new shaft is

(*a*) 90 kW ☐ (*b*) 180 kW ☐
(*c*) 360 kW ☐ (*d*) 720 kW. ☐

33. A cold rolled steel shaft is designed on the basis of maximum shear stress theory. The principal stresses induced at its critical section are 60 MPa and – 60 MPa respectively. If the yield stress for the shaft material is 360 MPa, the factor of safety of the design is

(*a*) 2 ☐ (*b*) 3 ☐
(*c*) 4 ☐ (*d*) 5. ☐

34. An eccentrically loaded riveted joint is shown in Fig. 8.35 with 4 rivets at *P*, *Q*, *R* and *S*. Which of the rivets are the most loaded?

(*a*) *P* and *Q* ☐ (*b*) *Q* and *R* ☐
(*c*) *R* and *S* ☐ (*d*) *S* and *P*. ☐

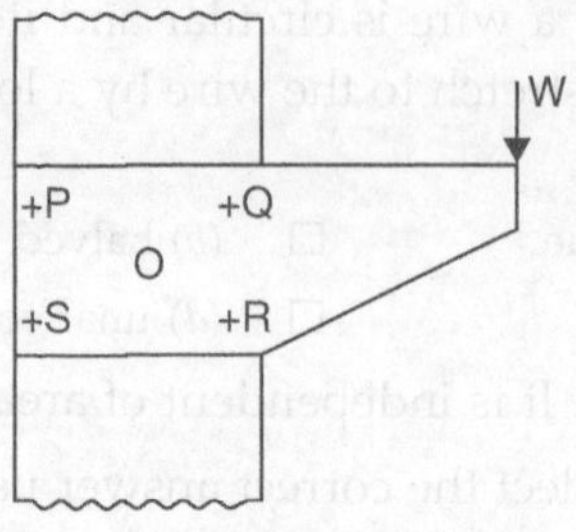

FIGURE 8.35

35. When a helical compression spring is cut into two equal halves, the stiffness of each of the resulting springs will be

(*a*) unaltered ☐ (*b*) double ☐

(*c*) one-half ☐ (*d*) one-fourth. ☐

36. While calculating the stress induced in a close-coiled helical spring, Wahl's factor must be considered to account for

(*a*) the curvature and stress concentration effect ☐

(*b*) shock loading ☐

(*c*) poor service conditions ☐

(*d*) fatigue loading. ☐

37. A straight bar is fixed at the edges *A* and *B* as shown in Fig. 8.36. Its elastic modulus is *E* and cross-section is *A*. There is a load *P* = 120 N acting at *C*. Determine the reactions at the ends

(*a*) 60 N at *A*, 60 N at *B* ☐

(*b*) 30 N at *A*, 90 N at *B* ☐

(*c*) 40 N at *A*, 80 N at *B* ☐

(*d*) 80 N at *A*, 40 N at *B*. ☐

FIGURE 8.36

38. For a material, the modulus of rigidity is 100 GPa and Poisson's ratio is 0.25. The value of modulus of elasticity in GPa is

(*a*) 125 ☐ (*b*) 150 ☐

(*c*) 200 ☐ (*d*) 250. ☐

39. A rigid beam of negligible weight is supported in a horizontal position by two rods of steel and aluminium 2 m and 1 m long having values of cross-sectional areas 1 cm^2 and 2 cm^2 and *E* of 200 GPa and 100 GPa respectively. A load *P* is applied as shown in Fig. 8.37.

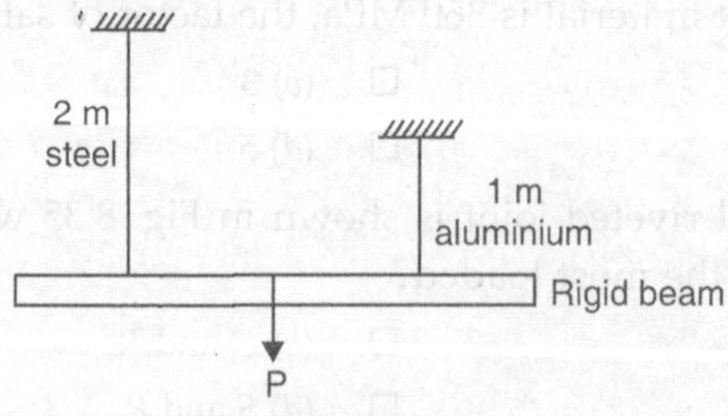

FIGURE 8.37

If the rigid beam is to remain horizontal, then

(a) the load on both rods should be equal ☐

(b) the load on aluminium rod should be twice the load on steel ☐

(c) the load on the steel rod should be twice the load on aluminium ☐

(d) the load P must be applied at the centre of the beam. ☐

40. A thin cylinder of radius r and thickness t when subjected to an internal hydrostatic pressure p causes a radial displacement u, then the tangential strain caused is

(a) $\frac{du}{dr}$ ☐ (b) $\frac{1}{r} \cdot \frac{du}{dr}$ ☐

(c) $\frac{u}{r}$ ☐ (d) $\frac{2u}{r}$. ☐

41. Determine the stiffness of the beam shown in Fig. 8.38 given below :

When $I = 375 \times 10^{-4}$ m^4

$L = 0.5$ m

$E = 200$ GPa

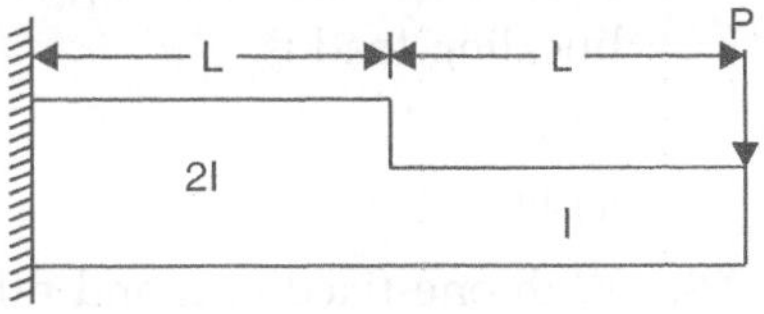

FIGURE 8.38

The stiffness is given by

(a) 12×10^{10} N/m ☐ (b) 10×10^{10} N/m ☐

(c) 4×10^{10} N/m ☐ (d) 8×10^{10} N/m. ☐

42. The strain energy stored in a body of volume V subjected to uniform stress σ is

(a) $\frac{\sigma \times E}{V}$ ☐ (b) $\frac{\sigma E^2}{V}$ ☐

(c) $\frac{\sigma \times V^2}{E}$ ☐ (d) $\frac{\sigma^2}{2E} \times V$. ☐

43. For the same internal diameter, wall thickness, material and internal pressure, the ratio of maximum stress, induced in a thin cylindrical and in a thin spherical vessel will be

(a) 2 ☐ (b) $\frac{1}{2}$ ☐

(c) 4 ☐ (d) $\frac{1}{4}$. ☐

44. Two identical springs labelled as 1 and 2 are arranged in series and subjected to force F as shown in Fig. 8.39. Assume that each spring constant is k. The strain energy stored in spring 1 is

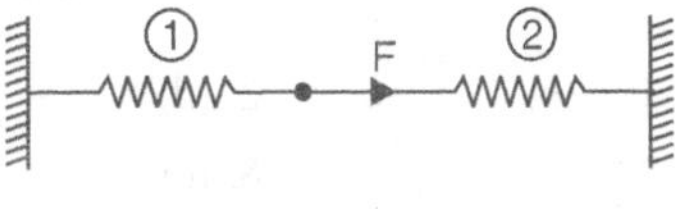

FIGURE 8.39

(a) $\frac{F^2}{2k}$ ☐ (b) $\frac{F^2}{4k}$ ☐

(c) $\frac{F^2}{8k}$ ☐ (d) $\frac{F^2}{16k}$. ☐

45. A rod having cross-sectional area 100×10^{-6} m^2 is subjected to a tensile load. Based on the Tresca failure criterion, if the uniaxial yield stress of the material is 200 MPa, the failure load is

(*a*) 10 kN ☐ (*b*) 20 kN ☐
(*c*) 100 kN ☐ (*d*) 200 kN. ☐

46. Wire diameter, mean coil diameter and number of turns of a closely-coiled steel spring are *d*, *D* and *N* respectively and stiffness of the spring is *k*. A second spring is made of the same steel but with wire diameter, mean coil diameter and number of turns as 2*d*, 2*D* and 2*N* respectively. The stiffness of the new spring is

(*a*) *k* ☐ (*b*) 2*k* ☐
(*c*) 4*k* ☐ (*d*) 8*k*. ☐

47. If the diameter of a long column is reduced by 20%, the percentage of reduction in Euler's buckling load is

(*a*) 4 ☐ (*b*) 36 ☐
(*c*) 49 ☐ (*d*) 59. ☐

48. With one fixed end and other free end, a column of length *L* buckles at load P_1. Another column of same length and same cross-section fixed at both ends buckles at load P_2. Then P_2/P_1 is

(*a*) 1 ☐ (*b*) 2 ☐
(*c*) 4 ☐ (*d*) 16. ☐

49. The principal stresses σ_1, σ_2 and σ_3 at a point respectively are 80 MPa, 30 MPa and – 40 MPa. The maximum shear stress is

(*a*) 25 MPa ☐ (*b*) 35 MPa ☐
(*c*) 55 MPa ☐ (*d*) 60 MPa. ☐

50. The Poisson's ratio of a material which has Young's modulus of 120 GPa and shear modulus of 50 GPa, is

(*a*) 0.1 ☐ (*b*) 0.2 ☐
(*c*) 0.3 ☐ (*d*) 0.4. ☐

ANSWERS

Answers to Objective Type Questions

1. (*b*)	**2.** (*d*)	**3.** (*a*)	**4.** (*c*)	**5.** (*b*)	**6.** (*a*)
7. (*c*)	**8.** (*c*)	**9.** (*b*)	**10.** (*a*)	**11.** (*a*)	**12.** (*a*)
13. (*b*)	**14.** (*c*)	**15.** (*d*)	**16.** (*b*)	**17.** (*a*)	**18.** (*c*)
19. (*b*)	**20.** (*a*)	**21.** (*b*)	**22.** (*b*)	**23.** (*a*)	**24.** (*c*)
25. (*c*)	**26.** (*a*)	**27.** (*b*)	**28.** (*d*)	**29.** (*b*)	**30.** (*b*)
31. (*b*)	**32.** (*c*)	**33.** (*c*)	**34.** (*c*)	**35.** (*a*)	**36.** (*b*)
37. (*c*)	**38.** (*c*)	**39.** (*d*)	**40.** (*d*)	**41.** (*c*)	**42.** (*c*)
43. (*c*)	**44.** (*b*)	**45.** (*c*)	**46.** (*a*)	**47.** (*b*)	**48.** (*b*)

49. (*c*)	**50.** (*c*)	**51.** (*a*)	**52.** (*c*)	**53.** (*c*)	**54.** (*a*)
55. (*b*)	**56.** (*b*)	**57.** (*b*)	**58.** (*c*)	**59.** (*b*)	**60.** (*b*)
61. (*d*)	**62.** (*c*)	**63.** (*b*)	**64.** (*c*)	**65.** (*d*)	**66.** (*b*)
67. (*b*)	**68.** (*c*)	**69.** (*d*)	**70.** (*d*)	**71.** (*c*)	**72.** (*e*)
73. (*b*)	**74.** (*b*)	**75.** (*c*)	**76.** (*b*)	**77.** (*a*)	**78.** (*d*)
79. (*b*)	**80.** (*a*)	**81.** (*c*)	**82.** (*c*)	**83.** (*d*)	**84.** (*b*)
85. (*a*)	**86.** (*c*)	**87.** (*a*)	**88.** (*a*)	**89.** (*d*)	**90.** (*b*)
91. (*d*)	**92.** (*c*)	**93.** (*a*)	**94.** (*b*)	**95.** (*b*)	**96.** (*a*)
97. (*a*)	**98.** (*a*)	**99.** (*c*)	**100.** (*a*)	**101.** (*b*)	**102.** (*c*)
103. (*a*)	**104.** (*b*)	**105.** (*c*)	**106.** (*d*)	**107.** (*b*)	**108.** (*d*)
109. (*c*)	**110.** (*a*)	**111.** (*b*)	**112.** (*c*)	**113.** (*c*)	**114.** (*b*)
115. (*a*)	**116.** (*c*)	**117.** (*d*)	**118.** (*a*)	**119.** (*a*)	**120.** (*b*)
121. (*b*)	**122.** (*d*)	**123.** (*b*)	**124.** (*d*)	**125.** (*c*)	**126.** (*a*)
127. (*d*)	**128.** (*b*)	**129.** (*c*)	**130.** (*d*)	**131.** (*b*)	**132.** (*d*)
133. (*e*)	**134.** (*d*)	**135.** (*c*)	**136.** (*d*)	**137.** (*a*)	**138.** (*b*)
139. (*d*)	**140.** (*a*)	**141.** (*d*)	**142.** (*b*)	**143.** (*d*)	**144.** (*d*)
145. (*a*)	**146.** (*b*)	**147.** (*d*)	**148.** (*d*)	**149.** (*b*)	**150.** (*a*)
151. (*c*)	**152.** (*c*)	**153.** (*c*)	**154.** (*b*)	**155.** (*b*)	**156.** (*b*)
157. (*a*)	**158.** (*e*)	**159.** (*d*)	**160.** (*c*)	**161.** (*c*)	**162.** (*c*)
163. (*b*)	**164.** (*b*)	**165.** (*b*)	**166.** (*c*)	**167.** (*a*)	**168.** (*b*)
169. (*b*)	**170.** (*b*)	**171.** (*a*)	**172.** (*a*)	**173.** (*d*)	**174.** (*d*)
175. (*c*)	**176.** (*c*)	**177.** (*c*)	**178.** (*b*)	**179.** (*c*)	**180.** (*c*)
181. (*a*)	**182.** (*c*)	**183.** (*d*)	**184.** (*a*)	**185.** (*c*)	**186.** (*d*)
187. (*d*)	**188.** (*b*)	**189.** (*c*)	**190.** (*d*)	**191.** (*d*)	**192.** (*c*)
193. (*b*)	**194.** (*c*)	**195.** (*b*)	**196.** (*c*)	**197.** (*c*)	**198.** (*b*)
199. (*c*)	**200.** (*b*)	**201.** (*c*)	**202.** (*d*)	**203.** (*d*)	**204.** (*c*)
205. (*c*)	**206.** (*b*)	**207.** (*a*)	**208.** (*b*)	**209.** (*c*)	**210.** (*c*)
211. (*b*)	**212.** (*c*)	**213.** (*d*)	**214.** (*b*)	**215.** (*c*)	**216.** (*d*)
217. (*a*)	**218.** (*b*)	**219.** (*b*)	**220.** (*d*)	**221.** (*b*)	**222.** (*c*)
223. (*a*)	**224.** (*c*)	**225.** (*a*)	**226.** (*b*)	**227.** (*c*)	**228.** (*a*)
229. (*b*)	**230.** (*b*)	**231.** (*c*)	**232.** (*d*)	**233.** (*d*)	**234.** (*b*)
235. (*c*)	**236.** (*b*)	**237.** (*c*)	**238.** (*c*)	**239.** (*d*)	**240.** (*b*)
241. (*a*)	**242.** (*c*)	**243.** (*b*)	**244.** (*a*)	**245.** (*c*)	**246.** (*c*)
247. (*d*)	**248.** (*c*)	**249.** (*b*)	**250.** (*a*)	**251.** (*b*)	**252.** (*c*)
253. (*c*)	**254.** (*b*)	**255.** (*b*)	**256.** (*a*)	**257.** (*b*)	**258.** (*d*)
259. (*c*)	**260.** (*a*)	**261.** (*d*)	**262.** (*a*)	**263.** (*b*)	**264.** (*a*)
265. (*d*)	**266.** (*c*)	**267.** (*c*)	**268.** (*d*)	**269.** (*a*)	**270.** (*b*)
271. (*b*)	**272.** (*b*)	**273.** (*a*)	**274.** (*b*)	**275.** (*a*)	**276.** (*c*)
277. (*c*)	**278.** (*a*)	**279.** (*c*)	**280.** (*a*)	**281.** (*b*)	**282.** (*a*)

283. (c)	284. (a)	285. (b)	286. (c)	287. (b)	288. (b)
289. (c)	290. (c)	291. (d)	292. (a)	293. (b)	294. (a)
295. (c)	296. (d)	297. (a)	298. (d)	299. (c)	300. (c)
301. (c)	302. (b)	303. (a)	304. (b)	305. (d)	306. (a)
307. (c)	308. (b)	309. (c)	310. (b)	311. (b)	312. (a)
313. (c)	314. (d)	315. (c)	316. (c)	317. (c)	318. (b)
319. (d)	320. (b)	321. (d)	322. (d)	323. (d)	324. (b)
325. (a)	326. (b)	327. (c)	328. (c)	329. (a)	330. (d)
331. (a)	332. (b)	333. (a)	334. (d)	335. (c)	336. (c)
337. (c)	338. (a)	339. (b)	340. (d)	341. (c)	342. (a)
343. (b)	344. (b)	345. (a)	346. (b)	347. (c)	348. (d)
349. (b)	350. (a)	351. (b)	352. (b)	353. (c)	354. (b)
355. (e)	356. (d)	357. (d)	358. (a)	359. (c)	360. (b)
361. (d)	362. (c)	363. (c)	364. (b)	365. (d)	366. (d)
367. (b)	368. (a)	369. (b)	370. (c)	371. (b)	372. (c)
373. (a)	374. (d)	375. (b)	376. (d)	377. (a)	378. (d)
379. (d)	380. (a)	381. (b)	382. (c)	383. (c)	384. (d)
385. (c)	386. (c)	387. (b)	388. (c)	389. (d)	390. (c)
391. (c)	392. (b)	393. (d)	394. (d)	395. (b)	396. (d)
397. (b)	398. (c)	399. (b)	400. (d)	401. (c)	402. (b)
403. (b)	404. (c)	405. (d)	406. (c)	407. (c)	408. (b)
409. (a)	410. (c)	411. (b)	412. (c)	413. (a)	414. (c).

True/False

415. (a)	416. (b)	417. (b)	418. (b)	419. (b)	420. (a)
421. (a)	422. (a)	423. (a)	424. (a)	425. (a)	426. (a)
427. (a)	428. (a)	429. (a)	430. (a)	431. (a)	432. (a)
433. (b)	434. (a)	435. (a)	436. (b)	437. (b)	438. (a)
439. (b)	440. (a)	441. (a)	442. (b)	443. (a)	444. (a)
445. (b)	446. (a)	447. (b)	448. (a)	449. (b)	450. (a)
451. (a)	452. (b)	453. (a)	454. (a)	455. (b)	456. (a)
457. (b)	458. (a)	459. (a)	460. (b)	461. (a)	462. (a)
463. (a)	464. (b)	465. (a)	466. (a)	467. (a)	468. (b)
469. (a)	470. (a)	471. (b)	472. (a)	473. (a)	474. (b)
475. (b)	476. (b)	477. (a)	478. (a)	479. (b)	480. (d)
481. (b)	482. (a)	483. (a)	484. (a)	485. (b)	486. (a)
487. (b)	488. (a)	489. (b)	490. (a)	491. (a)	492. (a)
493. (b)	494. (a)	495. (b)	496. (a)	497. (b)	498. (a)
499. (a)	500. (a)	501. (b)	502. (b)	503. (b)	504. (b)
505. (b)	506. (a)	507. (a)	508. (b)	509. (b)	510. (b)

Fill in the Blanks

511. (*b*)	512. (*b*)	513. (*b*)	514. (*b*)	515. (*b*)	516. (*a*)
517. (*b*)	518. (*b*)	519. (*b*)	520. (*b*)	521. (*b*)	522. (*b*)
523. (*a*)	524. (*b*)	525. (*a*)	526. (*b*)	527. (*b*)	528. (*b*)
529. (*b*)	530. (*b*)	531. (*a*)	532. (*b*)	533. (*a*)	534. (*b*)
535. (*b*)	536. (*b*)	537. (*b*)	538. (*b*)	539. (*b*)	540. (*b*)
541. (*b*).					

Answers to Objective Type Questions from Competitive Examinations with Explanations

1. (*d*)	2. (*a*)	3. (*a*)	4. (*a*)	5. (*b*)	6. (*a*)
7. (*a*)	8. (*b*)	9. (*c*)	10. (*d*)	11. (*c*)	12. (*c*)
13. (*b*)	14. (*b*)	15. (*c*)	16. (*d*)	17. (*c*)	18. (*b*)
19. (*d*)	20. (*c*)	21. (*c*)	22. (*d*)	23. (*d*)	24. (*b*)
25. (*b*)	26. (*b*)	27. (*c*)	28. (*b*)	29. (*d*)	30. (*c*)
31. (*b*)	32. (*c*)	33. (*b*)	34. (*d*)	35. (*b*)	36. (*a*)
37. (*d*)	38. (*d*)	39. (*b*)	40. (*c*)	41. (*c*)	42. (*d*)
43. (*a*)	44. (*c*)	45. (*b*)	46. (*a*)	47. (*d*)	48. (*d*)
49. (*d*)	50. (*b*).				

EXPLANATIONS

1. Here $\sigma_x = 50$ MPa

$\sigma_y = -10$ MPa

$\tau_{xy} = 40$ MPa

The principal stresses σ_1 and σ_2 are given as

$$\sigma_1, \sigma_2 = \frac{\sigma_x + \sigma_y}{2} \pm \sqrt{\left(\frac{\sigma_x - \sigma_y}{2}\right)^2 + \tau_{xy}^{\;2}}$$

$$= \frac{50 + (-10)}{2} \pm \sqrt{\left[\frac{50 - (-10)}{2}\right]^2 + 40^2}$$

$$= 20 \pm \sqrt{30^2 + 40^2}$$

$$= 20 \pm 50,$$

$\therefore$ $\sigma_1 = 20 + 50 = 70$ MPa

and $\sigma_2 = 20 - 50 = -30$ MPa $=$ **30 MPa. (compressive). Ans.**

2. μ = Poisson's ratio $= \dfrac{\text{Lateral strain}}{\text{Longitudinal strain}}$

The lateral strain is opposite to longitudinal strain. This means if longitudinal strain is tensile, then lateral strain is compressive. Hence μ is negative. For most of the material μ lies between -0.25 to -0.40.

4. Bending stress, $\sigma_b = \frac{M}{I} \times y$

$$\therefore \quad (\sigma_b)_{max} = \frac{M}{I} \times \frac{d}{2} \quad \text{where, } I = \frac{\pi}{64} d^4$$

$$= \frac{M}{\frac{\pi}{64} d^4} \times \frac{d}{2} = \frac{32\,M}{\pi d^3}$$

$$T = \frac{\pi}{16} \times d^3 \times \tau_{max}$$

$$\therefore \quad \tau_{max} = \frac{16\,T}{\pi d^3}$$

$\therefore$ If $\sigma_{max} = \tau_{max}$

Then $\frac{32\,M}{\pi d^3} = \frac{16\,T}{\pi d^3}$ or $M = \frac{T}{2}$. **Ans.**

5. Due to bending moment (M), the bending stress will be produced in the shaft. This bending stress (σ_b) is given by

$$\sigma_b = \frac{32\,M}{\pi d^3}$$

Due to torque (T), the shear stress will be produced. This shear stress (τ) is given by

$$\tau = \frac{16\,T}{\pi d^3}$$

$\therefore$ The principal stresses due to bending and shear stresses are

$$\sigma_1 \text{ and } \sigma_2 = \frac{\sigma_b}{2} \pm \sqrt{\left(\frac{\sigma_b}{2}\right)^2 + \tau^2}$$

$$= \frac{1}{2} \times \frac{32\,M}{\pi d^3} \pm \sqrt{\left(\frac{16M}{\pi d^3}\right)^2 + \left(\frac{16\,T}{\pi d^3}\right)^2}$$

For finding equivalent torque (T_e) when the shaft is subjected to bending moment and torque, we should determine the maximum shear produced by the principal stresses.

$\therefore$ Max. shear stress due to principal stresses is given by

$$\tau_{max} = \frac{\sigma_1 - \sigma_2}{2} \quad \text{where } \sigma_1 = \frac{16M}{\pi d^3} + \sqrt{\left(\frac{16M}{\pi d^3}\right)^2 + \left(\frac{16\,T}{\pi d^3}\right)^2}$$

$$= \sqrt{\left(\frac{16M}{\pi d^3}\right)^2 + \left(\frac{16\,T}{\pi d^3}\right)^2} \quad \text{and } \sigma_2 = \frac{16M}{\pi d^3} - \sqrt{\left(\frac{16M}{\pi d^3}\right)^2 + \left(\frac{16\,T}{\pi d^3}\right)^2}$$

$$= \frac{16}{\pi d^3} \sqrt{M^2 + T^2}$$

Equivalent torque is

$$T_e = \frac{\pi}{16} \times d^3 \times \tau_{max.}$$

$$= \frac{\pi}{16} \times d^3 \times \left[\frac{16}{\pi d^3} \times \sqrt{M^2 + T^2}\right] \qquad \left(\because \ \tau_{max} = \frac{16}{\pi d^3} \times \sqrt{M^2 + T^2}\right)$$

$$= \sqrt{\mathbf{M^2 + T^2}} \text{ . } \textbf{Ans.}$$

7. 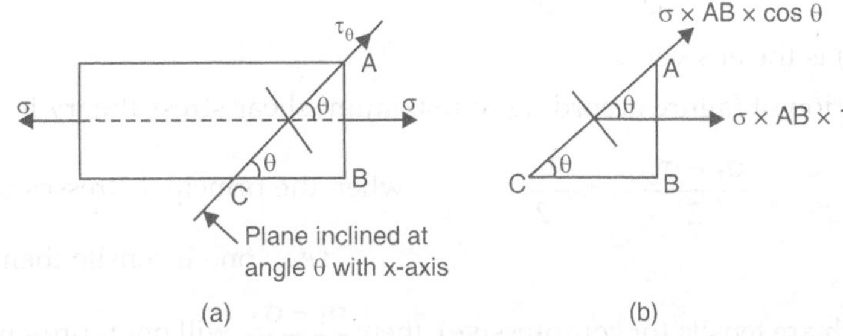

FIGURE 8.40

Force on $\quad AB = \sigma \times \text{Area}$

$\quad = \sigma \times AB \times 1$ (Thickness = unity)

Force on $\quad AC = (\sigma \times AB) \times \cos\theta$

$\therefore \quad \tau_\theta = \text{Shear stress on the plane } AC$

$$= \frac{\text{Shear force}}{\text{Area}} = \frac{\sigma \times AB \times \cos\theta}{AC \times 1} = \sigma \times \sin\theta \times \cos\theta \ \left(\because \ \frac{AB}{AC} = \sin\theta\right)$$

$$= \frac{\sigma}{2} \times 2\sin\theta\cos\theta = \frac{\boldsymbol{\sigma}}{\mathbf{2}} \times \mathbf{\sin 2\theta}. \quad \textbf{Ans.}$$

8. The tensile force in the bar at a distance y from the support

= Weight at the lower end + Weight of bar for a length $(L - y)$

= **W + w (L – y). Ans.**

9. The normal stress on the inclined plane in case of biaxial stress system is given by

$$\sigma_\theta = \frac{1}{2}(\sigma_x + \sigma_y) + \frac{1}{2}(\sigma_x - \sigma_y)\cos 2\theta$$

$$\therefore \quad \sigma_{45^\circ} = \frac{1}{2}(\sigma_x + \sigma_y) + \frac{1}{2}(\sigma_x - \sigma_y)\cos 90^\circ$$

$$= \frac{\mathbf{1}}{\mathbf{2}}(\boldsymbol{\sigma_x + \sigma_y}). \quad \textbf{Ans.}$$

10. Temperature stress = $\alpha \times (\Delta t) + E$

where $\quad \alpha$ = Co-efficient of linear expansion,

Δt = Temperature rise, and

E = Modulus of elasticity.

Hence temperature stress depends upon all the three and (*d*) is the answer.

12. The energy absorbed by a part subject to dynamic force is given by

$$U = \frac{\sigma^2}{2E} \times \text{volume}$$

when σ and E are constant then

$$U \propto \text{volume}$$

Hence (*c*) is the answer.

13. The criterion of failure according to maximum shear stress theory is

$$\frac{\sigma_1 - \sigma_2}{2} = \pm \frac{\sigma_{yp}}{2}$$ when the principal stresses σ_1 and σ_2 are opposite *i.e.,* one is tensile then other is compressive

But if both are tensile (or compressive), then $\frac{\sigma_1 - \sigma_2}{2}$ will not represent the maximum shear stress. It will represent the stress less than maximum shear stress. But $\frac{\sigma_1}{2}$ will represent the maximum shear stress. Hence criterion of failure is

$$\frac{\sigma_1}{2} = \pm \frac{\sigma_{yp}}{2}.$$ **Ans.**

14. Buckling load for column with different end condition are:

$$P_E = \frac{\pi^2 EI}{L^2}$$ ...both ends hinged

$$= \frac{\pi^2 EI}{4L^2}$$ ...one end is fixed (or clamped) and other is hinged

$$= \frac{2\pi^2 EI}{L^2}$$ ...one end is fixed (clamped) and other is hinged

$$= \frac{4\pi^2 EI}{L^2}$$ ...both ends fixed (or clamped)

$\therefore$ When both ends are clamped, the buckling load is maximum.

15. The spring stiffness (*k*) for a close-coiled helical spring in terms of dia. of wire (*d*), mean radius of coil (*R*), no. of turns (*n*) and modulus of rigidity (*C*) is given by

$$k = \frac{W}{\delta} = \frac{Cd^4}{64R^3 \times n}$$

When the dia. of wire (*d*) and material of coil is same, then

$$k \propto \frac{1}{R^3 \times n}$$ [$\because$ For the same material, *C* is constant]

$\therefore$ $k \times R^3 \times n = \text{constant}$

or $k_1 \times R_1^{\,3} \times n_1 = k_2 \times R_2^{\,3} \times n_2$

or $$k_2 = k_1 \times \left(\frac{R_1}{R_2}\right)^3 \times \left(\frac{n_1}{n_2}\right) = k_1 \times \left(\frac{D_1}{D_2}\right)^3 \times \left(\frac{n_1}{n_2}\right)$$

$$= k_1 \times \left(\frac{75}{60}\right)^3 \times \left(\frac{8}{10}\right) \qquad \left[\because \begin{array}{l} D_1 = 75 \text{ mm}, n_1 = 8 \\ D_2 = 60 \text{ mm}, n_2 = 10 \end{array}\right]$$

$$= \mathbf{1.56k_1}. \quad \textbf{Ans.}$$

∴ Answer is (*c*).

16. Here $d = 1 \text{ m} = 100 \text{ cm}$

$p = 100 \text{ N/cm}^2$

Max. permissible tensile stress = 2 kN/cm² = 2000 N/cm²

For thin cylinder, the maximum stress is circumferential (or hoop) stress.

Hence here $\sigma_c = 2000 \text{ N/cm}^2$

Let t = thickness.

Then $$\sigma_c = \frac{p \times d}{2t}$$ [Here p and σ_c should have same unit. Then d and t will have the same unit]

or $$t = \frac{p \times d}{2 \times \sigma_c} = \frac{100 \times 100}{2 \times 2000} = 2.5 \text{ cm} = \mathbf{25\ mm}. \quad \textbf{Ans.}$$

17. Original volume, $$V = \frac{\pi}{4} D^2 \times L$$

Change in volume, dV will be obtained by taking differential as

$$dV = \frac{\pi}{4} D^2 \times dL + \frac{\pi}{4} Ld(2D^2)$$

$$= \frac{\pi}{4} D^2 \times dL + \frac{\pi}{4} L \times 2D\, d(D)$$

∴ Volumetric strain, $$\frac{dV}{V} = \frac{\frac{\pi}{4}D^2 \times dL + \frac{\pi}{4}L \times 2D \times d(D)}{\frac{\pi}{4}D^2 \times L} = \frac{dL}{L} + \frac{2d(D)}{D}$$

$$\frac{dL}{L} = \text{Longitudinal strain} = e_2$$

$$\frac{d(D)}{D} = \text{Circumferential strain} = e_1 \qquad \left[\frac{\pi(D + dD) - \pi D}{\pi D} = \frac{dD}{D}\right]$$

∴ $$\frac{dV}{V} = e_2 + 2e_1 = 2e_1 + e_2 \qquad \left[\frac{dV}{V} = \text{Change in volume per unit volume}\right]$$

18. The part which is cheapest in overall cost and can be easily replaced when there is some damage, is made the weakest part. The key in comparison to pulley and shaft can be replaced easily.

19. The shear stress distribution in a beam of rectangular cross-section is parabolic and having maximum value at the neutral axis. Hence the answer is (*d*).

Shear stress is given by

$$\tau = \frac{FA \times \bar{y}}{I \times b}, \quad \text{where } A \times \bar{y} \text{ for rectangular section} = \frac{b}{2}\left(\frac{d^2}{4} - y^2\right)$$

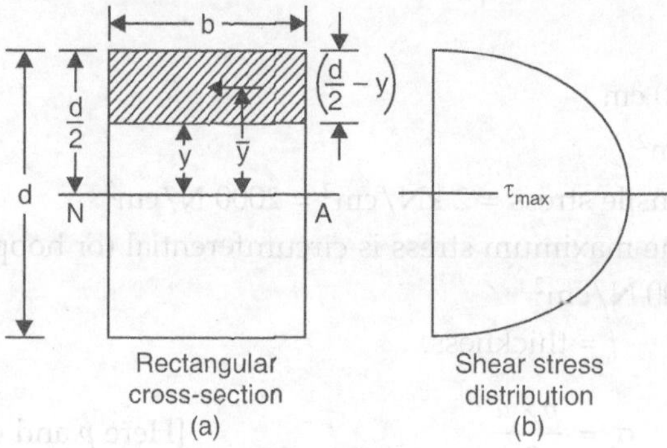

Rectangular cross-section (a) — Shear stress distribution (b)

FIGURE 8.41

where A = Shaded area = $b \times \left(\frac{d}{2} - y\right)$

y = Section where shear stress is τ

$\bar{y}$ = Distance of C.G. of shaded area from N.A. = $y + \frac{1}{2}\left(\frac{d}{2} - y\right)$

I = M.O. Inertia

b = Width at section y which is b here.

$$\therefore \quad \tau = \frac{F \times \frac{b}{2}\left(\frac{d^2}{4} - y^2\right)}{I \times b} = \frac{F}{2I}\left(\frac{d^2}{4} - y^2\right)$$

which is the equation of a parabola.

20. When beam is placed as shown in Fig. 8.42(*a*) 1st position, then bending stress is given by

$$\sigma_b = \frac{M}{I} \times y$$

(a) 1st position — (b) 2nd position

FIGURE 8.42

and $$(\sigma_b)_{max} = \frac{M}{I}\times\frac{d}{2} = \frac{M}{\frac{bd^3}{12}}\times\frac{d}{2} \qquad \left[\because\ I = \frac{bd^3}{12} = \frac{d\times d^3}{12} = \frac{d^4}{12}\right]$$

$$= \frac{6M}{bd^2} = \frac{6M}{d\times d^2} = \frac{6M}{d^3} \qquad (\because\ b = d \text{ being square}) \quad ...(i)$$

The bending stress when beam is placed as shown in Fig. 8.15(*b*) 2nd position is given by

$$(\sigma_b)^*_{max} = \frac{M}{I^*}\times y^*$$

where y^* = Distance of top layer from N.A. *i.e.*, distance OC

= Half of diagonal of the square = OC

$= d \times \sin 45° = \frac{d}{\sqrt{2}}$

I^* = M.O.I, of 2nd position about N.A.

$= 2\times\frac{bh^3}{12}$ where $b = BD = 2\times\frac{d}{\sqrt{2}}$

$h = \frac{d}{\sqrt{2}} = \sqrt{2}\times d$

$$= \frac{2\times\sqrt{2}d\times\left(\frac{d}{\sqrt{2}}\right)^3}{12} = \frac{1}{6}\times\frac{d^4}{2} = \frac{d^4}{12}$$

$$= \frac{M}{\left(\frac{d^4}{12}\right)}\times\frac{d}{\sqrt{2}}$$

$$= \frac{12M}{d^3\times\sqrt{2}}$$

$$\therefore \quad \frac{(\sigma_b)^*_{max}}{(\sigma_b)_{max}} = \frac{\left(\frac{12\,M}{d^3\times\sqrt{2}}\right)}{\left(\frac{6M}{d^3}\right)}$$

$$= \frac{2}{\sqrt{2}} = \sqrt{2}$$

$$\therefore \quad (\sigma_b)^*_{max} = \sqrt{2}\times(\sigma_b)_{max}$$

Hence the answer is (*c*).

21. $$\sigma_b = \frac{M}{I}\times y = \frac{M}{\frac{\pi}{64}\times d^4}\times\frac{d}{2} = \frac{32\,M}{\pi d^3}$$

$$\therefore \quad T = \frac{\pi}{16}\times d^3\times\tau$$

$$\therefore \quad \tau = \frac{16\,T}{\pi d^3}$$

$$\therefore \quad \frac{\sigma_b}{\tau} = \frac{32M}{\pi d^3}\times\frac{\pi d^3}{16T} = \frac{2M}{T} = 2. \qquad (\because\ M = T)$$

22. For a simply supported beam carrying a point load at the centre, the deflection (δ) is given by

$$\delta = \frac{PL^3}{48\,EI}$$

Here load P, span of beam L and E is same for both positions,

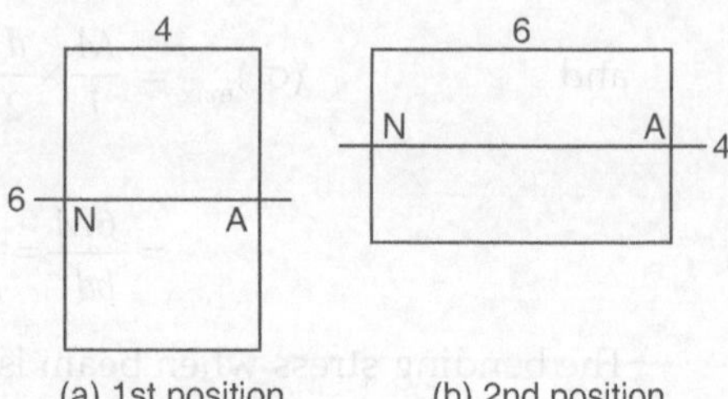

FIGURE 8.43

$\therefore \quad \delta \propto \frac{1}{I}$

or $\quad \delta \times I = \text{constant}$

or $\quad \delta_1 \times I_1 = \delta_2 \times I_2 \quad \text{or} \quad \delta_2 = \delta_1 \times \frac{I_1}{I_2}$

Now $\quad I_1 = \frac{4 \times 6^3}{12} = 72$

and $\quad I_2 = \frac{6 \times 4^3}{12} = 32$

$\therefore \quad \delta_2 = \delta_1 \times \frac{72}{32} = \mathbf{2.25\,\delta_1}$. **Ans.**

Hence answer is (*d*).

23. Couple acting on beam = $P \times C$ anticlockwise moment at any point should be zero

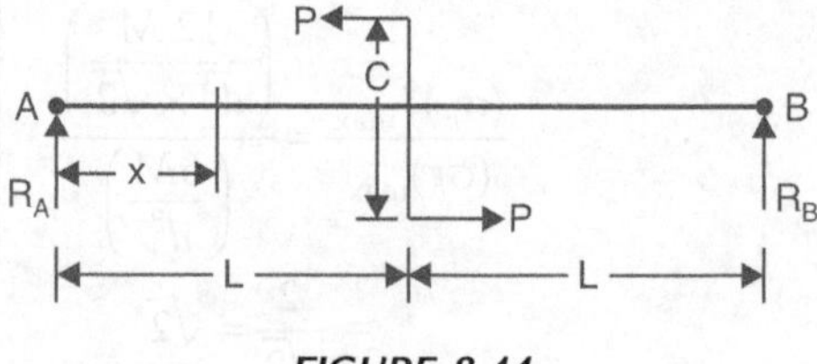

FIGURE 8.44

$\therefore \quad M_A = 0$

or $\quad R_B \times 2L + P \times C = 0$

or $\quad R_B = \frac{-P \times C}{2L}$

(–ve sign shows that reaction R_B is acting downwards)

There is no load on the beam

$\therefore \quad R_A + R_B = 0 \quad \text{or} \quad R_A = -R_B = \frac{P \times C}{2L}$

The reaction R_A is acting upwards.

Shear force at x, $\quad F_x = R_A = \frac{\mathbf{P \times C}}{\mathbf{2L}}$. **Ans.**

24. Resolving forces along the inclined plane and normal to the plane, we get

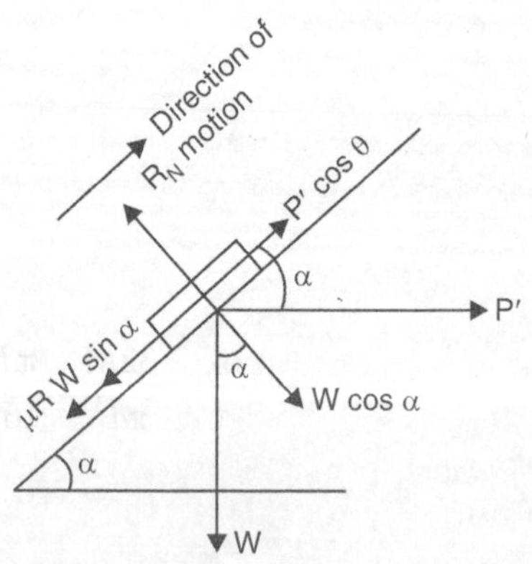

FIGURE 8.45

$$P' \cos\alpha = \mu R_N + W \sin\alpha \quad ...(1)$$

$$R_N = W\cos\alpha + P' \sin\alpha$$

Substituting the value of R_N in (1), we get

$$P' \cos\alpha = \mu\,[W\cos\alpha + P'\sin\alpha] + W\sin\alpha$$

Taking $\mu = \tan \phi = \dfrac{\sin \phi}{\cos \phi}$

and substituting in the above equation and simplifying, we get

$$P' = W \tan (\alpha + \phi)$$

$\therefore$ Frictional torque at mean radius (R) is $= R \times P' = R \times W \tan (\alpha + \phi)$.

26.
$$\sigma_\theta = \frac{\sigma_x + \sigma_y}{2} + \frac{\sigma_x - \sigma_y}{2} \times \cos 2\theta$$

$$\sigma_{45°} = \frac{\sigma_x + \sigma_y}{2} + \frac{\sigma_x - \sigma_y}{2} \times \cos 90°$$

$$= \frac{\sigma_x + \sigma_y}{2}$$

$$= \frac{5000 + 5000}{2} = \mathbf{5000\ N/cm^2.}\quad \textbf{Ans.}$$

27. There is no normal stress either in x-direction or in y-direction.

$\therefore$ $\sigma_x = 0$ and $\sigma_y = 0$. While drawing Mohr's circle of stresses, σ_x and σ_y are taken along x-axis from origin and shear stress (τ) along y-axis. Hence shear stress (τ) will be taken on the origin of axis along y-direction upward and downwards. Hence answer is (*c*).

28. $A = 1\ cm^2$; $L = 100$ cm; $E = 20 \times 10^6\ N/cm^2$, $P = 20$ kN $= 20 \times 1000$ N

$$\frac{\delta L}{L} = \frac{\sigma}{E} = \frac{\left(\frac{P}{A}\right)}{E}$$

$$\therefore \quad \delta L = \frac{P}{AE} \times L = \frac{20{,}000 \times 100}{1 \times 20 \times 10^6} = \mathbf{0.1\ cm.}\quad \textbf{Ans}$$

29.
$$A_1 = \pi r_1^2,\ A_2 = \pi \left(\frac{r_1}{2}\right)^2 = \frac{\pi r_1^2}{4}$$

The modulus of elasticity for a material is independent of area and load applied etc. Hence with the increase or decrease of area (*i.e.*, radius), the modulus of elasticity will be unaffected. **Ans.**

30. To find the correct answer proceed as given below

Ductility is determined by tension test (3)

Toughness is determined by Impact test (1)

Endurance limit is determined by Fatigue Test (2)

Resistance to penetration is determined by Hardness test (4)

Hence the correct code is which contains 3, 1, 2, 4.

Hence the correct code is (*c*).

31. $E = 21 \times 10^6\ N/cm^2$, $C = 8 \times 10^6\ N/cm^2$

$$E = 2C(1 + \mu)$$

$$\therefore \quad \mu = \frac{E}{2C} - 1$$

$$= \frac{21 \times 10^6}{2 \times 8 \times 10^6} - 1 = \frac{21}{16} - 1 = 1.3125 - 1 = \mathbf{0.3125.} \quad \textbf{Ans.}$$

32. $P = 2\pi NT$, where

$$T = \frac{\pi}{16} \times d^3 \times \tau \quad \therefore \quad T_1 = \frac{\pi}{16} \times d_1^{\,3} \times \tau$$

$$P_1 = 2\pi N_1 T_1 \qquad N_2 = \frac{N_1}{2} \qquad T_2 = \frac{\pi}{6} \times (2d_1)^3 \times \tau$$

$$P_2 = 2\pi N_1 T_1 \qquad\qquad = \frac{\pi}{16} \times 8 \times d_1^{\,3} \times \tau$$

$$= 2\pi \times \left(\frac{N_1}{2}\right) \times T_1 \qquad\qquad = 8 \times T_1$$

$$= 2\pi N_1 \times T_1 \times 4$$

$= P_1 \times 4 = 90 \times 4 =$ **360 kW. Ans.**

33. According to maximum shear stress theory for design purpose, we have the equation

$(\sigma_1 - \sigma_2) = \sigma_t$ where σ_t = Permissible stress in simple tension

or $60 - (-60) = \sigma_t$ or $60 + 60 = \sigma_t$

or $\sigma_t = 120$ and safety factor $= \dfrac{\sigma_t^*}{\sigma_t} = \dfrac{360}{120} = 3.$ **Ans.**

34. Here each rivet is subjected to direct stress due to load W and bending stress due to bending moment. Bending moment is equal to $W \times e$, where e = eccentricity.

The value of e is maximum for rivets P and S. The direct stress is same for all the rivets. Bending stress is maximum when e is maximum. Eccentricity is maximum for rivets P and S. Hence rivets P and S are having maximum bending stress. Hence rivets P and S are most loaded. **Ans.**

35. The stiffness of a helical compression spring is given by,

$$k = \frac{Cd^4}{64 R^3 \times n}$$

where C = Modulus of rigidity

d = Dia. of wire

n = No. of turns

R = Mean radius of coil.

When spring is cut into two equal halves, only no. of turns will be effected it will become half. Other value such as C, d and R will be same.

$$\therefore \quad k_2 = \frac{Cd^4}{64 \times R^3 \times \left(\frac{n}{2}\right)} = \frac{2Cd^4}{64R^3 \times n} = \mathbf{2 \times k.} \quad \textbf{Ans.}$$

36. For close-coiled helical spring,

$$\tau = \frac{16\,WR}{\pi d^3}$$

While deriving this equation, the effect of curvature of spring and stress concentration effect are neglected. Hence the correct expression for shear stress will be

$$\tau = \frac{16\,WR}{\pi d^3} \times K$$

where K = Wahl's correction factor

$$= \frac{4S-1}{4S-4} + \frac{0.615}{S}$$

where $S = \text{Spring index} = \frac{D}{d} = \frac{\text{Mean dia. of coil}}{\text{Dia. of wire}}.$

37. Free body diagram through C

$$R_1 + R_2 = 120 \text{ N} \quad \text{...}(i)$$

FIGURE 8.46

Extension in AC = Compression in CB

For $AC: \frac{\delta L_1}{L_1} = \frac{\sigma_1}{E}$ or $\delta L_1 = \frac{\sigma_1}{E} \times L_1 = \frac{R_1}{AE} \times L_1$ $\left(\because \ \sigma_1 = \frac{R_1}{A}\right)$

For $CB: \frac{\delta L_2}{L_2} = \frac{\sigma_2}{E}$ or $\delta L_2 = \frac{\sigma_2}{E} \times L_2 = \frac{R_2}{AE} \times L_2$ $\left(\because \ \sigma_2 = \frac{R_2}{A}\right)$

As $\delta L_1 = \delta L_2$

$\therefore$ $\frac{R_1 \times L_1}{AE} = \frac{R_2 \times L_2}{AE}$

or $R_1 \times L_1 = R_2 \times L_2$ But $L_1 = L$ and $L_2 = 2L$

$\therefore$ $R_1 \times L = R_2 \times 2L$

or $R_1 = 2R_2$ ...(*ii*)

From (*i*) and (*ii*), $3R_2 = 120 \text{ N} \ \therefore \ R_2 = \mathbf{40\ N}$. **Ans.**

and $R_1 = 120 - 40 = \mathbf{80\ N}$. **Ans.**

38. C = 100 G Pa = 100×10^9 N/m²

$$\mu = 0.25$$

$$E = 2C(1 + \mu)$$

$$= 2 \times 100 \times 10^9 (1 + 0.25) = 200 \times 1.25 \times 10^9$$

$$= 250 \times 10^9 \text{ N/m}^2 = \mathbf{250\ GPa.}\ \ \mathbf{Ans.}$$

39. Steel rod | Aluminium rod

$L_1 = 2$ m | $L_2 = 1$ m

$A_1 = 1$ cm² | $A_2 = 2$ cm²

$E_1 = 200$ GPa | $E_1 = 100$ GPa

The rigid beam will be horizontal if:

Extension of steel rod = Extension of aluminium rod

$$\delta L_1 = \delta L_2$$

$$\frac{P_1 \times L_1}{A_1 \times E_1} = \frac{P_2 \times L_2}{A_2 \times E_2}$$

or $$P_1 = P_2 \times \frac{A_1}{A_2} \times \frac{E_1}{E_2} \times \frac{L_2}{L_1} = P_2 \times \frac{1}{2} \times \frac{200}{100} \times \frac{1}{2} = \frac{P_2}{2}$$

or $$2P_1 = 2P_2 \quad \text{or} \quad P_2 = \mathbf{2P_1}.\ \ \mathbf{Ans.}$$

or load on aluminium rod = 2 times the load on steel rod.

40. Radial displacement $= u$

Initial radius $= r$

Final radius $= r + u$

$$\text{Tangential strain} = \text{Circumferential strain} = \frac{\text{Final circumference} - \text{Initial}}{\text{Initial circumference}}$$

$$= \frac{2\pi(r+u) - 2\pi r}{2\pi r} = \frac{2\pi u}{2\pi r} = \frac{\mathbf{u}}{\mathbf{r}}.\ \ \mathbf{Ans.}$$

41. $$\text{Stiffness} = \frac{\text{Load}}{\text{Deflection}} = \frac{P}{\text{Deflection under } P}$$

Let us find deflection under load *P*. This can be done by conjugate Beam Method. In this method, the beam carries the $\frac{M}{EI}$ load corresponding to actual load. The deflection at any section will be equal to B. M. at that section due the load carried by conjugate beam. Refer to Fig. 8.47 (*b*).

Load for conjugate beam is $\frac{M}{EI}$.

B.M. at $A = 0$, hence value of $\frac{M}{EI}$ at $A = 0$

B.M. at $C = P \times 2L$, hence value of $\frac{M}{EI}$ at $C = \frac{P \times 2L}{E \times (2I)} = \frac{PL}{EI}$

B.M. at $B = P \times L$, hence value of $\frac{M}{EI}$ at B for $AB = \frac{P \times L}{EI}$, for $BC = \frac{P \times L}{EI} = \frac{PL}{2EI}$

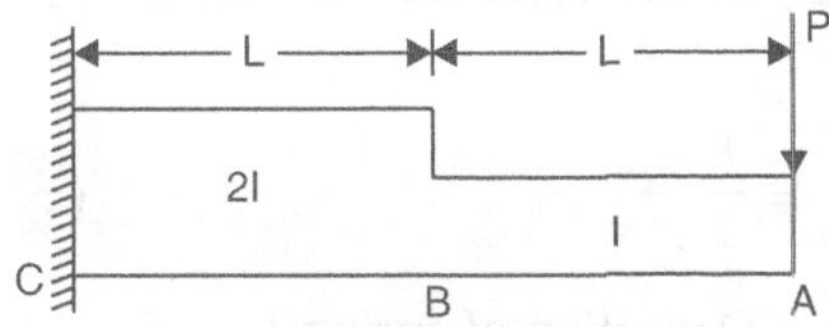

(a) Actual beam

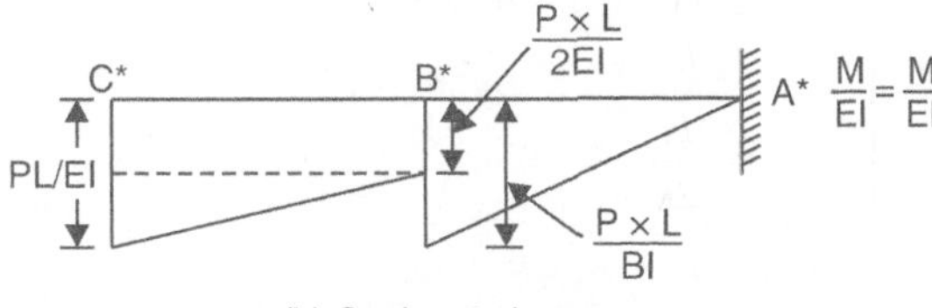

(b) Conjugate beam

FIGURE 8.47

Deflection (δ) at A = B.M. at A^* due to load carried by conjugate beam

$$= \left(\frac{1}{2} \times \frac{PL}{EI} \times L\right) \times \frac{2L}{3} + \left(\frac{PL}{2EI} \times L\right) \times 1.5L + \left(\frac{1}{2} \times \frac{PL}{2EI} \times L\right) \times \frac{5L}{3}$$

$$= \frac{PL^3}{EI}\left(\frac{1}{3} + 0.75 + \frac{5}{12}\right) = \frac{PL^3}{EI} \times \frac{18}{12} = \frac{1.5 \times PL^3}{EI}$$

$$\therefore \quad \text{Stiffness} = \frac{P}{\delta} = \frac{P}{\left(\frac{1.5 \times PL^3}{EI}\right)} = \frac{EI}{1.5 \times L^3} = \frac{(200 \times 10^9) \times (375 \times 10^{-6})}{1.5 \times (0.5)^3}$$

$$= \frac{200 \times 10^9 \times 375 \times 10^{-4} \times 10^3}{1.5 \times 125} = \mathbf{4 \times 10^{10}\ N/m.} \quad \textbf{Ans.}$$

42. The maximum stress induced in a thin cylinder is hoop stress (σ_c). It is given by

$$\sigma_c = \frac{P \times d}{2t}$$

The maximum hoop stress produced in spherical vessel $= \frac{P \times d}{4t}$

$$\therefore \quad \frac{\text{Max. stress in cylindrical vessel}}{\text{Max. stress in spherical vessel}} = \frac{\frac{pd}{2t}}{\frac{p \times d}{4t}} = 2.0.$$

44.

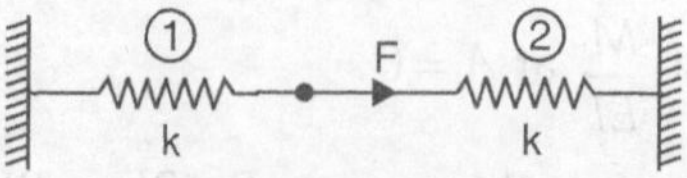

FIGURE 8.48

Strain energy stored in spring 1,

$U_1 = \frac{1}{2} \times F_1 \times \delta_1$ where F_1 = force carried by spring 1

$= \frac{1}{2} \times \frac{F}{2} \times \frac{F}{2k}$ | $= \frac{F}{2}$ ($\because$ both springs are identical)

δ_1 = Deflection of spring 1.

$= \frac{\mathbf{F^2}}{\mathbf{8k}}$. **Ans.** | $= \frac{F_1}{k}$

$= \frac{F_1}{2k}$ $\left(\because\ F_1 = \frac{F}{2}\right)$

45. $A = 100 \times 10^{-6}$ m²

Let P = tensile load at failure $\therefore$ For one-dimensional stress system,

we have stresses as $\left(\frac{P}{A}, 0, 0\right)$

Tensile stress due to load P, $\sigma_1 = \frac{P}{A}$

This stress is in one-direction only *i.e.*, $(\sigma, 0, 0)$

Max. shear stress due to stress system $(\sigma_1, 0, 0) = \frac{1}{2}(\sigma_1 - 0) = \frac{\sigma_1}{2}$

Uniaxial yield stress, $\sigma_t^* = 200$ MPa $= 200 \times 10^6$ N/m².

$\therefore$ For uniaxial yield stress, we have stress system as $(\sigma_t^*, 0, 0)$

Max. shear stress due to uniaxial yield stress $= \frac{1}{2}(\sigma_t^* - 0) = \frac{\sigma_t^*}{2}$

According to Tresca failure criterion,

Max. shear stress developed = Max. shear due to yield stress

i.e., $\frac{\sigma_1}{2} = \frac{\sigma_t^*}{2}$ or $\sigma_1 = \sigma_t^*$

or $\frac{P}{A} = 200 \times 10^6$ N/m² $\left(\because\ \sigma_t^* = 200 \times 10^6 \text{ N/m}^2 \text{ and } \sigma_1 = \frac{P}{A}\right)$

or $P = 200 \times 10^6 \times A$

$= (200 \times 10^6) \times (100 \times 10^{-6}) = 20000 =$ **20 kN. Ans.**

46. We know that stiffness of a close-coiled helical spring is given by

$$k = \frac{Cd^4}{64\,R^3 \times n}$$

1st case, $$k_1 = \frac{Cd^4}{64\left(\frac{D}{2}\right)^3 \times N} \qquad \left(\text{here } R = \frac{D}{2} \text{ and } n = N\right)$$

2nd case, $$k_2 = \frac{C(2d)^4}{64\left(\frac{2D}{2}\right)^3 \times 2N}$$

(here dia. of wire = $2d$; Mean coil dia. = $2D$ and number of turns = $2N$)

$$= \frac{Cd^4 \times 16}{64 \times \left(\frac{D}{2}\right)^3 \times 8 \times 2N} = \frac{Cd^4}{64\left(\frac{D}{2}\right) \times N} = \mathbf{k_1} \qquad \left(\because\ k_1 = \frac{Cd^4}{64\left(\frac{D}{2}\right)^3 \times N}\right)$$ **Ans.**

47. Euler's buckling load,

$$P = \frac{\pi^2 EI}{L^2} \quad \text{where} \quad I = \frac{\pi}{64}\,d^4$$

For circular column when diameter is reduced by 20%, then dia. of new column = 0.8 d

$\therefore$ New moment of inertia, $I^* = \frac{\pi}{4} \times (0.8\,d)^4 = \frac{\pi}{4} \times d^4 \times (0.8)^4$

Initial buckling load, $$P = \frac{\pi^2 E}{L^2} \times \frac{\pi}{64}\,d^4$$

New buckling load, $$P^* = \frac{\pi^2 E}{L^2} \times \frac{\pi}{64} \times d^4 \times 0.8^4$$

$\therefore$ % reduction in load $= \frac{P - P^*}{P} \times 100$

$$= \frac{\left(\frac{\pi^2 E}{L^2} \times \frac{\pi}{64} d^4 - \frac{\pi^2 E}{L^2} \times \frac{\pi}{64}\, d^4 \times 0.8^4\right)}{\frac{\pi^2 E}{L^2} \times \frac{\pi}{4} d^4} \times 100$$

$$= \frac{1 - 0.8^4}{1} = (1 - 0.4096) \times 100 = \mathbf{59\%}.$$ **Ans.**

48. For a column of one end fixed and other free, the bucking load is

$$P_1 = \frac{\pi^2 EI}{4L^2}$$

For a second column of same length and same cross-sectional area when both ends are fixed, the buckling load is

$$P_2 = \frac{4\pi^2 EI}{L^2}$$

$$\therefore \quad \frac{P_2}{P_1} = \frac{\left(\dfrac{4\pi^2 EI}{L^2}\right)}{\left(\dfrac{\pi^2 EI}{4L^2}\right)} = 4 \times 4 = 16. \quad \textbf{Ans.}$$

49. $\sigma_{max} = \frac{1}{2}[\sigma_1 - \sigma_3] = \frac{1}{2}[80 - (-40)] = \mathbf{60\ MPa.}$ **Ans.**

50. $E = 2C(1 + \mu)$

$$\therefore \quad \mu = \frac{E}{2C} - 1 = \frac{120}{2 \times 50} - 1 = 1.2 - 1.0 = \mathbf{0.2.} \quad \textbf{Ans.}$$

Chapter 9 THEORY OF MACHINES

I. THEORY

9.1. INTRODUCTION

Theory of machine is that branch of science, which deals with the study of the relative motion between the parts of a machine and study of the forces which act on these parts. Each part of a machine which has motion relative to some other part is known as *link or element.* A link should be capable of transmitting the required force with negligible deformation *i.e.,* a link must be a resistant body.

9.2. KINEMATIC PAIR

Kinematic pair is a combination of two links which are joined in such a way that their relative motion is completely constrained or successfully constrained. They are *classified* as :

(*a*) *According to the nature of relative motion between the elements, i.e.,*

(*i*) Sliding pair (*ii*) Turning pair

(*iii*) Rolling pair (*iv*) Spherical pair

(*v*) Cylindrical pair.

(*b*) *According to the nature of contact between the elements i.e.,*

(*i*) Lower pairs, and (*ii*) Higher pairs.

(*c*) *According to the nature of mechanical constraint i.e.,*

(*i*) Closed pair, and (*ii*) Unclosed pair.

9.3. KINEMATIC CHAIN

Kinematic chain is a combination of kinematic pairs in which each element of link forms part of two pairs and in which the relative motion is completely constrained. A chain may be locked, constrained or unconstrained. The nature of the chain is determined by the following relation,

$$J + \frac{H}{2} = \frac{3}{2}L - 2 \quad \text{...(1)}$$

where J = Number of binary joints in the chain,

H = Number of unclosed pairs, and

L = Number of links in the chain.

If in equation (1), R.H.S. < L.H.S., the chain is locked. If R.H.S. = L.H.S., then chain is a kinematic chain. If R.H.S. > L.H.S., the chain is unconstrained.

Ternary joint is a joint to which three links are connected. A ternary joint is equivalent to two binary joints. When n links are connected at the same joint, the joint is equivalent to $(n - 1)$ binary joints.

9.4. MECHANISM AND INVERSION

When one link of a kinematic chain is fixed, the arrangement is known as mechanism. This arrangement may be used for transmitting or transforming motion. If a different link of the chain is made the fixed link, a different mechanism will be obtained. Hence if a kinematic chain has 'l' links then 'l' different mechanism are obtained by fixing each of the links in turn. Each mechanism so obtained is known as the *inversion* of the original kinematic chain.

9.4.1. Machine

A mechanism, in which the elements are made in such a way that they are able to withstand the forces which act on them when mechanism is put to work, is called a machine.

9.5. KINEMATIC CHAINS WITH FOUR LOWER PAIRS

Most of the machines are based on the kinematic chains with four lower pairs (which are sliding and turning pairs). The different mechanism obtained from four lower pairs are :

1. Four bar chain which consists of four turning pairs.
2. Slider crank chain which consists of one sliding pair and three turning pairs.
3. Double slider crank chain which consists of two sliding pairs and two turning pairs. Also the two similar pairs are adjacent.
4. Crossed slider crank chain which consists of two sliding pairs and two turning pairs. And two similar pairs are not adjacent.

9.5.1. Four-bar Chain

Figure 9.1 shows the four-bar chain which consists of four turning pairs. By fixing any one of links, four different inversions can be obtained.

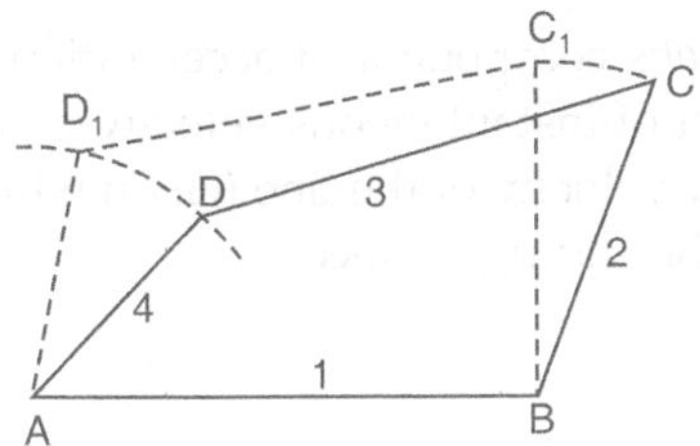

FIGURE 9.1. Four-bar chain.

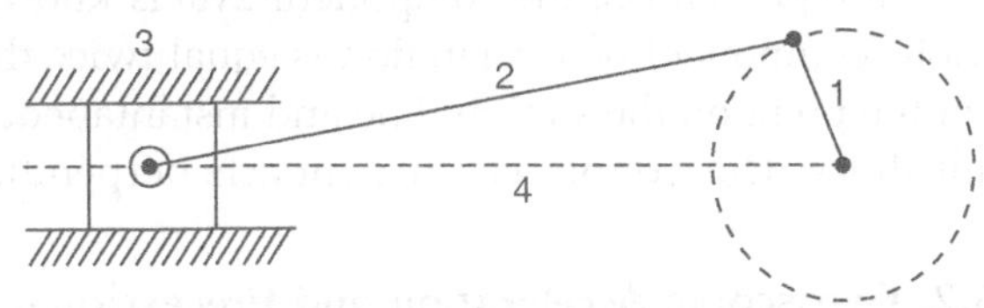

FIGURE 9.2. Reciprocating mechanism.

9.5.2. Slider Crank Chain

Slider crank chain consists of one sliding pair and three turning pairs. One of its inversion is shown in Fig. 9.2. This inversion is used in reciprocating engines, reciprocating pumps and reciprocating compressors.

9.5.3. Double Slider Crank Chain

The different inversions obtained from double slider crank chain are elliptical trammel and Oldham's Coupling.

9.6. MOTION IN A CURVED PATH

When a particle is moving in a curved path, the particle has got normal acceleration and tangential accelerations. The tangential acceleration is equal to the rate of change of velocity. The normal acceleration is equal to the square of velocity divided by the radius of curvature of the path. Mathematically,

Tangential acceleration, $a_t = \dfrac{dv}{dt}$

and normal acceleration, $a_n = \dfrac{v^2}{r}$.

9.6.1. Acceleration of a Point Moving Along a Rotating Straight Line

If a point is moving along a straight line and the straight line is rotating, then the normal component of acceleration (a_n) is given by

$$a_n = \frac{dv}{dt} - r\omega^2$$

and the tangential component of acceleration (a_t) is given by

$$a_t = 2v\omega + r\alpha \qquad \ldots(2)$$

where, v = Velocity of the point along the straight line,

r = Radius of the point,

ω = Angular velocity of the straight line, and

α = Angular acceleration of the straight line.

In equation (2), the component $2v\omega$ is known as *Coriolis* component of acceleration. Thus Coriolis component of acceleration is equal twice the product of instantaneous velocity of sliding (v) of the point on the straight line and instantaneous angular velocity of the line (ω) on which the point slides. The Coriolis component acts perpendicular to the sliding surface.

9.6.2. Gyroscopic Acceleration and Precessional Motion

If a disc is rotating about an axis with an angular velocity (ω) and axis of rotation is changing its position then the rate at which the axis of rotation changes its position is known as angular velocity of precession. The change in direction of the plane of rotation of the disc is called precessional motion. The product of angular velocity of disc (ω) and angular velocity of precession is known as gyroscopic acceleration.

9.7. VELOCITY OF A POINT IN A MECHANISM

The velocity of a point in a mechanism can be determined by the following methods:

(*i*) by means of instantaneous centres

(*ii*) by means of velocity diagrams

(*iii*) by drawing a displacement time graph

(*iv*) by analytical method.

9.8. CLUTCHES AND BRAKES

A clutch connects a moving member to another moving member whereas a brake connects a moving member to a stationary frame. The main two types of clutches are:

1. Positive clutch, and
2. Friction clutch.

The jaw clutch and toothed clutch are the main types of positive clutch whereas the disc clutch, cone clutch, block clutch and band clutch belong to friction clutch.

9.9. INSTANTANEOUS CENTRE

It is the point about which one body rotates relative to another body for the configuration being considered. The bodies do not have any linear velocity relative to each other at this point. If a mechanism is having 'n' links, then the number of instantaneous centres would be

$$= \frac{n(n-1)}{2}.$$

9.9.1. Kennedy's Theorem

It states that any three bodies moving relatively to each other have three instantaneous centres and these centres lie on a straight line.

9.9.2. Relative Velocity Method

For a rigid link, the velocity of one end of the link relative to the other is at right angles to the link. Relative velocity method is used for determining acceleration at any point of a link. If *AB* is a rigid link and point B is moving with respect to point A, then acceleration of B will be equal to the vector sum of acceleration of A and acceleration of B relative to A.

9.9.3. Analytical Method for Determining the Velocity and Acceleration of the Piston of a Reciprocating Engine

The displacement of the piston of a reciprocating engine from inner dead centre is given by

$$x = r\,(1 + n - \cos\theta - \sqrt{n^2 - \sin^2\theta})$$

where r = Crank radius,

θ = Angle turned by crank from I.D.C., and

n = Ratio of length of connecting rod to crank radius.

If $\sin^2\theta$ is small in comparison with n^2, then velocity (v) and acceleration (a) of the piston are given by

$$v = \omega r\left[\sin\theta + \frac{\sin 2\theta}{2n}\right] \text{ and } a = \omega^2 r\left[\cos\theta + \frac{\cos 2\theta}{n}\right].$$

9.10. SIMPLE HARMONIC MOTION

If a body oscillates about an equilibrium position in such a way that its acceleration is always directed towards the fixed equilibrium position, and is proportional to its displacement from the equilibrium position, the body is said to be having simple harmonic motion. The equation of motion of a simple harmonic motion is given by

$$\frac{d^? x}{dt^2} = -kx$$

where $\frac{d^2x}{dt^2}$ = Acceleration of the body,

x = Displacement from mean position, and

k = A constant.

9.10.1. Simple Pendulum

A simple pendulum consists of a bob suspended from a fixed point by a light string. In equilibrium position, the string is vertical. If bob is displaced by a small angle and released, it will perform simple harmonic motion. The time period of the simple pendulum is given by

$$T = 2\pi\sqrt{\frac{L}{g}}.$$

The time period (T) is independent of the size of the bob. It depends only on the length and acceleration due to gravity.

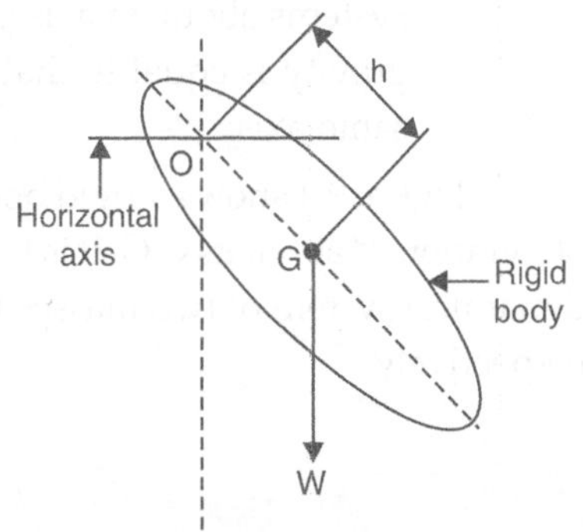

FIGURE 9.3

9.10.2. Compound Pendulum

Compound Pendulum is a rigid body oscillating in a vertical plane about a horizontal axis, which passes through *O* as shown in Fig. 9.3. The time period of the compound pendulum is given by

$$T = 2\pi\sqrt{\frac{k^2 + h^2}{h \times g}}$$

where k = Radius of gyration of the rigid body about an axis through *G* (*i.e.*, C.G. of the body), and

h = Distance of the C.G. of the body from the point of suspension.

The simple pendulum, which has the same periodic time of oscillation as a given compound pendulum, is called the *equivalent simple pendulum.* The length of the equivalent simple pendulum is given by

$$l_c = \frac{k^2 + h^2}{h}.$$

When the distance of the C.G. of the compound pendulum from the point of suspension is equal to the radius of gyration of the body about its centre of gravity, the periodic time of the compound pendulum will be minimum.

The minimum period of a compound pendulum is given by

$$T_{\min} = 2\pi\sqrt{\frac{2k}{g}}$$

where k = Radius of gyration of compound pendulum about its C.G.

9.11. EQUIVALENT DYNAMICAL SYSTEM

A rigid body can be replaced by two masses assumed to be concentrated at points and connected rigidly together. The two masses will be dynamically equivalent to the given rigid body, if

(*i*) the total mass of the two masses is equal to the total mass of the given body;

(*ii*) the centre of gravity of the two mass system coincides with the centre of gravity of the rigid body; and

(*iii*) the total mass moment of inertia of the two mass systems about an axis passing through the centre of gravity is equal to that of the rigid body about the same axis.

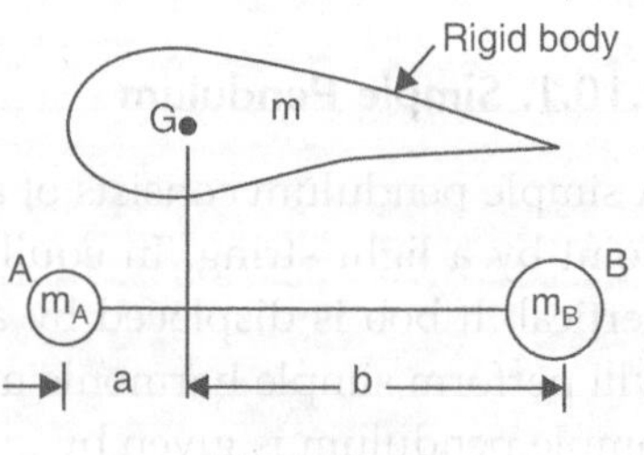

FIGURE 9.4

Figure 9.4 shows a rigid body of mass '*m*' having radius of gyration '*k*' about its C.G. It is to be replaced by an equivalent dynamical system of two masses m_A and m_B placed at *A* and *B* respectively.

Then by applying the above three conditions, the mass at A and B should be equal to,

$$m_A = \frac{b \times m}{a+b}$$

and
$$m_B = \frac{a \times m}{a+b}$$

The distances a and b are related to the radius of gyration by $k^2 = ab$.

9.12. GEARS

Gears are used for transmitting power from one shaft to another shafts, when the shafts are at a very small distance apart. Gears are having advantages over belts and ropes as there is no slip and creep in case of power transmitted by gears. Also gear drive is a positive and smooth.

9.13. DEFINITION OF THE TERMS USED IN GEARS

Figure 9.5 shows the profile of a gear along with important terms, which are used in the study of the gears. The terms are defined as :

(*i*) **The pitch circle diameter** is the diameter of a circle which by pure rolling action would produce the same motion as the toothed gear wheel.

(*ii*) **Pitch point** is the point of contact of two pitch circles of mating gears.

(*iii*) **Circular pitch (p_c)** is the distance, measured along the circumference of the pitch circle from a point on one tooth to a corresponding point on the adjacent tooth. It will be equal to the pitch circle circumference divided by number of teeth on the wheel. Hence

$$\text{Circular pitch} = \frac{\pi D}{T} \quad \text{or} \quad p_c = \frac{\pi D}{T} \qquad \text{...(3)}$$

where D = Diameter of pitch circle, and T = Number of teeth.

(*iv*) **The diametral pitch (p_d)** is equal to the number of teeth per unit length of pitch circle diameter.

$$\therefore \quad p_d = \frac{T}{D} \qquad \text{...(4)}$$

If we multiply equations (3) and (4), we get

$$p \cdot p_d = \pi$$

(*v*) **Module (*m*)** is defined as the length of the pitch circle diameter per tooth. Hence

$$m = \frac{D}{T}$$

m is generally expressed in millimetre. Module is the reciprocal of the diametral pitch.

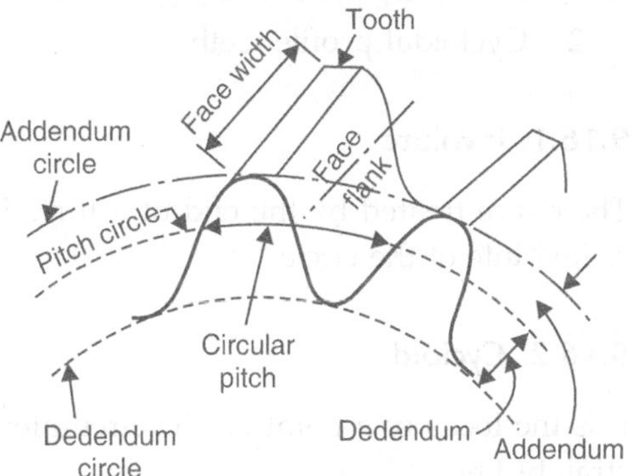

FIGURE 9.5

(*vi*) **The addendum** is the radial distance of the tooth above pitch circle. Its value is generally one module.

(*vii*) **The dedendum** is the radial distance of the tooth below the pitch circle. Its value of generally 1.157 module.

(*viii*) **Addendum circle** is the circle which passes through the top of the teeth.

(*ix*) **Dedendum circle** is the circle which passes through the bottom of the teeth.

(*x*) **Face of the tooth** is that part of tooth surface which is above the pitch circle.

(*xi*) **Flank of the tooth** is that part of the tooth surface which is below the pitch circle.

(*xii*) **Path of contact** is the curve traced by the point of contact of the two mating teeth from the beginning to the end of engagement of the two teeth.

(*xiii*) **Path of approach** is the path of contact from the beginning of the engagement to the pitch point of the two mating teeth.

(*xiv*) **Path of recess** is the path of contact from the pitch point to the end of engagement of the two mating gears.

(*xv*) **Pressure angle** is the angle which the common normal to the two teeth at the point of contact makes with the common tangent to the two pitch circles at the pitch point.

9.14. LAW OF GEARING

It states that the common normal at the point of contact between a pair of teeth always passes through the pitch point.

9.15. VELOCITY OF SLIDING

Velocity of sliding of the mating teeth of the two gears is equal to the product of the sum of the angular velocities of the two gears and distance of the point of contact from the point.

9.16. FORMS OF TEETH

The followings are the forms of the teeth which are commonly used and satisfy the law of gearing.

1. Involute profile teeth.
2. Cycloidal profile teeth.

9.16.1. Involute

The curve treated by the end of a thread as it is unwound from a stationary cylinder is known as involute of the circle.

9.16.2. Cycloid

It is the locus of a point on the circumference of a circle which rolls without slipping on a fixed straight line.

9.16.3. Epicycloid

It is the locus of a point of the circumference of a circle which rolls without slipping on the outside of another circle.

9.16.4. Hypo-cycloid

It is the locus of a point on the circumference of a circle which rolls without slipping on the inside of another circle.

9.16.5. Involute Gear Teeth

(*i*) The centre distance between the two involute gears = $\dfrac{R+r}{\cos\phi}$

where R and r = Radii of the base circles of the meshing gears, and ϕ = Pressure angle.

(*ii*) The angular velocity ratio of the two involute gears in mash, is inversely proportional to the size of the base circles.

(*iii*) The involute function [$INV(\phi)$] in terms of pressure angle (ϕ) is given by

$$INV(\phi) = \tan\phi - \phi.$$

(*iv*) The path of contact in involute gears is a straight line.

9.17. SPIRAL GEARS

To connect non-parallel and non-intersecting shafts, spiral gears are used. The spiral gears are having maximum efficiency if the spiral angle (α) is equal to half the sum of the shaft angle (θ) and friction angle (ϕ), *i.e.*, if

$$\alpha = \frac{\theta+\phi}{2}.$$

And the maximum efficiency is given by $\eta_{max} = \dfrac{1+\cos(\theta+\phi)}{1+\cos(\theta-\phi)}$.

9.17.1. Helical Gears

(*i*) The normal circular pitch = $p_c \times \cos\alpha$

where p = Circular pitch, and α = Helix angle.

(*ii*) Axial pitch for helical gears = $p_c/\tan\alpha$.

9.17.2. Bevel Gears

They are used to connect two shafts, whose axes intersect each other.

9.17.3. Worm Gears

They are the same as spiral gears except that the teeth of the worm gears are having line contact instead of point contact. Hence load-carrying capacity of worm gears are much higher. The maximum efficiency of the worm gear is given by

$$\eta_{max} = \frac{1-\sin\phi}{1+\sin\phi}$$

where ϕ = Friction angle.

9.18. GEAR TRAINS

Gear trains means more than two gears in mesh between the driving shaft and driven shaft. Hence the combination of more than two gears, by means of which power is transmitted from the driving shaft to driven shaft, is known as gear train. The gear train may be a *single gear train* or a *compound gear train.* If each gear is having separate shaft, the train is said to be simple gear train. And if except the first and last, two gears arc mounted on the same shaft, the train is known as compound gear train.

9.18.1. Simple Gear Train

Figure 9.6 shows a simple gear train having three gear wheels. Let the gear wheel number 1 is the driving wheel and gearwheel number 3 is the driven whccl. Thc intermediate gearwheel number 2, is called idler.

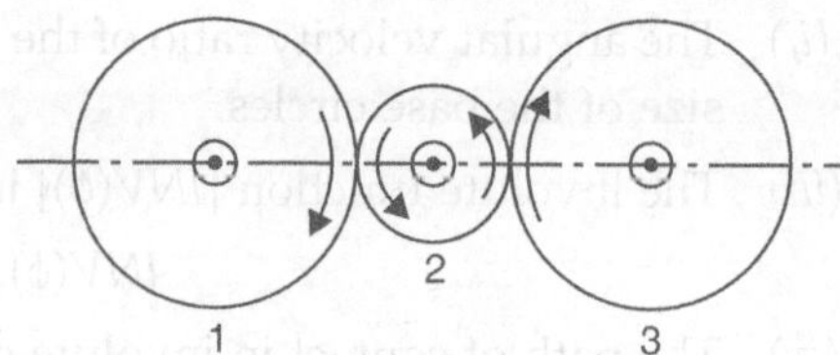

FIGURE 9.6. Simple gear train.

If N_1, N_2 and N_3 = Speed of wheels, 1, 2 and 3 respectively and T_1, T_2, T_3 = Number of teeth on wheels 1, 2, 3 respectively.

Then $$\frac{N_1}{N_2} = \frac{T_3}{T_1}$$

The ratio $\left(\frac{N_1}{N_3}\right)$ is called speed ratio and the ratio $\left(\frac{N_3}{N_1}\right)$ is called train value. The speed ratio is equal to the ratio of number of teeth on the driven wheel to the number of teeth on the driving wheel. Thus the speed ratio is independent of the number of teeth on the intermediate gear wheel (which is known as idler). The intermediate wheel only helps to change the direction of rotation. When there are odd number of gear wheels connected by a simple gear train, the first and the last gear wheel rotate in the same direction. But if there are even number of gear wheels then the first and last gear wheel rotate in the opposite direction.

9.18.2. Compound Gear Train

Figure 9.7 shows a compound gear train in which gear wheels 2 and 3 are mounted on the same shaft and they rotate together. The ratio of the speeds of the fourth wheel to that of first wheel is given by

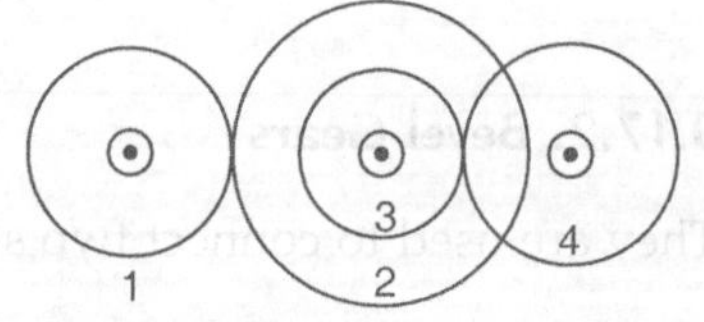

FIGURE 9.7. Compound gear train.

$$\frac{N_4}{N_1} = \frac{T_1 \times T_3}{T_2 \times T_4}$$

where T_1, T_2, T_3, T_4 are the number of teeth on gear wheels 1, 2, 3 and 4 respectively.

9.18.3. Reverted Gear Train

When the axes of the first and last wheels of a compound gear train are co-axial, then that gear train is known as reverted gear train.

9.19. SQUARE THREADED SCREW

The efficiency of the screw is given by, $\eta = \dfrac{\tan \alpha}{\tan(\alpha + \phi)}$

where α = Helix angle of the screw = $\tan^{-1}\left(\dfrac{\text{Lead of screw}}{\pi \times \text{Diameter of screw}}\right)$, and

ϕ = Angle of friction.

For a single start threads, lead and pitch are equal and for n start threads,

$$\text{Lead} = n \times \text{pitch}.$$

The efficiency of the screw is independent of the load. The efficiency will be maximum when

$$\alpha = \frac{\pi}{4} - \frac{\phi}{2}$$

Maximum efficiency is given by, $\eta_{max} = \dfrac{1 - \sin \phi}{1 + \sin \phi}$.

The screw is *self-locking* if its efficiency is less than 50%. Self-locking of a machine is a desired property.

9.19.1. Friction in V-threads

The equivalent coefficient of friction (μ_1) for V-threads is given by

$$\mu_1 = \frac{\mu}{\cos \beta}$$

where β = Semi angle of the V-threads

μ = Actual co-efficient of friction.

The equivalent coefficient to friction is always greater than the actual coefficiency of friction.

9.19.2. Friction Circle

When a journal is rotating in its bearing the resultant of the normal force (R_N) and frictional force (μR_N) does not pass through the centre of the journal, but it is tangent to a small circle of radius = $R \times \sin \phi$ in which R is the radius of the journal and ϕ is the angle of friction. This circle of radius $R \times \sin \phi$ is called friction circle. As ϕ is very small angle, then we have $\sin \phi \simeq \tan \phi \simeq \mu$. Hence radius of friction circle = $R \times \mu$.

9.19.3. Frictional Horse Power Lost at Pivot and Collar Bearings

For calculating the friction horse power lost at the pivot and collar bearings, the two assumptions made in practice are:

(*i*) the intensity of pressure is uniform, and

(*ii*) the rate of wearing at the surface is uniform.

9.19.4. Flat Pivot Bearing

The frictional torque (T) transmitted in case of flat pivot bearing for uniform pressure and uniform wear is given by

$$T = \frac{2}{3}\mu WR \quad \text{For uniform pressure}$$

$$= \frac{1}{2}\mu WR \quad \text{For uniform wear}$$

where W = Total axial load carried by the bearing,

R = Radius of bearing surface, and

μ = Co-efficient of friction.

9.19.5. Flat Collar Bearing

The frictional torque transmitted in case of flat collar bearing is given by

$$T = \frac{2}{3}\mu W\left[\frac{R_1^3 - R_2^3}{R_1^2 - R_2^2}\right] \quad \text{For uniform pressure}$$

$$= \frac{\mu W}{2}(R_1 + R_2) \quad \text{For uniform wear}$$

where R_1 = External radius of the collar, and

R_2 = Internal radius of the collar.

9.19.6. Conical Pivot Bearing

The frictional torque transmitted in case of conical pivot bearing having semi angle of the cone as θ; is given by

$$T = \frac{2}{3}\frac{\mu WR}{\sin\theta} \quad \text{For uniform pressure}$$

$$= \frac{\mu WR}{2\sin\theta} \quad \text{For uniform wear}$$

where R = Radius of the shaft.

9.19.7. Conical Collar Bearing

The frictional torque transmitted is given by

$$T = \frac{2}{3}\frac{\mu W}{\sin\theta}\left[\frac{R_1^3 - R_2^3}{R_1^2 - R_2^2}\right] \quad \text{For uniform pressure}$$

$$= \frac{\mu W}{2\sin\theta}[R_1 + R_2] \quad \text{For uniform wear}$$

For the safe design of a bearing, uniform pressure is assumed if horse power lost in friction is to be determined and uniform wear is assumed if horse power transmitted is to be determined.

9.20. BELTS AND ROPES

The power can be transmitted from one shaft to another shaft by means of belts and ropes. The shafts are fitted with pulleys. The belts or ropes pass over the pulleys and power is transmitted due to friction between pulleys and belts (or ropes).

9.20.1. Velocity Ratio

The ratio of the speed of the driven pulley to the speed of driving pulley is called velocity ratio.

If N_1 = Speed of the driving pulley in r.p.m.

N_2 = Speed of the driven pulley in r.p.m.

d_1 = Diameter of the driving pulley and

d_2 = Diameter of the driven pulley.

Then velocity ratio is given by $\frac{N_2}{N_1} = \frac{d_1}{d_2}$

This shows that the speed of the pulleys are inversely proportional to their diameters.

If t = Thickness of the belt then velocity ratio becomes as $\frac{N_2}{N_1} = \frac{d_1 + t}{d_2 + t}$.

9.20.2. Open and Crossed Belts

Two shafts may be connected by an open belt or a crossed belt as shown in Figs. 9.8 and 9.9 respectively. When the shafts are connected by an open belt, the two shafts rotate in the same direction but when the shafts are connected by a crossed belt, they rotate in the opposite direction. The length of open belt (L) connecting two pulleys of radii r_1 and r_2 and at a centre distance D apart, is given by

$$L = \pi(r_1 + r_2) + \frac{(r_1 - r_2)^2}{D} + 2D$$

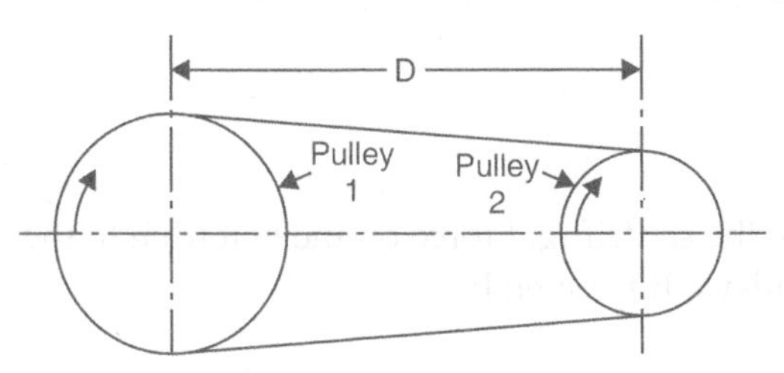

FIGURE 9.8. Open belt.

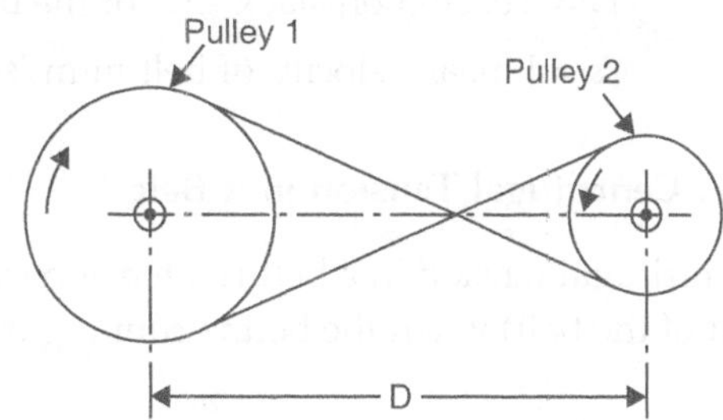

FIGURE 9.9. Crossed belt.

The length of the belt, when the shafts are connected by means of a crossed belt is given by

$$L = \pi(r_1 + r_2) + \frac{(r_1 + r_2)^2}{D} + 2D$$

where r_1 = Radius of pulleys No. 1, r_2 = Radius of pulley No. 2, and D = Centre distance between the two shafts.

The length of an open belt depends upon the sum and difference of the radii whereas the length of the crossed belt depends only on the sum of the radii of the two pulleys.

9.20.3. Slip

Slip is defined as the difference in speeds of the rim of the pulley and the belt on the pulley.

9.20.4. Angle of Contact (θ)

The minimum angle of lap of contact, *i.e.*, the angle of contact on the smaller pulley is termed as angle of contact.

9.20.5. Ratio of Belt Tensions

The ratio of belt tensions is given by

$$\frac{T_1}{T_2} = e^{\mu\theta}$$

where T_1 = Tension on tight side of the belt,
T_2 = Tension on slack side of the belt,
μ = Co-efficient of friction between belt and pulley rim, and
θ = Angle of contact in radians.

9.20.6. Power Transmitted by Belts

The horse power transmitted by a belt is given by

$$\text{H.P.} = \frac{(T_1 - T_2)v}{75}$$

where T_1 = Tension on tight side of the belt in kgf,
T_2 = Tension on slack side of the belt in kgf, and
v = Linear velocity of belt in m/sec.

9.20.7. Centrifugal Tension in a Belt

The centrifugal tension in a belt is a tension caused by the centrifugal force on the belt (due to the weight of the belt) when the belt is running on the pulleys. It is given by

$$T_c = \frac{w}{g} \times v^2$$

where w = Weight of belt per metre length, and v = Velocity of belt in m/sec.

The centrifugal tension is independent of the tight and slack side tensions of the belt. It depends upon the velocity of the belt and weight per unit length of the belt.

9.20.8. Effect of Centrifugal Tension on the Power Transmitted

The power transmitted by a belt, when centrifugal tension is to be considered, is given by

$$\text{H.P.} = (T_1 - T_s)\left(1 - \frac{1}{e^{\mu\theta}}\right)\frac{v}{75} = \left(T_1 - \frac{w}{g}V^2\right)\left(1 - \frac{1}{e^{\mu\theta}}\right) \times \frac{v}{75}.$$

The above power transmitted is zero, when v is zero. Also horse power will be zero when

$$T_1 = \frac{w}{g}V^2 \text{ or } v = \left(\frac{T_1 \times g}{w}\right)^{1/2}.$$

The horse power increases with the increase of speed upto a certain point, after that the horse power decreases and becomes zero. The velocity of the belt at which maximum horse power is transmitted is given by

$$v = \sqrt{\frac{T_1 \times g}{3w}}.$$

When maximum horse power is transmitted, the centrifugal tension in the belt is equal to one-third of the maximum tension allowed in the belt.

9.20.9. Brakes and Dynamometer

A brake is a device which uses frictional resistance either to regulate the speed of a machine or to stop the machine, whereas a dynamometer is a device used for measuring the effort or torque exerted by or on a machine.

9.21. CAMS

A cam is defined as a mechanical member which is used to transmit a desired motion to a follower by direct contact. The cam is the driving member and the follower is the driven member.

The followers are *classified* according to their shape, movement and location of their line of contact. Common shapes of the followers used are :

1. Knife-edge follower [Fig. 9.10 (*a*)];
2. Roller follower [Fig. 9.10 (*b*)]; and
3. Flat faced follower or mushroom follower [Fig. 9.10 (*c*)].

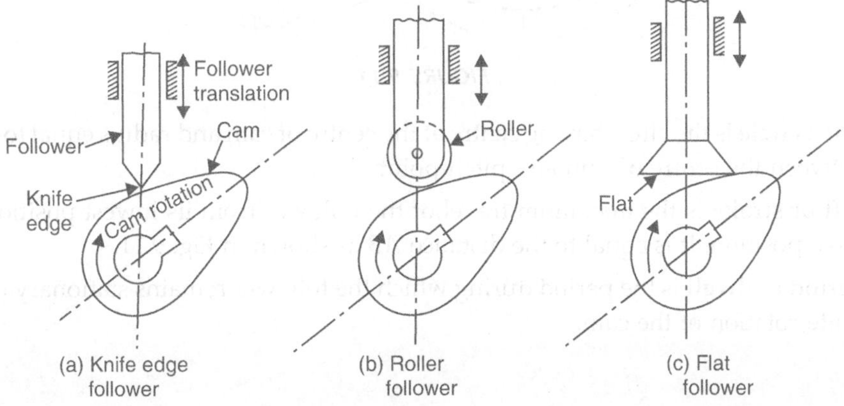

FIGURE 9.10. Type of followers.

The knife-edge follower is not a practical one. It is used only for theoretical analysis.

Definitions. Figure 9.11 shows the cam profile along with certain terms which are defined as:

1. **Cam profile.** The surface in contact with the follower is known as cam profile. This is the actual working curve of the cam.
2. **Trace point.** It is the reference point on the follower for the purpose of tracing the cam profile. It is situated at the knife-edge and knife-edge follower and at the centre in a roller follower.
3. **Base circle** is the smallest circle drawn from the cam centre to the cam profile. The size of the cam circle decides the cam size.
4. **Prime circle** is the smallest circle drawn from the cam centre to the pitch curve.
5. **The pitch curve** is the curve traced by the trace point if it is assumed that follower is rotating round the cam (instead of the cam rotation).
6. **The pitch point** is the point on the pitch curve of the cam having maximum pressure angle.
7. **Cam angle** is the angle of rotation of the cam for a definite displacement of the follower.
8. **The pressure angle** is the angle between the direction of motion of follower and a normal to the pitch curve.

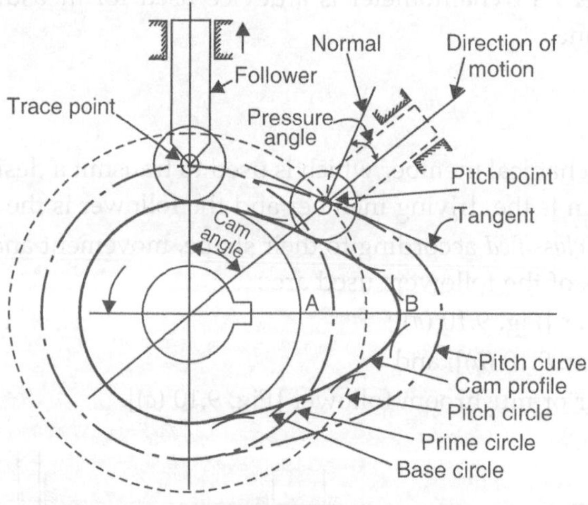

FIGURE 9.11

9. **Pitch circle** is the circle having centre at the centre of cam and radius equal to the distance between the centre of cam and pitch-point.
10. **Lift or stroke** is the maximum travel of the follower from its lowest position to the top most position. It is equal to the distance *AB* as shown in Fig. 9.11.
11. **Period of dwell** is the period during which the follower remains stationary during some finite rotation of the cam.

9.21.1. Follower Motions

A follower may have the following types of motions:

1. Simple harmonic motion,
2. Uniform velocity,
3. Uniform acceleration and retardation,
4. Cycloidal motion, and
5. Any type of desired motion.

Governors. A governor is a device which maintains the mean speed of a machine constant over long period during which the load on engine varies. A governor is different than a flywheel, which reduces the fluctuation of speed during a cycle for constant output load.

9.21.2. Types of Governors

Governors are of the following two types:

1. Fly-ball governors
2. Shaft governors.

Fly-ball governors are gravity loaded or spring loaded governors. Fly-ball gravity loaded governors are : Watt governor, Porter governor and Proell governor. Spring loaded fly-ball governors are : Hartnell, Hartung, Wilson and Pickering governors. Shaft governors are of two types: centrifugal governors and inertia governors.

Watt governor. The height (h) of a Watt's governor is given by

$$h = \frac{g}{\omega^2}$$

where ω = Angular speed of the governor.

The height of the governor is independent of the weight of the balls and length of the supporting arms of the governor. It varies inversely as the square of the speed.

9.21.3. Porter Governor

The height (h) of a Porter governor, when the upper and lower arms of the governor are equal and are pivoted either on the governor axis or at equal distance from the axis, is given by

$$h = \left(\frac{w + W}{w}\right) \times \frac{g}{\omega^2}$$

where w = Weight of the balls of governor, and

W = Weight of the sleeve.

If friction at the sleeve is taken into account, then height of Porter governor is given by

$$h = \left(\frac{w + W + F}{w}\right) \times \frac{g}{\omega^2}$$

where F = Frictional force at the sleeve +ve sign is used when speed is increasing and –ve sign is used when speed is decreasing.

9.21.4. Hartnell Governor (Spring Loaded Governor)

The spring force (S) exerted on sleeve if the effect of the pull of gravity on governor balls and arms is neglected is given by

$$S = 2F\frac{a}{b}$$

where F = Centrifugal force on the ball of a Hartnell governor,

a = Length of vertical arm of bell crank lever, and

b = Length of horizontal arm of bell crank lever.

The *lift* (h) of the sleeve in the Hartnell governor is given by

$$h = \frac{b}{a}(r_1 - r_2)$$

where r_1 = Maximum radius of rotation, and

r_2 = Minimum radius of rotation.

The *stiffness* (k) of the spring in the Hartnell governor is given by

$$k = 2\left(\frac{a}{b}\right)^2\left(\frac{F_1 - F_2}{r_1 - r_2}\right) \text{ or } \left(\frac{S_1 - S_2}{h}\right)$$

where F_1 = Centrifugal force at the maximum radius of rotation, and

F_2 = Centrifugal force at the minimum radius of rotation.

S_1 = Spring force exerted on the sleeve at maximum radius of rotation, and

S_2 = Spring force exerted on the sleeve at minimum radius of rotation.

9.21.5. Important Terms Used in Governors

(*i*) Sensitiveness of a governor. A governor is said to be sensitive when it readily responds to a very small change in speed. A governor generally operates between maximum speed and minimum speed. Maximum speed occurs when the sleeve touches the top stop whereas minimum speed occurs, when the sleeve touches the bottom stop. The difference between the maximum and minimum speed is known as *range of speed.* Sometimes the sensitiveness is defined as the ratio of the range of speed to the mean speed.

(*ii*) Effort and power of a governor. The mean force exerted by governor on the sleeve to raise it or lower it for a given change in speed is known as effort of governor. The change in speed is generally taken as *one per cent.* The power of a governor is the product of governor effort and the displacement of the sleeve. The power of a governor is also defined as the work done at the sleeve for a given percentage change in speed. The effort and power of a Porter governor (when the arms of the governor are equal and have equal inclination with the axis of the governor spindle) are given by

$$\text{Effort} = c(W + H), \text{ and Power} = \left(\frac{4c^2}{1+2c}\right)(w + W)\times h$$

where c = Percentage increase in speed, w = Weight of ball of governor,

W = Weight on sleeve, and h = Life of the governor.

The effort of a Hartnell governor (when the moments due to weight of the arms and ball are neglected) is given by

$$\text{Effort} = c \times S$$

where S = Spring force exerted on the sleeve.

(*iii*) Isochronism. When the equilibrium speed of a governor is constant for all radii of rotation of the balls within the working range, the governor is said to be isochronous.

(*iv*) Hunting. When a governor is oversensitive, the sleeve will oscillate between two extreme positions when there is a slightest change in speed. This phenomenon is known as hunting.

9.21.6. Controlling Force of a Governor

It is the radially inward force acting on each rotating ball. This is equal and opposite to the outward centrifugal force acting on the balls. The graph of variation of controlling force with the radius of rotation is known as *controlling force curve.*

9.22. BALANCING

Balancing is required to avoid the vibrations produced by the mass of the rotating machine parts on the bearings. If the centre of gravity of the rotating machine parts does not lie on the axis of rotation, but at a distance away from it, there is a centrifugal force which acts on the bearing in a constantly changing directions and result in a vibrating load. This unbalanced centrifugal forces is given by

$$F = \frac{W}{g}\omega^2 r$$

where r = The distance of centre of gravity from the axis of rotation,

W = Weight of rotating component, and

ω = Angular velocity of the rotating component.

The unbalanced centrifugal forces may be due to rotating masses or due to reciprocating masses.

9.22.1. Balancing or Rotating Masses

The rotating masses may be in (*i*) static balance or (*ii*) dynamic balance. If the rotating masses are in equilibrium among themselves when not running, the masses are said to in static balance. If the inertia forces and couples exerted by rotating masses are in equilibrium among themselves, the masses are said to be in dynamic balance.

9.22.2. Balancing of a Single Rotating Mass

If the balance weight is attached to a shaft in such a way that it revolves in the same plane as the disturbing weight as shown in Fig. 9.12, then necessary condition for balancing is that the centrifugal force due to rotating disturbing weight must be equal to the centrifugal force due to rotating balance weight.

If W = Disturbing weight,

B = Balance weight,

r = Distance of C.G. of the weight W from the axis of rotation,

h = Distance of C.G. of the weight B from the axis of rotation, and

ω = Angular speed of the shaft.

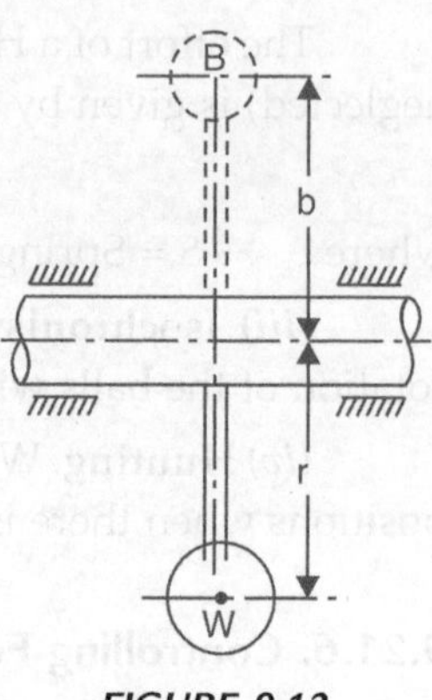

FIGURE 9.12

Then for balance $\frac{W}{g}\omega^2 r = \frac{B}{g}\omega^2 b$ or $W \times r = B \times b$.

Generally, the balance radius b is made as large as possible so as to keep down the magnitude of the balance weight B. The condition for balance is independent of speed.

9.22.3. Balancing of Several Masses Rotating in the Same Plane

Let a number of masses of weight W_1, W_2, W_3 etc. are rigidly attached to a shaft at a distance r_1, r_2, r_3 etc. from the axis of rotation respectively as shown in Fig. 9.13(*a*). They revolve in the same plane. The centrifugal force acting radially outwards on each weight is proportional to the product of weight and radius of rotation (*i.e.*, $W \times r$). The product of balance weight and its radius of rotation is determined either by graphical method or by analytical method.

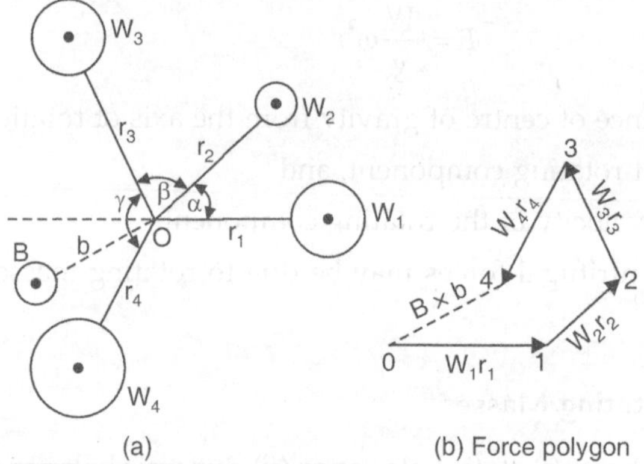

FIGURE 9.13

9.22.4. Graphical Method

Draw 01 parallel to the direction of W_1 and its magnitude equal to $W_1 r_1$ as shown in Fig. 9.13(*b*). From 1, draw 12 parallel to the direction of W_2 and its magnitude equal to $W_2 r_2$. Similarly, draw 23 parallel to the direction W_3 and 34 parallel to the direction of W_4. Take 23 equal to $W_3 r_3$ and 34 equal to $W_4 r_4$. Then 40 represents the direction and magnitude of $B \times b$, *i.e.*, product of balance weight and its radius of rotation.

Analytical method. The centrifugal forces acting on the rotating masses are proportional to the product of W and r. Resolve each force horizontally and vertically. Then resultant, horizontal and vertical components are

$$F_H = W_1r_4 + W_2r_2 \cos \alpha + W_3r_3 \cos (\alpha + \beta) + W_4r_4 \cos (\alpha + \beta + \gamma)$$

and $$F_V = W_2r_2 \sin \alpha + W_3r_3 \sin (\alpha + \beta) + W_4r_4 \sin (\alpha + \beta + \gamma)$$

Then $$B \times b = \sqrt{F_H^2 + F_V^2}$$

and the direction of the balance weight is given by $\tan \theta = \dfrac{F_V}{F_H}$.

9.22.5. Balancing of Several Masses Rotating in Different Planes

Let there are a number of weights W_1, W_2, W_3 at radii r_1, r_2, r_3 rotating in different planes as shown in Fig. 9.14. The weights W_1, W_2 and W_3 are at a distance of l_1, l_2 and l_3 respectively from the reference plane. Then for balancing:

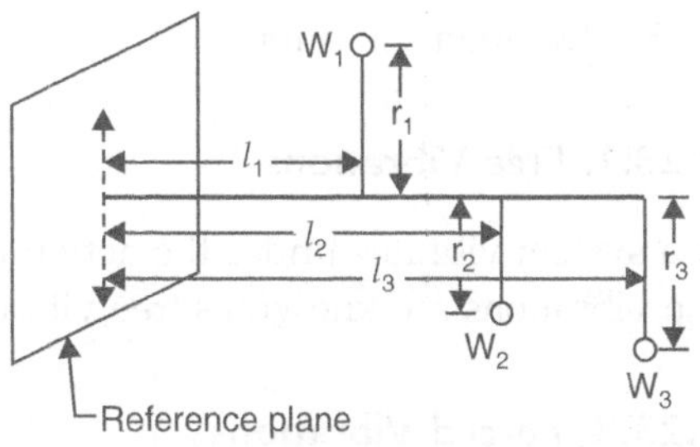

FIGURE 9.14

(*i*) A force polygon is drawn whose sides are proportional to W_1r_1, W_2r_2 and W_3r_3. The closing side of the polygon will give the product of balance weight and its radius (*i.e.*, $B \times b$).

(*ii*) A couple polygon is drawn whose sides are proportional to $W_1r_1l_1$, $W_2r_2l_2$ and $W_3r_3l_3$. The closing side of the couple polygon will represent $B \times b \times l$ where l is the distance of the balance weight from the reference plane.

Balancing of reciprocating masses. The force (F) required to accelerate the reciprocating mass of weight W of reciprocating engine is given by

$$F = \frac{W}{g}\omega^2 r\left(\cos\theta + \frac{\cos 2\theta}{n}\right)$$

where θ = Inclination of crank with inner dead centre

$n = \dfrac{l}{r}$ = Ratio of connecting rod length of crank radius

ω = Angular speed of the rotation of crank.

The force applied to the frame of the engine is equal and opposite to the force required to accelerate the reciprocating mass.

$\therefore$ Unbalanced reciprocating force $= F = \dfrac{W}{g}\omega^2 r\cos\theta + \dfrac{W}{g}\omega^2 r\dfrac{\cos 2\theta}{n} = F_P + F_S$

where F_P = Primary unbalanced force, and

F_S = Secondary unbalanced force.

Hence for complete balancing the reciprocating masses, we must balance:

1. Primary forces, *i.e.*, $W \times r = 0$,
2. Primary couples, *i.e.*, $W \times r \times l = 0$,
3. Secondary forces, and
4. Secondary couples.

9.23. VIBRATIONS

When a body is displaced from its equilibrium position by an external force and released, the body starts vibrating. The external forces may be due to the dynamically unbalanced masses in the rotating machines. Vibrations are of the following types:

1. Free vibration,
2. Forced vibration, and
3. Damped vibration.

9.23.1. Free Vibrations

If a system vibrates under the action of forces inherent in the system without any external force, the vibrations are known as free vibrations.

9.23.2. Forced Vibrations

If a system vibrates under the influence of external forces, the vibrations are known as forced vibrations. The external force applied to the system is periodic in nature.

9.23.3. Damped Vibrations

The energy possessed by a vibrating system is gradually dissipated in overcoming friction and other resistances. Thus the amplitude of the vibrations will go on decreasing and finally the body will come to rest. These types of vibrations are called damped vibrations.

9.23.4. Natural Frequencies

They are the frequencies of a system which is having free vibrations. For the free longitudinal vibration, the natural frequency

$$= \frac{1}{2\pi}\sqrt{\frac{g}{\delta}}$$

where δ = Static deflection.

9.23.5. Resonance

It is a phenomenon in which the frequency of the external exciting force coincides with the natural frequency of the system.

9.23.6. Equation of Motions of a Vibration System

The following are the equations of motion of a vibration system:

$$\frac{d^2x}{dt^2} + \frac{k}{m}x = 0 \qquad \text{Free from vibrations}$$

$$\frac{d^2x}{dt^2} + \frac{c}{m}\frac{dx}{dt} + \frac{k}{m}x = 0 \qquad \text{For free vibration with viscous damping}$$

$$\frac{d^2x}{dt^2} + \frac{c}{m}\frac{dx}{dt} + \frac{k}{m}x = F \sin \omega t \qquad \text{Forced vibration with viscous damping}$$

where c = Co-efficient of viscous damping,

m = Mass of the vibrating body, and

k = Stiffness of the vibrating body,

$F \sin \omega t$ = Periodic external force acting on the body.

9.23.7. Critical Speed of a Shaft

It is the speed of the shaft, at which the rotating shaft tends to vibrate violently in transverse direction. Critical speed is also called whirling speed or whipping speed. For a shaft, carrying a weight (W) at the centre, the critical speed

$$= \sqrt{\frac{k \times g}{W}}$$

where k = Stiffness of the shaft.

II. OBJECTIVE TYPE QUESTIONS

Tick mark the most appropriate statement of the multiple choice answers:

Basic Definitions

1. Theory of machines is the branch of science, which deals with the study of
   - (*a*) the relative motion between the parts of a machine and the study of forces which acts on those parts ☐
   - (*b*) the relative motion between the parts of a machine ☐
   - (*c*) the forces acting on the parts of the machine ☐
   - (*d*) none of the above. ☐
2. Link or element is a
   - (*a*) part of a machine ☐
   - (*b*) stationary part of a machine ☐
   - (*c*) art of a machine which has motion relative to some other part ☐
   - (*d*) all of the above. ☐

3. When the relative motion between two elements is completely or successfully constrained, then these two elements from a
   (*a*) mechanism ☐ (*b*) machine ☐
   (*c*) kinematic pair ☐ (*d*) kinematic chain. ☐
4. Kinematic pairs are classified as
   (*a*) sliding pair ☐ (*b*) rolling pair ☐
   (*c*) turning pair ☐ (*d*) spherical pair ☐
   (*e*) all of the above ☐ (*f*) none of the above. ☐
5. Piston and cylinder of a reciprocating steam engine forms a
   (*a*) turning pair ☐ (*b*) rolling pair ☐
   (*c*) sliding pair ☐ (*d*) spherical pair. ☐
   (*e*) none of the above. ☐
6. A ball and a socket joint form a
   (*a*) turning pair ☐ (*b*) rolling pair ☐
   (*c*) sliding pair ☐ (*d*) spherical pair. ☐
7. Shaft with collars at both ends fitted into a circular hole forms a
   (*a*) turning pair ☐ (*b*) rolling pair ☐
   (*c*) sliding pair ☐ (*d*) spherical pair. ☐
8. Ball-bearing and roller bearing form a
   (*a*) turning pair ☐ (*b*) rolling pair ☐
   (*c*) sliding pair ☐ (*d*) spherical pair. ☐
9. A bolt and nut form a
   (*a*) turning pair ☐ (*b*) rolling pair ☐
   (*c*) screw pair ☐ (*d*) spherical pair. ☐
10. When the two elements of a pair have a surface contact when in motion and the surface of one element slides over the surface of the other element, the pair formed is called a
    (*a*) higher pair ☐ (*b*) lower pair ☐
    (*c*) force-closed pair ☐ (*d*) none of the above. ☐
11. When the two elements of a pair have line or point contact when in motion, the pair formed is called a
    (*a*) higher pair ☐ (*b*) lower pair ☐
    (*c*) unclosed pair ☐ (*d*) none of the above. ☐
12. The lower pair is
    (*a*) close pair ☐ (*b*) unclosed pair ☐
    (*c*) point contact pair ☐ (*d*) surface contact pair ☐
    (*e*) both (*a*) and (*b*) ☐ (*f*) both (*a*) and (*c*). ☐
13. The higher pair is a
    (*a*) closed pair ☐ (*b*) unclosed pair ☐
    (*c*) point contact pair ☐ (*d*) surface contact pair ☐
    (*e*) both (*a*) and (*b*) ☐ (*f*) both (*a*) and (*c*). ☐

14. Which one of the following is a lower pair?

(*a*) ball and roller bearing ☐ (*b*) automobiles steering gear ☐
(*c*) cam and follower ☐ (*d*) belt and chain drives. ☐

15. Which of the following is a higher pair?

(*a*) ball and roller bearing ☐ (*b*) automobile steering gear ☐
(*c*) cam and follower ☐ (*d*) belt and chain drives ☐
(*e*) all of the above except ☐ (*f*) none of the above. ☐

16. Choose the correct statement

(*a*) A sliding pair has incompletely constrained motion. ☐
(*b*) A pair of friction discs constitutes a lower pair. ☐
(*c*) Rectilinear motion of a piston is converted into rotary motion by slider crank. ☐
(*d*) Automobile steering gear is an example of higher pair. ☐

17. Choose the wrong statement

(*a*) An element needs not be a rigid body, but it must be a resistant body. ☐
(*b*) Cam and follower is an example of lower pair. ☐
(*c*) A sliding pair has a completely constrained motion. ☐
(*d*) The motion of a shaft in a circular hole is an example of incompletely constrained motion. ☐

18. The function of an element is to

(*a*) transmit motion ☐ (*b*) to serve as a support ☐
(*c*) to guide other elements ☐ (*d*) all of the above ☐
(*e*) none of the above. ☐

19. The motion of a shaft between a foot-step bearing is an example of

(*a*) incompletely constrained motion ☐ (*b*) completely constrained motion ☐
(*c*) successfully constrained motion ☐ (*d*) none of the above. ☐

20. The motion of a shaft with collars at each end in a circular hole is an example of

(*a*) incompletely constrained motion ☐ (*b*) completely constrained motion ☐
(*c*) successfully constrained motion ☐ (*d*) none of the above. ☐

21. The motion of a shaft in a circular hole is an example of

(*a*) incompletely constrained motion ☐ (*b*) completely constrained motion ☐
(*c*) successfully constrained motion ☐ (*d*) none of the above. ☐

22. A kinetic chain is

(*a*) a combination of kinetic pair in which one link is fixed ☐
(*b*) a combination of kinetic pair in which each link forms part of two pairs and the relative motion between the links is completely constrained ☐
(*c*) the same as mechanism ☐
(*d*) also called inversion. ☐

23. In a kinetic chain, if the specification of one co-ordinate, or dimension or position of a single link is sufficient to define the position of all other links, then the chain is called a kinematic chain of

(*a*) two degree of freedom ☐ (*b*) one degree of freedom ☐
(*c*) three degrees of freedom ☐ (*d*) none of the above. ☐

24. The differential mechanism of an automobile is having

(*a*) one degree of freedom ☐ (*b*) two degrees of freedom ☐
(*c*) three degrees of freedom ☐ (*d*) zero degree of freedom. ☐

Kinematic Chain

25. If J = Number of binary joints in a chain, H = Number of unclosed pairs, and L = Number of links, then to determine whether a chain is a locked, constrained or constrained, the relation used is given by

(a) $J + H = \frac{3}{2}\,[L - 2]$ ☐ (b) $J + H = 3L - 2$ ☐

(c) $J + \frac{H}{2} = \frac{3}{2}\,L - 2$ ☐ (d) $\frac{J}{2} + H = \frac{3}{2}\,L - 2.$ ☐

26. If the right-hand side of the equation (which is used to determine whether a chain is locked, constrained or unconstrained) is greater than the left-hand side, then the chain is

(a) locked ☐ (b) constrained ☐

(c) unconstrained ☐ (d) none of the above. ☐

27. If R.H.S. = L.H.S. of the equation (which is used to determine whether a chain is locked, constrained or unconstrained), then the chain is

(a) locked ☐ (b) constrained ☐

(c) unconstrained ☐ (d) none of the above. ☐

28. If R.H.S. is less than L.H.S. of the equation (used for determining whether a chain is locked, constrained or unconstrained), then the chain is

(a) locked ☐ (b) constrained ☐

(c) unconstrained ☐ (d) none of the above. ☐

29. Ternary joint is a joint at which three links are joined at the same connection. It is equivalent to

(a) one binary joint ☐ (b) two binary joints ☐

(c) three binary joints ☐ (d) combination of (a) and (b). ☐

30. Quaternary joint is a joint at which four links are joined at the same connection and it is equivalent to

(a) one binary joint ☐ (b) two binary joints ☐

(c) three binary joints ☐ (d) combination of (a) and (b). ☐

31. If 'n' links are connected at the same joint, the joint is equivalent to

(a) $(n - 1)$ binary joints ☐ (b) $(2n - 1)$ binary joints ☐

(c) $(n - 2)$ binary joints ☐ (d) $(n - 3)$ binary joints. ☐

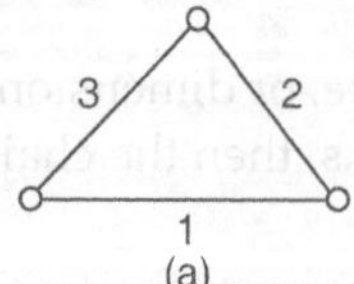

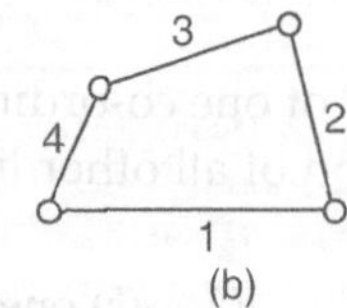

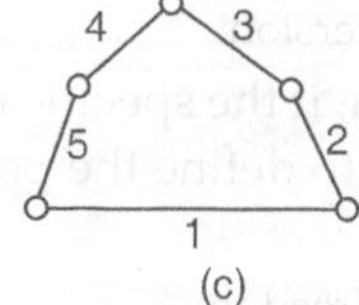

FIGURE 9.15

32. Figure 9.15 (a) is an example of a

(a) locked chain ☐ (b) unconstrained chain ☐

(c) constrained chain ☐ (d) none of the above. ☐

33. Figure 9.15(*b*) is an example of a

(*a*) locked chain ☐ (*b*) unconstrained chain ☐

(*c*) constrained chain ☐ (*d*) none of the above. ☐

34. Figure 9.15(*c*) is an example of a

(*a*) locked chain ☐ (*b*) unconstrained chain ☐

(*c*) constrained chain ☐ (*d*) none of the above. ☐

35. For a kinematic chain, the relationship between number of pairs (P) and number of links (L) is

(*a*) $L = P - 2$ ☐ (*b*) $L = 2P - 4$ ☐

(*c*) $L = 4P - 2$ ☐ (*d*) $L = P - 4$. ☐

36. For a kinematic chain, the relationship between number of joints (J) and number of links (L) is

(*a*) $L = \frac{2}{3}(J + 2)$ ☐ (*b*) $L = \frac{2}{3}(J + 3)$ ☐

(*c*) $L = \frac{3}{2}(J + 2)$ ☐ (*d*) $L = \frac{3}{2}(J + 3)$. ☐

37. If a kinematic chain has 'l' links, then the number of mechanism obtained are

(*a*) $l - 1$ ☐ (*b*) $l - 2$ ☐

(*c*) $l + 1$ ☐ (*d*) l. ☐

38. Choose the wrong statement

(*a*) A chain consisting of three links and three joints is known as locked chain. ☐

(*b*) A chain consisting of four links and four joints is known as kinematic chain. ☐

(*c*) Quorternary joint is equivalent to three binary joints. ☐

(*d*) Rectangular bar in a rectangular hole is the type of partially constrained motion. ☐

39. Choose the correct statement

(*a*) If a kinematic chain has 'l' links, then 'l' different mechanism are obtained by fixing each of the links in turn. ☐

(*b*) A machine serves to modify and transmit energy (or force and motion) while the structure modifies and transmits force only. ☐

(*c*) When one of the links of a kinematic chain is fixed, the chain is known as a mechanism. ☐

(*d*) All of the above ☐

(*e*) None of the above. ☐

40. In a kinematic chain with four lower pairs, if all the four lower pairs are turning pairs, the mechanism is classified into the chain known as

(*a*) crossed slider crank chain ☐ (*b*) four bar chain ☐

(*c*) slider crank chain ☐ (*d*) double slider crank chain. ☐

41. In a kinematic chain with four lower pairs, if one is sliding pair and three turning pairs, the mechanism is classified into the chain known as

(*a*) crossed slider crank chain ☐ (*b*) four bar chain ☐

(*c*) slider crank chain ☐ (*d*) double slider crank chain. ☐

42. In a kinematic chain with four lower pairs, if there are two sliding pairs and two turning pairs and two similar pairs are adjacent then the mechanism is classified into the chain known as

(a) crossed slider crank chain ☐ (b) four bar chain ☐
(c) slider crank chain ☐ (d) double slider crank chain. ☐

43. In a kinematic chain with four lower pairs, if there are two sliding pairs and two turning pairs and two similar pairs are not adjacent, then the mechanism is classified into the chain known as

(a) crossed slider crank chain ☐ (b) four bar chain ☐
(c) slider crank chain ☐ (d) double slider crank chain. ☐

Mechanism and Inversion

44. The Ackermann steering gear is the inversion of a

(a) slider crank chain ☐ (b) four bar chain ☐
(c) double slider crank chain ☐ (d) crossed slider crank chain. ☐

45. Whitworth quick return mechanism is an inversion of

(a) double slider crank chain ☐ (b) single slider crank chain ☐
(c) four bar chain ☐ (d) crossed slider crank chain. ☐

46. Oldham's coupling and elliptic trammels are the inversion of

(a) double slider crank chain ☐ (b) single slider crank chain ☐
(c) four bar chain ☐ (d) crossed slider crank chain. ☐

47. Rapson's slide is an inversion of

(a) double slider crank chain ☐ (b) single slider crank chain ☐
(c) four bar chain ☐ (d) crossed slider crank chain. ☐

48. Choose the correct statement

(a) The ordinary reciprocating steam engine is an inversion of double slider crank chain. ☐
(b) Scotch Yoke mechanism is an inversion of slider crank chain. ☐
(c) Quick return mechanism is an inversion of slider crank chain. ☐
(d) The number of inversions in a mechanism having number of links equal to l, will be $(l + 1)$. ☐

Dynamically Equivalent System

49. A rigid body can be replaced by two masses, connected rigidly together. The system of two masses will be dynamically equivalent to the rigid body if

(a) the mass of the two systems are the same ☐
(b) the centre of gravity of the two mass system coincides with the centre of gravity of the rigid body ☐
(c) the total mass moment of inertia of the two mass system about an axis through the centre of gravity is equal to that of rigid body about the same axis ☐
(d) all of the above ☐
(e) none of the above. ☐

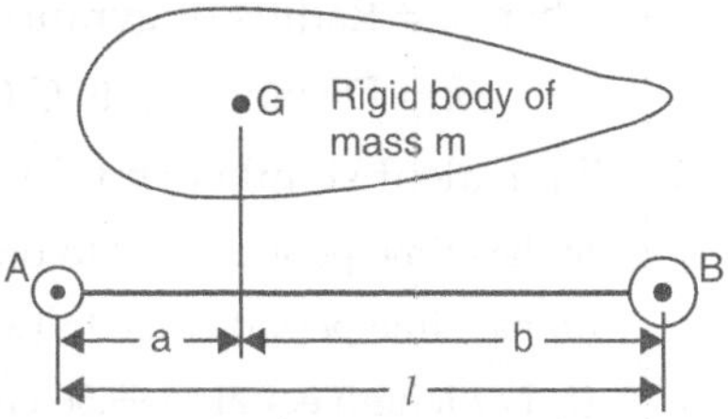

50. Figure 9.16 shows a rigid body of mass m having radius of gyration k about its centre of gravity. It is to be replaced by an equivalent dynamical system of two masses placed at A and B. The mass at A should be

(a) $\frac{a \times m}{a+b}$ ☐

(b) $\frac{b \times m}{a+b}$ ☐

(c) $\frac{m}{2}$ ☐

(d) $\frac{m}{3}$. ☐

FIGURE 9.16

51. In question 50, the mass at B should be

(a) $\frac{a \times m}{a+b}$ ☐ (b) $\frac{b \times m}{a+b}$ ☐

(c) $\frac{m}{2}$ ☐ (d) $\frac{m}{3}$. ☐

52. In question 50, the relation among k, a and b is given by

(a) $k = \sqrt{ab}$ ☐ (b) $k = \frac{a \times b}{2}$ ☐

(c) $k = \frac{a+b}{2}$ ☐ (d) $k = \frac{a}{b}$. ☐

where k = Radius of gyration of the rigid body about its C.G.

Simple Harmonic Motion

53. In simple harmonic motion, acceleration is proportional to

(a) displacement ☐ (b) velocity ☐

(c) square of displacement ☐ (d) square of velocity ☐

(e) none of the above. ☐

54. A body moving with simple harmonic motion has an amplitude of 30.0 cm. The maximum velocity of the body is 3 m/sec. The period of vibration will be equal to

(a) 0.314 sec ☐ (b) 0.628 sec ☐

(c) 1.59 sec ☐ (d) 3.18 sec. ☐

55. In question 54, the frequency of vibration will be equal to

(a) 2.0 vibrations/sec ☐ (b) 1.8 vibrations/sec ☐

(c) 1.59 vibrations/sec ☐ (d) 1.30 vibrations/sec. ☐

56. The time period of a compound pendulum is equal to

(a) $2\pi\sqrt{\frac{h}{g}}$ ☐ (b) $2\pi\sqrt{\frac{k^2+h^2}{h \times g}}$ ☐

(c) $2\pi\sqrt{\frac{k \times h}{g}}$ ☐ (d) $2\pi\sqrt{\frac{k^2 \times h^2}{g}}$. ☐

where k = Radius of gyration of the pendulum about an axis through its C.G.

h = Distance of its C.G. from the axis of suspension.

57. Equivalent simple pendulum means a simple pendulum which

(*a*) has time period equal to one second ☐

(*b*) has time period equal to two seconds ☐

(*c*) has length equal to 99.56 cm ☐

(*d*) has the same periodic time of oscillation as a given compound pendulum. ☐

58. The length of the equivalent simple pendulum is given by

(*a*) $\frac{k^2}{h} + h$ ☐ (*b*) $\frac{k^2}{h^2} + h$ ☐

(*c*) $\frac{k}{h^2} + h$ ☐ (*d*) $\frac{k^2}{h^2} + h^2$. ☐

where k = Radius of gyration of compound pendulum about its C.G.

h = Distance of the C.G. of the compound pendulum from the axis of suspension.

59. When the distance of suspension of a compound pendulum from its centre of gravity is equal to the radius of gyration about its centre of gravity, then the time period of compound pendulum is

(*a*) 1 sec ☐ (*b*) maximum ☐

(*c*) minimum ☐ (*d*) zero. ☐

60. The minimum period of a compound pendulum is equal to

(*a*) $2\pi\sqrt{\frac{h}{g}}$ ☐ (*b*) $2\pi\sqrt{\frac{k}{g}}$ ☐

(*c*) $2\pi\sqrt{\frac{2k}{g}}$ ☐ (*d*) $2\pi\sqrt{\frac{k}{2g}}$. ☐

where k = Radius of gyration of the compound pendulum about its C.G.

61. Choose the wrong statement

(*a*) The distance between the centre of suspension and centre of percussion is equal to the equivalent length of a simple pendulum. ☐

(*b*) The centre of suspension and centre of percussion are interchangeable. ☐

(*c*) The periodic time of a compound pendulum when suspended from the point of suspension is the same as when suspended from the centre of percussion. ☐

(*d*) The equivalent length of a simple pendulum is equal to the distance of its C.G. from the point of suspension. ☐

62. Bifilar suspension is an experimental method of determining

(*a*) periodic time of a body ☐ (*b*) moment of inertia of the body ☐

(*c*) centre of gravity of a body ☐ (*d*) centre of percussion of the body. ☐

63. Three springs having stiffness k_1, k_2, and k_3 are connected in series. The stiffness of a single spring equivalent to these three springs, would be equal to

(a) $k_1 + k_2 + k_3$ ☐ (b) $\frac{k_1 + k_2 + k_3}{3}$ ☐

(c) $\frac{k_1 \times k_2 \times k_3}{k_1k_2 + k_2k_3 + k_3k_1}$ ☐ (d) $\frac{k_1k_2 + k_2k_3 + k_3k_1}{k_1k_2k_3}$ ☐

64. In question 63, if the three springs are connected in parallel, then the stiffness of a single spring which is equivalent to these three springs will be equal to

(a) $k_1 + k_2 + k_3$ ☐ (b) $\frac{k_1 + k_2 + k_3}{3}$ ☐

(c) $\frac{k_1 \times k_2 \times k_3}{k_1k_2 + k_2k_3 + k_3k_1}$ ☐ (d) $\frac{k_1k_2 + k_2k_3 + k_3k_1}{k_1k_2k_3}$ ☐

65. Choose the wrong statement

(a) The instantaneous centre is the point, about which one body rotates relative to another body for the configuration being considered. ☐

(b) At the instantaneous centre the two bodies do not have any linear velocity relative to each other. ☐

(c) At the instantaneous centre the two bodies have the same linear velocity to the third body. ☐

(d) All of the above. ☐

Instantaneous Centre

66. A mechanism is having n links, then the number of instantaneous centres would be equal to

(a) n ☐ (b) $\frac{n(n-1)}{2}$ ☐

(c) $\frac{n}{2}$ ☐ (d) $(n-1)$. ☐

67. The total number of instantaneous centre of a mechanism having 8 links, is equal to

(a) 8 ☐ (b) 28 ☐

(c) 4 ☐ (d) 7. ☐

68. Kinnedy's theorem states that any three bodies moving relatively to each other

(a) having three instantaneous centres and these three centres lie on a curved line ☐

(b) have three instantaneous centres and these three centres lie on a straight line ☐

(c) have only one instantaneous centre ☐

(d) have two instantaneous centre and these two centres lie on a circle. ☐

69. Figure 9.17, show a four-bar chain mechanism in which a link 1 is fixed. The number of instantaneous centres would be

(a) 2 ☐ (b) 4 ☐

(c) 6 ☐ (d) 3. ☐

70. In Fig. 9.17, the instantaneous centres 41 and 21 are known as
(a) permanent instantaneous centres ☐
(b) fixed instantaneous centres ☐
(c) neither fixed nor permanent instantaneous centres ☐
(d) none of the above. ☐

FIGURE 9.17

71. In Fig. 9.17, the centres 32 and 34 are known as
(a) permanent instantaneous centres ☐
(b) fixed instantaneous centres ☐
(c) neither fixed nor permanent instantaneous centres ☐
(d) none of the above. ☐

72. In Fig. 9.17, the centres 31 and 24 are known as
(a) permanent instantaneous centres ☐
(b) fixed instantaneous centres ☐
(c) neither fixed nor permanent instantaneous centres ☐
(d) none of the above. ☐

73. Choose the wrong statement
(a) When two links are connected by a pin joint, their instantaneous centre will lie on the centre of the pin joint. ☐
(b) When a link slides over another fixed link which is having flat surface, then the instantaneous centre of the two links lies at infinity. ☐
(c) When a link slides over another fixed link which is having curved surface, then the instantaneous centre of the two links lies at the centre of curvature of the curved surface. ☐
(d) All of the above ☐
(e) None of the above. ☐

74. Choose the correct statement
(a) Bililar suspension-method is used to determine the moment of inertia of the body. ☐
(b) The instantaneous centres, which remain in the same place for all configuration of the mechanism, are called fixed instantaneous centres. ☐
(c) The instantaneous centres, which move when the mechanism moves but the joints are of permanent nature, are called permanent instantaneous centres. ☐
(d) All of the above ☐
(e) None of the above. ☐

Velocity and Acceleration

75. For a rigid link, the velocity of one end of the link relative to other end will be
(a) at 45° to the link ☐ (b) at right angles to the link ☐
(c) parallel to the link ☐ (d) none of the above. ☐

76. In the design of many machines, a knowledge of methods of determining acceleration is essential. The acceleration of any point in a mechanism is determined by

(a) instantaneous centre method ☐ (b) analytical method ☐
(c) acceleration diagram method ☐ (d) only (b) and (c) ☐
(e) none of the above. ☐

77. In a rigid link AB, the point B is moving with respect to A. Then the acceleration of B will be equal to

(a) acceleration of A × distance AB ☐
(b) (acceleration of A) ÷ distance AB ☐
(c) vector sum of acceleration of A and acceleration of B, relative to A ☐
(d) acceleration of A × square of distance AB. ☐

78. In question 77, if ω is the angular velocity of the link AB about A, then normal component of acceleration of B relative to A will be equal to

(a) V_{BA}^2/AB ☐ (b) $V_{BA} \times AB$ ☐
(c) $V_{BA}^2 \times AB$ ☐ (d) V_{BA}/AB ☐

where V_{BA} = Linear velocity of B relative to A.

79. In question 77, the angular acceleration of the link AB will be equal to

(a) (tangential acceleration) ÷ AB ☐ (b) (normal acceleration) ÷ AB ☐
(c) (total acceleration) ÷ AB ☐ (d) (tangential acceleration) × AB. ☐

80. The velocity of the piston of reciprocating engine is equal to

(a) $\omega^2 r\left(\cos\theta + \dfrac{\cos 2\theta}{n}\right)$ ☐ (b) $\omega^2 r\left(\sin\theta + \dfrac{\sin 2\theta}{n}\right)$ ☐

(c) $\omega r\left(\sin\theta + \dfrac{\sin 2\theta}{n}\right)$ ☐ (d) $\omega r\left(\cos\theta + \dfrac{\cos 2\theta}{n}\right)$ ☐

where ω = Angular velocity of crank, θ = Angle turned by crank from inner dead centre

r = Radius of crank, $n = \dfrac{l}{r}$ Ratio of length of connecting rod to crank radius.

81. The displacement of the piston of a reciprocating engine is equal to

(a) $r\,(1 + n - \cos\theta - \sqrt{n^2 - \sin^2\theta}\,)$ ☐ (b) $r\,(1 - \cos\theta - n)$ ☐

(c) $r\,(1 - \cos\theta - \sqrt{n^2 - \sin^2\theta}\,)$ ☐ (d) $r\,(1 + \sin\theta - n)$. ☐

82. The angular velocity of the connecting rod of a reciprocating engine is equal to

(a) $\dfrac{\omega\sin\theta}{\sqrt{n^2 + \sin^2\theta}}$ ☐ (b) $\dfrac{\omega\cos\theta}{\sqrt{n^2 - \sin^2\theta}}$ ☐

(c) $\dfrac{\omega\cos\theta}{\sqrt{n^2 - \cos^2\theta}}$ ☐ (d) $\dfrac{\omega\sin\theta}{\sqrt{n^2 - \sin^2\theta}}$ ☐

83. The angular acceleration of the connecting rod of a reciprocating engine is equal to

(a) $\dfrac{(n^2 - \sin^2\theta)}{\omega^2(1 - n^2)}$ ☐ (b) $\dfrac{-\omega^2(n^2 - 1)\sin^2\theta}{(n^2 - \sin^2\theta)^{3/2}}$ ☐

(c) $\dfrac{n^2 + \sin^2\theta}{\omega^2(1 + n^2)}$ ☐ (d) none of the above. ☐

84. Klein's construction is a graphical method of determining
(*a*) acceleration of various parts ☐ (*b*) velocity of various parts ☐
(*c*) displacement of various parts ☐ (*d*) all of the above ☐
(*e*) none of the above. ☐

85. Klein's construction is used when
(*a*) crank has uniform acceleration ☐ (*b*) crank has uniform angular velocity ☐
(*c*) crank has non-uniform acceleration ☐ (*d*) crank has non-uniform angular velocity. ☐

86. When a point is moving in a curved path, the tangential acceleration is equal to
(*a*) rate of change in speed ☐ (*b*) rate of change of square of speed ☐
(*c*) square of speed/radius of curvature ☐ (*d*) speed/radius of curvature. ☐

87. When a point is moving in a curved path, the normal acceleration is equal to
(*a*) rate of change in speed ☐ (*b*) rate of change of square of speed ☐
(*c*) square of speed/radius of curvature ☐ (*d*) speed/radius of curvature. ☐

88. Which one of the following statement is wrong?
(*a*) The normal component of acceleration of a particle moving in a straight path is zero. ☐
(*b*) The tangential component of acceleration of a particle moving with constant speed is zero. ☐
(*c*) The normal component of acceleration of a particle moving in a circular path with constant velocity (v) is v^2/r. ☐
(*d*) The tangential component of acceleration of a particle moving in a circular path with constant velocity (v) is v^2/r. ☐

89. If a point moves along a straight line which is rotating, then the normal component of acceleration is equal to
(*a*) $\frac{v^2}{r}$ ☐ (*b*) $\frac{dv}{dt} - r\omega^2$ ☐
(*c*) $\frac{dv}{dt}$ ☐ (*d*) $2v\omega + r\alpha$ ☐

where v = Velocity of the point along the straight line
r = Radius of the point, ω = Angular velocity of the line
α = Angular acceleration of the line.

90. In question 89, the tangential component of acceleration is equal to
(*a*) $\frac{v^2}{r}$ ☐ (*b*) $\frac{dv}{dt} - r\omega^2$ ☐
(*c*) $\frac{dv}{dt}$ ☐ (*d*) $2wv + r\alpha$. ☐

91. In question 89, the Coriolis component of acceleration is equal to
(*a*) $r\omega^2$ ☐ (*b*) $\frac{dv}{dt}$ ☐
(*c*) $2v\omega$ ☐ (*d*) $r\alpha$. ☐
(*e*) none of the above. ☐

92. The Coriolis component of acceleration exists only whenever a point

(*a*) moves along a circular path ☐

(*b*) moves in a straight line ☐

(*c*) moves along a straight line which has rotational motion ☐

(*d*) none of the above. ☐

93. The Coriolis law states that if a point moves along a path that has rotation, the absolute acceleration of the point is

(*a*) equal to the absolute acceleration of the point relative to coincident point in the path ☐

(*b*) equal to the absolute acceleration of the coincident point in the path ☐

(*c*) equal to the vector sum of (*a*), (*b*) and a third component known as Coriolis' component acceleration ☐

(*d*) zero. ☐

94. Which of the following statement concerning the Coriolis component is correct ?

(*a*) It is equal to twice the product of the instantaneous velocity of sliding of the point on the link and instantaneous angular velocity of the link on which the point slides. ☐

(*b*) It acts perpendicular to the sliding surface. ☐

(*c*) The direction of the Coriolis component of acceleration is the direction of the relative velocity vector when it has been rotated 90° in the direction of the angular velocity of the link. ☐

(*d*) All of the above ☐

(*e*) None of the above. ☐

95. In which of the following mechanism, the Coriolis component of acceleration exists ?

(*a*) Shaper mechanism ☐ (*b*) Whitworth Quick Return mechanism ☐

(*c*) Tangent cam mechanism ☐ (*d*) All of the above ☐

(*e*) None of the above. ☐

96. A disc is rotating about an axis with an angular velocity ω (also known as angular velocity of spin) and the axis of spin is changing its position, then the tangential component of acceleration is equal to

(*a*) $\omega \dfrac{d\theta}{dt}$ ☐ (*b*) $\dfrac{dw}{dt}$ ☐

(*c*) $r\,\omega^2$ ☐ (*d*) $r\dfrac{dw}{dt}$ ☐

where $\dfrac{d\theta}{dt} = \Omega$ *i.e.*, the rate at which the axis of rotation changes its position

r = Radius.

97. In question 96, the normal component of acceleration is equal to

(*a*) $\omega \dfrac{d\theta}{dt}$ ☐ (*b*) $\dfrac{dw}{dt}$ ☐

(*c*) $r\,\omega^2$ ☐ (*d*) $r\dfrac{dw}{dt}$. ☐

98. A disc is rotating about an axis with an angular velocity and the axis of rotation is changing its position. The change in direction of the plane of rotation of the disc is called

(*a*) rotational motion ☐ (*b*) precessional motion ☐
(*c*) rectilinear motion ☐ (*d*) none of the above. ☐

99. Gyroscopic acceleration is the

(*a*) tangential component of the acceleration of a disc rotating with an angular velocity about an axis and the axis of rotation is changing its position. ☐

(*b*) normal component of the acceleration of a disc rotating with an angular velocity about an axis and axis of rotation is changing its position. ☐

(*c*) tangential component of the acceleration of a point moving in a straight line which has rotational motion. ☐

(*d*) normal component of the acceleration of a point moving in a straight line which has rotational motion. ☐

100. The rate at which the axis of rotation (about which a disc is rotating) changes its position is known as

(*a*) angular velocity ☐ (*b*) angular velocity of precession ☐
(*c*) linear velocity ☐ (*d*) sliding velocity. ☐

101. The gyroscopic acceleration depends upon

(*a*) instantaneous value of ω ☐
(*b*) the rate at which the axis of spin changes its speed ☐
(*c*) both (*a*) and (*b*) ☐
(*d*) none of the above. ☐

102. The slotted lever of a quick return mechanism at a certain instant has an angular velocity of 5 rad/sec clockwise and angular acceleration of 2 rad/sec^2 clockwise. The slider on the lever at this instant is 0.5 m from the axis of oscillation of the lever and it moves away from the axis of rotation at a velocity of 1.5 m/sec and its acceleration is 0.5 m/sec^2 outwards. Then the Coriolis component of acceleration of the slider will be

(*a*) zero ☐ (*b*) 7.5 m/sec^2 ☐
(*c*) 10 m/sec^2 ☐ (*d*) 15 m/sec^2. ☐

103. In question 102, the tangential component of acceleration of slider will be

(*a*) 16 m/sec^2 ☐ (*b*) 11.0 m/sec^2 ☐
(*c*) 8.5 m/sec^2 ☐ (*d*) zero. ☐

104. In question 102, the normal component of acceleration of slider will be

(*a*) 12.5 m/sec^2 ☐ (*b*) 13 m/sec^2 ☐
(*c*) – 12.0 m/sec^2 ☐ (*d*) – 10 m/sec^2. ☐

105. In question 102, the total acceleration of the slider will be

(*a*) 10 m/sec^2 ☐ (*b*) 20 m/sec^2 ☐
(*c*) 25 m/sec^2 ☐ (*d*) 30 m/sec^2. ☐

106. The flywheel of a motor car rotates about a horizontal axis at 2100 r.p.m., while the car moves in a horizontal circle of radius 40 m at a speed of 72 km/hour. Then the angular velocity of precession will be

(*a*) 1 rad/sec ☐ (*b*) 0.5 rad/sec ☐
(*c*) 1.5 rad/sec ☐ (*d*) 2.0 rad/sec. ☐

107. In the equation 106, the gyroscopic acceleration will be

(*a*) 100 rad/sec^2 ☐ (*b*) 105 rad/sec^2 ☐
(*c*) 110 rad/sec^2 ☐ (*d*) 120 rad/sec^2. ☐

108. The total kinetic energy of a body whose centre of gravity is moving with a linear velocity 'v' and body is also rotating about its centre of gravity with angular velocity 'ω', is equal to

(*a*) $\frac{v^2}{2g} + \frac{1}{2} I\omega^2$ ☐ (*b*) $\frac{v^2}{2g} + I\omega^2$ ☐
(*c*) $\frac{1}{2}\frac{W}{g} v^2 + \frac{1}{2} I\omega^2$ ☐ (*d*) $\frac{1}{2}\frac{W}{g} v^2 + I\omega^2$. ☐

where W = Weight of body, and
I = Moment of inertia of the body about the axis of rotation.

109. The product of moment of inertia and angular velocity is known as

(*a*) kinetic energy ☐ (*b*) angular momentum ☐
(*c*) angular torque ☐ (*d*) none of the above. ☐

110. Two co-axial rotors having moments of inertia, I_1, I_2 and angular speeds ω_1 and ω_2 respectively are engaged together. The common speed of the two rotors alter engagement will be

(*a*) $\frac{\omega_1 + \omega_2}{2}$ ☐ (*b*) $\frac{I_1\omega_1 + I_2\omega_2}{I_1 + I_2}$ ☐
(*c*) $\frac{I_1\omega_1 - I_2\omega_2}{I_1}$ ☐ (*d*) $\frac{\omega_1 + \omega_2}{I_1 + I_2}$. ☐

111. In question 110, the loss of energy during engagement is equal to

(*a*) $\frac{I_1 I_2(\omega_1 - \omega_2)^2}{2(I_1 + I_2)}$ ☐ (*b*) $\frac{I_1 I_2(\omega_1^2 - \omega_2^2)}{2(I_1 + I_2)}$ ☐
(*c*) $\frac{2I_1 I_2(\omega_1 - \omega_2)^2}{(I_1 + I_2)}$ ☐ (*d*) $\frac{I_1\omega_1^2 + I_2\omega_2^2}{(I_1 + I_2)}$. ☐

112. A machine is said to be self-locking if efficiency of the machine is

(*a*) more than 50% ☐ (*b*) equal to 50% ☐
(*c*) less than 50% ☐ (*d*) equal to 100%. ☐

113. Choose the correct statement

(*a*) If the angle of inclined plane is less than the friction angle, the arrangement is known as self-locking. ☐
(*b*) The efficiency of a square threaded screw is maximum when the helix angle of the screw is equal to 45° minus half of friction angle. ☐

(*c*) Self-locking is a desired property of a machine. ☐
(*d*) All the above. ☐
(*e*) None of the above. ☐

Friction and Bearing

114. The equivalent co-efficient of friction (μ_1) for angular threads is equal to

(*a*) $\mu \cos \beta$ ☐ (*b*) $\mu \sin \beta$ ☐
(*c*) $\frac{\mu}{\cos \beta}$ ☐ (*d*) $\frac{\mu}{\sin \beta}$ ☐

where μ = Actual co-efficient of friction, and β = Semi-angle of the threads.

115. The equivalent co-efficient of friction for *V*-threads is

(*a*) equal to actual co-efficient of friction ☐
(*b*) less than actual co-efficient of friction ☐
(*c*) greater than the actual co-efficient of friction ☐
(*d*) none of the above. ☐

116. The total angle of metric threads is 60°. If the actual co-efficient of friction between the thread and the nut is 0.15, then the value of equivalent co-efficient of friction will be equal to

(*a*) 0.12 ☐ (*b*) 0.1730 ☐
(*c*) 0.225 ☐ (*d*) 0.245. ☐

117. In question 116, if the screw is self-locking, then the value of helix angle will be equal to approximately

(*a*) 8° ☐ (*b*) 12° ☐
(*c*) 10° ☐ (*d*) 6°. ☐

118. Which one is a correct statement ?

(*a*) The rubbing surface in case of pivot and collar bearing is always flat. ☐
(*b*) The rubbing surface in case of pivot and collar bearing is always conical. ☐
(*c*) The rubbing surface in case of pivot bearing is flat and in case of collar bearing it is conical. ☐
(*d*) The rubbing surface in case of pivot bearing is either flat or conical but in case of collar bearing it is generally flat. ☐

119. In case of flat pivot bearing, the rubbing velocity is

(*a*) maximum at the centre of the contact area ☐
(*b*) zero at the centre and maximum at the outer radius ☐
(*c*) uniform throughout the contact area ☐
(*d*) zero at the outer radius. ☐

120. The rate of wearing in case of flat pivot bearing is

(*a*) inversely proportional to the rubbing velocity ☐
(*b*) inversely proportional to the pressure intensity ☐
(*c*) directly proportional to the rubbing velocity ☐
(*d*) directly proportional to the pressure intensity ☐
(*e*) both (*c*) and (*d*). ☐

121. For calculating the friction power for bearing, the assumptions made in practice are

(*a*) the rate of wearing at the surface is uniform ☐

(*b*) the rubbing velocity is constant ☐

(*c*) the pressure intensity is uniform ☐

(*d*) (*a*) and (*b*) only ☐

(*e*) (*a*) and (*c*) only. ☐

122. In case of flat pivot bearing, the frictional torque transmitted for uniform pressure as compared to the frictional torque transmitted for uniform wear

(*a*) is more ☐ (*b*) is less ☐

(*c*) is constant ☐ (*d*) may be more or less. ☐

123. The frictional torque, transmitted in case of flat pivot bearing for uniform pressure, is equal to

(*a*) μWR ☐ (*b*) $\frac{2}{3}\mu WR$ ☐

(*c*) $\frac{1}{3}\mu WR$ ☐ (*d*) $\frac{1}{2}\mu WR$. ☐

where W = Total axial load carried by pivot, μ = Co-efficient of friction, and

R = Radius of bearing surface.

124. The frictional torque, transmitted in case of flat pivot bearing for uniform wear, is equal to

(*a*) μWR ☐ (*b*) $\frac{2}{3}\mu WR$ ☐

(*c*) $\frac{1}{3}\mu WR$ ☐ (*d*) $\frac{1}{2}\mu WR$. ☐

125. The ratio of frictional torque transmitted for uniform pressure to the frictional torque for uniform wear in case of flat pivot, is equal to

(*a*) 1/3 ☐ (*b*) 2/3 ☐

(*c*) 1.0 ☐ (*d*) 4/3. ☐

126. The frictional torque transmitted in case of a flat collar bearing for uniform pressure, is equal to

(*a*) $\mu W\left(\frac{R_1^3 - R_2^3}{R_1^2 - R_2^2}\right)$ ☐ (*b*) $\frac{2}{3}\mu W\left(\frac{R_1^3 - R_2^3}{R_1^2 - R_2^2}\right)$ ☐

(*c*) $\frac{\mu W}{2}(R_1 + R_2)$ ☐ (*d*) $\frac{\mu W}{2}(R_1 - R_2)$. ☐

where R_1 = External radius of the collar, R_2 = Internal radius of the collar, and

W = Total axial load carried by the shaft.

127. The frictional torque transmitted for uniform wear, in case of flat collar bearing, is equal to

(*a*) $\mu W\left(\frac{R_1^3 - R_2^3}{R_1^2 - R_2^2}\right)$ ☐ (*b*) $\frac{2}{3}\mu W\left(\frac{R_1^3 - R_2^3}{R_1^2 - R_2^2}\right)$ ☐

(*c*) $\frac{\mu W}{2}(R_1 + R_2)$ ☐ (*d*) $\frac{\mu W}{2}(R_1 - R_2)$. ☐

128. When the axial load carried by the shaft in case of flat collar bearing, is large; a number of collars are provided on the shaft. This is done

(*a*) to reduce the frictional torque ☐ (*b*) to increase the frictional torque ☐
(*c*) to reduce the intensity of pressure ☐ (*d*) to increase the intensity of pressure. ☐

129. The value of limiting bearing pressure on a collar is about

(*a*) 1 kgf/cm^2 ☐ (*b*) 5 kgf/cm^2 ☐
(*c*) 10 kgf/cm^2 ☐ (*d*) 15 kgf/cm^2. ☐

130. Choose the correct statement regarding the safe design of a bearing

(*a*) when horse power lost in friction is to be determined, uniform rate of wear is assumed ☐
(*b*) when horse power transmitted is to be determined, uniform pressure is assumed ☐
(*c*) when horse power lost in friction is to be determined, uniform pressure is assumed and when horse power transmitted is to be determined, uniform wear is assumed ☐
(*d*) none of the above. ☐

131. A flat foot step bearing 20 cm in diameter supports a load of 2500 kgf. The shaft is running at 420 r.p.m. and the co-efficient of friction is 0.09. The frictional torque, for uniform pressure, will be equal to

(*a*) 11.25 kgf-m ☐ (*b*) 15 kgf-m ☐
(*c*) 18.50 kgf-m ☐ (*d*) 20.0 kgf-m. ☐

132. In question 131, the frictional torque, for uniform wear, will be equal to

(*a*) 11.25 kgf-m ☐ (*b*) 15 kgf-m ☐
(*c*) 18.50 kgf-m ☐ (*d*) 20.0 kgf-m. ☐

133. In question 131, the horse power lost in overcoming friction at the bearing, will be equal to

(*a*) 6.6 h.p. ☐ (*b*) 8.8 h.p. ☐
(*c*) 4.4 h.p. ☐ (*d*) 2.2 h.p. ☐

134. In question 131, the horse power transmitted by the friction between the surface, will be equal to

(*a*) 6.6 h.p. ☐ (*b*) 8.8 h.p. ☐
(*c*) 4.4 h.p. ☐ (*d*) 2.2 h.p. ☐

135. The frictional torque transmitted for uniform pressure, in case of a conical pivot bearing having semi-angle of the cone as θ, is equal to

(*a*) $\frac{\mu WR}{\sin\theta}$ ☐ (*b*) $\frac{\mu WR}{2\sin\theta}$ ☐

(*c*) $\frac{2\mu WR}{\sin\theta}$ ☐ (*d*) $\frac{2\mu WR}{3\sin\theta}$ ☐

where W = Total axial load, and R = Radius of shaft.

136. The frictional torque transmitted for uniform wear, in case of a conical pivot bearing having semi-angle of the cone as θ, is equal to

(*a*) $\frac{\mu WR}{\sin\theta}$ ☐ (*b*) $\frac{\mu WR}{2\sin\theta}$ ☐

(*c*) $\frac{2\mu WR}{\sin\theta}$ ☐ (*d*) $\frac{2\mu WR}{3\sin\theta}$ ☐

137. For uniform pressure the ratio of the frictional torque for a flat collar bearing to the frictional torque for a conical collar bearing, is equal to

(*a*) 1.33 ☐ (*b*) 1.0 ☐

(*c*) sin θ ☐ (*d*) cos θ ☐

where θ = Semi-angle of the cone.

138. For uniform wear, the ratio of the frictional torque for a flat collar bearing to the frictional torque for a conical collar bearing, is equal to

(*a*) 1.33 ☐ (*b*) 1.0 ☐

(*c*) sin θ ☐ (*d*) cos θ ☐

139. The frictional torque transmitted by a conical collar bearing is

(*a*) less than that of flat collar bearing ☐ (*b*) equal to that of flat collar bearing ☐

(*c*) more than that of flat collar bearing ☐ (*d*) none of the above. ☐

140. Choose the wrong statement

(*a*) The frictional torque transmitted by a shaft Fitted with a large number of collars is the same as transmitted by a shaft with one collar. ☐

(*b*) The frictional force for uniform pressure, in case of a flat pivot bearing, can be assumed to be acting at a distance of two-thirds of the radius of the shaft from the axis of shaft. ☐

(*c*) When horse power lost in friction is to be determined, uniform pressure is assumed for safe design of bearing. ☐

(*d*) The frictional force for uniform pressure, in case of a flat collar pivot bearing, can be assumed to be acting at a distance of half of the sum of internal and external radii of the shaft. ☐

141. A thrust bearing, having collars of external and internal diameters as 40 cm and 20 cm respectively, carries an axial load of 3000 kgf. The co-efficient of friction is 0.09 and shaft runs at 420 r.p.m. The horse power lost in friction is to be determined. The corresponding value of friction torque would be

(*a*) 40.5 kgf-m ☐ (*b*) 42 kgf-m ☐

(*c*) 45.0 kgf-m ☐ (*d*) 48.5 kgf-m. ☐

142. In question 141, the horse power lost in friction would be equal to

(*a*) 22.8 ☐ (*b*) 24.8 ☐

(*c*) 26.2 ☐ (*d*) 28.1. ☐

143. Which of the following is an example of friction clutch?

(*a*) disc clutch ☐ (*b*) cone clutch ☐

(*c*) centrifugal clutch ☐ (*d*) plate clutch ☐

(*e*) all of the above ☐ (*f*) none of the above. ☐

144. When a journal is rotating in its bearing, the resultant of the normal force (R_N) and frictional force (μR_N)

(*a*) passes through the centre of the journal ☐

(*b*) is tangent to the circle of the journal at the point of contact ☐

(*c*) is tangent to a small circle of radius $r \times \sin \phi$ where $\phi = \tan^{-1} \mu$. ☐

(*d*) none of the above. ☐

145. Friction circle in case of journal bearing is

(*a*) the circle of the journal ☐

(*b*) the circle to which resultant force F is tangent ☐

(*c*) the circle which is having radius equal to half of the radius of the journal ☐

(*d*) the circle which is having radius equal to $\mu \times R$ where R is the radius of the journal ☐

(*e*) both (*b*) and (*d*) ☐

(*f*) none of the above. ☐

146. The radius of friction circle increases

(*a*) with the increase of load ☐

(*b*) with the increase of radius of journal ☐

(*c*) with the increase of speed ☐

(*d*) with the increase of co-efficient of sliding friction ☐

(*e*) all of the above , ☐

(*f*) (*b*) and (*d*) only. ☐

Screw Jack

147. The screw jack is having maximum efficiency, when the helix angle (α) is equal to

(*a*) $\frac{\pi}{4} + \phi$ ☐ (*b*) $\frac{\pi}{4} + \frac{\phi}{2}$ ☐

(*c*) $\frac{\pi}{4} - \phi$ ☐ (*d*) $\frac{\pi}{4} - \frac{\phi}{2}$ ☐

where ϕ = Angle of friction.

148. The maximum efficiency of a screw jack is a function of

(*a*) helix angle ☐ (*b*) angle of friction ☐

(*c*) load lifted ☐ (*d*) effort. ☐

149. The maximum efficiency of a screw jack is given by

(*a*) $\frac{1+\cos\phi}{1-\cos\phi}$ ☐ (*b*) $\frac{1-\sin\phi}{1+\sin\phi}$ ☐

(*c*) $\frac{1-\cos\phi}{1+\cos\phi}$ ☐ (*d*) $\frac{1-\tan\phi}{1+\tan\phi}$. ☐

where ϕ = Angle of friction.

150. Choose the wrong statement

(*a*) Angle of friction is the angle made by the normal reaction with the resultant of normal reaction and force of friction. ☐

(*b*) Static frictions is more than dynamic friction. ☐

(*c*) Limiting friction is the maximum frictional force, which comes into play when a body just begins to slide over the surface of other body. ☐

(*d*) The radius of friction circle for a shaft rotating inside a bearing is $r \sin\phi$ ☐

(*e*) None of the above. ☐

151. Choose the correct statement

(a) When the helix angle is equal to $\left(45° - \frac{\phi}{2}\right)$, the efficiency of a screw jack is maximum. ☐

(b) Angle of repose is the angle of inclination of the plane, at which the body begins to move down the plane. ☐

(c) The ratio of the limiting friction to the normal reaction is called co-efficient friction. ☐

(d) If the angle of friction is less than the angle of inclination of the plane, the body will begin to move down the inclined plane. ☐

(e) All of the above. ☐

Belt and Chain Drives

152. The length of an open belt, connecting two pulleys of radii r_1 and r_2 and at a centre distance D apart, is equal to

(a) $\pi(r_1 - r_2) + \frac{(r_1 + r_2)^2}{D} + 2D$ ☐ (b) $\pi(r_1 - r_2) + \frac{(r_1 - r_2)^2}{D} + 2D$ ☐

(c) $\pi(r_1 + r_2) + \frac{(r_1 - r_2)^2}{D} + 2D$ ☐ (d) $\pi(r_1 + r_2) + \frac{(r_1 + r_2)^2}{D} + 2D.$ ☐

153. The length of a crossed belt, connecting two pulleys of radii r_1 and r_2 and at a centre distance D apart, is equal to

(a) $\pi(r_1 + r_2) + \frac{(r_1 + r_2)^2}{D} + 2D$ ☐ (b) $\pi(r_1 - r_2) + \frac{(r_1 - r_2)^2}{D} + 2D$ ☐

(c) $\pi(r_1 + r_2) + \frac{(r_1 - r_2)^2}{D} + 2D$ ☐ (d) $\pi(r_1 + r_2)^2 + \frac{(r_1 + r_2)^2}{D} + 2D.$ ☐

154. The angular velocities of the two pulleys, connected either by an open belt or a crossed belt, are

(a) directly proportional to their diameters ☐

(b) inversely proportional to their diameters ☐

(c) proportional to the square of their diameters ☐

(d) proportional to the square root of their diameters. ☐

155. Choose the wrong statement

(a) With an open belt the pulleys rotate in the same direction. ☐

(b) The angle of contact in crossed belt drive is more than that in an open belt drive. ☐

(c) With a crossed belt the pulleys rotate in opposite direction. ☐

(d) The length of crossed belt depends on the sum and difference of the radii of the pulleys. ☐

156. When two pulleys, of unequal sizes, the connected by a belt drive the angle θ is taken as

(a) angle of contact on the bigger pulley ☐

(b) angle of contact on the smaller pulley ☐

(c) average angle of the contact on the two pulleys ☐

(d) none of the above. ☐

157. If T_1 and T_2 are the tensions on the tight and slack side of a belt and θ is the angle of contact, then ratio of tension is given by

(a) $\frac{T_1}{T_2} = \mu\theta$ ☐ (b) $\frac{T_1}{T_2} = e^{\mu\theta}$ ☐

(c) $\frac{T_1}{T_2} = e^{1/\mu\theta}$ ☐ (d) $\frac{T_1}{T_2} = \mu e^{\theta}$. ☐

where μ = Co-efficient of friction between the belt and pulley.

158. The effect of slip on velocity ratio of the belt is to

(a) increase the velocity ratio ☐ (b) decrease the velocity ratio ☐

(c) keep the velocity ratio constant ☐ (d) none of the above. ☐

159. The horse power transmitted by a belt is equal to

(a) $\frac{T_1 \times v}{75}$ ☐ (b) $\frac{T_2 \times v}{75}$ ☐

(c) $\frac{(T_1 + T_2)}{2} \times \frac{v}{75}$ ☐ (d) $\frac{(T_1 + T_2) \times v}{75}$. ☐

where T_1 = Tension to tight side, T_2 = Tension on slack side, and v = Linear velocity of belt.

160. The tension caused by centrifugal force on the belt is equal to

(a) $\frac{w}{g}\frac{V^2}{r}$ ☐ (b) $\frac{w}{g} V^2$ ☐

(c) $w \times \frac{V^2}{r}$ ☐ (d) $\frac{1}{g} \times \frac{V^2}{r}$. ☐

where w = Weight of belt per unit length, V = Velocity of belt, and

r = Radius of bigger pulley.

161. The tension caused by centrifugal force on the belts

(a) increases power transmission ☐ (b) decreases power transmission ☐

(c) does not affect power transmission ☐ (d) none of the above. ☐

162. If the effect of centrifugal tension in the belt is considered, the horse power transmitted is zero when the velocity of the belt is zero. The horse power will be again zero when the velocity of belt is equal to

(a) $\sqrt{\frac{w}{T_1 \times g}}$ ☐ (b) $\sqrt{\frac{T_1}{w \times g}}$ ☐

(c) $\sqrt{\frac{T_1 \times g}{w}}$ ☐ (d) $\sqrt{\frac{w \times g}{T}}$. ☐

where T_1 = Total tension on the tight side of the belt, and

w = Weight of belt per unit length.

163. If the effect of centrifugal tension in the belt is considered then the velocity of the belt, at which the maximum horse power is transmitted, is equal to

(*a*) $\sqrt{\dfrac{T_1 \times g}{3w}}$ ☐ (*b*) $\sqrt{\dfrac{3w}{T_1 \times g}}$ ☐

(*c*) $\sqrt{\dfrac{w}{3T_1 \times g}}$ ☐ (*d*) $\sqrt{\dfrac{T_1}{3w \times g}}$. ☐

where T_1 = Total tension on the tight side of the belt, and

w = Weight of belt per unit length.

164. When maximum horse power is transmitted by a belt, the centrifugal tension in the belt is

(*a*) half the maximum tension allowed in the belt ☐

(*b*) one-third the maximum tension allowed in the belt ☐

(*c*) two-thirds the maximum tension allowed in the belt ☐

(*d*) one-fourth the maximum tension allowed in the belt. ☐

165. When maximum horse power is transmitted, the effective tension in the tight side of the belt is equal to

(*a*) twice the centrifugal tension ☐ (*b*) three times the centrifugal tension ☐

(*c*) half the centrifugal tension ☐ (*d*) one-third the centrifugal tension. ☐

166. The belts are installed with initial tension, which is generally equal to

(*a*) sum of the tensions on the tight side and slack side of the belt ☐

(*b*) average tension of the tight side and slack side of the belt ☐

(*c*) half of tension on the tight side of the belt ☐

(*d*) half of tension on the slack side of the belt. ☐

167. Creep in belt drive is due to

(*a*) weak material of the belt ☐

(*b*) weak material of the pulley ☐

(*c*) uneven extensions and contractions of the belt when it passes from tight side to slack side ☐

(*d*) none of the above. ☐

168. Crowning of pulley is done

(*a*) to avoid the slipping of the belt ☐

(*b*) to make them more study ☐

(*c*) to enable pulley rigidly fixed to the shaft ☐

(*d*) to make pulley look more pleasant in appearance. ☐

169. A shaft running at 100 r.p.m. is to drive a parallel shaft at 200 r.p.m. If the diameter of the pulley on the driving shaft is 80 cm, then the diameter of the pulley on the driven shaft will be

(*a*) 160 cm ☐ (*b*) 80 cm ☐

(*c*) 40 cm ☐ (*d*) 80 cm. ☐

170. If the question 169, the thickness of belt is 4 mm, then the diameter of the pulley on the driven shaft will be

(*a*) 40.40 cm ☐ (*b*) 80.40 cm ☐

(*c*) 39.60 cm ☐ (*d*) 29.80 cm. ☐

171. Choose the wrong statement

(*a*) When two pulleys are connected by means of a cross-belt drive, then both the pulleys will rotate in opposite directions. ☐

(*b*) The centrifugal tension in belts reduces power transmission. ☐

(*c*) Abbreviation P.I.V, drive stands for positive infinitely variable drive. ☐

(*d*) The power transmitted by a belt is maximum when the maximum tension in the belt is one-third of centrifugal tension. ☐

172. Choose the correct statement

(*a*) In a belt drive, the speeds are directly proportional to the diameters of the pulley. ☐

(*b*) Considering the effect of centrifugal tension in the belt, the horse power transmitted increase directly with the increase of speed. ☐

(*c*) The length of open belt depends upon the sum and difference of the radii of the pulleys. ☐

(*d*) Crowning of pulley is done to make them more sturdy. ☐

173. Two parallel shafts, having centre distance 5 metres, are connected by an open belt. The diameters of the pulley fixed to the shafts are 40 cm and 30 cm respectively. The length of the belt will be

(*a*) 1000 cm ☐ (*b*) 1110.05 cm ☐

(*c*) 1112.45 cm ☐ (*d*) 1120 cm. ☐

174. If in the question 173, the shafts are connected by a cross belt, then the length of the belt would be

(*a*) 1000 cm ☐ (*b*) 110.05 cm ☐

(*c*) 1112.45 cm ☐ (*d*) 1120 cm. ☐

175. The effective tensions on the tight side and slack in a belt drive are 60 kgf and 30 kgf respectively. If the linear velocity of the belt is 300 metre/sec, then the horse power transmitted by the belt would be

(*a*) 120 h.p. ☐ (*b*) 0.5 h.p. ☐

(*c*) 4 h.p. ☐ (*d*) 2 h.p. ☐

176. If in question 175, the tension on the tight side only is increased to 1.5 times, then the horse power transmitted will be

(*a*) 120 h.p. ☐ (*b*) 0.5 h.p. ☐

(*c*) 4 h.p. ☐ (*d*) 2 h.p. ☐

177. If in question 175, tension on the slack side only is increased to 1.5 times then the horse power transmitted will be

(*a*) 120 h.p. ☐ (*b*) 0.5 h.p. ☐

(*c*) 4 h.p. ☐ (*d*) 2 h.p. ☐

178. If in question 175, the tensions on the tight and slack side are increased to 1.5 times then the horse power transmitted will be

(*a*) 180 h.p. ☐ (*b*) 0.75 h.p. ☐
(*c*) 6 h.p. ☐ (*d*) 3 h.p. ☐

179. The centrifugal tension in the belt depends upon

(*a*) tight side tension of the belt ☐ (*b*) slack side tension of the belt ☐
(*c*) velocity of the belt ☐ (*d*) weight per unit length of the belt ☐
(*e*) (*a*) and (*b*) ☐ (*f*) (*c*) and (*d*). ☐

180. When centrifugal tension (T_c) is considered, the belt is designed for the tension

(*a*) T_1 ☐ (*b*) $T_1 - T_c$ ☐
(*c*) $T_1 + T_c$ ☐ (*d*) none of the above. ☐

181. The horse power transmitted by a belt, when the effect of centrifugal tension is considered and belt is running with a velocity equal to $\left(\frac{T_1 \times g}{3w}\right)^{1/2}$, will be

(*a*) zero ☐ (*b*) maximum ☐
(*c*) minimum ☐ (*d*) none of the above. ☐

182. Choose the wrong statement

(*a*) When the shafts are at long centre distances, the power is transmitted with the help of belts, ropes and chains. ☐
(*b*) When the shafts are at short centre distances, the power is transmitted by gears. ☐
(*c*) Gear drive is a positive and smooth drive. ☐
(*d*) The belts and rope drives are not positive drive. ☐
(*e*) none of the above. ☐

Gears

183. Two intersecting and co-planar shafts are connected by gears. This type of gear is called

(*a*) helical gear ☐ (*b*) spur gear ☐
(*c*) bevel gear ☐ (*d*) spiral gear. ☐

184. Two parallel and co-planar shafts are connected by gears. This type of gear is called

(*a*) helical gear ☐ (*b*) spur gear ☐
(*c*) bevel gear ☐ (*d*) spiral gear. ☐

185. If the peripheral velocity of the gears are more than 15 m/sec, the gears are known as

(*a*) low velocity type ☐ (*b*) medium velocity type ☐
(*c*) high velocity type ☐ (*d*) none of the above. ☐

186. Figure 9.18 shows the various terms used in the study of gears. Pitch circle is a circle whose radius is equal to

(*a*) R_1 ☐ (*b*) R_2 ☐
(*c*) R_3 ☐ (*d*) $\frac{R_1 + R_2}{2}$ ☐

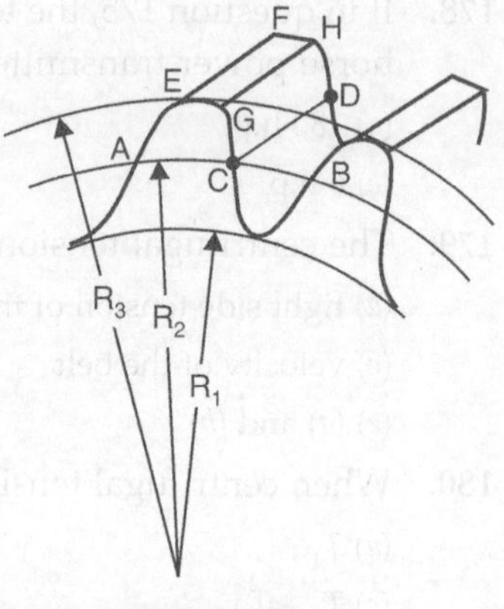

187. In Fig. 9.18, the dedendum circle is a circle whose radius is equal to

(*a*) R_1 ☐
(*b*) R_2 ☐
(*c*) R_3 ☐
(*d*) $\frac{R_1 + R_2}{2}$ ☐

188. In Fig. 9.18, the addendum circle is a circle whose radius is equal to

(*a*) R_1 ☐
(*b*) R_2 ☐
(*c*) R_3 ☐
(*d*) $\frac{R_1 + R_2}{2}$ ☐

FIGURE 9.18

189. In Fig. 9.18, dedendum is equal to

(*a*) $(R_3 - R_2)$ ☐ (*b*) $(R_2 - R_1)$. ☐
(*c*) $(R_3 - R_1)$ ☐ (*d*) $(R_2 - R_1)/2.0$. ☐

190. In Fig. 9.18, addendum is equal to

(*a*) $(R_3 - R_2)$ ☐ (*b*) $(R_2 - R_1)$ ☐
(*c*) $(R_3 - R_1)$ ☐ (*d*) $(R_2 - R_1)/2.0$. ☐

191. If Fig. 9.18, the surface of the tooth above line *CD* is known as

(*a*) top land ☐ (*b*) face ☐
(*c*) flank ☐ (*d*) face width. ☐

192. In Fig. 9.18, the surface of the tooth below the line *CD* is known as

(*a*) top land ☐ (*b*) face ☐
(*c*) flank ☐ (*d*) face width. ☐

193. In Fig. 9.18, the length *EF* represents

(*a*) circular pitch ☐ (*b*) face width ☐
(*c*) diametral pitch ☐ (*d*) addendum. ☐

194. In Fig. 9.18, the distance *AS* measured along the circumference of the circle of radius R_2 is known as

(*a*) circular pitch ☐ (*b*) face width ☐
(*c*) diametral pitch ☐ (*d*) dedendum. ☐

195. Choose the wrong statement

(*a*) The gears are termed as medium velocity gears if their peripheral velocity lies in the range of 3 to 14 m/sec. ☐
(*b*) When two intersecting and coplanar shafts are connected by gears, the arrangement is called bevel gearing. ☐
(*c*) The radial distance of a tooth from the pitch circle to the bottom of tooth is called addendum. ☐
(*d*) The face of the tooth is the surface of the tooth above the pitch surface. ☐

196. Choose the correct statement

(*a*) The gears are termed as high velocity gears if their peripheral velocity lies in the range of 3 to 10 m/sec. ☐

(*b*) The flank of the tooth is the surface of the tooth above the pitch surface. ☐

(*c*) The distance measured along the circumference of the pitch circle from a point on one tooth to a corresponding point on the adjacent tooth is known as circular pitch. ☐

(*d*) The dedendum is the radial distance of the tooth from the pitch circle to the top of the tooth. ☐

197. Diametral pitch is defined as the ratio of

(*a*) number of teeth to pitch circle, diameter ☐

(*b*) pitch circle diameter to number of teeth ☐

(*c*) circumference of pitch circle to number of teeth ☐

(*d*) none of the above. ☐

198. For a spur gear, the circular pitch is equal to the ratio of

(*a*) number of teeth to pitch circle diameter ☐

(*b*) pitch circle diameter to number of teeth ☐

(*c*) circumference of pitch circle to number of teeth ☐

(*d*) none of the above. ☐

199. The module (m) is defined as equal to the ratio of

(*a*) number of teeth to pitch circle diameter ☐

(*b*) pitch circle diameter to number of teeth ☐

(*c*) circumference of the pitch circle to number of teeth ☐

(*d*) none of the above. ☐

200. The product of the circular pitch and diametral pitch is equal to

(*a*) 2π ☐ (*b*) π ☐

(*c*) $\frac{\pi}{2}$ ☐ (*d*) 1.0. ☐

201. The product of the module and diametral pitch is equal to

(*a*) 2π ☐ (*b*) π ☐

(*c*) $\frac{\pi}{2}$ ☐ (*d*) 1.0. ☐

202. The pressure angles varies from

(*a*) 40° to 50° ☐ (*b*) 30° to 40° ☐

(*c*) 20° to 30° ☐ (*d*) 14° to 20°. ☐

203. Law of gearing states that

(*a*) the normal at the point of contact between a pair of teeth divides the line joining the centre of rotation directly in the ratio of angular velocities ☐

(*b*) the common normal at the point of contact between a pair of teeth does not pass, always through the pitch point ☐

(*c*) the common normal at the point of contact between a pair of teeth always passes through the pitch point ☐

(*d*) none of the above. ☐

204. The velocity of sliding of the mating teeth of the two gears is equal to
(*a*) product of angular velocity of first gear and the distance of the point of contact from the pitch point ☐
(*b*) product of angular velocity of second gear and the distance of the point of contact from the pitch point ☐
(*c*) product of the sum of the angular velocities of the two gears and the distance of the point of contact from the pitch point ☐
(*d*) none of the above. ☐

205. The teeth, which are normally used and satisfy the law of gearing are
(*a*) conjugate teeth ☐ (*b*) cycloidal teeth ☐
(*c*) involute teeth ☐ (*d*) only (*b*) and (*c*) ☐
(*e*) none of the above. ☐

206. The involute function [*INV* (ϕ)] in terms of pressure angle (ϕ) is expressed as
(*a*) $INV(\phi) = (\phi) - \tan(\phi)$ ☐ (*b*) $INV(\phi) = 1 - \tan(\phi)$ ☐
(*c*) $INV(\phi) = \tan(\phi) - (\phi)$ ☐ (*d*) $INV(\phi) = 1 + \tan(\phi)$ ☐

207. If the pressure angle is 20, the involute function will be equal to
(*a*) 0.702 ☐ (*b*) 0.636 ☐
(*c*) 0.0149 ☐ (*d*) 1.3639. ☐

208. In involute teeth, the tangent to the base circle is a
(*a*) tangent to the involute ☐ (*b*) normal to the involute ☐
(*c*) normal to the pitch circle ☐ (*d*) none of the above. ☐

209. The path of contact in involute gears is a
(*a*) circle ☐ (*b*) straight line ☐
(*c*) curved line ☐ (*d*) none of the above. ☐

210. Choose the wrong statement
(*a*) The product of the circular pitch and diametral pitch is equal to π. ☐
(*b*) The pressure angle for involute gears is always constant. ☐
(*c*) The angular velocity ratio when two involutes are in mesh, is inversely proportional to the radius of base circles. ☐
(*d*) The involute function in terms of pressure angle (ϕ) is equal to (ϕ) - tan ϕ). ☐

211. Choose the correct statement
(*a*) The range of pressure angle is usually from 20° to 30°. ☐
(*b*) The involute of circle is the locus of a point on the circumference of a circle which rolls without slipping on a fixed straight line. ☐
(*c*) The module is the reciprocal of diametral pitch. ☐
(*d*) The pressure angle for involute gears always depends upon the size of teeth. ☐

212. The centre distance between two meshing involute gears is
(*a*) directly proportional to the sum of base circle radii of the meshing gears ☐
(*b*) inversely proportional to the cosine of pressure angle ☐
(*c*) directly proportional to the cosine of pressure angle ☐
(*d*) both (*a*) and (*b*) ☐
(*e*) none of the above. ☐

213. The locus of a point on the circumference of a circle, which rolls without slipping on a fixed straight line, is known as

(*a*) involute ☐ (*b*) cycloid ☐
(*c*) hypo-cycloid' ☐ (*d*) epicycloid. ☐

214. The locus of a point on the circumference of a circle, which rolls without slipping on the outside of another circle, is known as

(*a*) involute ☐ (*b*) cycloid ☐
(*c*) hypo-cycloid ☐ (*d*) epicycloid. ☐

215. The locus of a point on the circumference of a circle, which rolls without slipping on the inside of another circle, is known as

(*a*) involute ☐ (*b*) cycloid ☐
(*c*) hypo-cycloid ☐ (*d*) epicycloid. ☐

216. Which is correct statement about cycloidal gears?

(*a*) The pressure angle is constant from the commencement of the engagement to the end of engagement. ☐
(*b*) The gears are manufactured with simplicity. ☐
(*c*) There is no interference in these gears. ☐
(*d*) Variation in centre distance within limits does not affect the velocity ratio of the two mating gears. ☐

217. The work wasted in friction, for a given total arc of action between gear teeth, will be minimum when arc of approach is

(*a*) half of arc of recess ☐ (*b*) one-fourth of arc of recess ☐
(*c*) equal to arc of recess ☐ (*d*) twice the arc of recess. ☐

218. For helical gears, the normal circular pitch is equal to

(*a*) $\dfrac{p_c}{\cos\alpha}$ ☐ (*b*) $p_c \times \cos\alpha$ ☐
(*c*) $\dfrac{\cos\alpha}{p_c}$ ☐ (*d*) $\dfrac{1}{p_c \times \cos\alpha}$. ☐

where p_c = Circular pitch, and α = Helix angle.

219. For helical gears, the axial pitch is equal to

(*a*) $p_c \times \tan\alpha$ ☐ (*b*) $\dfrac{\tan\alpha}{p_c}$ ☐
(*c*) $\dfrac{p_c}{\tan\alpha}$ ☐ (*d*) $\dfrac{1}{p_c \times \tan\alpha}$. ☐

where p_c = Circular pitch, and α = Helix angle.

220. The efficiency of spiral gears will be maximum, if

(*a*) $\alpha = \dfrac{\theta - \phi}{2}$ ☐ (*b*) $\alpha = \dfrac{\theta + \phi}{2}$ ☐
(*c*) $\alpha = \theta - 2\phi$ ☐ (*d*) $x = 2\theta - \phi$. ☐

where α = Spiral angle, and θ = Shaft angle and ϕ = Friction angle.

221. The maximum efficiency of spiral gears is equal to

(a) $\frac{1-\cos(\theta+\phi)}{1+\cos(\theta-\phi)}$ ☐ (b) $\frac{1+\cos(\theta+\phi)}{1+\cos(\theta-\phi)}$ ☐

(c) $\frac{1+\cos(\theta-\phi)}{1+\cos(\theta+\phi)}$ ☐ (d) $\frac{1-\cos(\theta-\phi)}{1+\cos(\theta+\phi)}$. ☐

where θ = Shaft angle, and ϕ = Friction angle.

222. In case of worm and worm gear, the maximum efficiency is equal to

(a) $\frac{1+\sin\phi}{1-\sin\phi}$ ☐ (b) $\frac{1-\sin\phi}{1+\sin\phi}$ ☐

(c) $\frac{1+\sin\phi}{\sin\phi}$ ☐ (d) $\frac{1-\sin\phi}{\sin\phi}$. ☐

where ϕ = Friction angle.

223. Two gears having involute teeth are in mesh. If the path of approach and path of recess are half of their maximum possible values, then the length of path of contact is equal to

(a) $\frac{(R-r)\sin\phi}{2}$ ☐ (b) $\frac{(R+r)\sin\phi}{2}$ ☐

(c) $\frac{(R-r)\cos\phi}{2}$ ☐ (d) $\frac{(R+r)\cos\phi}{2}$. ☐

where R = Pitch circle radius of wheel, r = Pitch circle radius of pinion, and ϕ = Pressure angle.

224. In question 223, the number of involute teeth on the pinion and wheels are 20 and 40 respectively. If the pressure angle is equal to 20° and module pitch 10 mm, then pitch circle radius of wheel would be equal to

(a) 100 mm ☐ (b) 150 mm ☐

(c) 175 mm ☐ (d) 200 mm. ☐

225. In equation 224, the pitch circle radius of the pinion would be equal to

(a) 100 mm ☐ (b) 150 mm ☐

(c) 175 mm ☐ (d) 200 mm. ☐

226. In equation 224, the required path of approach would be equal to

(a) 88.2 mm ☐ (b) 44.1 mm ☐

(c) 34.2 mm ☐ (d) 17.1 mm. ☐

227. In question 224, the required path of recess would be equal to

(a) 88.2 mm ☐ (b) 44.1 mm ☐

(c) 34.2 mm ☐ (d) 17.1 mm. ☐

228. In question 224, the path of contact would be

(a) 88.2 mm ☐ (b) 51.3 mm ☐

(c) 34.2 mm ☐ (d) 17.1 mm. ☐

229. In question 224, the arc of contact would be equal to

(*a*) 88.2 mm ☐ (*b*) 51.3 mm ☐

(*c*) 34.2 mm ☐ (*d*) 58.73 mm. ☐

230. The addendum as compared to dedendum of a toothed gear wheel is

(*a*) more ☐ (*b*) less ☐

(*c*) same ☐ (*d*) can be more or less. ☐

231. The centre distance between two involute teeth gears is equal to

(*a*) $(R + r) \cos \phi$ ☐ (*b*) $(R + r) \sin \phi$ ☐

(*c*) $\frac{(R+r)}{\cos \phi}$ ☐ (*d*) $\frac{(R+r)}{\sin \phi}$. ☐

where R and r = Radii of the base circle of two gears and ϕ = Pressure angle.

Gear Train

232. In a simple trains of three wheels, the third wheel will rotate

(*a*) in the opposite direction to the first wheel ☐

(*b*) in the same direction as the first wheel ☐

(*c*) in any direction ☐

(*d*) none of the above. ☐

233. In question 232, the second wheel is known as ideal wheel (or the idler). This wheel

(*a*) has no influence on the velocity ratio. It only affects the direction of rotation ☐

(*b*) changes the velocity ratio and also direction of rotation ☐

(*c*) may change the velocity ratio but may not affect the direction of rotation ☐

(*d*) none of the above. ☐

234. In question 232, the ratio of the speed of the third wheel to that of first wheel is equal to the ratio of

(*a*) number of teeth on the first wheel to the number of teeth on the third wheel ☐

(*b*) number of teeth on the third wheel to the number of teeth on the first wheel ☐

(*c*) product of number of teeth on the first and third wheels to the number of wheel ☐

(*d*) none of the above. ☐

235. Figure 9.19 shows compound gears train in which wheels 2 and 3 are mounted on the same shaft and they rotate together. The ratio of speed of the fourth wheel to that of first wheel is given by

(*a*) $\frac{N_4}{T_1} = \frac{T_1}{T_4}$ ☐ (*b*) $\frac{N_4}{N_1} = \frac{T_4}{T_1}$ ☐

(*c*) $\frac{N_4}{N_1} = \frac{T_1 \times T_3}{T_2 \times T_4}$ ☐ (*d*) $\frac{N_4}{N_1} = \frac{T_1 \times T_2}{T_3 \times T_4}$. ☐

where T_1, T_2, T_3 and T_4 = Number of teeth on first, second, third and fourth wheel respectively.

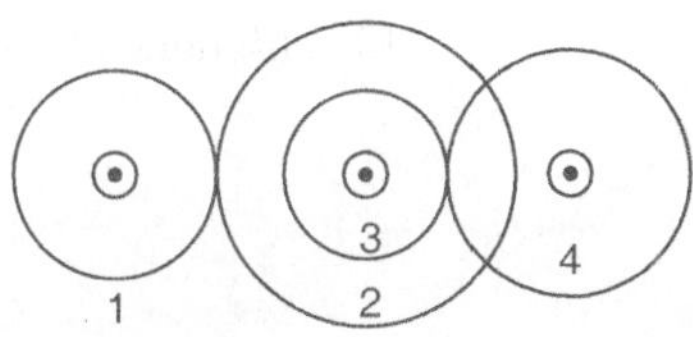

FIGURE 9.19

236. When the axes of the first and last wheels of a compound gear train are co-axial, then the train is known as

(*a*) non-reverted gear train ☐ (*b*) reverted gear train ☐

(*c*) epicyclic gear train ☐ (*d*) none of the above. ☐

237. If in a gear trains, the first wheel is a driver and the last wheel is a driven wheel, then train value is defined as the

(*a*) product of the speeds of the driver and driven wheels ☐

(*b*) ratio of the speed of the driver wheel to the speed of the driven wheel ☐

(*c*) ratio of the speed of the driven wheel to the speed of the driver wheel ☐

(*d*) products of number of teeth on driver and driven wheels. ☐

238. In question 237, the speed ratio of the gear train is defined as

(*a*) product of the speeds of the driver and driven wheels ☐

(*b*) ratio of the speed of the driver wheel to the speed of the driven wheel ☐

(*c*) ratio of the speed of the driven wheel to the speed of the driver wheel ☐

(*d*) products of number of teeth on driver and driven wheels. ☐

239. Train value of a gear train is

(*a*) equal to the speed ratio ☐ (*b*) half of the speed ratio ☐

(*c*) reciprocal of the speed ratio ☐ (*d*) double the speed ratio. ☐

240. The train values and speed ratios are specified as their plus or minus. The negative sign of speed ratio indicates that

(*a*) speed of driver wheel is less than the speed of driven wheel ☐

(*b*) speed of driven wheel is less than the speed of driver wheel ☐

(*c*) driver and driven wheels are rotating in opposite direction ☐

(*d*) driver and driven wheels are rotating in the same direction. ☐

241. A gear train, in which at least one of the gear axes is in motion relative to the frame, is known as

(*a*) reverted gear train ☐ (*b*) non-reverted gear train ☐

(*c*) epicyclic gear ☐ (*d*) none of the above. ☐

242. Figure 9.20 shows a wheel *B* fixed to the frame and an arm *C*, with wheel *A*, revolves along wheel *B*. This gear train is an example of

(*a*) reverted gear train ☐ (*b*) non-reverted gear train ☐

(*c*) epicyclic gear ☐ (*d*) none of the above. ☐

243. With reference to Fig. 9.20, the ratio of the speed of wheel *A* to the speed of arm *C* is equal to

(*a*) number of teeth on wheel *B* divided by number of teeth on wheel *A* ☐

(*b*) one plus (number of teeth on wheel *B* divided by number of teeth on wheel *A*) ☐

(*c*) one plus (number of teeth on wheel *A* divided by number of teeth on wheel *B*) ☐

(*d*) none of the above. ☐

FIGURE 9.20

244. In Fig. 9.20, if the arm C is rotated at 60 r.p.m. and number of teeth on wheels and *B* are 25 and 100, respectively then speed of wheel *A* would be

(*a*) 125 r.p.m. ☐ (*b*) 240 r.p.m. ☐

(*c*) 300 r.p.m. ☐ (*d*) 350 r.p.m. ☐

245. In question 244, if the wheel *A* is Fixed instead of wheel *B* (refer to Fig. 9.20) then the speed of the wheel *B* would be

(*a*) 125 r.p.m ☐ (*b*) 240 r.p.m ☐

(*c*) 300 r.m.p ☐ (*d*) 350 r.p.m ☐

Dynamometer

246. A dynamometer is a device used for measuring

(*a*) speed of a machine ☐ (*b*) torque exerted by or on a machine ☐

(*c*) kinetic energy of the machine ☐ (*d*) all of the above ☐

(*e*) none of the above. ☐

247. Which of the followings are the correct statement for clutch and brake ?

(*a*) clutch connects a moving member of stationary frame ☐

(*b*) clutch connects a moving member to another moving member ☐

(*c*) brake connects a moving member to another moving member ☐

(*d*) brake connects a moving members to stationary frame. ☐

248. Which one of the followings is an absorption dynamometer ?

(*a*) gear dynamometer ☐ (*b*) torsion dynamometer ☐

(*c*) belt dynamometer ☐ (*d*) prony brake dynamometer. ☐

249. Which one of the followings is an transmission dynamometer ?

(*a*) prony brake dynamometer ☐ (*b*) torsion dynamometer ☐

(*c*) fluid friction dynamometer ☐ (*d*) band brake dynamometer. ☐

250. The abbreviation FMR in brakes and dynamometer stands for

(*a*) friction motion ratio ☐ (*b*) friction multiplication ratio ☐

(*c*) force multiplication ratio ☐ (*d*) force mean rate. ☐

251. Figure 9.21 shows a band brake in which the drum is rotating clock-wise. The length *OA* is greater than length *OC*. Force *P* is applied at *B*. In order to apply, the brake, the force *P* should

(*a*) act in upward direction ☐ (*b*) act in downward direction ☐

(*c*) be zero ☐ (*d*) none of the above. ☐

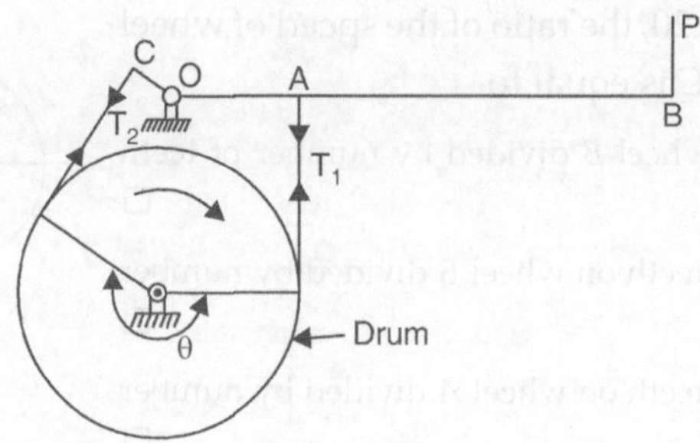

FIGURE 9.21

252. With reference to Fig. 9.21, the band will be loosened if the force P

(a) act in upward direction ☐ (b) act in downward direction ☐
(c) be zero ☐ (d) none of the above. ☐

253. With reference to Fig. 9.21, if force P is acting upward the system will not work if

(a) $OC = OA$ ☐ (b) $OC < OA$ ☐
(c) $OC > OA'$ ☐ (d) none of the above. ☐

254. If the drum is rotating anticlockwise and length OC is greater than length OA in Fig. 9.21, then in order to apply the brake, the force P at B should

(a) act in upward direction ☐ (b) acts in downward direction ☐
(c) is zero ☐ (d) none of the above. ☐

255. In question 254, the band will be loosened if the force P at B

(a) acts in upward direction ☐ (b) act in downward direction ☐
(c) be zero ☐ (d) none of the above. ☐

256. In question 254, for the brake to be self-locking, the force P at B should

(a) act in upward direction ☐ (b) act in downward direction ☐
(c) be zero ☐ (d) none of the above. ☐

257. In question 254, the condition for self-locking of the brake is

(a) $\frac{OC}{OA} = \frac{T_1}{T_2}$ ☐ (b) $\frac{OC}{OA} < \frac{T_1}{T_2}$ ☐
(c) $\frac{OC}{OA} > \frac{T_1}{T_2}$ ☐ (d) none of above. ☐

where T_1 = Tension on tight side, and T_2 = Tension on slack side.

Cams

258. The smallest circle drawn to the cam profile from the cam centre is known as

(a) prime circle ☐ (b) base circle ☐
(c) pitch circle ☐ (d) pitch curve. ☐

259. The smallest circle drawn to the pitch curve from the centre of rotation of the cam is known as

(a) prime circle ☐ (b) base circle ☐
(c) pitch circle ☐ (d) none of the above. ☐

260. The circle, with centre as the centre of the cam axis and radius such that it passes through the pitch point, is known as

(*a*) prime circle ☐ (*b*) base circle ☐
(*c*) pitch circle ☐ (*d*) pitch curve. ☐

261. The path, followed by the trace point if the follower moved about the cam, is known as

(*a*) prime circle ☐ (*b*) base circle ☐
(*c*) pitch circle ☐ (*d*) pitch curve. ☐

262. For the knife edge follower, the cam profile and pitch curve

(*a*) are different ☐ (*b*) are same ☐
(*c*) may be same or different ☐ (*d*) none of the above. ☐

263. On which circle or curve the size of the cam depends?

(*a*) prime circle ☐ (*b*) base circle ☐
(*c*) pitch circle ☐ (*d*) pitch curve. ☐

264. The reference point, on the follower for tracing the cam profile, is called

(*a*) the pitch point ☐ (*b*) the trace point ☐
(*c*) the centre of cam ☐ (*d*) none of the above. ☐

265. The point, on the cam pitch curve having the maximum pressure angle, is called

(*a*) the pitch point ☐ (*b*) the trace point ☐
(*c*) the centre of cam ☐ (*d*) none of the above. ☐

266. The pressure angle of a cam is the angle at any point on the pitch curve.

(*a*) made by line of motion of the follower with horizontal axis ☐
(*b*) made by line of motion of the follower with vertical axis ☐
(*c*) between the normal to that point on the curve and line of motion of follower at that instant ☐
(*d*) between the tangent to that point on the curve and line of motion of follower at that instant. ☐

267. The angle, which represents the cam profile and is most important in cam design, is

(*a*) cam angle ☐ (*b*) pressure angle ☐
(*c*) angle of dwell ☐ (*d*) angle of descent. ☐

268. The stroke of the follower is equal to

(*a*) half of the maximum travel of the follower from the lowest position to the topmost position ☐
(*b*) the maximum travel of the follower from the lowest position of the topmost position ☐
(*c*) half of the diameter of base circle ☐
(*d*) half of the diameter of the prime circle. ☐

269. Choose the wrong statement

(*a*) The reference point for a knife edge follower is the edge of the knife. ☐
(*b*) The reference point for a roller follower is the centre of the roller. ☐
(*c*) The pitch curve and cam profile, for a knife edge follower, are the same ☐
(*d*) The pressure angle is the angle between the direction of the follower motion and a tangent to the pitch curve. ☐

270. A follower of a cam has

(*a*) uniform velocity ☐ (*b*) simple harmonic motion ☐
(*c*) cyclodial motion ☐ (*d*) uniform acceleration ad retardation ☐
(*e*) any one of the above ☐ (*f*) none of the above. ☐

271. Which factor affects the size of a cam?

(*a*) base circle ☐ (*b*) pressure angle ☐
(*c*) hub size ☐ (*d*) undercutting ☐
(*e*) all of the above ☐ (*f*) none of the above. ☐

272. The pressure angle of a cam

(*a*) decreases if radius of base circle decreases ☐
(*b*) decreases if radius of base circle increase ☐
(*c*) is independent of the radius of base circle ☐
(*d*) first decreases then increases if the radius of base circle decreases. ☐

273. The lift of a roller follower, when it is an contact with the straight flank of a tangent cam, is equal to

(*a*) $(r + r_1)(1 - \cos\theta)$ ☐ (*b*) $(r + r_1)\left(\frac{1 - \cos\theta}{\cos\theta}\right)$ ☐
(*c*) $(r - r_1)(1 + \cos\theta)$ ☐ (*d*) $(r - r_1)\left(\frac{1 - \cos\theta}{\cos\theta}\right)$. ☐

where θ = Any angle turned by the cam from the beginning of the lift of the follower

r = Radius of the base circle of the cam, and r_1 = Radius of the roller follower.

274. For the question 273, the velocity of the follower is equal to

(*a*) $\omega(r + r_1)\sin\theta\sec^2\theta$ ☐ (*b*) $\omega(r + r_1)\sin\theta\cos\theta$ ☐
(*c*) $\omega(r - r_1)\sin\theta\sec^2\theta$ ☐ (*d*) $\omega^2(r - r_1)\cos\theta\sin^2\theta$. ☐

275. For the question 273, the acceleration of the follower is equal to

(*a*) $\omega^2(r - r_1)\left(\frac{2 - \cos^2\theta}{\cos^3\theta}\right)$ ☐ (*b*) $\omega^2(r + r_1)\left(\frac{2 - \cos^2\theta}{\cos^3\theta}\right)$ ☐
(*c*) $\omega^2(r - r_1)(1 + \cos^2 r)$ ☐ (*d*) $\omega^2(r - r_1)(1 + \sin^2\theta)$. ☐

276. A roller follower, the diameter 30 mm, is in contact with the straight flask of a tangent cam of base circle diameter 70 mm. The cam has rotated through 30° when the roller is just to leave contact of the flank on its ascent. If the angular speed of the cam is 10 rad/sec, then left the roller, when it is just to leave contact of the flank will be

(*a*) 10 mm ☐ (*b*) 7.73 mm ☐
(*c*) 5 mm ☐ (*d*) 4 mm. ☐

277. In question 275, the velocity of the roller, when the roller is just to leave contact of the flank, will be

(*a*) 10 cm/sec ☐ (*b*) 77.3 cm/sec ☐
(*c*) 33.33 cm/sec ☐ (*d*) 50 cm/sec. ☐

278. In question 276, the acceleration of the roller, when the roller is just to leave contact of the flank, will be

(*a*) 50.8 m/sec^2 ☐ (*b*) 25.3 m/sec^2 ☐
(*c*) 15.4 m/sec^2 ☐ (*d*) 9.615 m/sec^2. ☐

279. The term $\omega^2 \frac{d^3y}{d\theta^3}$, in a cam follower motion, represents

(*a*) acceleration of follower ☐ (*b*) velocity of follower ☐
(*c*) jerk ☐ (*d*) displacement. ☐

280. The left of a flat-faced follower, when it is in contact with the flank of a circular arc cam, is equal to

(*a*) $(R + r)(1 - \cos\theta)$ ☐ (*b*) $(R - r)(1 - \cos\theta)$ ☐
(*c*) $(R - r)(1 + \cos\theta)$ ☐ (*d*) $(R + r)(1 + \cos\theta)$ ☐

where R = Flank radius, r = Minimum radius of cam, and
θ = Angle turned by cam from beginning of the left of follower.

281. The acceleration of a flat-faced follower, when it is in contact with the flank of a circular arc cam, is equal to

(*a*) $\omega^2 (R - r)\sin\theta$ ☐ (*b*) $\omega^2 (R + r)\sin\theta$ ☐
(*c*) $\omega^2 (R - r)\cos\theta$ ☐ (*d*) $\omega^2 (R + r)\cos\theta$. ☐

282. The velocity of a flat-faced follower, when it is in contact with the flank of a circular arc cam, is equal to

(*a*) $\omega (R - r)\sin\theta$ ☐ (*b*) $\omega (R - r)\sin\theta$ ☐
(*c*) $\omega (R + r)\cos\theta$ ☐ (*d*) $\omega (R + r)\cos\theta$. ☐

283. Period of dwell (in case of cam and follower) is the period

(*a*) during which cam rotates ☐
(*b*) during which follower moves from its lower position to highest position ☐
(*c*) during which follower remains stationary during some finite rotation of the cam ☐
(*d*) none of the above. ☐

284. Figure 9.22 shows a cam with a roller follower. The circle given by curve A is called

(*a*) prime circle ☐ (*b*) base circle ☐
(*c*) pitch circle ☐ (*d*) pitch curve. ☐

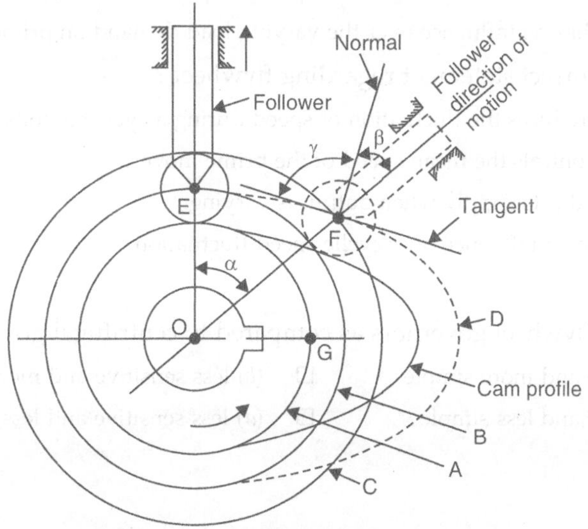

FIGURE 9.22

285. In Fig. 9.22, the circle given by curve *B* is called

(*a*) prime circle ☐ (*b*) base circle ☐

(*c*) pitch circle ☐ (*d*) pitch curve. ☐

286. In Fig. 9.22, the circle given by curve *C* is called

(*a*) prime circle ☐ (*b*) base circle ☐

(*c*) pitch circle ☐ (*d*) pitch curve. ☐

287. In Fig. 9.22, curve *D* is called

(*a*) prime circle ☐ (*b*) base circle ☐

(*c*) pitch circle ☐ (*d*) pitch curve. ☐

288. In Fig. 9.22, cam angle is given by

(*a*) angle α ☐ (*b*) angle β ☐

(*c*) angle γ ☐ (*d*) none of the above. ☐

289. In Fig. 9.22, pressure angle is given by

(*a*) angle α ☐ (*b*) angle β ☐

(*c*) angle γ ☐ (*d*) none of the above. ☐

290. In Fig. 9.22, pitch point is given by

(*a*) point *O* ☐ (*b*) point *E* ☐

(*c*) point *F* ☐ (*d*) point *G*. ☐

291. In Fig. 9.22, trace point is given by

(*a*) point *O* ☐ (*b*) point *E* ☐

(*c*) point *F* ☐ (*d*) point *G*. ☐

292. Which one is a correct statement regarding governor ?

(*a*) The governor reduces the fluctuation of speed during a cycle for constant output load. ☐

(*b*) The governor controls the mean speed over a period for output load variations. ☐

(*c*) The governor has no influence on mean speed for prime mover. ☐

(*d*) The governor has no influence over the varying load demand on prime mover. ☐

293. Which one is a correct statement regarding flywheel ?

(*a*) The fly-wheel reduces the fluctuation of speed during a cycle for constant output load. ☐

(*b*) The flywheel controls the mean speed of the prime mover. ☐

(*c*) The flywheel adjusts supply when output is varying. ☐

(*d*) The flywheel is no influence over cyclic speed fluctuation. ☐

Governors

294. The inertia and flywheel governors as compared to centrifugal governors are

(*a*) more sensitive and more simple ☐ (*b*) less sensitive and more simple ☐

(*c*) more sensitive and less simple ☐ (*d*) less sensitive and less simple. ☐

295. Which one of the following is a spring loaded type governor ?

(*a*) Watt governor ☐ (*b*) Hartnell governor ☐

(*c*) Porter governor ☐ (*d*) Proell governor. ☐

296. Which one of the following is a gravity controlled type governor ?

(*a*) Hartung governor ☐ (*b*) Hartnell governor ☐

(*c*) Pickering governor ☐ (*d*) Watt governor. ☐

297. The height (h) of a watt's governor is given by

(*a*) $h = g\omega^2$ ☐ (*b*) $h = \frac{\omega^2}{g}$ ☐

(*c*) $h = \frac{g}{\omega^2}$ ☐ (*d*) $h = \frac{1}{g\omega^2}$. ☐

298. The height (h) of a Porter governor (when the upper and lower arms are equal and pivoted either on the governor axis or at equal distance from it) is given by

(*a*) $h = \left(\frac{w+W}{w}\right) \times \frac{g}{\omega^2}$ ☐ (*b*) $h = \left(\frac{w+W}{w}\right) \times \frac{\omega^2}{g}$ ☐

(*c*) $h = \left(\frac{w}{w+W}\right) \times \frac{g}{\omega^2}$ ☐ (*d*) $h = \left(\frac{w}{w+W}\right) \times g\omega^2$. ☐

where w = Weight of the balls of governor, and W = Weight of the sleeve.

299. The height of a Watt's governor is

(*a*) inversely proportional to speed ☐ (*b*) inversely proportional to square of speed ☐

(*c*) directly proportional to speed ☐ (*d*) directly proportional to square of speed. ☐

300. The angular speed of a Watt's governor, when its height is 20 cm, will be equal to

(*a*) 20 rad/sec ☐ (*b*) 10 rad/sec ☐

(*c*) 6 rad/sec ☐ (*d*) 7 rad/sec. ☐

301. If, in question 300, the height of governor is 5 cm, then the angular speed of the governor will be

(*a*) 40 rad/sec ☐ (*b*) 5 rad/sec ☐

(*c*) 14 rad/sec ☐ (*d*) 70 rad/sec. ☐

302. If the upper and lower arms of a Porter governor are equal and are pivoted either on the governor axis or at equal distance from it, then the ratio of height of Porter governor to that of Watt's governor is equal to

(*a*) $\frac{W}{W-w}$ ☐ (*b*) $\frac{W-w}{W}$ ☐

(*c*) $\frac{W}{w+W}$ ☐ (*d*) $\frac{W+w}{w}$. ☐

where W = Weight of sleeve, and w = Weight of the balls of governor.

303. When the speed of governor increases

(*a*) height of governor and radius of rotation increases ☐

(*b*) height of governor and radius of rotation decreases ☐

(c) height of governor decreases but radius of rotation increases ☐
(d) height of governor increases but radius of rotation decreases. ☐

304. When the speed of governor decreases
(a) height of governor and radius of rotation increases ☐
(b) height of governor an radius of rotation decreases ☐
(c) height of governor decreases but radius of rotation increases ☐
(d) height of governor increases but radius of rotation decreases. ☐

305. For a Porter governor (which the upper and lower arms arc equal and pivoted either on the governor axis or at equal distance from it) the height of the governor, when speed is increasing and effect of friction is considered, is given by

(a) $\left(\frac{W+R+w}{w}\right)\times\omega^2\times g$ ☐ (b) $\left(\frac{W+R+w}{w}\right)\times\frac{g}{\omega^2}$ ☐

(c) $\left(\frac{W-R+w}{w}\right)\times\frac{g}{\omega^2}$ ☐ (d) $\left(\frac{W-R+w}{w}\right)\times\omega^2\times g$. ☐

where W = Weight of sleeve, w = Weight of balls of governor, and
R = Frictional resistance of sleeve and gears etc.

306. For question 305, when speed is decreasing and effect of friction is considered, then height of governor is given by

(a) $\left(\frac{W+R+w}{w}\right)\times\omega^2\times g$ ☐ (b) $\left(\frac{W+R+w}{w}\right)\times\frac{g}{\omega^2}$ ☐

(c) $\left(\frac{W-R+w}{w}\right)\times\frac{g}{\omega^2}$ ☐ (d) $\left(\frac{W-R+w}{w}\right)\times\omega^2\times g$. ☐

307. The friction, at the sleeve of a governor
(a) increases the sensitiveness of the governor ☐
(b) decreases the sensitiveness of the governor ☐
(c) has no effect on sensitiveness of the governor ☐
(d) none of the above. ☐

308. Proell governor as compared to Porter governor
(a) is more sensitive ☐ (b) is less sensitive ☐
(c) requires weights of smaller size ☐ (d) requires weights of bigger size ☐
(e) both (a) and (c). ☐

309. The spring loaded governors as compared to gravity controlled governors
(a) may be operated at very high speeds ☐
(b) are much smaller in overall size ☐
(c) may revolve about a vertical or inclined axis ☐
(d) all of the above ☐
(e) none of the above. ☐

310. If F = centrifugal force on the ball of a Hartnell governor, S = spring force exerted on the sleeve, a and b are the lengths of vertical and horizontal arms of the bell crank levers, then spring force exerted on the sleeve (neglecting the effect of the pull of gravity on governor balls and arms) is given by

(a) $S = \dfrac{F \times a}{2b}$ ☐ (b) $S = \dfrac{F \times b}{2a}$ ☐

(c) $S = 2F \times \dfrac{a}{b}$ ☐ (d) $S = \dfrac{b}{2F \times a}$. ☐

311. The lift of the sleeve in a Hartnell governor of question 310 is equal to

(a) $\dfrac{a}{b}(r_1 - r_2)$ ☐ (b) $\dfrac{b}{a}(r_1 - r_2)$ ☐

(c) $\dfrac{a}{b}(r_1 + r_2)$ ☐ (d) $\dfrac{b}{a}(r_1 + r_2)$. ☐

where r_1 = Maximum radius of rotation, and r_2 = Minimum radius of rotation.

312. The stiffness of the spring in Hartnell governor of question 310 is equal to

(a) $2\left(\dfrac{a}{b}\right)^2\left(\dfrac{F_1 + F_2}{r_1 - r_2}\right)$ ☐ (b) $2\left(\dfrac{a}{b}\right)^2\left(\dfrac{F_1 + F_2}{r_1 - r_2}\right)$ ☐

(c) $2\left(\dfrac{a}{b}\right)^2\left(\dfrac{F_1 - F_2}{r_1 + r_2}\right)$ ☐ (d) $2\left(\dfrac{a}{b}\right)^2\left(\dfrac{F_1 + F_2}{r_1 + r_2}\right)$. ☐

where F_1 = Centrifugal force at maximum radius of rotation, and

F_2 = Centrifugal force at minimum radius of rotation.

313. The stiffness of the spring in a Hartnell governor is also equal to

(a) $\dfrac{S_1 - S_2}{h}$ ☐ (b) $\dfrac{2(S_1 - S_2)}{h}$ ☐

(c) $\dfrac{S_1 - S_2}{2h}$ ☐ (d) $\dfrac{S_1 + S_2}{h}$. ☐

where S_2 = Spring force exerted on the sleeve at maximum radius of rotation,

S_2 = Spring force exerted on the sleeve at minimum radius of rotation, and

h = Lift of sleeve.

314. In a Hartnell governor, the weight of each ball is 2.45 kgf. The maximum and minimum speed of rotations are 10 rad/sec and 8 rad/sec. The maximum and minimum radii of rotations are 20 cm and 12 cm respectively. The lengths of vertical and horizontal arms of bell crank levers are 20 cm and 10 cm respectively. If the effect of pull of gravity is neglected then the lift of the sleeve will be equal to

(a) 10 cm ☐ (b) 8 cm ☐

(c) 6 cm ☐ (d) 4 cm. ☐

315. For the question 314, the spring force exerted on the sleeve at maximum radius of rotation will be

(a) 20 kgf ☐ (b) 10 kgf ☐

(c) 7.68 kgf ☐ (d) 3.08 kgf. ☐

316. For the question 314, the spring force exerted on the sleeve at minimum radius of rotation will be

(*a*) 20 kgf ☐ (*b*) 10 kgf ☐
(*c*) 7.68 kgf ☐ (*d*) 3.08 kgf. ☐

317. For the question 314, the stiffness of the spring will be

(*a*) 20 kgf ☐ (*b*) 10 kgf ☐
(*c*) 7.68 kgf ☐ (*d*) 3.08 kgf. ☐

318. The range of speed a governor is equal to

(*a*) sum of maximum and minimum equilibrium speeds ☐
(*b*) difference of maximum and minimum equilibrium speeds ☐
(*c*) average of sum of maximum and minimum equilibrium speeds ☐
(*d*) average of difference of maximum and minimum equilibrium speeds. ☐

319. Sensitiveness of a governor is defined as the ratio of

(*a*) mean speed to range of speed ☐
(*b*) range of speed to mean speed ☐
(*c*) maximum equilibrium speed to mean speed ☐
(*d*) minimum equilibrium speed to mean speed. ☐

320. A governor is said to be sensitive when

(*a*) for a small change of speed, the displacement of sleeve is maximum ☐
(*b*) the governor readily responds to a small alternation of speed ☐
(*c*) the ratio of range of speed to mean speed is maximum ☐
(*d*) any one of the above ☐
(*e*) none of the above. ☐

321. Which one of the following is a correct statement regarding the effect of friction on governors ?

(*a*) Due to friction, there is a range of speed for any sleeve position over which the governor is sensitive. ☐
(*b*) Due to friction, the governor will have same configuration whether it approaches equilibrium from the higher or lower speed side. ☐
(*c*) Due to friction, the governor will have different configuration whether it approaches equilibrium from the higher or lower speed side. ☐
(*d*) Due to fiction, the load on sleeve decreases when sleeve is rising. ☐

322. A governor is said to isochronous when

(*a*) the equilibrium speed is constant for all radii of rotation of the balls within working range ☐
(*b*) the range of speed is zero for all radii of rotation of the balls within working range ☐
(*c*) any one of the above ☐
(*d*) none of the above. ☐

323. The mean force exerted by the governor on the sleeve for a given change of speed is known as

(*a*) sensitiveness of governor ☐ (*b*) effort of governor ☐
(*c*) stability of governor ☐ (*d*) frictional force of governor. ☐

324. If for each speed within working range, there is only one radius of rotation of the governor's balls, then the governor is said to be

(*a*) hunting ☐ (*b*) isochronous ☐

(*c*) stable ☐ (*d*) sensitive. ☐

325. When a governor is over sensitive, the sleeve will oscillate between two extreme positions when there is a slightest change of speed. Then the governor is said to be

(*a*) hunting ☐ (*b*) isochronous ☐

(*c*) stable ☐ (*d*) none of the above. ☐

326. The power of governor is

(*a*) the work done at the sleeve for a given percentage change of speed ☐

(*b*) equal to the product of governor effort and maximum equilibrium speed ☐

(*c*) equal to product of governor effort and the displacement of sleeve ☐

(*d*) both (*a*) and (*c*) ☐

(*e*) none of the above. ☐

327. The effort of a Porter governor (when the arms of the governor are equal and have equal inclination with the axis of the governor spindle) is equal to

(*a*) $C(w + W)$ ☐ (*b*) $\frac{(w+W)}{C}$ ☐

(*c*) $\frac{(w+W \times C)}{C}$ ☐ (*d*) $\frac{(C \times w+W)}{C}$. ☐

where w = Weight of ball of governor, W = Weight on sleeve, and

C = Percentage increase in speed.

328. The effort of a Hartnell governor (where the moments due to weight of the arms and ball are neglected) is equal to

(*a*) $\frac{C}{S}$ ☐ (*b*) CS ☐

(*c*) $\frac{1}{CS}$ ☐ (*d*) $\frac{S}{C}$. ☐

where C = Percentage increase in speed, and S = Spring force exerted on the sleeve.

329. The power of a Porter governor (when the arms of the governor are equal and have equal inclination with the axis of the governor spindle) is equal to

(*a*) $\left(\frac{1+2C}{4C}\right)(w \times W) \times h$ ☐ (*b*) $\left(\frac{1-2C}{4C^2}\right)(w+W) \times h$ ☐

(*c*) $\left(\frac{4C^2}{1+2C}\right)(w \times W) \times h$ ☐ (*d*) $\left(\frac{4C^2}{1-2C}\right)(w+W) \times h$. ☐

where h = Height of governor, and C = Percentage increase in speed.

330. The controlling force in case of Porter governor is provided by

(*a*) weight of balls ☐ (*b*) springs ☐

(*c*) both (*a*) and (*c*) ☐ (*d*) none of the above. ☐

331. The controlling force, in case of Hartnell governor, is provided by

(*a*) weight of balls ☐ (*b*) springs ☐

(*c*) both (*a*) and (*b*) ☐ (*d*) none of the above. ☐

332. The controlling force curve is a graph between controlling force and

(*a*) speed of rotation ☐ (*b*) radius of rotation ☐

(*c*) range of speed ☐ (*d*) lift of governor. ☐

333. The controlling force curve is used to find out

(*a*) stability of a governor ☐ (*b*) sensitiveness of a governor ☐

(*c*) hunting of a governor ☐ (*d*) (*a*) and (*b*) only. ☐

334. Figure 9.23 shows that variation of controlling force with the radius of rotation of the governors. Curve *A* holds good for

(*a*) porter governor ☐

(*b*) stable Hartnell governor ☐

(*c*) unstable Hartnell governor ☐

(*d*) isochronous Hartnell governor. ☐

FIGURE 9.23

335. The curve *B* in Fig. 9.23 holds good for

(*a*) porter governor ☐

(*b*) stable Hartnell governor ☐

(*c*) unstable Hartnell governor ☐

(*d*) isochronous Hartnell governor. ☐

336. The curve C in Fig. 9.23 holds good for

(*a*) porter governor ☐ (*b*) stable Hartnell governor ☐

(*c*) unstable Hartnell governor ☐ (*d*) isochronous Hartnell governor. ☐

337. The curve *D* in Fig. 9.23 holds good for

(*a*) porter governor ☐ (*b*) stable Hartnell governor ☐

(*c*) unstable Hartnell governor ☐ (*d*) isochronous Hartnell governor. ☐

338. The condition for stability of a governor is stated as

(*a*) the rate of increase of controlling force is equal to the rate of increase of radius ☐

(*b*) the rate of increase of controlling force must be greater than the rate of increase of radius ☐

(*c*) the rate of increase of controlling force must be less than the rate of increase of radius ☐

(*d*) none of the above. ☐

339. The equation for controlling curve for a stable Hartnell governor is given by

(*a*) $F_c = mr + C$ ☐ (*b*) $F_c = mr$ ☐

(*c*) $F_c = mr - C$ ☐ (*d*) none of the above ☐

where F_c = Controlling force, and *m* and C = Arc constants.

340. Choose the wrong statement

(*a*) Static balance between revolving masses exists if the masses are in equilibrium when not running. ☐

(b) Dynamic balance exists when the inertia forces and couples exerted by the moving masses are in equilibrium among themselves. ☐

(c) All revolving and reciprocating masses should be balanced as completely as possible. ☐

(d) Balancing of rotating and reciprocating parts of an engine is necessary when it runs at slow speed. ☐

Balancing

341. A weight attached to a rotating shaft is to be balanced by a weight *B* attached to the shaft and revolving in the same plane as the weight *W*. If '*r*' is the distance of centre of gravity of the revolving weight *W* from the axis of shaft and '*b*' is the distance of centre of gravity of balancing weight *B* from the axis of shaft, then the necessary condition for balancing if

(a) $W \times b = B \times r$ ☐ (b) $W \times B = b \times r$ ☐

(c) $W \times r = B \times b$ ☐ (d) $W = B \times b \times r$. ☐

342. Which of the following is a correct statement for balancing a revolving disturbing weight by a weight revolving in the same plane ?

(a) The condition for balancing is independent of speed. ☐

(b) The shall is in static and dynamic balance. ☐

(c) The centre of gravity is on the axis of the shaft and remains in a fixed position relative to the frame whether the shaft is rotating or standing still. ☐

(d) All of the above. ☐

(e) None of the above. ☐

343. A weight *W* attached to a rotating shaft is to be balanced. But it is not possible to introduce the single balance weight in the plane of rotation of the disturbing weight. Then the single balance weight is introduced in a plane parallel to the plane of rotation of the disturbing weight. The system will be in

(a) static balance ☐ (b) dynamic balance ☐

(c) complete balance ☐ (d) none of the above. ☐

344. When the balance weights are introduced in a plane parallel to the plane of rotation of the disturbing weight, then the minimum number of balance weights, for balancing a single revolving disturbing weight, will be

(a) one ☐ (b) two ☐

(c) three ☐ (d) four. ☐

345. A weight of 10 kg is attached to a rotating shaft at a radius of 100 cm as shown in Fig. 9.24. This weight is to be balanced by two weights B_1 and B_2 at radii 50 cm and 125 cm respectively. If the planes of rotation of the balancing weights and disturbing weights are parallel, then for complete balancing, the weight B_1 will be equal to

(a) 15 kg ☐

(b) 10 kg ☐

(c) 5 kg ☐

(d) 2 kg. ☐

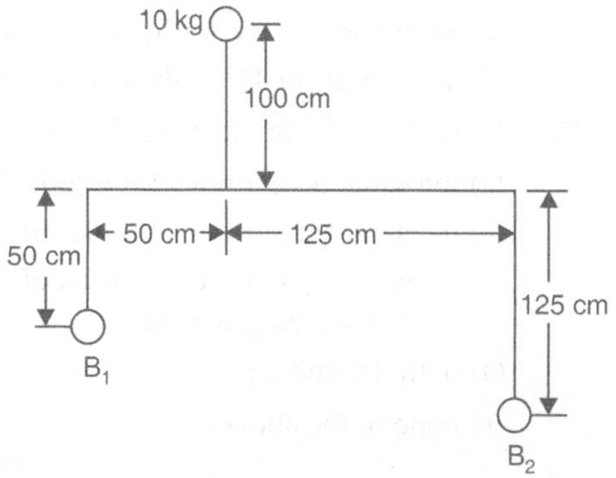

FIGURE 9.24

346. In question 345, for complete balancing, the weight B_2 will be equal to (refer to Fig. 9.24)

(*a*) 15 kg (*b*) 10 kg
(*c*) 5 kg (*d*) 2 kg.

347. Figure 9.25 shows the three weights *A*, *B*, *C* revolving in the same plane with radii r_1, r_2 and r_3 respectively. The angular positions of *B* and *C* are θ_1° and θ_2° respectively from *A*. The system will be balanced by θ weight whose magnitude is equal to

(*a*) $A + B + C$ (*b*) $A + B \cos \theta_1 + C \cos \theta_2$
(*c*) $A + B \sin \theta_1 + C \sin \theta_2$ (*d*) $Ar_1 + Br_2 + Cr_3$
(*e*) none of the above.

348. If in question 347, the balance weight is placed at an angle of θ° from *A* and at a radius *r*, then the magnitude of balance weight would be equal to (refer to Fig. 9.25)

(*a*) $\dfrac{A + B\cos\theta_1 + C\cos\theta_2}{\cos\theta}$

(*b*) $\dfrac{A + B\sin\theta_1 + C\sin\theta_2}{\sin\theta}$

(*c*) $\dfrac{Ar_1 + Br_2\cos\theta_1 + Cr_3\cos\theta_2}{r\cos\theta}$

(*d*) $\dfrac{Ar_1 + Br_2\cos\theta_1 + Cr_3\cos\theta_2}{r\sin\theta}$

(*e*) none of the above
(*f*) both (*c*) and (*d*).

FIGURE 9.25

349. In question 348, the value of θ (*i.e.*, the angle made by the balance weight with *A*) is given by

(*a*) $\sin^{-1}\dfrac{F_V}{F_H}$ (*b*) $\cos^{-1}\dfrac{F_V}{F_H}$

(*c*) $\tan^{-1}\dfrac{F_V}{F_H}$ (*d*) $\tan^{-1}\dfrac{F_H}{F_V}$

where $F_H = Ar_1 + Br_2 \cos \theta_1 + Cr_3 \cos \theta_2$, and $F_V = Ar_1 + Br_2 \sin \theta_1 + Cr_3 \sin \theta_2$.

350. Reference plane is a plane which is

(*a*) passing through the plane of rotation of the rotating weight
(*b*) passing through the plane of rotation of the balancing weight
(*c*) at an angle of 45° to the rotating weight
(*d*) at an angle of 45° to the balancing weight.

351. The effect of transferring the rotating weight in the reference plane is to produce

(*a*) the same magnitude of the centrifugal force as produced by the rotating weight
(*b*) the same magnitude of the centrifugal force as produced by the balancing weight
(*c*) a couple equal to the product of centrifugal force and distance of the reference plane of rotation from disturbing weight
(*d*) both (*a*) and (*c*)
(*e*) none of the above.

352. For complete balance of the several revolving weights in different planes, the conditions is

(a) the vector sum of the forces must be zero ☐

(b) the vector sum of all the couples must be zero ☐

(c) both (a) and (b) ☐

(d) none of the above. ☐

353. The acceleration of the piston of a reciprocating engine is equal to

(a) $\omega^2 r\left(\cos\theta + \frac{\sin 2\theta}{n}\right)$ ☐ (b) $\omega^2 r\,(\cos\theta + \cos 2\theta)$ ☐

(c) $\omega^2 r\left(\cos\theta + \frac{\cos 2\theta}{n}\right)$ ☐ (d) $\omega^2 r\left(\sin\theta + \frac{\sin 2\theta}{n}\right)$ ☐

where ω = Angular speed of crank, r = Crank radius,

θ = Inclination of crank with inner dead centre, and

n = Ratio of the length of connecting rod to crank radius = $\frac{l}{r}$.

354. The inertia force due to weight of reciprocating parts in a reciprocating engine is equal to

(a) $\frac{W}{g}\omega^2 r\left(\cos\theta + \frac{\sin 2\theta}{n}\right)$ ☐ (b) $\frac{W}{g}\omega^2 r\,(\cos\theta + \cos 2\theta)$ ☐

(c) $\frac{W}{g}\omega^2 r\left(\cos\theta + \frac{\cos 2\theta}{n}\right)$ ☐ (d) $\frac{W}{g}\omega^2 r\left(\sin\theta + \frac{\sin 2\theta}{n}\right)$ ☐

where W = Weight of reciprocating parts.

355. The term $\left(\frac{W}{g}\omega^2 r\cos 2\theta\right)$ in reciprocating engine is called

(a) primary couple ☐ (b) secondary couple ☐

(c) primary disturbing force ☐ (d) secondary disturbing force. ☐

356. The term $\left(\frac{W}{g}\omega^2 r\frac{\cos 2\theta}{n}\right)$ in reciprocating engine is called

(a) primary couple ☐ (b) secondary couple ☐

(c) primary disturbing force ☐ (d) secondary disturbing force. ☐

357. The frequency of the secondary force as compared to primary force is

(a) One-half ☐ (b) double ☐

(c) One-fourth ☐ (d) One-third. ☐

358. The condition of balancing the reciprocating masses of a reciprocating engine is to balance

(a) primary forces couple ☐ (b) secondary forces only ☐

(c) primary couples only ☐ (d) secondary couples only. ☐

(e) all of the above ☐ (f) none of the above, ☐

359. When the angle of inclination of crank with inner dead centre of a reciprocating engine is 0°, then

(a) primary force is maximum ☐ (b) secondary force is maximum ☐

(c) primary force is minimum ☐ (d) secondary force is minimum. ☐

(e) both (a) and (b) ☐ (f) both (c) and (d). ☐

360. When the angle of inclination of crank with inner dead centre of a reciprocating engine is 90°, then

(*a*) primary force is maximum ☐ (*b*) secondary force is maximum ☐
(*c*) primary force is zero ☐ (*d*) secondary force is zero ☐
(*e*) both (*b*) and (*c*) ☐ (*f*) both (*a*) and (*d*). ☐

361. When the angle of inclination of crank with inner dead centre of a reciprocating engine is 45°, then

(*a*) primary force is maximum ☐ (*b*) secondary force is maximum ☐
(*c*) primary force is zero ☐ (*d*) secondary force is zero ☐
(*e*) none of the above. ☐

362. Choose the correct statement

(*a*) The secondary disturbing force is more than primary disturbine force, ☐
(*b*) For one revolution of crank, the secondary force becomes maximum two times, ☐
(*c*) For one revolution of crank, the primary force becomes maximum four times. ☐
(*d*) The primary disturbing force due to inertia of reciprocating parts in an engine is equal to $\frac{W}{g}\omega^2 r\cos\theta$. ☐

363. The secondary disturbing force due to inertia of reciprocating parts of an engine is equal to the centrifugal produced by the equivalent crank

(*a*) having an equivalent radius of $\frac{r^2}{4l}$ ☐ (*b*) running at twice the speed ☐
(*c*) having a weight attached to its end ☐ (*d*) all of the above ☐
(*e*) none of the above. ☐

364. If a weight equal to the weights of reciprocating parts is attached diametrically opposite to crank at crank radius, then

(*a*) primary disturbing force is completely balanced ☐
(*b*) primary disturbing force is not balanced ☐
(*c*) primary disturbing force is completely balanced but an unbalanced vertical force is introduced ☐
(*d*) none of the above. ☐

365. The primary forces are balanced partially. The balance weight placed diametrically opposite to the crank is cW, where W is the weight of reciprocating parts. The usual value of c for locomotive is

(*a*) one-fourth ☐ (*b*) one-half ☐
(*c*) one-third ☐ (*d*) two-thirds. ☐

366. If the balance weight placed diametrically opposite to the crank at crank radius (r) is cW, then unbalanced force along the line of stroke of the reciprocating engine, is equal to

(*a*) $\frac{cW}{g}\omega^2 r\cos\theta$ ☐ (*b*) $(1-c)\frac{W}{g}\omega^2 r\cos\theta$ ☐
(*c*) $\frac{cW}{g}\omega^2 r\sin\theta$ ☐ (*d*) $(1-c)\frac{W}{g}\omega^2 r\sin\theta$. ☐

367. In question 366, the unbalanced force normal to the line of stroke of the reciprocating engine, is equal to

(a) $\frac{cW}{g}\omega^2 r\cos\theta$ ☐ (b) $(1-c)\frac{W}{g}\omega^2 r\cos\theta$ ☐

(c) $\frac{cW}{g}\omega^2 r\sin\theta$ ☐ (d) $(1-c)\frac{W}{g}\omega^2 r\sin\theta$. ☐

368. In question 366, the resultant unbalanced force at any instant, is equal to

(a) $\frac{cW}{g}\omega^2 r$ ☐ (b) $(1-c)\frac{W}{g}\omega^2 r$ ☐

(c) $\frac{W}{g}\omega^2 r\sqrt{(1-c)^2\cos^2\theta + c^2\sin^2\theta}$ ☐ (d) $\frac{W}{g}\omega^2 r\sqrt{(1-c)^2\cos^2\theta + \sin^2\theta}$. ☐

369. If the ratio of the length of connecting rod to the crank radius is very large, then

(a) primary forces are negligible ☐ (b) primary forces are maximum ☐

(c) secondary forces are negligible ☐ (d) secondary forces are maximum. ☐

370. In locomotives, with two cylinders, the cranks of two cylinders are placed at right angles to each other in order to

(a) start the locomotive in any position ☐ (b) reduce the effect of secondary forces ☐

(c) reduced the effect of primary forces ☐ (d) all of the above ☐

(e) (a) and (h) only. ☐

371. The swaying couple and variation in the tractive effort is due to

(a) primary disturbing force ☐ (b) secondary disturbing force ☐

(c) both (a) and (c) ☐ (d) none of the above. ☐

372. The unbalanced force acting vertically upward or downward due to balancing of the reciprocating parts in a locomotive varies

(a) directly with the speed ☐ (b) directly with the square of the speed ☐

(c) inversely with the speed ☐ (d) inversely with the square of the speed. ☐

373. The dead weight of the wheel and locomotive is

(a) a function of speed ☐ (b) a function of square of the speed ☐

(c) independent of the speed ☐ (d) none of the above. ☐

374. The lifting of the wheels from the rails and striking back on the rails take place at

(a) low speed ☐ (b) normal speed ☐

(c) high speed ☐ (d) none of the above. ☐

375. When the centre of gravity of the balance weights is either directly below the wheel centre or directly above the wheel centre, then

(a) unbalanced horizontal force is maximum ☐

(b) unbalanced vertical force is maximum ☐

(c) unbalanced vertical force is minimum ☐

(d) none of the above. ☐

376. Hammer blow is

(*a*) maximum value of unbalanced force along the line of stroke □

(*b*) maximum value of unbalanced force perpendicular to the line of stroke □

(*c*) resultant value of the unbalanced force □

(*d*) resultant value of unbalanced force perpendicular to the line of stroke. □

377. If B is the balance weight for reciprocating parts placed at a radius (r) in a locomotive for which W is the dead load acting on each wheel, then the force trying to lift to the wheels from the rails will be

(*a*) $W + \frac{B}{g}\omega^3 r \sin\theta$ □ (*b*) $W - \frac{B}{g}\omega^1 r^2 \sin\theta$ □

(*c*) $W - \frac{B}{g}\omega^2 r$ □ (*d*) $W - \frac{B}{g}\omega^2 r \cos\theta$. □

378. The lifting of wheel can be prevented if the angular speed is less than

(*a*) $\sqrt{\frac{W}{g}\frac{r}{B}}$ □ (*b*) $\sqrt{\frac{Wg}{B \times r}}$ □

(*c*) $\sqrt{\frac{g \times B}{W \times r}}$ □ (*d*) $\sqrt{\frac{W}{g \times B \times r}}$. □

where W = Dead weight acting on a wheel, r = Radius at which balance weight is placed, and B = Balance weight for reciprocating masses.

379. For two cylinder locomotives having their cranks at right angles to each other, the tractive force is equal to

(*a*) $(1-c)\frac{W}{g}\omega^2 r \cos\theta$ □ (*b*) $(1-c)\frac{W}{g}\omega^2 r \sin\theta$ □

(*c*) $(1-c)\frac{W}{g}\omega^2 r (\cos\theta - \sin\theta)$ □ (*d*) $(1-c)\frac{W}{g}\omega^2 r (\cos\theta - 1)$. □

where W = Weight of reciprocating parts, c = Fraction of the reciprocating parts, r = Radius of crank.

380. The tractive force will be maximum or minimum when angle of inclination of crank with the line of stroke is equal to

(*a*) 45° and 90° □ (*b*) 90° and 180° □

(*c*) 135° and 315° □ (*d*) 0° and 180°. □

381. If C = 2/3, the maximum value of tractive force is equal to

(*a*) $\sqrt{2}\frac{W}{g}\omega^2 r$ □ (*b*) $\frac{\sqrt{2}}{3}\frac{W}{g}\omega^2 r$ □

(*c*) $\frac{3}{\sqrt{2}}\frac{W}{g}\omega^2 r$ □ (*d*) $3\sqrt{2}\frac{W}{g}\omega^2 r$. □

382. Choose the wrong statement

(*a*) Swaying couple is maximum or minimum when angle of inclination of crank to the line of stroke is 45° or 225. □

(b) Tractive force is maximum or minimum when angle of inclination of crank to the line of stroke is 135° and 315°. ☐

(c) For starting the locomotive in any position, the cranks of the two cylinders of a locomotive are placed at right angle to each other. ☐

(d) The sway couple is due to secondary unbalanced force. ☐

383. Choose the correct statement

(a) The effect of hammer blow can be reduced by decreasing the speed and using two or three pairs of wheels coupled together. ☐

(b) Partial balancing in locomotive results in variation in tractive force, hammer blow and swaying couple. ☐

(c) In locomotives, the effect of secondary forces is minimised by making the ratio of the connecting rod length to the crank radius very large. ☐

(d) All of the above. ☐

(e) None of the above. ☐

Vibrations

384. The number of cycles of motion completed in a unit interval of time is known as

(a) period ☐ (b) frequency ☐

(c) resonance ☐ (d) damping. ☐

385. The time taken, by a vibrating body to complete one cycle, is known as

(a) period ☐ (b) frequency ☐

(c) resonance ☐ (d) damping. ☐

386. Natural frequency of a system is due to

(a) free vibration ☐ (b) forced vibration ☐

(c) resonance ☐ (d) damping. ☐

387. Resonance is a phenomenon in which the frequency of the external exciting force

(a) is half the natural frequency of the system ☐

(b) is double the natural frequency of the system ☐

(c) coincides with the natural frequency of the system ☐

(d) does not coincide with the natural frequency of the system. ☐

388. A vibrating body is said to be number the action of damping when

(a) the amplitude of the body gradually increases with time ☐

(b) the amplitude of the body is constant with respect to time ☐

(c) the amplitude of the body gradually decreases with time ☐

(d) none of the above. ☐

389. If the particles of a body vibrate parallel to the axis of the body, then the body is said to have

(a) transverse vibration ☐ (b) longitudinal vibration ☐

(c) torsional vibration ☐ (d) none of the above. ☐

390. If the particles of a body vibrate perpendicular to the axis of the body, then the body is said to have

(a) transverse vibration ☐ (b) longitudinal vibration ☐

(c) torsional vibration ☐ (d) none of the above. ☐

391. If the particles of a body vibrates along a circular arc, whose centre lies on the axis of the shaft, then the body is said to have

(*a*) transverse vibration ☐ (*b*) longitudinal vibration ☐

(*c*) torsional vibration ☐ (*d*) none of the above. ☐

392. Choose the wrong statement

(*a*) Bending stresses are induced when a body is having transverse vibrations. ☐

(*b*) Shear stresses are induced when a body is having torsional vibrations. ☐

(*c*) Tensile or compressive stresses are induced when a body is having longitudinal vibrations. ☐

(*d*) All of the above. ☐

(*e*) None of the above. ☐

393. The equation of motion for the free vibration of the spring mass system without viscous damping is given by

(*a*) $\frac{d^2x}{dt^2}+\frac{c}{m}+\frac{dx}{dt}+\frac{k}{m}x=0$ ☐ (*b*) $\frac{d^2x}{dt^2}+\frac{k}{m}x=0$ ☐

(*c*) $\frac{d^2x}{dt^2}+\frac{c}{m}\cdot\frac{dx}{dt}+\frac{k}{m}x=F\sin\omega t$ ☐ (*d*) $\frac{d^2x}{dt^2}+\frac{k}{m}x=F\sin\omega t.$ ☐

where c = Co-efficient of viscous damping, m = Mass of the vibrating body, k = Stiffness of the spring, $F\sin\omega t$ = Harmonic force acting on the body.

394. The equation of motion for the free vibration of the spring mass system with viscous damping is given by

(*a*) $\frac{d^2x}{dt^2}+\frac{c}{m}\frac{dx}{dt}+\frac{k}{m}x=0$ ☐ (*b*) $\frac{d^2x}{dt^2}+\frac{k}{m}x=0$ ☐

(*c*) $\frac{d^2x}{dt^2}+\frac{c}{m}\cdot\frac{dx}{dt}+\frac{k}{m}x=F\sin\omega t$ ☐ (*d*) $\frac{d^2x}{dt^2}+\frac{k}{m}x=F\sin\omega t.$ ☐

395. The equation of motion for the forced vibration of spring mass system with viscous damping is given by

(*a*) $\frac{d^2x}{dt^2}+\frac{c}{m}+\frac{dx}{dt}+\frac{k}{m}x=0$ ☐ (*b*) $\frac{d^2x}{dt^2}+\frac{k}{m}x=0$ ☐

(*c*) $\frac{d^2x}{dt^2}+\frac{c}{m}\cdot\frac{dx}{dt}+\frac{k}{m}x=F\sin\omega t$ ☐ (*d*) $\frac{d^2x}{dt^2}+\frac{k}{m}x=F\sin\omega t.$ ☐

396. The natural frequency of the free longitudinal vibrations is given by

(*a*) $2\pi\sqrt{\frac{m}{k}}$ ☐ (*b*) $\frac{1}{2\pi}\sqrt{\frac{k}{m}}$ ☐

(*c*) $2\pi\sqrt{\frac{g}{\delta}}$ ☐ (*d*) $2\pi\sqrt{\frac{\delta}{g}}$ ☐

(*e*) both (*b*) and (*c*) ☐ (*f*) both (*a*) and (*d*). ☐

where m = Mass of the body, k = Stiffness of spring, and δ = Static deflection of the spring.

397. The natural period of the free longitudinal vibrations is given by

(*a*) $2\pi\sqrt{\frac{m}{k}}$ ☐ (*b*) $\frac{1}{2\pi}\sqrt{\frac{k}{m}}$ ☐

(*c*) $\frac{1}{2\pi}\sqrt{\frac{g}{\delta}}$ ☐ (*d*) $2\pi\sqrt{\frac{\delta}{g}}$ ☐

(*e*) both (*b*) and (*c*) ☐ (*f*) both (*a*) and (*d*) ☐

398. The natural frequency of the free longitudinal vibrations is the

(*a*) same as the natural period of vibrations ☐

(*b*) reciprocal of the natural period of vibrations ☐

(*c*) twice the natural period of vibrations ☐

(*d*) half the natural period of vibrations. ☐

399. Two springs of stiffness k_1 and k_2 are connected in series. The resultant stiffness will be equal to

(*a*) $k_1 + k_2$ ☐ (*b*) $\frac{k_1 + k_2}{2}$ ☐

(*c*) $\frac{k_1 + k_2}{k_1 k_2}$ ☐ (*d*) $\frac{k_1 k_2}{k_1 + k_2}$. ☐

400. The natural frequency of a system is a function of

(*a*) stiffness of the system ☐ (*b*) mass of the system ☐

(*c*) amplitude of oscillation ☐ (*d*) both (*a*) and (*b*) ☐

(*e*) none of the above. ☐

401. A mass m is attached to one end of a spring, whose upper end is fixed. The mass and stiffness of the spring are m_1 and k respectively. If the mass oscillates longitudinally, then natural frequency of the mass will be equal to

(*a*) $\frac{1}{2\pi}\sqrt{\frac{k}{(m + m_1)}}$ ☐ (*b*) $\frac{1}{2\pi}\sqrt{\frac{2k}{(m + m_1)}}$ ☐

(*c*) $\frac{1}{2\pi}\sqrt{\frac{k}{\left(m + \frac{m_1}{2}\right)}}$ ☐ (*d*) $\frac{1}{2\pi}\sqrt{\frac{2k}{\left(m + \frac{m_1}{3}\right)}}$. ☐

402. If in question 401, the mass of spring is negligible then the natural frequency of the mass (attached to the spring) as compared to that of question 401, will be

(*a*) less ☐ (*b*) more ☐

(*c*) same ☐ (*d*) may be more or less ☐

403. At certain speed, the rotating shaft tends to vibrate violently in transverse direction. Then speed is called

(*a*) whirling speed ☐ (*b*) critical speed ☐

(*c*) whipping speed ☐ (*d*) all of the above ☐

(*e*) none of the above. ☐

404. The critical speed of a shaft carrying a weight (W) at the centre is given by

(a) $\sqrt{\dfrac{k \times W}{g}}$ ☐ (b) $\sqrt{\dfrac{k \times g}{W}}$ ☐

(c) $\sqrt{\dfrac{g}{kW}}$ ☐ (d) $\sqrt{\dfrac{W}{k \times g}}$. ☐

where k = Stiffness of the shaft.

405. The natural frequency of shaft carrying a weight (W) at the centre and vibrating laterally is given by

(a) $\dfrac{5.623}{\sqrt{\delta}}$ ☐ (b) $\dfrac{4.987}{\sqrt{\delta}}$ ☐

(c) $\dfrac{4.5}{\sqrt{\delta}}$ ☐ (d) $\dfrac{3.25}{\sqrt{\delta}}$. ☐

where δ = Static deflection in cm due to load W.

406. A shaft carries a weight (W) at the centre. Let the C.G. of the weight is displaced by an amount 'e' from the axis of the rotation. When the shaft is rotating, the C.G. will be subjected to centrifugal force and it will be displaced further. If 'y' is the additional displacement of C.G. from the axis of rotation due to centrifugal force, then the ratio of 'y' to 'e' is given by

(a) $\dfrac{1}{\left(\dfrac{\omega_c}{\omega}\right)^2 + 1}$ ☐ (b) $\dfrac{1}{\left(\dfrac{\omega_c}{\omega}\right)^2 - 1}$ ☐

(c) $\left(\dfrac{\omega_c}{\omega}\right)^2 + 1$ ☐ (d) $\dfrac{\omega}{\left(\dfrac{\omega_c}{\omega}\right)^2 - 1}$. ☐

where ω_c = Critical speed of the shaft and ω = Angular speed of the shaft.

407. In question 406, the additional displacement of C.G., from the axis of rotation due to centrifugal force, will be positive when the shaft is rotating with a speed which is

(a) equal to critical speed ☐ (b) less than critical speed ☐

(c) more than critical speed ☐ (d) none of the above. ☐

408. In question 406, the additional displacement of C.G. from the axis of rotation due to centrifugal force will be negative where the shaft is rotation with a speed which is

(a) equal to critical speed ☐ (b) less than critical speed ☐

(c) more than critical speed ☐ (d) none of the above. ☐

409. When a shaft is rotating at a speed which is less than critical speed the phase difference between displacement and centrifugal force is

(a) 180° ☐ (b) 90° ☐

(c) 45° ☐ (d) 0°. ☐

410. When a shaft is rotating at a speed which is more than critical speed, the phase difference between the displacement and centrifugal force is

(a) 180° ☐ (b) 90° ☐

(c) 45° ☐ (d) 0°. ☐

411. In rotating horizontal shafts, there are large vibrations when the shafts are rotating at a speed which is

(*a*) one-fourth of critical speed ☐ (*b*) one-third of critical speed ☐

(*c*) one-half of critical speed ☐ (*d*) two-third of critical speed. ☐

412. The natural frequency of free transverse vibrations of a uniformly distributed loaded shaft, having ends simply supported, is given by

(*a*) $\frac{5.623}{\sqrt{\delta}}$ ☐ (*b*) $\frac{4.987}{\sqrt{\delta}}$ ☐

(*c*) $\frac{4.5}{\sqrt{\delta}}$ ☐ (*d*) $\frac{3.25}{\sqrt{\delta}}$. ☐

where δ = Maximum static deflection of the shaft due to uniformly distributed load in cm.

413. Two heavy rotating masses are connected by shafts of lengths l_1, l_2, l_3 and corresponding diameters d_1, d_2 and d_3. This system is reduced to a torsionally equivalent system having uniform diameter 'd_1' of the shaft. The equivalent length of the shaft is equal to

(*a*) $l_1 + l_2 + l_3$ ☐ (*b*) $\frac{l_1 + l_2 + l_3}{3}$ ☐

(*c*) $l_1 + l_2\left(\frac{d_1}{d_2}\right)^3 + l_3\left(\frac{d_1}{d_3}\right)^3$ ☐ (*d*) $l_1 + l_2\left(\frac{d_1}{d_2}\right)^4 + l_3\left(\frac{d_1}{d_3}\right)^4$. ☐

414. In question 413, the system will become torsionally equivalent when equal and opposite torque are applied to the two rotors and the system twists through an angle which is

(*a*) half of the angle of twist of actual shafts ☐

(*b*) one-third of the angle twist of actual shafts ☐

(*c*) equal to the angle of twist of actual shafts ☐

(*d*) three times of the angle of twist of actual shafts. ☐

415. The angle of twist of a shaft is given by

(*a*) $\frac{T}{J} \times \frac{C}{l}$ ☐ (*b*) $\frac{T}{J} \times \frac{l}{C}$ ☐

(*c*) $\frac{T \times J}{C \times l}$ ☐ (*d*) $\frac{T}{J \times l \times C}$. ☐

where T = Torque, l = Length of shaft, C = Modulus of rigidity, and J = Polar moment of inertia of the shaft.

416. In question 413, the twist of one rotating mass with respect to the second rotating mass is equal to

(*a*) $\frac{32}{\pi C}\left(\frac{d_1^4}{l_1} + \frac{d_2^4}{l_2} + \frac{d_3^4}{l_3}\right)$ ☐ (*b*) $\frac{32}{\pi C}\left(\frac{d_1^3}{l_1} + \frac{d_2^3}{l_2} + \frac{d_3^3}{l_3}\right)$ ☐

(*c*) $\frac{32}{\pi C}\left(\frac{l_1}{d_1^4} + \frac{l_2}{d_2^4} + \frac{l_3}{d_3^4}\right)$ ☐ (*d*) none of the above. ☐

417. A shaft with two rotors at its ends will have

(a) three nodes ☐ (b) two nodes ☐
(c) one node ☐ (d) zero node. ☐

418. A shaft with three rotors will have

(a) three nodes ☐ (b) two nodes ☐
(c) one node ☐ (d) zero node. ☐

419. A node means a section where the amplitude of vibration is

(a) maximum ☐ (b) half of maximum ☐
(c) zero ☐ (d) one-fourth of the maximum. ☐

420. The number of node points is equal to

(a) number of rotors ☐ (b) one more than the number of rotors ☐
(c) one less than the number of rotors ☐ (d) half the number of rotors. ☐

421. Transmissibility is defined as

(a) the ratio of the force transmitted through rigid support to the force transmitted through elastic support ☐
(b) the ratio of the force transmitted through elastic support to the force transmitted through rigid support ☐
(c) the product of the force transmitted through the elastic support to the force transmitted through rigid support ☐
(d) none of the above. ☐

422. When damping is negligible and frequency ratio $\frac{\omega}{\omega_n} > \sqrt{2}$, the transmissibility is given by

(a) $\dfrac{1}{1-\left(\dfrac{\omega}{\omega_n}\right)^2}$ ☐ (b) $\dfrac{1}{\left(\dfrac{\omega}{\omega_n}\right)^2 - 1}$ ☐

(c) $1-\left(\dfrac{\omega}{\omega_n}\right)^2$ ☐ (d) $\left(\dfrac{\omega}{\omega_n}\right)^2 - 1$. ☐

423. Figure 9.26 shows the graph of transmissibility against frequency ratio $\left(\dfrac{\omega}{\omega_n}\right)$ for various values of damping factor (say damping factor equal to 0.0, 0.25, 0.50 and 1.0). The curve A holds good when damping factor is equal to

(a) 0.0 ☐
(b) 0.25 ☐
(c) 0.50 ☐
(d) 1.0. ☐

FIGURE 9.26

424. In Fig. 9.26, the curve *B* holds good when damping factor is equal to

(*a*) 0.0 ☐ (*b*) 0.25 ☐

(*c*) 0.50 ☐ (*d*) 1.0. ☐

425. In Fig. 9.26, the curve *C* holds good when damping factor is equal to

(*a*) 0.0 ☐ (*b*) 0.25 ☐

(*c*) 0.50 ☐ (*d*) 1.0. ☐

426. In Fig. 9.26, the curve *D* holds good when damping factor is equal to

(*a*) 1.0 ☐ (*b*) 0.25 ☐

(*c*) 0.50 ☐ (*d*) 1.0. ☐

427. In Fig. 9.26, all the four curves pass through the point *E*, at which transmissibility is equal to 1.0 and frequency ratio $\left(\frac{\omega}{\omega_n}\right)$ is equal to

(*a*) 1.0 ☐ (*b*) $\sqrt{2}$ ☐

(*c*) 2.0 ☐ (*d*) 0.5. ☐

428. With reference to Fig. 9.26, the damping is beneficial only when the value of frequency ratio $\left(\frac{\omega}{\omega_n}\right)$ is

(*a*) more than $\sqrt{2}$ ☐ (*b*) less than $\sqrt{2}$ ☐

(*c*) equal to 1.0 ☐ (*d*) less than 1.0. ☐

429. With reference to Fig. 9.26, the vibration isolation is possible only when the value of frequency ratio $\left(\frac{\omega}{\omega_n}\right)$ is

(*a*) more than $\sqrt{2}$ ☐ (*b*) less than $\sqrt{2}$ ☐

(*c*) equal to 1.0 ☐ (*d*) less than 1.0. ☐

430. In vibration isolation systems, the transmissibility (*i.e.*, the ratio of the force transmitted through elastic support to the force transmitted through rigid support) will be less than unity, for all values of damping factors, if the frequency ratio $\left(\frac{\omega}{\omega_n}\right)$ is

(*a*) more than $\sqrt{2}$ ☐ (*b*) less than $\sqrt{2}$ ☐

(*c*) equal to, 1.0 ☐ (*d*) less than 1.0. ☐

431. The transmitted force will be greater than that transmitted through rigid supports for all values of damping factors, if the frequency ratio $\left(\frac{\omega}{\omega_n}\right)$ is

(*a*) more than $\sqrt{2}$ ☐ (*b*) less than $\sqrt{2}$ ☐

(*c*) equal to, 1.0 ☐ (*d*) less than 1.0. ☐

432. Choose the wrong statement

(a) For vibration isolation, dampers need not be provided. ☐

(b) When frequency ratio $\left(\frac{\omega}{\omega_n}\right)$ is equal to $\sqrt{2}$, transmissibility for all damping factors is equal to unity. ☐

(c) When frequency ratio $\left(\frac{\omega}{\omega_n}\right)$ is less than $\sqrt{2}$, the transmissibility increases as damping factor decreases. ☐

(d) When frequency ratio $\left(\frac{\omega}{\omega_n}\right)$ is more than $\sqrt{2}$, the transmissibility decreases as damping factor increases. ☐

(e) All of the above ☐

(f) None of the above. ☐

433. A system is said to be critically damped if the damping factor for a vibrating system is

(a) more than one ☐ (b) equal to one ☐

(c) less than one ☐ (d) equal to zero. ☐

434. The amount of damping, necessary for a system to be critically damped, is known as

(a) damping factor ☐ (b) magnification factor ☐

(c) critical damping co-efficient ☐ (d) logarithmic decrement. ☐

435. The ratio of the actual damping co-efficient to the critical damping co-efficient is known as

(a) damping factor ☐ (b) magnification factor ☐

(c) critical damping co-efficient ☐ (d) logarithmic decrement. ☐

436. In a light or under damped vibrating system, the natural logarithm of the ratio of any two successive amplitudes is known as

(a) damping factor ☐ (b) magnification factor ☐

(c) critical damping co-efficient ☐ (d) logarithmic decrement. ☐

437. A system is said to be overdamped if the damping factor for the system is

(a) more than one ☐ (b) equal to one ☐

(c) less than one ☐ (d) equal to zero. ☐

438. An overdamped system is disturbed from equilibrium position with some initial velocity. The system will

(a) not cross the equilibrium position ☐

(b) vibrate about equilibrium position ☐

(c) immediately return to the equilibrium position ☐

(d) none of the above. ☐

439. Choose the wrong statement

(a) Critical damping is a function of mass and stiffness. ☐

(b) The amplitude in steady state forced vibration at resonance is directly proportional to the damping co-efficient. ☐

(*c*) For a vibrating system if the damping factor is equal to one, the system is known as critically damped. ☐

(*d*) For an overdamped system, motion is a periodic. ☐

440. The shafts (*i.e.*, driving and driven shafts) are connected by a Hooke's joint. The relation between the angular displacements of the driver (θ) and driven (ϕ) shafts and the inclination (α) between the axes of the driver and driven shafts is given by

(*a*) $\tan \theta = \tan \alpha \times \tan \phi$ ☐ (*b*) $\tan \theta = \tan \alpha \times \cos \phi$ ☐

(*c*) $\tan \theta = \cos \alpha \times \tan \phi$ ☐ (*d*) $\tan \theta = \sin \alpha \times \cos \phi$. ☐

441. If the inclination (α) between the axes of the driver and driven shafts of a Hooke's joint is constant, then the velocity ratio of the driven shaft to the driving shaft is given by

(*a*) $\dfrac{\cos \alpha}{1 + \cos^2 \theta \sin^2 \alpha}$ ☐ (*b*) $\dfrac{\cos \alpha}{1 - \cos^2 \theta \cos^2 \alpha}$ ☐

(*c*) $\dfrac{\cos \alpha}{1 - \cos^2 \theta \sin^2 \alpha}$ ☐ (*d*) $\dfrac{\sin \alpha}{1 - \cos^2 \theta \sin^2 \alpha}$. ☐

where θ = Angular displacement of the driving shaft.

442. The velocity of the driven shaft of a Hooke's joint is minimum when θ is equal to

(*a*) 0° ☐ (*b*) 90° ☐

(*c*) 180° ☐ (*d*) 270° ☐

(*e*) (*b*) and (*d*) ☐ (*f*) (*a*) and (*c*). ☐

443. The velocity of the driven shaft of a Hooke's joint is maximum when the value of angular displacement (θ) of the driving shaft is equal to

(*a*) 0° ☐ (*b*) 90° ☐

(*c*) 180° ☐ (*d*) 270° ☐

(*e*) (*b*) and (*d*) ☐ (*f*) (*a*) and (*c*). ☐

444. The value of angular displacement (θ) of the driving shaft, for which is velocity of the driven shaft is equal to the driver shaft, is given by the relation

(*a*) $\tan \theta = \cos \alpha$ ☐ (*b*) $\tan \theta = \pm\sqrt{\cos \alpha}$ ☐

(*c*) $\tan \theta = \sqrt{\sin \alpha}$ ☐ (*d*) $\tan \theta = \pm\sqrt{\cos \alpha \times \sin \alpha}$ ☐

where α = Inclination between the two shafts.

445. The co-efficient of fluctuation of speed of the driven shaft of a Hooke's joint, when the angle between the two shafts is small and is measured in radian

(*a*) is proportional to the angle between the two shafts ☐

(*b*) is proportional to the square of the angle between the two shafts ☐

(*c*) is inversely proportional to the angle between the two shafts ☐

(*d*) is inversely proportional to the square of the angle between the two shafts. ☐

446. The acceleration on the driven shaft of a Hooke's joint is maximum for a value of θ. This approximate value of θ is given by

(*a*) $\cos 2\alpha = \frac{2\sin^2\alpha}{2+\sin^2\alpha}$ ☐ (*b*) $\cos 2\theta = \frac{\sin^2\alpha}{1-2\sin^2\alpha}$ ☐

(*c*) $\cos 2\theta = \frac{2\sin^2\alpha}{1-\sin^2\alpha}$ ☐ (*d*) $\cos 2\theta = \frac{2\sin^2\alpha}{2-\sin^2\alpha}$. ☐

447. Choose the wrong statement

(*a*) The minimum value of the angular speed of a driven shaft of a Hooke's joint is equal to angular speed of the driving shaft multiplied by cos α. ☐

(*b*) The maximum angular speed of the driven shaft of a Hooke's joint is equal to the product of angular speed of driving shaft and cos α. ☐

(*c*) Hooke's joint is also known as universal joint. ☐

(*d*) The velocity of driven shaft and driving shaft is equal if tan θ = cos α. ☐

448. The centre lines of the axes of the four wheels of an automobile, for correct steering, should meet at a common point. This condition is satisfied if

(*a*) $\cos\theta - \cos\phi = \frac{a}{b}$ ☐ (*b*) $\cos\theta + \cos\phi = \frac{a}{b}$ ☐

(*c*) $\cot\theta - \cos\phi = \frac{b}{a}$ ☐ (*d*) $\cot\theta - \cot\phi = \frac{a}{b}$. ☐

where θ = Angle through which the axis of the outer forward wheel turns

φ = Angle through which the axis of the inner forward wheel turns

a = Distance between the pivots of front axle

b = Wheel base.

Tick mark the true or false statement:

449. The Coriolis component of acceleration acts perpendicular to the sliding surface.

(*a*) True ☐ (*b*) False. ☐

450. The term 2 × ω × v represents the Coriolis component of acceleration in which ω is angular velocity of a link in rad/sec and v is the velocity of the slider on the link.

(*a*) True ☐ (*b*) False. ☐

451. The Coriolis component of acceleration leads the sliding velocity by 45°.

(*a*) True ☐ (*b*) False. ☐

452. If three bodies move relative to each other, then according to Kennedy's theorem, their instantaneous centres will lie on a parabolic curve.

(*a*) True ☐ (*b*) False. ☐

453. For a mechanism consisting of *n* links, the total number of instantaneous centres will be $n(n-1)/2$.

(*a*) True ☐ (*b*) False. ☐

454. Incompletely constrained motion is the motion between a pair when limited to a definite direction, irrespective of the direction of force applied.

(*a*) True ☐ (*b*) False. ☐

455. A sliding pair has a completely constrained motion.

(*a*) True ☐ (*b*) False. ☐

456. The piston and cylinder of a reciprocating steam engine is an example of sliding pair.

(*a*) True ☐ (*b*) False. ☐

457. The motion of the shaft between a foot-stip bearing is an example of incompletely constrained motion.

(*a*) True ☐ (*b*) False. ☐

458. The motion of a shaft in a circular hole is an example of completely constrained motion.

(*a*) True ☐ (*b*) False. ☐

459. An assemblage of three links is known as a mechanism.

(*a*) True ☐ (*b*) False. ☐

460. For a kinematric chain, $L = \frac{1}{3}(j+2)$ where L = number of links and j = number of joints.

(*a*) True ☐ (*b*) False. ☐

461. Kinematric chain is a chain which consists of four links and four joints.

(*a*) True ☐ (*b*) False. ☐

462. The expression $(1 - \sin \phi)/(1 + \sin \phi)$ represents the maximum efficiency of a screw jack where ϕ = angle of friction.

(*a*) True ☐ (*b*) False. ☐

463. The efficiency of a screw jack does not depend upon the load lifted or load raised.

(*a*) True ☐ (*b*) False. ☐

464. If α = helix angle and ϕ = angle of friction, then the efficiency of a screw jack will be maximum when $\alpha = 45° + \phi$.

(*a*) True ☐ (*b*) False. ☐

465. The ratio of the limiting friction to the normal reaction between the two bodies is known as co-efficient of friction.

(*a*) True ☐ (*b*) False. ☐

466. The expression $P = W \tan (\alpha + \phi)$ in a screw jack, represents the effort required to lift the load W where P = effort, α = helix angle and ϕ = angle of friction.

(*a*) True ☐ (*b*) False. ☐

467. The effort (P) required to lower the load W in a screw jack is given by $P = W \tan (\phi - \alpha)$.

(*a*) True ☐ (*b*) False. ☐

468. The expression $\tan \alpha/\tan (\alpha + \phi)$ gives the efficiency of a screw jack.

(*a*) True ☐ (*b*) False. ☐

469. In angle of friction for a screw jack is 30°, then the maximum efficiency will be equal to 33.33%.

(*a*) True ☐ (*b*) False. ☐

470. A body is placed on an inclined plane. The angle of the plane is such that body just starts moving downwards without applying any force. This angle is known as angle of friction.

(*a*) True ☐ (*b*) False. ☐

471. The force of friction always acts in the opposite direction of motion of the body.

(*a*) True ☐ (*b*) False. ☐

472. The friction experienced by a body, when it is in motion is known as dynamic friction.

(*a*) True ☐ (*b*) False. ☐

473. The two pulleys will rotate in the same direction when they are connected by means of an open belt drive.

(*a*) True ☐ (*b*) False. ☐

474. The length of the open belt depends upon the sum and difference of the radii of the two pulleys, over which belt is passing.

(*a*) True ☐ (*b*) False. ☐

475. The length of a crossed belt depends upon only the sum of the radii of the two pulleys.

(*a*) True ☐ (*b*) False. ☐

476. When the maximum tension in the belt is three times of centrifugal tension, then the power transmitted by the belt, will be maximum.

(*a*) True ☐ (*b*) False. ☐

477. The angle of contact is taken on the smaller pulley.

(*a*) True ☐ (*b*) False. ☐

478. If D = centre distance between two shafts and r_1 and r_2 are the radii of pulley 1 and 2, then the length L of the belt when the shafts are connected by a crossed belt is $L = \pi(r_1 + r_2) + (r_1 + r_2)^2/D + 2D$.

(*a*) True ☐ (*b*) False. ☐

479. The centrifugal tension on the belt has no effect on power transmitted.

(*a*) True ☐ (*b*) False. ☐

480. For maximum power, the velocity of belt should be equal to $\sqrt{\dfrac{T \times g}{3w}}$ where T = maximum tension and w = weight of belt per metre length.

(*a*) True ☐ (*b*) False. ☐

481. A shaft will have two nodes if it carries three rotors.

(*a*) True ☐ (*b*) False. ☐

482. For forced vibrations, the body should vibrate under the influence of external force.

(*a*) True ☐ (*b*) False. ☐

483. For free vibrations, the external force acts on the body only to give the body initial displacement after that force is removed.

(*a*) True ☐ (*b*) False. ☐

484. When a body is subjected to longitudinal vibrations, the stresses induced in the body will be tensile or compressive.

(*a*) True ☐ (*b*) False. ☐

485. In torsional vibrations, the bending stress is produced in the body.

(*a*) True ☐ (*b*) False. ☐

486. In transverse vibrations, shear stress is produced in the body.

(*a*) True ☐ (*b*) False. ☐

487. For longitudinal vibrations, the particles of the body are moving parallel to the axis of the body.

(*a*) True ☐ (*b*) False. ☐

488. For free longitudinal vibrations, the natural frequency is equal to $\frac{1}{2\pi}\sqrt{\frac{g}{\delta}}$ where δ = static deflection of the body.

(*a*) True ☐ (*b*) False. ☐

489. The critical speed of a shaft depends upon mass and stiffness.

(*a*) True ☐ (*b*) False. ☐

490. A simply supported shaft carries a point load at the centre. The natural frequency of free transverse vibrations due to the point load will be equal to $\frac{4.985}{\sqrt{\delta}}$ where δ = static deflection of the simply supported shaft due to point load.

(*a*) True ☐ (*b*) False. ☐

491. When belt is stationary, then initial tension will be $(T_1 + T_2)/2$.

(*a*) True ☐ (*b*) False. ☐

492. The displacement, under the condition of resonance for a vibrating body, would lag behind the disturbing force by 90°.

(*a*) True ☐ (*b*) False. ☐

493. If the damping factor is unity, for a vibrating system, then maximum magnification factor will occur for $\omega/\omega_n = 1$ where ω = angular speed of the system and ω_n = natural frequency of vibration of the system.

(*a*) True ☐ (*b*) False. ☐

494. Transmissibility is the ratio of force transmitted to the disturbing force.

(*a*) True ☐ (*b*) False. ☐

495. The product of periodic time and frequency in simple harmonic motion is equal to one.

(*a*) True ☐ (*b*) False. ☐

496. In a continuous system, the number of degree of freedom would to be two.

(*a*) True ☐ (*b*) False. ☐

497. For a critically damped system, the damping factor of a vibrating system is unity.

(*a*) True ☐ (*b*) False. ☐

498. The range of speed to the mean speed in known as governor sensitivity.

(*a*) True ☐ (*b*) False. ☐

499. The term $895/N^2$ represents the height of a Watt's governor in meter where N = speed of the arm and ball about the spindle axis.

(*a*) True ☐ (*b*) False. ☐

500. The governor speed decreases when the sleeve of a Porter governor moves upwards.

(*a*) True ☐ (*b*) False. ☐

501. Watt governor is a pendulum type governor.

(*a*) True ☐ (*b*) False. ☐

502. As the speed range decreases, the sensitiveness of the governor increases.

(*a*) True ☐ (*b*) False. ☐

503. Porter governor is a dead weight governor.

(*a*) True ☐ (*b*) False. ☐

504. The lift of the sleeve in a Hartnell governor is given by $(r_1 + r_2) \times y/x$ where x = length of ball arm of the lever, y = length of sleeve arm of the lever and r_1, r_2 = maximum and minimum radii of rotation

(*a*) True ☐ (*b*) False. ☐

505. The height of a Watt's governor is inversely proportional to speed.

(*a*) True ☐ (*b*) False. ☐

506. The compression of the spring is equal to the lift of the sleeve in a Hartnell governor.

(*a*) True ☐ (*b*) False. ☐

507. If w = weight of the ball and W = weight of the sleeve, then the ratio of height of Porter governor (if length of arms and links are equal) to the height of Watt's governor is $(w + W)/w$.

(*a*) True ☐ (*b*) False. ☐

508. If S_1 = spring force exerted on the sleeve at maximum radius of rotation, S_2 = spring force exerted on the sleeve at minimum radius of rotation and h = compression of the spring, then stiffness of the spring is equal to $(S_1 - S_2)/h$.

(*a*) True ☐ (*b*) False. ☐

509. If the controlling force increases as the radius of rotation increases in a spring controlled governor, then governor is said to be stable governor.

(*a*) True ☐ (*b*) False. ☐

510. If the controlling force remains constant for all radii of rotation of a spring controlled governor, then the governor is known as isochronous.

(*a*) True ☐ (*b*) False. ☐

511. If W = Weight on the sleeve, w = Weight of ball and C = percentage increase in speed, then effort of a Porter governor is equal to $C/(w + W)$.

(*a*) True ☐ (*b*) False. ☐

512. A very sensitive governor will cause hunting.

(*a*) True ☐ (*b*) False. ☐

513. The work done at the sleeve for maximum equilibrium speed is known as effort of a governor.

(*a*) True ☐ (*b*) False. ☐

514. To find the periodic time of a body, Bifilar suspension method is used.

(*a*) True ☐ (*b*) False. ☐

515. The equivalent length of a simple pendulum is equal to the distance between centre of suspension and centre of percussion.

(*a*) True ☐ (*b*) False. ☐

516. The periodic time and frequency of a body will be same if it is suspended at the point of suspension or at the point of percussion.

(*a*) True ☐ (*b*) False. ☐

517. The frequency of motion of a helical spring which is fixed at one end and carries a load *W* at the other end is equal to $\frac{1}{2\pi}\sqrt{\frac{g}{\delta}}$.

(*a*) True ☐ (*b*) False. ☐

518. If the length of the string of a simple pendulum is made four times the original length, the time period will be doubled.

(*a*) True ☐ (*b*) False. ☐

519. At mean position, the velocity of a body moving with simple harmonic motion, is minimum.

(*a*) True ☐ (*b*) False. ☐

520. The number of cycles per second is known as frequency.

(*a*) True ☐ (*b*) False. ☐

521. Periodic time is the time taken by a particle for one complete oscillation.

(*a*) True ☐ (*b*) False. ☐

522. The product of frequency and time period of a body moving with simple harmonic motion is equal to one.

(*a*) True ☐ (*b*) False. ☐

523. The term $\omega^2 \times y$ represents the acceleration of a body moving with simple harmonic motion at any time (where y is the displacement from mean position).

(*a*) True ☐ (*b*) False. ☐

524. The term $\omega \times r$ represents the maximum velocity of a body moving with simple harmonic motion,

(*a*) True ☐ (*b*) False. ☐

525. When two intersecting and coplanar shafts are connected by gears, the arrangement is known as spur gearing.

(*a*) True ☐ (*b*) False. ☐

526. In case of bevel gearing, the two parallel and coplanar shafts are connected by gears having teeth parallel to the axis of the shaft.

(*a*) True ☐ (*b*) False. ☐

527. The size of a gear is usually specified by pitch circle diameter.

(*a*) True ☐ (*b*) False. ☐

528. Clearance is the radial distance from the top of a tooth to the bottom of the tooth.

(*a*) True ☐ (*b*) False. ☐

529. Diametral pitch is the ratio of number of teeth to the pitch circle diameter in millimetres.

(*a*) True ☐ (*b*) False. ☐

530. The product of diametral pitch and circular pitch is equal to one.

(*a*) True ☐ (*b*) False. ☐

531. The product of the diametral pitch and module is equal to π.

(*a*) True ☐ (*b*) False. ☐

532. The ratio of length of arc of contact to circular pitch is known as contact ratio.

(*a*) True ☐ (*b*) False. ☐

533. The contact ratio for gears is greater than one.

(*a*) True ☐ (*b*) False. ☐

534. If ϕ = angle of friction and θ = shaft angle, then the term $[\cos(\theta + \phi) + 1]/[\cos \theta - \phi) + 1]$ represents the maximum efficiency of spiral gears.

(*a*) True ☐ (*b*) False. ☐

535. The ratio of the driving tensions for *V*-belt is cosec α times that of flat belts.

(*a*) True ☐ (*b*) False. ☐

536. The relative velocity of *B* with respect to *A* in a rigid link *AB* is perpendicular to *AB*.

(*a*) True ☐ (*b*) False. ☐

537. Module is the reciprocal of diametrical pitch.

(*a*) True ☐ (*b*) False. ☐

538. The velocity of rubbing surface increases with the distance from the axis of the bearing.

(*a*) True ☐ (*b*) False. ☐

539. For a uniform pressure, the frictional torque trangential in a flat pivot bearing is $\mu WR/5$ where W = Load over the bearing, μ = Co-efficient of friction and R = Radius of bearing surface.

(*a*) True ☐ (*b*) False. ☐

540. For a uniform wear, the frictional torque transmitted in a flat pivot bearing is $\mu WR/2$.

(*a*) True ☐ (*b*) False. ☐

541. The frictional torque transmitted by a cone clutch is same as that of conical pivot bearing.

(*a*) True ☐ (*b*) False. ☐

542. The inertia force and accelerating forces are equal in magnitude but opposite in direction.

(*a*) True ☐ (*b*) False. ☐

543. To find the acceleration of various parts, Klein's diagram is used.

(*a*) True ☐ (*b*) False. ☐

544. When crank has uniform angular acceleration Klein's diagram is used.

(*a*) True ☐ (*b*) False. ☐

545. The difference of the maximum and minimum speeds is known as maximum fluctuations of speed.

(*a*) True ☐ (*b*) False. ☐

546. The ratio of the maximum energy to the minimum energy is known as co-efficient of fluctuation of energy.

(*a*) True ☐ (*b*) False. ☐

547. The time period of a compound pendulum is given by $T = 2\pi\sqrt{k^2 + h^2} / \sqrt{h \times g}$ where k = radius of gyration of the rigid body about an axis through the C.G. of the body and h = distance of the C.G. of the body from the point of suspension.

(*a*) True ☐ (*b*) False. ☐

548. The minimum period of a compound pendulum is equal to $T_{min} = 2\pi\sqrt{\frac{2k}{g}}$.

(*a*) True ☐ (*b*) False. ☐

549. A body of mass m and radius of gyration k about its C.G. is to be replaced by an equivalent dynamical system of two masses m_A and m_B placed at a distance of a and b from the C.G. of the body respectively. Then the relation between a, b and k are given by $k = \frac{a+b}{2}$.

(*a*) True ☐ (*b*) False. ☐

550. In the above equation 549, the mass m_A should be equal to $(a + b)/b \times m$.

(*a*) True ☐ (*b*) False. ☐

551. In the above equation 549, the mass m_B should be equal to $a \times m/(a + b)$.

(*a*) True ☐ (*b*) False. ☐

552. Gear are used for transmitting power from one shaft to another shaft, when the shafts are at a very small distance apart.

(*a*) True ☐ (*b*) False. ☐

553. In case of gears, there is no slip or creep as in case of belts or ropes,

(*a*) True ☐ (*b*) False. ☐

554. Gear drive is a positive and smooth,

(*a*) True ☐ (*b*) False. ☐

555. The point of contact of two pitch circles of mating gears is known as pitch point.

(*a*) True ☐ (*b*) False. ☐

556. The pitch circle circumference divided by the number of teeth on the wheel gives diametrical pitch,

(*a*) True ☐ (*b*) False. ☐

557. The number of teeth per unit length of pitch circle diameter is known as circular pitch.

(*a*) True ☐ (*b*) False. ☐

558. The length of pitch circle diameter per tooth, is known as module.

(*a*) True ☐ (*b*) False. ☐

559. Law of gearing states that the common normal at the point of contact between a pair of teeth always passes through the pitch point.

(*a*) True ☐ (*b*) False. ☐

560. Cycloid is the locus of a point on the circumference of a circle which rolls without slipping on a fixed straight line.

(*a*) True ☐ (*b*) False. ☐

561. Involute of a circle is the curve traced by the end of a thread as it is unwound from a stationary cylinder,

(*a*) True ☐ (*b*) False. ☐

562. The locus of a point on the circumference of a circle which rolls without slipping on the outside of another circle is known as Hypo-cycloid.

(*a*) True ☐ (*b*) False. ☐

563. The locus of a point on the circumference of a circle which rolls without slipping on the inside of another circle is known as Epicycloid.

(*a*) True ☐ (*b*) False. ☐

564. The involute function in terms of pressure angle (ϕ) is given by

$INV(\phi) = \tan \phi - \phi$ where $INV(\phi)$ = involute function,

(*a*) True ☐ (*b*) False. ☐

565. The path of contact in involute gears is a straight line.

(*a*) True ☐ (*b*) False. ☐

566. The angular velocity ratio of two involute gears in mesh is directly proportional to the size of the base circle.

(*a*) True ☐ (*b*) False. ☐

567. The spiral gears will have maximum efficiency if spiral angle (α) is equal to half the sum of the shaft angle (θ) and friction angle (ϕ).

(*a*) True ☐ (*b*) False. ☐

568. Static balance means the rotating masses are in equilibrium among themselves when not running,

(*a*) True ☐ (*b*) False. ☐

569. If the inertia forces and couples exerted by rotating masses are in equilibrium among themselves, the masses are said to be in dynamic balance.

(*a*) True ☐ (*b*) False. ☐

570. Natural frequencies are the frequencies of a system which is having free vibrations.

(*a*) True ☐ (*b*) False. ☐

Fill in the blanks:

571. The height of Watt's governor is ...... of the weight of the balls.

(*a*) directly proportional ☐ (*b*) independent. ☐

572. A governor is said to be ...... when it readily responds to a very small change in speed.

(*a*) isochronous ☐ (*b*) sensitive. ☐

573. The number of pairs of contact surfaces in a disc clutch will be ...... where n_1 = number of discs on the driving shaft and n_2 = number of discs on the driven shaft.

(*a*) $n_1 + n_2$ ☐ (*b*) $n_1 + n_2 - 1$. ☐

574. The uniform pressure theory gives a ...... frictional torque than the uniform wear theory.

(*a*) higher ☐ (*b*) lower. ☐

575. If the helix angle, α = ..., the efficiency of screw jack will be maximum

(*a*) $45° + \phi/2$ ☐ (*b*) $45° - \phi/2$. ☐

where ϕ = Angle of friction.

576. The efficiency of screw jack is ...... upon the load lifted.

(*a*) independent ☐ (*b*) dependent. ☐

577. The effort required is ...... to lift a load *W* by a screw jack.

(*a*) $W \tan(\alpha - \phi)$ ☐ (*b*) $W \tan(\alpha + \phi)$. ☐

where α = Helix angle, and ϕ = Angle of friction.

578. The two pulleys will rotate in the ...... direction, if they are connected by an open belt drive.

(*a*) opposite ☐ (*b*) same. ☐

579. When the maximum tension in the belt is ...... of centrifugal tension, the power transmitted by the belt will be maximum.

(*a*) two times ☐ (*b*) three times. ☐

580. The ratio of the driving tensions for *V*-belt is ...... times that of flat belts.

(*a*) $\sin \alpha$ ☐ (*b*) $\text{cosec}\,\alpha$. ☐

581. The initial tension in the belt is equal to ...... tension of the tight side and slack side of the belt.

(*a*) product of ☐ (*b*) average. ☐

582. The product of diametrical pitch and ...... is equal to one.

(*a*) circular pitch ☐ (*b*) module. ☐

583. The pitch circle in case of gears is always ...... than the base circle.

(*a*) less ☐ (*b*) greater. ☐

584. The contact ratio is ...... for gears.

(*a*) less than one ☐ (*b*) greater than one. ☐

585. The expression *L* = ... gives the relation between number of pairs (*p*) forming a kinematic chain and number of links (*L*).

(*a*) $2p - 3$ ☐ (*b*) $2p - 4$. ☐

586. A chain consists of four links and ...... joints is called a kinematic chain.
(*a*) three ☐ (*b*) four. ☐

587. A simple mechanism consists of ...... links.
(*a*) three ☐ (*b*) four. ☐

588. If ...... the links is fixed, then a kinematic chain is known as mechanism.
(*a*) two of ☐ (*b*) one of. ☐

589. A body is moving with simple harmonic motion, then its velocity will be ...... at the mean position.
(*a*) zero ☐ (*b*) maximum. ☐

590. A body is moving with simple harmonic motion. The product of its periodic time and ...... will be equal to one.
(*a*) amplitude ☐ (*b*) frequency. ☐

591. The total number of instantaneous centres are ...... for a mechanism consisting of n links.
(*a*) $n(n + 1)$ ☐ (*b*) $n(n - 1)/2$. ☐

592. The coriolis component of acceleration acts ...... to the sliding surface.
(*a*) parallel to ☐ (*b*) perpendicular to. ☐

593. In a reciprocating engine, usually ...... of the reciprocating masses are balanced.
(*a*) two-third ☐ (*b*) one-half. ☐

III. OBJECTIVE TYPE QUESTIONS FROM COMPETITIVE EXAMINATIONS

Tick mark the most appropriate answer:

1. In a kinematic chain with four lower pairs, if all the four lower pairs are turning pairs, the mechanism is classified into
(*a*) four bar chain ☐ (*b*) crossed slider crank chain ☐
(*c*) slider crank chain ☐ (*d*) double slider crank chain. ☐

2. In a kinematic chain with four lower pairs, if one is sliding pair and three turning pairs, the mechanism is classified into
(*a*) crossed slider crank chain ☐ (*b*) four bar chain ☐
(*c*) slider crank chain ☐ (*d*) double slider crank chain. ☐

3. If in the above question, there are two sliding pairs and two turning pairs and two similar pairs are adjacent then mechanism is classified into
(*a*) crossed slider crank chain ☐ (*b*) four bar chain ☐
(*c*) slider crank chain ☐ (*d*) double slider crank chain. ☐

4. Which is a false statement?
(*a*) The resultant force acting on a body together with the reversed effective force, are in equilibrium ☐

(b) The magnitude of inertia force is given by the expression $F_I = m_r\,\omega^2\,r\left(\cos\theta + \dfrac{\cos 2\theta}{n}\right)$

where m_r = Mass of reciprocating parts, r = Crank radius, $n = \dfrac{L}{r}$ ☐

(c) The inertia force is equal in magnitude and opposite in direction to accelerating force ☐

(d) D-Alembert's principle is used to reduce a dynamic problem into an equivalent static problem ☐

(e) None of the above. ☐

5. Choose the correct statement. A flywheel is

(a) an essential element of every prime mover ☐

(b) used in storing up energy and gives up whenever required, during a cycle ☐

(c) a device for coordination between the prime mover and the external resistance ☐

(d) used for all the above purposes. ☐

6. A solid disc of radius r rolls without slipping on a horizontal floor with angular velocity ω and angular acceleration α. The magnitude of the acceleration of the point of contact on the disc is

(a) zero ☐ (b) $r\alpha$ ☐

(c) $\sqrt{(r\alpha)^2 + (r\omega)^2}$ ☐ (d) $r\omega^2$. ☐

(ME-GATE-2012)

7. The speed of an engine varies from 210 rad/s to 190 rad/s. During cycle the change in kinetic energy is found to be 400 Nm. The inertia of the flywheel in kgm² is

(a) 0.10 ☐ (b) 0.20 ☐

(c) 0.30 ☐ (d) 0.40. ☐

(ME-GATE-2007)

8. A planar mechanism has 8 links and 10 rotary joints. The number of degrees of freedom of the mechanism, using Gruebler's criterion, is

(a) 0 ☐ (b) 1 ☐

(c) 2 ☐ (d) 3. ☐

(ME-GATE-2008)

9. In certain air-craft engines balancing masses for balancing secondary force produced by reciprocating parts are provided on

(a) secondary shaft running at half the engine speed ☐

(b) secondary shaft running at twice the engine speed ☐

(c) primary shaft running at twice the engine speed ☐

(d) secondary shaft running at same speed as that of the engine. ☐

10. Which of the following is a false statement ?

(a) Shoe brake is used in trains ☐

(b) Internal expanding brake is used for motor cars ☐

(c) Prony brake is an absorption type dynamometer ☐

(d) Watt governor is a spring load type governor. ☐

11. The planes of spin, precession and applied gyroscopic couple are in

(*a*) the same one plane ☐

(*b*) two planes perpendicular to one another ☐

(*c*) three planes perpendicular to one another ☐

(*d*) none of the above. ☐

12. The engine of an aeroplane rotates in clockwise direction when seen from the tail end and the aeroplane takes a turn to the left. The effect of the gyroscopic couple on the aeroplane will be

(*a*) to raise the nose and lower the tail ☐ (*b*) to raise the nose and tail ☐

(*c*) to lower the nose and raise the tail ☐ (*d*) to lower the nose and tail. ☐

13. When the pitching of a ship is upward, the effect of gyroscopic couple acting on it will be

(*a*) to raise the bow and lower the stern ☐ (*b*) to move the ship towards star-board ☐

(*c*) to move the ship towards port-side ☐ (*d*) to raise the stern and lower the bow. ☐

14. Four governors are given below. Choose a governor which can not be isochronous

(*a*) Watt ☐ (*b*) Porter ☐

(*c*) Hartnell ☐ (*d*) Hartung. ☐

15. The equation $F_c = a \cdot r \pm b$ shows the relation between the controlling force F_c and radius of rotation (r) of a governor in which a and b are constant. If b is negative then the governor will be

(*a*) isochronous ☐ (*b*) stable ☐

(*c*) unstable ☐ (*d*) hunt. ☐

16. The main objective of controlling force diagram is to

(*a*) determine the sensitiveness of governor ☐ (*b*) determine the stability of governor ☐

(*c*) determine the both (*a*) and (*b*) ☐ (*d*) none of the above. ☐

17. In dynamic balancing, the following condition will hold

(*a*) only the force polygon will be closed ☐

(*b*) only the couple polygon will be closed ☐

(*c*) both force and couple polygon's will be closed ☐

(*d*) none of the above. ☐

18. The resultant unbalanced force is minimum in reciprocating engines

(*a*) When one-third of the reciprocating masses are balanced by rotating masses ☐

(*b*) When half the reciprocating masses are balanced by rotating masses ☐

(*c*) When three-fourth of the reciprocating masses are balanced by rotating masses ☐

(*d*) None of the above. ☐

19. Instantaneous centre of a body rolling with sliding on a stationary curved surface lies

(*a*) at the point of contact ☐

(*b*) on the common normal at the point of contact ☐

(*c*) on the common tangent at the point of contact ☐

(*d*) at the center of curvature of the stationary surface. ☐

(ME-GATE-1992)

20. A flywheel connected to a punching machine has to supply energy of 400 Nm while running at a mean angular speed of 20 rad/s. If the total fluctuation of speed is not to exceed ±2%, the mass moment of inertia of the flywheel in kgm^2 is

(*a*) 25 ☐ (*b*) 50 ☐

(*c*) 100 ☐ (*d*) 125. ☐

(ME-GATE-2013)

21. Tooth interference in an external involute spur gear pair can be reduced by

(*a*) decreasing centre distance between gear pair ☐

(*b*) decreasing module ☐

(*c*) decreasing pressure angle ☐

(*d*) increasing number of gear teeth. ☐

(ME-GATE-2010)

22. By distributing balancing of reciprocating parts between coupled wheels, the hammer blow is

(*a*) completely eliminated ☐ (*b*) reduced to half ☐

(*c*) increased considerably ☐ (*d*) constant. ☐

23. Two cylinders uncoupled locomotives have their cranks at

(*a*) 45° ☐ (*b*) 90° ☐

(*c*) 135° ☐ (*d*) 20°. ☐

24. The secondary force produced by the reciprocating parts of a certain cylinder of a given engine can be considered as equal to primary force produced by the same weight having an equivalent crank of radius

(*a*) $\frac{r^2}{4l}$ and rotating at twice the engine speed ☐

(*b*) $\frac{r^2}{4l}$ and rotating at engine speed ☐

(*c*) $\frac{r^2}{4l}$ and rotating at half the engine speed ☐

(*d*) none of the above. ☐

25. A uniform rigid rod of mass $m = 1$ kg and length $L = 1$ m is hinged at its centre and laterally supported at one end by a spring of spring constant $k = 300$ N/m. The natural frequency ω_n in rad/s is

(*a*) 10 ☐ (*b*) 20 ☐

(*c*) 30 ☐ (*d*) 40. ☐

(ME-GATE-2008)

26. A link OB is rotating with a constant angular velocity of 2 rad/s in counter clockwise direction and a block is sliding radially outward on it with an uniform velocity of 0.75 m/s with respect to the rod, as shown in Fig. 9.27. If OA = 1 m, the magnitude of the absolute acceleration of the block at location A in m/s^2 is

(a) 3 ☐
(b) 4 ☐
(c) 5 ☐
(d) 6. ☐

FIGURE 9.27

(ME-GATE-2013)

27. A 1.5 kW motor is running at 1440 rev/min. Its is to be connected to a stirrer running at 36 rev/min. The gearing arrangement suitable for this application is

(a) differential gear ☐ (b) helical gear ☐
(c) spur gear ☐ (d) worm gear. ☐

(ME-GATE-2000)

28. The natural frequency of a spring mass system on earth is ω_n. The natural frequency of this system on the moon ($g_{moon} = g_{earth}/6$) is

(a) ω_n ☐ (b) $0.408\omega_n$ ☐
(c) $0.204\omega_n$ ☐ (d) $0.167\omega_n$. ☐

(ME-GATE-2010)

29. Governor effort is defined as

(a) force applied for 0% change in speed ☐ (b) force applied for 1% change in speed ☐
(c) force applied for 1% change in speed ☐ (d) none of the above. ☐

30. An isochronous governor is

(a) very stable ☐ (b) less sensitive ☐
(c) infinitely sensitive ☐ (d) none of the above. ☐

31. For spring controlled governors, the controlling force curve will generally be

(a) Circle ☐ (b) Parabola ☐
(c) Hyperbola ☐ (d) None of the above. ☐

32. For the same lift of sleeve, the range of speed of Proell governor as compared to Porter governor is

(a) less ☐ (b) same ☐
(c) more ☐ (d) may be more or less depending on other data. ☐

33. The air screw of an aeroplane is rotating clockwise when looking from the front. It makes a left turn. The gyroscopic effects will

(a) tend to depress the nose and raise the tail ☐ (b) tend to raise the nose and depress the tail ☐
(c) it will tilt the aeroplane ☐ (d) none of the above. ☐

34. The gyroscopic effects due to rotating parts of a turbo jet engine of an aircraft on a curved course depend on

(a) flight velocity ☐ (b) flight altitude ☐
(c) radius of the curve ☐ (d) flight velocity and radius of the curve. ☐

35. For a gyrometer with spin vector $\vec{\omega_s}$ and a perpendicular precession vector $\vec{\omega_p}$ the external torque needed is in the direction of

(a) $\vec{\omega_p}$ ☐ (b) $\vec{\omega_s}$ ☐

(c) $\vec{\omega_p} \times \vec{\omega_s}$ ☐ (d) $\vec{\omega_s} \times \vec{\omega_p}$. ☐

36. A flywheel absorbs energy during those periods of crank rotation when

(a) the turning moment is greater than the resisting moment ☐

(b) the turning moment is equal to the resisting moment ☐

(c) the turning moment is less than the resisting moment ☐

(d) absorbs energy during all periods of crank rotation. ☐

37. Absorption of energy is accompanied

(a) by decrease of speed ☐ (b) at all speeds ☐

(c) an increase of speed ☐ (d) no relation with speed. ☐

38. Which is the correct statement?

(a) the flywheel influences the mean speed of prime mover ☐

(b) the flywheel influences the variation of load demand on prime mover ☐

(c) the flywheel influences the cyclic variation of turning moment ☐

(d) the flywheel influences the mean torque developed by the prime mover. ☐

39. In a flywheel safe stress is 7×10^6 N/m^2 and density of flywheel material 700 kg/m^3, the maximum peripheral velocity will be

(a) 100 m/s ☐ (b) 50 m/s ☐

(c) 25 m /s ☐ (d) 10 m/s. ☐

40. Choose the wrong statement

(a) A chain consisting of three links and three joints is known as locked chain ☐

(b) A chain consisting of four links with four joints is known as kinematic chain ☐

(c) Quaternary joint is equivalent to three binary joints ☐

(d) Rectangular bar in a rectangular hole is the example of partially constrained motion. ☐

41. If 'n' links are connected at the same joint, the joint is equivalent to

(a) (n –1) binary joints ☐ (b) (n – 2) binary joints ☐

(c) (n – 3) binary joints ☐ (d) (2n – 1) binary joints. ☐

42. If a kinematic chain has 'n' links then number of mechanism obtained are

(a) (n – 1) ☐ (b) (n – 2) ☐

(c) (n + 1) ☐ (d) n. ☐

43. An automotive engine weighting 240 kg is supported on four springs with linear characteristics. Each of the front two springs have a stiffness of 16 MN/m while the stiffness of each rear spring is 32 MN/m. The engine speed (in rpm), at which resonance is likely to occur, is

(a) 6040 ☐ (b) 3020 ☐

(c) 1424 ☐ (d) 955. ☐

(ME-GATE-2009)

44. A circular solid disc of uniform thickness 20 mm, radius 200 mm and mass 20 kg, is used as a fly wheel. If it rotates at 600 rpm, the kinetic energy of the flywheel, in joules is

(*a*) 395 ☐ (*b*) 790 ☐

(*c*) 1580 ☐ (*d*) 3160. ☐

(ME-GATE-2012)

45. A vibratory system consists of a mass 12.5 kg, a spring of stiffness 1000 N/m, and a dashpot with damping coefficient of 15 Ns/m. The value of critical damping of the system is:

(*a*) 0.223 Ns/m ☐ (*b*) 17.88 Ns/m ☐

(*c*) 71.4 Ns/m ☐ (*d*) 223.6 Ns/m. ☐

(ME-GATE-2006)

46. If there are several unbalanced masses in a rotor in different planes, the minimum number of balancing masses required, is

(*a*) one ☐ (*b*) two ☐

(*c*) three ☐ (*d*) four. ☐

47. Which of the following is a wrong statement ?

(*a*) When brakes are applied to all the four wheels, the vehicle moves minimum distance ☐

(*b*) When vehicle moves up on an inclined plane, it experiences retardation ☐

(*c*) Dynamometer is used to measure frictional resistance ☐

(*d*) Brake lining material is Ferodo ☐

(*e*) None of the above. ☐

48. With the decrease of governor speed

(*a*) radius of rotation decreases but height of governor increases ☐

(*b*) radius of rotation and height of governor decrease ☐

(*c*) radius of rotation and height of governor increase ☐

(*d*) radius of rotation increases but height of governor decreases. ☐

49. The acceleration of the piston in a reciprocating steam engine is given by

(*a*) $\omega \cdot r\left(\sin\theta + \dfrac{\sin 2\theta}{n}\right)$ ☐ (*b*) $\omega^2 \cdot r\left(\cos\theta + \dfrac{\cos 2\theta}{n}\right)$ ☐

(*c*) $\omega^2 \cdot r\left(\sin\theta + \dfrac{\sin 2\theta}{2n}\right)$ ☐ (*d*) none of the above. ☐

where ω = Angular velocity of crank, r = Crank radius, θ = Angle of crank, $n = \dfrac{L}{r}$ = ratio of length of connecting rod to length of crank radius.

50. Which one of the following conditions is satisfied for a system to be dynamically equivalent

(*a*) $a_1 \cdot a_2 = k^2$ ☐ (*b*) $a_1 + a_2 = k^2$ ☐

(*c*) $a_1 - a_2 = k^2$ ☐ (*d*) $a_1 \cdot a_2 = k$. ☐

where a_1 and a_2 = Distance of two masses from C.G. of the body, and k = Radius of gyration of the body.

51. While considering balancing of coupled locomotives, one has to consider

(*a*) two planes, one of the cylinders and the other of the driving wheels ☐

(*b*) four planes, two of the cylinders and two of driving wheels ☐

(*c*) six planes, two of the cylinders, two of the coupling rods and two of the driving wheels containing balancing weights ☐

(*d*) none of the above. ☐

52. For complete balancing of the reciprocating parts the condition arrived at, is

(*a*) primary force polygon must close ☐ (*b*) secondary force polygon must close ☐

(*c*) primary couple polygon must close ☐ (*d*) secondary couple polygon must close ☐

(*e*) all of the above. ☐

53. Sensitivity of governor is expressed as

(*a*) $\frac{N_1 + N_2}{N_1 + N_2}$ ☐ (*b*) $\frac{2(N_2 - N_2)}{N_1 + N_2}$ ☐

(*c*) $\frac{2(N_1 + N_2)}{N_1 + N_2}$ ☐ (*d*) $\frac{N_1 - N_2}{2(N_1 + N_2)}$. ☐

54. A hunting governor is

(*a*) more stable ☐ (*b*) less sensitive ☐

(*c*) more sensitive ☐ (*d*) none of the above. ☐

55. A planar closed kinematic chain is formed with rigid links PQ = 2.0 m, QR = 3.0 m, RS = 2.5 m and SP = 2.7 m with all revolute joints. The link to be fixed to obtain a double rocker (rocker–rocker) mechanism is

(*a*) PQ ☐ (*b*) QR ☐

(*c*) RS ☐ (*d*) SP. ☐

(ME-GATE-2013)

56. A vibrating machine is isolated from the floor using springs. If the ratio of excitation frequency of vibration of machine to the natural frequency of the isolation system is equal to 0.5, the transmissibility ratio of isolation is

(*a*) $\frac{1}{2}$ ☐ (*b*) $\frac{3}{4}$ ☐

(*c*) $\frac{4}{3}$ ☐ (*d*) 2. ☐

(ME-GATE-2004)

57. The maximum fluctuation of energy of flywheel

(*a*) is directly proportional to coefficient of fluctuation of speed ☐

(*b*) is directly proportional to square of angular velocity of flywheel ☐

(*c*) is directly proportional to moment of inertia of flywheel ☐

(*d*) all of the above. ☐

58. The ratio of maximum fluctuation of energy to the work done per cycle is called

(*a*) coefficient of fluctuation of energy ☐ (*b*) coefficient of fluctuation of speed ☐

(*c*) all of the above. ☐ (*d*) none of the above. ☐

59. The mass of the flywheel is concentrated in the rims because then it will

(*a*) store much energy ☐ (*b*) store less energy ☐

(*c*) store zero energy ☐ (*d*) make the flywheel stronger. ☐

60. The partial balancing means

(*a*) best balancing of engines ☐

(*b*) balancing of engines ☐

(*c*) balancing partially the reciprocating masses ☐

(*d*) all of the above. ☐

61. In a locomotive, the ratio of length of connecting rod to the crank radius is kept very large in order to

(*a*) start the locomotive quickly ☐ (*b*) minimise the effects of primary forces ☐

(*c*) minimise the effect of secondary forces ☐ (*d*) none of the above. ☐

62. In a Hartnell governor, if a spring of lower stiffness is used, then the governor will be

(*a*) isochronous ☐ (*b*) more sensitive ☐

(*c*) less sensitive ☐ (*d*) none of the above. ☐

63. For Hartnell governor with equal arms the lift of the sleeve is given by

(*a*) $h = r_2 + r_1$ ☐ (*b*) $h = \frac{r_2 - r_1}{2}$ ☐

(*c*) $h = r_2 - r_1$ ☐ (*d*) $h = \frac{r_2 + r_1}{2}$. ☐

64. During taking a turn a cyclist inclines at an angle with the normal to the road. The equilibrium is maintained due to

(*a*) weight of the cyclist ☐

(*b*) centrifugal force alone ☐

(*c*) centrifugal force and gyrocouple ☐

(*d*) weight of the cyclist, centrifugal force and gyro-couple. ☐

65. Which of the following statement is correct?

(*a*) Flywheel reduces speed fluctuations during a cycle for a constant load, but flywheel does not control the mean speed of the engine if the load changes. ☐

(*b*) Flywheel does not reduce speed fluctuations during a cycle for a constant load, but flywheel does control the mean speed of the engine if the load changes ☐

(*c*) Governor control a speed fluctuations during a cycle for a constant load, but governor does not control the mean speed of the engine if the load changes. ☐

(*d*) Governor controls speed fluctuations during a cycle for a constant load, and governor also controls the mean speed of the engine if the load changes. ☐

(ME-GATE-2001)

66. Match the items in columns I and II

Column I	Column II
P. Addendum	1. Cam
Q. Instantaneous center of velocity	2. Beam
R. Section modulus	3. Linkage
S. Prime circle	4. Gear

Codes:

	P	*Q*	*R*	*S*	
(*a*)	4	2	3	1	☐
(*b*)	4	3	2	1	☐
(*c*)	3	2	1	4	☐
(*d*)	3	4	1	2	☐

(ME-GATE-2006)

67. The following are the data for two crossed helical gears used for speed reduction:

Gear I: Pitch circle diameter in the plane of rotation 80 mm and helix angle 30°

Gear II: Pitch circle diameter in the plane of rotation 120 mm and helix angle 22.5°

If the input speed is 1440 rpm. The output speed in rpm is

(*a*) 1200 ☐ (*b*) 900 ☐

(*c*) 875 ☐ (*d*) 720. ☐

(ME-GATE-2012)

68. The rotor shaft of a large electric motor supported between short bearing at both the ends shows a deflection of 1.8 mm in the middle of the rotor. Assuming the rotor to be perfectly balanced and supported at knife edges at both the ends, the likely critical speed (in rpm) of the shaft is

(*a*) 350 ☐ (*b*) 705 ☐

(*c*) 2810 ☐ (*d*) 4430. ☐

(ME-GATE-2009)

69. The number of inversions for a slider crank mechanism is

(*a*) 6 ☐ (*b*) 5 ☐

(*c*) 4 ☐ (*d*) 3. ☐

(ME-GATE-2006)

70. The function of an element is to

(*a*) transmit motion ☐ (*b*) to serve as a support ☐

(*c*) to guide other elements ☐ (*d*) all of the above ☐

(*e*) none of the above. ☐

71. Oldham's coupling and elliptic trammels are the inversions of

(*a*) double slider crank chain ☐ (*b*) single slider crank chain ☐

(*c*) four bar chain ☐ (*d*) none of the above. ☐

72. Whitworth quick return mechanism is an inversion of

(*a*) double slider crank chain ☐ (*b*) single slider crank chain ☐

(*c*) four bar chain ☐ (*d*) none of the above. ☐

73. The displacement of the piston in a reciprocating steam engine is given by

(a) $\omega \cdot r\left(\sin\theta + \dfrac{\cos 2\theta}{n}\right)$ ☐ (b) $r(1 - \cos\theta) + l(1 - \cos\phi)$ ☐

(c) $\omega^2 \cdot r\left(\sin\theta + \dfrac{\cos 2\theta}{n}\right)$ ☐ (d) none of the above. ☐

74. The velocity of the piston in a reciprocating steam engine is given by

(a) $\omega \cdot r\left(\sin\theta + \dfrac{\sin 2\theta}{2n}\right)$ ☐ (b) $\omega^2 \cdot r\left(\cos\theta + \dfrac{\cos 2\theta}{n}\right)$ ☐

(c) $\omega^2 r\left(\sin\theta + \dfrac{\sin 2\theta}{2n}\right)$ ☐ (d) none of above. ☐

75. There are two points P and Q on a planar rigid body. The relative velocity between the two points

(a) should always be along PQ ☐

(b) can be oriented along any direction ☐

(c) should always be perpendicular to PQ ☐

(d) should be along QP when the body undergoes pure translation. ☐

(GATE-ME-2010)

76. If two nodes are observed at a frequency of 1800 rpm during whirling of a simply supported long slender rotating shaft, the first critical speed of the shaft in rpm is

(a) 200 ☐ (b) 450 ☐

(c) 600 ☐ (d) 900. ☐

(GATE-ME-2013)

77. A single degree of freedom system having mass 1 kg and stiffness 10 kN/m initially at rest is subjected to an impulse force of magnitude 5 kN for 10^{-4} seconds. The amplitude in mm of the resulting free vibration is

(a) 0.5 ☐ (b) 1.0 ☐

(c) 5.0 ☐ (d) 10.0. ☐

(GATE-ME-2013)

78. The height of a Watt's governor is expressed as

(a) $h = \dfrac{\omega^2}{g}$ ☐ (b) $h = g\omega^2$ ☐

(c) $h = g \cdot \omega$ ☐ (d) $h = \dfrac{g}{\omega^2}$. ☐

79. Tick the correct statement

(a) Hartnell governor is spring loaded type governor. ☐

(b) Watt governor is spring load type governor. ☐

(c) Porter governor is spring loaded type governor. ☐

(d) Proel governor is spring loaded type governor. ☐

80. Tick the correct statement

(a) The governor does not have any control on the varying load on the engine ☐

(b) The governor reduces the speed fluctuation during a cycle of engine ☐

(c) The governor does not have any control on the speed of engine ☐

(d) The governor maintains the speed of the engine within prescribed limits for varying torque output conditions. ☐

ANSWERS

Answers to Objective Type Questions

1. (*a*)	2. (*c*)	3. (*c*)	4. (*e*)	5. (*c*)	6. (*d*)
7. (*a*)	8. (*b*)	9. (*c*)	10. (*b*)	11. (*a*)	12. (*e*)
13. (*f*)	14. (*b*)	15. (*e*)	16. (*c*)	17. (*b*)	18. (*d*)
19. (*c*)	20. (*b*)	21. (*a*)	22. (*b*)	23. (*b*)	24. (*b*)
25. (*c*)	26. (*c*)	27. (*b*)	28. (*a*)	29. (*b*)	30. (*c*)
31. (*a*)	32. (*a*)	33. (*c*)	34. (*b*)	35. (*b*)	36. (*a*)
37. (*d*)	38. (*d*)	39. (*d*)	40. (*b*)	41. (*c*)	42. (*d*)
43. (*a*)	44. (*b*)	45. (*b*)	46. (*a*)	47. (*d*)	48. (*c*)
49. (*d*)	50. (*b*)	51. (*a*)	52. (*a*)	53. (*a*)	54. (*b*)
55. (*c*)	56. (*b*)	57. (*d*)	58. (*a*)	59. (*c*)	60. (*c*)
61. (*d*)	62. (*b*)	63. (*c*)	64. (*a*)	65. (*d*)	66. (*b*)
67. (*b*)	68. (*b*)	69. (*c*)	70. (*b*)	71. (*a*)	72. (*c*)
73. (*e*)	74. (*d*)	75. (*b*)	76. (*d*)	77. (*c*)	78. (*a*)
79. (*c*)	80. (*c*)	81. (*a*)	82. (*b*)	83. (*b*)	84. (*a*)
85. (*b*)	86. (*a*)	87. (*c*)	88. (*d*)	89. (*b*)	90. (*d*)
91. (*c*)	92. (*c*)	93. (*c*)	94. (*d*)	95. (*d*)	96. (*b*)
97. (*a*)	98. (*b*)	99. (*b*)	100. (*b*)	101. (*c*)	102. (*d*)
103. (*a*)	104. (*c*)	105. (*b*)	106. (*b*)	107. (*c*)	108. (*c*)
109. (*b*)	110. (*b*)	111. (*a*)	112. (*c*)	113. (*d*)	114. (*c*)
115. (*c*)	116. (*b*)	117. (*c*)	118. (*d*)	119. (*b*)	120. (*e*)
121. (*e*)	122. (*a*)	123. (*b*)	124. (*d*)	125. (*d*)	126. (*b*)
127. (*c*)	128. (*c*)	129. (*b*)	130. (*c*)	131. (*b*)	132. (*a*)
133. (*b*)	134. (*a*)	135. (*d*)	136. (*b*)	137. (*c*)	138. (*c*)
139. (*c*)	140. (*d*)	141. (*b*)	142. (*b*)	143. (*e*)	144. (*c*)
145. (*e*)	146. (*f*)	147. (*d*)	148. (*b*)	149. (*b*)	150. (*e*)
151. (*e*)	152. (*c*)	153. (*d*)	154. (*b*)	155. (*d*)	156. (*b*)
157. (*b*)	158. (*b*)	159. (*d*)	160. (*b*)	161. (*b*)	162. (*c*)
163. (*a*)	164. (*b*)	165. (*a*)	166. (*b*)	167. (*c*)	168. (*a*)
169. (*c*)	170. (*d*)	171. (*d*)	172. (*c*)	173. (*b*)	174. (*c*)
175. (*d*)	176. (*c*)	177. (*b*)	178. (*d*)	179. (*f*)	180. (*c*)
181. (*b*)	182. (*e*)	183. (*c*)	184. (*b*)	185. (*c*)	186. (*b*)
187. (*a*)	188. (*c*)	189. (*b*)	190. (*a*)	191. (*b*)	192. (*c*)
193. (*b*)	194. (*a*)	195. (*c*)	196. (*c*)	197. (*a*)	198. (*c*)
199. (*b*)	200. (*b*)	201. (*d*)	202. (*d*)	203. (*c*)	204. (*c*)
205. (*d*)	206. (*c*)	207. (*c*)	208. (*b*)	209. (*b*)	210. (*d*)
211. (*c*)	212. (*d*)	213. (*b*)	214. (*d*)	215. (*c*)	216. (*c*)

217. (*c*)	**218.** (*b*)	**219.** (*c*)	**220.** (*b*)	**221.** (*b*)	**222.** (*b*)
223. (*b*)	**224.** (*d*)	**225.** (*a*)	**226.** (*d*)	**227.** (*c*)	**228.** (*b*)
229. (*d*)	**230.** (*b*)	**231.** (*c*)	**232.** (*b*)	**233.** (*a*)	**234.** (*a*)
235. (*c*)	**236.** (*b*)	**237.** (*c*)	**238.** (*b*)	**239.** (*c*)	**240.** (*c*)
241. (*c*)	**242.** (*c*)	**243.** (*b*)	**244.** (*c*)	**245.** (*a*)	**246.** (*b*)
247. (*b*), (*d*)	**248.** (*b*)	**249.** (*b*)	**250.** (*c*)	**251.** (*a*)	**252.** (*b*)
253. (*c*)	**254.** (*b*)	**255.** (*a*)	**256.** (*c*)	**257.** (*a*)	**258.** (*b*)
259. (*a*)	**260.** (*c*)	**261.** (*d*)	**262.** (*b*)	**263.** (*b*)	**264.** (*b*)
265. (*a*)	**266.** (*c*)	**267.** (*b*)	**268.** (*b*)	**269.** (*d*)	**270.** (*e*)
271. (*e*)	**272.** (*b*)	**273.** (*b*)	**274.** (*a*)	**275.** (*b*)	**276.** (*b*)
277. (*c*)	**278.** (*d*)	**279.** (*c*)	**280.** (*b*)	**281.** (*c*)	**282.** (*a*)
283. (*c*)	**284.** (*b*)	**285.** (*a*)	**286.** (*c*)	**287.** (*d*)	**288.** (*a*)
289. (*b*)	**290.** (*c*)	**291.** (*b*)	**292.** (*b*)	**293.** (*a*)	**294.** (*c*)
295. (*b*)	**296.** (*d*)	**297.** (*c*)	**298.** (*a*)	**299.** (*b*)	**300.** (*d*)
301. (*c*)	**302.** (*d*)	**303.** (*c*)	**304.** (*d*)	**305.** (*b*)	**306.** (*c*)
307. (*b*)	**308.** (*e*)	**309.** (*d*)	**310.** (*c*)	**311.** (*b*)	**312.** (*a*)
313. (*a*)	**314.** (*d*)	**315.** (*a*)	**316.** (*c*)	**317.** (*d*)	**318.** (*b*)
319. (*b*)	**320.** (*d*)	**321.** (*c*)	**322.** (*c*)	**323.** (*b*)	**324.** (*c*)
325. (*a*)	**326.** (*d*)	**327.** (*a*)	**328.** (*b*)	**329.** (*c*)	**330.** (*a*)
331. (*b*)	**332.** (*b*)	**333.** (*d*)	**334.** (*c*)	**335.** (*d*)	**336.** (*b*)
337. (*a*)	**338.** (*b*)	**339.** (*c*)	**340.** (*d*)	**341.** (*c*)	**342.** (*d*)
343. (*a*)	**344.** (*b*)	**345.** (*a*)	**346.** (*d*)	**347.** (*e*)	**348.** (*f*)
349. (*c*)	**350.** (*b*)	**351.** (*d*)	**352.** (*c*)	**353.** (*c*)	**354.** (*c*)
355. (*c*)	**356.** (*d*)	**357.** (*d*)	**358.** (*e*)	**359.** (*e*)	**360.** (*e*)
361. (*d*)	**362.** (*d*)	**363.** (*d*)	**364.** (*c*)	**365.** (*d*)	**366.** (*b*)
367. (*c*)	**368.** (*c*)	**369.** (*c*)	**370.** (*e*)	**371.** (*c*)	**372.** (*b*)
373. (*c*)	**374.** (*c*)	**375.** (*b*)	**376.** (*b*)	**377.** (*c*)	**378.** (*b*)
379. (*c*)	**380.** (*c*)	**381.** (*b*)	**382.** (*d*)	**383.** (*d*)	**384.** (*b*)
385. (*a*)	**386.** (*a*)	**387.** (*c*)	**388.** (*c*)	**389.** (*b*)	**390.** (*a*)
391. (*c*)	**392.** (*e*)	**393.** (*b*)	**394.** (*a*)	**395.** (*a*)	**396.** (*e*)
397. (*f*)	**398.** (*b*)	**399.** (*d*)	**400.** (*d*)	**401.** (*d*)	**402.** (*b*)
403. (*d*)	**404.** (*b*)	**405.** (*b*)	**406.** (*b*)	**407.** (*b*)	**408.** (*c*)
409. (*d*)	**410.** (*a*)	**411.** (*c*)	**412.** (*a*)	**413.** (*d*)	**414.** (*c*)
415. (*b*)	**416.** (*c*)	**417.** (*c*)	**418.** (*b*)	**419.** (*c*)	**420.** (*c*)
421. (*b*)	**422.** (*b*)	**423.** (*d*)	**424.** (*c*)	**425.** (*b*)	**426.** (*a*)
427. (*b*)	**428.** (*b*)	**429.** (*a*)	**430.** (*a*)	**431.** (*b*)	**432.** (*f*)
433. (*b*)	**434.** (*c*)	**435.** (*a*)	**436.** (*d*)	**437.** (*a*)	**438.** (*a*)
439. (*b*)	**440.** (*c*)	**441.** (*c*)	**442.** (*e*)	**443.** (*e*)	**444.** (*b*)
445. (*b*)	**446.** (*d*)	**447.** (*d*)	**448.** (*d*).		

True/False

449. (*a*) 450. (*a*) 451. (*b*) 452. (*b*) 453. (*a*) 454. (*b*)
455. (*a*) 456. (*a*) 457. (*b*) 458. (*b*) 459. (*b*) 460. (*b*)
461. (*a*) 462. (*a*) 463. (*a*) 464. (*b*) 465. (*a*) 466. (*a*)
467. (*a*) 468. (*a*) 469. (*a*) 470. (*b*) 471. (*a*) 472. (*a*)
473. (*a*) 474. (*a*) 475. (*a*) 476. (*a*) 477. (*a*) 478. (*a*)
479. (*a*) 480. (*a*) 481. (*a*) 482. (*a*) 483. (*a*) 484. (*a*)
485. (*b*) 486. (*b*) 487. (*b*) 488. (*a*) 489. (*a*) 490. (*a*)
491. (*a*) 492. (*a*) 493. (*b*) 494. (*a*) 495. (*a*) 496. (*b*)
497. (*a*) 498. (*a*) 499. (*a*) 500. (*a*) 501. (*a*) 502. (*a*)
503. (*a*) 504. (*b*) 505. (*b*) 506. (*a*) 507. (*a*) 508. (*a*)
509. (*a*) 510. (*a*) 511. (*b*) 512. (*a*) 513. (*b*) 514. (*b*)
515. (*a*) 516. (*a*) 517. (*a*) 518. (*a*) 519. (*b*) 520. (*a*)
521. (*a*) 522. (*a*) 523. (*a*) 524. (*a*) 525. (*b*) 526. (*b*)
527. (*a*) 528. (*a*) 529. (*a*) 530. (*b*) 531. (*b*) 532. (*a*)
533. (*a*) 534. (*a*) 535. (*a*) 536. (*a*) 537. (*a*) 538. (*a*)
539. (*b*) 540. (*a*) 541. (*b*) 542. (*a*) 543. (*a*) 544. (*a*)
545. (*a*) 546. (*b*) 547. (*a*) 548. (*a*) 549. (*b*) 550. (*b*)
551. (*a*) 552. (*a*) 553. (*a*) 554. (*a*) 555. (*a*) 556. (*b*)
557. (*a*) 558. (*a*) 559. (*a*) 560. (*a*) 561. (*a*) 562. (*b*)
563. (*b*) 564. (*a*) 565. (*a*) 566. (*b*) 567. (*a*) 568. (*a*)
569. (*a*) 570. (*a*).

Fill in the Blanks

571. (*b*) 572. (*b*) 573. (*b*) 574. (*a*) 575. (*b*) 576. (*a*)
577. (*b*) 578. (*b*) 579. (*b*) 580. (*b*) 581. (*b*) 582. (*b*)
583. (*b*) 584. (*b*) 585. (*b*) 586. (*b*) 587. (*b*) 588. (*b*)
589. (*b*) 590. (*b*) 591. (*b*) 592. (*b*) 593. (*a*).

Answers to Objective Type Questions from Competitive Examinations

1. (*a*) 2. (*c*) 3. (*d*) 4. (*e*) 5. (*b*) 6. (*d*)
7. (*a*) 8. (*b*) 9. (*b*) 10. (*d*) 11. (*c*) 12. (*a*)
13. (*b*) 14. (*b*) 15. (*b*) 16. (*c*) 17. (*c*) 18. (*b*)
19. (*b*) 20. (*a*) 21. (*d*) 22. (*b*) 23. (*d*) 24. (*a*)
25. (*c*) 26. (*c*) 27. (*d*) 28. (*a*) 29. (*c*) 30. (*a*)
31. (*d*) 32. (*d*) 33. (*a*) 34. (*d*) 35. (*c*) 36. (*a*)
37. (*c*) 38. (*c*) 39. (*a*) 40. (*d*) 41. (*a*) 42. (*d*)
43. (*a*) 44. (*b*) 45. (*d*) 46. (*b*) 47. (*e*) 48. (*a*)
49. (*b*) 50. (*a*) 51. (*c*) 52. (*e*) 53. (*b*) 54. (*c*)
55. (*c*) 56. (*c*) 57. (*d*) 58. (*a*) 59. (*a*) 60. (*c*)
61. (*c*) 62. (*b*) 63. (*c*) 64. (*d*) 65. (*a*) 66. (*b*)
67. (*b*) 68. (*b*) 69. (*c*) 70. (*d*) 71. (*a*) 72. (*b*)
73. (*b*) 74. (*a*) 75. (*c*) 76. (*a*) 77. (*c*) 78. (*d*)
79. (*a*) 80. (*a*).

Chapter 10 MACHINE DESIGN

I. THEORY

10.1. INTRODUCTION

Machine Design is defined as the practical application of mechanics of machinery to the design and construction of machines and structures. For designing any machine, the knowledge of materials and their properties is of great significance. The common materials used by design engineers are (*i*) iron such as cast iron, steel, malleable iron and wrought iron, (*ii*) copper and its alloys, (*iii*) brass and bronze, (*iv*) light alloys of aluminium, and (*v*) materials like wood, fibre, rubber, leather and mica.

In preparing a design, the following points should be remembered:

(*i*) Selection of materials from which the machine is to be constructed.

(*ii*) Study of cost of production and cost of operation.

(*iii*) Study of facilities of fabrication.

(*iv*) Specifications and drawings.

(*v*) Determination of machine mechanism which will give the required motion.

(*vi*) Method of lubrication.

(*vii*) Safety of the operator of the machine.

10.2. INTERCHANGEABILITY AND LIMIT SYSTEM

In a mass production, to produce a large number of parts of the exactly same dimension is not possible. Hence a small permissible variation in size is allowed, which will not prevent the proper functioning of the parts. This phenomenon is known as **interchangeability.** For controlling the small variation in the size of finished parts **a limit system** is used. The various **important terms** used in limit system are:

10.2.1. Nominal Size

It is the size of a part generally specified in the drawing.

10.2.2. Basic Size

It is the exact theoretical size to which all limits of variations are determined.

10.2.3. Allowance

It is the minimum clearance space, provided between the two meeting parts.

10.2.4. Interference

It is amount by which the dimensions of the meeting parts overlap.

10.2.5. Limits

They are the extreme permissible dimension of any part. There are two extreme permissible size for a dimension. The largest permissible size for a dimension is called **upper or higher limit** whereas the smallest size is known as **lower limit.**

10.2.6. Tolerance

It is the difference between the upper limit and lower limit of a dimension. The tolerance may be unilateral or bilateral. If the tolerance is allowed on one side of the nominal size such as $15^{+0.003}_{-0.000}$, the system is known as **unilateral.** If the tolerance is allowed on both sides of the nominal size such as $15^{+0.003}_{-0.003}$, the system is known as **bilateral.** Here +0.003 represents the upper limit and –0.003 is the lower limit.

10.2.7. Fit

It is defined as the degree of tightness and looseness between the two mating parts. The type of fits depend upon the size of clearance and interference. According to Indian Standards, the **fits are classified** into the following **three groups:**

(*i*) Clearance fit, (*ii*) Interference fit, and (*iii*) Transition fit.

10.2.8. Basis of Limit System

There are following two bases for limit system:

(*i*) Hole basis, and (*ii*) Shaft basis.

If the size of the hole is kept constant and different fits are obtained by varying the shaft size, then the limit system is said to be on a **hole basis.** On the other hand, if the size of the shaft is kept constant and different fits are obtained by varying the hole size, then the limit system is said to be a **shaft basis.**

10.2.9. The Indian Standard System of Limits and Fits

The system of limits and fits, according to Indian Standard (IS : 1919–1963), comprises 16 grades of fundamental tolerance (*i.e.*, grades of accuracy of manufacture) and 25 types of fundamental deviations indicated by letter symbols for both holes and shafts (capital letter *A* to *ZC* for holes and small letters *a* to *zc* for shafts) in diameter steps ranging from 1 mm to 500 mm.

10.3. POWER SCREWS AND THREADS

For transmission of power in screw operated machines, such as lathe, screw jack and press, the power screws are used. In some cases, the nut has axial motion and screw rotates in its bearing whereas in other cases, the screw rotates and moves axially while the nut is stationary. The standard forms of threads are :

(*i*) B.S.W. Threads, (*ii*) Metric Threads,
(*iii*) Square Threads, (*iv*) Acme Threads, and
(*v*) Buttress Threads.

The *B.S. W. thread* is a British Standard Whitworth thread having angle between the flanks as 55°. The *metric thread* is an Indian Standard thread having angle between the flanks as 60°. The square, Acme and buttress threads are used for power transmission. The square and Acme threads can transmit power in any direction whereas the buttress threads are used to transmit power in one direction only.

10.3.1. Bolt

It is a cylindrical rod, having a head at one end and a nut fitted to the other end.

10.3.2. Tap Bolt

It is a cylindrical rod, having a head at one end and other end fitting into a tapped hole in the other part to be joined.

10.3.3. Stud

It is a cylindrical rod with both ends threaded.

10.3.4. Set-Screw

It is a cylindrical rod having head at one end and is threaded practically throughout its length.

Screw threads according to IS : 1362–1962, are designated by a letter *'M'* followed by the diameter and pitch, the two being separated by the sign ×. The tolerance grade on the screw threads is indicated by words 7 or 8 or 9 showing fine or normal or coarse grade respectively. The symbol $M\ 24 \times 2 - 8d$ represents a bolt with metric threads of 24 mm size and 2 mm pitch with allowance on the threads and having normal tolerance grade.

10.4. STRESSES IN SCREWED FASTENING DUE TO STATIC LOADING

The stresses are:

1. Tightening-up stresses
2. Stresses due to external forces
3. Stresses due to tightening and due to external forces.

10.4.1. Tightening-up Stresses

The tightening-up stresses, induced in the bolt, screw or stud are:

(*i*) tensile stress, (*ii*) shear stress,

(*iii*) crushing stress on the threads, and (*iv*) bending stress.

(*i*) Tensile Stress due to Tightening-up. During tightening, the bolts of small diameter will fail. Hence bolts of diameter less than *M* 16 or *M* 18 are not used for making fluid tight joint. The initial tension in a bolt due to tightening is given by the relation

$$P = 2840\, d \text{ N} \quad \text{For a fluid tight joint}$$

$$= 1420\, d \text{ N} \quad \text{For an ordinary joint}$$

where d = Nominal diameter of bolt in mm.

10.4.2. Stresses Due to External Forces

They are:

(*i*) Tensile stress, (*ii*) Shear stress, and

(*iii*) Combined tensile and shear stress.

(*i*) Tensile Stress due to External Load. If P is the external load applied to a bolt of core diameter (or root diameter) equal to d_c, the tensile stress induced in the bolt material is given by

$$\sigma_t = \frac{P}{\frac{\pi}{4} d_c^2} \quad \text{If load } P \text{ is taken up by one bolt}$$

$$= \frac{P}{\frac{\pi}{4} d_c^2 \times n} \quad \text{If load } P \text{ is taken up by '}n\text{' bolts}$$

(*ii*) Shear Stress due to External Load. Shear stress is induced in the bolts, when bolts are used to prevent the relative movement of two parts as in the case of flange coupling. The value of shear stress is given by

$$\tau = \frac{P_s}{\frac{\pi}{4} d^2 \times n}$$

where P_s = Shearing load on the bolts, d = Major diameter of the bolt

n = Number of bolts.

(*iii*) **Combined Tensile and Shear Stress.** For a bolt, subjected to both tension and shear loads, the diameter of the shank of bolt is obtained from shear load and that of threaded part from the tensile load. A diameter slightly large than the two values is adopted.

10.4.3. Stresses Due to Initial Tension and Due to External Load

The resultant axial load on a bolt due to initial tension and due to external load depends upon the relative elastic yielding of the bolt and the connecting member. The resultant load is given by

$$P = P_1 + \left(\frac{a}{1+a}\right)P_2 = P_1 + KP_2$$

where P_1 = Initial tightening load on the bolt,

P_2 = External load on the bolt, and

a = Ratio of elasticity of connected parts to the elasticity of bolt.

(*i*) If gasket material is very soft, the value of a is high and $\frac{a}{1+a}$ becomes equal to one and then resultant load is equal to sum of initial tightening load and external load.

(*ii*) If gasket material is very hard, the value of a is very small and $\frac{a}{1+a}$ becomes negligible and then resultant load is equal to the initial tightening load.

10.4.4. Eccentric Load on Bolted Joint

The eccentric load acting on a bolted joint may be: (*a*) parallel to the axis of the bolt, (*b*) perpendicular to the axis of the bolt, and (*c*) in plane containing the bolts.

10.5. DESIGN OF SHAFTS

The shaft may be a transmission shaft (such as counter shafts, line shafts, overhead shafts etc.) or a machine shaft (such as crank shaft). They may be a hollow shafts or solids shafts. The stresses induced in the shafts are:

1. Shear stresses due to transmission of torque.
2. Bending stresses due to weight of the shafts and due to forces acting on machine elements like gears and pulleys etc.
3. Complex stresses which is due to torsion and bending.

10.5.1. Shafts in Pure Torsion

When a shaft is subjected to **pure torsion,** the stress induced is shear stress only. The shaft is designed for pure torsion only as given by equations.

(*i*) $$\frac{T}{J} = \frac{\tau}{R} = \frac{C\theta}{l}$$

(ii) $$T = \frac{\pi D^3}{16}\tau \quad \text{For a solid shaft}$$

$$= \frac{\pi(D_o^4 - D_i^4)}{16 D_o}\tau \quad \text{For a hallow shaft}$$

where T = Torque acting on the shaft,

J = Polar moment of inertia of cross-sectional area,

τ = Shear stress, R = Radius of the shaft,

D_o = Outer diameter of the hollow shaft,

D_i = Inner diameter of the shaft, C = Modulus of rigidity,

θ = Angle of twist, l = Length of the shaft.

10.5.2. Shaft in Pure Bending

When the shaft is subjected to **pure bending,** the stress induced will be tension and compression. The bending equations are

(i) $$\frac{M}{I} = \frac{\sigma_b}{y} \quad \text{and}$$

(ii) $$M = \frac{\pi D^3}{32}\sigma_b \quad \text{For a solid shaft}$$

$$= \frac{\pi(D_o^4 - D_i^4)}{32 D_o}\sigma_b \quad \text{For a hallow shaft}$$

where M = Bending moment,

σ_b = Bending stress,

I = Moment of inertia of the cross-sectional area,

y = Distance from neutral axis to be outer-most fibre,

D = Diameter of solid shaft,

D_o = Outer diameter of hollow shaft, and

D_i = Inner diameter of hollow shaft.

10.5.3. Shafts in Torsion and Bending

When the shaft is subjected to **both twisting moment and bending moment,** then the shaft is designed on the basis of the resultant stress which is determined by the various theories of failure which are given in next article.

10.6. VARIOUS THEORIES OF FAILURE

The various theories of failure are:

10.6.1. Maximum Shear Stress Theory or Guest Theory

According to this theory the maximum shear stress induced is given by the equation

$$\frac{\pi}{16} D^3 \times \tau = \sqrt{M^2 + T^2}\,.$$

10.6.2. Maximum Normal Stress Theory or Rankine's Theory

According to this theory the maximum normal stress-in the shaft is given by the equation

$$\frac{\pi}{32} D^3 \times \sigma_b = \frac{1}{2}[M + \sqrt{M^2 + T^2}]\,.$$

10.6.3. Maximum Strain Theory

According to this theory, we have

$$\frac{\pi}{32} D^3 \times \sigma_b - \frac{1}{2}(1-m)M + \frac{1}{2}(1+m)\sqrt{M^2 + T^2}$$

where m = Poisson's ratio and if it taken as 0.3, then the above equation becomes as

$$\frac{\pi}{32} D^3 \times \sigma_b = 0.35M + 0.65\sqrt{M^2 + T^2}\,.$$

10.6.4. Materials for Shafts

Generally, the material for the shaft is mild steel. But when high strength is required, an alloy steel such as nickel, nickel chromium steel is used.

10.6.5. Effect of Key Ways

The effect of key ways is to reduce the strength of the shaft. The strength of the keyed shaft is usually 75% of the solid shaft. The ratio of the strength of the shaft with key way to the strength of shaft without key way is known as strength factor. The shaft strength factor

$$= 1.0 - 0.2\frac{b}{d} - 1.1\frac{h}{d}$$

where b = Width of key way,

h = Depth of key way, and

d = Diameter of shaft.

10.7. HOLLOW SHAFTS

Large shafts are generally made hollow as hollow shafts have greater strength than solid shafts of equal weight. If a hollow shaft is equal in torsional strength to a solid shaft, then

$$d_o^3 = \frac{d^3}{1-k^4}$$

where d = Diameter of solid shaft,

d_o = Outer diameter of hollow shaft, and

$$k = \frac{d_i}{d_o} = \frac{\text{Inner diameter of hollow shaft}}{\text{Outer diameter of hollow shaft}}.$$

10.7.1. Fatigue of Metals and Endurance Limit

When a material is subjected to varying stresses, the material fails under stresses considerably lower than the ultimate stress. Such type of failure of material is known as fatigue and magnitude of such stresses depend upon the range of stresses, the mean stress and number of applications. The value of maximum stress, which a material can withstand without failure for infinite number of cycles when subjected to completely reversed load, is called **endurance limit.** The ratio of endurance limit stress to working stress is known as factor of safety for fatigue loading. The endurance limit of a material subjected to fatigue loading increases with the increases of ultimate tensile strength and with the increase of surface finish.

10.8. STRESS CONCENTRATION FACTOR

It is the ratio of the maximum stress in a machine component (at a notch or fillet) to the nominal stress at the same section. The stress concentration factor depends upon the geometry and material of the component. When there is a sudden change of cross-section, stress concentration is caused. Figure 10.1 shows a plate with an elliptical hole in the centre with semi-major axis = a and semi-minor axis = b. The plate is subjected to tensile load P. The maximum stress induced at the edge of the hole (*i.e.*, at the point A) is given by

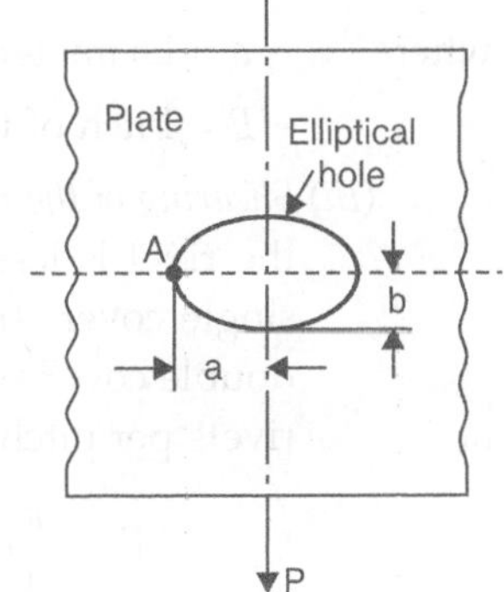

FIGURE 10.1

$$\sigma_{max} = \sigma\left(1+\frac{2a}{b}\right)$$

where σ = Nominal stress, *i.e.*, stress in the plate far away from the hole.

The stress concentration factor is equal to

$$\frac{\sigma_{max}}{\sigma} = \left(1+\frac{2a}{b}\right)$$

When $a = 2b$, stress concentration factor = 5.0,

When $a = b$, stress concentration factor = 3.0, and

When $a = \frac{b}{2}$, stress concentration factor = 2.0.

10.9. RIVETED JOINTS

Riveting is a permanent fastening in which two parts are joined together permanently. There are two types of riveted joints, namely (*i*) Lap joint, and (*ii*) Butt joint.

10.9.1. Lap Joint

In lap joints, the ends of the plates (to be joined) overlap each other and they are riveted together by a single row of rivets or by multiple rows.

10.9.2. Butt Joint

In butt joints, the ends of the plates (to be joined) are placed face to face and a separate plate known as strap, overlaps the two. The strap is riveted to the plates. The joint may be made with a single strap or with two straps one on either side of the plates.

10.9.3. Types of Failures in Riveted Joints

A riveted joint may fail in the following ways:

(i) *Tearing of the plate at an edge.* This can be avoided by keeping the distance between the centre of the rivet hole to the nearest edge of the plate (*i.e.*, margin, m) equal to $1.5d$ where d is the diameter of the rivet.

(ii) *Tearing of the plate across a row of rivets.* The pull required to tear off the plate across a row of rivets per pitch length is given by

$$P_t = (P - d)t \times \sigma_t$$

where d = Diameter of the rivet, t = Thickness of the plate

P = Pitch of the rivets, σ_t = Permissible tensile stress.

(iii) *Shearing of the rivets.* A riveted joint will fail by shearing off the rivets, if the diameter of the rivet is less than the required diameter. In case of lap joints and butt joints with single cover strap, the rivet will be in single shear, whereas in case of butt joints with double cover straps, the rivet will be in double shear. The pull required to shear off the rivets per pitch length is given by

$$P_s = \frac{\pi}{4} d^2 \times \tau \times n \quad \text{For single shear}$$

$$= 2 \times \frac{\pi}{4} d^2 \times \tau \times n \quad \text{For double shear (theoretically)}$$

$$= 1.875 \times \frac{\pi}{4} d^2 \times \tau \times n \quad \text{For double shear according to Indian Boiler Regulation}$$

where n = Number of rivets per pitch length, and

τ = Permissible shear stress.

(iv) *Crushing of the rivets.* The pull required to crush the rivets per pitch length is given by

$$P_c = n \times d \times t \times \sigma_c$$

where n = Number of rivets per pitch length, and

σ_c = Permissible crushing stress.

10.9.4. Strength and Efficiency of a Riveted Joint

The strength of a riveted joint is equal to the minimum of the pulley required to shear off the rivet or to tear off the plate or to crush the rivets. The efficiency of a riveted joint is the ratio of the strength of the joint to the strength of unriveted plate. Mathematically, efficiency of the riveted joint is given by

$$\eta = \frac{\text{Least of } P_s, P_t \text{ and } P_c}{\text{Strength of unriveted (or solid) plate}}$$

where P_s = Pull required to shear off the rivets,
P_t = Pull required to tear off the plate, and
P_c = Pull required to crush the rivets.

10.9.5. Design of Riveted Joints

(*i*) The diameter '*d*' of the rivet for the given thickness '*t*' of the plate is given by

$$d = 6\sqrt{t} \text{ to } 6.1\sqrt{t}\text{, where } t \text{ and } d \text{ are in mm}$$

$$= 1.91\sqrt{t} \text{ to } 2\sqrt{t}\text{, where } t \text{ and } d \text{ are in cm.}$$

(*ii*) The pitch of the rivets should not be less than 2*d* where there are equal number of rivets in each row. The maximum pitch = 3*d*.

(*iii*) The distance between the centre of the rivet hole and the edge of the plate (*i.e.*, the margin), according to Indian Boiler Code, should not be less than 1.5*d* and not more than 1.75*d*.

(*iv*) The thickness (t_1) for the butt straps, in a butt joint, in no case shall be less than 1 cm. It is generally taken as

$$t_1 = 1.125t\text{, for a single strap}$$
$$= 0.625t\text{, for double butt strap of equal width.}$$

(*v*) In a butt joint, when two unequal width of straps are used, then thickness of butt straps is given by

$$t_1 = 0.75t\text{, for wide strap on the inside}$$
$$= 0.625t\text{, for narrow strap on the outside.}$$

(*vi*) Length of the rivet = 1.25*d* to 1.75*d*.

10.10. WELDED JOINTS

The important types of welded joints are:

(*i*) Butt joint which may be a single butt joint or double butt joint,

(*ii*) Lap joint,

(*iii*) Edge joint,

(*iv*) *T*-joint, and

(*v*) Corner joint.

10.10.1. Strengths of Welds

For calculating the strength of the welds, the following design stresses may be used :

For shear, $\tau = 0.55$ N/mm^2

For tension, $\sigma_t = 0.91$ N/mm^2

For compression, $\sigma_c = 1.05$ N/mm^2

The maximum stress in bending should not be more than the working stresses given for tension or compression.

For a butt joint with a single *V*-groove weld loaded by a tensile force *F*, the tensile stress in the throat is given by

$$\sigma_t = \frac{F}{h \times l}$$

where h = Depth of the weld, and l = Length of the weld.

The shear stress in the butt weld, due to shear loading is given by

$$\tau = \frac{F}{h \times l}$$

But if the butt weld is subjected to a bending moment *M*, the bending stress is given by

$$\sigma_b = \frac{6M}{h^2 \times l}.$$

10.10.2. Transverse Filled Weld

(*i*) A 45° double fillet weld lap joint is shown in Fig. 10.2. This weld is subjected to tensile stress as well as shear stress. The tensile force acts on the vertical face *AC* and shear force acts on the horizontal face *AB* of the weld. At the throat *AD*, the minimum area of the weld is obtained. The tensile stress for the weld shown in Fig. 10.2 is given by

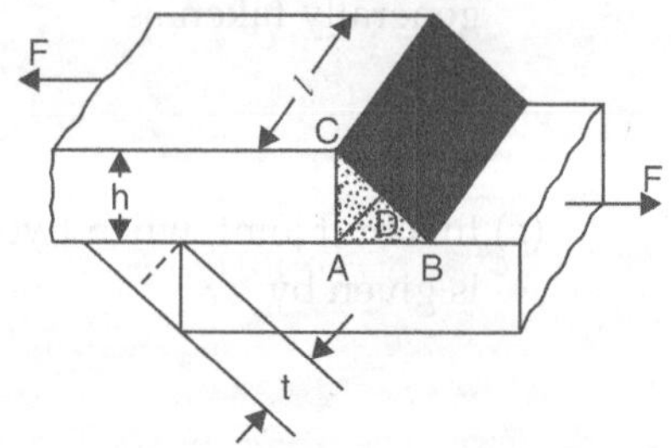

FIGURE 10.2

$$\sigma_t = \frac{F}{2 \times l \times t} = \frac{F}{2 \times l \times h \times \sin 45^\circ}$$

$$(\because \quad t = AD = AC \times \sin 45^\circ = h \sin 45^\circ = 0.707h)$$

$$= \frac{F}{2 \times l \times h \times 0.707} = \frac{F}{1.414 \times l \times h} = \frac{0.707F}{h \times l}$$

(*ii*) If the fillet welds are used on one side of the plate, due to eccentricity of forces, a bending moment is introduced. The maximum normal stress due to bending moment is equal to

$$\sigma_{t_1} = \frac{3F}{h \times l}$$

and tensile stress, $$\sigma_{t_2} = \frac{F}{l \times t} = \frac{F}{0.707h \times l} \times \frac{1.414F}{h \times l}$$

The resultant stress will be equal to

$$\sigma_t = \sigma_{t_1} + \sigma_{t_2} = \frac{3F}{hl} + \frac{1.414F}{h \times l} = \frac{4.414F}{h \times l}$$

(*iii*) In case of parallel fillet welded joint, the shear stress is given by

$$\tau = \frac{0.707F}{h \times l}.$$

10.11. KEYS AND COUPLINGS

A piece of mild steel, inserted between the shaft and hub or boss of a pulley to connect them together in order to prevent relative motion between them, is known as a key. The key is subjected to shearing and crushing stresses. The important type of keys are:

(*i*) Sunk keys, (*ii*) Saddle keys,
(*iii*) Tangent keys, (*iv*) Round keys, and
(*v*) Spline.

10.11.1. Sunk Keys

The keys, which fit half in the key way of the hub and half in the way of shaft, are known as sunk keys. The sunk keys are of the following types:

(*i*) Rectangular sunk key, (*ii*) Square sunk key,
(*iii*) Gib head key, (*iv*) Feather key, and
(*v*) Wood ruff key.

(*i*) **Rectangular sunk key.** The width and thickness of a rectangular sunk key are:

Width of key, $w = \frac{d}{4}$ and

Thickness of key, $t = \frac{2}{3}w = \frac{2}{3} \times \frac{d}{4} = \frac{d}{6}$, where d = Diameter of the shaft.

(*ii*) **Square sunk key.** The width and thickness of a square key are equal.

Hence $w = t = \frac{d}{4}$.

(*iii*) **Gib head key.** A rectangular sunk key with a head on one end is known as gib head key. The dimensions of a gib head key are the same as of a rectangular sunk key.

(*iv*) **Feather key.** A key attached to one member of a pair and which permits relative axial movement is known as feather key.

(*v*) **Wood ruff key.** A key, which is easily adjustable and is made from a cylindrical disc having segmental cross-section, is known as would ruff key.

10.11.2. Splines Shafts

They are the shafts on which keys are made integral with the shaft which fits in the key ways provided in the hub. The splined shafts are relatively stronger than shafts having key ways.

10.11.3. Design of the Key

The key is designed on the basis of power to the transmitted. When the torque is transmitted from the shaft to the hub or rotor by using a key, the following two types of failure of the key can take place :

(*i*) Shear failure of the key, and (*ii*) Crushing failure of the key.

(*i*) **Shear failure of the key.** When the key fails due to shear, the shearing force (F_s) is given by

$$F_s = \text{Shearing stress} \times \text{Area resisting shear} = \tau \times w \times l$$

where w = Width of key, and l = Length of key.

Torque transmitted by the shaft due to shearing force

$$T_s = F_s \times \frac{d}{2} = \tau \times w \times l \times \frac{d}{2} \quad \ldots(i)$$

(*ii*) **Crushing failure of the key.** When the key fails due to crushing, the crushing force (F_c) is given by

$$F_c = \text{Crushing stress} \times \text{Area resisting crushing}$$

$$= \sigma_c \times \left(l \times \frac{t}{2} \right)$$

where t = Thickness of the key.

Torque transmitted by the shaft due to crushing force

$$T_c = F_c \times \frac{d}{2} = \sigma_c \times l \times \frac{t}{2} \times \frac{d}{2} \quad \ldots(ii)$$

The key will be equally strong in shearing and crushing if,

$$T_s = T_c$$

or

$$\tau \times w \times l \times \frac{d}{2} = \sigma_c \times l \times \frac{t}{2} \times \frac{d}{2} \text{ or } \frac{w}{t} = \frac{\sigma_c}{2\tau}.$$

If $\sigma_c = 2\tau$ *i.e.*, crushing stress for the key material is two times the shearing stress, then width of the key will be equal to thickness of the key. This means a square key will be equally strong in shearing and crushing.

10.11.4. Length of Key

If the material of the key is different than the material of the shaft and width of the key is equal to one-fourth the diameter of the shaft, then length of the key is given by

$$l = 1.571 d \times \frac{\tau^*}{\tau}$$

where τ^* = Shear stress for the shaft material, and

τ = Shear stress for the key material.

If the material of the key and the shaft is the same, then $\tau^* = \tau$ and hence length of the key is given by

$$l = 1.571d.$$

10.11.5. Couplings

A coupling is a device which is used to connect two shafts. A shaft coupling should be easy to connect or disconnect, should transmit the full power of the shaft, should hold the shafts in perfect alignment and should have no projecting parts. They are of two main types:

(*i*) Rigid couplings, and (*ii*) Flexible coupling.

(*i*) **Rigid couplings.** The coupling, which is used to connect two shafts which are perfectly aligned, is known as rigid coupling. The examples of rigid couplings are:

(*a*) Sleeve or muff coupling,

(*b*) Clamp or split or compression coupling, and

(*c*) Flange coupling.

(*ii*) **Flexible coupling.** The coupling, which is used to connect two shafts which have both lateral and angular misalignment, is known as flexible coupling. The examples of flexible couplings are:

(*a*) Bushed pin type coupling,

(*b*) Universal coupling and Oldham's coupling.

10.11.6. Design Procedure for Flange Coupling

(*i*) The hub is designed by considering it a hollow shaft, transmitting the same torque as that of a solid shaft.

$$\therefore \qquad T = \frac{\pi}{16}\tau\left(\frac{d_o^4 - D_i^4}{D_o}\right) \qquad \ldots(i)$$

where D_o = Outer diameter of the hub,

D_i = Inner diameter of the hub or diameter of the shaft, and

τ = Permissible shear stress induced in the shaft.

Generally $D_o = 2D_i$ and hence from equation (*i*), the shear stress induced in the hub may be checked.

(*ii*) The length of the hub is taken equal to 1.5 times the diameter of the shaft and thickness of the flange is taken equal to half the diameter of the shaft.

The thickness of the projected portion of the flange is taken as one-eight of the diameter of the shaft.

(*iii*) The number of bolts is given by empirical formula, $n = 3 + \dfrac{\text{Diameter of shaft}}{5}$.

(*iv*) The pitch circle diameter is equal to three times the diameter of the shaft.

10.12. DESIGN OF COTTER JOINT AND KNUCKLE JOINT

The dimensions of a cotter joint is obtained by considering the various modes of failures as given below:

(*i*) **Diameter of the Rod (*d*)** is obtained by considering tension failure of the rods as

$$P = \frac{\pi}{4} d^2 \times \sigma_t$$

$$\therefore \quad d = \sqrt{\frac{4P}{\pi \times \sigma_t}}$$

(*ii*) **Inside Diameter of the Socket (d_1) and Thickness of Cotter (*t*)** are obtained by considering the tension failure of the socket across slot and crushing strength of the cotter as

$$P = \left(\frac{\pi}{4} d_1^2 - d_1 \times t\right) \times \sigma_t \text{, and } P = d_2 \times t \times \sigma_c.$$

From these two equations, two unknowns d_1 and t can be obtained. In actual practice, the thickness of cotter is given by $t = 0.25\, d_1$ or $0.31d$.

(*iii*) **Outside Diameter of the Socket (d_2)** is obtained by considering the failure of the socket in tension or compression across the slot as

$$P = \left[\frac{\pi}{4}(d_2^2 - d_1^2) - (d_2 - d_1)t\right] \times \sigma_t$$

(*iv*) **Width (*b*) of the Cotter** is obtained by considering the failure of the cotter in shear as

$$P = 2b \times t \times \tau.$$

(*v*) **Diameter of Socket Cotter (d_4)** is obtained by considering crushing failure of socket failure as

$$P = (d_4 - d_1) \times t \times \sigma_c.$$

(*vi*) **Thickness Socket Collar (*c*)** is obtained by considering failure of the socket end in shear as

$$P = 2(d_4 - d_1) \times c \times \tau.$$

(*vii*) **Diameter of Spigot Collar (d_3)** is obtained by considering the failure of spigot collar in crushing as

$$P = \frac{\pi}{4}(d_3^2 - d_1^2) \times \tau$$

(*viii*) **Thickness of the Spigot Collar (t_1)** is obtained by considering the failure of the collar in shearing as

$$P = \pi d_1 \times t_1 \times \tau$$

10.12.1. Design of Knuckle Joint

The dimensions of a knuckle joint are obtained as mentioned below:

(*i*) **Diameter of the Rod (*d*)** is obtained by considering tensile failure of the rod as

$$P = \frac{\pi}{4} d^2 \times \sigma_t$$

(*ii*) **Diameter of the Pin (d_1)** is obtained by considering the shear failure of the pin as

$$P = 2\times\frac{\pi}{4}d_1^2\times\tau$$

If the diameter of the pin is less than diameter of the rod from the above relation, then diameter of the pin is taken equal to the diameter of the rod.

(*iii*) The other dimensions of the joint are fixed by empirical relations as :

(*a*) Outer diameter of the eye, $d_2 = 2d$

(*b*) Diameter of the knuckle pin head and collar, $d_3 = 1.5d$

(*c*) Thickness of single eye, $t = 1.25d$

(*d*) Thickness of fork, $t_1 = 0.15d$

(*e*) Thickness of pin head, $t_2 = 0.5d$.

10.13. BELTS, ROPES AND CHAIN DRIVES

The belts, ropes and chains are used to transmit power from one shaft to another shaft when the two shafts are at a considerable distance apart.

The belts may be flat, *V*-belt or circular. The power is transmitted by open-belt drive or cross-belt drive.

10.13.1. Open-Belt Drive

Let r_1 = Radius of larger pulley, r_2 = Radius of smaller pulley,

D = Distance between the centres of the two shafts, and

L = Total length of the belt.

Then the value of L is given by, $L = \pi(r_1 + r_2) + \dfrac{(r_1 - r_2)^2}{D} + 2D$

The total length of the belt, required to connect driver and follower, depends upon the sum and difference of the radii of the two pulleys.

10.13.2. Cross-Belt Drive

In cross-belt drive, the two shafts are parallel and rotating in opposite directions. The total length of the belt required to connect the driver and follower is given by,

$$L = \pi(r_1 + r_2) + \frac{(r_1 + r_2)^2}{D} + 2D$$

where D = Distance between the centres of the two shafts

r_1 and r_2 = Radii of larger and smaller pulley.

In this case, the total length of the belt is a function of $(r_1 + r_2)$ or in otherwords, the total length depends upon the sum of the radii of two pulleys.

10.13.3. Velocity Ratio

Let N_1 = Speed of driver, N_2 = Speed of follower,
d_1 = Diameter of driver, and d_2 = Diameter of follower

then velocity ratio is given by $\frac{N_2}{N_1} = \frac{d_1}{d_2}$

Hence ratio of velocity of follower to the velocity of driver is known as velocity ratio.

If thickness of the belt is also considered, then velocity ratio is given by,

$$\frac{N_2}{N_1} = \frac{d_1 + t}{d_2 + t}$$

where t = Thickness of the belt.

The velocity ratio of a compound belt-drive is given by,

$$\frac{\text{Speed of last follower}}{\text{Speed of first driver}} = \frac{\text{Product of diameters of drivers}}{\text{Product of diameters of followers}}$$

10.13.4. Slip of the Belt

If the frictional grip between the belts and shafts is not sufficient, the driver may move slightly forward without carrying the belt with it. Also there may be some forward motion of the belt without carrying the driver pulley with it. This is known as slip of the belt and expressed generally as a percentage. Due to slip, the velocity ratio reduces. The velocity ratio in terms of slip of the belt is given by,

$$\frac{N_2}{N_1} = \frac{d_1}{d_2}\left(1 - \frac{S}{100}\right).$$

where S = Total percentage of slip = $S_1 + S_2$

where S_1 = Percentage slip between driver and the belt, and

S_2 = Percentage slip between follower and the belt.

If thickness of belt is also considered, then velocity ratio is given by,

$$\frac{N_2}{N_1} = \frac{d_1 + t}{d_2 + t}\left(1 - \frac{S}{100}\right)$$

10.13.5. Power Transmitted by a Belt

The power transmitted by a belt or rope is given by, Power = $\frac{(T_1 - T_2) \times v}{1000}$

where $(T_1 - T_2)$ = Difference between the tensions in the tight and slack sides in N

v = Linear velocity of belt or rope in m/sec.

When a belt is passed over a pulley, **the relation** between the tension (T_1) in the belt on the tight side to the tension (T_2) on the slack side is given by the relation

$$\frac{T_1}{T_2} = e^{\mu\theta}$$

where μ = Co-efficient of friction between the belt and the pulley

θ = Angle of lap in radians.

10.13.6. Centrifugal Tension

Centrifugal tension caused by centrifugal force is given by

$$T_c = \frac{w \times v^2}{g}$$

where w = Weight of belt per unit length, and

v = Linear velocity of the belt.

The ratio of driving tensions with centrifugal tension is given by $\frac{(T_1 - T_c)}{(T_2 - T_c)} = e^{\mu\theta}$.

10.13.7. Maximum Power Transmitted by a Belt

(*i*) The speed at which the belt transmits maximum power is given by

$$V = \sqrt{\frac{T_1 \times g}{3w}}$$

where T_1 = Maximum tension to which belt can be subjected, and

w = Weight of belt per unit length.

(*ii*) The centrifugal tension, when maximum power is transmitted, is given as

$$T_c = \frac{1}{3}T_1.$$

(*iii*) Maximum power is given by the relation

$$\text{Max. power.} = \frac{2}{3} \times \frac{K}{75}\sqrt{\frac{g}{3w}} \times T_1^{3/2}$$

where $K = 1 - e^{-\mu\theta}$.

10.13.8. Rope Drive

For a rope running in a suitable grooved pulley, if α is the angle of the groove,

$$\frac{T_1}{T_2} = e^{\mu_1\theta}$$

where $\mu_1 = \mu \operatorname{cosec}\left(\frac{\alpha}{2}\right)$.

The groove angle α is usually equal to 45° and ratio of $\frac{T_1}{T_2}$ is much greater than that of the belt. The horse power transmitted by n number of ropes is given by

$$\text{H.P.} = \frac{n \times v(T_1 - T_2)}{75}$$

where v = Linear speed of the ropes in m/sec.

10.13.9. Chain Drive

It is positive non-slip drive which is used to transmit power between parallel shafts comparatively at longer distance. Slip may be avoided by this type of drive but the velocity ratio is not constant.

The effective pitch line of a chain wheel is a polygon and not a circle. The velocity ratio varies from $\frac{R}{l}$ to $\frac{L}{r}$

where R and r = Radii of the circles passing through the centres of the large and small wheels respectively, and

L and l = Perpendicular from the wheel centres to the sides of the pitch polygons.

For better load distribution and for reducing wear and noise, the **silent** or **inverted**—tooth type of chain is used instead of roller type.

10.14. LUBRICANT

There is always a loss of power due to friction which should be avoided by providing proper lubrication of all surfaces which move in contact with other, whether in rotating, sliding or rolling bearings. The lubricants are used in bearings to reduce friction between the rubbing surfaces and to carry away the heat generated by friction. The lubricants also protect the bearing against corrosion. The lubricants may be a :

1. Liquid,
2. Semi-liquid, and
3. Solid.

The liquid lubricants which are mostly used, are mineral oils and synthetic oils. The semi-liquid lubricant is a grease. Where the heavy pressure and slow speed exist and where the oil drip from the bearing is undesirable, the semi-liquid lubricant (*i.e.* grease) is used. The solid lubricants are useful in reducing friction where oil films cannot be maintained because of pressure and temperature. A graphite is the most common of the solid lubricants.

10.15. BEARINGS

The device for supporting the rotating shaft is called a bearing. The factors which are considered in the design of a bearing are:

1. Specific pressure,
2. Viscosity of lubricant,
3. Rubbing speed, and
4. Heat dissipation.

Design of a bearing includes choosing the type of bearing for the given requirement, design of bearing, its geometry, tolerances selection of material, lubricant and provision of cooling requirements.

10.15.1. Design of Journal Bearing

Knowing the bearing load and speed, the following design procedure is adopted.

1. Assume a suitable ratio of L/D from the table.

2. Choose a suitable bearing pressure p and the values of D and L from the table and from the following equation.

$$P = p \times D \times L$$

where P = Load on bearing, and p = Bearing pressure.

3. Assume a suitable clearance ratio C/D.

4. Select a good lubricant, that is the best value of Z. Here Z gives the absolute viscosity in centipoise.

5. Determine the value of ZN/p in which

Z = Absolute viscosity in centipoise, N = Speed, and

p = Pressure in kgf/cm^2.

Compare the above value of ZN/p with the corresponding values for the type of bearing from the table.

6. Find the co-efficient of friction f.

7. Determine the heat generated (H_g) by using

$$H_g = fPV \text{ kgf-m/minute or } \frac{fPV}{J} \text{ kcal/min}$$

where f = Co-efficient of friction, P = Load on bearing in kgf,

V = Rubbing velocity in metre/minute, and

J = Mechanical equivalent of heat = 427 kgf m/kcal.

8. Determine the heat dissipated (H_d) by using

$$H_d = CA(t_b - t_a) \text{ kcal/minute,}$$

where C = Heat dissipation co-efficient is kcal/min/cm^2/°C

A = Projected area of bearing in cm^2

$= D \times L$

t_b = Temperature of bearing surface in °C, and

t_a = Temperature of the surrounding air in °C.

9. Check whether heat dissipated is greater than heat generated for satisfactory working of the bearing. If the heat generated is more than the heat dissipated, then either the baring is redesigned or it is artificially cooled by water.

10. The mass of the oil to remove the heat generated at the bearing may be obtained by equating the heat generated to the heat taken away by the oil. The heat taken away by the oil is given by

$$\text{Heat taken} = m \times S \times t \text{ kcal/min}$$

where m = Mass of oil in kg/min,

S = Specific heat of the oil which varies from 0.44 to 0.49 kcal/kg°C.

t = Difference between outlet and inlet temperature of the oil in °C.

10.15.2. Bearing Characteristic Number

The quantity *ZN/p* is known as bearing characteristic number and is a dimensionless number.

10.15.3. Bearing Modulus (K)

The value of *ZN/p*, at which the amount of friction is minimum, is known as bearing modulus. This is represented by *K*. The bearing should not be operated at this value of bearing modulus, because a slight decrease in speed or slight increase in pressure will make the journal to operate with metal to metal contact, which will result in high friction, wear and heating. In order to prevent such conditions, the bearing should be designed for a value of *ZN/p* at least three times the value of *K*.

For heavy impacts and for large fluctuations of load the value of $\frac{ZN}{p} = 15K$, may be used.

10.15.4. Rolling Contact Bearing

If the contact between the bearing surfaces is rolling instead of sliding, the bearing is known as rolling contact bearing. The rolling contact bearing is having a low starting friction. Due to low friction, the rolling contact bearings are also known as *antifriction bearings.* The rolling contact bearings are having low cost of maintenance as no lubrication is required while in service, are having small overall dimensions and are easy to mount and erect. But they are having more noise at very high speeds, more initial cost and the design of bearing housing is complicated.

They are two types of rolling contact bearings:

1. Ball bearings, and 2. Roller bearings.

The ball and roller bearings consists of an inner race which is mounted on the shaft and an outer race which is mounted on casing. A number of balls or rollers are used and these are held at proper distances by retainers so that they donot touch each other. For light loads, the ball bearings are used and for heavier loads, the roller bearings are used.

10.15.5. Classification of Ball Bearings

The ball bearings have been standardized into four classes known as extra light, light, medium and heavy series. They are designated by a number. In general, the number consists of at least three digits. Additional digits or letters are used to indicate special features. The last three digits give the series and the bore of the bearing. The bore of the bearing is given in millimetres. The last two digits from 04 onwards, when multiplied by five, give the bore diameter in millimetres. The third from the last designates the series of the bearing. The four most common series of ball bearings are:

1. Extra light series designated by (100)
2. Light series designated by (200)
3. Medium series designated by (300)
4. Heavy series designated by (400)

The light series (*i.e.* 200 series) is used generally for moderate loads. The medium series (*i.e.*, 300 series) is wider, has larger outside diameter and is capable of carrying 50 per cent greater load than the light series. This series is mostly used. The heavy series (*i.e.*, 400 series) is wider, has a larger outside diameter and is capable of carrying 30 per cent greater load than the medium series.

If a bearing is designated by the number 206, it means that the bearing is of light series whose bore is 06 × 5 *i.e.*, 30 mm.

10.16. FRICTION CLUTCHES

The friction clutch is used in automobiles to connect the engine to the driver shaft. The following are the important types of friction clutches :

1. Disc or plate clutches,
2. Cone clutches, and
3. Centrifugal clutches.

10.16.1. Disc Clutches

(*a*) *For a single disc (or plate) clutch* for uniform pressure, total frictional torque is given by,

$$T = \frac{2}{3}\mu W \frac{(r_1^2 - r_2^2)}{(r_1^2 - r_2^2)}$$

where r_1 = External radius of disc, r_2 = Internal radius of disc,

μ = Co-efficient of friction, and W = Load.

And for uniform wear, the torque is given by,

$$T = \frac{1}{2}\mu W(r_1 + r_2) = \mu WR$$

where R = Mean radius of friction surface = $\frac{r_1 + r_2}{2}$.

(*b*) *For a multiple disc clutch,* the total torque transmitted is given by

$$T = \frac{2}{3}\mu W \frac{(r_1^3 - r_2^3)}{(r_1^2 - r_2^2)} \quad \text{...For uniform pressure}$$

$$= n \times \mu WR \quad \text{... For uniform wear}$$

where n = No. of pairs of contact surfaces = $n_1 + n_2 - 1$

where n_1 = No. of discs on the driving shaft, and

n_2 = No. of discs on the driver shaft.

For a *new clutch* intensity of pressure is approximately uniform whereas for *an old clutch,* the uniform wear theory is more approximate.

10.16.2. Cone Clutch

The cone clutches are extensively used in automobiles and have a conical friction surface. The total frictional torque is given by,

$$T = \frac{2}{3}\mu W \frac{(r_1^3 - r_2^3)}{(r_1^2 - r_2^2)} \times \cos\alpha \quad \text{...For uniform pressure}$$

$$= \mu W\left(\frac{(r_1 + r_2)}{2}\right) \times \operatorname{cosec}\alpha \quad \text{...For uniform wear.}$$

10.16.3. Centrifugal Clutch

The centrifugal clutches are usually incorporated into motor pulleys. They usually consist of a number of shoes on the inside of the rim of the pulley. The outer surfaces of the shoes are covered with a frictional material.

10.16.4. Pivot and Collar Bearing

The pivot bearing may be flat or conical. Similarly, the collar bearings are flat and conical. For the study of bearings, two assumptions are made, which are:

(*i*) It is assumed that the pressure is uniformly distributed throughout the bearing surface, and

(*ii*) The wear is uniform throughout the bearing surface.

(*a*) **Flat Pivot Bearing:** The total frictional torque (T) is given by,

$$T = \frac{2}{3}\mu WR \quad \text{...For uniform pressure}$$

$$= \frac{1}{2}\mu WR \quad \text{...For uniform wear}$$

where W = Load transmitted over bearing surface,

μ = Co-efficient of friction, and

R = Radius of bearing surface.

The power lost in friction is given by,

$$P = \frac{2\pi NT}{60,000} \text{ kW}$$

$$= T \times \infty$$

The frictional torque for uniform pressure is more than for uniform wear.

(*b*) **Conical Pivot Bearing:** The total frictional torque is given by,

$$T = \frac{2}{3}\mu WR \operatorname{cosec}\alpha \quad \text{...For uniform pressure}$$

$$= \frac{1}{2}\mu WR \operatorname{cosec}\alpha \quad \text{...For uniform wear}$$

where R = Radius of shaft,

W = Axial load, and α = Semi-angle of the cone.

The total frictional torque in terms of slant length of the cone (*i.e.*, in terms of L) is given by

$$T = \frac{2}{3}\mu WL \quad \text{...For uniform pressure}$$

$$= \frac{1}{2}\mu WL \quad \text{...For uniform wear}$$

where L = Slant length of the cone = $R \operatorname{cosec}\alpha$.

(*c*) **Truncanted Conical Pivot Bearing:**

$$T = \frac{2}{3}\mu W \frac{(r_1^3 - r_2^3)}{(r_1^2 - r_2^2)} \times \text{cosec}\ \alpha \quad \text{...For uniform pressure}$$

$$= \frac{1}{2}\mu W (r_1 + r_2) \times \text{cosec}\ \alpha \quad \text{...For uniform wear}$$

where r_1 = External radius of shaft, and r_2 = Internal radius of shaft.

(*d*) **Flat Collar Bearing:**

$$T = \frac{2}{3}\mu W \frac{(r_1^3 - r_2^3)}{(r_1^2 - r_2^2)} \quad \text{...For uniform pressure}$$

$$= \frac{1}{2}\mu W (r_1 + r_2) \quad \text{...For uniform wear}$$

where r_1 = External radius of collar, r_2 = Internal radius of collar,
W = Axial load, and μ = Co-efficient of friction.

To reduce the intensity of pressure on the bearing, two or more collars are used. Then intensity of pressure is given by,

$$p = \frac{W}{n \times \pi(r_1^2 - r_2^2)}$$

where n = Number of collars.

But the total torque transmitted by a multi-collared shaft remains contact *i.e.*,

$$T = \frac{2}{3}\mu W \frac{(r_1^3 - r_2^3)}{(r_1^2 - r_2^2)}.$$

II. OBJECTIVE TYPE QUESTIONS

Tick mark the most appropriate statement of the multiple choice answers:

Basic Definition

1. When a material is subjected to varying stresses, it fails under stresses considerably lower than the ultimate stress. Such type of failure of the material is known as
   (*a*) creep ☐ (*b*) fatigue ☐
   (*c*) stress concentration ☐ (*d*) overstrain. ☐
2. When a ductile material is loaded in excess of a certain value, a gradual increase in elongation takes place with time. This phenomenon is known as
   (*a*) creep ☐ (*b*) fatigue ☐
   (*c*) stress concentration ☐ (*d*) overstrain. ☐
3. When a material is subjected to varying stresses, it fails under stresses considerably lower than the ultimate stress. The magnitude of such stresses, under which the material fails, depends on
   (*a*) range of stress ☐ (*b*) mean stress ☐

(c) number of applications ☐ (d) all of the above ☐
(e) none of the above. ☐

4. Endurance limit is the value of maximum stress which a material can withstand without failure for infinite number of cycles, when it is subjected to

(a) dynamic load ☐ (b) static load ☐
(c) bending load ☐ (d) completely reversed load. ☐

5. Figure 10.3 shows the range of stress causing failure against the number of cycles, when a material is subjected to varying stresses. The correct curve is given by

(a) curve A ☐
(b) curve B ☐
(c) curve C ☐
(d) curve D. ☐

FIGURE 10.3

6. In Fig. 10.3, the endurance limit (or fatigue limit) is represented by

(a) curve A ☐ (b) curve B ☐
(c) curve C ☐ (d) curve D. ☐

7. The factor of safety, for a component subjected to fatigue loading, is given by

(a) $\dfrac{\text{Ultimate stress}}{\text{Working stress}}$ ☐ (b) $\dfrac{\text{Ultimate stress}}{\text{Endurance limit stress}}$ ☐

(c) $\dfrac{\text{Endurance limit stress}}{\text{Working stress}}$ ☐ (d) None of the above. ☐

8. If the size of a standard specimen is increased, the endurance limit of the material will

(a) increase ☐ (b) decrease ☐
(c) remain same ☐ (d) increase first and then decrease. ☐

9. The endurance limit of a material, subjected to fatigue loading

(a) increases with the increase of ultimate tensile strength ☐
(b) increases with the decrease of ultimate tensile strength ☐
(c) is independent of ultimate tensile strength ☐
(d) none of the above. ☐

10. Which of the following materials is having maximum ratio of the endurance limit to the ultimate tensile strength?

(a) cast steel ☐ (b) cast iron ☐
(c) steel ☐ (d) non-ferrous metals. ☐

11. Which of the following materials is having minimum ratio of the endurance limit to the ultimate tensile strength?

(a) cast steel ☐ (b) cast iron ☐
(c) steel ☐ (d) non-ferrous metals. ☐

12. The endurance limit of a material, subjected to fatigue loading

(*a*) increases with the increase of surface finish

(*b*) increases with the decrease of surface finish

(*c*) is independent of surface finish

(*d*) none of the above.

Stress Concentration Factor

13. The ratio of maximum stress in a machine component (at a notch or a fillet) to the nominal stress at the same section, is known as

(*a*) endurance limit

(*b*) stress concentration factor

(*c*) surface finish factor

(*d*) factor of safety.

14. Stress concentration factor is a function of

(*a*) geometry of the machine component

(*b*) material of the machine component

(*c*) geometry and material of the component

(*d*) none of the above.

15. Choose the wrong statement

(*a*) Stress concentration is caused due to sudden change of cross-section.

(*b*) When a material is subjected to varying stresses, the material fails under stresses considerably lower than the ultimate stress. This type of failure is known as fatigue.

(*c*) The ratio of endurance limit stress to working stress is known as factor of safety for a component subjected to fatigue loading.

(*d*) All of the above.

(*e*) None of the above.

16. Choose the correct statement

(*a*) Fatigue is the failure of a material above the yield point.

(*b*) For a mirror polished material, the surface finish factor is 0.5.

(*c*) Stress concentration factor is independent of the geometry of the machine component.

(*d*) Endurance limit of a material subjected to fatigue loading depends on the surface finish and ultimate tensile strength of the material.

17. Figure 10.4 shows a plate with an elliptical hole in the centre with semi-major axis = a and semi-minor axis = b. The plate is subjected to a tensile load P. The maximum stress induced at the edge of the hole (*i.e.*, at point A) will be equal to

(*a*) $f\left(1-\frac{2a}{b}\right)$

(*b*) $f\left(1+\frac{a}{2b}\right)$

(*c*) $f\left(1-\frac{a}{2}\right)$

(*d*) $f\left(1+\frac{2a}{b}\right)$

where f = Nominal stress, *i.e.*, stress in the plate far away from hole.

FIGURE 10.4

18. If in question 17, the ratio of *a* and *b* are changed, the stress induced at the edge of the hole (*i.e.,* at point *A*) will be maximum when

(*a*) $\frac{a}{b} = 0.5$ ☐ (*b*) $\frac{a}{b} = 1.0$ ☐

(*c*) $\frac{a}{b} = 1.5$ ☐ (*d*) $\frac{a}{b} = 2.0$ ☐

19. With reference to Fig. 10.4, the stress induced at the edge of the hole (*i.e.,* at point *A*) will be minimum when

(*a*) $\frac{a}{b} = 0.5$ ☐ (*b*) $\frac{a}{b} = 1.0$ ☐

(*c*) $\frac{a}{b} = 1.5$ ☐ (*d*) $\frac{a}{b} = 2.0$ ☐

20. If in Fig. 10.4, $\frac{a}{b} = 2.0$, then theoretical stress concentration factor will be equal to

(*a*) 2.0 ☐ (*b*) 3.0 ☐

(*c*) 4.0 ☐ (*d*) 5.0 ☐

21. If in Fig. 10.4, $\frac{a}{b} = 1.0$, then theoretical stress concentration factor will be equal to

(*a*) 2.0 ☐ (*b*) 3.0 ☐

(*c*) 4.0 ☐ (*d*) 5.0 ☐

22. If in Fig. 10.4, $\frac{a}{b} = 0.5$, then theoretical stress concentration factor will be equal to

(*a*) 2.0 ☐ (*b*) 3.0 ☐

(*c*) 4.0 ☐ (*d*) 5.0 ☐

23. Which one of the following is a wrong statement?

(*a*) Sudden change in cross-section should be avoided as they cause stress concentration. ☐

(*b*) In brittle material, stress concentration is more serious for static loading. ☐

(*c*) In cyclic loading, stress concentration is more serious in ductile materials. ☐

(*d*) Stress concentration factor is the ratio of ultimate stress to working stress. ☐

24. When the deformation of a machine number is due to bending only, the maximum bending moment is given by $M = \tau Z$ in which

(*a*) τ is the maximum shear stress and *Z* is polar modulus of section ☐

(*b*) τ is the maximum tensile stress and *Z* is modulus of section ☐

(*c*) τ is the maximum compressive stress and *Z* is modulus of section ☐

(*d*) τ is the maximum tensile or compressive stress and *Z* modulus of section. ☐

25. When the deformation of a machine member is due to pure torsion only, the maximum torque is given by $T = \tau Z$ in which

(*a*) τ is the maximum shear stress and *Z* is polar modulus of section ☐

(*b*) τ is the maximum tensile stress and *Z* modulus of section. ☐

(c) τ is the maximum compressive stress and Z is modulus of section ☐

(d) τ is the maximum tensile or compressive stress and Z modulus of section. ☐

Polar Modulus

26. For a solid shaft of diameter D, the polar modulus of section is equal to

(a) $\frac{\pi D^3}{64}$ ☐ (b) $\frac{\pi D^3}{32}$ ☐

(c) $\frac{\pi D^3}{16}$ ☐ (d) $\frac{\pi D^4}{64}$. ☐

27. For a solid shaft of diameter D, the modulus of section is equal to

(a) $\frac{\pi D^3}{64}$ ☐ (b) $\frac{\pi D^3}{32}$ ☐

(c) $\frac{\pi D^3}{16}$ ☐ (d) $\frac{\pi D^4}{64}$. ☐

28. For a solid shaft of diameter D, the moment of inertia of the section about the neutral axis is equal to

(a) $\frac{\pi D^3}{64}$ ☐ (b) $\frac{\pi D^3}{32}$ ☐

(c) $\frac{\pi D^3}{16}$ ☐ (d) $\frac{\pi D^4}{64}$. ☐

29. For a hollow shaft of external diameter D and internal diameter d, the modulus of section is equal to

(a) $\frac{\pi(D^4 - d^4)}{64D}$ ☐ (b) $\frac{\pi(D^4 - d^4)}{32D}$ ☐

(c) $\frac{\pi(D^4 - d^4)}{16D}$ ☐ (d) $\frac{\pi(D^4 - d^4)}{32D}$. ☐

30. For a hollow shaft of external diameter D and internal diameter d the polar modulus of section is equal to

(a) $\frac{\pi(D^4 - d^4)}{64D}$ ☐ (b) $\frac{\pi(D^4 - d^4)}{32D}$ ☐

(c) $\frac{\pi(D^4 - d^4)}{16D}$ ☐ (d) $\frac{\pi(D^4 - d^4)}{32D}$. ☐

31. Two direct stresses σ_1 and σ_2 are acting on a body in two mutually perpendicular planes accompanied by shear stress τ as shown in Fig. 10.5. The maximum principal stresses are given by

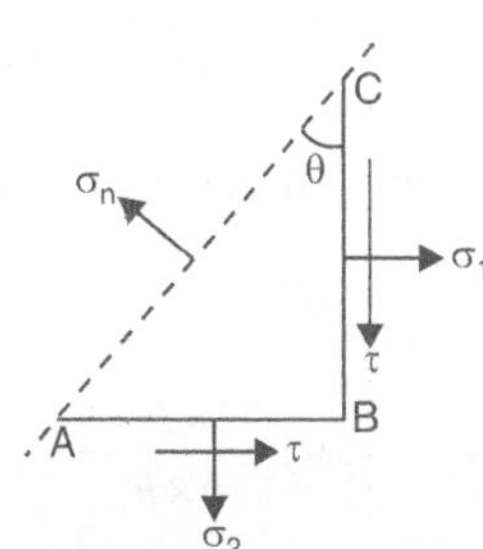

FIGURE 10.5

(a) $\sigma_n = \frac{\sigma_1 + \sigma_2}{2} \pm \frac{1}{2}\sqrt{(\sigma_1 - \sigma_2)^2 + 4\tau^2}$ ☐

(b) $\sigma_n = \frac{\sigma_1 - \sigma_2}{2} \pm \frac{1}{2}\sqrt{(\sigma_1 + \sigma_2)^2 + 4\tau^2}$ ☐

(c) $\sigma_n = \frac{\sigma_1 - \sigma_2}{2} \pm \frac{1}{2}\sqrt{(\sigma_1 - \sigma_2)^2 + 4\tau^2}$ ☐ (d) $\sigma_n = \frac{\sigma_1 + \sigma_2}{2} \pm \frac{1}{2}\sqrt{(\sigma_1 + \sigma_2)^2 + 4\tau^2}$. ☐

32. In question 31, the maximum principal stresses are acting on the place *AC*, which is inclined at an angle θ with the plane *BC*. The value of θ is given by the equation

(a) $\tan 2\theta = \frac{\sigma_1 - \sigma_2}{2\tau}$ ☐ (b) $\tan 2\theta = \frac{\tau}{2(\sigma_1 - \sigma_2)}$ ☐

(c) $\tan 2\theta = \frac{2\tau}{(\sigma_1 - \sigma_2)}$ ☐ (d) $\tan 2\theta = \frac{2\tau}{\sigma_1 + \sigma_2}$ ☐

33. In question 31, the maximum shear stress is given by

(a) $\pm\frac{1}{2}\sqrt{(\sigma_1 + \sigma_2)^2 + 4\tau^2}$ ☐ (b) $\pm\frac{1}{2}\sqrt{(\sigma_1 - \sigma_2)^2 + 4\tau^2}$ ☐

(c) $\pm\sqrt{\left(\frac{\sigma_1 + \sigma_2}{2}\right)^2 + 4\tau^2}$ ☐ (d) $\pm\sqrt{\left(\frac{\sigma_1 - \sigma_2}{2}\right)^2 + 4\tau^2}$ ☐

34. The planes on which maximum shear stress act is given by

(a) $\tan 2\theta_1 = \frac{\sigma_1 - \sigma_2}{2\tau}$ ☐ (b) $\tan 2\theta_1 = \frac{\tau}{2(\sigma_1 - \sigma_2)}$ ☐

(c) $\tan 2\theta_1 = \frac{2\tau}{(\sigma_1 - \sigma_2)}$ ☐ (d) $\tan 2\theta_1 = \frac{2\tau}{\sigma_1 + \sigma_2}$ ☐

35. Figure 10.6 shows a member subjected to a force *P* which is parallel to the axis and at a distance *x* from the axis. The member is rectangular in section. The resultant stress at any point is equal to

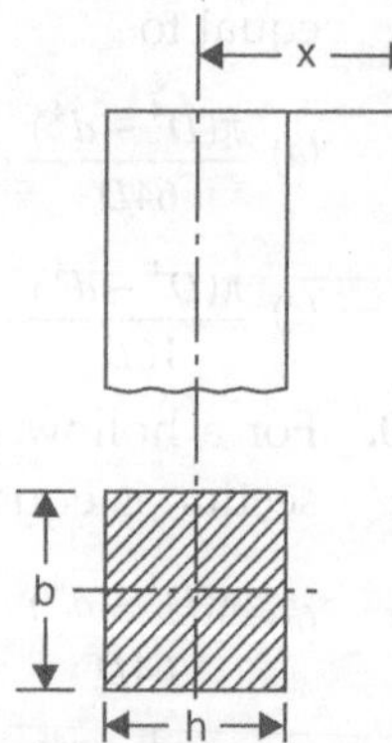

FIGURE 10.6

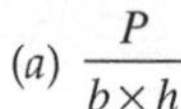

(a) $\frac{P}{b \times h}$ ☐

(b) $\frac{P}{b \times h} \pm P \times x$ ☐

(c) $\frac{P}{b \times h} \pm \frac{P \times x \times 6}{bh^2}$ ☐

(d) $\frac{P}{b \times h} \pm \frac{P \times x \times 12}{bh^3}$ ☐

36. In question 35, maximum stress due to bending will be equal to

(a) $P \times x$ ☐ (b) $\frac{6P \times x}{bh^2}$ ☐

(c) $\frac{12 \times P \times x}{b \times h^2}$ ☐ (d) $\frac{12P \times x}{b^2 \times h}$ ☐

37. In question 35, the direct stress and maximum bending stress will be equal if

(*a*) $x = \frac{h}{2}$ ☐ (*b*) $x = \frac{h}{3}$ ☐

(*c*) $x = \frac{h}{4}$ ☐ (*d*) $x = \frac{h}{6}$ ☐

38. A shaft is subjected to twisting moment (*T*) and bending moment (*M*). The equivalent bending moment is equal to

(*a*) $\frac{1}{2} \cdot [M + \sqrt{(M^2 + T^2)}\,]$ ☐ (*b*) $\frac{1}{2}\sqrt{(M^2 + T^2)}$ ☐

(*c*) $\sqrt{(M^2 + T^2)}$ ☐ (*d*) $M + \sqrt{M^2 + T^2}$ ☐

39. In the above question, the equivalent twisting moment is equal to

(*a*) $\frac{1}{2} \cdot [M + \sqrt{(M^2 + T^2)}\,]$ ☐ (*b*) $\frac{1}{2}\sqrt{(M^2 + T^2)}$ ☐

(*c*) $\sqrt{(M^2 + T^2)}$ ☐ (*d*) $M + \sqrt{M^2 + T^2}$ ☐

Torsion of Shaft

40. The torque transmitted by a rotating hollow shaft is equal to

(*a*) $\frac{\pi}{16}\tau d_o^3$ ☐ (*b*) $\frac{\pi}{16}\tau (d_o^3 - d_i^3)$ ☐

(*c*) $\frac{\pi}{16}\tau d_o^3 (1 - k^4)$ ☐ (*d*) $\frac{\pi}{16}\tau d_o^3 (1 - k^3)$ ☐

where τ = Torsional shear stress, d_o = Outside diameter of the shaft,

d_i = Inside diameter of the shaft, and $k = \frac{d_i}{d_o}$ = Ratio of inside diameter to outside diameter.

41. The relation between the diameter of the solid shaft and the diameters of a hollow shaft, which is having the same material and equal in strength to the solid shaft, is given by

(*a*) $d = d_o (1 - k^4)$ ☐ (*b*) $d = d_o (1 - k^4)^{1/3}$ ☐

(*c*) $d = d_o (1 + k^2)$ ☐ (*d*) $d = d_o (1 + k^3)$ ☐

where d = Diameter of the solid shaft, d_o = Outer diameter of the hollow shaft, and

k = Ratio of inside diameter to outside diameter of a hollow shaft.

42. A shaft is designed on the basis of

(*a*) strength ☐ (*b*) rigidity ☐

(*c*) strength and rigidity ☐ (*d*) none of the above. ☐

43. Choose the wrong statement

(*a*) Ductile materials are designed on the basis of maximum shear stress theory. ☐

(*b*) Brittle materials are designed on the basis of maximum normal stress theory. ☐

(c) If the twisting moments of the two shafts are equal, they will have equal strength. ☐
(d) A shafts is subjected to shear stresses only. ☐

44. Torsional rigidity of a shaft is equal to
(a) product of modulus of rigidity and polar moment of inertia ☐
(b) sum of modulus of rigidity and polar moment of inertia ☐
(c) difference of modulus of rigidity and polar moment of inertia ☐
(d) ratio of modulus of rigidity and polar moment of inertia. ☐

45. Two shafts, one solid and other hollow, are made of the same materials and are having same length and weight. The hollow shaft as compared to solid shaft will be
(a) more strong ☐ (b) less strong ☐
(c) having same strength ☐ (d) none of the above. ☐

46. The torque required to produce a twist of one radian per unit length of the shaft is known as
(a) polar modulus ☐ (b) torsional rigidity ☐
(c) flexural rigidity ☐ (d) maximum twisting moment. ☐

47. The ratio of torque transmitted by a hollow shaft (of inside diameter equal to the half of the outer diameter) to the torque transmitted by a solid shaft of the same material and of the same weight is equal to
(a) 1/2 ☐ (b) 1/4 ☐
(c) 3/4 ☐ (d) 15/16. ☐

48. If the diameter of a solid shaft is increased two times, the torque transmitted will be
(a) two times ☐ (b) four times ☐
(c) eight times ☐ (d) sixteen times. ☐

49. A solid shaft of diameter 100 mm is subjected to a maximum bending moment of 4,000 Nm and maximum torque of 3,000 Nm. The equivalent bending moment is
(a) 5000 Nm ☐ (b) 4500 Nm ☐
(c) 4000 Nm ☐ (d) 3500 Nm. ☐

50. In question 49, the equivalent torque will be equal to
(a) 5000 Nm ☐ (b) 4500 Nm ☐
(c) 4000 Nm ☐ (d) 3500 Nm. ☐

51. In question 49, the maximum shear stress will be
(a) $80/\pi$ N/mm^2 ☐ (b) $160/\pi$ N/mm^2 ☐
(c) $144/\pi$ N/mm^2 ☐ (d) $100/\pi$ N/mm^2. ☐

52. Two shafts, one solid and the other hollow, of the same material will have the same strength if they are having
(a) same length and same weight ☐ (b) same length and same polar modulus ☐
(c) same weight and same polar modulus ☐ (d) same length, weight and polar modulus. ☐

53. For calculating stresses in a hollow shaft subjected to torsion, the radius taken into consideration is
(a) outer radius ☐ (b) inner radius ☐
(c) mean radius ☐ (d) none of the above. ☐

54. Which of the following is a wrong statement?

(*a*) From a number of shafts of the same length and material, the shaft with greatest polar modulus will resist the maximum twisting moment. ☐

(*b*) For a shaft of a given material, the magnitude of polar modulus is a measure of its strength in resisting torsion. ☐

(*c*) Shaft of the same material, and length having the same polar modulus have the same strength. ☐

(*d*) The torsional rigidity of a shaft = C/J. ☐

55. Which one of a correct statement?

(*a*) Hot rolling produces a stronger shaft than cold rolling. ☐

(*b*) Cold rolling produces a stronger shaft than hot rolling. ☐

(*c*) Shaft are not made by rolling process. ☐

(*d*) Angle of twist of shaft is inversely proportional to shaft diameter. ☐

56. Yield point in fatigue loading as compared to static loading is

(*a*) higher ☐ (*b*) lower ☐

(*c*) same ☐ (*d*) none of the above. ☐

57. Fatigue strength

(*a*) increases by cold working ☐ (*b*) decreases by cold working ☐

(*c*) increases by hot working ☐ (*d*) none of the above. ☐

58. Choose the wrong statement

(*a*) The fatigue life of a part can be improved by shot peening. ☐

(*b*) The notch angle of the Izod impact test specimen is 45°. ☐

(*c*) The procedure of increasing fatigue limit by overstressing the metal by successively increasing loadings is known as Coaxing. ☐

(*d*) In Vicker's hardness testing, the pyramid indentor apex is 60°. ☐

59. The technique 'brittle coating' is used for

(*a*) experimental stress analysis ☐

(*b*) protecting metal against corrosion ☐

(*c*) non-destructive testing of metals ☐

(*d*) determining brittleness. ☐

60. Choose the correct statement

(*a*) Resilience of a material is important, when it is subjected shock loading. ☐

(*b*) The ultimate strength of steel in tension in comparison to shear is in the ratio of 3 : 2. ☐

(*c*) Stress concentration is caused due to abrupt change of section. ☐

(*d*) Stress concentration is cyclic loading is more serious in ductile materials. ☐

(*e*) All of the above. ☐

(*f*) None of the above. ☐

System of Limits

61. The largest permissible size for a dimension is known as

(*a*) lower limit (*b*) upper limit

(*c*) basic size (*d*) actual size.

62. The smallest permissible size for a dimension is known as

(*a*) lower limit (*b*) upper limit

(*c*) basic size (*d*) actual size.

63. The difference between the upper limit and lower limit of a dimension is known as

(*a*) basic size (*b*) nominal size

(*c*) tolerance (*d*) actual size.

64. A system is said to unilateral, if

(*a*) tolerance is zero (*b*) tolerance is one direction

(*c*) tolerance is in two direction (*d*) none of the above.

65. A system is said to bilateral, if

(*a*) tolerance is zero (*b*) tolerance is one direction

(*c*) tolerance is in two direction (*d*) none of the above.

66. The size of a shaft is given as $30^{+0.003}_{-0.003}$, it means

(*a*) the system is bilateral (*b*) upper limit is +0.003

(*c*) lower limit is –0.003 (*d*) all of the above

(*e*) none of the above.

67. The size of a part, to which all limits of variation are determined, is called

(*a*) actual size (*b*) basic size

(*c*) tolerance (*d*) zone of tolerance.

68. Choose the correct statement

(*a*) The algebraic difference between the lower limit and basic size is called lower deviation.

(*b*) The algebraic difference between an actual size and corresponding basic size is known as actual deviation.

(*c*) The algebraic difference between the upper limit and the basic size is called upper deviation.

(*d*) All of the above.

(*e*) None of the above.

69. Figure 10.7 shows a shaft and a hole. The distance *D* is known as

(*a*) upper limit

(*b*) lower limit

(*c*) tolerance

(*d*) allowance.

FIGURE 10.7

70. In Fig. 10.7, the distance C is known as

(*a*) upper limit ☐ (*b*) lower limit ☐
(*c*) tolerance ☐ (*d*) allowance. ☐

71. In Fig. 10.7, the distance A is known as

(*a*) upper limit ☐ (*b*) lower limit ☐
(*c*) tolerance ☐ (*d*) allowance. ☐

72. In Fig. 10.7, the distance B is known as

(*a*) upper limit ☐ (*b*) lower limit ☐
(*c*) tolerance ☐ (*d*) allowance. ☐

73. In Fig. 10.7, the distance $(B - A)$ is known as

(*a*) upper limit ☐ (*b*) lower limit ☐
(*c*) tolerance ☐ (*d*) allowance. ☐

74. Choose the wrong statement

(*a*) Fit is defined as the degree of tightness or looseness between two mating parts. ☐

(*b*) The interference is the amount by which the actual size of a shaft is larger than the actual finished size of the mating hole. ☐

(*c*) The clearance is the amount by which the actual size of the shaft is less than the actual size of the mating hole. ☐

(*d*) First are of two types namely hole basis fit and shaft basis fit. ☐

75. Figure 10.8 shows the different types of fits. Fig. 10.8 (*a*), represents

(*a*) transition fit ☐ (*b*) clearance fit ☐
(*c*) interference fit ☐ (*d*) none of the above. ☐

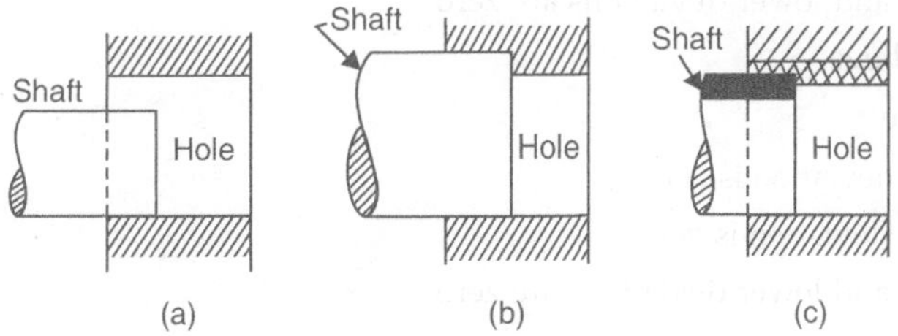

FIGURE 10.8

76. Figure 10.8 (*b*) represents

(*a*) transition fit ☐ (*b*) clearance fit ☐
(*c*) interference fit ☐ (*d*) none of the above. ☐

77. Figure 10.8 (*c*) represents

(*a*) transition fit ☐ (*b*) clearance fit ☐
(*c*) interference fit ☐ (*d*) none of the above. ☐

78. The system of limits and fits, according to Indian Standard (IS : 1919–1963) consists of

(*a*) fundamental tolerances = 25 and fundamental deviations = 16 ☐

(*b*) fundamental tolerances =16 and fundamental deviations = 25 ☐

(*c*) fundamental tolerances = 10 and fundamental deviations = 25 ☐

(*d*) fundamental tolerances =16 and fundamental deviations = 20. ☐

79. Choose the correct statement

(*a*) A limit system is said to be on a shaft basis if the hole is kept constant and shaft size is varied. ☐

(*b*) A limit system is said to be on a hole basis if the shaft is kept constant and the hole size is varied. ☐

(*c*) The standard tolerances are determined in terms of standard tolerances unit '*i*' in microns. ☐

(*d*) The standard tolerance unit (*i*), for the first fundamental tolerance (IT 1), is equal to 0.45 + 0.000 *D* where *D* is diameter in mm. ☐

80. S.A.E. 3120 means

(*a*) Carbon steel with 0.15 to 0.25 per cent carbon ☐

(*b*) Nickel steel with 1 to 1.5 per cent carbon ☐

(*c*) Chrome nickel steel with 1 to 1.5% nickel and 0.15 to 0.25% carbon ☐

(*d*) Chromium steel with 0.15 to 0.25% carbon. ☐

81. Choose the wrong statement

(*a*) S.A.E. stands for Society of Automobile Engineers. ☐

(*b*) A.S.T.M. stands for American Society for Testing Materials. ☐

(*c*) A.S.A. stands for American Standard Association. ☐

(*d*) Preferred numbers are of a series of numbers in arithmetic progression. ☐

82. Basic shaft is one

(*a*) whose lower deviation is zero ☐

(*b*) whose upper deviation is zero ☐

(*c*) whose upper and lower deviations are zero ☐

(*d*) none of the above. ☐

83. Basic hole is one

(*a*) whose lower deviation is zero ☐

(*b*) whose upper deviation is zero ☐

(*c*) whose upper and lower deviations are zero ☐

(*d*) none of the above. ☐

84. The notation 70 H 6/g 5 means

(*a*) basic size is 70 mm ☐

(*b*) basic size is 70 mm and tolerance grade for shaft is 6 and for the hole is 5 ☐

(*c*) basic size is 70 mm and tolerance grade for a hole is 6 and for the shall is 5 ☐

(*d*) basic size is 30 mm and tolerance grade is one. ☐

85. The fatigue duration of parts can be increased by a process, known as

(*a*) electroplating ☐ (*b*) shot-peening ☐

(*c*) finishing and polishing ☐ (*d*) decraburisation. ☐

Bolts, Nuts and Threads

86. A cylindrical rod, with both ends threaded, is known as a

(*a*) bolt ☐ (*b*) stud ☐
(*c*) tap bolt ☐ (*d*) set-screw. ☐

87. A cylinderial rod, having a head at one end and a nut fitted to the other end, is known as a

(*a*) bolt ☐ (*b*) stud ☐
(*c*) tap bolt ☐ (*d*) set-screw. ☐

88. A cylindrical rod, having a head at one end and other end lilting into a tapped hole in the other parts to be joined, is known as a

(*a*) bolt ☐ (*b*) stud ☐
(*c*) tap bolt ☐ (*d*) set-screw. ☐

89. A cylindrical rod, having a head at one end and is threaded practically throughout its length, is known as a

(*a*) bolt ☐ (*b*) stud ☐
(*c*) tap bolt ☐ (*d*) set-screw. ☐

90. Which one of the following threads is not used for transmission of power?

(*a*) Square thread ☐ (*b*) Buttress thread ☐
(*c*) B.S.W. thread ☐ (*d*) Acme thread. ☐

91. Which of the following threads is used when the force acts entirely in one direction?

(*a*) Square thread ☐ (*b*) Buttress thread ☐
(*c*) B.S.W. thread ☐ (*d*) Acme thread. ☐

92. Choose the correct statement

(*a*) Acme thread and knuckle thread are the modification of the square thread. ☐
(*b*) Buttress thread is a combination of triangular and square threads. ☐
(*c*) *M* 20 means metric thread with diameter of screw as 20 mm. ☐
(*d*) B.S.W. threads are used for power transmission. ☐

93. The diameter of a cylindrical rod, on which threads are cut, is known as

(*a*) major diameter ☐ (*b*) minor diameter ☐
(*c*) nominal diameter ☐ (*d*) pitch diameter. ☐

94. The diameter, at the root of the thread, is known as

(*a*) major diameter ☐ (*b*) minor diameter ☐
(*c*) nominal diameter ☐ (*d*) pitch diameter ☐
(*e*) core diameter ☐ (*f*) (*b*) or (*e*). ☐

95. The diameter, at the crest of the thread, is known as

(*a*) major diameter ☐ (*b*) minor diameter ☐
(*c*) nominal diameter ☐ (*d*) pitch diameter ☐

96. Which one of the following threads is having smallest included angle?

(*a*) Acme thread ☐ (*b*) B.S.W. thread ☐
(*c*) Buttress thread ☐ (*d*) Unified thread. ☐

97. Which one of the following threads is having largest included angle?

(*a*) Acme thread ☐ (*b*) B.S.W. thread ☐
(*c*) Buttress thread ☐ (*d*) Unified thread. ☐

98. The included angle for Buttress threads is

(*a*) 60° ☐ (*b*) 55° ☐
(*c*) 45° ☐ (*d*) 29°. ☐

99. The included angle for Acme threads is

(*a*) 60° ☐ (*b*) 55° ☐
(*c*) 45° ☐ (*d*) 29°. ☐

100. A screw, which is similar to a tap bolt and is used to prevent relative movement between the two parts, is known as

(*a*) machine screw ☐ (*b*) cap screw ☐
(*c*) set-screw ☐ (*d*) none of the above. ☐

101. A screw, which is slotted for a screw driver and are generally used with a nut, is known as

(*a*) machine screw ☐ (*b*) cap screw ☐
(*c*) set-screw ☐ (*d*) none of the above. ☐

102. A screw, which is similar to a tap bolt except that they are of small size and a greater variety of shapes of heads are available, is known as

(*a*) machine screw ☐ (*b*) cap screw ☐
(*c*) set-screw ☐ (*d*) none of the above. ☐

103. $M\ 30 \times 2$ for a bolt represents

(*a*) metric threads of 30 mm outside diameter and 2 mm pitch ☐
(*b*) metric threads of 60 mm^2 cross-sectional area ☐
(*c*) metric threads on a bolt of length 30 cm ☐
(*d*) none of the above. ☐

104. A coarse screw of major diameter 6 mm and pitch 1 mm according to IS : 1362–1962 is designated as

(*a*) $M \times 6 \times 1$ ☐ (*b*) $M\ 6 \times 1$ ☐
(*c*) $6\ M \times 1$ ☐ (*d*) $M\ 6$. ☐

105. A fine screw of major diameter 36 mm and pitch 3 mm according to IS : 1362–1962 is designated as

(*a*) $M \times 36 \times 3$ ☐ (*b*) $M\ 36 \times 3$ ☐
(*c*) $36\ M \times 3$ ☐ (*d*) $M\ 36$. ☐

106. The pitch of 24 mm diameter metric find threads would be

(*a*) 3 mm ☐ (*b*) 2 mm ☐
(*c*) 1.5 mm ☐ (*d*) 1 mm. ☐

107. If in the above question, the threads are coarse then pitch would be

(*a*) 3 mm ☐ (*b*) 2 mm ☐
(*c*) 1.5 mm ☐ (*d*) 1 mm. ☐

108. Choose the wrong statement

(*a*) For a double start threads, lead is equal to twice the pitch. ☐

(*b*) If the threads on a bolt are right-handed, then the threads on the nut will be also right-handed. ☐

(*c*) A stud is having threads on both ends. ☐

(*d*) The included angle for the acme thread is 55°. ☐

109. Eye bolts are used for

(*a*) transmission of power ☐ (*b*) locking devices ☐

(*c*) lifting and transporting heavy pieces ☐ (*d*) absorption of shocks and vibrations. ☐

110. The basis on which bolts are generally designed for initial tension due to screwing up with a large factor of safety is

(*a*) direct bending stress ☐ (*b*) direct tensile stress ☐

(*c*) direction shear stress ☐ (*d*) direct compressive stress. ☐

111. The initial tension in a bolt (in M.K.S. units) used for making fluid tight joint is equal to

(*a*) $284 \times d$ ☐ (*b*) $142\, d$ ☐

(*c*) $81 \times d$ ☐ (*d*) $568 \times d$ ☐

where d = Nominal diameter of bolt in mm.

112. The stresses included in a bolt, when it is subjected to an external load, will be

(*a*) shear stress ☐ (*b*) tensile stress ☐

(*c*) both shear and tensile stress ☐ (*d*) any one of the above. ☐

113. On which the resultant axial load on a bolt depends

(*a*) the initial tension due to tightening of the bolt ☐

(*b*) the external load applied ☐

(*c*) the relative elastic yielding of the bolt and the connected member ☐

(*d*) all of the above ☐

(*e*) none of the above. ☐

114. The resultant axial load (P) on a bolt is given by

(*a*) $P = P_1 + \frac{a}{a+1} P_2$ ☐ (*b*) $P = P_1 - \frac{a}{a+1} P_2$ ☐

(*c*) $P = P_1 + \frac{a}{a-1} P_2$ ☐ (*d*) $P = P_1 + \frac{a+1}{a} P_2$. ☐

where P_1 = Initial tension in the bolt due to tightening up,

P_2 = External load on the bolt, and

a = Ratio of elasticity of connected parts to the elasticity of the bolt.

115. When a soft gaskets are connected by large bolts, then the resultant axis load on the bolt is equal to

(*a*) sum of initial tension and external load applied ☐

(*b*) mean of the initial tension and external load applied ☐

(*c*) initial tension only ☐

(*d*) external load only ☐

(*e*) initial tension or external load whichever is greater. ☐

116. For marking a leak proof joint, the bolts or studs are screwed up tightly along with metal gasket or asbestos packing. The minimum diameter of the bolt or stud (used for this purpose) should be at least

(a) 5 mm ☐ (b) 10 mm ☐
(c) 16 mm ☐ (d) 24 mm. ☐

117. A bolt can be made of uniform strength by

(a) drilling an axial hole through the head upto threaded portion ☐
(b) turning down the diameter of the shank of the bolt ☐
(c) increasing the diameter of the shank of the bolt ☐
(d) (a) and (b) only. ☐

118. The diameter of the hole, that must be drilled in the head of the bolt to make the bolt of uniform strength, is equal to

(a) $D_o - D_c$ ☐ (b) $\frac{D_o + D_c}{2}$ ☐
(c) $\sqrt{D_o^2 - D_c^2}$ ☐ (d) $\frac{D_o - D_c}{4}$. ☐

where D_o = Outside diameter of thread, and D_c = Core diameter of thread.

119. If the nut is made of a weaker material than the bolt, then the height of the nut should be

(a) equal to nominal diameter of the bolt ☐
(b) less than the nominal diameter of the bolt ☐
(c) more than the nominal diameter of the bolt ☐
(d) none of the above. ☐

120. Choose the wrong statement

(a) For leak-proof joint, the circumferential pitch of the studs should be between $20\sqrt{d_1}$ and $30\sqrt{d_1}$ where d_1 is the diameter of stud hole in mm. ☐
(b) The effective height of the nut is made equal to the nominal diameter of the bolt, when a nut and bolt are made of mild steel. ☐
(c) If an axial hole is drilled through the head of a bolt upto threaded portion such that the area of the shank becomes equal to the root area of the thread, the bolt of uniform strength is obtained. ☐
(d) Turn buckle has left hand threads on both ends. ☐

121. Figure 10.9 shows a bracket having a rectangular base and bolted to a wall by means of four bolts. The bracket supports a load W. Two bolts are at a distance L_1 from O and other two are at a distance L_2 from O. If all the bolts are having equal cross-sectional area and w be the load per unit distance due to turning effect, then the total tensile load, on each bolt at distance L_1 from O, will be

O
L1
L2
L
W

FIGURE 10.9

(a) $\frac{W}{4}$ ☐ (b) $\frac{w \times L_1}{2}$ ☐
(c) $\frac{W}{4} + wL_1$ ☐ (d) $\frac{W}{4} + \frac{wL_1}{2}$. ☐

122. In the question 121 (refer to Fig. 10.9), the total tensile load on each bolt at a distance L_2 from O, will be

(a) $\frac{W}{4}$ ☐ (b) $\frac{w \times L_2}{2}$ ☐

(c) $\frac{W}{4} + wL_2$ ☐ (d) $\frac{W}{4} + \frac{wL_3}{2}$. ☐

123. In the question 121 (refer to Fig. 10.9), the value of the load in the bolt per unit distance due to turning effect (*i.e.*, value of w) is equal to

(a) $\frac{2WL}{L_1^2 + L_2^2}$ ☐ (b) $\frac{WL}{2(L_1^2 + L_2^2)}$ ☐

(c) $\frac{WL}{4(L_1^2 + L_2^2)}$ ☐ (d) $\frac{4WL}{(L_1^2 + L_2^2)}$. ☐

124. In the question 121 (refer to Fig. 10.9) all the bolts are

(a) equally loaded ☐

(b) two bolt at a distance L_1 are heavily loaded ☐

(c) two bolts at a distance L_2 are heavily loaded ☐

(d) none of the above. ☐

125. If in Fig. 10.9, the bracket supports a load of 4000 N and the distances L_1 L_2 and L are equal to 10 cm, 30 cm and 50 cm respectively then the load in the bolt per unit distance due to turning effect will be

(a) 50 N ☐ (b) 100 N ☐

(c) 125 N ☐ (d) 150 N. ☐

126. For the question 125, the total tensile load carried by each bolt at a distance 10 cm from O (refer to Fig. 10.9) will be

(a) 4000 N ☐ (b) 3000 N ☐

(c) 2000 N ☐ (d) 1000 N. ☐

127. For the question 125, the total tensile load carried by each bolt at a distance 30 cm from O (refer to Fig. 10.9) will be

(a) 4000 N ☐ (b) 3000 N ☐

(c) 2000 N ☐ (d) 1000 N. ☐

128. Figure 10.10 shows a wall bracket carrying an eccentric load of 800 N perpendicular to the axis of four bolts. Two bolts are at a distance of 5 cm from A and the other two are at a distance of 15 cm from A. If the bolts are of the same size, then each bolt will be subjected to

(a) direct shear stress ☐

(b) direct tensile stress ☐

(c) direct compressive stress ☐

(d) none of the above. ☐

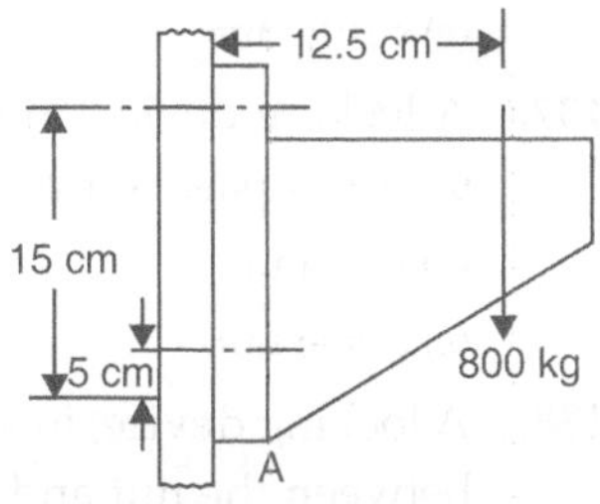

FIGURE 10.10

129. In Fig. 10.10, the load 800 kg will try to tilt the bracket about *A*. The stresses induced in each bolt due to turning moment, will be

(*a*) tensile ☐ (*b*) compressive ☐
(*c*) shear ☐ (*d*) none of the above. ☐

130. In question 128, the shear load carried by each bolt is equal to

(*a*) 100 N ☐ (*b*) 200 N ☐
(*c*) 300 N ☐ (*d*) 400 N. ☐

131. In question 128, the maximum tensile load carried by bolts which are at a distance 15 cm from *A*, will be equal to

(*a*) 100 N ☐ (*b*) 200 N ☐
(*c*) 300 N ☐ (*d*) 400 N. ☐

132. In question 128, the equivalent shear load will be equal to

(*a*) 100 N ☐ (*b*) 200 N ☐
(*c*) 250 N ☐ (*d*) 400 N. ☐

133. In question 128, the equivalent tensile load will be equal to

(*a*) 100 N ☐ (*b*) 200 N ☐
(*c*) 250 N ☐ (*d*) 400 N. ☐

134. The rotation of a bolt, while screwing a nut on or off the bolt, is prevented by providing on the bolt

(*a*) a square neck ☐ (*b*) a pip ☐
(*c*) a snug ☐ (*d*) any one of the above ☐
(*e*) none of the above. ☐

135. The locking arrangement for the nut is done by

(*a*) a split pin ☐ (*b*) a slotted nut ☐
(*c*) a castle nut ☐ (*d*) a sawn nut ☐
(*e*) a ring nut ☐ (*f*) any one of the above. ☐

136. A locking device, in which slots are cut in a cylindrical collar provided on the top of the nut on opposite faces and a split pin passes through the two slots in the nut and a hole in the bolt, is known as

(*a*) ring nut ☐ (*b*) slotted nut ☐
(*c*) castle nut ☐ (*d*) sawn nut. ☐

137. A locking device, in which slots are cut in the upper end of the nut on opposite faces and a split pin passes through the two slots in the nut and a hole in the bolt, is known as

(*a*) ring nut ☐ (*b*) slotted nut ☐
(*c*) castle nut ☐ (*d*) sawn nut. ☐

138. A locking device, in which a slot is cut half way across the nut and a small screw is tightened between the nut and the bolt, is known as

(*a*) ring nut ☐ (*b*) slotted nut ☐
(*c*) castle nut ☐ (*d*) sawn nut. ☐

139. A locking device, which is widely used in automobile and locomotive engines, is a

(*a*) ring nut ☐ (*b*) slotted nut ☐

(*c*) castle nut ☐ (*d*) sawn nut. ☐

140. Which forms of rivet heads of the following are mostly used for general work?

(*a*) conical ☐ (*b*) ellipsoid ☐

(*c*) snap ☐ (*d*) countersunk ☐

(*e*) (*a*) and (*b*) ☐ (*f*) (*c*) and (*d*). ☐

141. Which forms of the rivet heads of the following are generally used in boiler work?

(*a*) conical ☐ (*b*) ellipsoid ☐

(*c*) snap ☐ (*d*) countersunk ☐

(*e*) (*a*) and (*b*) ☐ (*f*) (*c*) and (*d*). ☐

142. Which of the following screw thread is used for power transmission in either direction?

(*a*) Buttress threads ☐ (*b*) Acme threads ☐

(*c*) Square threads ☐ (*d*) All of the above. ☐

143. Efficiency of a square threaded screw, neglecting collar friction is given by

(*a*) $\frac{\tan\alpha}{\tan\phi}$ ☐ (*b*) $\frac{\tan\alpha}{\tan(\alpha+\phi)}$ ☐

(*c*) $\frac{\tan\phi}{\tan\alpha}$ ☐ (*d*) $\frac{\tan\phi}{\tan(\alpha+\phi)}$. ☐

where α = Helix angle, and ϕ = Angle of friction.

144. The efficiency of a square threaded screw will be maximum, if the helix angle is equal to

(*a*) $\frac{\pi}{4}+\frac{\phi}{2}$ ☐

(*b*) $\frac{\pi}{4}-\phi$ ☐

(*c*) $\frac{\pi}{4}-\frac{\phi}{2}$ ☐

(*d*) $\frac{\pi}{4}+\phi$ ☐

where ϕ = Angle of friction.

FIGURE 10.11

145. Figure 10.11 shows the variation of efficiency of a square threaded screw for raising the load with the helix angle (α) for four different values of μ equal to 0.15, 0.125, 0.10 and 0.05. Curve *A* holds good for the values of μ, equal to

(*a*) 0.15 ☐ (*b*) 0.125 ☐

(*c*) 0.10 ☐ (*d*) 0.05. ☐

146. In Fig. 10.11, curve *B* holds good for the value of μ equal to

(*a*) 0.15 ☐ (*b*) 0.125 ☐

(*c*) 0.10 ☐ (*d*) 0.05. ☐

147. In Fig. 10.11, curve *C* holds good for the value of μ equal to

(*a*) 0.15 ☐ (*b*) 0.125 ☐
(*c*) 0.10 ☐ (*d*) 0.05. ☐

148. In Fig. 10.11, curve *D* holds good for the value of μ equal to

(*a*) 0.15 ☐ (*b*) 0.125 ☐
(*c*) 0.10 ☐ (*d*) 0.05. ☐

149. Factor of safety is the ratio of

(*a*) tensile stress to working stress ☐ (*b*) compressive stress to working stress ☐
(*c*) yield stress to working stress ☐ (*d*) bearing stress to working stress. ☐

150. The nut length, when shearing stress in nut is half the tensile stress in a bolt, would be equal to

(*a*) half the diameter of bolt ☐
(*b*) diameter of bolt ☐
(*c*) one and a half times the diameter of bolt ☐
(*d*) two times the diameter of the bolt. ☐

Riveted Joints

151. The two plates are to be joined by rivets. The plates are touching each other and two cover plates are placed on both sides of the main plates and then riveted. This type of joint is called

(*a*) lap joint ☐ (*b*) butt joint ☐
(*c*) single strap joint ☐ (*d*) double strap butt joint. ☐

152. Which of the following is a permanent fastening?

(*a*) screws ☐ (*b*) bolts and nuts ☐
(*c*) riveting ☐ (*d*) cotter ☐
(*e*) keys ☐ (*f*) none of the above. ☐

153. Which of the following is a temporary fastenings?

(*a*) welding ☐ (*b*) bolts and nuts ☐
(*c*) riveting ☐ (*d*) forging ☐
(*e*) soldering ☐ (*f*) none of the above. ☐

154. The distance between the centre of the rivet hole to the nearest edge of the plate is known as

(*a*) pitch ☐ (*b*) diagonal pitch ☐
(*c*) margin ☐ (*d*) back pitch. ☐

155. Two plates are joined by rivets. The tearing of the plate at an edge may be avoided if the margin is made at least equal to

(*a*) 0.25 *d* ☐ (*b*) 0.5 *d* ☐
(*c*) 1.0 *d* ☐ (*d*) 1.5 *d* ☐

where *d* = Diameter of the rivet.

156. Two plates are joined by rivets. The pull required to tear off the plate across a row of rivets per pitch length is equal to

(*a*) $(p - d)t \times \sigma_t$ ☐ (*b*) $p \times t \times \sigma_t$ ☐

(*c*) $d \times t \times \sigma_t$ ☐ (*d*) $(p + d) \times t \times \sigma_t$ ☐

where d = Diameter of rivet, p = Pitch of rivets,

t = Thickness of plate, and σ_t = Permissible tensile stress.

157. The plates are joined by a lap joint. The pull required to shear off the rivet per pitch length is equal to

(*a*) $n \times \frac{\pi}{4} d^2 \times \tau$ ☐ (*b*) $n \times 2 \times \frac{\pi}{4} d^2 \times \tau$ ☐

(*c*) $n \times 3 \times \frac{\pi}{4} d^2 \times \tau$ ☐ (*d*) $n \times 1.875 \times \frac{\pi}{4} d^2 \times \tau$. ☐

where n = Number of rivets per pitch length, d = Diameter of rivet, and

τ = Permissible shear stress for the rivet material.

158. Two plates are joined by a simple cover butt joint. The pull required to shear off the rivet per pitch length is equal to

(*a*) $n \times \frac{\pi}{4} d^2 \times \tau$ ☐ (*b*) $n \times 2 \times \frac{\pi}{4} d^2 \times \tau$ ☐

(*c*) $n \times 3 \times \frac{\pi}{4} d^2 \times \tau$ ☐ (*d*) $n \times 1.875 \times \frac{\pi}{4} d^2 \times \tau$. ☐

159. Two plates are joined by a double cover butt joint. The theoretical pull required to shear off the rivet per pitch length is equal to

(*a*) $n \times \frac{\pi}{4} d^2 \times \tau$ ☐ (*b*) $n \times 2 \times \frac{\pi}{4} d^2 \times \tau$ ☐

(*c*) $n \times 3 \times \frac{\pi}{4} d^2 \times \tau$ ☐ (*d*) $n \times 1.875 \times \frac{\pi}{4} d^2 \times \tau$. ☐

160. In question 159, according to Indian Boiler Regulation (I.B.R.), the pull required to shear off the rivets per pitch length is equal to

(*a*) $n \times \frac{\pi}{4} d^2 \times \tau$ ☐ (*b*) $n \times 2 \times \frac{\pi}{4} d^2 \times \tau$ ☐

(*c*) $n \times 3 \times \frac{\pi}{4} d^2 \times \tau$ ☐ (*d*) $n \times 1.875 \times \frac{\pi}{4} d^2 \times \tau$. ☐

161. Two plates are joined by a lap joint. The pull required to crush the rivet per pitch length is equal to

(*a*) $n \times d \times t \times \sigma_c$ ☐ (*b*) $\pi(d - t) \times t \times \sigma_c$ ☐

(*c*) $n \times \left(\frac{\pi}{4} d^2 - dt\right) \sigma_c$ ☐ (*d*) $n \times \frac{\pi}{4} d^2 \times \sigma_c$. ☐

where σ_c = Permissible crushing stress, n = Number of rivets per pitch length under crushing,

and d = Diameter of rivet.

162. The strength of a riveted joint is equal to

(*a*) the pull required to shear off the rivet ☐ (*b*) the pull required to crush the rivet ☐

(*c*) the pull required to tear off the plate ☐ (*d*) minimum of the above three values ☐

(*e*) maximum of the above three values. ☐

163. The efficiency of a riveted joint is the ratio of

(*a*) the pull required to shear off the rivets to the strength of unriveted plate ☐

(*b*) the pull required to crush the rivet to the strength of unriveted plate ☐

(*c*) the pull required to tear off the plate to the strength of unriveted plate ☐

(*d*) the strength of the riveted joint to the strength of unriveted plate. ☐

164. Two plates of thickness 5 mm are joined by a double riveted double cover butt joints. The diameter of rivets is 7 mm and pitch = 28 mm. The permissible tensile, shear and crushing stresses are 120 N/mm^2, 100 N/mm^2 and 150 N/mm^2 respectively. The pull required to tear off the plate across the row of rivets per pitch length is equal to

(*a*) 15400 N ☐ (*b*) 16800 N ☐

(*c*) 12600 N ☐ (*d*) 9800 N. ☐

165. In question 164, the theoretical pull required to shear off the rivets per pitch length is equal to

(*a*) 15400 N ☐ (*b*) 16800 N ☐

(*c*) 12600 N ☐ (*d*) 9800 N. ☐

166. In question 164, the pull required to crush the rivet per pitch length is equal to

(*a*) 15400 N ☐ (*b*) 16800 N ☐

(*c*) 12600 N ☐ (*d*) 9800 N. ☐

167. In question 164, the strength of the joint is equal to

(*a*) 15400 N ☐ (*b*) 16800 N ☐

(*c*) 12600 N ☐ (*d*) 9800 N. ☐

168. In question 164, the efficiency of the joint is equal to

(*a*) 80% ☐ (*b*) 70% ☐

(*c*) 58.33% ☐ (*d*) 50%. ☐

169. Unwin's empirical formula is used to find

(*a*) pitch of a riveted joint ☐

(*b*) diameter of the rivet hole, when thickness of the plates is less than 8 mm ☐

(*c*) diameter of the rivet hole, when thickness of the plate is more than 8 mm ☐

(*d*) none of the above. ☐

170. In Lozenge joint

(*a*) diamond riveting is used ☐

(*b*) number of rivets in each row is the same ☐

(*c*) number of rivets in each row goes no decreasing as we proceed from the outermost row to innermost row ☐

(*d*) none of the above. ☐

171. Lozenge joint is called a joint of

(*a*) maximum strength ☐ (*b*) minimum strength ☐

(*c*) uniform strength ☐ (*d*) none of the above. ☐

172. Diamond riveting is used for

(*a*) structural work ☐ (*b*) boiler work ☐

(*c*) structural and boiler ☐ (*d*) none of the above. ☐

173. If the diameter of the rivets varies from 12 mm to 24 mm, then according to I.S.I., the diameter of the rivet hole is equal to the diameter of rivet plus

(*a*) 0.5 mm ☐ (*b*) 1 mm ☐

(*c*) 1.5 mm ☐ (*d*) 2 mm. ☐

174. Which one is a wrong statement ?

(*a*) A butt joint with a single strap is in single shear. ☐

(*b*) A butt joint with double strap is in double shear. ☐

(*c*) The tearing of the plate at an edge can be avoided if the margin is equal to 1.5 times the diameter of the rivet bole. ☐

(*d*) The thickness of the boiler shell, according to I.B.R. (Indian Boiler Regulation) should not be more than 7 mm. ☐

175. Which one is a correct statement?

(*a*) The ratio of pitch to the diameter of the rivet is equal to one by two if the tearing efficiency of the riveted joint is 50%. ☐

(*b*) In chain riveting joints, minimum distance between the rows of the rivets should be equal to the diameter of rivet hole. ☐

(*c*) A butt joint with two cover plates should be used for longitudinal joint in boilers. ☐

(*d*) All of the above. ☐

(*e*) None of the above. ☐

176. The diameter '*d*' of the rivet for the given thickness '*t*' of the plate and to take up a given load *F* is given by

(*a*) $d = 1.91\sqrt{t}$ when *d* and *t* are in mm ☐ (*b*) $d = 6\sqrt{t}$ when *d* and *t* are in cm ☐

(*c*) both (*a*) and (*b*) ☐ (*d*) none of the above. ☐

177. In a butt joint, the thickness for the butt straps according to I.B.R. should not be less than

(*a*) 7 mm ☐ (*b*) 5 mm ☐

(*c*) 10 mm ☐ (*d*) 20 mm. ☐

178. In a butt joint if two unequal widths of straps are used, then the thickness of butt straps is

(*a*) 6.625 × *t* and 0.5 × *t* for wide strap and narrow strap ☐

(*b*) 0.75*t* and 0.625*t* for wide strap and narrow strap ☐

(*c*) 0.75*t* and 0.5*t* for wide strap and narrow strap ☐

(*d*) 0.5*t* and 0.35*t* for wide strap and narrow strap ☐

where *t* = Thickness of main plates.

179. For equal number of rivets is more than one row for lap joint or butt joint, the pitch of the rivets should not be less than

(a) $\frac{d}{2}$ ☐ (b) d ☐

(c) $1.5\ d$ ☐ (d) $2\ d$. ☐

where d = Diameter of rivet hole.

180. The pitch of the rivets is calculated by assuming the efficiency of the joint given by

(a) $\eta = \frac{p}{p-d}$ ☐ (b) $\eta = \frac{p}{p+d}$ ☐

(c) $\eta = \frac{p-d}{p}$ ☐ (d) $\eta = \frac{p+2}{p+2d}$. ☐

181. According to I.B.R., the maximum value of the pitch of rivets for a longitudinal joint of a boiler is given by

(a) 4.128 cm ☐ (b) t + 4.128 cm ☐

(c) $C \times t$ + 4.128 cm ☐ (d) $\frac{C \times t}{4.128}$ ☐

where t = Thickness of the shell plate in cm, and C = Constant.

182. A lap joint connecting two plates of thickness 16 mm is to be designed for an axial load of 64000 N. The permissible stresses are tensile stress = 40 N/mm^2, shear stress = 30 N/mm^2 and crushing stress = 70 N/mm^2. The diameter of the rivet would be equal to

(a) 244 mm ☐ (b) 24.4 mm ☐

(c) 2.44 mm ☐ (d) 10 mm. ☐

183. The width of the plates in question 182, would be equal to

(a) 124.4 mm ☐ (b) 12.44 mm ☐

(c) 200 mm ☐ (d) 250 mm. ☐

184. Choose the correct statement

(a) The diameter of the rivet, for the given thickness of the plate and to take up a given load, is obtained by Unwin's formula. ☐

(b) The pitch of the rivets is calculated by equating the shearing strength to the tearing strength of the joint. ☐

(c) The thickness of the boiler shell should not be less than 7 mm. ☐

(d) All of the above. ☐

(e) None of the above. ☐

Welded Joints

185. In a transverse double fillet welded lap joint, the allowable tensile stress for the well metal is equal to

(a) $\frac{1.414\ F}{h \times l}$ ☐ (b) $\frac{0.707\ F}{h \times l}$ ☐

(c) $\frac{4.414\ F}{h \times l}$ ☐ (d) $\frac{3F}{h \times l}$. ☐

where h = Thickness of the plate, l = Length of weld.

186. In a transverse single fillet welded lap joint, the allowable tensile stress for the weld metal is equal to

(a) $\frac{1.414\ F}{h \times l}$ ☐ (b) $\frac{0.707\ F}{h \times l}$ ☐

(c) $\frac{4.414\ F}{h \times l}$ ☐ (d) $\frac{3F}{h \times l}$. ☐

187. When fillet welds are used on one side of the lap joint, a bending moment is introduced because of the eccentricity of the forces. The maximum normal stress due to bending moment is equal to

(a) $\frac{1.414\ F}{h \times l}$ ☐ (b) $\frac{0.707\ F}{h \times l}$ ☐

(c) $\frac{4.414\ F}{h \times l}$ ☐ (d) $\frac{3F}{h \times l}$. ☐

188. In question 187, the resultant stress at the root of the weld is equal to

(a) $\frac{1.414\ F}{h \times l}$ ☐ (b) $\frac{0.707\ F}{h \times l}$ ☐

(c) $\frac{4.414\ F}{h \times l}$ ☐ (d) $\frac{3F}{h \times l}$. ☐

189. In case of parallel fillet welded joint, the shear stress is given by

(a) $\frac{0.707\ F}{h \times l}$ ☐ (b) $\frac{1.414\ F}{h \times l}$ ☐

(c) $\frac{0.5\ F}{h \times l}$ ☐ (d) $\frac{0.75\ F}{h \times l}$. ☐

190. The size of the weld, in case of a transverse fillet welded joint, is equal to

(a) one-third of the throat of weld ☐ (b) half of the throat of weld ☐

(c) 1.414 times the throat of weld ☐ (d) the throat of the weld. ☐

Keys and Couplings

191. A key is a piece of mild steel inserted between the shaft and hub to connect them together. The key

(a) prevents relative motion between the shaft and hub ☐

(b) is always inserted parallel to the axis of the shaft ☐

(c) is subjected to crashing and shearing stresses ☐

(d) has all the above function. ☐

192. A sunk key is a key

(a) made from a cylindrical disc having segmental cross-section ☐

(b) which fits half in the key way of the hub and half in the key way of shaft ☐

(c) which is flat on the shaft and fits in a key way in the hub ☐

(d) none of the above. ☐

193. A wood-ruff key is a key
(*a*) made from a cylindrical disc having segmental cross-section ☐
(*b*) which fits half in the key way of the hub and half in the key way of the shaft ☐
(*c*) which is flat on the shaft and fits in a key way in the hub ☐
(*d*) none of the above. ☐

194. A saddle key is a key
(*a*) made from a cylindrical disc having segmental cross-section ☐
(*b*) which fits half in the key way of the hub and half in the key way of the shaft ☐
(*c*) which is flat on the shaft and fits in a key way in the hub ☐
(*d*) none of the above. ☐

195. The most commonly used cross-section for a key is
(*a*) circular ☐ (*b*) square ☐
(*c*) rectangular ☐ (*d*) conical. ☐

196. A rectangular sunk key is having taper on top side only. The suitable taper is
(*a*) 1 in 100 ☐ (*b*) 1 in 60 ☐
(*c*) 1 in 30 ☐ (*d*) 1 in 10. ☐

197. The dimensions of a parallel sunk key are
(*a*) width = $\frac{d}{4}$ and thickness = $\frac{d}{10}$ ☐ (*b*) width = $\frac{d}{6}$ and thickness = $\frac{d}{10}$ ☐
(*c*) width = $\frac{d}{4}$ and thickness = $\frac{d}{6}$ ☐ (*d*) width = $\frac{d}{10}$ and thickness = $\frac{d}{6}$ ☐
where d = Diameter of the shaft.

198. A sunk key is provided with a gib head in order
(*a*) to fit the key easily ☐ (*b*) to remove the key easily ☐
(*c*) to transmit the torque ☐ (*d*) to connect shaft and hub. ☐

199. When relative axial movement of the hub is required
(*a*) a wood ruff key is used ☐ (*b*) a leather key is used ☐
(*c*) a round key used ☐ (*d*) flat saddle key is used. ☐

200. In automobile and allied industries mostly
(*a*) shaft with key ways are used ☐ (*b*) shafts having splines are used ☐
(*c*) shafts with serrations are used ☐ (*d*) (*b*) and (*c*) only. ☐

201. Which one of the following is a wrong statement?
(*a*) Shafts having splines are relatively stronger than those having key ways. ☐
(*b*) Shape of wood-ruff key is like semicircular. ☐
(*c*) A tangent key is provided in pairs at right angles and each key withstands torsion in one direction only. ☐
(*d*) For heavy duty, hollow or flat saddle keys are used. ☐

202. Choose the correct statement
(*a*) Feather keys are generally tight in the shaft and loose in the hub. ☐
(*b*) The distribution of the stresses along the length of the key fitted in a shaft is of exponential shape, being more at the torque input end. ☐

(*c*) In designing a key, the distribution of the stresses along the length of key is assumed uniform throughout. ☐

(*d*) All of the above. ☐

(*e*) None of the above. ☐

203. If the permissible crushing stress for the key material is double the permissible shearing stress, then a sunk key will be equally strong in shearing and crushing if the key is a

(*a*) rectangular key with width equal to twice the thickness ☐

(*b*) rectangular key with width equal to half the thickness ☐

(*c*) rectangular key with width equal to one-fourth the thickness ☐

(*d*) square key. ☐

204. The power transmitted from the shaft by the key will be 100% if the shearing strength of the key is

(*a*) half the torsional shear strength of the shaft ☐

(*b*) 80% of the torsional shear strength of the shaft ☐

(*c*) 90% of the torsional shear strength of the shaft ☐

(*d*) equal to the torsional shear strength of the shaft. ☐

205. For transmitting full power of the shaft, the length of the key (if the material of the key and shaft is same and width of key is one-fourth of the diameter of the shaft) should be equal to

(*a*) D ☐ (*b*) $1.25D$ ☐

(*c*) $1.571D$ ☐ (*d*) $1.75D$. ☐

206. Which of the following couplings is used to connect two shafts which are parallel and in line?

(*a*) muff-coupling ☐ (*b*) split-coupling ☐

(*c*) flange-coupling ☐ (*d*) compression-coupling ☐

(*e*) any one of the above ☐ (*f*) none of the above. ☐

207. Which of the following couplings is used to connect two shafts which are not parallel (or which are out of line)?

(*a*) muff-coupling ☐ (*b*) flange-coupling ☐

(*c*) Oldham-coupling ☐ (*d*) universal-coupling ☐

(*e*) both (*a*) and (*b*) ☐ (*f*) both (*c*) and (*d*). ☐

208. Which one of the following couplings is used to connect two shafts when the shafts are having relative speed?

(*a*) muff-coupling ☐ (*b*) flange-coupling ☐

(*c*) fluid-coupling ☐ (*d*) compression-coupling. ☐

209. In designing a flange-coupling, the hub diameter and pitch circle diameter is taken as

(*a*) Hub dia = $2D$ and pitch circle dia = $4D$ ☐

(*b*) Hub dia = $3D$ and pitch circle dia = $2D$ ☐

(*c*) Hub dia = $1.5D$ to $2D$ and pitch circle dia = $3D$ ☐

(*d*) Hub dia = $2.5D$ and pitch circle dia = $5D$. ☐

where D = Diameter of solid shaft.

210. The hub length of the flange coupling should not be less than

(a) 3D ☐ (b) 2D ☐

(c) 1.5D ☐ (d) D ☐

where D = Diameter of solid shaft.

211. The thickness of the projected portion of the flange may vary from

(a) 10 mm to 15 mm ☐ (b) 15 mm to 20 mm ☐

(c) 20 mm to 25 mm ☐ (d) 5 mm to 8 mm. ☐

212. The number of bolts by empirical formula in a flange coupling is given by

(a) $n = \frac{D}{2} + 3$ ☐ (b) $n = \frac{D}{5} + 3$ ☐

(c) $n = \frac{3D}{4} + 3$ ☐ (d) $n = \frac{2D}{7} + 3$ ☐

where D = Diameter of solid shaft in cm.

213. The number of bolts in a flange coupling should not be less than

(a) 2 ☐ (b) 3 ☐

(c) 4 ☐ (d) 5. ☐

214. The diameter of the bolt for a flange coupling is given by

(a) $\frac{0.75\,D}{n}$ ☐ (b) $\frac{0.5\,D}{n}$ ☐

(c) $\frac{0.5\,D}{\sqrt{n}}$ ☐ (d) $\frac{D}{\sqrt{n}}$. ☐

where n = Number of bolts, and D = Dia. of solid shaft.

215. The diameter of a bolt for a flange coupling should not be less than

(a) 12 mm ☐ (b) 25 mm ☐

(c) 8 mm ☐ (d) 5 mm. ☐

216. In case of a flange coupling the width of the flange varies from

(a) D to D + 1.25 ☐ (b) 1.5 D to 1.5 D + 1.25 ☐

(c) $\frac{D}{2}$ to $\frac{D}{2} + 1.25$ ☐ (d) $\frac{D}{4}$ to $\frac{D}{4} + 1.25$. ☐

217. In case of a flange coupling usually a rectangular key is used. The length of the key should not be less than

(a) 0.5 D ☐ (b) 1.0 D ☐

(c) 1.5 D ☐ (d) 3.0 D. ☐

218. In question 217, the width (b) and thickness (t) of the rectangular key are taken as

(a) $b = \frac{D}{4}$ and $t = \frac{2D}{3}$ ☐ (b) $b = \frac{D}{3}$ and $t = \frac{2D}{3}$ ☐

(c) $b = \frac{D}{4}$ and $t = \frac{D}{6}$ ☐ (d) $t = \frac{D}{2}$ and $t = \frac{D}{6}$. ☐

219. In case of marine flange couplings, the number of bolts are given by

(a) $\frac{D}{5} + 2$ ☐ (b) $\frac{D}{4} + 2$ ☐

(c) $\frac{D}{7.5} - 2$ ☐ (d) $\frac{D}{7.5} + 2.$ ☐

220. In case of marine flange coupling, the pitch circle diameter of the bolt is taken as equal to

(a) D ☐ (b) 1.6 D ☐

(c) 2D ☐ (d) 2.5 D. ☐

Belt Drive

221. The power transmitted by a belt or rope is given by

(a) $\frac{(T_1 + T_2) \times v}{1000}$ ☐ (b) $\frac{(T_1 + T_2) \times v}{1000}$ ☐

(c) $\frac{T_1 \times v}{1000}$ ☐ (d) $\frac{T_2 \times v}{1000}$. ☐

where T_1 = Tension in the tight side, T_2 = Tension in the slack side, and

v = Linear speed of the belt in m/sec.

222. The power transmitted by a belt depends upon

(a) centre distance between the two pulleys over which belt is passing ☐

(b) velocity of belt ☐

(c) difference between the tensions in the tight and slack sides of the belt ☐

(d) all of the above. ☐

(e) (b) and (c) only. ☐

223. The relation between the tension (T_1) in the belt on the tight side to the tension (T_2) on the slack side is given by

(a) $\frac{T_1}{T_2} = \mu\theta$ ☐ (b) $\frac{T_1}{T_2} = \mu\theta$ ☐

(c) $\frac{T_1}{T_2} = e^{-\mu\theta}$ ☐ (d) $\frac{T_1}{T_2} = e^{\mu\theta}$. ☐

where μ = Co-efficient of friction between the bolt and the pulley, and

θ = Angle of lap in radians.

224. The centrifugal tension caused by centrifugal force is given by

(a) $\frac{1}{2} mv^2$ ☐ (b) $\frac{mv^2}{2g}$ ☐

(c) $\frac{mv^2}{g}$ ☐ (d) $\frac{2wv^2}{g}$. ☐

where v = Linear velocity of the belt, w = Weight per unit length of the belt, and

m = Mass of the belt.

225. Taking centrifugal tension into account, the speed of the belt at which the belt transmits maximum horse power is equal to

(a) $\sqrt{\frac{3T \times g}{w}}$ ☐ (b) $\sqrt{\frac{3T}{w \times g}}$ ☐

(c) $\sqrt{\frac{T \times g}{3 \times w}}$ ☐ (d) $\sqrt{\frac{3 \times w}{T \times g}}$ ☐

where T = Maximum tension to which belt can be subjected, and

w = Weight per unit length of the belt.

226. When the horse power transmitted is maximum, the centrifugal tension is equal to

(a) one-half of the maximum tension ☐ (b) one-third to the maximum tension ☐

(c) one-fourth of the maximum tension ☐ (d) three times the maximum tension. ☐

227. If α is the angle of groove of a grooved pulley in which a rope is running, then ratio of tensions of the two sides of the rope is given by

(a) $\frac{T_1}{T_2} = \mu\theta$ ☐ (b) $\frac{T_1}{T_2} = e^{\mu\alpha}$ ☐

(c) $\frac{T_1}{T_2} = e^{\mu_1 \times \theta}$ ☐ (d) $\frac{T_1}{T_2} = \mu \times \theta \times \alpha$. ☐

where $\mu_1 = \mu \operatorname{cosec}\left(\frac{\alpha}{2}\right)$, μ = Co-efficient of friction between the rope and the grooved pulley, and θ = Angle of lap in radians.

228. Choose the wrong statement

(a) In a multiple V-belt drive, if one of the belts break, then the entire set of belts is replaced. ☐

(b) When the shafts are arranged in parallel and rotate in the same direction, an open belt drive is used. ☐

(c) The ratio of driving tensions in V-belt drive is more than flat bell drives. ☐

(d) For rope drive, the groove angle of the pulley is usually 20 to 30°. ☐

229. The included angle for the V-belt varies from

(a) 10° to 20° ☐ (b) 20° to 30° ☐

(c) 30° to 40° ☐ (d) 40° to 50°. ☐

230. Angle of contact, if both the pulleys are of the same material, is the

(a) angle of contact at the bigger pulley ☐

(b) angle of contact at the smaller pulley ☐

(c) mean angle or contact at both the pulley ☐

(d) difference of the angle of contact at the bigger and smaller pulley. ☐

231. When the pulleys are made of different material, then design will refer to the pulley for which

(a) $\mu \times \theta$ is maximum ☐ (b) $\mu \times \theta$ is minimum ☐

(c) $\mu \times \theta = 1$ ☐ (d) $\mu \times \theta > 1$. ☐

232. The initial tension in the perfectly elastic material of the belt neglecting centrifugal tension is given by

(*a*) $\frac{T_1 - T_2}{2}$ ☐ (*b*) $\frac{T_1 + T_2}{2}$ ☐

(*c*) $\frac{2T_1 - T_2}{2}$ ☐ (*d*) $\frac{T_1 - 2T_2}{2}$. ☐

where T_1 = Tension on tight side, and T_2 = Tension on slack side.

233. The initial tension in the perfectly elastic material of the belt considering the centrifugal tension is given by

(*a*) $\frac{T_1 - T_2 - T_c}{2}$ ☐ (*b*) $\frac{T_1 + T_2 + T_c}{2}$ ☐

(*c*) $\frac{T_1 + T_2 + 2T_c}{2}$ ☐ (*d*) $\frac{T_1 + T_2 - 2T_c}{2}$. ☐

where T_c = Centrifugal tension in the belt.

234. The relation between initial tension (T_0), tension on tight side (T_1) and tension on slack side (T_2) (if the material of the belt is not perfectly elastic) is given by

(*a*) $T_0 = \sqrt{T_1} + \sqrt{T_2}$ ☐ (*b*) $T_0 = \frac{\sqrt{T_1} + \sqrt{T_2}}{2}$ ☐

(*c*) $\sqrt{T_0} = \frac{\sqrt{T_1} + \sqrt{T_2}}{2}$ ☐ (*d*) $T_0 = \sqrt{T_1} + \sqrt{T_2}$. ☐

235. The power transmitted by a belt is maximum when initial tension in the belt is

(*a*) half of centrifugal tension ☐ (*b*) one-third of centrifugal tension ☐

(*c*) twice of centrifugal tension ☐ (*d*) three times the centrifugal tension. ☐

236. The angle of lap on the smaller pulley for open belt is given by

(*a*) $180° - 2\sin^{-1}\left(\frac{r_1 - r_2}{d}\right)$ ☐ (*b*) $180° - 2\cos^{-1}\left(\frac{r_1 - r_2}{d}\right)$ ☐

(*c*) $180° - 2\sin^{-1}\left(\frac{r_1 + r_2}{d}\right)$ ☐ (*d*) $180° - 2\cos^{-1}\left(\frac{r_1 + r_2}{d}\right)$ ☐

where r_1 = Radius of large pulley, r_2 = Radius of small pulley, and

d = Centre distance between the shafts.

237. The centre distance between two parallel shafts is 5 m. They are connected by an open belt drive. The diameters of large and small pulley are 1.5 m and 1.0 respectively. The weight of the belt is 0.98 kgf/m length and initial tension in the belt, when belt is stationary is 250 kgf. If the smaller pulley rotates at 420 r.p.m., the linear velocity of the belt would be

(*a*) 11 m/sec ☐ (*b*) 22 m/sec ☐

(*c*) 33 m/sec ☐ (*d*) 44 m/sec. ☐

238. In question 237, the centrifugal tension would be equal to

(*a*) 46.4 kgf ☐ (*b*) 23.2 kgf ☐

(*c*) 69.6 kgf ☐ (*d*) 11.6 kgf. ☐

239. In question 237, the sum of tension in the tight side and slack side would be equal to

(*a*) 592.8 kgf ☐ (*b*) 407.2 kgf ☐
(*c*) 92.8 kgf ☐ (*d*) 500 kgf. ☐

240. For the same tension and co-efficient of friction, the type of belt which can transmit more power, is

(*a*) V-belt ☐ (*b*) flat belt ☐
(*c*) timing belt ☐ (*d*) none of the above. ☐

241. The type of belt used for obtaining clutch action by shifting the belt from a loose to a tight pulley, would be

(*a*) V-belt ☐ (*b*) flat belt ☐
(*c*) timing belt ☐ (*d*) none of the above. ☐

Chain and Rope Drives

242. Chain drives are usually put out of the commission due to

(*a*) failure of joints and plates ☐ (*b*) chain elongation ☐
(*c*) water of sprocket teeth ☐ (*d*) all of the above ☐
(*e*) none of the above. ☐

243. In chain drive, it is usually preferable to use

(*a*) odd number of teeth on sprocket ☐ (*b*) even number of teeth on sprocket ☐
(*c*) fifty number of teeth on sprocket ☐ (*d*) none of the above. ☐

244. Choose the correct statement

(*a*) In motor cycle silent type of chain is used. ☐
(*b*) Silent chain is made of three or more roller chains. ☐
(*c*) Friction radius for new clutches compared to worn-out would be more. ☐
(*d*) Unit of dynamic viscosity are N m/s^2. ☐

245. Which one of the following ropes is most flexible?

(*a*) 6 by 38 ☐ (*b*) 6 by 30 ☐
(*c*) 5 by 19 ☐ (*d*) 6 by 9. ☐

Design of Cotter Joint and Knuckle Joint

246. Cotter joints is used to connect two co-axial rods which are subjected to

(*a*) bending ☐ (*b*) twisting ☐
(*c*) axial loading ☐ (*d*) all of the above ☐
(*e*) none of the above. ☐

247. If P is the axial load and σ_t is the permissible tensile stress in the material of the two co-axial rods connected by cotter joint, then diameter of the rods is given by

(*a*) $\sqrt{\dfrac{\pi P}{4\sigma_t}}$ ☐ (*b*) $\sqrt{\dfrac{4P}{\pi \times \sigma_t}}$ ☐

(*c*) $\sqrt{\dfrac{4\sigma_t}{\pi P}}$ ☐ (*d*) $\sqrt{\dfrac{\pi \times \sigma_t}{4P}}$. ☐

248. The mean width *(b)* of the collar in case of cotter joint is given by

(*a*) $\frac{P}{2t \times \tau}$ ☐ (*b*) $\frac{2P}{t \times \tau}$ ☐

(*c*) $\frac{P}{t \times \tau}$ ☐ (*d*) $\frac{4P}{t \times \tau}$. ☐

where P = Axial load, t = Thickness of cotter, and

τ = Permissible shear stress in the cotter material.

249. The diameter of the spigot (or inside diameter of the socket) of a cotter joint is obtained from

(*a*) $P = \frac{\pi}{4} d_1^2 \times \sigma_t$ ☐ (*b*) $P = d_1 \times t \times \sigma_t$ ☐

(*c*) $P = \left(\frac{\pi}{4} d_1^2 - d_1 \times t\right) \sigma_t$ ☐ (*d*) $P = 2d_1 \times t \times \tau$ ☐

where P = Axial load, t = Thickness of cotter,

d_1 = Diameter of spigot, σ_t = Permissible tensile stress, and

τ = Permissible shear stress.

250. The cotter, in case of cotter joint, is having

(*a*) taper on one side ☐ (*b*) taper on both sides ☐

(*c*) no taper ☐ (*d*) none of the above. ☐

251. A cotter joint is having maximum draw of

(*a*) 10 mm ☐ (*b*) 8 mm ☐

(*c*) 5 cm ☐ (*d*) 3 mm. ☐

252. A cotter is used to

(*a*) rigidly connect rods ☐ (*b*) transmit motion ☐

(*c*) prevent rotation of rods ☐ (*d*) all of the above ☐

(*e*) none of the above. ☐

253. For connecting piston rod and cross-head of an engine

(*a*) a cotter joint is used ☐ (*b*) a knuckle joint is used ☐

(*c*) both are used ☐ (*d*) none is used. ☐

254. In the links of suspension chains

(*a*) a cotter joint is used ☐ (*b*) a knuckle joint is used ☐

(*c*) both are used ☐ (*d*) none is used. ☐

255. A cotter joint may fail due to

(*a*) shearing ☐ (*b*) bending ☐

(*c*) crushing ☐ (*d*) all of the above. ☐

256. A knuckle joint may fail due to

(*a*) shearing ☐ (*b*) bending ☐

(*c*) crushing ☐ (*d*) all of the above. ☐

257. The loose knuckle pin in the fork will be subjected to

(*a*) shearing ☐ (*b*) bending ☐

(*c*) crushing ☐ (*d*) tearing. ☐

258. When threading a pipe, the cutting oil is used to

(*a*) lubricate the dies ☐ (*b*) provide cooling action ☐

(*c*) remove chips ☐ (*d*) all of the above. ☐

259. The type of threads best suited for tight leakage joints would be

(*a*) acme ☐ (*b*) buttress ☐

(*c*) square ☐ (*d*) national pipe threads (NPT). ☐

260. The diameter of the pin of a knuckle joint is obtained by considering the failure of the pin due to

(*a*) shearing ☐ (*b*) tension ☐

(*c*) bending ☐ (*d*) crushing. ☐

261. The thickness of the cotter is generally taken equal to

(*a*) 0.5*d* ☐ (*b*) 0.4*d* ☐

(*c*) 0.31*d* ☐ (*d*) 0.10*d*. ☐

Thin and Thick Cylinders

262. When a thin-walled cylinder is subjected to internal fluid pressure, two types of stresses are developed. These stresses are

(*a*) both tensile ☐ (*b*) both compressive ☐

(*c*) both shear ☐ (*d*) one tensile and other compressive ☐

(*e*) none of the above. ☐

263. In the above question, the two stresses developed in the material of the cylinder are called longitudinal stress and circumferential stress. The longitudinal stress is given by

(*a*) $\frac{p \times D}{2t}$ ☐ (*b*) $\frac{p \times D}{4 \times t}$ ☐

(*c*) $\frac{p \times D}{8 \times t}$ ☐ (*d*) $\frac{2p \times D}{t}$ ☐

264. The circumferential stress in question 263, is given by

(*a*) $\frac{p \times D}{2t}$ ☐ (*b*) $\frac{p \times D}{4 \times t}$ ☐

(*c*) $\frac{p \times D}{8 \times t}$ ☐ (*d*) $\frac{2p \times D}{t}$ ☐

265. The stress induced in a thin cylinder (subjected to internal fluid pressure) parallel to the axis of the cylinder is

(*a*) two times the circumferential stress ☐ (*b*) equal to the circumferential stress ☐

(*c*) one-half of the circumferential stress ☐ (*d*) one-third of the circumferential stress. ☐

266. In the design of steam boilers, according to Boilers Code, the factor of safety should be at least equal to

(*a*) 2 ☐ (*b*) 3 ☐
(*c*) 5 ☐ (*d*) 8. ☐

267. The design of a pressure vessel, having thin walls, is based on

(*a*) longitudinal stress ☐ (*b*) hoop stress ☐
(*c*) both (*a*) and (*b*) ☐ (*d*) none of the above. ☐

268. A spherical vessel of 2 m diameter is subjected to an internal pressure of 16 kg/cm^2. If the maximum stress in the material of the vessel is limited to 800 kg/cm^2 and efficiency of joint is 80% then the thickness of the vessel will be

(*a*) 2.0 cm ☐ (*b*) 1.75 cm ☐
(*c*) 1.50 cm ☐ (*d*) 1.25 cm. ☐

269. Which one of the following equations is used in the design of thick cylinders?

(*a*) Lame's equation ☐ (*b*) Clavarino's equation ☐
(*c*) Barlow's equation ☐ (*d*) Birnei's equation ☐
(*e*) any one of the above ☐ (*f*) none of the above. ☐

270. The circumferential stress (hoop stress) induced in the material of a thin cylinder, subjected to an internal pressure of intensity p, is assume to be

(*a*) having maximum stress at the outer radius and minimum stress at inner radius ☐
(*b*) having maximum stress at the inner radius and minimum stress at outer radius ☐
(*c*) uniformly distributed across the thickness of the wall ☐
(*d*) none of the above. ☐

271. The circumferential stress induced in the material of a thick cylinder, subjected to an internal pressure of intensity p, is

(*a*) having maximum stress at the outer radius and minimum stress at the inner radius ☐
(*b*) having maximum stress at the inner radius and minimum stress at the outer radius ☐
(*c*) uniformly distributed across the thickness of the wall ☐
(*d*) none of the above. ☐

272. The thickness (t) of a thick cylinder according to Lame's theory is equal to

(*a*) $\frac{d}{2}\left[\sqrt{\left(\frac{f+p}{f-p}\right)}-1\right]$ ☐ (*b*) $\frac{d}{2}[\sqrt{(f+p)}-1]$ ☐

(*c*) $\frac{d}{2}\left[\sqrt{\frac{f-p}{f+p}}-1\right]$ ☐ (*d*) $\frac{d}{2}\sqrt{(f-p)}$. ☐

where f = Hoop stress, d = Internal diameter of the cylinder, and

p = Internal fluid pressure.

273. The maximum hoop stress in case of a thick cylinder is equal to

(*a*) $\frac{p(d_o^2-d_i^2)}{d_o^2+d_i^2}$ ☐ (*b*) $\frac{p(d_o^2+d_i^2)}{d_o^2-d_i^2}$ ☐

(*c*) $\frac{p(d_o^2-d_i^2)}{d_o^2}$ ☐ (*d*) $\frac{p(d_o^2+d_i^2)}{d_i^2}$. ☐

where d_o = Outer diameter of cylinder, d_i = Inner diameter of cylinder, and

p = Internal fluid pressure.

274. The thickness (t) of a thick cylinder according to Grashof formula is equal to

(a) $\frac{d}{2}\left[\sqrt{\left(\frac{3f+2p}{3f-4p}\right)}+1\right]$ ☐ (b) $\frac{d}{2}\left[\sqrt{\left(\frac{3f-2p}{3f+2p}\right)}-1\right]$ ☐

(c) $\frac{d}{2}\left[\sqrt{\left(\frac{3f+2p}{3f-4p}\right)}-1\right]$ ☐ (d) $\frac{d}{2}\left[\sqrt{\frac{3f-4p}{3f+2p}}+1\right]$. ☐

275. The minimum thickness of the flat plates, used for the head of thick cylinders is given by

(a) $t = d\sqrt{\frac{kp}{f}}$ ☐ (b) $t = \sqrt{\frac{dkp}{f}}$ ☐

(c) $t = k\sqrt{\frac{dp}{f}}$ ☐ (d) $t = p\sqrt{\frac{k \times d}{f}}$ ☐

where d = Internal diameter of the cylinder, p = Internal fluid pressure,
f = Permissible tensile stress for the material of the plate, and k = A constant = 0.162 for flat heads.

276. A liquid under a pressure of 140 kg/cm^2 is stored in a thick cast steel cylinder of inside diameter 40 cm. Flat cover plates are used for closing the two ends of the cylinder. If the maximum hoop stress in the material of the cylinder is limited to 560 kg/cm^2, the wall thickness of the cylinder would be approximately

(a) 4 cm ☐ (b) 5 cm ☐

(c) 6 cm ☐ (d) 7 cm. ☐

277. If in question 276, the permissible tensile stress for the material of the plate is 600 kg/cm^2, the minimum necessary thickness of the cover plates would be

(a) 4 cm ☐ (b) 6 cm ☐

(c) 8 cm ☐ (d) 10 cm. ☐

278. The design of a pipe means the determination of the inside diameter of the pipe and its wall thickness. The inside diameter of the pipe is given by

(a) $\sqrt{\frac{Q}{v}}$ ☐ (b) $1.13\sqrt{\frac{Q}{v}}$ ☐

(c) $2.125\sqrt{\frac{Q}{v}}$ ☐ (d) $0.755\sqrt{\frac{Q}{v}}$. ☐

where Q = Volume of fluid carried per sec, and v = Velocity of the following fluid per sec.

279. If the ratio of allowable stress to the fluid pressure inside the pipe is more than 6.0, then thickness of the pipe is given by

(a) $\frac{p \times d}{2f} + C$ ☐ (b) $\frac{p \times d}{4 \times f} + C$ ☐

(c) $\frac{2p \times d}{f} + C$ ☐ (d) $\frac{2p + d}{3f} + C$. ☐

where p = Internal fluid pressure, d = Internal diameter of pipe
f = Allowable stress in the pipe material, and
C = A constant which is equal to 0.9 for a cast iron pipe and = 0.3 for a steel pipe.

280. Socket joint is mostly used for pipes which

(*a*) carry steam at high pressure ☐ (*b*) carry water at low pressure ☐
(*c*) are burried in the earth ☐ (*d*) carry fluid at high pressure. ☐

281. Expansion joints are mostly used for pipes which

(*a*) carry steam at high pressure ☐ (*b*) carry water at low pressure ☐
(*c*) are burried in the earth ☐ (*d*) carry fluid at high pressure. ☐

282. The spigot and socket joint is mostly used for pipes which

(*a*) carry steam at high pressure ☐ (*b*) carry water at low pressure ☐
(*c*) are burried in the earth ☐ (*d*) carry fluid at high pressure. ☐

283. The hydraulic pipe joint is mostly used for pipes which

(*a*) carry steam at high pressure ☐ (*b*) carry water at low pressure ☐
(*c*) are hurried in the earth ☐ (*d*) carry fluid at high pressure. ☐

284. In case of flanged pipe joint to make the joint leak proof the circumferential pitch of the bolts should be

(*a*) more than $30\sqrt{d}$ and diameter of bolt less than 16 mm ☐

(*b*) less than $20\sqrt{d}$ and diameter of bolt more than 16 mm ☐

(*c*) more than $20\sqrt{d}$ and diameter of bolt less than 16 mm ☐

(*d*) between $20\sqrt{d}$ and $30\sqrt{d}$ and diameter of bolt more than 16 mm ☐

where d = Diameter of the bolt hole.

Springs

285. Springs are used mainly

(*a*) to store strain energy ☐
(*b*) to absorb shocks and vibrations ☐
(*c*) to return an element to its original position ☐
(*b*) to act as a reservoirs of energy ☐
(*e*) all of the above ☐
(*f*) none of the above. ☐

286. A spring material should have the following properties

(*a*) high yield strength ☐ (*b*) high breaking strength ☐
(*c*) more hardness ☐ (*d*) all of the above. ☐
(*e*) none of the above. ☐

287. In a helical spring if the pitch angle is less than 10° (or the pitch of the helix is small compared with the mean diameter of the coils) and the spring is subjected to an axial compressive or tensile load, the stress induced in the wire would be

(*a*) compressive only ☐ (*b*) tensile only ☐
(*c*) bending only ☐ (*d*) shear only ☐
(*e*) none of the above. ☐

288. In a helical spring if the pitch angle is more than 10° and spring is subjected to an axial compressive load, the stresses induced in the wire would be

(*a*) compressive ☐ (*b*) tensile ☐
(*c*) bending ☐ (*d*) shear ☐
(*e*) all of the above ☐ (*f*) (*c*) and (*d*) only. ☐

289. In close-coiled helical spring, the maximum shear stress induced in a wire of the circular section is equal to

(*a*) $\frac{16WD}{\pi d^3}$ ☐ (*b*) $\frac{8WD}{\pi d^3}$ ☐
(*c*) $\frac{4WD}{\pi d^3}$ ☐ (*d*) $\frac{32WD}{\pi d^3}$ ☐

where W = Applied load, D = Mean diameter of coils, and
d = Diameter of spring wire.

290. In a close-coiled helical spring, the deflection of spring coils is equal to

(*a*) $\frac{16WD^3n}{Cd^4}$ ☐ (*b*) $\frac{32WD^3n}{Cd^4}$ ☐
(*c*) $\frac{8WD^3n}{Cd^4}$ ☐ (*d*) $\frac{4WD^3n}{Cd^4}$ ☐

where C = Modulus of rigidity, and n = Number of effective coils.

291. The maximum shear stress induced in the wire of a circular section of a helical spring depends on

(*a*) the material of wire ☐ (*b*) size of cross-section ☐
(*c*) the ratio D/d ☐ (*d*) all of the above ☐
(*e*) none of the above. ☐

292. The ratio of the mean diameter of the coil to the diameter of the wire (*i.e.*, D/d) is known as

(*a*) solid length of spring ☐ (*b*) free length of spring ☐
(*c*) spring index ☐ (*d*) spring rate. ☐

293. The wire of a close-coiled helical spring on compression, is subjected to

(*a*) compression ☐ (*b*) tension ☐
(*c*) shear ☐ (*d*) all of the above. ☐

294. Two springs (each having stiffness *K*) are in parallel. The overall stiffness of the two springs would be equal to

(*a*) 2*K* ☐ (*b*) *K* ☐
(*c*) *K*/2 ☐ (*d*) *K*/4. ☐

295. Two springs (each having stiffness *K*) are in series. The overall stiffness of the two springs would be equal to

(*a*) 2*K* ☐ (*b*) *K* ☐
(*c*) *K*/2 ☐ (*d*) *K*/4. ☐

296. A closely coiled helical spring is made out of 1 cm diameter steel rod. The spring carries an axial pull of 20 kg. The coil consists of 10 complete turns with a mean radius of 5 cm. If modulus of rigidity of spring material is 8×10^5 kg/cm^2, the maximum shear stress induced in the section of the rod would be

(*a*) $\frac{3200}{\pi}$ kg/cm^2 ☐ (*b*) $\frac{1600}{\pi}$ kg/cm^2 ☐

(*c*) $\frac{800}{\pi}$ kg/cm^2 ☐ (*d*) $\frac{400}{\pi}$ kg/cm^2. ☐

297. In the above question, the deflection of the spring would be

(*a*) 2 cm ☐ (*b*) 1 cm ☐

(*c*) 0.5 cm ☐ (*d*) 0.4 cm. ☐

298. In question 296, stiffness of the spring would be

(*a*) 20 kg per cm ☐ (*b*) 15 kg per cm ☐

(*c*) 10 kg per cm ☐ (*d*) 8 kg per cm. ☐

299. In an open-coiled helical spring, the maximum shear stress induced in a wire of circular section taking into consideration the effect of direct shear stress and effect of curvature of the wire, is equal to

(*a*) $\frac{8WD}{\pi d^3}$ ☐ (*b*) $\frac{8WD}{\pi d^3} \times K$ ☐

(*c*) $\frac{16WD}{\pi d^3} \times K$ ☐ (*d*) $\frac{8W}{\pi D^2} \times K$. ☐

where W = Applied axial load, D = Mean diameter of coils,

d = Diameter of spring wire, and K – Wahl's shear stress factor.

300. Figure 10.12 shows the variation of Wahl's shear stress factor versus spring index of a helical spring. Which one of the following is a correct curve?

(*a*) curve A ☐

(*b*) curve *B* ☐

(*c*) curve *C* ☐

(*d*) curve *D*. ☐

FIGURE 10.12

301. Choose the correct statement

(*a*) The ratio of the mean diameter of the coil to the diameter of the spring wire is known as spring index. ☐

(*b*) The product of total number of coils and the diameter of the spring wire gives the solid length of the spring. ☐

(*c*) If the helix angle is more than 10°, the helical spring is called open-coiled helical spring. ☐

(*d*) All of the above. ☐

(*e*) None of the above. ☐

302. The type of springs, mostly used in railway wagon coach and road vehicles, are

(a) spiral torsion spring ☐ (b) helical spring ☐
(c) leaf spring ☐ (d) bellevile spring. ☐

303. The type of springs used, where the space is relatively small and large amount of energy must be stored, are

(a) spiral torsion spring ☐ (b) helical spring ☐
(c) leaf spring ☐ (d) bellevile spring. ☐

304. The Wahl's stress factor is used mostly for the design of

(a) spiral torsion spring ☐ (b) helical spring ☐
(c) leaf spring ☐ (d) bellevile spring. ☐

305. The Wahl's stress factor is equal to

(a) $\frac{4C-1}{4C-4}+\frac{0.615}{C}$ ☐ (b) $\frac{C-1}{C-4}+\frac{0.615}{C}$ ☐

(c) $\frac{4C-1}{C-4}+\frac{0.615}{C}$ ☐ (d) $\frac{C-1}{4C-4}+\frac{0.615}{C}$. ☐

where C = Spring index.

306. Which one of the followings is a wrong statement?

(a) The value of Wahl's stress factor (*K*) for springs decreases exponentially with the increase of value of spring index (C). ☐
(b) The spring mostly used in gramophones is flat spiral spring. ☐
(c) The deflection of the helical spring is directly proportional to the cube of mean diameter of coil. ☐
(d) With the increase of the size of the wire of the compression springs, the allowable stress increases. ☐

307. The cracks due to fatigue spreads only by

(a) tensile stress and in direction along the tensile stress ☐
(b) compressive stress and in direction along the compressive stress ☐
(c) shear stress ☐
(d) tensile stress and in direction perpendicular to the tensile stress. ☐

Friction Clutch

308. Which of following is a friction clutch?

(a) cone clutch ☐ (b) band clutch ☐
(c) disc clutch ☐ (d) block clutch ☐
(e) all of the above ☐ (f) none of the above. ☐

309. The frictional torque transmitted by a disc clutch, considering uniform pressure for a single friction surface is equal to

(a) $\frac{2}{3}\mu W\frac{(r_1^2-r_2^2)}{(r_1^3-r_2^3)}$ ☐ (b) $\frac{2}{3}\mu W\frac{(r_1^3-r_2^3)}{(r_1^2-r_2^2)}$ ☐

(c) $\frac{1}{2}\pi W(r_1-r_2)$ ☐ (d) $\frac{1}{2}\pi W(r_1+r_2)$. ☐

where W = Axial thrust with which the contact surfaces are held together,

r_1 = Outer radius of the plate, and r_2 = Inner radius of the plate.

310. The frictional torque transmitted by a disc clutch for a single friction surface considering uniform wear is equal to

(a) $\frac{2}{3}\ \mu W \frac{(r_1^2 - r_2^2)}{(r_1^3 - r_2^3)}$ ☐ (b) $\frac{2}{3}\ \mu W \frac{(r_1^3 - r_2^3)}{(r_1^2 - r_2^2)}$ ☐

(c) $\frac{1}{2}\ \pi W (r_1 - r_2)$ ☐ (d) $\frac{1}{2}\ \pi W (r_1 + r_2)$. ☐

311. If there are n_1 number of discs on the driving shaft and n_2 number of discs on the driven shaft, then number of pairs of contact surfaces would be equal to

(a) $n_1 + n_2$ ☐ (b) $2(n_1 + n_2)$ ☐

(c) $n_1 + n_2 - 1$ ☐ (d) $n_1 + n_2 - 2$. ☐

312. The frictional radius for new clutches is equal to

(a) $\frac{2}{3}\left(\frac{r_1^3 - r_2^3}{r_1^2 - r_2^2}\right)$ ☐ (b) $\frac{3}{2}\left(\frac{r_1^3 - r_2^3}{r_1^2 - r_2^2}\right)$ ☐

(c) $\frac{r_1 - r_2}{2}$ ☐ (d) $\frac{r_1 + r_2}{2}$. ☐

313. The friction radius for old clutches is equal to

(a) $\frac{2}{3}\frac{(r_1^3 - r_2^3)}{(r_1^2 - r_2^2)}$ ☐ (b) $\frac{3}{2}\left(\frac{r_1^3 - r_2^3}{r_1^2 - r_2^2}\right)$ ☐

(c) $\frac{r_1 - r_2}{2}$ ☐ (d) $\frac{r_1 + r_2}{2}$. ☐

314. The outside and inside diameters of contact surfaces of a multidisc clutch are 24 cm and 16 cm respectively. The clutch has three discs on the driving shaft and two on the driven shaft. The number of pairs of contact surfaces would be

(a) 5 ☐ (b) 6 ☐

(c) 4 ☐ (d) 3. ☐

315. If in question 314, the wear is uniform, the friction radius will be equal to

(a) 10 cm ☐ (b) 6 cm ☐

(c) 4 cm ☐ (d) 3 cm. ☐

316. If in question 315, co-efficient of friction = 0.25 and power transmitted = 22 kW at 1680 r.p.m., then torque transmitted will be equal to

(a) 150 cm ☐ (b) 125 cm ☐

(c) 100 cm ☐ (d) 80 cm. ☐

317. In the above question, the axial force on the friction faces would be equal to

(a) 1000 N ☐ (b) 1200 N ☐

(c) 1250 N ☐ (d) 1500 N. ☐

318. The uniform pressure theory gives

(a) same frictional torque as uniform wear theory ☐

(b) higher frictional torque than the uniform wear theory ☐

(c) lesser frictional torque than the uniform wear theory ☐

(d) none of the above. ☐

319. For a block brake, the equivalent co-efficient of friction is equal to

(a) $\frac{\mu \sin \theta}{2\theta + \sin 2\theta}$ ☐ (b) $\frac{4\mu \sin \theta}{2\theta + \sin 2\theta}$ ☐

(c) $\frac{4\mu \sin \theta}{\theta + 4 \sin 2\theta}$ ☐ (d) $\frac{\mu \sin \theta}{\theta + 2 \sin 2\theta}$. ☐

where θ = Semiblock angle, and μ = Actual co-efficient of friction.

320. In motor cars, commonly used type of brake is

(a) internal expanding brake ☐ (b) shoe brake ☐

(c) band brake ☐ (d) any one of the above. ☐

321. The ratio of shoe width to wheel diameter in block brakes is kept between

(a) 0.75 to 1.0 ☐ (b) 0.5 to 0.75 ☐

(c) 0.25 to 0.5 ☐ (d) 0.1 to 0.25. ☐

322. A zero film bearing is one in which

(a) there is no lubricant between the journal and bearing ☐

(b) there is a thin film of lubricant between the journal and the bearing ☐

(c) there is a thick film of lubricant between the journal and the bearing ☐

(d) steady load is supported without any relative motion between the journal and the bearing. ☐

323. A hydrodynamic lubricated bearing is one in which

(a) there is no lubricant between the journal and bearing ☐

(b) there is a thin film of lubricant between the journal and the bearing ☐

(c) there is a thick film of lubricant between the journal and the bearing ☐

(d) steady load is supported without any relative motion between the journal and the bearing. ☐

324. Figure 10.13 shows a shaft rotating in a fixed bearing subjected to steady load *W*. If the shaft is rotating in anticlockwise direction at slow speed then the correct position of shaft and bearing is given by

Bearing
Shaft
W W W W
(a) (b) (c) (d)

FIGURE 10.13

(a) Fig. 10.13 (a) ☐ (b) Fig. 10.13 (b) ☐

(c) Fig. 10.13 (c) ☐ (d) Fig. 10.13 (d). ☐

325. The correct position of shaft and bearing, which shaft is rotating in anticlockwise direction at high speed will be (refer to Fig. 10.13).

(a) Fig. 10.13 (a) ☐ (b) Fig. 10.13 (b) ☐

(c) Fig. 10.13 (c) ☐ (d) Fig. 10.13 (d). ☐

326. The correct position of shaft and bearing, when shaft is rotating in clockwise direction at high speed will be (refer to Fig. 10.13).

(*a*) Fig. 10.13 (*a*) ☐ (*b*) Fig. 10.13 (*b*) ☐

(*c*) Fig. 10.13 (*c*) ☐ (*d*) Fig. 10.13 (*d*). ☐

327. The correct position of shaft and bearing, when shaft is rotating in clockwise direction at slow speed will be (refer to Fig. 10.13)

(*a*) Fig. 10.13 (*a*) ☐ (*b*) Fig. 10.13 (*b*) ☐

(*c*) Fig. 10.13 (*c*) ☐ (*d*) Fig. 10.13 (*d*). ☐

328. Which one of the following is a solid lubricant?

(*a*) mineral oil ☐ (*b*) grease ☐

(*c*) graphite ☐ (*d*) synthetic oil. ☐

329. Which one of the following is a semi-liquid lubricant?

(*a*) mineral oil ☐ (*b*) grease ☐

(*c*) graphite ☐ (*d*) synthetic oil. ☐

Bearings

330. The bearing characteristic number relating the absolute viscosity (Z) of lubricant, speed (*N*) of journal and bearing pressure (*p*) on the projected bearing area is given by

(*a*) $\frac{Zp}{N}$ ☐ (*b*) $\frac{Z}{Np}$ ☐

(*c*) $\frac{ZN}{p}$ ☐ (*d*) $\frac{p}{NZ}$. ☐

331. Bearing modulus (*K*) is equal to bearing characteristic number when for a full lubricated journal bearing the amount of friction is

(*a*) maximum ☐ (*b*) minimum ☐

(*c*) zero ☐ (*d*) half of maximum. ☐

332. Choose the wrong statement

(*a*) The bearing should be designed for a value of $\frac{ZN}{p}$ at least three times the value of bearing modulus (*K*). ☐

(*b*) In a boundary lubricated bearing, there is a thin film of lubricant between the journal and the bearing. ☐

(*c*) The bearing characteristic number is a dimensionless number. ☐

(*d*) A shaft rotating in anticlockwise direction as slow speed inside a bearing will be towards right side of the bearing and making no metal to metal contact. ☐

333. If the bearing is subjected to large fluctuations of load and heavy impacts, the bearing is designed for bearing characteristic number equal to

(*a*) bearing modulus ☐ (*b*) 4 times the bearing modulus ☐

(*c*) 10 times the bearing modulus ☐ (*d*) 15 times the hearing modulus. ☐

334. The heat generated in a bearing due to fluid friction and due to friction of parts having relative motion is given by

(a) $\frac{\mu WV}{2J}$ ☐ (b) $\frac{\mu WV}{4J}$ ☐

(c) $\frac{\mu WV}{J}$ ☐ (d) $\frac{4\mu WV}{J}$ ☐

where μ = Co-efficient of friction, W = Load on bearing,

V = Rubbing velocity, and J = Mechanical equivalent of heat.

335. A journal diameter 15 cm is rotating at a speed of 900 r.p.m. in a bearing of length 20 cm. The absolute viscosity of lubricating oil is 16 centipoise and load on the bearing is 3000 kg. The bearing pressure (p) will be equal to

(a) 20 kg/cm^2 ☐ (b) 10 kg/cm^2 ☐

(c) 5 kg/cm^2 ☐ (d) 4 kg/cm^2. ☐

336. In question 335, the bearing characteristic number will be equal to

(a) 2000 ☐ (b) 1600 ☐

(c) 1440 ☐ (d) 1000. ☐

337. In journal bearing, the load acts

(a) perpendicular to the axis of shaft ☐ (b) along the axis of the shaft ☐

(c) parallel to the axis of the shaft ☐ (d) none of the above. ☐

338. In thrust bearing, the load acts

(a) perpendicular to the axis of shaft ☐ (b) along the axis of the shaft ☐

(c) parallel to the axis of the shaft ☐ (d) none of the above. ☐

339. In case of thrust bearing, the shaft is generally counter-bored at the end. For a counter-bored shaft, the uniform bearing pressure is equal to

(a) $\frac{W}{\pi R^2}$ ☐ (b) $\frac{W}{\pi(R^2 - r^2)}$ ☐

(c) $\frac{W}{\pi r^2}$ ☐ (d) $\frac{W}{2\pi R \times r}$. ☐

where W = Loading supported by bearing, R = Radius of the bearing surface, and

r = Radius of counter-bore.

340. For a counter-bored shaft at the end and supported by a footstep bearing, the torque transmitted is equal to (assuming uniform bearing pressure)

(a) $\frac{2}{3}\,\mu WR$ ☐ (b) $\frac{2}{3}\,\mu W\frac{(R^2 - r^2)}{(R - r)}$ ☐

(c) $\frac{2}{3}\,\mu W\frac{(R^3 - r^3)}{(R^2 - r^2)}$ ☐ (d) $\frac{2}{3}\,\mu W\frac{(R^4 - r^4)}{(R^3 - r^3)}$. ☐

341. A shaft of diameter 20 cm is supported in a foot-step bearing. The load transmitted to the bearing is 2860 kg and the shaft is counter-bored with a hole of 6 cm, at the end. The value of the pressure, which is assumed to be uniformly distributed, will be equal to

(*a*) 20 kg/cm^2 ☐ (*b*) 10 kg/cm^2 ☐
(*c*) 8 kg/cm^2 ☐ (*d*) 5 kg/cm^2. ☐

342. In the above question, if $\mu = 0.015$, then fictional torque will be equal to

(*a*) 10 kg-m ☐ (*b*) 5 kg-m ☐
(*c*) 4 kg-m ☐ (*d*) 3.06 kg-m. ☐

343. Which of the following is an anti-friction bearing?

(*a*) gas lubricated bearings ☐ (*b*) ball and roller bearings ☐
(*c*) plastic bearings ☐ (*d*) collar bearing. ☐

344. The bearings are designated by a number, generally consisting of at least three digits such as 205, 305, 405. The last three digits of the bearing number represents

(*a*) bore of the bearing ☐ (*b*) series of the bearing ☐
(*c*) width of bearing ☐ (*d*) bore and series of the bearing. ☐

345. In question 344, the third from the last digit represents

(*a*) bore of the bearing ☐ (*b*) series of the bearing ☐
(*c*) width of bearing ☐ (*d*) bore and series of the bearing. ☐

346. In question 344, the last two digits give bore diameter in millimetre when

(*a*) multiplied by 4 ☐ (*b*) multiplied by 5 ☐
(*c*) added to 10 ☐ (*d*) added to 20. ☐

347. The bearing number 406 means that the bearing is of

(*a*) heavy series whose bore is 6 mm ☐ (*b*) heavy series whose bore is 30 mm ☐
(*c*) light series whose bore is 30 mm ☐ (*d*) medium series whose bore is 30 mm. ☐

348. Which part is made harder than others in case of ball bearings?

(*a*) outer race ☐ (*b*) inner race ☐
(*c*) ball ☐ (*d*) all are equally hard. ☐

349. Which one is a wrong statement?

(*a*) The bearings of medium series have capacity 30 to 40 per cent over the light series. ☐
(*b*) The bearings of heavy series have capacity 20 to 30 per cent over medium series. ☐
(*c*) The bearing number 205 means that the bearing is of light series whose bore is 25 mm. ☐
(*d*) In thrust bearing load acts perpendicular to the axis of the shaft. ☐

350. In journal bearing, the oil should be applied at point where load is

(*a*) maximum ☐ (*b*) minimum ☐
(*c*) average ☐ (*d*) no need of oil. ☐

351. A connecting rod having length 32 cm is to be designed for a petrol engine. If the diameter of piston = 14 cm, weight of reciprocating parts = 2 kg, crank radius = 8 cm, angular speed of crank = 100 rad/sec and maximum explosive pressure is 200 N/cm^2, then maximum force on the piston due to pressure will be

(*a*) 3080 N ☐ (*b*) 30800 N ☐
(*c*) 15400 N ☐ (*d*) 7700 N. ☐

352. In question 351, the inertia force of the reciprocating parts when the crank is at inner dead centre (*i.e.*, when $\theta = 0°$) will be equal to

(*a*) 2000 N ☐ (*b*) 2500 N ☐
(*c*) 3000 N ☐ (*d*) 1600 N. ☐

353. In question 351, the net force acting on the gudgeon pin will be equal to

(*a*) 1080 N ☐ (*b*) 5080 N ☐
(*c*) 32800 N ☐ (*d*) 28800 N. ☐

354. In question 351, the force acting along the connecting rod at the inner dead centre will be equal to

(*a*) 30800 N ☐ (*b*) 2000 N ☐
(*c*) 32800 N ☐ (*d*) 28800 N. ☐

355. In question 351, the connecting rod will be designed for an axial force of

(*a*) 30800 N ☐ (*b*) 2000 N ☐
(*c*) 32800 N ☐ (*d*) 28800 N. ☐

356. A connecting rod is designed as a

(*a*) short column ☐ (*b*) long column ☐
(*c*) strut ☐ (*d*) none of the above. ☐

357. Due to centripetal action, bending stress will be induced in a connecting rod. The resultant normal force acting on the connecting rod is equal to (if connecting rod is assumed of uniform cross-section)

(*a*) $\frac{wl}{2g}\,\omega^2 r$ ☐ (*b*) $\frac{wl}{g}\,\omega^2 r$ ☐
(*c*) $\frac{2wl}{g}\,\omega^2 r$ ☐ (*d*) $\frac{4wl}{g}\,\omega^2 r$. ☐

where w = Weight of connecting rod per unit length, l = Length of connecting rod, ω = Angular velocity of crank, and r = Radius of crank.

358. In question 357, the resultant normal force acts at a distance of

(*a*) $\frac{l}{3}$ from the gudgeon pin ☐ (*b*) $\frac{2l}{3}$ from the gudgeon pin ☐
(*c*) $\frac{l}{2}$ from the gudgeon pin ☐ (*d*) $\frac{2l}{5}$ from the gudgeon pin ☐

359. In question 357, the maximum bending moment on the connecting rod is equal to

(*a*) $\frac{W}{g}\,\omega^2 r \times \frac{l}{9}$ ☐ (*b*) $\frac{W}{g}\,\omega^2 r \times \frac{1}{9\sqrt{3}}$ ☐
(*c*) $\frac{W}{2g}\,\omega^2 r \times \frac{1}{9\sqrt{3}}$ ☐ (*d*) $\frac{W}{2g}\,\omega^2 r \times \frac{l}{9}$ ☐

where $W = w \times l$.

360. In question 359, the maximum bending moment acts at a distance of

(*a*) $\frac{l}{3}$ from gudgeon pin ☐ (*b*) $\frac{l}{\sqrt{3}}$ from the gudgeon pin ☐

(*c*) $\frac{2l}{3}$ from gudgeon pin ☐ (*d*) $\frac{l}{2}$ from the gudgeon pin. ☐

361. A connecting rod is shown in Fig. 10.14. For buckling in the plane of motion of the rod (*i.e.*, buckling about the neutral axis *X-X*), the rod is considered as

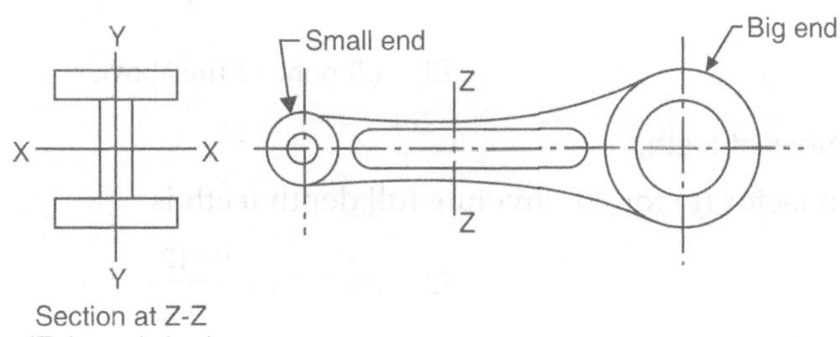

FIGURE 10.14

(*a*) fixed at both ends ☐ (*b*) hinged at both ends ☐

(*c*) fixed at small end and hinged at big end ☐ (*d*) fixed at big end and hinged at small end. ☐

362. In questions 361, for buckling of the connecting rod in the plane perpendicular to the plane of motion (*i.e.*, about the axis *Y*-Y), the rod is considered as

(*a*) fixed at both ends ☐ (*b*) hinged at both ends ☐

(*c*) fixed at small end and hinged at big end ☐ (*d*) fixed at big end and hinged at small end. ☐

363. The connecting rod will be equally strong in buckling about *X*-axis and *Y*-axis, if

(*a*) $I_{XX} = I_{YY}$ ☐ (*b*) $I_{XX} = 2I_{YY}$ ☐

(*c*) $I_{XX} = 3I_{YY}$ ☐ (*d*) $I_{XX} = 4I_{YY}$. ☐

where I_{XX} = Moment of inertia about *X*-axis, and I_{YY} = Moment of inertia about *Y*-axis.

364. If in question 363, moment of inertia about *X*-axis is more than four times the moment of inertia about *Y*-axis, then buckling will

(*a*) not occur about any axis ☐ (*b*) occur about *X*-axis ☐

(*c*) occur about *Y*-axis ☐ (*d*) none of the above. ☐

365. If in question 363, moment of inertia about *X*-axis is less than four limes the moment of inertia about *Y*-axis, then buckling will

(*a*) not occur about any axis ☐ (*b*) occur about *X*-axis ☐

(*c*) occur about *Y*-axis ☐ (*d*) none of the above. ☐

366. The beam strength of the tooth is given by Lewis equation, which is equal to

(*a*) $b \times p_c \times f_1 \times y$ ☐ (*b*) $\frac{b \times p_c \times y}{f_1}$ ☐

(*c*) $\frac{p_c \times b \times y}{f_1}$ ☐ (*d*) $\frac{p_c \times f_1}{b \times y}$. ☐

where p_c = Circular pitch, b = Width of face of the tooth,

y = Lewis form factor or tooth form factor, and f_1 = Permissible working stress.

367. The Lewis form factor (y) depends upon

(*a*) size of the tooth ☐ (*b*) number of teeth of a gear ☐

(*c*) pressure angle ☐ (*d*) all of the above ☐

(*e*) (*b*) and (*c*) only. ☐

368. The Lewis form factor (y) for 20° stub involute teeth is

(*a*) $0.124 - \frac{0.648}{n}$ ☐ (*b*) $0.154 - \frac{0.912}{n}$ ☐

(*c*) $0.175 - \frac{0.841}{n}$ ☐ (*d*) none of the above. ☐

where n = Number of teeth.

369. The Lewis from factor (y) for 20° involute full depth teeth is

(*a*) $0.124 - \frac{0.648}{n}$ ☐ (*b*) $0.154 - \frac{0.912}{n}$ ☐

(*c*) $0.175 - \frac{0.841}{n}$ ☐ (*d*) none of the above. ☐

370. The permissible working stress (f_1) in the Lewis equation depends upon

(*a*) the material of the teeth ☐ (*b*) pitch line velocity ☐

(*c*) load condition ☐ (*d*) all of the above. ☐

371. The permissible working stress (f_1) in the Lewis equation

(*a*) is increased with the increase of pitch line velocity ☐

(*b*) is decreased with the increase of pitch line velocity ☐

(*c*) is independent of pitch line velocity ☐

(*d*) none of the above. ☐

Gears

372. The velocity factor (C_v) for ordinary commercial gears operating at velocities upto 760 m/min is equal to

(*a*) $\frac{183}{183 + V}$ ☐ (*b*) $\frac{3}{3 + V}$ ☐

(*c*) $\frac{4.5}{4.5 + V}$ ☐ (*d*) $\frac{366}{366 + V}$. ☐

where V = Pitch line velocity in m/min.

373. The velocity factor (C_v) for carefully cut gears operating at velocities upto 760 m/min, is equal to

(*a*) $\frac{183}{183 + V}$ ☐ (*b*) $\frac{3}{3 + V}$ ☐

(*c*) $\frac{4.5}{4.5 + V}$ ☐ (*d*) $\frac{275}{275 + V}$. ☐

374. The velocity factor (C_v) for very accurately cut and ground metallic gears operating at velocities upto 1220 m/min, is equal to

(a) $\frac{6}{6+V}$ ☐ (b) $\frac{366}{366+V}$ ☐

(c) $\frac{43}{43+\sqrt{V}}$ ☐ (d) $\frac{0.75}{0.75+\sqrt{V}}$. ☐

375. The velocity factor (C_v) for hardened steel ground and high procession gears operating at velocities upto 1220 m/min, is equal to

(a) $\frac{6}{6+V}$ ☐ (b) $\frac{366}{366+V}$ ☐

(c) $\frac{43}{43+\sqrt{V}}$ ☐ (d) $\frac{0.75}{0.75+\sqrt{V}}$. ☐

where V = Pitch line velocity in m/min.

376. The velocity factor (C_v) for non-metallic gears operating upto 760 m/min, is equal to

(a) $\frac{45.75}{61+V}+0.25$ ☐ (b) $\frac{0.75}{1+V}+0.25$ ☐

(c) $\frac{43}{43+\sqrt{V}}$ ☐ (d) $\frac{0.75}{0.75+\sqrt{V}}$. ☐

where W = Pitch line velocity in m/min.

377. The allowable static stress for steel gears is approximately

(a) one-hall of the ultimate tensile stress ☐ (b) one-third of the ultimate tensile stress ☐

(c) two-third of the ultimate tensile stress ☐ (d) one-fourth of the ultimate tensile stress. ☐

378. The power transmitted by a tooth in kW is given by

(a) $\frac{W \times V}{1000}$ ☐ (b) $\frac{W \times 75}{V}$ ☐

(c) $\frac{W \times V}{4500}$ ☐ (d) $\frac{W \times V}{550}$. ☐

where W = Tangential tooth load, V = Pitch line velocity in m/sec.

379. When the pinion and gear are made of the same material, then

(a) pinion is stronger ☐ (b) pinion is weaker ☐

(c) both are of equal strength ☐ (d) none of the above. ☐

380. The Lewis equation is applied to

(a) pinion ☐ (b) gear ☐

(c) stronger of pinion or gear ☐ (d) weaker of pinion or gear. ☐

381. When the pinion and the gear are made of different materials then the Lewis equation is applied to

(a) pinion or gear ☐

(b) the wheel which has the product of working stress and tooth factor more ☐

(c) the wheel which has the product of working stress and tooth factor less ☐

(d) none of the above. ☐

382. Best profile of gear to withstand resistance to wear is

(*a*) 20° full depth involute teeth ☐ (*b*) 20° involute stub teeth ☐

(*c*) $14\frac{1}{2}$° full depth involute teeth ☐ (*d*) none of the above. ☐

383. The backlash for spur gears depend upon

(*a*) tooth profile ☐ (*b*) module ☐

(*c*) pitch line velocity ☐ (*d*) all of the above ☐

(*e*) both (*a*) and (*b*) ☐ (*f*) both (*b*) and (*c*). ☐

384. Choose the correct statement

(*a*) The product of the diametral pitch and circular pitch is equal to one. ☐

(*b*) The pressure angle for involute gears depends upon the size of teeth. ☐

(*c*) The normal to the involute, in a gear having involute teeth, is a tangent to base circle. ☐

(*d*) For commercially cut gears, the limiting pitch line velocity is 60 m/min. ☐

385. Choose the wrong statement

(*a*) Low pressure angle gears result in weaker teeth. ☐

(*b*) The form factor used in design of gear depends upon the number of teeth and system of teeth ☐

(*c*) Gear teeth are made harder to avoid pitting. ☐

(*d*) Stub tooth in gears is shorter than standard tooth. ☐

386. In cycloidal teeth, the reference

(*a*) is maximum ☐

(*b*) is minimum ☐

(*c*) is absent completely ☐

(*d*) depends upon pressure angle and number of teeth. ☐

387. The surface of the tooth above the pitch surface is known as

(*a*) dedendum ☐ (*b*) addendum ☐

(*c*) face ☐ (*d*) flank. ☐

388. The surface of the tooth below the pitch surface is known as

(*a*) dedendum ☐ (*b*) addendum ☐

(*c*) face ☐ (*d*) flank. ☐

389. The radial distance from the pitch circle to the bottom of a tooth is known as

(*a*) dedendum ☐ (*b*) addendum ☐

(*c*) face ☐ (*d*) flank. ☐

390. The radial distance from the pitch circle to the top of a tooth is known as

(*a*) dedendum ☐ (*b*) addendum ☐

(*c*) face ☐ (*d*) flank. ☐

391. The gears, used to connect two non-intersecting and non-coplanar shafts, are known as

(*a*) bevel gears ☐ (*b*) spiral gears ☐

(*c*) helical gears ☐ (*d*) spur gears. ☐

392. In watches and clocks, the type of tooth profile used for gears, is

(*a*) cycloidal ☐ (*b*) involute ☐
(c) epicycloid ☐ (*d*) hypocycloid. ☐

393. The type of gear used to keep noise minimum, should be

(*a*) cycloidal ☐ (*b*) involute ☐
(*c*) bevel ☐ (*d*) helical. ☐

394. The materials used for gears to run quietly at high speed, should be

(*a*) non-ferrous ☐ (*b*) harder steel ☐
(*c*) stainless steel ☐ (*d*) non-metallic. ☐

395. The amount by which a tooth space exceeds the thickness of an engaging teeth, is called

(*a*) clearance ☐ (*b*) backlash ☐
(*c*) module ☐ (*d*) working depth. ☐

396. In the Lewis equation, $W_t = f \times p_c \times y \times b$, the strength factor is given by the product of

(*a*) $f \times y$ ☐ (*b*) $f \times b$ ☐
(*c*) $p_c \times y$ ☐ (*d*) $p_c \times b$ ☐

where f = Permissible working stress, y = Lewis form factor,
b = Face width, and p_c = Circular pitch.

397. In a corrosive atmosphere, the material for the gear should be

(*a*) brass and bronze ☐ (*b*) steel ☐
(*c*) cast iron ☐ (*d*) none of the above. ☐

398. The power from one shaft to another shaft is transmitted by toothed gears under the condition of

(*a*) constant velocity ratio ☐ (*b*) short distance between the two shafts ☐
(*c*) low speed and high torque ☐ (*d*) all of the above ☐
(*e*) none of the above. ☐

399. Stub tooth is

(*a*) longer than standard tooth depth ☐ (*b*) shorter than standard tooth depth ☐
(*c*) standard gear tooth ☐ (*d*) none of the above. ☐

400. In an automobile, the function of gear box is to

(*a*) increase speed ☐ (*b*) decrease speed ☐
(*c*) provided, variable speed ☐ (*d*) produce torque. ☐

401. For 20° pressure angle teeth, the interference can be avoided if minimum number of teeth is

(*a*) 4 ☐ (*b*) 8 ☐
(*c*) 16 ☐ (*d*) more than 24. ☐

402. Stress concentration factor in case of gears depends upon

(*a*) the thickness of tooth at the root ☐ (*b*) the material of the gear ☐
(*c*) the load position on the tooth ☐ (*d*) the pressure angle ☐
(*e*) fillet radius ☐ (*f*) all of the above. ☐

403. Choose the wrong statement

(*a*) In a corrosive atmosphere, the material for the gear should be brass and bronze. ☐

(*b*) Surface endurance limit of gear material is dependent upon its brinell hardness number. ☐

(*c*) To keep noise minimum, helical gear should be used. ☐

(*d*) In helical gears, the right hand helix will mesh right hand helix. ☐

404. Bevel gears are used for transmitting motion from one shaft to the other when the two shafts are

(*a*) parallel and non-coplanar ☐

(*b*) parallel and coplanar ☐

(*c*) non-parallel and non-coplanar ☐

(*d*) non-parallel and coplanar. ☐

405. In case of bevel gearing, the pitch angle of the pinion is given by

(*a*) $\tan^{-1}\dfrac{\cos\theta}{\left(\dfrac{N_g}{N_p}+\sin\theta\right)}$ ☐ (*b*) $\tan^{-1}\dfrac{\sin\theta}{\left(\dfrac{N_g}{N_p}\right)+\cos\theta}$ ☐

(*c*) $\tan^{-1}\dfrac{\sin\theta}{\left(\dfrac{N_p}{N_g}+\cos\theta\right)}$ ☐ (*d*) $\tan^{-1}\dfrac{\cos\theta}{\left(\dfrac{N_g}{N_p}+1\right)}$ ☐

where θ = Shaft angle, and N_g and N_p = Number of teeth on gears and pinions.

406. In case of bevel gearing, the pitch angle of the gear is given by

(*a*) $\tan^{-1}\dfrac{\cos\theta}{\left(\dfrac{N_g}{N_p}+\sin\theta\right)}$ ☐ (*b*) $\tan^{-1}\dfrac{\sin\theta}{\left(\dfrac{N_g}{N_p}\right)+\cos\theta}$ ☐

(*c*) $\tan^{-1}\dfrac{\sin\theta}{\left(\dfrac{N_p}{N_g}+\cos\theta\right)}$ ☐ (*d*) $\tan^{-1}\dfrac{\cos\theta}{\left(\dfrac{N_g}{N_p}+1\right)}$ ☐

407. Formative number of teeth on a bevel gear are the number of teeth on the gear whose radius is

(*a*) more than the back cone radius ☐ (*b*) half the back cone radius ☐

(*c*) one-fourth of the cone radius ☐ (*d*) equal to the back cone radius. ☐

408. The formative (or virtual) number of teeth on a bevel gear is given by

(*a*) $\dfrac{N_p \times L}{r_p}$ ☐ (*b*) $\dfrac{N_g \times L}{r_g}$ ☐

(*c*) $\dfrac{N_p \times L}{r_g}$ ☐ (*d*) $\dfrac{N_g \times L}{r_p}$ ☐

(*e*) both (*a*) and (*b*) ☐ (*f*) both (*c*) and (*d*). ☐

where N_p and N_g = Number of teeth on pinion and gear respectively, and

r_p and r_g = Radius of pinion and gear respectively.

409. The bevel factor in terms of cone distance (L) and face width (b) of a bevel gear is given by

(*a*) $\frac{L}{b}$ ☐ (*b*) $\frac{L-b}{L}$ ☐

(*c*) $\frac{L-b}{b}$ ☐ (*d*) $\frac{b}{L}$. ☐

410. The general value of bevel factor is

(*a*) 2.0 ☐ (*b*) 1.5 ☐

(*c*) 1.0 ☐ (*d*) 0.75. ☐

411. If the shaft angle = 90° for a bevel gear arrangement, then cone distance (L) is equal to

(*a*) $\sqrt{D_p^2 + D_g^2}$ ☐ (*b*) $\frac{1}{2}\sqrt{D_p^2 + D_g^2}$ ☐

(*c*) $\frac{D_p + D_g}{2}$ ☐ (*d*) $\frac{D_p - D_g}{2}$. ☐

412. Worm gearing is used to obtain considerable speed reduction between shafts whose axes are

(*a*) parallel ☐ (*b*) perpendicular and intersect ☐

(*c*) perpendicular and do not intersect ☐ (*d*) inclined. ☐

413. Crown bevel gears are the bevel gears which connect two shafts whose axes intersect

(*a*) at right angles ☐

(*b*) at an angle greater than 90° and one of the bevel gears has at pitch angle of 90° ☐

(*c*) at 45° angles ☐

(*d*) none of the above. ☐

414. Mitre gears arc the bevel gears which connect two shafts whose axes intersect

(*a*) at right angles ☐

(*b*) at right angles and gears have equal teeth and equal pitch angle ☐

(*c*) at 45° angles ☐

(*d*) none of the above. ☐

415. The cutting angle or root angle of the pinion of a bevel gear is equal to

(*a*) $\alpha - \tan^{-1}\frac{\text{dedendum}}{\text{cone distance}}$ ☐ (*b*) $\alpha + \tan^{-1}\frac{\text{dedendum}}{\text{cone distance}}$ ☐

(*c*) $\beta - \tan^{-1}\frac{\text{dedendum}}{\text{cone distance}}$ ☐ (*d*) $\beta + \tan^{-1}\frac{\text{dedendum}}{\text{cone distance}}$. ☐

where α = Pitch angle of pinion, and β = Pitch angle of gear.

416. The root angle of the gear of the bevel gears is equal to

(*a*) $\alpha - \tan^{-1}\frac{\text{dedendum}}{\text{cone distance}}$ ☐ (*b*) $\alpha + \tan^{-1}\frac{\text{dedendum}}{\text{cone distance}}$ ☐

(*c*) $\beta - \tan^{-1}\frac{\text{dedendum}}{\text{cone distance}}$ ☐ (*d*) $\beta + \tan^{-1}\frac{\text{dedendum}}{\text{cone distance}}$. ☐

417. The face angle of the pinion of a bevel gear is equal to

(a) $\alpha - \tan^{-1} \dfrac{\text{dedendum}}{\text{cone distance}}$ ☐ (b) $\alpha + \tan^{-1} \dfrac{\text{dedendum}}{\text{cone distance}}$ ☐

(c) $\beta - \tan^{-1} \dfrac{\text{dedendum}}{\text{cone distance}}$ ☐ (d) $\beta + \tan^{-1} \dfrac{\text{dedendum}}{\text{cone distance}}$. ☐

418. The face angle of the gear of the bevel gears is equal to

(a) $\alpha - \tan^{-1} \dfrac{\text{dedendum}}{\text{cone distance}}$ ☐ (b) $\alpha + \tan^{-1} \dfrac{\text{dedendum}}{\text{cone distance}}$ ☐

(c) $\beta - \tan^{-1} \dfrac{\text{dedendum}}{\text{cone distance}}$ ☐ (d) $\beta + \tan^{-1} \dfrac{\text{dedendum}}{\text{cone distance}}$. ☐

419. Lead of a worm is the distance from any point on one thread to the corresponding point on

(a) the adjacent thread measured parallel to the worm axis ☐

(b) the next turn of the same thread measured parallel to the worm axis ☐

(c) the adjacent thread measured parallel to the worm axis divided by two ☐

(d) none of the above. ☐

420. The ratio of the lead of the worm to the pitch circumference of the worm is equal to the

(a) sine of lead angle ☐ (b) tangent of lead angle ☐

(c) tangent of root angle ☐ (d) tangent of face angle. ☐

421. The relationship between minimum pitch diameter for worms (d_w) cut on a shaft and centre distance (C) is given by

(a) $d_w = \dfrac{C^{0.875}}{1.96}$ ☐ (b) $d_w = 1.96C^{0.875}$ ☐

(c) $d_w = \dfrac{1.96}{C^{0.876}}$ ☐ (d) $\dfrac{1}{1.96C^{0.876}}$. ☐

422. The relationship between centre distance between two shafts (connected by worm gears), diameter of worm wheel and diameter of gear wheel is given by

(a) $C = 2(d_w + d_g)$ ☐ (b) $C = \dfrac{d_w + d_g}{2}$ ☐

(c) $C = d_w + d_g$ ☐ (d) $C = \dfrac{d_w + d_g}{4}$ ☐

where d_w = Diameter of worm wheel, and d_g = Diameter of gear wheel.

423. In the velocity ratio for the worm gear is 20 or more than 20, then the threads on the worm should be

(a) single start ☐ (b) double start ☐

(c) triple start ☐ (d) could be any start. ☐

424. A three-threaded worm rotating at 1000 r.p.m. drives a 30 tooth-worm gear and transmits 10 H.P. The worm has $14\frac{1}{2}$ degree teeth with 5 mm module and 5 cm pich diameter. The velocity ratio would be equal to

(a) 20 ☐ (b) 10 ☐

(c) 15 ☐ (d) 8. ☐

425. In question 424, the diameter of the worm gear would be equal to

(*a*) 20 cm ☐ (*b*) 10 cm ☐
(*c*) 15 cm ☐ (*d*) 8 cm. ☐

426. In question 424, the centre distance between the shafts would be

(*a*) 20 cm ☐ (*b*) 10 cm ☐
(*c*) 15 cm ☐ (*d*) 8 cm. ☐

427. In question 424, the tangential velocity of the worm gear would be

(*a*) 10π m/min ☐ (*b*) 15π m/min ☐
(*c*) 5π m/min ☐ (*d*) 2π m/min. ☐

428. In question 424, the force transmitted would be equal to

(*a*) $\frac{3000}{\pi}$ kgf ☐ (*b*) $\frac{2000}{\pi}$ kgf ☐
(*c*) $\frac{1500}{\pi}$ kgf ☐ (*d*) $\frac{1000}{\pi}$ kgf. ☐

429. For high speed engine, the type of cam used should be

(*a*) flat ☐ (*b*) involute ☐
(*c*) cycloidal ☐ (*d*) simple harmonic ☐
(*e*) uniform acceleration and retardation. ☐

430. For low and moderate speed engines, the type of cam used should be

(*a*) flat ☐ (*b*) involute ☐
(*c*) cycloidal ☐ (*d*) simple harmonic ☐
(*e*) uniform acceleration and retardation. ☐

431. The type of cam used, in order to minimise jerks, is

(*a*) flat ☐ (*b*) involute ☐
(*c*) cycloidal ☐ (*d*) simple harmonic ☐

432. The type of cam follower used, in automobiles, is

(*a*) roller type ☐ (*b*) spherical type ☐
(*c*) knife-edge type ☐ (*d*) flat-type. ☐

433. A point on the pitch curve of a cam, having maximum pressure angle is known as

(*a*) trace point ☐ (*b*) centre of rotation of cam ☐
(*c*) pitch point ☐ (*d*) none of the above. ☐

434. The period of dwell is the period during which

(*a*) cam is at rest ☐ (*b*) cam is moving ☐
(*c*) cam is moving but cam follower is at rest ☐ (*d*) none of the above. ☐

435. Choose the correct statement

(*a*) The tangent of lead angle is equal to the ratio of the lead of the worm to the pitch diameter of the worm. ☐
(*b*) The size of the cam depends upon pitch circle. ☐
(*c*) Non-metallic materials should be used for gears to run quietly at high speed. ☐
(*d*) To keep noise minimum, cycloidal spur gear should be used. ☐

436. A worm gear set with worm gear of 100 teeth and worm of double lead thread would have gear reduction ratio of

(*a*) 25 : 1 ☐ (*b*) 50 : 1 ☐
(*c*) 75 : 1 ☐ (*d*) 100 : 1. ☐

437. Two bevel gears are in mesh. They have the number of teeth as 40 and 80, then the cone pitch angle of the pinion would be

(*a*) $\sin^{-1} 1.0$ ☐ (*b*) $\sin^{-1} 0.5$ ☐
(*c*) $\tan^{-1} 0.5$ ☐ (*d*) $\cos^{-1} 0.5$. ☐

438. In question 437, the cone pitch angle of the gear would be

(*a*) $\frac{\pi}{4} - \tan^{-1} 0.5$ ☐ (*b*) $\frac{\pi}{2} - \tan^{-1} 0.5$ ☐
(*c*) $\frac{\pi}{2} - \sin^{-1} 0.5$ ☐ (*d*) $\frac{\pi}{4} - \cos^{-1} 0.5$. ☐

439. For minimum axial thrust, the type of gear used should be

(*a*) worm gears ☐ (*b*) helical gears ☐
(*c*) herringbone gears ☐ (*d*) hypoid gears. ☐

440. The hunting phenomenon is observed in

(*a*) d.c. shunt motors ☐ (*b*) synchronous motor ☐
(*c*) induction motor ☐ (*d*) d.c. series motor. ☐

Lubricants

441. Lubricants are used in bearings

(*a*) to protect bearings against corrosion ☐
(*b*) to reduce friction between the rubbing surfaces ☐
(*c*) to carry away the heat generated ☐
(*d*) all of the above ☐
(*e*) none of the above. ☐

442. The lubricants are only

(*a*) a liquid ☐ (*b*) a semi-liquid ☐
(*c*) a solid ☐ (*d*) any one of the above. ☐
(*e*) none of the above. ☐

443. Grease is a

(*a*) liquid lubricant ☐ (*b*) solid lubricant ☐
(*c*) semi-liquid lubricant ☐ (*d*) none of the above. ☐

444. Where slow speed and heavy pressure exist, the lubricant used should be a

(*a*) liquid lubricant ☐ (*b*) solid lubricant ☐
(*c*) semi-liquid lubricant ☐ (*d*) any one of the above. ☐

445. A graphite is a

(*a*) solid lubricant ☐ (*b*) liquid lubricant ☐
(*c*) semi-liquid lubricant ☐ (*d*) none of the above. ☐

446. The factor which is considered in the design of bearing is

(*a*) specific pressure ☐ (*b*) rubbing speed ☐
(*c*) viscosity of lubricant ☐ (*d*) heat dissipation ☐
(*e*) all of the above ☐ (*f*) none of the above. ☐

447. The load on the bearing is equal to

(*a*) $\frac{\pi}{4} d^2 \times p$ ☐ (*b*) $\pi d \times l \times p$ ☐
(*c*) $p \times d \times l$ ☐ (*d*) $2 \times p \times d \times l$ ☐

where p = Bearing pressure, l = Length of bearing, and d = Diameter of the journal.

448. The heat generated in a journal bearing in kcal/min is equal to

(*a*) $\frac{f \times p \times V}{J}$ ☐ (*b*) $\frac{f \times P \times V}{J}$ ☐
(*c*) $\frac{f \times l \times d \times V}{J}$ ☐ (*d*) $\frac{f \times V \times d}{J}$ ☐

where p = Bearing pressure, P = Load on the bearing in kgf,
V = Rubbing in velocity in m/min, l = Length of bearing,
d = Diameter of journal, and J = Mechanical equivalent of heat.

449. The heat dissipated by the bearing is equal to

(*a*) $C \times A \times t_b$ ☐ (*b*) $C \times A \times t_a$ ☐
(*c*) $C \times A \times (t_b + t_a)$ ☐ (*d*) $C \times A \times (t_b - t_a)$ ☐

where C = Heat dissipation co-efficient, A = Projected area of the bearing,
t_b = Temperature of the bearing surfaces in °C, and
t_a = Temperature of the surrounding air in °C.

450. For satisfactory working of the bearing, the heat dissipated (in comparison to heat generated) should be

(*a*) less ☐ (*b*) more ☐
(*c*) equal ☐ (*d*) none of the above. ☐

451. The bearing should be cooled artificially by water if heat generated (in comparison to heat dissipated) is

(*a*) less ☐ (*b*) more ☐
(*c*) equal ☐ (*d*) none of the above. ☐

452. Heat taken away by the oil is equal to

(*a*) $m \times t$ ☐ (*b*) $m \times s \times t$ ☐
(*c*) $m \times (s + 1) \times t$ ☐ (*d*) $m \times (s - 1) \times t$ ☐

where m = Mass of oil, s = Specific heat of oil, and
t = Difference between outlet and inlet temperature of the oil in °C.

453. The bearing characteristic number is given by

(a) $\frac{Z \times p}{N}$ ☐ (b) $\frac{Z \times N}{p}$ ☐

(c) $\frac{p}{Z \times N}$ ☐ (d) $Z \times N \times p$ ☐

where Z = Absolute viscosity of lubricant in centipoise,

N = Speed, and p = Pressure in W/mm^2

454. Bearing modulus is the value of $\frac{ZN}{p}$ at which the amount of friction is

(a) maximum ☐ (b) minimum ☐

(c) average ☐ (d) none of the above. ☐

455. The bearing should be designed for a value of $\frac{ZN}{p}$ which is having a value

(a) equal to the value of bearing modulus ☐ (b) half the value of bearing modulus ☐

(c) three times the value of bearing modulus ☐ (d) none of the above. ☐

456. For heavy impacts and for large fluctuations of load on the bearing, the value of $\frac{ZN}{p}$ should be

(a) K ☐ (b) $3K$ ☐

(c) $15K$ ☐ (d) $30K$ ☐

where K = Bearing modulus.

457. The Sommerfield number is equal to

(a) $\frac{ZN}{p}$ ☐ (b) $\frac{ZN}{p} \times \left(\frac{d}{c}\right)$ ☐

(c) $\frac{ZN}{p} \times \left(\frac{d}{c}\right)^2$ ☐ (d) $\frac{ZN}{p} \times \left(\frac{d}{c}\right)^3$ ☐

where d = Diameter of the journal, and

c = Difference between the diameter of bushing and diameter of journal.

458. For design purposes, the value of Sommerfield number should be equal to

(a) 10^5 ☐ (b) 1.43×10^7 ☐

(c) 1.43×10^9 ☐ (d) 1.43×10^{15}. ☐

459. The diametral clearance ratio is equal to

(a) $\frac{l}{d}$ ☐ (b) $\frac{c}{d}$ ☐

(c) $\frac{l}{\pi d}$ ☐ (d) $\frac{c}{\pi d}$ ☐

where c = Difference between the diameter of bushing and diameter of journal, and

d = Diameter of the journal.

460. The minimum thickness of the oil film in bearing should be

(*a*) c ☐ (*b*) $\frac{c}{2}$ ☐

(*c*) $\frac{c}{4}$ ☐ (*d*) $\frac{c}{8}$ ☐

where c = Clearance between journal and bush.

461. If the length of the journal is equal to the diameter of the journal, then bearing is known as

(*a*) short bearing ☐ (*b*) long bearing ☐

(*c*) square bearing ☐ (*d*) none of the above. ☐

462. The load on a journal bearing is given 4500 kgf. If the diameter of the journal is 15 cm and length of bearing is 30 cm, then the bearing pressure will be equal to

(*a*) 30 kgf/cm^2 ☐ (*b*) 25 kgf/cm^2 ☐

(*c*) 15 kgf/cm^2 ☐ (*d*) 10 kgf/cm^2. ☐

463. The rolling contact bearing is having a

(*a*) high starting friction ☐ (*b*) low starting friction ☐

(*c*) medium starting friction ☐ (*d*) none of the above. ☐

464. Antifriction bearing is another name given to

(*a*) sliding friction ☐ (*b*) rolling contact bearing ☐

(*c*) both of them ☐ (*d*) none of the above. ☐

465. The maintenance cost of rolling contact bearings are

(*a*) low ☐ (*b*) high ☐

(*c*) average ☐ (*d*) none of the above. ☐

466. The initial cost of rolling contact bearings are

(*a*) low ☐ (*b*) high ☐

(*c*) average ☐ (*d*) none of the above. ☐

467. The bearings used for light loads are

(*a*) roller bearings ☐ (*b*) ball bearings ☐

(*c*) any one of the above ☐ (*d*) none of the above. ☐

468. The bearings used for heavier loads are

(*a*) roller bearings ☐ (*b*) ball bearings ☐

(*c*) any one of the above ☐ (*d*) none of the above. ☐

469. A ball bearing is designated by a number 206. The series of the bearings is

(*a*) heavy ☐ (*b*) light ☐

(*c*) medium ☐ (*d*) extra-light. ☐

470. For the above question, the bore of bearing is

(*a*) 2 cm ☐ (*b*) 6 cm ☐

(*c*) 30 cm ☐ (*d*) 30 mm. ☐

471. Which series of the bearing is mostly used?

(*a*) light ☐ (*b*) extra light ☐
(*c*) medium ☐ (*d*) heavy. ☐

472. A ball bearing is designated by a number which consists of at least

(*a*) one digit ☐ (*b*) two digits ☐
(*c*) three digits ☐ (*d*) ten digits. ☐

473. The relation between angle of repose (α), angle of friction (ϕ) and co-efficient of friction (μ) is

(*a*) $\tan \alpha = \sin \phi = \mu$ ☐ (*b*) $\tan \alpha = \tan \phi = \tan \mu$ ☐
(*c*) $\tan \alpha = \tan \phi = \mu$ ☐ (*d*) $\alpha = \phi = \tan \mu$. ☐

474. The angle of helix in a screw jack is given by

(*a*) $\tan \alpha = \dfrac{\pi d}{p}$ ☐ (*b*) $\tan \alpha = \dfrac{p}{\pi d}$ ☐
(*c*) $\cos \alpha = \dfrac{p}{\pi d}$ ☐ (*d*) $\sin \alpha = \dfrac{p}{\pi d}$. ☐

where α = Helix angle, p = Pitch, and d = Mean dia.

475. The expression $\tan \alpha / \tan (\alpha + \phi)$ for a screw jack represents

(*a*) effort to lift a given load W ☐ (*b*) effort required to lower a load ☐
(*c*) efficiency of a screw jack ☐ (*d*) none of the above. ☐

476. The maximum efficiency of a screw jack is given by

(*a*) $\dfrac{1 - \tan \phi}{1 + \tan \phi}$ ☐ (*b*) $\dfrac{1 - \cos \phi}{1 + \cos \phi}$ ☐
(*c*) $\dfrac{1 - \sin \phi}{1 + \sin \phi}$ ☐ (*d*) $\dfrac{1 + \sin \phi}{1 - \sin \phi}$. ☐

where ϕ = Angle of friction.

477. The efficiency of a screw jack

(*a*) increases with the increase of load lifted ☐
(*b*) decreases with the increase of load lifted ☐
(*c*) decreases with the increase of load and effort applied ☐
(*d*) is independent of load lilted and effort applied. ☐

478. The friction circle of a bearing is having radius equal to

(*a*) the radius of shaft ☐ (*b*) $r \cos \phi$ ☐
(*c*) $r \sin \phi$ ☐ (*d*) $r \tan \phi$ ☐

where ϕ = Angle of friction, and r = Radius of shaft.

479. In a flat pivot bearing (considering uniform pressure), the frictional torque transmitted is given by

(*a*) $0.50\ \mu WR$ ☐ (*b*) $0.67\ \mu WR$ ☐
(*c*) $0.75\ \mu WR$ ☐ (*d*) $\mu W \times R$ ☐

where μ = Co-efficient of friction, R = Radius of bearing surface, and

W = Load transmitted over the bearing surface.

480. In a flat pivot bearing (considering uniform wear), the frictional torque transmitted is given by

(a) 0.50 μWR ☐ (b) 0.67 μWR ☐
(c) 0.75 μWR ☐ (d) μW × R. ☐

481. A vertical shaft of dia. 150 mm carries a vertical load of 20 kN and rests on a flat end of a foot-step bearing (*i.e.* flat pivot bearing). The shaft is rotating at 100 r.p.m. and co-efficient of friction is 0.045. Assuming uniform pressure distribution, the frictional torque will be

(a) 25 Nm ☐ (b) 3375 Nm ☐
(c) 45 Nm ☐ (d) 675 Nm. ☐

482. In question 481, the power lost in friction will be

(a) 0.8 π h.p ☐ (b) 0.20 π h.p ☐
(c) 0.15 π h.p ☐ (d) 0.1 π h.p. ☐

483. In question 481, if uniform wear is assumed then the frictional torque will be equal to

(a) 25 Nm ☐ (b) 3375 Nm ☐
(c) 45 Nm ☐ (d) 675 Nm. ☐

484. In a flat pivot bearing with the assumption of uniform pressure, the frictional torque transmitted as compared to be uniform wear will be

(a) more ☐ (b) less ☐
(c) same ☐ (d) none of the above. ☐

485. For a flat collar bearing, assuming uniform pressure, the intensity of pressure is given by

(a) $p = \frac{W}{\pi r_1^2}$ ☐ (b) $p = \frac{W}{\pi r_2^2}$ ☐

(c) $p = \frac{W}{\pi(r_1^2 - r_2^2)}$ ☐ (d) $p = \frac{W}{\pi(r_1^2 + r_2^2)}$. ☐

where W = Vertical load on the bearing, r_1 = External radius of the collar, and
r_2 = Internal radius of the collar.

486. The intensity of pressure, in any bearing, will be reduced by

(a) decreasing the area of rubbing surface ☐
(b) increasing the area of rubbing surface ☐
(c) making area of rubbing surface as constant ☐
(d) none of the above. ☐

487. The intensity of uniform pressure in case of a multi-collared bearing is given by

(a) $p = \frac{W}{n \times \pi r_1^2}$ ☐ (b) $p = \frac{W}{n \times \pi r_2^2}$ ☐

(c) $p = \frac{n \times W}{\pi(r_1^2 - r_2^2)}$ ☐ (d) $p = \frac{W}{n \times \pi(r_1^2 + r_2^2)}$. ☐

488. The total torque transmitted in a multi-collared shaft having *n* collars is equal to

(a) *n* times the frictional torque of a single shaft collared ☐
(b) frictional torque of a single collared shaft divided by number of collars ☐

(c) frictional torque of a single collared shaft ☐
(d) none of the above. ☐

489. For a conical pivot bearing, the frictional torque transmitted (assuming uniform pressure) is equal to

(a) μWL ☐ (b) $\frac{3}{4}\mu WL$ ☐
(c) $\frac{2}{3}\mu WL$ ☐ (d) $\frac{1}{2}\mu WL$. ☐

where L = Slant length of the cone.

490. A conical pivot bearing is subjected to an axial load of 300 kgf. The diameter of vertical shaft is 30 cm. The co-efficient of friction is 0.05 and slant length of cone is 0.3 m. The shaft is rotating at 450 r.p.m. Then for uniform pressure, the total frictional torque transmitted will be

(a) 60 kgf-m ☐ (b) 30 kgf-m ☐
(c) 22.5 kgf-m ☐ (d) 11.25 kgf-m. ☐

491. For question 490, if the uniform wear is assumed then the total frictional torque transmitted will be

(a) 60 kgf-m ☐ (b) 30 kgf-m ☐
(c) 22.5 kgf-m ☐ (d) 11.25 kgf-m. ☐

492. For the question 490, the total horse power lost in friction will be equal to

(a) 6π h.p. ☐ (b) 4.5π h.p. ☐
(c) 3.5π h.p. ☐ (d) 2.25π h.p. ☐

493. For the question 490, the total horse power lost in friction for uniform wear, will be equal to

(a) 6π h.p. ☐ (b) 4.5π h.p. ☐
(c) 3.5π h.p. ☐ (d) 2.25π h.p. ☐

494. The number of pairs of contact surfaces in a disc clutch, in which there are n_1 number of discs on the driving shaft and n_2 number of discs on the driver shaft, will be

(a) $n_1 \times n_2$ ☐ (b) $n_1 + n_2$ ☐
(c) $n_1 + n_2 + 1$ ☐ (d) $n_1 + n_2 - 2$. ☐

495. For a new clutch, generally

(a) pressure is assumed uniform ☐ (b) wear is assumed uniform ☐
(c) pressure and wear are assumed uniform ☐ (d) none of the above. ☐

496. For an old clutch, generally

(a) pressure is assumed uniform ☐ (b) wear is assumed uniform ☐
(c) pressure and wear are assumed uniform ☐ (d) none of the above. ☐

497. The length of a belt in open-belt drive is a function of

(a) difference of the radii of the two pulleys only ☐
(b) sum of the radii of the two pulleys only ☐
(c) sum as well as difference of the radii of the two pulleys ☐
(d) none of the above. ☐

498. The length of a belt in cross-belt drive is a function of

(*a*) difference of the radii of the two pulleys only ☐

(*b*) sum of the radii of the two pulleys only ☐

(*c*) sum as well as difference of the radii of the two pulleys ☐

(*d*) none of the above. ☐

499. Power transmitted by a belt in S.I. unit is given by

(*a*) $T_1 \times V$ Watt ☐ (*b*) $T_2 \times V$ ☐

(*c*) $\frac{(T_1 - T_2) \times V}{1000}$ kW ☐ (*d*) $\frac{(T_1 + T_2) \times V}{1000}$. ☐

where T_1 = Tension on tight side, T_2 = Tension on slack side, V = Velocity in m/sec.

500. The centrifugal tension acting on a rotating bell is given by

(*a*) $T_c = \frac{1}{2} mV^2$ ☐ (*b*) $T_c = mV^2$ ☐

(*c*) $T_c = \frac{2W}{g} V^2$ ☐ (*d*) $T_c = \frac{3W}{g} \times V^2$ ☐

where W= Weight of belt per unit length, V = Linear velocity of belt, and

T_c = Centrifugal tension.

Tick mark the true or false statement:

501. The common normal at the point of contact between a pair of teeth, according to law of gearing, must always pass through the pitch point.

(*a*) True ☐ (*b*) False. ☐

502. The product of the diametrical pitch and circular pitch is equal to one.

(*a*) True ☐ (*b*) False. ☐

503. The module is the reciprocal of diametrical pitch.

(*a*) True ☐ (*b*) False. ☐

504. The load acts perpendicular to the axis of rotation is radial bearings.

(*a*) True ☐ (*b*) False. ☐

505. Zero film bearing is a sliding bearing which operates without any lubricant present.

(*a*) True ☐ (*b*) False. ☐

506. The angle of contact, in a full journal bearing with the journal is 180°.

(*a*) True ☐ (*b*) False. ☐

507. The expression pZ/N represents the bearing characteristic number where p = Bearing pressure on projected bearing area, Z = Absolute viscosity of the lubricant, and N = speed of the journal.

(*a*) True ☐ (*b*) False. ☐

508. There is a thick film of lubricant between the journal and the bearing, in a boundary lubricated bearing.

(*a*) True ☐ (*b*) False. ☐

509. The bearing number 305 means that the bearing is of medium series whose bore is 25 mm.

(a) True ☐ (b) False. ☐

510. The load acts along the axis of rotation in thrust bearing.

(a) True ☐ (b) False. ☐

511. In a hydrodynamic lubricated bearing, there is a thick film of lubricant between the journal and the bearing.

(a) True ☐ (b) False. ☐

512. To store strain energy and to absorb shock and vibrations, a spring is used.

(a) True ☐ (b) False. ☐

513. The stress induced in the spring wire of an open coiled helical compression spring which is subjected to an axial compressive load is shear stress.

(a) True ☐ (b) False. ☐

514. The maximum shear stress induced in the wire of an open coiled helical compression spring is $16WR \times K/\pi d^3$

where R = Mean radius of spring, d = Diameter of wire

W = Axial compressive load, K = Wahl's stress factor.

(a) True ☐ (b) False. ☐

515. The ratio of mean dia. of spring to dia. of wire is known as spring index.

(a) True ☐ (b) False. ☐

516. The expression $(4C - 1)/(4C + 1) + 0.615/C$ represents the Wahl's stress factor for spring where C is spring index.

(a) True ☐ (b) False. ☐

517. On the bases of rigidity and stiffness, a shaft is designed.

(a) True ☐ (b) False. ☐

518. Compression members tends to buckle in the direction of greatest radius of gyration.

(a) True ☐ (b) False. ☐

519. The ratio of column width to column depth is known as slenderness ratio.

(a) True ☐ (b) False. ☐

520. In the design of nut, the most important dimension is its inside diameter.

(a) True ☐ (b) False. ☐

521. To connect two shafts which have lateral misalignment. Oldham's coupling is used.

(a) True ☐ (b) False. ☐

522. Saddle key transmits power through frictional resistance only.

(a) True ☐ (b) False. ☐

523. A beam of circular cross-section is subjected to a shearing force F. The maximum shear stress induced will be equal to $2F/bt$.

(a) True ☐ (b) False. ☐

524. The stress induced by a suddenly applied load W on a beam of cross-sectional area A is W/A.

(*a*) True ☐ (*b*) False. ☐

525. For experimental stress analysis, brittle coating technique is used.

(*a*) True ☐ (*b*) False. ☐

526. The notch angle of the Izod impact test specimen is usually 30°.

(*a*) True ☐ (*b*) False. ☐

527. For brittle material, maximum principal stress theory is used.

(*a*) True ☐ (*b*) False. ☐

528. If the twisting moment of two shafts is same, then the two shafts will have equal strength.

(*a*) True ☐ (*b*) False. ☐

529. For ductile materials, the maximum shear stress theory is used.

(*a*) True ☐ (*b*) False. ☐

530. Eutectoid steel means a steel containing 0.85% carbon.

(*a*) True ☐ (*b*) False. ☐

531. The ductility of steel increases, when percentage of carbon in steel is increased.

(*a*) True ☐ (*b*) False. ☐

532. By sand mould casting, the gears are casted.

(*a*) True ☐ (*b*) False. ☐

533. Cold working decreases fatigue strength.

(*a*) True ☐ (*b*) False. ☐

534. The boiler plates edges are bevelled to an angle of 45°.

(*a*) True ☐ (*b*) False. ☐

535. A lap joint is always in single shear.

(*a*) True ☐ (*b*) False. ☐

536. The ratio of least value of tearing resistance, crushing resistance or shearing resistance to the strength of solid plate, is known as efficiency of the joint.

(*a*) True ☐ (*b*) False. ☐

537. To connect two co-axial rods, a cotter joint is used.

(*a*) True ☐ (*b*) False. ☐

538. The included angle for the acme thread is 55°.

(*a*) True ☐ (*b*) False. ☐

539. When shafts are arranged parallel and rotate in the same directions, then an open belt drive is used.

(*a*) True ☐ (*b*) False. ☐

540. To secure high efficiency, multiple threads are used.

(*a*) True ☐ (*b*) False. ☐

541. If the efficiency of a machine is less than 50%, the machine will be self-locking.

(*a*) True ☐ (*b*) False. ☐

Fill in the blanks:

542. For ......, Rankine's theory is used.

(*a*) ductile materials ☐ (*b*) brittle materials. ☐

543. For ......, Guest's theory is used.

(*a*) ductile materials ☐ (*b*) brittle materials. ☐

544. The cast iron will be of ......, when carbon in the cast iron is principally in the form of graphite.

(*a*) white colour ☐ (*b*) grey colour. ☐

545. The module is the ...... of diametral pitch.

(*a*) square ☐ (*b*) reciprocal. ☐

546. There is a ...... of lubricant between the journal and the bearing in a boundary lubricated bearing.

(*a*) thin film ☐ (*b*) thick film. ☐

547. Screws used for power transmission should have ......

(*a*) high efficiency ☐ (*b*) low efficiency. ☐

548. A feather key is generally ...... in shaft and loose in hub.

(*a*) loose ☐ (*b*) tight. ☐

549. The ...... coupling is designed as a hollow shaft.

(*a*) Oldham ☐ (*b*) Muff. ☐

550. A cotter joint is used to connect two ...... rods.

(*a*) parallel ☐ (*b*) co-axial. ☐

551. The ...... threads arc commonly used in railway carriage couplings.

(*a*) square ☐ (*b*) buttress. ☐

552. As rivetted joint is a ...... fastening.

(*a*) temporary ☐ (*b*) permanent. ☐

553. A ...... is always in single shear.

(*a*) lap joint ☐ (*b*) butt joint. ☐

554. For pipes which carry steam at high pressure, mostly ...... joint is used.

(*a*) riveted ☐ (*b*) expansion. ☐

555. The strength of rivetted joint is equal to the ...... value of P_t, P_s and P_c, where P_t = tearing resistance, P_s = shearing resistance and P_c = crushing resistance.

(*a*) least ☐ (*b*) maximum. ☐

556. The circumferential stress in thin cylinders is ...... times the longitudinal stress.

(*a*) two ☐ (*b*) three. ☐

557. The size of weld is ...... the throat of weld in a butt welded joint.
(*a*) 0.707 times ☐ (*b*) 0.5 times. ☐

558. The stress concentration is more serious in ...... in static loading.
(*a*) ductile materials ☐ (*b*) brittle material. ☐

559. The ratio of maximum stress to the nominal stress is known as ......
(*a*) fatigue stress ☐ (*b*) stress concentration factor. ☐

560. In cyclic loading in ...... the stress concentration is more serious.
(*a*) ductile material ☐ (*b*) brittle material. ☐

561. Total number of fundamental deviations are ...... according to I.S.I.
(*a*) 30 ☐ (*b*) 25. ☐

562. For a basic hole, ... deviation is zero.
(*a*) lower ☐ (*b*) upper. ☐

563. For a basic shaft, ... deviation is zero.
(*a*) lower ☐ (*b*) upper. ☐

564. In static loading, the yield point is ...... as compared to fatigue loading.
(*a*) lower ☐ (*b*) higher. ☐

565. Due to ...... change of cross-section, stress concentration is caused.
(*a*) gradually ☐ (*b*) suddenly. ☐

566. In the design of a nut, the ...... is the most important dimensions.
(*a*) pitch diameter ☐ (*b*) height. ☐

567. The ...... of cotter should not be more than 3 mm.
(*a*) taper ☐ (*b*) draw. ☐

568. The ...... on cotter varies from 1 in 48 to 1 in 24.
(*a*) taper ☐ (*b*) draw. ☐

III. OBJECTIVE TYPE QUESTIONS FROM COMPETITIVE EXAMINATIONS

Tick mark the most appropriate answer:

1. The maximum principal stress for the stress state shown in Fig. 10.15 is
(*a*) σ ☐
(*b*) 2σ ☐
(*c*) 3σ ☐
(*d*) 1.5σ. ☐

(GATE 2001)

FIGURE 10.15

2. The friction circle of a bearing is having radius equal to
(*a*) r ☐ (*b*) $r \cos \phi$ ☐
(*c*) $r \sin \phi$ ☐ (*d*) $r \tan \phi$. ☐

where r = Radius of shaft, and ϕ = Angle of friction.

3. If failure in shear along 45° planes is to be avoided, then a material subjected to uniaxial tension should have its shear strength equal to at least
   (*a*) tensile strength ☐
   (*b*) compressive strength ☐
   (*c*) half the difference between the tensile and compressive strengths ☐
   (*d*) half the tensile strength. ☐

(IES 1994)

4. The bearing characteristic number is given by
   (*a*) $\frac{Z \times p}{N}$ ☐ (*b*) $\frac{Z \times N}{p}$ ☐
   (*c*) $\frac{p}{Z \times N}$ ☐ (*d*) $\frac{Z}{p \times N}$. ☐
   where Z = Absolute viscosity of lubricant in centipoise, N = Speed, and
   p = Pressure in N/mm^2.

5. If the rotating mass of a rim type flywheel is distributed on another rim type flywheel whose mean radius is half mean radius of the former, the energy stored in the latter at the same speed will be
   (*a*) four times the first one ☐ (*b*) same as the first one ☐
   (*c*) one-fourth of the first one ☐ (*d*) one and half times the first one. ☐

(IES 1993)

6. A metal pipe of 1.0 m diameter is filled with a fluid having a pressure of 10 kg/cm^2. If the permissible tensile stress in the metal is 200 kg/cm^2, then the thickness of the metal required for making the pipe would be
   (*a*) 5 mm ☐ (*b*) 10 mm ☐
   (*c*) 20 mm ☐ (*d*) 25 mm. ☐

(IES 1993)

7. A vertical shaft of dia. 150 mm carries a vertical load of 20 kN and rests flat foot-step bearing. The shaft is rotating at 100 r.p.m. and μ = 0.045. For uniform pressure distribution, the frictional torque will be
   (*a*) 25 Nm ☐ (*b*) 35 Nm ☐
   (*c*) 45 Nm ☐ (*d*) 55 Nm. ☐

8. If the tearing efficiency of a riveted joint is 75 per cent, then the ratio of rivet hole diameter to the pitch is equal to
   (*a*) 0.2 ☐ (*b*) 0.25 ☐
   (*c*) 0.50 ☐ (*d*) 0.60. ☐

(GATE 1996)

9. For design purposes, the value of sommerfield number should be
   (*a*) 10^5 ☐ (*b*) 1.43×10^7 ☐
   (*c*) 1.43×10^9 ☐ (*d*) 1.43×10^{15}. ☐

10. A double fillet welded joint with parallel weld of length L and leg B is subjected to a tensile force P. Assuming uniform stress distribution, the shear stress in the weld is given by

(a) $\frac{\sqrt{2}P}{BL}$ ☐ (b) $\frac{P}{2BL}$ ☐

(c) $\frac{P}{\sqrt{2}BL}$ ☐ (d) $\frac{2P}{BL}$. ☐

(IES 1996)

11. The length of the belt in the case of a cross-belt drive is given in terms of centre distance between pulleys (C), diameters of the pulleys D and d as

(a) $2C+\frac{\pi}{2}(D+d)+\frac{(D+d)^2}{4C}$ ☐ (b) $2C+\frac{\pi}{2}(D-d)+\frac{(D+d)^2}{4C}$ ☐

(c) $2C+\frac{\pi}{2}(D+d)+\frac{(D-d)^2}{4C}$ ☐ (d) $2C+\frac{\pi}{2}(D-d)+\frac{(D-d)^2}{4C}$. ☐

12. The type of gear used to keep noise minimum, should be

(a) cycloidal ☐ (b) involute ☐

(c) bevel ☐ (d) helical. ☐

13. The cracks due to fatigue spreads only by

(a) tensile stress along its direction ☐

(b) compressive stress along its direction ☐

(c) shear stress ☐

(d) tensile stress and in direction perpendicular to the tensile stress. ☐

14. A spring material should have the following properties

(a) high yield strength ☐ (b) more hardness ☐

(c) high breaking strength ☐ (d) all of the above ☐

(e) none of the above. ☐

15. Which of the following statements is correct?

(a) Flywheel reduces speed fluctuations during a cycle for a constant load, but flywheel does not control the mean speed of the engine if the load changes. ☐

(b) Flywheel does not reduce speed fluctuations during a cycle for a constant load, but flywheel does control the mean speed of the engine if the load changes. ☐

(c) Governor controls speed fluctuations during a cycle for a constant load, but governor does not control the mean speed of the engine if the load changes. ☐

(d) Governor controls speed fluctuations during a cycle for constant load, and governor also controls the mean speed of the engine if the load changes. ☐

(GATE 2001)

16. The endurance limit of a material subject to fatigue loading

(a) increases with the increase of ultimate strength ☐

(b) increases with the decrease of ultimate strength ☐

(c) in independent of ultimate strength ☐

(d) none of the above. ☐

17. Maximum shear stress developed on the surface of a solid circular shaft under pure torsion is 240 MPa. If the shaft diameter is doubled, then the maximum shear stress developed corresponding to the same torque is

(*a*) 120 MPa ☐ (*b*) 60 MPa ☐

(*c*) 30 MPa ☐ (*d*) 15 MPa. ☐

(GATE 2003)

18. The ratio of maximum stress (at a notch or a fillet) in a machine to the nominal stress at the same section, is known as

(*a*) endurance limit ☐ (*b*) stress concentration factor ☐

(*c*) surface finish factor ☐ (*d*) factor of safety. ☐

19. The design calculations for members subjected to fluctuating loads with the same factor of safety yield the most conservative estimates when using

(*a*) Gerber relation ☐ (*b*) Soderberg relation ☐

(*c*) Goodman relation ☐ (*d*) None of the above. ☐

(IES 1995)

20. If a shaft made from ductile material is subjected to combined bending and twisting moments, calculations based on which one of the following failure theories would give the most conservative value?

(*a*) Maximum principal stress theory ☐ (*b*) Maximum shear stress theory ☐

(*c*) Maximum strain energy theory ☐ (*d*) Maximum distortion energy theory. ☐

(IES 1996)

21. In the case of a flywheel, the maximum fluctuation of energy is the

(*a*) sum of maximum and minimum energies ☐

(*b*) difference between the maximum and minimum energies ☐

(*c*) ratio of maximum and minimum energy ☐

(*d*) ratio of minimum and maximum energy. ☐

(IES 1998)

22. Yield point in fatigue loading as compared to static loading, is

(*a*) higher ☐ (*b*) lower ☐

(*c*) same ☐ (*d*) none of the above. ☐

23. The basic load rating of a ball bearing is

(*a*) the maximum static radial load that can be applied without causing any plastic deformation of bearing components. ☐

(*b*) the radial load at which 90% of the group of apparently identical bearings run for 1 million revolutions before the first evidence of failure. ☐

(*c*) the maximum radial load that can be applied during operation without any plastic deformation of bearing components. ☐

(*d*) a combination of radial and axial loads that can be applied without any plastic deformation. ☐

24. Two shafts, one solid and the other hollow, of the same material will have the same strength if they are having

(*a*) same length and weight ☐

(*b*) same length and same polar modulus ☐

(*c*) same weight and same polar modulus ☐

(*d*) same length, weight and polar modulus. ☐

25. If principal stresses corresponding to a two dimensional state of stress are σ_1 and σ_2 and σ_1 is greater than σ_2 and both are tensile, then which one of the following would be the correct criterion for failure by yielding, according to the maximum shear stress criterion?

(*a*) $\frac{(\sigma_1 - \sigma_2)}{2} = \pm \frac{\sigma_{yp}}{2}$ ☐ (*b*) $\frac{\sigma_1}{2} = \pm \frac{\sigma_{yp}}{2}$ ☐

(*c*) $\frac{\sigma_2}{2} = \pm \frac{\sigma_{yp}}{2}$ ☐ (*d*) $\sigma_1 = \pm 2\sigma_{yp}$. ☐

(IES 1993)

26. A solid circular shaft is subjected to a maximum shear stress of 140 MPa. The magnitude of the maximum normal stress developed in the shaft is

(*a*) 140 MPa ☐ (*b*) 80 MPa ☐

(*c*) 70 MPa ☐ (*d*) 60 MPa. ☐

(IAS 1995)

27. Match list I with list II and select the correct answer using the codes given below the lists:

List I	List II
A. Crankshaft	1. Supports the revolving parts and transmits torque
B. Wire shaft	2. Transmits motion between shafts where it is not possible to effect a rigid coupling between them
C. Axle	3. Converts linear motion into rotary motion
D. Plain shaft	4. Supports only the revolving parts

Codes:

	A	*B*	*C*	*D*	
(*a*)	3	2	1	4	☐
(*b*)	4	2	3	1	☐
(*c*)	3	2	4	1	☐
(*d*)	1	4	2	3	☐

(IES 1995)

28. The power transmitted by a belt is dependent on the centrifugal effect in the belt. The maximum power can be transmitted when the centrifugal tension is

(*a*) $\frac{1}{3}$ of tension (T_1) on the tight side. ☐

(*b*) $\frac{1}{3}$ of total tension (T_t) on the tight side. ☐

(c) $\frac{1}{3}$ of the tension (T_2) on the slack side. ☐

(d) $\frac{1}{3}$ of sum of tensions T_1 and T_2 *i.e.*, $\frac{1}{3}(T_1 + T_2)$. ☐

29. A transmission shaft subjected to bending loads must be designed on the basis of

(a) maximum normal stress theory ☐

(b) maximum shear stress theory ☐

(c) maximum normal stress and maximum shear stress theories ☐

(d) fatigue strength. ☐

(IES 1995)

30. A shaft can safety transmit 90 kW while rotating at a given speed. If this shaft is replaced by a shaft of diameter double of the previous one and rotated at half the speed of the previous, the power that can be transmitted by the new shaft is

(a) 90 kW ☐ (b) 180 kW ☐

(c) 360 kW ☐ (d) 720 kW. ☐

31. When compared to a rod of same diameter and material, a wire rope

(a) is less flexible ☐

(b) has a much smaller load carrying capacity ☐

(c) does not provided much warning before failure ☐

(d) provides much greater time for remedial action before failure. ☐

(IES 1994)

32. A cold rolled steel shaft is designed on the basis of maximum shear stress theory. The principal stresses induced at its critical section are 60 MPa and –60 MPa respectively. If the yield stress for the shift material is 360 MPa, the factor of safety of the design is

(a) 2 ☐ (b) 3 ☐

(c) 4 ☐ (d) 6. ☐

33. A 50 kW motor using six *V*-belts is used in a pulp mill. If one of the belts breaks after a month of continuous running, then

(a) the broken belt is to be replaced by a similar belt ☐

(b) all the belts are to be replaced ☐

(c) the broken belt and two adjacent belts are to be replaced ☐

(d) the broken belt and one adjacent belt is to be replaced ☐

(IAS 1994)

34. The bolts in a rigid flanged coupling connecting two shafts transmitting power are subjected to

(a) shear force and bending moment ☐

(b) axial force ☐

(c) torsion and bending moment ☐

(d) torsion. ☐

35. Which one of the following pairs is not correctly matched?

(*a*) Positive drive ...... Belt drive ☐

(*b*) High velocity ratio ...... Worm gearing ☐

(*c*) To connect non-parallel and non-intersecting shafts ...... spiral gearing ☐

(*d*) Diminished noise and smooth operation ...... Helical gears. ☐

(IES 1995)

36. If the area of cross-section of a wire is circular and if the radius of the circle decreases to half its original value due to the stretch of the wire by a load, then the modulus of elasticity of the wire will be

(*a*) one-fourth of its original value ☐ (*b*) Halved ☐

(*c*) doubled ☐ (*d*) Unaffected. ☐

(IES 1993)

37. Two shafts one solid and other hollow, are made of the same materials and are having same length and weight. The hollow shaft as compared to solid shaft will be

(*a*) more strong ☐ (*b*) less strong ☐

(*c*) of same strength ☐ (*d*) none of the above. ☐

38. Which of the following pairs are correctly matched?

1. Resilience—Resistance to deformation
2. Malleability—Shape change
3. Creep—Progressive deformation
4. Plasticity—Permanent deformation

Select the correct answer using the codes given below:

(*a*) 2, 3, 4 ☐ (*b*) 1, 2, 3 ☐

(*c*) 1, 2, 4 ☐ (*d*) 1, 3, 4. ☐

(IES 1994)

39. During a tensile testing of a specimen using a UTM, the parameters actually measured include

(*a*) True stress and True strain ☐

(*b*) Poisson's ratio and young's modulus ☐

(*c*) Engineering stress and engineering strain ☐

(*d*) Load and elongation. ☐

(IES 1996)

40. A static load is mounted at the centre of a shaft rotating at uniform angular velocity. The shaft will be designed for

(*a*) maximum compressive stress (static) ☐ (*b*) the maximum tensile stress (static) ☐

(*c*) the maximum bending moment (static) ☐ (*d*) fatigue loading. ☐

41. Match list I with list II and select the correct answer using the codes given below the lists:

List I (Type of joint)	**List II (Mode of joining members)**
A. Cotter joint	1. Connects two rods or bars permitting small amount of flexibility
B. Knuckle joint	2. Rigidly connects two members
C. Turn buckle	3. Connects two rods having threaded ends
D. Riveted joint	4. Permanent fluid tight joint between two flat pieces.
	5. Connects two shafts and transmits torque.

Codes:

	A	*B*	*C*	*D*	
(*a*)	5	1	3	2	☐
(*b*)	2	1	3	4	☐
(*c*)	5	3	2	4	☐
(*d*)	2	3	1	4.	☐

(IES 1993)

42. When two shafts are neither parallel nor intersecting, power can be transmitted by using

(*a*) a pair of spur gears ☐ (*b*) a pair of helical gears ☐

(*c*) an Oldham's coupling ☐ (*d*) a pair of spiral gears. ☐

(IES 1998)

43. Match list I with list II and select the correct answer using the codes given below in the list:

List I (Types of joint)	**List II (Use of these joints)**
A. Cotter joint	1. Tie rod of a wall crane
B. Knuckle joint	2. Suspension bridges
C. Suspension link joint	3. Diagonal stays in boiler
D. Turn buckle	4. Cross-head of a steam engine

Codes:

	A	*B*	*C*	*D*	
(*a*)	4	2	3	1	☐
(*b*)	4	3	2	1	☐
(*c*)	3	2	1	4	☐
(*d*)	2	1	4	3.	☐

(IES 1995)

44. Which of the following stresses are associated with the design of pins in bushed pin type flexible coupling?

1. Bearing stress 2. Bending stress

3. Axial tensile stress 4. Transverse shear stress.

Codes:

(*a*) 1, 3 and 4 ☐ (*b*) 2, 3 and 4 ☐

(*c*) 1, 2 and 3 ☐ (*d*) 1, 2 and 4. ☐

(IES 1998)

45. Consider the following design considerations

1. Tensile failure
2. Creep failure
3. Bearing failure
4. Shearing failure
5. Bending failure

The design of the pin of a rocker arm of an I.C. engine is based on

(*a*) 1, 2 and 4 ☐ (*b*) 1, 3 and 4 ☐

(*c*) 2, 3 and 5 ☐ (*d*) 3, 4 and 5. ☐

(IES 1995)

46. The coupling used to connect two shafts with large angular misalignment is

(*a*) a flange coupling ☐ (*b*) an Oldham's coupling ☐

(*c*) a flexible bush coupling ☐ (*d*) a hook's joint. ☐

(GATE 2002)

47. A friction circle is drawn when a journal rotates in a bearing. Its radius depends on the co-efficient of friction and the

(*a*) magnitudes of the forces on the journal ☐

(*b*) angular velocity of the journal ☐

(*c*) clearance between the journal and the bearing ☐

(*d*) radius of the journal. ☐

(IES 1993)

48. The most suitable bearing for carrying very heavy loads with slow speed is

(*a*) hydro-dynamic bearing ☐ (*b*) ball bearing ☐

(*c*) roller bearing ☐ (*d*) hydrostatic bearing. ☐

(IAS 1994)

49. A flywheel of moment of inertia 9.8 kg-m^2 fluctuates by 30 r.p.m. for a fluctuation of energy of 19360 J. The mean speed of the flywheel is (r.p.m.)

(*a*) 600 ☐ (*b*) 900 ☐

(*c*) 968 ☐ (*d*) 2940. ☐

(GATE 1998)

50. Thrust bearings of sliding type are often provided with multiple sector shaped bearing pads of the tilting type instead of a continuous annular bearing surface in order to

(*a*) distribute the thrust load more non-uniformly ☐

(*b*) provide limited adjustments to shaft misalignments ☐

(*c*) enable the formation of a wedge-shaped oil film ☐

(*d*) enabling lubricating oil to come into contact with the total bearing area. ☐

(IAS 1994)

51. The most suitable theory of failure for brittle materials is

(*a*) maximum normal stress theory ☐ (*b*) maximum strain energy theory ☐

(*c*) maximum distortion energy theory ☐ (*d*) maximum shear stress theory. ☐

52. Deep groove ball bearings are used for

(*a*) heavy thrust load only ☐
(*b*) small angular displacements of shafts ☐
(*c*) radial load at high speed ☐
(*d*) combined thrust and radial loads at high speed. ☐

(IAS 1995)

53. The endurance limit of a material can be improved by

(*a*) polishing ☐ (*b*) coating ☐
(*c*) heat treatment ☐ (*d*) shot peening ☐
(*e*) (*a*), (*c*) and (*d*). ☐

54. In an oil lubricated journal bearing, co-efficient of friction between the journal and the bearing

(*a*) remains constant at all speeds ☐
(*b*) is minimum at zero speed and decreases monotonically with increase in speed ☐
(*c*) is maximum at zero speed and decreases monotonically with increase in speed ☐
(*d*) becomes minimum at an optimum speed and then increases with further increase in speed. ☐

(IES 1995)

55. Which one of the following materials is highly elastic?

(*a*) Rubber ☐ (*b*) Brass ☐
(*c*) Steel ☐ (*d*) Glass. ☐

(IAS 1995)

56. Match list I with list II and select the correct answer using the codes given below the lists:

List I (Property)	**List II (Testing machine)**
A. Tensile strength	1. Rotating bending machine
B. Impact strength	2. Three point loading machine
C. Bending strength	3. Universal testing machine
D. Fatigue strength	4. Izod testing machine

Codes:

	A	*B*	*C*	*D*	
(*a*)	4	3	2	1	☐
(*b*)	3	2	1	4	☐
(*c*)	2	1	4	3	☐
(*d*)	3	4	2	1.	☐

(IAS 1995)

57. The highest stress that a material can withstand for a specified length of time without excessive deformation is called

(*a*) fatigue strength ☐ (*b*) endurance strength ☐
(*c*) creep strength ☐ (*d*) creep rupture strength. ☐

(IES 1997)

58. The unit of elastic modulus is same as those of

(*a*) stress, shear modulus and pressure ☐ (*b*) strain, shear modulus and force ☐
(*c*) shear modulus, stress and force ☐ (*d*) stress, strain and pressure. ☐

(IAS 1994)

59. Which theory of failure will you use for aluminium components under steady loading?

(*a*) Principal stress theory ☐ (*b*) Principal strain theory ☐
(*c*) Strain energy theory ☐ (*d*) Maximum shear stress theory. ☐

(GATE 1999)

60. A test specimen is stressed slightly beyond the yield point and then unloaded. Its yield strength will

(*a*) decrease ☐ (*b*) increase ☐
(*c*) remain same ☐ (*d*) become equal to ultimate tensile strength. ☐

(GATE 1995)

61. Fatigue strength of a rod subjected to cyclic axial force is less than that of a rotating beam of the same dimensions subjected to steady lateral force because

(*a*) axial stiffness is less than bending stiffness ☐
(*b*) of absence of centrifugal effects in rod ☐
(*c*) the number of discontinuities vulnerable to fatigue are more in the rod ☐
(*d*) at a particular time, the rod has only one type of stress whereas the beam has both the tensile and compressive stresses. ☐

(GATE 1992)

62. **Assertion (A):** Endurance limits for all materials are always less than the ultimate strength of the corresponding materials.

Reason (R): Stress concentration in a machine part due to any dislocation is very damaging when the part is subjected to variable loading.

(*a*) *A* is true and *R* is false. ☐ (*b*) *A* is false and *R* is true. ☐
(*c*) both *A* and *R* are true. ☐ (*d*) both *A* and *R* are wrong. ☐

(IAS 1994)

ANSWERS

Answers to Objective Type Questions

1. (*b*)	**2.** (*a*)	**3.** (*d*)	**4.** (*d*)	**5.** (*c*)	**6.** (*d*)
7. (*c*)	**8.** (*b*)	**9.** (*a*)	**10.** (*c*)	**11.** (*d*)	**12.** (*a*)
13. (*b*)	**14.** (*c*)	**15.** (*e*)	**16.** (*d*)	**17.** (*d*)	**18.** (*d*)
19. (*a*)	**20.** (*d*)	**21.** (*b*)	**22.** (*a*)	**23.** (*d*)	**24.** (*d*)
25. (*a*)	**26.** (*c*)	**27.** (*b*)	**28.** (*d*)	**29.** (*b*)	**30.** (*c*)
31. (*a*)	**32.** (*c*)	**33.** (*b*)	**34.** (*a*)	**35.** (*c*)	**36.** (*b*)
37. (*d*)	**38.** (*a*)	**39.** (*c*)	**40.** (*c*)	**41.** (*b*)	**42.** (*c*)

43. (*d*) **44.** (*a*) **45.** (*a*) **46.** (*b*) **47.** (*d*) **48.** (*c*)
49. (*b*) **50.** (*a*) **51.** (*a*) **52.** (*b*) **53.** (*a*) **54.** (*d*)
55. (*b*) **56.** (*b*) **57.** (*a*) **58.** (*d*) **59.** (*a*) **60.** (*e*)
61. (*b*) **62.** (*a*) **63.** (*c*) **64.** (*b*) **65.** (*c*) **66.** (*d*)
67. (*b*) **68.** (*d*) **69.** (*c*) **70.** (*d*) **71.** (*b*) **72.** (*a*)
73. (*c*) **74.** (*d*) **75.** (*b*) **76.** (*c*) **77.** (*a*) **78.** (*b*)
79. (*c*) **80.** (*c*) **81.** (*d*) **82.** (*a*) **83.** (*a*) **84.** (*c*)
85. (*b*) **86.** (*b*) **87.** (*a*) **88.** (*c*) **89.** (*d*) **90.** (*c*)
91. (*b*) **92.** (*d*) **93.** (*c*) **94.** (*f*) **95.** (*a*) **96.** (*a*)
97. (*d*) **98.** (*c*) **99.** (*d*) **100.** (*c*) **101.** (*a*) **102.** (*b*)
103. (*a*) **104.** (*d*) **105.** (*b*) **106.** (*b*) **107.** (*a*) **108.** (*d*)
109. (*c*) **110.** (*b*) **111.** (*a*) **112.** (*d*) **113.** (*d*) **114.** (*a*)
115. (*a*) **116.** (*c*) **117.** (*d*) **118.** (*c*) **119.** (*c*) **120.** (*d*)
121. (*c*) **122.** (*c*) **123.** (*b*) **124.** (*c*) **125.** (*b*) **126.** (*c*)
127. (*a*) **128.** (*a*) **129.** (*a*) **130.** (*b*) **131.** (*c*) **132.** (*c*)
133. (*b*) **134.** (*c*) **135.** (*d*) **136.** (*c*) **137.** (*b*) **138.** (*d*)
139. (*c*) **140.** (*f*) **141.** (*e*) **142.** (*c*) **143.** (*b*) **144.** (*c*)
145. (*d*) **146.** (*c*) **147.** (*b*) **148.** (*a*) **149.** (*c*) **150.** (*b*)
151. (*d*) **152.** (*c*) **153.** (*b*) **154.** (*c*) **155.** (*d*) **156.** (*a*)
157. (*a*) **158.** (*a*) **159.** (*b*) **160.** (*d*) **161.** (*a*) **162.** (*d*)
163. (*d*) **164.** (*c*) **165.** (*a*) **166.** (*d*) **167.** (*d*) **168.** (*c*)
169. (*c*) **170.** (*a*) **171.** (*c*) **172.** (*a*) **173.** (*c*) **174.** (*d*)
175. (*c*) **176.** (*d*) **177.** (*c*) **178.** (*b*) **179.** (*c*) **180.** (*c*)
181. (*c*) **182.** (*b*) **183.** (*a*) **184.** (*d*) **185.** (*b*) **186.** (*a*)
187. (*d*) **188.** (*c*) **189.** (*a*) **190.** (*c*) **191.** (*d*) **192.** (*b*)
193. (*a*) **194.** (*c*) **195.** (*c*) **196.** (*a*) **197.** (*c*) **198.** (*b*)
199. (*b*) **200.** (*d*) **201.** (*d*) **202.** (*d*) **203.** (*d*) **204.** (*d*)
205. (*c*) **206.** (*e*) **207.** (*f*) **208.** (*c*) **209.** (*c*) **210.** (*c*)
211. (*d*) **212.** (*b*) **213.** (*b*) **214.** (*c*) **215.** (*a*) **216.** (*c*)
217. (*c*) **218.** (*c*) **219.** (*d*) **220.** (*b*) **221.** (*b*) **222.** (*e*)
223. (*d*) **224.** (*c*) **225.** (*c*) **226.** (*b*) **227.** (*c*) **228.** (*d*)
229. (*c*) **230.** (*b*) **231.** (*b*) **232.** (*b*) **233.** (*c*) **234.** (*b*)
235. (*d*) **236.** (*a*) **237.** (*b*) **238.** (*a*) **239.** (*b*) **240.** (*a*)
241. (*b*) **242.** (*d*) **243.** (*a*) **244.** (*c*) **245.** (*a*) **246.** (*c*)
247. (*b*) **248.** (*a*) **249.** (*c*) **250.** (*a*) **251.** (*d*) **252.** (*a*)
253. (*a*) **254.** (*b*) **255.** (*d*) **256.** (*d*) **257.** (*b*) **258.** (*d*)
259. (*d*) **260.** (*a*) **261.** (*c*) **262.** (*a*) **263.** (*b*) **264.** (*a*)
265. (*c*) **266.** (*c*) **267.** (*b*) **268.** (*d*) **269.** (*e*) **270.** (*c*)
271. (*b*) **272.** (*a*) **273.** (*b*) **274.** (*c*) **275.** (*a*) **276.** (*c*)

277. (*c*)	**278.** (*b*)	**279.** (*a*)	**280.** (*b*)	**281.** (*a*)	**282.** (*c*)
283. (*d*)	**284.** (*d*)	**285.** (*e*)	**286.** (*d*)	**287.** (*d*)	**288.** (*f*)
289. (*b*)	**290.** (*a*)	**291.** (*d*)	**292.** (*c*)	**293.** (*b*)	**294.** (*a*)
295. (*c*)	**296.** (*d*)	**297.** (*a*)	**298.** (*b*)	**299.** (*b*)	**300.** (*a*)
301. (*d*)	**302.** (*c*)	**303.** (*a*)	**304.** (*b*)	**305.** (*a*)	**306.** (*d*)
307. (*d*)	**308.** (*e*)	**309.** (*b*)	**310.** (*d*)	**311.** (*c*)	**312.** (*a*)
313. (*d*)	**314.** (*c*)	**315.** (*a*)	**316.** (*b*)	**317.** (*c*)	**318.** (*b*)
319. (*b*)	**320.** (*a*)	**321.** (*c*)	**322.** (*a*)	**323.** (*c*)	**324.** (*c*)
325. (*d*)	**326.** (*b*)	**327.** (*a*)	**328.** (*c*)	**329.** (*b*)	**330.** (*c*)
331. (*b*)	**332.** (*d*)	**333.** (*d*)	**334.** (*c*)	**335.** (*b*)	**336.** (*c*)
337. (*a*)	**338.** (*b*)	**339.** (*b*)	**340.** (*c*)	**341.** (*a*)	**342.** (*d*)
343. (*b*)	**344.** (*d*)	**345.** (*b*)	**346.** (*b*)	**347.** (*b*)	**348.** (*d*)
349. (*d*)	**350.** (*b*)	**351.** (*b*)	**352.** (*a*)	**353.** (*d*)	**354.** (*d*)
355. (*b*)	**356.** (*c*)	**357.** (*a*)	**358.** (*b*)	**359.** (*b*)	**360.** (*b*)
361. (*b*)	**362.** (*a*)	**363.** (*d*)	**364.** (*c*)	**365.** (*b*)	**366.** (*a*)
367. (*e*)	**368.** (*c*)	**369.** (*b*)	**370.** (*d*)	**371.** (*b*)	**372.** (*a*)
373. (*d*)	**374.** (*b*)	**375.** (*c*)	**376.** (*c*)	**377.** (*b*)	**378.** (*a*)
379. (*b*)	**380.** (*d*)	**381.** (*b*)	**382.** (*c*)	**383.** (*f*)	**384.** (*c*)
385. (*c*)	**386.** (*c*)	**387.** (*c*)	**388.** (*d*)	**389.** (*a*)	**390.** (*b*)
391. (*b*)	**392.** (*a*)	**393.** (*d*)	**394.** (*d*)	**395.** (*b*)	**396.** (*a*)
397. (*a*)	**398.** (*d*)	**399.** (*b*)	**400.** (*c*)	**401.** (*d*)	**402.** (*f*)
403. (*d*)	**404.** (*d*)	**405.** (*b*)	**406.** (*c*)	**407.** (*d*)	**408.** (*f*)
409. (*b*)	**410.** (*d*)	**411.** (*b*)	**412.** (*c*)	**413.** (*b*)	**414.** (*a*)
415. (*a*)	**416.** (*c*)	**417.** (*b*)	**418.** (*d*)	**419.** (*b*)	**420.** (*b*)
421. (*a*)	**422.** (*c*)	**423.** (*a*)	**424.** (*b*)	**425.** (*c*)	**426.** (*b*)
427. (*b*)	**428.** (*a*)	**429.** (*c*)	**430.** (*d*)	**431.** (*c*)	**432.** (*b*)
433. (*c*)	**434.** (*c*)	**435.** (*c*)	**436.** (*b*)	**437.** (*b*)	**438.** (*b*)
439. (*c*)	**440.** (*b*)	**441.** (*d*)	**442.** (*d*)	**443.** (*c*)	**444.** (*c*)
445. (*a*)	**446.** (*e*)	**447.** (*c*)	**448.** (*b*)	**449.** (*d*)	**450.** (*b*)
451. (*b*)	**452.** (*b*)	**453.** (*b*)	**454.** (*b*)	**455.** (*c*)	**456.** (*c*)
457. (*c*)	**458.** (*c*)	**459.** (*b*)	**460.** (*c*)	**461.** (*c*)	**462.** (*d*)
463. (*b*)	**464.** (*b*)	**465.** (*a*)	**466.** (*b*)	**467.** (*b*)	**468.** (*a*)
469. (*b*)	**470.** (*d*)	**471.** (*c*)	**472.** (*c*)	**473.** (*c*)	**474.** (*b*)
475. (*c*)	**476.** (*c*)	**477.** (*d*)	**478.** (*c*)	**479.** (*b*)	**480.** (*a*)
481. (*c*)	**482.** (*b*)	**483.** (*b*)	**484.** (*a*)	**485.** (*c*)	**486.** (*b*)
487. (*d*)	**488.** (*c*)	**489.** (*c*)	**490.** (*b*)	**491.** (*c*)	**492.** (*a*)
493. (*b*)	**494.** (*c*)	**495.** (*a*)	**496.** (*b*)	**497.** (*c*)	**498.** (*b*)

499. (*c*)	**500.** (*b*).				

True/False

501. (*a*)	**502.** (*b*)	**503.** (*a*)	**504.** (*a*)	**505.** (*a*)	**506.** (*b*)
507. (*b*)	**508.** (*b*)	**509.** (*a*)	**510.** (*a*)	**511.** (*a*)	**512.** (*a*)
513. (*a*)	**514.** (*a*)	**515.** (*a*)	**516.** (*a*)	**517.** (*a*)	**518.** (*b*)
519. (*b*)	**520.** (*b*)	**521.** (*a*)	**522.** (*a*)	**523.** (*a*)	**524.** (*b*)
525. (*a*)	**526.** (*b*)	**527.** (*a*)	**528.** (*a*)	**529.** (*a*)	**530.** (*a*)
531. (*b*)	**532.** (*b*)	**533.** (*b*)	**534.** (*b*)	**535.** (*a*)	**536.** (*a*)
537. (*a*)	**538.** (*b*)	**539.** (*a*)	**540.** (*a*)	**541.** (*a*).	

Fill in the Blanks

542. (*a*)	**543.** (*a*)	**544.** (*b*)	**545.** (*b*)	**546.** (*a*)	**547.** (*a*)
548. (*b*)	**549.** (*b*)	**550.** (*b*)	**551.** (*b*)	**552.** (*b*)	**553.** (*a*)
554. (*b*)	**555.** (*a*)	**556.** (*a*)	**557.** (*b*)	**558.** (*b*)	**559.** (*b*)
560. (*a*)	**561.** (*b*)	**562.** (*a*)	**563.** (*b*)	**564.** (*b*)	**565.** (*b*)
566. (*b*)	**567.** (*b*)	**568.** (*a*).			

Answers to Objective Type Questions from Competitive Examinations

1. (*b*)	**2.** (*c*)	**3.** (*c*)	**4.** (*b*)	**5.** (*c*)	**6.** (*d*)
7. (*c*)	**8.** (*b*)	**9.** (*c*)	**10.** (*c*)	**11.** (*b*)	**12.** (*d*)
13. (*d*)	**14.** (*d*)	**15.** (*a*)	**16.** (*a*)	**17.** (*c*)	**18.** (*b*)
19. (*b*)	**20.** (*d*)	**21.** (*b*)	**22.** (*b*)	**23.** (*b*)	**24.** (*b*)
25. (*b*)	**26.** (*a*)	**27.** (*c*)	**28.** (*b*)	**29.** (*c*)	**30.** (*a*)
31. (*d*)	**32.** (*b*)	**33.** (*b*)	**34.** (*a*)	**35.** (*a*)	**36.** (*d*)
37. (*a*)	**38.** (*d*)	**39.** (*d*)	**40.** (*d*)	**41.** (*b*)	**42.** (*c*)
43. (*b*)	**44.** (*d*)	**45.** (*d*)	**46.** (*d*)	**47.** (*d*)	**48.** (*d*)
49. (*a*)	**50.** (*c*)	**51.** (*a*)	**52.** (*d*)	**53.** (*e*)	**54.** (*d*)
55. (*c*)	**56.** (*d*)	**57.** (*c*)	**58.** (*a*)	**59.** (*d*)	**60.** (*b*)
61. (*d*)	**62.** (*c*).				

Chapter 11

ENGINEERING MATERIALS

I. THEORY

11.1. INTRODUCTION

The engineering materials are mostly classified into three types namely (*a*) Metals, (*b*) Ceramic materials, and (*c*) Organic materials. Metals are further classified into (*i*) Ferrous materials, and (*ii*) Non-ferrous materials.

11.1.1. Ferrous Materials

Ferrous materials are those materials, which contain iron. The ferrous materials are classified as

1. Pig iron,
2. Wrought iron,
3. Cast iron, and
4. Steels.

11.1.2. Cast Iron

It is divided into four groups namely (*i*) Grey cast iron, (*ii*) White cast iron, (*iii*) Malleable cast iron, and (*iv*) Ductile cast iron. *Steels* are divided into plain carbon steels, low alloy steels and high alloy steels. *Plain* carbon steels are divided into low carbon steels, medium carbon steels and high carbon steels.

11.1.3. Non-Ferrous Materials

Non-ferrous materials are those materials which do not contain iron. The examples of non-ferrous materials are: copper, lead, aluminium, tin, nickel, zinc and their alloys.

11.1.4. Ceramic Materials

The oxides, carbides, nitrides and silicates of various metals usually are known as ceramic materials. The examples of ceramic materials are: glass, brick, sand, concrete, cement, silicon carbide, insulators, Tungsten carbide etc.

11.1.5. Organic Materials

The polymeric materials composed of carbon-compounds are called organic materials. The examples of organic materials are: paper, rubber, fuel, lubricant, wood, textiles, paints and finishes.

11.2. PIG IRON

It is the first product in the process of converting iron ore into useful metals. It is produced in a blast furnace. Pig iron is the castic materials from which cast iron, wrought iron and steels are obtained. If pig iron is refined in a cupola, *cast iron* is produced. If pig iron is put in a puddling furnace, *wrought iron* is produced. And if pig iron is refined in furnaces such as Bessemer converter, open-hearth furnace, electric furnace etc., *steel* is produced.

11.2.1. Wrought Iron

It is produced from pig iron by puddling process or Aston's process. It is never cast. All shaping is accomplished by hammering, pressing or forging. It possesses high ductility and high resistance towards corrosion.

11.2.2. Cast Iron

When the pig iron is refined in a cupola, cast iron is produced. It is brittle and grey in colour. It contains carbon more than 2%. It can be cast into a desired shape in a sand muld. By varying the carbon and silicon contents, several types of cast irons are produced.

11.2.3. Steel

If the pig iron is refined in furnaces such as Bessemer converter, Open-Hearth furnaces, electric furnaces etc., steel is produced. Actually steel is an alloy of iron and carbon in which carbon content is not more than 1.7%.

11.3. NON-FERROUS MATERIALS

The metals, which do not contain iron, are called non-ferrous metals. Examples of non-ferrous metals are: copper, zinc, nickel, lead, tin, aluminium, etc.

11.3.1. Copper

Copper possesses high thermal and electrical conductivity, high resistance to corrosion and high ductility and malleability. With a large number of metals, it forms very useful alloys. The followings are the important copper alloys:

1. Brass, and
2. Bronze.

11.3.2. Brass

If zinc is the principal alloying material with copper, the resultant alloy is known as brass. In brass, the zinc contents varies from 25 to 40 per cent.

11.3.3. Bronze

If tin is the principal alloying metal with copper, the resultant alloy is known as bronze.

11.4. ALLOTROPIC FORMS OF IRON

Iron is allotropic metal and hence exists is more than one type depending upon the temperature. The allotropic forms of iron are:

1. Alpha iron (α),
2. Beta iron (β),
3. Gamma iron (γ), and
4. Delta iron (δ).

11.4.1. Alpha Iron (α)

It exists below 768°C. It is magnetic in nature.

11.4.2. Beta Iron (β)

It exists between 768°C and 910°C. It is non-magnetic in nature.

11.4.3. Gamma Iron (γ)

It exists between 910°C and 1400°C.

11.4.4. Delta Iron (δ)

It exists between 1400°C and 1539°C.

11.5. IRON-CARBON EQUILIBRIUM DIAGRAM

It is obtained by plotting temperature along the y-axis and percentage composition of the carbon along the x-axis, when the iron-carbon alloy is heated or cooled slowly. This diagram covers the changes from pure iron to cementite (Fe_3C), *i.e.*, iron with carbon contents of 6.67%. If the percentage of carbon in the iron-carbon alloy is less than 2%, it is known as **steel**. If the percentage of carbon in the steel is less than 0.8%, it is known as **hypoeutectoid steel.** If the percentage of carbon in the steel is more than 0.8%, then it is known as **hypereutectoid steel.** But if the percentage of carbon in the steel is 0.8%, it is known as **eutectoid steel.**

If the percentage of carbon in the iron-carbon alloy is more than 2%, it is known as **cast iron.** If the percentage of carbon in the cast iron lies between 2%, and 4.3%, the cast iron is known as **hypoeutectic cast iron.** If the percentage of carbon in the cast iron in 4.3%, the cast iron is known as **eutectic cast iron.** If the percentage of carbon in the cast iron lies between 4.3% to 6.67%, the cast iron is known as **hypereutectic cast iron.**

11.6. ENGINEERING MATERIAL PROPERTIES

The following are the important properties of materials:

1. Mechanical properties,
2. Electrical properties,

3. Thermal properties,
4. Magnetic properties, and
5. Optical properties.

11.6.1. Mechanical Properties

The following are the important mechanical properties of the materials:

(*a*) Strength,
(*b*) Ductility,
(*c*) Stiffness,
(*d*) Hardness.
(*e*) Fatigue,
(*f*) Creep,
(*g*) Toughness,
(*h*) Impact strength, and
(*i*) Malleability.

The ability of a material to sustain loads without undue failure or distortion is known as **strength** of a material. The capacity of a material to undergo deformation under tension without rupture as in the wire drawing process, is known as **ductility.** The ability of a material to resist elastic deflection is known as **stiffness.** The property of a material to resist penetration by another material is known as **hardness.** When a material is subjected to repeated or fluctuating stresses, fracture takes place under a stress whose maximum value is less than the tensile strength of the material. This phenomenon is known as **fatigue.** When a material sustains steady loads for long periods of time, the material may continue to deform until they may tend to fracture under the same load. This phenomenon is known as **creep.** The ability of a material to withstand bending without fracture is known **toughness.** The capacity of a material to withstand deformation under compression without rupture as in forging and rolling processes is known as **malleability.**

11.6.2. Electrical Properties

The following are the electrical properties of the material:

(*a*) Conductivity,
(*b*) Resistivity,
(*c*) Dielectric strength, and
(*d*) Temperature co-efficient of resistance.

11.6.3. Thermal Properties

They are:

(*a*) Specific heat,
(*b*) Thermal expansion,
(*c*) Melting point, and
(*d*) Thermal conductivity.

11.6.4. Magnetic Properties

They are:

(*a*) Hysteresis,
(*b*) Permeability, and
(*c*) Coercive force.

11.6.5. Optical Properties

They are:

(*a*) Reflectivity, (*b*) Refractive index, and

(*c*) Absorptivity.

11.7. HEAT TREATMENT PROCESS

Heat treatment is defined as a process involving heating and cooling of a metal (or alloy) in a solid state to obtain certain desirable properties (such as homogeneous structure, better machinability, improved ductility etc.). The following are the important heat treatment processes:

1. Annealing,
2. Normalising,
3. Hardening (by quenching),
4. Tempering,
5. Martempering, and
6. Austempering.

11.7.1. Annealing

It is the process of heating a metal which is in a distorted structural state to a temperature which will remove distortion and then cooling the metal at a slow rate to room temperature so that the metal is stable. Annealing may be **process annealing or full annealing.** The process of heating steel to a temperature below or close to lower critical temperature and then cooling is at a desired rate is known as process annealing. This is done to soften the steel partially and to relieve internal stresses. The process of heating the steel to 40°C above the upper critical temperature and keeping the steel at that temperature for a definite period of time and then cooling it very slowly in the furnace, is known as full annealing. Full annealing softens the steel, refines the grains and removes strains.

11.7.2. Normalising

It is process of heating steel to approximately 40 to 50°C above the upper critical temperature of steel and then cooling it in still air at room temperature. In case of normalising, the rate of cooling is more rapid as compared to full annealing. Hence normalising produces a harder and stronger steel than full annealing.

11.7.3. Hardening (by Quenching)

It is process of heating steel to about 40°C above the critical temperature and holding the steel at that temperature for a definite period of time and then quenching in a medium such as water, oil or brine etc. If the cooling rate is equal to or greater than the critical cooling rate, the hardness produced will be maximum.

11.7.4. Tempering

It is process of heating the hardened steel below the lower critical temperature and holding the steel at that temperature for three to five minutes for each mm of thickness or diameter and then cooling the steel in water, oil or air either rapidly or slowly. Tempering is done to reduce the hardness and to relieve the internal stresses of a quenched steel.

11.7.5. Martempering

It is the process of heating the steel above the critical range and quenching into a salt bath and then cooling in air.

11.7.6. Austempering

It is the process of heating the steel above the critical range and then quenching into a salt bath or lead bath (which is maintained at 200 to 420°C) and keeping the steel in the bath for a definite period of time and then cooling it to room temperature in air.

11.8. CASE HARDENING

It is the process of adding carbon, nitrogen or both to the surface of the steel parts in order to provide a hardened layer of a definite depth on the surface. The following are the important processes of case hardening:

1. Carburizing,
2. Nitriding,
3. Cyaniding, and
4. Carbonitriding.

11.8.1. Carburizing

It is the process of adding carbon into solid iron-base alloys such as low carbon steels in order to produce a hard surface. The low carbon steel is heated at 870 to 930°C in contact with gaseous, solid or liquid carbon containing substances for several hours and then quenched to obtain hardened surface. The following are the important methods of carburizing:

(*a*) Pack carburizing which uses solid carburizing medium,

(*b*) Gas carburizing which uses suitable hydro-carbon gas, and

(*c*) Liquid carburizing which uses fused baths of carburizing salts.

11.8.2. Nitriding

It is the process of introducing nitrogen into the surface of certain types of steel (*i.e.*, the steel containing Al and Cr) by heating the surface and keeping the surface at a suitable temperature in contact with partially dissociated ammonia.

11.8.3. Cyaniding

It is the process of introducing carbon and nitrogen into the surfaces of steel by heating the surface to a suitable temperature and holding it in contact with molten cyanide. The low carbon steel is heated between 800 to 875°C in a molten sodium cyanide bath for a definite period of time and then quenched in oil or water.

11.8.4. Carbonitriding

It is the process of introducing carbon and nitrogen into the solid ferrous alloy which is kept in contact with suitable gases such as hydrocarbon, carbon monoxide and ammonia.

11.8.5. Flame Hardening

It is the process of introducing local hardness at the surface of the steel by oxyactylene flame. The surface of the steel is heated upto 1500°F by the flame and then quenched.

11.8.6. Induction Hardening

It is the process of heating the surface of carbon steel upto 1500°F by an alternating magnetic field and then quenching immediately.

11.9. SURFACE TREATMENTS

Surface treatments are the processes in which the coatings are applied on the surface of metals or alloys in order to improve corrosion or water resistance, to lengthen the useful life of a part and to serve as an finish. Coatings are of the following types:

1. Metallic coating, and
2. Non-metallic coatings.

11.9.1. Metallic Coatings

They are applied by the following methods:

(*i*) Hot dipping,
(*ii*) Metal spraying,
(*iii*) Electroplating,
(*iv*) Cladding, and
(*v*) Facing.

11.9.2. Non-Metallic Coatings

They consist of:

(*i*) Oxide coating,
(*ii*) Phosphate coating, and
(*iii*) Enamel coating.

11.10. ATOMIC STRUCTURE

An atom consists of a nucleus containing protons and neutrons and with electrons moving about the nucleus in definite paths called electron shells. The nucleus is an extremely dense structure and constitutes about 99.98% of the weight of the atom. The protons carries a positive electrical charge of fixed value whereas the neutron carries no charge. The protons and neutrons have approximately the same mass, and their total number indicates the atomic weight.

The electron is a very light particle only about 1/2000 the mass of a proton, but it carries an electrical charge of equal magnitude and opposite sign. The charge on a proton is denoted by $+e$ and that on electron by $-e$, where $e = 1.602 \times 10^{-19}$ coulomb. The number of either electrons or protons is the atomic number.

11.10.1. Electronic Configurations

A system is the most stable when it has minimum energy. Therefore, the electrons occupy the low energy levels first. This will make the atom more stable. The electrons will fill first of all *K* shell, then *L* shell and so on. This arrangement of electrons in various shells of an atom of an electrons is known as *electronic configuration* Or the distribution of element in different shells and sub-shells of an atom in accordance with the Pauli's Principal is known as electronic configuration.

11.10.2. Ionic Bond

In this ionic bond atoms of different elements transfer electrons one to the other so that both have stable outer shells and at the same time become ions, one positively and other negatively charged. The attractive force is electrostatic.

11.10.3. Covalent Bond

A covalent bond results from sharing the pairs of valence electrons by two or more atoms.

11.10.4. Metallic Bond

In the metallic bond atoms of the same or different elements give up their valence electrons to form an electron cloud (generally referred to as an electron gas) throughout the space occupied by the atoms.

11.10.5. van der Waal's Bond

A van der Waal's bond arises because there are at any one time a few more electrons on one side of the nucleus than on the other side, the centres of positive and negative charge donot coincide at this moment and thus a weak dipole is produced. A force then exists between opposite ends of dipoles in adjacent atoms and tends to draw them together. The bond produced by these fluctuating dipoles is non-directional and is weaker than the hydrogen bond.

11.10.6. Space Lattices and Crystal Classes

A crystal is formed whenever atoms arrange themselves in an orderly three-dimensional pattern, in which rows can be identified running in various directions, along which the atoms are regularly spaced. The locations of the atoms and their particular arrangement in a given crystal are described by means of a *space lattice*, which is a three-dimensional network of straight lines and acts as a coordinate system in space.

In crystal structure, a certain fundamental grouping of atoms is repeated indefinitely in all three dimensions. This fundamental grouping is called a *unit cell.* In case of a unit cell, the three axes may be at right angles or the axes may be inclined to each other at different angles.

11.10.7. Indices of Planes

In every crystal-system, there exist planes which have maximum concentration of atoms. Such planes are called rich planes. To describe rich planes and direction in a crystal, a system of crystallographic indices called *Miller*-indices is used. These indices are based on the intercepts of a plane with three crystal axes *i.e.* three edges of the unit cell.

11.11. IMPERFECTIONS IN METAL CRYSTALS

The imperfections or defects in crystals are of three types. They are:

1. Point defects,
2. Line defects, and
3. Surface defects.

11.11.1. Point Defects

A point defect comes about because of the absence of matrix atoms, the presence of an impurity or a matrix atom in the wrong place (a site not occupied by the atom in the normal lattice). Point defects are classified as:

(*a*) Vacancies,
(*b*) Impurities,
(*c*) Interstitialcies, and
(*d*) Electric defect.

The absence of an atom from a normally occupied site is called *vacancy*. Such defects arise either due to imperfect packing during crystallisation or from thermal vibrations of atoms at high temperature.

The presence of a foreign atom in the crystal structure is called an *impurity*.

The number of point defects in a crystal lattice at a given temperature at equilibrium is given by

$$n = Ne^{-E_d/k.T}$$

where n = Number of defects,

N = Total number of atomic sites per cubic metre,

E_d = Energy of activation necessary to form the defect,

k = Boltzmann constant, and

T = Absolute temperature.

11.11.2. Line Defects

These defects are one dimensional defects and extend along some directions. These defects are also called dislocations. A dislocation may be defined as a disturbed region between two substantially perfect parts of a crystal. The dislocation is responsible for the phenomenon of slip, by which most metals deform plastically. The line defects are of two types:

(*a*) Edge dislocation, and
(*b*) Screw dislocation.

(*a*) **Edge dislocation.** An edge dislocation involves an extra row of atoms, either above or below the slip plane. The edge dislocation lies perpendicular to the Burger's Vector. The edge dislocation moves (in its slip plane) in the direction of the Burger's Vector (slip direction).

(*b*) **Screw dislocation.** In the screw dislocation, the distortion follows a helical or screw path. A screw dislocation lies parallel to its Burger's Vector. A screw dislocation moves in a direction perpendicular to the Burger's Vector. Screw dislocation is useful in explaining crystal growth as well as slip in plastic deformation.

11.11.3. Surface Defects

Surface defects includes; grain boundaries, tilt boundaries and twin boundaries.

(*a*) **Grain boundaries** are those surface defects that separate crystal (grain) of different orientation. A grain boundary is formed when two growing grain surfaces meet.

(*b*) **Tilt boundaries.** If the angle between two grain boundaries is less than 10°, then the boundaries are called tilt boundaries.

(*c*) **Twin boundaries.** A boundary, which separates two parts of a crystal having the same orientation and the two parts look like mirror image of each other, is called a twin boundary.

11.11.4. Burger's Vector

The magnitude and direction of the displacement of atoms is determined by a vector, which is known as Burger's Vector. Thus the Burger's Vector can completely describe the dislocation if the orientation of dislocation line is known. The Burger's Vector is determined as follows:

The step-by-step circuit (in counterclockwise or clockwise) is made around a dislocation in an imperfect crystal. The end-point will not coincide with the starting point. The vector connecting the end point with the starting point is known as the Burger's Vector of the dislocation.

11.12. DIFFUSION IN SOLIDS

Diffusion is the movement of atoms and molecules to new sites within a material, tending to make the composition of all parts uniform. Diffusion takes place due to thermal agitation.

11.12.1. Laws of Diffusion

The atomic movement in diffusion is generally in a direction from higher to lower concentration. The rate of diffusion, given by Fick, is expressed as

$$J = \frac{F}{R} \qquad \text{...}(i)$$

where J = Rate of diffusion, F = Driving force, and

R = Resistance to movement of atoms through lattice

$$= d \cdot \frac{L}{A} \qquad \text{...}(ii)$$

where d = A constant, L = Distance of atomic movement, and

A = Area of plane of diffusion.

Substituting equation (*ii*) in equation (*i*), we get

$$J = \frac{1}{d} A \cdot \left(\frac{F}{L}\right) = D \cdot A \cdot \left(\frac{F}{L}\right) \quad \left(\text{where } D = \frac{1}{d}\right) \qquad \text{...}(iii)$$

Here D is known as **co-efficient of diffusion.** It is defined as the amount of substance diffusing in unit time across a unit area through a unit concentration gradient and is generally expressed in cm²/sec (or m²/sec).

11.12.2. Fick's First Law of Diffusion

In equation (*iii*), the rate of diffusion J and driving force F are represented as

$$J = \frac{dm}{dt} = \text{Rate of mass flow or rate of diffusion}$$

$$\frac{F}{L} = -\frac{dc}{dx} \text{ (Expressing in terms of concentration gradient)}$$

(–ve sign indicates that concentration decreases in the direction of diffusion)

then equation (*iii*) becomes as $\dfrac{dm}{dt} = -D \cdot A \cdot \dfrac{dc}{dx}$ ...(*iv*)

The equation (*iv*) represents Fick's first law of diffusion, which states that the rate of diffusion is directly proportional to the concentration gradient and area of plane of diffusion.

11.12.3. Fick's Second Law of Diffusion

Fick's first law of diffusion deals with stationary states of diffusion only whereas the second law deals with those states of diffusion in which the concentration in a fixed region changes with time. According to Fick's second law of diffusion,

$$\frac{dc}{dt} = \frac{d}{dx} \cdot \left(D \cdot \frac{dc}{dx} \right)$$

where $\dfrac{dc}{dt}$ = Rate of accumulation of diffusing material at a point where the concentration gradient is $\dfrac{dc}{dx}$.

D = Co-efficient of diffusion.

11.12.4. Factors Affecting Diffusions

The following factors influences the diffusion co-efficients:

(*i*) Temperature, (*ii*) Concentration,

(*iii*) Pressure, (*iv*) Crystal structure, and (*v*) Grain size.

11.12.5. Solid Solution

A solid solution is simply a solution in the solid state and consists of two kinds of atoms combined in one type of space lattice. A solid solution is the result of metals dissolving in each other's crystal lattice. Solid solutions are conductors, but not so good as the pure metals on which they are based.

11.13. UNIT CELL

It is the smallest group of atoms possessing symmetry of the crystal which when repeated in all directions, will develop the crystal lattice. The unit-cell may be:

1. Body-centred cubic structure or B.C.C.
2. Face-centred cubic structure of F.C.C.

11.13.1. Body-Centred Cubic Structure

If a unit-cell of a structure has one atom in the centre of the cube and one atom each at all the corners, then the structure is known as body-centred cubic structure. The unit-cell of body-centred cubic structure contains two atoms. The relation between atomic radius (r) and length of the cube edge (a) for body-centred cubic structure is given by

$$r = \frac{a\sqrt{3}}{4}.$$

11.13.2. Face-Centred Cubic Structure

If a unit-cell of a structure has one atom at each corner of the cube and one atom at the intersection of the diagonals of each of the six faces of the cube, the structure is known as face centred cubic structure. The unit-cell of the face-centred cubic structure contains four atoms. The relation between atomic radius (r) and length of the cube edge (a) for face-centred cubic structure is given by

$$r = \frac{a\sqrt{2}}{4}.$$

11.13.3. Atomic Packing Factor

It is defined as the ratio of the volume of the atoms per unit-cell to the total volume occupied by the unit cell. The atomic packing factor

$$= 0.74 \quad \text{...in a face-centred cube}$$
$$= 0.68 \quad \text{...in a body-centred cube.}$$

11.14. THERMAL PROPERTIES OF MATERIALS

The following are the important thermal properties of the material:

1. Specific heat,
2. Thermal conductivity, and
3. Thermal expansion.

11.14.1. Specific Heat

The quantity of heat, that must be added to a unit mass of the solid to raise its temperature by one degree, is known as specific heat. If it is assumed that no structural or chemical changes are involved, then specific heat (C) is expressed mathematically as

$$C = \frac{1}{m}\frac{dE}{dT}$$

where m = Mass, E = Total energy content, and T = Temperature.

The unit of specific heat in Metric System is given as (cal/g) K (or °C).

11.14.2. Thermal Conductivity

It is defined as the amount of heat in calories which flows in one second through a cm cube of material with its opposite faces maintained at a temperature difference of 1°C. The heat flow through an area A, perpendicular to the direction of flow is given by

$$Q = -kA\frac{dT}{dx}$$

where k = Thermal conductivity,

x = Then distance in the direction of heat flow, and

$\frac{dT}{dx}$ = Thermal gradient.

The –ve sign is necessary as heat flows in the direction of a negative thermal gradient.

11.14.3. Thermal Expansion

The dimensions of a material changes when thermal energy is added to a material. The increase in the dimensions of the material due to the addition of thermal energy is known as thermal expansion. The property of a material, responsible for this increase in dimensions, is known as co-efficient of thermal expansion.

The co-efficient of (linear) thermal expansion is defined as the rate of change of length with respect to temperature, per unit length. Mathematically, the co-efficient of linear thermal expansion (α) is written as

$$\alpha = \frac{1}{L}\frac{dL}{dT} = \frac{d\varepsilon}{dT}, \quad \text{where } L = \text{Length}$$

$$\frac{dL}{L} = d\varepsilon, \text{ where } \varepsilon = \text{Strain.}$$

11.15. ELECTRICAL PROPERTIES OF SOLIDS

The following are the important electrical properties of solids:

(*i*) Conductivity,
(*ii*) Dielectric strength,
(*iii*) Resistivity, and
(*iv*) Thermoelectricity.

11.15.1. Conductivity

The property of a material owing to which the electrical current flows easily through the solid material is known as electrical conductivity.

Mathematically conductivity is given by $\sigma = \frac{L}{RA}$

where L = Length of conductor, R = Resistance in ohms of a conductor

A = Area of conductor.

The unit of conductivity is ohm^{-1} cm^{-1}.

Electrical conductivity allows the movement of electrical charge from the location to another. Charge is carried by ions or electrons.

(*i*) **Electronic conductivity.** In electronic conductivity, the charge carriers are electrons.

(*ii*) **Ionic conductivity.** In ionic conductivity, the charge carriers are negative or positive ions. On applying external electrical field, the ions receive more energy and are accelerated in the direction of field resulting in the net flow of ions.

Mathematically conductivity is given by $\sigma = nq\mu$

where n = The number of charge carriers in a material,

q = Charge carried by each carrier, and

μ = The net velocity of carrier per unit voltage gradient.

(*iii*) **Super conductivity.** Super conductivity is the phenomenon of a bright drop in electrical resistance of certain metals at temperature approaching absolute zero. Metals, which are comparatively poor conductor at room temperature tend to become good conductors at very low temperature.

11.15.2. Dielectric Strength

The maximum voltage per unit thickness of a dielectric material which a dielectric can withstand before failure occurs, is known as dielectric strength. Hence a material having high dielectric strength can withstand sufficiently high voltage field across it before it will break down and conduct.

11.15.3. Resistivity

The property of a material owing to which, the material resists the flow of electricity through it, is known as resistivity. It is the reciprocal of electrical conductivity. Mathematically resistivity is given by

$$\rho = \frac{R_0 A}{L}$$

Unit of ρ are 'ohm cm'.

11.15.4. Thermoelectricity

If two dissimilar metals are joined and this junction is then heated, a small voltage in the millivolt range is produced. This effect is known as thermoelectric effect.

11.16. DIELECTRIC LOSS

If an alternating voltage V is applied across a capacitor C having vacuum as dielectric, then the resulting current I leads the voltage by 90°. But when a real dielectric is present, the current leads the voltage by (90° – δ), where δ is the measure of the dielectric power loss.

The dielectric power loss = $VI \tan \delta$

where $\tan \delta$ = Loss tangent or dissipation factor.

11.16.1. Permittivity or Dielectric Constant

When a thin slice of dielectric of thickness *t*, is placed in an electric field *E* of the capacitor plates, the static charges, on the surface of the thin slice due to polarization, is developed. The charge per unit area of the dielectric surface is known as electric flux density (*D*) and is given by

$$D = \frac{Q}{A} C/m^2$$

where Q = Charge, and A = Area of dielectric surface.

The ratio of electric flux density and the electric field (*i.e.*, D/E) is known as permittivity of the dielectric (or dielectric constant) and is expressed as

$$\varepsilon = \frac{D}{E} \text{ Farad/meter } \left(\frac{F}{m}\right)$$

or,

$$D = \varepsilon E.$$

Relative permittivity of the dielectric medium is defined as the ratio of permittivity of the dielectric to the permittivity of vacuum. It is a dimensionless number.

11.16.2. Frequency and Temperature Dependence of Dielectric Constant

The dielectric constant (or permittivity) increases more rapidly with rising temperature. But as the frequency increases, there is a reduction of polarization and hence the dielectric constant decreases. When the alternating electric field is not producing any polarization due to high frequency, the process is known as relaxation and the frequency beyond which the polarization no longer follows the applied field, is known as the relaxation frequency.

11.16.3. Piezoelectricity

The phenomena of the appearance of the charges on the surface of a crystal when it is subjected to the mechanical deformation, is known as piezoelectricity. This is due to the relative displacement of the ions. Piezoelectric crystals are used in transducers which convert electrical energy to mechanical energy.

11.16.4. Ferroelectricity

The phenomena of the spontaneous polarization of a dielectric due to the presence of permanent electric dipoles, is known as ferroelectricity. When a dielectric possessing permanent dipoles is subjected to an electric field, the dipoles line up with the applied field. The dipoles lengths are changed with voltage gradients. Thus the material either elongates or contracts in the electric field.

Like ferromagnetic materials, ferroelectrics have a *Curie temperature,* at temperature higher than the Curie point thermal agitation prevents spontaneous alignments of dipoles, while at lower temperatures the material becomes ferroelectric.

11.17. BAND THEORY OF SOLIDS

The band theory (or zone theory) of solids is very useful for understanding the structures and properties of metals, alloy and non-metallic solids. According to this theory, the electrons move in a periodic field provided by the lattice. When the electrons (being waves) are subjected to oscillating potential field, they undergo diffraction in accordance with the Bragg's diffraction law, given by

$$n\lambda = 2d \sin \theta$$

where θ = Angle of incidence of the electron beam,

λ = Wave, length of electron,

d = Spacing between the planes of crystal, and

n = Any integer.

And the wave number (k) is given by $k = \dfrac{2\pi}{\lambda} = \dfrac{n\pi}{d \sin \theta}$ $\left(\text{Substituting } \lambda = \dfrac{2d \sin \theta}{n}\right)$

The above equation shows that there is at least one series of values of k corresponding to the integer n, for which electrons are diffracted and they to not pass freely through the crystal. Hence corresponding to these values of k, the energies given by $E = \dfrac{h^2 k^2}{8\pi^2 m}$ do not exist. This means that there can be no electrons which have such energies or we can say that certain energy levels are forbidden. These forbidden energy levels from forbidden energy gaps.

11.17.1. Intrinsic and Extrinsic Semi-conductors

Some non-metallic elements have an electron structure which allows for more valence electrons to be made available by thermal energy than in the majority of insulators. These elements are known as semi-conductors. They are insulator at absolute zero but develop significant conductivities, at room temperature.

The semi-conduction produced by thermal energy alone is known as *intrinsic semi-conduction* whereas the conduction resulting from electrons are from-holes provided by impurity atoms is known as extrinsic semi-conduction.

11.17.2. Uses of Semi-conductors

Semi-conductors are used in fabricating devices such as *p–n* rectifiers, transistors, photo cells, integrated circuits and thermistors etc. The mobility of the charge carriers, their concentration and conductivity of a semi-conductor are all affected by temperature. The different elements and compounds such as sulphides and oxides are widely employed as semi-conducing materials.

11.17.3. The p–n Junction (Rectifier)

A *p–n* junction is formed if donor impurities are introduced into one side and acceptor into the other side of a single crystal of silicon or germanium. The semi-conductor diode and transistor depends upon the properties of this junction. The donor ions are represented by plus signs and the acceptor ions are indicated by minus signs. The *p*-side of the junction has holes as the majority carriers and the *n*-side contains many conduction band electrons.

11.17.4. Transistor

Transistor is a device which amplifies signals when two *p–n* junctions are arranged back to back in a single crystal. There are two types of transistors (*i*) *p–n–p* type and (*ii*) *n–p–n* type. In the *p–n–p* type, the *p* regions are separated by a *n* region and the *n–p–n* transistor has two *n* region separated by *p*-region. There are two junction J_1 and J_2 the transistor. The *p*-region of the junction J_1 is called *emitter* and *n*-region the base and the *p*-region of the junction J_2 is known as collector. For amplifier action, junction J_2 is forward biased and J_2 is reverse biased.

11.17.5. Photo Cells

Photo cells are the devices which are used for converting radiation energy into electrical energy. The followings are the different types of photo cells which are mostly used:

(*i*) Photo emissive cell, (*ii*) Photo conductive cell, and

(*iii*) Photo voltaic cell.

Photo emissive cell depends on the emission of electrons from a metallic cathode when it is exposed to light or other radiations.

11.18. MAGNETIC PROPERTIES OF SOLIDS

The following are the important magnetic properties of solids :

1. Permeability,
2. Coercive force, and
3. Hyteresis.

11.18.1. Permeability

The ratio of magnetic induction (*B*) to the intensity of the magnetizing field (*H*) is known as magnetic permeability. Mathematically, it is expressed as

$$\mu = \frac{B}{H}$$

A vacuum is assigned a permeability of 1. The diamagnetic materials have a permeability of less than 1 whereas paramagnetic materials have a permeability greater than 1.

11.18.2. Coercive Force

The opposing magnetizing force which is necessary to remove previous magnetization or residual magnetization is known as coercive force. It is small in magnetically soft materials.

11.18.3. Hysteresis

Magnetic hysteresis is defined as the lagging of flux density behind magnetizing force. The *B–H* curve is a closed loop and is known as hysteresis loop.

Consider a piece of material placed in a magnetic field of strength H. A force will act on the piece of material and it will be magnetised. If intensity of magnetisation is J then the susceptibility (k), which is the property of the material, is given by

$$k = \frac{J}{H}.$$

11.19. DIAMAGNETIC MATERIALS

The materials, for which the susceptibility (k) is small and negative, are known diamagnetic materials.

Examples are: copper, gold, silver, bismuth. They are weakly repelled by the field.

11.19.1. Paramagnetic Materials

The materials, for which the susceptibility (k) is small and positive, are known paramagnetic materials.

Examples are: the alkalies, the alkaline, earth etc. They are weakly attracted.

11.19.2. Ferromagnetic Materials

The materials, for which the susceptibility (k) is large and positive, are known ferromagnetic materials.

Examples are: iron, nickel, cobalt etc. They are strongly attracted by the magnetic field.

11.19.3. Soft Magnetic Materials

The materials with higher permeability low hysteresis loss and low coercive force are called soft magnetic materials.

11.19.4. Hard Magnetic Materials

The materials with high coercive and high saturation magnetisation are known as hard magnetic materials.

II. OBJECTIVE TYPE QUESTIONS

Tick mark the most appropriate statement of the multiple choice answers:

Basic Definitions

1. The total energy E, of an electron in term of principal quantum (n) is given by

(*a*) $E \propto \frac{1}{n}$ ☐ (*b*) $E \propto \frac{1}{n^2}$ ☐

(*c*) $E \propto n$ ☐ (*d*) $E \propto n^2$. ☐

2. The total energy E, of the electron in any fixed orbit in term of the atomic number (Z) is given by

(a) $E \propto \frac{1}{Z}$ ☐ (b) $E \propto \frac{1}{Z^2}$ ☐

(c) $E \propto Z$ ☐ (d) $E \propto Z^2$. ☐

3. In the periodical table, elements are arranged in order of increasing

(a) atomic number ☐ (b) atomic weight ☐

(c) molecular weight ☐ (d) none of the above. ☐

4. The atomic number of an element is given by the number of

(a) protons ☐ (b) electrons ☐

(c) sum of protons and electrons ☐ (d) either (a) or (b). ☐

5. The extra electrons, in the outermost incomplete main shell, are called

(a) valence electrons ☐ (b) extra electrons ☐

(c) surplus electrons ☐ (d) none of the above. ☐

6. The valence of an element with structure $1s^2, 2s^1, 2p^2$ (carbon) is

(a) 1 ☐ (b) 2 ☐

(c) 3 ☐ (d) 4. ☐

7. The electronic configuration of sodium with atomic number ($Z = 11$) is

(a) $1s^2, 2s^2, 2p^4, 3s^3$ ☐ (b) $1s^2, 2s^2, 2p^6, 3s^1$ ☐

(c) $1s^2, 2s^2, 3s^2, 3p^5$ ☐ (d) none of the above. ☐

8. The maximum number of electrons present in a sub-shell is equal to

(a) $2(2l + 1)$ ☐ (b) $2l(l + 1)$ ☐

(c) $l(2l + 1)$ ☐ (d) $2(l + 1)$. ☐

where l = Orbital quantum number.

9. The maximum number of electrons in any main shell is equal to

(a) $3n^2$ ☐ (b) n^2 ☐

(c) $2n^2$ ☐ (d) $4(n^2 + 1)$. ☐

10. The chemical bonds are formed due to the forces between atoms. These forces are

(a) electromagnetic ☐ (b) electrostatic ☐

(c) gravitational ☐ (d) none of the above. ☐

11. The nature of the atomic bond found in diamond is

(a) metallic ☐ (b) ionic ☐

(c) covalent ☐ (d) none of the above. ☐

12. When the atoms in a molecule are at their equilibrium distance, their

(a) potential energy is minimum ☐

(b) potential energy is maximum ☐

(c) potential energy is average of minimum and maximum ☐

(d) none of the above. ☐

13. The bond between two hydrogen atom is
(a) ionic bond ☐ (b) covalent bond ☐
(c) hydrogen bond ☐ (d) van der Waal's bond. ☐

14. The type of atomic bond found in CH_4 is
(a) covalent ☐ (b) ionic ☐
(c) metallic ☐ (d) van der Wall's. ☐

15. Which one of the following is a secondary bond ?
(a) covalent bond ☐ (b) ionic bond ☐
(c) metallic bond ☐ (d) van der Waal's bond. ☐

16. Primary bond as compared to secondary bond is
(a) stronger but less stable ☐ (b) stronger and more stable ☐
(c) weaker and less stable ☐ (d) weaker but more stable. ☐

17. The bond which exists between a metallic atom (sodium) and a non-metallic atom (chlorine) is
(a) covalent bond ☐ (b) ionic bond ☐
(c) metallic bond ☐ (d) van der Waal's bond. ☐

18. The number of covalent bonds formed by an element is equal to
(a) $4 - N$ ☐ (b) $6 - N$ ☐
(c) $8 - N$ ☐ (d) $W - N$. ☐
where N = Number of electrons outside the full shell.

19. The numbers of the atom consists of
(a) electrons and protons ☐ (b) electrons and neutrons ☐
(c) protons and neutrons ☐ (d) none of the above. ☐

20. The ratio of the weight of nucleus of the atom to the total weight of the atom is approximately equal to
(a) 0.5 ☐ (b) 0.667 ☐
(c) 0.8 ☐ (d) 0.999. ☐

21. The atomic weight of an atom is equal to the sum of the number of
(a) electron and protons ☐ (b) electrons and neutrons ☐
(c) protons and neutrons ☐ (d) none of the above. ☐

22. The locations of the atoms and their particular arrangement in a given crystal are described by means of
(a) potential energy ☐ (b) space lattice ☐
(c) intermolecular bond ☐ (d) Diffusion. ☐

23. Which one of the following is a correct statement? In space lattice
(a) the axes are at right angles ☐
(b) the spacing of the lines differ along different axes ☐
(c) the axes are inclined to each other at different angles ☐
(d) all of the above ☐
(e) none of the above. ☐

24. In each type of crystal structure a certain fundamental grouping of atoms is repeated indefinitely in all three dimensions. This fundamental grouping is called

(*a*) a unit molecule ☐ (*b*) a bond ☐

(*c*) a unit cell ☐ (*d*) atomic space. ☐

25. If the axes of a unit cell are at right angles and the axial lengths are equal, then the unit cell is known as

(*a*) cubic ☐ (*b*) tetragonal ☐

(*c*) monoclinic ☐ (*d*) ortho-rhombic. ☐

26. If the axes of a unit cell are at right angles and only two axial lengths are equal, then the unit cell is known as

(*a*) cubic ☐ (*b*) tetragonal ☐

(*c*) monoclinic ☐ (*d*) ortho-rhombic. ☐

27. If the axes of a unit cell are at right angles and all the three axial lengths are different, then unit cell is known as

(*a*) cubic ☐ (*b*) tetragonal ☐

(*c*) monoclinic ☐ (*d*) ortho-rhombic. ☐

28. If only the two axes of a unit cell are at right angles and all the three axial lengths are different, then the unit cell is known as

(*a*) cubic ☐ (*b*) tetragonal ☐

(*c*) monoclinic ☐ (*d*) ortho-rhombic. ☐

29. Vacancies are the

(*a*) line defects ☐ (*b*) point defects ☐

(*c*) surface defects ☐ (*d*) volume defects. ☐

30. Edge dislocation is a

(*a*) line defects ☐ (*b*) point defects ☐

(*c*) surface defects ☐ (*d*) volume defects. ☐

31. Cracks are

(*a*) line defects ☐ (*b*) point defects ☐

(*c*) surface defects ☐ (*d*) volume defects ☐

32. Twin boundaries are

(*a*) line defects ☐ (*b*) point defects ☐

(*c*) surface defects ☐ (*d*) volume defects. ☐

33. In a point defect is introduced in a crystal, then the internal energy of the crystal as compared to that of the perfect crystal

(*a*) decreases ☐ (*b*) increases ☐

(*c*) remains constant ☐ (*d*) none of the above. ☐

34. The Burger's Vector lies at an angle to the line of dislocation, along the axis of rows of atoms in the same plane, in

(*a*) screw dislocation ☐ (*b*) edge dislocation ☐

(*c*) cracks ☐ (*d*) vacancies. ☐

35. The Burger's Vector lies parallel to the dislocation, line along the axis of a line of atoms in the same plane, in
(*a*) screw dislocation ☐ (*b*) edge dislocation ☐
(*c*) cracks ☐ (*d*) vacancies. ☐

36. Line imperfection in a crystal is called
(*a*) Frenkel defect ☐ (*b*) Schottky defect ☐
(*c*) edge dislocation ☐ (*d*) none of the above. ☐

37. The missing of an atom from its normal site is
(*a*) Frenkel defect ☐ (*b*) Schottky defect ☐
(*c*) interstial impurity ☐ (*d*) none of the above. ☐

38. Line defects play an important role in
(*a*) deformation ☐ (*b*) self diffusion ☐
(*c*) climb ☐ (*d*) none of the above. ☐

39. The missing of an atom from its normal site is called
(*a*) edge dislocation ☐ (*b*) vacancy ☐
(*c*) screw dislocation ☐ (*d*) none of the above. ☐

40. The number of point defects in a crystal can be increased by
(*a*) hammering ☐ (*b*) rolling ☐
(*c*) quenching ☐ (*d*) all of the above ☐
(*e*) none of the above. ☐

41. The number of point defects at equilibrium at a given temperature is given by
(*a*) $n = Ne^{E_d/kT}$ ☐ (*b*) $n = Ne^{-E_d/kT}$ ☐
(*c*) $n = N \cdot E_d \cdot T$ ☐ (*d*) $n = N \cdot k \cdot T$ ☐
where n = Number of defects, N = Total number of atomic sites per cubic metre,
E_d = Energy of activation necessary to form defect,
k = Boltzmann constant, T = Absolute temperature.

42. The line defects are formed in the process of
(*a*) solidification of metals ☐ (*b*) recrystallisation of metals ☐
(*c*) deformation of metals ☐ (*d*) all of the above ☐
(*e*) none of the above. ☐

43. The magnitude and direction of the displacement of atom is determined by
(*a*) axis of symmetry ☐ (*b*) plane of symmetry ☐
(*c*) Burger Vector ☐ (*d*) none of the above. ☐

44. Burger Vector in case of edge dislocation is
(*a*) inclined to the dislocation line ☐ (*b*) parallel to dislocation line ☐
(*c*) perpendicular to the dislocation line ☐ (*d*) none of the above. ☐

45. Burger Vector in case of screw dislocation is

(*a*) inclined to the dislocation line ☐ (*b*) parallel to dislocation line ☐

(*c*) perpendicular to the dislocation line ☐ (*d*) none of the above. ☐

46. Dislocation play an important role in the process of

(*a*) elastic deformation ☐ (*b*) plastic deformation ☐

(*c*) self diffusion ☐ (*d*) none of the above. ☐

47. If the angle between two grain boundaries is less than 10°, then the boundaries are called

(*a*) high angle grain boundary ☐ (*b*) tilt boundary ☐

(*c*) twin boundary ☐ (*d*) none of the above. ☐

48. If the angle between two grain boundaries is more than 10 to 15°, then the boundaries are called

(*a*) high angle grain boundary ☐ (*b*) tilt boundary ☐

(*c*) twin boundary ☐ (*d*) none of the above. ☐

49. If the two parts of crystal having the same orientation and look-like the mirror image of each other, are separated by a boundary, then the boundary is known as

(*a*) high angle grain boundary ☐ (*b*) tilt boundary ☐

(*c*) twin boundary ☐ (*d*) none of the above. ☐

Diffusion in Solids

50. The diffusion is generally in a direction

(*a*) from lower concentration to higher concentration ☐

(*b*) from higher concentration to lower concentration ☐

(*c*) of constant concentration ☐

(*d*) none of the above. ☐

51. The rate of diffusion is expressed as

(*a*) $J = F \times R$ ☐ (*b*) $J = \dfrac{R}{F}$ ☐

(*c*) $J = \dfrac{F}{R}$ ☐ (*d*) $J = \dfrac{1}{R \times F}$. ☐

where J = Rate of diffusion, F = Driving force, and R = Resistance to movement.

52. Resistance R to diffusion

(*a*) decreases with increase of temperature ☐

(*b*) decreases with decrease of temperature ☐

(*c*) remains constant with the increase of temperature ☐

(*d*) none of the above. ☐

53. The resistance R to diffusion is given by

(*a*) $R = d \cdot \dfrac{A}{L}$ ☐ (*b*) $R = d \cdot \dfrac{L}{A}$ ☐

(*c*) $R = d \cdot A \cdot L$ ☐ (*d*) $R = \dfrac{1}{d \cdot A \cdot L}$. ☐

where A = Area of plane of diffusion, L = Distance of atomic movement, and d = A constant.

54. The amount of substance diffusing in unit time across a unit area through a unit concentration gradient it known as

(*a*) first law of diffusion ☐ (*b*) second law of diffusion ☐

(*c*) co-efficient of diffusion ☐ (*d*) co-efficient of viscosity. ☐

55. The co-efficient of diffusion is generally expressed in

(*a*) m/sec ☐ (*b*) m^2/sec ☐

(*c*) m^3/sec ☐ (*d*) m-sec. ☐

56. The Fick's first law of diffusion is given by

(*a*) $\frac{dm}{dt} = \frac{d}{dx}\left(D \cdot \frac{dc}{dx}\right)$ ☐ (*b*) $\frac{dm}{dt} = D \cdot A \cdot \frac{dc}{dx}$ ☐

(*c*) $\frac{dm}{dt} = -D \cdot A \cdot \frac{dc}{dx}$ ☐ (*d*) $\frac{dm}{dt} = D \cdot \frac{dc}{dx}$. ☐

where $\frac{dm}{dt}$ = Rate of mass flow or rate of diffusion, $\frac{dc}{dx}$ = Concentration gradient

D = Co-efficient of diffusion, and A = Area of plane of diffusion.

57. Fick's second law of diffusion is given by

(*a*) $\frac{dc}{dt} = \frac{d}{dx}\left(D \cdot \frac{dc}{dx}\right)$ ☐ (*b*) $\frac{dm}{dt} = D \cdot A \cdot \frac{dc}{dx}$ ☐

(*c*) $\frac{dm}{dt} = -D \cdot A \cdot \frac{dc}{dx}$ ☐ (*d*) $\frac{dm}{dt} = D \cdot \frac{dc}{dx}$. ☐

where $\frac{dc}{dt}$ = Rate of accumulation of the diffusing material at a point where the concentration gradient is $\frac{dc}{dx}$.

58. The diffusion co-efficient is influenced by

(*a*) pressure ☐ (*b*) temperature ☐

(*c*) grain size ☐ (*d*) concentration ☐

(*e*) all of the above ☐ (*f*) none of the above. ☐

59. The diffusion co-efficient is related to the absolute temperature by

(*a*) $D = A \cdot QR \cdot T$ ☐ (*b*) $D = A \cdot e^{-Q \cdot R \cdot T}$ ☐

(*c*) $D = A \cdot e^{-Q/RT}$ ☐ (*d*) $D = A \cdot e^{Q/RT}$ ☐

where A = Frequency factor, R = Gas constant,

T = Absolute temperature, and Q = Activation energy for diffusion to occur.

60. In case of diffusive processes, the entropy

(*a*) decreases ☐ (*b*) increases ☐

(*c*) remains constant ☐ (*d*) none of the above. ☐

61. The diffusion increases when temperature

(*a*) decreases ☐ (*b*) increases ☐

(*c*) remains constant ☐ (*d*) none of the above. ☐

62. In a perfect crystal, diffusion
(*a*) is maximum ☐ (*b*) is normal ☐
(*c*) does not take place ☐ (*d*) none of the above. ☐

63. The activation energy for grain boundary diffusion is
(*a*) more than the volume diffusion ☐ (*b*) equal to that of volume diffusion ☐
(*c*) lower than the volume diffusion ☐ (*d*) none of the above. ☐

64. According to Fick's first law of diffusion, rate of diffusion
(*a*) is directly proportional to the square of concentration gradient ☐
(*b*) is directly proportional to the concentration gradient ☐
(*c*) is inversely proportional to the concentration gradient ☐
(*d*) none of the above. ☐

65. The solid solution of copper and zinc is
(*a*) steel ☐ (*b*) brass ☐
(*c*) monal metal ☐ (*d*) sterling silver. ☐

66. The solid solution of Ni—Cu alloys is
(*a*) steel ☐ (*b*) brass ☐
(*c*) monal metal ☐ (*d*) sterling silver. ☐

67. Solid solutions form most readily when the solvent and the solute atoms have
(*a*) similar sizes but different electron structure ☐
(*b*) different sizes but similar electron structure ☐
(*c*) different sizes and different electron structure ☐
(*d*) similar sizes and similar electron structure. ☐

68. In the solid solution of copper and zinc
(*a*) copper atoms are solvent atoms ☐ (*b*) copper atoms are solute atoms ☐
(*c*) zinc atoms are solute atoms ☐ (*d*) zinc atoms are solvent atoms ☐
(*e*) both (*a*) and (*c*). ☐

69. A phase diagram is also known as
(*a*) equilibrium diagram ☐ (*b*) constitutional diagram ☐
(*c*) indicator diagram ☐ (*d*) both (*a*) and (*h*) ☐
(*e*) none of the above. ☐

70. A phase diagram has
(*a*) Alloy composition are ordinate and temperature as its abscissa ☐
(*b*) temperature as its ordinate and alloy composition as its abscissa ☐
(*c*) temperature as its ordinate and time as its abscissa ☐
(*d*) none of the above. ☐

71. A phase diagram permits to study and control processes such as
(*a*) purification of materials ☐
(*b*) phase separation ☐

(c) solidification of metals and alloys ☐
(d) the structural changes produced by heat treatment, casting etc. ☐
(e) all of the above ☐
(f) none of the above. ☐

72. A solid solution consists of two kinds of atoms combined in one type of space lattice and is simply a solution in the
(a) liquid state of the two elements ☐ (b) solid state ☐
(c) either (a) or (b) ☐ (d) none of the above. ☐

73. The phase diagram shows two different and distinct phases. The phases are
(a) solid solution and gas ☐ (b) liquid metal solution and gas ☐
(c) solid solution and liquid metal solution ☐ (d) none of the above. ☐

Thermal Properties of Materials

74. Mathematically, the specific heat (C) is expressed as
(a) $C = m\frac{dE}{dt}$ ☐ (b) $C = \frac{dE}{dt}$ ☐
(c) $C = \frac{1}{m}\frac{dE}{dt}$ ☐ (d) $C = \frac{1}{m^2}\frac{dE}{dt}$. ☐
where m = Mass, E = Total energy, and t = Temperature.

75. The line, below which the alloy is in solid state and where solidification completes, is known as
(a) liquidus ☐ (b) solidus ☐
(c) gaseous ☐ (d) none of the above. ☐

76. The line, above which the alloy is in liquid state and where solidification starts, is known as
(a) liquidus ☐ (b) solidus ☐
(c) gaseous ☐ (d) none of the above. ☐

77. The Lever Rule is used to determine
(a) the chemical composition of each phase ☐
(b) the phases that are present ☐
(c) the amount of each phase at any given temperature ☐
(d) all of the above. ☐
(e) none of the above. ☐

78. According to Lever-rule
(a) the relative lengths of the lever arms must balance ☐
(b) the amount of phase present mast balance ☐
(c) the relative lengths of the level arms multiplied by the amounts of the phase present must balance ☐
(d) none of the above. ☐

79. The phase rule establishes the relationship between
(a) number of degrees of freedom and number of phases ☐
(b) number of degrees of freedom and number of components ☐

(*c*) number of degrees of freedom, number of components and number of phases ☐
(*d*) none of the above. ☐

80. The phase rule is expressed mathematically as
(*a*) $P = F + C + 2$ ☐ (*b*) $P + F = C + 2$ ☐
(*c*) $P + F = C$ ☐ (*d*) $P + F = C - 2$. ☐
where P = Number of phases, F = Number of degrees of freedom
C = Number of components in the system.

81. The atoms and molecules of gases
(*a*) have same freedom of motion as of solids ☐
(*b*) have lesser freedom of motion than those in solids ☐
(*c*) have greater freedom of motion than those in solids ☐
(*d*) none of the above. ☐

82. Heat is transmitted through solids by
(*a*) transfer of energy by free electrons ☐ (*b*) elastic vibrations of atoms and molecules ☐
(*c*) both (*a*) and (*b*) ☐ (*d*) none of the above. ☐

83. The metals have a plentiful supply of free electrons, and hence they have
(*a*) high conductivity ☐ (*b*) low conductivity ☐
(*c*) zero conductivity ☐ (*d*) none of the above. ☐

84. The lower conductivity in the insulators is due to the presence of
(*a*) maximum free electrons ☐ (*b*) vibrations of atoms and molecules ☐
(*c*) both (*a*) and (*b*) ☐ (*d*) none of the above. ☐

85. The driving force in case of electronic thermal conduction is provided by
(*a*) an electric field ☐ (*b*) potential gradient ☐
(*c*) thermal gradient ☐ (*d*) none of the above. ☐

86. Heat always flows in the direction of a
(*a*) positive thermal gradient ☐ (*b*) negative thermal gradient ☐
(*c*) zero thermal gradient ☐ (*d*) none of the above. ☐

87. The heat flow through an area A, perpendicular to the direction of flow is
(*a*) directly proportional to the area ☐ (*b*) directly proportional to the thermal gradient ☐
(*c*) both (*a*) and (*b*) ☐ (*d*) none of the above. ☐

88. The presence of pores, impurities, and other defects tends to shorten the mean free path and hence conductivity
(*a*) increases ☐ (*b*) decreases ☐
(*c*) remains constant ☐ (*d*) none of the above. ☐

89. The quantity of heat that must be added to a unit mass of a solid to raise its temperature by one degree is known as
(*a*) specific gravity ☐ (*b*) specific heat ☐
(*c*) specific energy ☐ (*d*) none of the above. ☐

90. The ratio of the heat capacity of a material to that of water is known as
(*a*) specific gravity ☐ (*b*) specific heat ☐
(*c*) specific energy ☐ (*d*) specific force. ☐

91. Molar heat is defined as the specific heat capacity
(*a*) per mole ☐ (*b*) per unit mass ☐
(*c*) per unit volume ☐ (*d*) per unit weight. ☐

92. For most elements at room temperature, the product of the specific heat and molecular weight is approximately equal to
(*a*) 1 ☐ (*b*) 6 ☐
(*c*) 12 ☐ (*d*) 18. ☐

93. If the atoms always vibrate symmetrically about their equilibrium positions, their average spacing would be constant and the thermal expansion will be
(*a*) maximum ☐ (*b*) zero ☐
(*c*) average ☐ (*d*) none of the above, ☐

94. The actual energy curve for atoms of a solid is
(*a*) symmetrical about the vertical line ☐
(*b*) not symmetrical about the vertical line ☐
(*c*) symmetrical about a line passing through origin and making an angle 45° with x-axis ☐
(*d*) none of the above. ☐

95. In case of actual energy curve, at the energy level increases, the average spacing between the atoms
(*a*) decreases ☐ (*b*) increases ☐
(*c*) remains constant ☐ (*d*) none of the above. ☐

96. The rate of change of length with respect to temperature per unit length is known as
(*a*) specific heat ☐ (*b*) co-efficient of thermal expansion ☐
(*c*) co-efficient of viscosity ☐ (*d*) co-efficient of friction. ☐

97. For body-centred cubic crystals, γ is generally various from
(*a*) 0.5 to 1.0 ☐ (*b*) 1.17 to 1.75 ☐
(*c*) 2.1 to 3.4 ☐ (*d*) 4 to 6. ☐

98. The co-efficient of linear expansion (α) for covalent and ionic crystals as compared to that of metallic crystals is
(*a*) higher ☐ (*b*) same ☐
(*c*) lower ☐ (*d*) none of the above. ☐

99. Specific heat, with the increase of temperature
(*a*) increases ☐ (*b*) decreases ☐
(*c*) remains constant ☐ (*d*) none of the above. ☐

100. The amount of heat in calories, which flows in one second through a cm cube of material when its opposite faces are maintained at a temperature difference of 1°C is known as
(*a*) thermal expansion ☐ (*b*) thermal conductivity ☐
(*c*) specific heat ☐ (*d*) specific energy. ☐

Engineering Materials

101. Which one of the following is a ceramic material?

(*a*) zinc ☐ (*b*) iron ☐
(*c*) silicon carbide ☐ (*d*) wood. ☐

102. Which one of the following is a organic material?

(*a*) zinc ☐ (*b*) iron ☐
(*c*) silicon carbide ☐ (*d*) wood. ☐

103. Which one of the following is a ferrous material?

(*a*) zinc ☐ (*b*) iron ☐
(*c*) silicon carbide ☐ (*d*) wood. ☐

104. Which one of the following is a non-ferrous material?

(*a*) zinc ☐ (*b*) iron ☐
(*c*) silicon carbide ☐ (*d*) wood. ☐

105. Which of the following are the mechanical properties of materials?

(*a*) fatigue ☐ (*b*) specific heat ☐
(*c*) creep ☐ (*d*) thermoelectricity ☐
(*e*) all of the above. ☐ (*f*) both (*a*) and (*c*). ☐

106. Which of the following are the electrical properties of materials?

(*a*) fatigue ☐ (*b*) conductivity ☐
(*c*) hysteresis ☐ (*d*) thermoelectricity ☐
(*e*) none of the above. ☐ (*f*) both (*b*) and (*d*). ☐

107. Which of the following are the thermal properties of materials?

(*a*) fatigue ☐ (*b*) specific heat ☐
(*c*) thermal conductivity ☐ (*d*) conductivity ☐
(*e*) both (*b*) and (*c*). ☐

108. Refractive index and reflectivity are the properties of materials known as

(*a*) mechanical properties ☐ (*b*) thermal properties ☐
(*c*) chemical properties ☐ (*d*) optical properties. ☐

109. The ability of a material to sustain loads without failure to known as

(*a*) mechanical strength ☐ (*b*) stiffness ☐
(*c*) toughness ☐ (*d*) ductility. ☐

110. The ability of a material to withstand bending without fracture is known as

(*a*) mechanical strength ☐ (*b*) stiffness ☐
(*c*) toughness ☐ (*d*) ductility. ☐

111. The ability of a material to resist elastic deflection is known as

(*a*) mechanical strength ☐ (*b*) stiffness ☐
(*c*) toughness ☐ (*d*) ductility. ☐

112. The capacity of a material to undergo deformation under tension without rupture is known as

(*a*) mechanical strength ☐ (*b*) stiffness ☐
(*c*) toughness ☐ (*d*) ductility. ☐

113. Choose the wrong statement

(*a*) Hardness is measured by Vicker's Brinell and Rockwell tests. ☐
(*b*) Copper is more tough than cast iron. ☐
(*c*) Silicon carbide is a ceramic material. ☐
(*d*) The capacity of material to withstand deformation under compression without rupture is known as hardness. ☐

114. When a material is subjected to fluctuating or repeated stresses, fracture takes place under a stress whose maximum value is less than the tensile strength of the material. This phenomenon is known as

(*a*) creep ☐ (*b*) fatigue ☐
(*c*) impact ☐ (*d*) malleability. ☐

115. When a material sustains steady loads for long periods of time, the material may continue to deform until they may tend to fracture under the same load. This phenomenon is known as

(*a*) creep ☐ (*b*) fatigue ☐
(*c*) impact ☐ (*d*) malleability. ☐

116. The melting point of mild steel is approximately

(*a*) 600°C ☐ (*b*) 800°C ☐
(*c*) 1000°C ☐ (*d*) 1500°C. ☐

117. The capacity of a material to absorb energy in the elastic range is known as

(*a*) creep ☐ (*b*) fatigue ☐
(*c*) resilience ☐ (*d*) impact. ☐

118. The property of a material to resist penetration by another material is known as

(*a*) toughness ☐ (*b*) hardness ☐
(*c*) stiffness ☐ (*d*) resilience. ☐

119. The capacity of a metal to exhibit considerable elastic recovery upon release of load, is known as

(*a*) toughness ☐ (*b*) hardness ☐
(*c*) stiffness ☐ (*d*) resilience. ☐

120. Which of the following are used as tool materials?

(*a*) tool steels ☐ (*b*) carbon steels ☐
(*c*) cemented carbides ☐ (*d*) ceramic tools ☐
(*e*) diamond tools ☐ (*f*) all of the above. ☐

121. Which of the following are the cast non-ferrous materials?

(*a*) cobalt ☐ (*b*) gormet ☐
(*c*) stellite ☐ (*d*) tantung ☐
(*e*) all of the above ☐ (*f*) none of the above. ☐

122. Stellite is used for machining

(*a*) steel ☐ (*b*) cast steel ☐
(*c*) cast iron ☐ (*d*) stainless steel ☐
(*e*) brass ☐ (*f*) all of the above. ☐

123. Which tool material is the hardest?

(*a*) ceremic tool ☐ (*b*) steel tool ☐
(*c*) diamond tools ☐ (*d*) cemented carbide tools. ☐

124. Choose the wrong statement

(*a*) Ceramic tools are brittle and tend to chip easily. ☐
(*b*) Carbon steel is less wear resistant as compared to high speed steel. ☐
(*c*) Diamond is the costly but hardest known material. ☐
(*d*) Cemented carbide tools are manufactured primarily from aluminium oxide. ☐

125. Choose the correct statement

(*a*) The ultimate strength, yield strength and elastic modulus of steel decreases with the increase of temperature. ☐
(*b*) The ductile properties of steel increases with the increase of temperature. ☐
(*c*) The steel used, for low temperature use, should have resistance to brittle fracture. ☐
(*d*) All of the above. ☐
(*e*) None of the above. ☐

126. Which of the following are the nuclear energy metals?

(*a*) plutonium ☐ (*b*) uranium ☐
(*c*) thorium ☐ (*d*) beryllium ☐
(*e*) all of the above ☐ (*f*) none of the above. ☐

127. The oxides, nitrides, carbides and silicates of metals are known as

(*a*) organic materials ☐ (*b*) ceramic materials ☐
(*c*) ferrous materials ☐ (*d*) non-ferrous materials. ☐

128. The polymeric materials composed of carbon compounds are known as

(*a*) organic materials ☐ (*b*) ceramic materials ☐
(*c*) ferrous materials ☐ (*d*) non-ferrous materials. ☐

129. If a solution has pH value less than seven, the solution is known as

(*a*) basic solution ☐ (*b*) acidic solution ☐
(*c*) neutral solution ☐ (*d*) none of the above. ☐

130. If a solution has pH value greater than seven, the solution is known as

(*a*) basic solution ☐ (*b*) acidic solution ☐
(*c*) neutral solution ☐ (*d*) none of the above. ☐

131. White metal is best suited as a bearing material for

(*a*) lighter duty bearings ☐ (*b*) heavier duty bearings ☐
(*c*) medium duty bearings ☐ (*d*) none of the above. ☐

132. Phosphor bronze is best suited as a bearing material for

(*a*) lighter duty bearings ☐ (*b*) heavier duty bearings ☐
(*c*) medium duty bearings ☐ (*d*) none of the above. ☐

133. Pyrosilicates is in example of

(*a*) chain structure ☐ (*b*) double tetrahedral structure ☐
(*c*) sheet structure ☐ (*d*) frame work structure. ☐

134. Quartz is an example of

(*a*) chain structure ☐ (*b*) double tetrahedral structure ☐
(*c*) sheet structure ☐ (*d*) frame work structure. ☐

135. The property of a material to exist in more than one type of space lattice in the solid state, is known as

(*a*) polymerization ☐ (*b*) polymorphism ☐
(*c*) hysteresis ☐ (*d*) allotropy. ☐

136. The process of growing large molecules from small molecules is known as

(*a*) polymerization ☐ (*b*) polymorphism ☐
(*c*) hysteresis ☐ (*d*) allotropy. ☐

137. Silicon tetrahedral units have hexagonal pattern upto

(*a*) 1000°C ☐ (*b*) 870°C ☐
(*c*) 1200°C ☐ (*d*) 1400°C. ☐

138. Silicon tetrahedral units changes to cubic pattern above

(*a*) 500°C ☐ (*b*) 600°C ☐
(*c*) 870°C ☐ (*d*) 1000°C. ☐

139. Certain ceramic materials are used for tape-recorder heads because they have

(*a*) low tensile strength ☐ (*b*) high compressive strength ☐
(*c*) great hardness and resistance to wear ☐ (*d*) low fracture strength. ☐

140. Silicon carbide is a

(*a*) natural abrasive ☐ (*b*) synthetic abrasive ☐
(*c*) artificial abrasive ☐ (*d*) none of the above. ☐

141. Choose the wrong statement

(*a*) Creamic materials are often used as electrical insulators. ☐
(*b*) Where resistance to attack from acids, bases and salt solutions is required, glass is used. ☐
(*c*) An abrasive is commonly made of a ceramic material. ☐
(*d*) Quartz is a natural abrasive. ☐
(*e*) All of the above. ☐
(*f*) None of the above. ☐

142. The abbreviation T.T.T. diagrams stand for

(*a*) tensile temperature time diagrams ☐ (*b*) time temperature transformation diagrams ☐
(*c*) temperature time testing diagrams ☐ (*d*) time transformation testing diagrams. ☐

143. Figure 11.1 shows stress-strain curves for low carbon steel, medium carbon steel and high carbon steel. For high carbon steel, the stress-strain curve is represented by

(*a*) curve *A* ☐
(*b*) curve *B* ☐
(*c*) curve *C* ☐
(*d*) curve *D*. ☐

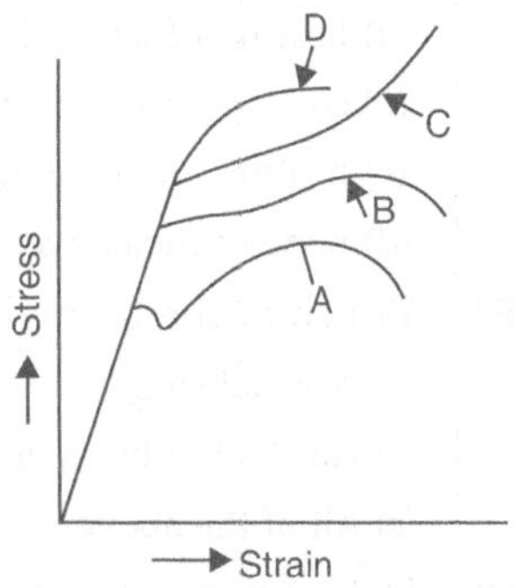

144. In Fig. 11.1, for medium carbon steel, the stress-strain curve is represented by

(*a*) curve *A* ☐
(*b*) curve *B* ☐
(*c*) curve *C* ☐
(*d*) curve *D*. ☐

FIGURE 11.1

145. In Fig. 11.1, for low carbon steel, the stress-strain curve is represented by

(*a*) curve *A* ☐ (*b*) curve *B* ☐
(*c*) curve *C* ☐ (*d*) curve *D*. ☐

146. Pearlite is a combination of

(*a*) 6.67% carbon and 93.33% iron ☐ (*b*) 13% ferrite and 87% cementite ☐
(*c*) 13% cementite and 87% ferrite ☐ (*d*) 13% carbon and 87% ferrite. ☐

147. Cementite is a combination of

(*a*) 6.67% carbon and 93.33% iron ☐ (*b*) 13% carbon and 87% ferrite ☐
(*c*) 13% cementite and 87% ferrite ☐ (*d*) 87% cementite and 13% ferrite. ☐

148. Which one of the following statements is a wrong?

(*a*) The body-centred cubic space lattice is found in alpha-iron. ☐
(*b*) A steel containing ferrite and pearlite is tough. ☐
(*c*) A steel with 0.8% carbon is wholly pearlite. ☐
(*d*) The lower critical temperature is different for all steels. ☐

149. Which one of the following statements is correct?

(*a*) The hardness of steel depends upon the amount of carbon it contains. ☐
(*b*) The process in which steel is coated with a thin layer of phosphate is known as anodising. ☐
(*c*) The imperfection in the crystal structure of a metal is known as dislocation. ☐
(*d*) Lime stone is added in blast furnace to flux carbon. ☐

150. The hardness of steel primarily depends on

(*a*) method of manufacture ☐
(*b*) heat treatment employed ☐
(*c*) shape of carbide and their distribution in iron ☐
(*d*) percentage of carbon. ☐

151. Heat treatment is a process involving
(*a*) heating of a metal in a solid state to obtain certain desirable properties ☐
(*b*) cooling of a metal in a solid state to obtain certain desirable properties ☐
(*c*) heating and cooling of a metal in a solid state to obtain certain desirable properties ☐
(*d*) none of the above. ☐

152. Which of the following are the heat treatment processes?
(*a*) Normalising ☐ (*b*) Annealing ☐
(*c*) Hardening by quenching ☐ (*d*) Tempering ☐
(*e*) All of the above ☐ (*f*) None of the above. ☐

153. Heat treatment is done to
(*a*) change grain size ☐
(*b*) soften the metal ☐
(*c*) improve electrical and magnetic properties ☐
(*d*) relieve internal stresses set up during cold and hot working ☐
(*e*) all of the above ☐
(*f*) none of the above. ☐

154. Choose the wrong statement
(*a*) Heat treatment is effective only with certain alloys. ☐
(*b*) Cooling rate is an important factor in heat treatment process. ☐
(*c*) The temperature at which the change starts on heating the steel is called lower critical temperature. ☐
(*d*) The lower critical temperature varies according to the carbon content in steel. ☐

155. The process involving the heating of steel to approximately 40°C above the critical temperature and then cooling the steel in still air is called
(*a*) annealing ☐ (*b*) normalising ☐
(*c*) tempering ☐ (*d*) hardening. ☐

156. The process involving the heating of steel above upper critical temperature and then cooling it in the furnace is known as
(*a*) annealing ☐ (*b*) normalising ☐
(*c*) tempering ☐ (*d*) hardening. ☐

157. The process involving the heating of steel above upper critical temperature and then quenching in a medium, such as brine, water or oil, as known as
(*a*) annealing ☐ (*b*) normalising ☐
(*c*) tempering ☐ (*d*) hardening. ☐

158. The process of heating hardened steel to any temperature below the lower critical temperature and then cooling at a desired rate is known as
(*a*) annealing ☐ (*b*) normalising ☐
(*c*) tempering ☐ (*d*) hardening. ☐

159. The process of heating the steel above the critical range, plunging it into a bath of molten salt of molten lead (maintained at 320°C) for several minutes and then cooling it in air, is known as

(*a*) annealing ☐ (*b*) normalising ☐
(*c*) tempering ☐ (*d*) austempering. ☐

160. Eutectoid steel is a steel containing carbon

(*a*) less than 0.8% ☐ (*b*) equal to 0.8% ☐
(*c*) from 0.8 to 2.0% ☐ (*d*) zero per cent. ☐

161. Hypoeutectoid steel is a steel containing carbon

(*a*) less than 0.8% ☐ (*b*) equal to 0.8% ☐
(*c*) from 0.8 to 2.0% ☐ (*d*) zero per cent. ☐

162. Hyper-eutectoid steel is a steel containing carbon

(*a*) less than 0.8% ☐ (*b*) equal to 0.8% ☐
(*c*) from 0.8 to 2.0% ☐ (*d*) zero per cent. ☐

163. The upper critical temperature for a steel

(*a*) is constant ☐
(*b*) depends upon the rate of heating ☐
(*c*) varies according to the carbon content in steel ☐
(*d*) none of the above. ☐

164. The lower critical temperature for a steel

(*a*) is constant ☐
(*b*) depends upon the rate of heating ☐
(*c*) varies according to the carbon content in steel ☐
(*d*) none of the above. ☐

Allotropic Forms of Iron

165. Alpha iron (*i.e.*, α-iron) exists

(*a*) between 900°C and 1404°C ☐ (*b*) between 1404°C and 1535°C ☐
(*c*) below 768°C ☐ (*d*) between 768°C and 900°C. ☐

166. Beta iron (*i.e.*, β-iron) exists

(*a*) between 900°C and 1404°C ☐ (*b*) between 1404°C and 1535°C ☐
(*c*) below 768°C ☐ (*d*) between 768°C and 900°C. ☐

167. Gamma iron (*i.e.*, γ-iron) exists

(*a*) between 900°C and 1404°C ☐ (*b*) between 1404°C and 1535°C ☐
(*c*) between 768°C ☐ (*d*) between 768°C and 900°C. ☐

168. Delta iron (*i.e.*, δ-iron) exists

(*a*) between 900°C and 1404°C ☐ (*b*) between 1404°C and 1535°C ☐
(*c*) between 768°C ☐ (*d*) between 768°C and 900°C. ☐

Iron-Carbon Equilibrium Diagram

169. Iron-carbon equilibrium diagram

(*a*) is constructed by plotting temperature along *y*-axis and carbon percentage along *x*-axis ☐
(*b*) establishes a correlation between the microstructure and properties of steel and cast iron ☐
(*c*) indicates the phase changes that occur during heating and cooling ☐
(*d*) all of the above. ☐
(*e*) none of the above. ☐

170. The limit of carbon solubility in austenite on iron-carbon equilibrium diagram is represented by

(*a*) line A_1 ☐ (*b*) line A_{cm} ☐
(*c*) line A_3 ☐ (*d*) none of the above. ☐

171. The beginning of transition from austenite to ferrite on iron-carbon equilibrium diagram is represented by

(*a*) line A_1 ☐ (*b*) line A_{cm} ☐
(*c*) line A_3 ☐ (*d*) none of the above. ☐

172. The completion of austenite transition to ferrite and pearlite on iron-carbon equilibrium diagram is represented by

(*a*) line A_1 ☐ (*b*) line A_{cm} ☐
(*c*) line A_3 ☐ (*d*) none of the above. ☐

173. The body-centred cube space lattice is formed in

(*a*) alpha iron ☐ (*b*) beta iron ☐
(*c*) gamma iron ☐ (*d*) delta iron ☐
(*e*) both (*a*) and (*b*) ☐ (*f*) none of the above. ☐

174. The face-centred cubic space lattice is formed in

(*a*) alpha iron ☐ (*b*) beta iron ☐
(*c*) gamma iron ☐ (*d*) delta iron ☐
(*e*) both (*a*) and (*b*) ☐ (*f*) none of the above. ☐

175. The carbon content in steel is

(*a*) above 2% ☐ (*b*) upto 2% ☐
(*c*) below 0.8% ☐ (*d*) above 6.3%. ☐

176. The carbon content in cast iron is

(*a*) above 2% ☐ (*b*) upto 2% ☐
(*c*) below 0.8% ☐ (*d*) above 6.3%. ☐

177. An allotropic material has

(*a*) different crystal structures at the same temperature ☐
(*b*) different crystal structure at different temperature ☐
(*c*) fixed structure at different temperature ☐
(*d*) none of the above. ☐

178. When a pure iron is cooled from its melting temperature of 1539°C, it has non-magnetic properties between the temperatures of

(*a*) 1400°C and 910°C ☐ (*b*) 910°C and 768°C ☐
(*c*) 768°C and room temperature ☐ (*d*) 1400°C and 768°C. ☐

179. When a pure iron is cooled from its melting temperature of 1539°C, it has magnetic properties between the temperatures of

(*a*) 1400°C and 910°C ☐ (*b*) 910°C and 768°C ☐
(*c*) 768°C and room temperature ☐ (*d*) 1400°C and 768°C. ☐

180. Choose the wrong statement.

(*a*) Austenite is non-magnetic and soft. ☐
(*b*) The solid solution of carbon is gamma iron in known as austenite. ☐
(*c*) Cementite contains 6.67% carbon by weight. ☐
(*d*) Pearlite is the hardest structure. ☐

181. The abbrevitation ASTM stands for

(*a*) Area surface temperature measurement ☐ (*b*) Area stiffness testing method ☐
(*c*) American society for testing materials ☐ (*d*) none of the above. ☐

182. According to ASTM, the grain size number, *N*, is obtained by the formula

(*a*) $n = 2^N + 1$ ☐ (*b*) $n = 2^{N+1}$ ☐
(*c*) $n = 2^{N-1}$ ☐ (*d*) $n = 2^{N-1}$ ☐

where n = Number of grains per square inch as viewed at 100 magnification

N = ASTM grain size number.

183. Which one of the following is the method for grain size measurement?

(*a*) comparison method ☐ (*b*) intercept method ☐
(*c*) planimetric method ☐ (*d*) all of the above ☐
(*e*) none of the above. ☐

184. Which of the following factors govern the grain size?

(*a*) nature and extent of deoxidizers ☐ (*b*) chemical composition of steel ☐
(*c*) alloying elements ☐ (*d*) heat treatment processes ☐
(*e*) all of the above ☐ (*f*) none of the above. ☐

185. Which of the following is an amorphous material?

(*a*) lead ☐ (*b*) glass ☐
(*c*) brass ☐ (*d*) zinc. ☐

186. Tin when added in lead increases

(*a*) strength ☐ (*b*) hardness ☐
(*c*) strength and hardness ☐ (*d*) strength and ductility. ☐

187. Nickel when added to copper increases

(*a*) strength ☐ (*b*) hardness ☐
(*c*) strength and hardness ☐ (*d*) strength and ductility. ☐

188. Silicon when added to copper increases

(*a*) strength ☐ (*b*) hardness ☐

(*c*) strength and hardness ☐ (*d*) strength and ductility. ☐

189. Figure 11.2 shows the growth of grain size with the annealing time and annealing temperatures of 300°C, 500°C and 700°C. Curve A shows the growth of grain size at

(*a*) 300°C ☐

(*b*) 500°C ☐

(*c*) 700°C ☐

(*d*) none of the above. ☐

FIGURE 11.2

190. Curve *B* in Fig. 11.2 shows the growth of grain size at

(*a*) 300°C ☐

(*b*) 500°C ☐

(*c*) 700°C ☐

(*d*) none of the above. ☐

191. Curve *C* in Fig. 11.2 shows the growth of grain size at

(*a*) 300°C ☐ (*b*) 500°C ☐

(*c*) 700°C ☐ (*d*) none of the above. ☐

192. Curve *D* in Fig. 11.2 shows the growth of grain size at

(*a*) 300°C ☐ (*b*) 500°C ☐

(*c*) 700°C ☐ (*d*) none of the above. ☐

193. Which one of the following statements is wrong?

(*a*) Formation and propagation of cracks in a metal is caused due to excessive cold working. ☐

(*b*) The shaping of metals by impact or by a hammer is known as forging. ☐

(*c*) The mechanism of deformation wherein one part of the crystal moves over another part along certain planes, is known as slip. ☐

(*d*) The slip plane is the plane of least atomic density. ☐

194. The process of reducing the ore with carbon in the presence of a flux is known as

(*a*) hardening ☐ (*b*) smelting ☐

(*c*) quenching ☐ (*d*) shaping. ☐

195. The furnace is lined with silica bricks in

(*a*) basic bessemer process ☐ (*b*) acidic bessemer process ☐

(*c*) duplex process ☐ (*d*) L.D. process. ☐

196. The furnace is lined with a mixture of tar and burnt dolomite bricks in

(*a*) basic bessemer process ☐ (*b*) acidic bessemer process ☐

(*c*) duplex process ☐ (*d*) L.D. process. ☐

197. The steel-making process used at Tata Iron and Steel Works, Jamshedpur is

(*a*) L.D. process ☐ (*b*) open hearth process ☐

(*c*) duplex process ☐ (*d*) electric process. ☐

198. The steel-making process used at Rourkela Steel Plant is
(*a*) L.D. process ☐ (*b*) open hearth process ☐
(*c*) duplex process ☐ (*d*) electric process. ☐

199. The process of steel-making which is a combination of acidic bessemer and basic open hearth process is known as
(*a*) L.D. process ☐ (*b*) open hearth process ☐
(*c*) duplex process ☐ (*d*) electric process. ☐

200. Which one of the following is a wrong statement?
(*a*) In low carbon steels, phosphorus in added to raise its yield point. ☐
(*b*) Lining of open hearth furnace controls impurities in steel. ☐
(*c*) Blast furnace uses coke as fuel. ☐
(*d*) Manganese is added in blast furnace for better fluidity. ☐

201. Which one of the following is a correct statement?
(*a*) Lime stone is added in blast furnace of flux carbon. ☐
(*b*) Lead is an amorphous material. ☐
(*c*) Beta iron exists below 768°C. ☐
(*d*) The product of cupola is known as cast iron. ☐

202. Under microscope, cementite in white cast iron appears as
(*a*) white ☐ (*b*) dark ☐
(*c*) light ☐ (*d*) finger print. ☐

203. Under microscope, ferrite appears as
(*a*) white ☐ (*b*) dark ☐
(*c*) light ☐ (*d*) finger print. ☐

204. Under microscope, pearlite appears as
(*a*) white ☐ (*b*) dark ☐
(*c*) light ☐ (*d*) finger print. ☐

205. Under microscope, cementite in the form of lamellar pearlite appears as
(*a*) white ☐ (*b*) dark ☐
(*c*) light ☐ (*d*) finger print. ☐

206. Macro-structure of a material is, generally, examined by
(*a*) X-ray technique ☐ (*b*) naked eye ☐
(*c*) optical microscope ☐ (*d*) spectroscope technique. ☐

207. Crystal structure of material is generally, examined by
(*a*) X-ray technique ☐ (*b*) naked eye ☐
(*c*) optical microscope ☐ (*d*) spectroscope technique. ☐

208. Figure 11.3 shows the fatigue curves for various metals and alloys. Curve *A* holds good for
(*a*) cast iron ☐ (*b*) mild steel ☐
(*c*) aluminium alloy ☐ (*d*) none of the above. ☐

209. In Fig. 11.3, curve *B* holds good for

(*a*) cast iron ☐
(*b*) mild steel ☐
(*c*) aluminium alloy ☐
(*d*) none of the above. ☐

210. In Fig. 11.3, curve *C* holds good for

(*a*) cast iron ☐
(*b*) mild steel ☐
(*c*) aluminium alloy ☐
(*d*) none of the above. ☐

Stress kg/cm²: 4200, 3500, 2800, 2100, 1400, 700, 0
A, B, C, D
10^5 10^6 10^7 10^8 10^9
No. of cycles to failure

FIGURE 11.3

211. In Fig. 11.3, curve *D* holds good for

(*a*) cast iron ☐ (*b*) mild steel ☐
(*c*) aluminium alloy ☐ (*d*) none of the above. ☐

Atomic Structure

212. The smallest group of atoms possessing the symmetry of the crystal which when repeated in all directions, will develop the crystal lattice is known as

(*a*) space lattice ☐ (*b*) unit-cell ☐
(*c*) body-centred cubic ☐ (*d*) face-centered cubic. ☐

213. If a unit-cell of a structure has one atom in the centre of the cube and one atom each at all the corners, then the structure is known as

(*a*) face-centred cubic structure ☐ (*b*) body-centred cubic structure ☐
(*c*) hexagonal close packed structure ☐ (*d*) none of the above. ☐

214. If a unit cell of a structure has an atom at each corner of the cube and one atom at the intersection of the diagonals of each of the six faces of the cube, the structure is known as

(*a*) face-centred cubic structure ☐ (*b*) body-centred cubic structure ☐
(*c*) hexagonal close packed structure ☐ (*d*) none of the above. ☐

215. The unit-cell of body-centred cubic structure contains number of atoms as

(*a*) 1 ☐ (*b*) 2 ☐
(*c*) 4 ☐ (*d*) 8. ☐

216. The unit cell of face-centred cubic structure contains number of atoms as

(*a*) 1 ☐ (*b*) 2 ☐
(*c*) 4 ☐ (*d*) 8. ☐

217. The atomic radius (*r*) in terms of lattice parameter *i.e.*, length of the cube edge (*a*) for body-centred cubic lattice is given by

(*a*) $r = \frac{a}{2}$ ☐ (*b*) $r = \frac{a\sqrt{2}}{4}$ ☐

(*c*) $r = \frac{a\sqrt{3}}{4}$ ☐ (*d*) $r = \frac{a\sqrt{5}}{4}$. ☐

218. The relation between atomic radius (r) and length of the cube edge 'a' for face-centred cubic lattice is given by

(*a*) $r = \frac{a}{2}$ ☐ (*b*) $r = \frac{a\sqrt{2}}{4}$ ☐

(*c*) $r = \frac{a\sqrt{3}}{4}$ ☐ (*d*) $r = \frac{a\sqrt{5}}{4}$. ☐

219. The atomic packing factor is defined as the

(*a*) ratio of the volume occupied by the unit-cell to the volumes of the atoms per unit-cell ☐

(*b*) ratio of the volume of the atoms per unit cell to the total volume occupied by the unit-cell ☐

(*c*) sum of the volume of the atoms per unit-cell and the total volume occupied by the unit-cell ☐

(*d*) difference of the volume of the atoms per unit-cell and the total volume occupied by the unit-cell. ☐

220. The atomic packing factor in a face-centred cube is equal to

(*a*) 0.68 ☐ (*b*) 0.74 ☐

(*c*) 0.90 ☐ (*d*) 1.0. ☐

221. The atomic packing factor in a body-centred cube is equal to

(*a*) 0.68 ☐ (*b*) 0.74 ☐

(*c*) 0.90 ☐ (*d*) 1.0. ☐

222. The atomic packing factor for hexagonal close packed structure is equal to

(*a*) 0.68 ☐ (*b*) 0.74 ☐

(*c*) 0.90 ☐ (*d*) 1.0. ☐

223. The small molecules, which combine end-to-end to form a large molecule, is known as

(*a*) polymers ☐ (*b*) manomers ☐

(*c*) ceramic ☐ (*d*) stellite. ☐

224. The small molecules combine end-to-end to form a large molecule. The large molecule is known as

(*a*) polymer ☐ (*b*) manomers ☐

(*c*) ceramic ☐ (*d*) stellite. ☐

225. Which of the following are examples of polymers?

(*a*) starch ☐ (*b*) nylon ☐

(*c*) resin ☐ (*d*) wood ☐

(*e*) all of the above ☐ (*f*) none of the above. ☐

226. The electron theory of metal is used in understanding the concept of

(*a*) energy levels in metals ☐ (*b*) cohesive and repulsive forces in metals ☐

(*c*) binding in solids ☐ (*d*) behaviour of conductors and insulators ☐

(*e*) all of the above ☐ (*f*) none of the above. ☐

227. The bond between atoms which results from the sharing of pairs of valence electrons by two or more atoms, is known as

(*a*) ionic bond ☐ (*b*) metallic bond ☐

(*c*) covalent bond ☐ (*d*) none of the above. ☐

228. The bond, in which the atoms of different elements transfer electrons one to the other so that both have stable outer shells and at the same time become ions, one positively and other negatively charged, is known as

(a) ionic bond ☐ (b) metallic bond ☐
(c) covalent bond ☐ (d) none of the above. ☐

229. The bond, in which atoms of the same or different elements give up their valence electrons to form an electron cloud (or electron gas) throughout the space occupied by the atoms, is known as

(a) ionic bond ☐ (b) metallic bond ☐
(c) covalent bond ☐ (d) none of the above. ☐

230. The bonding force (F) between atoms is represented by the equation

(a) $F = \frac{A}{r^M} + B$ ☐ (b) $F = \frac{A}{r^M} + B \times r$ ☐
(c) $F = \frac{A}{r^M} - \frac{B}{r^N}$ ☐ (d) $F = A \times r - Br^N$. ☐

where F = Centre-to-centre spacing between atoms. A, B, M and N are constants depending upon the form of bond and $N > M$.

231. The force of attraction, force of repulsion and the resultant force between atoms versus atomic spacing r is shown in Fig. 11.4. The variation of force of attraction with r is given by

(a) curve A ☐ (b) curve B ☐
(c) curve C ☐ (d) none of the above. ☐

232. In Fig. 11.4, the variation of force of repulsion is given by

(a) curve A ☐
(b) curve B ☐
(c) curve C ☐
(d) none of the above. ☐

233. In Fig. 11.4, the variation of resultant force is given by

(a) curve A ☐
(b) curve B ☐
(c) curve C ☐
(d) none of the above. ☐

FIGURE 11.4

234. In Fig. 11.4, the distance r_0 represents

(a) maximum attractive force ☐ (b) maximum repulsive force ☐
(c) maximum resultant force ☐ (d) zero resultant force. ☐

235. In Fig. 11.4, the distance F represents

(a) maximum attractive force ☐ (b) maximum repulsive force ☐
(c) maximum resultant force ☐ (d) zero resultant force. ☐

236. The equilibrium spacing between the atoms generally varies form

(a) 10^{-4} cm to 10^{-5} cm ☐ (b) 10^{-5} cm to 10^{-6} cm ☐

(c) 10^{-6} cm to 10^{-7} cm ☐ (d) 10^{-8} cm to 4×10^{-8} cm. ☐

237. Figure 11.5 shows variation of attractive energy, repulsive energy and resultant potential energy versus atomic spacing (r). Curve A represents variation of

(a) attractive energy ☐

(b) repulsive energy ☐

(c) resultant potential energy ☐

(d) none of the above. ☐

FIGURE 11.5

238. In Fig. 11.5, curve B represent variation of

(a) attractive energy ☐

(b) repulsive energy ☐

(c) resultant energy ☐

(d) none of the above. ☐

239. In Fig. 11.5, curve C represents the variation of

(a) attractive energy ☐ (b) repulsive energy ☐

(c) resultant energy ☐ (d) none of the above. ☐

240. In Fig. 11.5, U^* represents

(a) maximum attractive energy ☐ (b) maximum repulsive energy ☐

(c) minimum attractive energy ☐ (d) minimum potential energy. ☐

241. Choose the wrong statement

(a) The bonding energy is defined as the energy to separate the atoms completely. ☐

(b) The repulsive forces between atoms are caused by the interaction of their electron shells. ☐

(c) The equilibrium spacing between the atoms is of the older of 10^{-8} cm. ☐

(d) High cohesive forces produce small elastic constants, small melting point and small co-efficients of thermal expansion. ☐

242. The total energy E is given by

(a) $E = \frac{b}{mv}$ ☐ (b) $E = \frac{k^2h^2}{\pi^2 m}$ ☐

(c) $E = \frac{k^2h^2}{8k^2 m}$ ☐ (d) $E = \frac{8\pi^2 m}{k^2h^2}$ ☐

where h = Planck's constant, $k = \frac{2\pi}{\lambda}$, λ = Wavelength that describes the motion of electron

m = Mass of electron.

243. The property of material to be strongly attracted to a magnetic field and to become a powerful magnet, is known as

(a) hysteresis ☐ (b) polymerization ☐

(c) ferromagnetism ☐ (d) polymorphism. ☐

244. A piece of material is said to be magnetised if it is placed in a magnetic field. The intensity of magnetisation in the material is given by

(a) $k = JH$ ☐ (b) $k = \frac{H}{J}$ ☐

(c) $k = \frac{J}{H}$ ☐ (d) $k = \frac{1}{JH}$. ☐

where J = Intensity of magnetisation, k = Susceptibility, H = Magnetic field strength.

245. Diamagnetic materials are those materials for which the value of susceptibility (k) is

(a) large and positive ☐ (b) large and negative ☐

(c) small and positive ☐ (d) small and negative. ☐

246. Ferromagnetic materials are those materials for which the value of susceptibility (k) is

(a) large and positive ☐ (b) large and negative ☐

(c) small and positive ☐ (d) small and negative. ☐

247. Paramagnetic materials are those materials for which the value of susceptibility (k) is

(a) large and positive ☐ (b) large and negative ☐

(c) small and positive ☐ (d) small and negative. ☐

248. The materials, which retain their magnetism when the magnetising field has been removed, are called

(a) diamagnetic materials ☐ (b) paramagnetic materials ☐

(c) ferromagnetic materials ☐ (d) none of the above. ☐

249. Which one of the following is an example of diamagnetic material?

(a) iron ☐ (b) copper ☐

(c) alkaline earth ☐ (d) cobalt. ☐

250. Which one of the following is an example of paramagnetic material?

(a) iron ☐ (b) copper ☐

(c) alkaline earth ☐ (d) cobalt. ☐

251. Which one of the following is an example of ferromagnetic material?

(a) silver ☐ (b) copper ☐

(c) gold ☐ (d) cobalt. ☐

252. The resistance of a metal to plastic deformation by indentation is known as

(a) stiffness ☐ (b) toughness ☐

(c) hardness ☐ (d) creep. ☐

253. The resistance to permanent deformation can be expressed in terms of

(a) load only ☐ (b) area only ☐

(c) load and corresponding area ☐ (d) none of the above. ☐

254. Which of the following geometrical shapes are used for indenters?

(a) cones ☐ (b) spheres ☐

(c) pyramids ☐ (d) all of the above. ☐

(e) none of the above. ☐

255. The Brinell hardness number is given by

(*a*) $\frac{2W}{\pi D(D+\sqrt{D^2-d^2})}$ ☐ (*b*) $\frac{2W}{\pi D(D-\sqrt{D^2-d^2})}$ ☐

(*c*) $\frac{4W}{\pi D^2}$ ☐ (*d*) $\frac{4W}{\pi(D^2-d^2)}$. ☐

where W = Load on indenter in kg, D = Diameter of steel ball
d = Average measured diameter of indentation in mm.

256. According to ASTM specifications, the diameter of the steel ball used for Brinell Test is

(*a*) 5 mm ☐ (*b*) 10 mm ☐
(*c*) 15 mm ☐ (*d*) 20 mm. ☐

257. In case of Rockwell hardness test as compared to Brinell hardness test

(*a*) indenters and loads are larger ☐
(*b*) indenters and loads are smaller ☐
(*c*) indenters are larger but loads are smaller ☐
(*d*) indenters are smaller but loads are larger. ☐

258. Which of the following tests are known as destructive tests?

(*a*) radiography ☐ (*b*) ultrasonic inspection ☐
(*c*) tensile test ☐ (*d*) bend test ☐
(*e*) only (*a*) and (*b*) ☐ (*f*) only (*c*) and (*d*). ☐

259. Which of the following tests are known as non-destructive tests?

(*a*) radiography ☐ (*b*) ultrasonic inspection ☐
(*c*) tensile test ☐ (*d*) bend test ☐
(*e*) only (*a*) and (*b*) ☐ (*f*) only (*c*) and (*d*). ☐

260. A test used to determine yield strength, modulus of elasticity, % elongation and % reduction in area, is known as

(*a*) hardness test ☐ (*b*) impact test ☐
(*c*) tensile test ☐ (*d*) torsion test ☐
(*e*) bend test ☐ (*f*) fatigue test. ☐

261. Which one of the following tests is known as semi-destructive test?

(*a*) hardness test ☐ (*b*) impact test ☐
(*c*) tensile test ☐ (*d*) torsion test ☐
(*e*) bend test ☐ (*f*) fatigue test. ☐

262. A test, used to determine the endurance strength and endurance limit for a metal, is known as

(*a*) hardness test ☐ (*b*) impact test ☐
(*c*) tensile test ☐ (*d*) torsion test ☐
(*e*) bend test ☐ (*f*) fatigue test. ☐

263. A test, used to determine the behaviour of materials when subjected to high rates of loading, is known as

(*a*) hardness test ☐ (*b*) impact test ☐
(*c*) fatigue test ☐ (*d*) torsion test ☐

264. The Brinell hardness number for mild steel approximately lies in the range of

(*a*) 50 to 70 ☐ (*b*) 70 to 100 ☐
(*c*) 110 to 150 ☐ (*d*) 150 to 300. ☐

265. The Brinell hardness number for soft brass approximately lies in the range of

(*a*) 50 to 70 ☐ (*b*) 70 to 100 ☐
(*c*) 110 to 150 ☐ (*d*) 150 to 300. ☐

266. To increase the corrosion resistance of steel,

(*a*) vanadium is added as an alloying element ☐
(*b*) chromium is added as an alloying element ☐
(*c*) nickel is added as an alloying element ☐
(*d*) copper is added as an alloying element. ☐

267. To modify yield and tensile strength properties of steel,

(*a*) vanadium is added as an alloying element ☐
(*b*) chromium is added as an alloying element ☐
(*c*) nickel is added as an alloying element ☐
(*d*) copper is added as an alloying element. ☐

268. The imperfection in the crystal structure of a metal is called

(*a*) slip ☐ (*b*) impurity ☐
(*c*) dislocation ☐ (*d*) fracture. ☐

269. On a Rockwell testing machine, there are two scales, *i.e.*, *B* scale and C scale. These scale use

(*a*) steel ball indenter ☐
(*b*) diamond cone penetrator ☐
(*c*) steel ball indenter for *B* scale and diamond cone penetrator for C scale ☐
(*d*) steel ball indenter for C scale and diamond cone penetrator for *B* scale. ☐

270. In a rockwell test, for testing hardness of low and medium carbon steels in the annealed condition

(*a*) scale *B* is used ☐ (*b*) scale C is used ☐
(*c*) scale *B* and scale C are used ☐ (*d*) none of the above. ☐

271. In a rockwell test, for testing hardness of alloy cast iron generally

(*a*) scale *B* is used ☐ (*b*) scale C is used ☐
(*c*) both scales are used ☐ (*d*) none of the above. ☐

272. In rockwell hardness test, minor load for all cases is

(*a*) 5 kg ☐ (*b*) 10 kg ☐
(*c*) 15 kg ☐ (*d*) 20 kg. ☐

273. In rockwell hardness test, major load for scale *B* (including minor load) is

(*a*) 90 kg ☐ (*b*) 100 kg ☐
(*c*) 140 kg ☐ (*d*) 150 kg. ☐

274. In rockwell hardness test, major load for scale *C* (including minor load) is

(*a*) 90 kg ☐ (*b*) 100 kg ☐
(*c*) 140 kg ☐ (*d*) 150 kg. ☐

275. The change of ferromagnetic alpha iron to paramagnetic alpha iron takes place at

(*a*) 1539°C ☐ (*b*) 910°C ☐
(*c*) 770°C ☐ (*d*) 400°C. ☐

276. The allotropic forms of iron are

(*a*) alpha iron and gamma iron ☐ (*b*) alpha iron and beta iron ☐
(*c*) alpha iron, beta iron and gamma iron ☐ (*d*) alpha iron, gamma iron and delta iron. ☐

277. Hypoeutectic cast irons are the iron-carbon alloys containing

(*a*) 0.8% carbon ☐ (*b*) 1.7 to 4.3% carbon ☐
(*c*) more than 4.3% carbon ☐ (*d*) less than 0.8% carbon. ☐

278. Hypereutectic cast irons are the iron-carbon alloys containing

(*a*) 0.8% carbon ☐ (*b*) 1.7 to 4.3% carbon ☐
(*c*) more than 4.3% carbon ☐ (*d*) less than 0.8% carbon. ☐

279. Hypoeutectoid steel contains

(*a*) 0.8% carbon ☐ (*b*) 1.7 to 4.3% carbon ☐
(*c*) more than 4.3% carbon ☐ (*d*) less than 0.8% carbon. ☐

280. Eutectoid steel contains

(*a*) 0.8% carbon ☐ (*b*) 1.7 to 4.3% carbon ☐
(*c*) more than 4.3% carbon ☐ (*d*) less than 0.8% carbon. ☐

281. Hypereutectoid steel contains

(*a*) 0.8% carbon ☐ (*b*) more than 0.8% carbon ☐
(*c*) less than 0.8% carbon ☐ (*d*) none of the above. ☐

282. On heating a low carbon steel upto lower critical temperature, the grain size

(*a*) is a minimum ☐
(*b*) increases very rapidly ☐
(*c*) remains same ☐
(*d*) first decreases and then increases very rapidly. ☐

283. The first product in the process of converting iron ore into useful metal from a blast furnace is known as

(*a*) cast iron ☐ (*b*) wrought iron ☐
(*c*) pig iron ☐ (*d*) steel. ☐

284. Which one of the following is the raw material for all iron and steel products?

(*a*) cast iron ☐ (*b*) wrought iron ☐
(*c*) pig iron ☐ (*d*) steel. ☐

285. Pig iron refined in a cupola produces

(*a*) cast iron ☐ (*b*) wrought iron ☐
(*c*) pig iron ☐ (*d*) none of the above. ☐

286. Pig iron refined in a puddling furnace produces

(*a*) cast iron ☐ (*b*) wrought iron ☐
(*c*) pig iron ☐ (*d*) none of the above. ☐

287. Pig iron refined in a Bessemer converter, open hearth furnace or electric furnace produces

(*a*) cast iron ☐ (*b*) wrought iron ☐
(*c*) pig iron ☐ (*d*) none of the above. ☐

288. The height of blast furnace varies from

(*a*) 10 to 15 m ☐ (*b*) 15 to 20 m ☐
(*c*) 25 to 32 m ☐ (*d*) 35 to 45 m. ☐

289. Choose the wrong statement

(*a*) The bosh is the smelting zone and the hottest part of the blast furnace. ☐
(*b*) The maximum diameter of a blast furnace is approximately 9 m. ☐
(*c*) The cupola is used to manufacture pig iron. ☐
(*d*) The portion of the blast furnace below its widest cross-section is called bosh. ☐

290. The pig iron contains approximately

(*a*) 95% carbon, 4% iron and remainder sulphur, silicon, phosphorus ☐
(*b*) 4% carbon, 95% iron and remainder sulphur, silicon, phosphorus ☐
(*c*) 2% carbon, 97% iron and remainder sulphur, silicon, phosphorus ☐
(*d*) 8% carbon, 91% iron and remainder sulphur, silicon, phosphorus. ☐

291. Cast iron is a

(*a*) brittle material ☐ (*b*) ductile material ☐
(*c*) malleable material ☐ (*d*) none of the above. ☐

292. When the carbon in the cast iron is mostly in tree state, the cast iron is known as

(*a*) molted cast iron ☐ (*b*) white cast iron ☐
(*c*) grey cast iron ☐ (*d*) black cast iron. ☐

293. When the carbon in cast iron is mostly in the combined state, the cast iron is known as

(*a*) molted cast iron ☐ (*b*) white cast iron ☐
(*c*) grey cast iron ☐ (*d*) black cast iron. ☐

294. When the carbon in cast iron is partly in free and partly in combined state, the cast iron is called

(*a*) molted cast iron ☐ (*b*) white cast iron ☐
(*c*) grey cast iron ☐ (*d*) black cast iron. ☐

295. Low carbon steel contains carbon from

(*a*) 0.05 to 0.1% ☐ (*b*) 0.1 to 0.2% ☐
(*c*) 0.3 to 0.85% ☐ (*d*) 0.85 to 1.3%. ☐

296. High carbon steel contains carbon from

(*a*) 0.05 to 0.1% ☐ (*b*) 0.1 to 0.2% ☐
(*c*) 0.3 to 0.85% ☐ (*d*) 0.85 to 1.3%. ☐

297. Carbon steel is

(*a*) refined form of cast iron ☐

(*b*) an alloy of iron and carbon with varying quantities of phosphorus and sulphur ☐

(*c*) made by adding carbon is steel ☐

(*d*) none of the above. ☐

298. Alloy steels are

(*a*) the steels in which the properties are primarily derived from the presence of carbon ☐

(*b*) the steels that contain elements added for the purpose of modifying the mechanical properties of plain carbon steels ☐

(*c*) refined form of cast iron ☐

(*d*) none of the above. ☐

299. The wrought iron is having

(*a*) tensile strength minimum and ductility maximum in the longitudinal direction ☐

(*b*) tensile strength maximum and ductility minimum in the longitudinal direction ☐

(*c*) tensile strength and ductility maximum in the longitudinal direction ☐

(*d*) tensile strength and ductility minimum in the longitudinal direction. ☐

300. Carbon steel contains carbon from

(*a*) 0.1 to 0.8% ☐ (*b*) 0.1 to 1.1% ☐

(*c*) 1.8 to 4.2% ☐ (*d*) 0.1 to 1.5%. ☐

301. Cast iron contains carbon from

(*a*) 0.1 to 0.8% ☐ (*b*) 0.1 to 1.1% ☐

(*c*) 1.8 to 4.2% ☐ (*d*) 0.1 to 1.5%. ☐

302. Carbon steel castings are

(*a*) ductile ☐ (*b*) tough ☐

(*c*) easily weldable ☐ (*d*) brittle ☐

(*e*) all of the above except (*d*). ☐

303. Choose the wrong statement

(*a*) A plain carbon steel is an alloy if iron and carbon. ☐

(*b*) An alloy steel contains more than 0.5% Mn and 0.5% Si. ☐

(*c*) The alloy steels are more stronger, tougher and fatigue resistant than carbon steels. ☐

(*d*) White cast iron has a low tensile strength and a high compressive strength. ☐

304. Choose the correct statement

(*a*) Grey cast iron possesses lowest melting point of the ferrous alloys. ☐

(*b*) Wrought iron possesses a high resistance towards corrosion. ☐

(*c*) White cast iron under normal circumstances is brittle and not machinable. ☐

(*d*) All of the above ☐

(*e*) None of the above. ☐

305. Killed steels are the steels
(*a*) in which carbon is completely burnt ☐
(*b*) which are destroyed by burning ☐
(*c*) which are deoxidised in the ladle with silicon and aluminium ☐
(*d*) which after their destruction are recycled to produce fresh steel. ☐

306. According to I.S.I. specification, the symbol for killed steel is
(*a*) *KR* ☐ (*b*) *K* ☐
(*c*) *R* ☐ (*d*) no symbol. ☐

307. According to I.S.I. specification, P_{35} means
(*a*) phosphorus content is maximum and equal to 0.35% ☐
(*b*) phosphorus and sulphur contents are maximum and each equal to 0.35% ☐
(*c*) phosphorus content is minimum and equal to 0.35% ☐
(*d*) phosphorus percentage is equal to 35%. ☐

308. According to I.S.I, specification *SP* symbol means
(*a*) sulphur and phosphorus contents are maximum ☐
(*b*) sulphur and phosphorus contents are equal ☐
(*c*) sulphur and phosphorus contents are not equal ☐
(*d*) both (*a*) and (*c*). ☐

309. According to I.S.I, specification, $F_e\, E_{400}\, P_{40}$ means
(*a*) it is a semi-killed steel ☐
(*b*) minimum yield strength = 400 N/mm^2 ☐
(*c*) sulphur and phosphorus are maximum and equal to 0.04% ☐
(*d*) all of the above. ☐

310. The copper-zinc alloys are known as
(*a*) bronzes ☐ (*b*) brasses ☐
(*c*) lead ☐ (*d*) zinc. ☐

311. The copper-tin alloys are known as
(*a*) bronzes ☐ (*b*) brasses ☐
(*c*) lead ☐ (*d*) zinc. ☐

312. Which one of the following is an example of a brass?
(*a*) gun metal ☐ (*b*) bell metal ☐
(*c*) muntz metal ☐ (*d*) lead. ☐

313. Which one of the following is an example of a bronze?
(*a*) gun metal ☐ (*b*) delta metal ☐
(*c*) muntz metal ☐ (*d*) lead. ☐

314. Gun-metal contains
(*a*) copper = 80%, tin = 15% and zinc = 5% ☐ (*b*) copper = 70% and zinc = 30% ☐
(*c*) copper = 88%, tin = 10% and zinc =2% ☐ (*d*) copper = 60% and zinc = 40%. ☐

315. Muntz metal contains

(*a*) 50% copper and 50% zinc ☐

(*b*) 60.45% copper, 35.2% zinc and 4.35% nickel ☐

(*c*) 60% copper and 40% zinc ☐

(*d*) 70% copper and 30% zinc. ☐

316. German silver contains

(*a*) 50% copper and 50% zinc ☐

(*b*) 60.45% copper, 35.2% zinc and 4.35% nickel ☐

(*c*) 60% copper and 40% zinc ☐

(*d*) 70% copper and 30% zinc. ☐

317. Which one of the followings is a correct statement?

(*a*) German silver contains 5% silver. ☐

(*b*) An alloy of copper, tin and zinc is known as brass. ☐

(*c*) An alloy of copper and zinc is known as bronze. ☐

(*d*) Bronzes are superior to brasses in corrosion resistant properties. ☐

318. Which of the following processes are used for case hardening?

(*a*) carburizing ☐ (*b*) nitriding ☐

(*c*) spraying ☐ (*d*) electroplating ☐

(*e*) oxide coating ☐ (*f*) (*a*) and (*b*) only. ☐

319. Which of the following processes are used for surface treatments?

(*a*) carburizing ☐ (*b*) nitriding ☐

(*c*) spraying ☐ (*d*) electroplating ☐

(*e*) oxide coating ☐ (*f*) all except (*a*) and (*b*). ☐

320. Case hardening involves

(*a*) applying coatings to the surface of metals or alloys ☐

(*b*) penetrating carbon to the surface of the steel part upto a definite depth ☐

(*c*) penetrating nitrogen to the surface of the steel part upto a definite depth ☐

(*d*) penetrating carbon, nitrogen or both to the surface of the steel part upto a definite depth. ☐

321. Surface treatments involve

(*a*) applying coatings to the surface of metals or alloys ☐

(*b*) penetrating carbon to the surface to the steel part upto a definite depth ☐

(*c*) penetrating nitrogen to the surface of the steel part upto a definite depth ☐

(*d*) penetrating carbon, nitrogen or both to the surface of the steel part upto a definite depth. ☐

322. The process of introducing carbon into low carbon steels in order to produce a hard surface is known as

(*a*) carbonitriding ☐ (*b*) nitriding ☐

(*c*) carburizing ☐ (*d*) cyaniding ☐

(*e*) induction hardening ☐ (*f*) none of the above. ☐

323. The process of introducing nitrogen into the surface of certain types of steel (*i.e.*, the steels containing Al and Cr) by heating surface and holding it at a suitable temperature in contact with partially dissociated ammonia, is known as

(*a*) carbonitriding ☐ (*b*) nitriding ☐
(*c*) carburizing ☐ (*d*) cyaniding ☐
(*e*) induction hardening ☐ (*f*) none of the above. ☐

324. The process of introducing carbon and nitrogen into the surface of steel by heating the surface to a suitable temperature and holding it in contact with molten cyanide, is known as

(*a*) carbonitriding ☐ (*b*) nitriding ☐
(*c*) carburizing ☐ (*d*) cyaniding ☐
(*e*) induction hardening ☐ (*f*) none of the above. ☐

325. The process of introducing carbon and nitrogen into a solid ferrous alloy is known as

(*a*) carbonitriding ☐ (*b*) nitriding ☐
(*c*) carburizing ☐ (*d*) cyaniding ☐
(*e*) induction hardening ☐ (*f*) none of the above. ☐

326. The process of heating the surface of the steel upto 1500°F by oxyacetylene flame and then quenching is known as

(*a*) carburizing ☐ (*b*) nitriding ☐
(*c*) flame hardening ☐ (*d*) induction hardening. ☐

327. The process of healing the surface of the carbon steel upto 1500°F by an alternating magnetic field and then quenching is known as

(*a*) carburizing ☐ (*b*) nitriding ☐
(*c*) flame hardening ☐ (*d*) induction hardening. ☐

328. In case of flame hardening, depth of the hardened zone depends upon

(*a*) flame intensity ☐ (*b*) heating time ☐
(*c*) carbon contents ☐ (*d*) all of the above ☐
(*e*) only (*a*) and (*b*). ☐

329. Choose the wrong statement

(*a*) In flame hardening it is difficult to produce hardened zones less than 1.5 mm in depth. ☐
(*b*) In case of induction hardening, the depth of hardening is controlled by controlling voltage. ☐
(*c*) Induction hardening cannot be applied to steels having carbon less than 0.4%. ☐
(*d*) In case of nitriding, the distortion or crack formation is maximum. ☐

330. Choose the correct statement

(*a*) A technique of measuring temperature generally above 500°C is known as pyrometry. ☐
(*b*) Cladding is the method by which layers of another material becomes an integral part of the base metal. ☐
(*c*) A gas fired furnace is much better than an oil fired furnace. ☐
(*d*) Cementation is also known as carburizing. ☐
(*e*) All of the above ☐
(*f*) None of the above. ☐

331. Fine grains of austenite
(*a*) increases hardenability in steel ☐
(*b*) decreases hardenability in steel ☐
(*c*) has no effect on hardenability in steel ☐
(*d*) first increases then decreases hardenability in steel. ☐

332. Dissolved elements in austenite
(*a*) increases hardenability in steel ☐
(*b*) decreases hardenability in steel ☐
(*c*) has no effect on hardenability in steel ☐
(*d*) first increases then decreases hardenability in steel. ☐

333. Steel balls for ball bearings are hardened to
(*a*) 700 to 800 VPN ☐ (*b*) 600 to 700 VPN ☐
(*c*) 400 to 600 VPN ☐ (*d*) 100 to 200 VPN. ☐

334. Pearlite is obtained when steel is
(*a*) quenched in oil ☐ (*b*) cooled in still air ☐
(*c*) slowly cooled in furnace ☐ (*d*) quenched in water. ☐

335. Sorbite is obtained when steel is
(*a*) quenched in oil ☐ (*b*) cooled in still air ☐
(*c*) slowly cooled in furnace ☐ (*d*) quenched in water. ☐

336. A circular is obtained when steel is
(*a*) quenched in oil ☐ (*b*) cooled in still air ☐
(*c*) slowly cooled in furnace ☐ (*d*) quenched in water. ☐

337. Troosite is obtained when steel is
(*a*) quenched in oil ☐ (*b*) cooled in still air ☐
(*c*) slowly cooled in furnace ☐ (*d*) quenched in water. ☐

338. A temperature, at which the first tiny new grain appears, is known as
(*a*) boiling temperature ☐ (*b*) melting temperature ☐
(*c*) recrystallization temperature ☐ (*d*) none of the above. ☐

339. The recrystallization temperature depends upon
(*a*) grain size ☐ (*b*) type of metal ☐
(*c*) annealing time ☐ (*d*) purity of metal ☐
(*e*) all of the above. ☐ (*f*) none of the above. ☐

340. Recrystallization temperature for pure metals is approximately equal to
(*a*) $0.2T_m$ ☐ (*b*) $0.3T_m$ ☐
(*c*) $0.5T_m$ ☐ (*d*) $0.8T_m$ ☐
where T_m = Melting temperature.

341. Recrystallization temperature for alloys is approximately equal to
(*a*) $0.2T_m$ ☐ (*b*) $0.3T_m$ ☐
(*c*) $0.5T_m$ ☐ (*d*) $0.8T_m$ ☐

342. Conductivity (σ) is given by

(a) $\sigma = \frac{RA}{L}$ ☐ (b) $\sigma = \frac{L}{RA}$ ☐

(c) $\sigma = \frac{AL}{R}$ ☐ (d) $\sigma = \frac{LR}{A}$. ☐

343. The dimension of conductivity is given by

(a) ohm cm ☐ (b) ohm ☐

(c) $ohm^{-1}\ cm^{-1}$ ☐ (d) none of the above. ☐

344. The dimensions of resistivity is given by

(a) ohm cm ☐ (b) ohm ☐

(c) $ohm^{-1}\ cm^{-1}$ ☐ (d) none of the above. ☐

345. The resistivity is given by

(a) $\sigma = \frac{RA}{L}$ ☐ (b) $\sigma = \frac{L}{RA}$ ☐

(c) $\sigma = \frac{AL}{R}$ ☐ (d) $\sigma = \frac{LR}{A}$. ☐

346. In which conductivity, the charge carriers are electrons?

(a) super conductivity ☐ (b) ionic conductivity ☐

(c) electronic conductivity ☐ (d) none of the above. ☐

347. In which conductivity, the charge carriers are ions?

(a) super conductivity ☐ (b) ionic conductivity ☐

(c) electronic conductivity ☐ (d) none of the above. ☐

348. Metals, which are comparatively poor conductors at room temperature, tend to become good conductors at very low temperature. This phenomen is known as

(a) super conductivity ☐ (b) ionic conductivity ☐

(c) electronic conductivity ☐ (d) none of the above. ☐

349. Bragg's diffraction law is given by

(a) $n\lambda = d \sin \theta$ ☐ (b) $n\lambda = d/2 \sin \theta$ ☐

(c) $n\lambda = 2d \sin \theta$ ☐ (d) $n\lambda = 4d \sin \theta$. ☐

where θ = Angle of incidence, λ = Wavelength of electron,

d = Spacing between the plane of crystal, n = Any integer.

350. The wave number (k) is given by

(a) $k = \frac{n\pi}{\sin \theta}$ ☐ (b) $k = \frac{n\pi}{d \sin \theta}$ ☐

(c) $k = \frac{d \sin \theta}{n\pi}$ ☐ (d) $k = \frac{dn\pi}{\sin \theta}$. ☐

351. The semi-conduction produced by thermal energy alone is known as

(a) extrinsic semi-conduction ☐ (b) intrinsic semi-conduction ☐

(c) both (a) and (b) ☐ (d) none of the above. ☐

352. The semi-conduction produced by holes due to impurity atoms is known as

(*a*) extrinsic semi-conduction ☐ (*b*) intrinsic semi-conduction ☐
(*c*) both (*a*) and (*b*) ☐ (*d*) none of the above. ☐

353. A *p–n* junction is

(*a*) an amplifier ☐ (*b*) a rectifier ☐
(*c*) an insulator ☐ (*d*) none of the above. ☐

354. A three layer *p–n–p* device can be used as

(*a*) an amplifier ☐ (*b*) a rectifier ☐
(*c*) zener diode ☐ (*d*) none of the above. ☐

355. The most important factor which can influence the operation of a *p–n* junction is

(*a*) mobility of charge carriers ☐ (*b*) temperature ☐
(*c*) impurity atom concentration ☐ (*d*) none of the above. ☐

356. No external battery is required in

(*a*) photo conductive cell ☐ (*b*) photo voltage cell ☐
(*c*) photo emissive cell ☐ (*d*) all of the above. ☐

357. The mobility of holes is

(*a*) more than that of electrons ☐ (*b*) equal to that of electrons ☐
(*c*) less than that of electrons ☐ (*d*) none of the above. ☐

358. The conductivity of an extrinsic semi-conductor

(*a*) decreases with temperature ☐ (*b*) increases with temperature ☐
(*c*) remains constant ☐ (*d*) first decreases then increases. ☐

359. Semi-conductors have electrical conductivities of the order of (units $ohm^{-1}\ cm^{-1}$)

(*a*) 10^{-15} ☐ (*b*) 10^{-2} ☐
(*c*) 10^{2} ☐ (*d*) 10^{5}. ☐

360. It is possible to obtain *p*-type or *n*-type semi-conductor from a single compound by adding Ge. That compound is

(*a*) GaAs ☐ (*b*) GaP ☐
(*c*) InSb ☐ (*d*) none of the above. ☐

The p–n Junction

361. In donor impurities are introduced into one side and acceptor into the other side of a single crystal of silicon or germanium,

(*a*) a transistor is formed ☐ (*b*) a *p–n* junction is formed ☐
(*c*) a photo cell is formed ☐ (*d*) none of the above. ☐

362. In case of *p–n* junction (rectifier), donor ions are represented by

(*a*) minus sign ☐ (*b*) plus sign ☐
(*c*) no sign ☐ (*d*) both plus and minus sign. ☐

363. In case of *p–n* junction, as the number of dipoles builds up, a thin layer on either side of the junction which is virtually empty of charge carriers is formed. This layer is known as

(*a*) boundary layer ☐ (*b*) depletion layer ☐
(*c*) sub-layer ☐ (*d*) none of the above. ☐

364. The layer in the above question is of the order of

(a) 10^{-1} m ☐ (b) 10^{-3} m ☐
(c) 10^{-6} m ☐ (d) 10^{-30} m. ☐

365. When the *p*-side of the *p–n* junction is positive with respect to the *n*-side and the applied voltage is more than the barrier voltage, the junction behaves like a

(a) closed switch ☐ (b) open switch ☐
(c) open or closed switch ☐ (d) none of the above. ☐

366. The *p–n* junction has

(a) a very large resistance in the forward and opposite directions ☐
(b) a very small resistance in the forward and opposite direction ☐
(c) a very large resistance in forward direction and a very small resistance in opposite direction ☐
(d) a very small resistance in the forward direction and a very large resistance in the opposite direction. ☐

367. The device in which two *p–n* junctions are arranged back to back in a single crystal is known as

(a) an rectifier ☐ (b) a transistor ☐
(c) a photo cell ☐ (d) none of the above. ☐

368. In case of a transistor, for amplifying action, the junction

(a) J_1 is reverse biased and J_2 is forward biased ☐
(b) J_1 is forward biased and J_2 is reverse biased ☐
(c) J_1 and J_2 are forward biased ☐
(d) J_1 and J_2 are reverse biased. ☐

369. Photo cells are employed for converting

(a) mechanical energy into electrical energy ☐ (b) electrical energy into mechanical energy ☐
(c) electrical energy into radiation energy ☐ (d) radiation energy into electrical energy. ☐

370. The photo-emission starts only when the frequency of the incident radiations

(a) is less than the threshold frequency ☐ (b) is equal to the threshold frequency ☐
(c) is more than the threshold frequency ☐ (d) none of the above. ☐

371. The resistivity of a material is equal to

(a) two lines the conductivity ☐ (b) conductivity ☐
(c) reciprocal of conductivity ☐ (d) half times the conductivity. ☐

372. The unit of resistivity is

(a) ohm/cm ☐ (b) cm/ohm ☐
(c) ohm/cm ☐ (d) $(\text{ohm cm})^{-1}$. ☐

373. The relation between ρ (resistivity), *A* (Area), *L* (length of conductor) and *R* (resistance) is given by

(a) $\rho = \dfrac{L}{RA}$ ☐ (b) $\rho = \dfrac{A}{RL}$ ☐

(c) $\rho = \dfrac{R}{AL}$ ☐ (d) $\rho = \dfrac{RA}{L}$. ☐

374. If two dissimilar metals are joined and this junction is then heated, a small voltage in the millivolt range is produced. This effect is known as

(*a*) ferro-electric effect ☐ (*b*) piezo-electric effect ☐
(*c*) pyroelectric effect ☐ (*d*) thermo-electric effect. ☐

375. A natural crystal of tourmaline acquires an electric charge when heated. This effect is known as

(*a*) ferro-electric effect ☐ (*b*) piezo-electric effect ☐
(*c*) pyroelectric effect ☐ (*d*) thermo-electric effect. ☐

376. Barium titanate is

(*a*) ferro-electric ☐ (*b*) piezo-electric ☐
(*c*) ferro-electric as well as piezo-electric ☐ (*d*) none of the above. ☐

377. If an alternating voltage *V* is applied across a capacitor *C* having vacuum as dielectric, then the resulting current *I* leads the voltage by

(*a*) 30° ☐ (*b*) 60° ☐
(*c*) 90° ☐ (*d*) 135°. ☐

378. If in the above question, a real dielectric is present then the current leads the voltage by

(*a*) $(30° - \delta)$ ☐ (*b*) $(60° - \delta)$ ☐
(*c*) $(75° - \delta)$ ☐ (*d*) $(90° - \delta)$. ☐

where δ is the measure of dielectric power loss.

379. The expression '$VI \tan \delta$' represents

(*a*) loss factor ☐ (*b*) dissipation factor ☐
(*c*) dielectric power loss ☐ (*d*) none of the above. ☐

380. In the above expression '$\tan \delta$' is known as

(*a*) loss factor ☐ (*b*) dissipation factor ☐
(*c*) dielectric power loss ☐ (*d*) none of the above. ☐

381. The dielectric constant (*i.e.*, permittivity)

(*a*) increase with decrease of temperature ☐
(*b*) increases with increase of temperature ☐
(*c*) remains constant with the change of temperature ☐
(*d*) none of the above. ☐

382. The dielectric constant

(*a*) increases with the decrease of frequency ☐
(*b*) increases with the increase of frequency ☐
(*c*) remains constant with the change of frequency ☐
(*d*) none of the above. ☐

383. Dielectrics are used

(*a*) capacitors ☐ (*b*) as an electrical insulators ☐
(*c*) in transistor ☐ (*d*) in all the above ☐
(*e*) only (*a*) and (*b*). ☐

384. The valence band of the energy diagram of a dielectric is separated from a conduction band by

(a) a small energy gap ☐ (b) a large energy gap ☐
(c) an average energy gap ☐ (d) none of the above. ☐

385. Which one of the following is a dielectrics?

(a) Copper ☐ (b) Wood ☐
(c) Zinc ☐ (d) Aluminium. ☐

386. The main function of a dielectric is to store static electric charges. The phenomena of development of static charges on the surface of the dielectric under the influence of the electric field is known as

(a) dielectric strength ☐ (b) dielectric loss ☐
(c) dielectric polarization ☐ (d) permittivity. ☐

387. The performance of a dielectric depends upon

(a) dielectric loss ☐ (b) dielectric strength ☐
(c) permittivity ☐ (d) all of the above ☐
(e) none of the above. ☐

388. The electric flux density (D) in terms of area (A) of the dielectric surface and charge (Q) on the dielectric surface is given by

(a) $D = Q \times A$ ☐ (b) $D = \frac{A}{Q}$ ☐
(c) $D = \frac{Q}{A}$ ☐ (d) $D = \frac{1}{A \times Q}$. ☐

389. The permittivity of the dielectric (ε) is given by

(a) $\varepsilon = DE$ ☐ (b) $\varepsilon = \frac{E}{D}$ ☐
(c) $\varepsilon = \frac{1}{DE}$ ☐ (d) $\varepsilon = \frac{D}{E}$. ☐

where D = Electric flux density, and E = Electric field.

390. The permittivity of any dielectric material is influenced by

(a) temperature only ☐ (b) frequency only ☐
(c) both temperature and frequency ☐ (d) none of the above. ☐

391. The phenomenon of the appearance of the charges on the surface of a crystal when it is subjected to mechanical deformation, is known as

(a) ferroelectricity ☐ (b) piezoelectricity ☐
(c) thermoelectricity ☐ (d) none of the above. ☐

392. The phenomena of the spontaneous polarization of a dielectric due to the presence of permanent electric dipoles, is known as

(a) ferroelectricity ☐ (b) piezoelectricity ☐
(c) thermoelectricity ☐ (d) none of the above. ☐

393. The ferroelectric curie temperature is the temperature at which spontaneous polarisation in ferroelectric material

(*a*) is maximum ☐ (*b*) is average ☐
(*c*) vanishes ☐ (*d*) none of the above. ☐

394. Below the ferroelectric curie temperature, the material is

(*a*) Ferroelectric ☐ (*b*) piezoelectric ☐
(*c*) both ☐ (*d*) none of the above. ☐

395. Barium titanate has the curie temperature as equal to

(*a*) room temperature ☐ (*b*) about 260°F ± ☐
(*c*) lower than room temperature ☐ (*d*) none of the above. ☐

396. Rochelle salt has the curie temperature equal to

(*a*) room temperature ☐ (*b*) about 260°F ± ☐
(*c*) lower than room temperature ☐ (*d*) none of the above. ☐

397. Most of the ferro-electrics have the curie temperature equal to

(*a*) room temperature ☐ (*b*) about 260°F ± ☐
(*c*) lower than room temperature ☐ (*d*) none of the above. ☐

398. Which one of the followings is a magnetic property?

(*a*) Resistivity ☐ (*b*) Conductivity ☐
(*c*) Permeability ☐ (*d*) Ductility. ☐

399. The relation between permeability (μ), magnetic induction (*B*) and intensity of magnetising Held (*H*) is expressed as

(*a*) $\mu = \frac{H}{B}$ ☐ (*b*) $\mu = \frac{B}{H}$ ☐
(*c*) $\mu = \frac{H+1}{B}$ ☐ (*d*) $\mu = \frac{B+1}{H}$. ☐

400. Diamagnetic materials have a permeability

(*a*) equal to one ☐ (*b*) greater than one ☐
(*c*) less than one ☐ (*d*) none of the above. ☐

401. Paramagnetic materials have a permeability

(*a*) equal to one ☐ (*b*) greater than one ☐
(*c*) less than one ☐ (*d*) none of the above. ☐

402. Vacuum is assigned a permeability

(*a*) equal to one ☐ (*b*) greater than one ☐
(*c*) less than one ☐ (*d*) none of the above. ☐

403. X-ray diffraction is used

(*a*) to determine the defects in the crystals ☐ (*b*) to study the structure of variety of crystals ☐
(*c*) in the field of medicine ☐ (*d*) none of the above. ☐

404. The science of describing the arrangements of atoms of molecules with the help of X-ray diffraction is known as

(*a*) pyrometry ☐ (*b*) refractory ☐
(*c*) crystallography ☐ (*d*) cladding. ☐

Tick mark true and false statement:

405. In the periodical table, elements are arranged in order of increasing atomic weight.

(*a*) True ☐ (*b*) False. ☐

406. The total energy E of the electron in any fixed orbit is directly proportional to the square of the atomic number.

(*a*) True ☐ (*b*) False. ☐

407. Valence electrons are the extra electrons in the outermost incomplete main shell.

(*a*) True ☐ (*b*) False. ☐

408. The maximum number of electrons present in a sub-shell is equal to $2(L + 1)$ where L = Orbital quantum number.

(*a*) True ☐ (*b*) False. ☐

409. The maximum number of electrons in any main shell is equal to $2n^2$.

(*a*) True ☐ (*b*) False. ☐

410. Vacancies are the point defect.

(*a*) True ☐ (*b*) False. ☐

411. Edge dislocation is a line defect.

(*a*) True ☐ (*b*) False. ☐

412. In space lattice, the axes are at right angles.

(*a*) True ☐ (*b*) False. ☐

413. If N = number of electrons outside the full shell, then number of covalent bonds formed by an element is $8 - N$.

(*a*) True ☐ (*b*) False. ☐

414. The sum of the number of electron and protons gives the atomic weight of an atom.

(*a*) True ☐ (*b*) False. ☐

415. The unit cell is known as cubic, if the axes of the unit cell are at right angles and only two axial lengths are equal.

(*a*) True ☐ (*b*) False. ☐

416. In the process of solidification of metals, the line defects are formed.

(*a*) True ☐ (*b*) False. ☐

417. Non-ferrous materials are those materials which do not contain iron.

(*a*) True ☐ (*b*) False. ☐

418. Copper, lead and aluminium are ferrous materials.

(*a*) True ☐ (*b*) False. ☐

419. The oxides, carbides and silicates of various metals are known as ceramic materials.

(*a*) True ☐ (*b*) False. ☐

420. The polymeric materials composed of carbon-compounds are called organic materials.

(*a*) True ☐ (*b*) False. ☐

421. Paper, rubber and fuel are non-organic materials.

(*a*) True ☐ (*b*) False. ☐

422. Cast iron, wrought iron and steel are obtained from pig iron.

(*a*) True ☐ (*b*) False. ☐

423. If pig iron is put in a puddling furnace, wrought iron is produced.

(*a*) True ☐ (*b*) False. ☐

424. If pig iron is refined in furnaces such as Bessemer converter, electric furnace etc., cast iron is produced,

(*a*) True ☐ (*b*) False. ☐

425. Steel is an alloy of iron and carbon is which carbon content is not more than 1.7%.

(*a*) True ☐ (*b*) False. ☐

426. Brass and bronze are the important copper alloys.

(*a*) True ☐ (*b*) False. ☐

427. If tin is the principal alloying metal with copper, the resultant alloy is known as bronze.

(*a*) True ☐ (*b*) False. ☐

428. Beta iron (β) is magnetic in nature.

(*a*) True ☐ (*b*) False. ☐

429. Alpha iron (α) exists below 768°C.

(*a*) True ☐ (*b*) False. ☐

430. Delta iron (δ) exists between 1400°C and 1539°C.

(*a*) True ☐ (*b*) False. ☐

431. In iron-carbon equilibrium diagram, temperature is plotted along y-axis and % composition of carbon along x-axis.

(*a*) True ☐ (*b*) False. ☐

432. Eutectoid steel is a steel in which percentage of carbon is more than 0.8%.

(*a*) True ☐ (*b*) False. ☐

433. If percentage of carbon in iron-carbon alloy is more than 2%, it is known as cast iron.

(*a*) True ☐ (*b*) False. ☐

434. The capacity of a material to undergo deformation under tension without rupture as in wire drawing process, is known as ductility.

(*a*) True ☐ (*b*) False. ☐

435. The ability of a material to withstand bending without fracture is known as toughness.

(*a*) True ☐ (*b*) False. ☐

436. The ability of a material to resist elastic deflection is known as stiffness.

(*a*) True ☐ (*b*) False. ☐

437. When a material sustains steady loads for long periods of time, the material may continue to deform until they may tend to fracture under the same load. This phenomenon is known as fatigue.

(*a*) True ☐ (*b*) False. ☐

438. Creep is the phenomenon in which a material is subjected to repeated or fluctuating stresses and fracture takes place under a stress whose maximum value is less than the tensile strength of the material.

(*a*) True ☐ (*b*) False. ☐

439. Strength of material means the ability of the material to sustain loads without undue failure or distortion.

(*a*) True ☐ (*b*) False. ☐

440. Heating and cooling of a metal in a solid state to obtain certain desirable properties, is known as heat treatment.

(*a*) True ☐ (*b*) False. ☐

441. The process of heating a metal to a temperature which will remove distortion and then cooling the metal at a slow rate to room temperature so that metal is stable, is known as normalising.

(*a*) True ☐ (*b*) False. ☐

442. The relation between atomic radius (r) and length of the cube edge (a) for body-centred cubic structure is given by $r = a \times \sqrt{2}/4$.

(*a*) True ☐ (*b*) False. ☐

443. The relation $r = a \times \sqrt{3}/4$ is between atomic radius (r) and length of the cube edge (a) for face-centred cubic structure.

(*a*) True ☐ (*b*) False. ☐

444. The ratio of the volume of the atoms per unit-cell to the total volume occupied by the unit cell is known as atomic packing factor.

(*a*) True ☐ (*b*) False. ☐

445. The atomic packing factor for a face-centred cube is equal to 0.74.

(*a*) True ☐ (*b*) False. ☐

446. The atomic packing factor for a body-centred cube is equal to 0.68

(*a*) True ☐ (*b*) False. ☐

447. The quantity of heat, that must be added to a unit mass of the solid to raise its temperature by one degree, is known as specific heat.

(*a*) True ☐ (*b*) False. ☐

448. The amount of heat in calories which flows in one second through a cm cube of material with its opposite faces maintained at a temperature difference of 1°C, is known as thermal conductivity.

(*a*) True ☐ (*b*) False. ☐

449. Thermal expansion is the increase in the dimensions of a material due to the addition of thermal energy.

(*a*) True ☐ (*b*) False. ☐

450. The rate of change of length with respect to temperature per unit length is known as co-efficient of linear thermal expansion.

(*a*) True ☐ (*b*) False. ☐

451. Conductivity is expressed mathematically as $\sigma = L/RA$ where L = length of conductor, R = resistance in ohms of a conductor and A = area of conductor.

(*a*) True ☐ (*b*) False. ☐

452. The unit of conductivity is ohm per cm.

(*a*) True ☐ (*b*) False. ☐

453. The ionic conductivity is given by $\sigma = n \times q \times \mu$ where n = number of charge carriers in a material q = charge carried by each carrier.

(*a*) True ☐ (*b*) False. ☐

454. Metals which are comparatively poor conductor at room temperature tend to become good conductor at very low temperature.

(*a*) True ☐ (*b*) False. ☐

455. Resistivity is the reciprocal of electrical conductivity.

(*a*) True ☐ (*b*) False. ☐

456. The unit of resistivity is ohm cm.

(*a*) True ☐ (*b*) False. ☐

457. If two dissimilar metals are joined and this junction is then heated, a small voltage in millivolt range is produced. This effect is known as thermoelectric effect.

(*a*) True ☐ (*b*) False. ☐

458. If an alternating voltage V is applied across a capacitor C having vacuum on dielectric, then the resulting current I leads the voltage by 90°.

(*a*) True ☐ (*b*) False. ☐

459. If in the above question, a real dielectric is present, the current leads the voltage by $(180° - \delta)$ where δ is the measure of dielectric power loss.

(*a*) True ☐ (*b*) False. ☐

460. In the term $VI \tan \delta$, the $\tan \delta$ is known as dissipation factor.

(*a*) True ☐ (*b*) False. ☐

461. The dielectric power loss is given by $VI \tan \delta$.

(*a*) True ☐ (*b*) False. ☐

462. Electric flux density is the charge per unit area of dielectric surface.

(*a*) True ☐ (*b*) False. ☐

463. The electric flux density (D) is given by $Q \times A$ where Q = Charge and A = Area of dielectric surface.

(*a*) True ☐ (*b*) False. ☐

464. Transistor is a device which amplifies signals when two *p–n* junctions are arranged back to back in a single crystal.

(*a*) True ☐ (*b*) False. ☐

465. There are two types of transistors (*i*) *p–n–p* type and (*ii*) *n–p–n* type.

(*a*) True ☐ (*b*) False. ☐

466. The devices which are used for converting radiation energy into electric energy are known as photo cells.

(*a*) True ☐ (*b*) False. ☐

467. A vacuum is assigned a permeability of one.

(*a*) True ☐ (*b*) False. ☐

468. The diamagnetic materials have a permeability of more than 1.

(*a*) True ☐ (*b*) False. ☐

469. Magnetic permeability is the ratio of magnetic induction to the intensity of the magnetizing field.

(*a*) True ☐ (*b*) False. ☐

470. Copper, gold and silver are diamagnetic materials.

(*a*) True ☐ (*b*) False. ☐

471. Alkalies, alkaline and earth are the examples of paramagnetic materials.

(*a*) True ☐ (*b*) False. ☐

472. Burger vector in case of edge dislocation is parallel to dislocation line.

(*a*) True ☐ (*b*) False. ☐

473. Burger vector in case of screw dislocation is perpendicular to the dislocation line.

(*a*) True ☐ (*b*) False. ☐

474. Grey cast iron is softer than white cast iron.

(*a*) True ☐ (*b*) False. ☐

475. Sulphur in cast iron increases hardness and brittleness.

(*a*) True ☐ (*b*) False. ☐

476. The tensile strength of cast iron is less than that of its compressive strength.

(*a*) True ☐ (*b*) False. ☐

477. By quick cooling of molten cast iron, chilled cast iron is produced.

(*a*) True ☐ (*b*) False. ☐

478. Vanadium when added to steel decreases tensile strength.

(*a*) True ☐ (*b*) False. ☐

479. The critical temperature increases when tungsten is added to steel.

(*a*) True ☐ (*b*) False. ☐

480. By adding sulphur, lead and phosphorus, hardness of steel is increased.

(*a*) True ☐ (*b*) False. ☐

481. By adding nickel and chromium, corrosion resistance of steel is increased

(*a*) True ☐ (*b*) False. ☐

482. Glass is a mesomorphous material.
(*a*) True ☐ (*b*) False. ☐

483. Pearlite is a combination of 87% fessite and 13% cementite.
(*a*) True ☐ (*b*) False. ☐

484. The hardness of steel increases if it contains cementite.
(*a*) True ☐ (*b*) False. ☐

485. According to the carbon content in steel the upper critical point varies.
(*a*) True ☐ (*b*) False. ☐

486. Case hardening involves penetrating carbon, nitrogen or both to the surface of the steel part upto a definite depth.
(*a*) True ☐ (*b*) False. ☐

487. Surface treatments involves penetrating nitrogen to the surface of the steel part upto a definite depth.
(*a*) True ☐ (*b*) False. ☐

488. Killed steels are the steels which are destroyed by burning.
(*a*) True ☐ (*b*) False. ☐

489. According to I.S.I. specifications, the symbol for killed steel is K.R.
(*a*) True ☐ (*b*) False. ☐

490. The phenomena of the spontaneous polarization of a dielectric due to the presence of permanent electric dipoles, is known as piezoelectricity.
(*a*) True ☐ (*b*) False. ☐

491. The permittivity of any dielectrical material is influenced by temperature only.
(*a*) True ☐ (*b*) False. ☐

492. The main function of a dielectric is to store static electric charges. The phenomena of development of static charges on the surface of the dielectric under the influence of the electric field is known as dielectric polarization.
(*a*) True ☐ (*b*) False. ☐

493. The performance of a dielectric depends upon dielectric loss and dielectric strength.
(*a*) True ☐ (*b*) False. ☐

494. The unit-cell of body-centred cubic structure contains number of atoms as 4.
(*a*) True ☐ (*b*) False. ☐

495. The process of reducing the ore with carbon in the presence of a flux is known as smelting.
(*a*) True ☐ (*b*) False. ☐

496. The furnace is lined with silica bricks in acidic bessemer process.
(*a*) True ☐ (*b*) False. ☐

497. This slip plane is the plane of least atomic density.
(*a*) True ☐ (*b*) False. ☐

498. The shaping of metals by impact or by a hammer is known as forging
(*a*) True ☐ (*b*) False. ☐

499. Formation and propagation of cracks in a metal is caused due to excessive cold working.

(*a*) True ☐ (*b*) False. ☐

500. In low carbon steels, phosphorus is added to raise its yield point.

(*a*) True ☐ (*b*) False. ☐

Fill in the blanks:

501. A steel with carbon ...... is known as hypereutectoid steel.

(*a*) below 0.8% ☐ (*b*) above 0.8%. ☐

502. Between the temperature range of ......, gamma-iron occurs.

(*a*) 400 to 600°C ☐ (*b*) 900 to 1400°C. ☐

503. There is only ...... critical point, for a steel containing 0.8% carbon.

(*a*) one ☐ (*b*) two. ☐

504. At 910°C...... iron exists.

(*a*) α-iron ☐ (*b*) β-iron. ☐

505. 18/8 steel consists of 18% ...... and 8% nickel.

(*a*) carbon ☐ (*b*) chromium. ☐

506. The high speed steel has ...... percentage of tungsten.

(*a*) maximum ☐ (*b*) minimum. ☐

507. A permeability ...... is assigned to vacuum.

(*a*) equal to one ☐ (*b*) greater than one. ☐

508. The expression ...... represents dielectric power loss.

(*a*) $VI/\tan\delta$ ☐ (*b*) $VI \tan\delta$. ☐

509. The phase rule is ...... mathematically.

(*a*) $P + F = C$ ☐ (*b*) $P + F = C + 2$ ☐

where P = Number of phases, F = Number of degrees of freedom, and
C = Number of components in the system.

510. When ...... is added to aluminium, the machinibility of aluminium increases.

(*a*) silicon ☐ (*b*) lead and bismuth. ☐

511. When ...... is added to aluminium, the casting ability of aluminium increases.

(*a*) silicon ☐ (*b*) lead and bismuth. ☐

512. An alloy of copper, ... and zinc is known as bronze.

(*a*) tin ☐ (*b*) brass. ☐

513. Bronzes are ...... to brasses in corrosion resistant properties.

(*a*) superior ☐ (*b*) inferior. ☐

514. The ...... is manufactured by the use of cupola.

(*a*) wrought iron ☐ (*b*) cast iron. ☐

515. In a blast furnace ...... is used as a fuel.

(*a*) coal ☐ (*b*) coke. ☐

516. In iron, combined carbon makes the metal ...... and gives a fine grained crystalline structure.
(*a*) hard ☐ (*b*) soft. ☐

517. In iron, free carbon makes the metal ...... and gives a coarse grained crystalline structure.
(*a*) hard ☐ (*b*) soft. ☐

518. In iron, the presence of ...... in the free form is called graphite.
(*a*) carbon ☐ (*b*) silicon. ☐

519. Cast iron is a ......
(*a*) ductile material ☐ (*b*) brittle material. ☐

520. The ...... is the property of a material due to which it can be drawn into wires.
(*a*) stiffness ☐ (*b*) ductility. ☐

521. The ...... is the property of a material due to which it can be rolled or hammered into thin sheets.
(*a*) ductility ☐ (*b*) malleability. ☐

522. The ...... is the property of a material which enables it to retain the deformation permanently.
(*a*) elasticity ☐ (*b*) plasticity. ☐

523. The ...... is a process of introducing local hardness at the surface of the steel by oxyacetylene flame.
(*a*) nitriding ☐ (*b*) flame hardening. ☐

524. The ...... is the process of introducing nitrogen into the surface of certain types of steels by heating the surface.
(*a*) nitriding ☐ (*b*) cyaniding. ☐

525. Carburizing is the process of adding carbon into solid iron base alloys such as low carbon steels in order to produce a ...... surface.
(*a*) hard ☐ (*b*) soft. ☐

526. In the periodical table, elements are arranged in order of ...... atomic number.
(*a*) increasing ☐ (*b*) decreasing. ☐

527. The ...... is the line below which the alloy is in solid state and where solidification completes.
(*a*) liquidus ☐ (*b*) solidus. ☐

528. The ...... rule is used to determine the amount of each phase at any given temperature.
(*a*) phase ☐ (*b*) lever. ☐

529. According to Lever-rule, the relative lengths of the lever arms ...... by the amounts of phase present must balance.
(*a*) divided ☐ (*b*) multiplied. ☐

530. Mathematically ...... is expressed as $\frac{1}{m}\frac{dE}{dt}$

where m = Mass, E = Total energy, and T = Temperature.
(*a*) specific volume ☐ (*b*) specific heat. ☐

531. A ...... diagram is also known as constitutional diagram.

(*a*) phase ☐ (*b*) indicator. ☐

532. Monal metal is the ...... solution of Ni-Cu alloys.

(*a*) solid ☐ (*b*) liquid. ☐

533. Brass is the ...... solution of copper and zinc.

(*a*) solid ☐ (*b*) liquid. ☐

534. According to Fick's first law of diffusion, the rate of diffusion is ...... proportional to the concentration gradient.

(*a*) directly ☐ (*b*) indirectly. ☐

535. Diffusion ...... in a perfect crystal.

(*a*) is maximum ☐ (*b*) does not take place. ☐

536. Fick's ...... law of diffusion is given by $\frac{dc}{dt} = \frac{d}{dx}\left(D \cdot \frac{dc}{dx}\right)$

where $\frac{dc}{dt}$ = Rate of accumulation, $\frac{dc}{dx}$ = Concentration gradient, and D = Co-efficient of diffusion.

(*a*) first ☐ (*b*) second. ☐

537. Resistance R to diffusion ...... with increase of temperature.

(*a*) increases ☐ (*b*) decreases. ☐

538. The ...... is defined as the amount of substance diffusing in unit time across a unit area through a unit concentration gradient.

(*a*) first law of diffusion ☐ (*b*) co-efficient of diffusion. ☐

539. The ...... is the missing of an atom from its normal site.

(*a*) edge dislocation ☐ (*b*) vacancy. ☐

540. The ...... are volume defects.

(*a*) twin boundaries ☐ (*b*) cracks. ☐

541. Vacancies are the ......

(*a*) line defects ☐ (*b*) point defects. ☐

542. In diamond ...... type of atomic bond is found.

(*a*) ionic ☐ (*b*) covalent. ☐

543. In CH_4 ...... type of atomic bond is found.

(*a*) ionic ☐ (*b*) covalent. ☐

544. Primary bond is ...... as compared to secondary bond.

(*a*) weaker and less stable ☐ (*b*) stronger and more stable. ☐

545. The ...... of an atom is equal to the sum of the number of protons and neutrons.

(*a*) atomic weight ☐ (*b*) specific weight. ☐

546. The axes are ...... in space lattice.

(*a*) parallel ☐ (*b*) at right angles. ☐

547. Burger vector is ...... in case of edge dislocation.

(*a*) parallel to dislocation line ☐ (*b*) perpendicular to the dislocation line. ☐

548. Burger vector is ...... in case of screw dislocation.

(*a*) parallel to dislocation line ☐ (*b*) perpendicular to the dislocation line. ☐

549. The entropy ...... in case of diffusive processes.

(*a*) decreases ☐ (*b*) increases. ☐

550. The ...... is produced when pig iron is refined in a Bessemer converter.

(*a*) steel ☐ (*b*) cast iron. ☐

551. The imperfection is known as ...... in the crystal structure of a metal.

(*a*) dislocation ☐ (*b*) slip. ☐

552. The minor load for all cases is ...... in rockwell hardness test.

(*a*) 5 kg ☐ (*b*) 10 kg. ☐

553. The major load for scale *B* (including minor load) is ...... in rockwell hardness test.

(*a*) 100 kg ☐ (*b*) 90 kg. ☐

554. The major load for scale *C* (including minor load) is ...... in rockwell hardness test.

(*a*) 150 kg ☐ (*b*) 160 kg. ☐

555. At a temperature of ......, the change of ferromagnetic alpha-iron to paramagnetic alpha iron takes place.

(*a*) 400°C ☐ (*b*) 770°C. ☐

556. In a rockwell test, the scale ...... is used for testing hardness of low and medium carbon steels in the annealed condition.

(*a*) *B* ☐ (*b*) *C*. ☐

557. The ...... is a semi-destructive test.

(*a*) impact test ☐ (*b*) hardness test. ☐

558. The ...... test is used to determine the endurance strength and endurance limit for a metal.

(*a*) impact test ☐ (*b*) fatigue test. ☐

559. The Brinell hardness number lies approximately in the range of ...... for soft brass.

(*a*) 70 to 100°C ☐ (*b*) 50 to 70°C. ☐

560. The Brinell hardness number lies approximately in the range of ...... for mild steel.

(*a*) 70 to 100°C ☐ (*b*) 110 to 150°C. ☐

561. The diameter of the steel ball used for Brinell Test is ...... according to ASTM specifications.

(*a*) 10 mm ☐ (*b*) 5 mm. ☐

562. In case of Rockwell hardness test indenters and loads are ...... as compared to Brinell hardness test.

(*a*) larger ☐ (*b*) smaller. ☐

563. The ...... is a dielectrics.

(*a*) copper ☐ (*b*) wood. ☐

564. The dielectric constant .... with increase of temperature.

(*a*) increases ☐ (*b*) decreases. ☐

565. The donor ions are represented by ...... in case of *p–n* junction (rectifier).

(*a*) minus sign ☐ (*b*) plus sign. ☐

566. In ...... no external battery is required.

(*a*) photo emissive cell ☐ (*b*) photo voltaic cell. ☐

567. The mobility of holes is ...... than that of electrons.

(*a*) more ☐ (*b*) less. ☐

568. A *p–n* junction is

(*a*) an amplifier ☐ (*b*) a rectifier. ☐

569. The conductivity of an extrinsic semi-conductor ...... with temperature.

(*a*) decreases ☐ (*b*) increases. ☐

570. The equation ...... is known as Bragg's diffraction law.

(*a*) $n\lambda = d \sin \theta/2$ ☐ (*b*) $n\lambda = d \sin \theta$. ☐

where θ = Angle of incidence, λ = Wavelength of electron,

d = Spacing between the plane of crystal, n = Any integer.

571. In ...... conductivity, the charge carriers are ions.

(*a*) super ☐ (*b*) ionic. ☐

572. In ...... conductivity, the charge carriers are electrons.

(*a*) ionic ☐ (*b*) electronic. ☐

573. Recrystallization temperature is ...... approximately for pure metals.

(*a*) $0.2T_m$ ☐ (*b*) $0.3T_m$ ☐

where T_m = Melting temperature.

574. Recrystallization temperature is approximately ...... for alloys.

(*a*) $0.3T_m$ ☐ (*b*) $0.5T_m$. ☐

575. When steel is ...... troosite is obtained.

(*a*) cooled in still air ☐ (*b*) quenched in oil. ☐

576. When steel is ...... sorbite is obtained.

(*a*) cooled in still air ☐ (*b*) quenched in oil. ☐

577. Steel balls are hardened to ...... for ball bearings.

(*a*) 400–600 VPN ☐ (*b*) 700–800 VPN. ☐

578. The depth of the hardened zone depends upon ...... in case of flame hardening.

(*a*) flame intensity ☐ (*b*) carbon contents. ☐

III. OBJECTIVE TYPE QUESTIONS FROM COMPETITIVE EXAMINATIONS

Tick mark the most appropriate answer:

1. Carburized machine components have high endurance limit because carburization
   (*a*) raise the yield point of the material ☐
   (*b*) produces a better surface finish ☐
   (*c*) introduces a compressive layer on the surface ☐
   (*d*) suppresses any stress concentration produced in the component ☐
   (GATE 1992)
2. The alloying element mainly used to improve the endurance strength of steel materials is
   (*a*) Nickel ☐ (*b*) Vanadium ☐
   (*c*) Molybdenum ☐ (*d*) Tungsten ☐
   (GATE 1997)
3. Which one of the followings is a correct statement?
   (*a*) German silver contains 5% silver. ☐
   (*b*) An alloy of copper, tin and zinc is known as brass. ☐
   (*c*) An alloy of copper and zinc is known as bronze. ☐
   (*d*) Bronzes are superior to brasses in corrosion resistant properties. ☐
4. Which of the following processes are used for case hardening?
   (*a*) Carburizing ☐ (*b*) Nitriding ☐
   (*c*) Spraying ☐ (*d*) Electroplating ☐
   (*e*) Oxide coating ☐ (*f*) (*a*) and (*b*) only. ☐
5. The process involving the heating of steel to approximately 40°C above the critical temperature and then cooling the steel in still air is called
   (*a*) annealing ☐ (*b*) normalising ☐
   (*c*) tempering ☐ (*d*) hardening. ☐
6. Match list I with list II and select the correct answer using the codes given below the list:

List I (Heat treatment)	**List II (Effect on the properties)**
A. Annealing	1. Refined grain structure
B. Nitriding	2. Improves the hardness of the whole mass
C. Martempering	3. Increases surface hardness
D. Normalising	4. Improved ductility

Codes:

	A	*B*	*C*	*D*	
(*a*)	4	3	2	1	☐
(*b*)	1	3	4	2	☐
(*c*)	4	2	1	3	☐
(*d*)	2	1	3	4	☐

(IES 1995)

7. Choose the wrong statement
   (a) Heat treatment is effective only with certain alloys. □
   (b) Cooling rate is an important factor in heal treatment process. □
   (c) The temperature at which the change starts on heating the steel is called lower critical temperature. □
   (d) The lower critical temperature varies according to the carbon content in steel. □

8. Heat treatment is done to
   (a) change grain size □
   (b) soften the metal □
   (c) improve electrical and magnetic properties □
   (d) relieve internal stresses set up during cold and hot working □
   (e) all of the above □
   (f) none of the above. □

9. Match list I with list II and select the correct answer using the codes given below the list:

List I (Materials)	List II (Applications)
A. Tungsten carbide	1. Abrasive wheels
B. Silicon nitride	2. Heating elements
C. Aluminium oxide	3. Pipes for carrying liquid metals.
D. Silicon carbide	4. Drawing dies

Codes:

	A	B	C	D	
(a)	3	4	1	2	□
(b)	4	3	2	1	□
(c)	3	4	2	1	□
(d)	4	3	1	2	□

(IES 1999)

10. Match list I with list II and select the correct answer using the codes given below the list:

List I (Steel type)	List II (Product)
A. Mild steel	1. Screw driver
B. Tool steel	2. Commercial beams
C. Medium carbon steel	3. Crane hooks
D. High carbon steel	4. Blanking dies

Codes:

	A	B	C	D	
(a)	1	4	3	2	□
(b)	2	4	1	3	□
(c)	1	3	4	2	□
(d)	2	4	3	1	□

(IES 1993)

11. Which of the following statements are true of annealing of steels?

1. Steels are heated to 500 to 700°C
2. Cooling is done slowly and steadily
3. Internal stresses are relieved
4. Ductility of steel is reduced.

Codes:

(*a*) 2, 3 and 4 ☐ (*b*) 1, 3 and 4 ☐
(*c*) 1, 2 and 4 ☐ (*d*) 1, 2 and 3 ☐

12. Consider the following work materials

1. Titanium 2. Mild steel
3. Stainless steel 4. Grey cast iron

The correct sequence of these materials in terms of increasing order of difficulty in machining is

(*a*) 4, 2, 3, 1 ☐ (*b*) 4, 2, 1, 3 ☐
(*c*) 2, 4, 3, 1 ☐ (*d*) 2, 4, 1, 3 ☐

(IES 1995)

13. Which of the following processes are used for surface treatments?

(*a*) carburizing ☐ (*b*) nitriding ☐
(*c*) spraying ☐ (*d*) electroplating ☐
(*e*) oxide coating ☐ (*f*) all except (*a*) and (*b*). ☐

14. Case hardening involves

(*a*) applying coatings to the surface of metals or alloys ☐
(*b*) penetrating carbon to the surface of the steel part upto a definite depth ☐
(*c*) penetrating nitrogen to the surface of the steel part upto a definite depth ☐
(*d*) penetrating carbon, nitrogen or both to the surface of the steel part upto a definite depth. ☐

15. Heating the steels to 30°C above the critical temperature line, soaking at that temperature and then cooling slowly to room temperature to form a pearlite and ferrite structure is known as

(*a*) hardening ☐ (*b*) normalising ☐
(*c*) tempering ☐ (*d*) annealing ☐

(IES 1999)

16. Cast iron is used for machine beds because of its high

(*a*) tensile strength ☐ (*b*) endurance strength ☐
(*c*) damping capacity ☐ (*d*) compressive strength ☐

(IES 1999)

17. The total area under the stress strain curve of a mild steel specimen tested upto failure under tension is a measure of

(*a*) ductility ☐ (*b*) ultimate strength ☐
(*c*) stiffness ☐ (*d*) toughness ☐

(GATE 2002)

18. Surface treatments involve

(*a*) applying coatings to the surface of metals or alloys ☐

(*b*) penetrating carbon to the surface to the steel part upto a definite depth ☐

(*c*) penetrating nitrogen to the surface of the steel part upto a definite depth ☐

(*d*) penetrating carbon, nitrogen or both to the surface of the steel part upto a definite depth. ☐

19. The process of introducing carbon into low carbon steels in order to produce a hard surface is known as

(*a*) carbonitriding ☐ (*b*) nitriding ☐

(*c*) carburizing ☐ (*d*) cyaniding ☐

(*e*) induction hardening ☐ (*f*) none of the above ☐

20. Match list I with list II and select the correct answer using the codes given below the list:

List I (Material properties)	**List II (Tests to determine material properties)**
A. Ductility	1. Impact test
B. Toughness	2. Fatigue test
C. Endurance limit	3. Tension test
D. Resistance to penetration	4. Hardness test

Codes:

	A	*B*	*C*	*D*	
(*a*)	3	2	1	4	☐
(*b*)	4	2	1	3	☐
(*c*)	3	1	2	4	☐
(*d*)	4	1	2	3	☐

21. For a linearly elastic, isotropic and homogeneous material, the number of elastic constants required to relate stress and strain is

(*a*) two ☐ (*b*) three ☐

(*c*) four ☐ (*d*) six ☐

(IAS 1994)

22. Which one of the following materials is used for car tyres as a standard material?

(*a*) Styrene-butadiene rubber (SBR) ☐ (*b*) Butyl rubber ☐

(*c*) Nitrile rubber ☐ (*d*) Any of the above depending on need ☐

(IES 1997)

23. Heat treatment is a process involving

(*a*) heating of a metal in a solid state to obtain certain desirable properties ☐

(*b*) cooling of a metal in a solid state to obtain certain desirable properties ☐

(*c*) heating and cooling of a metal in a solid state to obtain certain desirable properties ☐

(*d*) none of the above. ☐

24. Match list I (alloying element in steel) with list II (property conferred on steel by the element) and select the correct answer using the codes given below the list.

List I	**List II**
A. Nickel	1. Corrosion resistance
B. Chromium	2. Magnetic permeability
C. Tungsten	3. Heat resistance
D. Silicon	4. Hardenability

Codes:

	A	*B*	*C*	*D*	
(*a*)	4	1	3	2	□
(*b*)	4	1	2	3	□
(*c*)	1	4	3	2	□
(*d*)	1	4	2	3	□

(IES 1998)

25. German silver contains

(*a*) 50% copper and 50% zinc □

(*b*) 60.45% copper, 35.2% zinc and 435% nickel □

(*c*) 60% copper and 40% zinc □

(*d*) 70% copper and 30% zinc. □

26. Duralumin alloy contains aluminium and copper in the ratio of

	Al (Percentage)	**Cu (Percentage)**	
(*a*)	94	4	□
(*b*)	90	8	□
(*c*)	88	10	□
(*d*)	86	12	□

(IES 1993)

27. The blade of power saw is made of

(*a*) Boron steel □ (*b*) High speed steel □

(*c*) Stainless steel □ (*d*) Malleable cast iron □

(IES 1993)

28. Quartz is a

(*a*) ferroelectric material □ (*b*) ferromagnetic material □

(*c*) piezoelectric material □ (*d*) diamagnetic material □

(IES 1993)

29. Which one of the following statements is correct?

(*a*) The hardness of steel depends upon the amount of carbon it contains. □

(*b*) The process in which steel is coated with a thin layer of phosphate is known as anodising. □

(*c*) The imperfection in the crystal structure of a metal is known as dislocation. □

(*d*) Lime stone is added in blast furnace to flux carbon. □

30. The hardness of steel primarily depends on

(*a*) method of manufacture ☐
(*b*) heat treatment employed ☐
(*c*) shape of carbide and their distribution in iron ☐
(*d*) percentage of carbon. ☐

31. Which of the following pairs regarding the effect of alloying elements in steel are correctly matched?

1. **Molybdenum:** Forms abrasion resisting particles
2. **Phosphorous:** Improves machinability in free cutting steels
3. **Cobalt:** Contributes to red hardness by hardening ferrite
4. **Silicon:** Reduces oxidation resistance

Codes:

(*a*) 2, 3 and 4 ☐ (*b*) 1, 3 and 4 ☐
(*c*) 1, 2 and 4 ☐ (*d*) 1, 2 and 3 ☐

(IES 1996)

32. Match list I (alloys) with list II (applications) and select the correct answer using the codes given below the list:

List I	**List II**
A. Chromel	1. Journal bearing
B. Babbit alloy	2. Milling cutter
C. Nimonic alloy	3. Thermocouple wire
D. High speed steel	4. Gas turbine blades

Codes:

	A	*B*	*C*	*D*	
(*a*)	3	1	4	2	☐
(*b*)	3	4	1	2	☐
(*c*)	2	4	1	3	☐
(*d*)	2	1	4	3	☐

(IES 1998)

33. Cementite is a combination of

(*a*) 6.67% carbon and 93.33% iron ☐ (*b*) 13% carbon and 87% ferrite ☐
(*c*) 13% cementite and 87% ferrite ☐ (*d*) 87% cementite and 13% ferrite. ☐

34. Which one of the following statements is a wrong?

(*a*) The body-centred cubic space lattice is found in alpha-iron. ☐
(*b*) A steel containing ferrite and pearlite is tough. ☐
(*c*) A steel with 0.8% carbon is wholly pearlite. ☐
(*d*) The lower critical temperature is different for all steels. ☐

35. Match list I with list II and select the correct answer using the codes given below the lists:

List I (Material/part)	**List II (Techniques)**
A. Ductile iron	1. Inoculation
B. Malleable iron	2. Chilled
C. Rail steel joints	3. Annealing
D. White cast iron	4. Thermit welding
	5. Isothermal annealing

Codes:

	A	*B*	*C*	*D*	
(*a*)	1	3	4	2	☐
(*b*)	5	3	2	1	☐
(*c*)	2	1	4	3	☐
(*d*)	1	4	2	3	☐

(IAS 1995)

36. Addition of magnesium to cast iron increases its

(*a*) hardness ☐ (*b*) ductility and strength in tension ☐

(*c*) corrosion resistance ☐ (*d*) creep strength. ☐

(IES 1995)

37. Small amounts of which one of the following elements/pair of elements is added to steel to increase its machinability?

(*a*) Nickel ☐ (*b*) Sulphur and phosphorus ☐

(*c*) All of the above ☐ (*d*) None of the above. ☐

38. The atomic packing factor is defined as the

(*a*) ratio of the volume occupied by the unit-cell to the volumes of the atoms per unit-cell ☐

(*b*) ratio of the volume of the atoms per unit-cell to the total volume occupied by the unit-cell ☐

(*c*) sum of the volume of the atoms per unit-cell and the total volume occupied by the unit-cell ☐

(*d*) difference of the volume of the atoms per unit-cell and the total volume occupied by the unit-cell. ☐

39. Consider the following treatments:

1. Normalising 2. Hardening
3. Martempering 4. Cold working

Hardness and tensile strength in austenictic stainless steel can be increased by

(*a*) 1, 2 and 3 ☐ (*b*) 1 and 3 ☐

(*c*) 2 and 4 ☐ (*d*) 4 alone. ☐

(IES 1994)

40. Killed steels

(*a*) have minimum impurity level ☐

(*b*) have almost zero percentage of phosphorus and sulphur ☐

(*c*) are produced by L.D. process ☐
(*d*) are free from oxygen. ☐

(IES 1994)

41. 18/8 stainless steel contains

(*a*) 18% steel, 8% chromium ☐ (*b*) 18% chromium, 8% nickel ☐
(*c*) 18% tungsten, 8% nickel ☐ (*d*) 18% tungsten, 8% chromium. ☐

(IES 1996)

42. Tin base white metals are used where the bearings are subjected to

(*a*) large surface wear ☐ (*b*) elevated temperature ☐
(*c*) light load and pressure ☐ (*d*) high pressure and load. ☐

(IES 1996)

43. Which one of the following is an example of a bronze?

(*a*) gun metal ☐ (*b*) delta metal ☐
(*c*) muntz metal ☐ (*d*) lead. ☐

44. Gun-metal contains

(*a*) copper = 80%, tin = 15% and zinc = 5% ☐ (*b*) copper = 70% and zinc = 30% ☐
(*c*) copper = 88%, tin = 10% and zinc = 2% ☐ (*d*) copper = 60% and zinc = 40%. ☐

45. Which of the following pairs are correctly matched?

1. Silicon steels-transformer stampings
2. Duralumin-Cooking utensils
3. Gun metal-bearings

Codes:

(*a*) 1, 2 and 3 ☐ (*b*) 1 and 2 ☐
(*c*) 1 and 3 ☐ (*d*) 2 and 3 ☐

(IES 1994)

46. Which of the following are correctly matched?

1. Lead screw nut-phosphor bronze
2. Piston-cast iron
3. Cam-EN-31 steel
4. Lead screw-wrought iron

Codes:

(*a*) 2, 3 and 4 ☐ (*b*) 1, 3 and 4 ☐
(*c*) 1, 2 and 4 ☐ (*d*) 1, 2 and 3 ☐

(IES 1994)

47. Choose the wrong statement

(*a*) A plain carton steel is an alloy if iron and carbon. ☐
(*b*) An alloy steel contains more than 0.5% Mn and 0.5% Si. ☐

(*c*) The alloy steels are more stronger, tougher and fatigue resistant than carbon steels. ☐

(*d*) White cast iron has a low tensile strength and a high compressive strength. ☐

48. Choose the correct statement

(*a*) Grey cast iron possesses lowest melting point of the ferrous alloys. ☐

(*b*) Wrought iron possesses a high resistance towards corrosion. ☐

(*c*) White cast iron under normal circumstances is brittle and not machinable. ☐

(*d*) All of the above. ☐

(*e*) None of the above. ☐

49. Alloy steel which is work handenable and which is used to make the blades of bulldozers, buckets, bucket wheel excavators and other earth moving equipment contain iron, carbon and

(*a*) chromium ☐ (*b*) silicon ☐

(*c*) manganese ☐ (*d*) magnesium. ☐

(IES 1996)

50. Match list I with list II and select the correct answer using the codes given below the list:

List I	**List II**
A. Toughness	1. Moment area method
B. Endurance strength	2. Hardness
C. Resistance to abrasion	3. Energy absorbed before fracture in a tension test
D. Deflection in a beam	4. Fatigue loading

Codes:

	A	*B*	*C*	*D*	
(*a*)	4	3	1	2	☐
(*b*)	4	3	2	1	☐
(*c*)	3	4	2	1	☐
(*d*)	3	4	1	2	☐

(IES 1997)

51. The upper critical temperature for a steel

(*a*) is constant ☐

(*b*) depends upon the rate of heating ☐

(*c*) varies according to the carbon content in steel ☐

(*d*) none of the above. ☐

52. The lowest critical temperature for a steel

(*a*) is constant ☐

(*b*) depends upon the rate of heating ☐

(*c*) varies according to the carbon content in steel ☐

(*d*) none of the above. ☐

53. Assertion (*A*): In high speed steels, alloying elements tungsten, Chromium, and Vanadium are added to make them suitable to work at higher cutting speeds than tool steel or low alloy steels.

Reason (*R*): Vanadium adds to the property of red hardness and tungsten and chromium add to high wear resistance

(*a*) *A* is false and *R* is true ☐ (*b*) *A* is true and *R* is false ☐
(*c*) both *A* and *R* are true ☐ (*d*) both *A* and *R* are false. ☐

(IAS 1995)

54. Assertion (*A*): Non-ferrous materials are best machined with diamond tools.

Reason (*R*): Diamond tools are suitable for high speed machining

(*a*) *A* is false and *R* is true ☐ (*b*) *A* is true and *R* is false ☐
(*c*) both *A* and *R* are true ☐ (*d*) both *A* and *R* are false. ☐

(IES 1995)

55. Assertion (*A*): An aluminium alloy with 11 per cent silicon is used for making engine pistons by die casting technique.

Reason (*R*): Aluminium has low density and addition of silicon improves its fluidity and therefore its castability

(*a*) *A* is false and *R* is true ☐ (*b*) *A* is true and *R* is false ☐
(*c*) both *A* and *R* are true ☐ (*d*) both *A* and *R* are false. ☐

(IES 1995)

56. Decreasing grain size in a polycrystalline material

(*a*) increases yield strength and Corrosion resistance. ☐
(*b*) decreases yieid strength and Corrosion resistance. ☐
(*c*) decreases yield strength but increases Corrosion resistance ☐
(*d*) Increases yield strength but decreases Corrosion resistance. ☐

(GATE 1998)

ANSWERS

Answers to Objective Type Questions

1. (*b*)	**2.** (*d*)	**3.** (*a*)	**4.** (*d*)	**5.** (*a*)	**6.** (*d*)
7. (*b*)	**8.** (*a*)	**9.** (*c*)	**10.** (*b*)	**11.** (*c*)	**12.** (*a*)
13. (*b*)	**14.** (*a*)	**15.** (*d*)	**16.** (*b*)	**17.** (*b*)	**18.** (*c*)
19. (*c*)	**20.** (*d*)	**21.** (*c*)	**22.** (*b*)	**23.** (*d*)	**24.** (*c*)
25. (*a*)	**26.** (*b*)	**27.** (*d*)	**28.** (*c*)	**29.** (*b*)	**30.** (*a*)
31. (*d*)	**32.** (*c*)	**33.** (*b*)	**34.** (*b*)	**35.** (*a*)	**36.** (*c*)
37. (*b*)	**38.** (*a*)	**39.** (*b*)	**40.** (*d*)	**41.** (*b*)	**42.** (*d*)
43. (*c*)	**44.** (*c*)	**45.** (*b*)	**46.** (*b*)	**47.** (*b*)	**48.** (*a*)
49. (*c*)	**50.** (*b*)	**51.** (*c*)	**52.** (*a*)	**53.** (*b*)	**54.** (*c*)
55. (*b*)	**56.** (*c*)	**57.** (*a*)	**58.** (*e*)	**59.** (*c*)	**60.** (*b*)

61. (*b*)	62. (*c*)	63. (*c*)	64. (*b*)	65. (*b*)	66. (*c*)
67. (*d*)	68. (*e*)	69. (*d*)	70. (*b*)	71. (*e*)	72. (*b*)
73. (*c*)	74. (*c*)	75. (*b*)	76. (*a*)	77. (*c*)	78. (*c*)
79. (*c*)	80. (*b*)	81. (*c*)	82. (*c*)	83. (*a*)	84. (*b*)
85. (*c*)	86. (*b*)	87. (*c*)	88. (*a*)	89. (*c*)	90. (*b*)
91. (*a*)	92. (*b*)	93. (*b*)	94. (*b*)	95. (*b*)	96. (*b*)
97. (*b*)	98. (*c*)	99. (*a*)	100. (*b*)	101. (*c*)	102. (*d*)
103. (*b*)	104. (*a*)	105. (*f*)	106. (*f*)	107. (*e*)	108. (*d*)
109. (*a*)	110. (*c*)	111. (*b*)	112. (*d*)	113. (*d*)	114. (*d*)
115. (*a*)	116. (*d*)	117. (*c*)	118. (*b*)	119. (*d*)	120. (*f*)
121. (*e*)	122. (*f*)	123. (*c*)	124. (*d*)	125. (*d*)	126. (*e*)
127. (*b*)	128. (*a*)	129. (*d*)	130. (*a*)	131. (*b*)	132. (*a*)
133. (*b*)	134. (*d*)	135. (*b*)	136. (*a*)	137. (*b*)	138. (*c*)
139. (*c*)	140. (*b*)	141. (*f*)	142. (*b*)	143. (*d*)	144. (*b*)
145. (*a*)	146. (*c*)	147. (*a*)	148. (*b*)	149. (*c*)	150. (*c*)
151. (*c*)	152. (*e*)	153. (*e*)	154. (*d*)	155. (*b*)	156. (*a*)
157. (*d*)	158. (*c*)	159. (*d*)	160. (*b*)	161. (*a*)	162. (*c*)
163. (*c*)	164. (*a*)	165. (*c*)	166. (*d*)	167. (*a*)	168. (*b*)
169. (*d*)	170. (*b*)	171. (*c*)	172. (*a*)	173. (*e*)	174. (*c*)
175. (*b*)	176. (*a*)	177. (*b*)	178. (*d*)	179. (*c*)	180. (*d*)
181. (*c*)	182. (*c*)	183. (*d*)	184. (*e*)	185. (*b*)	186. (*c*)
187. (*b*)	188. (*c*)	189. (*d*)	190. (*a*)	191. (*b*)	192. (*c*)
193. (*d*)	194. (*b*)	195. (*b*)	196. (*a*)	197. (*c*)	198. (*a*)
199. (*c*)	200. (*b*)	201. (*d*)	202. (*a*)	203. (*c*)	204. (*d*)
205. (*b*)	206. (*b*)	207. (*a*)	208. (*b*)	209. (*c*)	210. (*d*)
211. (*a*)	212. (*b*)	213. (*b*)	214. (*a*)	215. (*b*)	216. (*c*)
217. (*c*)	218. (*b*)	219. (*b*)	220. (*b*)	221. (*a*)	222. (*b*)
223. (*b*)	224. (*a*)	225. (*e*)	226. (*e*)	227. (*c*)	228. (*a*)
229. (*b*)	230. (*c*)	231. (*a*)	232. (*b*)	233. (*c*)	234. (*d*)
235. (*c*)	236. (*d*)	237. (*b*)	238. (*c*)	239. (*a*)	240. (*d*)
241. (*d*)	242. (*c*)	243. (*c*)	244. (*c*)	245. (*d*)	246. (*a*)
247. (*c*)	248. (*c*)	249. (*b*)	250. (*c*)	251. (*d*)	252. (*c*)
253. (*c*)	254. (*d*)	255. (*b*)	256. (*b*)	257. (*b*)	258. (*f*)
259. (*e*)	260. (*c*)	261. (*a*)	262. (*f*)	263. (*b*)	264. (*c*)
265. (*a*)	266. (*b*)	267. (*a*)	268. (*c*)	269. (*c*)	270. (*a*)
271. (*b*)	272. (*b*)	273. (*b*)	274. (*d*)	275. (*c*)	276. (*d*)
277. (*b*)	278. (*c*)	279. (*d*)	280. (*a*)	281. (*b*)	282. (*c*)
283. (*c*)	284. (*c*)	285. (*a*)	286. (*b*)	287. (*c*)	288. (*c*)
289. (*c*)	290. (*b*)	291. (*a*)	292. (*c*)	293. (*b*)	294. (*a*)

295. (*b*)	**296.** (*d*)	**297.** (*b*)	**298.** (*b*)	**299.** (*c*)	**300.** (*d*)
301. (*c*)	**302.** (*e*)	**303.** (*d*)	**304.** (*d*)	**305.** (*b*)	**306.** (*b*)
307. (*b*)	**308.** (*d*)	**309.** (*d*)	**310.** (*b*)	**311.** (*a*)	**312.** (*c*)
313. (*a*)	**314.** (*c*)	**315.** (*c*)	**316.** (*b*)	**317.** (*d*)	**318.** (*f*)
319. (*f*)	**320.** (*d*)	**321.** (*a*)	**322.** (*c*)	**323.** (*b*)	**324.** (*d*)
325. (*a*)	**326.** (*c*)	**327.** (*d*)	**328.** (*e*)	**329.** (*d*)	**330.** (*e*)
331. (*b*)	**332.** (*a*)	**333.** (*a*)	**334.** (*c*)	**335.** (*b*)	**336.** (*d*)
337. (*a*)	**338.** (*c*)	**339.** (*e*)	**340.** (*b*)	**341.** (*c*)	**342.** (*b*)
343. (*c*)	**344.** (*a*)	**345.** (*a*)	**346.** (*c*)	**347.** (*b*)	**348.** (*a*)
349. (*a*)	**350.** (*b*)	**351.** (*b*)	**352.** (*a*)	**353.** (*b*)	**354.** (*a*)
355. (*b*)	**356.** (*b*)	**357.** (*c*)	**358.** (*b*)	**359.** (*a*)	**360.** (*a*)
361. (*b*)	**362.** (*b*)	**363.** (*b*)	**364.** (*c*)	**365.** (*a*)	**366.** (*d*)
367. (*b*)	**368.** (*b*)	**369.** (*d*)	**370.** (*d*)	**371.** (*c*)	**372.** (*c*)
373. (*d*)	**374.** (*d*)	**375.** (*b*)	**376.** (*b*)	**377.** (*c*)	**378.** (*d*)
379. (*c*)	**380.** (*f*)	**381.** (*b*)	**382.** (*a*)	**383.** (*e*)	**384.** (*b*)
385. (*b*)	**386.** (*c*)	**387.** (*d*)	**388.** (*c*)	**389.** (*d*)	**390.** (*d*)
391. (*b*)	**392.** (*a*)	**393.** (*c*)	**394.** (*a*)	**395.** (*b*)	**396.** (*a*)
397. (*c*)	**398.** (*c*)	**399.** (*b*)	**400.** (*c*)	**401.** (*b*)	**402.** (*a*)
403. (*b*)	**404.** (*c*).				

True/False

405. (*b*)	**406.** (*a*)	**407.** (*a*)	**408.** (*b*)	**409.** (*a*)	**410.** (*a*)
411. (*a*)	**412.** (*a*)	**413.** (*a*)	**414.** (*b*)	**415.** (*b*)	**416.** (*a*)
417. (*a*)	**418.** (*d*)	**419.** (*a*)	**420.** (*a*)	**421.** (*b*)	**422.** (*a*)
423. (*a*)	**424.** (*b*)	**425.** (*a*)	**426.** (*a*)	**427.** (*a*)	**428.** (*b*)
429. (*a*)	**430.** (*a*)	**431.** (*a*)	**432.** (*b*)	**433.** (*a*)	**434.** (*a*)
435. (*a*)	**436.** (*a*)	**437.** (*b*)	**438.** (*b*)	**439.** (*a*)	**440.** (*a*)
441. (*b*)	**442.** (*b*)	**443.** (*b*)	**444.** (*a*)	**445.** (*a*)	**446.** (*a*)
447. (*a*)	**448.** (*a*)	**449.** (*a*)	**450.** (*a*)	**451.** (*a*)	**452.** (*b*)
453. (*a*)	**454.** (*a*)	**455.** (*a*)	**456.** (*a*)	**457.** (*a*)	**458.** (*a*)
459. (*b*)	**460.** (*a*)	**461.** (*a*)	**462.** (*a*)	**463.** (*b*)	**464.** (*a*)
465. (*a*)	**466.** (*a*)	**467.** (*a*)	**468.** (*b*)	**469.** (*a*)	**470.** (*a*)
471. (*a*)	**472.** (*b*)	**473.** (*b*)	**474.** (*a*)	**475.** (*a*)	**476.** (*a*)
477. (*a*)	**478.** (*a*)	**479.** (*a*)	**480.** (*b*)	**481.** (*a*)	**482.** (*b*)
483. (*a*)	**484.** (*a*)	**485.** (*a*)	**486.** (*a*)	**487.** (*b*)	**488.** (*a*)
489. (*b*)	**490.** (*b*)	**491.** (*b*)	**492.** (*a*)	**493.** (*a*)	**494.** (*b*)
495. (*a*)	**496.** (*a*)	**497.** (*b*)	**498.** (*a*)	**499.** (*a*)	**500.** (*a*).

Fill in the Blanks

501. (*b*)	502. (*b*)	503. (*a*)	504. (*a*)	505. (*b*)	506. (*a*)
507. (*a*)	508. (*b*)	509. (*b*)	510. (*b*)	511. (*a*)	512. (*a*)
513. (*a*)	514. (*b*)	515. (*b*)	516. (*a*)	517. (*b*)	518. (*a*)
519. (*b*)	520. (*b*)	521. (*b*)	522. (*b*)	523. (*b*)	524. (*a*)
525. (*a*)	526. (*a*)	527. (*b*)	528. (*b*)	529. (*b*)	530. (*b*)
531. (*a*)	532. (*a*)	533. (*a*)	534. (*a*)	535. (*b*)	536. (*b*)
537. (*b*)	538. (*b*)	539. (*b*)	540. (*b*)	541. (*b*)	542. (*b*)
543. (*b*)	544. (*b*)	545. (*a*)	546. (*b*)	547. (*b*)	548. (*a*)
549. (*b*)	550. (*a*)	551. (*a*)	552. (*b*)	553. (*a*)	554. (*a*)
555. (*b*)	556. (*a*)	557. (*b*)	558. (*b*)	559. (*a*)	560. (*b*)
561. (*a*)	562. (*b*)	563. (*b*)	564. (*a*)	565. (*b*)	566. (*b*)
567. (*b*)	568. (*b*)	569. (*b*)	570. (*b*)	571. (*b*)	572. (*b*)
573. (*b*)	574. (*b*)	575. (*b*)	576. (*a*)	577. (*b*)	578. (*a*).

Answers to Objective Type Questions from Competitive Examinations

1. (*c*)	2. (*b*)	3. (*d*)	4. (*f*)	5. (*b*)	6. (*a*)
7. (*d*)	8. (*e*)	9. (*d*)	10. (*d*)	11. (*a*)	12. (*a*)
13. (*f*)	14. (*d*)	15. (*d*)	16. (*b*)	17. (*d*)	18. (*a*)
19. (*c*)	20. (*c*)	21. (*c*)	22. (*a*)	23. (*c*)	24. (*a*)
25. (*b*)	26. (*a*)	27. (*b*)	28. (*c*)	29. (*c*)	30. (*c*)
31. (*a*)	32. (*a*)	33. (*a*)	34. (*d*)	35. (*a*)	36. (*b*)
37. (*b*)	38. (*b*)	39. (*d*)	40. (*d*)	41. (*b*)	42. (*d*)
43. (*a*)	44. (*c*)	45. (*a*)	46. (*d*)	47. (*d*)	48. (*d*)
49. (*c*)	50. (*c*)	51. (*c*)	52. (*a*)	53. (*c*)	54. (*a*)
55. (*c*)	56. (*a*).				

Chapter 12 PRODUCTION ENGINEERING

I. THEORY

12.1. INTRODUCTION

Production engineering is the science which deals with knowledge of raw materials, selection of material, manufacturing process, planning process layouts, plant layouts, equipment specification, tool design, method developments, work standards, value analysis, cost control, final assembly, inspection, packing and dispatch.

The tools and machines used for production should be such that they can produce the goods economically as well as accurately. Economy depends upon to a large extent on the proper selection of the machine or process for the job that will give a satisfactory finished product. The selection depends upon the quantity of items to be produced.

12.2. CASTING

Casting is one of the most commonly used processes for producing parts. In case of casting, the molten metal is poured into the mould. The molten metal flows into the corners and fills all the voids. When metal solidifies it takes the shape of the mould. But the casting is slightly smaller and hence the provision should be made for shrinkage of metal. This casting is now ready for further machining operations.

12.2.1. Pattern

The first step in making a casting is to prepare a model known as **pattern** whose shape and size is determined from the finished part drawing. Allowances for draft, shrinkage, finish etc. are provided in the pattern. **Materials** used in the patterns are wood, metal, plastic and plaster. The pattern must be able to withstand wear, rough handling, rapping, ramming and moisture.

12.2.2. Types of Pattern

The important types of pattern are:

(*i*) Solid or one-piece pattern,
(*ii*) Split or two-piece pattern,
(*iii*) Cope and drag pattern,
(*iv*) Follow-board pattern,
(*v*) Match plate pattern,
(*vi*) Skeleton pattern, and
(*vii*) Loose-piece pattern.

12.3. CENTRIFUGAL CASTING

If the mould is fitted to a revolving disc which is rotated at about 1500 r.p.m., then the casting is known as centrifugal casting. The centrifugal forces causes the molten metal to be distributed to all parts of the mould. In this casting, impurities collect at the inner wall of the casting. Centrifugal casting may be:

(*i*) True centrifugal casting or
(*ii*) Semi-centrifugal casting.

12.3.1. True Centrifugal Casting

True centrifugal casting is used to cast symmetrical objects such as pipes. The metal is forced against the mould wall by the centrifugal force until it solidifies. No core is needed to form the hole in the casting.

12.3.2. Semi-Centrifugal Casting

Semi-centrifugal casting is used to ensure purity and density at the extremitives of a casting such as a cast wheel. The sand mould is rotated at a lower speed than in the centrifugal casting process.

12.3.3. Defects in Castings

The important **defects in castings** are:

1. Blow-holes,
2. Roughness of surfaces,
3. Scales or buckles
4. Fins,
5. Sand spots
6. Shrinkage, and
7. Cold shots.

12.4. METAL FORMING

The process of shaping the metals into required form is known as metal forming. Metal forming techniques occupy an important position in engineering industries because of many advantages. Casting is not feasible in all cases. The shaping of metals is done by mechanical means either in cold working state or hotworking state. Cold working takes place at a temperature below the recrystallization temperature whereas hot working takes place at a temperature well above the recrystallization temperature.

12.4.1. Important Metal Forming Processes

Important metal forming processes are:

(*a*) Rolling (*b*) Hot spinning
(*c*) Seaming (*d*) Shear forming
(*e*) Stretch forming.

(*a*) **Rolling:** Rolling is the most important of all the processes employed for shaping the metals. It is the simplest and most economical way of producing large tonnages of metal products. Rolling or forcing a piece of metal through a pair of revolving rolls with plain or grooved barrels, is widely used in engineering and metal industry these days. Rolling may be hot or cold.

Plug rolling: It is used to reduce wall thickness and increase the diameter of tubes already made. This process cannot be used to make seamless tubing from solid billets.

(*b*) **Hot spinning:** The process which uses pressure and plastic flow to shape material, is known as hot spinning. It may be done hot or cold. In either case the metal is forced to flow over a rotating shape by pressure of a blunt tool or rollers. The pressure of tool against disc generates heat. Hot spinning is restricted to thick wall forming. The temperature is maintained by heat generated by the spinning operation.

(*c*) **Seaming:** This is accomplished with rollers so that the bends lap the material back on itself to close the form. This type of seam may be made in various types. They may be single or multiple seam laps.

(*d*) **Shear forming:** It is a metal forming process used for production of hollow rotationally symmetric sheet metal components.

(*e*) **Stretch forming:** It is a metal forming process in which a piece of metal sheet is stretched and bent simultaneously over a die in order to form large contoured parts.

12.5. FORGING PROCESSES

The process forging is defined as the controlled plastic deformation of metals at elevated temperatures into a predetermined size or shape using compressive forces existed through some type of die by a hammer, press or an upsetting machine. The important type of forgings are:

(*a*) Smith forging, (*b*) Drop forging,
(*c*) Hot press forging, (*d*) Upset forging,
(*e*) Swing forging, (*f*) Roll forging, and
(*g*) Cold forging.

12.5.1. Smith Forging

It is one of the oldest types of forging processes. In this forging, hand tools are required to shape desired parts. This process is used when low production is required. The process is carried out by striking the heated part repeatedly by a hammer until it takes on the desired shape and size.

12.5.2. Drop Forging

It is a process used to shape metal in closed dies. Forms are machined into two halves of a closed die set. The metal is heated and under impact the metal becomes plastic. Under repeated blows the metal fills the die cavity. Excess metal (flash) is trimmed off.

12.5.3. Hot Press Forging

Here a steadily pressure is applied instead of an impact force. Under a steady high pressure the entire work-price is subjected to the deforming pressure whereas under impact, energy is transmitted through the surface of the work-piece. Multiple die cavities may be machined into one die set and usually one stroke is needed to complete a particular forging operations. As the pressure is applied, the metal flows into and fills the cavity.

12.5.4. Upset Forging

Here the end of a bar is heated and this heated bar end is forced into a desired shape. Actually the bar is heated, clamped and upset into the die-opening. The pressure builds rapidly and becomes a maximum at the end of the stroke.

12.5.5. Swing Forging

It employs two pairs of rotating forging tools which close and open rapidly such that they provide squeezing action at a fast rate to the work-piece. The forging operation may reduce the diameter of the tube which is fed between the two pairs of the forging tools.

12.6. EXTRUSION

The process of pushing the heated billet or slug of metal through an orifice provided into a die, thus forming an elongated part of uniform cross-section is known as extrusion. The pressure is applied either hydraulically or mechanically. The extrusion pressure to the stroke is given by

$$P = \rho \times \log_e \delta \times \exp.\frac{2fL}{R}$$

where δ = Extrusion ratio = $\dfrac{A}{a}$,

A = Cross-section area of upset billet in the container,

a = Cross-section area of the hole in the die,

R = Container radius,

f = Coefficient of friction,

L = Billet length after upsetting,

P = Specific resistance to deformation of metal at a specific temperature.

The two basic methods for forcing hot metals through a die opening are:

1. The direct pressure method, and 2. The indirect pressure method.

12.7. GRINDING

The process of metal removal comparatively in smaller volume, by a rotating abrasive wheel that acts as a cutting tool, is known as grinding. This is used to finish work-pieces which must show a high surface quality, accuracy of shape and dimension. Hence grinding is mainly used for the following purposes.

1. To remove a very small amount of metal from the work-piece to bring its dimensions within very close tolerances after all rough finishing and heat treatment operations have been carried out.
2. It is some times used to obtain belter finish.
3. It is also used for sharpening the cutting tools. Grinding wheels are made of abrasive particles bonded together by means some suitable bond. The abrasive used for grinding is divided into two groups *i.e.*,

   (*i*) Natural abrasive, and

   (*ii*) Artificial or manufactured abrasive.

The natural abrasive are produced by uncontrolled forces of nature. They are:

(*a*) Sandstone or solid quartz, (*b*) Emergy,

(*c*) Corrundum, and (*d*) Diamond.

Grindstones are made of naturally occurring sandstone. Emergy or corrundum is naturally occurring aluminium oxide with many impurities.

The artificial or manufactured abrasives include chiefly:

(*a*) Simcon carbide (SiC)-trade name is carborandum, crystolon, electron etc.

(*b*) Aluminium oxide.

The silicon carbide is manufactured from 56 parts of silica sand, 34 parts of powdered coke, 12 parts of saw dust and 2 parts of salt.

By heating mineral bauxite, mixed with ground coke and iron borings in an arc-type electric furnace, aluminium oxide is manufactured.

12.7.1. Grinding Wheels

They are classified as:

(*a*) Straight-side grinding wheels, (*b*) Cylinder wheels,

(*c*) Cup wheels, and (*d*) Dish wheels.

Hence the grinding wheels are made in many different shapes and sizes to adopt them for use in different types of grinding machines and on different classes of work. The selection of a particular type of grinding wheel depends upon the following factors:

(*i*) The material to be ground (*ii*) Amount of the stock to be removed
(*iii*) Area of contact (*iv*) Type of grinding machine.

Classification

The grinding may be classified broadly into two groups:

(*a*) Rough or non-precision grinding and (*b*) Precision grinding.

The rough grinding includes snagging and off-hand grinding. In this type of grinding the accuracy and surface finish obtained are of secondary importance.

The precision grinding is concerned with producing good surface finish and high degree of accuracy. The precision grinding includes

(*a*) External cylindrical grinding, (*b*) Internal cylindrical grinding,
(*c*) Surface grinding, and (*d*) Form grinding.

12.8. WELDING

The process of joining two pieces of metals by application of heat and pressure is known as welding. The welding may be classified as

(*a*) Pressure welding, (*b*) Fusion welding, and
(*c*) Brazing and soldering.

In welding, the type of joint used depends upon:

(*i*) The type of load to be carried by the weld (*i.e.*, load by tensile or compressive)
(*ii*) The nature of the load (*i.e.*, load may be steady or impact etc.) and
(*iii*) The cost of preparation prior to or after welding.

Pressure welding requires heat and pressure to join the two pieces of metals. The two metals are heated to the plastic state and pressure is applied before, during or after the plastic state is reached. The resistance to electric flow, supplies the heat. The pressure required for welding may be mechanical, hydraulic, spring or air. The following processes require heat and pressure.

1. **Forge Welding:** Here pressure is applied by hammer blow. This process is slow, costly and not suitable for welding structural shapes.
2. **Spot Welding:** Here the two pieces to be joined are overlapped and placed between two electrodes one stationary and other movable. The movable electrode places a load on the work.
3. **Seam Welding:** It is a continuous spot welding process. It is used for drawn and rolled products or sheet welding of metal containers, stove pipe gas tanks etc.
4. **Projection Welding:** It uses projections to concentrate the current. The localized current and the multiple projections create a multiple spot-weld effect.
5. **Butt Welding**
6. **Percussion Welding**

Numerically Controlled Machine Tools

To increase production at a lower cost without decreasing the degree of precision, is the purpose of any automatic system.

II. OBJECTIVE TYPE QUESTIONS

Tick mark the most appropriate statement of the multiple choice answers:

Cutting Tools

1. Which of the followings are the elements of machinery?
(*a*) cutting fluid ☐ (*b*) chip ☐
(*c*) tool ☐ (*d*) work-piece ☐
(*e*) none of the above ☐ (*f*) all of the above. ☐

2. In orthogonal cutting of metals
(*a*) three components of the cutting forces, which are mutually perpendicular are acting at the cutting edge ☐
(*b*) two components of the cutting forces, which are perpendicular, are acting on the cutting tool ☐
(*c*) one component of the cutting force, is acting on the cutting tool ☐
(*d*) none of the above. ☐

3. In oblique cutting of metals
(*a*) three components of the cutting forces, which are mutually perpendicular are acting at the cutting edge ☐
(*b*) two components of the cutting forces, which are perpendicular, are acting on the cutting tool ☐
(*c*) one component of the cutting force, is acting on the cutting tool ☐
(*d*) none of the above. ☐

4. The cutting edge of the tool is perpendicular to the direction of tool travel in case of
(*a*) orthogonal cutting system ☐ (*b*) oblique cutting system ☐
(*c*) orthogonal and oblique cutting systems ☐ (*d*) none of the above. ☐

5. The cutting edge is inclined at angle '*i*' with the normal to the cutting velocity in case of
(*a*) orthogonal cutting system ☐ (*b*) oblique cutting system ☐
(*c*) orthogonal and oblique cutting systems ☐ (*d*) none of the above. ☐

6. Orthogonal cutting system is also known as
(*a*) one-dimensional cutting system ☐ (*b*) two-dimensional cutting system ☐
(*c*) three-dimensional cutting system ☐ (*d*) none of the above. ☐

7. Oblique cutting system is also known as
(*a*) one-dimensional cutting system ☐ (*b*) two-dimensional cutting system ☐
(*c*) three-dimensional cutting system ☐ (*d*) none of the above. ☐

8. In orthogonal cutting system, the cutting tool prepares a surface which is
(*a*) normal to the work surface ☐ (*b*) parallel to the work surface ☐
(*c*) inclined to the work surface ☐ (*d*) none of the above. ☐

9. In orthogonal cutting system, the chip flows over the tool face and the direction of the chip flow-velocity is
(*a*) normal to the cutting edge ☐
(*b*) parallel to the cutting edge ☐
(*c*) inclined with the normal to the cutting edge ☐
(*d*) none of the above. ☐

10. In oblique cutting system, the chip flows over the tool face and the direction of the chip flow-velocity is

(*a*) normal to the cutting edge ☐
(*b*) parallel to the cutting edge ☐
(*c*) inclined with the normal to the cutting edge ☐
(*d*) none of the above. ☐

11. In orthogonal cutting system, the chip thickness is

(*a*) minimum at the middle ☐ (*b*) maximum at the middle ☐
(*c*) maximum at the ends ☐ (*d*) none of the above. ☐

12. The formation of the chip is due to

(*a*) linear deformation ☐ (*b*) shear deformation ☐
(*c*) linear translation ☐ (*d*) none of the above. ☐

13. If the metals are brittle and non-ductile, then

(*a*) continuous chips are formed ☐
(*b*) discontinuous chips are formed ☐
(*c*) continuous chips with built-up edge are formed ☐
(*d*) none of the above. ☐

14. If the metals are ductile and cutting speed is high, then

(*a*) continuous chips are formed ☐
(*b*) discontinuous chips are formed ☐
(*c*) continuous chips with built-up edge are formed ☐
(*d*) none of the above. ☐

15. If the metals are ductile and cutting speed is low, then

(*a*) continuous chips are formed ☐
(*b*) discontinuous chips are formed ☐
(*c*) continuous chips with built-up edge are formed ☐
(*d*) none of the above. ☐

16. Choose the wrong statement

(*a*) When cutting cast iron, the type of chip produced is discontinuous. ☐
(*b*) Best coolant and lubricant for steel and wrought iron is water soluble oils. ☐
(*c*) The cutting edge of the tool is perpendicular to the direction of tool travel in orthogonal cutting system. ☐
(*d*) Mild steel during machining produces discontinuous chips. ☐

17. Choose the correct statement

(*a*) In oblique cutting system, the maximum chip thickness occurs at the middle. ☐
(*b*) In orthogonal cutting system, three components of the cutting forces, which are mutually perpendiculare are acting at the cutting edge. ☐
(*c*) Orthogonal cutting system is also known as three-dimensional cutting system. ☐
(*d*) The maximum chip thickness occurs at the middle in case of orthogonal cutting system. ☐

18. The angle between the tool face and the ground end surface of flank, is known as

(*a*) clearance angle ☐ (*b*) rake angle ☐
(*c*) cutting angle ☐ (*d*) lip angle. ☐

19. The angle between the face of tool and line tangent to the machined surface at the cutting point, is known as

(*a*) clearance angle ☐ (*b*) rake angle ☐
(*c*) cutting angle ☐ (*d*) lip angle. ☐

20. The angle made by the face of the tool and the plane parallel to the base of the cutting tool, is known as

(*a*) clearance angle ☐ (*b*) rake angle ☐
(*c*) cutting angle ☐ (*d*) lip angle. ☐

21. The only angle on which the strength of the tool depends, is

(*a*) clearance angle ☐ (*b*) rake angle ☐
(*c*) cutting angle ☐ (*d*) lip angle. ☐

22. For brittle materials like brass

(*a*) positive rake angle is provided ☐ (*b*) negative rake angle is provided ☐
(*c*) zero rake angle is provided ☐ (*d*) any rake angle is provided. ☐

23. For tougher materials like copper

(*a*) positive rake angle is provided ☐ (*b*) negative rake angle is provided ☐
(*c*) zero rake angle is provided ☐ (*d*) any rake angle is provided. ☐

24. The negative rake angles are used to

(*a*) increase the strength of cutting tool point ☐
(*b*) decrease the temperature rise at the tool-tip ☐
(*c*) give better finish ☐
(*d*) all of the above ☐
(*e*) none of the above. ☐

25. Choose the wrong statement

(*a*) The lip angle of a single point tool is usually between 60° to 80°. ☐
(*b*) The negative rake angle is usually provided on carbide tipped tools. ☐
(*c*) The relief or clearance angles are never made zero or negative. ☐
(*d*) The relief angles on carbide tools are kept as large as possible. ☐

26. The commonly used value of point angle for a standard twist drill is

(*a*) 12° ☐ (*b*) 29° ☐
(*c*) 60° ☐ (*d*) 118°. ☐

27. The commonly used value of helix for a standard twist drill is

(*a*) 12° ☐ (*b*) 29° ☐
(*c*) 60° ☐ (*d*) 118°. ☐

28. The commonly used value of lip relief angle for a standard twist drill is

(*a*) 12° ☐ (*b*) 29° ☐

(*c*) 60° ☐ (*d*) 118°. ☐

29. The point angle of a drill for drilling Bakelite, hard rubber etc. is

(*a*) 60° ☐ (*b*) 90° ☐

(*c*) 135° ☐ (*d*) 150°. ☐

30. The point angle of a drill for drilling hard steel is

(*a*) 60° ☐ (*b*) 90° ☐

(*c*) 135° ☐ (*d*) 150°. ☐

31. For drilling brass of bronze, a drill with

(*a*) zero helix angle is used ☐ (*b*) negative helix angle is used ☐

(*c*) low helix angle is used ☐ (*d*) high helix angle is used. ☐

32. For drilling zinc alloys or magnesium, a drill with

(*a*) zero helix angle is used ☐ (*b*) negative helix angle is used ☐

(*c*) low helix angle is used ☐ (*d*) high helix angle is used. ☐

33. The chip thickness ratio (r) is given by the relation

(*a*) $r = \dfrac{t}{t_c}$ ☐ (*b*) $r = \dfrac{t_c}{t}$ ☐

(*c*) $r = \dfrac{t+1}{t_c+1}$ ☐ (*d*) $r = t \times t_c$. ☐

where t = Depth of cut, and t_c = Chip thickness.

34. The chip compression factor is

(*a*) equal to chip thickness ratio ☐

(*b*) reciprocal of chip thickness ratio ☐

(*c*) equal to the product of chip thickness ratio and shear angle ϕ ☐

(*d*) none of the above. ☐

35. Cutting ratio is also defined as

(*a*) the ratio of cutting speed to the chip velocity ☐

(*b*) the ratio of chip velocity to the cutting speed ☐

(*c*) the ratio of depth of cut to chip thickness ☐

(*d*) the ratio of chip thickness to depth of cut. ☐

36. Choose the wrong statement

(*a*) The commonly used value of margin for a standard twist drill is from 0.4 to 1.6. mm. ☐

(*b*) The commonly used material for a standard twist drill is H.S.S. ☐

(*c*) If the lip relief angle on drills is zero, then drill would not cut. ☐

(*d*) Slow or low helix drills are used for drilling materials like zinc alloys, magnesium etc. ☐

(*e*) Chip thickness ratio is also known as cutting ratio. ☐

37. Velocity of chip relative to the tool is acting

(*a*) along the shear plane ☐ (*b*) along the tool face ☐

(*c*) normal to the tool face ☐ (*d*) normal to the shear plane. ☐

38. Velocity of chip relative to the work-piece is acting

(*a*) along the shear plane ☐ (*b*) along the tool face ☐

(*c*) normal to the tool face ☐ (*d*) normal to the shear plane. ☐

39. The shear angle (ϕ) in terms of chip thickness ratio (r) and rake angle of the tool (α) is given by

(*a*) $\tan\phi = \dfrac{r\sin\alpha}{1 + r\cos\alpha}$ ☐ (*b*) $\tan\phi = \dfrac{r\cos\alpha}{1 + r\sin\alpha}$ ☐

(*c*) $\tan\phi = \dfrac{r\cos\alpha}{1 - r\sin\alpha}$ ☐ (*d*) $\tan\phi = r\tan\alpha$. ☐

40. When turning mild steel, if the area of cross-section of the cut remains constant, the cutting force will be minimum if depth of cut is

(*a*) half the feed per revolution ☐

(*b*) approximately equal to the feed per revolution ☐

(*c*) one and a half times the feed per revolution ☐

(*d*) two times the feed per revolution. ☐

41. In a lathe operation, the machining time is calculated by the relation

(*a*) $T = \dfrac{\text{Length of cut}}{\text{R.P.M.}}$ ☐ (*b*) $T = \dfrac{\text{Length of cut} \times \text{Feed}}{\text{R.P.M.}}$ ☐

(*c*) $T = \dfrac{\text{Length of cut}}{\text{Feed} \times \text{R.P.M.}}$ ☐ (*d*) $T = \dfrac{\text{R.P.M.}}{\text{Length of cut} \times \text{Feed}}$. ☐

where T = Time for one cut.

42. A mild steel rod 300 mm long, 25 mm diameter is to be turned to reduce its diameter to 24 mm in one cut. If the rod is rotating at 510 r.p.m., the cutting speed would be equal to

(*a*) 20 m/min ☐ (*b*) 40 m/min ☐

(*c*) 60 m/min ☐ (*d*) 80 m/min. ☐

43. In question 42, if the feed = 0.5 mm/min then the machining time would be equal to

(*a*) 2 min ☐ (*b*) 1 min 30 sec ☐

(*c*) 1 min 10.6 sec ☐ (*d*) 1 min 5 sec. ☐

44. Which is the correct statement about tool life?

(*a*) Tool life is the time between tool resharpenings. ☐

(*b*) Tool life is the time for which the tool is in contact with the work-piece. ☐

(*c*) Tool life is measured by the volume of material removed between tool sharpenings. ☐

(*d*) All of the above. ☐

(*e*) None of the above. ☐

45. The relationship between tool life (T) and cutting speed (V) is expressed as

(a) $V^nT = C$ ☐ (b) $\frac{V}{T} = C$ ☐

(c) $VT^n = C$ ☐ (d) $\frac{T}{V} = C$. ☐

where n and C are constants.

46. Figure 12.1 shows the plots of log of tool life versus log of cutting speed for different tool materials. Curve A holds good for

(a) H.S.S. tools ☐

(b) cemented carbide tools ☐

(c) ceramic tools ☐

(d) ideal material tools. ☐

FIGURE 12.1

47. In Fig. 12.1, the curve B holds good for

(a) H.S.S. tools ☐

(b) cemented carbide tools ☐

(c) ceramic tools ☐

(d) ideal material tools. ☐

48. In Fig. 12.1, the curve C holds good for

(a) H.S.S. tools ☐ (b) cemented carbide tools ☐

(c) ceramic tools ☐ (d) ideal material tools. ☐

49. In Fig. 12.1, the curve D holds good for

(a) H.S.S. tools ☐ (b) cemented carbide tools ☐

(c) ceramic tools ☐ (d) ideal material tools. ☐

50. The equation $VT^n = C$ is known as Taylor's equation. The value of n for H.S.S. tools is

(a) 0.1 to 0.15 ☐ (b) 0.2 to 0.25 ☐

(c) 0.3 to 0.4 ☐ (d) 0.4 to 0.55. ☐

51. In question 50, the value of n for ceramic tools is

(a) 0.1 to 0.15 ☐ (b) 0.2 to 0.25 ☐

(c) 0.3 to 0.4 ☐ (d) 0.4 to 0.55. ☐

52. In question 50, the value of n for carbide tools is

(a) 0.1 to 0.15 ☐ (b) 0.2 to 0.25 ☐

(c) 0.3 to 0.4 ☐ (d) 0.4 to 0.55. ☐

53. Choose the wrong statement

(a) Ceramic tools have greater tool life than carbide tools. ☐

(b) Tool signature is a numerical method of indentification of tool. ☐

(c) The tool life depends upon cutting speed. ☐

(d) Tool signature consists of four elements. ☐

54. A carbide tool (having $n = 0.25$) with a mild steel work-piece was found to give life of 1 hour 21 minutes while cutting at 60 mpm. The value of C in Taylor's equation $VT^n = C$ would be equal to

(a) 200 ☐ (b) 180 ☐

(c) 150 ☐ (d) 100. ☐

55. If in question 54, the speed is increased to 25% higher, then the tool life will be

(*a*) 1 hour ☐ (*b*) 45 minutes ☐
(*c*) 33.17 minutes ☐ (*d*) 24.5 minutes. ☐

56. If in question 54, the tool life is 2 hours, then the cutting speed of the tool should be

(*a*) 70 mpm ☐ (*b*) 65 mpm ☐
(*c*) 58 mpm ☐ (*d*) 54.4. mpm. ☐

57. Increase in cutting speed

(*a*) deteriorates the surface finish ☐
(*b*) improves the surface finish ☐
(*c*) has no effect on surface finish ☐
(*d*) first improves the surface finish and then deteriorates. ☐

58. Increase in true rake angle

(*a*) deteriorates the surface finish ☐
(*b*) improves the surface finish ☐
(*c*) has no effect on surface finish ☐
(*d*) first improves the surface finish and then deteriorates. ☐

59. Increase in feed rate

(*a*) deteriorates the surface finish ☐
(*b*) improves the surface finish ☐
(*c*) has no effect on surface finish ☐
(*d*) first improves the surface finish and then deteriorates. ☐

60. Increase in depth of cut

(*a*) deteriorates the surface finish ☐
(*b*) improves the surface finish ☐
(*c*) has no effect on surface finish ☐
(*d*) first improves the surface finish and then deteriorates. ☐

61. Increase in nose radius

(*a*) deteriorates the surface finish ☐
(*b*) improves the surface finish ☐
(*c*) has no effect on surface finish ☐
(*d*) first improves the surface finish and then deteriorates. ☐

Machine Tool Operations

62. The operation of removal of excess metal from the edge of a strip to make it suitable for drawing without wrinkling, is known as

(*a*) tumbling ☐ (*b*) slugging ☐
(*c*) lancing ☐ (*d*) notching. ☐

63. The operation of cutting in a single line across a part of the metal strip to allow bending or forming in progressive die operation while the part remains attached to the strip, is known as

(*a*) tumbling ☐ (*b*) slugging ☐
(*c*) lancing ☐ (*d*) notching. ☐

64. The operation of punching in which punch is stopped as soon as the metal fracture is completed and metal is not removed but held in hole, is known as

(a) tumbling ☐ (b) slugging ☐

(c) lancing ☐ (d) notching. ☐

65. The operation of cutting of the excess metal at edge which was required for gripping purposes during press working operation, is known as

(a) tumbling ☐ (b) trimming ☐

(c) hemming ☐ (d) bulging. ☐

66. The operation in which the edges of the sheet are turned over to provide stiffness and a smooth edge, is known as

(a) tumbling ☐ (b) trimming ☐

(c) hemming ☐ (d) bulging. ☐

67. The operation employed to expand a tubular or cylindrical part, is known as

(a) tumbling ☐ (b) trimming ☐

(c) hemming ☐ (d) bulging. ☐

68. Choose the wrong statement

(a) The metal lows due to plasticity in drawing operation. ☐

(b) The process of cleaning the surface of small parts is known as tumbling. ☐

(c) The cutting force at the cutting edge is measured by a pyrometer. ☐

(d) Crater wear occurs mainly on the nose part, front relief face and side relief face of the cutting tool. ☐

69. A process of removing metal, by pushing or pulling a cutting tool, is known as

(a) forming ☐ (b) hemming ☐

(c) broaching ☐ (d) bulging. ☐

70. Choose the correct statement

(a) A jig does not guide the tool. ☐

(b) The process of improving the cutting action of the grinding wheel is known as truing. ☐

(c) In machining cast iron no cutting fluid is required. ☐

(d) Laser is produced by graphite. ☐

71. The ability of a tool material, by virtue of which it retains its hardness while working at elevated temperature, is known as

(a) wear resistance ☐ (b) toughness ☐

(c) red hardness ☐ (d) machinability. ☐

72. The ability of a tool material to resist shock to impact forces is known as

(a) wear resistance ☐ (b) toughness ☐

(c) red hardness ☐ (d) machinability. ☐

73. Which one of the following cutting tool materials has low hardness at elevated temperature?

(a) cemented carbides ☐ (b) high speed steel ☐

(c) high carbon steel ☐ (d) ceramics. ☐

74. Which one of the following cutting tool materials is the hardest?
(*a*) cemented carbides ☐ (*b*) high speed steel ☐
(*c*) high carbon steel ☐ (*d*) diamond. ☐

75. The most popular and standard type for all purpose tool steels is 18 : 4 : 1 high speed steel (H.S.S.), which contains
(*a*) 18% chromium, 4% tungsten and 1% vanadium ☐
(*b*) 18% tungsten, 4% vanadium and 1% chromium ☐
(*c*) 18% tungsten, 4% chromium and 1% vanadium ☐
(*d*) 18% vanadium, 4% chromium and 1% tungsten. ☐

76. The device, which holds and locates a workpiece and guides and controls one or more cutting tools, is known as
(*a*) fixture ☐ (*b*) jig ☐
(*c*) lathe ☐ (*d*) templates. ☐

77. The device, which holds and locates a work-piece during an inspection or for a manufacturing operation, is known as
(*a*) fixture ☐ (*b*) jig ☐
(*c*) lathe ☐ (*d*) templates. ☐

78. The purpose of jigs and fixtures is to
(*a*) facilitate interchangeable manufacture ☐ (*b*) increase production rate ☐
(*c*) increase machining accuracy ☐ (*d*) enable employ less skilled operators ☐
(*e*) all of the above ☐ (*f*) none of the above. ☐

Questions on Lathe

79. The size of a lathe is specified by the
(*a*) maximum job length in mm that may be held between centres ☐
(*b*) the height of centres measured over the lathe bed ☐
(*c*) maximum diameter that can be rotated over the bed ways ☐
(*d*) any one of the above ☐
(*e*) none of the above. ☐

80. If D = large diameter, d = small diameter, and l = length of a job, then tangent of half of the taper angle is equal to
(*a*) $\frac{D-d}{l}$ ☐ (*b*) $\frac{2(D-d)}{l}$ ☐
(*c*) $\frac{D-d}{2}$ ☐ (*d*) $\frac{(D+d)}{2l}$. ☐

81. Which of the following methods is used for taper turning?
(*a*) compound rest method ☐ (*b*) taper attachment method ☐
(*c*) forming tool method ☐ (*d*) tail stock set over method ☐
(*e*) all of the above ☐ (*f*) none of the above. ☐

82. The angle at which the compound rest should be set to turn taper on the work-piece having a length of 400 mm, larger diameter 100 mm and smaller diameter 60 mm, would be

(*a*) $\tan^{-1} 0.1$ ☐ (*b*) $\tan^{-1} 0.01$ ☐
(*c*) $\tan^{-1} 0.05$ ☐ (*d*) $\tan^{-1} 0.025$. ☐

83. When a part of length of job, of diameters D and d, is to be given taper, then the tailstock set over is equal to

(*a*) $\frac{D-d}{2L}$ ☐ (*b*) $\frac{D-d}{L}$ ☐
(*c*) $\frac{D-d}{2} \times \frac{L}{l}$ ☐ (*d*) $\frac{D-d}{2}$. ☐

where L = Total length of job, and l = Length of the taper.

84. When the entire length of job (of diameters D and d) is to be given taper, then the tailstock set over is equal to

(*a*) $\frac{D-d}{2L}$ ☐ (*b*) $\frac{D-d}{L}$ ☐
(*c*) $\frac{D-d}{2} \times \frac{L}{l}$ ☐ (*d*) $\frac{D-d}{2}$. ☐

85. The tailstock set over for a job [having D = 35 mm, d = 27 mm, l = 75 mm and L = 225 mm] would be equal to

(*a*) 4 mm ☐ (*b*) 10 mm ☐
(*c*) 12 mm ☐ (*d*) 15 mm. ☐

86. If in question 85, the taper is given to the entire length of the job then the tailstock set over would be equal to

(*a*) 4 mm ☐ (*b*) 10 mm ☐
(*c*) 12 mm ☐ (*d*) 15 mm. ☐

87. Choose the wrong statement

(*a*) The different spindle speeds on a lathe form geometrical progression. ☐
(*b*) The lead screw of a lathe has single start threads. ☐
(*c*) The lathe centres are provided with standard tapers known as Morse taper. ☐
(*d*) Tumbler gears in lathe are used to reduce the spindle speed. ☐

88. Feed normally varies from

(*a*) 2 mm to 3 mm ☐ (*b*) 1 mm to 2 mm ☐
(*c*) 0.1 mm to 1.5 mm ☐ (*d*) 0.01 mm to 0.1 mm. ☐

89. A high rate of feed will get the job done in

(*a*) less time and will give a better finish ☐ (*b*) less time and will give a rough finish ☐
(*c*) longer time and will give a better finish ☐ (*d*) longer time and will give a rough finish. ☐

90. Depth of cut for roughing operation normally varies from

(*a*) 5 mm to 10 mm ☐ (*b*) 1 mm to 5 mm ☐
(*c*) 0.2 mm to 1 mm ☐ (*d*) 0.01 mm to 0.2 mm. ☐

91. Depth of cut for finishing operation normally varies from
(*a*) 5 mm to 10 mm ☐ (*b*) 1 mm to 5 mm ☐
(*c*) 0.2 mm to 1 mm ☐ (*d*) 0.01 mm to 0.2 mm. ☐

92. The recommended average cutting speed in metres per minute for a high speed steel tool for cutting aluminium is
(*a*) 15 ☐ (*b*) 30 ☐
(*c*) 60 ☐ (*d*) 90. ☐

93. The recommended average cutting speed in metres per minute for a high speed steel tool for cutting mild steel is
(*a*) 15 ☐ (*b*) 30 ☐
(*c*) 60 ☐ (*d*) 90. ☐

94. The recommended average cutting speed in metres per minute for a high speed steel tool for cutting cast steel is
(*a*) 15 ☐ (*b*) 30 ☐
(*c*) 60 ☐ (*d*) 90. ☐

95. The recommended average cutting speed in metres per minute for a high speed steel tool or cutting brass is
(*a*) 15 ☐ (*b*) 30 ☐
(*c*) 60 ☐ (*d*) 90. ☐

96. The cutting pressure on a single point tool is given by $P = K \times d \times f$ in which d = depth of cut, f = feed and K is a constant and is equal to the pressure on the tool in kgf per square cm of cut area. The value of K for cutting aluminium is
(*a*) 16000 ☐ (*b*) 12000 ☐
(*c*) 6700 ☐ (*d*) 8600. ☐

97. In question 96, the value of K for cutting bronze is
(*a*) 16000 ☐ (*b*) 12000 ☐
(*c*) 6700 ☐ (*d*) 8600. ☐

98. In question 96, the value of K for cutting cast iron is
(*a*) 16000 ☐ (*b*) 12000 ☐
(*c*) 6700 ☐ (*d*) 8600. ☐

99. In question 96, the value of K for cutting brass is
(*a*) 16000 ☐ (*b*) 12000 ☐
(*c*) 6700 ☐ (*d*) 8600. ☐

100. A lathe is to turn a metallic bar at 0.5 mm feed with 0.6 mm depth of cut at a cutting speed of 20 metres per minute. If K for the metal being cut = 16000, the ideal power required would be
(*a*) 1.5 h.p. ☐ (*b*) 1.0 h.p. ☐
(*c*) 0.5 h.p. ☐ (*d*) 0.21 h.p. ☐

101. A metal cutting test on a lathe gave a following observations: cutting speed = 20 m/minute, depth of cut = 2.00 mm, feed = 5 cuts/mm and vertical force on the tool = 2400 N. The feed per revolution would be

(*a*) 5 mm ☐ (*b*) 2 mm ☐
(*c*) 0.5 mm ☐ (*d*) 0.2 mm. ☐

102. In question 101, cross-sectional area of the chip would be

(*a*) 10×10^{-6} m^2 ☐ (*b*) 2×10^{-6} m^2 ☐
(*c*) 0.4×10^{-6} m^2 ☐ (*d*) 0.2×10^{-6} m^2. ☐

103. In question 101, pressure on the tool would be

(*a*) 8000 MN/m^2 ☐ (*b*) 6000 MN/m^2 ☐
(*c*) 4000 MN/m^2 ☐ (*d*) 2000 MN/m^2. ☐

104. In question 101, power required at the tool point would be

(*a*) 1000 watts ☐ (*b*) 800 watts ☐
(*c*) 600 watts ☐ (*d*) 400 watts. ☐

105. If in question 101, the overall efficiency of the machine is 80%, then power required at motor would be

(*a*) 1000 watts ☐ (*b*) 800 watts ☐
(*c*) 600 watts ☐ (*d*) 400 watts. ☐

106. The process of removing metal, by feeding the work past a rotating multipoint cutter, is known as

(*a*) broaching ☐ (*b*) sawing ☐
(*c*) milling ☐ (*d*) grinding. ☐

107. Which is the correct statement about milling operation?

(*a*) The rate of metal removal is rapid. ☐ (*b*) The jobs are machined at a faster rate. ☐
(*c*) Surface finish is better. ☐ (*d*) All of the above. ☐
(*e*) None of the above. ☐

108. The cutting tool in a milling machine is held in position by

(*a*) arbor ☐ (*b*) spindle ☐
(*c*) column ☐ (*d*) knee. ☐

109. The cutting tool in a milling machine is having

(*a*) longitudinal motion ☐ (*b*) vertical motion ☐
(*c*) rotational motion ☐ (*d*) all of the above motions. ☐

110. The milling machine is used for machining

(*a*) flat surfaces ☐ (*b*) contoured surfaces ☐
(*c*) surface of revolution ☐ (*d*) all of the above. ☐

111. The process of removing metal by a cutter, which is rotated in a direction opposite to the feed of the work-piece, is known as

(*a*) conventional milling ☐ (*b*) climb milling ☐
(*c*) up milling ☐ (*d*) down milling ☐
(*e*) (*a*) or (*c*) ☐ (*f*) (*b*) or (*d*). ☐

112. The process of removing metal by a cutter, which is rotated in the same direction as the feed of the work-piece, is known as

(*a*) conventional milling ☐ (*b*) climb milling ☐
(*c*) up milling ☐ (*d*) down milling ☐
(*e*) (*a*) or (*c*) ☐ (*f*) (*b*) or (*d*). ☐

113. The thickness of the chip is minimum at the beginning of the cut and maximum at the end of the cut in case of

(*a*) climb milling ☐ (*b*) up milling ☐
(*c*) down milling ☐ (*d*) face milling. ☐

114. The thickness of the chip is maximum at the beginning of the cut and minimum at the end of the cut in case of

(*a*) climb milling ☐ (*b*) up milling ☐
(*c*) conventional milling ☐ (*d*) face milling. ☐

115. In which milling operation, the cutting force tends to lift the work-piece?

(*a*) climb ☐ (*b*) down ☐
(*c*) conventional ☐ (*d*) face. ☐

116. In which milling operation, the surface finish is better?

(*a*) climb ☐ (*b*) up ☐
(*c*) conventional ☐ (*d*) face. ☐

117. In which milling operation, there is a greater tendency for the chip to weld to the cutter?

(*a*) climb ☐ (*b*) down ☐
(*c*) conventional ☐ (*d*) face. ☐

118. In which milling operation, the cutting force tends to push the work-piece into the fixture?

(*a*) climb ☐ (*b*) up ☐
(*c*) conventional ☐ (*d*) face. ☐

119. Choose the wrong statement

(*a*) In conventional milling the cutter is somewhat guided by the portion already machined. ☐
(*b*) In climb milling the chips are deposited behind the work zone and out of the way of cutter. ☐
(*c*) The cutting tool is mounted on arbor in a milling machine. ☐
(*d*) The conventional milling is also called climb milling. ☐

120. Which of the following machines utilize fly cutter?

(*a*) Planer ☐ (*b*) Shaper ☐
(*c*) Broaching ☐ (*d*) Milling machine. ☐

121. Which one of the following statements about cutting fluids is wrong?

(*a*) A cutting fluid permits the use of higher cutting speeds. ☐
(*b*) A cutting fluid improves the tool life. ☐
(*c*) A cutting fluid provide lubrication and washes away the chips. ☐
(*d*) A cutting fluid increases the kinetic co-efficient of friction and keeps down the cutting forces. ☐

122. Which one of the following statements about the requirements of a good cutting fluid is wrong?

(*a*) A cutting fluid should not gum the moving parts of the machine tools. ☐
(*b*) A cutting fluid should possess high specific heat and heat conductivity. ☐
(*c*) A cutting fluid should possess low flash point. ☐
(*d*) A cutting fluid should have low cost. ☐

123. The process of rotating the job correct to the friction on minutes between the two successive cuts is known as

(*a*) milling ☐ (*b*) indexing ☐
(*c*) broaching ☐ (*d*) sawing. ☐

124. A milling cutter having 18 teeth is to operate at a cutting speed of 22 m/min. The diameter of the cutter is 10 cm. If the feed is 0.20 mm per tooth per revolution, then r.p.m. of the cutter would be

(*a*) 100 r.p.m. ☐ (*b*) 80 r.p.m. ☐
(*c*) 70 r.p.m. ☐ (*d*) 50 r.p.m. ☐

125. In question 124, the feed per revolution would be equal to

(*a*) 3.5 mm ☐ (*b*) 3.6 mm ☐
(*c*) 4.0 mm ☐ (*d*) 5.0 mm. ☐

126. In question 124, the feed per minute would be equal to

(*a*) 35 cm ☐ (*b*) 25.20 cm ☐
(*c*) 16.4 cm ☐ (*d*) 10.8 cm. ☐

127. If in milling operation, depth and width of the cut is constant and feed rate is doubled, the power consumption will increase by

(*a*) 100% ☐ (*b*) 90% ☐
(*c*) 50% ☐ (*d*) 30%. ☐

128. If in milling operation, depth is doubled keeping other factors constant, then power consumption will increase by

(*a*) 100% ☐ (*b*) 90% ☐
(*c*) 50% ☐ (*d*) 30%. ☐

129. If in the milling operation, width of cut is doubled keeping feed and depth constant, then power consumption will increase by

(*a*) 100% ☐ (*b*) 90% ☐
(*c*) 50% ☐ (*d*) 30%. ☐

130. If in milling operation, feed is halved and speed doubled, then power consumption will increase by

(*a*) 100% ☐ (*b*) 90% ☐
(*c*) 50% ☐ (*d*) 30%. ☐

131. In milling operation, if positive rake angle is increased

(*a*) power consumption increases ☐
(*b*) power consumption decreases ☐

(c) power consumption remains constant ☐
(d) power consumption first increases, then decreases. ☐

132. Choose the wrong statement
(a) The coarse teeth of a milling cutter are more efficient for removing metal than fine teeth. ☐
(b) Soft materials having high ductility offer greater resistance to the formation of the chip. ☐
(c) For high-speed steel and carbide-tipped cutters, negative rake angles are generally used. ☐
(d) The action of a milling cutter is the same as that of a drill of lathe tool. ☐

133. For milling the sides of a hexagonal nut, the indexing movement required would be
(a) 4 spaces of a 40-space circle ☐ (b) 10 spaces of a 40-space circle ☐
(c) 4 spaces of a 24-space circle ☐ (d) 6 spaces of a 24-space circle. ☐

134. For milling the sides of a square nut, the indexing movement required would be
(a) 4 spaces of a 40-space circle ☐ (b) 8 spaces of a 40-space circle ☐
(c) 10 spaces of a 40-space circle ☐ (d) 6 spaces of a 24-space circle. ☐

135. For milling 14 teeth on gear blank, the indexing movement required would be
(a) 2 complete turns and 8 spaces on 21-space circle ☐
(b) 2 complete turns and 12 spaces on 21-space circle ☐
(c) 2 complete turns and 18 spaces on 21-space circle ☐
(d) 2 complete turns and 42 spaces on 49-space circle ☐
(e) (c) or (d). ☐

136. Shear angle is the angle between
(a) shear plane and vertical ☐ (b) shear plane and tool face ☐
(c) shear plane and job surface ☐ (d) shear plane and horizontal. ☐

137. Choose the wrong statement
(a) The cross-section of a chisel is usually octagonal. ☐
(b) A hacksaw blade cuts on the forward stroke. ☐
(c) The tool life is most affected by cutting speed. ☐
(d) A zinc diffusion process is known as galvanizing. ☐

138. Which of the following process is used for making cold chisels?
(a) rolling ☐ (b) forging ☐
(c) piercing ☐ (d) drawing. ☐

139. A casting is to be machined on a lathe. The casting should be held in
(a) magnetic chuck ☐ (b) collect chuck ☐
(c) four-jaw chuck ☐ (d) three-jaw chuck. ☐

140. When large number of components are machined from a bar on a automatic machine, the bar is held in
(a) magnetic chuck ☐ (b) collect chuck ☐
(c) four-jaw chuck ☐ (d) three-jaw chuck. ☐

141. A diamond pointed chisel is used for cutting

(*a*) grooves ☐ (*b*) keyways ☐

(*c*) V-shaped grooves ☐ (*d*) flat surfaces. ☐

142. Sanding is a process of removing metal surfaces or wood fibre by

(*a*) sawing ☐ (*b*) cutting ☐

(*c*) filing ☐ (*d*) planing. ☐

143. Which of the following type of gear should be selected when the gear is subjected to shock and vibrations?

(*a*) bevel gears ☐ (*b*) gear with stub teeth ☐

(*c*) hardened gears ☐ (*d*) hybrid gears. ☐

144. Which of the following parameter is used for specifying a hacksaw?

(*a*) length ☐ (*b*) width ☐

(*c*) material ☐ (*d*) number of teeth. ☐

145. Which of the following operation is required for making a chamfer on the edge of a hole?

(*a*) spot facing ☐ (*b*) counter sinking ☐

(*c*) counter boring ☐ (*d*) reaming. ☐

146. Which of the following device is used for holding a reamer?

(*a*) mandrel ☐ (*b*) collect chuck ☐

(*c*) universal chuck ☐ (*d*) floating chuck. ☐

147. A file with 20 teeth in 25 mm is known as

(*a*) bastard file ☐ (*b*) rough file ☐

(*c*) smooth file ☐ (*d*) second cut file. ☐

148. Seamless tubes are made by

(*a*) plug rolling ☐ (*b*) cold rolling ☐

(*c*) piercing operations ☐ (*d*) none of the above. ☐

149. Which material of the following is mostly used for drills, taps and reamers?

(*a*) low alloy carbon steel ☐ (*b*) high speed steel ☐

(*c*) ceramics ☐ (*d*) cemented carbide. ☐

150. Which material of the following is added in low carbon tool steel for improving the wear resistance?

(*a*) chromium ☐ (*b*) tungsten ☐

(*c*) copper ☐ (*d*) molybdenum. ☐

151. Mechanical working of metals is the shaping of metals. This is carried by

(*a*) hot working process ☐ (*b*) cold working process ☐

(*c*) casting ☐ (*d*) grinding ☐

(*e*) machining ☐ (*f*) (*a*) or (*e*). ☐

152. The process which takes place below recrystallization temperature is known as

(*a*) not working process ☐ (*b*) cold working process ☐

(*c*) casting ☐ (*d*) grinding. ☐

153. Hot working process takes place

(*a*) at recrystallization temperature ☐ (*b*) below recrystallization temperature ☐

(*c*) above recrystallization temperature ☐ (*d*) none of the above. ☐

154. The recrystallization temperature of steel is

(*a*) 800°C ☐ (*b*) 400°C ☐

(*c*) 200°C ☐ (*d*) room temperature. ☐

155. The recrystallization temperature of lead, tin or zinc is

(*a*) 800°C ☐ (*b*) 400°C ☐

(*c*) 200°C ☐ (*d*) room temperature. ☐

156. Which one is the correct statement about hot working process?

(*a*) The surface finish of the product is fine. ☐

(*b*) The life of the tools, used for hot working process, is high. ☐

(*c*) It is a rapid and economical process. ☐

(*d*) Tooling and handling costs are less. ☐

157. Which one is the wrong statement about hot working process?

(*a*) In hot working process, the metal becomes plastic and causes the growth of grains. ☐

(*b*) The mechanical properties such as ductility, toughness, elongation and reduction in area are improved. ☐

(*c*) The power required to finish the part from ingot is less. ☐

(*d*) The surface finish of the product is fine. ☐

Forging

158. The plastic deformation of metals at elevated temperature into a predetermined size or shape using compressive forces exerted through some type of die by a hammer, a press or an upsetting machine, is known as

(*a*) extrusion ☐ (*b*) forging ☐

(*c*) casting ☐ (*d*) piercing. ☐

159. The process of pushing the heated billet or slug of metal through an orifice, provided into a die and forming an elongated part of uniform cross-section, is known as

(*a*) extrusion ☐ (*b*) forging ☐

(*c*) casting ☐ (*d*) piercing. ☐

160. The process which is employed for the production of seamless tubes, is known as

(*a*) extrusion ☐ (*b*) forging ☐

(*c*) casting ☐ (*d*) piercing. ☐

161. The process, in which the cross-sectional area of bars, rods or uses in the desired area is reduced by repeated blows, is known as

(*a*) extrusion ☐ (*b*) forging ☐

(*c*) swaging ☐ (*d*) piercing. ☐

162. The process of shaping thin metal sheets by pressing them against a form is known as

(*a*) swaging ☐ (*b*) spinning ☐

(*c*) reaming ☐ (*d*) sawing. ☐

163. The process of enlarging a machined hole to proper size with a smooth finish is known as

(*a*) swaging ☐ (*b*) spinning ☐

(*c*) reaming ☐ (*d*) sawing. ☐

164. Choose the wrong statement

(*a*) Boring is a process of enlarging the hole that has already been drilled. ☐

(*b*) Drilling is a process of making hole in a work-piece by rotating tool called the drill. ☐

(*c*) Reaming is the process of enlarging a machined hole to proper size with a smooth finish. ☐

(*d*) Swaging is the process of shaping thin metal sheets by pressing them against a form. ☐

165. The forging of steel specimen is done at a temperature of

(*a*) 400°C ☐ (*b*) 600°C ☐

(*c*) 800°C ☐ (*d*)1000°C. ☐

166. For plain carbon steel, if carbon content is 0.2%, the forging should be done at a temperature of

(*a*) 800 to 900°C ☐ (*b*) 900 to 1000°C ☐

(*c*) 1000 to 1150°C ☐ (*d*) 1150 to 1300°C. ☐

167. For plain carbon steel if carbon content is 0.7%, the forging should be done at a temperature of

(*a*) 800 to 900°C ☐ (*b*) 900 to 1000°C ☐

(*c*) 1000 to 1150°C ☐ (*d*) 1150 to 1300°C. ☐

168. Which one is the wrong statement about forging?

(*a*) The forgings have high strength and ductility. ☐

(*b*) The forgings offer great resistance to impact and fatigue loads. ☐

(*c*) Forging renders the parts uniform in density as well as dimensions. ☐

(*d*) The work-price after forging has a less reliability. ☐

169. For conventional extrusion, the capacities of horizontal hydraulic press varies from

(*a*) 10 to 20 tonnes ☐ (*b*) 20 to 100 tonnes ☐

(*c*) 100 to 250 tonnes ☐ (*d*) 250 to 5500 tonnes. ☐

170. For conventional extrusion, the pressure acting on the metal varies from

(*a*) 100 to 200 kg/cm^2 ☐ (*b*) 200 to 1000 kg/cm^2 ☐

(*c*) 1000 to 5000 kg/cm^2 ☐ (*d*) 5000 to 7000 kg/cm^2. ☐

171. For extrusion, the temperature of magnesium metal in the die should be

(*a*) 1100 to 1250°C ☐ (*b*) 650 to 900°C ☐

(*c*) 425 to 480°C ☐ (*d*) 350 to 425°C. ☐

172. For extrusion, the temperature of copper alloys in the die should be

(*a*) 1100 to 1250°C ☐ (*b*) 650 to 900°C ☐

(*c*) 425 to 480°C ☐ (*d*) 350 to 425°C. ☐

173. For extrusion, the temperature of aluminium in the die should be

(*a*) 1100 to 1250°C ☐ (*b*) 650 to 900°C ☐

(*c*) 425 to 480°C ☐ (*d*) 350 to 425°C. ☐

174. The extrusion chamber, die and ram are generally lubricated by

(*a*) vegetable oils ☐ (*b*) grease ☐
(*c*) water ☐ (*d*) solid metal. ☐

175. The grain structure is refined in

(*a*) cold working process ☐ (*b*) hot working process ☐
(*c*) both (*a*) and (*b*) ☐ (*d*) none of the above. ☐

176. Choose the wrong statement

(*a*) Cold working process requires greater force than hot working process. ☐
(*b*) The deformation of metals takes place by the method of slip of planes in cold working process. ☐
(*c*) Cold working produces an improved surface finish and scale-free surface. ☐
(*d*) Cold working results in increases of ductility and loss of strength and hardness of metal. ☐

177. The cold working process can be performed on the components having diameter upto

(*a*) 10 mm ☐ (*b*) 25 mm ☐
(*c*) 40 mm ☐ (*d*) 60 mm. ☐

178. The process of removing metal by an elongated tool having a number of successive teeth of increasing size which cut in a fixed path, is known as

(*a*) reaming ☐ (*b*) boring ☐
(*c*) broaching ☐ (*d*) honing. ☐

179. The process of grinder or abrading in which very little material is removed, is known as

(*a*) reaming ☐ (*b*) honing ☐
(*c*) broaching ☐ (*d*) lapping. ☐

180. The process, which is used to produce geometrically true surfaces, is known as

(*a*) reaming ☐ (*b*) honing ☐
(*c*) broaching ☐ (*d*) lapping. ☐

181. EDM stands for

(*a*) Energy discharge method ☐ (*b*) Electro-discharge machining ☐
(*c*) Energy direct method ☐ (*d*) Efficient direction method. ☐

182. EBM stands for

(*a*) Energy best method ☐ (*b*) Electron beam method ☐
(*c*) Electron beam machining ☐ (*d*) Explosive beam method. ☐

183. PAM stands for

(*a*) Plastic arc method ☐ (*b*) Plasma arc method ☐
(*c*) Plasma arc machining ☐ (*d*) Plastic abrasive method. ☐

184. In EDM process, a suitable gap known as spark gap is maintained between the tool and work-piece. This gap should be of the order of

(*a*) 0.5 cm to 1.0 cm ☐ (*b*) 0.1 cm to 0.5 cm ☐
(*c*) 0.01 cm to 0.1 cm ☐ (*d*) 0.002 cm to 0.008 cm. ☐

185. In EDM process, if both the electrodes (one tool and the other work-piece) are made of the same material, the erosion of the metal

(*a*) is greatest at positive electrode (anode) ☐ (*b*) is greatest at negative electrode (cathode) ☐
(*c*) is same at anode and cathode ☐ (*d*) is minimum at anode and cathode. ☐

186. In electro-discharge machining process, in order to remove maximum metal and have minimum wear on the tool

(*a*) the tool is made cathode and work-piece as anode ☐
(*b*) the tool is made anode and work-piece as cathode ☐
(*c*) the tool and work-piece should be of different metals ☐
(*d*) none of the above. ☐

187. The dielectric, in electro-discharge machining process, is used to

(*a*) help in cooling the work and tool ☐
(*b*) carry away the eroded metal along with it ☐
(*c*) control the spark discharge ☐
(*d*) all of the above. ☐

188. In electro-discharge machining, the tool is made of

(*a*) cast iron ☐ (*b*) copper ☐
(*c*) brass ☐ (*d*) alloy of copper ☐
(*e*) all of the above ☐ (*f*) none of the above. ☐

189. The wear ratio (*i.e.*, ratio of electrode wear rate to metal removal rate) for bross work is

(*a*) 0.5 ☐ (*b*) 1.0 ☐
(*c*) 2.0 ☐ (*d*) 3.0. ☐

190. The wear ratio for tungsten carbide work is

(*a*) 0.5 ☐ (*b*) 1.0 ☐
(*c*) 2.0 ☐ (*d*) 3.0. ☐

191. The wear ratio for hardened plain carbon steel work is

(*a*) 0.5 ☐ (*b*) 1.0 ☐
(*c*) 2.0 ☐ (*d*) 3.0. ☐

192. Choose the wrong statement

(*a*) In electro-discharge machining process, tool does not come into contact with the work-piece. ☐
(*b*) Life of tool is long in EDM process. ☐
(*c*) The dielectric fluids generally used in EDM process are oil, white spirit or transformer oil. ☐
(*d*) In EDM process, the tool is connected to anode. ☐

193. Choose the correct statement

(*a*) EDM process can be used for any hard material. ☐
(*b*) In EDM process, metal removal rate is slow. ☐
(*c*) In EDM process, tool material need not be harder than work material. ☐
(*d*) All of the above. ☐
(*e*) None of the above. ☐

194. Which one of the following processes is the reverse of electroplating process?

(*a*) EDM ☐ (*b*) ECM ☐

(*c*) ECG ☐ (*d*) LBM. ☐

Electrochemical Milling Process

195. The electrolyte generally used in electrochemical milling (ECM) process, is

(*a*) white spirit ☐ (*b*) transformer oil ☐

(*c*) aqueous solution of common salt ☐ (*d*) vegetable oil. ☐

196. In ECM process, the metal removal rate (M.R.R.) is given by

(*a*) M.R.R. = $\frac{F \times I}{\rho A v}$ ☐ (*b*) M.R.R. = $\frac{I \times At}{F \times \rho \times A \times v}$ ☐

(*c*) M.R.R. = $\frac{F \times \rho \times A \times v}{I \times At}$ ☐ (*d*) M.R.R. = $\frac{R \times I \times At}{\rho \times A \times v}$ ☐

where F = Faraday's constant, I = Current amperes,

At = Atomic weight of material in gm

v = Valency of metal dissolved, A = Machined area in cm^2,

ρ = Density of work-piece in gm/cm^2.

197. The current required in an ECM process for machining iron (of At. wt. 56 gm, valency = 2, density = 7.8 gm/cm^2 and Faraday's constant = 1609 amp-minute) is 224 amp. Then the metal removal rate will be approximately equal to

(*a*) 1 cm^3/min ☐ (*b*) 0.5 cm^3/min ☐

(*c*) 0.25 cm^3/min ☐ (*d*) 0.1 cm^3/min. ☐

198. In electrochemical grinding (ECG) process, the metal is removed by

(*a*) abrasion of the metal only ☐

(*b*) electrochemical decomposition only ☐

(*c*) 10% electrochemical decomposition and 90% abrasion of the metal ☐

(*d*) 90% electrochemical decomposition and 10% abrasion of the metal. ☐

199. Choose the wrong statement

(*a*) In ECM process, no spark is produced. ☐

(*b*) In ECM process, machining is done at low voltage. ☐

(*c*) In ECM process, amount of energy consumed is low. ☐

(*d*) The ECM process is independent of hardness, toughness or brittleness of the material to be machined. ☐

200. During ECG process, a continuous steam of non-corrosive salt solution is passed through work and tool. This solution acts as

(*a*) an electrolyte ☐ (*b*) a coolant ☐

(*c*) an electrolyte and coolant ☐ (*d*) none of the above. ☐

201. In ECG process

(*a*) grinding pressure is low ☐

(*b*) temperature is very low ☐

(*c*) pressure is less but temperature is high ☐

(*d*) both grinding pressure and temperature are low. ☐

202. In chemically milling, a coating is applied

(*a*) on that portion of the job which is to be machined ☐

(*b*) on that portion of the job which is not to be machined ☐

(*c*) on the complete portion of the job ☐

(*d*) none of the above. ☐

203. In chemical milling process, chemical etching reagent is used. The chemical reagent for aluminium work-piece is

(*a*) nitric acid ☐ (*b*) sulphuric acid ☐

(*c*) caustic soda ☐ (*d*) sodium chloride. ☐

204. In chemical milling process, the chemical reagent for steel work-piece is

(*a*) nitric acid ☐ (*b*) sodium sulphate ☐

(*c*) caustic soda ☐ (*d*) sodium chloride. ☐

205. LBM stands for

(*a*) laser best method ☐ (*b*) light beam method ☐

(*c*) laser beam machining ☐ (*d*) light beam machining. ☐

206. The cutting rate in mm/min for LBM process is equal to

(*a*) $\frac{C \times P \times t}{EA}$ ☐ (*b*) $\frac{P \times t}{CEA}$ ☐

(*c*) $\frac{E \times A \times t}{C \times P}$ ☐ (*d*) $\frac{C \times P}{E \times A \times t}$ ☐

where P = Laser power incident on surface in watts,

A = Area of laser beam at focal point in mm^2,

t = Thickness of material, mm,

E = Vaporisation energy of material, in W/mm^2

C = Constant.

207. The cutting rate by LBM process

(*a*) is maximum for thin materials ☐

(*b*) is minimum for thin materials ☐

(*c*) is independent of thickness of the material ☐

(*d*) none of the above. ☐

208. In abrasive jet machining (AJM) process, a high pressure gas or air is used. The velocity of the gas or air is of the order of

(*a*) 20 to 50 m/sec ☐ (*b*) 50 to 100 m/sec ☐

(*c*) 100 to 150 m/sec ☐ (*d*) 200 to 400 m/sec. ☐

209. Choose the correct statement

(*a*) In chemical milling process, the metal is removed by the chemical conversion of the metal into metallic salt. □

(*b*) Chemical milling process is mostly used in aircraft industry. □

(*c*) Laser beam is the light amplification by stimulated emission radiation. □

(*d*) The tool is made of brass in EDM process. □

(*e*) All of the above. □

(*f*) None of the above. □

210. In AJM process, inside diameter of the nozzle through which abrasive particle flow is about

(*a*) 10 mm □ (*b*) 5 mm □

(*c*) 1 mm □ (*d*) 0.04 mm. □

211. In AJM process, the distance between nozzle tip and work-piece is kept about

(*a*) 10 to 5 mm □ (*b*) 5 to 1 mm □

(*c*) 1 to 0.7 mm □ (*d*) 0.5 to 0.1 mm. □

212. In AJM process, the material removal rate

(*a*) increases with the increase of tip distance from work-piece □

(*b*) decreases with the increase of tip distance from work-piece □

(*c*) first increases with the increase of tip distance from work-piece upto a certain limit after which it remains unchanged for a certain tip distance and then falls gradually □

(*d*) is independent of the tip distance from work-piece. □

213. USM stands for

(*a*) ultimate strength of metal □ (*b*) universal sound method □

(*c*) universal sonic machining □ (*d*) ultrasonic machining. □

214. USM process is used for machining

(*a*) hard metals □ (*b*) brittle metals □

(*c*) conducting metals □ (*d*) non-conducting metals □

(*e*) all of the above □ (*f*) none of the above. □

215. In USM process, the cutting rate will be factor if amplitude of vibration is

(*a*) minimum □ (*b*) maximum □

(*c*) constant □ (*d*) none of the above. □

216. Which of the following machining methods require electrolyte?

(*a*) LBM □ (*b*) UTM □

(*c*) USM □ (*d*) EDM. □

217. The cutting angle of a chisel for cutting mild steel should be

(*a*) 90° □ (*b*) 60° □

(*c*) 45° □ (*d*) 30° □

218. Which metal of the following is used for making cold chisel?

(*a*) mild steel □ (*b*) stainless steel □

(*c*) cast tool steel □ (*d*) cast iron. □

219. A chisel should have cutting edge

(*a*) tempered ☐ (*b*) hardened ☐
(*c*) tempered and hardened ☐ (*d*) annealed. ☐

220. The band saws cut best under conditions of

(*a*) slow speed and light feed ☐ (*b*) slow speed and heavy feed ☐
(*c*) high speed and light feed ☐ (*d*) high speed and heavy feed. ☐

221. In a single V-butt welds, and angle between edges is kept about

(*a*) 10 to 20° ☐ (*b*) 20 to 40° ☐
(*c*) 40 to 60° ☐ (*d*) 70 to 90°. ☐

222. The single V and single U-butt welds are used for sheets of about

(*a*) 1 to 5 mm thickness ☐ (*b*) 5 to 15 mm thickness ☐
(*c*) 15 to 25 mm thickness ☐ (*d*) more than 25 mm thickness. ☐

223. A flux is used in welding to

(*a*) remove the oxides of the metals formed at high temperature ☐
(*b*) permit perfect cohesion of the metals ☐
(*c*) coll the metals ☐
(*d*) all of the above ☐
(*e*) both (*a*) and (*b*). ☐

224. In case of arc welding, the flux

(*a*) is used from outside in the form of powder ☐
(*b*) is coated on the electrodes ☐
(*c*) is not used ☐
(*d*) none of the above. ☐

225. In case of gas welding, the flux

(*a*) is used from outside in the form of powder ☐
(*b*) is coated on the electrodes ☐
(*c*) is not used ☐
(*d*) none of the above. ☐

226. Which one of the following metals has the least weldability?

(*a*) carbon steel ☐ (*b*) iron ☐
(*c*) stainless steel ☐ (*d*) cast iron. ☐

227. Which one of the following metals has the best weldability?

(*a*) carbon steel ☐ (*b*) iron ☐
(*c*) stainless steel ☐ (*d*) cast iron. ☐

228. In a molten state, a weld metal has

(*a*) no capacity of dissolving gases like oxygen, nitrogen and hydrogen ☐
(*b*) last capacity of dissolving gases ☐
(*c*) good capacity of dissolving gases ☐
(*d*) none of the above. ☐

229. In case of plastic welding, the metals to be joined are heated

(*a*) to the plastic state and allowed to solidity ☐

(*b*) to the plastic state and then forced together by external pressure without the addition of filter material ☐

(*c*) to a molten state and allowed to cool in presence of a filter material ☐

(*d*) none of the above. ☐

230. Choose the wrong statement

(*a*) Impurities, in case of welded joint, make the joint weaker as the welded portion is filled with gas and slag inclusions and metal becomes brittle. ☐

(*b*) In square butt welds, the distance between two faces is kept about 3 mm. ☐

(*c*) In case of welding due to high temperature, oxides of the metals are formed which weaken the weld. ☐

(*d*) The material, used for coating the electrode in case of welding, is called slag. ☐

231. In case of fusion welding, the metals to be joined are heated to

(*a*) the plastic state and allowed to solidify ☐

(*b*) to the plastic state and then forced together by external pressure without the addition of filler material ☐

(*c*) to a molten state and allowed to solidify in presence of a filler material ☐

(*d*) none of the above. ☐

232. Which one of the following is an example of plastic welding?

(*a*) gas welding ☐ (*b*) arc-welding ☐

(*c*) thermit welding ☐ (*d*) forge welding. ☐

233. Which one of the following is an example of fusion welding?

(*a*) gas welding ☐ (*b*) forge welding ☐

(*c*) resistance welding ☐ (*d*) thermit welding with pressure. ☐

234. If an electric current is passed through the metals to be joined and heated to the plastic state and then weld is completed by the application of pressure, the welding is known as

(*a*) forge welding ☐ (*b*) electric arc welding ☐

(*c*) resistance welding ☐ (*d*) thermit welding with pressure. ☐

235. In resistance welding, the voltage is stepped down to about

(*a*) 100 to 50 volts ☐ (*b*) 50 to 20 volts ☐

(*c*) 20 to 10 volts ☐ (*d*) 4 to 12 volts. ☐

236. The amount of current needed for resistance welding is

(*a*) directly proportional to the duration of time ☐

(*b*) inversely proportional to the duration of time ☐

(*c*) directly proportional to the area in contact ☐

(*d*) inversely proportional to the area in contact ☐

(*e*) both (*a*) and (*d*) ☐

(*f*) both (*b*) and (*c*). ☐

237. The pressure required for resistance welding varies from

(*a*) 50 to 100 kgf/cm^2 ☐ (*b*) 100 to 200 kgf/cm^2 ☐

(*c*) 250 to 550 kgf/cm^2 ☐ (*d*) 500 to 800 kgf/cm^2 ☐

238. In resistance welding, the heat generated is given by

(*a*) $H = \dfrac{I^2 \times R}{T}$ ☐ (*b*) $\dfrac{I^2 \times T}{R}$ ☐

(*c*) $H = I^2 \times R \times T$ ☐ (*d*) $H = \dfrac{R \times T}{I^2}$ ☐

where H = Heat generated,

R = Resistance of metal being welded,

T = Duration of current flow, and

I = Current in amperes through weld.

239. In resistance welding two copper electrodes are used. The electrodes are cooled by

(*a*) air ☐ (*b*) water ☐

(*c*) either air or water ☐ (*d*) both air and water. ☐

240. Choose the correct statement

(*a*) Low carbon steel and wrought iron are generally joined by forge welding. ☐

(*b*) If the ends of the metals to be joined by forge welding are not heated enough, they will not stick together. ☐

(*c*) The flux forms a protective coating of slag over the weld metal and creates a non-oxidising atmosphere. ☐

(*d*) All of the above. ☐

(*e*) None of the above. ☐

241. Spot welding, projection welding and seam welding are the classification of

(*a*) forge welding ☐ (*b*) resistance welding ☐

(*c*) thermit welding ☐ (*d*) arc welding. ☐

242. The number of zones of heat generation in spot welding are

(*a*) two ☐ (*b*) three ☐

(*c*) four ☐ (*d*) five. ☐

243. In resistance welding between the electrodes a current of

(*a*) high voltage and high ampere is passed ☐

(*b*) high voltage and low ampere is passed ☐

(*c*) low voltage and low ampere is passed ☐

(*d*) low voltage and high ampere is passed. ☐

244. In spot welding, the weld time is about

(*a*) 45 to 30 seconds ☐ (*b*) 30 to 50 seconds ☐

(*c*) 10 to 20 seconds ☐ (*d*) 3 to 10 seconds ☐

245. Spot welding is most suitable for joining parts having thickness upto

(*a*) 50 mm ☐ (*b*) 30 mm ☐

(*c*) 20 mm ☐ (*d*) 10 mm. ☐

246. In case of projection welding, the diameter of the projection is approximately equal to
(*a*) thickness of the sheet ☐
(*b*) 1.5 times the thickness of the sheet ☐
(*c*) twice the thickness of the sheet ☐
(*d*) half the thickness of the sheet. ☐

247. Seam welding is
(*a*) an arc welding process ☐
(*b*) a continuous spot welding process ☐
(*c*) a multi-spot welding process ☐
(*d*) a process used for joining round bars. ☐

248. Projection welding is
(*a*) an arc welding process ☐
(*b*) a continuous spot welding process ☐
(*c*) a multi-spot welding process ☐
(*d*) a process used for joining round bars. ☐

249. Which one is the correct statement about inert gas-arc welding?
(*a*) Inert arc gas-welding is faster ☐
(*b*) Inert arc gas-welding can weld metals considered impossible to weld ☐
(*c*) Inert arc gas-welding produces cleaner weld ☐
(*d*) All of the above ☐
(*e*) None of the above. ☐

250. Choose the wrong statement
(*a*) Seam welding is continuous spot welding process. ☐
(*b*) Projection welding is multi-spot welding process. ☐
(*c*) In resistance welding, the pressure is released after the weld cools. ☐
(*d*) The resistance welding is best suited for lead and tin. ☐

251. Which of the following welding processes uses non-consumable electrodes?
(*a*) MIG welding ☐ (*b*) TIG welding ☐
(*c*) CIG welding ☐ (*d*) Sub-merged arc welding. ☐

252. Which of the following welding processes used consumable electrodes?
(*a*) MIG welding ☐ (*b*) TIG welding ☐
(*c*) CIG welding ☐ (*d*) Sub-merged arc welding. ☐

253. Arc-voltage is the voltage drop across the arc between the two electrodes. The arc-voltage
(*a*) increases with the decrease of arc length ☐
(*b*) increases with the increase of arc length ☐
(*c*) is independent of arc length ☐
(*d*) none of the above. ☐

254. In arc-welding processes, penetration is deepest for
(*a*) DCRP ☐ (*b*) DCSP ☐
(*c*) A.C. ☐ (*d*) none of the above. ☐

255. In arc-welding processes, penetration is least for

(*a*) DCRP ☐ (*b*) DCSP ☐

(*c*) A.C. ☐ (*d*) none of the above. ☐

256. In an arc-welding process, using direct current, the amount of useful arc heat

(*a*) at the anode and cathode are equal ☐

(*b*) at the anode is one-third and at the cathode two-third ☐

(*c*) at the anode is two-third and at the cathode one-third ☐

(*d*) at the anode is 80% and at the cathode 20%. ☐

257. In an arc welding process using A.C, the amount of useful arc heat

(*a*) at the anode and cathode are equal ☐

(*b*) at the anode is one-third and at the cathode two-third ☐

(*c*) at the anode is two-third and at the cathode one-third ☐

(*d*) at the anode is 80% and at the cathode 20%. ☐

258. In TIG arc welding, an arc is produced between the work-piece and tungsten electrode in an atmosphere of

(*a*) helium gas ☐ (*b*) hydrogen gas ☐

(*c*) oxygen gas ☐ (*d*) argon gas ☐

(*e*) either (*a*) or (*d*) ☐ (*f*) either (*b*) or (*c*). ☐

259. The welding process in which heat is produced for welding by chemical reaction is known as

(*a*) forge welding ☐ (*b*) resistance welding ☐

(*c*) gas welding ☐ (*d*) thermit welding. ☐

260. Choose the wrong statement

(*a*) In case of direct current straight polarity (DCSP), the work-piece is connected to the positive terminal of the welding machine. ☐

(*b*) In case of direct current reverse polarity (DCRP), the work-piece is connected to the negative terminal of the welding machine. ☐

(*c*) In tungsten-inert gas arc welding a consumable electrode is used. ☐

(*d*) Direct current with straight polarity is used in welding copper alloys with TIG arc welding. ☐

261. In thermit welding, the aluminium and iron oxide are mixed in the proportion of

(*a*) 1 : 3 ☐ (*b*) 1 : 2 ☐

(*c*) 1 : 1 ☐ (*d*) 2 : 1. ☐

262. In thermit welding, a very high temperature is developed. This temperature is of the order of

(*a*) 1000°C ☐ (*b*) 2000°C ☐

(*c*) 2700°C ☐ (*d*) 3600°C. ☐

263. For welding heavy parts such as trucks, broken motor castings, connecting rods etc., the welding process used is generally

(*a*) resistance welding ☐ (*b*) gas welding ☐

(*c*) thermit welding ☐ (*d*) spot welding. ☐

264. The process of joining two pieces of metals in which a non-ferrous alloy is introduced in a liquid state between the pieces of metals to be joined and allowed to solidify, is known as

(*a*) welding ☐ (*b*) soldering ☐
(*c*) brazing ☐ (*d*) lancing. ☐

265. The process of joining two or more pieces of metal-sheets, by means of a fusible alloy or metal applied in the molten state, is known as

(*a*) welding ☐ (*b*) wiping ☐
(*c*) GMAW ☐ (*d*) lancing. ☐

266. Soldering of lead-pipe is known as

(*a*) lancing ☐ (*b*) soldering ☐
(*c*) brazing ☐ (*d*) GTAW. ☐

267. After the welding operation, the residual flux must be removed from the metal surface immediately otherwise the presence of residual flux will promote

(*a*) crack formation ☐ (*b*) corrosion ☐
(*c*) diffusion ☐ (*d*) undercutting. ☐

268. Inadequate penetration in case of welding operation will lead to

(*a*) crack formation ☐ (*b*) corrosion ☐
(*c*) diffusion ☐ (*d*) undercutting. ☐

269. For micro welding applications, the most suitable welding technique is

(*a*) electron beam-welding ☐ (*b*) laser welding ☐
(*c*) thermit welding ☐ (*d*) arc-welding. ☐

270. For welding steel by MIG process, the gas used is

(*a*) pure argon gas ☐ (*b*) CO_2 ☐
(*c*) argon-oxygen mixture ☐ (*d*) nitrogen. ☐

271. Choose the correct statement

(*a*) Hot cracking is influenced by the sulphur and carbon content of mild steel weld metals. ☐
(*b*) Fluxes are to be used in welding all types of metals except mild steel. ☐
(*c*) The temperature produced by oxy-hydrogen flame is about 2400°C. ☐
(*d*) The temperature of oxy-hydrogen flame is less than oxy-acetylene flame. ☐
(*e*) All of the above. ☐
(*f*) None of the above. ☐

272. For welding titanium by MIG process, the gas used is

(*a*) pure argon gas ☐ (*b*) CO_2 ☐
(*c*) argon-oxygen mixture ☐ (*d*) nitrogen. ☐

273. For welding stainless steel by MIG process, the gas used is

(*a*) pure argon gas ☐ (*b*) CO_2 ☐
(*c*) argon-oxygen mixture ☐ (*d*) nitrogen. ☐

274. Which is the correct statement about MIG welding process?

(*a*) MIG welding does not require any flux. ☐
(*b*) In MIG welding, corrosion resistance is high. ☐
(*c*) MIG welding can weld aluminium and stainless steel metals which are difficult to weld. ☐
(*d*) All of the above ☐
(*e*) None of the above. ☐

275. The laser welding can have the thickness and width of the weld in the ratio of

(*a*) 50 : 1 ☐ (*b*) 100 : 1 ☐
(*c*) 125 : 1 ☐ (*d*) 200 : 1. ☐

276. The pressure applied in case of ultrasonic welding is of the order of

(*a*) 2 to 3 kg/cm^2 ☐ (*b*) 10 to 20 kg/cm^2 ☐
(*c*) 20 to 40 kg/cm^2 ☐ (*d*) 40 to 100 kg/cm^2. ☐

277. In case of sub-merged arc welding, the electrodes up to

(*a*) 30 mm diameter may be used ☐ (*b*) 20 mm diameter may be used ☐
(*c*) 15 mm diameter may be used ☐ (*d*) 12 mm diameter may be used. ☐

278. The ratio of oxygen to acetylene for complete combustion is

(*a*) 1 : 1 ☐ (*b*) 2 : 1 ☐
(*c*) 2.5 : 1 ☐ (*d*) 3 : 1. ☐

279. Neutral flame is a flame which is obtained by supplying

(*a*) more volume of acetylene and less volume of oxygen ☐
(*b*) less volume of acetylene and more volume of oxygen ☐
(*c*) equal volume of acetylene and oxygen ☐
(*d*) none of the above. ☐

280. Carburising flame is a flame which is obtained by supplying

(*a*) more volume of acetylene and less volume of oxygen ☐
(*b*) less volume of acetylene and more volume of oxygen ☐
(*c*) equal volume of acetylene and oxygen ☐
(*d*) none of the above. ☐

281. Oxidising flame is a flame which is obtained by supplying

(*a*) more volume of acetylene and less volume of oxygen ☐
(*b*) less volume of acetylene and more volume of oxygen ☐
(*c*) equal volume of acetylene and oxygen ☐
(*d*) none of the above. ☐

282. Neutral flame is used to weld

(*a*) copper ☐ (*b*) cast iron ☐
(*c*) steel ☐ (*d*) brass ☐
(*e*) all above except (*d*). ☐

283. Choose the wrong statement

(*a*) The temperature of the oxy-acetylene flame depends upon the relative proportion of the oxygen and acetylene gases. ☐

(*b*) The maximum temperature of the flame occurs at the inner cone of the flame. ☐

(*c*) The temperature of neutral flame at the inner luminous cone is about 3200°C. ☐

(*d*) Most of the oxy-acetylene welding is done by oxidising flame. ☐

284. Rate of welding steel by carburising flame as compared to that of neutral flame is

(*a*) slow ☐ (*b*) fast ☐

(*c*) same ☐ (*d*) none of the above. ☐

285. For cutting operation, the most suitable flame is

(*a*) neutral ☐ (*b*) oxidising ☐

(*c*) carburising ☐ (*d*) none of the above. ☐

286. For welding non-ferrous metal such as brass and bronze, the most suitable flame is

(*a*) neutral ☐ (*b*) oxidising ☐

(*c*) carburising ☐ (*d*) none of the above. ☐

287. Back hand oxy-acetylene welding is normally used for welding metals of thickness up to

(*a*) 20 mm ☐ (*b*) 15 mm ☐

(*c*) 10 mm ☐ (*d*) 5 mm. ☐

288. Flux is not used for welding

(*a*) cast iron ☐ (*b*) brass ☐

(*c*) bronze ☐ (*d*) carbon steel. ☐

289. The oxygen cutting process can cut the following materials

(*a*) brass ☐ (*b*) bronze ☐

(*c*) aluminium ☐ (*d*) mild steel ☐

(*e*) all of the above ☐ (*f*) none of the above. ☐

290. Black colour is usually painted on a

(*a*) acetylene cylinder ☐ (*b*) oxygen cylinder ☐

(*c*) hydrogen cylinder ☐ (*d*) nitrogen cylinder. ☐

291. Maroon colour is usually painted on a

(*a*) acetylene cylinder ☐ (*b*) oxygen cylinder ☐

(*c*) hydrogen cylinder ☐ (*d*) nitrogen cylinder. ☐

292. Weld spatter refers to

(*a*) flux ☐ (*b*) welding defect ☐

(*c*) filler material ☐ (*d*) shield. ☐

293. Filler material is used in

(*a*) seam welding ☐ (*b*) spot welding ☐

(*c*) projection welding ☐ (*d*) all of the above ☐

(*e*) none of the above. ☐

294. Acetylene gas is produced from

(*a*) calcium chloride ☐ (*b*) calcium carbide ☐

(*c*) carbon ☐ (*d*) calcium carbonate. ☐

295. For brazing aluminium and magnesium, the following flux is used

(*a*) chlorides and fluorides mixed with water ☐

(*b*) mixture of boric acid, borax and wetting agent ☐

(*c*) boric acid, borax and fluoride with a wetting agent ☐

(*d*) none of the above. ☐

296. Choose the correct statement

(*a*) Neutral flame is used to weld all metals. ☐

(*b*) By supplying equal volumes of oxygen and acetylene, an oxidising flame is obtained. ☐

(*c*) The joint should be cleaned before brazing to avoid scale effect. ☐

(*d*) Soldering iron is made of wedge shape in order to retain heat. ☐

297. For oxidising flame, the ratio of oxygen to acetylene is

(*a*) 1 : 3 ☐ (*b*) 1 : 2 ☐

(*c*) 1 : 1 ☐ (*d*) 1.5 : 1. ☐

298. For carburizing flame, the ratio of oxygen to acetylene is

(*a*) 1 : 3 ☐ (*b*) 1 : 2 ☐

(*c*) 1 : 1 ☐ (*d*) 0.9 : 1. ☐

299. Choose the wrong statement

(*a*) The temperature produced in oxy-hydrogen flame is compared to oxy-acetylene flame is less. ☐

(*b*) For arc welding direct current is used. ☐

(*c*) Thermit welding is a process which uses a mixture of iron oxide and aluminium. ☐

(*d*) The temperature of the inner luminous core of neutral flame is of the order of 2000°C. ☐

300. The temperature of the plasma torch is of the order of

(*a*) 80000°C ☐ (*b*) 40000°C ☐

(*c*) 33000°C ☐ (*d*) 1000°C. ☐

301. In brazing, the melting point of the filler metal should be

(*a*) above 1000°C ☐ (*b*) above 800°C ☐

(*c*) above 420°C ☐ (*d*) above 300°C. ☐

302. In arc welding, the arc length should be approximately equal to

(*a*) twice the diameter of electrode rod ☐ (*b*) 1.5 times the rod diameter ☐

(*c*) diameter of the rod ☐ (*d*) half the diameter of the rod. ☐

303. Choose the correct statement

(*a*) Forge welding is best suited for wrought iron. ☐

(*b*) Electrode gets consumed in arc welding process. ☐

(*c*) Carbon arc welding is used to weld carbon rods. ☐

(*d*) Helium or argon is used in order to act as shielding medium in MIG welding process. ☐

(e) All of the above. ☐
(f) All of the above except (c). ☐

304. Choose the wrong statement
(a) When measuring the cutting forces, the force measuring equipment will undergo same deflection. ☐
(b) Dynamometer is used to measure the cutting forces of the tool. ☐
(c) If the dynamometer is rigid enough, the cutting operation will not be influenced by the accompanying deflections. ☐
(d) Increase in feed rate, improves the surface finish. ☐

305. Which of the following is not a heat treatment process?
(a) cyaniding ☐ (b) martempering ☐
(c) tempering ☐ (d) parkerizing. ☐

306. The operation of cutting a hole (other than cylindrical) in a sheet of metal by the punch and the die is known as
(a) slitting ☐ (b) blanking ☐
(c) piercing ☐ (d) notching. ☐

307. The operation of removal of metal to the desired shape from the edge of a plate is known as
(a) slitting ☐ (b) blanking ☐
(c) piercing ☐ (d) notching. ☐

308. The operation of cutting a sheet of metal in a straight line along the length, is known as
(a) slitting ☐ (b) blanking ☐
(c) piercing ☐ (d) notching. ☐

309. The operation of cutting of flat sheet to the desired shape is known as
(a) slitting ☐ (b) blanking ☐
(c) piercing ☐ (d) notching. ☐

310. Choose the wrong statement
(a) Wood for pattern is considered dry when moisture content is less than 15%. ☐
(b) The coarser-grain sand is better for steel castings. ☐
(c) Slick is used to make or repair corners in a mould. ☐
(d) The purpose of sprue is to act as a reservoir for molten metal. ☐

311. Choose the wrong statement
(a) An abrasive is a hard material which is used to cut or wear away other materials. ☐
(b) Grinding is the only method of removing material after hardening. ☐
(c) Grinding is basically a finishing process used for producing close dimensional and geometrical accuracies. ☐
(d) Natural abrasives are more efficient than artificial abrasives. ☐

312. Which one of the following is a natural abrasive?
(a) silicon carbide ☐ (b) boron carbide ☐
(c) aluminium oxide ☐ (d) diamond. ☐

313. Which one of the following is an artificial abrasive?

(*a*) silicon carbide ☐ (*b*) boron carbide ☐
(*c*) aluminium oxide ☐ (*d*) diamond ☐
(*e*) all of the above ☐ (*f*) all of the above except (*d*). ☐

314. Which of the following materials are grounded by aluminium oxide abrasive?

(*a*) tough bronze ☐ (*b*) touch copper ☐
(*c*) high speed steel ☐ (*d*) cast iron ☐
(*e*) all of the above ☐ (*f*) all of the above except (*d*). ☐

315. Which of the following materials are grounded by silicon carbide abrasive?

(*a*) cast iron ☐ (*b*) carbide tips ☐
(*c*) tungsten tools ☐ (*d*) non-metallic materials ☐
(*e*) all of the above ☐ (*f*) none of the above. ☐

316. Choose the correct statement

(*a*) Silicon carbide is more tough than aluminium oxide. ☐
(*b*) Silicon carbide as compared to aluminium oxide is less hard and less brittle. ☐
(*c*) Sand stone or solid quartz is a neutral abrasive. ☐
(*d*) The trade name of aluminium oxide is crystolon. ☐

317. Grain depth of cut is equal to

(*a*) half the maximum cut depth ☐ (*b*) the maximum cut depth ☐
(*c*) 1.5 times the maximum cut depth ☐ (*d*) twice the maximum cut depth. ☐

318. If the speed of work-piece is increased, the grain depth of cut in case of grinding

(*a*) decreases ☐ (*b*) increases ☐
(*c*) remains constant ☐ (*d*) first decreases then increases. ☐

319. If the speed of grinding wheel is increased, the grain depth of cut

(*a*) decreases ☐ (*b*) increases ☐
(*c*) remains constant ☐ (*d*) first decreases then increases. ☐

320. The maximum chip thickness per grain depth of cut is given by

(*a*) $t = \dfrac{v \times \sin(\alpha + \beta)}{N}$ ☐ (*b*) $t = \dfrac{v \times \sin(\alpha + \beta)}{N \times V}$ ☐

(*c*) $t = \dfrac{V \times \sin(\alpha + \beta)}{N \times V}$ ☐ (*d*) $t = \dfrac{N \times V \times \sin(\alpha + \beta)}{v}$. ☐

where v = Linear velocity of work, V = Linear velocity of grinding wheel,

α = Angle subtended by arc of contact at the centre of wheel,

β = Angle subtended by arc of contact at the centre of work, and

N = Number of grains (or grift) per unit length of the wheel circumference.

321. If the speed of the work-piece is increased, the wheel will appear

(*a*) harder ☐ (*b*) softer ☐
(*c*) neither harder nor softer ☐ (*d*) none of the above. ☐

322. If the depth of cut is decreased the wheel will appear

(*a*) harder ☐ (*b*) softer ☐

(*c*) neither harder nor softer ☐ (*d*) none of the above. ☐

323. If the speed of the grinding wheel is reduced, the wheel will appear

(*a*) harder ☐ (*b*) softer ☐

(*c*) neither harder nor softer ☐ (*d*) none of the above. ☐

324. For internal grinding

(*a*) hard wheel is used ☐ (*b*) soft wheel is used ☐

(*c*) any wheel is used ☐ (*d*) none of the above. ☐

325. The temperature in fine grinding can be reduced by

(*a*) decreasing wheel speed and increasing chip thickness ☐

(*b*) decreasing wheel speed and chip thickness ☐

(*c*) increasing wheel speed and decreasing chip thickness ☐

(*d*) increasing wheel speed and chip thickness. ☐

326. The temperature in normal grinding can be reduced by

(*a*) decreasing wheel speed ☐ (*b*) decreasing wheel speed and chip thickness ☐

(*c*) increasing wheel speed ☐ (*d*) increasing wheel speed and chip thickness. ☐

327. Which of the following factors affects the selection of the grinding wheel?

(*a*) work speed ☐ (*b*) wheel speed ☐

(*c*) work material ☐ (*d*) area of contact ☐

(*e*) all of the above ☐ (*f*) none of the above. ☐

328. The force on individual grain of grinding wheel is

(*a*) proportional to the work speed ☐ (*b*) proportional to the square of work speed ☐

(*c*) proportional to the cube of work speed ☐ (*d*) inversely proportional to the work speed. ☐

329. Silicon carbide abrasive is best suited for

(*a*) brittle materials ☐ (*b*) hard materials ☐

(*c*) low tensile strength material ☐ (*d*) tough materials ☐

(*e*) all of the above except (*d*) ☐ (*f*) none of the above. ☐

330. Aluminium oxide is best suited abrasive for

(*a*) brittle materials ☐ (*b*) hard materials ☐

(*c*) high tensile strength material ☐ (*d*) tough materials ☐

(*e*) both (*a*) and (*b*) ☐ (*f*) both (*c*) and (*d*). ☐

331. The excess wheel wear indicates that the

(*a*) wheel is running too slow ☐ (*b*) work-piece is running too fast ☐

(*c*) wheel is running too fast ☐ (*d*) work-piece is running too slow ☐

(*e*) both (*a*) and (*b*) ☐ (*f*) both (*c*) and (*d*). ☐

332. In dry condition of grinding

(*a*) soft wheel is required ☐ (*b*) hard wheel is required ☐

(*c*) any wheel is required ☐ (*d*) none of the above. ☐

333. The soft grade grinding wheels are denoted by the letters

(*a*) *A* to *K* ☐ (*b*) *L* to *O* ☐

(*c*) *P* to *Z* ☐ (*d*) *A* to *Z*. ☐

334. The hard grade grinding wheels are denoted by the letters

(*a*) *A* to *K* ☐ (*b*) *L* to *O* ☐

(*c*) *P* to *Z* ☐ (*d*) *A* to *Z*. ☐

335. The silicon carbide abrasive is denoted by

(*a*) *S* ☐ (*b*) *C* ☐

(*c*) *SC* ☐ (*d*) *SB*. ☐

336. The white aluminium oxide is denoted by

(*a*) *Al* ☐ (*b*) *A* ☐

(*c*) *WA* ☐ (*d*) *WAl*. ☐

337. The dense structure of the grinding wheel is denoted by numbers

(*a*) 1 to 8 ☐ (*b*) 1 to 6 ☐

(*c*) 9 to 15 ☐ (*d*) 15 to 30. ☐

338. The act of restoring the cutting face of a grinding wheel is known as

(*a*) wheel dressing ☐ (*b*) wheel truing ☐

(*c*) wheel grinding ☐ (*d*) wheel grading. ☐

339. The act of improving the cutting action of a grinding wheel is known as

(*a*) wheel dressing ☐ (*b*) wheel truing ☐

(*c*) wheel grinding ☐ (*d*) wheel grading. ☐

340. Wheel truing is carried by

(*a*) abrasive sticks ☐ (*b*) abrasive wheel ☐

(*c*) stainless steel ☐ (*d*) diamond. ☐

341. Grinding ratio is defined as the ratio between

(*a*) Volume of the metal removed from the work-piece to the total volume of the work-piece ☐

(*b*) Volume of the work-piece to the volume of the metal removed from the work-piece ☐

(*c*) Volume of the metal removed from the work-piece to the wear of grinding wheel ☐

(*b*) Volume of the wear of grinding wheel to the volume of the metal removed from the work-piece. ☐

342. The grinding ratio varies from

(*a*) 10 to 20 ☐ (*b*) 30 to 60 ☐

(*c*) 75 to 125 ☐ (*d*) 150 to 200. ☐

343. Choose the wrong statement

(*a*) Green grain silicon carbide is represented by *GC*. ☐

(*b*) The number 8 to 24 represents coarse grain. ☐

(*c*) Very hard grinding wheels are denoted by the letters *L* to *O*. ☐

(*d*) For resinoid bond, the notation used is letter *B*. ☐

344. A lot of heat is generated between the cutting tool and work-piece during the process of grinding. This heat is distributed

(*a*) equally among work-piece and grinding wheel ☐
(*b*) more in work-piece and less in grinding wheel ☐
(*c*) less in work-piece and more in grinding wheel ☐
(*d*) none of the above. ☐

345. The cracks, which are developed due to generation of heat on the grinding wheel, are

(*a*) in the direction of grinding marks ☐
(*b*) inclined to the direction of grinding marks ☐
(*c*) perpendicular to the direction of grinding marks ☐
(*d*) none of the above. ☐

346. In case of cylindrical grinder, the depth of cut for roughing cut in normally

(*a*) 0.05 mm ☐ (*b*) 0.01 mm ☐
(*c*) 0.005 mm ☐ (*d*) 0.001 mm. ☐

347. In case of cylindrical grinder, the depth of cut for finishing cut is normally

(*a*) 0.05 mm ☐ (*b*) 0.01 mm ☐
(*c*) 0.005 mm ☐ (*d*) 0.001 mm. ☐

348. The process of removing surface roughness, tool marks, surface cracks from grinding, slight distortions and other minor defects from previous operation is known as

(*a*) honing ☐ (*b*) lapping ☐
(*c*) brazing ☐ (*d*) milling. ☐

349. The process of removing stock from metallic and non-metallic surfaces is known as

(*a*) honing ☐ (*b*) lapping ☐
(*c*) brazing ☐ (*d*) milling. ☐

350. The stock, left on diameter for honing, varies from

(*a*) 1 mm to 3 mm ☐ (*b*) 0.5 mm to 1.0 mm ☐
(*c*) 0.005 mm to 0.5 mm ☐ (*d*) 0.0001 mm to 0.001 mm. ☐

351. The minimum stock removed by honing is

(*a*) 0.2 mm ☐ (*b*) 0.02 mm ☐
(*c*) 0.002 mm ☐ (*d*) 0.0002 mm. ☐

352. The maximum stock removed by honing is

(*a*) 10 mm ☐ (*b*) 5 mm ☐
(*c*) 3 mm ☐ (*d*) 1 mm. ☐

353. Which of the following materials can be honed?

(*a*) brass ☐ (*b*) hardened steel ☐
(*c*) aluminium ☐ (*d*) glass ☐
(*e*) carbides ☐ (*f*) all of the above. ☐

354. Choose the wrong statement

(*a*) A grinding wheel is considered better if it has greater grinding ratio. ☐

(*b*) The tangent of the curve between the wear grinding wheel and volume of the metal removed represents the grinding ratio. ☐

(*c*) The property by which the metal can be removed easily by grinding wheel is known grindability. ☐

(*d*) The material removed by lapping is approximately equal to 0.05 mm. ☐

355. The stock left for lapping varies from

(*a*) 1 to 3 mm ☐ (*b*) 0.5 to 1.0 mm ☐

(*c*) 0.01 to 0.1 mm ☐ (*d*) 0.001 to 0.01 mm. ☐

356. Honing can correct irregularities but grinding cannot because honing uses

(*a*) small contact area at slow speed ☐ (*b*) large contact area at high speed ☐

(*c*) large contact area at slow speed ☐ (*d*) small contact area at high speed. ☐

357. When grinding with diamond wheel, the coolant used should be

(*a*) water ☐ (*b*) paraffin ☐

(*c*) paraffin in equal quantity with water ☐ (*d*) soluble oil. ☐

358. When grinding with silicon carbide wheel, the coolant should be

(*a*) water ☐ (*b*) paraffin ☐

(*c*) paraffin in equal quantity with water ☐ (*d*) soluble oil. ☐

359. The process of improving the cutting action of the grinding wheel is known as

(*a*) honing ☐ (*b*) truing ☐

(*c*) lapping ☐ (*d*) dressing. ☐

360. Which of the following abrasive should be used for grinding materials of high tensile strength?

(*a*) diamond ☐ (*b*) silicon carbide ☐

(*c*) aluminium oxide ☐ (*d*) sand stone. ☐

361. Which of the following abrasive should be used for grinding materials of low tensile strength?

(*a*) diamond ☐ (*b*) silicon carbide ☐

(*c*) aluminium oxide ☐ (*d*) sand stone. ☐

362. Which of the following statement is correct?

(*a*) In wet condition hard wheel is required. ☐ (*b*) In wet condition soft wheel is required. ☐

(*c*) In dry condition hard wheel is required. ☐ (*d*) In dry condition soft wheel is required. ☐

(*e*) both (*a*) and (*c*). ☐ (*f*) both (*b*) and (*d*). ☐

363. The hardness of a grinding wheel is specified by

(*a*) Rockweel hardness number ☐ (*b*) Brinell hardness number ☐

(*c*) Vickers pyramid number ☐ (*d*) Letter of alphabet. ☐

364. Which of the following statement about honing is correct?

(*a*) Honing operation can change the location of hole. ☐

(*b*) Honing operation can correct a sloped condition of a hole. ☐

(c) Honing operation cannot change the location of hole. ☐
(d) Honing operation cannot correct a slopped condition of a hole. ☐
(e) Both (a) and (b). ☐
(f) Both (c) and (d). ☐

365. The grinder used for grinding irregular, curved, tapered, concave and convex surfaces should be
(a) internal grinder ☐ (b) surface grinder ☐
(c) external grinder ☐ (d) all of the above ☐
(e) none of the above. ☐

366. According to Indian Standard specifications, a grinding wheel is specified by *'WA* 36 K 8 *VBE'*. The first letters *WA* stands for
(a) type of grade ☐ (b) type of abrasive ☐
(c) type of bond ☐ (d) grain size. ☐

367. In question 366, the figure 36 stands for
(a) type of grade ☐ (b) type of abrasive ☐
(c) type of bond ☐ (d) grain size. ☐

368. In question 366, the letter *K* stands for
(a) type of grade ☐ (b) type of abrasive ☐
(c) type of bond ☐ (d) grain size. ☐

369. In question 366, the figure 8 stands for
(a) type of grade ☐ (b) type of abrasive ☐
(c) type of bond ☐ (d) grain size. ☐

370. In question 366, the letter *V stands* for
(a) type of grade ☐ (b) type of abrasive ☐
(c) type of bond ☐ (d) grain size. ☐

371. In case of casting, the molten metal is introduced into a mould of the desired shape and the molten metal is allowed to solidify. The shape of the casting from the mould
(a) is same but slightly bigger ☐ (b) is same but slightly smaller ☐
(c) is same and size is also same ☐ (d) none of the above. ☐

372. Choose the wrong statement
(a) For moulding, pure silica sand is not suitable. ☐
(b) Sand can be easily packed to any shape and is somewhat porous and resist high temperature. ☐
(c) The fine sand is used for small and intricate castings. ☐
(d) Synthetic sand is also called green sand. ☐

373. The binder is case of natural sand is
(a) clay ☐ (b) molasses ☐
(c) water ☐ (d) bentonite. ☐

374. The binder in case of synthetic sand is
(a) clay ☐ (b) molasses ☐
(c) water ☐ (d) bentonite and water. ☐

375. For large castings, the sand grains should be

(a) fine ☐ (b) medium ☐
(c) coarse ☐ (d) rounded. ☐

376. For small and intricate castings, the sand grains should be

(a) fine ☐ (b) medium ☐
(c) coarse ☐ (d) rounded. ☐

377. For bench work and light floor work, the sand grains should be

(a) fine ☐ (b) medium ☐
(c) coarse ☐ (d) rounded. ☐

378. Permeability is poor for

(a) fine grains ☐ (b) medium grains ☐
(c) coarse grains ☐ (d) rounded grains. ☐

379. Permeability is high for

(a) fine grains ☐ (b) medium grains ☐
(c) coarse grains ☐ (d) rounded grains. ☐

380. Choose the wrong statement

(a) For large castings, coarse sand gains are used to permit gases to escape. ☐
(b) Angular grains are having more strength and less permeability. ☐
(c) The impurities in a good moulding sand is below 2%. ☐
(d) In silica sand the amount of clay is generally 20 to 30%. ☐

381. Which of the following properties should a good moulding sand have?

(a) A good moulding sand must be porous enough ☐
(b) A good moulding sand must have a good strength ☐
(c) A good moulding sand must have chemical resistivity with molten metal ☐
(d) A good moulding sand must have adhesiveness ☐
(e) All of the above. ☐

382. Choose the correct statement

(a) Natural sand maintains moisture content for a long time. ☐
(b) Synthetic sand is an artificial sand. ☐
(c) The two chief constituents of the moulding sand are silica and clay. ☐
(d) For fine sand grains, permeability is poor. ☐
(e) All of the above. ☐
(f) None of the above. ☐

383. Figure 12.2 shows the variation of fineness of the grain with permeability by four different curves. The correct relationship between fineness and permeability is given by

(a) curve *A* ☐
(b) curve *B* ☐
(c) curve *C* ☐
(d) curve *D*. ☐

FIGURE 12.2

384. The grain-fineness number according to American Foundry Men's Society (A.F.S.) is equal to

(*a*) $\dfrac{\text{Total product}}{\text{Total weight of sand retained}}$ ☐

(*b*) $\dfrac{\text{Total product}}{\text{Total \% sand retained by different sieves}}$ ☐

(*c*) $\dfrac{\text{Total weight of sand retained}}{\text{Total product}}$ ☐

(*d*) $\dfrac{\text{Total \% sand retained by different sieves}}{\text{Total product}}$. ☐

385. Figure 12.3 shows the variation of permeability, compression strength and density with moisture content. The correct relationship between permeability and moisture content is given by

(*a*) curve *A* ☐

(*b*) curve *B* ☐

(*c*) curve *C* ☐

(*d*) curve *D*. ☐

FIGURE 12.3

386. In Fig. 12.3, the correct relationship between compression strength and moisture content is given by

(*a*) curve *A* ☐ (*b*) curve *B* ☐

(*c*) curve *C* ☐ (*d*) curve *D*. ☐

387. In Fig. 12.3, the correct relationship between density and moisture content is given by

(*a*) curve *A* ☐ (*b*) curve *B* ☐

(*c*) curve *C* ☐ (*d*) curve *D*. ☐

388. In sand moulding, drag is the

(*a*) uppermost part of the flask ☐ (*b*) bottom most part of the flask ☐

(*c*) middle part of the flask ☐ (*d*) none of the above. ☐

389. In sand moulding, cope is the

(*a*) uppermost part of the flask ☐ (*b*) bottom most part of the flask ☐

(*c*) middle part of the flask ☐ (*d*) none of the above. ☐

390. In sand moulding, check is the

(*a*) uppermost part of the flask ☐ (*b*) bottom most part of the flask ☐

(*c*) middle part of the flask ☐ (*d*) none of the above. ☐

391. The purpose of sprue is to

(*a*) act as a reservoir for molten metal ☐

(*b*) feed molten metal from pouring basin to gate ☐

(*c*) removing pattern from the mould ☐

(*d*) split the pattern in two parts. ☐

392. A core is used in the mould to

(*a*) obtain hole or desired cavities ☐
(*b*) reduce metal erosion in gate and runners ☐
(*c*) retard foreign matter in the melt ☐
(*d*) all of the above. ☐
(*e*) none of the above. ☐

393. Which of the following are casting defects?

(*a*) blow-holes ☐
(*b*) cold-cracks ☐
(*c*) sand spots ☐
(*d*) scabs ☐
(*e*) all of the above ☐
(*f*) none of the above. ☐

394. The shrinkage rule for cast iron pattern is

(*a*) 30 mm per metre longer than the actual object ☐
(*b*) 10 mm per metre longer than the actual object ☐
(*c*) 10 mm per metre shorter than the actual object ☐
(*d*) 30 mm per metre shorter than the actual object. ☐

395. The tapering of the sides of the pattern in a direction parallel to which the pattern is drawn out from a mould, is known as

(*a*) shake ☐
(*b*) drag ☐
(*c*) draft ☐
(*d*) sprue. ☐

396. Aluminium is the best of the metals for making pattern, because it is

(*a*) easy to work ☐
(*b*) light in weight ☐
(*c*) resistant to corrosion ☐
(*d*) all of the above ☐
(*e*) none of the above. ☐

397. The process of making hollow castings of desired thickness by permanent mould without the use of cores is known as

(*a*) die casting ☐
(*b*) slush casting ☐
(*c*) pressed casting ☐
(*d*) centrifugal casting. ☐

398. The process of making hollow casting from permanent mould by a close fitting core is known as

(*a*) die casting ☐
(*b*) slush casting ☐
(*c*) pressed casting ☐
(*d*) centrifugal casting. ☐

399. Choose the correct statement

(*a*) A core is more permeable than the mould itself. ☐
(*b*) Core sand should be highly refractory. ☐
(*c*) Double shrinkage should be allowed when metal patterns are to be cast from the original patterns. ☐
(*d*) A natural binder is linseed oil. ☐
(*e*) All of the above. ☐
(*f*) None of the above. ☐

400. The process of pouring molten metal in the gravity of a metallic mould by gravity, is known as

(a) die casting ☐ (b) slush casting ☐

(c) pressed casting ☐ (d) permanent mould casting. ☐

401. The process of pouring molten metal under high pressure into mould, is known as

(a) die casting ☐ (b) slush casting ☐

(c) pressed casting ☐ (d) permanent mould casting. ☐

402. The process of pouring molten metal into mould and allowed to solidify, when the mould is revolving, is known as

(a) die casting ☐ (b) slush casting ☐

(c) pressed casting ☐ (d) centrifugal casting. ☐

403. Choose the correct statement

(a) The permeability of a fine sand will be high. ☐

(b) Adhesiveness is the property of the sand due to which the sand grains stick together. ☐

(c) A riser is used to deliver molten metal from pouring basin to gate. ☐

(d) The bottom part of a pattern is known as cope. ☐

(e) all of the above. ☐

(f) none of the above. ☐

404. For ornamental parts, and toys of non-ferrous alloys, the casting method used is known as

(a) die casting ☐ (b) slush casting ☐

(c) centrifugal casting ☐ (d) permanent mould casting. ☐

405. The tolerance obtained by investment casting is of the order of

(a) ± 0.5 mm ☐ (b) ± 0.05 mm ☐

(c) ± 0.005 mm ☐ (d) ± 0.0005 mm. ☐

406. Which one of the following casting processes is an expensive process and is used only where small number of intricate and highly accurate parts are to be manufactured?

(a) die casting ☐ (b) true centrifugal casting ☐

(c) slush casting ☐ (d) investment casting. ☐

407. Pipes of large length and diameters are made by

(a) pressed casting ☐ (b) semi-centrifugal casting ☐

(c) slush casting ☐ (d) extrusion process. ☐

408. Pipes subjected to very high pressure of the order 120 kg/cm^2 are made by

(a) pressed casting ☐ (b) semi-centrifugal casting ☐

(c) slush casting ☐ (d) extrusion process. ☐

409. Choose the wrong statement

(a) The molten metal is forced into mould under high pressure. ☐

(b) In case of slush casting, the thickness of casting depends upon the time for which the metal is allowed to solidify into the permanent mould. ☐

(c) Slush casting is mostly used for ornaments and toys of non-ferrous alloys. ☐

(d) Riser acts as a reservoir for the molten metal. ☐

410. Choose the wrong statement
- (*a*) In centrifugal casting, the metal solidifies under the pressure of centrifugal force. ☐
- (*b*) The unfinished surfaces of a pattern are marked by blue colour. ☐
- (*c*) Loam sand comprises of 85% clay and 15% moisture. ☐
- (*d*) Stag inclusion in casting is a surface defect. ☐

411. The impurities in centrifugal casting
- (*a*) are forced outside the surface ☐
- (*b*) are collected at the centre of the casting ☐
- (*c*) are uniformly distributed ☐
- (*d*) none of the above. ☐

412. The low melting point metals can be cast by
- (*a*) cold chamber method of die casting ☐
- (*b*) hot chamber method of die casting ☐
- (*c*) cold or hot chamber method of die casting ☐
- (*d*) none of the above. ☐

413. The high melting point metals can be cast by
- (*a*) cold chamber method of die casting ☐
- (*b*) hot chamber method of die casting ☐
- (*c*) cold or hot chamber method of die casting ☐
- (*d*) none of the above. ☐

414. The melting pot is integral with die casting machine in
- (*a*) cold chamber method of die casting ☐
- (*b*) hot chamber method of die casting ☐
- (*c*) both the above methods ☐
- (*d*) none of the above. ☐

415. The melting pot is separate from die casting machine in
- (*a*) cold chamber method of die casting ☐
- (*b*) hot chamber method of die casting ☐
- (*c*) both the above methods ☐
- (*d*) none of the above. ☐

416. Which of the following are casting defects?
- (*a*) blow holes ☐
- (*b*) slag ☐
- (*c*) scabs ☐
- (*d*) hot tears ☐
- (*e*) all of the above ☐
- (*f*) all of the above except (*b*). ☐

417. Choose the wrong statement
- (*a*) In centrifugal casting no core is used. ☐
- (*b*) For casting cast iron pipes, true centrifugal casting is used. ☐
- (*c*) Antioch process is a welding process. ☐
- (*d*) Red colour is used on the surface of a pattern, which is to be machined. ☐

418. If molten metal is fed into the cavity of metallic mould by gravity, the casting method is known as
- (*a*) pressed casting ☐
- (*b*) die casting ☐
- (*c*) permanent mould casting ☐
- (*d*) slush casting. ☐

419. If the molten metal is poured into die cavity after a predetermined time, the mould is inverted to permit a part of metal still in molten stale to flow out of cavity, the casting method is known as

(a) pressed casting ☐ (b) die casting ☐

(c) permanent mould casting ☐ (d) slush casting. ☐

420. To obtain high density and pure casting, which of the following method is used?

(a) die casting ☐ (b) centrifugal casting ☐

(c) slush casting ☐ (d) permanent mould casting. ☐

421. To cast symmetrical objects, which of the following casting method is used?

(a) die casting ☐ (b) true centrifugal casting ☐

(c) investment casting ☐ (d) none of the above. ☐

422. The colour marked on the surface of a pattern to be machined, is

(a) black ☐ (b) green ☐

(c) red ☐ (d) blue. ☐

423. The colour marked on the surface of a pattern not to be machined, is

(a) black ☐ (b) green ☐

(c) red ☐ (d) blue. ☐

424. For repairing and finishing the mould

(a) riddle is used ☐ (b) slick is used ☐

(c) swab is used ☐ (d) gagger is used. ☐

425. For moisturing the sand around the edge before removing pattern

(a) riddle is used ☐ (b) slick is used ☐

(c) swab is used ☐ (d) lifter is used. ☐

426. For reinforcement of sand in the top part of moulding box

(a) riddle is used ☐ (b) slick is used ☐

(c) swab is used ☐ (d) lifter is used. ☐

427. Choose the wrong statement

(a) Green sand is a mixture of 70% sand and 30% clay. ☐

(b) Loam sand is a sand employed on the faces of the pattern before moulding. ☐

(c) A pouring basin acts as a reservoir for molten metal. ☐

(d) Cast iron and steel pipes are produced by true centrifugal casting method. ☐

428. The glazing of the grinding wheel is due to

(a) cracks on grinding wheel ☐ (b) wear of bond ☐

(c) wear of abrasive grains ☐ (d) breaking up of abrasive grains. ☐

429. The stroke length of a planer varies from

(a) 1 to 4 m ☐ (b) 1 to 12 m ☐

(c) 12 to 16 m ☐ (d) 16 to 20 m. ☐

430. The lift the tool on the return stroke, air cylinder is employed in case of a

(*a*) shaper ☐ (*b*) planer ☐
(*c*) both shaper and planer ☐ (*d*) none of the above. ☐

431. The tool is stationary and the work reciprocates in case of a

(*a*) milling machine ☐ (*b*) shaper ☐
(*c*) planer ☐ (*d*) none of the above. ☐

432. The work is stationary and the tool reciprocates in case of a

(*a*) milling machine ☐ (*b*) shaper ☐
(*c*) planer ☐ (*d*) none of the above. ☐

433. For metal removal, single-point cutting tools are used in case of

(*a*) planers ☐ (*b*) shapers ☐
(*c*) both planers and shapers ☐ (*d*) none of the above. ☐

434. Comparatively small work-pieces are machined in case of

(*a*) planers ☐ (*b*) shapers ☐
(*c*) both planers and shapers ☐ (*d*) none of the above. ☐

435. The shaping process is very slow but it is mostly used because of

(*a*) ease of operation ☐ (*b*) inexpensive tooling ☐
(*c*) short set-up time ☐ (*d*) all of the above ☐
(*e*) none of the above. ☐

436. The maximum stroke length upto which a shaper works satisfactorily is

(*a*) 50 mm ☐ (*b*) 500 mm ☐
(*c*) 750 mm ☐ (*d*) 1200 mm. ☐

437. The stroke length of planer varies from

(*a*) 50 to 1000 mm ☐ (*b*) 1 to 3 m ☐
(*c*) 1 to 12 m ☐ (*d*) 1 to 30 m. ☐

438. In case of a shaper, the ram moves

(*a*) faster during forward stroke ☐
(*b*) slower during forward stroke ☐
(*c*) at the same speed during forward and return stroke ☐
(*d*) none of the above. ☐

439. The size of a shaper is given by

(*a*) maximum length of the machine ☐ (*b*) maximum length of the stroke ☐
(*c*) maximum height of the machine ☐ (*d*) all of the above. ☐

440. The push-cut type shaper removes the metal during

(*a*) forward stroke ☐ (*b*) backward stroke ☐
(*c*) both the forward and backward strokes ☐ (*d*) none of the above strokes. ☐

441. The draw-cut type shaper removes the metal during

(*a*) forward stroke ☐ (*b*) backward stroke ☐
(*c*) both the forward and backward strokes ☐ (*d*) none of the above strokes. ☐

442. Choose the wrong statement

(*a*) The shaper in comparison to planer is easier to operate and about three times quicker in action. ☐

(*b*) In a shaper, the metal is removed during forward stroke. ☐

(*c*) The shaper is best suited for cutting keyways and splines on shafts. ☐

(*d*) In case of planer, reciprocating motion is given to the cutting tool. ☐

443. The term sheaper means the

(*a*) push-cut type ☐ (*b*) draw-cut type ☐

(*c*) any one of the above types ☐ (*d*) none of the above. ☐

444. In case of a shaper, the ram should move

(*a*) faster during return stroke ☐ (*b*) slower during return stroke ☐

(*c*) at the same speed during return stroke ☐ (*d*) none of the above. ☐

445. In case of a shaper, the surface cutting speed depends upon

(*a*) the metal to be machined ☐ (*b*) the cutting tool material ☐

(*c*) size of the machine ☐ (*d*) all of the above ☐

(*e*) only (*a*) and (*b*). ☐

446. The length cut in one minute by a shaper in given by

(*a*) $\frac{N}{1000 \times L}$ ☐ (*b*) $\frac{L}{N \times 1000}$ ☐

(*c*) $\frac{N \times L}{1000}$ ☐ (*d*) $N \times L$ ☐

where L = Length of stroke in mm; and N = Number of stroke per minute.

447. The cutting speed for a shaper is equal to

(*a*) $\left(\frac{N \times L}{1000}\right) \times \alpha$ ☐ (*b*) $\left(\frac{N \times L}{1000}\right)\left(\frac{\alpha + \beta}{\alpha}\right)$ ☐

(*c*) $\left(\frac{N \times L}{1000}\right) \times \left(\frac{\alpha}{\alpha + \beta}\right)$ ☐ (*d*) $\left(\frac{N \times L}{1000}\right) \times \left(\frac{\beta}{\alpha + \beta}\right)$ ☐

where L = Length of stroke in mm, N = Number of stroke per minute,

α = Actual cutting time for machining, and β = Time for idle stroke.

448. In case of shaper, for roughing cuts, the practice is to use

(*a*) maximum feeds at higher speeds ☐ (*b*) maximum feeds at slow speeds ☐

(*c*) minimum feeds at slow speeds ☐ (*d*) minimum feeds at higher speeds. ☐

449. In case of shaper, for finish machining, the practice is to use

(*a*) maximum feeds at higher speeds ☐ (*b*) maximum feeds at slow speeds ☐

(*c*) minimum feeds at slow speeds ☐ (*d*) minimum feeds at higher speeds. ☐

450. The depth of cut in case of shaper should not be less than

(*a*) 0.1 mm ☐ (*b*) 0.2 m ☐

(*c*) 0.01 mm ☐ (*d*) 0.4 mm. ☐

451. Choose the correct statement

(*a*) The ram is heavy in case of draw-cut type shaper. ☐

(*b*) The shaper as compared to planer occupies less floor area. ☐

(*c*) In case of hydraulic shaper, the operation is very smooth. ☐

(*d*) The velocity diagram of hydraulically operated shaper, as compared to crank operated shaper, is nearly constant. ☐

(*e*) All of the above. ☐

(*f*) None of the above. ☐

452. The distance between tool bit and tool post in case of a shaper should not be more than

(*a*) 200 mm ☐ (*b*) 150 mm ☐

(*c*) 100 mm ☐ (*d*) 50 mm. ☐

453. Which of the following is the cause for chattering on the shaper?

(*a*) suspension of tool far from the tool holder ☐

(*b*) not holding the work-piece rigidly in the vice ☐

(*c*) improper adjustment of the ram jib ☐

(*d*) all of the above ☐

(*e*) none of the above. ☐

454. The overhang of the tool slide in case of a shaper should be

(*a*) maximum ☐ (*b*) minimum ☐

(*c*) zero ☐ (*d*) none of the above. ☐

455. The front clearance angle is case of a shaper when the tool is in the cutting position should be

(*a*) 10 to 12° ☐ (*b*) 8 to 10° ☐

(*c*) 4 to 6° ☐ (*d*) 3 to 4°. ☐

456. A slotter can be considered as a

(*a*) vertical shaper, having horizontal and vertical movement of ram ☐

(*b*) vertical shaper, having only vertical movement of ram ☐

(*c*) horizontal shaper, having horizontal and vertical movement of ram ☐

(*d*) horizontal shaper, having only horizontal movement of ram. ☐

457. The stroke of slotting machines ranges from

(*a*) 100 to 200 mm ☐ (*b*) 200 to 300 mm ☐

(*c*) 300 to 1800 mm ☐ (*d*) 1800 to 3000 mm. ☐

458. In case of slotting machine, cutting action takes place in

(*a*) upward stroke ☐ (*b*) forward stroke ☐

(*c*) downward stroke ☐ (*d*) backward stroke. ☐

459. The longitudinal and transverse power feeds in case of a slotting machine varies from

(*a*) 2 to 4 mm/stroke ☐ (*b*) 0.05 to 2.5 mm/stroke ☐

(*c*) 1 to 2 mm/stroke ☐ (*d*) 0.5 to 1.0 mm/stroke. ☐

460. The setting time in case of planers as compared to that of a shaper is of the order of
(*a*) two times ☐ (*b*) five or six times ☐
(*c*) half time ☐ (*d*) one-fourth time. ☐

461. Portable drilling machine is specified by
(*a*) the maximum diameter of the drill which can be held ☐
(*b*) the length of arm and column diameter ☐
(*c*) the diameter of the largest work-piece that can be drilled ☐
(*d*) the drilling area, the size and number of holes a machine can drill. ☐

462. Multiple spindle drilling machine is specified by
(*a*) the maximum diameter of the drill which can be held ☐
(*b*) the length of the arm and column diameter ☐
(*c*) the diameter of the largest work-piece that can be drilled ☐
(*d*) the drilling area, the size and number of holes a machine can drill. ☐

463. The radial drilling machine is specified by
(*a*) the maximum diameter of the drill which can be held ☐
(*b*) the length of the arm and column diameter ☐
(*c*) the diameter of the largest work-piece that can be drilled ☐
(*d*) the drilling area, the size and number of holes a machine can drill. ☐

464. Which of the following operations can be performed by a drilling machine?
(*a*) spot facing ☐ (*b*) reaming ☐
(*c*) tapping ☐ (*d*) boring ☐
(*e*) all of the above ☐ (*f*) none of the above. ☐

465. The cutting speed for high speed drills is approximately equal to
(*a*) four times the cutting speed for carbon steel drills ☐
(*b*) three times the cutting speed for carbon steel drills ☐
(*c*) two times the cutting speed for carbon steel drills ☐
(*d*) one and a half times the cutting speed for carbon steel drills. ☐

466. Drilling is an example of
(*a*) simple cutting ☐ (*b*) orthogonal cutting ☐
(*c*) oblique cutting ☐ (*d*) none of the above. ☐

467. Feed of drill is
(*a*) equal to half the radius of the drill ☐
(*b*) the distance moved by the drill into the work-piece on every revolution of the drill ☐
(*c*) the distance moved by the drill per second ☐
(*d*) equal to one-fourth the radius of the drill. ☐

468. The feed of the drill varies from
(*a*) 0.5 to 1 mm per revolution ☐ (*b*) 0.1 to 0.5 mm per revolution ☐
(*c*) 0.05 to 0.35 mm per revolution ☐ (*d*) 0.005 to 0.05 mm per revolution. ☐

469. Which of the following parameters affects the amount of metal removed?

(*a*) depth of hole ☐ (*b*) cutting speed ☐

(*c*) feed ☐ (*d*) diameter of hole ☐

(*e*) all of the above ☐ (*f*) none of the above. ☐

470. Choose the wrong statement

(*a*) The cutting speed for a drill depends upon the material to be drilled. This varies from 10 to 90 m/min. ☐

(*b*) Deep-hole drilling machines are used for drilling such holes whose length exceeds three times the drill size. ☐

(*c*) The Gang drilling machine is used for drilling holes of several different sizes on a work-piece. ☐

(*d*) Sensitive and upright drilling machines are specified by the length of the arm and column diameter. ☐

471. If f = feed per revolution of drill, L = length of hole to be drilled and N = r.p.m. of drill then drilling time (T) in minutes is given by

(*a*) $T = N \times L \times f$ ☐ (*b*) $T = \frac{N \times L}{f}$ ☐

(*c*) $T = \frac{f}{N \times L}$ ☐ (*d*) $T = \frac{L}{N \times f}$. ☐

472. The length of approach in case of drilling operation is equal to

(*a*) 0.8 D ☐ (*b*) 0.6 D ☐

(*c*) 0.29 D ☐ (*d*) 0.14 D ☐

where D = Diameter of drill.

473. The total length (L) in drilling an open hole is equal to

(*a*) $H + 0.29\,D$ ☐ (*b*) $H + 0.29\,D \times 2$ ☐

(*c*) $H + D$ ☐ (*d*) H. ☐

where D = Diameter of drill, and H = Depth of plate.

474. The total length (L) in drilling a blind hole is equal to

(*a*) $H + 0.29\,D$ ☐ (*b*) $H + 0.29\,D \times 2$ ☐

(*c*) $H + D$ ☐ (*d*) H. ☐

475. A hole of diameter 10 mm is to be drilled in a work-piece of thickness 20 mm. The length of approach will be equal to

(*a*) 0.29 mm ☐ (*b*) 2.9 mm ☐

(*c*) 4.0 mm ☐ (*d*) 4.50 mm. ☐

476. If in question 475, the r.p.m. of drill is 420 and length of approach in neglected then cutting speed in metre per minute will be equal to

(*a*) 10 m/min ☐ (*b*) 13.2 m/min ☐

(*c*) 20 m/min ☐ (*d*) 25 m/min. ☐

477. The time required for drilling a 20 mm hole in a work-piece of thickness 40 mm at a drill speed of 200 r.p.m. and feed = 0.2 mm/revolution when length of approach is neglected, will be equal to

(*a*) 2.50 min ☐ (*b*) 1.5 rain ☐

(*c*) 1.0 rain ☐ (*d*) 0.5 min. ☐

478. For a constant feed, the tool life increases as the cutting speed

(*a*) is increased ☐ (*b*) is decreased ☐

(*c*) is constant ☐ (*d*) none of the above. ☐

479. For a given rate of metal removal, the best tool life will be obtained by using the

(*a*) lowest possible feed ☐

(*b*) highest possible feed ☐

(*c*) average of the highest and lowest possible feed ☐

(*d*) none of the above. ☐

480. The intermediate speeds in case of drilling are chosen in

(*a*) Geometrical progression series ☐ (*b*) Arithmetical progression series ☐

(*c*) Logarithmic series ☐ (*d*) none of the above. ☐

481. The recommended values of speed for drilling magnesium alloys are

(*a*) 12 to 15 m/min ☐ (*b*) 20 to 23 m/min ☐

(*c*) 35 to 55 m/min ☐ (*d*) 60 to 105 m/min. ☐

482. The recommended values of speed for drilling stainless steels are

(*a*) 12 to 15 m/min ☐ (*b*) 20 to 23 m/min ☐

(*c*) 35 to 55 m/min ☐ (*d*) 60 to 105 m/min. ☐

483. The recommended values of speed for drilling aluminium alloys are

(*a*) 12 to 15 m/min ☐ (*b*) 20 to 23 m/min ☐

(*c*) 35 to 55 m/min ☐ (*d*) 60 to 105 m/min. ☐

484. The recommended values of speed for drilling grey cast iron are

(*a*) 12 to 15 m/min ☐ (*b*) 20 to 23 m/min ☐

(*c*) 35 to 55 m/min ☐ (*d*) 60 to 105 m/min. ☐

485. The relation between torque, diameter of drill and feed is given by

(*a*) $T = C \times f^{0.5} \times d^{1.5}$ ☐ (*b*) $T = C \times f^{0.75} \times d^{1.5}$ ☐

(*c*) $T = C \times f \times d^2$ ☐ (*d*) $T = C \times f^{0.75} \times d^{1.8}$ ☐

where C = Constant depending upon the material being drilled.

486. If in the relation between torque, diameter of drill and feed, the diameter of drill is in mm, feed in mm per revolution and torque in Newton metre, then the value of constant C for aluminium will be equal to

(*a*) 0.4 ☐ (*b*) 0.11 ☐

(*c*) 0.07 ☐ (*d*) 1.0. ☐

487. In question 486, the value of constant C for cast is equal to

(*a*) 0.4 ☐ (*b*) 0.11 ☐

(*c*) 0.007 ☐ (*d*) 1.0. ☐

488. In question 486, the value of constant C for carbon tool steel is equal

(*a*) 0.4 ☐ (*b*) 0.11 ☐

(*c*) 0.07 ☐ (*d*) 1.0. ☐

489. A hole of diameter 20 mm is to be drilled at a drill speed of 250 r.p.m. and a feed of 0.03 mm per revolution in a mild steel plate for which value of constant C = 0.36. Then the torque required in Newton metre will be equal to approximately

(*a*) 10.2 ☐ (*b*) 8.0 ☐
(*c*) 5.7 ☐ (*d*) 3.0. ☐

490. The drill spindles are provided with standard taper known as

(*a*) Seller's taper ☐ (*b*) Sharp taper ☐
(*c*) Morse taper ☐ (*d*) None of the above. ☐

491. The operation of smoothening and squaring the surface around a hole is known as

(*a*) reaming ☐ (*b*) spot facing ☐
(*c*) tapping ☐ (*d*) counter sinking. ☐

492. The operation of sizing and finishing a hole is known as

(*a*) reaming ☐ (*b*) spot facing ☐
(*c*) tapping ☐ (*d*) counter sinking. ☐

493. The operation of a cone-shaped enlargement of the end of a hole is known as

(*a*) reaming ☐ (*b*) spot facing ☐
(*c*) tapping ☐ (*d*) counter sinking. ☐

494. The operation of forming internal threads is known as

(*a*) reaming ☐ (*b*) spot facing ☐
(*c*) tapping ☐ (*d*) counter sinking. ☐

495. Which of the following operation is first performed?

(*a*) tapping ☐ (*b*) drilling ☐
(*c*) boring ☐ (*d*) spot facing. ☐

496. Choose the wrong statement

(*a*) Drilling is an example of oblique cutting. ☐
(*b*) The lead screw of a lathe has multi-start threads. ☐
(*c*) In drilling operation, the metal is removed by shearing and extrusion. ☐
(*d*) A 20 mm drilling machine means that it can drill a hole of maximum diameter 20 mm. ☐

497. A drill consists of a cylindrical piece of steel with special grooves. These grooves are generally called

(*a*) margins ☐ (*b*) lips ☐
(*c*) flutes ☐ (*d*) clearances. ☐

498. High helix drills are having helix angles of

(*a*) 10 to 15° ☐ (*b*) 20 to 30° ☐
(*c*) 35 to 40° ☐ (*d*) 50 to 70°. ☐

499. High helix drills are used for deep-hole drilling in

(*a*) hard rubber ☐ (*b*) bakelite ☐
(*c*) fibre asbestos ☐ (*d*) aluminium. ☐

500. The drills are made in

(*a*) number sizes ☐ (*b*) letter sizes ☐

(*c*) metric sizes ☐ (*d*) all of the above ☐

(*e*) none of the above. ☐

501. The main cutting edges of a drill that are formed by the intersection of the flank and flute surfaces are known as

(*a*) shank ☐ (*b*) lips ☐

(*c*) webs ☐ (*d*) none of the above. ☐

502. For efficient cutting, the lips should be

(*a*) straight ☐ (*b*) equal to length ☐

(*c*) symmetrical with the axis of the drill ☐ (*d*) all of the above ☐

(*e*) none of the above. ☐

503. The best angle to grind a drill for work on steel or cast iron is

(*a*) 70° ☐ (*b*) 59° ☐

(*c*) 40° ☐ (*d*) 30°. ☐

504. The best angle to grind a drill for work on extremely hard material is

(*a*) 70° ☐ (*b*) 59° ☐

(*c*) 40° ☐ (*d*) 30°. ☐

505. The best angle to grind a drill for work on soft material like fibre is

(*a*) 70° ☐ (*b*) 59° ☐

(*c*) 40° ☐ (*d*) 30°. ☐

506. With the increase of helix angle, the cutting edge becomes

(*a*) weak ☐ (*b*) strong ☐

(*c*) normal ☐ (*d*) none of the above. ☐

507. For harder material, helix angles are of the order of

(*a*) 50 to 60° ☐ (*b*) 35 to 45° ☐

(*c*) 20 to 30° ☐ (*d*) 10 to 13°. ☐

508. For softer material, the helix angles are of the order of

(*a*) 50 to 60° ☐ (*b*) 35 to 45° ☐

(*c*) 20 to 30° ☐ (*d*) 10 to 13°. ☐

509. The point angle for most of the material is taken as

(*a*) 140° ☐ (*b*) 118° ☐

(*c*) 80° ☐ (*d*) 60°. ☐

510. The point angle for hard and tough material is taken as

(*a*) 140° ☐ (*b*) 118° ☐

(*c*) 80° ☐ (*d*) 60°. ☐

511. The point angle for marble is taken as

(*a*) 140° ☐ (*b*) 118° ☐

(*c*) 80° ☐ (*d*) 60°. ☐

512. Which of the following material is used for making drills?

(*a*) high speed steel ☐ (*b*) carbon steel ☐
(*c*) carbide-tipped ☐ (*d*) any one of the above ☐
(*e*) none of the above. ☐

513. While grinding carbon-steel drill, if the drill is allowed to get hot enough, the drill will loose temper. The temper is indicated by the cutting edge showing

(*a*) red colour ☐ (*b*) blue colour ☐
(*c*) green colour ☐ (*d*) yellow colour. ☐

514. In grinding carbon-steel drill,

(*a*) no water should be used ☐ (*b*) plenty of water should be used ☐
(*c*) cold air should be used ☐ (*d*) none of the above. ☐

515. The average cutting speed with drill of high speed steel is

(*a*) half the cutting speed with drill of carbon steel ☐
(*b*) equal to the cutting speed with drill of carbon steel ☐
(*c*) twice the cutting speed with drill of carbon steel ☐
(*d*) thrice the cutting speed with drill of carbon steel. ☐

516. For drilling cast iron, the lubricant used

(*a*) kerosene ☐ (*b*) soda water ☐
(*c*) turpentine ☐ (*d*) dry. ☐

517. For drilling aluminium and copper, the lubricant used is

(*a*) kerosene ☐ (*b*) soda water ☐
(*c*) turpentine ☐ (*d*) dry. ☐

518. For drilling malleable carbon, the lubricant used is

(*a*) kerosene ☐ (*b*) soda water ☐
(*c*) turpentine ☐ (*d*) dry. ☐

519. For drilling brass and bronze, the lubricant used is

(*a*) kerosene ☐ (*b*) a flood of soluble oil ☐
(*c*) dry ☐ (*d*) soda water ☐
(*e*) (*b*) or (*c*) ☐ (*f*) none of the above. ☐

520. Choose the wrong statement

(*a*) If helix angle decreases, rake angle also decreases. ☐
(*b*) A very large rake angle makes a tightly rolled chips. ☐
(*c*) The point angle less than 118° is used for brittle materials. ☐
(*d*) Higher point angle reduces width of cut and produces thicker chips for same feed rate. ☐
(*e*) All of the above. ☐
(*f*) None of the above. ☐

521. For drilling brass, a drill with

(*a*) zero helix angle is required ☐ (*b*) low helix angle is required ☐
(*c*) high helix angle is required ☐ (*d*) any helix angle can be used. ☐

522. A twist drill is specified by its shank, material and

(*a*) diameter ☐ (*b*) lip angle ☐
(*c*) size of flute ☐ (*d*) length of body. ☐

523. For drilling aluminium, a drill with

(*a*) zero helix angle is required ☐ (*b*) low helix angle is required ☐
(*c*) high helix angle is required ☐ (*d*) any helix angle can be used. ☐

524. The tool used to withdraw a drill from its sleeve, is known as

(*a*) drift ☐ (*b*) drill puller ☐
(*c*) key ☐ (*d*) drill remover. ☐

525. For drilling aluminium with drill of carbon steel, the average cutting speed should be

(*a*) 10 m/min ☐ (*b*) 15 m/min ☐
(*c*) 30 m/min ☐ (*d*) 45 m/min. ☐

526. For drilling brass and copper with drill of carbon steel, the average cutting speed should be

(*a*) 10 m/min ☐ (*b*) 15 m/min ☐
(*c*) 30 m/min ☐ (*d*) 45 m/min. ☐

527. For drilling high carbon steel with drill of carbon steel, the average cutting speed should be

(*a*) 10 m/min ☐ (*b*) 15 m/min ☐
(*c*) 30 m/min ☐ (*d*) 45 m/min. ☐

528. For drilling very soft steel and soft grey iron with drill of carbon steel, the average cutting speed should be

(*a*) 10 m/min ☐ (*b*) 15 m/min ☐
(*c*) 30 m/min ☐ (*d*) 45 m/min. ☐

529. A dimension is expressed as $20.5^{+00.0}_{-00.2}$. This is a case of

(*a*) bilateral tolerance ☐ (*b*) unilateral tolerance ☐
(*c*) tri-lateral tolerance ☐ (*d*) none of the above. ☐

530. In grinding operation, for grinding harder material

(*a*) fine grain size is used ☐ (*b*) coarser grain size is used ☐
(*c*) medium grain size is used ☐ (*d*) none of the above. ☐

531. The method of removing metal by a tool that has successively higher cutting edges in a fixed path, is called

(*a*) drilling ☐ (*b*) broaching ☐
(*c*) milling ☐ (*d*) sawing. ☐

532. Which of the following processes are used for completing job in one stroke of the machine?

(*a*) milling ☐ (*b*) drilling ☐
(*c*) broaching ☐ (*d*) boring. ☐

533. The accuracy of surface finish in case of broaching is of the order of

(*a*) 8.0 microns ☐ (*b*) 0.8 microns ☐
(*c*) 0.08 microns ☐ (*d*) 0.008 microns. ☐

534. On micron is equal to

(*a*) 0.1 mm ☐ (*b*) 0.01 mm ☐

(*c*) 0.001 mm ☐ (*d*) 1×10^{-6} mm. ☐

535. The maximum cut per tooth in case of broaching is

(*a*) 1 mm ☐ (*b*) 0.5 mm ☐

(*c*) 0.15 mm ☐ (*d*) 0.01 mm. ☐

536. A broach has

(*a*) finishing teeth ☐ (*b*) roughing teeth ☐

(*c*) semi-finishing teeth ☐ (*d*) all of the above ☐

(*e*) none of the above. ☐

537. One which of the following parameters, the face angle of the tooth of a broaching tool depends?

(*a*) material to be cut ☐ (*b*) toughness of the material ☐

(*c*) hardness of the material ☐ (*d*) ductility of the material ☐

(*e*) all of the above ☐ (*f*) none of the above. ☐

538. Which of the following pitch leads to chatler in case of broaching?

(*a*) fine pitch ☐ (*b*) large pitch ☐

(*c*) uniform pitch ☐ (*d*) all of the above. ☐

539. The face-angles of the teeth of a broaching tool for cast iron is approximately equal to

(*a*) 6 to 8° ☐ (*b*) 10° ☐

(*c*) 15° ☐ (*d*) – 5 to + 5°. ☐

540. The face angles of the teeth of a broaching tool for brass is approximately equal to

(*a*) 6 to 8° ☐ (*b*) 10° ☐

(*c*) 15° ☐ (*d*) – 5 to + 5°. ☐

541. The pitch for internal plain broaches is equal to

(*a*) (1.5 to 2.0) $\sqrt{\text{Length of cut in mm}}$ ☐ (*b*) (1.24 to 1.50) $\sqrt{\text{Length of cut in mm}}$ ☐

(*c*) 3 $\sqrt{\text{Length of cut in mm}}$ ☐ (*d*) 4.5 $\sqrt{\text{Length of cut in mm}}$. ☐

542. Choose the wrong statement

(*a*) In case of broaching, the job is completed in one stroke of the machine. ☐

(*b*) Broaching is used for straight or irregular surfaces either externally or internally. ☐

(*c*) Broaching is a metal cutting operation in which rate of production is high. ☐

(*d*) The operation of broaching is simple and tooling cost is low and hence it is used for mass production. ☐

543. Choose the correct statement

(*a*) In case of broaching, clearance angle is chosen independent of the work material. ☐

(*b*) For proper alignment between the work-piece and broach, at least two teeth and preferably three should be in contact at all times. ☐

(*c*) The length of cut and chip thickness in case of broaching depends upon the pitch of the broaching tool. ☐

(d) All of the above. ☐ (e) None of the above. ☐

544. The pitch for rotary cut broaches is equal to

(a) (1.5 to 2.0) $\sqrt{\text{Length of cut in mm}}$ ☐ (b) (1.25 to 1.5) $\sqrt{\text{Length of cut in mm}}$ ☐

(c) 3.0 $\sqrt{\text{Length of cut in mm}}$ ☐ (d) 4.5 $\sqrt{\text{Length of cut in mm}}$. ☐

545. The finishing teeth of a broaching tool are provided with

(a) large amount of land ☐ (b) small amount of land ☐

(c) average amount of land ☐ (d) none of the above. ☐

546. The broaching operation in which either the work or the tool moves across the other, is known as

(a) push broaching ☐ (b) pull broaching ☐

(c) continuous broaching ☐ (d) surface broaching. ☐

547. The ratio of gullet area to chip cross-section for surface broach for roughing is taken as

(a) 2.0 ☐ (b) 3 to 5.0 ☐

(c) 8.0 ☐ (d) 12.0. ☐

548. The ratio of gullet area to chip cross-section for surface broach for finishing is taken as

(a) 2.0 ☐ (b) 3 to 5.0 ☐

(c) 8.0 ☐ (d) 12.0. ☐

549. If the teeth of the broach are at an angle of 5 to 20 degrees to the direction of travel of the broach, then

(a) smooth shear cutting is produced ☐ (b) finer finish is produced ☐

(c) vibrations are eliminated ☐ (d) all of the above ☐

(e) none of the above. ☐

550. The range of hardness of a material upto which broaching can be employed is

(a) 8 to 10 Rockwell C ☐ (b) 12 to 22 Rockwell C ☐

(c) 25 to 40 Rockwell C ☐ (d) 40 to 50 Rockwell C. ☐

551. The primary aim of broaching is

(a) high finish ☐ (b) economy in production ☐

(c) removal of maximum amount of stock ☐ (d) none of the above. ☐

552. Choose the wrong statement

(a) The length of the broach changes with the cut per tooth and the pitch. ☐

(b) The cost of broaching operation is very high as compared to other machine tool operations. ☐

(c) In surface broaching, the job and the broaching tool move across each other. ☐

(d) The process of broaching is best suited for the short ran jobs and for the removal of large amount of stock. ☐

553. The effective length of a broach is given by

(a) $E_l = \dfrac{C_t \times C_d}{p}$ ☐ (b) $E_l = \dfrac{C_d}{C_t \times p}$ ☐

(c) $E_l = \dfrac{p}{C_t \times C_d}$ ☐ (d) $E_l = \dfrac{p \times C_t}{C_d}$. ☐

where E_l = Effective length of broach, C_t = Average cut per tooth,

C_d = Depth of cut, and p = Pitch of broach tooth.

554. The effective length of a broach (having average amount of stock removed per tooth = 0.05 mm, pitch = 12 mm and material to be removed = 3 mm) will be equal to

(*a*) 4.0 mm ☐ (*b*) 5.0 mm ☐
(*c*) 5.5 mm ☐ (*d*) 6.0 mm. ☐

555. The operation of cutting a metal to some specified length before the metal is presented to a machine tool, is known as

(*a*) broaching ☐ (*b*) milling ☐
(*c*) sawing ☐ (*d*) boring. ☐

556. The average thickness of the cut in one travel of the circular saw is

(*a*) 1.5 mm ☐ (*b*) 3.0 mm ☐
(*c*) 6.25 mm ☐ (*d*) 8.0 mm. ☐

557. The average thickness of the cut in one travel of the band saw is

(*a*) 1.5 mm ☐ (*b*) 3.0 mm ☐
(*c*) 6.25 mm ☐ (*d*) 8.0 mm. ☐

558. The average thickness of the cut in one travel of the reciprocating saw is

(*a*) 1.5 mm ☐ (*b*) 3.0 mm ☐
(*c*) 6.25 mm ☐ (*d*) 8.0 mm. ☐

559. Which one of the following saw is having average thickness of cut in one travel as 6.25 mm?

(*a*) band saw ☐ (*b*) reciprocating saw ☐
(*c*) circular saw ☐ (*d*) none of the above. ☐

560. Which one of the following saw is having average thickness of the cut in one travel as 1.5 mm?

(*a*) band saw ☐ (*b*) circular saw ☐
(*c*) reciprocating saw ☐ (*d*) none of the above. ☐

561. The height of each tooth of a broach is

(*a*) in progressively decreasing order ☐ (*b*) in progressively increasing order ☐
(*c*) same throughout ☐ (*d*) none of the above. ☐

562. The rear teeth of a broach

(*a*) remove maximum metal ☐ (*b*) remove minimum metal ☐
(*c*) remove no metal ☐ (*d*) none of the above. ☐

563. Which of the following process is used for making a complicated contour in a carbide piece?

(*a*) plasma-arc machining ☐ (*b*) electrochemical milling ☐
(*c*) electro-discharge machining ☐ (*d*) laser machining. ☐

564. If the dimension is expressed as $30^{+0.045}_{-0.035}$, then tolerance is

(*a*) 0.045 mm ☐ (*b*) 0.035 mm ☐
(*c*) 0.08 mm ☐ (*d*) 0.01 mm. ☐

565. In the above question, the basic size is

(*a*) 30.045 mm ☐ (*b*) 30.035 mm ☐
(*c*) 30.00 mm ☐ (*d*) 29.90 mm. ☐

566. Choose the wrong statement

(*a*) Cast iron cannot be fabricated by welding ☐

(*b*) Mild steel and other alloys can be fabricated by welding ☐

(*c*) The bed, column and frame of machine tools are made of cast iron ☐

(*d*) The shock absorbing capacity of cast iron is very low. ☐

567. In order to obtain a high surface finish the machine should

(*a*) have vibrations of maximum amplitude ☐ (*b*) have vibration of minimum amplitude ☐

(*c*) be free from vibrations ☐ (*d*) none of the above. ☐

568. For selecting guideway materials, which of the following factors should be considered?

(*a*) damping capacity ☐ (*b*) wear resistance ☐

(*c*) strength ☐ (*d*) all of the above ☐

(*e*) none of the above. ☐

569. If the cutting speed is constant, then with the increase of diameter of the job, the speed of spindle

(*a*) increases ☐ (*b*) decreases ☐

(*c*) remains constant ☐ (*d*) first increases then decreases. ☐

570. For a constant speed of spindle, the relation between cutting speed and diameter of the job is

(*a*) parabolic ☐ (*b*) a straight line ☐

(*c*) hyperbolic ☐ (*d*) none of the above. ☐

571. Ray diagram is a graph showing the relation between

(*a*) cutting speed and speed of spindle ☐

(*b*) cutting speed and diameter of work-piece ☐

(*c*) diameter of work-piece and spindle speed ☐

(*d*) cutting speed, diameter of job and spindle speed. ☐

572. Figure 12.4 shows the variations of spindle speed and cutting speed with change of job diameter. The variation of cutting speed at constant spindle speed with the change of job diameter is given by

(*a*) curve *A* ☐

(*b*) curve *B* ☐

(*c*) curve *C* ☐

(*d*) curve *D*. ☐

573. In Fig. 12.4, the variation of spindle speed at constant cutting speed with the change of job diameter is given by

(*a*) curve *A* ☐

(*b*) curve *B* ☐

(*c*) curve *C* ☐

(*d*) curve *D*. ☐

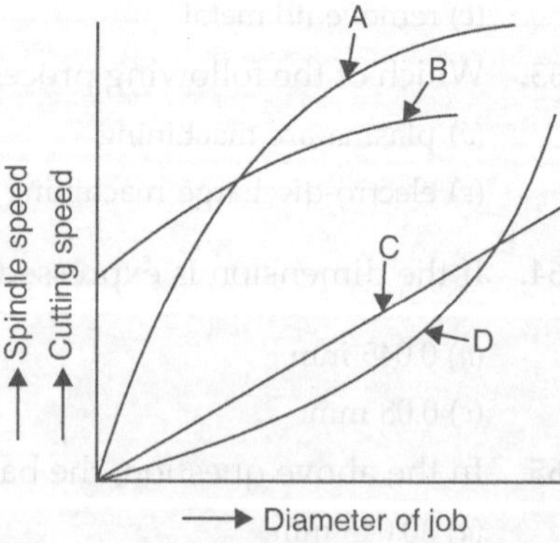

FIGURE 12.4

574. Choose the wrong statement

(*a*) Concrete is a good vibration damping material. ☐

(*b*) The thrust bearings are located close to the spindle nose to avoid the possibility of buckling. ☐

(*c*) The minimum diameter of the job is taken = $h/8$, where h being the height of the centre above the bed. ☐

(*d*) Cast iron has high tensile strength. ☐

575. The variation of spindle speed with diameter of job at constant cutting speed form a series which is in

(*a*) arithmetical progression ☐ (*b*) geometrical progression ☐

(*c*) logarithmic progression ☐ (*d*) none of the above. ☐

576. A series of numbers in a geometric progression, which is used for standardisation in preference to any other random number, is known as

(*a*) best number ☐ (*b*) preferred numbers ☐

(*c*) basic number ☐ (*d*) none of the above. ☐

577. Which two of the following parameters need standardisation for main drives?

(*a*) power ☐ (*b*) speed range of the main drive ☐

(*c*) diameter of job ☐ (*d*) length of job ☐

(*e*) only (*a*) and (*b*) ☐ (*f*) only (*c*) and (*d*). ☐

578. Which of the following factors are effected by vibrations?

(*a*) work-piece ☐ (*b*) machine tool ☐

(*c*) tool-life ☐ (*d*) cutting conditions ☐

(*e*) all of the above ☐ (*f*) none of the above. ☐

579. The common ratio ϕ for spindle speeds is geometrical progression lies between

(*a*) 2.0 to 4.0 ☐ (*b*) 1.0 to 2.0 ☐

(*c*) 0.5 to 1.0 ☐ (*d*) 0.1 to 0.5. ☐

580. The most commonly used ratio, according to Renard series is

(*a*) 2.0 ☐ (*b*) 1.56 ☐

(*c*) 1.41 ☐ (*d*) 1.26 ☐

(*e*) 1.12 ☐ (*f*) all of the above. ☐

581. The series according to 'Androin Progression Ratio' is given by

(*a*) $\sqrt{2}, \sqrt{4}, \sqrt{8}, \sqrt{16}$ ☐ (*b*) $\sqrt{3}, \sqrt{9}, \sqrt{27}, \sqrt{81}$ ☐

(*c*) $10, 10^2, 10^3, 10^4$ ☐ (*d*) $\sqrt[5]{10}, \sqrt[10]{10}, \sqrt[20]{10}, \sqrt[40]{10}$. ☐

582. For a cutting tool, the spindle speeds are 160, 229, 328, 469 ... The next higher speed will be

(*a*) 800 ☐ (*b*) 704 ☐

(*c*) 671 ☐ (*d*) 643. ☐

583. In case of turning, as the machining proceeds, with the decrease in diameter of work, the spindle speed should

(*a*) increase ☐ (*b*) decrease ☐

(*c*) remain constant ☐ (*d*) increase first and then decrease. ☐

584. Choose the wrong statement

(*a*) Due to vibration, the surface finish obtained will be very poor. ☐

(*b*) Due to vibrations, the tool life is decreased by about 70 to 80% of the normal value. ☐

(*c*) Due to vibrations in machine tools, the chip thickness removed by the cutting tool is constant. ☐

(*d*) In dynamically stable system, the amplitude of vibration keeps on decaying with time. ☐

585. In electro-forming process, first a negative image of the part is prepared. This negative image of the part is called

(*a*) mould ☐ (*b*) matrix ☐

(*c*) pattern ☐ (*d*) any one of the above ☐

(*e*) none of the above. ☐

586. The substance used in case of expendable type of moulds is

(*a*) conducting ☐ (*b*) non-conducting ☐

(*c*) any one of the above ☐ (*d*) none of the above. ☐

587. Which one of the following metals is best suited for electro-forming?

(*a*) zinc ☐ (*b*) lead ☐

(*c*) gold ☐ (*d*) silver. ☐

588. The best suited metal for electro-forming is one which has the property of

(*a*) good bearing surface ☐ (*b*) good reproducibility ☐

(*c*) resistance to corrosion ☐ (*d*) adequate strength for small thickness ☐

(*e*) all of the above ☐ (*f*) none of the above. ☐

589. Which one is the correct statement about electro-forming?

(*a*) This is an economical method of production. ☐

(*b*) Rate of production is high. ☐

(*c*) High surface finish can be obtained. ☐

(*d*) Process is limited only to very thick parts. ☐

590. Which one is the wrong statement about electro-forming?

(*a*) High dimensional accuracy in complicated shaped parts can be obtained. ☐

(*b*) Parts of any thickness can be easily made. ☐

(*c*) Any amount of purity can be achieved. ☐

(*d*) Cost of production is very low. ☐

591. The feed in drilling operation is expressed as

(*a*) mm/revolution ☐ (*b*) mm ☐

(*c*) mm/sec ☐ (*d*) mm/min. ☐

592. Steels, in which no gas evolution occurs on solidification, are known as

(*a*) stainless steels ☐ (*b*) killed steels ☐

(*c*) carbon steels ☐ (*d*) alloy steels. ☐

593. A steel, in which an element like lead or sulphur is added to promote rapid machining, is known as

(*a*) killed steel ☐ (*b*) stainless steel ☐

(*c*) free-cutting steel ☐ (*d*) none of the above. ☐

594. Gears which work under low specific pressure and low velocities (*i.e.*, gears of relatively small power transmitting capacity) are made of

(*a*) alloy steel ☐ (*b*) carburised steel ☐
(*c*) carbon steel ☐ (*d*) mild steel. ☐

595. Gears which work under heavier service conditions (*i.e.*, gears working under high unit stresses and high velocities at the pitch diameter) are made of

(*a*) alloy steel ☐ (*b*) carburised steel ☐
(*c*) carbon steel ☐ (*d*) mild steel. ☐

596. Gears which work under severe conditions and also likely to be subjected to shocks, are made of

(*a*) alloy steel ☐ (*b*) carburised steel ☐
(*c*) carbon steel ☐ (*d*) mild steel. ☐

597. Drills should be hardened

(*a*) at the tip only ☐ (*b*) upto one-fourth of its length ☐
(*c*) upto half of its length ☐ (*d*) throughout its length. ☐

598. Plain high carbon steel are tempered in order to

(*a*) reduce brittleness in hardened steel ☐ (*b*) improve toughness ☐
(*c*) reduce the internal stresses ☐ (*d*) all of the above ☐
(*e*) none of the above. ☐

599. The tool force measuring apparatus is called

(*a*) pyrometer ☐ (*b*) dynamometer ☐
(*c*) radiator ☐ (*d*) reflector. ☐

600. The natural frequency of dynamometer, so that the recorded force for actual cut may not be influenced by any vibrating motion of the dynamometer, should be at least

(*a*) equal to the frequency of the exciting vibrations ☐
(*b*) half of the frequency of the exciting vibrations ☐
(*c*) two times the frequency of the exciting vibrations ☐
(*d*) four times the frequency of the exciting vibrations. ☐

601. A dynamometer should be stable with respect to

(*a*) temperature only ☐ (*b*) time only ☐
(*c*) humidity only ☐ (*d*) all of the above ☐
(*e*) none of the above. ☐

602. Which of the following are the devices to measure small deflections in dynamometers?

(*a*) electric transducers ☐ (*b*) piezoelectric crystals ☐
(*c*) dial indicator ☐ (*d*) optical devices ☐
(*e*) hydraulic pressure cells ☐ (*f*) all of the above. ☐

603. Choose the correct statement

(*a*) The object becomes more tough if tempering is done at low temperature. ☐
(*b*) The object retains its hardness if tempering is done at high temperature. ☐
(*c*) The combination of hardness and toughness is provided by case-hardening. ☐
(*d*) The process of measuring high temperature is known as dynamometry. ☐

604. Choose the wrong statement

(*a*) The heat treatment of screw taps is done throughout its length. ☐

(*b*) The carbon, that must be present in a steel article which may be hardened sufficiently by heat treatment, must be at least 0.2%. ☐

(*c*) Arrest points are temperatures at which a change in phase occurs. ☐

(*d*) By carburising process, the components can be hardened up to a depth of 2 mm. ☐

605. Dial indicator can measure the deflections up to

(*a*) 25×10^{-10} mm ☐ (*b*) 25×10^{-8} mm ☐

(*c*) 25×10^{-6} mm ☐ (*d*) 25×10^{-4} mm. ☐

606. The main actions of cutting fluids are

(*a*) friction reduction ☐

(*b*) cooling ☐

(*c*) washing away the chip ☐

(*d*) reduction of the shear-strength of work material ☐

(*e*) all of the above ☐

(*f*) none of the above. ☐

607. The cutting fluid used for providing cooling action to the work-piece and tool and to wash away the chips, is

(*a*) mineral oil ☐ (*b*) chlorinated cutting oils ☐

(*c*) aqueous solution ☐ (*d*) water. ☐

608. The advantages of using cutting fluids are

(*a*) low power consumption ☐ (*b*) better dimensional accuracy ☐

(*c*) increased tool life ☐ (*d*) better surface finish ☐

(*e*) all of the above ☐ (*f*) none of the above. ☐

609. In die cutting, when the punch presses the steel metal, the bottom-most layers of the sheet in die are subjected to

(*a*) compressive stresses ☐ (*b*) tensile stresses ☐

(*c*) shear stresses ☐ (*d*) bending stresses. ☐

610. In question 609, the layers below the punch are subjected to

(*a*) compressive stresses ☐ (*b*) tensile stresses ☐

(*c*) shear stresses ☐ (*d*) bending stresses. ☐

611. The magnitude of the cutting pressure on the cutting dies is equal to

(*a*) $\frac{f_s}{L \times t}$ ☐ (*b*) $\frac{f_s}{L}$ ☐

(*c*) $f_s \times L \times t$ ☐ (*d*) $\frac{L \times t}{f_s}$ ☐

where f_s = Shear strength of work-piece, t = Thickness of work-piece,

L = Periphery of work-piece.

612. The total cutting forces in case of cutting dies depends upon
(*a*) angle of shear on die and punch ☐ (*b*) sharpness of cutting edge ☐
(*c*) clearance between die and punch ☐ (*d*) hardness of the material to be cut ☐
(*e*) all of the above ☐ (*f*) none of the above. ☐

613. The instantaneous fracture of the metal after its plastic deformation in case of die cutting operation is known as
(*a*) depth of cut ☐ (*b*) penetration ☐
(*c*) creep ☐ (*d*) feed. ☐

614. The value of penetration for mild steel in case of die cutting operation is
(*a*) 70% of punch stroke ☐ (*b*) 50% of punch stroke ☐
(*c*) 33.33% of punch stroke ☐ (*d*) 20% of punch stroke. ☐

615. The force acting on the punch goes on increasing if angle of shear
(*a*) increases ☐ (*b*) decreases ☐
(*c*) remains constant ☐ (*d*) none of the above. ☐

616. The average force to cut any work-piece in case of die cutting operation is equal to
(*a*) one-fourth of the maximum force ☐ (*b*) one-third of the maximum force ☐
(*c*) one-half of the maximum force ☐ (*d*) two-third of the maximum force. ☐

617. Which of the following methods are used for manufacturing gears?
(*a*) casting ☐ (*b*) stamping ☐
(*c*) extruding ☐ (*d*) power-metallurgy ☐
(*e*) all of the above ☐ (*f*) none of the above. ☐

618. For making heavy gears of cast iron and steel, which of the following methods is particularly used?
(*a*) sand casting ☐ (*b*) stamping ☐
(*c*) extruding ☐ (*d*) power-metallurgy ☐

619. Gears made by sand casting
(*a*) have poor accuracy ☐
(*b*) are mostly used for slow speed drive ☐
(*c*) are not very efficient in power transmission ☐
(*d*) all of the above ☐
(*e*) none of the above. ☐

620. Die-casting gears
(*a*) are used for transmission of light loads ☐
(*b*) are made from low melting temperature metals ☐
(*c*) are used for high speed drives ☐
(*d*) are very efficient in power transmission ☐
(*e*) both (*a*) and (*b*) ☐
(*f*) both (*c*) and (*d*). ☐

621. The gears used in watches, clocks, toys etc. are made by
(*a*) casting ☐ (*b*) stamping ☐
(*c*) extruding ☐ (*d*) coining. ☐

622. For making thin gears from sheet metals up to thickness 3 mm, the method used is
(*a*) casting ☐ (*b*) stamping ☐
(*c*) extruding ☐ (*d*) coining. ☐

623. Which of the following material is used for manufacturing gears by power-metallurgy?
(*a*) cast iron ☐ (*b*) steel ☐
(*c*) bakelite ☐ (*d*) plastic materials ☐
(*e*) both (*a*) and (*b*) ☐ (*f*) both (*c*) and (*d*). ☐

624. The method of manufacturing gears by extrusion is used for
(*a*) bevel gears ☐ (*b*) worm gears ☐
(*c*) spur gears ☐ (*d*) helical gears. ☐

625. Which one of the following process is not used for manufacturing gears?
(*a*) casting ☐ (*b*) extruding ☐
(*c*) milling ☐ (*d*) stamping. ☐

626. Which of the following process is used for machining gears from the blanks?
(*a*) milling ☐ (*b*) broaching ☐
(*c*) formed tooth process ☐ (*d*) all of the above ☐
(*e*) none of the above. ☐

627. Which of the following methods are considered as generating type of gear manufacturing methods?
(*a*) milling ☐ (*b*) broaching ☐
(*c*) hobbing ☐ (*d*) shear cutting ☐
(*e*) both (*a*) and (*b*) ☐ (*f*) both (*c*) and (*d*). ☐

628. Which of the following methods are considered as forming type of gear manufacturing methods?
(*a*) milling ☐ (*b*) broaching ☐
(*c*) hobbing ☐ (*d*) shear cutting ☐
(*e*) both (*a*) and (*b*) ☐ (*f*) both (*c*) and (*d*). ☐

629. Milling machines can cut
(*a*) spur gear ☐ (*b*) helical gear ☐
(*c*) worm gear ☐ (*d*) all of the above ☐
(*e*) none of the above. ☐

630. The hobbing process of cutting gears as compared to that of shaping is
(*a*) slower but more accurate ☐ (*b*) slower and less accurate ☐
(*c*) faster but less accurate ☐ (*d*) faster and more accurate. ☐

631. A process of coating zinc by hot dipping is known as
(*a*) anodising ☐ (*b*) galvanising ☐
(*c*) sheradising ☐ (*d*) brazing. ☐

632. A zinc diffusion process is known as
(*a*) anodising ☐ (*b*) galvanising ☐
(*c*) sheradising ☐ (*d*) brazing. ☐

633. An oxidising process, used for aluminium and magnesium articles, is known as

(*a*) anodising ☐ (*b*) galvanising ☐
(*c*) sheradising ☐ (*d*) brazing. ☐

634. The process of coating a thin layer of phosphate on steel is known as

(*a*) anodising ☐ (*b*) parkerising ☐
(*c*) galvanising ☐ (*d*) brazing. ☐

635. Choose the wrong statement

(*a*) Milling is not a production process. ☐
(*b*) The initial cost of the cutter of the milling machines is lower than cutters for hobbing and shaping processes. ☐
(*c*) Cutter No. 4 is suitable for cutting gears with 35 to 54 teeth. ☐
(*d*) Internal gears can be very conveniently cut by a broach tool. ☐

636. Numerical control can be applied to

(*a*) drilling machines ☐ (*b*) milling machines ☐
(*c*) lathes ☐ (*d*) all of the above machines. ☐

637. Machining centre is a

(*a*) automatic tool changing unit ☐
(*b*) group of automatic machine tools ☐
(*c*) next logical step beyond NC (numerical controlled) machine ☐
(*d*) NC machine tool. ☐

638. The numerical control machines are controlled by the tape whose width is

(*a*) 50 mm ☐ (*b*) 40 mm ☐
(*c*) 30 mm ☐ (*d*) 20 mm. ☐

639. *NC* machine tool is operated by

(*a*) output-input modules ☐ (*b*) a series of coded instructions ☐
(*c*) feed back system ☐ (*d*) none of the above. ☐

640. Choose the wrong statement.

(*a*) The hobbing process of cutting gears as compared to that of shaping is faster and more accurate. ☐
(*b*) The gears used in watches, clocks, toys, etc. are made by stamping. ☐
(*c*) Galvanising is a zinc diffusion process. ☐
(*d*) The job is completed in one stroke of the machine, in case of broaching. ☐

Tick mark the true or false statement:

641. Ultrasonic machining is best suited for machining hard and brittle materials.

(*a*) True ☐ (*b*) False. ☐

642. In electro-discharge machining, dielectric is used.

(*a*) True ☐ (*b*) False. ☐

643. The tool is connected to anode, in electro-discharge machining.

(*a*) True ☐ (*b*) False. ☐

644. The word EDM stands for electro-discharge machining.

(*a*) True ☐ (*b*) False. ☐

645. The word ECM stands for electro-cathode machining.

(*a*) True ☐ (*b*) False. ☐

646. EDM machining produces high degree of surface finish.

(*a*) True ☐ (*b*) False. ☐

647. Chemical milling is used for preparing parts having large curved surfaces and thin sections.

(*a*) True ☐ (*b*) False. ☐

648. Chemical milling operation is performed in a tank containing an etching solution.

(*a*) True ☐ (*b*) False. ☐

649. The metal is removed by maintaining an electrolyte between the work and tool in a very small gap between the two in electrochemical machining.

(*a*) True ☐ (*b*) False. ☐

650. Galvanising is a zinc diffusion process.

(*a*) True ☐ (*b*) False. ☐

651. The gears are best produced on mass production by casting.

(*a*) True ☐ (*b*) False. ☐

652. During machining of hard metals, discontinuous chips are formed.

(*a*) True ☐ (*b*) False. ☐

653. During machining of ductile metals, continuous chips with built-up edges are formed.

(*a*) True ☐ (*b*) False. ☐

654. The low cutting speed and small rake angle is responsible for the formation of discontinuous chips.

(*a*) True ☐ (*b*) False. ☐

655. The formation of chip is due to shear deformation.

(*a*) True ☐ (*b*) False. ☐

656. Three-dimensional cutting system is also known as oblique cutting system.

(*a*) True ☐ (*b*) False. ☐

657. The chip flows over the tool face and the direction of the chip flow velocity is inclined with the normal to the cutting edge in orthogonal cutting system.

(*a*) True ☐ (*b*) False. ☐

658. If the metals are ductile and cutting speed is high, then continuous chips are formed.

(*a*) True ☐ (*b*) False. ☐

659. The relief angles on carbide tools are kept as large as possible.

(*a*) True ☐ (*b*) False. ☐

660. The relief or clearance angles are never made zero or negative.

(*a*) True ☐ (*b*) False. ☐

661. The lip angle of a single point tool is usually between 60° to 80°.

(*a*) True ☐ (*b*) False. ☐

662. The negative rake angle is usually provided on carbide tipped tools.
(*a*) True ☐ (*b*) False. ☐

663. For drilling brass or bronze, a drill with low helix angle is used.
(*a*) True ☐ (*b*) False. ☐

664. For drilling hard steel, the point angle of a drill is 90°.
(*a*) True ☐ (*b*) False. ☐

665. The commonly used valve of point angle for a standard twist drill is 180°.
(*a*) True ☐ (*b*) False. ☐

666. The only angle on which the strength of the tool depends as rake angle.
(*a*) True ☐ (*b*) False. ☐

667. The negative rake angles are used to increase the strength of the cutting tool point, to decrease the temperature rise at the tool-tip and to give better finish.
(*a*) True ☐ (*b*) False. ☐

668. For tougher materials like copper positive rake angle is provided.
(*a*) True ☐ (*b*) False. ☐

669. For brittle materials, zero rake angle is provided.
(*a*) True ☐ (*b*) False. ☐

670. Mild steel during machining produces discontinuous chips.
(*a*) True ☐ (*b*) False. ☐

671. When cutting cast iron, the type of chip produced is discontinuous.
(*a*) True ☐ (*b*) False. ☐

672. Best coolant and lubricant for steel and wrought iron is water soluble oils.
(*a*) True ☐ (*b*) False. ☐

673. The cutting edge of the tool is perpendicular to the direction of tool travel in orthogonal cutting system.
(*a*) True ☐ (*b*) False. ☐

674. In oblique cutting system, the maximum chip thickness occurs at the middle.
(*a*) True ☐ (*b*) False. ☐

675. In orthogonal cutting system, three components of the cutting forces, which are mutually perpendicular are acting at the cutting edge.
(*a*) True ☐ (*b*) False. ☐

676. The maximum chip thickness occurs at the middle in case of orthogonal cutting system.
(*a*) True ☐ (*b*) False. ☐

677. For drilling zinc alloys or magnesium, a drill with zero helix angle is used.
(*a*) True ☐ (*b*) False. ☐

678. Chip thickness ratio is also known as cutting ratio.
(*a*) True ☐ (*b*) False. ☐

679. If the lip relief angle on drill is zero, then drill would not cut.
(*a*) True ☐ (*b*) False. ☐

680. The commonly used material for a standard twist drill is H.S.S.

(*a*) True ☐ (*b*) False. ☐

681. Cutting ratio is defined as the ratio of depth of cut to chip thickness.

(*a*) True ☐ (*b*) False. ☐

682. The chip compression factor is the reciprocal of chip thickness ratio.

(*a*) True ☐ (*b*) False. ☐

683. The shear angle ϕ in terms of chip thickness ratio r and rake angle of the tool a is given by

$$\tan \phi = (r \cos \alpha)/(1 - r \sin \alpha).$$

(*a*) True ☐ (*b*) False. ☐

684. Tool life is the time between tool re-sharpenings.

(*a*) True ☐ (*b*) False. ☐

685. The equation $V/T = C$ represents the relationship between tool life (T) and cutting speed (V)

(*a*) True ☐ (*b*) False. ☐

686. Tumbler gears in lathe are used to reduce the spindle speed.

(*a*) True ☐ (*b*) False. ☐

687. The different spindle speeds on a lathe form geometrical progression.

(*a*) True ☐ (*b*) False. ☐

688. The lead screw of *A* lathe has single start threads.

(*a*) True ☐ (*b*) False. ☐

689. The lathe centres are provided with standard tapers known as Morse taper.

(*a*) True ☐ (*b*) False. ☐

690. The conventional milling is also called climb nulling.

(*a*) True ☐ (*b*) False. ☐

691. The cutting tool is mounted on arbor in a milling machine.

(*a*) True ☐ (*b*) False. ☐

692. In climb milling the chips are deposited behind the work zone and out of the way of cutter.

(*a*) True ☐ (*b*) False. ☐

693. The surface finish improves, if cutting speed in increased.

(*a*) True ☐ (*b*) False. ☐

694. Increase in true rake angle, deteriorates the surface finish.

(*a*) True ☐ (*b*) False. ☐

695. Increase in feed rate, improves the surface finish.

(*a*) True ☐ (*b*) False. ☐

696. There is no effect on surface finish, if the depth of cut is increased.

(*a*) True ☐ (*b*) False. ☐

697. Increase in nose radius improves the surface finish.

(*a*) True ☐ (*b*) False. ☐

698. Notching is the operation of removal of excess metal from the edge of a strip to make it suitable for drawing without wrinkling.

(*a*) True ☐ (*b*) False. ☐

699. Slugging is the operation of cutting in a single line across a part of the metal strip to allow bending or forming in progressive die operation while the part remains attached to the strip.

(*a*) True ☐ (*b*) False. ☐

700. The operation of cutting of the excess metal at edge which was required for gripping purposes during press operation, is known at trimming.

(*a*) True ☐ (*b*) False. ☐

701. Crater wear occurs mainly on the nose part, front relief face and side relief face of the cutting tool.

(*a*) True ☐ (*b*) False. ☐

702. The metal flows due to plasticity in drawing operation.

(*a*) True ☐ (*b*) False. ☐

703. The process of cleaning the surface of small parts is known as tumbling.

(*a*) True ☐ (*b*) False. ☐

704. The cutting force at the cutting edge is measured by pyrometer.

(*a*) True ☐ (*b*) False. ☐

705. In machining cast iron, no cutting fluid is required.

(*a*) True ☐ (*b*) False. ☐

706. A jig does not guide the tool.

(*a*) True ☐ (*b*) False. ☐

707. The process of improving the cutting action of the grinding wheel is known as truing.

(*a*) True ☐ (*b*) False. ☐

708. Laser is produced by graphite.

(*a*) True ☐ (*b*) False. ☐

709. The size of the lathe is specified by the maximum job length in mm that may be held between centres.

(*a*) True ☐ (*b*) False. ☐

710. Tumbler gears in lathe are used to reduce the spindle speed.

(*a*) True ☐ (*b*) False. ☐

711. The different spindle speeds on a lathe form geometrical progression.

(*a*) True ☐ (*b*) False. ☐

712. The lead screw of a lathe has single start threads.

(*a*) True ☐ (*b*) False. ☐

713. The lathe centres are provided with standard tapers known as Morse taper.

(*a*) True ☐ (*b*) False. ☐

714. A high rate of feed will get the job done in less time and will give a rough finish.

(*a*) True ☐ (*b*) False. ☐

715. Shaper utilize fly cutter.

(*a*) True ☐ (*b*) False. ☐

716. A cutting fluid provides lubrication and washes away the chips.

(*a*) True ☐ (*b*) False. ☐

717. A good cutting fluid should possess low flash point.

(*a*) True ☐ (*b*) False. ☐

718. The angle between the shear plane and job surface is known as shear angle.

(*a*) True ☐ (*b*) False. ☐

719. A zinc diffusion process is known as galvanizing.

(*a*) True ☐ (*b*) False. ☐

720. The cross-section of a chisel is usually octagonal.

(*a*) True ☐ (*b*) False. ☐

721. A hacksaw blade cuts on the forward stroke.

(*a*) True ☐ (*b*) False. ☐

722. The tool life is most affected by cutting speed.

(*a*) True ☐ (*b*) False. ☐

723. Forging is used for making cold chisels.

(*a*) True ☐ (*b*) False. ☐

724. Cold working results in increase of ductility and loss of strength and hardness of metal.

(*a*) True ☐ (*b*) False. ☐

725. Cold working process requires greater force than hot working process.

(*a*) True ☐ (*b*) False. ☐

726. The deformation of metals takes place by the method of slip of planes in cold working process.

(*a*) True ☐ (*b*) False. ☐

727. Cold working produces an improved surface finish and scale-free surface.

(*a*) True ☐ (*b*) False. ☐

728. The tool is made of brass in EDM process.

(*a*) True ☐ (*b*) False. ☐

729. Laser beam is the light amplification by stimulated emission radiation.

(*a*) True ☐ (*b*) False. ☐

730. Chemical milling process is mostly used in aircraft industry.

(*a*) True ☐ (*b*) False. ☐

731. The material, used for coating the electrode in case of welding, is called slag.

(*a*) True ☐ (*b*) False. ☐

732. Impurities, in case of welded joint, make the joint weaker as the welded portion is filled with gas and slag inclusions and metal becomes brittle.

(*a*) True ☐ (*b*) False. ☐

733. In square butt welds, the distance between two faces is kept about 3 mm.

(*a*) True ☐ (*b*) False. ☐

734. In case of welding due to high temperature, oxides of the metals are formed which weaken the weld.

(*a*) True ☐ (*b*) False. ☐

735. Low carbon steel and wrought iron are generally joined by forge welding.

(*a*) True ☐ (*b*) False. ☐

736. If the ends of the metals to be joined by forge welding are not heated enough, they will not stick together.

(*a*) True ☐ (*b*) False. ☐

737. The flux forms a protective coating of the slag over the weld metal and creates a non-oxidising atmosphere.

(*a*) True ☐ (*b*) False. ☐

738. Seam welding is continuous spot welding process.

(*a*) True ☐ (*b*) False. ☐

739. Projection welding is multi-spot welding process.

(*a*) True ☐ (*b*) False. ☐

740. In resistance welding, the pressure is released alter the weld cools.

(*a*) True ☐ (*b*) False. ☐

741. The resistance welding is best suited for lead and tin.

(*a*) True ☐ (*b*) False. ☐

742. In tungsten-inert gas arc welding a consumable electrode is used.

(*a*) True ☐ (*b*) False. ☐

743. In case of direct current straight polarity (DCSP), the work-piece is connected to the positive terminal of the welding.

(*a*) True ☐ (*b*) False. ☐

744. Fluxes are to be used in welding all types of metals except mild steel.

(*a*) True ☐ (*b*) False. ☐

745. The temperature produced by oxy-hydrogen flame is about 2400°C.

(*a*) True ☐ (*b*) False. ☐

746. The temperature of oxy-hydrogen flame is less than oxy-acetylene flame.

(*a*) True ☐ (*b*) False. ☐

747. MIG welding does not required any flux.

(*a*) True ☐ (*b*) False. ☐

748. Galvanising is a zinc diffusion process.

(*a*) True ☐ (*b*) False. ☐

749. The gears used in watches, clocks, toys etc. are made by stamping.
(*a*) True ☐ (*b*) False. ☐

750. The hobbing process of cutting gears as compared to that of shaping is faster and more accurate.
(*a*) True ☐ (*b*) False. ☐

751. In case of broaching, the job is completed in one stroke of the machine.
(*a*) True ☐ (*b*) False. ☐

752. Cutter No. 4 is suitable for cutting gears with 35 to 54 teeth.
(*a*) True ☐ (*b*) False. ☐

753. Milling is not a production process.
(*a*) True ☐ (*b*) False. ☐

754. Internal gears can be very conveniently cut by a broach tool.
(*a*) True ☐ (*b*) False. ☐

755. The combination of hardness and toughness is provided by case-hardening.
(*a*) True ☐ (*b*) False. ☐

756. The object becomes more tough if tempering is done at low temperature.
(*a*) True ☐ (*b*) False. ☐

757. The process of measuring high temperature is known as dynometry.
(*a*) True ☐ (*b*) False. ☐

758. The object retains its hardness if tempering is done at high temperature.
(*a*) True ☐ (*b*) False. ☐

759. Arrest points are temperature at which a change in phase occurs.
(*a*) True ☐ (*b*) False. ☐

760. The heat treatment of screw taps is done throughout its length.
(*a*) True ☐ (*b*) False. ☐

761. By carburising process, the components can be hardened up to a depth of 2 mm.
(*a*) True ☐ (*b*) False. ☐

762. The carbon, that must be present in a steel article which may he hardened sufficiently by heat treatment, must be at least 0.2%.
(*a*) True ☐ (*b*) False. ☐

763. Due to vibrations in machine tools, the chip thickness removed by the cutting tool is constant.
(*a*) True ☐ (*b*) False. ☐

764. Due to vibration, the surface finish obtained will be very poor.
(*a*) True ☐ (*b*) False. ☐

765. Due to vibrations, the tool life is decreased by about 70 to 80% of normal life.
(*a*) True ☐ (*b*) False. ☐

766. In dynamically stable system, the amplitude of vibration keeps on decaying with time.
(*a*) True ☐ (*b*) False. ☐

767. Cast iron has high tensile strength.
(*a*) True ☐ (*b*) False. ☐

768. Concrete is a good vibration damping material.
(*a*) True ☐ (*b*) False. ☐

769. The minimum diameter of job is taken as $h/8$, where h is the height of the centre above the bed.
(*a*) True ☐ (*b*) False. ☐

770. The thrust bearing are located close to the spindle nose to avoid the possibility of buckling.
(*a*) True ☐ (*b*) False. ☐

771. Cast iron cannot be fabricated by welding.
(*a*) True ☐ (*b*) False. ☐

772. Mild steel and other alloys can be fabricated by welding.
(*a*) True ☐ (*b*) False. ☐

773. The bed, column and frame of machine tools are made of cast iron.
(*a*) True ☐ (*b*) False. ☐

774. The shock absorbing capacity of cast iron is very low.
(*a*) True ☐ (*b*) False. ☐

775. For short run jobs and for the removal of large amount of stock, the process of broaching is best suited.
(*a*) True ☐ (*b*) False. ☐

776. The length of the broach changes with the cut per tooth and the pitch.
(*a*) True ☐ (*b*) False. ☐

777. The cost of broaching operation is very high as compared to other machine tool operations.
(*a*) True ☐ (*b*) False. ☐

778. In surface-broaching, the job and the broaching tool move across each other.
(*a*) True ☐ (*b*) False. ☐

779. The operation of broaching is simple and tooling cost is low and hence it is used for mass production.
(*a*) True ☐ (*b*) False. ☐

780. In case of broaching, the job is completed in one stroke of the machine.
(*a*) True ☐ (*b*) False. ☐

781. Broaching is used for straight or irregular surfaces either externally or internally.
(*a*) True ☐ (*b*) False. ☐

782. Broaching is a metal cutting operation in which rate of production is very high.
(*a*) True ☐ (*b*) False. ☐

783. The length of cut and chip thickness in case of broaching depends upon the pitch of the broaching tool.
(*a*) True ☐ (*b*) False. ☐

784. For proper alignment between the work-piece and broach, at least two teeth and preferably three should be in contact at all times.
(*a*) True ☐ (*b*) False. ☐

785. If helix angle decreases, rake angle also decreases.
(*a*) True ☐ (*b*) False. ☐

786. A very large ranke angle makes a tightly rolled chips.
(*a*) True ☐ (*b*) False. ☐

787. The point angle less than 118° is used for brittle materials.
(*a*) True ☐ (*b*) False. ☐

788. Higher point angle reduces width of cut and produces thicker chips for same feed rate.
(*a*) True ☐ (*b*) False. ☐

789. The lead screw of a lathe has multi-start threads.
(*a*) True ☐ (*b*) False. ☐

790. Drilling in an example of oblique cutting.
(*a*) True ☐ (*b*) False. ☐

791. In drilling operation, the metal is removed by shearing and extrusion.
(*a*) True ☐ (*b*) False. ☐

792. A 20 mm drilling machine means that it can drill a hole of maximum diameter 20 mm.
(*a*) True ☐ (*b*) False. ☐

793. Sensitive and upright drilling machines are specified by the length of the ami and column diameter.
(*a*) True ☐ (*b*) False. ☐

794. The Gang drilling machine is used for drilling holes of several different sizes on a work-piece.
(*a*) True ☐ (*b*) False. ☐

795. Deep-hole drilling machines are used for drilling such holes whose length exceeds three times the drill size.
(*a*) True ☐ (*b*) False. ☐

796. The ram is heavy in case of draw-cut type shaper.
(*a*) True ☐ (*b*) False. ☐

797. The shaper as compared to planer occupies less floor area.
(*a*) True ☐ (*b*) False. ☐

798. In case of hydraulic shaper, the operation is very smooth.
(*a*) True ☐ (*b*) False. ☐

799. In centrifugal casting no core is used.
(*a*) True ☐ (*b*) False. ☐

800. Red colour is used on the surface of a pattern, which is to be machined.
(*a*) True ☐ (*b*) False. ☐

Fill in the blanks:

801. The wear ratio (*i.e.*, ratio of electrode wear rate to metal removal rate) is ...... for brass work.
(*a*) 1.0 ☐ (*b*) 0.5. ☐

802. Life of tool is ...... in electro-discharge machining (EDM) process.
(*a*) long ☐ (*b*) short. ☐

803. In EDM process, the tool is made cathode and work-piece as anode, in order to remove ...... metal and have minimum wear on the tool.
(*a*) minimum ☐ (*b*) maximum. ☐

804. Oblique cutting system is also known as ...... cutting system.
(*a*) three-dimensional ☐ (*b*) two-dimensional. ☐

805. In orthogonal cutting system, the chip thickness is ...... at the middle.
(*a*) minimum ☐ (*b*) maximum. ☐

806. The formation of the chip is due to ...... deformation.
(*a*) linear ☐ (*b*) shear. ☐

807. If the metals are ductile and cutting speed is high, then ...... chips are formed.
(*a*) continuous ☐ (*b*) discontinuous. ☐

808. The deflections up to ...... can be measured by dial indicator.
(*a*) 25×10^{-8} mm ☐ (*b*) 25×10^{-4} mm. ☐

809. For providing cooling action to the work-piece and tool and to wash away the chips, ... is used as the cutting fluid.
(*a*) chlorinated cutting oils ☐ (*b*) aqueous solution. ☐

810. In die-cutting, the bottom-most layers of the sheet in die are subjected to ...... when the punch presses the steel metal.
(*a*) compressive stress ☐ (*b*) tensile stress. ☐

811. In the above question, the layer below the punch is subjected to ......
(*a*) compressive stress ☐ (*b*) tensile stress. ☐

812. The instantaneous fracture of the metal after its plastic deformation is known as ...... in case of die cutting operation.
(*a*) depth of cut ☐ (*b*) penetration. ☐

813. If the angle of shear ......, the force acting on the punch goes on increasing.
(*a*) increases ☐ (*b*) decreases. ☐

814. For making thin gears from sheet metals up to thickness 3 mm, ... is used.
(*a*) casting ☐ (*b*) stamping. ☐

815. In ......, the molten metal under high pressure is poured into mold.
(*a*) slush casting ☐ (*b*) die casting. ☐

816. For ornamental parts and toys of non-ferrous alloys, ... is used.
(*a*) slush casting ☐ (*b*) die casting. ☐

817. Pipes are made by ...... if they are subjected to very high pressure of the order of 120 kgf/cm^2.

(*a*) pressed casting ☐ (*b*) extrusion process. ☐

818. The ...... is the casting defect.

(*a*) slag ☐ (*b*) blow holes. ☐

819. To obtain high density and pure casting, ... is used.

(*a*) die casting ☐ (*b*) centrifugal casting. ☐

820. A mixture of 70% sand and 30% clay is known as ...... sand.

(*a*) red ☐ (*b*) green. ☐

821. In case of ......, the work is stationary and tool reciprocates.

(*a*) planner ☐ (*b*) shaper. ☐

822. In case of ......, the tool is stationary and work reciprocates,

(*a*) planner ☐ (*b*) shaper. ☐

823. In case of a shaper, the ram moves ...... during forward stroke.

(*a*) faster ☐ (*b*) slower. ☐

824. The push-cut type shaper removes the metal during ...... stroke.

(*a*) forward ☐ (*b*) backward. ☐

825. The draw-but type shaper removes the metal during ...... stroke.

(*a*) forward ☐ (*b*) backward. ☐

826. The term shaper means the ...... type.

(*a*) push-cut ☐ (*b*) draw-cut. ☐

827. In case of a shaper, for rough cuts, the practice is to use ...... feeds at slow speeds.

(*a*) minimum ☐ (*b*) maximum. ☐

828. The overhang of the tool slide should be ...... in case of a shaper.

(*a*) minimum ☐ (*b*) maximum. ☐

829. The setting time in case of planers is ...... as compared to that of a shaper.

(*a*) two times ☐ (*b*) five times. ☐

830. Portable drilling machine is specified by ...... diameter of the drill which can be held.

(*a*) minimum ☐ (*b*) maximum. ☐

831. The...... drilling machine is specified by the drilling area, the size and number of holes a machine can drill.

(*a*) radial ☐ (*b*) multiple spindle. ☐

832. The cutting speed for high speed drills is approximately equal to ...... times the cutting speed for carbon steel drills.

(*a*) four ☐ (*b*) two. ☐

833. The drilling (T) in minute is equal to ...... where N = r.p.m. of drill, L = length of hole to be drilled, and f = feed per revolution of drill.

(*a*) $T = f/N \times L$ ☐ (*b*) $T = L/N \times f$. ☐

834. The total length L = ... in drilling a blind hole.

(*a*) $H + 0.29\ D$ ☐ (*b*) $H + 0.29\ D \times 2$ ☐

where H = Depth of plate, and D = Diameter of drill.

835. The total life ...... as the cutting speed is decreased for a constant feed.

(*a*) increases ☐ (*b*) decreases. ☐

836. For a given rate of metal removal, the best tool life will be obtained by using the ...... possible feed.

(*a*) lowest ☐ (*b*) highest. ☐

837. The intermediate speeds in case of drilling are chosen in ...... progression series.

(*a*) arithmetical ☐ (*b*) geometrical. ☐

838. For deep-hole drilling in ......, high helix drills are used.

(*a*) hard rubber ☐ (*b*) aluminium. ☐

839. The best angle is ...... to grind a drill for work on soft materials like fibre.

(*a*) 70° ☐ (*b*) 40°. ☐

840. The cutting edge becomes ...... with the increase of helix angle.

(*a*) strong ☐ (*b*) weak. ☐

841. The helix angles are of the order of ...... for harder material.

(*a*) 50 to 60° ☐ (*b*) 10 to 13°. ☐

842. The point angle is ...... for hard and tough materials.

(*a*) 60° ☐ (*b*) 140°. ☐

843. In grinding carbon-steel drill ...... should be used.

(*a*) no water ☐ (*b*) plenty of water. ☐

844. For drilling cast iron ...... is used.

(*a*) dry lubricant ☐ (*b*) turpentine. ☐

845. A drill with ...... helix angle is required for drilling brass.

(*a*) zero ☐ (*b*) low. ☐

846. Glazing in grinding wheels ...... cutting capacity.

(*a*) increases ☐ (*b*) decreases. ☐

847. A grinding wheel is said to be of ...... if the abrassive grains can be easily dislodged.

(*a*) soft grade ☐ (*b*) hard grade. ☐

848. A fixture ...... the tool.

(*a*) guides ☐ (*b*) does not guide. ☐

849. The vector sum of cutting velocity and chip velocity is ...... shear velocity.

(*a*) half of ☐ (*b*) equal to. ☐

850. The velocity of tool along the tool face is known as ...... velocity.

(*a*) shear ☐ (*b*) chip. ☐

851. The angle between the shear plane and ...... is called shear angle.

(*a*) tool face ☐ (*b*) work surface. ☐

852. The strength of the tool depends on ...... angle.

(*a*) rake ☐ (*b*) cutting. ☐

853. On cemented carbide tools usually, the ...... rake is provided.

(*a*) positive ☐ (*b*) negative. ☐

854. The rate of removal metal is ...... by increasing the depth of cut.

(*a*) increased ☐ (*b*) decreased. ☐

855. In order to prevent tool from rubbing the work, ... on tools are provided.

(*a*) rake angles ☐ (*b*) relief angles. ☐

856. If the grain size is ...... the tool life is better.

(*a*) larger ☐ (*b*) smaller. ☐

857. The increase in depth of cut and feed rate ...... surface finish.

(*a*) improves ☐ (*b*) deteriorates. ☐

858. Tool signature consists of ...... elements.

(*a*) four ☐ (*b*) seven. ☐

859. To weld ...... neutral flame is used.

(*a*) brass ☐ (*b*) cast iron. ☐

860. Rate of welding steel by carburising flame is ...... as compared to that of neutral flame.

(*a*) slow ☐ (*b*) fast. ☐

861. The most suitable flame is ...... for cutting operation.

(*a*) oxidising ☐ (*b*) carburising. ☐

862. Flux ...... used for welding carbon steel.

(*a*) is ☐ (*b*) is not. ☐

863. On a oxygen cylinder ...... colour is usually painted.

(*a*) red ☐ (*b*) black. ☐

864. From ......, acetylene gas is produced.

(*a*) calcium chloride ☐ (*b*) calcium carbide. ☐

865. For brazing aluminium and magnesium ...... is used as a flux.

(*a*) chloride and fluorides mixed with water ☐

(*b*) mixture of boric acid, borax and wetting agent. ☐

866. The ratio of oxygen to acetylene is ...... for oxidising flame.

(*a*) 1 : 1 ☐ (*b*) 1.5 : 1. ☐

867. For carburizing flame ...... is the ratio of oxygen to acetylene.

(*a*) 1 : 1 ☐ (*b*) 0.9 : 1. ☐

868. The arc length should be approximately equal to ...... in arc welding.

(*a*) diameter of the rod ☐ (*b*) half the diameter of the rod. ☐

869. The ...... is a natural abrasive.

(*a*) silicon carbide ☐ (*b*) diamond. ☐

870. The ...... is an artificial abrasive.

(*a*) silicon carbide ☐ (*b*) diamond. ☐

871. Grain depth of cut is equal to ......

(*a*) half the maximum cut depth ☐ (*b*) the maximum cut depth. ☐

872. The grain depth of cut is case of grinding ...... if the speed of work-piece is increased.

(*a*) decreases ☐ (*b*) increases. ☐

873. The wheel will appear .... if the depth of cut is decreased.

(*a*) harder ☐ (*b*) softer. ☐

Machine Tools

874. To remove the excess material in the form of chips, which one of the following is used?

(*a*) shaper ☐ (*b*) planer ☐

(*c*) machine tool ☐ (*d*) all of the above. ☐

875. The tool life is affected by

(*a*) chip thickness ☐ (*b*) depth of cut ☐

(*c*) cutting speed ☐ (*d*) feed ☐

(*e*) all of the above ☐ (*f*) none of the above. ☐

876. The relation between the tool life in minute and cutting speed in m/min is given by

(*a*) $V \times T = C$ ☐ (*b*) $V^n \times T = C$ ☐

(*d*) $V \times T^n = C$ ☐ (*d*) $V/T^n = C$ ☐

where V = Cutting speed in meter/minute, T = Tool life in minute,

C = A constant, n = An exponent, which depends upon the tool and work-piece.

877. The machine tools, which are used for performing all metal cutting operations within their range, are known as

(*a*) production machine tools ☐ (*b*) basic machine tools ☐

(*c*) special purpose machine tools ☐ (*d*) none of the above. ☐

878. The machines tools, which are used to reduce the manufacturing cost and to increase the rate of production, are known as

(*a*) production machine tools ☐ (*b*) basic machine tools ☐

(*c*) special purpose machine tools ☐ (*d*) none of the above. ☐

879. The machine tools, which are used for mass production are known as

(*a*) production machine tools ☐ (*b*) basic machine tools ☐

(*c*) special purpose machine tools ☐ (*d*) none of the above. ☐

880. The process of chip removal is affected by

(*a*) motion of machine tool ☐

(*b*) motion of work piece ☐

(*c*) relative motion between tool and work-piece ☐

(*d*) none of the above. ☐

881. Process capability of a machine tool is defined as its ability

(*a*) to produce jobs quickly ☐

(*b*) to produce jobs accurately ☐

(*c*) to produce jobs within specified degree of accuracy consistently over a long period of time ☐

(*d*) none of the above. ☐

882. Compliance of a machine tool is

(*a*) equal to rigidity ☐ (*b*) reciprocal of rigidity ☐

(*c*) square of rigidity ☐ (*d*) square root of rigidity. ☐

883. Which one of the following is an objective of a machine tool?

(*a*) to hold the job ☐ (*b*) to hold the cutter ☐

(*c*) to produce a relative motion ☐ (*d*) all of the above ☐

(*e*) none of the above. ☐

884. Which one is responsible for tool failure?

(*a*) plastic deformation of the cutting edge ☐

(*b*) cracking at the cutting edge due to thermal stresses ☐

(*c*) chipping of the cutting edge ☐

(*d*) flank and crater wear ☐

(*e*) all of the above. ☐

885. With the increase of cutting speed, the tool life

(*a*) increases ☐ (*b*) decreases ☐

(*c*) is not affected ☐ (*d*) increases first and then becomes constant. ☐

886. The cutting fluid required for machining cast iron is

(*a*) water ☐ (*b*) soluble oil ☐

(*c*) heavy oil ☐ (*d*) none of the above. ☐

887. The cutting fluid mostly used for machining alloy steels is

(*a*) soluble oil ☐ (*b*) water ☐

(*c*) heavy oil ☐ (*d*) sulphurised mineral oil. ☐

888. Which one of the following is a correct statement?

(*a*) compliance of a machine tool is equal to the square of rigidity. ☐

(*b*) crater wear occurs mainly due to oxidation. ☐

(*c*) continuous chips are produced when cutting cast iron. ☐

(*d*) soluble oils are best coolant and lubricant for brass, copper, bronze and monel metal. ☐

889. Which one of the following is a wrong statement?

(*a*) Tool signature is a numerical method of indentification of tool. ☐

(*b*) Carbon tool steels have low-heat and wear-resistance. ☐

(*c*) In machining soft materials, a tool with negative relief angle is used. ☐

(*d*) With the same tool life, the maximum material per minute is removed by increasing the speed. ☐

890. The discontinuous chips arc formed due to

(*a*) high cutting speed and small rake angle ☐ (*b*) low cutting speed and large rake angle ☐
(*c*) low cutting speed and small rake angle ☐ (*d*) high cutting speed and large rake angle. ☐

891. During machining process, cast iron produces

(*a*) discontinuous chips ☐ (*b*) continuous chips. ☐
(*c*) continuous chips with built-up-edge ☐ (*d*) none of the above. ☐

892. High speed steel tools retain their hardness up to a temperature of

(*a*) 1400°C ☐ (*b*) 1200°C ☐
(*c*) 900°C ☐ (*d*) 500°C. ☐

893. The surface finish deteriorates when

(*a*) depth of cut and feed rate is increased ☐
(*b*) depth of cut is decreased but feed rate is increased ☐
(*c*) depth of cut and feed rate is decreased ☐
(*d*) depth of cut is increased but feed rate is decreased. ☐

894. The maximum material per minute is removed by (keeping the same tool life)

(*a*) increasing the feed rate ☐ (*b*) increasing the depth of cut ☐
(*c*) increasing the cutting speed ☐ (*d*) decreasing the depth of cut. ☐

895. Soft materials are machined by a tool which has

(*a*) positive relief angle ☐ (*b*) negative relief angle ☐
(*c*) zero relief angle ☐ (*d*) none of the above. ☐

896. In oblique cutting of metals, the cutting edge of the tool is

(*a*) parallel to the direction of tool travel ☐
(*b*) perpendicular to the work-piece. ☐
(*c*) perpendicular to the direction of tool travel ☐
(*d*) inclined at an angle less than 90° to the direction of the tool travel. ☐

897. In orthogonal cutting of metals, the cutting edge of the tool is

(*a*) parallel to the direction of tool travel ☐
(*b*) perpendicular to the work-piece ☐
(*c*) perpendicular to the direction of tool travel ☐
(*d*) inclined at an angle less than 90° to the direction of the tool travel. ☐

898. On lathe, shaper and planer, the type of tool used is

(*a*) multi-point cutting tool ☐ (*b*) three point cutting tool ☐
(*c*) two point cutting tool ☐ (*d*) single point cutting tool. ☐

899. On milling machine and broaching machine, the type of tool used is

(*a*) multi-point cutting tool ☐ (*b*) three point cutting tool ☐
(*c*) two point cutting tool ☐ (*d*) single point cutting tool. ☐

900. A single point tool has

(*a*) cutting angle ☐ (*b*) rake angle ☐
(*c*) lip angle ☐ (*d*) clearance angle ☐
(*e*) all of the above ☐ (*f*) none of the above. ☐

901. The angle between the tool face and the ground end surface of flank is known as

(*a*) rake angle ☐ (*b*) lip angle ☐
(*c*) clearance angle ☐ (*d*) cutting angle. ☐

902. The angle made by the face of the tool and the plane parallel to the base of cutting tool is known as

(*a*) rake angle ☐ (*b*) lip angle ☐
(*c*) clearance angle ☐ (*d*) cutting angle. ☐

903. Shear angle is the angle between the shear plane and

(*a*) tool face ☐ (*b*) the plane parallel to tool face ☐
(*c*) work surface ☐ (*d*) the plane normal to tool face ☐

904. Lip angle is usually

(*a*) 45° to 50° ☐ (*b*) 20° to 25° ☐
(*c*) 10° to 15° ☐ (*d*) 3° to 5°. ☐

905. Relief angle on carbide tipped tools is

(*a*) 45° ☐ (*b*) 30° ☐
(*c*) 10° ☐ (*d*) 0°. ☐

906. Relief angle on high speed steel tools vary from

(*a*) 60° to 80° ☐ (*b*) 40° to 60° ☐
(*c*) 20° to 40° ☐ (*d*) 0° to 20°. ☐

907. The cutting fluid, for machining cast iron, is

(*a*) water ☐ (*b*) soluble oils ☐
(*c*) heavy oils ☐ (*d*) no fluid. ☐

908. The number of elements in tool signatures are

(*a*) 3 elements ☐ (*b*) 5 elements ☐
(*c*) 7 elements ☐ (*d*) 9 elements. ☐

909. The numerical method of identification of tool is known as

(*a*) rake angle ☐ (*b*) relief angle ☐
(*c*) tool signature ☐ (*d*) pressure angle. ☐

910. Chip breakers are used to

(*a*) minimize heat generation ☐ (*b*) break with chips into short segments ☐
(*c*) increase the tool life ☐ (*d*) remove chips from bed. ☐

911. The tool cutting forces decreases as the cutting speed.

(*a*) decreases ☐ (*b*) increases ☐
(*c*) remains constant ☐ (*d*) none of the above. ☐

912. The ceramic tools can withstand temperatures up to 600°C only.

(*a*) True ☐ (*b*) False ☐

913. The diamond is the hardest tool material and can run at cutting speeds about 45 times that of high speed steel tool.

(*a*) True ☐ (*b*) False ☐

914. Tool life is said to be over if poor surface finish is obtained or overheating and fuming due to friction start or sudden increase in power and cutting force with chattering take place.

(*a*) True ☐ (*b*) False ☐

915. The ceramic tools can be used at cutting speeds 40 times that of high speed steel tools.

(*a*) True ☐ (*b*) False ☐

916. The life of tool is most affected by

(*a*) feed and depth ☐ (*b*) tool geometry ☐

(*c*) cutting speed ☐ (*d*) none of the above. ☐

917. Flank wear occurs mainly on

(*a*) nose part and top face ☐ (*b*) nose part, front relief face and side relief face ☐

(*c*) front face ☐ (*d*) cutting edges. ☐

III. OBJECTIVE TYPE QUESTIONS FROM COMPETITIVE EXAMINATIONS

Tick mark the most appropriate answer:

1. A cube shaped casting solidifies in 5 min. The solidification time in min. for a cube of the same material, which is 8 times heavier than the original casting, will be

(*a*) 10 ☐ (*b*) 20 ☐

(*c*) 24 ☐ (*d*) 40. ☐

(ME & PI-GATE-2013)

2. An expandable pattern is used in

(*a*) slush casting ☐ (*b*) squeeze casting ☐

(*c*) centrifugal casting ☐ (*d*) investment casting. ☐

(PI-GATE-2005)

3. As tool and work are not in contact in EDM process

(*a*) no relative motion occurs between them ☐

(*b*) no wear of tool occurs ☐

(*c*) no power is consumed during metal cutting ☐

(*d*) no force between tool and work occurs. ☐

(ME-GATE-2003)

4. In ultrasonic machining process the material removal rate will be higher for material with

(*a*) higher toughness ☐ (*b*) higher ductility ☐

(*c*) lower toughness ☐ (*d*) higher fracture strain. ☐

(ME-GATE-2010)

5. Diamond cutting tools are not recommended for machining of ferrous metals due to

(*a*) high tool hardness ☐

(*b*) high thermal conductivity of the work material ☐

(*c*) poor tool roughness ☐

(*d*) chemical affinity of tool material with iron. ☐

(PI-GATE-2009)

6. The formation of the chip is due to
   (a) linear deformation ☐ (b) shear deformation ☐
   (c) linear translation ☐ (d) none of the above. ☐
7. Choose the wrong statement
   (a) When cutting cast iron, the type of chip produced is discontinuous. ☐
   (b) Best coolant and lubricant for steel and wrought iron is water soluble oil. ☐
   (c) The cutting edge of the tool is perpendicular to the direction of tool travel in orthogonal cutting system. ☐
   (d) Mild steel during machining produces discontinuous chips. ☐
8. Friction at the tool-chip interface can be reduced by
   (a) decreasing the rake angle ☐ (b) increasing the depth of cut ☐
   (c) decreasing the cutting speed ☐ (d) increasing the cutting speed. ☐

(ME-GATE-2009)

9. Match the following:

List I	List II
A. Welding of aluminium alloy	1. Submerged are welding
B. Ship welding	2. Electron beam welding
C. Joining of HSS drill bit to shank	3. TIG welding
D. Deep penetration precision welds	4. Friction welding
	5. Gas welding

Codes:

(a)	A – 3	B – 1	C – 4	D – 2	☐
(b)	A – 2	B – 1	C – 5	D – 3	☐
(c)	A – 1	B – 2	C – 3	D – 4	☐
(d)	A – 4	B – 3	C – 2	D – 5	☐

(ME-GATE-2009)

10. Match the following:

Work material	Type of joining
P. Aluminum	1. Submerged arc welding
Q. Die steel	2. Soldering
R. Copper wire	3. Thermit welding
S. Titanium sheet	4. Atomic hydrogen welding
	5. Gas tungsten arc welding
	6. Laser beam welding

Codes:

(a)	P – 2	Q – 5	R – 1	S – 3	☐
(b)	P – 6	Q – 3	R – 4	S – 4	☐
(c)	P – 4	Q – 1	R – 6	S – 2	☐
(d)	P – 5	Q – 4	R – 2	S – 6	☐

(ME-GATE-2003)

11. Match the following:

NC code	Definition
P. M05	1. Absolute co-ordinate system
Q. G01	2. Dwell
R. G04	3. Spindle stop
S. G90	4. Linear interpolation

(*a*) *P*–2, *Q*–3, *R*–4, *S*–1 ☐
(*b*) *P*–3, *Q*–4, *R*–1, *S*–2 ☐
(*c*) *P*–3, *Q*–4, *R*–2, *S*–1 ☐
(*d*) *P*–4, *Q*–3, *R*–2, *S*–1. ☐

(ME-GATE-2009)

12. Consider the following characteristics:

1. Single machine tool
2. Manual materials handling system
3. Computer control
4. Random sequencing of part of machines

Which of the above characteristics are associated with flexible manufacturing system?

(*a*) 1, 2 and 3 ☐ (*b*) 1 and 2 ☐
(*c*) 3 and 4 ☐ (*d*) 2, 3 and 4. ☐

(PI-GATE-2005)

13. Choose the wrong statement

(*a*) The cross-section of a chiesel is usually octagonal. ☐
(*b*) A hacksaw blade cuts on the forward stroke. ☐
(*c*) The tool life is most affected by cutting speed. ☐
(*d*) A zinc diffusion process is known as galvanizing. ☐

14. In hallow cylindrical parts made by centrifugal casting, the density of the part is

(*a*) maximum at the outer region ☐
(*b*) maximum at the inner region ☐
(*c*) maximum at the mid point between outer and inner surfaces ☐
(*d*) uniform throughout ☐

(PI-GATE-2008)

15. In oxy-acetylene gas welding, temperature at the inner cone of the flame is around

(*a*) 3500°C ☐ (*b*) 3200°C ☐
(*c*) 2900°C ☐ (*d*) 2550°C ☐

(ME-GATE-2003)

16. In deep drawing of sheets, the values of limiting draw ratio depends on

(*a*) percentage elongation of the sheet ☐ (*b*) yield strength of the sheet ☐
(*c*) type of press used ☐ (*d*) thickness of sheet ☐

(ME-GATE-1994)

17. Which one is not a method of reducing cutting forces to prevent the overloading of press?

(*a*) providing shear on die ☐ (*b*) providing shear on punch ☐
(*c*) increasing the clearance ☐ (*d*) stepping punches. ☐

(PI-GATE-2003)

18. In electrochemical grinding (ECG) process

(*a*) grinding pressure is low ☐
(*b*) temperature is very low ☐
(*c*) pressure is less but temperature is high ☐
(*d*) both grinding pressure and temperature are low. ☐

19. Instrument principle of inspection

P. Dial indicator	1. Non-contact
Q. Pneumatic gage	2. Limit of size
R. GO/NO GO gage	3. Comparator

(*a*) *P*–2, *Q*–3, *R*–1 ☐ (*b*) *P*–3, *Q*–1, *R*–2 ☐
(*c*) *P*–1, *Q*–2, *R*–3 ☐ (*d*) *P*–2, *Q*–1, *R*–3 ☐

(PI-GATE-2003)

20. Machining of complex shapes on CNC machines requires

(*a*) simultaneous control of x, y and z-axes ☐ (*b*) simultaneous control of x, and y-axes ☐
(*c*) independent control of x and y-axes ☐ (*d*) independent control of x, y and z-axes ☐

(ME-GATE-2002)

21. Deep hole drilling of small diameter, say 0.2 mm is done with EDM by selecting the tool material as

(*a*) copper wire ☐ (*b*) tungsten wire ☐
(*c*) brass wire ☐ (*d*) tungsten carbide. ☐

(ME-GATE-2000)

22. The cutting rate in mm/min for LBM process is equal to

(*a*) $\frac{C \times P \times t}{E \times A}$ ☐ (*b*) $\frac{P \times t}{CEA}$ ☐

(*c*) $\frac{EA \times t}{C \times P}$ ☐ (*d*) $\frac{C \times P}{EA \times t}$. ☐

where P = Laser power incident on surface in watts,

A = Area of laser beam at focal point in mm^2,

t = Thickness of material in mm,

E = Vaporisation energy of material in W/mm^2, and

C = Constant.

23. In an interchangeable assembly, shafts of size $25{,}000^{+0.040}_{-0.010}$ mm mate with holes of size $25{,}000^{+0.030}_{+0.020}$ mm. The maximum interference (in microns) in the assembly is

(*a*) 40 ☐ (*b*) 30 ☐
(*c*) 20 ☐ (*d*) 10. ☐

(ME & PI-GATE-2012)

24. Which type of motor is used in axial or spindle drives of CNC m/C tools?

(*a*) Induction motors ☐ (*b*) DC servomotors ☐
(*c*) Stepper motor ☐ (*d*) Linear servo motor. ☐

(ME-GATE-2007)

25. In sand casting fluidity of the molten metal increases with

(*a*) increase in degree of superheat ☐
(*b*) decrease in pouring rate ☐
(*c*) increase in thermal conductivity of the mold ☐
(*d*) increase in sand grain size. ☐

(PI-GATE-2011)

26. The feed in drilling operation is expressed as

(*a*) mm/revolution ☐ (*b*) mm ☐
(*c*) mm/s ☐ (*d*) mm/min. ☐

27. Gray cast iron blocks 200 × 100 × 10 mm are to be cast in sand moulds. Shrinkage allowance for pattern making is 1%. The ratio of the volume of pattern to that of the casting will be

(*a*) 0.97 ☐ (*b*) 0.99 ☐
(*c*) 1.01 ☐ (*d*) 1.03. ☐

(ME-GATE-2004)

28. Numerical control can be applied to

(*a*) drilling machines ☐ (*b*) lathes ☐
(*c*) milling machines ☐ (*d*) all of the above machines. ☐

29. In blanking operation the clearance is provided on

(*a*) the die ☐ (*b*) the punch ☐
(*c*) both die and punch equally ☐ (*d*) neither the punch nor the die ☐

(ME-GATE-2002)

30. A built up edge is formed while machining

(*a*) ductile materials at high speed ☐ (*b*) ductile materials at low speed ☐
(*c*) brittle materials at high speed ☐ (*d*) brittle materials at low speed. ☐

(ME-GATE-2000)

31. Which of the following is a solid welding process?

(*a*) GTAW ☐ (*b*) Resistance spot welding ☐
(*c*) Friction welding ☐ (*d*) SAW. ☐

(ME-GATE-2000)

32. The material most commonly used manufacturing of machine tool because

(*a*) MS ☐ (*b*) gray CI ☐

(*c*) white CI ☐ (*d*) galvanized. ☐

(PI-GATE-2011)

33. Two plates of the same metal having equal thickness are to be butt welded with electric arc. When the plate thickness changes, welding is achieved by

(*a*) adjusting the current ☐ (*b*) adjusting the duration of the current ☐

(*c*) changing the electrode size ☐ (*d*) changing the electrode coating. ☐

(ME-GATE-2001)

34. The cutting tool material normally used for turning steel of very high hardness is

(*a*) HSS ☐ (*b*) tungsten carbide ☐

(*c*) CBN ☐ (*d*) diamond. ☐

(PI-GATE-2006)

35. The gears used in watches, clocks, toys etc. are made by

(*a*) casting ☐ (*b*) stamping ☐

(*c*) extruding ☐ (*d*) coining. ☐

36. A grinding wheel is said to be glazed if

(*a*) grains have become blunt, but do not fall out ☐

(*b*) gap between the grains is filled by swarf ☐

(*c*) it becomes black due to burning ☐

(*d*) part of the wheel is chipped off. ☐

(PI-GATE-1995)

37. Grinding ratio is defined as

(*a*) $\dfrac{\text{Volume of wheel wear}}{\text{Volume of work material removed}}$ ☐ (*b*) $\dfrac{\text{Volume of work material removed}}{\text{Volume of wheel wear}}$ ☐

(*c*) $\dfrac{\text{Cutting speed}}{\text{feed}}$ ☐ (*d*) $\dfrac{\text{Longitudinal feed}}{\text{Transverse feed}}$. ☐

(PI-GATE-2011)

38. In resistance seam welding, the electrode is in the form of a

(*a*) cylinder ☐ (*b*) flat plate ☐

(*c*) coil of wire ☐ (*d*) circular disc. ☐

(PI-GATE-2011)

39. The temp of a carburizing flame in gas welding .... that of a neutral or an oxidizing flame

(*a*) Lower than ☐ (*b*) Higher than ☐

(*c*) Equal to ☐ (*d*) Unrelated to. ☐

(ME-GATE-2002)

40. Cold shut is a defect in casting due to

(*a*) sand sliding from the cope surface ☐
(*b*) internal voids or surface depression due to excessive gas trapped ☐
(*c*) discontinuity resulting from hindered contraction ☐
(*d*) two streams of material that are too cold to fuse properly. ☐

(PI-GATE-2006)

41. Increase in feed rate, improves the surface finish.

(*a*) True ☐ (*b*) False. ☐

42. Which one of the following is not a synthetic abrasive material?

(*a*) silican carbide ☐ (*b*) aluminum oxide ☐
(*c*) titanium nitride ☐ (*d*) cubic boron nitride. ☐

(PI-GATE-2003)

43. Which one of the Instrument is comparator?

(*a*) tool makers microscope ☐ (*b*) GO/NOGO gage ☐
(*c*) optical interferometer ☐ (*d*) dial gage. ☐

(PI-GATE-2007)

44. For generating coon's surface we require

(*a*) a set of grid points on the surface ☐
(*b*) a set of grid control points ☐
(*c*) four bounding curves defining the surface ☐
(*d*) two bounding curves and a set of grid control points. ☐

(ME-GATE-2008)

45. In ultrasonic machining process, the material removal rate will be higher for material with

(*a*) higher toughness ☐ (*b*) higher ductility ☐
(*c*) lower toughness ☐ (*d*) higher fracture strain. ☐

(ME-GATE-2010)

46. The process which is employed for production seamless tubes, is known as

(*a*) extrusion ☐ (*b*) forging ☐
(*c*) casting ☐ (*d*) piercing. ☐

47. The operation in which oil is permeated into the pores of a powder metallurgy product is known as

(*a*) mixing ☐ (*b*) sintering ☐
(*c*) impregnation ☐ (*d*) infiltration. ☐

(ME-GATE-2011)

48. The cutting force in punching and blanking operations mainly depends on

(*a*) the modulus of elasticity of the material ☐ (*b*) the shear strength of the material ☐
(*c*) the bulk modulus of the material ☐ (*d*) the yield strength of the material. ☐

(ME-GATE-2001)

49. Circular blanks of 10 mm diameter are punched from an aluminum sheet of 2 mm thickness. The shear strength of aluminum is 80 MPa. The minimum punching force required in kN is

(*a*) 2.57 ☐ (*b*) 3.29 ☐
(*c*) 5.03 ☐ (*d*) 6.33. ☐

(PI-GATE-2013)

50. In blanking operation, the best way to improve the smoothness and squarness of the edges is to

(*a*) have reduced gap between punch and die ☐
(*b*) increase the ductility of the sheet ☐
(*c*) decrease the speed of blanking ☐
(*d*) provide shear on the punch. ☐

(PI-GATE-1994)

51. The crystal structure of austenite is

(*a*) body centered cubic ☐ (*b*) face centered cubic ☐
(*c*) hexagonal closed packed ☐ (*d*) body centered tetragonal. ☐

(ME-GATE-2011)

52. Solidification time of a metallic alloy casting is

(*a*) directly proportional to its surface area ☐
(*b*) inversely proportional to the specific heat of the cast material ☐
(*c*) inversely proportional to the thermal diffusivity of the mould material ☐
(*d*) inversely proportional to the pouring temp. ☐

(PI-GATE-2010)

53. With a solidification factor of 0.97×10^6 s/m^2, the solidification time (in seconds) for a spherical casting of 200 mm diameter is

(*a*) 539 ☐ (*b*) 1078 ☐
(*c*) 4311 ☐ (*d*) none. ☐

(ME-GATE-2010)

54. The strength of a brazed joint

(*a*) decreases with increase in gap between the two joining surface ☐
(*b*) increases with increase in gap between the two joining surface ☐
(*c*) decreases up to certain gap between the two joining surface beyond which it increases ☐
(*d*) increases up to certain gap between the two joining surfaces beyond which it decreases. ☐

(ME-GATE-2010)

55. The cutting rate by LBM process for thin materials is

(*a*) maximum ☐ (*b*) minimum ☐
(*c*) independent of thickness ☐ (*d*) none of the above. ☐

56. Friction at the tool–chip interface can be reduced by

(*a*) decreasing the rake angle ☐ (*b*) increasing the depth of cut ☐
(*c*) decreasing the cutting speed ☐ (*d*) increasing the cutting speed. ☐

(ME-GATE-2009)

57. Which of the following casting processes uses expandable pattern and expandable mould

(*a*) shell mod casting ☐ (*b*) investment casting ☐
(*c*) pressure die casting ☐ (*d*) centrifugal casting. ☐

(PI-GATE-2011)

58. In spot welding, the weld time is about

(*a*) 45 to 30 seconds ☐ (*b*) 30 to 25 seconds ☐
(*c*) 25 to 20 seconds ☐ (*d*) 3 to 10 seconds. ☐

59. Match List–I (product) with List–II (casting process) and select the correct answer using the codes given below the lists:

List–I	**List–II**
A. Hollow status	1. Centrifugal casting
B. Dentures	2. Investment casting
C. Aluminum alloy pistons	3. Slush casting
D. Rocker arm	4. Shell moulding
	5. Gravity die casting

(*a*) *A*–3, *B*–2, *C*–4, *D*–5 ☐ (*b*) *A*–1, *B*–3, *C*–4, *D*–5 ☐
(*c*) *A*–1, *B*–2, *C*–3, *D*–4 ☐ (*d*) *A*–3, *B*–2, *C*–5, *D*–4. ☐

(ME-GATE-2005)

60. Which one among the following welding processes uses non-consumable electrode?

(*a*) Gas metal arc welding ☐ (*b*) Submerged arc welding ☐
(*c*) Gas tungsten arc welding ☐ (*d*) Flux coated arc welding. ☐

(ME-GATE-2011)

61. In resistance welding , heat is generated due to the resistance between

(*a*) electrode and work-piece ☐ (*b*) asperities between touching plates ☐
(*c*) two dissimilar metals being in contact ☐ (*d*) inter-atomic forces. ☐

(PI-GATE-2003)

62. In case of cylindrical grinder, the depth of cut for finishing cut is normally

(*a*) 0.05 mm ☐ (*b*) 0.01 mm ☐
(*c*) 0.005 mm ☐ (*d*) 0.001 mm. ☐

63. The effects of setting a boring tool above centre height leads to a/an.

(*a*) increase in the effective rake angle and a decrease in the effective clearance angle ☐
(*b*) increase in both effective rake angle and effective clearance angle ☐
(*c*) decrease in effective rake angle and an increase in the effective clearance angle ☐
(*d*) decrease in both effective rake angle and effective clearance angle. ☐

(PI-GATE-2005)

64. Formation of build-up edge during machining can be avoided by using

(*a*) tool with low positive rake angle ☐ (*b*) high feed rate ☐
(*c*) high cutting speed ☐ (*d*) large depth of cut ☐

(ME-GATE-2003)

65. A steel bar 200 mm in diameter is turned at a feed of 0.25 mm/rev with a depth of cut of 4 mm. The rotational speed of the work-piece is 160 rpm. The material removal rate in mm^3/s is

(*a*) 160 ☐ (*b*) 167.6 ☐
(*c*) 1600 ☐ (*d*) 1675.5 ☐

(ME & PI-GATE-2013)

66. Choose the wrong statement

(*a*) If helix angle decreases, rake angle also decreases. ☐
(*b*) A very large rake angle makes a tightly rolled chips. ☐
(*c*) The point angle less than 118° is used for brittle materials. ☐
(*d*) Higher point angle reduced width of cut and produces thicker chips for same feed rate . ☐
(*e*) All of the above. ☐
(*f*) None of the above. ☐

67. In with-worth quick return mechanism (used in shapers) the velocity of the ram is maximum at

(*a*) middle of the forward stroke ☐ (*b*) beginning of return stroke ☐
(*c*) end of return stroke ☐ (*d*) middle of the return stroke ☐

(PI-GATE-2001)

68. Match the following metal forming processing with their associated stresses in the work-piece

Metal forming process	**Type of stress**
1. Coining	*P*. Tensile
2. Wire drawing	*Q*. Shear
3. Blanking	*R*. Tensile and compressive
4. Deep drawing	*S*. Compressive

Codes:

(*a*)	1 – *S*	2 – *P*	3 – *Q*	4 – *R*	☐
(*b*)	1 – *P*	2 – *Q*	3 – *S*	4 – *R*	☐
(*c*)	1 – *S*	2 – *P*	3 – *R*	4 – *Q*	☐
(*d*)	1 – *P*	2 – *R*	3 – *Q*	4 – *S*	☐

(ME & PI-GATE-2012)

69. A dimension is expressed as $20.5^{+00.0}_{-00.2}$. This is a case of

(*a*) bilateral tolerance ☐ (*b*) unilateral tolerance ☐
(*c*) trilateral tolerance ☐ (*d*) none of the above. ☐

70. In a bending operation, if the modulus of elasticity *E* is increased keeping all other parameters unchanged, the spring back will

(*a*) increase ☐ (*b*) decrease ☐
(*c*) remains unchanged ☐ (*d*) be independent of *E* ☐

(PI-GATE-1994)

71. The face angles of the teeth of a broaching tool for brass is approximately equal to

(*a*) 6 to 8° ☐ (*b*) 15° ☐

(*c*) 70° ☐ (*d*) – 5 to + 5°. ☐

ANSWERS

Answers to Objective Type Questions

1. (*f*)	**2.** (*b*)	**3.** (*a*)	**4.** (*a*)	**5.** (*b*)	**6.** (*b*)
7. (*c*)	**8.** (*b*)	**9.** (*a*)	**10.** (*c*)	**11.** (*b*)	**12.** (*b*)
13. (*b*)	**14.** (*a*)	**15.** (*c*)	**16.** (*d*)	**17.** (*d*)	**18.** (*d*)
19. (*c*)	**20.** (*b*)	**21.** (*b*)	**22.** (*c*)	**23.** (*b*)	**24.** (*d*)
25. (*d*)	**26.** (*d*)	**27.** (*b*)	**28.** (*a*)	**29.** (*b*)	**30.** (*c*)
31. (*c*)	**32.** (*d*)	**33.** (*a*)	**34.** (*b*)	**35.** (*b*), (*c*)	**36.** (*d*)
37. (*b*)	**38.** (*a*)	**39.** (*c*)	**40.** (*b*)	**41.** (*c*)	**42.** (*b*)
43. (*c*)	**44.** (*d*)	**45.** (*c*)	**46.** (*d*)	**47.** (*c*)	**48.** (*b*)
49. (*a*)	**50.** (*a*)	**51.** (*d*)	**52.** (*b*)	**53.** (*d*)	**54.** (*b*)
55. (*c*)	**56.** (*d*)	**57.** (*b*)	**58.** (*b*)	**59.** (*a*)	**60.** (*a*)
61. (*b*)	**62.** (*d*)	**63.** (*c*)	**64.** (*b*)	**65.** (*b*)	**66.** (*c*)
67. (*d*)	**68.** (*d*)	**69.** (*c*)	**70.** (*c*)	**71.** (*c*)	**72.** (*b*)
73. (*c*)	**74.** (*d*)	**75.** (*c*)	**76.** (*b*)	**77.** (*a*)	**78.** (*e*)
79. (*d*)	**80.** (*c*)	**81.** (*e*)	**82.** (*c*)	**83.** (*c*)	**84.** (*d*)
85. (*c*)	**86.** (*a*)	**87.** (*d*)	**88.** (*c*)	**89.** (*b*)	**90.** (*b*)
91. (*c*)	**92.** (*d*)	**93.** (*b*)	**94.** (*a*)	**95.** (*c*)	**96.** (*c*)
97. (*a*)	**98.** (*d*)	**99.** (*b*)	**100.** (*d*)	**101.** (*d*)	**102.** (*c*)
103. (*b*)	**104.** (*b*)	**105.** (*a*)	**106.** (*c*)	**107.** (*d*)	**108.** (*a*)
109. (*c*)	**110.** (*d*)	**111.** (*e*)	**112.** (*f*)	**113.** (*b*)	**114.** (*a*)
115. (*c*)	**116.** (*a*)	**117.** (*c*)	**118.** (*a*)	**119.** (*d*)	**120.** (*d*)
121. (*d*)	**122.** (*c*)	**123.** (*b*)	**124.** (*c*)	**125.** (*b*)	**126.** (*b*)
127. (*c*)	**128.** (*b*)	**129.** (*a*)	**130.** (*d*)	**131.** (*b*)	**132.** (*d*)
133. (*c*)	**134.** (*c*)	**135.** (*f*)	**136.** (*c*)	**137.** (*d*)	**138.** (*b*)
139. (*c*)	**140.** (*b*)	**141.** (*c*)	**142.** (*b*)	**143.** (*b*)	**144.** (*a*)
145. (*c*)	**146.** (*a*)	**147.** (*b*)	**148.** (*c*)	**149.** (*a*)	**150.** (*b*)
151. (*f*)	**152.** (*b*)	**153.** (*c*)	**154.** (*a*)	**155.** (*d*)	**156.** (*c*)
157. (*d*)	**158.** (*b*)	**159.** (*a*)	**160.** (*d*)	**161.** (*c*)	**162.** (*b*)
163. (*c*)	**164.** (*d*)	**165.** (*d*)	**166.** (*d*)	**167.** (*c*)	**168.** (*d*)
169. (*d*)	**170.** (*d*)	**171.** (*d*)	**172.** (*b*)	**173.** (*c*)	**174.** (*a*)
175. (*b*)	**176.** (*d*)	**177.** (*b*)	**178.** (*c*)	**179.** (*b*)	**180.** (*d*)
181. (*b*)	**182.** (*c*)	**183.** (*c*)	**184.** (*d*)	**185.** (*a*)	**186.** (*a*)
187. (*d*)	**188.** (*e*)	**189.** (*a*)	**190.** (*d*)	**191.** (*b*)	**192.** (*d*)
193. (*d*)	**194.** (*b*)	**195.** (*c*)	**196.** (*b*)	**197.** (*b*)	**198.** (*d*)

199. (*c*)	**200.** (*c*)	**201.** (*d*)	**202.** (*b*)	**203.** (*c*)	**204.** (*a*)
205. (*c*)	**206.** (*d*)	**207.** (*a*)	**208.** (*d*)	**209.** (*e*)	**210.** (*d*)
211. (*c*)	**212.** (*c*)	**213.** (*d*)	**214.** (*e*)	**215.** (*b*)	**216.** (*d*)
217. (*b*)	**218.** (*c*)	**219.** (*c*)	**220.** (*b*)	**221.** (*d*)	**222.** (*b*)
223. (*e*)	**224.** (*b*)	**225.** (*a*)	**226.** (*c*)	**227.** (*b*)	**228.** (*c*)
229. (*b*)	**230.** (*d*)	**231.** (*c*)	**232.** (*d*)	**233.** (*a*)	**234.** (*c*)
235. (*d*)	**236.** (*f*)	**237.** (*c*)	**238.** (*c*)	**239.** (*b*)	**240.** (*d*)
241. (*b*)	**242.** (*d*)	**243.** (*d*)	**244.** (*d*)	**245.** (*d*)	**246.** (*a*)
247. (*b*)	**248.** (*c*)	**249.** (*d*)	**250.** (*d*)	**251.** (*b*)	**252.** (*a*)
253. (*b*)	**254.** (*b*)	**255.** (*a*)	**256.** (*c*)	**257.** (*a*)	**258.** (*e*)
259. (*d*)	**260.** (*c*)	**261.** (*a*)	**262.** (*c*)	**263.** (*c*)	**264.** (*c*)
265. (*b*)	**266.** (*b*)	**267.** (*b*)	**268.** (*a*)	**269.** (*b*)	**270.** (*b*)
271. (*e*)	**272.** (*a*)	**273.** (*c*)	**274.** (*d*)	**275.** (*d*)	**276.** (*a*)
277. (*d*)	**278.** (*c*)	**279.** (*c*)	**280.** (*a*)	**281.** (*b*)	**282.** (*e*)
283. (*d*)	**284.** (*b*)	**285.** (*b*)	**286.** (*b*)	**287.** (*d*)	**288.** (*d*)
289. (*d*)	**290.** (*b*)	**291.** (*b*)	**292.** (*b*)	**293.** (*e*)	**294.** (*b*)
295. (*a*)	**296.** (*d*)	**297.** (*d*)	**298.** (*d*)	**299.** (*d*)	**300.** (*c*)
301. (*c*)	**302.** (*c*)	**303.** (*f*)	**304.** (*d*)	**305.** (*d*)	**306.** (*c*)
307. (*d*)	**308.** (*a*)	**309.** (*b*)	**310.** (*d*)	**311.** (*d*)	**312.** (*d*)
313. (*f*)	**314.** (*d*)	**315.** (*e*)	**316.** (*c*)	**317.** (*b*)	**318.** (*b*)
319. (*a*)	**320.** (*b*)	**321.** (*b*)	**322.** (*a*)	**323.** (*b*)	**324.** (*c*)
325. (*c*)	**326.** (*a*)	**327.** (*e*)	**328.** (*b*)	**329.** (*e*)	**330.** (*f*)
331. (*e*)	**332.** (*a*)	**333.** (*a*)	**334.** (*c*)	**335.** (*d*)	**336.** (*b*)
337. (*a*)	**338.** (*b*)	**339.** (*a*)	**340.** (*d*)	**341.** (*c*)	**342.** (*c*)
343. (*c*)	**344.** (*b*)	**345.** (*c*)	**346.** (*a*)	**347.** (*e*)	**348.** (*b*)
349. (*a*)	**350.** (*c*)	**351.** (*c*)	**352.** (*c*)	**353.** (*f*)	**354.** (*d*)
355. (*c*)	**356.** (*c*)	**357.** (*c*)	**358.** (*d*)	**359.** (*b*)	**360.** (*c*)
361. (*b*)	**362.** (*e*)	**363.** (*d*)	**364.** (*f*)	**365.** (*b*)	**366.** (*b*)
367. (*d*)	**368.** (*a*)	**369.** (*b*)	**370.** (*c*)	**371.** (*b*)	**372.** (*d*)
373. (*c*)	**374.** (*d*)	**375.** (*c*)	**376.** (*a*)	**377.** (*b*)	**378.** (*a*)
379. (*d*)	**380.** (*d*)	**381.** (*d*)	**382.** (*e*)	**383.** (*c*)	**384.** (*b*)
385. (*c*)	**386.** (*b*)	**387.** (*d*)	**388.** (*b*)	**389.** (*a*)	**390.** (*c*)
391. (*b*)	**392.** (*d*)	**393.** (*e*)	**394.** (*b*)	**395.** (*c*)	**396.** (*d*)
397. (*b*)	**398.** (*c*)	**399.** (*e*)	**400.** (*d*)	**401.** (*a*)	**402.** (*d*)
403. (*f*)	**404.** (*b*)	**405.** (*c*)	**406.** (*d*)	**407.** (*b*)	**408.** (*d*)
409. (*d*)	**410.** (*a*)	**411.** (*b*)	**412.** (*b*)	**413.** (*a*)	**414.** (*b*)
415. (*a*)	**416.** (*f*)	**417.** (*c*)	**418.** (*c*)	**419.** (*d*)	**420.** (*b*)
421. (*b*)	**422.** (*c*)	**423.** (*a*)	**424.** (*b*)	**425.** (*c*)	**426.** (*a*)
427. (*b*)	**428.** (*c*)	**429.** (*d*)	**430.** (*b*)	**431.** (*c*)	**432.** (*b*)

433. (*c*)	434. (*b*)	435. (*d*)	436. (*c*)	437. (*c*)	438. (*b*)
439. (*b*)	440. (*a*)	441. (*b*)	442. (*d*)	443. (*a*)	444. (*a*)
445. (*e*)	446. (*c*)	447. (*b*)	448. (*b*)	449. (*b*)	450. (*d*)
451. (*e*)	452. (*d*)	453. (*d*)	454. (*b*)	455. (*d*)	456. (*b*)
457. (*c*)	458. (*c*)	459. (*b*)	460. (*b*)	461. (*a*)	462. (*d*)
463. (*b*)	464. (*e*)	465. (*c*)	466. (*c*)	467. (*b*)	468. (*c*)
469. (*e*)	470. (*d*)	471. (*d*)	472. (*c*)	473. (*b*)	474. (*a*)
475. (*b*)	476. (*b*)	477. (*c*)	478. (*b*)	479. (*b*)	480. (*a*)
481. (*d*)	482. (*a*)	483. (*c*)	484. (*b*)	485. (*d*)	486. (*b*)
487. (*c*)	488. (*a*)	489. (*c*)	490. (*c*)	491. (*b*)	492. (*a*)
493. (*d*)	494. (*c*)	495. (*c*)	496. (*b*)	497. (*c*)	498. (*c*)
499. (*d*)	500. (*d*)	501. (*b*)	502. (*d*)	503. (*b*)	504. (*a*)
505. (*c*)	506. (*a*)	507. (*d*)	508. (*b*)	509. (*b*)	510. (*a*)
511. (*c*)	512. (*d*)	513. (*b*)	514. (*b*)	515. (*c*)	516. (*d*)
517. (*a*)	518. (*b*)	519. (*c*)	520. (*f*)	521. (*b*)	522. (*a*)
523. (*c*)	524. (*a*)	525. (*d*)	526. (*c*)	527. (*a*)	528. (*b*)
529. (*b*)	530. (*a*)	531. (*b*)	532. (*a*)	533. (*b*)	534. (*c*)
535. (*c*)	536. (*d*)	537. (*e*)	538. (*c*)	539. (*a*)	540. (*d*)
541. (*b*)	542. (*d*)	543. (*d*)	544. (*a*)	545. (*b*)	546. (*d*)
547. (*b*)	548. (*c*)	549. (*d*)	550. (*c*)	551. (*b*)	552. (*d*)
553. (*b*)	554. (*b*)	555. (*c*)	556. (*c*)	557. (*a*)	558. (*b*)
559. (*c*)	560. (*a*)	561. (*b*)	562. (*a*)	563. (*c*)	564. (*c*)
565. (*c*)	566. (*d*)	567. (*c*)	568. (*d*)	569. (*b*)	570. (*b*)
571. (*d*)	572. (*c*)	573. (*d*)	574. (*d*)	575. (*b*)	576. (*b*)
577. (*e*)	578. (*e*)	579. (*b*)	580. (*f*)	581. (*d*)	582. (*c*)
583. (*a*)	584. (*c*)	585. (*d*)	586. (*b*)	587. (*d*)	588. (*e*)
589. (*c*)	590. (*d*)	591. (*a*)	592. (*b*)	593. (*c*)	594. (*c*)
595. (*a*)	596. (*b*)	597. (*d*)	598. (*d*)	599. (*b*)	600. (*d*)
601. (*d*)	602. (*f*)	603. (*c*)	604. (*a*)	605. (*d*)	606. (*e*)
607. (*c*)	608. (*e*)	609. (*b*)	610. (*a*)	611. (*c*)	612. (*e*)
613. (*b*)	614. (*c*)	615. (*b*)	616. (*b*)	617. (*e*)	618. (*a*)
619. (*d*)	620. (*e*)	621. (*b*)	622. (*b*)	623. (*f*)	624. (*e*)
625. (*c*)	626. (*d*)	627. (*f*)	628. (*e*)	629. (*d*)	630. (*d*)
631. (*b*)	632. (*c*)	633. (*a*)	634. (*b*)	635. (*c*)	636. (*d*)
637. (*c*)	638. (*d*)	639. (*b*)	640. (*c*).		

True/False

641. (*a*)	642. (*a*)	643. (*b*)	644. (*a*)	645. (*b*)	646. (*a*)
647. (*a*)	648. (*a*)	649. (*a*)	650. (*b*)	651. (*b*)	652. (*b*)
653. (*a*)	654. (*a*)	655. (*a*)	656. (*a*)	657. (*b*)	658. (*a*)
659. (*b*)	660. (*a*)	661. (*b*)	662. (*a*)	663. (*c*)	664. (*b*)

665. (*a*) **666.** (*a*) **667.** (*a*) **668.** (*b*) **669.** (*a*) **670.** (*b*)
671. (*a*) **672.** (*a*) **673.** (*a*) **674.** (*b*) **675.** (*b*) **676.** (*a*)
677. (*b*) **678.** (*a*) **679.** (*a*) **680.** (*a*) **681.** (*a*) **682.** (*a*)
683. (*a*) **684.** (*a*) **685.** (*b*) **686.** (*b*) **687.** (*a*) **688.** (*a*)
689. (*a*) **690.** (*b*) **691.** (*a*) **692.** (*a*) **693.** (*a*) **694.** (*b*)
695. (*b*) **696.** (*b*) **697.** (*a*) **698.** (*a*) **699.** (*b*) **700.** (*a*)
701. (*b*) **702.** (*a*) **703.** (*a*) **704.** (*a*) **705.** (*a*) **706.** (*b*)
707. (*b*) **708.** (*b*) **709.** (*a*) **710.** (*b*) **711.** (*a*) **712.** (*a*)
713. (*a*) **714.** (*a*) **715.** (*b*) **716.** (*a*) **717.** (*b*) **718.** (*a*)
719. (*b*) **720.** (*a*) **721.** (*a*) **722.** (*a*) **723.** (*a*) **724.** (*b*)
725. (*a*) **726.** (*a*) **727.** (*a*) **728.** (*a*) **729.** (*a*) **730.** (*a*)
731. (*b*) **732.** (*a*) **733.** (*a*) **734.** (*a*) **735.** (*a*) **736.** (*a*)
737. (*a*) **738.** (*a*) **739.** (*a*) **740.** (*a*) **741.** (*b*) **742.** (*b*)
743. (*a*) **744.** (*a*) **745.** (*a*) **746.** (*a*) **747.** (*a*) **748.** (*b*)
749. (*a*) **750.** (*a*) **751.** (*a*) **752.** (*b*) **753.** (*a*) **754.** (*a*)
755. (*a*) **756.** (*b*) **757.** (*b*) **758.** (*a*) **759.** (*a*) **760.** (*b*)
761. (*a*) **762.** (*a*) **763.** (*b*) **764.** (*a*) **765.** (*a*) **766.** (*a*)
767. (*b*) **768.** (*a*) **769.** (*a*) **770.** (*a*) **771.** (*a*) **772.** (*a*)
773. (*a*) **774.** (*b*) **775.** (*b*) **776.** (*a*) **777.** (*a*) **778.** (*a*)
779. (*b*) **780.** (*a*) **781.** (*a*) **782.** (*a*) **783.** (*a*) **784.** (*a*)
785. (*a*) **786.** (*a*) **787.** (*a*) **788.** (*a*) **789.** (*b*) **790.** (*a*)
791. (*a*) **792.** (*a*) **793.** (*b*) **794.** (*a*) **795.** (*a*) **796.** (*a*)
797. (*a*) **798.** (*a*) **799.** (*a*) **800.** (*a*).

Fill in the Blanks

801. (*b*) **802.** (*a*) **803.** (*b*) **804.** (*a*) **805.** (*b*) **806.** (*b*)
807. (*a*) **808.** (*b*) **809.** (*b*) **810.** (*b*) **811.** (*a*) **812.** (*b*)
813. (*b*) **814.** (*b*) **815.** (*b*) **816.** (*a*) **817.** (*b*) **818.** (*b*)
819. (*b*) **820.** (*b*) **821.** (*b*) **822.** (*a*) **823.** (*b*) **824.** (*a*)
825. (*b*) **826.** (*a*) **827.** (*b*) **828.** (*a*) **829.** (*b*) **830.** (*b*)
831. (*b*) **832.** (*b*) **833.** (*b*) **834.** (*a*) **835.** (*a*) **836.** (*b*)
837. (*b*) **838.** (*b*) **839.** (*b*) **840.** (*b*) **841.** (*b*) **842.** (*b*)
843. (*b*) **844.** (*a*) **845.** (*b*) **846.** (*b*) **847.** (*a*) **848.** (*b*)
849. (*b*) **850.** (*b*) **851.** (*b*) **852.** (*a*) **853.** (*b*) **854.** (*a*)
855. (*b*) **856.** (*a*) **857.** (*b*) **858.** (*b*) **859.** (*b*) **860.** (*b*)
861. (*a*) **862.** (*b*) **863.** (*b*) **864.** (*b*) **865.** (*a*) **866.** (*b*)
867. (*b*) **868.** (*a*) **869.** (*b*) **870.** (*a*) **871.** (*b*) **872.** (*b*)
873. (*a*) **874.** (*c*) **875.** (*e*) **876.** (*c*) **877.** (*b*) **878.** (*a*)
879. (*c*) **880.** (*c*) **881.** (*c*) **882.** (*b*) **883.** (*d*) **884.** (*e*)
885. (*b*) **886.** (*d*) **887.** (*d*) **888.** (*d*) **889.** (*d*) **890.** (*c*)

891. (*a*)	**892.** (*c*)	**893.** (*a*)	**894.** (*b*)	**895.** (*b*)	**896.** (*d*)
897. (*c*)	**898.** (*d*)	**899.** (*a*)	**900.** (*c*)	**901.** (*b*)	**902.** (*a*)
903. (*c*)	**904.** (*d*)	**905.** (*d*)	**906.** (*a*)	**907.** (*d*)	**908.** (*c*)
909. (*c*)	**910.** (*c*)	**911.** (*b*)	**912.** (*b*)	**913.** (*a*)	**914.** (*a*)
915. (*a*)	**916.** (*c*)	**917.** (*b*).			

Answers to Objective Type Questions from Competitive Examinations

1. (*b*)	**2.** (*d*)	**3.** (*d*)	**4.** (*c*)	**5.** (*d*)	**6.** (*b*)
7. (*d*)	**8.** (*d*)	**9.** (*a*)	**10.** (*d*)	**11.** (*c*)	**12.** (*c*)
13. (*d*)	**14.** (*a*)	**15.** (*b*)	**16.** (*b*)	**17.** (*c*)	**18.** (*d*)
19. (*b*)	**20.** (*a*)	**21.** (*b*)	**22.** (*d*)	**23.** (*c*)	**24.** (*b*)
25. (*a*)	**26.** (*a*)	**27.** (*a*)	**28.** (*d*)	**29.** (*b*)	**30.** (*b*)
31. (*c*)	**32.** (*b*)	**33.** (*a*)	**34.** (*c*)	**35.** (*b*)	**36.** (*a*)
37. (*b*)	**38.** (*d*)	**39.** (*a*)	**40.** (*c*)	**41.** (*b*)	**42.** (*a*)
43. (*d*)	**44.** (*b*)	**45.** (*c*)	**46.** (*d*)	**47.** (*d*)	**48.** (*b*)
49. (*c*)	**50.** (*a*)	**51.** (*b*)	**52.** (*c*)	**53.** (*b*)	**54.** (*d*)
55. (*a*)	**56.** (*d*)	**57.** (*b*)	**58.** (*d*)	**59.** (*d*)	**60.** (*c*)
61. (*b*)	**62.** (*c*)	**63.** (*c*)	**64.** (*c*)	**65.** (*d*)	**66.** (*f*)
67. (*d*)	**68.** (*a*)	**69.** (*b*)	**70.** (*b*)	**71.** (*a*)	

Chapter 13

Industrial Engineering and Production Management

I. THEORY

13.1. FUNCTION OF MANAGEMENT

The general functions of management are:

(*i*) Undertaking risk and handling uncertainty.

(*ii*) Planning and innovation.

(*iii*) Co-ordination, administration and control.

(*iv*) Routine supervision.

13.1.1. Undertaking Risk and Handling Uncertainty

This aspect is very prominent when a new industry is established.

13.1.2. Planning and Innovation

This aspect is very prominent while establishing a new industry. In very big industries mass producing similar products, careful planning is done in the initial stage. Thereafter the work becomes routine. Planning is generally confined to taking note of scientific developments and better methods of manufacture. In smaller job works planning had to be done at every stage and continuously.

13.1.3. Co-ordination, Administration and Control

The activities of the various personnel in an industry has to be planned and co-ordinated in such a way that work is done smoothly without wastage. Control consists in telling people what to do and how to do and to see that they carry out the instructions.

13.1.4. Routine Supervision

It is necessary to see that in the industry, correct working hours and correct working habits are observed. It is also necessary to see that materials and equipment and tools are well-kept and proper security arrangements are provided for their safety. A routine check up is necessary to look after this aspect of the industry.

13.2. DECISION-MAKING

These are systematic procedures to help choose the best course of action under given circumstances. In a typical decision-making situation, the decision-maker has two or more courses of action open to him, and he is faced with the problem of selecting one of them—one that best fulfils his objectives. Such problems frequently arise in choosing one of the many alternative design concepts, materials of construction, processes and machinery for production etc. In every such case, an inferior decision would increase costs, reduce profits or slow down production. It is, therefore, necessary that all decisions should be arrived at in a rational manner, and not taken on the heat of the moment. The decision problems can be classified in two categories:

(*a*) Decision-making under certainty (*b*) Decision-making under uncertainty.

13.2.1. Decision-Making Under Certainty

When circumstances are such that outcomes resulting from all possible actions of the decision-maker can be estimated with certainty, the chance plays little part. Decision-making under such conditions is called decision-making under certainty. Many problems of decision-making in production systems engineering more or less fall under this category. For example, choosing between alternative construction materials for specific parts and components, and choice between alternative production processes, plant layouts and product mix are problems which frequently belong to this category.

13.2.2. Decision-Making Under Uncertainty

Many situations are encountered in production systems engineering where the outcomes resulting from an action are not certain because of factors over which the company has little or no control. For example, demand for a new product or service introduced by the company in a new market, is controlled by several external factors, like consumer preferences, competition, substitutes etc. which are beyond the control of the company and therefore make any estimate of demand quite uncertain. Decision-making under such circumstances is called decision-making under uncertainty.

13.3. INTERNAL ORGANISATION

Internal Organisation refers to allocation of authority and responsibility to each employee of the production concern.

The main terms used in internal organisational planning are defined below:

1. **Department.** It is a grouping of facilities and employees for carrying out similar activities.
2. **Line Functions.** Functions connected with primary goals of production are known as line functions, *i.e.,* production, sales etc.

3. **Staff-Functions.** Functions which advise and support the line functions, *i.e.,* Quality control. Production planning and control.
4. **Span of Control.** It refers to number of subordinates who take orders and report to a superior directly.
5. **Unity of Commands.** In this a subordinate takes orders from one superior directly.

13.4. FORMS OF OWNERSHIP

Business ownership can be classified as follows :

1. Sole proprietorship
2. Partnership
3. Joint stock companies
4. Co-operative societies
5. State enterprises.

13.4.1. Sole Proprietorship

It is the earliest form of business unit. One man or family owns the enterprise. The entire capital is provided by him. He knows his business well. He knows his employees intimately. He can take quick decisions. He can control the business efficiently. But the capital available is rather limited. The control tends to be autocratic.

13.4.2. Partnership

Concerns are those in which two or more partners provide the capital and divide the profit in an agreed manner. All the partners may work or one may be the active partner and other sleeping partners. At least one of the partners should have unlimited liability. Partnerships enable collection of greater capital and sharing of the management work.

13.4.3. Joint Stock Companies

When very large capital is required, joint stock companies are formed. The capital is divided into small units called shares and these are sold to the public. The company should be registered before formation.

13.4.4. Co-operative Societies

They are similar to joint stock companies as far as shares are concerned. The difference is in the process of electing the directors. Each member is entitled to only one vote irrespective of the shares held. This makes the election democratic. In actual practice, this system may not be conducive for risk-taking and initiative as often very little stakes are involved.

13.4.5. State Enterprises

They are sometimes owned by the state. Posts, telegraphs, railways etc., are some of the common services owned and operated by the state. In such enterprises service and not profit is generally the guiding factor. The absence of profit motive leads to inefficiency and sometimes loss which are borne by the public. The risk of loss is borne by the tax-payers and control is vested in persons who are appointed. The control becomes very complicated. Elaborate rules and regulations are framed which are not conductive to quick and efficient work and decisions.

13.5. ELEMENTARY ECONOMICS (DEMAND, SUPPLY AND PRICES: ELASTICITY)

The price of articles will depend on the following factors:

1. The demand for the article,
2. The supply of the article, and
3. The marginal utility of the article.

The demand and supply are interlinked. The greater the price, the greater will be the supply. The less the price, the greater will be the demand. This is known as the law of supply and demand. This may not be true always. But, in general, the law can be takes to be true.

When prices change, the demand may also change to a large extent. In such cases, the demand is said to be *elastic*. If changes in prices do not cause corresponding changes in demand then the demand is *inelastic*.

The prices that articles will fetch, will depend on the elasticity of demand.

If the demand is elastic, it means that

1. alternate products are available,
2. the available supply is large,
3. necessity is not intense.

In such cases, the chances of getting higher prices are remote. Manufactures must be careful in increasing their prices as their sale may be affected if the demand for their products is elastic.

If the demand is inelastic, it means that

1. the articles are necessities,
2. changes of alternate products remote, and
3. availability is limited.

In such cases, the manufacturers can easily get the prices they fix. But they cannot fix the prices too high as they will attract competition. Then the prices will be forced down.

The ultimate price the manufacturer will get is called the **equilibrium price.**

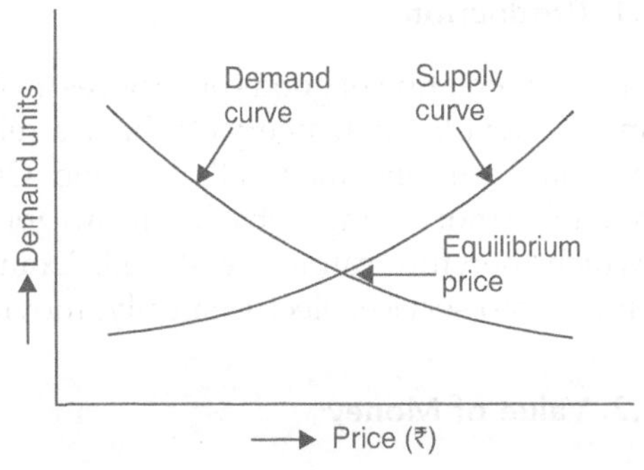

FIGURE 13.1

The higher the price the greater will be the supply and *vice versa*. At a particular price, the supplier's price and the consumer's price will coincide. This is known as the equilibrium price. This can be graphically illustrated as shown in Fig. 13.1.

At low prices the demand will be high. As prices increase the demand will fall down. This is indicated by the demand curve. At high prices, the supply is high and at low prices, the supply is small. This is indicated by the supply curve. The price at the intersection of these curves is known as the equilibrium price.

At this price customers are willing to purchase and suppliers are willing to sell as much as they like.

Besides the law of supply and demand there is another factor that influences the price. This is known as the marginal utility of the product.

13.6. COMPETITION

The perfect competition is that which fulfils the following requirements:

1. Larger number of producers, retailers and customers.
2. Production of 'identical items'.
3. Free market-entry.
4. Perfect mobility in production factors.
5. Knowledge of prices to customers.

Perfect competition practically never exists. However, pure competition is there in foodgrains, vegetables and fruits. Here first three conditions are satisfied. Prices in the pure competition depends on the particular industry producing it.

Imperfect competition is a more practical case. In this the producers produce same goods but not identical. Due to special quality in goods some customers continue to purchase that even if there is hike in price. If prices increase beyond a particular level, the customer may switch over to a substitute.

In true monopoly, the total income of the customers is taken away by one supplier. A monopolist has control over production, supply and prices of an item.

13.7. PRODUCTION AND CONSUMPTION

13.7.1. Production

Production is creation of goods or services (*a*) Goods (also called products or Durables), for example automobile, textiles, machinery etc. have a relatively long life and can be stored. The goods have a life-cycle and are consumed in that period. Creation of goods is called manufacturing, (*b*) Services are intangible (not having a physical form) goods which are consumed as soon as they are produced, and which therefore cannot be stored. Examples of services are—repair of an automobile, dry cleaning, cargo service, electric supply, movie show, entertainment etc.

13.7.2. Value of Money

It means that a sum of money can always be put to some productive use, thereby earning profit and increasing the amount or value of money. Larger the period and larger the investment, greater are the profits. On the other hand, if a sum of money is not put to any positive use, *i.e.*, it is just as idle cash, no profits can be earned.

13.7.3. Job and Work Situation

Job and work situation refer to (*a*) the physical conditions surrounding the workplace, (*b*) physical and mental stress caused while performing the job, and (*c*) the psychological environment on the shop floor. Following conditions are normally conducive or employee morale.

(*i*) The worker is protected against discomfort or harm arising from heat and cold, vibration and noise, fumes and radiation etc.

(*ii*) The machine is provided with safety guards to minimize the risk of accident.

(*iii*) Every effort is made to minimize physical exertion and mental drudgery of the employee while on the job.

(*iv*) If there is any risk of accident, the worker is given a thorough training in safe procedures.

(*v*) Some workers prefer variety in their job to prevent monotory and boredom. Such workers are allowed (*a*) to periodically switch over from one job to another (job rotation), (*b*) to participate in planning, organizing and controlling their job (job enrichment), or (*c*) to work on larger jobs *i.e.*, those involving greater number of operations and longer cycle times (job enlargement).

(*vi*) The pace of work is not tiresome or fatiguing.

(*vii*) The worker is allowed as far as possible to set his own working pace and schedule his own rest periods.

(*viii*) The shop and its surroundings are regularly cleaned. Dirt and refuse is not allowed to accumulate.

(*ix*) The workers are provided adequate wash-up, toilet, locker and canteen facilities.

(*x*) It is the policy of management to promote courteous behaviour of superiors with their subordinates. Harsh dealings are discouraged. No brawls or quarrels are allowed between one employee and another within the plant premises.

13.7.4. Motivational Aspects

The fundamental in motivation is that a person desires to exist and survive for this he requires basic necessities of life, *e.g.*, food, shelters, cloth, education and medical aid. The another aspects of motivation is the desire to achieve satisfaction and bliss a goal. Motivation performs the following functions:

(*i*) It originates action.

(*ii*) It directs activities towards a goal.

(*iii*) It helps to continue the activities until the goal is achieved. Motivation is needed to create interest, initiative, enthusiasm and loyalty. It increases the production and proves quality of the products. Motivation helps in maintaining better relations between the employee and the employees.

13.7.5. Kinds of Motivation

Motivation is either negative or positive, punishments, reprimands, fear of loss of jobs form negative motivation. They tend to create psychological fear. Positive motivation makes people do their work in the best manner and improves their performance.

13.7.6. Safety Measures

Following are the usual safety measures adopted to enhance an employee's morale:

(*i*) The m/c is provided with safety guards to minimize the risk of accident.

(*ii*) The worker is given through training in safe procedures.

(*iii*) Medical services to take care of accidental injuries and occupational diseases.

(*iv*) Medical assistance either in the form of reimbursement of medical bills or medical attention to the employees and their families.

(*v*) Insurance for health, and against accidents and death.

(*vi*) Provision of housing or assistance in finding houses.

(*vii*) Adequate eating facilities for the employees, especially if the plant is located away from urban areas. Clean, wholesome and subsidized metals may be arranged.

(*viii*) Transportation between the place of work and places of employee residence.

13.8. INVESTMENT DECISIONS

Capital costs affect investment decisions. Whenever a physical asset or expenditure is involved that provides a continuing benefit or return. In management science economic choice are required when (*a*) phasing-in of new products and services, and phasing-out of existing products and services, (*b*) making a choice between alternative production technologies, (*c*) choosing plant location and layout, and (*d*) when deciding about the questions of equipment replacement etc.

13.9. BREAK-EVEN ANALYSIS

This is a graphical or mathematical representation of the relationship between cost, income and quantity of production, in order to find the minimum quantity of production, called the Break-Even Quantity, at which cost of production equals income from production.

Suppose F is annual indirect costs (including capital-recovery, costs of land, buildings and equipment as well as other indirect costs) chargeable to a given product (or service). It is called the Fixed Cost. Also let V be the direct costs, called the Variable Cost, per unit of that product or service. If Q is the quantity of production and T_C is the corresponding total annual cost of production, then we have

$$T_C = F + VQ$$

If S is the unit selling price and I_S the total sales incomes,

$$I_S = SQ$$

In Fig. 13.2 lines ADB and OC represent T_c and I_S as function of Q.

When $$Q = Q_{BE}$$

$$T_C = I_S$$

$$\therefore \quad F + V.\, Q_{BE} = SQ_{BE}$$

$$\therefore \quad Q_{BE} = \frac{F}{S - V}$$

It would be observed as Q is less than Q_{BE}, we incur loss. Also greater is the difference Q–Q_{BE}, greater is the profit. In this discussion, we have assumed both V and S as constant.

13.9.1. Some Definitions for Break-Even Analysis

(*i*) **Break-Even Chart.** It is a graphic device required to determine the break-even point.

(*ii*) **Break-Even Point** (*i.e.*, Point *B* in Fig. 13.2). The point *B* at which the total cost of production line (*i.e.*, line *AB*) and income line (*i.e.*, line *OC*) intersect, is known as Break-even point. At this point, for a given volume of production, there will be no profit and no loss.

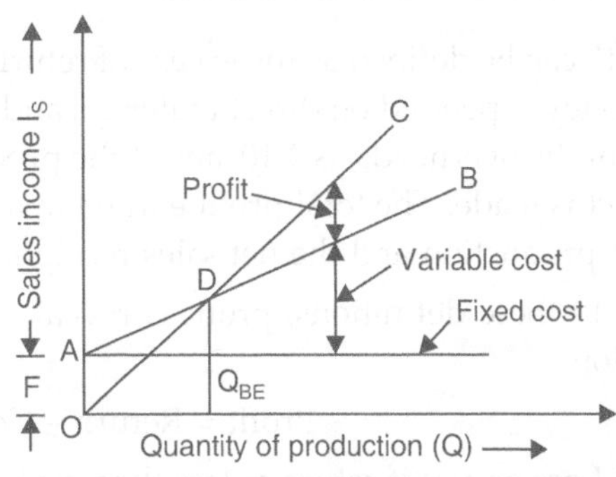

FIGURE 13.2

(*iii*) **Angle of Incidence.** It is the included angle between the income line and the total cost-line. The large angle indicates high profits and small angle indicates profits under less favourable conditions. Angle *CDB* is the angle of incidence in Fig. 13.2.

(*iv*) **Safety Margin.** It refers to the extent to which the concern can afford to decline in sales upto the break-even point. It can be represented by the percentage ratio of sales income over break-even point volume to the total sales income.

$$\therefore \quad \text{Safety margin} = \frac{\text{Total sales income} - \text{Sales income at B.E.P.}}{\text{Total sales income}} \times 100.$$

13.9.2. Application of Break-Even Chart

The application of B.E. chart are to find out:

(*i*) Safety margin.

(*ii*) Quantity needed for desired profit.

(*iii*) Effect of change in sale price.

(*iv*) Effect of change in costs (fixed and variable).

(*v*) Whether to accept or reject an order.

(*vi*) Whether to add or subtract a product.

(*vii*) Whether to make or buy a product.

13.9.3. Depreciation

The wear and tear of any equipment in use can not be totally prevented. However this can be minimized to some extent with proper care and proper maintenance. The equipment become less efficient with the lapse of time and therefore its value is reduced. The loss of value of an equipment is known is Depreciation. This amount however is deducted from the yearly gross profits of a company. This amount is kept separate and is used for replacement at end of useful life of that equipment. Depreciation can also be defined as systematic procedure of recovering every year a portion of investment made on an asset.

13.9.4. Profits and Loss Accounts

"Profit" can be defined as the excess of return over cost. To take a simple example, if the sum of all the money expended on direct materials and direct labour to produce a product, plus an equitable share of the overheads, is ₹ 10, and if the product is then sold for a return of ₹ 15, a profit of ₹ 5 per product is made. The total profit earned in a business will then be the difference between the total cost of production and the net sales receipts or "return".

In these definitions, profit is revealed as a simple factor related to cost and return by the equation.

$$\text{Profit} = \text{Return} - \text{Cost}$$

Loss occurs if return is less than cost.

13.10. PLANT LOCATION

Following are the main factors influencing the choice of location for a proposed plant.

1. Availability of land
2. Construction Cost.
3. Availability of Utilities.
4. Proximity to Sources of Raw Materials.
5. Proximity to the Market.
6. Availability of Skilled Personnel.
7. Government Policy.
8. Community Acceptability.
9. Local and State Laws.
10. Climate.

13.10.1. Steps in Plant location

The major steps in plant location are shown in :

1. Listing of Alternative Locations. A careful survey is made of all possible geographical locations within and outside the country. This is usually done by a site selection committee composed of knowledgeable people. The main considerations while preparing this list should be not to omit any likely location without proper analysis.

2. Identification of Feasible Locations. For each of the location listed in the forgoing step, detailed date is collected regarding the relevant factors, *e.g.*, market, raw materials, labour, transportation, utilities, climate, government policies, availability of land etc. Any location that seriously falls short of requirements with respect to one or more of the above factors is considered as infeasible.

3. Choice of the Best Location. Out of the feasible locations, the best one has to be selected. There are several methods of making such of choice, some of which are:

(*i*) Equivalent Annual Cost Method.

(*ii*) Transportation Method of Linear Programming.

(*iii*) Force Analogy Method.

13.10.2. Plant Layout

Plant layouts may be classified into four types as described below:

1. Fixed Position Layout. In this type of layout, the work (part or product being manufactured or serviced) remains in a fixed position. The machines, materials, tools and workers needed for production are brought to it. In all these cases, the product is so big in size or has some other characteristic that it is not convenient or desirable to move it.

2. Process Layout. In process layout (or Functional layout), the machines, equipment and workcentres performing similar functions are grouped together and arranged according to any particular sequence of operations.

3. Product Layout. In product layout (also known as Flowline layout, line layout or production line layout), the machines, equipment and workcentres are arranged in a straight or curved line according to the particular sequence of operations required for producing the given product (or service). The work flows from one workcentre to another in the line without back-tracking or cross-movements.

4. Group Technology (G.T.) Layout. In group technology (also known as cellular layout), a number of line layouts are established, one for each family of G.T. parts. The G.T. layout is being increasingly used for batch production machine shops in place of the coventional process layout.

13.10.3. Importance of Layout Planning

A well-planned layout of a production plant is essential for the following reasons:

1. Interdepartmental materials handlings cost depend on the relative location of the departments within the site.

2. The size and complexity of buildings and therefore the first cost of the plant depends on the plant layout.

3. Good layout avoids congestion and under utilization, and promotes safety.

4. Good layout minimizes workers fatigue and improves quality of production.

In actual fact, costs and effort of careful planning pay for itself many times over in the form of reduced operating cost and other benefits. On the other hand, the costs of inadequate layout planning are incurred during every working day of the plant.

13.10.4. Steps in Layout

1. Determination of space requirements.

2. Determination of flow of materials.

3. Synthesis of alternative overall layouts.
4. Choice of the best overall layout.
5. Detailed layout planning.

13.11. MATERIALS HANDLING

Materials handling within a plant is a non-productive operation because it does not add any value to the product or service. Nevertheless it is unavoidable. It is required to move materials, parts and products: (*i*) from receiving area to incoming inspection area, (*ii*) from incoming inspection area to stores, (*iii*) from stores to shop floor, (*iv*) between operations on the shop floor, and (*v*) from shop floor to warehouse. It has been estimated that the average cast of material handling in a manufacturing plant may be as much as 20% of the total production cost.

13.11.1. Principles of Materials Handling

(*i*) Eliminate materials handling wherever possible, *e.g.*, (*a*) by carrying out several operations at the same workstation instead of materials, and (*b*) by avoiding transfers of materials from one container to another, etc.

(*ii*) Minimize materials handling as much as possible, *e.g.*, (*a*) by devising optimum layout of plant and equipment, (*b*) by moving the loads in largest units possible, and (*c*) by correctly selecting materials handling equipment, etc.

(*iii*) Minimize handling effort by using gravity feed and by designing parts and products for easy handling.

(*iv*) All handling activities should be planned.

(*v*) Standardize methods as well as types and sizes of handling equipment. Handling methods should be safe.

13.11.2. Materials Handling Equipment

The usual types of materials handling equipment may be classified as under:

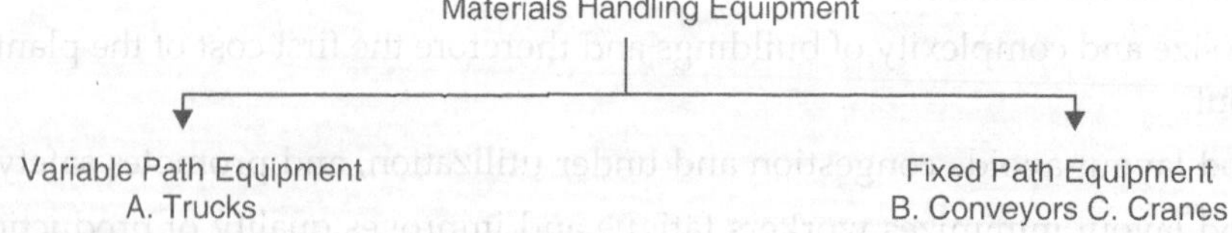

13.12. WORK-STUDY

Work-study includes methods study and work measurement.

13.12.1. Methods-Study

Methods-study is a systematic procedure of methods improvement. It involves (*a*) identification of the objectives of a given task, (*b*) through study of the existing method of performing the task, (*c*) critical examination of the existing method, (*d*) synthesis of alternative methods of achieving the desired objectives, and (*e*) choice of the best alternative. It may be usefully employed where the existing method appears to be inefficient, wasteful of human or machine effort, dangerous, unduly slow, slowing down other operations in the line or causing quality problems. The procedure of methods-study described below may be applied to both production or non-production operations (like office procedures, equipment maintenance etc.).

1. Choosing the Operation to be Studied. It is profitable to employ methods-study for operations (*a*) whose proportion in total man-hours or machine-hours is high, (*b*) which are production bottleneck, or (*c*) which are fatiguing, unpleasant or dangerous. The reasons for carrying out the study should in any case be recorded.

2. Identifying Objectives of the Operation. The need and the technological requirements of the operation, characteristics of the work-piece, and the place of the given operation in relation to other operations required for the production of the given part (or product etc.) are identified. This would normally require a careful study of product design as well.

3. Studying the Existing Method. The existing method of performing the operation is studied in detail including identification of the different tasks (steps) involved, characteristics of the machine, tools, work-place layout, garage of the operator and the times required for carrying out the different tasks of the operation. Various charting techniques, *e.g.*, man-machine chart, left and right-hand chart or flow diagram may be used to record details of the method.

4. Critically Examining the Existing Method. A systematic questioning technique may be adopted to examine the existing method in all its aspects. The questioning technique may be applied not only to the operation but also to the product design and work material.

5. Synthesizing Alternatives. Having analyzed all aspects of the existing method, alternative methods (solutions) are formulated. (The best alternative, of course, is to altogether eliminate the need of the operation if that is practicable). Attempt is made to synthesize as many alternative solutions as possible, some of which may be very different from the existing method and the others may merely by modifications if the existing method. While synthesizing alternative solution, principles of motion economy should be made use of.

6. Choosing the Best Alternative. The proposed solutions are first subjected to initial screening to reject those which give only marginal improvement or which fail to meet technological requirements concerning rate and quality of production. The remaining feasible solutions are compared amongst themselves and with the existing solutions of the basis of speed of operation, cost, quality of production, safety etc. The best solution is tried out, developed and put into practice.

Methods study is usually carried out by a team headed by a Methods Engineer. The members of the team are drawn from various 'interested' departments like design, research and development, production, purchasing etc.

13.13. WORK MEASUREMENT

Work measurement is concerned with establishing a time standard for specific task of job. Time standards are needed for the following reasons:

(*i*) To determine the number of machines required for a given capacity,

(*ii*) For performance evaluation of individual workers,

(*iii*) For incentive payments,

(*iv*) For evaluating alternative methods and equipment,

(*v*) For production scheduling, and

(*vi*) For cost estimating.

13.14. MAINTENANCE AND REPLACEMENT DECISIONS

13.14.1. Maintenance

Machine breakdowns are very costly. They not only interrupt production on the affected machines but sometimes on other machines also to which this machine feeds work. This multiplying effect is maximum in flow production where breakdown of one machine in the line may stop production on all other machines in the line. It is highly desirable to adopt a maintenance system in which the machines are periodically inspected if possible, without stopping them and those parts and components which are approaching the end of their service life are replaced before they cause expensive breakdown. This system of maintenance is called Preventive Maintenance. Because preventive maintenance is pre-planned, it can even be carried out in off time such as holidays and non-working hours to minimize disruption of normal production. Although the cost of preventive maintenance is higher than breakdown repairs, it pays for itself many times over in reduced downtime of the plant, especially in plants when flow production is carried out.

13.14.2. Breakdown Repairs

Even if there exists an effective preventive maintenance routine for an equipment, there would be occasions when the equipment breaks down inspite of preventive maintenance. The reason of such catastrophic failures is overloading or misuse of the equipment. When such sudden breakdowns occur, it calls for emergency repairs. The main objective during breakdown repairs is to put it back into operation as quickly as possible through speedy repairs. Reserve to highly skilled technicians, ample stock of spare parts, efficient filing system of catalogues and availability of stand by equipment—all serve to minimize ill-effects of equipment breakdowns.

13.14.3. Maintenance Economics

Preventive maintenance should not become uneconomical. To this effect the cost of breakdown and cost of maintenance schedule is studied. The cost should include expenditure on idle men, idle machines and spoilage. When the cost of maintenance schedule increases, the breakdown cost decreases. Increase in the expenditure on maintenance schedule becomes uneconomical after some extent. Two costs are plotted on a graph in Fig. 13.3. An optimum point 'O' gives the economical maintenance required.

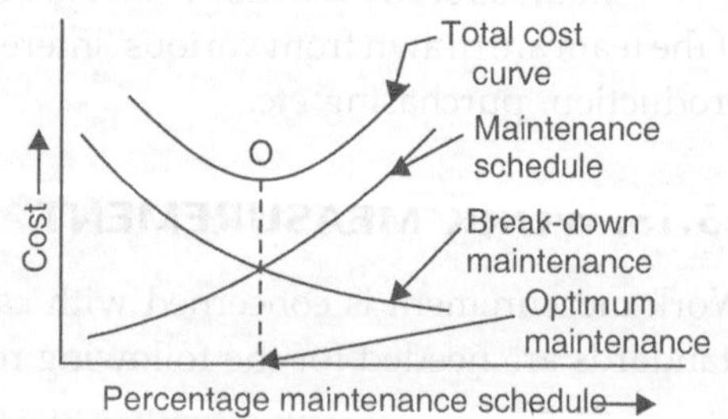

FIGURE 13.3

It is necessary to decide the amount of insection required on the basis of part records from an economic point of view. Over inspection results in large production time loss while under inspection causes more breakdowns. Schedule for inspection and lubrication should coincide for less production distruption. These schedules should be communicated properly to the production department.

13.14.4. Replacement Decisions

Replacement of an existing asset, may be necessitated by the following reasons:

(*i*) End of technological service repairs of the existing asset,

(*ii*) The existing asset is expected to fail catastrophically in the near future, and

(*iii*) Obsolescence of the asset because of design innovation such that a greatly superior equipment becomes available.

Two kinds of problems arise during replacement. The first problem is concerned with the determination of Economic Life of an Equipment, say a company car, which is expected to be replaced by a similar equipment. The second problem arises when it is to be explored whether it is economical to replace an existing asset by a superior asset.

1. Economic Life of Equipment. This problem may be solved by the undiscounted cash flow method or the discounted cash flow method.

2. Choice between Existing and New Equipment. There are three important replacement models for making a choice between an existing equipment and a new equipment of better design.

(*a*) **Engineering Economy Model.** In this method, either the equivalent annual costs or the expected rates of return for both the existing equipment and its proposed replacement are calculated and compared. Whilst making the economic analysis, the capital recovery costs in case of the existing equipment should be based on its present salvage value. Past investment made on the existing equipment are sunk costs. They are only relevant in so far as they affect depreciation and income-tax.

(*b*) **Dynamic Equipment Policy Model (Terbogh's Model).** In this method, the so-called Challenger's Adverse Minimum is compared with the cost of extending the service of the defender (existing equipment) by one year more.

(*c*) **MAPI Plan (Machinery and Allied Products Institute Plan).** Like Terbogh's plan, this plan also compares the costs of extending the service of the defender by another year with the challenger's adverse minimum. However, while calculating the challenger's adverse minimum, it takes into the **account** the challenger's salvage value and income tax.

Simulation in Replacement Studies

When the life on an equipment is likely to come to a sudden end, for example fluorescent tubes, electronic values etc., it is said to be statistical. The optimum replacement policy in such cases can be found by using simulation.

13.15. LINEAR PROGRAMMING

It is the mathematical technique used to select the best choice among several alternatives. When applied to an industry linear programming is used to find the best use of the limited resources of the industry is an optimum manner (*e.g.*, maximum or minimum) to achieve the industry objectives which may be maximum overall profit or minimum overall cost.

13.15.1. Applications

Such optimization problems arise many a time in production system engineering, *e.g.*, while (*a*) deciding about the optimum product-mix during resource allocation in a plant where demand exceeds capacity (Product Mix Problem), (*b*) deciding about the optimum location of a proposed plant (Transportation Problem), and (*c*) deciding about the optimum assignment of jobs to machines (Assignment Problem) etc. The linear programming problems are solved by the following methods:

(*i*) Graphical method, (*ii*) Simplex method, (*iii*) Index distribution, (*iv*) Modified distribution (MODI). The first two are commonly used.

13.16. PROJECT MANAGEMENT CPM AND PERT

13.16.1. CPM (Critical Path Method)

CPM is a management-network technique required to schedule and control special projects which are not continuous (Complicated). With its help management can know the progress operations required for completing the project work. CPM helps to find the earliest completion date for the whole project and the scheduling times for different activities. When the method is used to schedule and control the project, the following steps are required to plan and watch the progress:

1. Statement of the problem.
2. Determination of activities, events and duration times.
3. Drawing of the network diagram.
4. Check and supervise the schedule if desirable.
5. Progress of the project.

13.16.2. Advantage of CPM

1. It helps in better and detailed planning.
2. Most critical elements are identified and attended to at proper time.
3. It is standard method for communicating project plans, schedules etc.
4. Control of project becomes easier.

13.16.3. PERT (Programme Evaluation and Review Technique)

This is also a network technique used for scheduling and controlling complex projects, where the activities are subject to considerable degree of uncertainly in performance time. Timely project execution is very essential and is done by PERT. The main difference between CPM and PERT is that, with CPM a single estimate of time for each activity is made while with PERT three estimates of duration time for an activity are made. The three time estimate are:

(*a*) Pessimistic time. (*b*) Optimistic time. (*c*) Most likely time.

CPM is activity-oriented technique while PERT is event-oriented technique. For network diagram line and arrows are used to connect events. These events are represented by circles. Usually computers are necessary for PERT technique while for CPM computers may be only required when the activities are large in number.

13.16.4. Job Evaluation

Job Evaluation, or Job Appraisal, refers to the analysis **of** relative difficulty of jobs in an organization, and determination of a rational wage structure.

The important methods of job evaluation are:

1. Ranking method.
2. Factor comparison method.
3. Weighted points method.

13.16.5. Incentives

The purpose of incentives is to reward a more productive employee in proportion to his achievements of output or his own efforts. Incentives are given in addition to the guaranteed wages on base rates. A good incentive scheme motivatives workers to produce better and more.

13.16.6. Wage Payment

The wage payment is mainly divided into three systems. They are:

1. Time rate system
2. Piece rate system
3. Combination of time and piece rate system.

1. Time Rate System. According to this system, the workers (or employees) are paid at hourly, daily, weekly and monthly rate. In this system, there is no distinction between efficient workers as they are not paid according to their merit. This is the oldest system of wage payment used in industry.

2. Piece Rate System. According to **this** system, the workers (or employees) are paid a fixed rate for each unit produced or job competed.

In this system, there is no consideration to the time taken by the workers to complete the work but the payment is made according to the quantity of the work done.

3. Combination of the Time and Piece Rate System. This is similar to the piece rate system. In addition, a basic minimum rate guaranteed.

A standard time per unit of output is fixed. A minimum guaranteed hourly or daily rate is also fixed.

If a worker finishes a work within the standard time, he is allowed the basic rate for the full standard time. If his output is lower, he is allowed the basic rate for the actual time spent.

13.16.7. The Financial Incentive Plans

They are:

(*i*) The Halsey plan (*ii*) The Rowan plan

(*iii*) Emerson plan (*iv*) Taylor's differential piece rate system

(*v*) Merric's multiple piece rate system.

II. OBJECTIVE TYPE QUESTIONS

Tick mark the most appropriate statement of the multiple choice answers:

Work Study

1. Which one of the following is the basic tool in work study?

(*a*) stop watch ☐ (*b*) process chart ☐

(*c*) bar chart ☐ (*d*) planning chart. ☐

2. The engineering approach of detailed analysis and search for the most efficient way of producing products under certain constraints such as material, man power, machines, money etc. is known as

(*a*) production engineering ☐ (*b*) industrial engineering ☐

(*c*) mechanical engineering ☐ (*d*) thermal engineering. ☐

3. The study, which is used to find to a simpler, easier and better way of performing a job, is known as

(*a*) time study ☐ (*b*) motion study ☐

(*c*) motion and time study ☐ (*d*) none of the above. ☐

4. The study, which is used for symmetrically recording, analysing and synthesizing the times required to perform a motion or series of motion, is known as

(*a*) time study ☐ (*b*) motion study ☐

(*c*) motion and time study ☐ (*d*) none of the above. ☐

5. The study, which is used for determining the most economical and effective method of performing a job and also for determining time required by trained employees working at a normal pace to perform the job, known as

(*a*) time study ☐ (*b*) motion study ☐

(*c*) motion and time study ☐ (*d*) none of the above. ☐

6. Work study is used in

(*a*) industries ☐ (*b*) transport ☐

(*c*) hospital ☐ (*d*) design ☐

(*e*) all of the above ☐ (*f*) none of the above. ☐

7. The science which deals with the systematic investigation of the existing method of doing a job in order to develop and install an easy, rapid, efficient and effective procedure for doing the same job and at lower costs, is known as

(*a*) motion study ☐ (*b*) method study ☐
(*c*) time study ☐ (*d*) none of the above. ☐

8. The objective of the work measurement is to
(*a*) formulate a proper incentive scheme ☐
(*b*) plan and schedule of production ☐
(*c*) estimate the selling prices and delivery dates ☐
(*d*) all of the above ☐
(*e*) none of the above. ☐

9. Which one of the following is known as father of industrial engineering?
(*a*) Newton ☐ (*b*) Gilberth ☐
(*c*) Gnatt ☐ (*d*) Taylor. ☐

10. The Fig. 13.4 shows some symbols as recommended by the A.S.M.E. in work study. The symbol for transportation is given by
(*a*) Fig. 13.4 (*a*) ☐ (*b*) Fig. 13.4 (*b*) ☐
(*c*) Fig. 13.4 (*c*) ☐ (*d*) Fig. 13.4 (*d*) ☐
(*e*) Fig. 13.4 (*e*) ☐ (*f*) none of the above. ☐

(a) (b) (c) (d) (e)

FIGURE 13.4

11. In Fig. 13.4, the symbol for storage is given by
(*a*) Fig. 13.4 (*a*) ☐ (*b*) Fig. 13.4 (*b*) ☐
(*c*) Fig. 13.4 (*c*) ☐ (*d*) Fig. 13.4 (*d*) ☐
(*e*) Fig. 13.4 (*e*) ☐ (*f*) none of the above. ☐

12. In Fig. 13.4, the symbol for inspection is given by
(*a*) Fig. 13.4 (*a*) ☐ (*b*) Fig. 13.4 (*b*) ☐
(*c*) Fig. 13.4 (*c*) ☐ (*d*) Fig. 13.4 (*d*) ☐
(*e*) Fig. 13.4 (*e*) ☐ (*f*) none of the above. ☐

13. In Fig. 13.4, the symbol for operation is given by
(*a*) Fig. 13.4 (*a*) ☐ (*b*) Fig. 13.4 (*b*) ☐
(*c*) Fig. 13.4 (*c*) ☐ (*d*) Fig. 13.4 (*d*) ☐
(*e*) Fig. 13.4 (*e*) ☐ (*f*) none of the above. ☐

14. In Fig. 13.4, the symbol for delay is given by
(*a*) Fig. 13.4 (*a*) ☐ (*b*) Fig. 13.4 (*b*) ☐
(*c*) Fig. 13.4 (*c*) ☐ (*d*) Fig. 13.4 (*d*) ☐
(*e*) Fig. 13.4 (*e*) ☐ (*f*) none of the above. ☐

15. Which two of the following are used to make a time study?

(*a*) stop watch ☐ (*b*) bar chart ☐
(*c*) time study sheet ☐ (*d*) planning chart ☐
(*e*) (*a*) and (*c*) ☐ (*f*) (*b*) and (*d*). ☐

16. Standard times is equal to

(*a*) normal time + idle time ☐ (*b*) normal time + allowance ☐
(*c*) normal time + idle time + allowance ☐ (*d*) normal time – allowances. ☐

17. Choose the wrong statement

(*a*) Stop watch is the basic tool in work study. ☐
(*b*) In time study, normal time is more than the standard time. ☐
(*c*) Material handling in automobile industry is done by overhead cranes ☐
(*d*) An event is a function of two or more activities. ☐

Break-Even Analysis

18. Figure 13.5 shows the break-even chart. In this chart, the fixed expenses are given by

(*a*) line *AB* ☐
(*b*) line *CD* ☐
(*c*) line *CE* ☐
(*d*) line *AC*. ☐

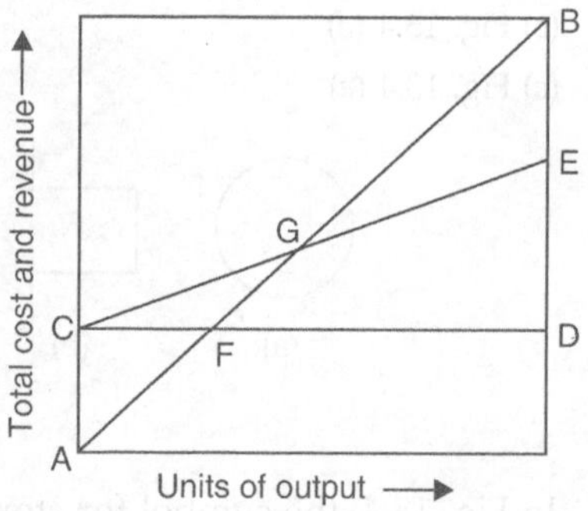

FIGURE 13.5. Break-even chart.

19. In Fig. 13.5, total cost is represented by

(*a*) line *AB* ☐
(*b*) line *CD* ☐
(*c*) line *CE* ☐
(*d*) line *AC*. ☐

20. In Fig. 13.5, total sales revenue is represented by

(*a*) line *AB* ☐ (*b*) line *CD* ☐
(*c*) line *CE* ☐ (*d*) line *AC*. ☐

21. In Fig. 13.5, the break-even point is given by

(*a*) point *A* ☐ (*b*) point *C* ☐
(*c*) point *F* ☐ (*d*) point *E*. ☐

22. In Fig. 13.5, the profit zone is represented by

(*a*) *ACF* ☐ (*b*) *ACG* ☐
(*c*) *EGB* ☐ (*d*) *CFG*. ☐

23. In Fig. 13.5, the loss zone is represented by

(*a*) *ACF* ☐ (*b*) *ACG* ☐
(*c*) *EGB* ☐ (*d*) *CFG*. ☐

24. In break-even analysis, the total cost consists of
(*a*) fixed cost + variable cost ☐ (*b*) fixed cost only ☐
(*c*) fixed cost + sales revenue ☐ (*d*) variable cost only. ☐

25. At the break-even point
(*a*) fixed cost and variable cost are equal ☐ (*b*) sales revenue and total cost are equal ☐
(*c*) sales revenue is more than total cost ☐ (*d*) sales revenue is less than total cost. ☐

26. If the sales revenue is more than total cost, then break-even analysis shows
(*a*) loss ☐ (*b*) profit ☐
(*c*) no loss no profit ☐ (*d*) none of the above. ☐

27. Choose the wrong statement
(*a*) Break-even analysis consists of fixed and variable costs. ☐
(*b*) Break-even analysis represents the relationship between cost and volume. ☐
(*c*) At the break-even point, total cost is equal to sales revenue. ☐
(*d*) If variable cost is less than fixed cost, then break-even analysis shows profit. ☐

Linear Programming

28. The mathematical technique, for finding the best use of limited resources of a concern in an optimum manner, is known as
(*a*) queuing theory ☐ (*b*) network analysis ☐
(*c*) value analysis ☐ (*d*) linear programming. ☐

29. The problem of optimisation in operation research is solved by the technique of
(*a*) queuing theory ☐ (*b*) game theory ☐
(*c*) linear programming ☐ (*d*) any of the above method. ☐

30. The simplex method is the basic method for
(*a*) queuing theory ☐ (*b*) network analysis ☐
(*c*) value analysis ☐ (*d*) linear programming. ☐

31. To which of the following industries, the linear programming technique is applied successfully?
(*a*) food processing ☐ (*b*) banking ☐
(*c*) oil and chemical ☐ (*d*) iron and steel ☐
(*e*) all of the above. ☐

32. The engineering, which aims at minimising the cost without change in quality of the product, is known as
(*a*) queuing theory ☐ (*b*) network analysis ☐
(*c*) value engineering ☐ (*d*) game theory. ☐

33. The term 'value' in value engineering refers to
(*a*) selling piece of the product ☐ (*b*) total cost of the product ☐
(*c*) utility of the product ☐ (*d*) none of the above. ☐

34. Which of the following items are analysed by the technique of value analysis?

(*a*) simple items ☐ (*b*) complicated items ☐
(*c*) crash-programme items ☐ (*d*) cost consciousness item ☐
(*e*) any item. ☐

35. The theory which deals with the problem of reducing the waiting time, is known as

(*a*) queuing theory ☐ (*b*) game theory ☐
(*c*) value engineering ☐ (*d*) network analysis. ☐

36. The process, which prescribes the sequence of operation to be followed, is known as

(*a*) despatching ☐ (*b*) scheduling ☐
(*c*) routing ☐ (*d*) loading. ☐

37. The process, which determines the programme for the operations, is known as

(*a*) despatching ☐ (*b*) scheduling ☐
(*c*) routing ☐ (*d*) loading. ☐

38. The process, which is concerned with the starting of processes, is known as

(*a*) despatching ☐ (*b*) scheduling ☐
(*c*) routing ☐ (*d*) loading. ☐

39. MIS stands for

(*a*) management inspection scheme ☐ (*b*) management information service ☐
(*c*) management information system ☐ (*d*) management inspection system. ☐

40. The utilization factor in case of queuing theory is defined as

(*a*) the ratio of mean arrival rate to mean service rate ☐
(*b*) the ratio of mean service rate to mean arrival rate ☐
(*c*) the product of mean arrival rate and mean service rate ☐
(*d*) the sum of mean arrival rate and mean service rate. ☐

41. If utilization factor is less than one and service facility is idle, then probability is given by

(*a*) $p_0 = 1 + \frac{\lambda}{\mu}$ ☐ (*b*) $p_0 = 1 - \frac{\lambda}{\mu}$ ☐
(*c*) $p_0 = 1 - \frac{\mu}{\lambda}$ ☐ (*d*) $p_0 = 1 + \frac{\mu}{\lambda}$. ☐

where $\frac{\mu}{\lambda}$ = Utilization factor.

42. The objective of the value analysis is to increase the value by

(*a*) decreasing the cost for the same worth ☐ (*b*) increasing the worth for the same cost ☐
(*c*) both of the above ways ☐ (*d*) none of the above. ☐

43. The information about when work should start and how much work should be completed during a certain period is obtained from

(*a*) routing ☐ (*b*) scheduling ☐
(*c*) despatching ☐ (*d*) loading. ☐

44. The information about the production schedule is obtained from
(*a*) starting diagram ☐ (*b*) Gnatt chart ☐
(*c*) distribution curve ☐ (*d*) travel chart. ☐

45. A compact estimate of the handling which must be done between various work sections, is obtained from
(*a*) starting diagram ☐ (*b*) Gnatt chart ☐
(*c*) distribution curve ☐ (*d*) travel chart. ☐

46. Which of the following methods is used for choosing the best location of a plant between alternative sites?
(*a*) Force analogy method ☐
(*b*) Equivalent annual cost method ☐
(*c*) Dimensional analysis method ☐
(*d*) Transportation method of linear programming ☐
(*e*) Any one of the above ☐
(*f*) None of the above. ☐

47. The continuous production of products of the same design on a line of machines arranged according to the required sequence of operations, is known as
(*a*) flow production ☐ (*b*) mass production ☐
(*c*) line production ☐ (*d*) continuous production ☐
(*e*) any one of the above ☐ (*f*) none of the above. ☐

48. In case of jobbing production
(*a*) highly skilled workers are needed ☐ (*b*) the unit cost is low ☐
(*c*) the operations are capital intensive ☐ (*d*) the operations are labour-intensive ☐
(*e*) the unit costs are high ☐ (*f*) only (*a*), (*b*) and (*e*). ☐

49. In case of mass production
(*a*) highly skilled workers are needed ☐ (*b*) the unit cost is low ☐
(*c*) the operations are capital-intensive ☐ (*d*) the operations are labour-intensive ☐
(*e*) the unit costs are high ☐ (*f*) only (*b*) and (*c*). ☐

50. The unit cost in case of batch production is
(*a*) higher than in jobbing production ☐ (*b*) lower than in jobbing production ☐
(*c*) the same as in jobbing production ☐ (*d*) lower than in mass production. ☐

51. Gnatt charts are used for
(*a*) machine utilization ☐ (*b*) production schedules ☐
(*c*) inventory control ☐ (*d*) none of the above. ☐

52. Bin cards are used for
(*a*) machine loading ☐ (*b*) stores ☐
(*c*) accounts ☐ (*d*) inventory control. ☐

53. The work flows from one work centre to another in the line without back-tracking or cross-movement. This type of plant layout, is known as
(*a*) product layout ☐ (*b*) process layout ☐
(*c*) flow-line layout ☐ (*d*) production line layout ☐
(*e*) any one except (*b*) ☐ (*f*) none of the above. ☐

54. In a process layout, the flow of work

(*a*) many involve back-tracking ☐ (*b*) does not involve back-tracking ☐

(*c*) remains in a fixed position ☐ (*d*) none of the above. ☐

55. Choose the wrong statement

(*a*) Process layout is also known as functional layout ☐

(*b*) In fixed position layout, the materials, tools, workers and machines remain in a fixed position ☐

(*c*) The abbreviation G.T. stands for group technology ☐

(*d*) In mass production, the operations are capital intensive and the unit cost is low. ☐

56. Choose the correct statement

(*a*) Good plant layout minimizes worker fatigue and improves quality of production ☐

(*b*) The plant layout, for which the total materials handling cost is minimum, should be chosen ☐

(*c*) From the various alternative layouts, the one having the smallest equivalent annual cost, should be chosen. ☐

(*d*) All of the above ☐

(*e*) None of the above. ☐

Work Management

57. Work measurement is concerned with establishing

(*a*) a standard job ☐ (*b*) a time standard for a specific task ☐

(*c*) the number of jobs for a given time ☐ (*d*) none of the above. ☐

58. Time standards are used for

(*a*) performance evaluation of individual workers ☐

(*b*) incentive payments ☐

(*c*) production scheduling ☐

(*d*) cost estimating ☐

(*e*) any one of the above ☐

(*f*) none of the above. ☐

59. Which of the following are the techniques of work measurement?

(*a*) work sampling ☐

(*b*) time study ☐

(*c*) synthesis of predetermined time standards ☐

(*d*) all of the above ☐

(*e*) none of the above. ☐

60. The standard time of an operation is defined as the time taken by a

(*a*) faster workers to perform that operation ☐

(*b*) slow worker to perform that operation ☐

(*c*) trained worker to perform that operation while working at a steady pace ☐

(*d*) none of the above. ☐

61. Which of the following allowance are added to the basic time of a given work element to obtain elemental standard tome?

(*a*) relaxation allowance ☐ (*b*) tool allowance ☐
(*c*) reject allowance ☐ (*d*) all of the above ☐
(*e*) none of the above. ☐

62. Choose the wrong statement

(*a*) Actual progress of an activity in a Gnatt chart is shown by drawing a thick bar. ☐
(*b*) From network analysis, the earliest and latest completion dates of all activities are known. ☐
(*c*) Merit rating is the procedure by which the value of individuals can be measured for selection for or promotion to a given job. ☐
(*d*) Standard time is equal to normal time. ☐

63. The objective of materials planning is to

(*a*) minimise overall production cost ☐
(*b*) minimise production delays owing to non-availability of raw materials and parts ☐
(*c*) to stabilize the production level ☐
(*d*) all of the above ☐
(*e*) none of the above. ☐

64. The ratio of total assets to the total liabilities of a company is known as

(*a*) acid test liquidity ratio ☐ (*b*) liquidity ratio ☐
(*c*) primary profitability ratio ☐ (*d*) secondary profitability ratio. ☐

65. The ratio of net profit to total assets of a company is known as

(*a*) acid test liquidity ratio ☐ (*b*) liquidity ratio ☐
(*c*) primary profitability ratio ☐ (*d*) secondary profitability ratio. ☐

66. The ratio of the sum of cash in hand and accounts receivable to the total liabilities of the company is known as

(*a*) acid test liquidity ratio ☐ (*b*) liquidity ratio ☐
(*c*) primary profitability ratio ☐ (*d*) secondary profitability ratio. ☐

67. EOQ stands for

(*a*) elimination of quality inspection ☐ (*b*) economic order quantity ☐
(*c*) elements of quality control ☐ (*d*) end of quality inspection stage. ☐

68. Choose the wrong statement

(*a*) SPT rule means the shortest processing time rule. ☐
(*b*) The SPT rule is the most satisfactory of all in minimizing the average waiting time as well as the number of jobs waiting in the queue. ☐
(*c*) Labour utilization is measured as the ratio of the total number of direct labour hours scheduled on standard operations to the total number of direct labour hours available. ☐
(*d*) Machine utilization is measured as the ratio of the total number of hours the production machines are scheduled to the productivity employed to the number of available machine hours. ☐
(*e*) None of the above. ☐

69. The theory, which states that the average industrial worker dislikes work and wishes to avoid it so long he can, is known as McGregor's
(a) Theory Z ☐ (b) Theory Y ☐
(c) Theory X ☐ (d) Theory S. ☐

70. The theory, which states that physical and mental effort in work is as much desired by the average industrial worker as play or rest, is known as McGregor's
(a) Theory Z ☐ (b) Theory Y ☐
(c) Theory X ☐ (d) Theory S. ☐

71. As the frequency of inspection increases, the frequency of breakdown
(a) increases ☐ (b) decreases ☐
(c) remains constant ☐ (d) first increases alter that decreases. ☐

72. The optimum inspection frequency is given by
(a) $f_i = \sqrt{\dfrac{k \times T_i}{T_r}}$ ☐ (b) $f_i = \sqrt{\dfrac{k \times T_r}{T_i}}$ ☐
(c) $f_i = \sqrt{k \times T_i \times T_r}$ ☐ (d) $f_i = \sqrt{T_i \times T_r}$. ☐

where k = Constant depending upon the specific equipment,
T_i = Equipment downtime due to each inspection, and
T_r = Equipment downtime due to each breakdown repairs.

73. If the equipment downtime due to each inspection = 3 hrs, the equipment downtime due to each breakdown = 16 hrs and k = 3 per month, then the optimum inspection frequency would be equal to
(a) 8 per month ☐ (b) 4 per month ☐
(c) 2 per month ☐ (d) one per month. ☐

Material Handling Equipment

74. During manufacture of cement, the material handling is done by
(a) overhead crane ☐ (b) belt conveyor ☐
(c) bucket conveyor ☐ (d) fork lift truck. ☐

75. Coal from the coal handling plant, in a thermal power plant, is moved to the boiler bunker through a
(a) overhead crane ☐ (b) belt conveyor ☐
(c) bucket conveyor ☐ (d) fork lift truck. ☐

76. The most economic order quantity in terms of total item consumed per year (A), procurement cost (P) per order and the annual inventory carrying cost (C) per item is given by
(a) $\dfrac{AP}{2C}$ ☐ (b) $\dfrac{2AP}{C}$ ☐
(c) APC ☐ (d) $\sqrt{\dfrac{2AP}{C}}$ ☐

77. If total item consumed per year = 2000, procurement cost per order = ₹ 10 and annual inventory carrying cost per item = ₹ 1.0, then the most economic order quantity would be equal to

(*a*) 100 ☐ (*b*) 200 ☐
(*c*) 300 ☐ (*d*) 400. ☐

78. In automobile industry, material handling is done by
(*a*) belt conveyor ☐ (*b*) overhead crane ☐
(*c*) bucket conveyor ☐ (*d*) trolley. ☐

79. The sub-division of an operation into therblings and their analysis is known as
(*a*) work study ☐ (*b*) time study ☐
(*c*) micromotion study ☐ (*d*) none of the above. ☐

80. Choose the wrong statement
(*a*) Queuing theory deals with the problems of reducing the waiting time. ☐
(*b*) Work study is done with the help of stop watch. ☐
(*c*) At the break-even point, the sales revenue is more than total cost. ☐
(*d*) Gnatt charts are used for production schedules. ☐

81. Which one is the correct statement about time wage system?
(*a*) Workers are paid according to their efficiency. ☐
(*b*) Workers are paid a fixed rate for each unit produced. ☐
(*c*) No consideration is given to the time taken by the workers to perform the work. ☐
(*d*) The workers are paid at hourly, daily, weekly or monthly rate. ☐

Financial Incentive Plans

82. The wages according to Halsey 50–50 plan is equal to
(*a*) $HA + \frac{(S-A)}{S} HA$ ☐ (*b*) HA ☐
(*c*) $HA + (S-A)H$ ☐ (*d*) $HA + \frac{(S-A)H}{2}$. ☐

where H = Hourly rate, A = Actual time, and S = Standard time.

83. The wages according to Rowan Plan is equal to
(*a*) $HA + \frac{(S-A)}{S} HA$ ☐ (*b*) HA ☐
(*c*) $HA + (S-A)H$ ☐ (*d*) $HA + \frac{(S-A)H}{2}$. ☐

84. According to Emerson plan, a worker gets a bonus whenever the efficiency of a worker is more than
(*a*) 50% ☐ (*b*) 55% ☐
(*c*) 60% ☐ (*d*) 66.67%. ☐

85. The incentive plan which works on bonus based on standard time is
(*a*) Rowan plan ☐ (*b*) Halsey premium bonus plan ☐
(*c*) Bedaux plan ☐ (*d*) Gnatt task bonus plan. ☐

86. The incentive wage plan, in which savings are expressed as a percentage of the standard time, is

(a) Rowan plan ☐ (b) Halsey premium bonus plan ☐
(c) Bedaux plan ☐ (d) Gnatt task bonus plan. ☐

87. Which of the following wage incentive plan guarantee minimum wage to a worker and bonus is paid for the fixed percentage of time saved?

(a) Rowan plan ☐ (b) Halsey plan ☐
(c) Bedaux plan ☐ (d) Gnatt plan. ☐

88. The wage incentive plan, which is applied to all level of workers, is

(a) Rowan plan ☐ (b) Halsey plan ☐
(c) Bedaux plan ☐ (d) Gnatt plan. ☐

89. The incentive plan which ensures a part of the saving to the worker and rest to the employer is

(a) Taylor plan ☐ (b) Emerson efficiency plan ☐
(c) Halsey premium plan ☐ (d) none of the above. ☐

90. Choose the wrong statement

(a) The bonus is paid to a worker whose efficiency is more than 66.67%, in Emerson's efficiency plan of wage incentive plan. ☐

(b) The savings are expressed as a percentage of standard time in a Rowan incentive plan. ☐

(c) In Halsey plan, a minimum wage is guaranteed to a worker and bonus is paid for the fixed percentage of time saved. ☐

(d) According to Gnatt plan, the bonus increases in proportion to the increase in efficiency. ☐

91. Acceptance sampling is used in

(a) batch production ☐ (b) mass production ☐
(c) job production ☐ (d) none of the above. ☐

92. *ABC* analysis deals with

(a) flow of material ☐ (b) analysis of process chart ☐
(c) controlling inventory costs money ☐ (d) none of the above. ☐

Functional Organisation

93. A system of working known as functional organisation was introduced by

(a) Newton ☐ (b) Gnatt ☐
(c) F.W. Taylor ☐ (d) Gilberth. ☐

94. Line organisation is also known as

(a) functional organisation ☐ (b) military organisation ☐
(c) staff and functional organisation ☐ (d) none of the above. ☐

95. Which of the following statement is correct about line organisation ? In a line organisation

(a) discipline is strong ☐ (b) quick decisions are taken ☐
(c) responsibility of each individual is fixed ☐ (d) all of the above ☐
(e) none of the above. ☐

96. Which of the following statement is correct about functional organisation?

(*a*) Wastage of material is minimum. ☐
(*b*) Quality of work is better. ☐
(*c*) Specialised knowledge and guidance to individual worker is provided. ☐
(*d*) All of the above. ☐
(*e*) None of the above. ☐

97. For steel industry, which of the following type of organisation is preferred?

(*a*) functional organisation ☐ (*b*) line organisation ☐
(*c*) line and staff organisation ☐ (*d*) none of the above. ☐

98. For civil engineering construction, which of the following type of organisation is most popular?

(*a*) functional organisation ☐ (*b*) line organisation ☐
(*c*) line and staff organisation ☐ (*d*) none of the above. ☐

99. For an automobile industry, which of the following type of organisation is preferred?

(*a*) functional organisation ☐ (*b*) line organisation ☐
(*c*) line and staff organisation ☐ (*d*) none of the above. ☐

100. Choose the wrong statement

(*a*) Product layout is used for continuous production. ☐
(*b*) A process layout is generally used for batch production. ☐
(*c*) Line organisation is best suited for big organisation. ☐
(*d*) Military organisation is known as line organisation. ☐

Project Management CPM and PERT

101. CPM stands for

(*a*) Common Project Method ☐ (*b*) Common Path Method ☐
(*c*) Critical Project Method ☐ (*d*) Critical Path Method. ☐

102. PERT stands for

(*a*) Planning Estimation and Review Technique ☐
(*b*) Planning Evaluation and Reporting Technique ☐
(*c*) Programme Evaluation and Review Technique ☐
(*d*) Process Evaluation and Reporting Technique. ☐

103. Choose the wrong statement

(*a*) Network analysis helps in designing, planning, controlling and in decision-making in order to accomplish the project. ☐
(*b*) Network analysis is synthesis of two techniques namely CPM and PERT. ☐
(*c*) The first method invented for planning project was critical path method. ☐
(*d*) Bar charts are suitable for minor works. ☐

104. The mathematical technique, employed to pick the best choice from among several alternatives, is known as

(a) production planning ☐ (b) production control ☐
(c) linear programming ☐ (d) value analysis. ☐

105. Which of the following methods are used to solve linear programming problems?
(a) simplex method ☐ (b) graphical method ☐
(c) transportation method ☐ (d) all of the above ☐
(e) none of the above. ☐

106. Figure 13.6 shows some symbols, which are used on a PERT/CPM chart. Figure 13.6 (a) represents
(a) an event to be transferred to other network chart ☐
(b) an ordinary event ☐
(c) a significant event representing some mile-stone ☐
(d) a dummy event. ☐

(a) (b) (c)

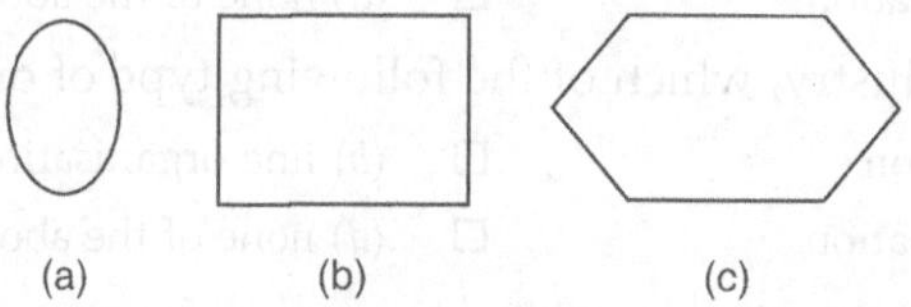

FIGURE 13.6

107. Figure 13.6 (b) on a PERT/CPM chart represents
(a) an event to be transferred to other network chart ☐
(b) an ordinary event ☐
(c) a significant event representing some mile-stone ☐
(d) a dummy event. ☐

108. Figure 13.6 (c) on a PERT/CPM chart represents
(a) an event to be transferred to other network chart ☐
(b) an ordinary event ☐
(c) a significant event representing some mile-stone ☐
(d) a dummy event. ☐

109. An activity, which consumes neither time nor resources on a PERT/CPM chart, is known as
(a) an event to lie transferred to other network chart ☐
(b) an ordinary event ☐
(c) a significant event representing same mile-stone ☐
(d) a dummy event. ☐

110. If the project is ahead of schedule, it is represented on a PERT chart by
(a) positive slack ☐ (b) negative slack ☐
(c) zero slack ☐ (d) none of the above. ☐

111. Slack represents the difference between the
(a) normal expected time and earliest completion time ☐
(b) latest allowable time and normal expected time ☐

(*c*) latest allowable time and earliest expected time ☐
(*d*) earliest competition time and normal expected time. ☐

112. Choose the wrong statement
(*a*) Actual performance of a task is known as activity. ☐
(*b*) The start or completion of a task is known as an event. ☐
(*c*) An event is indicated on the network by a number enclosed in a circle or a square. ☐
(*d*) In a PERT chart only critical activities are numbered. ☐

113. Slack is always
(*a*) positive ☐ (*b*) negative ☐
(*c*) zero ☐ (*d*) any one of the above. ☐

114. CPM is the
(*a*) target-oriented technique ☐ (*b*) time-oriented technique ☐
(*c*) event oriented technique ☐ (*d*) activity-oriented technique. ☐

115. PERT is the
(*a*) target-oriented technique ☐ (*b*) time-oriented technique ☐
(*c*) event-oriented technique ☐ (*d*) activity-oriented technique. ☐

116. Single-time estimate is required for
(*a*) PERT ☐ (*b*) CPM ☐
(*c*) MIS ☐ (*d*) none of the above. ☐

117. Tripple-time estimate is required for
(*a*) PERT ☐ (*b*) CPM ☐
(*c*) MIS ☐ (*d*) none of the above. ☐

118. In PERT/CPM chart, slack of various events on the critical path
(*a*) remains constant ☐ (*b*) increases continuously ☐
(*c*) decreases continuously ☐ (*d*) none of the above. ☐

119. Critical path is obtained in PERT analysis by joining events having
(*a*) maximum slack ☐ (*b*) minimum slack ☐
(*c*) negative slack ☐ (*d*) zero slack. ☐

120. Choose the wrong statement
(*a*) The bar chart represents the activities while a mile-stone chart represents the events which mark either the beginning or the end of an activity. ☐
(*b*) The project whether big or small should be completed with a minimum of capital investment without delay. ☐
(*c*) PERT uses event-oriented network diagram. ☐
(*d*) PERT is preferred in those projects where time can be estimated fairly well and when costs can be calculated in advance. ☐

Project Management

121. Which of the following names are used for the network techniques?
(*a*) PERT ☐ (*b*) CPM ☐
(*c*) TOPS ☐ (*d*) SCANS ☐
(*e*) all of the above ☐ (*f*) none of the above. ☐

122. In the network diagram

(*a*) an activity and an event are represented by a circle ☐
(*b*) an activity and an event are represented by an arrow ☐
(*c*) an activity is represented by an arrow and an event by a circle ☐
(*d*) an activity is represented by a circle and an event by an arrow. ☐

123. It is always possible to break up the entire project into a number of distinct and well defined jobs or task. Each task is called

(*a*) an event ☐ (*b*) an activity ☐
(*c*) any one of the above ☐ (*d*) none of the above. ☐

124. The beginning or end of a task is called

(*a*) an event ☐ (*b*) an activity ☐
(*c*) any one of the above ☐ (*d*) none of the above. ☐

125. Which of the following networks are the techniques of the project management?

(*a*) CPM ☐ (*b*) PERT ☐
(*c*) MIS ☐ (*d*) only (*a*) and (*b*) ☐
(*e*) all of the above. ☐

126. Which of the following are the basic elements of a project network?

(*a*) an event ☐ (*b*) an activity ☐
(*c*) a duration ☐ (*d*) all of the above ☐
(*e*) only (*a*) and (*b*). ☐

127. Choose the correct statement

(*a*) The commencement or completion of an activity is called an event. ☐
(*b*) Events acts as control points in a project. ☐
(*c*) Events does not require either time or resources. ☐
(*d*) A significant event is always positive. ☐
(*e*) All of the above. ☐
(*f*) None of the above. ☐

128. The event, which marks the beginning of an activity, is known as

(*a*) dual role event ☐ (*b*) tail event ☐
(*c*) head event ☐ (*d*) initial event ☐
(*e*) final event ☐ (*f*) none of the above. ☐

129. The event, which marks the completion of an activity, is known as

(*a*) dual role event ☐ (*b*) tail event ☐
(*c*) head event ☐ (*d*) initial event ☐
(*e*) final event ☐ (*f*) none of the above. ☐

130. The event, which represents commencement of the project, is known as

(*a*) dual role event ☐ (*b*) tail event ☐
(*c*) head event ☐ (*d*) initial event ☐
(*e*) final event ☐ (*f*) none of the above. ☐

131. The event, which represents the completion of the project, is known as

(*a*) dual role event ☐ (*b*) tail event ☐
(*c*) head event ☐ (*d*) initial event ☐
(*e*) final event ☐ (*f*) none of the above. ☐

132. Choose the wrong statement

(*a*) All events except initial and final events are dual role events. ☐
(*b*) The events are represented by a geometrical figure which may be circular, square, rectangular or oval in shape. ☐
(*c*) The events are numbered and the number of the events are written inside the circle, square or rectangle. ☐
(*d*) The actual performance of a task is known as an event. ☐

133. An activity requires

(*a*) time only ☐ (*b*) resources only ☐
(*c*) both time and resources ☐ (*d*) none of the above. ☐

134. The activities, in a network diagram, are represented by

(*a*) a circle ☐ (*b*) a square ☐
(*c*) a rectangle ☐ (*d*) a simple arrow, drawn from left to right ☐
(*e*) none of the above. ☐

135. A dummy in a network requires

(*a*) time only ☐ (*b*) resources only ☐
(*c*) both (*a*) and (*b*) ☐ (*d*) neither time nor resources. ☐

136. Choose the correct statement

(*a*) An event cannot occur until all the activities leading to it are completed. ☐
(*b*) No activity can start until its tail end event has occurred. ☐
(*c*) An event cannot occur twice. ☐
(*d*) In a network, there must be any dead end left except the final node. ☐
(*e*) All of the above ☐
(*f*) None of the above. ☐

137. Which of the following kinds of time estimates are made in case of PERT analysis?

(*a*) the optimistic time estimate ☐ (*b*) the pessimistic time estimate ☐
(*c*) the most likely time estimate ☐ (*d*) all of the above ☐
(*e*) none of the above. ☐

138. The shortest possible time, in which an activity can be completed under ideal conditions, is known as

(*a*) the pessimistic time estimate ☐ (*b*) the optimistic time estimate ☐
(*c*) the most likely time estimate ☐ (*d*) none of the above. ☐

139. The maximum time, in which an activity may be completed when abnormal situations prevail and every thing wrong, is known as

(a) the pessimistic time estimate ☐ (b) the optimistic time estimate ☐

(c) the most likely time estimate ☐ (d) none of the above. ☐

140. The time, in which an activity can be completed under normal condition, is known as

(a) the pessimistic time estimate ☐ (b) the optimistic time estimate ☐

(c) the most likely time estimate ☐ (d) none of the above. ☐

141. The three time estimates are combined into a single time which is called

(a) the pessimistic time ☐ (b) the optimistic time ☐

(c) the most likely time ☐ (d) the expected time. ☐

142. The expected time (t_E) of a PERT activity in terms of optimistic time (t_O), pessimistic time (t_P) and most likely time (t_L) is given by

(a) $t_E = \frac{t_O + t_L + t_P}{3}$ ☐ (b) $t_E = \frac{4t_O + t_L + t_P}{3}$ ☐

(c) $t_E = \frac{t_O + 4t_L + t_P}{6}$ ☐ (d) $t_E = \frac{t_O + 4t_L + t_P}{3}$. ☐

(GATE–ME–2009)

143. The deviation is the difference between

(a) mean time and the time under consideration ☐

(b) the time under consideration and the mean time ☐

(c) the time under consideration and optimistic time ☐

(d) pessimistic time and optimistic time. ☐

144. Variance is defined as the

(a) square of the deviation ☐ (b) mean of the squared deviation ☐

(c) square root of the deviation ☐ (d) none of the above. ☐

145. Standard deviation is defined as

(a) square of the variance ☐ (b) square root of the variance ☐

(c) cube root of the variance ☐ (d) cube of the variance. ☐

146. The beta distribution is one which is

(a) symmetrical about its apex ☐ (b) not symmetrical about its apex ☐

(c) increasing linearly ☐ (d) increasing hyperbolic. ☐

147. In PERT analysis

(a) beta distribution is used ☐ (b) normal distribution is used ☐

(c) alpha distribution is used ☐ (d) gamma distribution is used. ☐

148. For beta distribution, the standard deviation is given by

(a) $\frac{t_P - t_O}{2}$ ☐ (b) $\frac{t_P - t_O}{6}$ ☐

(c) $\left(\frac{t_P - t_O}{6}\right)^2$ ☐ (d) $\frac{t_P + t_O}{2}$. ☐

where t_P = Most pessimistic time estimate, and t_O = Most optimistic time estimate.

149. For beta distribution, the variance is given by

(*a*) $\frac{t_P - t_O}{2}$ ☐ (*b*) $\frac{t_P - t_O}{6}$ ☐

(*c*) $\left(\frac{t_P - t_O}{6}\right)^2$ ☐ (*d*) $\frac{t_P + t_O}{2}$. ☐

150. Critical path is that sequence of activities between the start and finish of the project which requires the

(*a*) shortest time ☐ (*b*) longest time ☐

(*c*) normal time ☐ (*d*) none of the above. ☐

151. Choose the wrong statement

(*a*) Variance is the measure of uncertainty. ☐

(*b*) The beta distribution has a small probability of reaching the most optimistic time. ☐

(*c*) The curve with its height so standardised that the area under the curve is equal to unity is known as probability distribution. ☐

(*d*) Greater the variance, lesser will be uncertainty. ☐

152. A zero slack is an indication of

(*a*) an ahead of schedule condition ☐ (*b*) a behind of scheduled condition ☐

(*c*) an on schedule condition ☐ (*d*) none of the above. ☐

153. The probability factor is given by the relation

(*a*) $Z = \frac{T_S + T_E}{2}$ ☐ (*b*) $Z = \frac{T_S - T_E}{6}$ ☐

(*c*) $Z = \frac{T_S + T_E}{2}$ ☐ (*d*) $Z = \frac{4T_S + T_E}{6}$. ☐

where T_S = Scheduled time of completion, and T_E = Earliest expected time of completion.

154. The probability factor is

(*a*) always positive ☐ (*b*) always zero ☐

(*c*) always negative ☐ (*d*) any one of the above ☐

(*e*) none of the above. ☐

155. If the probability factor is positive, the changes of completing the project in time are

(*a*) less than 50% ☐ (*b*) more than 50% ☐

(*c*) 50% ☐ (*d*) none of the above. ☐

156. If the probability factor is zero, the chances of completing the project in time are

(*a*) less than 50% ☐ (*b*) more than 50% ☐

(*c*) 50% ☐ (*d*) none of the above. ☐

157. If the probability factor is negative, the chances of completing the project in time are

(*a*) less than 50% ☐ (*b*) more than 50% ☐

(*c*) 50% ☐ (*d*) none of the above. ☐

158. Choose the correct statement

(*a*) In PERT system, time estimates are not so accurate and definite as in CPM. ☐

(*b*) In CPM, cost is the direct controlling factor while in PERT, time is controlling factor. ☐

(*c*) In PERT, the path that joins the critical events, is known as critical path. ☐

(*d*) In CPM, the path that passes through critical activities, is known as critical path. ☐

(*e*) All of the above ☐

(*f*) None of the above. ☐

159. The critical path in PERT is determined on the basis of

(*a*) minimum float for each activity ☐ (*b*) maximum float for each activity ☐

(*c*) slack at each event ☐ (*d*) none of the above. ☐

160. The critical path in CPM is determined on the basis of

(*a*) minimum float for each activity ☐ (*b*) maximum float for each activity ☐

(*c*) slack at each event ☐ (*d*) none of the above. ☐

161. The term which is associated with the activity times is known as

(*a*) slack ☐ (*b*) float ☐

(*c*) dummy ☐ (*d*) all of the above ☐

(*e*) none of the above. ☐

162. The term which is associated with the event times is known as

(*a*) slack ☐ (*b*) float ☐

(*c*) dummy ☐ (*d*) all of the above ☐

(*e*) none of the above. ☐

163. The amount of time by which an activity can be delayed without increasing the completion time of a project, is known as

(*a*) slack ☐ (*b*) float ☐

(*c*) dummy ☐ (*d*) constraint. ☐

164. The time span by which the starting (or finishing) of an activity can be delayed without delaying the completion of the project, is known as

(*a*) independent float ☐ (*b*) free float ☐

(*c*) total float ☐ (*d*) interfering float. ☐

165. The time by which the activity completion time can be delayed without affecting the start of succeeding activities, is known as

(*a*) independent float ☐ (*b*) free float ☐

(*c*) total float ☐ (*d*) interfering float. ☐

166. The difference between total float and free float is known as

(*a*) independent float ☐ (*b*) free float ☐

(*c*) total float ☐ (*d*) interfering float. ☐

167. Total float is always

(*a*) positive ☐ (*b*) negative ☐

(*c*) zero ☐ (*d*) any one of the above ☐

(*e*) none of the above. ☐

168. If the total float is negative, the activity is called

(*a*) sub-critical ☐ (*b*) critical ☐

(*c*) super-critical ☐ (*d*) none of the above. ☐

169. If the total float is positive, the activity is called

(*a*) sub-critical ☐ (*b*) critical ☐

(*c*) super-critical ☐ (*d*) none of the above. ☐

170. If the total float is zero, then the activity is called

(*a*) sub-critical ☐ (*b*) critical ☐

(*c*) super-critical ☐ (*d*) none of the above. ☐

171. The activities on the critical path are those activities that have total float

(*a*) positive ☐ (*b*) negative ☐

(*c*) equal to zero ☐ (*d*) none of the above. ☐

172. Critical path passes through those events where slack is

(*a*) positive ☐ (*b*) negative ☐

(*c*) equal to zero ☐ (*d*) none of the above. ☐

173. The standard time that an estimater would usually allow for an activity is known as

(*a*) crash time ☐ (*b*) normal time ☐

(*c*) finish time ☐ (*d*) none of the above. ☐

174. The minimum possible time, in which an activity can be completed by employing extra resources, is known as

(*a*) crash time ☐ (*b*) normal time ☐

(*c*) finish time ☐ (*d*) none of the above. ☐

175. The direct cost required to complete the activity in normal lime duration, is known as

(*a*) normal cost ☐ (*b*) crash cost ☐

(*c*) minimum cost ☐ (*d*) none of the above. ☐

176. The direct cost corresponding to the completion of the activity within minimum possible time, is known as

(*a*) normal cost ☐ (*b*) crash cost ☐

(*c*) minimum cost ☐ (*d*) none of the above. ☐

177. The slope of the direct cost curve is equal to

(*a*) the ratio of increase in cost to decrease in time ☐

(*b*) the ratio of decrease in time to increase in cost ☐

(*c*) one minus the ratio of increase in cost to decrease in time ☐

(*d*) one plus the ratio of increase in cost to decrease in time. ☐

178. Choose the wrong statement

(*a*) The longest path through the network is known as critical path and the time along the critical path gives the project duration. ☐

(*b*) PERT is an event-oriented technique. ☐

(c) PERT requires double time estimate. ☐
(d) Slack may be positive, zero or negative. ☐

179. Choose the correct statement
(a) Total float calculations help in determining critical path. ☐
(b) The critical path for the network is shown by thick lines. ☐
(c) All events and activities lying along the critical path are critical. ☐
(d) Positive slack on a PERT indicates that project is ahead of schedule. ☐
(e) All of the above ☐
(f) None of the above. ☐

180. Choose the statement which represents an event
(a) Plastering of the walls ☐ (b) Fixing of windows, doors and ventilators ☐
(c) Fixing of electrical wiring ☐ (d) Curing of concrete ☐
(e) Concrete cured. ☐

181. Merit rating is the method of determining the
(a) worth of a machine ☐ (b) relative values of a worker ☐
(c) worker's performance on a job ☐ (d) none of the above. ☐

182. Job evaluation is the method of determining the
(a) worth of a machine ☐ (b) relative values of a worker ☐
(c) worker's performance on a job ☐ (d) none of the above. ☐

183. Statistical Quantity Control (SQC) is based on
(a) probability ☐ (b) sampling ☐
(c) statistical inference ☐ (d) all of the above factors ☐
(e) none of the above. ☐

184. *ABC* analysis
(a) is meant for relative inventory control ☐
(b) is a basic technique of materials management ☐
(c) does not depend upon the unit cost of the item but on its annual consumption ☐
(d) all of the above. ☐

185. A graphic representation of work performed by the left and right hands in accomplishing a job, is called
(a) SIMQ chart ☐ (b) Operation chart ☐
(c) Gnatt chart ☐ (d) Process chart. ☐

186. A chart in which time values are recorded and motions are classified by therblings is called
(a) SIMQ chart ☐ (b) Operation chart ☐
(c) Gnatt chart ☐ (d) Process chart. ☐

187. The material in perpetual inventory control is checked as the material reaches its
(a) minimum value ☐ (b) maximum value ☐
(c) average value ☐ (d) none of the above. ☐

188. Work cost is equal to

(*a*) factory cost ☐ (*b*) primary cost ☐
(*c*) factory cost + primary cost ☐ (*d*) factory cost + primary cost. ☐

189. The lines must be balanced in

(*a*) product layout ☐ (*b*) process layout ☐
(*c*) plant layout ☐ (*d*) functional layout ☐
(*e*) all of the above except (*a*) ☐ (*f*) none of the above. ☐

190. Functional layout is suited for

(*a*) mass production ☐ (*b*) job production ☐
(*c*) batch production ☐ (*d*) none of the above. ☐

191. Product layout is suited for

(*a*) mass production ☐ (*b*) job production ☐
(*c*) batch production ☐ (*d*) none of the above. ☐

192. Choose the wrong statement.

(*a*) Gnatt charts are used for production schedule. ☐
(*b*) Bar charts are suitable for minor work. ☐
(*c*) A critical activity is having maximum slack. ☐
(*d*) Standard time is greater than normal time. ☐

193. Choose the correct statement

(*a*) For value analysis, simplex method is used. ☐
(*b*) In batch production, acceptance sampling is used. ☐
(*c*) Military organisation is known as functional organisation. ☐
(*d*) An activity of the project is represented graphically by a circle on the network. ☐
(*e*) All of the above. ☐
(*f*) None of the above. ☐

194. The deductions for employees provident fund start

(*a*) after one year of joining the service ☐ (*b*) after 240 days of joining the service ☐
(*c*) after 120 days of joining the service ☐ (*d*) after 60 days of joining the service ☐
(*e*) immediately on joining the service. ☐

195. For medium and large scale computers, the fastest internal storage is of the type of

(*a*) magnetic tape ☐ (*b*) magnetic core storage ☐
(*c*) punched cards ☐ (*d*) none of the above. ☐

196. The unit which retains the information read by the computer so that it can be used in subsequent calculations, is known as

(*a*) control unit ☐ (*b*) input unit ☐
(*c*) output unit ☐ (*d*) storage unit. ☐

197. For search therbligs colour is

(*a*) red ☐ (*b*) brown ☐
(*c*) black ☐ (*d*) blue. ☐

198. Cash discounts are reduction in price of goods

(*a*) which depends on prompt cash payment ☐

(*b*) which depends on assurance of payment ☐

(*c*) sold on credit ☐

(*d*) none of the above. ☐

199. Shadow prices in linear programming are

(*a*) maximum cost per item ☐

(*b*) lowest sales price ☐

(*c*) the values assigned to one unit of capacity ☐

(*d*) none of the above. ☐

200. A public sector undertaking is

(*a*) fully owned by government ☐

(*b*) fully owned by public through shareholders ☐

(*c*) jointly owned by private parties ☐

(*d*) jointly owned by private parties and government. ☐

201. Standing orders contain

(*a*) welfare measures adopted by management ☐ (*b*) list of workers in the factory ☐

(*c*) details of shift working attendance rules etc. ☐ (*d*) number of vacancies in the factory. ☐

202. Standing orders are applicable to

(*a*) only major industries ☐

(*b*) all industries employing more than 500 workers ☐

(*c*) all industries employing more than 100 workers ☐

(*d*) none of the above. ☐

203. The workman, for interpretation of Standing Orders, can refer to

(*a*) management ☐ (*b*) labour court ☐

(*c*) shop supervisor ☐ (*d*) trade union. ☐

204. For an industrial dispute, the proper authority is

(*a*) management ☐ (*b*) labour court ☐

(*c*) labour minister state/centre ☐ (*d*) High Court/Supreme Court. ☐

205. Which of the following are the methods of optimisation?

(*a*) differential calculus ☐ (*b*) search ☐

(*c*) statistical method ☐ (*d*) queuing theory ☐

(*e*) linear programming ☐ (*f*) dynamic programming ☐

(*g*) all of the above ☐ (*h*) none of the above. ☐

Tick mark the true and false statement:

206. In time study, normal time is more than the standard time.

(*a*) True ☐ (*b*) False. ☐

207. Stop watch is the basic tool in work study.
(*a*) True ☐ (*b*) False. ☐

208. Material handling in automobile industry is done by overhead cranes.
(*a*) True ☐ (*b*) False. ☐

209. An event is a function of two or more activities.
(*a*) True ☐ (*b*) False. ☐

210. If variable cost is less than fixed cost than break-even analysis shows profit.
(*a*) True ☐ (*b*) False. ☐

211. At the break-even point, total cost is equal to sales revenue.
(*a*) True ☐ (*b*) False. ☐

212. Break-even analysis consists of fixed and variable costs.
(*a*) True ☐ (*b*) False. ☐

213. MIS stands for management inspection system.
(*a*) True ☐ (*b*) False. ☐

214. Process layout is also known as functional layout.
(*a*) True ☐ (*b*) False. ☐

215. In fixed position layout the materials, tools, workers and machines remain in a fixed position.
(*a*) True ☐ (*b*) False. ☐

216. The abbreviation G.T. stands for group technology.
(*a*) True ☐ (*b*) False. ☐

217. In mass production, the operations are capital intensive and unit cost is low.
(*a*) True ☐ (*b*) False. ☐

218. Good plant layout minimises worker fatigue and improves quality of production.
(*a*) True ☐ (*b*) False. ☐

219. The plant layout, for which the total materials handling cost is minimum, should be chosen.
(*a*) True ☐ (*b*) False. ☐

220. From the various alternative layouts, the one having the smallest equivalent annual cost, should be chosen.
(*a*) True ☐ (*b*) False. ☐

221. Standard time is equal to normal time.
(*a*) True ☐ (*b*) False. ☐

222. Actual progress of an activity in a Gnatt chart is shown by drawing a thick bar.
(*a*) True ☐ (*b*) False. ☐

223. From network analysis, the earliest and latest completion dates of all activities are known.
(*a*) True ☐ (*b*) False. ☐

224. Merit rating is the procedure by which the value of individuals can be measured for solution for or promotion to a given job.
(*a*) True ☐ (*b*) False. ☐

225. The word EOQ stands for Economic order quantity.

(*a*) True ☐ (*b*) False. ☐

226. SPT rule means the shortest processing time rule.

(*a*) True ☐ (*b*) False. ☐

227. The SPT rule is the most satisfactory of all in minimizing the average waiting time as well as the number of jobs waiting in the queue.

(*a*) True ☐ (*b*) False. ☐

228. Labour utilization is measured as the ratio of the total number of direct labour hours scheduled on standard operations to the total number of direct labour hours available.

(*a*) True ☐ (*b*) False. ☐

229. All the break-even point, the sales revenue is more than total cost.

(*a*) True ☐ (*b*) False. ☐

230. Queuing theory deals with the problems of reducing the waiting time.

(*a*) True ☐ (*b*) False. ☐

231. Work study is done with the help of stop watch.

(*a*) True ☐ (*b*) False. ☐

232. Gnatt charts are used for production schedules.

(*a*) True ☐ (*b*) False. ☐

233. In a time wage system, workers are paid according to their efficiency.

(*a*) True ☐ (*b*) False. ☐

234. The bonus is paid to a worker whose efficiency is more than 66.67% in Emerson's efficiency plan of wage incentive plan.

(*a*) True ☐ (*b*) False. ☐

235. The savings as expressed as a percentage of standard time in a Rowan incentive plan.

(*a*) True ☐ (*b*) False. ☐

236. In Halsey plan, a minimum wage is guaranteed to a worker and bonus is paid for the fixed percentage of time saved.

(*a*) True ☐ (*b*) False. ☐

237. In a line organisation discipline is strong, quick decisions are taken and responsibility of each individual is fixed.

(*a*) True ☐ (*b*) False. ☐

238. The first method invented for planning project was critical path method.

(*a*) True ☐ (*b*) False. ☐

239. Network analysis helps in designing, planning, controlling and in decision-making in order to accomplish the project.

(*a*) True ☐ (*b*) False. ☐

240. Network analysis is synthesis of two techniques namely CPM and PERT.

(*a*) True ☐ (*b*) False. ☐

241. Bar charts are suitable for minor works.
(*a*) True ☐ (*b*) False. ☐

242. Line organisation is best suited for big organisation.
(*a*) True ☐ (*b*) False. ☐

243. Military organisation is known as line organisation.
(*a*) True ☐ (*b*) False. ☐

244. Product layout is used for continuous production.
(*a*) True ☐ (*b*) False. ☐

245. For batch production, generally a process layout is used.
(*a*) True ☐ (*b*) False. ☐

246. PERT stands for Planning Estimation and Review Technique.
(*a*) True ☐ (*b*) False. ☐

247. The ratio of amount used in the cause of production to the amount produced is known as productivity,
(*a*) True ☐ (*b*) False. ☐

248. Shortage of raw materials, spare parts, cutting tools etc. results in low productivity.
(*a*) True ☐ (*b*) False. ☐

249. In product layout, the machine and processing equipments are arranged according to the sequence of operations of a product.
(*a*) True ☐ (*b*) False. ☐

250. In case of process layout, the production of the whole section stops in the event of breakdown of even one machine.
(*a*) True ☐ (*b*) False. ☐

251. In operation process chart, only three symbols are used.
(*a*) True ☐ (*b*) False. ☐

252. In string diagram the actual distances between the work-stations and the distance travelled by the work-piece can be measured.
(*a*) True ☐ (*b*) False. ☐

253. In flow diagram, no symbol is used.
(*a*) True ☐ (*b*) False. ☐

254. Material handling is done to reduce unnecessary movement of the workers, to reduce idle time of workers and machines and to reduce the overall cost of manufacture.
(*a*) True ☐ (*b*) False. ☐

255. The estimate of level of demand to be expected for a product or products for some period of time in future is known an forecasting.
(*a*) True ☐ (*b*) False. ☐

256. The inventory of finished goods is necessary to meet the demand rate in the event of failure of any one of the equipments or to meet the demand rate if there is any delay in receiving the raw materials.
(*a*) True ☐ (*b*) False. ☐

257. The level of inventory should be less.

(a) True ☐ (b) False. ☐

258. The expression $(2NA/CI)^{1/2}$ represents the economy lot size

where N = Yearly requirements, A = Cost of procurement per Q quantity,

C = Cost of unit material, and I = Inventory carrying rate,

(a) True ☐ (b) False. ☐

259. The fundamental management principle used for inventory management problem is *ABC* analysis,

(a) True ☐ (b) False. ☐

260. Work samplying is not work-measurement technique.

(a) True ☐ (b) False. ☐

261. Time study is the process of conducting experiments on a model instead of attempting the experiments with the real system.

(a) True ☐ (b) False. ☐

262. In case of work-study PMT stands for pre-determined motion time study.

(a) True ☐ (b) False. ☐

263. S.Q.C. stands for statistical quality control.

(a) True ☐ (b) False. ☐

264. M.T.A. in case of work-study stands for method of time analysis.

(a) True ☐ (b) False. ☐

265. PERT has two-times estimates.

(a) True ☐ (b) False. ☐

266. CPM terminology employs word like events.

(a) True ☐ (b) False. ☐

267. The value engineering application and implementation in an organisation is through the principle of value analysis.

(a) True ☐ (b) False. ☐

268. System of functional organisation was introduced by Gilbreath.

(a) True ☐ (b) False. ☐

269. According to Rowan wage incentive plan, a worker is guaranted a minimum wage and bonus is paid for the fixed percentage of time saved.

(a) True ☐ (b) False. ☐

270. The incentive wage plan in which savings are expressed as a percentage of the standard time is Halsey plan.

(a) True ☐ (b) False. ☐

271. The bonus over the minimum guaranteed wage to a worker when there is 100% time saving, according to Rowan Plan will be minimum.

(a) True ☐ (b) False. ☐

272. The bonus, according to Rowan Plan when time saved is 50%, will be maximum.

(*a*) True ☐ (*b*) False. ☐

273. The Rowan Plan pays more bonus to the workers in comparison to Halsey Plan when lime saved is 60%.

(*a*) True ☐ (*b*) False. ☐

274. In Emerson's efficiency plan of wage incentive, bonus is paid to a Worker, whose output exceeds 67% efficiency.

(*a*) True ☐ (*b*) False. ☐

275. In Emerson's efficiency plan, the bonus increases in proportion to the increased in efficiency.

(*a*) True ☐ (*b*) False. ☐

276. For determination of standard time for long cycles, sampling method is useful.

(*a*) True ☐ (*b*) False. ☐

277. Micro motion study is used to analyse rapid and short-duration movements.

(*a*) True ☐ (*b*) False. ☐

278. Halsey wage incentive plan guarantees minimum wage for a worker and in this plan bonus is paid for the fixed percentage of time saved.

(*a*) True ☐ (*b*) False. ☐

279. For value analysis, simplex method is used.

(*a*) True ☐ (*b*) False. ☐

280. Gnatt charts are used for production schedule.

(*a*) True ☐ (*b*) False. ☐

281. In batch production, acceptance sampling is used.

(*a*) True ☐ (*b*) False. ☐

282. A critical activity is having maximum slack.

(*a*) True ☐ (*b*) False. ☐

283. Standard time is greater than normal time.

(*a*) True ☐ (*b*) False. ☐

284. Military organisation is known as functional organisation.

(*a*) True ☐ (*b*) False. ☐

285. An activity of the project is represented graphically be a circle on the network.

(*a*) True ☐ (*b*) False. ☐

286. The statement 'Fixing of windows, doors and ventilators' represents an event.

(*a*) True ☐ (*b*) False. ☐

287. Slack may be positive, zero or negative.

(*a*) True ☐ (*b*) False. ☐

288. PERT requires double time estimate.

(*a*) True ☐ (*b*) False. ☐

289. PERT is an event-oriented technique.
(*a*) True ☐ (*b*) False. ☐

290. The longest path through the network is known as critical path; and the time along the critical path gives the project duration.
(*a*) True ☐ (*b*) False. ☐

291. Total float calculations help in determining critical path.
(*a*) True ☐ (*b*) False. ☐

292. The critical path for the network is shown by thick lines.
(*a*) True ☐ (*b*) False. ☐

293. Positive slack on a PERT indicates that project is ahead of schedule.
(*a*) True ☐ (*b*) False. ☐

294. All events and activities lying along the critical path are critical.
(*a*) True ☐ (*b*) False. ☐

295. In PERT system, time estimates are not so accurate and definite as in CPM.
(*a*) True ☐ (*b*) False. ☐

296. In CPM, cost is the direct controlling factor while in PERT, time is the controlling factor.
(*a*) True ☐ (*b*) False. ☐

297. In CPM, the path that passes through critical activities, is known as critical path.
(*a*) True ☐ (*b*) False. ☐

298. Greater the variance, lesser will be uncertainty.
(*a*) True ☐ (*b*) False. ☐

299. Variance is the measures of uncertainty.
(*a*) True ☐ (*b*) False. ☐

300. The β (beta) distribution has a small probability of reaching the most optimistic times.
(*a*) True ☐ (*b*) False. ☐

301. Probability distribution is the curve with its height so standardised that the area under the curve is equal to unity.
(*a*) True ☐ (*b*) False. ☐

302. In PERT analysis, normal distribution is used.
(*a*) True ☐ (*b*) False. ☐

303. Actual performance of a task is known as activity.
(*a*) True ☐ (*b*) False. ☐

304. The start or completion of a task is known as event.
(*a*) True ☐ (*b*) False. ☐

305. An event is indicated on the network by a number enclosed in a circle or a square.
(*a*) True ☐ (*b*) False. ☐

306. In a PERT chart only critical activities are numbered.
(*a*) True ☐ (*b*) False. ☐

307. The bar chart represents the activities while a milestone chart represents the events which mark either the beginning or the end of an activity.

(*a*) True ☐ (*b*) False. ☐

308. The project whether big or small should be completed with a minimum of capital investment without delay.

(*a*) True ☐ (*b*) False. ☐

309. PERT is preferred in those projects where time can be estimated fairly well and when costs can be calculated in advance.

(*a*) True ☐ (*b*) False. ☐

310. PERT uses event-oriented network diagram.

(*a*) True ☐ (*b*) False. ☐

311. The commencement or completion of an activity is called an event.

(*a*) True ☐ (*b*) False. ☐

312. Events do not require either time or resources.

(*a*) True ☐ (*b*) False. ☐

313. No activity can start until its tail end event has occurred.

(*a*) True ☐ (*b*) False. ☐

314. An event cannot occur twice.

(*a*) True ☐ (*b*) False. ☐

315. An event cannot occur until all the activities leading to it are completed.

(*a*) True ☐ (*b*) False. ☐

316. The three time estimates are combined into a single time which is called the expected time.

(*a*) True ☐ (*b*) False. ☐

317. A graphical device used to determine the break-even point and profit potential under varying conditions of output and costs, is known as PERT chart.

(*a*) True ☐ (*b*) False. ☐

318. The selling price of a product is determined by the factors such as buyers capability to pay, various elements of the cost and sales turn over.

(*a*) True ☐ (*b*) False. ☐

319. Scheduling gives information about when work should start and how much work should be completed during a certain period.

(*a*) True ☐ (*b*) False. ☐

320. Time study is used to determine overhead expenses.

(*a*) True ☐ (*b*) False. ☐

Fill in the blanks:

321. At......, the first free trade zone in India was established.

(*a*) Delhi ☐ (*b*) Mumbai. ☐

322. Subsidy and .... are included in shadow prices.

(*a*) taxes ☐ (*b*) not taxes. ☐

323. The theory of transactional analysis (TA) is applied to determine the ......, in works management.

(*a*) best layout ☐ (*b*) cause of behaviour of personnel. ☐

324. The trade between ...... is known as free trade zone.

(*a*) manufacturer and wholesaler. ☐ (*b*) two countries without tariff. ☐

325. Along the activities having total float of ......, critical path moves.

(*a*) negative valve ☐ (*b*) zero valve. ☐

326. The expected time of activity is ...... in terms of optimistic time '*a*', pessimistic time '*b*' and most likely time '*m*'.

(*a*) $(a + 4m + b)/6$ ☐ (*b*) $(a + m + b)/6$. ☐

327. In ...... industry, *p* routing is essential.

(*a*) assembly ☐ (*b*) process. ☐

328. CPM has ...... time estimate.

(*a*) one ☐ (*b*) two. ☐

329. PERT has...... time estimate.

(*a*) two ☐ (*b*) three. ☐

330. A worker receives only his daily wage and no bonus is paid till his efficiency reaches ...... is the Emerson efficiency plan.

(*a*) 50% ☐ (*b*) 66.67%. ☐

331. Maximum attention is given to ...... in *A*—*B*—*C* control policy.

(*a*) those items which are not readily available ☐

(*b*) those items which consume more money. ☐

332. For......, the simplex method is the basic method.

(*a*) operational research ☐ (*b*) linear programming. ☐

333. In ...., Bin card is used.

(*a*) stores ☐ (*b*) workshop. ☐

334. CPM is the ...... oriented technique.

(*a*) time ☐ (*b*) activity. ☐

335. The problems of ...... are dealt with Queuing theory.

(*a*) reducing the waiting time ☐ (*b*) effective use of machines. ☐

336. The old machine should be replaced by new one when .... according to MAPI formula.

(*a*) CAM > DAM ☐ (*b*) CAM < DAM ☐

where CAM = Challenger's Adverse Minimum, DAM = Defender's Adverse Minimum.

337. The ...... is the difference between the time available to do the job and the time required to do the job.

(*a*) float ☐ (*b*) event. ☐

338. The critical activity has ...... float.

(*a*) maximum ☐ (*b*) zero. ☐

339. Therblig colour is ...... for search.

(*a*) red ☐ (*b*) black. ☐

340. For value analysis ...... is used.

(*a*) simplex method ☐ (*b*) linear programming. ☐

341. Shadow prices are ...... in linear programming.

(*a*) maximum cost per item ☐ (*b*) the valves assigned to one unit of capacity. ☐

342. The workman can refer to ...... for interpretation of standing orders.

(*a*) management ☐ (*b*) labour court. ☐

343. Bar charts are suitable for .... work.

(*a*) minor ☐ (*b*) major. ☐

344. An event ...... occur twice.

(*a*) can ☐ (*b*) cannot ☐

345. The ...... is the basic tool in work study.

(*a*) bar chart ☐ (*b*) stop watch. ☐

346. To find a simpler, easier and better way of performing a job ...... is done.

(*a*) time study ☐ (*b*) motion study. ☐

347. When ......, break-even analysis shows profit.

(*a*) sales revenue > total cost ☐ (*b*) sales revenue < total cost. ☐

348. The symbol O is used for ...... in work study.

(*a*) inspection ☐ (*b*) operation. ☐

349. The symbol V is used for ...... in work study.

(*a*) permanent storage ☐ (*b*) inspection. ☐

350. The number of fundamental hand motions are ...... in micromotion study.

(*a*) 8 ☐ (*b*) 16. ☐

351. With the help of ...... work study is done.

(*a*) process chart ☐ (*b*) stop watch. ☐

352. To ...... time study is done.

(*a*) determine standard costs ☐ (*b*) determine overhead expenses. ☐

353. For ......, motion study is done.

(*a*) reducing inventory costs ☐ (*b*) improving a work method. ☐

354. For ......, templates are used.

(*a*) flow of material ☐ (*b*) a planning layout. ☐

355. The ...... is independent of sales forecast

(*a*) productivity ☐ (*b*) inventory control. ☐

356. The economic order quantity is the ...... in inventory control.

(*a*) highest level of inventory ☐ (*b*) optimum lot size. ☐

357. On the theory of ...... statistical quality control techniques are based.

(*a*) probability ☐ (*b*) quality. ☐

358. With ...... the two-bin system is concerned.

(*a*) forecasting sales ☐ (*b*) ordinary procedure. ☐

359. An event is indicated by ...... on the network.

(*a*) a straight line ☐ (*b*) a number enclosed in a circle or a square. ☐

360. According to ......, the bonus increases in proportions to the increase in efficiency.

(*a*) Halsey plan ☐ (*b*) Emerson's efficiency plan. ☐

361. Monte Carlo solution in ...... is extremely useful in queuing problems that cannot be analyses mathematically.

(*a*) queuing theory ☐ (*b*) simplex method. ☐

III. OBJECTIVE TYPE QUESTIONS FROM COMPETITIVE EXAMINATIONS

Tick mark the most appropriate answer:

1. A manufacturer produces two types of products, 1 and 2, at production levels of x_1 and x_2 respectively. The profit is given is $2x_1 + 5x_2$. The production constraints are

$$x_1 + 3x_2 \le 40$$
$$3x_1 + x_2 \le 24$$
$$x_1 + x_2 \le 10$$
$$x_1 > 0,\ x_2 > 0$$

The maximum profit which can meet the constraints is

(*a*) 29 ☐ (*b*) 38 ☐

(*c*) 44 ☐ (*d*) 75 ☐

(GATE-ME-2003)

2. The expected time (t_e) of a PERT activity in terms of optimistic time (t_0), pessimistic (t_p) and most likely time (t_L) is given by

(*a*) $t_e = \dfrac{t_0 + 4t_L + t_p}{6}$ ☐ (*b*) $t_e = \dfrac{t_0 + 4t_p + t_L}{6}$ ☐

(*c*) $t_e = \dfrac{t_0 + 4t_L + t_p}{3}$ ☐ (*d*) $t_e = \dfrac{t_0 + 4t_p + t_L}{3}$ ☐

(GATE-ME-2009)

3. Which one of the following is NOT a decision taken during the aggregate production planning stage?

(*a*) Scheduling of machines ☐ (*b*) Amount of labour to be committed ☐

(*c*) Rate at which production should happen ☐ (*d*) Inventory to be carried forward ☐

(GATE-ME-2012)

4. A company has two factories S_1, S_2 and two warehouses D_1, D_2. The supplies from S_1 and S_2 are 50 and 40 units respectively. Warehouse D_1 requires a minimum of 20 units and a maximum of 40 units. Warehouse D_2 requires a minimum of 20 units, and, over and above, it can take as much as can be supplied. A balanced transportation problem is to be formulated for the above situation. The number of supply points, the number of demand points, and the total supply (or total demand) in the balanced transportation problem respectively are

(*a*) 2, 4, 90 ☐ (*b*) 2, 4, 110 ☐

(*c*) 3, 4, 90 ☐ (*d*) 3, 4, 110. ☐

(GATE-ME-2005)

5. Standard time is equal to

(*a*) normal time + idle time ☐ (*b*) normal time + allowances ☐

(*c*) normal time + idle time + allowance ☐ (*d*) normal time – allowance ☐

6. The maximum level of inventory of an item is 100 and it is achieved with infinite replenishment rate. The inventory becomes zero over one and half month due to consumption at a uniform rate. This cycle continues throughout the year. Ordering cost is ₹ 100 per order and inventory carrying cost is ₹ 10 per item per month. Annual cost (in ₹) of the plan, neglecting material cost, is

(*a*) 800 ☐ (*b*) 2800 ☐

(*c*) 4800 ☐ (*d*) 6800. ☐

(GATE-ME-2007)

7. In a single serve infinite population queuing model, arrivals follow a Poisson distribution with mean $\lambda = 4$ per hour. The service times are exponential with mean service time equal to 12 minutes. The expected length of the queue will be

(*a*) 4 ☐ (*b*) 3.2 ☐

(*c*) 1.25 ☐ (*d*) 24.3 ☐

(GATE-ME-2000)

8. An electronic equipment manufacturer has decided to add a component sub-assembly operation that can produce 80 units during a regular 8-hour shift. This operation consists of three activities as below.

Activity	**Standard time (min)**
M. Mechanical assembly	12
E. Electric wiring	16
T. Test	3

For line balancing the number of work stations required for the activities *M*, *E* and *T* would respectively be

(*a*) 2, 3, 1 ☐ (*b*) 3, 2, 1 ☐

(*c*) 2, 4, 2 ☐ (*d*) 2, 1, 3. ☐

(GATE-ME-2004)

9. At the break-even point

(*a*) fixed cost and variable cost are equal ☐ (*b*) sales revenue and total cost are equal ☐

(*c*) sales revenue is more than total cost ☐ (*d*) sales revenue is less than total cost. ☐

10. The word **kanban** is most appropriately associated with

(*a*) economic order quantity ☐ (*b*) just-in-time production ☐
(*c*) capacity planning ☐ (*d*) product design. ☐

(GATE-ME-2011)

11. The number of customers arriving at a railway reservation counter is Poisson distributed with an arrival rate of eight customers per hour. The reservation clerk at this counter takes six minutes per customer on an average with an exponentially distributed service time. The average number of the customers in the queue will be

(*a*) 3 ☐ (*b*) 3.2 ☐
(*c*) 4 ☐ (*d*) 4.2 ☐

(GATE-ME-2006)

12. A set of 5 jobs is to be processed on a single machine. The processing time (in days) is given in the table below. The holding cost for cash job is ₹ *K* per day.

Job	*Processing time*
P	5
Q	2
R	3
S	2
T	1

A schedule that minimizes the total inventory cost is

(*a*) *T-S-Q-R-P* ☐ (*b*) *P-R-S-Q-T* ☐
(*c*) *T-R-S-Q-P* ☐ (*d*) *P-Q-R-S-T*. ☐

(GATE-ME-2008)

13. The product structure of an assembly *P* is shown in the figure.
Estimated demand for end product *P* is as follows:

Week	1	2	3	4	5	6
Demand	1000	1000	1000	1000	1200	1200

Ignore lead times for assembly and sub-assembly. Production capacity (per week) for component *R* is the bottleneck operation. Starting with zero inventory, the smallest capacity that will ensure a feasible production plan up to week 6 is

(*a*) 1000 ☐
(*b*) 1200 ☐
(*c*) 2200 ☐
(*d*) 2400 ☐

FIGURE 13.7

(GATE-ME-2008)

14. A manufacturing shop processes sheet metal jobs, where in each job must pass through two machines (M_1 and M_2, in that order). The processing time (in hours) for these jobs is

Machine	*Jobs*					
	P	*Q*	*R*	*S*	*T*	*U*
M_1	15	32	8	27	11	16
M_2	6	19	13	20	14	7

The optimal make-span (in hours) of the shop is

(*a*) 120 ☐ (*b*) 115 ☐
(*c*) 109 ☐ (*d*) 79. ☐

(GATE-ME-2006)

15. The cost of providing service in a queuing system increases with

(*a*) increased mean time in the queue ☐ (*b*) increased arrival rate ☐
(*c*) decreased mean time in the queue ☐ (*d*) decreased arrival rate ☐

(GATE-ME-1997)

16. A public sector undertaking is

(*a*) fully owned by Government ☐
(*b*) fully owned by public through shareholders ☐
(*c*) jointly owned by private parties ☐
(*d*) jointly owned by private parties and Government. ☐

17. Which of the following are the method of optimisation?

(*a*) Differential calcular ☐ (*b*) Linear programming ☐
(*c*) Queuing theory ☐ (*d*) Satistical method ☐
(*e*) All of the above ☐ (*f*) None of the above ☐

18. A firm is required to procure three items (*P*, *Q* and *R*). The prices quoted for these items (in ₹) by suppliers S_1, S_2 and S_3 are given in table. The management policy requires that each item has to be supplied by only one supplier and one supplier supply only one item. The minimum total cost (in ₹) of procurement to the firm is

Item	*Suppliers*		
	S_1	S_2	S_3
P	110	120	130
Q	115	140	140
R	125	145	165

(*a*) 350 ☐ (*b*) 360 ☐
(*c*) 385 ☐ (*d*) 395. ☐

(GATE-ME-2006)

19. A project consists of three parallel paths with durations and variances of (10, 4), (12, 4) and (12, 9) respectively. According to the standard PERT assumptions, the distribution of the project duration is

(*a*) Beta with mean 10 and standard deviation 2 ☐

(*b*) Beta with mean 12 and standard deviation 2 ☐

(*c*) Beta with mean 10 and standard deviation 3 ☐

(*d*) Beta with mean 12 and standard deviation 3. ☐

(GATE-ME-2002)

20. In an assembly line for assembling toys, five workers are assigned tasks which take times of 10, 8, 6, 9 and 10 minutes respectively. The balance delay for line is

(*a*) 43.3% ☐ (*b*) 14.8% ☐

(*c*) 14.0% ☐ (*d*) 16.3%. ☐

(GATE-ME-1996)

21. Six jobs arrived in a sequence as given below:

Job	*Completion time*
I	4
II	9
III	5
IV	10
V	6
VI	8

Average flow time (in days) for the above jobs using shortest processing time rule is

(*a*) 20.83 ☐ (*b*) 23.16 ☐

(*c*) 125.00 ☐ (*d*) 139.00. ☐

(GATE-ME-2009)

22. Little's law is a relationship between

(*a*) stock level and lead time in an inventory system ☐

(*b*) waiting time and length of the queue in a queuing system ☐

(*c*) number of machines and job due dates in a scheduling problem ☐

(*d*) uncertainly in the activity time and project completion time. ☐

(GATE-ME-2010)

23. In simple exponential smoothing forecasting, to give higher weightage to recent demand information, the smoothing constant must be close to

(*a*) – 1 ☐ (*b*) zero ☐

(*c*) 0.5 ☐ (*d*) 1.0. ☐

(GATE-ME-2013)

24. In an MRP system, component demand is
(*a*) forecasted ☐
(*b*) established by the master production schedule ☐
(*c*) calculated by the MRP system from the master production schedule ☐
(*d*) ignored. ☐

(GATE-ME-2006)

25. For planning the procurement or production of dependent demand items, the technique most suitable is
(*a*) *MPR* ☐ (*b*) *MRP* ☐
(*c*) *RPM* ☐ (*d*) *PRM.* ☐

(GATE-ME-1995)

26. Which of the following forecasting methods takes a fraction of forecast error into account for the next period forecast?
(*a*) simple average method ☐ (*b*) moving average method ☐
(*c*) weighted moving average method ☐ (*d*) exponential smoothing method ☐

(GATE-ME-2009)

27. Production flow analysis (PFA) is a method of identifying part families that uses data from
(*a*) engineering drawings ☐ (*b*) production schedule ☐
(*c*) bill of materials ☐ (*d*) route sheets. ☐

(GATE-ME-2001)

28. The supply at three sources is 50, 40 and 60 units respectively while the demand at the four destinations is 20, 30, 10 and 50 units. In solving this transportation problem
(*a*) a dummy source of capacity 40 units is needed ☐
(*b*) a dummy destination of capacity 40 units is needed ☐
(*c*) no solution exists as the problem is infeasible ☐
(*d*) none solution exists as the problem is degenerate. ☐

(GATE-ME-2002)

29. A stockiest wishes to optimize the number of perishable items he needs to stock in any month in his store. The demand distribution for this perishable item is

Demand (in units)	2	3	4	5
Probability	0.10	0.35	0.35	0.20

The stockiest pays ₹ 70 for each item and he sells each at ₹ 90. If the stock is left unsold in any month, he can sell the item at ₹ 500 each. There is no penalty for unfulfilled demand. To maximize the expected profit, the optimal stock level is
(*a*) 5 units ☐ (*b*) 4 units ☐
(*c*) 3 units ☐ (*d*) 2 units. ☐

(GATE-ME-2006)

30. The mathematical technique, employed to pick the best choice from among several alternatives, is known as
(*a*) production planning ☐ (*b*) production control ☐
(*c*) linear programming ☐ (*d*) value analysis. ☐

31. An item can be purchased for ₹ 100. The ordering cost is ₹ 200 and the inventory carrying cost is 10% of the item cost per annum. If the annual demand is 4000 units, the economic order quantity (in units) is

(*a*) 50 ☐ (*b*) 100 ☐
(*c*) 200 ☐ (*d*) 400. ☐

(GATE-ME-2002)

32. The manufacturing area of a plant is divided into four quadrants. Four machines have to located one in each quadrant. The total number of possible layouts is

(*a*) 4 ☐ (*b*) 8 ☐
(*c*) 16 ☐ (*d*) 24. ☐

(GATE-ME-1995)

33. At a production machine, parts arrive according to a Poisson process at the rate of 0.35 parts per minute. Processing time for parts have exponential distribution with mean of 2 minutes. What is the probability that a random part arrival finds that there are already 8 parts in the system (in machine + in queue)?

(*a*) 0.0247 ☐ (*b*) 0.0576 ☐
(*c*) 0.0173 ☐ (*d*) 0.082. ☐

(GATE-ME-1999)

34. CPM stands for

(*a*) common project method ☐ (*b*) common path method ☐
(*c*) critical project method ☐ (*d*) critical path method ☐

35. In PERT analysis a critical activity has

(*a*) maximum float ☐ (*b*) zero float ☐
(*c*) maximum cost ☐ (*d*) minimum cost ☐

(GATE-ME-2004)

36. Consider a single server queuing model with Poisson arrivals (λ = 4/hour) and exponential service (μ = 4/hour). The number in the system is restricted to a maximum of 10. The probability that a person who comes in leaves without joining the queue is

(*a*) $\frac{1}{11}$ ☐ (*b*) $\frac{1}{10}$ ☐
(*c*) $\frac{1}{9}$ ☐ (*d*) $\frac{1}{2}$. ☐

(GATE-ME-2005)

37. In PERT/CPM chart, slack of various events on the critical path

(*a*) remains constant ☐ (*b*) increases continuously ☐
(*c*) decreases continuously ☐ (*d*) none of the above. ☐

38. Simplex method of solving linear programming problem uses

(*a*) all the points in the feasible region ☐
(*b*) only the corner points of the feasible region ☐

(c) intermediate points within the infeasible region ☐
(d) only the interior points in the feasible region. ☐

(GATE-ME-2010)

39. A regression model is used to express a variable Y as a function of another variable X this implies that

(a) there is a casual relationship between Y and X ☐
(b) a value of X may be used to estimate a value of Y ☐
(c) values of X exactly determine values of Y ☐
(d) there is no casual relationship between Y and X. ☐

(GATE-ME-2002)

40. The table gives details of an assembly line.

Work station	*I*	*II*	*III*	*IV*	*V*	*VI*
Total task time at the workstation (in min)	7	9	7	10	9	6

What is the line efficiency of the assembly line?

(a) 70% ☐ (b) 75% ☐
(c) 80% ☐ (d) 85%. ☐

(GATE-ME-2006)

41. Cars arrive at a service station according to Poisson's distribution with a mean rate of 5 per hour. The service time per car is exponential with a mean of 10 minutes. At state, the average waiting time in the queue is

(a) 10 min ☐ (b) 20 min ☐
(c) 25 min ☐ (d) 50 min. ☐

(GATE-ME-2011)

42. Capacities of production of an item over 3 consecutive months in regular time are 100, 100 and 80 and in overtime are 20, 20 and 40. The demands over those 3 months are 90, 130 and 110. The cost of production in regular time and overtime are respectively ₹ 20 per item and ₹ 24 per item. Inventory carrying cost is ₹ 2 per item per month. The levels of starting and final inventory are nil. Backorder is not permitted. For minimum cost of plan, the level of planned production in overtime in the third month is

(a) 40 ☐ (b) 30 ☐
(c) 20 ☐ (d) 0. ☐

(GATE-ME-2007)

43. For the standard transportation linear programme with m sources and n destinations and total supply equaling total demand, an optimal solution (lowest cost) with the smallest number of non-zero x_{ij} values (amounts from source i to destination j) is desired. The best upper bound for this number is

(a) mn ☐ (b) $2(m + n)$ ☐
(c) $m + n$ ☐ (d) $m + n - 1$. ☐

(GATE-ME-2008)

44. Annual demand for window frames is 10000. Each costs ₹ 200 and ordering cost is ₹ 300 per order. Inventory holding cost is ₹ 40 per frame per year. The supplier is willing to offer 2% discount if the order quantity is 1000 or more, and 4% if order quantity is 2000 or more. If the total cost is to be minimized, the retailer should

(*a*) order 200 frames every time ☐ (*b*) accept 2% discount ☐
(*c*) accept 4% discount ☐ (*d*) order economic order quantity. ☐

(GATE-ME-2010)

45. Customers arrive at a ticket counter at a rate of 50 per hr and tickets are issued in the order of their arrival. The average time taken for issuing a ticket is 1 min. Assuming that customer arrivals form a Poisson process and service times are exponentially distributed, the average waiting time in queue in min is

(*a*) 3 ☐ (*b*) 4 ☐
(*c*) 5 ☐ (*d*) 6. ☐

(GATE-ME-2013)

46. The project activities, precedence relationships and durations are described in the table. The critical path of the project is

Activity	*Precedence*	*Duration (in days)*
P	–	3
Q	–	4
R	*P*	5
S	*Q*	5
T	*R, S*	7
U	*R, S*	5
V	*T*	2
W	*U*	10

(*a*) *P-R-T-V* ☐ (*b*) *Q-S-T-V* ☐
(*c*) *P-R-U-W* ☐ (*d*) *Q-S-U-W*. ☐

(GATE-ME-2010)

47. In inventory planning, extra inventory is unnecessarily carried to the end of the planning period when using one of the following lot size decision policies

(*a*) Lot-for-lot production ☐
(*b*) Economic order quantity (*EOQ*) lot size ☐
(*c*) Period order quantity (*POQ*) lot size ☐
(*d*) Part period total cost balancing ☐

(GATE-ME-1998)

48. If at the optimum in a linear programming problem, a dual variable corresponding to a particular primal constraint is zero, then it means that

(*a*) right hand side of the primal constraint can be altered without affecting the optimum solution ☐

(*b*) changing the right hand side of the primal constraint will disturb the optimum solution ☐

(*c*) the objective function is unbounded ☐

(*d*) the problem is degenerate. ☐

(GATE-ME-1996)

49. In the construction of networks, dummy activities are introduced in order to

(*a*) compute the slack on all events ☐

(*b*) transfer resources, if necessary, during monitoring ☐

(*c*) clearly designate a precedence relationship ☐

(*d*) simplify the crashing plan. ☐

(GATE-ME-1990)

50. The distribution of lead time demand for an item is as follows:

Lead time demand	*Probability*
80	0.20
100	0.25
120	0.30
140	0.25

The reorder level is 1.25 times the expected value of the lead demand. The service level is

(*a*) 25% ☐ (*b*) 50% ☐

(*c*) 75% ☐ (*d*) 100%. ☐

(GATE-ME-2005)

ANSWERS

Answers to Objective Type Questions

1. (*a*)	**2.** (*b*)	**3.** (*b*)	**4.** (*a*)	**5.** (*c*)	**6.** (*e*)
7. (*b*)	**8.** (*d*)	**9.** (*c*)	**10.** (*b*)	**11.** (*e*)	**12.** (*c*)
13. (*a*)	**14.** (*d*)	**15.** (*e*)	**16.** (*b*)	**17.** (*b*)	**18.** (*b*)
19. (*c*)	**20.** (*a*)	**21.** (*b*)	**22.** (*c*)	**23.** (*b*)	**24.** (*a*)
25. (*b*)	**26.** (*b*)	**27.** (*d*)	**28.** (*d*)	**29.** (*d*)	**30.** (*d*)
31. (*e*)	**32.** (*c*)	**33.** (*c*)	**34.** (*e*)	**35.** (*a*)	**36.** (*c*)
37. (*b*)	**38.** (*a*)	**39.** (*c*)	**40.** (*a*)	**41.** (*b*)	**42.** (*b*)
43. (*b*)	**44.** (*b*)	**45.** (*d*)	**46.** (*e*)	**47.** (*e*)	**48.** (*f*)
49. (*f*)	**50.** (*b*)	**51.** (*b*)	**52.** (*b*)	**53.** (*e*)	**54.** (*a*)

55. (*b*)	**56.** (*d*)	**57.** (*b*)	**58.** (*e*)	**59.** (*d*)	**60.** (*c*)
61. (*d*)	**62.** (*d*)	**63.** (*d*)	**64.** (*b*)	**65.** (*c*)	**66.** (*a*)
67. (*b*)	**68.** (*e*)	**69.** (*c*)	**70.** (*b*)	**71.** (*b*)	**72.** (*b*)
73. (*b*)	**74.** (*c*)	**75.** (*b*)	**76.** (*d*)	**77.** (*b*)	**78.** (*b*)
79. (*c*)	**80.** (*c*)	**81.** (*d*)	**82.** (*d*)	**83.** (*a*)	**84.** (*d*)
85. (*d*)	**86.** (*a*)	**87.** (*b*)	**88.** (*a*)	**89.** (*c*)	**90.** (*d*)
91. (*b*)	**92.** (*c*)	**93.** (*c*)	**94.** (*b*)	**95.** (*d*)	**96.** (*d*)
97. (*b*)	**98.** (*b*)	**99.** (*c*)	**100.** (*c*)	**101.** (*d*)	**102.** (*c*)
103. (*c*)	**104.** (*c*)	**105.** (*d*)	**106.** (*b*)	**107.** (*c*)	**108.** (*a*)
109. (*d*)	**110.** (*a*)	**111.** (*c*)	**112.** (*d*)	**113.** (*d*)	**114.** (*d*)
115. (*c*)	**116.** (*b*)	**117.** (*a*)	**118.** (*a*)	**119.** (*d*)	**120.** (*d*)
121. (*e*)	**122.** (*c*)	**123.** (*b*)	**124.** (*a*)	**125.** (*d*)	**126.** (*e*)
127. (*e*)	**128.** (*b*)	**129.** (*c*)	**130.** (*d*)	**131.** (*e*)	**132.** (*d*)
133. (*c*)	**134.** (*d*)	**135.** (*d*)	**136.** (*e*)	**137.** (*d*)	**138.** (*b*)
139. (*a*)	**140.** (*c*)	**141.** (*d*)	**142.** (*c*)	**143.** (*b*)	**144.** (*b*)
145. (*b*)	**146.** (*b*)	**147.** (*a*)	**148.** (*b*)	**149.** (*c*)	**150.** (*b*)
151. (*d*)	**152.** (*c*)	**153.** (*b*)	**154.** (*d*)	**155.** (*b*)	**156.** (*c*)
157. (*a*)	**158.** (*e*)	**159.** (*c*)	**160.** (*a*)	**161.** (*b*)	**162.** (*a*)
163. (*b*)	**164.** (*c*)	**165.** (*b*)	**166.** (*d*)	**167.** (*d*)	**168.** (*c*)
169. (*a*)	**170.** (*b*)	**171.** (*c*)	**172.** (*c*)	**173.** (*b*)	**174.** (*a*)
175. (*a*)	**176.** (*b*)	**177.** (*a*)	**178.** (*c*)	**179.** (*e*)	**180.** (*e*)
181. (*c*)	**182.** (*b*)	**183.** (*a*)	**184.** (*d*)	**185.** (*b*)	**186.** (*a*)
187. (*a*)	**188.** (*c*)	**189.** (*e*)	**190.** (*b*)	**191.** (*a*)	**192.** (*c*)
193. (*f*)	**194.** (*e*)	**195.** (*b*)	**196.** (*d*)	**197.** (*c*)	**198.** (*a*)
199. (*c*)	**200.** (*a*)	**201.** (*c*)	**202.** (*c*)	**203.** (*b*)	**204.** (*c*)
205. (*g*).					

True/False

206. (*b*)	**207.** (*a*)	**208.** (*a*)	**209.** (*a*)	**210.** (*b*)	**211.** (*a*)
212. (*a*)	**213.** (*b*)	**214.** (*a*)	**215.** (*b*)	**216.** (*a*)	**217.** (*a*)
218. (*a*)	**219.** (*a*)	**220.** (*a*)	**221.** (*b*)	**222.** (*a*)	**223.** (*a*)
224. (*a*)	**225.** (*a*)	**226.** (*a*)	**227.** (*a*)	**228.** (*a*)	**229.** (*b*)
230. (*a*)	**231.** (*a*)	**232.** (*b*)	**233.** (*a*)	**234.** (*a*)	**235.** (*a*)
236. (*a*)	**237.** (*a*)	**238.** (*b*)	**239.** (*a*)	**240.** (*a*)	**241.** (*a*)
242. (*b*)	**243.** (*a*)	**244.** (*a*)	**245.** (*a*)	**246.** (*b*)	**247.** (*b*)
248. (*a*)	**249.** (*a*)	**250.** (*b*)	**251.** (*b*)	**252.** (*b*)	**253.** (*b*)
254. (*a*)	**255.** (*a*)	**256.** (*a*)	**257.** (*b*)	**258.** (*a*)	**259.** (*a*)
260. (*b*)	**261.** (*b*)	**262.** (*a*)	**263.** (*a*)	**264.** (*b*)	**265.** (*b*)

266. (*b*)	**267.** (*b*)	**268.** (*b*)	**269.** (*b*)	**270.** (*b*)	**271.** (*b*)
272. (*a*)	**273.** (*a*)	**274.** (*a*)	**275.** (*a*)	**276.** (*a*)	**277.** (*a*)
278. (*a*)	**279.** (*b*)	**280.** (*a*)	**281.** (*b*)	**282.** (*b*)	**283.** (*a*)
284. (*b*)	**285.** (*b*)	**286.** (*b*)	**287.** (*a*)	**288.** (*b*)	**289.** (*a*)
290. (*a*)	**291.** (*a*)	**292.** (*a*)	**293.** (*a*)	**294.** (*a*)	**295.** (*a*)
296. (*a*)	**297.** (*a*)	**298.** (*b*)	**299.** (*a*)	**300.** (*a*)	**301.** (*a*)
302. (*b*)	**303.** (*a*)	**304.** (*a*)	**305.** (*a*)	**306.** (*b*)	**307.** (*a*)
308. (*a*)	**309.** (*b*)	**310.** (*a*)	**311.** (*a*)	**312.** (*a*)	**313.** (*a*)
314. (*a*)	**315.** (*a*)	**316.** (*a*)	**317.** (*b*)	**318.** (*a*)	**319.** (*a*)
320. (*b*).					

Fill in the Blanks

321. (*b*)	**322.** (*b*)	**323.** (*b*)	**324.** (*b*)	**325.** (*b*)	**326.** (*a*)
327. (*a*)	**328.** (*a*)	**329.** (*b*)	**330.** (*b*)	**331.** (*b*)	**332.** (*b*)
333. (*a*)	**334.** (*b*)	**335.** (*a*)	**336.** (*b*)	**337.** (*a*)	**338.** (*b*)
339. (*b*)	**340.** (*b*)	**341.** (*b*)	**342.** (*b*)	**343.** (*a*)	**344.** (*b*)
345. (*b*)	**346.** (*b*)	**347.** (*b*)	**348.** (*b*)	**349.** (*b*)	**350.** (*b*)
351. (*b*)	**352.** (*a*)	**353.** (*b*)	**354.** (*b*)	**355.** (*a*)	**356.** (*b*)
357. (*a*)	**358.** (*b*)	**359.** (*b*)	**360.** (*b*)	**361.** (*a*).	

Answers to Objective Type Questions from Competitive Examinations

1. (*a*)	**2.** (*a*)	**3.** (*a*)	**4.** (*a*)	**5.** (*b*)	**6.** (*d*)
7. (*b*)	**8.** (*a*)	**9.** (*b*)	**10.** (*b*)	**11.** (*b*)	**12.** (*a*)
13. (*c*)	**14.** (*b*)	**15.** (*c*)	**16.** (*a*)	**17.** (*e*)	**18.** (*c*)
19. (*d*)	**20.** (*c*)	**21.** (*a*)	**22.** (*b*)	**23.** (*d*)	**24.** (*c*)
25. (*b*)	**26.** (*d*)	**27.** (*b*), (*c*)	**28.** (*b*)	**29.** (*b*)	**30.** (*c*)
31. (*d*)	**32.** (*d*)	**33.** (*c*)	**34.** (*d*)	**35.** (*b*)	**36.** (*a*)
37. (*a*)	**38.** (*b*)	**39.** (*b*)	**40.** (*c*)	**41.** (*d*)	**42.** (*b*)
43. (*d*)	**44.** (*c*)	**45.** (*c*)	**46.** (*d*)	**47.** (*b*)	**48.** (*a*)
49. (*c*)	**50.** (*c*).				

Chapter 14

GENERAL ENGINEERING AND GENERAL APTITUDE

I. THEORY

MATHEMATICS

14.1. FUNCTIONS

(*i*) **Function of a Single Variable.** A variable y is said to be a function of another variable x if corresponding to some value of x there exists a value of y depending upon x. Here y is called **dependent variable** and x **independent variable.** This relationship is denoted by $y = f(x)$.

(*ii*) **Function of Several Variables.** If y depends on several variables, say $x_1, x_2, x_3, ..., x_n$, we say that y is a function of several variables $x_1, x_2, x_3, ..., x_n$. We denote it as $y = f(x_1, x_2, x_3, ..., x_n)$.

(*iii*) **Some Important Limits**

(*a*) $\lim\limits_{x \to 0} \dfrac{\sin x}{x} = 1$ (*b*) $\lim\limits_{x \to 0} (1 + x)^{\frac{1}{x}} = e$

(*c*) $\lim\limits_{x \to 0} \left(\dfrac{e^x - 1}{x}\right) = 1$ (*d*) $\lim\limits_{x \to 0} \left(\dfrac{a^x - 1}{x}\right) = \log_e a.$

(*iv*) **Continuity.** A function $f(x)$ is said to be **continuous** at the point $x = a$, if $\lim\limits_{x \to a} f(x) = f(a)$.

A function is said to be continuous in a given interval (a, b), if it is continuous at every point of the interval (a, b).

The sum, product or quotient of two continuous functions is also continuous.

14.2. SEQUENCE AND SERIES

(*i*) **Sequence.** A set of real numbers in a definite order, formed according to some law, is called a **sequence.** A sequence is generally denoted by $\{u_n\} = u_1, u_2, u_3, ..., u_n, ...$ to ∞.

For example, $\left\{\frac{1}{n}\right\} = 1, \frac{1}{2}, \frac{1}{3}, ..., \frac{1}{n}, ...$ to ∞ is a sequence.

A sequence $\{u_n\}$ is **convergent**, if $\lim\limits_{n \to \infty} u_n$ is finite, *e.g.*, $\left\{\frac{1}{n}\right\}$.

A sequence $\{u_n\}$ is **divergent,** if $\lim\limits_{n \to \infty} u_n$ is not finite, *e.g.*, $\{n^2\}$.

Every convergent sequence is bounded.

(*ii*) **Series.** If $\{u_n\}$ is a sequence, then the expression

$$u_1 + u_2 + u_3 + ... + u_n + ... \infty$$

is called an **infinite series.**

(*iii*) Let $S_n = u_1 + u_2 + u_3 + ... + u_n$. If $\lim\limits_{n \to \infty} S_n$ is finite, the series is said to be **convergent.** But if $\lim\limits_{n \to \infty} S_n \to \infty$, the series is said to be **divergent.** If $\lim\limits_{n \to \infty} S_n$ does not tend to a definite sum, the series is said to be oscillatory.

(*iv*) **P-series.** The series $1 + \frac{1}{2^p} + \frac{1}{3^p} + \frac{1}{4^p} + ... + \frac{1}{n^p} + ...$ is convergent when $p > 1$ and divergent when $p \le 1$.

(*v*) **Ratio Test.** Let $\lim\limits_{n \to \infty} \frac{u_{n+1}}{u_n} = k$. If $k > 1$, the series is divergent and if $k < 1$, the series is convergent. If $k = 1$, the ratio test fails.

(*vi*) **Comparison Test.** If $\lim\limits_{n \to \infty} \frac{u_n}{v_n}$ is finite, then either both series Σu_n and Σv_n are convergent or both are divergent.

14.3. DIFFERENTIATION

(*i*) **Derivative** $f'(x) = \lim\limits_{h \to 0} \frac{f(x+h) - f(x)}{h}$.

(*ii*) **Physical Meaning of** $\frac{dy}{dx}$. If $y = f(x)$, then $\frac{dy}{dx} = f'(x)$ represents the rate of change of y with respect to x.

(*iii*) **Geometrical Meaning.** If $y = f(x)$ is the equation of a curve, then $\frac{dy}{dx}$ represents the slope or gradient of the curve at any point (x, y).

(*iv*) **Monotonocity of Functions.** A function $y = f(x)$ is an increasing function in the interval (a, b) when $\frac{dy}{dx} > 0$ in (a, b). If $\frac{dy}{dx} < 0$ in the interval (a, b), then $y = f(x)$ is a decreasing function in (a, b).

(*v*) **Taylor Series.** $f(x+h) = f(x) + hf'(x) + \frac{h^2}{2!} f''(x) + \frac{h^3}{3!} f'''(x) + \ldots\ldots$

(*vi*) **Maxima and Minima.** A function $y = f(x)$ is said to have a maximum at a point when $\frac{dy}{dx} = f'(x) = 0$ at that point and $\frac{d^2y}{dx^2} = f''(x)$ is negative at that point.

For minimum $\frac{dy}{dx} = f'(x) = 0$ but $\frac{d^2y}{dx^2} = f''(x)$ is positive at that point.

14.4. INTEGRATION

It is regarded as the inverse of differentiation. Since $\frac{d}{dx}(\tan x) = \sec^2 x$, we have $\int \sec^2 x \, dx = \tan x$.

We integrate functions with the help of standard formulae. In case of complicated functions, we simplify them by the method of substitution. In case of products of functions, we integrate by parts. We also use the method of partial fractions, reduction formulae etc. to evaluate integrals.

14.5. MULTIPLE INTEGRALS

(*i*) **Double Integral.** $\int_{y_1}^{y_2} \int_{x_1}^{x_2} f(x, y)\, dx\, dy$

(*ii*) **Triple Integral.** $\int_{z_1}^{z_2} \int_{y_1}^{y_2} \int_{x_1}^{x_2} f(x, y, z)\, dx\, dy\, dz.$

(*a*) If $x_1, x_2, y_1, y_2, z_1, z_2$ are constants, then the order of integration is immaterial, provided the limits of integration are changed accordingly. For example

$$\int_0^2 \boxed{\int_0^1 (x+y)\, dx}\, dy = \int_0^2 \left[\frac{x^2}{2} + xy\right]_0^1 dy = \int_0^2 \left(\tfrac{1}{2} + y\right) dy = 3$$

or $$\int_0^1 \boxed{\int_0^2 (x+y)\, dy}\, dx = \int_0^1 \left[xy + \frac{y^2}{2}\right]_0^2 dx = \int_0^1 (2x+2)\, dx = 3.$$

(*b*) If z_1, z_2 are functions of x and y, y_1 and y_2 are functions of x, while x_1, x_2 are constants, then integration is to performed first w.r.t. z, then w.r.t. y and finally w.r.t. x. Thus

$$\int_{x=a}^{x=b} \boxed{\int_{y=f_1(x)}^{y=f_2(x)} \boxed{\int_{z_1=\phi_1(x,y)}^{z_2=\phi_2(x,y)} F(x, y, z)\, dz}\, dy}\, dx.$$

14.6. FUNCTIONS OF A COMPLEX VARIABLE

(*i*) **Complex Variable.** A variable number of the form $x + iy$, where both x and y are real variables, is called a complex variable. We write $z = x + iy$.

(*ii*) **Function of a Complex Variable.** If corresponding to each value of a complex variable z, there are one or more than one values of another complex variable w, then w is said to be a function of z and it is expressed as $w = f(z)$.

Since $z = x + iy$, we can always write

$$w = f(z) = f(x + iy) = u(x, y) + iv(x, y)$$

or simply, $w = u + iv$

where u and v are functions of x and y.

Thus if $w = z^3 = (x + iy)^3 = x^3 - 3xy^2 + i(3x^2y - y^3)$

Then $u = x^3 - 3xy^2$ and $v = 3x^2y - y^3$.

(*iii*) **Continuity of f(z).** A single-valued function $w = f(z)$ is said to be continuous at $z = z_0$, if

$$\underset{z \to z_0}{\text{limit}}\ f(z) = f(z_0).$$

Further, $f(z)$ is said to be continuous in any region R of the z-plane, if it is continuous at every point of that region.

If $f(z)$ is continuous, then $u(x, y)$ and $v(x, y)$ are also continuous and *vice versa.*

(*iv*) **Derivative of $f(z)$.** Let $w = f(z)$ be a single-valued function of z. The derivative of $w = f(z)$ is defined as

$$\frac{dw}{dz} = f'(z) = \underset{\delta z \to 0}{\text{limit}} \frac{f(z + \delta z) - f(z)}{\delta z} \quad \text{provided the limit exists.}$$

The necessary and sufficient conditions for $\frac{dw}{dz} = f'(z)$ to exist for all values of z in a given region R, are

(*a*) $\frac{\partial u}{\partial x}, \frac{\partial u}{\partial y}, \frac{\partial v}{\partial x}, \frac{\partial v}{\partial y}$ are continuous functions of x and y in R.

(*b*) $\frac{\partial u}{\partial x} = \frac{\partial v}{\partial y}$ and $\frac{\partial u}{\partial y} = -\frac{\partial v}{\partial x}$. These relations are known as **Cauchy-Riemann Equations.**

(*v*) **Analytic Functions.** A single-valued function $f(z)$ is said to be **analytic** or **regular** in a region R if its derivative $f'(z)$ exists at all points of the region R.

Every analytic function $f(z) = u + iv$ satisfies the conditions mentioned in (*iv*) (*a* and *b*). The real part $u(x, y)$ and imaginary part $v(x, z)$ of $f(z)$ are called conjugate functions. They are also called **harmonic functions** since they satisfy the Laplace equations

$$\frac{\partial^2 u}{\partial x^2} + \frac{\partial^2 u}{\partial y^2} = 0, \quad \frac{\partial^2 v}{\partial x^2} + \frac{\partial^2 v}{\partial y^2} = 0.$$

(*vi*) **Conformal Transformation or Mapping.** For every point $z = x + iy = (x, y)$ in the z-plane the relation $w = f(z)$ defines a corresponding point $w = u + iv = (u, v)$ in the w-plane. We call this a mapping of the z-plane into the w-plane.

The relation $w = f(z)$ transforms curves and regions in z-plane into some other curves and regions in the w-plane.

Let two curves c_1 and c_2 intersecting at z_0 in the z-plane be mapped on two curves c_1' and c_2' intersecting at w_0 in the w-plane. If the angle between c_1 and c_2 is equal to the angle between c_1' and c_2', then the transformation is called **isogonal.** In case sense of transformation as well as the magnitude of the angle is preserved then the transformation is said to be **conformal.**

A transformation $w = f(z)$ is conformal if $f(z)$ is analytic and $f'(z) \neq 0$ at $z = z_0$.

(*vii*) **Bilinear Transformation.** The transformation $w = \dfrac{az + b}{cz + d}$, $ad - bc \neq 0$ is known as **bilinear transformation.** It maps lines into lines and circles into circles.

(*viii*) **Cauchy's Theorem.** If $f(z)$ is analytic and $f'(z)$ is continuous at each point within or on a closed curve C, then $\int_C f(z)\, dz = 0$.

Corollary. If $f(z)$ is analytic in the region R between two simple closed curves C and C_1, then

$$\int_C f(z)\, dz = \int_{C_1} f(z)\, dz \cdot$$

This shows that the integral of an analytic function $f(z)$ does not depend upon the contour.

(*ix*) **Cauchy's Integral Formula.** If $f(z)$ is analytic inside and on a simple closed curve C, and if z is in any point within C, then

$$f(z_0) = \frac{1}{2\pi i} \int_C \frac{f(z)}{z - z_0}\, dz,$$

the integration around C being taken in anti-clockwise direction.

(*x*) **Taylor's Series.** A function $f(z)$, analytic inside a circle C with centre z_0, can be expanded in the series

$$f(z) = f(z_0) + f'(z) \,.\, (z - z_0) + \frac{f''(z_0)}{2!}(z - z_0)^2 + \ldots\ldots + \frac{f^n(z_0)}{n!}(z - z_0)^n + \ldots\ldots,$$

which is convergent at every point inside C.

(*xi*) **Laurent's Series.** Let C_1 and C_2 be two concentric circles of radii R_1 and R_2 $(R_1 > R_2)$ with centre z_0, and let $f(z)$ be analytic on C_1 and C_2, and in the annular region R between them. Then for any point z in R

$$f(z) = \sum_{n=0}^{\infty} a_n (z - z_0)^n + \sum_{n=1}^{\infty} \frac{b_n}{(z - z_0)^n}$$

where $a_n = \dfrac{1}{2\pi i} \displaystyle\int_{C_1} \frac{f(t)}{(t - z_0)^{n+1}}\, dt$ and $b_n = \dfrac{1}{2\pi i} \displaystyle\int_{C_2} (t - z_0)^{n-1} f(t)\, dt$.

(*xii*) **Singular Points.** A point z_0 at which a function $f(z)$ is not analytic is known as a **singular point** or **singularity** of $f(z)$. If there is no other singular point in its neighbourhood, we call it an **isolated** singular point. The function $f(z) = \dfrac{z^2 + 1}{(z - 2)^3 (z^2 + 9)}$ has three isolated singular points.

They are $z = 2$, $z = \pm 3i$.

The second part of the Laurent's series is

$$\sum_{n=1}^{\infty} \frac{b_n}{(z-z_0)^n} = \frac{b_1}{(z-z_0)} + \frac{b_2}{(z-z_0)^2} + \ldots\ldots + \frac{b_m}{(z-z_0)^m} + \ldots\ldots .$$

If this infinite series terminates due to $b_{m+1} = b_{m+2} = \ldots\ldots = 0$ after some particular term b_m, then $f(z)$ is said to have a singularity, called **pole of order m.** A pole of order one is called a **simple pole.**

But if this series does not terminate then z_0 is called an essential singularity of $f(z)$.

(*xiii*) **Residue.** The co-efficient b_1 in Laurent's series is called residue of $f(z)$ at $z = z_0$. It is given by

$$b_1 = \frac{1}{2\pi i} \int_C f(z)\,dz$$

If z_0 is a simple pole, then $\quad b_1 = \lim_{z \to z_0} \{(z - z_0) f(z)\}$

If z_0 is a pole of order m, then $b_1 = \frac{1}{(m-1)!} \lim_{z \to z_0} \frac{d^{m-1}}{dz^{m-1}} \{(z-z_0)^m f(z)\}$.

(*xiv*) **Residue Theorem.** If $f(z)$ be analytic at all points inside and on a simple closed curve C, except for a finite number of isolated singular points $z_1, z_2, \ldots\ldots, z_n$ inside C, then

$$\int_C f(z)\,dz = 2\pi i \text{ (sum of the residues at } z_1, z_2, \ldots\ldots, z_n).$$

ELECTRICAL TECHNOLOGY

14.7. INTRODUCTION TO ELECTRICAL TECHNOLOGY

An alternating quantity is one which goes through a complete cycle of change repeating periodically and has positive and negative values. The average value of an a.c. quantity for a full cycle is zero. However, the average value is defined for a half cycle only $E_{av} = \frac{1}{\pi} \int_0^{\pi} E_m \sin wt\,dt = \frac{2E_m}{\pi}$

The root mean square value is $I_{rms} = \sqrt{\frac{1}{T} \int_0^T i^2\,dt} = \frac{I_m}{\sqrt{2}}$

where i is the instantaneous value.

In a series *R-L* circuit driven by an a.c. voltage v, let the current drawn is $i = I_m \sin wt$, then the *KVL* equation is

$$Ri + L\left(\frac{di}{dt}\right) = v$$

$$I_m \sqrt{R^2 + w^2L^2}\ \sin(wt + \phi) = v, \quad \text{where } \phi = \tan^{-1}\left(\frac{wL}{R}\right).$$

If the voltage is sinusoidal, the current is also sinusoidal. The current will lag behind the voltage by an angle ϕ. wL is the inductive reactance and $\sqrt{(R^2 + w^2L^2)}$ is the magnitude of impedance Z.

Similarly, in a series *R-C* circuit driven by an a.c. voltage v, the *KVL* equation is

$$Ri + \frac{1}{C}\int i\,dt = v$$

current is leading the voltage by an angle ϕ, where $\phi = \tan^{-1}\left(\frac{1}{wCR}\right)$

$$\text{Power} = V_{rms}\, I_{rms} \cos\phi$$

which is same for *R-L* circuit also. $\cos\phi$ is the power factor. The capacitive energy stored is $\frac{1}{2}CV_m^2$ and that of inductive circuit is $\frac{1}{2}LI^2{}_m$. In the same way, for a series *R-L-C* circuit driven by an a.c. voltage v, the v-i equation is $\left[R + j\left(wL - \frac{1}{wC}\right)\right]i = v.$

For $wL > \frac{1}{wC}$, the circuit is equivalent to series *R-L* circuit and for $wL < \frac{1}{wC}$, it is equivalent to series *R-C* circuit and at $wL = \frac{1}{wC}$, it is purely resistive.

The concept of impedance is useful for series circuit where $Z = R + jX$, R is resistance and X is reactance. Admittance Y is the inverse of impedance. It consists of susceptance B and conductance G.

$$Y = G + jB.$$

A high pass *RC* network with step voltage of V as input gives the output response as $Ve^{-t/RC}$. *RC* is the time constant. After one time constant the response decays to 63% of the applied input V and after $2T$, it reaches 84%. For very small value of time constant, this circuit behaves as a differentiator with a pulse input of *RC*. Differentiating circuit spikes are obtained at the output having zero average value. The low pass *RC* circuit with step voltage V as input, the output becomes $V(1 - e^{-t/T})$ which reaches 63% of the final value after one time constant $T = RC$. For very large T this acts as integrator.

The property of cancellation of reactances when inductive and capacitive reactances are in series or cancellation of susceptances when in parallel, has become known as resonance.

Resonance occurs at unity power factor. The figure of merit or Q of energy storing devices has been defined as

$$Q = 2\pi \times \frac{\text{Maximum energy stored per cycle}}{\text{Energy dissipated per cycle}}.$$

The definition of Q for an inductor is

$$Q = \frac{2\pi \times LI_m^2/2}{I_m^2\, R_s/2f} = \frac{wL}{R_s}$$

where R_s is the series resistance of inductor and I_m is the maximum value of current through the inductor and resistor.

The Q of a capacitor is $\quad Q = \frac{2\pi \times CE_m^2/2}{E_m^2/2R_p f} = wCR_p$

where R_p is the leakage resistance of capacitor in shunt and E_m is the madiman voltage across the capacitor. When Q of a resonant circuit is stated, the resonant frequency of the circuit is implied as that of the management.

For a series R-C-L circuit, resonance occurs at frequency $f_r = \dfrac{1}{2\pi\sqrt{LC}}$.

The current at resonance, $I_r = \dfrac{E}{R}$.

The voltage across L at resonance, $E_L = jQE$ and the voltage across C at resonance, $E_C = -jQE$. In R-L-C series circuit, the impedance is minimum and the current is maximum at resonance frequency.

The impedance, $Z = R[1 + j \,.\, Q\delta\,(2 - \delta)]$ is expressed in terms of Q and δ; where δ is the small fractional deviation of the actual frequency from the resonant frequency. The bandwidth of a resonant circuit is defined as the width of the resonant curve, in cycles, at the frequency at which the power in the circuit is one-half the maximum power. The bandwidth B in cycles is $B = \dfrac{f_r}{Q}$.

The bandwidth for matched condition is $\left(\dfrac{2f_r}{Q}\right)$.

For a parallel resonance or antiresonance, the antiresonant frequency

$$f_{ar} = \frac{1}{2\pi}\sqrt{\frac{1}{LC}}\,.\,\sqrt{1-\frac{1}{Q^2}}\,.$$

For Q of very large value, $\quad f_{ar} = f_r = \dfrac{1}{2\pi\sqrt{LC}}$

At unity power factor, the impedance $Z = R_{ar}$.

14.8. MUTUAL INDUCTANCE

If the current through an inductance varies with time, the flux through the coil likewise varies and an e.m.f. is induced in the coil. The direction of this self-induced e.m.f. is such as to oppose the circuit e.m.f. if the current is increasing or to add to the circuit e.m.f. if the current is decreasing

$e = -L\dfrac{di}{dt}$.

Faraday's law is $e = -N\dfrac{d\phi}{dt}$ hence $L = N\dfrac{d\phi}{di}$ Henry.

14.9. DEFINITIONS IN NETWORK ANALYSIS

It is desirable to introduce certain definitions in network analysis. (*i*) Lumped network—One in which physically separate resistors, inductors, capacitors can be represented, (*ii*) Passive network—A network containing circuit elements without energy sources, (*iii*) Active network—A network containing generators or energy sources as well as other elements, (*iv*) Linear elements—A circuit element is linear if the relation between current and voltage involves a constant coefficient as in

$$e = Ri; e = L\frac{di}{dt}, e = \frac{1}{C}\int i\,dt$$

Iron-cored reactors and incandescent lamps are examples of elements that are not linear or in which the coefficient is a function of current.

Linear networks are those in which the differential equation relating the instantaneous current and voltage is a linear equation which constant coefficients, (*v*) Mesh—A set of branches forming a closed path in a network, provided that if one branch is omitted, the remaining branches do not form a closed path, (*vi*) Node—A terminal of any branch of a network or a terminal common to two or more branches.

Kirchhoff's basic circuit laws provide two methods for the solution of networks. The potential law (*i.e., KVL*) states that the algebraic summation of potential around a closed traverse of a circuit is zero and this leads to the method of network solution known as mesh or loop analysis. The current law (*i.e., KCL*) states that the algebraic summation of the currents towards a junction point is zero and this leads to node or junction analysis.

In general, if N junction points and B branches exist in a circuit, $N - 1$ current summations are required by the *KCL* method. Likewise $B - (N - 1) = B - N + 1$ voltage summations will be required by *KVL* method of analysis. However, when $(N - 1) < (B - N + 1)$; the node-voltage method will yield fewer equations.

Again principle of duality is of importance because *KVL* and *KCL* are dual and advantage is obviously on the choice of method of solution provided one can find the dual network easily. A parallel *GLC* circuit driven by a current source is dual of a series *RLC* circuit excited by a voltage source. The mesh equations for the series case is

$$Ri + L\frac{di}{dt} + \frac{1}{C}\int i\,dt = e$$

and for the dual parallel circuit $Ge + C\frac{de}{dt} + \frac{1}{L}\int e\,dt = i$

illustrating the various dual relations.

Again the conversion of T to π or *vice versa* has the advantages in the analysis of electric circuit. Any linear, bilateral, passive electrical network can be represented at a single frequency by a T or π network. The network theorems lead to the advantage in the analysis of the network. The theorems are stated.

14.10. THE NETWORK THEOREMS

The network theorems are:

(*i*) Superposition theorem (*ii*) Reciprocating theorem
(*iii*) Thevenin's theorem (*iv*) Norton's theorem
(*v*) The compensation theorem (*vi*) The maximum power transfer theorem
(*vii*) Millman's theorem and (*viii*) Tellegen's theorem.

14.10.1. Superposition Theorem

In any linear network containing bilateral linear impedances and energy sources, the current flowing in any element is the vector sum of the currents that are separately caused to flow in that element by each energy source. This is the linearity test of the network.

14.10.2. Reciprocating Theorem

In any linear network containing bilateral linear impedances and energy sources, the ratio of a voltage E introduced in one mesh to the current I in any second mesh is the same as the ratio obtained if the positions of E and I are interchanged, other e.m.f.'s being removed.

A network in which reciprocity theorem is applicable is said to be bilateral. This reciprocity theorem leads to a definition of the transfer impedance Z_T, where if E_1 is the voltage to mesh 1 and I_2 is the current in any second mesh, then $Z_{T12} = \dfrac{E_1}{I_2}$.

14.10.3. Thevenin's Theorem

Any two-terminal linear network containing energy sources (generators) and impedances can be replaced with an equivalent circuit consisting of a voltage source E' in series with an impedance Z'. The value of E' is the open-circuit voltage between the terminals of the network, and Z' is the impedance measured between the terminals with all energy sources eliminated (but not their impedances).

Thevenin's theorem is the voltage source equivalent circuit.

14.10.4. Norton's Theorem

Any two terminal linear network containing energy sources (generators) and impedances can be replaced with an equivalent circuit consisting of a current source I' in parallel with an admittance Y'. The value of I' is the short circuit current between the terminals of the network and Y' is the admittance measured between the terminals with all energy sources eliminated (but not their admittances). Norton's theorem is the current source equivalent circuit.

Thevenin's and Norton's theorem are dual to each other.

14.10.5. The Compensation Theorem

Any impedance having an e.m.f. across its terminals may be replaced by a generator of zero internal impedance whose e.m.f. is at every instant equal to the e.m.f. across the impedance.

14.10.6. The Maximum Power Transfer Theorem

Maximum power will be delivered by a network to an impedance Z_R if the impedance of Z_R is the conjugate of the impedance Z' of the network, measured looking back into the terminals of the network.

A corollary to the above theorem states that if only the absolute magnitude and not the angle of Z_R may be varied, then the greatest power output be delivered from the network if the absolute magnitude of Z_R is made equal to the absolute magnitude of Z'.

14.10.7. Millman's Theorem

It is the generalisation of Thevenin and Norton's theorem for N number of networks. The substitution theorem states that a known voltage in a circuit can be replaced by an ideal voltage source and a known current can be replaced by an ideal current source.

14.10.8. Tellegen's Theorem

Let N_1 and N_2 be the networks having the same graph but possibly different types of elements between corresponding nodes. Then

$$\sum_{k=1}^{b} v_{1k} i_{2k} = 0 \quad \text{and} \quad \sum_{k=1}^{b} v_{2k} i_{2k} = 0$$

where b is the number of branches, v_{jk} and i_{jk} are the voltage and current of the kth branch of network N_1 for $j = 1$ and of network N_2 for $j = 2$ such that v_{jk} and i_{jk} are opposite in directions of all k. A special case of this theorem occurs for $N_1 = N_2$.

CHEMISTRY

14.11. BOHR'S MODEL OF ATOM

Atom consists of a small positively charged nucleus at the centres, containing most of the mass, around which negatively charged electrons are revolving in orbits.

The main **postulates of Bohr's theory** are

(*i*) The electrons continue revolving in their respective orbits without losing energy. Energy in fractions of a quantum cannot be lost. Orbits are also known as *energy levels* or *energy shells.* These are designated by *K, L, M, N* etc.

(*ii*) Energy is emitted or absorbed by an atom only when an electron moves from one level to another. The energy cannot change continuously. It changes abruptly as it jumps from one level to other.

(*iii*) The angular momentum of an electron moving round the nucleus is quantised and is given by

$$mvr = \frac{nh}{2\pi}$$

where m = Mass of an electron, n = Integer value (1, 2, 3), v = Velocity of electron, h = Planck's constant, and r = Radius of orbit.

14.12. DUAL NATURE OF MATTER

An electron has dual character of particle and wave. The wavelength (λ) of particle of mass (m) moving with velocity (ν) is given by de-Broglie equation, which can be easily derived using mass energy relationship.

$$E = mc^2 \quad \text{...}(i)$$

$$\text{Einstein} \quad E = h\nu \quad \text{...}(ii)$$

$$\text{Planck} \quad h\nu = mc^2 \quad \text{...}(iii)$$

$$\nu = \frac{c}{\lambda} \text{ (we know)}$$

Substituting this value of ν in equation (*iii*),

$$\frac{hc}{\lambda} = mc^2$$

$$\lambda = \frac{h}{mc} \quad \text{...}(iv)$$

Substituting c by velocity of electron (v),

$$\lambda = \frac{h}{mv} \quad \text{or} \quad \boxed{\lambda = \frac{h}{p}}$$

where p = Momentum = mv.

14.13. HEISENBERG'S UNCERTAINTY PRINCIPLE

The concept of probability in electron in motion follows from Heisenberg's uncertainty principle. According to this principle, we cannot determine the position and momentum of an electron exactly at a time, because when the light photon is used to see the position of electron, former disturbs the momentum (velocity).

It is expressed mathematically as

$$\Delta x \,.\, \Delta p = \frac{h}{2\pi}$$

where Δx = Uncertainty in position, Δp = Uncertainty in momentum.

If Δx is low Δp will be high and *vice versa,* but the product of these two remains constant. Therefore, it is not possible to know exactly the position of an electron as Bohr considered in his model. It is possible only to state or predict probability of locating an electron of a particular energy in a given region of space at a given time.

14.14. QUANTUM NUMBERS

The implementation of Schrödinger's wave equation to hydrogen atom and other, furnishes the idea of quantum numbers, namely

(*i*) Principal quantum number, n (*ii*) Azimuthal quantum number, l

(*iii*) Magnetic quantum number, m (*iv*) Spin quantum number, s

The fourth quantum number is followed from spectral evidence.

14.14.1. Principal Quantum Number (n)

It is the main quantum number which determines the energy of an electron as well as its distance from the nucleus. As it increases, energy and distance also increase.

14.14.2. Azimuthal Quantum Number (l)

It is also known as angular momentum quantum number or orbital quantum number. The value of l gives the sub-level and sub-shell in which electron is present. It also gives the shape of orbitals in which the electron is present. The value of l varies from 0 to $n - 1$, when n is the number of principal shell in which electron is present. The various sub-levels are designated as s, p, d and f depending upon the value of l.

Value of l	*Designation of sub-level*
0	s
1	p
2	d
3	f

If $n = 1$, the value of $l = n - 1 = 0$, *i.e.*, only s sub-shell is there. If $n = 2$ the values of $l = 0$ to $n - 1$ or 0 and 1. It shows two sub-shell (s and p) are present there.

14.14.3. Magnetic Quantum Number (m)

The motion of electron in orbital may be considered as the motion of charged particle in a loop. It will create a magnetic field. This field will interact with the external magnetic or electric field. As a result, the electrons in a given energy sub-level orient themselves in certain specific region in space around the nucleus. Its values varies from $- l$ to $+ l$ through zero, thus making total values $2l + 1$.

If $l = 0$ (*i.e.*, subshell s) m can have only one value zero. It means in s subshell only one orbital or only one possible orientation in space.

If $l = 1$ (*i.e.*, p sub-shell) m can have three values like $- 1, 0, + 1$. It mean in p sub-shell three orientation of electrons motions are possible. There are three orbitals similarly in case of $l = 2$, five values for m means d sub-shell has five orbitals and in case of $l = 3$ seven values for m mean f sub-shell has seven orbitals.

14.14.4. Spin Quantum Number (s)

The spinning of an electron adds to the angular momentum of the electron and therefore to its energy relationship. The electron can spin clockwise or anticlockwise. The values designed to these motions $+\frac{1}{2}$ and $-\frac{1}{2}$ respectively.

PHYSICS

14.15. DALTON'S LAW OF PARTIAL PRESSURE

If two or more gases which do not react with one another are present in an enclosed space then at a constant temperature the total pressure exerted by gaseous mixture is equal to the sum of the individual (partial) pressure which each gas would exert if it was present alone occupying the same total volume.

Thus, if p_1, p_2, p_3 are partial pressures of various gases present in a mixture then the total pressure of the gaseous mixture is given by $P = p_1 + p_2 + p_3 + \ldots\ldots$.

14.16. KINETIC MOLECULAR THEORY OF GASES

The kinetic theory consists of following postulates or assumptions. The summary is as follows:

(*i*) A gas consists of a vast number of very small particles (molecules) which are in rapid and continuous motion in all direction continually undergoing collision with each other and with the walls of the vessel but moving in straight lines between collisions.

(*ii*) The pressure exerted by a gas on the walls of its containing vessel is due to bombardment by moving particles.

(*iii*) The collisions between particles are perfectly elastic *i.e.*, the speed of the molecule is the same as after a collision as before only the direction of motion being charged.

(*iv*) Attractive forces between molecules are negligible as pressure on the gas is released, the gas expands.

(*v*) The molecules are so small as compared with the average distance travelled between consecutive collisions (which is known as the mean free path that the space actually occupied by the molecules is negligible as compared to the volume of the gas as a whole).

(*vi*) The average kinetic (motion) energy of the molecule is proportional to the absolute temperature. After applying all these points following kinetic gas equation was derived

$$PV = \frac{1}{3} mNU^2$$

where P = Pressure, V = Volume, m = Mass of a molecule

N = Number of molecules, U = Velocity of molecule.

14.17. TRANSLATIONAL, VIBRATIONAL AND ROTATIONAL ENERGIES

Molecules require different amount of energies for their translation, rotation and vibration. These energies are quantised say for example, vibrational energy is given by the equation

$$E_v = \frac{h}{2\pi}\sqrt{\frac{k}{\mu}}\left(v + \frac{1}{2}\right)$$

14.17.1. Heat Capacity

The mean heat capacity of a system between any two temperatures is defined as the quantity of heat required to raise the temperature of the system from the lower to the higher temperature divided by the temperature difference.

$$C_{(T_2, T_1)} = \frac{q}{T_2 - T_1}$$

q = Heat absorbed.

It is determined in two ways:

(*i*) Heat capacity at constant volume is given by $C_v = \left(\frac{q}{T_2 - T_1}\right)_v$.

(*ii*) Heat capacity at constant pressure is given by $C_p = \left(\frac{q}{T_2 - T_1}\right)_p$.

14.17.2. van der Waal's Equation

An ideal gas equation is $PV = RT$, van der Waal's introduced two corrections (*i*) b the volume of molecules themselves should be considered, the resulting volume $(V - b)$, (*ii*) attraction due to inter molecular forces lowers the exact pressure, a correction term $\frac{a}{V^2}$ should be added to P. Pressure term becomes $\left(P + \frac{a}{V^2}\right)$.

Corrected gas equation as follows $\left(P + \frac{a}{V^2}\right)(V - b) = RT$.

This equation is known as van der Waal's gas equation.

14.17.3. Vapour Pressure of Liquid

Vapour pressure of a liquid at a given temperature is defined as the pressure of the vapour in equilibrium with the liquid at the temperature. As the temperature of any liquid is raised, greater number of molecules acquire sufficient energy to escape as vapour. Thus with a rise in temperature the equilibrium concentration of vapour molecules increases. There is a relation between vapour pressure and temperature

$$\log \frac{P_2}{P_1} = \frac{\Delta H_v}{2.303\,R}\left[\frac{1}{T_1} - \frac{1}{T_2}\right]$$

where P_2 and P_1 are the vapour pressure of the liquid at temperature T_2 *K* and T_1 *K* respectively.

ΔH_v = Heat of vaporization per mole of the liquid.

14.17.4. Surface Tension of Liquid

Surface tension may be defined the force in dynes acting at right angles to the surface of a liquid along one centimetre length of the surface. Due to surface tension liquids tend to minimise their surface area. It can also be defined as the work (energy) required to expand the surface of a liquid by unit area. Mathematically,

$$\text{Surface tension, } (\gamma) = \frac{\text{Work done}}{\text{Change in area}}.$$

II. OBJECTIVE TYPE QUESTIONS

Tick mark the most appropriate of the multiple choice answers:

MATHEMATICS

1. Let $f(x) = |x| + |x-1| + |x-2| + |x-3|$. The value of $f(x)$ at $x = 0$ is
   (a) 0 ☐ (b) not defined ☐
   (c) – 6 ☐ (d) none of these. ☐
2. The period of the function ($\sin 3x + \cos 6x$) is
   (a) $\frac{\pi}{3}$ ☐ (b) $\frac{2\pi}{3}$ ☐
   (c) π ☐ (d) none of these. ☐
3. Which one of the following is not an even function?
   (a) $|x|$ ☐ (b) $3x^4 + 5x^2 + 1$ ☐
   (c) $\sin(x^2) + 4\cos(x^2)$ ☐ (d) $\tan(x^2) + \sin(x^3)$. ☐
4. A function of two variables $z = f(x, y)$ represents a
   (a) curve in a plane ☐ (b) curve in space ☐
   (c) surface in a plane ☐ (d) surface in space. ☐
5. $\lim_{\theta \to \pi/2} \left(\frac{2\theta - \pi}{\cos \theta} \right)$ is equal to
   (a) 0 ☐ (b) – 1 ☐
   (c) – 2 ☐ (d) none of these. ☐
6. Let $f(x) = (1 + \cos x)^{3 \sec x}$. Then $\lim_{x \to \pi/2} f(x)$ is
   (a) 1 ☐ (b) e ☐
   (c) e^2 ☐ (d) e^3. ☐
7. $\lim_{x \to b} \frac{|x-b|}{(x-b)}$
   (a) is – 1 ☐ (b) is + 1 ☐
   (c) does not exist ☐ (d) none of these. ☐
8. If $y = \frac{2^x - 1}{\sqrt{1+x-1}}$, then $\lim_{x \to 0} (y)$ is
   (a) log 2 ☐ (b) 2 log 2 ☐
   (c) zero ☐ (d) none of these. ☐
9. The function $f(x) = \frac{e^{1/x} - 1}{e^{1/x} + 1}$ when $x \neq 0$, and $f(0) = 0$ is
   (a) continuous at $x = 0$ ☐ (b) discontinuous at $x = 0$ ☐
   (c) none of these ☐ (d) continuous at $x = 1$. ☐

10. If $f(x) = \dfrac{\sin (a+1)x + \sin x}{x}$, for $x < 0$

$= c,$ for $x = 0$

$= \dfrac{\sqrt{(x + bx^2)} - \sqrt{x}}{bx^{3/2}}$, for $x >= 0$

is continuous at $x = 0$, then

(a) $a = \frac{3}{2}, c = \frac{1}{2}$ ☐ (b) $a = -\frac{3}{2}, c = -\frac{1}{2}$ ☐

(c) $a = \frac{3}{2}, c = -\frac{1}{2}$ ☐ (d) $a = -\frac{3}{2}, c = \frac{1}{2}$. ☐

11. If $f(x)$ is the integral of $\dfrac{2 \sin x - \sin 2x}{x^3}$, $x \neq 0$, then $\lim\limits_{x \to 0} f'(x)$ is

(a) zero ☐ (b) 1 ☐

(c) – 1 ☐ (d) none of these. ☐

12. If $f(x) = \dfrac{x-1}{2x^2 - 7x + 5}$, when $x \neq 1$ and

$f(x) = -\dfrac{1}{3}$, when $x = 1$, then $f'(x)$ at $x = 1$ is

(a) not defined ☐ (b) $-\frac{2}{9}$ ☐

(c) zero ☐ (d) none of these. ☐

13. If $\phi(x + y) = \phi(x)\,\phi(y)$ for all x and y, and $\phi(5) = 2$, $\phi'(0) = 3$, then $\phi'(5)$ is

(a) 6 ☐ (b) zero ☐

(c) not defined ☐ (d) none of these. ☐

14. If $\phi''(x) = -\phi(x)$, $\phi'(x) = \psi(x)$, $f(x) = [\phi(x)]^2 + [\psi(x)]^2$ and $f(10) = 11$, then $f(11)$ is

(a) 10 ☐ (b) 11 ☐

(c) 12 ☐ (d) none of these. ☐

15. If $f(x) = 1 + x$ when $x \leq 2$ and $f(x) = 5 - x$ when $x \geq 2$, then $f(x)$ is

(a) discontinuous at $x = 2$ ☐ (b) differentiable at $x = 2$ ☐

(c) not differentiable at $x = 2$ ☐ (d) none of these. ☐

16. Let $y = x^3 - 8x + 7$ and $x = f(t)$. Also let $x = 3$ when $t = 0$ and $\dfrac{df}{dt} = 2$, then at $t = 0$, $\dfrac{dy}{dt}$ is

(a) 19 ☐ (b) 38 ☐

(c) 57 ☐ (d) none of these. ☐

17. If $y = \sec^{-1}\left(\dfrac{x+1}{x-1}\right) + \sin^{-1}\left(\dfrac{x-1}{x+1}\right)$, then $\dfrac{dy}{dx}$ is

(a) zero ☐ (b) $\pi/2$ ☐

(c) 1 ☐ (d) none of these. ☐

18. The tangent to the curve $3xy^2 - 2x^2y = 1$ at (1, 1) meets the curve again at

(a) $\left(\frac{16}{5}, -\frac{1}{20}\right)$ ☐ (b) $\left(-\frac{16}{5}, \frac{1}{20}\right)$ ☐

(c) $\left(\frac{16}{5}, \frac{1}{20}\right)$ ☐ (d) $\left(-\frac{16}{5}, -\frac{1}{20}\right)$. ☐

19. If $x > 0$, then

(a) $\sin x > x$ ☐ (b) $\sin x = x$ ☐

(c) $\sin x < x$ ☐ (d) none of these. ☐

20. The maximum value of $f(x) = \frac{\log x}{x}$ is

(a) $\frac{1}{e}$, at $x = 1$ ☐ (b) $\frac{1}{e}$, at $x = e$ ☐

(c) 1, at $x = \frac{1}{e}$ ☐ (d) none of these. ☐

21. For the function $y = x^{\frac{1}{x}}$

(a) minimum y is 1 ☐ (b) maximum y is e^e ☐

(c) maximum y is $e^{\frac{1}{e}}$ ☐ (d) none of these. ☐

22. The extreme value of $\frac{a^2}{\cos^2 x} + \frac{b^2}{\sin^2 x}$ is

(a) $a - b$ ☐ (b) $a + b$ ☐

(c) $(a - b)^2$ ☐ (d) $(a + b)^2$. ☐

23. If $y = \frac{ax + b}{(x - 1)(x - 4)}$ has a turning point at (2, – 1), then

(a) $a = 0, b = 1$ ☐ (b) $a = -1, b = 0$ ☐

(c) $a = 1, b = 0$ ☐ (d) none of these. ☐

24. If $S = \frac{1}{4}t^4 - 2t^3 + 4t^2 - 7$, then $\frac{d^2S}{dt^2}$ is minimum at

(a) $t = 0$ ☐ (b) $t = 1$ ☐

(c) $t = 2$ ☐ (d) none of these. ☐

25. If $R = \frac{V^2}{54} - \frac{3(V - 12)}{V + 12}$, then R will have an extreme value at

(a) $V = 2$ ☐ (b) $V = 4$ ☐

(c) $V = 6$ ☐ (d) $V = 8$. ☐

26. The value of the function $\frac{(x+1)^2}{(x+3)^3}$ lies between

(*a*) 0 and $\frac{2}{27}$ ☐ (*b*) $-\frac{2}{27}$ and 0 ☐

(*c*) 1 and 3 ☐ (*d*) none of these. ☐

27. If h is small, then $\sin(x+h)$ is nearly equal to

(*a*) $\sin x + h\cos x + \frac{h^2}{2}\sin x$ ☐ (*b*) $\sin x - h\cos x - \frac{h^2}{2}\sin x$ ☐

(*c*) $\sin x - h\cos x + \frac{h^2}{2}\sin x$ ☐ (*d*) $\sin x + h\cos x - \frac{h^2}{2}\sin x.$ ☐

28. A circular plate expands under the influence of heat so that its radius increases from 5 cm to 5.06 cm. The approximate increase in the area is

(*a*) 18.8 cm^2 ☐ (*b*) 0.06 π cm^2 ☐

(*c*) 1.88 cm^2 ☐ (*d*) none of these. ☐

29. $Q = CH^{\frac{5}{2}}$ and error in the measurement of H committed of the order of 1.5 per cent, then error in the calculated value of Q will be of the order of

(*a*) 7.5% ☐ (*b*) 3.75% ☐

(*c*) 1.5% ☐ (*d*) none of these. ☐

30. In $pv = 20$ and p is measured as 5 ± 0.01, then v is

(*a*) 4 ± 0.002 ☐ (*b*) 4 ± 0.004 ☐

(*c*) 4 ± 0.006 ☐ (*d*) 4 ± 0.008. ☐

31. The power P required to propel a ship of length L moving with velocity V is given by $P = kV^3L^2$. An increase of 3% in V and 4% in L require

(*a*) 18.18% increase in P ☐ (*b*) 27% increase in P ☐

(*c*) 7% increase in P ☐ (*d*) none of these. ☐

32. If $R = \frac{E}{I}$ and possible errors in E and I are 20% and 10%, then error in R is

(*a*) 5% ☐ (*b*) 10% ☐

(*c*) 30% ☐ (*d*) none of these. ☐

33. The gradient of a curve is given by $3x^2 + 2bx + 5$. The curve passes through (0, 3) and (–1, – 1). The equation of the curve is

(*a*) $y = x^3 + 2x^2 + 5x + 1$ ☐ (*b*) $y = x^3 - 2x^2 + 5x + 3$ ☐

(*c*) $y = x^3 + 2x^2 + 5x + 3$ ☐ (*d*) none of these. ☐

34. The slope of a curve is given by $\frac{dy}{dx} = \frac{2y}{x}$, $x > 0$, $y > 0$. If the curve passes through the point (1, 1), then its equation is

(*a*) $y = x$ ☐ (*b*) $y = x^2$ ☐

(*c*) $y = x^3$ ☐ (*d*) none of these. ☐

35. A car moves on a straight road. Its velocity is given by $V(t) = 3\sqrt{t}$. In first 100 seconds, it will move

(*a*) 1000 ☐ (*b*) 1500 ☐

(*c*) 2000 ☐ (*d*) none of these. ☐

36. If $\frac{dV}{dt} = 3t^2 + 4t + 10$ and $S = 0$, $V = 5$ when $t = 0$, then at $t = 3$, V is

(*a*) 80 ☐ (*b*) 40 ☐

(*c*) 120 ☐ (*d*) none of these. ☐

37. $\int_0^{\pi} \cos mx \cos nx \, dx$, when $m \neq n$, is

(*a*) π ☐ (*b*) $\frac{\pi}{2}$ ☐

(*c*) zero ☐ (*d*) none of these. ☐

38. $\int_0^1 \frac{1-x^2}{1+x^2} dx$ is equal to

(*a*) $\frac{\pi}{2} + 1$ ☐ (*b*) $\frac{\pi}{2} - 1$ ☐

(*c*) $\pi + 1$ ☐ (*d*) $\pi - 1$. ☐

39. If $J = \int_1^2 \frac{1}{x^2} e^{-\frac{1}{x}} dx$, then J is equal to

(*a*) $\frac{1}{\sqrt{e}}$ ☐ (*b*) $\frac{\sqrt{e}+1}{e}$ ☐

(*c*) $\frac{1-\sqrt{e}}{e}$ ☐ (*d*) $\frac{\sqrt{e}-1}{e}$. ☐

40. $\int_0^{\pi/2} e^x (\sin x + \cos x) \, dx$ is equal to

(*a*) $e^{-\frac{\pi}{2}}$ ☐ (*b*) 1 ☐

(*c*) zero ☐ (*d*) none of these. ☐

41. $\int_0^{\pi/2} \frac{\sqrt{\sin x}}{\sqrt{\sin x} + \sqrt{\cos x}} dx$ is equal to

(*a*) $\frac{\pi}{4}$ ☐ (*b*) $\frac{\pi}{2}$ ☐

(*c*) π ☐ (*d*) none of these. ☐

42. $\int_0^{\pi} x f(\sin x) \, dx$ is equal to

(*a*) $\pi \int_0^{\pi} f(\sin x) \, dx$ ☐ (*b*) $\frac{\pi}{2} \int_0^{\pi} f(\sin x) \, dx$ ☐

(*c*) $\int_0^{\pi} f(\sin x) \, dx$ ☐ (*d*) none of these. ☐

43. Let $f(x) = a + bx + cx^2$, then $\int_0^1 f(x)\,dx$

(a) $\frac{1}{2}[f(0) + 2f\left(\frac{1}{2}\right) + f(1)]$ ☐ (b) $\frac{1}{3}[f(0) + f\left(\frac{1}{2}\right) + 4f(1)]$ ☐

(c) $\frac{1}{6}[f(0) + 4f\left(\frac{1}{2}\right) + f(1)]$ ☐ (d) none of these. ☐

44. Let $P = \int_0^1 \frac{a_1 a_2\,dx}{[(a_1 - a_2)x + a_2]^2}$, then P is equal to

(a) $\sin\frac{\pi}{2}$ ☐ (b) $\cos\frac{\pi}{2}$ ☐

(c) $\cos\pi$ ☐ (d) none of these. ☐

45. $\int_0^1 x^2(1 - x^2)^{3/2}\,dx$ is equal to

(a) $\frac{\pi}{8}$ ☐ (b) $\frac{\pi}{16}$ ☐

(c) $\frac{\pi}{32}$ ☐ (d) none of these. ☐

46. The double integral $\int_0^1 \int_0^1 (x^2 + y^2)\,dx\,dy$ reduces to

(a) $\frac{1}{3}$ ☐ (b) $\frac{2}{3}$ ☐

(c) 1 ☐ (d) none of these. ☐

47. The triple integral $\int_0^3 \int_0^2 \int_0^1 (x + y + z)\,dz\,dy\,dx$ is equal to

(a) 18 ☐ (b) 20 ☐

(c) 21 ☐ (d) none of these. ☐

48. Let $I = \int_{-1}^{2} \int_{x^2}^{x+2} dy\,dx$, then I is

(a) $\frac{7}{2}$ ☐ (b) $\frac{9}{2}$ ☐

(c) $\frac{11}{3}$ ☐ (d) none of these. ☐

49. Let $I = \iint_R x^2y^2\,dxdy$, where R is the region bounded by $x = 0$, $y = 0$ and $x^2 + y^2 = 1$. Then I is

(a) $\int_{-1}^{1} \int_0^{\sqrt{1-x^2}} x^2y^2\,dxdy$ ☐ (b) $\int_0^1 \int_{-\sqrt{1-y^2}}^{\sqrt{1-y^2}} x^2y^2\,dxdy$ ☐

(c) $\int_0^1 \int_0^{\sqrt{1-y^2}} x^2y^2\,dxdy$ ☐ (d) none of these. ☐

50. To evaluate $\iint dx\,dy$ over the area bounded by $x = 0$, $y = 0$, $x^2 + y^2 = 1$ and $5y = 3$, we must have the limits as indicated below

(a) $\int_0^{3/5} \int_0^{\sqrt{1-y^2}} dxdy$ ☐ (b) $\int_0^{\sqrt{1-y^2}} \int_0^{3/5} dxdy$ ☐

(c) $\int_0^{3/5} \int_0^{\sqrt{1-x^2}} dxdy$ ☐ (d) none of these. ☐

51. The volume of tetrahedron bounded by the co-ordinate planes and the plane $x + y + z = 1$ is given by the triple integral

(a) $\int_0^{1-x} \int_0^1 \int_0^{1-x-y} dz\, dx\, dy$ ☐ (b) $\int_0^1 \int_0^1 \int_0^1 dz\, dx\, dy$ ☐

(c) $\int_0^1 \int_0^{1-x} \int_0^{1-x-y} dz\, dy\, dx$ ☐ (d) none of these. ☐

52. The integral $\iint xy\, dx\, dy$ over the rectangle bounded by $x = 2, x = 5, y = 1$ and $y = 2$ is equal to

(a) 16 ☐ (b) $\frac{63}{2}$ ☐

(c) $\frac{63}{5}$ ☐ (d) none of these. ☐

53. If density at a point of a solid is $\rho(x, y, z)$, then the mass of the solid can be evaluated by the triple integral with appropriate limits of integration

(a) $\rho \iiint dx\, dy\, dz$ ☐ (b) $\iiint \rho\, dx\, dy\, dz$ ☐

(c) $\iiint dx\, dy\, dz$ ☐ (d) none of these. ☐

54. The C.G. of a plane lamina having variable density $\rho(x, y)$ at any point (x, y) is $(\bar{x}, \bar{y})$. The $\bar{x}$ is

(a) $\frac{\iint x\, dx\, dy}{\iint dx\, dy}$ ☐ (b) $\frac{\rho \iint x\, dx\, dy}{\iint dx\, dy}$ ☐

(c) $\frac{\iint \rho\, dx\, dy}{\iint \rho x\, dx\, dy}$ ☐ (d) $\frac{\iint x\rho\, dx\, dy}{\iint \rho\, dx\, dy}$. ☐

55. The moment of inertia of a plane lamina about y-axis is given by

(a) $\iint \rho\, x^2\, dx\, dy$ ☐ (b) $\iint \rho y^2\, dx\, dy$ ☐

(c) $\iint \rho xy\, dx\, dy$ ☐ (d) none of these. ☐

56. The product of inertia of a solid of volume V about x and y-axes is given by

(a) $\iiint_V \rho y^2\, dx\, dy\, dz$ ☐ (b) $\iiint_V (x^2 + y^2)\, \rho\, dx\, dy\, dz$ ☐

(c) $\iiint_V \rho\, xyz\, dx\, dy\, dz$ ☐ (d) none of these. ☐

where ρ is the density of the solid at any point (x, y, z) of the solid.

57. The sequence $\left\{\frac{1}{2^n}\right\}$ is

(a) convergent ☐ (b) divergent ☐

(c) oscillating ☐ (d) none of these. ☐

58. The sequence $\{(-1)^n n\}$ is

(*a*) bounded below ☐ (*b*) bounded above ☐

(*c*) bounded ☐ (*d*) none of these. ☐

59. Every convergent sequence is

(*a*) bounded ☐ (*b*) finite ☐

(*c*) monotonic ☐ (*d*) none of these. ☐

60. Every convergent sequence has

(*a*) a unique limit ☐ (*b*) two limits ☐

(*c*) finite value ☐ (*d*) none of these. ☐

61. The sequence $\{a_n\}$, where $a_n = 1 + \frac{1}{3} + \frac{1}{3^2} + \ldots.. + \frac{1}{3^n}$ is

(*a*) convergent ☐ (*b*) conditionally convergent ☐

(*c*) non-convergent ☐ (*d*) none of these. ☐

62. If 137 + 276 = 435, how much is 731 + 672 =?

(*a*) 534 ☐ (*b*) 1403 ☐

(*c*) 1623 ☐ (*d*) 1513 ☐

(GATE ALL BRANCHES-2010)

63. The series $1 + x + x^2 + x^3 + \ldots\ldots$ to ∞, converges if

(*a*) $x < 1$ ☐ (*b*) $-1 < x < 1$ ☐

(*c*) $-1 \le x \le 1$ ☐ (*d*) none of these. ☐

64. If a series Σu_n is convergent, then $\lim\limits_{n \to \infty} u_n$ is

(*a*) 1 ☐ (*b*) zero ☐

(*c*) finite converent ☐ (*d*) none of these. ☐

65. Which statement is not correct?

(*a*) $1 + \frac{1}{2} + \frac{1}{3} + \ldots.. + \frac{1}{n} + \ldots\ldots$, converges ☐

(*b*) A positive term series either converges or diverges to infinity ☐

(*c*) $1^2 + 2^2 + 3^2 + \ldots\ldots + n^2 + \ldots\ldots$, is not convergent ☐

(*d*) $\sum\limits_{n=0}^{\infty} (-1)^n$ oscillates finitely. ☐

66. The series $1 + \frac{1}{2^p} + \frac{1}{3^p} + \frac{1}{4^p} + \ldots\ldots + \frac{1}{n^p} + \ldots\ldots$ converges when

(*a*) $p \ge 1$ ☐ (*b*) $p < 1$ ☐

(*c*) $p > 1$ ☐ (*d*) none of these. ☐

67. The series $\frac{2^p}{1^q} + \frac{3^p}{2^q} + \frac{4^p}{3^q} + \ldots\ldots$, p and q being positive numbers is divergent if

(*a*) $a < p + 1$ ☐ (*b*) $q \le p + 1$ ☐

(*c*) $q > p + 1$ ☐ (*d*) $q \ge p + 1$. ☐

68. The series $\Sigma[\sqrt{(n^2+1)} - \sqrt{(n^2-1)}]$ is

(a) convergent ☐ (b) divergent ☐

(c) conditionally convergent ☐ (d) none of these. ☐

69. In the series $\sum \frac{1}{n}\sin\left(\frac{1}{n}\right)$ and $\sum \frac{1}{\sqrt{n}}\tan\left(\frac{1}{n}\right)$

(a) both are convergent ☐ (b) only one is convergent ☐

(c) none is convergent ☐ (d) none of these. ☐

70. To test the convergence of the series $\frac{2!}{3} + \frac{3!}{3^2} + \frac{4!}{3^3} + \ldots\ldots$, we apply

(a) comparison test ☐ (b) ratio test ☐

(c) root test ☐ (d) none of these. ☐

71. $\frac{1}{2\sqrt{1}} + \frac{x^2}{3\sqrt{2}} + \frac{x^4}{4\sqrt{3}} + \frac{x^6}{5\sqrt{4}} + \ldots\ldots$ is convergent when

(a) $x^2 \geq 1$ ☐ (b) $x^2 \leq 1$ ☐

(c) $x \leq 1$ ☐ (d) none of these. ☐

72. The series $1 - \frac{1}{2^p} + \frac{1}{3^p} - \frac{1}{4^p} + \ldots\ldots, p > 0$ is

(a) convergent for all values of p ☐ (b) conditionally convergent ☐

(c) divergent ☐ (d) none of these. ☐

73. The series $1 + \frac{1}{2} + \frac{1}{2^2} + \frac{1}{2^3} + \ldots\ldots$ is

(a) conditionally convergent ☐ (b) non-convergent ☐

(c) absolutely convergent ☐ (d) none of these. ☐

74. Let $f(z) = \bar{z} = x - iy$. Then $f(z)$ is

(a) continuous at $x = 0$ ☐ (b) differentiable at $x = 0$ ☐

(c) neither continuous nor differentiable ☐ (d) none of these. ☐

75. $f(z) = |z|^2 = x^2 + y^2$ is

(a) continuous at $z = 0$ only ☐

(b) differentiable at $z = 0$ only ☐

(c) continuous as well as differentiable for every value of z ☐

(d) none of these. ☐

76. If $w = f(z) = u(x, y) + iv(x, y)$ is analytic, then

(a) $\frac{\partial u}{\partial x} = \frac{\partial v}{\partial y}$ and $\frac{\partial u}{\partial y} = \frac{\partial v}{\partial x}$ ☐ (b) $\frac{\partial u}{\partial x} = -\frac{\partial v}{\partial y}$ and $\frac{\partial u}{\partial y} = \frac{\partial v}{\partial x}$ ☐

(c) $\frac{\partial u}{\partial x} = -\frac{\partial v}{\partial y}$ and $\frac{\partial u}{\partial y} = -\frac{\partial v}{\partial x}$ ☐ (d) $\frac{\partial u}{\partial x} = \frac{\partial v}{\partial y}$ and $\frac{\partial u}{\partial y} = -\frac{\partial v}{\partial x}$. ☐

77. If a function $f(z)$ be analytic in a domain D, then it must be
(a) differentiable in the domain D ☐
(b) continuous but not differentiable in the domain D ☐
(c) differentiable but not continuous in the domain D ☐
(d) none of these. ☐

78. If $f(z) = u(x, y) + iv(x, y)$, then u and v satisfy
(a) $\frac{\partial^2 u}{\partial x^2} + \frac{\partial^2 u}{\partial y^2} = 0$ but not $\frac{\partial^2 v}{\partial x^2} + \frac{\partial^2 v}{\partial y^2} = 0$ ☐
(b) $\frac{\partial^2 v}{\partial x^2} + \frac{\partial^2 v}{\partial y^2} = 0$ but not $\frac{\partial^2 u}{\partial x^2} + \frac{\partial^2 u}{\partial y^2} = 0$ ☐
(c) $\frac{\partial^2 u}{\partial x^2} + \frac{\partial^2 u}{\partial y^2} = 0$ as well as $\frac{\partial^2 v}{\partial x^2} + \frac{\partial^2 v}{\partial y^2} = 0$ ☐
(d) none of these. ☐

79. Let $f(z) = u(x, y) + iv(x, y)$. Then u and v are called
(a) analytic functions ☐ (b) harmonic functions ☐
(c) orthogonal functions ☐ (d) none of these. ☐

80. Let $f(z) = u(x, y) + iv(x, y)$. If $u = x^3 - 3xy^2$, then
(a) $f(z) = z^2$ ☐ (b) $f(z) = z^3$ ☐
(c) $f(z) = z^4$ ☐ (d) none of these. ☐

81. If real part of a function $f(z)$ is given by $u = \frac{1}{2} \log (x^2 + y^2)$, then its imaginary part v is
(a) $\tan^{-1} \frac{y}{x}$ ☐ (b) $\tan^{-1} \frac{x}{y}$ ☐
(c) indeterminate ☐ (d) none of these. ☐

82. At each point z of a domain where $f(z)$ is analytic and $f'(z) \neq 0$, mapping $w = f(z)$ is
(a) isogonal ☐ (b) bilinear ☐
(c) conformal ☐ (d) none of these. ☐

83. A bilinear transformation is defined by
(a) $w = \frac{az + b}{cz + d}$ ☐ (b) $w = \frac{az^2 + b}{cz + d}$ ☐
(c) $w = \frac{az + b}{c + dz^2}$ ☐ (d) none of these. ☐

84. In the transformation $w = f(z)$, a point at which $f'(z) = 0$ is called
(a) singular point ☐ (b) regular point ☐
(c) critical point ☐ (d) none of these. ☐

85. The transformation $w = z^2$ transform
(a) the upper half of the z-plane into the entire w-plane ☐
(b) the entire z-plane into the upper half of w-plane ☐

(c) the entire z-plane into the entire w-plane ☐

(d) none of these. ☐

86. If C is the circle $| z - a | = r$, then $\int_C \frac{dz}{z-a}$ is equal to

(a) πi ☐ (b) $2\pi i$ ☐

(c) zero ☐ (d) none of these. ☐

87. If C is the circle $|z - b| = R$ and n is any integer $\neq -1$, then $\int_C (z-b)^n\, dz$ is

(a) zero ☐ (b) πi ☐

(c) $2\pi i$ ☐ (d) none of these. ☐

88. $\int_C \frac{e^z}{z-1}\, dz$, where C is the circle $| z | = 2$ is equal to

(a) πei ☐ (b) $2\pi ei$ ☐

(c) zero ☐ (d) none of these. ☐

89. If $f(\xi) = \int_C \frac{3z^2+7z+1}{z-\xi}\, dz$, where C is the circle $x^2 + y^2 = 4$, then $f''(1-i)$ is

(a) πi ☐ (b) $6\pi i$ ☐

(c) $12\pi i$ ☐ (d) none of these. ☐

90. Out of all the 2-digit integers between 1 and 100, a 2-digit number has to be selected at random. What is the probability that the selected number is not divisible by 7?

(a) 13/90 ☐ (b) 12/90 ☐

(c) 78/90 ☐ (d) 77/90 ☐

(GATE-ME/CSE/PI-2013)

91. If C be the circle $| z | = 3$, then $\int_C \frac{\cos \pi z}{z-1}\, dz$ is

(a) zero ☐ (b) $-\pi i$ ☐

(c) $-2\pi i$ ☐ (d) none of these. ☐

92. If $\int_C \frac{\cos \pi z^2}{(z-1)(z-2)}\, dz$, where C is the circle $| z | = 13$, is equal to

(a) πi ☐ (b) $2\pi i$ ☐

(c) $3\pi i$ ☐ (d) $4\pi i$. ☐

93. Let C be the circle $| z | = 2.2$, then the complex integral $\int \frac{e^{2z}}{(z+1)^4}\, dz$ is equal to

(a) $\frac{4\pi i}{3e}$ ☐ (b) $\frac{8\pi i}{3e^2}$ ☐

(c) $\frac{4\pi i}{3e^2}$ ☐ (d) none of these. ☐

94. Let $f(z)$ be analytic inside a circle C with centre at a, then for z inside C, $f(z)$ can be expanded as a

(a) Fourier series ☐ (b) Laurent's series ☐
(c) Taylor's series ☐ (d) none of these. ☐

95. What will be the maximum sum of 44, 42, 40, ...?

(a) 502 ☐ (b) 504 ☐
(c) 506 ☐ (d) 500 ☐

(GATE-ME/CSE/PI-2013)

CHEMISTRY

96. de-Broglie equation is

(a) $h/mv = \lambda$ ☐ (b) $hv = E_1 - E_2$ ☐
(c) $n\lambda = 2d \sin\theta$ ☐ (d) $c - \lambda v$ ☐
(e) $E = mc^2$. ☐

97. Two isobars will have in their nuclei the same number of

(a) p ☐ (b) n ☐
(c) e^{-1} ☐ (d) $(p + n)$. ☐

98. Which one of the following isotopes would be most likely to undergo fission?

(a) ${}_6C^{14}$ ☐ (b) ${}_{27}Co^{59}$ ☐
(c) ${}_{94}Pu^{249}$ ☐ (d) ${}_1H^2$. ☐

99. Magnetic quantum number can take values

(a) $-n \ldots 0 \ldots +n$ ☐ (b) $-m \ldots 0 \ldots +m$ ☐
(c) $-l \ldots -1 \ldots 0$ ☐ (d) $0, 1, 2, 3 \ldots (l-1)$ ☐
(e) $-l \ldots 0 \ldots +l$. ☐

100. The electronic configuration of Cr is $3d^5\,4s^1$. How many unpaired electrons would there be Cr^{3+} ion?

(a) 1 ☐ (b) 2 ☐
(c) 3 ☐ (d) 4 ☐
(e) 5 ☐

101. The crystal co-ordination number in a body centred cubic type of packing is

(a) 6 ☐ (b) 12 ☐
(c) 10 ☐ (d) 8. ☐

102. Cations and anions are held together in a crystal by

(a) electrostatic force ☐ (b) nuclear forces ☐
(c) covalent bonds ☐ (d) electrons. ☐

103. Which of the following has a tetrahedral structure?

(a) NH_3 ☐ (b) NH_4^+ ☐
(c) $K_4[Fe(CN)_6]$ ☐ (d) $[Ni(CN)_4]^{2-}$. ☐

104. The shape of a *dsp*^{2} orbital is

(*a*) trigonal ☐ (*b*) tetrahedral ☐
(*c*) octahedral ☐ (*d*) square planar. ☐

105. Radioactivity was accidently discovered by

(*a*) M. Curie ☐ (*b*) H. Becquerel ☐
(*c*) J.J. Thomson ☐ (*d*) E. Rutherford ☐
(*e*) J.W. Aston. ☐

106. Radiation which has the least energy

(*a*) α-rays ☐ (*b*) β-rays ☐
(*c*) γ-rays ☐ (*d*) X-rays ☐
(*e*) electrical waves. ☐

107. The emission of an alpha particle from Ra^{226} will produce

(*a*) $_{88}Ra^{222}$ ☐ (*b*) $_{87}Fr^{222}$ ☐
(*c*) $_{87}Fr^{223}$ ☐ (*d*) $_{86}Rn^{222}$ ☐
(*e*) $_{86}Rn^{223}$. ☐

108. The radioactive isotope of hydrogen is called

(*a*) Radium ☐ (*b*) Protium ☐
(*c*) Deuterium ☐ (*d*) Tritium ☐
(*e*) Ortho-hydrogen. ☐

109. C^{14} has a half life of 5760 years. 100 milligrams of a sample of C^{14} will completely disintegrate in

(*a*) 23,040 yrs ☐ (*b*) 1440 yrs ☐
(*c*) 11,520 yrs ☐ (*d*) 2,880 yrs ☐
(*e*) infinite time. ☐

110. $_6C^{11}$ and $_5B^{11}$ are referred to as

(*a*) Nuclear isomers ☐ (*b*) Isotopes ☐
(*c*) Fission products ☐ (*d*) Isobars ☐
(*e*) None of these. ☐

111. Choose the one which is not used as a moderator of neutrons in controlled chain reactions

(*a*) water ☐ (*b*) heavy water ☐
(*c*) oxygen ☐ (*d*) graphite ☐
(*e*) paraffin. ☐

112. One electron volt is equal to

(*a*) 4.18 kcal/mole ☐ (*b*) 6.625 kcal/mole ☐
(*c*) 23.06 kcal/mole ☐ (*d*) 23.06 cal ☐
(*e*) 16.5 cal/mole. ☐

113. Isotopes have similar chemical properties because they have

(*a*) the same atomic number ☐ (*b*) different atomic weights ☐
(*c*) different number of neutrons ☐ (*d*) the same mass number. ☐

114. Isobars are produced as the result of the emission of

(a) α-particles ☐ (b) γ-rays ☐
(c) X-rays ☐ (d) β-particles. ☐

115. The rate of radioactive disintegration with time

(a) increases ☐ (b) decreases ☐
(c) is constant ☐ (d) fluctuates ☐
(e) decrease first then increases. ☐

116. $_6C^{14}$ in the upper atmosphere is produced by the action of neutron on

(a) $_8O^{16}$ ☐ (b) $_7N^{14}$ ☐
(c) $_6C^{12}$ ☐ (d) $_1H^1$ ☐
(e) CO_2. ☐

117. In a nuclear reactor thermal neutrons are produced by the use of

(a) cyclotron ☐ (b) U^{238} ☐
(c) boron ☐ (d) graphite ☐
(e) cadmium. ☐

118. The material used to control neutron flux in an atomic reactor is

(a) heavy water ☐ (b) paraffin ☐
(c) cadmium ☐ (d) graphite ☐
(e) thorium. ☐

119. Which among the following will be paramagnetic?

(a) $Ni(CO)_4$ ☐ (b) $Cr(CO)_6$ ☐
(c) $Fe(CO)_5$ ☐ (d) $Fe_2(CO)_9$ ☐
(e) $V(CO)_6$. ☐

120. Phosphorus has the oxidation state of + 3 in

(a) phosphorus acid ☐ (b) hypophosphorus acid ☐
(c) ortho-phosphoric acid ☐ (d) metaphosphoric acid ☐
(e) pyrophosphoric acid. ☐

121. The valency of Cr in the complex $[Cr(H_2O)_4Cl_2]^+$ is

(a) 3 ☐ (b) 1 ☐
(c) 6 ☐ (d) 5 ☐
(e) 2. ☐

122. AgCl is soluble in ammonia due to the formation of

(a) $(Ag(NH_3)_2]^{2+}$ ☐ (b) $[Ag(NH_3)_2]^+$ ☐
(c) $[Ag(NH_3)_4]^+$ ☐ (d) $AgNH_2$ ☐
(e) $[Ag(NH_4)_2]^+$. ☐

123. In K_2MnO_4 the oxidation state of Mn is

(a) + 3 ☐ (b) + 7 ☐
(c) + 1 ☐ (d) + 6 ☐
(e) + 4. ☐

124. C(graphite) + O_2 (g) ⟶ CO_2(g) ΔH°_{298} = – 94.05 kcal mol^{-1}; C(diamond) + O_2(g) ⟶ CO_2(g) ΔH°_{298} = – 94.50 kcal mol^{-1}. Therefore

(*a*) C(diamond) ⟶ C(graphite) ΔH°_{298} = 450 cal mol^{-1} □

(*b*) C(graphite) ⟶ C(diamond) ΔH°_{298} = – 450 cal mol^{-1} □

(*c*) diamond is harder than graphite □

(*d*) graphite is the stable allotrope. □

125. The standard heat of formation of 1 mole of graphite is

(*a*) 1 cal $mole^{-1}$ □ (*b*) the same as that for charcoal □

(*c*) more than that for diamond □ (*d*) assumed to be zero. □

126. For a particular reaction ΔC_p is found to be positive. It means that

(*a*) ΔH increases with increase in temperature □

(*b*) ΔH decreases with increase in temperature □

(*c*) ΔH is independent of temperature □

(*d*) ΔH and ΔE are identical □

(*e*) the reaction is slow. □

127. If the cell reaction is spontaneous

(*a*) E° is –ve □ (*b*) ΔG is +ve □

(*c*) E° is +ve □ (*d*) (ΔG + E°) is +ve □

(*e*) ΔG/E° is +ve. □

128. The voltage of a cell whose half cells are given below is

$$Mg^{2+} + 2e^- \longrightarrow Mg(s) \qquad E^\circ = -2.37\ v$$

$$Cu^{2+} + 2e^- \longrightarrow Cu(s) \qquad E^\circ = +0.34\ v$$

(*a*) – 2.03 *v* □ (*b*) 1.36 *v* □

(*c*) 2.71 *v* □ (*d*) 2.03 *v* □

(*e*) – 1.02. □

129. The value of E for the half cell in which the following reaction takes place is

$$Cu^{2+}\ (0.1a) + 2e^- \rightleftharpoons Cu(s)\ E^\circ = 0.34\ v$$

(*a*) 0.31 *v* □ (*b*) 0.37 *v* □

(*c*) 0.34 *v* □ (*d*) 0.27 *v* □

(*e*) 0.49 *v*. □

130. The half time of a reaction following zero order kinetics is ...... initial concentration

(*a*) independent □ (*b*) proportional to □

(*c*) inversely proportional to. □

131. For a second order reaction governed by the equation $\frac{dx}{dt} = k(a-x)^2$ the half life is given by

(*a*) *k* □ (*b*) $\frac{\ln}{2k}$ □

(*c*) $\frac{l}{ka}$ □ (*d*) *ka*. □

132. Arhenius equation is written as

(a) $\frac{dl\,nk}{dT} = -\frac{\Delta E^*}{RT}$ ☐ (b) $\frac{dl\,nk}{dT} = -\frac{\Delta E^*}{RT^2}$ ☐

(c) $\frac{dl\,nk}{dT} = -\Delta E^* . (RT)^2$. ☐

133. In the reaction $A_2(g) + 2B_2(g) \rightleftharpoons 2AB_2(g)$ + heat, the equilibrium is shifted to the left on

(a) decreasing the pressure and increasing the temperature ☐
(b) decreasing the pressure and decreasing the temperature ☐
(c) increasing the pressure and increasing the temperature ☐
(d) increasing the pressure and decreasing the temperature ☐
(e) adding more $A_2(g)$. ☐

134. Close packing is maximum in the lattice type

(a) body centred cubic ☐ (b) face centred cubic ☐
(c) simple cubic. ☐

135. A certain substance decomposes according to zero order kinetics. If the rate constant k and initial concentration a the half life is

(a) $\frac{1}{ka}$ ☐ (b) $\frac{k}{2}$ ☐

(c) $\frac{ka}{2}$ ☐ (d) $\frac{a}{2k}$ ☐

(e) $\frac{2.303 \log a}{k}$. ☐

136. An atom of mass number 23 is likely to have the following number of electrons (e), protons (p) and neutrons (n)

(a) 10e, 11p, 12n ☐ (b) 11e, 12p, 12n ☐
(c) 11e, 11p, 12n ☐ (d) 11e, 11p, 23n ☐
(e) 12e, 11p, 12n. ☐

137. The ratio of the radii of an atom and its nucleus is nearly

(a) 10^{-5} ☐ (b) 10^{5} ☐
(c) 10^{-8} ☐ (d) 10^{12} ☐
(e) 10^{8}. ☐

138. For a dumb bell shaped orbital the l value is

(a) 3 ☐ (b) 0 ☐
(c) 1 ☐ (d) + 1 or – 1 ☐
(e) 2. ☐

139. Which of the following orbital will have the highest energy in a hydrogen form?

(a) 4f ☐ (b) 5s ☐
(c) 5p ☐ (d) 6s ☐
(e) 6p. ☐

140. The number of possible values for the magnetic quantum number (m) for a d electron is

(*a*) 1 ☐ (*b*) 5 ☐
(*c*) 2 ☐ (*d*) 4 ☐
(*e*) 3. ☐

141. In sodium chloride crystal each chloride ion is surrounded by

(*a*) 8 Na^+ ☐ (*b*) 6 Na^+ ☐
(*c*) 4 Na^+ ☐ (*d*) 6 Cl^- ☐
(*e*) none of the above. ☐

142. The mass of a unit cell of Cs Cl corresponds to

(*a*) one Cs^+ and eight Cl^- ☐ (*b*) one Cl^- and eight Cs^+ ☐
(*c*) one Cs^+ and one Cl^- ☐ (*d*) four Cs^+ and four Cl^- ☐
(*e*) two Cs^+ and two Cl^-. ☐

143. Which of the following is correct?

(*a*) At absolute zero all lattice vibrations in a metal cease ☐
(*b*) At absolute zero all electron mobility is reduced to zero ☐
(*c*) At absolute zero electron remain mobile while lattice vibrations cease ☐
(*d*) At absolute zero electron mobility reduced to zero while lattice vibration are not reduced to zero ☐
(*e*) none of these. ☐

144. Properties which depend on the number rather than the nature of the dissolved particles in a solution are called

(*a*) general ☐ (*b*) colligative ☐
(*c*) isotropic ☐ (*d*) isotonic ☐
(*e*) isoelectronic. ☐

145. For the system $C(s) + O_2(g) = CO_2(g)$

(*a*) $\Delta H = \Delta E$ ☐ (*b*) $\Delta H < \Delta E$ ☐
(*c*) $\Delta H > \Delta E$ ☐ (*d*) $\Delta H = 0$ ☐
(*e*) $\Delta E = 0$. ☐

146. The condition for a gas to behave ideally is

(*a*) $\left(\frac{dV}{dT}\right)_T = 0$ ☐ (*b*) $\left(\frac{dP}{dV}\right)_T = 0$ ☐
(*c*) $\left(\frac{dE}{dV}\right)_T = 0$ ☐ (*d*) $\left(\frac{dP}{dn}\right)_T = 0$ ☐
(*e*) $dE = 0$. ☐

147. A compound to which no other element or group of elements can be directly added is said to be

(*a*) homologous ☐ (*b*) isomeric ☐
(*c*) olefinic ☐ (*d*) unsaturated ☐
(*e*) saturated. ☐

148. The number of isomers possible for C_5H_{12} is

(a) three ☐ (b) eight ☐
(c) two ☐ (d) five ☐
(e) four. ☐

149. Methane has

(a) sp hybridised carbon ☐ (b) sp^3 hybridised carbon ☐
(c) sp^2 hybridised carbon ☐ (d) linear structure ☐
(e) square planar structure. ☐

150. The formula of 3-methyl pent-2-ene is

(a) $CH_2 = CHCH(CH_3)CH_2CH_3$ ☐ (b) $CH_3CH_2CH(CH_3)CH_2CH_3$ ☐
(c) $CH_3CH = C(CH_3)CH_2CH_3$ ☐ (d) $CH_3CH = CHCH(CH_3)_2$ ☐
(e) $CH_3CH = C(CH_3)CH = CH_2$. ☐

151. Which is not a nucleophile?

(a) H_2 ☐ (b) H_2O ☐
(c) CH_3OH ☐ (d) NH_3 ☐
(e) CN^- ☐

152. Ca^{2+} is isoelectronic with

(a) Sr^{2+} ☐ (b) Mg^{2+} ☐
(c) Kr ☐ (d) Na^+ ☐
(e) Ar. ☐

153. The shape of s orbital is

(a) elliptical ☐ (b) pyramidal ☐
(c) dumb-bell shaped ☐ (d) tetrahedral ☐
(e) spherical. ☐

154. The olefinic bond is easily reduced by

(a) hydrogen gas ☐ (b) metal and acid ☐
(c) H_2/Ni ☐ (d) sodium and ethanol ☐
(e) none of the method listed. ☐

155. When ethylene is passed into cold dilute alkaline permagnate solution the product is

(a) glycol ☐ (b) methyl alcohol ☐
(c) acetylene ☐ (d) 2 molecules of CO_2 ☐
(e) epoxide. ☐

156. Polythene is a polymer of

(a) isobutane ☐ (b) ethylene ☐
(c) vinyl chloride ☐ (d) propylene ☐
(e) chloroprene. ☐

157. Olefins are converted to paraffins by

(a) halogenation ☐ (b) hydrolysis ☐
(c) dehydration ☐ (d) substitution ☐
(e) hydrogenation. ☐

158. Select the compound which has acidic hydrogen

(*a*) acetylene ☐ (*b*) methane ☐
(*c*) ethylene ☐ (*d*) butadiene ☐
(*e*) benzene. ☐

159. Select the compound which forms tetrabromide with excess liquid bromine at room temperature

(*a*) $CH_3CH_2CH_2OH$ ☐ (*b*) $CH_3C{\equiv}CH$ ☐
(*c*) CH_3OCH_3 ☐ (*d*) $CH_3CH = CHC{\equiv}C\ CH_3$ ☐
(*e*) $CH_3CH_2OCH_2CH_3$. ☐

160. An alkyl halide may be converted into alcohol by

(*a*) addition ☐ (*b*) elimination ☐
(*c*) hydrogenation ☐ (*d*) dehydration ☐
(*e*) substitution. ☐

161. The fundamental basis of the present day periodic table is that the elements are

(*a*) arranged in the order of increasing atomic weight ☐
(*b*) taken in groups of eight ☐
(*c*) grouped according to chemical properties ☐
(*d*) arranged in the order of increasing number of neutrons in the atomic nucleus ☐
(*e*) arranged in the order of increasing number of protons in the nucleus. ☐

162. The principal quantum number (n) can take values from

(*a*) 0 to 7 ☐ (*b*) – 1 to 7 ☐
(*c*) 1 to 7 ☐ (*d*) 0 to (n – 1) ☐
(*e*) – n to + n. ☐

163. The energy of an electron in its orbit is

(*a*) its kinetic energy ☐ (*b*) sum of potential and kinetic energy ☐
(*c*) potential energy only ☐ (*d*) due to rotation ☐
(*e*) determined from the nucleus. ☐

164. The average kinetic energy of molecule is

(*a*) directly proportional to its velocity ☐
(*b*) indirectly proportional to its velocity ☐
(*c*) directly proportional to its temperature ☐
(*d*) inversely proportional to its absolute temperature. ☐

165. In the ideal gas $PV = nRT$, the gas constant R has the dimension of

(*a*) ergs ☐ (*b*) litre mole^{-1} ☐
(*c*) ml atm deg^{-1} ☐ (*d*) cal mole^{-1} ☐
(*e*) litre atm deg^{-1} mole^{-1}. ☐

166. If each gas is at S.T.P. 2.016 gm of hydrogen gas contains the same number of molecules as

(*a*) 16 g of oxygen ☐ (*b*) 14 g of nitrogen ☐
(*c*) 2.016 g mole of helium ☐ (*d*) 22.4 litre of fluorine. ☐

167. A gas behaves most nearly like an ideal gas when under conditions of

(*a*) high pressure and low temperature ☐ (*b*) low pressure and high temperature ☐
(*c*) low pressure and low temperature ☐ (*d*) high pressure and high temperature. ☐

168. Gases deviate from ideal gas behaviour because their molecules

(*a*) are colourless ☐ (*b*) attract each other ☐
(*c*) contain covalent bonds ☐ (*d*) show Brownian motion. ☐

169. A weather balloon contains 12 litre of hydrogen at 740 mm of Hg pressure. At what pressure in mm of Hg will the volume become 24 litre (tem. const.)

(*a*) 370 ☐ (*b*) 444 ☐
(*c*) 760 ☐ (*d*) 1230. ☐

170. Dalton's law of partial pressure is applicable to mixture of

(*a*) HCl and NH_3 ☐ (*b*) H_2 and Cl_2 ☐
(*c*) SO_2 and O_2 ☐ (*d*) CO and CO_2. ☐

171. Kinetic expression for the pressure exerted by *n* moles of a gas is

(*a*) $PV = \frac{3}{4} nc^2$ ☐ (*b*) $PV = \frac{1}{3} mc^2$ ☐
(*c*) $PV = \frac{1}{3} n^2c^2$ ☐ (*d*) $PV = \frac{1}{3} mnc^2$. ☐

172. The general form of van der Waal's equation for one mole of a gas is as

(*a*) $\left(P - \frac{a^a}{V}\right)(V - b) = RT^2$ ☐ (*b*) $\left(P + \frac{a^2}{V}\right)(V + b^2) = RT$ ☐
(*c*) $\left(P + \frac{a}{V^2}\right)(V - b) = RT$ ☐ (*d*) $\left(P - \frac{a}{V^2}\right)(V - b) = RT$. ☐

173. For a van der Waal's fluid, *a* has the dimensions of

(*a*) atm/litre ☐ (*b*) atm/litre^2 ☐
(*c*) energy/mole ☐ (*d*) atm litre^2 mol^{-1}. ☐

174. For a van der Waal's fluid, *b* has the dimensions of

(*a*) litres ☐ (*b*) atmospheres ☐
(*c*) litre mole^{-1} ☐ (*d*) atmosphere mole^{-1}. ☐

175. In case of monoatomic gas the ratio of two specific heat is equal to

(*a*) 1.33 ☐ (*b*) 1.40 ☐
(*c*) 1.667 ☐ (*d*) 1.0. ☐

176. A gas has 1.4 as the C_p/C_v. Its atomic mass is *X* its molecules mass will be

(*a*) X ☐ (*b*) 2X ☐
(*c*) X/2 ☐ (*d*) 3X. ☐

177. At constant volume the pressure of a gas increases with increasing temperature because as the temperature increases

(*a*) molecules move faster ☐ (*b*) molecular volume increases ☐
(*c*) the mass of the molecules increases ☐ (*d*) collision number increases ☐
(*e*) molecules are closer together. ☐

178. If the pressure on a gas is trebled at const. temp. its volume is reduced to

(*a*) 50% ☐ (*b*) 25% ☐
(*c*) 33.3% ☐ (*d*) 10%. ☐

179. The molecules present in 0.1 mole of any gas are

(*a*) 6.023×10^{23} ☐ (*b*) 6.023×10^{24} ☐
(*c*) 6.023×10^{22} ☐ (*d*) 6.023×10^{13}. ☐

180. Kinetic energy per mole of a gas is equal to

(*a*) RT ☐ (*b*) $\frac{2}{3} RT$ ☐
(*c*) $\frac{3}{2} RT$ ☐ (*d*) $3\ RT$. ☐

181. The translational kinetic energy of an ideal gas depends upon

(*a*) the nature of the gas ☐ (*b*) temperature ☐
(*c*) the pressure ☐ (*d*) the pressure and the nature of the gas. ☐

182. In the gas equation $PV = nRT$. n represents

(*a*) Avogadro's number ☐ (*b*) the number of gms ☐
(*c*) the number of moles ☐ (*d*) the number of molecules. ☐

183. According to first law of thermodynamics

(*a*) $dE = dq - dw$ ☐ (*b*) $dq = dE - dw$ ☐
(*c*) $dw = dE - dq$ ☐ (*d*) $dw = dq - dE$. ☐

184. Heat change in a reaction at constant pressure is called

(*a*) internal energy change ☐ (*b*) enthalpy change ☐
(*c*) entropy change ☐ (*d*) none of these. ☐

185. The efficiency of a reversible engine working between the heat source and a sink temperatures of 1000 K and 300 K respectively is

(*a*) 1 ☐ (*b*) $\frac{1}{2}$ ☐
(*c*) 0.7 ☐ (*d*) 0.3. ☐

186. The entropy of the universe

(*a*) tends towards maximum ☐ (*b*) tends towards minimum ☐
(*c*) remains constant ☐ (*d*) tends to be zero. ☐

187. The value of Joule-Thomson coefficient is

(*a*) positive at the inversion temperature ☐ (*b*) zero above the inversion temperature ☐
(*c*) negative at the inversion temperature ☐ (*d*) positive at the inversion temperature. ☐

ELECTRICAL SCIENCE

188. AM transmitter radiates 50 kW of carrier power. The radiated power at 85% modulation is

(*a*) 60.1 kW ☐ (*b*) 80.1 kW ☐
(*c*) 68.1 kW ☐ (*d*) 70.1 kW. ☐

189. When a broadcast AM transmitter is 50% modulated, its antenna current is 12 A. What will be the current when modulation depth is increased to 0.9?

(*a*) 12.4 A ☐ (*b*) 13.4 A ☐

(*c*) 14.4 A ☐ (*d*) 10 A. ☐

190. *P*-side of a semi-conductor diode is applied a potential of 2.5 V whereas *N*-side is applied a potential of – 0.5 V, the diode will

(*a*) not conduct ☐ (*b*) conduct ☐

(*c*) breakdown ☐ (*d*) none of these. ☐

191. The capacitance appearing across a reverse biased semi-conductor junction

(*a*) increases with increase in the bias voltage ☐

(*b*) decreases with increase in the bias voltage ☐

(*c*) is independent of bias voltage ☐

(*d*) none of these. ☐

192. The input impedance of a common base transistor stage may be

(*a*) as small as 100 Ω ☐ (*b*) as large as several mega ohms ☐

(*c*) about 50 k Ω ☐ (*d*) none of these. ☐

193. A transistor amplifier consists of four cascaded stages common base configuration with 50 as the voltage gain per stage. The overall gain is

(*a*) $(50)^4$ ☐ (*b*) 200 ☐

(*c*) 50 ☐ (*d*) none of these. ☐

194. When the current through a Zener diode increases by a factor of 2, the voltage across its terminals

(*a*) gets doubled too ☐ (*b*) remains unaltered ☐

(*c*) is halved ☐ (*d*) none of these. ☐

195. There identical Zener diodes when connected in series were found to exhibit a breakdown voltage of 15 V. Each Zener diode has a breakdown potential of

(*a*) 5 V ☐ (*b*) 15 V ☐

(*c*) 45 V ☐ (*d*) none of these. ☐

196. Tunnel diode characteristics differ from ordinary junction diode characteristics in the respect that former's

(*a*) are linear ☐ (*b*) are non-linear ☐

(*c*) exhibit a negative resistance region ☐ (*d*) none of these. ☐

197. As regards a UJT, the peak point is the point

(*a*) where there is a transition from (+ve) resistance to negative resistance ☐

(*b*) where the current is maximum ☐

(*c*) where there is a transition from (–ve) resistance to (+ve) resistance ☐

(*d*) where voltage is minimum. ☐

198. Holding current, in a *p-n-p-n* diode, is the

(*a*) normal operating current ☐

(*b*) current corresponding to breakover voltage ☐

(c) minimum current to keep the device on ☐
(d) none of these. ☐

199. A silicon controlled switch with a breakover voltage of 12 V suffers from rate effect. When used in a circuit, the device may fire at a voltage

(a) slightly greater than 12 V ☐ (b) slightly less than 12 V ☐
(c) very much less than the firing voltage ☐ (d) none of the above. ☐

200. Pinch off voltage in a field effect transistor is

(a) the gate to source voltage that gives zero drain current ☐
(b) the drain voltage that gives zero drain current ☐
(c) the gate to source voltage for which there is no source current ☐
(d) none of these. ☐

201. For operating a *p*-channel MOSFET, the gate is applied a

(a) (+ve) potential (for enhancement mode) ☐ (b) (–ve) potential (for depletion mode) ☐
(c) (–ve) potential (for enhancement mode) ☐ (d) none of the above. ☐

202. An operational-amplifier can be used

(a) with D.C. signals only ☐ (b) with A.C. signals only ☐
(c) with A.C. as well as D.C. signals ☐ (d) none of these. ☐

203. Common Model Rejection Ratio (CMRR) in an op. amp. should be

(a) as large as possible ☐ (b) as small as possible ☐
(c) as close to unity as possible ☐ (d) none of the above. ☐

204. An operational amplifier has an open loop gain of 100 dB. Its gain starts falling at a frequency of 900 kHz with a rate of 6 dB per octave. The gain at a frequency of 3.6 MHz is

(a) 25 dB ☐ (b) 94 dB ☐
(c) 88 dB ☐ (d) none of the above. ☐

205. PSRR (Power Supply Rejection Ratio) should be

(a) as large as possible ☐ (b) as small as possible ☐
(c) unity ☐ (d) none of the above. ☐

206. A stationary pattern is obtained on C.R.O. only when

(a) the time period of the time-base signal is equal to the time period of the applied signal ☐
(b) time period of the time base signal is equal to the time occupied by whose number of cycles of applied signal ☐
(c) there is no proper synchronisation ☐
(d) none of these. ☐

207. While measuring a voltage of approximately 0.05 mV, we will prefer

(a) an amplifier-rectifier VTVM ☐ (b) a rectifier-amplifier VTVM ☐
(c) either of two ☐ (d) none of these. ☐

208. Quality factor (Q-factor), in general, is defined as

(a) $\frac{wL}{R}$ ☐ (b) $\frac{1}{wRC}$ ☐

(c) $\dfrac{\text{Powered stored}}{\text{Power lost}}$ ☐ (d) $\dfrac{\text{Power dissipated}}{\text{Power stored}}$. ☐

209. The conduction band

(a) is always located at the top of the crystal ☐

(b) is a range of energies corresponding to the energies of free electrons ☐

(c) is also called the forbidden energy gap ☐

(d) none of the above. ☐

210. In a semi-conductor diode, the barrier potential offers oppositions to only

(a) the majority carriers in both regions ☐ (b) the minority carriers in both regions ☐

(c) the free electrons in the *N*-region ☐ (d) the holes in the *P*-region. ☐

211. The heater filament of a vacuum tube is generally supplied with A.C. voltage (and not D.C. voltage) for heating because

(a) it results in a uniform heating of filament so that the electron emission is also uniform ☐

(b) it is very easy to obtain this voltage from the A.C. power mains ☐

(c) the D.C. voltage that would be required for heating has much greater magnitude than the A.C. voltage ☐

(d) when D.C. is used for heating, a different type of filament is required which is very expensive. ☐

212. The amplitude distortion is an amplifier mainly depends upon

(a) the active device (tube or transistor) itself ☐

(b) the associated circuit-elements ☐

(c) both the active device and the associated circuit-elements ☐

(d) none of the above. ☐

213. In an *RC* coupled amplifier, the rate change of voltage gain in the low and high frequency ranges is

(a) 3 dB/octave ☐ (b) 6 dB/decade ☐

(c) 10 dB/decade ☐ (d) none of the above. ☐

214. A transistor connected in common-base configuration has

(a) a low input resistance and a high output resistance ☐

(b) a high input resistance and a low output resistance ☐

(c) a low input resistance and a low output resistance ☐

(d) a high input resistance and a high output resistance. ☐

215. Compared to a CB amplifier, the CE amplifier has

(a) lower input resistance ☐ (b) higher output resistance ☐

(c) lower current amplification ☐ (d) higher current amplification. ☐

216. When a positive-voltage signal is applied to the base of a normally biased *n-p-n* common emitter transistor amplifier

(a) the emitter current decreases ☐ (b) the collector voltage goes less positive ☐

(c) the base current decreases ☐ (d) the collector current decreases. ☐

217. The output power of a power amplifier is several times its input power. It is possible because

(*a*) the power amplifier introduces negative resistance ☐

(*b*) the power amplifier converts a part of the input D.C. power into A.C. output power ☐

(*c*) positive feedback exists in the circuit ☐

(*d*) step-up transformer is used in the circuit. ☐

218. Heat sinks are used in the transistor working as power amplifiers as to

(*a*) increase the output power ☐

(*b*) to reduce the heat losses in the transistors ☐

(*c*) to increase the voltage gain of the amplifiers ☐

(*d*) to increase the collector dissipation rating of the transistors. ☐

219. As the temperature of a transistor increases, the base spreading resistance $r_{bb'}$, in the hybrid-pi model at high frequencies

(*a*) increases, because the concentration of the current-carriers in the base region increases ☐

(*b*) increases, because the mobility of the carriers reduces ☐

(*c*) decreases, because the base-width reduces ☐

(*d*) decreases, because the base-width increases. ☐

220. The capacitance between the gate and drain, $C_{gd.'}$ in the equivalent circuit of a JFET at high frequencies is of the order of

(*a*) a few microfarads ☐ (*b*) a few tens of picofarads ☐

(*c*) a few picofarads ☐ (*d*) a small fraction of a picofarad. ☐

221. The control grid in a triode is usually given a negative D.C. potential with respect to the cathode so as to

(*a*) reduce the space-charge ☐ (*b*) increase the space-charge ☐

(*c*) limit the plate current to a safe value ☐ (*d*) make the grid current zero. ☐

222. In a half-wave rectifier, the peak value of the A.C. voltage across the secondary of the transformer 20/2 V. If no filter circuit is used, the maximum D.C. voltage across the load will be

(*a*) 28.28 V ☐ (*b*) 14.14 V ☐

(*c*) 20 V ☐ (*d*) 9 V. ☐

223. In an unbiased *p-n* junction, the junction current at equilibrium is

(*a*) due to diffusion of minority carriers only ☐

(*b*) due to diffusion of majority carriers only ☐

(*c*) zero, because equal but opposite carriers are crossing the junction ☐

(*d*) zero, because no charge carriers are crossing the junction. ☐

224. When the temperature of an intrinsic semi-conductor increased

(*a*) the resistance of the semi-conductor increases ☐

(*b*) the heat energy decreases the atomic radius ☐

(*c*) the holes are created in the conduction band ☐

(*d*) the energy of the atoms is increased. ☐

225. At half power frequencies, the reduction in voltage gain of an amplifier equals

(*a*) 6 dB ☐ (*b*) 20 dB ☐
(*c*) 3 dB ☐ (*d*) 12 dB. ☐

226. The frequency of the ripple voltage at the output of a bridge rectifier operating from a 50 Hz supply is

(*a*) 25 Hz ☐ (*b*) 50 Hz ☐
(*c*) 100 Hz ☐ (*d*) 200 Hz. ☐

227. In an amplifier using a BJT, the emitter resistor R_E bypassed by a capacitor C_E

(*a*) reduces the voltage gain ☐ (*b*) causes thermal run away ☐
(*c*) increases the voltage gain ☐ (*d*) stabilizes the *Q*-point. ☐

228. The signal fed at the input of an idea push-pull amplifier has frequency components – 150 Hz, 300 Hz, 450 Hz and 600 Hz. The output signal will contain

(*a*) only 150 Hz frequency components ☐
(*b*) only 150 Hz and 450 Hz frequency components ☐
(*c*) only 320 Hz and 600 Hz frequency components ☐
(*d*) all the four frequency components. ☐

229. An important advantage of the RC coupling scheme over other schemes is

(*a*) economy ☐ (*b*) excellent frequency response ☐
(*c*) high efficiency ☐ (*d*) good impedance matching. ☐

230. The A.C. input to an oscillator is obtained from

(*a*) the previous stage ☐ (*b*) a signal generator ☐
(*c*) a battery ☐ (*d*) its own internal circuit. ☐

231. In an unsymmetrical *p-n* junction, *p*-region is more heavily dopped than the *n*-region. Then, the depletion layer will

(*a*) penetrate more into the *p*-region ☐ (*b*) penetrate more into the *n*-region ☐
(*c*) penetrate equally into both the regions ☐ (*d*) not be formed at all. ☐

232. In a VTVM, the a.c. voltage measurement corresponds to

(*a*) sine wave inputs ☐ (*b*) square wave inputs ☐
(*c*) saw-tooth wave inputs ☐ (*d*) any way form at the input. ☐

233. If the retrace is visible on the CRT display, then the trouble may be that

(*a*) the intensity is to high ☐
(*b*) the blanking control is not set properly ☐
(*c*) there is loss of SYNC signal ☐
(*d*) the flyback time of the time-base saw-tooth wave is not zero. ☐

234. In a Doppler Radar, the transmitted signal frequency was 100 MHz whereas that of the received echo was 110 MHz. This indicates

(*a*) the target is moving towards the radar receive ☐
(*b*) the target is moving away from the radar receiver ☐
(*c*) target is at an angle with the horizontal determined by change in frequency ☐
(*d*) none of the above. ☐

235. The aspect-ratio in a television system is defined as

(*a*) raster height/raster width ☐ (*b*) raster width/raster height ☐
(*c*) diagonal length of the raster ☐ (*d*) none of the above. ☐

236. In a radar system, the peak power transmitted is increased by a factor 18, the maximum range will be increased by a factor of

(*a*) 81 ☐ (*b*) 9 ☐
(*c*) 3 ☐ (*d*) none of the above. ☐

237. One of the following statements is false. What is it?

(*a*) A high PRR makes the target tracking easier. ☐
(*b*) A high PRR increases the maximum range. ☐
(*c*) Change in PRR has no effect. ☐
(*d*) A high PRR makes the returned echos easier to distinguish from noise. ☐

238. An antenna should have

(*a*) a very high SWR ☐ (*b*) very small SWR ☐
(*c*) moderate value of SWR ☐ (*d*) none of the above. ☐

239. The most commonly used filters in SSB generation are

(*a*) mechanical ☐ (*b*) RC ☐
(*c*) LC ☐ (*d*) low-pass. ☐

240. Indicate in which one of the following only one side band is transmitted

(*a*) A 3H ☐ (*b*) A 3 ☐
(*c*) A 3B ☐ (*d*) A 5C. ☐

241. Indicate which of the following circuits could not demodulate SSB

(*a*) balanced modulator ☐ (*b*) product detector ☐
(*c*) BFO ☐ (*d*) phase discriminator. ☐

242. For a transmission line load matching over a range of frequencies, it is best to use

(*a*) balun ☐ (*b*) broadband directional coupler ☐
(*c*) double stub ☐ (*d*) single stub of adjustable position. ☐

243. The main disadvantage of two directional couple is

(*a*) low directional coupling ☐ (*b*) poor directivity ☐
(*c*) high SWR ☐ (*d*) narrow bandwidth. ☐

244. To couple a coaxial line to a parallel-wire line, it is best to use a

(*a*) slotted line ☐ (*b*) balun ☐
(*c*) directional coupler ☐ (*d*) quarter-wave transformer. ☐

245. A carrier is simultaneously modulated by two sine waves with modulation indices of 0.3 and 0.4, the total modulation index is

(*a*) 1 ☐ (*b*) 0.7 ☐
(*c*) 0.5 ☐ (*d*) none of the above. ☐

246. An FM signal with a modulation index m_f is passed through a frequency tripler. The wave in the output of the tripler will have a modulation index of

(*a*) $m_f/3$ ☐ (*b*) m_f ☐
(*c*) 3 m_f ☐ (*d*) $9m_f$. ☐

247. When the modulating frequency is doubled, the modulation index is halved and the modulating voltage remains constant. The modulation system is

(*a*) amplitude modulation ☐ (*b*) phase modulation ☐
(*c*) frequency modulation ☐ (*d*) any one of them. ☐

248. One of the following is an indirect way of generating FM. This is the

(*a*) reactance FET modulator ☐ (*b*) varactor diode modulator ☐
(*c*) arm strong modulator ☐ (*d*) reactance bipolar transistor modulator. ☐

249. When the modulation index of an AM wave is double, the antenna current is also doubled. The AM system used is

(*a*) A 3H ☐ (*b*) A 5C ☐
(*c*) A 3J ☐ (*d*) A 3. ☐

250. The following crystal filter is used only at the higher frequencies

(*a*) half-lattice ☐ (*b*) full-lattice ☐
(*c*) ladder ☐ (*d*) crystal gate. ☐

251. The frequency at which the reactances of the two arms in a crystal gate are equal is the

(*a*) pole ☐ (*b*) zero ☐
(*c*) zero attenuation frequency ☐ (*d*) maximum attenuation frequency. ☐

252. If the plate supply voltage for a plate modulated class-*C* amplifier is *E*, the maximum plate cathode voltage could be almost as high as

(*a*) 4*E* ☐ (*b*) 3*E* ☐
(*c*) 2*E* ☐ (*d*) *E*. ☐

253. In a low-level AM system, amplifiers following the modulated stage must be

(*a*) linear devices ☐ (*b*) harmonic devices ☐
(*c*) class-C amplifier ☐ (*d*) non-linear devices. ☐

254. If the carrier of a 100% modulated AM wave is suppressed the % power saving will be

(*a*) 50 ☐ (*b*) 150 ☐
(*c*) 100 ☐ (*d*) 66.66. ☐

255. The modulation index of an AM wave is changed from 0 to 1. The transmitted power is

(*a*) unchanged ☐ (*b*) halved ☐
(*c*) doubled ☐ (*d*) increased by 50% ☐

256. In a communication system, noise is most likely to affect the signal

(*a*) at the transmitter ☐ (*b*) in the channel ☐
(*c*) in the formation source ☐ (*d*) at the destination. ☐

257. One of the following types of noise becomes of great importance at high frequencies. It is the

(*a*) shot noise ☐ (*b*) agitation noise ☐
(*c*) flicker noise ☐ (*d*) transit time noise. ☐

258. The value of a resistor creating thermal noise is doubled. The noise power generated is therefore

(*a*) halved ☐ (*b*) quadrupled ☐
(*c*) doubled ☐ (*d*) unchanged. ☐

259. Indicate which one of the following types of noise does not occur in transistors?

(*a*) shot noise ☐ (*b*) flicker noise ☐
(*c*) partition noise ☐ (*d*) resistance noise. ☐

260. One of the following is not a useful quantity for comparing the noise performance of the receivers

(*a*) input noise voltage ☐ (*b*) equivalent noise resistance ☐
(*c*) noise temperature ☐ (*d*) noise figure. ☐

261. Indicate the noise whose source is in a category different from that of the other three

(*a*) solar noise ☐ (*b*) cosmic noise ☐
(*c*) atmospheric noise ☐ (*d*) galactic noise. ☐

262. Indicate which one of the following terms applies to troposcatter propagation

(*a*) SIDs ☐ (*b*) Fading ☐
(*c*) Atmospheric storms ☐ (*d*) Faraday rotation. ☐

263. Indicate which of the following diodes does not use negative resistance in its operation

(*a*) Backward ☐ (*b*) Gunn ☐
(*c*) IMPATT ☐ (*d*) Tunnel. ☐

264. One of the following is not used as a microwave mixer or detector

(*a*) crystal diode ☐ (*b*) schottky-barrier diode ☐
(*c*) backward diode ☐ (*d*) PIN diode. ☐

265. One of the following microwave diodes is suitable for very low power oscillators only

(*a*) Tunnel ☐ (*b*) LSA ☐
(*c*) Gunn ☐ (*d*) IMPATT. ☐

266. The signals sent by the TV transmitter to ensure correct scanning in the receiver are called

(*a*) sync ☐ (*b*) chroma ☐
(*c*) luminance ☐ (*d*) video. ☐

267. Negative resistance channel is

(*a*) IMPATT ☐ (*b*) Tunnel ☐
(*c*) Gunn ☐ (*d*) None of these. ☐

268. Power dissipated in a pure capacitor is

(*a*) maximum ☐ (*b*) minimum ☐
(*c*) zero ☐ (*d*) depends on voltage and current. ☐

269. Number of electron in one cubic centimetre of a metal is of the order of

(*a*) 10^5 ☐ (*b*) 10^8 ☐
(*c*) 10^{12} ☐ (*d*) 10^{23} ☐

270. An RC differentiator circuit has a

(*a*) very long time constant ☐ (*b*) very short time constant ☐
(*c*) moderate value of time constant ☐ (*d*) none of these. ☐

271. Passage of current is effective due to the movement of

(*a*) atoms ☐ (*b*) electrons ☐
(*c*) molecules ☐ (*d*) protons. ☐

272. Negative feedback in amplifier stage introduces

(*a*) increase in gain ☐ (*b*) decrease in gain ☐
(*c*) decrease in bandwidth ☐ (*d*) none of these. ☐

273. Bolometer is used for measurement of

(*a*) microwave power ☐ (*b*) microwave current ☐
(*c*) RF power ☐ (*d*) audio power. ☐

274. Impedance of RLC series circuit is

(*a*) $R + X_L + X_C$ ☐ (*b*) $1/X_C$ ☐
(*c*) $R^2 + (X_L - X_C)^2$ ☐ (*d*) $R^2 + (X_L + X_C)^2$. ☐

275. An emitter follower is

(*a*) voltage shunt feedback ☐ (*b*) voltage series feedback ☐
(*c*) current series feedback ☐ (*d*) current shunt feedback. ☐

276. Lissajous attern of two waveforms having same amplitude but 90° phase difference is

(*a*) circle ☐ (*b*) straight line ☐
(*c*) ellipse ☐ (*d*) cycloid. ☐

277. Cavity is equivalent to filter as

(*a*) band-pass ☐ (*b*) LP ☐
(*c*) HP ☐ (*d*) band-elimination. ☐

278. In TV, frame frequency in India is

(*a*) 25 ☐ (*b*) 30 ☐
(*c*) 35 ☐ (*d*) 40. ☐

279. Thyratron is

(*a*) UJT ☐ (*b*) BJT ☐
(*c*) Gas filled triode ☐ (*d*) Gas filled pentode. ☐

280. Separation of sound and picture carrier in TV system is

(*a*) 5.5 MHz ☐ (*b*) 5.5 kHz ☐
(*c*) 6.5 MHz ☐ (*d*) 8.5 MHz. ☐

281. Which of the following is used in digital communication system?

(*a*) AM ☐ (*b*) FM ☐
(*c*) PCM ☐ (*d*) PAM. ☐

282. Path of an electron in uniform electric field is

(*a*) straight ☐ (*b*) circular ☐
(*c*) helical ☐ (*d*) elliptical. ☐

283. Trimmers are generally having

(*a*) air dielectric ☐ (*b*) paper dielectric ☐
(*c*) mica dielectric ☐ (*d*) waxed paper dielectric. ☐

284. Zener diode is used as

(*a*) voltage reference ☐ (*b*) relay ☐
(*c*) switching circuits ☐ (*d*) none of the above. ☐

285. VHF is

(*a*) 20 GHz ☐ (*b*) 300 MHz ☐
(*c*) 20 kHz ☐ (*d*) none of the above. ☐

286. Collector current increases by increasing

(*a*) V_{ee} ☐ (*b*) I_b ☐
(*c*) I_e ☐ (*d*) V_{EE}. ☐

287. I_{CBO} increases due to the increases of

(*a*) Temperature ☐ (*b*) V_{EE} ☐
(*c*) I_e ☐ (*d*) V_{ee}. ☐

288. Increasing the emitter current will

(*a*) decrease I_e ☐ (*b*) increase I_e ☐
(*c*) not influence I_e ☐ (*d*) decrease I_E. ☐

289. As a bypassing capacitor, which is better

(*a*) Mica ☐ (*b*) Ceramic ☐
(*c*) Electrolytic ☐ (*d*) None of the above. ☐

290. When four resistances are in parallel, the effective resistance is

(*a*) lower than the least ☐ (*b*) greater than the least but lesser greatest ☐
(*c*) depends on the values of the resistances ☐ (*d*) none of these. ☐

291. Integration of 150 gates in I.C. technology is called

(*a*) SSI ☐ (*b*) MSI ☐
(*c*) LSI ☐ (*d*) None of the above. ☐

292. Material changes polarization under mechanical strain

(*a*) di-electric ☐ (*b*) fero-electric ☐
(*c*) piezo-electric ☐ (*d*) none of the above. ☐

293. Interelectrode capacitance causes

(*a*) noise in amplifier ☐ (*b*) harmonic distortion ☐
(*c*) parasitic noise ☐ (*d*) none of the above. ☐

294. The third sub-harmonic of 60 cycles is

(*a*) 180 ☐ (*b*) 20 ☐
(*c*) 360 ☐ (*d*) 100. ☐

295. Resistance of 100 W, 200 V lamp is

(*a*) 100 ohms ☐ (*b*) 200 ohms ☐
(*c*) 400 ohms ☐ (*d*) 50 ohms. ☐

296. Non-conductors whose polarization is caused by an electric field are known as

(a) semi-conductor ☐ (b) super-conductor ☐

(c) di-electric. ☐ (d) none of the above. ☐

297. There are two d.c. currents, that is, 2 amperes and 45 amperes

(a) 4.5 A signal carries more intelligence ☐ (b) 2 A signal carries more intelligence ☐

(c) both carry same intelligence ☐ (d) none of these. ☐

298. A complex waveform is composed of frequency components 1 Hz, 3 Hz, 7 Hz and 9 Hz. Its fundamental frequency is

(a) 1 Hz ☐ (b) 9 Hz ☐

(c) 7 Hz ☐ (d) none of these. ☐

PHYSICS

299. When an elevator moves upward with uniform acceleration the apparent weight of a body kept in the elevator is

(a) decreased ☐ (b) increased ☐

(c) remains same ☐ (d) none of these. ☐

300. An object of mass 50 kg is hung by a rope from the ceiling of a lift. If the lift is moving upward with an acceleration of 4 m/s^2, the tension in the rope will be

(a) 500 N ☐ (b) 980 N ☐

(c) 690 N ☐ (d) none of these. ☐

301. In question 300, if the lift is moving downs with the same acceleration, the tension in the rope will be

(a) 290 N ☐ (b) 530 N ☐

(c) 980 N ☐ (d) none of these. ☐

302. In question 300, if the lift is moving upward with constant velocity, the tension in the rope will be

(a) 500 N ☐ (b) 490 N ☐

(c) 980 N ☐ (d) none of these. ☐

303. A car of mass M is required to be given an acceleration of a m/s^2 on a rough road whose coefficient of friction is μ. The required force will be

(a) Ma ☐ (b) $Ma + \mu Mg$ ☐

(c) $Ma - \mu Mg$ ☐ (d) none of these. ☐

304. In question 303, once the car has started moving, the force required to keep the car moving with constant speed will be

(a) Ma ☐ (b) Mg ☐

(c) μMg ☐ (d) none of these. ☐

305. An automobile of mass M has attained an constant speed of v m/s. The engine is put off. Its retardation will be (coefficient of friction = μ)

(a) $\mu M + Mg$ ☐ (b) $\mu M - Mg$ ☐

(c) $-\mu g$ ☐ (d) none of these. ☐

306. The radius of earth is 6.37×10^6 m. If the earth stops rotating about its axis, the change in the value of g at a place of latitude 45° will be

(*a*) 6.37×10^{-2} m/s ☐ (*b*) 1.18×10^{-2} m/s ☐
(*c*) 2.36×10^{-2} m/s ☐ (*d*) none of these. ☐

307. A rocket is moving upwards with acceleration $2g$. A man weighting 60 kg is sitting in the rocket. The effective weight of the man will be

(*a*) 60 kg ☐ (*b*) 180 kg ☐
(*c*) 120 kg ☐ (*d*) none of these. ☐

308. A satellite is moving in a circular orbit round the earth at a height $Re/2$ from earth's surface. (Re is radius of earth). Its period of revolution will be

(*a*) 9.8×10^3 sec ☐ (*b*) 9.3×10^3 sec ☐
(*c*) 19.8×10^3 sec ☐ (*d*) none of these. ☐

309. A rocket is moving upwards with acceleration $5g$. The effective weight of an observer weighing 100 kg sitting in the socket will be

(*a*) 200 kg ☐ (*b*) 600 kg ☐
(*c*) 500 kg ☐ (*d*) none of these. ☐

310. A particle of mass 'm' is moving along a circular path in a plane. The force acting on it is

(*a*) non-conservative ☐ (*b*) conservative ☐
(*c*) both conservative and non-conservative ☐ (*d*) none of these. ☐

311. The frequency of a mass 'm' oscillating on a spring having force constant k is given by

(*a*) $\sqrt{km}$ ☐ (*b*) $\frac{1}{2\pi}\sqrt{\frac{k}{m}}$ ☐
(*c*) $\frac{2\pi\sqrt{k}}{m}$ ☐ (*d*) none of these. ☐

312. A mass 'm' attached to a spring oscillates every 2 seconds. If the mass is increased by 2 kg, the period increases by one second. The initial mass 'm' is

(*a*) 9.8 kg ☐ (*b*) 1.6 kg ☐
(*c*) 3.2 kg ☐ (*d*) none of these. ☐

313. The total energy of a particle of mass 'm' is S.H.M. having amplitude 'a' and angular velocity 'w' is

(*a*) m^2w^2 ☐ (*b*) $\frac{1}{2}\ mw^2$ ☐
(*c*) $\frac{1}{2}\ mw^2 a^2$ ☐ (*d*) none of these. ☐

314. In question 313, the time period of the satellite will be

(*a*) 8.53×10^3 s ☐ (*b*) 9.8×10^3 s ☐
(*c*) 4.76×10^3 s ☐ (*d*) 16.98×10^3 s. ☐

315. The phenomena of bending of light rays about sharp edges of an object is called

(*a*) dispersion ☐ (*b*) diffraction ☐

(*c*) deviation ☐ (*d*) refraction. ☐

316. If the distance between two sources (interfering slits) is 2*d*, the fringe width at a distance *D* for wavelength λ is

(*a*) $\frac{D\lambda}{d}$ ☐ (*b*) $\frac{2d}{D\lambda}$ ☐

(*c*) $\frac{D\lambda}{2d}$ ☐ (*d*) $\frac{\lambda d}{2D}$. ☐

317. If the distance of the screen is increased from the slits producing interference, the fringe width will be

(*a*) remain constant ☐ (*b*) increase ☐

(*c*) decrease ☐ (*d*) none of the above. ☐

318. The condition for producing maxima for two interfering beams is that its path difference is

(*a*) $\frac{n\lambda}{2}$ ☐ (*b*) $n\lambda$ ☐

(*c*) $\frac{2n}{\lambda}$ ☐ (*d*) $\frac{2}{\lambda n}$. ☐

319. In a double slit experiment, the two slits are 0.25 mm apart and the screen in 2 m away. The linear distance between the two second order maxima is 20 mm. The wavelength of the incident light is

(*a*) 5000 Å ☐ (*b*) 5893 Å ☐

(*c*) 6250 Å ☐ (*d*) 7000 Å. ☐

320. Mercury light of wavelength 5461 Å is used to form fringes at a distance of 1 m from the double slit. The linear separation between two-fifth order fringes (one on each side of the central maxima) is 1.2 cm. The distance between two slits is

(*a*) 1 mm ☐ (*b*) 0.455 mm ☐

(*c*) 0.85 mm ☐ (*d*) none of the above. ☐

321. A glass surface of refractive index 1.50 has a coating of magnesium fluoride of refractive index 1.25. Reflected light of wavelength 5000 Å produces destructive interference. The thickness of the coating is

(*a*) 0.5 Å ☐ (*b*) 100 Å ☐

(*c*) 1 Å ☐ (*d*) 50 Å. ☐

322. When a thin sheet of glass having thickness 6×10^{-2} cm and refractive index 1.5 is introduced in one of the interfering beams the central fringe shifts to a position previously occupied by the 5th bright fringe. The wavelength of the light used is

(*a*) 6 Å ☐ (*b*) 60 Å ☐

(*c*) 600 Å ☐ (*d*) none of these. ☐

323. The additional path difference introduced due to a film of thickness 't' and refractive index μ is given by

(*a*) 2 (μ – 1) ☐ (*b*) 2 μt ☐

(*c*) t (μ – 1) ☐ (*d*) μt. ☐

324. The resolving power of a grating having *N* number of lines in 2nd order spectrum will be

(*a*) $N/2$ ☐ (*b*) $2N$ ☐

(*c*) $N/4$ ☐ (*d*) none of these. ☐

325. Monochromatic light of wavelength 5800 Å produces second order spectrum at 18° when it is used with a grating. The number of lines per cm on the grating is

(*a*) 2664 ☐ (*b*) 500 ☐

(*c*) 2500 ☐ (*d*) 1332. ☐

326. A film of oil of refractive index 1.40 is formed on a glass surface have refractive index 1.50. The thickness of the film to eliminate the light of wavelength 5000 Å from the reflected light should be

(*a*) 5.0×10^{-8} m ☐ (*b*) 4.46×10^{-8} m ☐

(*c*) 1.10×10^{-8} m ☐ (*d*) 8.93×10^{-8} m. ☐

327. Light of wavelength 5800 Å is incident normally on a grating having 4000 lines per cm. The angular deviation in the second order is

(*a*) 66° ☐ (*b*) 44° ☐

(*c*) 22° ☐ (*d*) 11°. ☐

328. The light reflected from a glass surface of refractive index 1.66 is linearly polarised. The angle of incidence is

(*a*) 30.0° ☐ (*b*) 58.9° ☐

(*c*) 29.8° ☐ (*d*) 60.0°. ☐

329. The light reflected from the surface of benzene having refractive index 1.50 is found to be linearly polarised. The angle of refraction is

(*a*) 56.3° ☐ (*b*) 42.5° ☐

(*c*) 33.77 ☐ (*d*) none of these. ☐

330. A grating 2.5 cm in size has 15000 lines. Light of wavelength 6000 Å is incident on it. The angular position of second order maxima is

(*a*) 46.5° ☐ (*b*) 36.8° ☐

(*c*) 21.3° ☐ (*d*) 17.4°. ☐

331. The intensity *I* of a wave is proportional to

(*a*) amplitude ☐ (*b*) amplitude^2 ☐

(*c*) amplitude^3 ☐ (*d*) none of these. ☐

332. The resolving power of a grating depends on

(*a*) the intensity of light ☐ (*b*) no. of lines in the grating ☐

(*c*) wavelength of light ☐ (*d*) none of these. ☐

333. A glass biprism (μ = 1.5) with refracting angle 2° produces fringes at 1.1 m distance from the slit. The slit is 10 cm from the biprism. If λ = 5900 Å, the fringe width is

(*a*) 1.86 cm ☐ (*b*) 0.186 cm ☐

(*c*) 0.372 cm ☐ (*d*) none of these. ☐

334. Light of wavelength 5893 Å is reflected at nearly normal incidence from a soap film of refractive index 1.42. For darkness the least thickness of the film will be

(*a*) 5893 Å ☐ (*b*) 2075 Å ☐
(*c*) 4000 Å ☐ (*d*) 4150 Å. ☐

335. In question 334, the thickness of the film for brightness will be

(*a*) 4000 Å ☐ (*b*) 4150 Å ☐
(*c*) 1037.5 Å ☐ (*d*) 5893 Å. ☐

336. Using sodium light of wavelength 5893 Å interference fringes are formed due to a thin wedge having air as medium. When viewed normally 10 fringes are observed at a distance of 1 cm. The angle of the wedge is

(*a*) 61 sec ☐ (*b*) 1° ☐
(*c*) 1°–2′ ☐ (*d*) none of these. ☐

337. In a Newton's rings experiment the diameters of 4th and 12th dark rings are 0.4 and 0.7 cm respectively. The diameter of 20th ring will be

(*a*) 1.00 cm ☐ (*b*) 1.42 cm ☐
(*c*) 0.894 cm ☐ (*d*) none of these. ☐

338. In an interference pattern the ratio between maximum and minimum intensity is 36 : 1. The ratio between the amplitudes of the two interfering waves is

(*a*) 6 : 1 ☐ (*b*) 7 : 5 ☐
(*c*) 18 : 5 ☐ (*d*) none of these. ☐

339. In question 338, the ratio between the intensities of the interfering waves is

(*a*) 49 : 25 ☐ (*b*) 6 : 1 ☐
(*c*) 18 : 5 ☐ (*d*) 36 : 1. ☐

340. At thin sheet of glass of refractive index 1.5 and thickness 6 microns, introduced in the path of one of the interfering beams in a double slit arrangement shifts the central fringe to a position previously occupied by the fifth bright fringe. The wavelength of light is

(*a*) 5893 Å ☐ (*b*) 5461 Å ☐
(*c*) 6000 Å ☐ (*d*) 4000 Å. ☐

341. The limit of angular resolution of telescope of aperture '*a*' is

(*a*) $\frac{1.22\,\lambda}{a}$ ☐ (*b*) $\frac{1.22\,a}{\lambda}$ ☐
(*c*) $\frac{\lambda a}{1.22}$ ☐ (*d*) none of these. ☐

342. A grating just resolves 5890 Å and 5876 Å in second order. The number of lines on the grating is

(*a*) 1000 ☐ (*b*) 589 ☐
(*c*) 992 ☐ (*d*) 5890. ☐

343. Two straight and narrow parallel slits 1 mm apart produce fringes of width 0.5 mm on a screen 1 m away from the slits. The wavelength of light is

(a) 4000 Å ☐ (b) 5893 Å ☐
(c) 6000 Å ☐ (d) 5000 Å. ☐

344. Light of wavelength 5100 Å from a narrow slit is incident on a double slit. If the overall separation of 10 fringes on a screen 2 m away is 20 mm, the slit separation is

(a) 1 mm ☐ (b) 0.51 mm ☐
(c) 5.1 mm ☐ (d) none of these. ☐

345. In an interference fringe pattern the fringe width are

(a) increasing order ☐ (b) equal thickness ☐
(c) decreasing order ☐ (d) none of these. ☐

346. In the case of interference phenomenon

(a) there is loss of energy ☐ (b) there is gain in energy ☐
(c) there is transfer of energy ☐ (d) none of these. ☐

347. In Young's double slit experiment, the separation of the slits is 1.9×10^{-3} m and the fringe width is 3.10^{-4} m at a distance of 1 m from the slits. The wavelengths of light used is

(a) 5890 Å ☐ (b) 6000 Å ☐
(c) 5500 Å ☐ (d) 5893 Å. ☐

348. If the refracting angle 'α' of a prism having refractive index μ is small, the deviation is given by

(a) $\mu\alpha$ ☐ (b) $\mu(\alpha - 1)$ ☐
(c) $(\mu - 1)\,\alpha$ ☐ (d) $\frac{\mu\alpha}{\mu - \alpha}$. ☐

349. A biprism of refracting angle 2° and μ = 1.5 is kept at 0.1 m from the source. The separation of the two sources will be

(a) 0.3×10^{-3} m ☐ (b) 7.5×10^{-2} m ☐
(c) 0.0035 m ☐ (d) none of these. ☐

350. In an interference fringe pattern, if the screen is drawn out, the fringe width will

(a) decrease ☐ (b) increase ☐
(c) remain same ☐ (d) none of these. ☐

351. In a biprism set up, if the biprism is moved towards the slit, the distance between two interfering sources will

(a) decrease ☐ (b) increase ☐
(c) remain same ☐ (d) none of these. ☐

352. Colours produced by thin films of oil on the surface of water is due to

(a) reflection of light ☐ (b) diffraction of light ☐
(c) interference phenomena ☐ (d) none of these. ☐

353. For the production of Newton's rings we need

(a) narrow slits ☐ (b) broad source ☐
(c) circular slits ☐ (d) none of these. ☐

354. Light of wavelength 6000 Å falls normally on a thin wedge shaped film of μ = 1.5 which forms fringes that are 2×10^{-3} m apart. The angle of the wedge is

(a) 3.0×10^{-4} radian ☐ (b) 0.75×10^{-4} radian ☐

(c) 1.07×10^{-4} radian ☐ (d) none of these. ☐

355. A plano-convex lens of radius 3 m is used to produce Newton rings. The diameter of eighth dark ring in the transmitted system is 7.2×10^{-3} m. The wavelength of light used is

(a) 6000 Å ☐ (b) 5890 Å ☐

(c) 5500 Å ☐ (d) 5760 Å. ☐

356. In the reflected system of Newton rings, the central spot is

(a) dark ☐ (b) bright ☐

(c) uniform intensity ☐ (d) none of these. ☐

357. In the case of Newton rings, the fringe width of the pattern with increasing number of rings

(a) remaining constant ☐ (b) increases ☐

(c) decreases ☐ (d) none of these. ☐

358. The visibility of interference fringes is given by

(a) $\dfrac{I_{max} - I_{min}}{I_{max} - I_{min}}$ ☐ (b) $I_{max} - I_{min}$ ☐

(c) $I^2_{max} - I^2_{min}$ ☐ (d) none of these. ☐

359. In the phenomena of diffraction, a zone plate acts as

(a) Diverging lens ☐ (b) Converging lens ☐

(c) Plano-achromate lens ☐ (d) none of these. ☐

360. In question 359, the zone plate acts as a lens having focal length

(a) $\dfrac{\gamma_n^2}{n\lambda}$ ☐ (b) $\gamma_n{}^2\lambda$ ☐

(c) $\dfrac{n\lambda}{\gamma_n}$ ☐ (d) none of these. ☐

361. A plane diffraction grating has 50×10^4 lines per metre. What is the highest order spectrum which may be seen with wavelength 6000 Å?

(a) 2 ☐ (b) 3 ☐

(c) 4 ☐ (d) none of these. ☐

362. In a plane transmission grating, the angle of diffraction for the second order principal maximum for wavelength 5×10^{-7} m is 30°. The number of lines in one metre of the grating surface is

(a) 50000 ☐ (b) 100000 ☐

(c) 25000 ☐ (d) none of these. ☐

363. The diffraction of the *n*th order principal maximum produced by a grating having grating element $(a + b)$ for a wavelength λ is given by

(a) $\dfrac{\sin\theta}{a+b} = n\lambda$ ☐ (b) $\dfrac{d\theta}{d\lambda}$ ☐

(c) $(a + b)\sin\theta = n\lambda$ ☐ (d) none of these. ☐

364. The dispersive power of a grating is given by

(*a*) $\frac{d\theta}{d\lambda}$ ☐ (*b*) nN ☐

(*c*) $(a + b) \sin \theta$ ☐ (*d*) none of these. ☐

365. If the number of lines on a grating is increased, it will produce

(*a*) higher intensity ☐ (*b*) higher dispersion ☐

(*c*) higher resolving power ☐ (*d*) none of these. ☐

366. A diffraction grating which has 400000 lines per metre is used for normal incidence with wavelengths 5000 Å. The dispersive power of the grating in the third order spectrum will be

(*a*) 10000 ☐ (*b*) 15000 ☐

(*c*) 20000 ☐ (*d*) 5000. ☐

367. The minimum number of lines per metre in a plane grating to resolve D_1(5896 Å) and D_2 (5890 Å) in the first order spectrum is

(*a*) 100000 ☐ (*b*) 50000 ☐

(*c*) 98330 ☐ (*d*) none of these. ☐

368. The number of blunt corners in a Nicol prism is

(*a*) one ☐ (*b*) two ☐

(*c*) three ☐ (*d*) four. ☐

369. The line joining the blunt corners in a crystal is known as

(*a*) crystallographic axis ☐ (*b*) double separation axis ☐

(*c*) optic axis ☐ (*d*) none of these. ☐

370. Uniaxial crystals are called negative crystals if

(*a*) $\mu_0 = \mu_e$ ☐ (*b*) $\mu_e > \mu_0$ ☐

(*c*) $\mu_0 > \mu_e$ ☐ (*d*) none of these. ☐

371. Nicol prism cannot be used with ultra-violet light because the Canada balsam

(*a*) reflects the ultra-violet ray ☐ (*b*) absorbs the ultra-violet ray ☐

(*c*) deviates the ultra-violet ray ☐ (*d*) none of these. ☐

372. A nicol prism can be used as

(*a*) polariser only ☐ (*b*) polariser and analyser ☐

(*c*) analyser only ☐ (*d*) none of these. ☐

373. A quarter wave plate is introduced to produce a path difference of

(*a*) λ ☐ (*b*) $\frac{\lambda}{2}$ ☐

(*c*) $\frac{\lambda}{4}$ ☐ (*d*) none of these. ☐

374. The phenomena of polarisation indicates that

(*a*) light waves are electromagnetic ☐ (*b*) light waves are longitudinal ☐

(*c*) light waves are standing ☐ (*d*) light waves are transverse. ☐

375. Einstein's photo-electric equation is given by

(*a*) $E = h\nu$ ☐ (*b*) $E = mc^2$ ☐

(*c*) $h\nu = h\nu_0 + \frac{1}{2}mv^2$ ☐ (*d*) none of these. ☐

376. If the intensity of incident light is increased for photo-electric emission, the photo-electrons emitted have

(*a*) more photons ☐ (*b*) more velocity ☐

(*c*) more energy ☐ (*d*) none of these. ☐

377. The work function of a photo-electric material depends on

(*a*) nature of the incident radiation ☐ (*b*) nature of the material ☐

(*c*) nature of the intensity of light ☐ (*d*) none of these. ☐

378. Light of wavelength 50000 Å falls on a metal whose work function is 1.9 eV. The energy of the incident photon will be

(*a*) 1.9 eV ☐ (*b*) 2.48 eV ☐

(*c*) 3.8 eV ☐ (*d*) none of these. ☐

379. Two polaroids are set to give maximum transmission of intensity I_0. If one of the polaroid is rotated by 30°, the intensity of transmission will be

(*a*) 0.75 ☐ (*b*) 0.50 ☐

(*c*) 0.414 ☐ (*d*) none of these. ☐

380. If the diameter of the objective glass of a telescope is increased, it resolving power will

(*a*) remain same ☐ (*b*) decrease ☐

(*c*) increase ☐ (*d*) none of these. ☐

381. The limit of resolution of a telescope having aperture '*a*' is given by

(*a*) $\frac{\lambda}{a}$ ☐ (*b*) λa ☐

(*c*) $\frac{1.22\,\lambda}{a}$ ☐ (*d*) none of these. ☐

382. The aperture of the objective of a telescope used to resolve stars separated by 4.88×10^{-6} radian with wavelength 6000 Å is

(*a*) 1.0 m ☐ (*b*) 1.5 m ☐

(*c*) 15×105^2 m ☐ (*d*) none of these. ☐

383. The resolving power of a grating is given by

(*a*) $\frac{d\lambda}{\lambda}$ ☐ (*b*) $\frac{\lambda}{d\lambda}$ ☐

(*c*) $\lambda \,.\, d\lambda$ ☐ (*d*) none of these. ☐

384. The equation of motion for a particular mass at the end of a spring is $x = 0.40 \cos (0.70\,t - 0.30)$ m. Its amplitude is

(*a*) 0.70 m ☐ (*b*) 0.30 m ☐

(*c*) 0.40 m ☐ (*d*) 0.20 m. ☐

385. In question 384, the frequency is

(*a*) 0.111 Hz ☐ (*b*) 10 Hz ☐

(*c*) 2.2 Hz ☐ (*d*) 0.4 Hz. ☐

386. In question 384, the ratio k/m is

(*a*) 9.8 ☐ (*b*) 0.49 ☐

(*c*) 0.89 ☐ (*d*) 1.47. ☐

387. In question 384, the period of motion is

(*a*) 9.87 sec ☐ (*b*) 0.89 sec ☐

(*c*) 1.47 sec ☐ (*d*) 8.97 sec. ☐

388. Sound waves in air are

(*a*) transverse ☐ (*b*) longitudinal ☐

(*c*) travelling ☐ (*d*) stationary. ☐

389. The waves propagated along the surface of water when a stone is dropped into it are

(*a*) longitudinal ☐ (*b*) transverse ☐

(*c*) stationary ☐ (*d*) none of these. ☐

390. In the propagation of sound wave, the distance between consecutive compression is

(*a*) λ ☐ (*b*) 2λ ☐

(*c*) $\frac{\lambda}{2}$ ☐ (*d*) $\frac{3\lambda}{2}$. ☐

391. The equation of a transverse travelling wave is given by $y = 0.05 \sin 2\pi (0.5x - 10t)$. Its amplitude is

(*a*) 10 m ☐ (*b*) 0.05 m ☐

(*c*) 0.05 m ☐ (*d*) none of these. ☐

392. Its wavelength is

(*a*) 5 m ☐ (*b*) 0.5 m ☐

(*c*) 2 m ☐ (*d*) 10 m. ☐

393. Sound waves in air are

(*a*) longitudinal ☐ (*b*) transverse ☐

(*c*) partially transverse ☐ (*d*) none of these. ☐

394. The energy is not carried by

(*a*) stationary waves ☐ (*b*) longitudinal waves ☐

(*c*) transverse waves ☐ (*d*) mechanical waves. ☐

395. An engine approaching towards a stationary observer sounds a whistle. The observer will hear a sound of

(*a*) slightly lower frequency ☐ (*b*) slightly higher frequency ☐

(*c*) the same frequency ☐ (*d*) none of these. ☐

396. An observer while moving towards a stationary source, observes the frequency of the source double of its real value. The observer is moving at a speed of

(a) 500 m/s ☐ (b) 660 m/s ☐
(c) 330 m/s ☐ (d) 750 m/s. ☐

397. When the source is moving towards the observer at rest, the apparent frequency is given by

(a) $\frac{v - v_s}{v}$ ☐ (b) $\frac{v + v_s}{v}$ ☐
(c) $\frac{v}{v - v_s}$ ☐ (d) $\frac{v + v_s}{v - v_s}$. ☐

398. When the source is in motion, it causes

(a) change in wavelength ☐ (b) change in frequency ☐
(c) change in velocity ☐ (d) change in medium. ☐

399. When the observer moves it causes change in

(a) frequency ☐ (b) velocity ☐
(c) length of wave train ☐ (d) none of these. ☐

400. The pitch of a sound wave is directly related to

(a) wavelength ☐ (b) frequency ☐
(c) velocity ☐ (d) amplitude. ☐

401. A spectrum line of wavelength 4000 Å from a star is found to be displaced from its normal position towards red end by 1 Å. The velocity of the star in the line of slight is

(a) 7.5×10^3 km/s ☐ (b) 4.0×10^3 km/s ☐
(c) 150×10^2 km/s ☐ (d) 8.0×10^3 km/s. ☐

402. The velocity of sound in a liquid is given by [k-Bulk modulus]

(a) $\sqrt{\frac{\rho}{k}}$ ☐ (b) $\sqrt{\frac{k}{\rho}}$ ☐
(c) $\sqrt{k\rho}$ ☐ (d) $\sqrt{\frac{k\rho}{v}}$. ☐

403. A tuning fork when sounded produces waves which are

(a) transverse ☐ (b) longitudinal ☐
(c) progressive ☐ (d) stationary. ☐

404. The velocity of transverse waves in stretched string is given by

(a) $\sqrt{mT}$ ☐ (b) $\sqrt{\frac{m}{T}}$ ☐
(c) $\sqrt{\frac{T}{m}}$ ☐ (d) $\sqrt{\frac{2T}{m}}$. ☐

405. Two nicols are first crossed and then one of them is rotated through 60°. The percentage of the incident light transmitted is

(a) 60% ☐ (b) 37.5% ☐
(c) 75% ☐ (d) 30%. ☐

406. The relation between phase diff. and path diff. is

(a) $\frac{\lambda}{2\pi}$ ☐ (b) $\frac{2\pi}{\lambda}$ ☐

(c) $2\pi\lambda$ ☐ (d) $\frac{\pi}{2\lambda}$. ☐

407. The distance between the two virtual images of double slit is given by

(a) $d_1 d_2$ ☐ (b) $\sqrt{d_1 d_2}$ ☐

(c) $d_1^2 d_2^2$ ☐ (d) $d_1 \sqrt{d_2}$. ☐

408. The velocity of the unpolarised transverse waves in a three dimensional medium having shear modulus G and P the density is given by

(a) $\frac{G}{P}$ ☐ (b) $\sqrt{\frac{G}{P}}$ ☐

(c) $\sqrt{\frac{P}{G}}$ ☐ (d) none of these. ☐

409. The velocity of longitudinal waves in a solid medium having Youngs modulus E and denotes P is given by

(a) $\sqrt{PE}$ ☐ (b) $\sqrt{\frac{P}{E}}$ ☐

(c) $\sqrt{\frac{E}{P}}$ ☐ (d) none of these. ☐

III. OBJECTIVE TYPE QUESTIONS FROM COMPETITIVE EXAMINATIONS

Tick mark the most appropriate answer:

1. Wanted temporary, part-time persons for the post of field interviewer to conduct personal interviews to collect and collate economic data. Requirements: high school-pass, must be available for day, evening and saturday work. Transportation paid, expanses reimbursed. Which one of the following is the best inference from the above advertisement?

   (a) Gender-discriminatory ☐ (b) Xenophobic ☐

   (c) Not designed to make the post attractive ☐ (d) Not gender-discriminatory. ☐

   (ME/CE/CSE/PI-2012)

2. Choose the most appropriate word from the options given below to complete the following sentence:

   If you are trying to make a strong impression on your audience, you cannot do so by being understated, tentative or ....

   (a) hyperbolic ☐ (b) restrained ☐

   (c) argumentative ☐ (d) indifferent. ☐

   (ME/CE/CSE/PI-2011)

3. Choose the most appropriate word from the options given below to complete the following sentence:

His rather casual remars on politics .... his lack of seriousness about the subject.

(*a*) masked ☐ (*b*) belied ☐

(*c*) betrayed ☐ (*d*) suppressed. ☐

(ALL BRANCHES-2010)

4. Few school curricula include a unit on how to deal with bereavement and grief, and yet all students at some point in their lives suffer from losses through death and parting.

Based on the above passage which topic would not be included in a unit on bereavement?

(*a*) How to write a letter of condolence? ☐

(*b*) What emotional stages are passed through in the healing process? ☐

(*c*) What are the leading causes of death? ☐

(*d*) How to give support to a grieving friend? ☐

(ME/CE/CSE/PI-2011)

5. 5 skilled workers can build a wall in 20 days, 8 semi-skilled workers can build a wall in 25 days; 10 unskilled workers can build a wall in 30 days. If a team has 2 skilled, 6 semi-skilled and 5 unskilled workers, how long will it take to build the wall?

(*a*) 20 days ☐ (*b*) 18 days ☐

(*c*) 16 days ☐ (*d*) 15 days. ☐

(ALL BRANCHES-2010)

6. After several defeats in wars, Robert Bruce went in exile and wanted to commit suicide. Just before committing suicide, he came across a spider attempting tirelessly to have its net. Time and again, the spider failed but that did not deter it to refrain from making attempts. Such attempts by the spider made Bruce curious. Thus, Bruce started observing the near-impossible goal of the spider to have the net. Ultimately, the spider succeeded in having its net despite several failures. Such act of the spider encouraged Bruce not to commit suicide. And then, Bruce went back again and won many a battle, and the rest is history.

(*a*) Failure is the pillar of success ☐ (*b*) Honesty is the best policy ☐

(*c*) Life begins and ends with adventures ☐ (*d*) No adversity justifies giving up hope. ☐

(ME/CSE/PI-2013)

7. The question below consists of a pair of related words followed by four pairs of words. Select the pair that best expresses the relation in the original pair:

Gladiator : Arena

(*a*) dancer : stage ☐ (*b*) commuter : train ☐

(*c*) teacher : classroom ☐ (*d*) lawyer : courtroom. ☐

(EC/EE/INST-2011)

8. Find the sum of the expression

$$\frac{1}{\sqrt{1}+\sqrt{2}}+\frac{1}{\sqrt{2}+\sqrt{3}}+\frac{1}{\sqrt{3}+\sqrt{4}}+\ldots\ldots+\frac{1}{\sqrt{80}+\sqrt{81}}$$

(*a*) 7 ☐ (*b*) 8 ☐

(*c*) 9 ☐ (*d*) 10. ☐

(ME/CSE/PI-2013)

9. Hari (H), Gita (G), Irfan (I) and Saira (S) are siblings (*i.e.* brothers and sisters). All were born on 1st January. The age difference between any two successive siblings (*i.e.* born one after another is less than 3 years).

Given following facts:

(*i*) Hari's age + Gita's age > Irfan's age + Saira's age

(*ii*) The age difference between Gita and Saira is 1 year. However, Gita is not the oldest and Saira is not the youngest.

(*iii*) There are no twins.

In what order were they born (oldest first)?

(*a*) HSIG ☐ (*b*) SGHI ☐
(*c*) IGSH ☐ (*d*) IHSG. ☐

(ALL BRANCHES-2010)

10. Statement: There were different streams of freedom movements in colonial India carried out by the moderates, liberals, radicals socialists, and so on.

Which one of the following is the best inference from the above statement?

(*a*) The emergence of nationalism in colonial India led to our independence. ☐
(*b*) Nationalism in India is homogeneous. ☐
(*c*) Nationalism in India is homogeneous. ☐
(*d*) Nationalism in India is heterogeneous. ☐

(EC/EE/INST-2013)

11. In the summer of 2012, in New Delhi, the mean temperature of Monday to Wednesday was 41°C and of Tuesday to Thursday was 43°C. If the temperature on Thursday was 15% higher than that of Monday, then the temperature in °C on Thursday was

(*a*) 40 ☐ (*b*) 43 ☐
(*c*) 46 ☐ (*d*) 49. ☐

(EC/EE/INST-2013)

12. The current erection cost of a structure is ₹ 13,200. If the labour wages per day increase by 1/5 of the current wages and the working hours decrease by 1/24 of the current period, then the new cost of erection in ₹ is

(*a*) 16,500 ☐ (*b*) 15,180 ☐
(*c*) 11,000 ☐ (*d*) 10,120. ☐

(ME/CSE/PI-2013)

13. The question below consists of a pair of related words followed by four pair of words. Select the pair that best expresses the relation in the original pair.

Unemployed : Worker

(*a*) follow : land ☐ (*b*) unaware : sleeper ☐
(*c*) wit : jester ☐ (*d*) renovated : house. ☐

(ALL BRANCHES-2010)

14. Choose the most appropriate alternative from the options given below to complete the following sentence:

Despite several ...... the mission succeeded in its attempt to resolve the conflict.

(*a*) attempts ☐ (*b*) setbacks ☐
(*c*) meetings ☐ (*d*) delegations. ☐

(ME/CE/CSE/PI-2012)

15. 25 persons are in a room, 15 of them play hockey, 17 of them play football and 10 of them play both hockey and football. Then the number of persons playing neither hockey nor football is

(*a*) 2 ☐ (*b*) 17 ☐

(*c*) 13 ☐ (*d*) 3. ☐

(ALL BRANCHES-2010)

16. The variable cost (V) of manufacturing a product varies according to the equation $V = 4q$, where q is the quantity produced. The fixed cost (F) of production of same product reduces with q according to the equation $F = 100/q$. How many units should be produced to minimize the total cost ($V + F$)?

(*a*) 5 ☐ (*b*) 4 ☐

(*c*) 7 ☐ (*d*) 6. ☐

(ME/CE/CSE/PI-2011)

17. Which of the following assertions are correct?

P : Adding 7 to each entry in a list adds 7 to the mean of the list.

Q : Adding 7 to each entry in a list adds 7 to the standard deviation of the list.

R : Doubling each entry in a list doubles the mean of the list.

S : Doubling each entry in a list leaves the standard deviation of the list unchanged.

(*a*) P, Q ☐ (*b*) Q, R ☐

(*c*) P, R ☐ (*d*) R, S. ☐

(ME/CE/CSE/PI-2012)

18. $Q = CH$ and error in the measurement of H committed of the order of 1.5%, then error in the calculated value of Q will be of the order of value of Q will be of the order of

(*a*) 7.5% ☐ (*b*) 3.75% ☐

(*c*) 1.5% ☐ (*d*) none of the above. ☐

19. The power P required to propel a ship of length L moving with velocity V is given by $P = KVL$. An increase of 3% of V and 4% in L require

(*a*) 18.18% increase of P ☐ (*b*) 27% increase in P ☐

(*c*) 7% increase in P ☐ (*d*) none of the above. ☐

20. A container originally contains 10 litres of pure spirit. From this container 1 litre of spirit is replaced with 1 litre of water. Subsequently, 1 litre of the mixture is again replaced with 1 litre of water and this process is repeated one more time. How much spirit in litre is now left in the container?

(*a*) 9 ☐ (*b*) 8 ☐

(*c*) 7.5 ☐ (*d*) 7.29. ☐

(ME/CE/CSE/PI-2011)

21. A tourist covers half of his journey by train at 60 km/h, half of the remainder by bus at 30 km/h and the rest by cycle at 10 km/h. The average speed of the tourist in km/h during his entire journey is

(*a*) 36 ☐ (*b*) 30 ☐

(*c*) 24 ☐ (*d*) 18. ☐

(ME/CSE/PI-2013)

22. Given the sequence of terms, *AD CG FK JP*, the next term is

(*a*) *OV* ☐ (*b*) *OW* ☐

(*c*) *PV* ☐ (*d*) *PW*. ☐

(ME/CE/CSE/PI-2012)

23. Choose the grammatically *INCORRECT* sentence

(*a*) He is of Asian origin. ☐ (*b*) They belonged to Africa. ☐

(*c*) She is an European. ☐ (*d*) They migrated from India to Australia. ☐

(ME/CSE/PI-2013)

24. Modern warfare has changed from large scale clashes of armies to suppression of civilian populations. Chemical agents that do their work silently appear to be suited to such warfare; and regretfully, there exist people in military establishments who think that chemical agents are useful tools for their cause.

Which of the following statements best sums up the meaning of the above passage?

(*a*) Modern warfare has resulted in civil strife. ☐

(*b*) Chemical agents are useful in modern warfare. ☐

(*c*) Use of chemical agents in warfare would be undesirable. ☐

(*d*) People in military establishments like to use chemical agents in war. ☐

(ALL BRANCHES-2010)

25. If Log (*P*) = (1/2) Log (*Q*) = (1/3) Log (*R*) then which of the following options is TRUE?

(*a*) $P^2 = Q^2R^2$ ☐ (*b*) $Q^2 = PR$ ☐

(*c*) $Q^2 = R^3P$ ☐ (*d*) $R = P^2Q^2$. ☐

(ME/CE/CSE/PI-2011)

26. Which one of the following options is the closest in meaning to the word given below?

Mitigate

(*a*) Diminish ☐ (*b*) Divulge ☐

(*c*) Dedicate ☐ (*d*) Denote. ☐

(ME/CE/CSE/PI-2012)

27. Choose the most appropriate word from the options given below to complete the following sentence:

If we manage to .... our natural resources, we would leave a better planet for our children.

(*a*) uphold ☐ (*b*) restrain ☐

(*c*) cherish ☐ (*d*) conserve. ☐

(ALL BRANCHES-2010)

28. If $R = \frac{V^2}{54} - \frac{5(V-12)}{V+12}$, then *R* will have an extreme value at

(*a*) $V = 2$ ☐ (*b*) $V = 4$ ☐

(*c*) $V = 6$ ☐ (*d*) $V = 8$. ☐

29. A circular plate expands under the influence of heat so that its radius increases from 5 cm to 5.06 cm. The approximate increase in the area is

(*a*) 18.8 cm^2 ☐ (*b*) $0.06 \times \pi$ cm^2 ☐

(*c*) 1.88 cm^2 ☐ (*d*) none of the above. ☐

ANSWERS

Answers to Objective Type Questions

Mathematics

1. (*d*) 2. (*b*) 3. (*d*) 4. (*d*) 5. (*c*) 6. (*d*)
7. (*c*) 8. (*b*) 9. (*b*) 10. (*d*) 11. (*b*) 12. (*b*)
13. (*a*) 14. (*b*) 15. (*c*) 16. (*b*) 17. (*a*) 18. (*d*)
19. (*c*) 20. (*a*) 21. (*c*) 22. (*d*) 23. (*c*) 24. (*c*)
25. (*c*) 26. (*a*) 27. (*d*) 28. (*c*) 29. (*b*) 30. (*d*)
31. (*a*) 32. (*b*) 33. (*c*) 34. (*b*) 35. (*c*) 36. (*a*)
37. (*c*) 38. (*b*) 39. (*c*) 40. (*d*) 41. (*a*) 42. (*b*)
43. (*c*) 44. (*a*) 45. (*c*) 46. (*b*) 47. (*a*) 48. (*b*)
49. (*c*) 50. (*a*) 51. (*c*) 52. (*d*) 53. (*b*) 54. (*d*)
55. (*a*) 56. (*c*) 57. (*a*) 58. (*c*) 59. (*a*) 60. (*a*)
61. (*a*) 62. (*c*) 63. (*b*) 64. (*b*) 65. (*a*) 66. (*c*)
67. (*b*) 68. (*b*) 69. (*a*) 70. (*b*) 71. (*b*) 72. (*a*)
73. (*c*) 74. (*a*) 75. (*b*) 76. (*d*) 77. (*a*) 78. (*c*)
79. (*b*) 80. (*b*) 81. (*a*) 82. (*c*) 83. (*a*) 84. (*c*)
85. (*a*) 86. (*b*) 87. (*a*) 88. (*b*) 89. (*c*) 90. (*d*)
91. (*c*) 92. (*d*) 93. (*b*) 94. (*c*) 95. (*c*).

Chemistry

96. (*a*) 97. (*d*) 98. (*c*) 99. (*e*) 100. (*c*) 101. (*d*)
102. (*a*) 103. (*b*) 104. (*d*) 105. (*b*) 106. (*e*) 107. (*d*)
108. (*d*) 109. (*e*) 110. (*d*) 111. (*c*) 112. (*c*) 113. (*a*)
114. (*d*) 115. (*b*) 116. (*d*) 117. (*d*) 118. (*c*) 119. (*e*)
120. (*a*) 121. (*a*) 122. (*b*) 123. (*d*) 124. (*d*) 125. (*d*)
126. (*a*) 127. (*c*) 128. (*c*) 129. (*a*) 130. (*b*) 131. (*c*)
132. (*b*) 133. (*a*) 134. (*b*) 135. (*d*) 136. (*c*) 137. (*b*)
138. (*c*) 139. (*e*) 140. (*b*) 141. (*b*) 142. (*c*) 143. (*e*)
144. (*b*) 145. (*a*) 146. (*c*) 147. (*e*) 148. (*a*) 149. (*b*)
150. (*c*) 151. (*a*) 152. (*e*) 153. (*e*) 154. (*c*) 155. (*a*)
156. (*b*) 157. (*e*) 158. (*a*) 159. (*b*) 160. (*e*) 161. (*e*)
162. (*c*) 163. (*b*) 164. (*c*) 165. (*e*) 166. (*d*) 167. (*b*)
168. (*b*) 169. (*b*) 170. (*d*) 171. (*d*) 172. (*c*) 173. (*d*)
174. (*c*) 175. (*c*) 176. (*b*) 177. (*a*) 178. (*c*) 179. (*c*)
180. (*b*) 181. (*b*) 182. (*c*) 183. (*a*) 184. (*b*) 185. (*b*)
186. (*a*) 187. (*d*).

Electrical Science

188. (*c*) 189. (*b*) 190. (*b*) 191. (*b*) 192. (*a*) 193. (*c*)
194. (*b*) 195. (*a*) 196. (*c*) 197. (*a*) 198. (*c*) 199. (*b*)
200. (*a*) 201. (*b*) 202. (*c*) 203. (*a*) 204. (*c*) 205. (*b*)
206. (*b*) 207. (*a*) 208. (*c*) 209. (*b*) 210. (*a*) 211. (*b*)
212. (*b*) 213. (*d*) 214. (*a*) 215. (*d*) 216. (*b*) 217. (*b*)

218. (*d*) 219. (*b*) 220. (*d*) 221. (*d*) 222. (*d*) 223. (*c*)
224. (*d*) 225. (*a*) 226. (*c*) 227. (*d*) 228. (*d*) 229. (*a*)
230. (*d*) 231. (*b*) 232. (*a*) 233. (*b*) 234. (*b*) 235. (*b*)
236. (*c*) 237. (*b*) 238. (*c*) 239. (*a*) 240. (*b*) 241. (*d*)
242. (*c*) 243. (*d*) 244. (*b*) 245. (*c*) 246. (*c*) 247. (*c*)
248. (*c*) 249. (*c*) 250. (*c*) 251. (*d*) 252. (*a*) 253. (*a*)
254. (*d*) 255. (*d*) 256. (*b*) 257. (*d*) 258. (*d*) 259. (*c*)
260. (*a*) 261. (*c*) 262. (*b*) 263. (*a*) 264. (*d*) 265. (*a*)
266. (*a*) 267. (*b*) 268. (*c*) 269. (*d*) 270. (*b*) 271. (*b*)
272. (*b*) 273. (*a*) 274. (*a*) 275. (*b*) 276. (*a*) 277. (*a*)
278. (*a*) 279. (*c*) 280. (*a*) 281. (*c*) 282. (*a*) 283. (*c*)
284. (*a*) 285. (*b*) 286. (*a*) 287. (*a*) 288. (*b*) 289. (*c*)
290. (*a*) 291. (*c*) 292. (*c*) 293. (*c*) 294. (*b*) 295. (*c*)
296. (*c*) 297. (*d*) 298. (*a*).

Physics

299. (*b*) 300. (*c*) 301. (*a*) 302. (*b*) 303. (*b*) 304. (*c*)
305. (*c*) 306. (*c*) 307. (*b*) 308. (*b*) 309. (*b*) 310. (*b*)
311. (*b*) 312. (*b*) 313. (*c*) 314. (*a*) 315. (*b*) 316. (*c*)
317. (*b*) 318. (*b*) 319. (*c*) 320. (*b*) 321. (*c*) 322. (*a*)
323. (*c*) 324. (*b*) 325. (*a*) 326. (*a*) 327. (*b*) 328. (*b*)
329. (*b*) 330. (*c*) 331. (*b*) 332. (*b*) 333. (*a*) 334. (*b*)
335. (*c*) 336. (*a*) 337. (*c*) 338. (*c*) 339. (*a*) 340. (*c*)
341. (*a*) 342. (*c*) 343. (*d*) 344. (*b*) 345. (*b*) 346. (*c*)
347. (*a*) 348. (*c*) 349. (*c*) 350. (*b*) 351. (*a*) 352. (*c*)
353. (*b*) 354. (*c*) 355. (*d*) 356. (*a*) 357. (*c*) 358. (*a*)
359. (*b*) 360. (*a*) 361. (*b*) 362. (*b*) 363. (*c*) 364. (*a*)
365. (*c*) 366. (*b*) 367. (*c*) 368. (*b*) 369. (*c*) 370. (*c*)
371. (*b*) 372. (*b*) 373. (*c*) 374. (*c*) 375. (*c*) 376. (*a*)
377. (*b*) 378. (*b*) 379. (*a*) 380. (*c*) 381. (*c*) 382. (*c*)
383. (*b*) 384. (*c*) 385. (*a*) 386. (*b*) 387. (*d*) 388. (*b*)
389. (*b*) 390. (*c*) 391. (*c*) 392. (*c*) 393. (*b*) 394. (*a*)
395. (*b*) 396. (*c*) 397. (*c*) 398. (*a*) 399. (*c*) 400. (*b*)
401. (*d*) 402. (*b*) 403. (*b*) 404. (*c*) 405. (*b*) 406. (*b*)
407. (*b*) 408. (*b*) 409. (*c*).

Answers to Objective Type Questions from Competitive Examinations

1. (*d*) 2. (*b*) 3. (*c*) 4. (*c*) 5. (*d*) 6. (*d*)
7. (*d*) 8. (*b*) 9. (*b*) 10. (*d*) 11. (*c*) 12. (*b*)
13. (*a*) 14. (*b*) 15. (*d*) 16. (*a*) 17. (*c*) 18. (*b*)
19. (*a*) 20. (*d*) 21. (*c*) 22. (*a*) 23. (*c*) 24. (*c*)
25. (*b*) 26. (*a*) 27. (*d*) 28. (*c*) 29. (*c*).

Chapter 15 TYPICAL OBJECTIVE TYPE TEST PAPERS

I. TEST PAPER NO. 1

Time : 3 hours *Max. Marks : 100*

Tick mark the most appropriate statement of the multiple choice answers :

1. The pressure, dry bulb temperature and relative humidity of air in a room are 1 bar, 30°C and 70%, respectively. If the saturated steam pressure at 30°C is 4.25 kPa, the specific humidity of the room air in kg water vapour/kg dry air is

(*a*) 0.0083 ☐ (*b*) 0.0101 ☐
(*c*) 0.0191 ☐ (*d*) 0.0232 ☐

(GATE-ME-2013)

2. A turbo-charged four-stroke direct injection diesel engine has a displacement volume of 0.0259 m^3 (25.9 litres). The engine has an output of 950 kW at 2200 rpm. The mean effective pressure (in MPa) is closest to

(*a*) 2 ☐ (*b*) 1 ☐
(*c*) 0.2 ☐ (*d*) 0.1 ☐

(GATE-ME-2010)

3. A frictionless piston-cylinder device contains a gas initially at 0.8 MPa and 0.015 m^3. It expands quasi-statically at constant temperature to a final volume of 0.030 m^3. The work output (in kJ) during this process will be

(*a*) 8.32 ☐ (*b*) 12.00 ☐
(*c*) 554.67 ☐ (*d*) 8320.00 ☐

(GATE-ME-2009)

4. A pump handling a liquid raises its pressure from 1 bar to 30 bar. Take the density of the liquid as 990 kg/m^3. The isentropic specific work done by the pump in kJ/kg is

(*a*) 0.10 ☐ (*b*) 0.30 ☐
(*c*) 2.50 ☐ (*d*) 2.93 ☐

(GATE-ME-2011)

5. The specific heats of an ideal gas depend on its
(*a*) temperature ☐ (*b*) pressure ☐
(*c*) volume ☐ (*d*) molecular weight and structure ☐
(GATE-ME-1996)

6. A small copper ball of 5 mm diameter at 500 K is dropped into an oil bath whose temperature is 300 K. The thermal conductivity of copper is 400 W/m. K, its density 9000 kg/m^3 and its specific heat 385 J/kg. K. If the heat transfer coefficient is 250 W/m^2.K and lumped analysis is assumed to be valid, the rate of fall of the temperature of the ball at the beginning of cooling will be, in K/s.
(*a*) 8.7 ☐ (*b*) 13.9 ☐
(*c*) 17.3 ☐ (*d*) 27.7 ☐
(GATE-ME-2005)

7. The law, which gives the basis for measuring the thermodynamic property known as temperature, is called
(*a*) first law of thermodynamics ☐ (*b*) second law of thermodynamics ☐
(*c*) third law of thermodynamics ☐ (*d*) zeroth law of thermodynamics. ☐

8. A two dimensional fluid element rotates like a rigid body. At a point with in the element, the pressure is 1 unit. Radius of the Mohr's circle, charactizing the state of stress at the point, is
(*a*) 0.5 unit ☐ (*b*) 0 unit ☐
(*c*) 1 unit ☐ (*d*) 2 units ☐
(GATE-ME-2008)

9. A cyclic heat engine does 50 kJ of work per cycle. If the efficiency of the heat engine is 75%.The heat rejected per cycle is
(*a*) $16\frac{2}{3}$ kJ ☐ (*b*) $33\frac{1}{3}$ kJ ☐
(*c*) $37\frac{1}{2}$ kJ ☐ (*d*) $66\frac{2}{3}$ kJ ☐
(GATE-ME-2001)

10. A perfect gas is heated at constant pressure. The final volume of the gas becomes 1.5 times the initial volume. If its initial temperature is 30°C, the final temperature will be
(*a*) 45°C ☐ (*b*) 100°C ☐
(*c*) 177°C ☐ (*d*) 330°C. ☐

11. A heat engine is supplied with 300 kcal/sec of heat at constant fixed temperature of 250°C. If 200 kcal/sec are rejected at 10°C, the cycle is
(*a*) reversible ☐ (*b*) irreversible ☐
(*c*) impossible ☐ (*d*) none of the above. ☐

12. The efficiency of a Carnot cycle may be increased by
(*a*) increasing the highest temperature ☐ (*b*) decreasing the highest temperature ☐
(*c*) increasing the lowest temperature ☐ (*d*) both (*a*) and (*b*). ☐

13. The gas constant (*R*) is equal to
(*a*) ratio of two specific heats ☐ (*b*) sum of two specific heats ☐
(*c*) difference of two specific heats ☐ (*d*) product of two specific heats. ☐

14. Air contains by weight

(*a*) 23 parts O_2 and 77 parts N_2 ☐ (*b*) 21 parts O_2 and 79 parts N_2 ☐

(*c*) 77 parts O_2 and 23 parts N_2 ☐ (*d*) 79 parts O_2 and 21 parts N_2 ☐

15. Which one is a spark ignition engine?

(*a*) petrol engine ☐ (*b*) diesel engine ☐

(*c*) steam engine ☐ (*d*) none of the above. ☐

16. The thermal efficiency of a two-stroke cycle engine as compared to four-stroke cycle engine is

(*a*) more ☐ (*b*) equal ☐

(*c*) less ☐ (*d*) none of the above. ☐

17. The process in which the molecules of a compound becomes larger is known as

(*a*) supercharging ☐ (*b*) scavenging ☐

(*c*) polymerisation ☐ (*d*) detonation. ☐

18. Morse test is used to determine

(*a*) indicated horse power for multi-cylinder engines ☐

(*b*) shaft horse power ☐

(*c*) mean effective pressure ☐

(*d*) temperature of the exhaust gases. ☐

19. Air fuel ratio theoretically for a petrol engine is approximately

(*a*) 25 : 1 ☐ (*b*) 20 : 1 ☐

(*c*) 10 : 1 ☐ (*d*) 15 : 1. ☐

20. The mechanism used for controlling air-fuel ratio in petrol engine is known as

(*a*) injector ☐ (*b*) governor ☐

(*c*) carburettor ☐ (*d*) none of the above. ☐

21. Advancing the spark timing in spark ignition, the possibility of knock will

(*a*) increase ☐ (*b*) decrease ☐

(*c*) not take place ☐ (*d*) none of the above. ☐

22. If a fuel has higher self-ignition temperature, the tendency for detonation would

(*a*) be more ☐ (*b*) be less ☐

(*c*) not take place ☐ (*d*) none of the above. ☐

23. The octane number of iso-octane is

(*a*) 40 ☐ (*b*) 30 ☐

(*c*) 10 ☐ (*d*) 100. ☐

24. For the diesel engine, the type of governing used is

(*a*) quality governing ☐ (*b*) hit and miss governing ☐

(*c*) quantity governing ☐ (*d*) none of the above. ☐

25. In an I.C. engine, firing order depends upon

(*a*) crank shaft design ☐ (*b*) arrangement of cylinder ☐

(*c*) number of cylinders ☐ (*d*) all of the above. ☐
(*e*) none of the above. ☐

26. Knocking in the S.I. engine can be prevented if the end gas has
(*a*) high density ☐ (*b*) high temperature ☐
(*c*) long ignition delay ☐ (*d*) none of the above. ☐

27. The energy produced by 4.5 tonnes of high grade coal is equivalent to the energy produced by
(*a*) one kg of uranium ☐ (*b*) one gram of uranium ☐
(*c*) 100 gram of uranium ☐ (*d*) 10 gram of uranium. ☐

28. The isotope of the uranium, which is mostly used is
(*a*) U_{235} ☐ (*b*) U_{234} ☐
(*c*) U_{238} ☐ (*d*) U_{240}. ☐

29. To protect against neutron and gamma rays
(*a*) reflector is used ☐ (*b*) moderator is used ☐
(*c*) shielding is done ☐ (*d*) control rod is used. ☐

30. The production of neutron for a critical reactor is
(*a*) zero ☐
(*b*) infinite ☐
(*c*) equal to the number of neutron lost by leakage ☐
(*d*) none of the above. ☐

31. La-Mont boiler is a
(*a*) low pressure water-tube boiler ☐ (*b*) forced circulation boiler ☐
(*c*) natural circulation boiler ☐ (*d*) none of the above. ☐

32. Consider two infinitely long thin concentric tubes of circular cross-section as shown in Fig. 15.1. If D_1 and D_2 are the diameters of the inner and outer tubes respectively, then the view factor F_{22} is given by

(*a*) $\left(\frac{D_2}{D_1}\right) - 1$ ☐

(*b*) Zero ☐

(*c*) $\left(\frac{D_1}{D_2}\right) + 1$ ☐

(*d*) $1 - \left(\frac{D_1}{D_2}\right)$. ☐

1
2

FIGURE 15.1

(ME & PI-GATE-2012)

33. In a counter flow heat exchanger, hot fluid enters at 65°C and cold fluid leaves at 30°C, mass flow rate of the hot fluid is 1 kg/s and that of cold fluid is 2 kg/s. Specific heat of the hot fluid is 10 kJ/kg K and that of cold fluid is 5 kJ/kg K. The LMTD for the heat exchanger is
(*a*) 15 ☐ (*b*) 30 ☐
(*c*) 35 ☐ (*d*) 45 ☐

(ME-GATE-2007)

34. The pressure ratio of a gas power plant cycle corresponding to maximum work output for the given temperature limits of T_{min} and T_{max} will be

(a) $\left(\frac{T_{max}}{T_{min}}\right)^{\frac{\gamma}{2(\gamma-1)}}$ ☐ (b) $\left(\frac{T_{min}}{T_{max}}\right)^{\frac{\gamma}{2(\gamma-1)}}$ ☐

(c) $\left(\frac{T_{max}}{T_{min}}\right)^{\frac{\gamma-1}{\gamma}}$ ☐ (d) $\left(\frac{T_{min}}{T_{max}}\right)^{\frac{\gamma-1}{\gamma}}$ ☐

(GATE-ME-2004)

35. At the critical point, the latent heat of vaporisation is

(a) maximum ☐ (b) minimum ☐

(c) zero ☐ (d) none of the above. ☐

36. The diagram factor is always

(a) more than 1.0 ☐ (b) equal to 1.0 ☐

(c) equal to zero ☐ (d) less than 1.0. ☐

37. The ratio of clearance volume of the swept volume is known as

(a) expansion ratio ☐ (b) cut off ratio ☐

(c) compression ratio ☐ (d) clearance ratio. ☐

38. Willan's line is a straight line graph between the rate of steam consumption and

(a) pressure of steam ☐ (b) temperature of steam ☐

(c) indicated horse power ☐ (d) none of the above. ☐

39. By compounding the steam engines, condensation is

(a) increased ☐ (b) reduced ☐

(c) constant ☐ (d) none of the above. ☐

40. Missing quantity is due to

(a) condensation of steam ☐ (b) leakage past the piston ☐

(c) both (a) and (b) ☐ (d) none of the above. ☐

41. A steam nozzle is a device used fro converting

(a) heat energy of steam into pressure energy ☐

(b) heat energy of steam into kinetic energy ☐

(c) pressure energy of steam into kinetic energy ☐

(d) pressure energy of steam into heat energy. ☐

42. If the exit pressure is less than the critical pressure, the nozzle used should be

(a) divergent ☐ (b) convergent ☐

(c) convergent-divergent ☐ (d) none of the above. ☐

43. When a nozzle operates with the maximum mass flow, it is said to be

(a) choked ☐ (b) under-expanding ☐

(c) over-expanding ☐ (d) none of the above. ☐

44. In case of reaction turbines, the heat drop takes place on

(a) fixed blade only ☐ (b) moving blade only ☐

(c) both fixed and moving blades ☐ (d) none of the above. ☐

45. For Parson's reaction turbine, degree of reaction is equal to

(a) 80% ☐ (b) 50% ☐

(c) 75% ☐ (d) 100%. ☐

46. The degree of reaction, for a turbine in which heat drop in moving blade is 8 kcal/kg and in fixed blade 12 kcal/kg, would be

(a) 80% ☐ (b) 60% ☐

(c) 40% ☐ (d) 20% ☐

47. The value of reheat factor varies normally between

(a) 1.0 to 1.5 ☐ (b) 1.5 to 2.0 ☐

(c) 1.02 to 1.06 ☐ (d) none of the above. ☐

48. The compressor capacity is equal to

(a) volume of air sucked ☐ (b) volume of air delivered ☐

(c) sum of (a) and (b) ☐ (d) none of the above. ☐

49. With the increase of pressure ratio, the volumetric efficiency of an air compressor

(a) increases ☐ (b) decrease ☐

(c) constant ☐ (d) first increases then decreases. ☐

50. For minimum work required, the intercooler pressure for a two-stage reciprocating air compressor, is given by

(a) $p_2 = \sqrt{p_1 p_3}$ ☐ (b) $p_2 = p_2 \times p_3$ ☐

(c) $p_2 = \dfrac{1}{\sqrt{p_1 \times p_3}}$ ☐ (d) $p_2 = \dfrac{1}{p_1 \times p_3}$. ☐

51. The compression ratio, for maximum efficiency for a three-stage compressor, is equal to

(a) square of the ratio of the final pressure to initial pressure ☐

(b) the ratio of final pressure to initial pressure ☐

(c) square root of the ratio of final pressure to initial pressure ☐

(d) cube root of the ratio of the final pressure to initial pressure. ☐

52. For the least work to be done on a compressor, the compression should be

(a) adiabatic ☐ (b) isothermal ☐

(c) polytropic ☐ (d) none of the above. ☐

53. At high altitudes, the efficiency of a screw propeller as compared to jet engine is

(a) more ☐ (b) less ☐

(c) same ☐ (d) none of the above. ☐

54. The basic law of heat conduction is called

(a) Newton's law of cooling ☐ (b) Fourier's law ☐

(c) Kirchhoff's law ☐ (d) Stefan's law. ☐

55. The co-efficient of thermal conductivity is defined as the heat flow per unit time

(a) through unit thickness ☐

(b) when temperature difference of unity is maintained between opposite faces ☐

(c) when temperature gradient is unity ☐

(d) across unit area when temperature gradient is unity. ☐

56. The rate of heat transfer through a hollow cylinder of inner and outer radii r_1 and r_2 respectively, depends on

(a) difference of radii $(r_2 - r_1)$ ☐ (b) ratio of (r_2/r_1) ☐
(c) product of r_2 and r_1 ☐ (d) sum of r_2 and r_1. ☐

57. Critical radius of a hollow cylinder is defined as

(a) inner radius which would give maximum heat flow ☐
(b) outer radius which would give minimum heat flow ☐
(c) outer radius which would give maximum heat flow ☐
(d) none of the above. ☐

58. In case of heat exchanger, the value of arithmetic mean temperature difference should be

(a) as small as possible ☐ (b) as large as possible ☐
(c) constant ☐ (d) none of the above. ☐

59. The term $(\mu \times C_p)/k$ is called

(a) Reynold number ☐ (b) Nusselt number ☐
(c) Prandtl number ☐ (d) Froude number ☐

where μ = Co-efficient of viscosity, C_p = Specific heat at constant pressure, and
k = Thermal conductivity.

60. Critical temperature is the temperature above which

(a) a gas will immediately liquify ☐ (b) a gas will never liquify ☐
(c) water will evaporate ☐ (d) none of the above. ☐

61. The process of adding heat of moist air at the same humidity ratio is known as

(a) sensible heating ☐ (b) sensible cooling ☐
(c) humidification ☐ (d) dehumidification. ☐

62. The viscosity (μ) has the dimensions as

(a) MLT^2 ☐ (b) $ML^{-1}T^{-1}$ ☐
(c) $ML^{-1}T^{-2}$ ☐ (d) $(MLT)^{-1}$ ☐

63. The hydrostatic pressure on an inclined plane surface is equal to

(a) $wA\bar{h}$ ☐ (b) $wA\bar{h}\ \sin^2\theta$ ☐
(c) $\frac{1}{2}wA\bar{h}$ ☐ (d) $wA\bar{h}\ \sin\theta$ ☐

where A = Area of plane surface,

$\bar{h}$ = Depth of centroid of the plane area below the liquid free surface, and

θ = Angle made by the inclined surface with free surface of liquid.

64. The resultant hydrostatic force acts through a point which is known as

(a) centre of gravity ☐ (b) centre of buoyancy ☐
(c) centre of pressure ☐ (d) none of the above. ☐

65. Pitot-tube is used for measurement of

(*a*) pressure ☐ (*b*) density ☐

(*c*) velocity at a point ☐ (*d*) rate of flow. ☐

66. The velocity distribution in laminar flow through a circular pipe follows

(*a*) parabolic law ☐ (*b*) linear law ☐

(*c*) logarithmic law ☐ (*d*) none of the above. ☐

67. Reynold's number is defined as the ratio of

(*a*) inertia force to gravity force ☐ (*b*) viscous force to gravity force ☐

(*c*) viscous force to elastic force ☐ (*d*) inertia force to viscous force. ☐

68. In sonic flow, the disturbances created by a projectile moves

(*a*) along the projectile ☐ (*b*) ahead of the projectile ☐

(*c*) behind the projectile ☐ (*d*) none of the above. ☐

69. For super-sonic flow, if the area of flow increases

(*a*) velocity decreases ☐ (*b*) velocity increases ☐

(*c*) velocity is constant ☐ (*d*) none of the above. ☐

70. Specific speed of a turbine is defined as the speed of the turbine which

(*a*) produces unit horse power at unit head ☐ (*b*) produces unit horse power at unit discharge ☐

(*c*) delivers unit discharge at unit head ☐ (*d*) delivers unit discharge at unit power. ☐

71. A Pelton turbine is

(*a*) a reaction turbine ☐ (*b*) a radial flow turbine ☐

(*c*) an impulse turbine ☐ (*d*) none of the above. ☐

72. Operating characteristic curves of a turbine means

(*a*) curves drawn at constant speed ☐ (*b*) curves drawn at constant efficiency ☐

(*c*) curves drawn at constant head ☐ (*d*) none of the above. ☐

73. Cavitation will take place if the pressure of the flowing fluid at any point is

(*a*) more than vapour pressure of the fluid ☐ (*b*) equal to vapour pressure of the fluid ☐

(*c*) is less than vapour pressure of the fluid ☐ (*d*) none of the above. ☐

74. For water, the limiting value of separation pressure head is

(*a*) 2.5 m (abs.) ☐ (*b*) 7.5 m (abs.) ☐

(*c*) 10.3 m (abs.) ☐ (*d*) 5.0 m (abs.). ☐

75. Rotameter is used for measuring

(*a*) density ☐ (*b*) velocity ☐

(*c*) discharge ☐ (*d*) viscosity. ☐

76. The number of members (n) and number of joints (j) in a perfect frame is given by

(*a*) $n = 3j - 2$ ☐ (*b*) $n = 2j - 3$ ☐

(*c*) $j = 2n - 3$ ☐ (*d*) $j = 3n - 2$. ☐

77. A body is moving with a velocity of 10 m/sec. The time required, to stop the body within a distance of 5 m, is equal to

(a) 3 sec ☐ (b) 5 sec ☐

(c) 1 sec ☐ (d) 0.5 sec. ☐

78. The angle of projection for maximum range of a projectile is

(a) 90° ☐ (b) 30° ☐

(c) 60° ☐ (d) 45°. ☐

79. The relation between co-efficient of friction (μ) and angle of friction (φ) is given by

(a) $\phi = \tan \mu$ ☐ (b) $\mu = \sin \phi$ ☐

(c) $\mu = \tan \phi$ ☐ (d) $\mu = \dfrac{1}{\tan \phi}$. ☐

80. The moment of inertia of a triangular section about an axis passing through its base is given by

(a) $\dfrac{bh^3}{12}$ ☐ (b) $\dfrac{bh^3}{32}$ ☐

(c) $\dfrac{bh^3}{36}$ ☐ (d) $\dfrac{bh^3}{48}$. ☐

where b = Width at a base, and h = Height of the triangle.

81. The angle of super-elevation (or banking) of roads is the process of

(a) raising the inner edge of the roads above the outer edge ☐

(b) raising the outer edge of the roads above the inner edge ☐

(c) keeping both the edges at the same level ☐

(d) none of the above. ☐

82. The time period of a simple pendulum will be doubled if

(a) its length is doubled ☐ (b) it length is halved ☐

(c) its length is increased four times ☐ (d) its length is increased eight times. ☐

83. The ratio of lateral strain to longitudinal strain is called

(a) Poisson's ratio ☐ (b) bulk modulus ☐

(c) modulus of rigidity ☐ (d) modulus of elasticity. ☐

84. The elongation produced in a rod (by its own weight) of length (l) and diameter (d) rigidity fixed at the upper end and hanging, is equal to

(a) $\dfrac{wl}{2E}$ ☐ (b) $\dfrac{wl^2}{2E}$ ☐

(c) $\dfrac{wl^3}{2E}$ ☐ (d) $\dfrac{wl^4}{2E}$. ☐

where w = Weight per unit volume of the rod, and E = Modulus of elasticity.

85. The ratio of modulus of rigidity to bulk modulus for a Poisson's ratio of 0.25 would be

(a) 1/3 ☐ (b) 2/3 ☐

(c) 3/5 ☐ (d) 1.0. ☐

86. Principal plane is a plane on which shear stress is

(*a*) maximum ☐ (*b*) minimum ☐

(*c*) average of maximum and minimum ☐ (*d*) zero. ☐

87. A body is subjected to a direct tensile stress of 300 kg/cm^2 in one plane accompanied by a simple shear stress of 200 kg/cm^2

(*a*) 250 kg/cm^2 ☐ (*b*) 400 kg/cm^2 ☐

(*c*) – 100 kg/cm^2 ☐ (*d*) 300 kg/cm^2. ☐

88. If a beam is fixed at both its ends, it is called a

(*a*) fixed beam ☐ (*b*) built in beam ☐

(*c*) encastered beam ☐ (*d*) any one of the above. ☐

(*e*) none of the above. ☐

89. The point of contra-flexure occurs only in

(*a*) continuous beams ☐ (*b*) cantilever beams ☐

(*c*) overhanging beams ☐ (*d*) simply supported beam ☐

(*e*) all of the above ☐ (*f*) none of the above. ☐

90. The shear force and bending moments are zero at the free end of a cantilever, if it carries a

(*a*) point load at the free end ☐

(*b*) uniformly distributed load over the whole length ☐

(*c*) point load in the middle of its length ☐

(*d*) none of the above. ☐

91. A simply supported beam carries a uniformly distributed load oven the whole span. The deflection at the centre is '*y*'. If the distributed load per unit length is doubled and also depth of the beam is doubled, then deflection at the centre would be

(*a*) $2y$ ☐ (*b*) $4y$ ☐

(*c*) $\frac{y}{2}$ ☐ (*d*) $\frac{y}{4}$. ☐

92. In a concrete beam, the steel bars are embedded

(*a*) near top section ☐ (*b*) near bottom section ☐

(*c*) in the centre ☐ (*d*) anywhere. ☐

93. Two shafts, one solid and the other hollow, are made of the same materials and are having same length and weight. The hollow shaft as compared to solid shaft is

(*a*) more strong ☐ (*b*) less strong ☐

(*c*) having same strength ☐ (*d*) none of the above. ☐

94. A coil is having stiffness *k*. It is cut into two halves, then the stiffness of the cut coils will be

(*a*) same ☐ (*b*) half ☐

(*c*) double ☐ (*d*) one-fourth. ☐

95. In case of thick cylinders, the difference of circumferential stress and radial stress

(*a*) is constant ☐ (*b*) varies directly with radius ☐

(*c*) varies inversely with radius ☐ (*d*) none of the above. ☐

96. The equivalent length is equal to actual length of a column with

(*a*) one end fixed and other end free ☐ (*b*) both ends fixed ☐

(*c*) one end fixed and other end hinged ☐ (*d*) both ends hinged. ☐

97. In a simply supported beam, the temperature variation produces

(*a*) large stresses ☐ (*b*) small stresses ☐

(*c*) zero stress ☐ (*d*) none of the above. ☐

98. Which one of the following is a lower pair?

(*a*) ball and roller bearing ☐ (*b*) automobile steering gear ☐

(*c*) cam and follower ☐ (*d*) belt and chain drives. ☐

99. If '*n*' links are connected at the same joint, the joint is equivalent to

(*a*) $(n-1)$ binary joints ☐ (*b*) $(2n-1)$ binary joints ☐

(*c*) $(n-2)$ binary joints ☐ (*d*) $(n-3)$ binary joints. ☐

100. The total number instantaneous centres for a mechanism having 8 links, equal to

(*a*) 8 ☐ (*b*) 28 ☐

(*c*) 4 ☐ (*d*) 7. ☐

101. For a rigid link, the velocity of one end of the link relative to other end will be

(*a*) at 45° to the link ☐ (*b*) at right angles of the link ☐

(*c*) parallel to the link ☐ (*d*) none of the above. ☐

102. Klein's construction is a graphical method of determining

(*a*) acceleration of various parts ☐ (*b*) velocity of various parts ☐

(*c*) displacement of various parts ☐ (*d*) all of the above ☐

(*e*) none of the above. ☐

103. The Coriolis component of acceleration exists only whenever a point

(*a*) moves along a circular path ☐

(*b*) moves in a straight line ☐

(*c*) moves along a straight line which has rotational motion ☐

(*d*) none of the above. ☐

104. The maximum efficiency of a screw jack is a function of

(*a*) helix angle ☐ (*b*) angle of friction ☐

(*c*) load lifted ☐ (*d*) effort applied. ☐

105. When maximum power is transmitted by a belt, the centrifugal tension in the belt is

(*a*) half the maximum tension allowed in the belt ☐

(*b*) one-third maximum tension allowed in the belt ☐

(*c*) two-third maximum tension allowed in the belt ☐

(*d*) one-fourth maximum tension allowed in the belt. ☐

106. The product of the circular pitch and diametral pitch is equal to

(*a*) 2π ☐ (*b*) π ☐

(*c*) $\frac{\pi}{2.0}$ ☐ (*d*) 1.0. ☐

107. The locus of a point on the circumference of a circle, which rolls without slipping on the inside of another circle, is known as

(*a*) involute ☐ (*b*) cycloid ☐

(*c*) hypo-cycloid ☐ (*d*) epicycloid. ☐

108. The pressure angle of a cam

(*a*) decreases if the radius of base circle decrease ☐

(*b*) decreases of the radius of base circle increases ☐

(*c*) is independent of the radius of the base circle ☐

(*d*) first decreases then increases if the radius of the base circle decreases. ☐

109. When the speed of governor increases

(*a*) height of governor and radius of rotation increases ☐

(*b*) height of governor and radius of rotation decreases ☐

(*c*) height of governor decreases but radius of rotation increases ☐

(*d*) height of governor increases but radius of rotation decreases. ☐

110. The controlling force curve is a graphed between controlling force and

(*a*) speed of rotation ☐ (*b*) radius of rotation ☐

(*c*) range of speed ☐ (*d*) lift of governor. ☐

111. The swaying couple and variation in the tractive effort is due to

(*a*) primary disturbing force ☐ (*b*) secondary disturbing force ☐

(*c*) both (*a*) and (*b*) ☐ (*d*) none of the above. ☐

112. Natural frequency of a system is due to

(*a*) free vibration ☐ (*b*) forced vibration ☐

(*c*) resonance ☐ (*d*) damping. ☐

113. At certain speed, the rotating shaft tends to vibrate violently in transverse direction. This speed is called

(*a*) whirling speed ☐ (*b*) critical speed ☐

(*c*) whipping speed ☐ (*d*) all of the above ☐

(*e*) none of the above. ☐

114. A system is said to be critically damped if the damping factor for a vibrating system, is

(*a*) more than one ☐ (*b*) equal to one ☐

(*c*) less than one ☐ (*d*) equal to none. ☐

115. Stress concentration factor is a function of

(*a*) geometry of the machine component ☐ (*b*) material of the machine component ☐

(*c*) both (*a*) and (*b*) ☐ (*d*) none of the above. ☐

116. A shaft is designed on the basis of

(*a*) strength ☐ (*b*) rigidity ☐

(*c*) strength and rigidity ☐ (*d*) none of the above. ☐

117. Eye bolts are used for

(*a*) transmission of power ☐ (*b*) locking device ☐

(*c*) lifting and transporting heavy pieces ☐ (*d*) absorption of shocks and vibrations. ☐

118. The distance between the centre of the rivet hole to the nearest edge of the plate is known as

(*a*) pitch ☐ (*b*) diagonal pitch ☐

(*c*) margin ☐ (*d*) back pitch. ☐

119. The strength of a riveted joint is equal to

(*a*) the pull required to shear oft the rivet ☐ (*b*) the pull required to crush the rivet ☐
(*c*) the pull required to tear off the rivet ☐ (*d*) minimum of the above three values ☐
(*e*) maximum of the above three values. ☐

120. The power transmitted from the shaft by the key will be 100% if the shearing strength of the key is

(*a*) hall the torsional shear strength of the shaft ☐
(*b*) 80% of the torsional shear strength of the shaft ☐
(*c*) 90% of the torsional shear strength of the shaft ☐
(*d*) equal to the torsional shear strength of the shaft. ☐

121. Cotter joints is used to connect two co-axial rods which are subjected to

(*a*) bending ☐ (*b*) twisting ☐
(*c*) axial loading ☐ (*d*) all of the above ☐
(*e*) none of the above. ☐

122. The ratio of the mean diameter of the coil to the diameter of the wire (*i.e.* D/d) is known as

(*a*) solid length of spring ☐ (*b*) free length of spring ☐
(*c*) spring index ☐ (*d*) spring rate. ☐

123. The cracks due to fatigue spreads only by

(*a*) tensile stress and in direction along the tensile stress ☐
(*b*) compressive stress and in direction along the compressive stress ☐
(*c*) shear stress ☐
(*d*) tensile stress and in direction perpendicular to the tensile stress. ☐

124. The bearing number 406 means that the bearing is of

(*a*) heavy series whose bore is 6 mm ☐ (*b*) heavy series whose bore is 30 mm ☐
(*c*) light series whose bore is 30 mm ☐ (*d*) medium series whose bore is 30 mm ☐

125. The Lewis form factor (y) depends upon

(*a*) size of the tooth ☐ (*b*) number of teeth a gear ☐
(*c*) pressure angle ☐ (*d*) all of the above ☐
(*e*) only (*b*) and (*c*). ☐

126. The materials used for gears to run quitely at high speed, should be

(*a*) non-ferrous ☐ (*b*) harder steel ☐
(*c*) stainless steel ☐ (*d*) non-metallic. ☐

127. The electron theory of metal is used in understanding the concept of

(*a*) energy levels in metals ☐ (*b*) cohesive and repulsive forces in metals ☐
(*c*) binding in solids ☐ (*d*) behaviour of conductors and insulators ☐
(*e*) none of the above. ☐

128. The oxides, nitrides, carbides and silicate of metals are known as

(*a*) organic materials ☐ (*b*) ceramic materials ☐
(*c*) ferrous materials ☐ (*d*) non-ferrous materials. ☐

129. Diamagnetic materials are those materials for which the value of susceptibility (*k*) is

(*a*) large and positive ☐ (*b*) large and negative ☐

(*c*) small and positive ☐ (*d*) small and negative. ☐

130. In the mechanism given below, if the angular velocity of the eccentric circular disc is 1 rad/s, the angular velocity (rad/s) of the follower link for the instant shown in Fig. 15.2 is

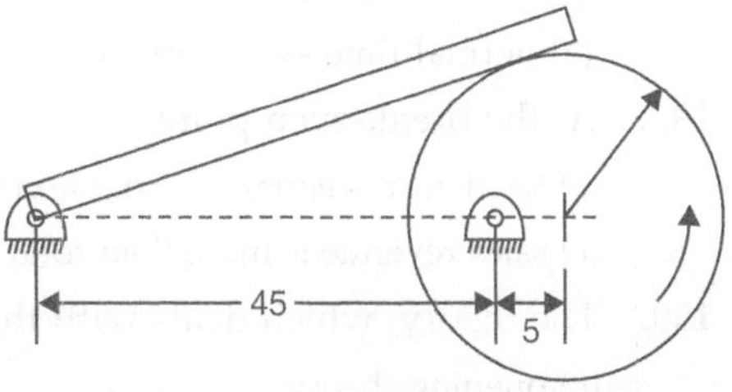

(*a*) 0.05 ☐

(*b*) 0.1 ☐

(*c*) 5.0 ☐

(*d*) 10.0 ☐

45
5

FIGURE 15.2

Note : All dimensions are in mm

(ME-GATE-2012)

131. If the metals are ductile and cutting speed is low, then

(*a*) continuous chips are formed ☐

(*b*) discontinuous chips are formed ☐

(*c*) continuous chips with built-up edge are formed ☐

(*d*) none of the above. ☐

132. EDM stands for

(*a*) energy discharge method ☐ (*b*) electro-discharge machining ☐

(*c*) energy direct method ☐ (*d*) efficient direct method. ☐

133. The relationship between tool life (*T*) and cutting speed (*V*) is expressed as

(*a*) $V^nT = C$ ☐ (*b*) $\frac{V}{T} = C$ ☐

(*c*) $VT^n = C$ ☐ (*d*) $\frac{T}{V} = C.$ ☐

where *n* and *C* are constants.

134. A milling cutter having 18 teeth is operate at a cutting speed of 22 m/min. The diameter of the cutter is 10 cm. If the feed is 0.20 mm per tooth per revolution, then r.p.m. of the cutter would be

(*a*) 100 r.p.m. ☐ (*b*) 80 r.p.m. ☐

(*c*) 70 r.p.m. ☐ (*d*) 50 r.p.m. ☐

135. For large castings, the sand grains should be?

(*a*) fine ☐ (*b*) medium ☐

(*c*) coarse ☐ (*d*) rounded. ☐

136. Which of the following are casting defects?

(*a*) blow holes ☐ (*b*) cold cracks ☐

(*c*) sand spots ☐ (*d*) scabs ☐

(*e*) all of the above ☐ (*f*) none of the above. ☐

137. Standard time is equal to

(*a*) normal time plus idle time ☐

(*b*) normal time plus allowance ☐

(*c*) normal time plus idle time plus allowance ☐

(*d*) normal time—allowances. ☐

138. At the break-even point

(*a*) fixed cost is equal to variable cost ☐ (*b*) sales revenue and total cost are equal ☐

(*c*) sales revenue is more than total cost ☐ (*d*) sales revenue is less than total cost. ☐

139. The theory, which deals with the problem of reducing the waiting time, is known as

(*a*) queuing theory ☐ (*b*) game theory ☐

(*c*) value engineering ☐ (*d*) net work analysis. ☐

140. According to Emerson plan, a worker gets a bonus whenever the efficiency of a worker is more than

(*a*) 50% ☐ (*b*) 55% ☐

(*c*) 60% ☐ (*d*) 66.67%. ☐

141. The period of the function (sin $3x$ + cos $6x$) is

(*a*) $\frac{\pi}{3}$ ☐ (*b*) $\frac{2\pi}{3}$ ☐

(*c*) π ☐ (*d*) none of the above. ☐

142. $\int_0^{\pi} \cos mx \cos nx \, dx$, when $m \neq n$, is

(*a*) π ☐ (*b*) $\frac{\pi}{2}$ ☐

(*c*) zero ☐ (*d*) none of these. ☐

143. An equation $A\frac{\partial^2 z}{\partial x^2} + B\frac{\partial^2 z}{\partial x \partial y} + C\frac{\partial^2 z}{\partial y^2} + f(x, y, z, p, q) = 0$ is said to be elliptic, parabolic or hyperbolic according as

(*a*) $B^2 - 4AC <, = \text{or} > 0$ ☐ (*b*) $B^2 - 4AC >, = \text{or} < 0$ ☐

(*c*) $B^2 - 4AC <, > \text{or} = 0$ ☐ (*d*) none of these. ☐

144. Let $A = \begin{pmatrix} 1 & -2 & 3 \\ -4 & 2 & 5 \end{pmatrix}$ and $B = \begin{pmatrix} 1 & 3 \\ -1 & 0 \\ 2 & 4 \end{pmatrix}$. Then the product

(*a*) *AB* only is conformal ☐ (*b*) *BA* only is conformal ☐

(*c*) Both *AB* and *BA* are conformal ☐ (*d*) none of these. ☐

145. The equation $xdy - ydx = \sqrt{(x^2 + y^2)}\, dx$ is

(*a*) exact ☐ (*b*) linear ☐

(c) homogeneous ☐ (d) none of these. ☐

146. In tossing 10 coins, the probability of having exactly 10 coins is

(a) $\frac{63}{256}$ ☐ (b) $\frac{63}{64}$ ☐

(c) $\frac{63}{128}$ ☐ (d) none of these. ☐

147. The triple integral $\int_0^3 \int_0^2 \int_0^1 (x + y + z)\, dz\, dy\, dx$ is equal to

(a) 18 ☐ (b) 20 ☐

(c) 21 ☐ (d) none of these. ☐

148. The product of matrices is

(a) commutative as well as associative ☐ (b) commutative as well as distributive ☐

(c) associative as well as distributive ☐ (d) none of these. ☐

149. Let $x\frac{dx}{dy} + \cot y = 0$, $y = \frac{\pi}{4}$ when $x = \sqrt{2}$. Then

(a) $y = 2 \cos x$ ☐ (b) $x = 2 \cos y$ ☐

(c) $x = \cos y$ ☐ (d) none of these. ☐

150. The gradient of curve is given by $3x^2 + 2bx + 5$. The curve passes through (0, 3) and (– 1, – 1). The equation of the curve is

(a) $y = x^3 + 2x^2 + 5x + 1$ ☐ (b) $y = x^3 - 2x^2 + 5x + 3$ ☐

(c) $y = x^3 + 2x^2 + 5x + 3$ ☐ (d) none of these. ☐

151. The total area under the normal probability curve is

(a) 1 ☐ (b) 2 ☐

(c) ∞ ☐ (d) none of these. ☐

152. The product of a solid of volume V about x and y-axis is given by

(a) $\int\int\int_V \rho y^2 \, dx\, dy\, dz$ ☐ (b) $\int\int\int_V (x^2 + y^2)\rho \, dx\, dy\, dz$ ☐

(c) $\iiint_V \rho\, xy\, dx\, dy\, dz$ ☐ (d) none of these. ☐

where ρ is the density of the solid at any point (x, y, z) of the solid.

153. In the nth electron level, the number of electron is

(a) $2n^2$ ☐ (b) $n + l$ ☐

(c) n^2 ☐ (d) $3n$ ☐

(e) $2n$. ☐

154. Isobars are produced as the result of the emission of

(a) α-particles ☐ (b) γ-rays ☐

(c) X-rays ☐ (d) β-particles. ☐

155. The energy of an electron in its orbit is

(a) its kinetic energy ☐ (b) sum of potential and kinetic energy ☐

(c) potential energy only ☐ (d) due to rotation ☐

(e) determined from the nucleus. ☐

156. The entropy of the universe

(*a*) tends towards maximum ☐ (*b*) tends towards minimum ☐

(*c*) remains constant ☐ (*d*) tends to be zero. ☐

157. Nitrogen tetraoxide dissociate as $N_2O_4 \rightleftharpoons 2NO_2(g)$. If one mole of N_2O_4 is brought in equilibrium and α is the degree of dissociation of N_2O_4, then the total number of moles of N_2O_4 and NO_2 present is

(*a*) 1 ☐ (*b*) $1 - \alpha$ ☐

(*c*) $1 + \alpha$ ☐ (*d*) 2 ☐

(*e*) $2 - \alpha$. ☐

158. The chemical reaction taking place at the anode is

(*a*) reduction ☐ (*b*) oxidation ☐

(*c*) ionisation ☐ (*d*) dissociation ☐

(*e*) hydrolysis. ☐

159. Hydrogen and deuterium are examples of

(*a*) isotopes ☐ (*b*) isobars ☐

(*c*) isotones ☐ (*d*) isomers. ☐

160. The strength of an acid depends upon

(*a*) its degree of dissociation ☐ (*b*) its basicity ☐

(*c*) its molecular weight ☐ (*d*) the number of H atoms in the molecule. ☐

161. Choose the one which is not used as a moderator of neutrons in controlled chain reactions

(*a*) water ☐ (*b*) heavy water ☐

(*c*) oxygen ☐ (*d*) graphite ☐

(*e*) paraffin. ☐

162. De-Broglie equation is

(*a*) $h/mv = \lambda$ ☐ (*b*) $h\nu = E_1 - E_2$ ☐

(*c*) $n\lambda = 2d \sin \theta$ ☐ (*d*) $c = \lambda\nu$ ☐

(*e*) $E = mc^2$. ☐

163. Ethylene may be produced by the dehydration of

(*a*) CH_3OH ☐ (*b*) C_2H_5OH ☐

(*c*) CH_3COCH_3 ☐ (*d*) CH_3COOH ☐

(*e*) HCOOH ☐ (*f*) CH_3CHO. ☐

164. The general formula of alkenes is

(*a*) C_nH_{2n} ☐ (*b*) C_nH_{2n+2} ☐

(*c*) C_nH_{2n+1} ☐ (*d*) C_nH_{2n-2} ☐

165. A company has an annual demand of 1000 units, ordering cost of ₹ 100 per order and carrying cost of ₹ 100 per unit-year. If the stock-out costs are estimated to be nearly ₹ 400 each time the company runs out-of-stock, then safety stock justified by the carrying cost will be

(*a*) 4 ☐ (*b*) 20 ☐

(*c*) 40 ☐ (*d*) 100. ☐

(GATE-ME-2004)

166. A linear programming problem is shown below.

Maximize: $3x + 7y$

Subject to: $3x + 7y \leq 10$

$4x + 6y \leq 8$

$x, y \geq 0$

It has

(*a*) an unbounded objective function ☐ (*b*) exactly one optimal solution ☐
(*c*) exactly two optimal solutions ☐ (*d*) infinitely many optimal solutions ☐

(GATE-ME-2013)

167. A company produces two types of toys: *P* and *Q*. Production time of *Q* is twice that of *P* and the company has a maximum of 2000 time units per day. The supply of raw material is just sufficient to produce 1500 toys (of any type) per day. Toy type *Q* requires an electric switch which is available @ 600 pieces per day only. The company makes a profit of ₹ 3 and ₹ 5 on type *P* and *Q* respectively. For maximization of profits, the daily production quantities of *P* and *Q* toys should respectively be

(*a*) 100, 500 ☐ (*b*) 500, 1000 ☐
(*c*) 800, 600 ☐ (*d*) 1000, 1000 ☐

(GATE-ME-2004)

168. A political party orders an arch for the entrance to the ground in which the annual convention is being held. The profile of the arch follows the equation $y = 2x - 0.1x^2$ where y is the height of the arch in metres. The maximum possible height of the arch is

(*a*) 8 metres ☐ (*b*) 10 metres ☐
(*c*) 12 metres ☐ (*d*) 14 metres ☐

(ME/CE/CSE/PI-2012)

169. Given digits 2, 2, 3, 3, 3, 4, 4, 4, 4 how many distinct 4 digit numbers greater than 3000 can be formed

(*a*) 50 ☐ (*b*) 51 ☐
(*c*) 52 ☐ (*d*) 54 ☐

(GATE ALL BRANCHES-2010)

170. The demand and forecast for February are 12000 and 10275, respectively. Using single exponential smoothening method (smoothening coefficient = 0.25), forecast for the month of March is

(*a*) 431 ☐ (*b*) 9587 ☐
(*c*) 10706 ☐ (*d*) 11000 ☐

(GATE-ME-2010)

171. When using a simple moving average to forecast demand, one would

(*a*) give equal weight to all demand data ☐
(*b*) assign more weight to the recent demand data ☐
(*c*) include new demand data in the average without discarding the earlier data ☐
(*d*) include new demand data in the average after discarding some of the earlier demand data ☐

(GATE-ME-2001)

172. The dimension of surface tension is

(*a*) N/m^2 ☐ (*b*) J/m ☐
(*c*) J/m^2 ☐ (*d*) W/m ☐

(GATE-PI-1997)

173. The 2-D flow with, velocity $\vec{v} = (x + 2y + 2)\vec{i} + (4 - y)\vec{j}$ is

(*a*) compressible and irrotational ☐ (*b*) compressible and not irrotational ☐
(*c*) incompressible and irrotational ☐ (*d*) incompressible and not irrotational ☐

(GATE- ME-2001)

174. The arm *OA* of an epicyllic gear train shown in Fig. 15.3 revolves counter clockwise about *O* with an angular velocity of 4 rad/s. Both gears are of same size. The angular velocity of gear *C*, if the sun gear *B* is fixed, is

(*a*) 4 rad/s ☐
(*b*) 8 rad/s ☐
(*c*) 10 rad/s ☐
(*d*) 12 rad/s ☐

A
B

FIGURE 15.3

(GATE-ME-1995)

175. The lengths of the links of a 4-bar linkage with revolute pairs only are *p*, *q*, *r* and *s* units. Given that $p < q < r < s$. Which of these links should be the fixed one, for obtaining a "double crank" mechanism?

(*a*) link of length *p* ☐ (*b*) link of length *q* ☐
(*c*) link of length *r* ☐ (*d*) link of length *s*. ☐

(ME-GATE-2003)

176. Consider the following Linear Programming problem (LPP)

Maximize: $Z = 3X_1 + 2X_2$

Subject to: $x_1 \le 4$

$x_2 \le 6$

$3x_1 + 2x_2 \le 18$

$x_1 \ge 0,\ x_2 \ge 0$

(*a*) The LPP has a unique optimal solution ☐ (*b*) The LPP is infeasible ☐
(*c*) The LPP is unbounded ☐ (*d*) The LPP has multiple optimal solutions. ☐

(GATE-ME-2009)

177. The unit of flux density is

(*a*) *AT* ☐ (*b*) weber ☐
(*c*) tesla ☐ (*d*) none of the above. ☐

178. In a since-wave *AC* circuit with a 90 Ω resistance is series with a 90 Ω capacitive reactance, the angle of phase difference between the applied voltage and the current is

(*a*) – 90° ☐ (*b*) – 45° ☐
(*c*) 0° ☐ (*d*) 90°. ☐

179. The equivalent admittance of 500 Ω resistor, 100 mH inductor and 0.05 μF capacitor, all connected in parallel to a 100 kHz source is

(*a*) (2 – *j* 1.28) Ω ☐ (*b*) (2 – *j* 1.28) × 10^3 Ω ☐
(*c*) (2 – *j* 1.28) × 10^{-3} Ω ☐ (*d*) none of the above. ☐

180. The voltage and current across a resistor are 5 V and 2 A respectively. Its resistance and power are

(*a*) 2.5 ohms 10 watts ☐ (*b*) 0.5 ohm, 10 watts ☐
(*c*) 3 ohms, 7 watts ☐ (*d*) 7 ohms, 3 watts. ☐

181. Universal gate is
(*a*) NAND ☐ (*b*) OR ☐
(*c*) AND ☐ (*d*) none of these. ☐

182. Resistance of 100 W, 200 V lamp is
(*a*) 100 ohms ☐ (*b*) 200 ohms ☐
(*c*) 400 ohms ☐ (*d*) 600 ohms. ☐

183. By Boolean algebra $A + AB \neq A + B$
(*a*) yes ☐ (*b*) no ☐
(*c*) none of the above ☐ (*d*) none of the above. ☐

184. The value of a resistor creating thermal noise is doubled. The noise power generated is therefore
(*a*) halved ☐ (*b*) quadrupled ☐
(*c*) doubled ☐ (*d*) unchanged. ☐

185. In a series RLC circuit resistance is doubled, then which of the following statements is true?
(*a*) Time constant of the circuit increases. ☐
(*b*) Damping ratio of the circuit decreases. ☐
(*c*) Natural frequency of the circuit increases. ☐
(*d*) None of the above. ☐

186. The signals sent by the TV transmitter to ensure correct scanning in the receiver are called
(*a*) sync ☐ (*b*) chroma ☐
(*c*) luminance ☐ (*d*) video. ☐

187. When the current through a Zener diode increases by a factor of 2, the voltage across its terminals
(*a*) gets doubled too ☐ (*b*) remains unaltered ☐
(*c*) is halved ☐ (*d*) none of these. ☐

188. A 47 K resistor with 10% tolerance reads in the ohmmeter as
(*a*) 143 K ☐ (*b*) 43 K ☐
(*c*) 52 K ☐ (*d*) none of these. ☐

189. Lami's theorem states that if
(*a*) three forces acting at a point arc in equilibrium, they can be represented by the three sides of a triangle ☐
(*b*) the three forces acting at a point can be represented in magnitude and direction by the sides of a triangle, the forces are in equilibrium ☐
(*c*) three forces acting at a point are in equilibrium, each force is proportional to the sine of the angle between the other two ☐
(*d*) none of the above. ☐

190. Co-efficient of friction is the ratio of
(*a*) force of friction to reaction between two bodies ☐
(*b*) force of friction to normal reaction between two bodies ☐

(c) force of limiting friction to reaction between two bodies ☐
(d) force of limiting friction to normal reaction between two bodies. ☐

191. The time taken by a ball (of weight 500 gm) to return back to earth, if it is thrown vertically upwards with a velocity 4.9 m/sec is equal to

(a) $\frac{1}{2}$ sec ☐ (b) 1 sec ☐
(c) 2 sec ☐ (d) 3 sec. ☐

192. On newton is a force acting on a mass of

(a) one gm to produce an acceleration of one m/sec^2 ☐
(b) one kg to produce an acceleration of one m/sec^2 ☐
(c) one kg to produce an acceleration of one cm/sec^2 ☐
(d) none of the above. ☐

193. Polar moment of inertia of an equilateral triangle of sides x is given by

(a) $\frac{x^2}{16}$ ☐ (b) $\frac{x^4}{16\sqrt{3}}$ ☐
(c) $\frac{x^4}{32}$ ☐ (d) $\frac{x^4}{64}$. ☐

194. The time period of a simple pendulum will be doubled if

(a) its length is doubled ☐ (b) its length is halved ☐
(c) its length is increased four times ☐ (d) its length is increased eight times. ☐

195. Dynamic viscosity (μ) has the dimensions are

(a) MLT^{-2} ☐ (b) $ML^{-1}T^{-1}$ ☐
(c) $ML^{-1}T^{-2}$ ☐ (d) $M^{-1}L^{-1}T^{-1}$. ☐

196. Continuity equation deals with the law of conservation of

(a) mass ☐ (b) momentum ☐
(c) energy ☐ (d) none of the above. ☐

197. The local acceleration in the direction of x is given by

(a) $u\frac{\partial u}{\partial x} + \frac{\partial u}{\partial t}$ ☐ (b) $\frac{\partial u}{\partial t}$ ☐
(c) $u\frac{\partial u}{\partial x}$ ☐ (d) none of the above. ☐

198. The value of the momentum correction factor (β) for the viscous flow through a circular pipe is

(a) 1.33 ☐ (b) 1.50 ☐
(c) 2.0 ☐ (d) 1.25. ☐

199. The velocity distribution in laminar flow through a circular pipe follows the

(a) parabolic law ☐ (b) linear law ☐
(c) logarithmic law ☐ (d) none of the above. ☐

200. For a floating body, if the meta-centre coincides with the centre of gravity, the equilibrium is called

(*a*) stable ☐ (*b*) unstable ☐
(*c*) neutral ☐ (*d*) none of the above. ☐

201. If the distance of the screen is increased from the slits producing interference, the fringe width will

(*a*) remain constant ☐ (*b*) increase ☐
(*c*) decrease ☐ (*d*) none of the above. ☐

202. Monochromatic light of wavelength 5800 Å produces second order spectrum at 18° when it is used with a grating. The number of lines per cm on the grating is

(*a*) 2664 ☐ (*b*) 5000 ☐
(*c*) 2500 ☐ (*d*) 1332. ☐

203. The pitch of a sound wave is directly related to

(*a*) wavelength ☐ (*b*) frequency ☐
(*c*) velocity ☐ (*d*) amplitude. ☐

204. Sound waves in air are

(*a*) transverse ☐ (*b*) longitudinal ☐
(*c*) travelling ☐ (*d*) stationary. ☐

205. The energy stored by a capacitor of capacitance C carrying a charge 'q' at a potential difference V is given by

(*a*) $\frac{V}{q}$ ☐ (*b*) $\frac{1}{2}qV$ ☐
(*c*) $\frac{q}{2V}$ ☐ (*d*) $\frac{q}{V}$. ☐

206. What is the wavelength associated with a beam of electrons having average kinetic energy as 100 eV?

(*a*) 10 Å ☐ (*b*) 2.4 Å ☐
(*c*) 1.2 Å ☐ (*d*) 12 Å ☐

207. A parallel plate mica capacitor is to be designed for voltage of 9000 volts. The minimum thickness of mica should be

(*a*) 1.8 mm ☐ (*b*) 18.0 mm ☐
(*c*) 3.6 mm ☐ (*d*) 0.18 mm ☐

208. 1 volt/metre is equal to

(*a*) 1 metre/coulomb ☐ (*b*) 1 joule/coulomb ☐
(*c*) 1 newton/coulomb ☐ (*d*) 1 newton metre. ☐

209. The momentum of a particle of mass m and kinetic energy k is given by

(*a*) $\frac{h}{\sqrt{2mk}}$ ☐ (*b*) $\sqrt{2mk}$ ☐
(*c*) $\frac{\sqrt{2mk}}{h}$ ☐ (*d*) none of these. ☐

210. An electron of charge 'e' and mass'm' is liberated from filament is attracted by the anode having V potential with respect to the filament. The speed of the electron when its strikes the anode is

(a) $\frac{1}{2}meV$ ☐ (b) $\sqrt{\frac{2eV}{m}}$ ☐

(c) $\frac{1}{2}eV^2$ ☐ (d) $\sqrt{\frac{2m}{eV}}$. ☐

211. The energy is not carried by

(a) stationary waves ☐ (b) longitudinal waves ☐

(c) transverse waves ☐ (d) mechanical waves. ☐

212. The equation of a transverse travelling waves is given by $y = 0.05 \sin 2\pi (0.5\,x - 10\,t)$. Its amplitude is

(a) 10 m ☐ (b) 0.5 m ☐

(c) 0.05 m ☐ (d) none of these. ☐

213. The law, which states that heat and work are mutually convertible, is known as

(a) zeroth law of thermodynamics ☐ (b) first law of thermodynamics ☐

(c) second law of thermodynamics ☐ (d) none of the above. ☐

214. Unit of power is

(a) Watt (W) ☐ (b) Joule (J) ☐

(c) Joule-metre (J m) ☐ (d) none of the above. ☐

215. The sum of internal energy and pressure volume product is called

(a) Entropy ☐ (b) Enthalpy ☐

(c) Heat supplied ☐ (d) none of the above. ☐

216. The relation between two specific heats of a gas (*i.e.*, C_p and C_v) and gas constant is given by

(a) $C_p - C_v = \frac{R+1}{J}$ ☐ (b) $C_v - C_p = \frac{R}{J}$ ☐

(c) $C_p - C_v = \frac{R}{J}$ ☐ (d) $C_p - C_v = \frac{J}{R}$. ☐

217. Compression ratio of defined as the ratio of

(a) $\frac{\text{total volume}}{\text{swept volume}}$ ☐ (b) $\frac{\text{swept volume}}{\text{total volume}}$ ☐

(c) $\frac{\text{total volume}}{\text{clearance volume}}$ ☐ (d) $\frac{\text{swept volume}}{\text{clearance volume}}$. ☐

218. Air standard efficiency of an Otto cycle is equal to

(a) $1 - \frac{1}{r^{\gamma+1}}$ ☐ (b) $1 - \frac{r}{r^{\gamma-1}}$ ☐

(c) $1 - \frac{1}{r^{\gamma-1}}$ ☐ (d) $1 + \frac{1}{r^{\gamma+1}}$. ☐

where r = Ratio of compression and expansion, γ = Ratio of two specific heats.

219. In a vapour compression refrigeration cycle, lowest temperature occurs in

(*a*) condenser ☐ (*b*) evaporator ☐

(*c*) compressor ☐ (*d*) throttle valve. ☐

220. If the cut off is decreased, the efficiency of Diesel cycle

(*a*) increases ☐ (*b*) decreases ☐

(*c*) is same ☐ (*d*) none of the above. ☐

221. The co-efficient of performance (C.O.P.) of a refrigerator working on a reversed Carnot cycle is mathematically equal to

(*a*) $\frac{(T_1 - T_2)}{T_1}$ ☐ (*b*) $\left(\frac{T_1}{T_1 - T_2}\right)$ ☐

(*c*) $\frac{T_2}{(T_1 - T_2)}$ ☐ (*d*) $\frac{(T_2 - T_1)}{T_1}$. ☐

where T_2 = Highest absolute temperature, T_1 = Lowest absolute temperature.

222. For flow of fluid over a heated plate, the following fluid properties are known: Viscosity = 0.001 Pa.s; Specific heat at constant pressure = 1 kJ/kg K; Thermal conductivity = 1 W/m K; The hydrodynamic boundary layer thickness at a specified location on the plate is 1 mm, thermal boundary layer thickness at the same location is

(*a*) 0.001 mm ☐ (*b*) 0.01 mm ☐

(*c*) 1 mm ☐ (*d*) 1000 mm. ☐

(GATE-ME-2008)

223. Consider one-dimensional steady state heat conduction along *x*-axis ($0 \leq x \leq L$), through a plane wall with the boundary surfaces ($x = 0$ and $x = L$) maintained at temperatures of 0°C and 100°C. Heat is generated uniformly throughout the wall. Choose the correct statement

(*a*) The direction of heat transfer will be from the surface at 100°C to the surface at 0°C. ☐

(*b*) The maximum temperature in side the wall must be greater than 100°C. ☐

(*c*) The temperature distribution is linear with in the wall. ☐

(*d*) The temperature distribution is symmetric about the mid-plane of the wall. ☐

(GATE-ME-2013)

224. Two rods, one of length *L* and the other of length 2*L* are made of the same material and have the same diameter. The two ends of the longer rod are maintained at 100°C. One end of the shorter rod is maintained at 100°C while the other end is insulated. Both the rods are exposed to the same environment at 40°C. The temperature at the insulated end of the shorter rod is measured to be 55°C. The temperature at the mid-point of the longer rod would

(*a*) 40°C ☐ (*b*) 50°C ☐

(*c*) 55°C ☐ (*d*) 100°C ☐

(GATE-ME-1992)

II. TEST PAPER NO. 2

Tick mark the most appropriate statement of the multiple choice answers:

1. A system, which permits the passage of energy across the boundary but does not permit the passage of matter, is known as
   (*a*) open system ☐ (*b*) closed system ☐
   (*c*) isolated system ☐ (*d*) none of the above. ☐
2. Joule's law states that
   (*a*) change of pressure is proportional of change of temperature ☐
   (*b*) change of volume is proportional to the change of temperature ☐
   (*c*) change of internal energy is proportional to the change of temperature ☐
   (*d*) none of the above. ☐
3. A Carnot cycle consists of
   (*a*) two constant pressure and two adiabatic processes ☐
   (*b*) two constant volume and two adiabatic processes ☐
   (*c*) two isothermal and two adiabatic processes ☐
   (*d*) one constant pressure, one constant volume and two adiabatic processes. ☐
4. Mean effective pressure is obtained if the work done is divided by
   (*a*) total volume ☐ (*b*) swept volume ☐
   (*c*) clearance volume ☐ (*d*) none of the above. ☐
5. The compression ratios for petrol engines lie in the range of
   (*a*) 5 to 8 ☐ (*b*) 15 to 20 ☐
   (*c*) 3 to 6 ☐ (*d*) 30 to 40. ☐
6. Bar is the unit of
   (*a*) power ☐ (*b*) energy ☐
   (*c*) pressure ☐ (*d*) entropy. ☐
7. The law, which states that heat and work are mutually convertible, is known as
   (*a*) zeroth law of thermodynamics ☐ (*b*) first law of thermodynamics ☐
   (*c*) second law of thermodynamics ☐ (*d*) third law of thermodynamics. ☐
8. A fin has 5 mm diameter and 100 mm long. The thermal conductivity of fin material is 400 W/m K. One end of the fin is maintained at 130°C and its remaining surface is exposed to ambient air at 30°C. If the convective heat transfer coefficient is 40 W/m^2 K, the heat loss (in W) from the fin is
   (*a*) 0.08 ☐ (*b*) 5.0 ☐
   (*c*) 7.0 ☐ (*d*) 7.8. ☐

   (GATE-ME-2010)
9. A cylinder contains 5 m^3 of an ideal gas at a pressure of 1 bar. This gas is compressed in a reversible isothermal process till its pressure increases to 5 bar. The work in kJ required for this process is
   (*a*) 804.7 ☐ (*b*) 953.2 ☐
   (*c*) 981.7 ☐ (*d*) 1012.2. ☐

   (GATE-ME-2013)

10. If a mass of moist air in an airtight vessel is heated to a higher temperature, then
(a) specific humidity of the air increases ☐ (b) specific humidity of the air decreases ☐
(c) relative humidity of the air increases ☐ (d) relative humidity of the air decreases. ☐

(GATE-ME-2011)

11. In diesel engine, the suction consists of
(a) air only ☐ (b) a mixture of air and fuel ☐
(c) fuel only ☐ (d) none of the above. ☐

12. The pressure at the end of compression in petrol engine as compared to that of diesel engine would be
(a) higher ☐ (b) lower ☐
(c) same ☐ (d) none of the above. ☐

13. The relative fuel-air ratio is equal to
(a) actual fuel-air ratio × chemically correct fuel-air ratio ☐
(b) chemically correct fuel-air ratio ÷ actual fuel-air ratio ☐
(c) actual fuel-air ratio + chemically correct fuel-air ratio ☐
(d) none of the above. ☐

14. In spark-ignition engines, the possibility of knocking can be reduced by
(a) advancing the spark timing ☐ (b) increasing the coolant temperature ☐
(c) reducing compression ratio ☐ (d) none of the above. ☐

15. For spark ignition engine fuels, the anti-knock agent is
(a) naphthene ☐ (b) amyl nitrate ☐
(c) tetraethyl lead ☐ (d) none of the above. ☐

16. For the petrol engines, the type of governing used is
(a) quality governing ☐ (b) hit and miss governing ☐
(c) quantity governing ☐ (d) none of the above. ☐

17. Knocking in C.I. engine can be prevented if the first element of the fuel and air has
(a) a short delay ☐ (b) a low density ☐
(c) a low temperature ☐ (d) all of the above. ☐

18. Which one of the following power plants was first located in India at Tarapur?
(a) steam power plant ☐ (b) hydro-electric power plant ☐
(c) nuclear power plant ☐ (d) none of the above. ☐

19. Fast breeder reactors produces
(a) less fuel than they consume ☐ (b) more fuel than they consume ☐
(c) same fuel as they consume ☐ (d) no fuel. ☐

20. The material for control rod is
(a) graphite ☐ (b) lead ☐
(c) boron or cadmium ☐ (d) zinc. ☐

21. Two large diffuse gray parallel plates, separated by a small distance, have surface temperatures of 400 K and 300 K. If the emissivities of the surfaces are 0.8 and the

Stefan-Boltzman constant is 5.67×10^{-8} W/m^{2}K^4, the net radiation heat exchange rate in kW/m^2 between the two plates is

(*a*) 0.66 ☐ (*b*) 0.79 ☐
(*c*) 0.99 ☐ (*d*) 3.96. ☐

(GATE-ME-2013)

22. If a closed system is undergoing an irreversible process, the entropy of the system

(*a*) must increase ☐ (*b*) always remains constant ☐
(*c*) must decrease ☐ (*d*) can increase, decrease or remain constant ☐

(GATE-ME-2009)

23. An ideal Brayton cycle, operating between the pressure limits of 1 bar and 6 bar, has minimum and maximum temperatures of 300 K and 1500 K. The ratio of specific heats of the working fluid is 1.4. The approximate final temperatures in Kelvin at the end of the compression and expansion processes are respectively

(*a*) 500 and 900 ☐ (*b*) 900 and 500 ☐
(*c*) 500 and 500 ☐ (*d*) 900 and 900. ☐

(GATE-ME-2011)

24. The LMTD of a counter flow heat exchanger is 20°C, the cold fluid enters at 20°C and the hot fluid enters at 100°C, mass flow rate of the cold fluid is twice that of the hot fluid specific heat at constant pressure of the fluid is twice that of the cold fluid. The exit temperature of the cold fluid is

(*a*) 40°C ☐ (*b*) 60°C ☐
(*c*) 80°C ☐ (*d*) cannot be determined ☐

(ME-GATE-2008)

25. The crank radius of a single-cylinder I.C. engine is 60 mm and the diameter of the cylinder is 80 mm. The swept volume of the cylinder in cm^3 us

(*a*) 48 ☐ (*b*) 96 ☐
(*c*) 302 ☐ (*d*) 603. ☐

(GATE-ME-2011)

26. A mono-atomic ideal gas (γ = 1.67, molecular weight = 40) is compressed adiabatically from 0.1 MPa, 300 K to 0.2 MPa. The universal gas constant is 8.314 kJ kmol^{-1} K^{-1}. The work of compression of the gas (in kJ kg^{-1}) is

(*a*) 29.7 ☐ (*b*) 19.9 ☐
(*c*) 13.3 ☐ (*d*) 0. ☐

(GATE-ME-2010)

27. Nozzle designed for

(*a*) maximum pressure at outlet ☐
(*b*) maximum discharge ☐
(*c*) maximum pressure and maximum discharge ☐
(*d*) none of the above. ☐

28. The maximum mass rate of flow of a gas through a convergent nozzle depends upon

(*a*) pressure and density at the inlet only ☐ (*b*) adiabatic exponent (γ) only ☐
(*c*) pressure ratio at the exit and inlet only ☐ (*d*) all of the above ☐
(*e*) only (*a*) and (*b*). ☐

29. Formation of shock wave front in case of convergent-divergent nozzle takes place

(*a*) at throat ☐ (*b*) in convergent portion ☐

(*c*) in divergent portion ☐ (*d*) none of the above. ☐

30. In case of reaction steam turbine

(*a*) the steam is expanded in nozzle only ☐

(*b*) the steam is expanded both in moving blades only ☐

(*c*) the steam is expanded both in fixed and moving blades continuously ☐

(*d*) none of the above. ☐

31. Consider one-dimensional steady state heat conduction, without heat generation, in a plane wall; with boundary conditions as shown in Fig. 15.4. The conductivity of the wall is given by $k = k_0 + bT$, where k_0 and b are positive constants, and T is temperature.

FIGURE 15.4

As x increases, the state temperature gradient (dT/dx) will

(*a*) remain constant ☐ (*b*) be zero ☐

(*c*) increase ☐ (*d*) decrease. ☐

(GATE-ME-2013)

32. With the increase of pressure ratio, the volumetric efficiency, of an air compressor

(*a*) increases ☐ (*b*) decreases ☐

(*c*) constant ☐ (*d*) first increases then decreases. ☐

33. The assumptions made for calculating the work done in multi-stage compressor are

(*a*) the index of compression is same in each stage ☐

(*b*) intercooling in each stage is at constant pressure ☐

(*c*) suction and delivery pressures remain constant during each stage ☐

(*d*) all of the above. ☐

34. For a three stage reciprocating at compressor, for perfect intercooling and minimum work required, the condition is

(*a*) $\frac{p_2}{p_1} = \frac{p_3}{p_2} = \frac{p_4}{p_3}$ ☐ (*b*) $p_1p_2 = p_3p_4$ ☐

(*c*) $p_1p_2 = \sqrt{p_3p_4}$ ☐ (*d*) $p_1p_4 = \frac{1}{\sqrt{p_3p_4}}$. ☐

35. The thermal efficiency of a gas turbine with regenerator is maximum when pressure ratio is

(*a*) less than 1.0 ☐ (*b*) more than 1.0 ☐

(*c*) equal to 1.0 ☐ (*d*) zero. ☐

36. Work ratio of a turbine plant is the ratio of

(*a*) $\frac{\text{net work output}}{\text{heat supplied}}$ ☐ (*b*) $\frac{\text{heat supplied}}{\text{net work output}}$ ☐

(*c*) $\frac{\text{net work output}}{\text{work from turbine}}$ ☐ (*d*) net work output × heat supplied. ☐

37. By employing intercooler in multi-stage compressor, compression obtained is

(*a*) adiabatic ☐ (*b*) isothermal ☐

(*c*) polytropic ☐ (*d*) none of the above. ☐

38. In a compressor, the cylinder clearance should be

(*a*) minimum ☐ (*b*) maximum ☐

(*c*) about 100% of swept volume ☐ (*d*) none of the above. ☐

39. A simple gas turbine cycle will have more efficiency and more work if

(*a*) multiple-stage compressor with intercoolers are used ☐

(*b*) heat exchangers are used ☐

(*c*) the air after partial expansion in the turbine is reheated ☐

(*d*) all of the above ☐

(*e*) none of the above. ☐

40. The rate of heat transfer from a solid surface to a fluid is obtained from

(*a*) Newton's law of cooling ☐ (*b*) Fourier's law ☐

(*c*) Kirchhoff's law ☐ (*d*) Stefan's law. ☐

41. For steady flow and constant value of conductivity, the temperature distribution for a hollow cylinder is

(*a*) parabolic ☐ (*b*) linear ☐

(*c*) logarithmic function of radii ☐ (*d*) cubic. ☐

42. Which one of the following is a correct statement?

(*a*) Fourier's law of heat conduction gives the heat flow for two-dimensional cases. ☐

(*b*) Thermal conductivity of air increases with decrease in temperature. ☐

(*c*) Thermal conductivity of solids increases with rise in temperature. ☐

(*d*) The unit of thermal conductivity is S.I. units is W/m K. ☐

43. A body, which absorbs all the radiations falling on it, is called

(*a*) opaque body ☐ (*b*) white body ☐

(*c*) black body ☐ (*d*) transparent body. ☐

44. One tonne refrigeration in S.I. units is equivalent to

(*a*) 50 kJ/min ☐ (*b*) 3.5 kW ☐

(*c*) 3000 J/min ☐ (*d*) 200 kJ/min. ☐

45. Kinematic viscosity is defined as equal to

(*a*) dynamic viscosity × density ☐ (*b*) dynamic viscosity/density ☐

(*c*) dynamic viscosity × pressure ☐ (*d*) pressure × density. ☐

46. Centre of pressure of a vertical plane surface immersed in a liquid is

(*a*) above the centre of gravity of the plane surface ☐

(*b*) at the centre of gravity of the plane surface ☐

(*c*) below the centre of gravity of the plane surface ☐

(*d*) none of the above. ☐

47. The flow in a pipe is laminar if

(*a*) Reynold number is equal to 5200 ☐ (*b*) Reynold number is equal to 4000 ☐
(*c*) Reynold number is more than 2500 ☐ (*d*) none of the above. ☐

48. The value of the kinetic energy correction factor (α) for the viscous flow through a circular pipe is

(*a*) 1.33 ☐ (*b*) 1.50 ☐
(*c*) 2.0 ☐ (*d*) 1.25. ☐

49. Froude number is defined as the ratio of

(*a*) inertia force to viscous force ☐ (*b*) inertia force to gravity force ☐
(*c*) inertia force to elastic force ☐ (*d*) inertia force to pressure force. ☐

50. In supersonic flow, the projectile (which creates disturbances) moves

(*a*) ahead of the disturbances ☐ (*b*) along the disturbances ☐
(*c*) behind the disturbance ☐ (*d*) none of the above. ☐

51. Francis turbine is

(*a*) an impulse turbine ☐ (*b*) a radial flow impulse turbine ☐
(*c*) an axial flow turbine ☐ (*d*) a radial flow reaction turbine. ☐

52. The speed ratio for Pelton wheel varies from

(*a*) 0.45 to 0.50 ☐ (*b*) 0.6 to 0.7 ☐
(*c*) 0.3 to 0.40 ☐ (*d*) 0.8 to 0.9. ☐

53. Muschel curves means

(*a*) curves at constant head ☐ (*b*) curves at constant speed ☐
(*c*) curves at constant efficiency ☐ (*d*) none of the above. ☐

54. Hydraulic ram is a pump which works on the principle of

(*a*) water-hammer ☐ (*b*) centrifugal action ☐
(*c*) reciprocating action ☐ (*d*) none of the above. ☐

55. A current metre is a device used for measuring

(*a*) velocity ☐ (*b*) viscosity ☐
(*c*) current ☐ (*d*) pressure. ☐

56. The time taken by a ball of weight 500 gm to return back to earth, if it is thrown vertically upwards with a velocity 4.9 m/sec is equal to

(*a*) 0.5 second ☐ (*b*) 1 second ☐
(*c*) 2 second ☐ (*d*) 3 second. ☐

57. One newton is equal to

(*a*) 10^3 dyne ☐ (*b*) 10^2 dyne ☐
(*c*) 10^5 dyne ☐ (*d*) 10^4 dyne. ☐

58. The C.G. of a semi-circular arc is at the central radius at a distance of

(*a*) $\frac{3r}{4}$ from base diameter ☐ (*b*) $\frac{3r}{4}$ from the base diameter ☐
(*c*) $\frac{2r}{\pi}$ above base diameter ☐ (*d*) 0.5 r from the base diameter. ☐

59. At a place, the clock will go slow, if

(a) length of pendulum is increased ☐ (b) length of pendulum is decreased ☐
(c) mass of its bob is increased ☐ (d) none of the above. ☐

60. A rod of steel of 1 cm^2 in cross-sectional area and 100 cm long is subjected to an axial pull of 2000 kgf. If $E = 2 \times 10^6$ kgf/cm^2 the elongation will be

(a) 1 cm ☐ (b) 0.2 cm ☐
(c) 0.1 cm ☐ (d) 0.15 cm. ☐

61. The ratio of modulus of rigidity to modulus of elasticity for a Poisson's ratio of 0.25 would be

(a) 0.5 ☐ (b) 0.4 ☐
(c) 0.3 ☐ (d) 1.0. ☐

62. Which of the following is a wrong statement?

(a) The thermal stresses are only set up in the bodies, when they are not allowed to expand or contract freely with the rise or fall of temperature. ☐
(b) If $\propto$ is the co-efficient of linear expansion and T is rise in temperature, then thermal stress is equal to '$E \propto T$'. ☐
(c) The thermal stress is independent of cross-sectional area of the bar. ☐
(d) Rubber is more elastic than steel. ☐

63. If a member is subjected to an axial tensile load, the plane inclined at 45° to the axis of loading carries

(a) minimum shear stress ☐ (b) maximum normal stress ☐
(c) maximum shear stress ☐ (d) none of the above. ☐

64. A strained element is subjected to principal stresses p_1 and p_2 in two mutually perpendicular direction. The strain (e_1) in the direction of principal stress (p_1) is equal to

(a) $\frac{p_1}{E} + \frac{p_2}{mE}$ ☐ (b) $\frac{p_1}{E} - \frac{p_2}{mE}$ ☐
(c) $\frac{p_2}{E} + \frac{p_1}{mE}$ ☐ (d) $\frac{p_2}{E} - \frac{p_1}{mE}$. ☐

where $\frac{1}{m}$ = Poisson's ratio.

65. A simply supported beam of span (l) carries a uniformly distributed load over the whole span. The shear force diagram will be

(a) a rectangle ☐ (b) a triangle ☐
(c) two equal and opposite rectangles ☐ (d) two equal and opposite triangles. ☐

66. A beam of uniform strength is one which has same

(a) bending stress at every section ☐ (b) deflection throughout the beam ☐
(c) bending moment throughout the beam ☐ (d) shear force throughout the beam. ☐

67. A fixed beam of length (l) carries a point load (W) at the centre. The deflection at the centre is

(a) same as for a simply supported beam ☐

(*b*) half of the deflection for a simply supported beam ☐
(*c*) one-fourth of the deflection for a simply supported beam ☐
(*d*) double the deflection for a simply supported beam. ☐

68. The load on a circular column of diameter (*d*) for keeping the stress wholly compressive may be applied anywhere within a concentric circle of diameter

(*a*) $\frac{d}{8}$ ☐ (*b*) $\frac{d}{4}$ ☐
(*c*) $\frac{d}{3}$ ☐ (*d*) $\frac{d}{4}$. ☐

69. Equivalent torque is the torque which alone produces

(*a*) maximum normal stress ☐ (*b*) minimum normal stress ☐
(*c*) maximum shear stress ☐ (*d*) minimum shear stress. ☐

70. In case of thick cylinders, at any point the three principal stresses (*i.e.*, radial, circumferential and logitudinal) are all

(*a*) tensile stresses ☐ (*b*) compressive stresses ☐
(*c*) shear stresses ☐ (*d*) none of the above. ☐

71. The differential mechanism of an automobile is having

(*a*) one degree of freedom ☐ (*b*) two degrees of freedom ☐
(*c*) three degrees of freedom ☐ (*d*) zero degree of freedom. ☐

72. Whitworth quick return mechanism is an inversion of

(*a*) double slider crank chain ☐ (*b*) single slider crank chain ☐
(*c*) four bar chain ☐ (*d*) crossed slider crank chain. ☐

73. Kennedy's theorem states that any three bodies moving relatively to each other, have

(*a*) three instantaneous centres and these three centres lie on a curved line ☐
(*b*) three instantaneous centres and these three centres lie on straight line ☐
(*c*) only one instantaneous centre ☐
(*d*) two instantaneous centres and these two centres lie on a circle. ☐

74. If a point moves along a straight line which is rotating, then the Coriolis component of acceleration is equal to

(*a*) $r\omega^2$ ☐ (*b*) $\frac{dv}{dt}$ ☐
(*c*) $22v\omega$ ☐ (*d*) $r\alpha$ ☐
(*e*) none of the above ☐

where v = Velocity of point along the straight line, r = Radius of the point

ω = Angular velocity of the line, α = Angular acceleration of the line.

75. A machine is said to be self-locking if efficiency of the machine is

(*a*) more than 50% ☐ (*b*) equal to 50% ☐
(*c*) less than 50% ☐ (*d*) equal to 100%. ☐

76. Two intersecting and co-planer shafts are connected by gears. This type of gear is called

(*a*) helical gear ☐ (*b*) spur gear ☐

(*c*) bevel gear ☐ (*d*) spiral gear. ☐

77. The path of contact in involute gears is a

(*a*) circle ☐ (*b*) straight line ☐

(*c*) curved line ☐ (*d*) none of the above. ☐

78. A dynamometer is a device use for measuring

(*a*) speed of a machine ☐ (*b*) torque exerted by or on a machine ☐

(*c*) kinetic energy of the machine ☐ (*d*) all of the above ☐

(*e*) none of the above. ☐

79. The height (h) of a Watt's governor is given by

(*a*) $h = g\omega^2$ ☐ (*b*) $h = \frac{\omega^2}{g}$ ☐

(*c*) $h = \frac{g}{\omega^2}$ ☐ (*d*) $h = \frac{1}{g\omega^2}$ ☐

80. A system is said to be overdamped if the damping factor for the system is

(*a*) more than one ☐ (*b*) equal to one ☐

(*c*) less than one ☐ (*d*) equal to zero. ☐

81. Endurance limit is the value of maximum stress which a material can withstand without failure for infinite number of cycles, when it is subjected to

(*a*) dynamic load ☐ (*b*) static load ☐

(*c*) bending load ☐ (*d*) completely reversed load. ☐

82. For a solid shaft of diameter D, the modulus of section is equal to

(*a*) $\frac{\pi D^3}{64}$ ☐ (*b*) $\frac{\pi D^3}{32}$ ☐

(*c*) $\frac{\pi D^3}{16}$ ☐ (*d*) $\frac{\pi D^4}{64}$. ☐

83. Two shafts, one solid and other hollow, are made of the same materials and are having same length and weight. The hollow shaft as compared to solid shaft will be

(*a*) more strong ☐ (*b*) less strong ☐

(*c*) of the same strength ☐ (*d*) none of the above. ☐

84. Basic shaft is one

(*a*) whose lower deviation is zero ☐

(*b*) whose upper deviation is zero ☐

(*c*) whose upper and lower deviations are zero ☐

(*d*) none of the above. ☐

85. A coarse screw of major diameter 6 mm and pitch 1 mm according to IS : 1362–1962 is represented as

(a) $M \times 6 \times 1$ ☐ (b) $M6 \times 1$ ☐
(c) $6M \times 1$ ☐ (d) $M6$. ☐

86. The stress induced in a thin cylinder (subjected to internal fluid pressure) parallel to the axis of the cylinder is

(a) two times the circumferential stress ☐ (b) equal to the circumferential stress ☐
(c) one-half of the circumferential stress ☐ (d) one-third of the circumferential stress. ☐

87. Two springs (each having stiffness k) are is parallel. The overall stiffness of the two springs would be equal to

(a) $2k$ ☐ (b) k ☐
(c) $\frac{k}{2}$ ☐ (d) $\frac{k}{4}$ ☐

88. The bearings are designated by a number, generally consisting of at least three digits such as 205, 305 and 405. The last three digits of the bearing number represents

(a) bore of the bearing ☐ (b) series of the bearing ☐
(c) width of the bearing ☐ (d) bore and series of the bearing. ☐

89. In cycloidal teeth, the interference

(a) is maximum ☐
(b) is minimum ☐
(c) is absent completely ☐
(d) depends upon pressure angle and number of teeth. ☐

90. Which one of the following is a ceramic material?

(a) zinc ☐ (b) iron ☐
(c) silicon carbide ☐ (d) wood. ☐

91. Delta iron exists

(a) between 900 and 1404°C ☐ (b) between 1404 and 1535°C ☐
(c) below 768°C ☐ (d) between 768 and 900°C. ☐

92. The carbon content in steel is

(a) above 2% ☐ (b) upto 2% ☐
(c) below 0.8% ☐ (d) above 6.3% ☐

93. The process of reducing the ore with carbon in the presence of a flux is known as

(a) hardening ☐ (b) smelting ☐
(c) quenching ☐ (d) shaping. ☐

94. The unit cell of face-centered cubic structure contains number of atoms as

(a) one ☐ (b) two ☐
(c) four ☐ (d) eight. ☐

95. Carbon steel castings ore

(a) ductile ☐ (b) tough ☐
(c) easily weldable ☐ (d) brittle ☐
(e) all of the above except (d). ☐

96. Slack is always

(*a*) positive ☐ (*b*) negative ☐

(*c*) zero ☐ (*d*) any one of the above. ☐

97. CPM is the

(*a*) target-oriented technique ☐ (*b*) time-oriented technique ☐

(*c*) event-oriented technique ☐ (*d*) activity-oriented technique. ☐

98. PERT is the

(*a*) target-oriented technique ☐ (*b*) time-oriented technique ☐

(*c*) event-oriented technique ☐ (*d*) activity-oriented technique. ☐

99. The three time estimates are combined into a single time which is called

(*a*) the pessimistic time ☐ (*b*) the optimistic time ☐

(*c*) the most likely time ☐ (*d*) the expected time. ☐

100. Critical path is that sequence of activities between the start and finish of the project which requires the

(*a*) shortest time ☐ (*b*) longest time ☐

(*c*) normal time ☐ (*d*) none of the above. ☐

101. The term, which is associated with the activity times, is known as

(*a*) slack ☐ (*b*) float ☐

(*c*) dummy ☐ (*d*) none of the above. ☐

102. If the total flat is negative, activity is called

(*a*) sub-critical ☐ (*b*) critical ☐

(*c*) super-critical ☐ (*d*) none of the above. ☐

103. The minimum possible time, in which an activity can be completed by employing extra-resources, is known as

(*a*) crash time ☐ (*b*) normal time ☐

(*c*) finish time ☐ (*d*) none of the above. ☐

104. A chart, in which time values are recorded and motions are classified by therblings, is called

(*a*) SIMQ chart ☐ (*b*) Operation chart ☐

(*c*) Gnatt chart ☐ (*d*) Process chart. ☐

105. A public sector undertaking is

(*a*) fully owned by Government ☐

(*b*) fully owned by public through shareholders ☐

(*c*) jointly owned by private parties ☐

(*d*) jointly owned by private parties and Government. ☐

106. The term 'value' in value engineering refers to

(*a*) selling price of the product ☐ (*b*) total cost of the product ☐

(*c*) utility of the product ☐ (*d*) none of the above. ☐

107. The simplex method is the basic method for

(*a*) queuing theory ☐ (*b*) network analysis ☐

(*c*) linear programming ☐ (*d*) value analysis. ☐

108. The information about the production schedule is obtained from

(*a*) string diagram ☐ (*b*) Gnatt chart ☐

(*c*) distribution curve ☐ (*d*) travel chart. ☐

109. Which one of the following is a natural abrasive ?

(*a*) silicon carbide ☐ (*b*) boron carbide ☐

(*c*) aluminium oxide ☐ (*d*) diamond. ☐

110. Grain depth of cut is equal to

(*a*) half the maximum cut depth ☐ (*b*) the maximum cut depth ☐

(*c*) 1.5 times the maximum cut depth ☐ (*d*) twice the maximum cut depth. ☐

111. Which of the following factors effects are section of the grinding wheels ?

(*a*) work speed ☐ (*b*) wheel speed ☐

(*c*) work material ☐ (*d*) area of contact ☐

(*e*) all of the above ☐ (*f*) none of the above. ☐

112. The act of improving the cutting action of a grinding wheel is known as

(*a*) wheel dressing ☐ (*b*) wheel truing ☐

(*c*) wheel grinding ☐ (*d*) wheel grading. ☐

113. For small and intricate castings, the sand grains should be

(*a*) fine ☐ (*b*) medium ☐

(*c*) coarse ☐ (*d*) rounded. ☐

114. Which one of the following is a wrong statement ?

(*a*) Natural sand maintains moisture content for a long time. ☐

(*b*) Synthetic sand in an artificial sand. ☐

(*c*) Silica and clay are the two chief constituents of the moulding sand. ☐

(*d*) For fine sand grains, permeability is high. ☐

115. In sand moulding cope is the

(*a*) upper most part of the flask ☐ (*b*) bottom most part of the flask ☐

(*c*) middle part of the flask ☐ (*d*) none of the above. ☐

116. For ornamental parts and toys of non-ferrous alloys, the casting method used is known as

(*a*) die casting ☐ (*b*) slush casting ☐

(*c*) centrifugal casting ☐ (*d*) permanent mould casting. ☐

117. In case of a shaper, the ram moves

(*a*) faster during forward stroke ☐

(*b*) slower during forward stroke ☐

(*c*) at the same speed during forward and return stroke ☐

(*d*) none of the above. ☐

118. If L = length of stroke in mm and N = number of stroke per minute of a shaper, then length cut in one minute is equal to

(a) $\frac{N}{1000\,L}$ ☐ (b) $\frac{L}{1000\,N}$ ☐

(c) $\frac{N \times L}{1000}$ ☐ (d) $N \times L$. ☐

119. Feed of the drill is

(a) equal to half the radius of the drill ☐

(b) equal to one-fourth the radius of the drill ☐

(c) the distance moved by the drill per second ☐

(d) the distance moved by the drill into the work-piece on every revolution of the drill. ☐

120. The time required for drilling a 20 mm hole in a work-piece of thickness 40 mm at a drill speed of 200 r.p.m. and feed = 0.2 mm/revolution when length of approach is neglected, will be equal to

(a) 2.50 min ☐ (b) 1.5 min ☐

(c) 1.0 min ☐ (d) 0.5 min. ☐

121. The relation between torque (T), diameter of drill (d) and feed (f) is given by

(a) $T = C \times f^{0.5} \times d^{1.5}$ ☐ (b) $T = C \times f^{0.75} \times d^{1.5}$ ☐

(c) $T = C \times f \times d^2$ ☐ (d) $T = C \times f^{0.75} \times d^{1.8}$ ☐

where C = Constant depending upon the material being drilled.

122. For drilling aluminium, a drill with

(a) zero helix angle is required ☐ (b) low helix angle is required ☐

(c) high helix angle is required ☐ (d) any helix angle can be used. ☐

123. Choose the wrong statement

(a) In case of broaching, the job is completed in one stroke of the machine. ☐

(b) Broaching is used for straight or irregular surface either externally or internally. ☐

(c) Broaching is a metal cutting operation in which rate of production is high. ☐

(d) The operation of broaching is simple and tooling cost is low and hence it is used for mass production. ☐

124. Which of the following process is used for making a complicated contour in a carbide piece ?

(a) plasma arc machining ☐ (b) electro-chemical milling ☐

(c) electro-discharge machining ☐ (d) laser machining. ☐

125. For a cutting tool, the spindle speeds are 160, 229, 328, 469, ... The next higher speed will be

(a) 800 ☐ (b) 704 ☐

(c) 671 ☐ (d) 643. ☐

126. Which one of the following metals is best suited for electro forming ?

(a) zinc ☐ (b) lead ☐

(c) gold ☐ (d) silver. ☐

127. Gears which work under low specific pressure and low velocities (*i.e.*, gears of relatively small power transmiting capacity) are made of

(*a*) alloy steel ☐ (*b*) carburised steel ☐

(*c*) carbon steel ☐ (*d*) mild steel. ☐

128. A process of coating zinc by hot dipping is known as

(*a*) anodising ☐ (*b*) galvanising ☐

(*c*) sheradising ☐ (*d*) brazing. ☐

129. In EDM process, if both the electrodes (one tool and the other work-piece) are made of the same material, the erosion of the metal is

(*a*) greatest at positive electrode (anode) ☐

(*b*) greatest at negative electrode (cathode) ☐

(*c*) same at anode and cathode ☐

(*d*) minimum at anode and cathode. ☐

130. PAM stands for

(*a*) Plastic arc method ☐ (*b*) Plasma arc method ☐

(*c*) Plasma arc machining ☐ (*d*) Plastic abrasive method. ☐

131. In electro-chemical grinding (ECG), the metals is removed by

(*a*) abrasion of the metal only ☐

(*b*) electro-chemical decomposition only ☐

(*c*) 10% electro-chemical decomposition and 90% abrasion of the metal ☐

(*d*) 90% electro-chemical decomposition and 10% abrasion of the metal. ☐

132. In case of plastic welding, the metals to be joined are heated

(*a*) to the plastic state and allowed to solidify ☐

(*b*) to the plastic state and then forced together by external pressure without the addition of filler material ☐

(*c*) to a molten state and allowed to cool in presence of a filler material ☐

(*d*) none of the above. ☐

133. An automobile plant contracted to buy shock absorbers from two suppliers *X* and *Y*, *X* supplies 60% and *Y* supplies 40% of the shock absorbers. All shock absorbers are subjected to a quality test. The ones that pass the quality test are considered reliable. Of *X*'s shock absorbers, 96% are reliable. Of *Y*'s shock absorbers, 72% are reliable.

The probability that a randomly chosen shock absorber, which is found to be reliable, is made by *Y* is

(*a*) 0.288 ☐ (*b*) 0.334 ☐

(*c*) 0.667 ☐ (*d*) 0.720 ☐

(ME/CE/CSE/PI-2012)

134. The welding process, in which heat is produced for welding by chemical reaction, is known as

(*a*) force welding ☐ (*b*) resistance welding ☐

(*c*) gas welding ☐ (*d*) thermit welding. ☐

135. In thermit welding, the aluminium and iron oxide are mixed in the proportion of

(*a*) 1 : 3 ☐ (*b*) 1 : 2 ☐

(*c*) 1 : 1 ☐ (*d*) 2 : 1. ☐

136. Filler material is used in

(*a*) seam welding ☐ (*b*) spot welding ☐

(*c*) projection welding ☐ (*d*) all of the above ☐

(*e*) none of the above. ☐

137. For brittle materials like brass

(*a*) positive rake angle is provided ☐ (*b*) negative rake angle is provided ☐

(*c*) zero rake angle is provided ☐ (*d*) any rake angle is provided. ☐

138. A carbide tool having ($n = 0.25$ in Taylor's equation $VT^m = C$) with a mild steel work-piece was found to give life of 1 hour 21 minutes while cutting at 60 metre per minute. The value of C would be equal to

(*a*) 200 ☐ (*b*) 180 ☐

(*c*) 150 ☐ (*d*) 100. ☐

139. If D = large diameter, d = small diameter, and L = length of a job then tangent of half of the taper angle is equal to

(*a*) $\frac{D-d}{L}$ ☐ (*b*) $\frac{2(D-d)}{L}$ ☐

(*c*) $\frac{D-d}{2L}$ ☐ (*d*) $\frac{D+d}{2L}$ ☐

140. Hot working process takes place

(*a*) at recrystallization temperature ☐ (*b*) below recrystallization temperature ☐

(*c*) above recrystallization temperature ☐ (*d*) none of the above. ☐

141. If $\frac{dV}{dt} = 3t^2 + 4t + 10$ and $S = 0$, $V = 5$ when $t = 0$, then at $t = 3$, V is

(*a*) 80 ☐ (*b*) 40 ☐

(*c*) 120 ☐ (*d*) none of these. ☐

142. To solve Laplace equation $\frac{\partial^2 u}{\partial x^2} + \frac{\partial^3 u}{\partial y^2} + \frac{\partial^2 u}{\partial z^2} = 0$, we make the assumption

(*a*) $u = X(x) \cdot Y(y)$ ☐ (*b*) $u = X(x) \cdot Z(z)$ ☐

(*c*) $u = Y(y) \cdot Z(z)$ ☐ (*d*) $u = X(x) \cdot Y(y) \cdot Z(z)$. ☐

143. If $A = \begin{pmatrix} 4 & 3 \\ 2 & 5 \end{pmatrix}$, then $A - 9I + 14I$ is equal to

(*a*) zero ☐ (*b*) unit matrix I ☐

(*c*) null matrix ☐ (*d*) none of these. ☐

144. A square matrix, all of whose elements below the leading diagonal are zero, is called

(*a*) a lower triangular matrix ☐ (*b*) an upper triangular matrix ☐

(*c*) a diagonal matrix ☐ (*d*) none of these. ☐

145. It is not possible to obtain series solution in ascending power of x when a differential equation has

(*a*) an ordinary point at $x = 0$ ☐ (*b*) a regular singularity at $x = 0$ ☐

(*c*) an essential singularity at $x = 0$ ☐ (*d*) none of these. ☐

146. On bag contains 3 white and 2 black balls and another contains 5 white and 3 black balls. If a bag is chosen at random and a ball is drawn from it, then the chance that it is white is

(*a*) $\frac{5}{8}$ ☐ (*b*) $\frac{49}{160}$ ☐

(*c*) $\frac{48}{80}$ ☐ (*d*) none of these. ☐

147. Let $I = \int_{-1}^{2} \int_{x^2}^{x+2} dy\, dx$, then I is

(*a*) $\frac{7}{2}$ ☐ (*b*) $\frac{9}{2}$ ☐

(*c*) $\frac{11}{3}$ ☐ (*d*) none of these. ☐

148. If $\Gamma\left(-\frac{1}{2}\right) = \sqrt{\pi}$, then $\Gamma\left(-\frac{3}{2}\right)$ is given by

(*a*) $\frac{\sqrt{\pi}}{3}$ ☐ (*b*) $\frac{2\sqrt{\pi}}{3}$ ☐

(*c*) $\frac{4\sqrt{\pi}}{3}$ ☐ (*d*) none of these. ☐

149. If $pv = 20$ and p is measured as $\pm$ 0.01, then v is

(*a*) 4 ± 0.002 ☐ (*b*) 4 ± 0.004 ☐

(*c*) 4 ± 0.006 ☐ (*d*) 4 ± 0.008. ☐

150. The equations $4x + y + 4z = 0$, $2x + \lambda y + z = 0$, $(\lambda - 1)x - y + 2z = 0$ have non-trivial solution if

(*a*) $\lambda = \frac{9}{2}$ ☐ (*b*) $\lambda = -1$ ☐

(*c*) $\lambda = 1, \frac{9}{4}$ ☐ (*d*) none of these. ☐

151. A car moves on a straight road. Its velocity is given by $V(t) = 3\sqrt{t}$. In first 100 seconds, it will move

(*a*) 1000 ☐ (*b*) 1500 ☐

(*c*) 2000 ☐ (*d*) none of these. ☐

152. If h is small, then $\sin(x + h)$ is nearly equal to

(*a*) $\sin x + h\cos x + \frac{h^2}{2}\sin x$ ☐ (*b*) $\sin x - h\cos x - \frac{h^2}{2}\sin x$ ☐

(*c*) $\sin x - h\cos x + \frac{h^2}{2}\sin x$ ☐ (*d*) $\sin x + h\cos x - \frac{h^2}{2}\sin x$. ☐

153. The principal quantum number (n) can take values from

(*a*) 0 to 7 ☐ (*b*) – 1 to 7 ☐

(*c*) 1 to 7 ☐ (*d*) 0 to $(n - 1)$ ☐

(*e*) $-n$ to $+ n$. ☐

154. The valency of Cr in the complex $[Cr(H_2O)_4Cl_2]^+$ is

(*a*) 3 ☐ (*b*) 1 ☐

(*c*) 6 ☐ (*d*) 5 ☐

(*e*) 2. ☐

155. Solutions I and II contain 1 g mole each of urea (mol. mass = 60) and cane sugar (mol. mass = 342) in 1000 of water respectively. Then

(*a*) solution I boils at higher temperature ☐

(*b*) solution II boils at higher temperature ☐

(*c*) both the solutions boil at a same temperature. ☐

156. When $\pm n = 2; l = 0$ it represents the sub-shell

(*a*) 2*s* ☐ (*b*) 2*p* ☐

(*c*) 1*s* ☐ (*d*) 3*p*. ☐

157. The half life of decomposition of gaseous CH_3CHO at constant temperature but at initial pressure of 300 mm and 150 mm of Hg were 400 and 800 sec respectively. The order of reaction is

(*a*) 2 ☐ (*b*) 0 ☐

(*c*) 1.5 ☐ (*d*) 1 ☐

(*e*) 3. ☐

158. A gas behaves most nearly like an ideal gas when under conditions of

(*a*) high pressure and low temperature ☐ (*b*) low pressure and high temperature ☐

(*c*) low pressure and low temperature ☐ (*d*) high pressure and high temperature. ☐

159. Which one of the following solution in water will have the highest freezing point

(*a*) 0.1 m NaCl ☐ (*b*) 0.1 m sugar ☐

(*c*) 0.1 m $BaCl_2$ ☐ (*d*) 0.1 m $FeCl_3$. ☐

160. Oxidation number of Cl is $KClO_4$ is

(*a*) – 7 ☐ (*b*) + 7 ☐

(*c*) + 1 ☐ (*d*) – 1. ☐

161. Loss of an α-particle means

(*a*) no change in atomic mass ☐

(*b*) no change in atomic number ☐

(*c*) increase in atomic mass by 4 units ☐

(*d*) decrease in atomic number by 2 units ☐

(*e*) decrease in atomic mass by 4 units and of atomic number by 2 units. ☐

162. C(graphite) + $O_2(g) \rightarrow CO_2(g) \Delta H^0_{198} = - 94.05$ kcal $mole^{-1}$

C(diamond) + $O_2(g) \rightarrow CO_2(g) \Delta H^0_{298} = - 94.50$ kcal $mole^{-1}$

Therefore,

(*a*) C(diamond) → C(graphite) Δ^0_{298} = 450 cal mole^{-1} ☐
(*b*) C(graphite) → C(diamond) Δ^0_{298} = – 450 cal mole^{-1} ☐
(*c*) diamond is harder than graphite ☐
(*d*) graphite is the stable allotrope. ☐

163. Chromic acid oxidises acetylene to

(*a*) COOH–COOH ☐ (*b*) CH_3COOH ☐
(*c*) CH_2OH–CH_2OH ☐ (*d*) HCOOH. ☐

164. Electro metric effect

(*a*) is a temporary effect ☐
(*b*) comes into play only at the demand of the attacking reagent ☐
(*c*) is shown by two atoms joined by double or triple bond ☐
(*d*) possess all the above. ☐

165. For a product, the forecast and the actual sales for December 2002 were 25, and 20 respectively. If the exponential smoothing constant (α) is taken as 0.2, then forecast sales for January 2003 would be

(*a*) 21 ☐ (*b*) 23 ☐
(*c*) 24 ☐ (*d*) 27. ☐

(GATE-ME-2004)

166. The number of degrees of freedom of a planar linkage with 8 links and 9 simple revolute joints is

(*a*) 1 ☐ (*b*) 2 ☐
(*c*) 3 ☐ (*d*) 4. ☐

(ME-GATE-2005)

167. The sale of cycle in a shop in four consecutive months are given as 70, 68, 82, 95. Exponentially smoothing average method with a smoothing factor of 0.4 is used in forecasting. The expected number of sales in the next month is

(*a*) 59 ☐ (*b*) 72 ☐
(*c*) 86 ☐ (*d*) 136. ☐

(GATE-ME-2003)

168. The differential equation governing the vibrating system shown in Fig. 15.5 is

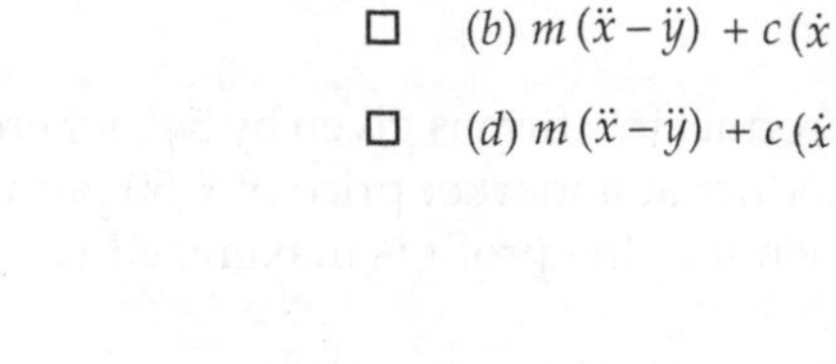

FIGURE 15.5

(*a*) $m\ddot{x} + c\ddot{x} + k(x - y) = 0$ ☐ (*b*) $m(\ddot{x} - \ddot{y}) + c(\dot{x} - \dot{y}) + kx = 0$ ☐

(*c*) $m\ddot{x} + c(\dot{x} - \dot{y}) + kx = 0$ ☐ (*d*) $m(\ddot{x} - \ddot{y}) + c(\dot{x} - \dot{y})k(x - y) = 0$ ☐

(ME-GATE-2006)

169. Choose the most appropriate word(s) from the options given below to complete the following sentence.

I contemplated .... Singapore for my vacation but decided against it

(*a*) to visit ☐ (*b*) having to visit ☐
(*c*) visting ☐ (*d*) for a visit. ☐

(ME/CE/CSE/PI-2011)

170. Were you a bird, you .... in the sky.

(*a*) would fly ☐ (*b*) shall fly ☐
(*c*) should fly ☐ (*d*) shall have flown ☐

(ME/CSE/PI-2013)

171. A model of a hydraulic turbine is tested at a head of $\frac{1}{4\text{th}}$ of that under which the full scale turbine works. The diameter of the model is half of that of the full scale turbine. If N is the RPM of the full scale turbine, then the RPM of the model will be

(*a*) $N/4$ ☐ (*b*) $N/2$ ☐
(*c*) N ☐ (*d*) $2N$ ☐

(GATE-ME-2007)

172. Consider the following data for an item:

Annual demand 2500 units per year, Ordering cost ₹ 100 per order, Inventory holding rate 25% of unit price.

Price quoted by a supplier

Order quantity (units)	Unit Price(₹)
< 500	10
≥ 500	9

The optimum order quantity (in units) is

(*a*) 447 ☐ (*b*) 471 ☐
(*c*) 500 ☐ (*d*) ≥ 600 ☐

(GATE-ME-2006)

173. The dual for the LP is

(*a*) Z_{min} $6u + 6v$ ☐
subject to:
$3u + 2v \geq 4$
$2u + 3v \geq 6$
$u, v \geq 0$

(*b*) Z_{max} $6u + 6v$ ☐
subject to:
$3u + 2v \leq 4$
$2u + 3v \leq 6$
$u, v \geq 0$

(*c*) Z_{max} $4u + 6v$ ☐
subject to:
$3u + 2v \geq 6$
$2u + 3v \geq 6$
$u, v \geq 0$

(*b*) Z_{max} $4u + 6v$ ☐
subject to:
$3u + 2v \leq 6$
$2u + 3v \leq 6$
$u, v \geq 0$

(GATE-ME-2008)

174. The cost function for a product in a firm is given by $5q^2$, where q is the amount of production. The firm can sell the product at a market price of ₹ 50 per unit. The number of units to be produced by the firm such that the profit is maximized is

(*a*) 5 ☐ (*b*) 10 ☐
(*c*) 15 ☐ (*d*) 25 ☐

(ME/CE/CSE/PI-2012)

175. Which one of the following options is the closest in meaning to the word given below?
Nadir
(*a*) highest ☐ (*b*) lowest ☐
(*c*) medium ☐ (*d*) integration ☐

(ME/CSE/PI-2013)

176. Which of the following options is the closest in the meaning to the word below?
Inexplicable
(*a*) incomprehensible ☐ (*b*) indelible ☐
(*c*) inextricable ☐ (*d*) infallible ☐

(ME/CE/CSE/PI-2011)

177. The conductance of a wire is directly proportional to
(*a*) area of cross-section ☐ (*b*) surface area of the conductor ☐
(*c*) length of the conductor ☐ (*d*) resistivity of the material. ☐

178. The unit of impedance is
(*a*) Mho ☐ (*b*) Ohm ☐
(*c*) Hertz ☐ (*d*) None of the above. ☐

179. *Q* of a circuit can be increased by
(*a*) decreasing the bandwidth ☐ (*b*) increasing the bandwidth ☐
(*c*) increasing *R* ☐ (*d*) none of the above. ☐

180. The Fourier series expansion of an odd function contains only
(*a*) sine terms ☐ (*b*) cosine terms ☐
(*c*) even harmonics only ☐ (*d*) odd harmonics only. ☐

181. The circuit elements absorbs 500 watts of power. Its current is 10 A. Its voltage is
(*a*) 510 volts ☐ (*b*) 490 volts ☐
(*c*) 500 volts ☐ (*d*) 50 volts. ☐

182. One of the following microwave diodes is suitable for very low power oscillators only
(*a*) Tunnel ☐ (*b*) LSA ☐
(*c*) Gunn ☐ (*d*) IMPATT. ☐

183. In the series RLC circuit if inductance of 1 H is replaced by 2 H then in the circuit
(*a*) time constant increases ☐ (*b*) damping ratio increases ☐
(*c*) natural frequency decreases ☐ (*d*) none of the above. ☐

184. An amplifier operating over the frequency range of 455 to 460 kHz has a 200 kΩ input resistor. Considering the ambient temperature to be 17°C. The r.m.s. noise voltage at the input to this amplifier is
(*a*) 4 μV ☐ (*b*) 4 mV ☐
(*c*) 40 μ V ☐ (*d*) none of the above. ☐

185. Volume control on the front panel of a TV set is

(*a*) variable coil ☐ (*b*) preset resistor ☐
(*c*) variable resistor ☐ (*d*) none of the above. ☐

186. Power consumption is least in microprocessor with the technology given in

(*a*) Bipolar junction transistor ☐ (*b*) FET ☐
(*c*) UJT ☐ (*d*) CMOS. ☐

187. In a VDR as voltage increases, the resistance

(*a*) decreases ☐ (*b*) increases ☐
(*c*) remains same ☐ (*d*) none of the above. ☐

188. Indicate which one of the following types of noise does not occur in transistors ?

(*a*) short noise ☐ (*b*) flicker noise ☐
(*c*) partition noise ☐ (*d*) resistance noise. ☐

189. Two couples will balance one another when they are in the same plane and

(*a*) have equal moments and their direction of rotation is same ☐
(*b*) have unequal moments and their direction of rotation is opposite ☐
(*c*) have equal moments and their direction of rotation is opposite ☐
(*d*) none of the above. ☐

190. Limiting force of friction is defined as the frictional force which exists when a boby

(*a*) is moving with maximum velocity ☐ (*b*) is stationary ☐
(*c*) just begins to slide over the surface ☐ (*d*) none of the above. ☐

191. If a body is moving with a uniform acceleration (*f*), then the distance travelled by the body in *n*th second is given by

(*a*) $\frac{u+f}{2}(1-2n)$ ☐ (*b*) $\frac{u+f}{2}(n-2)$ ☐
(*c*) $u+\frac{f}{2}(2n-1)$ ☐ (*d*) none of the above. ☐

192. A stone dropped into a well is heard to strike the water after 4 seconds. If the velocity of sound is 350 m/sec, the depth of well would be

(*a*) 150 m ☐ (*b*) 70.75 m ☐
(*c*) 100 m ☐ (*d*) 35.375 m. ☐

193. Dyne is the force acting on a mass of

(*a*) one kilogram to produce an acceleration of the metre per second square ☐
(*b*) one gm to produce an acceleration of one m/sec^2 ☐
(*c*) one kg to produce an acceleration of one cm/sec^2 ☐
(*d*) one gm to produce an acceleration of one cm/sec^2 ☐

194. Polar moment of inertia of a circle (I_0) is given by

(*a*) $\frac{\pi}{64}d^4$ ☐ (*b*) $\frac{\pi}{32}d^4$ ☐
(*c*) $\frac{\pi}{16}d^4$ ☐ (*d*) $\frac{\pi}{32}d^2$. ☐

195. Consider a linear programming problem with two variables and two constraints. The objective function is: Maximize $X_1 + X_2$. The corner points of the feasible region are (0, 0), (0, 2), (2, 0) and (4/3, 4/3).

If an additional constraint $X_1 + X_2 \le 5$ is added, the optimal solution is

(*a*) $\left(\frac{5}{3}, \frac{5}{3}\right)$ ☐ (*b*) $\left(\frac{4}{3}, \frac{4}{3}\right)$ ☐

(*c*) $\left(\frac{5}{2}, \frac{5}{2}\right)$ ☐ (*d*) (5, 0). ☐

(GATE-ME-2005)

196. The flow in a pipe is laminar, if

(*a*) Reynold number is equal to 2500 ☐ (*b*) Reynold number is equal to 4000 ☐

(*c*) Reynold number is more than 2500 ☐ (*d*) none of the above. ☐

197. Mach number is defined as the ratio of

(*a*) inertia force to viscous force ☐ (*b*) viscous force to surface tension force ☐

(*c*) viscous force to elastic force ☐ (*d*) inertia force to elastic force. ☐

198. The value of the kinetic energy correction factor (α) for the viscous flow through a circular pipe is

(*a*) 1.33 ☐ (*b*) 1.50 ☐

(*c*) 2.0 ☐ (*d*) 1.25. ☐

199. The less of pressure head for the laminar flow through pipes varies

(*a*) as the square of velocity ☐ (*b*) directly as the velocity ☐

(*c*) as the inverse of the velocity ☐ (*d*) none of the above. ☐

200. For a two-dimensional fluid flow in *x-y* plane, the rotational component is given as

(*a*) $\omega_z = \frac{1}{2}\left(\frac{\partial v}{\partial x} + \frac{\partial u}{\partial y}\right)$ ☐ (*b*) $\omega_z = \frac{1}{2}\left(\frac{\partial u}{\partial x} - \frac{\partial v}{\partial y}\right)$ ☐

(*c*) $\omega_z = \frac{1}{2}\left(\frac{\partial u}{\partial x} + \frac{\partial v}{\partial y}\right)$ ☐ (*d*) $\omega_z = \frac{1}{2}\left(\frac{\partial v}{\partial x} - \frac{\partial u}{\partial y}\right)$. ☐

201. In a double slit experiment the two slits are 0.25 mm apart and the screen is 2 m away. The linear distance between the two second order maxima is 20 mm. The wavelength of the incident light is

(*a*) 5500 Å ☐ (*b*) 5893 Å ☐

(*c*) 6250 Å ☐ (*d*) 7000 Å. ☐

202. A grating 2.5 cm size has 1500 lines. Light of wavelength 6000 Å is incident on it. The angular position of second order maxima is

(*a*) 46.5° ☐ (*b*) 36.8° ☐

(*c*) 21.3° ☐ (*d*) 17.4°. ☐

203. In the propagation of sound wave the distance between consecutive compression is

(*a*) λ ☐ (*b*) 2λ ☐

(*c*) $\frac{\lambda}{2}$ ☐ (*d*) $\frac{3\lambda}{2}$. ☐

204. If the potential difference across a parallel plate condenser is doubled, its capacity becomes

(*a*) half ☐ (*b*) double ☐

(*c*) four-times ☐ (*d*) no change. ☐

205. What is the electric potential at the surface of an atom having Z = 79? The radius of the atom is 6.6×10^{-15} m

(*a*) 10^5 volts ☐ (*b*) 3.4×10^7 volts ☐

(*c*) 1.7×10^7 volts ☐ (*d*) 1.7×10^5 volts. ☐

206. The form of the energy stored in a capacitor is

(*a*) dielectric ☐ (*b*) magnetic ☐

(*c*) electrostatic ☐ (*d*) electromagnetic. ☐

207. In a Netwon's rings experiment the diameters of 4th and 12th dark rings are 0.4 and 0.7 cm respectively. The diameter of 20th ring will be

(*a*) 1.00 cm ☐ (*b*) 1.43 cm ☐

(*c*) 0.894 cm ☐ (*d*) none of the above. ☐

208. The radius of earth is 6.37×10^6 m. If the earth stops rotating about its axis, the change in the value of '*g*' at a place of latitude 45° will be

(*a*) 6.37×10^{-2} m/s ☐ (*b*) 1.18×10^{-2} m/s ☐

(*c*) 2.36×10^{-2} m/s ☐ (*d*) none of these. ☐

209. A parallel plate mica capacitor is to be designed for a voltage of 9000 volts. The minimum thickness of mica should be

(*a*) 1.8 mm ☐ (*b*) 18.0 mm ☐

(*c*) 3.6 mm ☐ (*d*) 0.18 mm. ☐

210. An alloy wire has a diameter of 2 mm and carries a current of 2 amp. The drift velocity of the electrons will be

(*a*) 3.5×10^{-5} m/s ☐ (*b*) 4.0×10^{-5} m/s ☐

(*c*) 7×10^{-5} m/s ☐ (*d*) none of these. ☐

211. What is the wavelength associated with a beam of electrons having average kinetic energy as 100 eV?

(*a*) 10 Å ☐ (*b*) 2.4 Å ☐

(*c*) 1.2 Å ☐ (*d*) 12 Å. ☐

212. The limit of resolution of a telescope having aperture '*a*' is given by

(*a*) $\frac{\lambda}{2}$ ☐ (*b*) λa ☐

(*c*) $\frac{1.22\,\lambda}{a}$ ☐ (*d*) none of these. ☐

213. According to Charle's law

(*a*) $p_1V_2 = p_2V_1$ if T is constant ☐ (*b*) $\frac{p_1}{T_1} = \frac{p_2}{T_2}$ if V is constant ☐

(*c*) $\frac{V_1}{T_1} = \frac{V_2}{T_2}$ if p is constant ☐ (*d*) none of the above. ☐

214. If C_p is the specfic heat at constant pressure and C_v is the specific heat at constant volume, then

(a) $C_p < C_v$ ☐ (b) $C_p = C_v$ ☐
(c) $C_p > C_v$ ☐ (d) none of the above. ☐

215. The efficiency of the Carnot cycle may be increased by

(a) increasing the highest temperature ☐ (b) decreasing the lowest temperature ☐
(c) increasing the lowest temperature ☐ (d) (a) and (b) both. ☐

216. A Carnot cycle is having an efficiency of 0.75. If the temperature of the high temperature reservoir is 727°C, what is the temperature of low temperature reservoir?

(a) 23°C ☐ (b) – 23°C ☐
(c) 0°C ☐ (d) 250°C. ☐

(GATE-ME-2002)

217. Which one of the following configurations has the highest fin effectiveness?

(a) Thin, closely spaced fins ☐ (b) Thin, widely spaced fins ☐
(c) Thick, widely spaced fins ☐ (d) Thick, closely spaced fins. ☐

(GATE-ME&PI-2012)

218. A coolant fluid at 30°C flows over a heated flat plate maintained at a constant temperature of 100°C. The boundary layer temp distribution at a given location on the plate may be approximated as $T = 30 + 70 \exp(-y)$, where y(in m) is the distance normal to the plate and T is in °C. If thermal conductivity of the fluid is 1.0 W/mK, the local convective heat transfer (in W/m^{2}K) at that location will be

(a) 0.2 ☐ (b) 1 ☐
(c) 5 ☐ (d) 10. ☐

(GATE-ME-2009)

219. Heat is rejected by a refrigerant, during a refrigeration cycle in a

(a) condensor ☐ (b) evaporator ☐
(c) compressor ☐ (d) throttle value. ☐

220. The thermal efficiency of an engine, which is supplied heat at the rate of 15000 N m/s and gives an output of 4500 W, is

(a) 45% ☐ (b) 50% ☐
(c) 30% ☐ (d) 33%. ☐

221. The co-efficient of performance (C.O.P.) of a refrigerator working on the reversed Carnot cycle is 5. The ratio of highest absolute temperature to lowest absolute temperature (*i.e.*, T_2/T_1) would be

(a) 1.25 ☐ (b) 1.3 ☐
(c) 1.4 ☐ (d) 1.2. ☐

222. The statement that when any two bodies are in thermal equilibrium with third, they are also equal in thermal equilibrium with each other is

(a) first law of thermodynamics ☐ (b) zeroth law of thermodynamics ☐
(c) second law of thermodynamics ☐ (d) none of the above. ☐

223. A Carnot cycle consists of

(a) two constant pressure and two adiabatic processes ☐
(b) two constant volume and two adiabatic processes ☐

(*c*) two isothermal and two adiabatic processes ☐

(*d*) one constant pressure, one constant volume and two-adiabatic processes. ☐

224. If the exit pressure is equal to or more than critical pressure, the nozzle used should be

(*a*) Divergent ☐ (*b*) Convergent ☐

(*c*) Convergent-divergent ☐ (*d*) None of the above. ☐

ANSWERS TO TEST PAPER NO. 1

1. (*c*) **2.** (*a*) **3.** (*a*) **4.** (*d*) **5.** (*d*) **6.** (*c*)
7. (*d*) **8.** (*b*) **9.** (*a*) **10.** (*c*) **11.** (*b*) **12.** (*d*)
13. (*c*) **14.** (*a*) **15.** (*a*) **16.** (*c*) **17.** (*c*) **18.** (*a*)
19. (*d*) **20.** (*c*) **21.** (*a*) **22.** (*b*) **23.** (*d*) **24.** (*c*)
25. (*e*) **26.** (*c*) **27.** (*b*) **28.** (*a*) **29.** (*c*) **30.** (*c*)
31. (*b*) **32.** (*d*) **33.** (*c*) **34.** (*a*) **35.** (*c*) **36.** (*d*)
37. (*d*) **38.** (*c*) **39.** (*b*) **40.** (*c*) **41.** (*b*) **42.** (*c*)
43. (*a*) **44.** (*c*) **45.** (*b*) **46.** (*c*) **47.** (*c*) **48.** (*b*)
49. (*b*) **50.** (*a*) **51.** (*d*) **52.** (*b*) **53.** (*b*) **54.** (*b*)
55. (*b*) **56.** (*b*) **57.** (*c*) **58.** (*b*) **59.** (*c*) **60.** (*b*)
61. (*a*) **62.** (*b*) **63.** (*a*) **64.** (*c*) **65.** (*c*) **66.** (*a*)
67. (*d*) **68.** (*a*) **69.** (*b*) **70.** (*a*) **71.** (*c*) **72.** (*a*)
73. (*c*) **74.** (*a*) **75.** (*c*) **76.** (*b*) **77.** (*c*) **78.** (*d*)
79. (*c*) **80.** (*c*) **81.** (*b*) **82.** (*c*) **83.** (*a*) **84.** (*b*)
85. (*c*) **86.** (*d*) **87.** (*b*) **88.** (*d*) **89.** (*c*) **90.** (*b*)
91. (*d*) **92.** (*b*) **93.** (*a*) **94.** (*c*) **95.** (*a*) **96.** (*d*)
97. (*c*) **98.** (*b*) **99.** (*a*) **100.** (*b*) **101.** (*b*) **102.** (*a*)
103. (*c*) **104.** (*b*) **105.** (*b*) **106.** (*b*) **107.** (*d*) **108.** (*b*)
109. (*c*) **110.** (*b*) **111.** (*a*) **112.** (*a*) **113.** (*d*) **114.** (*b*)
115. (*c*) **116.** (*c*) **117.** (*c*) **118.** (*c*) **119.** (*d*) **120.** (*d*)
121. (*c*) **122.** (*c*) **123.** (*d*) **124.** (*b*) **125.** (*e*) **126.** (*d*)
127. (*e*) **128.** (*a*) **129.** (*d*) **130.** (*b*) **131.** (*c*) **132.** (*b*)
133. (*c*) **134.** (*c*) **135.** (*c*) **136.** (*e*) **137.** (*b*) **138.** (*b*)
139. (*c*) **140.** (*d*) **141.** (*b*) **142.** (*c*) **143.** (*a*) **144.** (*c*)
145. (*c*) **146.** (*a*) **147.** (*a*) **148.** (*b*) **149.** (*b*) **150.** (*c*)
151. (*a*) **152.** (*c*) **153.** (*a*) **154.** (*d*) **155.** (*b*) **156.** (*a*)
157. (*c*) **158.** (*b*) **159.** (*a*) **160.** (*a*) **161.** (*c*) **162.** (*a*)
163. (*b*) **164.** (*a*) **165.** (*a*) **166.** (*b*) **167.** (*c*) **168.** (*b*)
169. (*d*) **170.** (*c*) **171.** (*a*) **172.** (*c*) **173.** (*d*) **174.** (*b*)
175. (*a*) **176.** (*d*) **177.** (*b*) **178.** (*b*) **179.** (*c*) **180.** (*a*)
181. (*a*) **182.** (*c*) **183.** (*a*) **184.** (*d*) **185.** (*a*) **186.** (*a*)
187. (*b*) **188.** (*b*) **189.** (*c*) **190.** (*d*) **191.** (*b*) **192.** (*b*)
193. (*b*) **194.** (*c*) **195.** (*b*) **196.** (*a*) **197.** (*b*) **198.** (*a*)
199. (*a*) **200.** (*c*) **201.** (*b*) **202.** (*a*) **203.** (*b*) **204.** (*b*)
205. (*b*) **206.** (*c*) **207.** (*d*) **208.** (*c*) **209.** (*b*) **210.** (*b*)
211. (*a*) **212.** (*c*) **213.** (*b*) **214.** (*a*) **215.** (*d*) **216.** (*c*)

217. (*c*)	218. (*c*)	219. (*b*)	220. (*a*)	221. (*b*)	222. (*c*)
223. (*b*)	224. (*c*).				

ANSWERS TO TEST PAPER NO. 2

1. (*b*)	2. (*c*)	3. (*e*)	4. (*b*)	5. (*a*)	6. (*c*)
7. (*b*)	8. (*b*)	9. (*a*)	10. (*d*)	11. (*a*)	12. (*b*)
13. (*c*)	14. (*c*)	15. (*b*)	16. (*a*)	17. (*a*)	18. (*c*)
19. (*b*)	20. (*c*)	21. (*a*)	22. (*d*)	23. (*a*)	24. (*a*)
25. (*d*)	26. (*a*)	27. (*b*)	28. (*e*)	29. (*c*)	30. (*c*)
31. (*d*)	32. (*b*)	33. (*d*)	34. (*a*)	35. (*c*)	36. (*c*)
37. (*b*)	38. (*a*)	39. (*d*)	40. (*a*)	41. (*c*)	42. (*d*)
43. (*c*)	44. (*b*)	45. (*b*)	46. (*c*)	47. (*d*)	48. (*c*)
49. (*b*)	50. (*a*)	51. (*d*)	52. (*a*)	53. (*c*)	54. (*a*)
55. (*a*)	56. (*b*)	57. (*c*)	58. (*c*)	59. (*a*)	60. (*c*)
61. (*b*)	62. (*b*)	63. (*c*)	64. (*b*)	65. (*d*)	66. (*a*)
67. (*c*)	68. (*b*)	69. (*c*)	70. (*d*)	71. (*b*)	72. (*b*)
73. (*b*)	74. (*c*)	75. (*c*)	76. (*c*)	77. (*b*)	78. (*b*)
79. (*c*)	80. (*a*)	81. (*d*)	82. (*b*)	83. (*a*)	84. (*a*)
85. (*d*)	86. (*c*)	87. (*a*)	88. (*d*)	89. (*c*)	90. (*c*)
91. (*b*)	92. (*b*)	93. (*b*)	94. (*c*)	95. (*e*)	96. (*d*)
97. (*d*)	98. (*c*)	99. (*d*)	100. (*b*)	101. (*b*)	102. (*c*)
103. (*a*)	104. (*a*)	105. (*a*)	106. (*c*)	107. (*c*)	108. (*b*)
109. (*d*)	110. (*b*)	111. (*e*)	112. (*a*)	113. (*a*)	114. (*d*)
115. (*a*)	116. (*b*)	117. (*b*)	118. (*c*)	119. (*d*)	120. (*c*)
121. (*d*)	122. (*c*)	123. (*d*)	124. (*c*)	125. (*c*)	126. (*d*)
127. (*c*)	128. (*b*)	129. (*a*)	130. (*c*)	131. (*d*)	132. (*b*)
133. (*b*)	134. (*d*)	135. (*a*)	136. (*e*)	137. (*c*)	138. (*b*)
139. (*c*)	140. (*c*)	141. (*a*)	142. (*d*)	143. (*c*)	144. (*b*)
145. (*c*)	146. (*c*)	147. (*b*)	148. (*c*)	149. (*d*)	150. (*c*)
151. (*c*)	152. (*d*)	153. (*c*)	154. (*a*)	155. (*c*)	156. (*a*)
157. (*a*)	158. (*b*)	159. (*b*)	160. (*b*)	161. (*e*)	162. (*d*)
163. (*b*)	164. (*d*)	165. (*c*)	166. (*c*)	167. (*b*)	168. (*c*)
169. (*c*)	170. (*a*)	171. (*c*)	172. (*c*)	173. (*a*)	174. (*a*)
175. (*b*)	176. (*a*)	177. (*a*)	178. (*b*)	179. (*a*)	180. (*a*)
181. (*d*)	182. (*a*)	183. (*c*)	184. (*a*)	185. (*c*)	186. (*d*)
187. (*a*)	188. (*c*)	189. (*c*)	190. (*c*)	191. (*c*)	192. (*b*)
193. (*d*)	194. (*b*)	195. (*b*)	196. (*d*)	197. (*d*)	198. (*c*)
199. (*b*)	200. (*d*)	201. (*c*)	202. (*a*)	203. (*c*)	204. (*b*)
205. (*c*)	206. (*c*)	207. (*a*)	208. (*c*)	209. (*d*)	210. (*c*)
211. (*c*)	212. (*c*)	213. (*c*)	214. (*c*)	215. (*d*)	216. (*b*)
217. (*b*)	218. (*b*)	219. (*a*)	220. (*c*)	221. (*d*)	222. (*b*)
223. (*a*)	224. (*b*).				